LANZKOWSKY'S MANUAL OF PEDIATRIC HEMATOLOGY AND ONCOLOGY

SIXTH EDITION

LANZKOWSKY'S MANUAL OF PEDIATRIC HEMATOLOGY AND ONCOLOGY

SIXTH EDITION

Edited by

PHILIP LANZKOWSKY, MBCHB, MD, ScD (HONORIS CAUSA), FRCP, DCH, FAAP
Chief Emeritus, Pediatric Hematology-Oncology
Chairman Emeritus, Department of Pediatrics
Executive Director and Chief-of-Staff (Retired)
Steven and Alexandra Cohen Children's Medical Center of New York, New Hyde Park, New York
Vice President, Children's Health Network (Retired)
Northwell Health System
Professor of Pediatrics
Hofstra Northwell School of Medicine, Hempstead, New York

JEFFREY M. LIPTON, MD, PHD
Chief of Hematology-Oncology and Stem Cell Transplantation
Steven and Alexandra Cohen Children's Medical Center of New York, New Hyde Park, New York
Center Head, Patient Oriented Research, Feinstein Institute for Medical Research, Manhasset, New York
Professor of Pediatrics and Molecular Medicine
Hofstra Northwell School of Medicine, Hempstead, New York
Professor, Elmezzi Graduate School of Medicine, Manhasset, New York

JONATHAN D. FISH, MD
Director Survivors Facing Forward Program
Steven and Alexandra Cohen Children's Medical Center of New York, New Hyde Park, New York
Assistant Professor of Pediatrics
Hofstra Northwell School of Medicine, Hempstead, New York

AMSTERDAM • BOSTON • HEIDELBERG • LONDON
NEW YORK • OXFORD • PARIS • SAN DIEGO
SAN FRANCISCO • SINGAPORE • SYDNEY • TOKYO
Academic Press is an imprint of Elsevier

Academic Press is an imprint of Elsevier
125 London Wall, London EC2Y 5AS, UK
525 B Street, Suite 1800, San Diego, CA 92101-4495, USA
50 Hampshire Street, 5th Floor, Cambridge, MA 02139, USA
The Boulevard, Langford Lane, Kidlington, Oxford OX5 1GB, UK

British Library Cataloguing-in-Publication Data
A catalogue record for this book is available from the British Library.

Library of Congress Cataloging-in-Publication Data
A catalog record for this book is available from the Library of Congress.

ISBN: 978-0-12-801368-7

For Information on all Academic Press publications
visit our website at http://www.elsevier.com/

Working together
to grow libraries in
developing countries

www.elsevier.com • www.bookaid.org

Publisher: Mica Haley
Acquisition Editor: Tari Broderick
Editorial Project Manager: Jeffrey L. Rossetti
Production Project Manager: Lucía Pérez
Designer: Mark Rogers

Typeset by MPS Limited, Chennai, India

Dedication

This book is dedicated

To our parents,
Abe and Lily Lanzkowsky, Thelma and Al Lipton,
and Vicky and Lawrence Fish, who instilled in us the importance of integrity,
the rewards of industry and the primacy of being a mensch.

To our wives,
Rhona Lanzkowsky, Linda Lipton, and Leah Fish,
who understand that the study of medicine is a
lifelong and consuming process.

To our children and grandchildren,
our pride and joy.

And

To our patients,
students, pediatric house staff,
fellows in pediatric hematology and oncology,
and to our colleagues who have taught us so much over the years.

Today he can discover the errors of yesterday
and tomorrow he may obtain new light
on what he thinks himself sure of today
Moses Maimónides

Contents

List of Contributors

Hisham Abdel-Azim, MD, MS Division of Hematology, Oncology and Blood & Marrow Transplantation, Children's Hospital Los Angeles, University of Southern California Keck School of Medicine, Los Angeles, CA

Hematopoietic Stem Cell Transplantation

Suchitra S. Acharya, MD, MBBS, DCH Division of Pediatric Hematology-Oncology and Stem Cell Transplantation, Steven and Alexandra Cohen Children's Medical Center of New York, Hofstra Northwell School of Medicine, New Hyde Park, New York

Disorders of Coagulation

Anurag K. Agrawal, MD Pediatric Hematology and Oncology, Children's Hospital and Research Center Oakland, Oakland, CA

Supportive Care of Patients with Cancer

Robert J. Arceci, MD, PhD[†] Children's Center for Cancer and Blood Disorders, and Ronald A. Matricaria Institute of Molecular Medicine, Phoenix Children's Hospital, Department of Child Health, College of Medicine, University of Arizona, Phoenix, AZ

Histiocytosis Syndromes

Mark P. Atlas, MD Department of Pediatrics, Hofstra Northwell School of Medicine; Steven and Alexandra Cohen Children's Medical Center of New York, New Hyde Park, NY

Central Nervous System Malignancies

Rochelle Bagatell, MD Department of Pediatrics, Perelman School of Medicine, University of Pennsylvania, Children's Hospital of Philadelphia, Philadelphia, PA

Neuroblastoma

Nicholas Bernthal, MD Department of Orthopaedic Surgery, David Geffen School of Medicine, University of California, Los Angeles, Santa Monica, CA

Malignant Bone Tumors

Teena Bhatla, MD Division of Pediatric Hematology-Oncology, NYU Langone Medical Center, New York University School of Medicine, New York, NY

Acute Lymphoblastic Leukemia

Lionel Blanc, PhD Departments of Molecular Medicine and Pediatrics, Hofstra Northwell School of Medicine; Center for Oncology and Cell Biology, Laboratory of Developmental Erythropoiesis, The Feinstein Institute for Medical Research, Manhasset, NY

General Considerations of Hemolytic Diseases, Red Cell Membrane, and Enzyme Defects

Mary Ann Bonilla, MD Pediatric Hematology-Oncology, St. Joseph's Children's Hospital, Paterson, NJ; Division of Pediatric Hematology-Oncology, New York Medical College, Valhalla, NY

Disorders of White Blood Cells

James B. Bussel, MD Department of Pediatrics, Division of Hematology, Weill Cornell Medical College and New York-Presbyterian Hospital, New York, NY

Disorders of Platelets

William L. Carroll, MD Division of Pediatric Hematology-Oncology, NYU Langone Medical Center, New York University School of Medicine, New York, NY

Acute Lymphoblastic Leukemia

Jeffrey S. Dome, MD, PhD Department of Pediatrics, George Washington University School of Medicine; Divisions of Hematology and Oncology, Children's National Health System, Washington, DC

Renal Tumors

Elizabeth A. Van Dyne, MD Division of Pediatric Hematology-Oncology, University of California Los Angeles Mattel, Children's Hospital, Los Angeles, CA

Malignant Bone Tumors

[†]Deceased

Tarek M. Elghetany, MD Departments of Pathology and Immunology, and Pediatrics, Baylor College of Medicine, Texas Children's Hospital, Houston, TX

Myelodysplastic Syndromes and Myeloproliferative Disorders

Noah Federman, MD Department of Pediatrics and Orthopaedic Surgery, David Geffen School of Medicine, University of California; Division of Pediatric Hematology-Oncology, University of California Los Angeles Mattel, Children's Hospital, UCLA's Jonsson Comprehensive Cancer Center, Los Angeles, CA

Malignant Bone Tumors

Carolyn Fein Levy, MD Department of Pediatrics, Hofstra Northwell School of Medicine; Divisions of Pediatric Hematology-Oncology and Stem Cell Transplantation, Department of Pediatrics, Steven and Alexandra Cohen Children's Medical Center of New York, New Hyde Park, NY

Rhabdomyosarcoma and Other Soft-Tissue Sarcomas

James Feusner, MD Department of Pediatrics, University of California San Francisco Medical Center; Department of Oncology, Children's Hospital and Research Center Oakland, Oakland, CA

Supportive Care of Patients with Cancer

Jonathan D. Fish, MD Division of Pediatric Hematology-Oncology, Hofstra Northwell School of Medicine, Hempstead, New York; Steven and Alexandra Cohen Children's Medical Center of New York, New Hyde Park, NY

Evaluation, Investigations, and Management of Late Effects of Childhood Cancer

Jason L. Freedman, MD, MSCE Division of Oncology, Children's Hospital of Philadelphia, Philadelphia, PA

Management of Oncologic Emergencies

Debra L. Friedman, MD, MS Department of Pediatrics, Vanderbilt University School of Medicine; Division of Pediatric Hematology-Oncology, Vanderbilt-Ingram Cancer Center, Nashville, TN

Hodgkin Lymphoma

Derek R. Hanson, MD Pediatric Neuro-Oncology, Hackensack University Medical Center, Joseph M. Sanzari Children's Hospital, Hackensack, NJ

Central Nervous System Malignancies

Inga Hofmann, MD Deparmtent of Pediatrics, Harvard Medical School, Dana-Farber/Boston Children's Cancer and Blood Disorders Center; Department of Hematopathology, Boston Children's Hospital, Boston, MA

Myelodysplastic Syndromes and Myeloproliferative Disorders

Mary S. Huang, MD Department of Pediatrics, Harvard Medical School, Massachusetts General Hospital for Children, Boston, MA

Non-Hodgkin Lymphoma

Rachel Kessel, MD Department of Pediatrics, Hofstra Northwell School of Medicine; Division of Pediatric Hematology-Oncology and Stem Cell Transplantation, Steven and Alexandra Cohen Children's Medical Center of New York, New Hyde Park, NY, USA

Acute Myeloid Leukemia

Julie I. Krystal, MD, MPH Department of Pediatrics, Hofstra Northwell School of Medicine; Division of Pediatric Hematology-Oncology and Stem Cell Transplantation, Department of Pediatrics, Steven and Alexandra Cohen Children's Medical Center of New York, New Hyde Park, NY

Evaluation, Investigations, and Management of Late Effects of Childhood Cancer

Janet L. Kwiatkowski, MD, MSCE Division of Hematology, Children's Hospital of Philadelphia; Department of Pediatrics, Perelman School of Medicine at the University of Pennsylvania, Philadelphia, PA

Hemoglobinopathies

Philip. Lanzkowsky, MBChB, MD, ScD (Honoris Causa), FRCP, DCH, FAAP Department of Pediatrics, Hofstra Northwell School of Medicine, Hempstead, NY; Division of Pediatric Hematology-Oncology, Department of Pediatrics, Steven and Alexandra Cohen Children's Medical Center of New York, New Hyde Park, NY

Classification and Diagnosis of Anemia in Children; Lymphadenopathy and Diseases of the Spleen; Anemia During the Neonatal Period; Iron-Deficiency Anemia; Megaloblastic Anemia

Ann Leahey, MD Department of Pediatrics, Perelman School of Medicine at the University of Pennsylvania, Children's Hospital of Philadelphia, Philadelphia, PA

Retinoblastoma

Jeffrey M. Lipton, MD, PhD Department of Pediatrics and Molecular Medicine, Hofstra Northwell School of Medicine, Hempstead, NY; Division of Pediatric Hematology-Oncology and Stem Cell Transplantation, Steven and Alexandra Cohen Children's Medical Center of New York, New Hyde Park, NY; The Feinstein Institute for Medical Research, Manhasset, NY

Bone Marrow Failure

Naomi L.C. Luban, MD Edward J. Miller Blood Donor Center; Division of Laboratory Medicine and Pathology, Center for Cancer and Blood Disorders, Children's

National Medical Center, Sheikh Zayed Campus for Advanced Children's Medicine; Departments of Pediatrics and Pathology, George Washington School of Medicine and Health Sciences, Washington, DC

Blood Banking Principles and Practices

Catherine McGuinn, MD Department of Pediatrics, Division of Hematology, Weill Cornell Medical College and New York Presbyterian Hospital, New York, NY

Disorders of Platelets

Jill S. Menell, MD Pediatric Hematology-Oncology, St. Joseph's Children's Hospital, Paterson, NJ; Division of Pediatric Hematology-Oncology, New York Medical College, Valhalla, NY

Disorders of White Blood Cells

Thomas A. Olson, MD Department of Pediatrics, Emory University School of Medicine; Pediatric Hematology-Oncology, Aflac Cancer and Blood Disorders Center, Children's Healthcare of Atlanta, Atlanta, GA

Germ Cell Tumors; Hepatic Tumors

Julie R. Park, MD Department of Pediatrics, University of Washington School of Medicine; Department of Hematology-Oncology Education, Division of Pediatric Hematology-Oncology, Seattle Children's Hospital, Seattle, Washington, DC

Neuroblastoma

Josef T. Prchal, MD Division of Hematology and Hematologic Malignancies, Department of Internal Medicine at the University of Utah; Huntsman Cancer Hospital, Myeloproliferative Disorders Clinic, Salt Lake City, Utah; First Faculty of Medicine, Charles University, Prague, Czech Republic

Polycythemia

Louis B. Rapkin, MD Department of Pediatrics, Emory University School of Medicine; Pediatric Hematology-Oncology, Aflac Cancer and Blood Disorders Center, Children's Healthcare of Atlanta, Atlanta, GA

Hepatic Tumors

Arlene Redner, MD Department of Pediatrics, Hofstra Northwell School of Medicine; Department of Oncology, Steven and Alexandra Cohen Children's Medical Center of New York, New Hyde Park, NY, USA

Acute Myeloid Leukemia

Susan R. Rheingold, MD Children's Hospital of Philadelphia; Division of Oncology, Perelman School of Medicine at the University of Pennsylvania, Philadelphia, PA

Management of Oncologic Emergencies

Indira Sahdev, MD Department of Pediatrics, Hofstra Northwell School of Medicine; Division of Pediatric Hematology-Oncology and Stem Cell Transplantation, Steven and Alexandra Cohen Children's Medical Center of New York, New Hyde Park, NY

Hematopoietic Stem Cell Transplantation

Vijay G. Sankaran, MD, PhD Department of Pediatrics, Harvard Medical School, Division of Hematology-Oncology, Boston Children's Hospital & Department of Pediatric Oncology, Dana-Farber Cancer Institute, Boston, MA

Diagnostic Molecular and Genomic Methodologies for the Hematologist/Oncologist

Susmita N. Sarangi, MD, MBBS Division of Pediatric Hematology-Oncology, Steven and Alexandra Cohen Children's Medical Center of New York, New Hyde Park, NY

Disorders of Coagulation

Tsewang Tashi, MD Division of Hematology and Hematologic Malignacies, University of Utah School of Medicine, Salt Lake City, UT

Polycythemia

David T. Teachey, MD Department of Pediatrics, Perelman School of Medicine at the University of Pennsylvania, Children's Hospital of Philadelphia, Philadelphia, PA

Lymphoproliferative and Myeloproliferative Disorders

Meg Tippy, PsyD Mental Health and Psychosocial Services, Division of Pediatric Hematology-Oncology and Stem Cell Transplantation, Department of Pediatrics, Steven and Alexandra Cohen Children's Medical Center of New York, New Hyde Park, NY

Psychosocial Aspects of Cancer for Children and Their Families

Adrianna Vlachos, MD Department of Pediatrics, Hofstra Northwell School of Medicine; Division of Pediatric Hematology-Oncology and Stem Cell Transplantation, Steven and Alexandra Cohen Children's Medical Center of New York, New Hyde Park, NY

Bone Marrow Failure

Anne B. Warwick, MD, MPH Department of Pediatrics, F. Edward Hébert School of Medicine, Uniformed Services University of the Health Sciences, Bethesda, MD

Renal Tumors

Howard Weinstein, MD Department of Pediatrics, Harvard Medical School; Division of Pediatric Hematology and Oncology, Massachusetts General Hospital for Children, Boston, MA

Non-Hodgkin Lymphoma

Leonard H. Wexler, MD Department of Pediatrics, Memorial Sloan-Kettering Cancer Center, New York, NY

Rhabdomyosarcoma and Other Soft-Tissue Sarcomas

Lawrence C. Wolfe, MD Department of Pediatrics, Hofstra Northwell School of Medicine; Division of Pediatric Hematology-Oncology and Stem Cell Transplantation, Steven and Alexandra Cohen Children's Medical Center of New York, New Hyde Park, NY

Hematologic Manifestations of Systemic Illness; General Considerations of Hemolytic Diseases, Red Cell Membrane, and Enzyme Defects; Extracorpuscular Hemolytic Anemia

Edward C.C. Wong, MD Department of Transfusion Medicine, Division of Laboratory Medicine, Center for Cancer and Blood Disorders, Children's National Medical Center Sheikh Zayed Campus for Advanced Children's Medicine; Department of Pediatrics and Pathology, George Washington School of Medicine and Health Sciences, Washington, DC

Blood Banking Principles and Practices

About the Editors

DR PHILIP LANZKOWSKY

PHILIP LANZKOWSKY

Dr Philip Lanzkowsky was born in Cape Town on March 17, 1932 and graduated high school from the South African College and obtained his MBChB degree from the University of Cape Town School of Medicine in 1954 and his Doctorate degree in 1959 for his thesis on *Iron Deficiency Anemia In Children*. He completed a pediatric residency at the Red Cross War Memorial Children's Hospital in Cape Town in 1960. In the same year, he received the Diploma in Child Health (DCH) from the Royal College of Physicians and Surgeons of London, and in 1961 was made member of the prestigious Royal College of Physicians of Edinburgh (RCPE). After working in Pediatrics at the University of Edinburgh and at St Mary's Hospital of the University of London, Dr Lanzkowsky did a pediatric Hematology-Oncology fellowship at Duke University School of Medicine and at the University of Utah.

In 1963 he was appointed Consultant Pediatrician and Pediatric Hematologist to the Red Cross War Memorial Children's Hospital at the University of Cape Town and introduced Pediatric Hematology and Oncology as a distinct discipline. In 1965 he was appointed Director of Pediatric Hematology and Associate Professor of Pediatrics at the New York Hospital-Cornell University School of Medicine.

In 1970 he was appointed Professor of Pediatrics and Chairman of Pediatrics at Long Island Jewish Medical Center and established a division of Pediatric Hematology-Oncology which he directed until 2000. He was the founder of the Schneider Children's Hospital (presently named Steven and Alexandra Cohen Children's Medical Center of New York), which he developed, planned, and was the hospital's Executive Director and Chief of Staff from its inception in 1983 until 2010.

Dr Lanzkowsky has received numerous honors and awards and has lectured extensively at various institutions and medical schools in the United States and around the world. In 1973 he was appointed Fellow of the Royal College of Physicians of Edinburgh, and in 1994 he received a Doctor of Science Degree (Honoris Causa) from St Johns University in New York for "his notable contribution to the field of pediatric medicine and to the children of the world". Among many other awards he was the recipient of the Joseph Arenow Prize for original postgraduate research in the field of Science, Medicine and Applied Science from the University of Cape Town, the John Adams Memorial Traveling fellowship administered by the Nuffield Foundation, the Hill-Pattison-Struthers Bursary from the Royal College of Physicians of Edinburgh and the Sonia Mechanick Traveling Fellowship from the South African College of Medicine. In addition to having been the editor of five editions of the *Manual of Pediatric Hematology and Oncology* used by clinicians worldwide, he is the author of *How It All Began: The History of a Children's Hospital* and over 280 scientific papers, abstracts, monographs, and book chapters.

Dr Lanzkowsky's medical writings have been prodigious. His seminal contributions to the medical literature have included the first description of the relationship of pica to iron-deficiency anemia (*Arch. Dis Child.*, 1959), Effects of timing of clamping of umbilical cord on infant's hemoglobin level (*Br. Med. J.*, 1960), Normal oral D-xylose test values in children (*New Engl. J. Med.*, 1963), Normal coagulation factors in women in labor and in the newborn (*Thromboses at Diath. Hemorr.*, 1966), Erythrocyte abnormalities induced by malnutrition (*Br. J. Haemat.*, 1967), Radiologic features in iron deficiency anemia (*Am. J. Dis. Child.*, 1968), Isolated defect of folic acid absorption associated with mental retardation (*Blood*, 1969; *Am. J. Med.*, 1970), Disaccharidase levels in iron deficiency (*J. Pediat.*, 1981) and Hexokinase "New Hyde Park" in a Chinese kindred (*Am. J. Hematol.*, 1981).

JEFFREY M. LIPTON

Jeffrey M. Lipton, MD, PhD, is the Chief of Hematology-Oncology and Stem Cell Transplantation at the Steven and Alexandra Cohen Children's Medical Center of New York and Center Head, Patient Oriented Research, The Feinstein Institute for Medical Research. He is Professor, Elmezzi Graduate School of Molecular Medicine, The Feinstein Institute for Medical Research and Professor of Pediatrics and Molecular Medicine at the Hofstra Northwell School of Medicine. He holds a BA from Queens College, City University of New York, a PhD in Chemistry from Syracuse University and an MD degree, *magna cum laude*, from Saint Louis University Medical School. He did his Pediatric residency training at the Children's Hospital, Boston, MA and his Pediatric Hematology-Oncology fellowship training at the Children's Hospital and the Sidney Farber Cancer Institute (now DFCI) in Boston. Dr Lipton is a Past-President of the American Society of Pediatric Hematology-Oncology (ASPHO). His main interest is bone marrow failure.

JONATHAN D. FISH

Jonathan D. Fish, MD, is Assistant Professor of Pediatrics at the Hofstra Northwell School of Medicine. He is Director of the Survivors Facing Forward Program, a long-term follow-up program for survivors of childhood cancer, in the Division of Pediatric Hematology-Oncology and Stem Cell Transplantation at the Steven and Alexandra Cohen Children's Medical Center of New York. He holds a BA, *with distinction*, from the University of Western Ontario and received his MD degree, *magna cum laude*, from the Upstate Medical Center of the State University of New York. He did his Pediatric training at the Schneider Children's Hospital, New Hyde Park, NY (presently known as the Steven and Alexandra Cohen Children's Medical Center of New York) and completed his Pediatric Hematology-Oncology fellowship training at the Children's Hospital of Philadelphia. Dr Fish received the American Society of Pediatric Hematology-Oncology Young Investigator Award in 2008. His clinical and research interest is on the late effects of childhood cancer treatment and survivorship care.

Preface to the Sixth Edition

The sixth edition of *Lanzkowsky's Manual of Pediatric Hematology and Oncology* has significant changes from previous editions. The title of the book, the editors, and its content have changed but the objective has remained unchanged and every effort has been made to retain the original style and clarity which have become the hallmark of the previous editions.

The title has changed to include the name of the original and sole editor of the book in its various editions for the past 45 years. In addition, the list of editors has increased from single editorship to include two additional hematologist-oncologists to reflect advances in pediatric hematology and oncology over the years. Jeffrey Lipton MD, PhD and Jonathan Fish MD have been selected as coeditors.

The book has been expanded from 33 chapters in the fifth edition to 36 in the present edition. A chapter has been added on the burgeoning subject of diagnostic, molecular, and genomic methodologies for the hematologist-oncologist and a new chapter has been added on transfusion medicine. Lymphoproliferative disorders and myelodysplasia have been assigned separate chapters and lymphoid and myeloid leukemia have also been assigned distinct chapters.

A number of new experts in particular fields have been added to the contributing-author panel.

Despite these significant changes the book has retained the original objective, format, and clarity of the founding editor. It remains a practical, concise, up-to-date guide to all professional staff treating children with hematological and oncologic diseases. The book is replete with detailed tables, practical algorithms, and flow diagrams useful for teaching housestaff, fellows, nursing staff, and practicing physicians and essential for the day-to-day investigation and management of patients with hematologic and oncologic conditions.

I would like to pay tribute to Drs Gungor Karayalcin and Ashok Shende, my close associates for over 40 years who retired after a lifelong career in clinical practice and research in Pediatric Hematology-Oncology, for their major contribution to the first four editions of the book which formed the very foundation of all subsequent editions.

Philip Lanzkowsky
MBChB, MD, ScD (Honoris Causa), FRCP, DCH, FAAP

Royalty payments accrued to Dr Lanzkowsky for all future editions of this book will be donated to the Division of Pediatric Hematology-Oncology at the Cohen Children's Medical Center of New York for children with cancer whose families have financial difficulties.

Preface to the Fifth Edition

The fifth edition of the *Manual of Pediatric Hematology and Oncology* differs considerably from previous editions but has retained the original intent of the author to offer a concise manual of predominantly clinical material culled from personal experience and to be an immediate reference for the diagnosis and management of hematologic and oncologic diseases. I have resisted succumbing to the common tendency of writing a comprehensive tome which is not helpful to the practicing hematologist-oncologist at the bedside. The book has remained true to its original intent.

The information included at all times keeps "the eye on the ball" to ensure that pertinent, up-to-date, practical clinical advice is presented without extraneous information, however interesting or pertinent this information may be in a different context.

The book differs from previous editions in many respects. The number of contributors has been considerably expanded drawing on the expertise of leaders in different subjects from various institutions in the United States. Increased specialization within the field of hematology and oncology has necessitated including this large a number of contributors in order to bring to the reader balanced and up-to-date information for the care of patients. In addition, the number of chapters has increased from 27, in the previous edition, to 33. The reason for this is that many of the chapters, such as hemolytic anemia and coagulation, had become so large and the subject so extensive that they were better handled by subdividing the chapter into a number of smaller chapters. An additional chapter on the psychosocial aspects of cancer for children and their families, not present in previous editions, has been added.

Some chapters have been extensively revised and re-written where advancement in knowledge has dictated this approach, for example, Hodgkin lymphoma, neuroblastoma and rhabdomyosarcoma and other soft-tissue sarcomas, whereas other chapters have been only slightly modified. In nearly all the chapters there has been significant change in the management and treatment section reflecting advances that have occurred in these areas.

This edition has retained the essential format written and developed decades ago by the author and, with usage over the years, has proven to be highly effective as a concise, practical, up-to-date guide replete with detailed tables, algorithms and flow diagrams for investigation and management of hematologic and oncologic conditions. The tables and flow diagrams included in the book have been updated using the latest information and the most recent protocols of treatment, which have received general acceptance and have become the standard of care, have been included. In a book with so many details, errors inevitably occur. I do not know where they are because if I did they would have been corrected. I apologize in advance for any inaccuracies that may have crept in inadvertently.

The four previous editions of this book were published when the name of the hospital was the Schneider Children's Hospital. Effective April 1, 2010 the name of the hospital was changed to the Steven and Alexandra Cohen Children's Medical Center of New York.

I would like to acknowledge Morris Edelman, MB, BCh, BSc (Laboratory Medicine) for his contribution in reviewing the pathology on Hodgkin disease.

I thank Rose Grosso for her untiring efforts in the typing and coordination of the various phases of the development of this edition.

Philip Lanzkowsky
MBChB, MD, ScD (Honoris Causa), FRCP, DCH, FAAP

Preface to the Fourth Edition

This edition of the *Manual of Pediatric Hematology and Oncology* is the fourth edition and the sixth book written by the author on pediatric hematology and oncology. The first book written by the author 25 years ago was exclusively on pediatric hematology and its companion book, exclusively on pediatric oncology, was written 3 years later. The book reviewers at the time suggested that these two books be combined into a single book on pediatric hematology and oncology and the first edition of the *Manual of Pediatric Hematology and Oncology* was published by the author in 1989.

It is from these origins that this 4th edition arises—the original book written in its entirety by the author was 456 pages—has more than doubled in size. The basic format and content of the clinical manifestations, diagnosis and differential diagnosis has persisted with little change as originally written by the author. The management and treatment of various diseases have undergone profound changes over time and these aspects of the book have been brought up-to-date by the subspecialists in the various disease entities. The increase in the size of the book is reflective of the advances that have occurred in both hematology and oncology over the past 25 years. Despite the size of the book, the philosophy has remained unchanged over the past quarter century. The author and his contributors have retained this book as a concise manual of personal experiences on the subject over these decades rather than developing a comprehensive tome culled from the literature. Its central theme remains clinical as an immediate reference for the practicing pediatric hematologist-oncologist concerned with the diagnosis and management of hematologic and oncologic diseases. It is extremely useful for students, residents, fellows and pediatric hematologists and oncologists as a basic reference assembling in one place, essential knowledge required for clinical practice.

This edition has retained the essential format written and developed decades ago by the author and, with usage over the years, has proven to be highly effective as a concise, practical, up-to-date guide replete with detailed tables, algorithms and flow diagrams for investigation and management of hematologic and oncologic conditions. The tables and flow diagrams have been updated with the latest information and the most recent protocols of treatment, that have received general acceptance and have produced the best results, have been included in the book.

Since the previous edition, some 5 years ago, there have been considerable advances particularly in the management of oncologic disease in children and these sections of the book have been completely rewritten. In addition, advances in certain areas have required that other sections of the book be updated. There has been extensive revision of certain chapters such as on Diseases of the White Cells, Lymphoproliferative Disorders, Myeloproliferative Disorders and Myelodysplastic Syndromes and Bone Marrow Failure. Because of the extensive advances in thrombosis we have rewritten that entire section contained in the chapter on Disorders of Coagulation to encompass recent advances in that area. The book, like its previous editions, reflects the practical experience of the author and his colleagues based on half a century of clinical experience. The number of contributors has been expanded but consists essentially of the faculty of the Division of Hematology Oncology at the Schneider Children's Hospital, all working together to provide the readers of the manual with a practical guide to the management of the wide spectrum of diseases within the discipline of pediatric hematology-oncology.

I would like to thank Laurie Locastro for her editorial assistance, cover design, and for her untiring efforts in the coordination of the various phases of the production of this edition. I also appreciate the efforts of Lawrence Tavnier for his expert typing of parts of the manuscript and would like to thank Elizabeth Dowling and Patrician Mastrolembo for proof reading of the book to ensure its accuracy.

Philip Lanzkowsky
MBChB, MD, ScD (Honoris Causa), FRCP, DCH, FAAP

Preface to the Third Edition

This edition of the *Manual of Pediatric Hematology and Oncology*, published 5 years after the second edition, has been written with the original philosophy in mind. It presents the synthesis of experience of four decades of clinical practice in pediatric hematology and oncology and is designed to be of paramount use to the practicing hematologist and oncologist. The book, like its previous editions, contains the most recent information from the literature coupled with the practical experience of the author and his colleagues to provide a guide to the practicing clinician in the investigation and up-to-date treatment of hematologic and oncologic diseases in childhood.

The past 5 years have seen considerable advances in the management of oncologic diseases in children. Most of the advances have been designed to reduce the immediate and long-term toxicity of therapy without influencing the excellent results that have been achieved in the past. This has been accomplished by reducing dosages, varying the schedules of chemotherapy, and reducing the field and volume of radiation.

The book is designed to be a concise, practical, up-to-date guide and is replete with detailed tables, algorithms, and flow diagrams for investigation and management of hematologic and oncologic conditions. The tables and flow diagrams have been updated with the latest information, and the most recent protocols that have received general acceptance and have produced the best results have been included in the book.

Certain parts of the book have been totally rewritten because our understanding of the pathogenesis of various diseases has been altered in the light of modern biological investigations. Once again, we have included only those basic science advances that have been universally accepted and impinge on clinical practice.

I thank Ms Christine Grabowski, Ms Lisa Phelps, Ms Ellen Healy and Ms Patricia Mastrolembo for their untiring efforts in the coordination of the writing and various phases of the development of this edition. Additionally, I acknowledge our fellows, Drs Banu Aygun, Samuel Bangug, Mahmut Celiker, Naghma Husain, Youssef Khabbase, Stacey Rifkin-Zenenberg, and Rosa Ana Gonzalez, for their assistance in culling the literature.

I also thank Dr Bhoomi Mehrotra for reviewing the chapter on bone marrow transplantation, Dr Lorry Rubin for reviewing the sections of the book dealing with infection, and Dr Leonard Kahn for reviewing the pathology.

Philip Lanzkowsky
MBChB, MD, ScD (Honoris Causa), FRCP, DCH, FAAP

Preface to the Second Edition

This edition of the *Manual of Pediatric Hematology and Oncology*, published 5 years after the first edition, has been written with a similar philosophy in mind. The basic objective of the book is to present useful clinical information from the recent literature in pediatric hematology and oncology and to temper it with experience derived from an active clinical practice.

The manual is designed to be a concise, practical, up-to-date book for practitioners responsible for the care of children with hematologic and oncologic diseases by presenting them with detailed tables and flow diagrams for investigation and clinical management.

Since the publication of the first edition, major advances have occurred, particularly in the management of oncologic diseases in children, including major advances in recombinant human growth factors and bone marrow transplantation. We have included only those basic science advances that have been universally accepted and impinge on clinical practice.

I would like to thank Dr Raj Pahwa for his contributions on bone marrow transplantation, Drs Alan Diamond and Leora Lanzkowsky-Diamond for their assistance with the neuro-radiology section, and Christine Grabowski and Lisa Phelps for their expert typing of the manuscript and for their untiring assistance in the various phases of the development of this book.

Philip Lanzkowsky
MBChB, MD, ScD (Honoris Causa), FRCP, DCH, FAAP

Preface to the First Edition

The *Manual of Pediatric Hematology and Oncology* represents the synthesis of personal experience of three decades of active clinical and research endeavors in pediatric hematology and oncology. The basic orientation and intent of the book is clinical, and the book reflects a uniform systematic approach to the diagnosis and management of hematologic and oncologic diseases in children. The book is designed to cover the entire spectrum of these diseases, and although emphasis is placed on relatively common disorders, rare disorders are included for the sake of completion. Recent developments in hematology-oncology based on pertinent advances in molecular genetics, cytogenetics, immunology, transplantation, and biochemistry are included if the issues have proven of value and applicability to clinical practice.

Our aim in writing this manual was to cull pertinent and useful clinical information from the recent literature in pediatric hematology and oncology and to temper it with experience derived from active clinical practice. The result, we hope, is a concise, practical, readable, up-to-date book for practitioners responsible for the care of children with hematologic and oncologic diseases. It is specifically designed for the medical student and practitioner seeking more detailed information on the subject, the pediatric house officer responsible for the care of patients with these disorders, the fellow in pediatric hematology-oncology seeking a systemic approach to these diseases and a guide in preparation for the board examinations, and the practicing pediatric hematologist-oncologist seeking another opinion and approach to these disorders. As with all brief texts, some dogmatism and "matters of opinion" have been unavoidable in the interests of clarity. The opinions expressed on management are prudent clinical opinions; and although they may not be accepted by all, pediatric hematologists-oncologists will certainly find a consensus. The reader is presented with a consistency of approach and philosophy describing the management of various diseases rather than with different managements derived from various approaches described in the literature. Where there are divergent or currently unresolved views on the investigation or management of a particular disease, we have attempted to state our own opinion and practice so as to provide some guidance rather than to leave the reader perplexed.

The manual is not designed as a tome containing the minutiae of basic physiology, biochemistry, genetics, molecular biology, cellular kinetics, and other esoteric and abstruse detail. These subjects are covered extensively in larger works. Only those basic science advances that impinge on clinical practice have been included here. Each chapter stresses the pathogenesis, pathology, diagnosis, differential diagnosis, investigations, and detailed therapy of hematologic and oncologic diseases seen in children.

I would like to thank Ms Joan Dowdell and Ms Helen Witkowski for their expert typing and for their untiring assistance in the various phases of the development of this book.

Philip Lanzkowsky
MBChB, MD, ScD (Honoris Causa), FRCP, DCH, FAAP

Introduction: Historic Perspective

Philip Lanzkowsky

REFLECTION ON 60 YEARS OF PROGRESS IN PEDIATRIC HEMATOLOGY-ONCOLOGY

As the sixth edition of the *Manual of Pediatric Hematology-Oncology* is published, I have reflected on the advances that have occurred since I began practicing hematology-oncology almost 60 years ago and since my first book on the subject was published by McGraw Hill in 1980. The present edition is more than double the size of the original book.

Our understanding of hematologic conditions has advanced considerably with the explosion of molecular biology and the management of most hematologic conditions has kept pace with these scientific advances. Our understanding of the basic science of oncology, molecular biology, genetics, and the management of oncologic conditions has undergone a seismic change. The previous age of dismal and almost consistent fatal outcomes for most childhood cancers has been replaced by an era in which most childhood cancers are cured. This has been made possible not only because of advances in oncology but because of the parallel development of radiology, radiologic oncology, and surgery as well as supportive care such as the pre-emptive use of antibiotics and blood component therapy. It has been a privilege to be a witness and participant in this great evolution over the past 60 years. Yet we still have a long way to go as current advances are superseded by therapy based upon the application of knowledge garnered from an accurate understanding of the fundamental biology of cancer.

In the early days of hematology-oncology practice, hematology dominated and occupied most of the practitioner's time because most patients with cancer had a short lifespan and limited therapeutic modalities were available.

Automated electronic blood-counting equipment has enabled valuable red cell parameters such as mean corpuscular volume (MCV) and red cell distribution width (RDW) to be applied in routine clinical practice. This advance permitted the reclassification of anemias based on MCV and RDW. Previously these parameters were determined by microscopy with considerable observer variability. The attempt at a more accurate determination of any one of these parameters was a laborious, time-consuming enterprise relegated only as a demonstration in physiology laboratories.

Rh hemolytic disease of the newborn and its management by exchange transfusion, which occupied a major place in the hematologists' domain, has now become almost extinct in developed countries due to the use of Rh immunoglobulin.

The description of the various genetic differences in patients with vitamin B_{12} deficiency has opened up new vistas of our understanding of cobalamin transport and metabolism. Similar advances have occurred with reference to folate transport and metabolism.

Gaucher and other similar diseases have been converted from crippling and often disabling disorders to ones where patients can live a normal and productive life thanks to the advent of enzyme replacement therapy.

Aplastic anemia has been transformed from a near death sentence to a disease with hope and cure in 90% of patients thanks to immunosuppressive therapies, hematopoietic stem cell transplantation, and advanced supportive care. The emergence of clonal disease years later in patients treated medically with immunosuppressive therapy, however, does remain a challenge. The discovery of the various genes responsible for Fanconi anemia and other inherited bone marrow failure syndromes has revealed heretofore unimaginable advances in our understanding of DNA repair, telomere maintenance, ribosome biology and other new fields of biology. The relationship of these syndromes to the development of various cancers may hold the key to our better understanding of the etiology of cancer as well as birth defects.

The hemolytic anemias, previously lumped together as a group of congenital hemolytic anemias, can now be identified as separate and distinct enzyme defects of the Embden—Meyerhof and hexose monophosphate pathways in intracellular red cell metabolism as well as various well-defined defects of red cell skeletal proteins due to advances in molecular biology and genetics. With improvement in electrophoretic and other biochemical as well as molecular techniques, hemoglobinopathies are being identified which were not previously possible.

Diseases requiring a chronic transfusion program to maintain a hemoglobin level for hemodynamic stability, such as in thalassemia major, frequently had marked facial characteristics with broad cheekbones and developed what was called "bronze diabetes," a bronzing of the skin along with organ damage and failure, particularly of the heart, liver, beta cells of the pancreas, and other tissues due to secondary hemochromatosis because of excessive iron deposition. The clinical findings attributed to extramedullary hematopoiesis are essentially of historic interest because of the development and widespread use of proper transfusion and chelation regimens. However, the full potential of the role of intravenous and oral chelating agents is yet to be realized due to the problems of compliance with difficult treatment regimens and also due to failure of some patients to respond adequately. Advances in our understanding of the biology of iron absorption and transport at the molecular level hold out promise for further improvement in the management of these conditions. Curative therapy in thalassemia major and other conditions by hematopoietic stem cell transplantation in suitable cases is widely available today. Gene therapy looms on the horizon but will not, for some time, be available to patients in the developing world requiring the development of other approaches.

In the treatment of idiopathic thrombocytopenic purpura, intravenous gamma globulin and anti-D immunoglobulin as well as thrombopoietin mimetic agents have been added to the armamentarium of management and are useful in specific indications in patients with this disorder.

Major advances in the management of hemophilia have included the introduction of commercially available products for replacement therapy which has saved these patients from a life threatened by hemorrhage into joints, muscles, and vital organs. Surgery has become possible in hemophilia without the fear of being unable to control massive hemorrhage during or after surgery. The devastating clinical history of tragic hemophilia outcomes has been relegated to the pages of medical history. Patients with inhibitors, however, still remain a clinical challenge. The whole subject of factors associated with inherited thrombophilia such as mutations of factor V, prothrombin G20210A and 5,10-methylenetetrahydrofolate reductase as well as the roles of antithrombin, protein C and S deficiency, and antiphospholipid antibodies in the development of thrombosis has opened new vistas of understanding of thrombotic disorders. Notwithstanding these advances, the management of these patients still presents a clinical challenge.

There are few diseases in which advances in therapy have been as dramatic as in the treatment of childhood leukemia. In my early days as a medical student, the only available treatment for leukemia was blood transfusion. Patients never benefitted from a remission and died within a few months. Steroids and single-agent chemotherapy, first with aminopterin, demonstrated the first remissions in leukemia and raised hope of a potential cure; however, relapse ensued in almost all cases and most patients died within the first year of diagnosis. In most large pediatric oncology centers there were few patients with leukemia as the disease was like a revolving door—diagnosis and death. The development of multiple-agent chemotherapy for induction, consolidation, and maintenance, CNS prophylaxis and supportive care ushered in a new era of cure for patients with leukemia. These principles were refined over time by more accurate classification of acute leukemia using morphological, cytochemical, immunological, cytogenetic, and molecular criteria which replaced the crude microscopic and highly subjective characteristics previously utilized for the classification of leukemia cells. These advances paved the way for the development of specific protocols of treatment for different types of leukemia. The management of leukemia was further refined by risk stratification, response-based therapy, and identification of minimal residual disease, all of which have led to additional chemotherapy or different chemotherapy protocols, resulting in an enormous improvement in the cure rate of acute leukemia. The results have been enhanced by modern supportive care including antibiotic, antifungal, antiviral therapy, and blood component therapy. Those patients whose leukemia is resistant to treatment or who have recurrences can be successfully treated by advances that have occurred with the development of hematopoietic stem cell transplantation. The challenge of finding appropriate, unrelated transplantation donors has been ameliorated by molecular HLA-typing techniques and the development of large, international donor registries. Emerging targeted and pharmacogenetic therapies hold great promise for the future. Already dramatic results are being reported exploiting surface antigens targeted by engineered cytotoxic T-cells.

Hodgkin disease, originally defined as a "fatal illness of the lymphatics," is a disease that is cured in most cases today. Initially, Hodgkin disease was treated with high-dose radiation to the sites of identifiable disease resulting in some cures but with major life-long radiation damage to normal tissues because of the use of cobalt machines and higher doses of radiation than is currently used. The introduction of nitrogen mustard early on, as a single-agent chemotherapy, improved the prognosis somewhat. A major breakthrough occurred with the staging of Hodgkin disease and the use of radiation therapy coupled with multiple-agent chemotherapy (MOPP). With time this therapeutic approach was considerably refined to include a reduction in radiation dosage and field and a modification of the chemotherapy regimens designed to reduce toxicity of high-dose radiation and of some of the chemotherapeutic agents. These major advances in treatment ushered in a new era in the management and cure of most patients with this disease. The management of Hodgkin disease, however, did go through a phase of staging laparotomy and splenectomy with a great deal of unnecessary surgery and splenectomies being performed. There were considerable surgical morbidity and post-splenectomy sepsis, occasionally fatal, which occurred in some cases. With the advent of MRI and PET scans, surgical staging, splenectomy, and lymphangiography have become unnecessary.

Non-Hodgkin lymphoma, previously considered a dismal disease, is another success story. Improvements in histologic, immunologic, and cytogenetic techniques have made the diagnosis and classification more accurate. The development of a staging system and multiagent chemotherapy was a major step forward in the management of this disease. This, together with enhanced supportive care, including the successful management of tumor lysis syndrome, have all contributed to the excellent results that occur today.

Brain tumors were treated by surgery and radiation therapy with devastating results due to primitive neurosurgical techniques and radiation damage. The advent of MRI scans has made the diagnosis and the determination of the extent of disease more accurate. Major technical advances in neurosurgery such as image guidance, which allows 3D mapping of tumors, functional mapping, and electrocorticography, which allow pre- and intraoperative differentiation of normal and tumor tissue, the use of ultrasonic aspirators and neuroendoscopy, have all improved the results of neurosurgical intervention and has resulted in less surgical damage to normal brain tissue. These neurosurgical advances, coupled with the use of various chemotherapy regimens, have resulted in considerable improvements in outcome for some. This field, however, still remains an area begging for a better understanding of the optimum management of these devastating and often fatal tumors. Improved radiation techniques, including proton beam therapy, have led to more precise radiation fields, sparing normal brain.

In the early days of pediatric oncology Wilms tumor in its early stages was cured with surgery followed by radiation therapy. The diagnosis was made with an intravenous pyelogram and inferior venocavogram and chest radiography was employed to detect pulmonary metastases. The diagnosis and extent of disease were better defined when CT of the abdomen and chest became available. The development of the clinicopathological staging system and the more accurate definition of the histology into favorable and unfavorable histologic types, allowed for more focused treatment with radiation and multiple chemotherapy agents, for different stages and histology of Wilms tumor, resulting in the excellent outcomes observed today. The success of the National Wilms Tumor Study Group (NWTSG), more than any other effort, provided the model for cooperative group therapeutic cancer trials, which in large measure have been responsible for advances in the treatment of Wilms tumor.

The diagnosis of neuroblastoma and its differentiation histologically from other round blue cell tumors such as rhabdomyosarcoma, Ewing sarcoma, and non-Hodgkin lymphoma was difficult before neurone-specific enolase cytochemical staining, Shimada histopathology classification, N-myc gene status, VMA and HVA determinations, and MIBG scintigraphy were introduced. In the future, new molecular approaches will offer diagnostic tools to provide even greater precision for diagnosis. The existing markers coupled with a staging system have enabled neuroblastoma to be assigned to various risk group categories with specific multimodality treatment protocols for each risk group, which has improved the prognosis in this disease. Improvements in diagnostic radiology determining extent of disease and modern surgical techniques have enhanced the advances in chemotherapy in this condition. With these advances and the addition of targeted immunologic approaches and radiopharmaceutical-linked therapy to our armamentarium the progress for disseminated neuroblastoma appears to be improving.

Major advances have occurred in rhabdomyosarcoma treatment over the years. Early on treatment of this disease was characterized by mutilating surgery including amputation and a generally poor outcome. More accurate histologic diagnosis, careful staging, judicious surgery, combination chemotherapy, and radiotherapy have all contributed a great deal to the improved cure rates with significantly less disability.

Malignant bone tumors had a terrible prognosis. They were generally treated by amputation of the limb with the primary tumor; however, this was usually followed by pulmonary metastases and death. The major advance in the treatment of this disease came with the use of high-dose methotrexate and leukovorin rescue which, coupled with limb salvage treatment, has resulted in improved survival and quality-of-life outcomes.

Of note however, improvement in outcomes for pediatric sarcomas have not kept pace with those for leukemia, lymphoma, and other tumors. The advances in the treatment of hepatoblastoma were made possible by safer anesthesia, more radical surgery, intensive postoperative management together with multiagent chemotherapy and more recently the increased use of liver transplantation. These advances have allowed many patients to be cured compared to past years.

Histiocytosis is a disease that has undergone many name changes from Letter–Siwe disease, Hand–Schüller–Christian disease and eosinophilic granuloma to the realization that these entities are one disease, renamed histiocytosis X (to include all three entities) to its present name of Langerhans cell histiocytosis (LCH) due to the realization that these entities have one pathognomonic pathologic feature that is the immunohistochemical presence of Langerhans cells defined in part by expression of CD1a or langerin (CD207), which induces the formation of Birbeck granules. Advances have occurred in the management of this disease by an appreciation of risk stratification depending on number and type of organs involved in this disease process as well as by early response to therapy. Once this was established, systemic therapy was developed for the various risk groups which led to appropriate and improved primary and salvage therapy and the introduction of new agents with better overall results. Recent advances describing stereotypic mutations in LCH offer hope for new targeted approaches.

Until a final prevention or cure for cancer in children is at hand, hematopoietic stem cell transplantation must be viewed as a major advance. Improved methods for tissue typing, the use of umbilical and peripheral blood stem cells, improved preparative regimens, including intensity-reduced approaches and better management of graft-versus-host disease (GVHD) has made this an almost routine treatment modality for many metabolic disorders, hemoglobinopathies, and malignant diseases following ablative chemotherapy in chemotherapy-sensitive tumors. Post-transplantation support with antibiotic, antifungal, antiviral, hematopoietic growth factors and judicious use of blood component therapy has made this procedure safer than it was in years gone by.

The recognition of severe and often permanent damage to organs and life-threatening complications from chemotherapy and radiation therapy has, over the years, led to regimens consisting of combination chemotherapy at reduced doses and reduction in dose and field of radiation with improved outcome. An entire new scientific discipline, survivorship, has arisen because of the near 80% overall cure rate for childhood cancer. Focusing on the improvement of the quality of life of survivors coupled with research in this new discipline gives hope that many of the remaining long-term effects of cancer chemotherapy in children will be mitigated and possibly eliminated.

Major advances have occurred in the management of chemotherapy-induced vomiting and pain management because of the greater recognition and attention to these issues and the discovery of many new, effective drugs to deal with these symptoms. The availability of symptom control and palliative care has provided a degree of comfort for children undergoing chemotherapy, radiation, and surgery that did not exist only a few years ago.

Hematologist-oncologists today are privileged to practice their specialty in an era in which most oncologic and many hematologic diseases in children are curable and at a time when national and international cooperative groups are making major advances in the management of these diseases and when basic research is at the threshold of making major breakthroughs. The present practice is grounded in evidence-based research that has been and is still being performed by hematologist-oncologists and researchers that form the foundation for ongoing advances. Today we stand on the shoulders of others, which permits us to see future advances unfold to benefit generations of children. While we bask in the glory of past achievements, we should always be cognizant that much work remains to be done until the permanent cure of all childhood malignancies and blood diseases is at hand.

This book encompasses the advances in the management of childhood cancer which have been accomplished to date and which have become the standard of care.

1

Diagnostic Molecular and Genomic Methodologies for the Hematologist/Oncologist

Vijay G. Sankaran

Over the past several years, molecular diagnostic testing in patients with hematologic and oncologic disorders has become increasingly sophisticated and prevalent. While in the past focused genetic tests were performed, in recent years the widespread use of genomic and molecular approaches in both research and clinical settings has shown potential to refine our understanding of pediatric blood disorders and cancer. This chapter provides an overview of the currently used molecular and genomic methods. In that way, the format for this chapter differs from those that describe specific disease entities. Thus the chapter can be read in its entirety as essential background for the modern practice of pediatric hematology/oncology. We will primarily focus on those genetic methods that are currently in use in clinical settings. Undoubtedly, in the coming years, the use of certain methods will evolve and new methods will become available. With this in mind, there are two goals in this chapter. We first aim to provide an overview of the types of currently used clinical genetic testing methods and specifically attempt to examine their utility in detecting specific changes at the molecular level that underlie both congenital and acquired conditions that are commonly seen by pediatric hematologists and oncologists. This overview will be important for clinicians to better understand how newly developed methods could supplant the currently used approaches in the coming years. The second goal of this chapter is to provide a basic understanding of the limitations that exist for the most common molecular and genomic methods in use, so that clinicians who receive these results can be sufficiently versed in these methods and avoid misinterpreting the results obtained from these tests.

CLINICAL MOLECULAR AND GENOMIC METHODOLOGIES

Despite the large range of approaches that have been developed, all of the methods remain focused on the goal of identifying patients' molecular lesions that underlie their disease. The methods can be broadly classified into two categories:

1. Direct testing: These approaches look for the presence of genetic mutations that directly contribute to disease.
2. Indirect testing: This category includes approaches that compare genomic or molecular markers in multiple affected individuals to unaffected individuals. These approaches often identify markers that may segregate with a disease, but the markers themselves may not cause the disease itself.

There may be overlap between these categories and certain methods may identify causal genetic mutations in some instances (direct testing), while only identifying segregating markers (indirect testing) in other cases.

Table 1.1 lists the commonly used genetic testing methodologies, along with the types of molecular lesions that they are able to identify.

Linkage Analysis

While most methods are now focused on identifying the precise molecular cause of disease, indirect tests can be quite useful, particularly for mapping causes of a disease in a family. For example, testing for markers such as

Lanzkowsky's Manual of Pediatric Hematology and Oncology.
DOI: http://dx.doi.org/10.1016/B978-0-12-801368-7.00001-6

TABLE 1.1 Overview of Molecular and Genomic Diagnostic Methodologies

Method	Common point mutations	Rare point mutations	Copy number variants	Uniparental disomy	Balanced inversions or translocations	Repeat expansions	Examples of use in pediatric hematology/oncology
Linkage analysis (using markers such as short tandem repeats)	X		X				Family pedigree with history of hereditary spherocytosis and interest in identifying causal gene
Fluorescent *in situ* hybridization			X		X		Acquired monosomy in myelodysplastic syndrome
Array comparative genomic hybridization			X	X			Testing for microdeletion in patient with hematologic and syndromic phenotype
Genome-wide single nucleotide polymorphism microarrays	X		X				Testing for small copy number variants in pediatric leukemia
Targeted polymerase chain reaction analysis	X	X				X	Testing for JAK2 V617F mutation in patient with a myeloproliferative disorder
Sanger sequencing	X	X					Molecular diagnosis of a patient with pyruvate kinase deficiency
Multiplex ligation-dependent probe amplification			X			X	Deletions in α- or δβ- thalassemia cases
Gene panel sequencing	X	X					Severe congenital neutropenia
Whole-genome or -exome sequencing	X	X	X				Unknown bone marrow failure syndrome

single nucleotide polymorphisms (SNPs) or short tandem repeats (STRs, which are 2–5 base long repetitive elements with varying numbers of repeats) that are found throughout the genome can be extremely useful as a way to identify likely causal genes, particularly in diseases where multiple possible causal genes have been implicated. For example, in hereditary spherocytosis a number of genes including *ANK1, SPTB, SPTA, SLC4A1*, and *EBP42* are implicated in the disease. Many of these genes are quite large and while sequencing a panel of genes is certainly possible, in a large family, the use of either SNP-based methods or other markers such as STRs can identify the likely causal locus to help focus targeted sequencing efforts. These methods are commonly used for diagnostic mapping in resource-poor settings where whole-genome sequencing (WGS) methods may not be available and these methods can also be extremely useful in other settings. For example, if a family is being followed with a known disease, but no coding mutations are identified on targeted sequencing, these approaches can help validate that there is linkage to a specific gene and they may assist in the efforts to identify mutations in regulatory regions of the implicated gene. Specifically, segregation of markers that are in linkage with the causal mutation should only be found in affected family members and would suggest that the causal mutation is located nearby. These methods are also commonly used as an initial screen in families where possible cancer predisposition syndromes may exist and can help focus in-depth analysis on certain regions of the genome. Even in cases where whole-exome or -genome sequencing is performed, linkage can provide an excellent indirect approach to focus on marker genes that segregate appropriately in individuals who have a particular disease.

Fluorescent *In Situ* Hybridization

Fluorescent *in situ* hybridization (FISH) was developed in the 1980s and uses fluorescently labeled DNA probes to query whether entire chromosomes or parts of a chromosome may be duplicated or deleted in cells. The fluorescently labeled DNA probes are complementary to the region of interest on a chromosome and therefore specifically hybridize only to this region and not to others. FISH is commonly used to assess for gain or loss of chromosomes or large parts of chromosomes in patients with hematologic malignancies. Typically, a number of cells from the bone marrow are tested for the presence of such chromosomal aberrations, which can have important roles both in terms of disease diagnosis and prognosis. FISH has a benefit in that it is a cytogenetic method and therefore individual cells are assessed rather than a population of cells in aggregate, which is the case for other methods that examine copy number variation, including array comparative genomic hybridization (CGH) or genome-wide SNP microarrays. In addition, FISH remains the best clinically available method to detect classic cytogenetic changes that are diagnostic and implicated in a number of pediatric cancers, such as translocations that are frequently seen in leukemia and certain solid tumors.

Array CGH

FISH lacks the sensitivity to detect smaller chromosomal deletions or duplications, which often represent important DNA copy number variations found in disease. Array CGH takes advantage of microarrays that have oligonucleotide probes at varying densities to detect differences in DNA copy number by comparing a sample genome with a reference sample (or group of reference samples) and examining whether there is an increase or decrease in signals at a particular genomic region in comparison with the control, which would be indicative of duplications or deletions, respectively. This method has significant sensitivity to detect DNA copy number changes, particularly smaller changes, in a variety of different samples. This can be applied to congenital disorders, where copy number changes can cause disease when present in the germline. In addition, in acquired hematologic or other malignancies, there can be acquired copy number changes. This increased sensitivity to detect smaller copy number changes has led to an increased detection of copy number changes in genomes of unclear significance. A number of resources are cataloging such changes in humans, although nonuniform deposition of deletion information into such databases makes the ongoing interpretation of either germline or acquired somatic copy number changes difficult in some cases. It is likely that as these databases grow with more phenotype information available, there will be increased insight into whether a deletion or duplication may be pathogenic.

Genome-Wide or Focused SNP Arrays

Microarrays provide the opportunity to genotype SNPs in a large-scale and potentially genome-wide manner. These approaches can be used for several applications. Similar to array CGH, these methods can be used to detect copy number variation that is either found in the germline or that is acquired. The resolution of the deletions detected using such approaches can be as good or in some cases, depending upon SNP or probe density, better than the resolution achieved with array CGH methods. In many cases, SNP arrays are used in place of array CGH in many clinical labs to detect copy number changes routinely. Indeed, the use of these SNP arrays is particularly widespread in the diagnostic evaluation of hematologic malignancies. Another application of such arrays is to genotype common SNPs in the genome, either for linkage mapping, as discussed above, or for identification of a common variation that confers risk of having certain diseases. While this has proven useful in some diseases, it is important to bear in mind that most such associations are probabilistic and not deterministic of acquiring disease. The clinical utility of this application is not clear, although a number of direct-to-consumer services offer such genotyping and will report relative risk information to individuals who request such services.

As more disease-associated mutations are being identified, there have been efforts to develop focused SNP arrays for specific phenotypes or diseases. In the future, such approaches may have clinical utility. However this application may be surpassed by large-scale genome sequencing as it becomes affordable (discussed below). One limitation of this approach is that there is continuous discovery of new causal alleles and genes in many diseases, which limits the clinical utility of such arrays.

Multiplex Ligation-Dependent Probe Amplification

Multiplex ligation-dependent probe amplification (MLPA) is a molecular approach that involves annealing of two adjacent oligonucleotides to a segment of genomic DNA followed by quantitative polymerase chain reaction (PCR) amplification to characterize copy number or other changes in the DNA. A series of MLPA probes can together screen and map deletions that occur in a particular region. In contrast to array CGH or SNP arrays, this approach is best applied to detect DNA copy number alterations in specific focused regions and this approach can allow such alterations to be finely mapped. For example, MLPA is commonly used to map deletions that commonly occur in the α-globin gene locus in cases of α-thalassemia. Traditionally, this mapping was done using Southern blotting (to determine a specific DNA sequence), but this is now rarely done for clinical purposes and in most instances MLPA is used for such applications in clinical labs. When results from MLPA are reported, it is important to bear in mind that the resolution will depend upon the number of probes used and, in general, precise deletion coordinates will not be defined using MLPA alone. Often to better map deletion sites, PCR-based Sanger sequencing approaches are used to map breakpoints once general coordinates have been defined using MLPA (discussed below).

Targeted PCR Analysis

Often a single mutation confers significant disease risk or occurs in the majority of cases of a disease. In these cases, PCR approaches can be used for amplification and separation of different alleles. A number of approaches to separate different alleles using PCR have been developed and since these methods are largely specific to individual platforms, the details of specific methods will not be covered here. These approaches are commonly used for detection of mutations, such as factor V Leiden that confers an increased risk of venous thrombosis and is found in several percent of the general population. In certain hematologic malignancies, such as myeloproliferative diseases, there are common mutations such as JAK2 V617F that occurs in many cases and focused genotyping of this variant is often performed as a clinical test. These tests can be done at low cost and relatively rapidly because of their focused nature and output of the presence/absence of a single mutation. However, using this approach it is impossible to detect relevant variants that have not been genotyped in a gene of interest.

Focused Sanger and Gene Panel Sequencing

Traditional DNA sequencing has relied upon the chain terminator method developed by Frederick Sanger, where a chain-terminating dideoxynucleotide is coupled to a fluorescent dye and this can allow sequencing in a single reaction by detecting terminated DNA fragments of various sizes. A chromatogram obtained after capillary-based separation displays the sequentially elongated fragments that each end in a specific fluorescent terminating dideoxynucleotide, allowing identification of the sequence of DNA. This approach is often applied to a series of reactions using PCR that cover a gene or in some instances a panel of genes implicated in disease. A limitation is that a single reaction can only assess a single sequence of several hundred bases and therefore multiple reactions will need to be run for most genes. It should be kept in mind that while Sanger sequencing can be very sensitive to detect point mutations, copy number or structural changes in genes will often be missed using this approach. Therefore, targeted sequencing of a disease gene can often be complemented using array CGH or SNP array-based approaches to look for deletions that may be implicated in a subset of cases of a particular disease.

Whole-Genome or -Exome Sequencing

While Sanger sequencing methods were once the primary method to map DNA sequences, the development of high-throughput next-generation sequencing (NGS) platforms over the past several years has allowed rapid and low-cost sequencing of large portions of the genome. NGS takes advantage of various technologies to sequence millions to billions of DNA strands in parallel to yield substantially more throughput than Sanger sequencing approaches. Moreover, NGS approaches bypass fragment-cloning steps that were necessary for genome sequencing using traditional Sanger sequencing approaches. In NGS, DNA is broken into short fragments, a subset of fragments may be enriched (such as with exome sequencing where sequences that encode regions overlapping with exons are enriched), and the sequences of all the fragments are then read on NGS platforms. The details of these approaches vary depending upon the technology or platform used, but in general all NGS methods use similar principles.

The use of NGS platforms to sequence the entire genome or exome of a patient with a particular disorder for clinical purposes has only begun to emerge and these technologies are currently being primarily used in the research setting. Studies are beginning to address the diagnostic yield of these approaches. One important consideration in using these approaches is that since the entire genome (or a substantial portion) is sequenced, potentially pathogenic variants may be identified that may not contribute to the disease that initially motivated the

WGS. At the current time, there are varying opinions on the types of circumstances in which such incidental findings can be reported to patients and studies are exploring the impact of delivering such information to patients. There has been substantial debate in the community regarding the delivery and broad impact of the information derived from NGS on patients, physicians, and society. Undoubtedly, this is an area that will evolve considerably in the coming years.

Currently WGS or whole-exome sequencing (WES) are ordered in the clinical setting for detection of rare variants in patients with a phenotype that is suspected to be due to a single-gene disorder, after known single-gene candidates have either been eliminated or when a multigene testing approach is prohibitively expensive. Before such tests are ordered, it is important that clinicians gather a thorough family history, fully evaluate a patient's phenotype, and obtain appropriate informed consent. There is little doubt that as WGS and WES are routinely employed in clinical settings, specific guidelines for when it is best to use these tests for patients with particular groups of diseases will emerge.

It is important for clinicians to be aware of the significant limitations of WGS and WES. NGS approaches can currently only sequence nonrepetitive regions of the genome and some diseases occur in repetitive DNA (i.e., fragile X syndrome) and thus diagnoses will be missed using such approaches. In addition, NGS approaches cannot currently detect most copy number variants, insertion–deletion variants, or chromosomal translocations. It is likely that with improvements in technology to allow for longer reads in NGS and with better computational methods, these types of variants could be detected using these approaches in the future.

Since numerous potential causal variants may be identified using such broad-based sequencing approaches, the American College of Medical Geneticists has developed recommendations regarding categories to which variants should be assigned and these are helpful for clinicians to be aware of when interpreting results from WES or WGS (although they apply to other sequencing approaches as well).

- Disease-causing: A sequence variant that has previously been reported as a cause of a disorder
- Likely disease-causing: A sequence variant that has not been previously reported, but is of a type that would be expected to cause disease
- Possibly disease-causing: A sequence variant that has not been previously reported and is of the type that may or may not cause a particular disorder
- Likely not disease-causing: Sequence variation that has not been reported and is not likely of the type that would cause a particular disorder
- Not disease-causing: Sequence variation has previously been reported and is a recognized neutral variation
- Variant of unknown clinical significance: Sequence variation is not known or expected to be causative of disease, but is found to be associated with a clinical presentation.

While these categories are useful when results are reported, they are largely dependent upon databases of prior variants. As more sequencing data are being reported, it is important to bear in mind that some variants previously thought to be pathogenic are being reclassified as benign. It is very likely that the categorization for a particular variant may evolve over time and therefore it is useful to evaluate the prior reports of any genetic variants identified in such studies on an individual basis. In some cases, it is important to bear in mind that in many single-gene disorders, variable penetrance or expressivity (where patients with a particular mutation may or may not have the disease or may have varying severities of the disease) may have a significant and underestimated impact.

INTERPRETING AND EVALUATING THE RESULTS FROM CLINICAL GENETIC TESTING

A key role of any pediatric hematologist/oncologist will be interpreting and evaluating the results obtained from clinical diagnostic genetic testing. As newer methods are used, different limitations of each approach need to be addressed. While for each diagnostic approach we have addressed some specific limitations, we also want to present a general framework for evaluating the validity of such testing in general. Table 1.2 presents some considerations that are applicable for any test and provides examples of complications that may need to be kept in mind.

For each particular test, specific test sensitivities and specificities will exist. In addition, for a particular test and diagnosis there will also be certain clinical sensitivities and specificities that need to be accounted for.

TABLE 1.2 Overview of Terms Used to Evaluate the Validity of Genetic Tests

Term	Definition	Calculation	Examples of complications
Test sensitivity	Proportion of assays with the genetic change that have a positive test result	True positives/ (true positives + false negatives)	Impaired amplification of mutant allele in mosaic setting due to selection
Test specificity	Proportion of assays without the genetic change that have a negative test result	True negatives/ (true negatives + false positives)	Technical impairment in MLPA leading to false appearance of deletion will lower test specificity
Clinical sensitivity	Proportion of people with a disease who have a positive test result	True positives/ (true positives + false negatives)	Multiple genes involved in a disease may lead to some patients not having a particular mutation or set of mutations
Clinical specificity	Proportion of people without a disease who have a negative test result	True negatives/ (true negatives + false positives)	Variable penetrance of disease (i.e., carriers of Diamond–Blackfan anemia mutations without symptoms)
Positive predictive value	The likelihood that a patient has a disease, given a positive test result	True positives/ (true positives + false positives)	Variable penetrance or expressivity can complicate this
Negative predictive value	The likelihood that a patient does not have a disease, given a negative test result	True negatives/ (true negatives + false negatives)	Multiple genes involved in a disease may complicate this

Finally, the positive and negative predictive values for any test should also be considered. Depending upon the genetic lesion under consideration, this may alter what tests are performed. For example, while large chromosomal alterations or translocations are readily detected using FISH, smaller deletions may only be detected using array CGH or SNP arrays. Therefore, depending upon the type of lesion expected, the various tests may be used in different ways or combinations. Another example is that while WES or WGS can be useful for looking for many causes of a disease, they may miss deletions that could result in a subset of cases of a particular disease. Therefore, it may be useful to both run WES and a SNP array on a particular patient if the disease can be caused by deletions in some cases.

One limitation currently faced in interpreting the results of WGS is that even when a mutation is identified in a regulatory region of a gene, its affect on the gene implicated in that disease may not be immediately apparent. If other affected and unaffected family members are available, linkage information can help demonstrate that a particular region is associated with the disease and support the presence of a nearby causal allele. For example, in a few cases of sideroblastic anemia, mutations were identified in regulatory regions of the *ALAS2* gene, which normally harbors coding mutations in most other cases of this disease. Appropriate identification of these mutations required a combination of linkage analysis and functional follow-up tests. This work was primarily done as part of a research study, but it is possible that in the future as more such noncoding mutations are identified, that these may be found to be important contributors to a variety of disorders seen in patients.

Further Reading and References

Bisecker, L., Green, R.C., 2014. Diagnostic clinical genome and exome sequencing. N. Engl. J. Med. 370, 2418–2425.

Feero, W.G., Green, E.D., 2011. Genomics education for health care professionals in the 21st century. JAMA 306, 989–990.

Johnsen, J.M., Nickerson, D.A., Reiner, A.P., 2013. Massively parallel sequencing: the new frontier of hematologic genomics. Blood 122, 3268–3275.

Katsanis, S.H., Katsanis, N., 2013. Molecular genetic testing and the future of clinical genomics. Nat. Rev. Genet. 14, 415–426.

Lindsley, R.C., Ebert, B.L., 2013. The biology and clinical impact of genetic lesions in myeloid malignancies. Blood 122, 3741–3748.

MacArthur, D.G., Manolio, T.A., et al., 2014. Guidelines for investigating causality of sequence variants in human disease. Nature 508, 469–476.

Mullighan, C.G., 2013. Genome sequencing of lymphoid malignancies. Blood 122, 3899–3907.

Richards, C.S., Bale, S., et al., on behalf of the Molecular Subcommittee of the ACMG Laboratory Quality Assurance Committee, 2008. ACMG recommendations for standards for interpretation and reporting of sequence variations: revisions 2007. Genet. Med. 10, 294–300.

Sankaran, V.G., Gallagher, P.G., 2013. Applications of high-throughput DNA sequencing to benign hematology. Blood 122, 3575–3582.

Hematologic Manifestations of Systemic Illness

Lawrence C. Wolfe

A variety of systemic illnesses including acute and chronic infections, neoplastic diseases, connective tissue disorders, and storage diseases are associated with hematologic manifestations. The hematologic manifestations are the result of the following mechanisms:

- Bone marrow dysfunction
  - Anemia or polycythemia
  - Thrombocytopenia or thrombocytosis
  - Leukopenia or leukocytosis
- Hemolysis
- Immune cytopenias
- Alterations in hemostasis
  - Acquired inhibitors to coagulation factors
  - Acquired von Willebrand disease
  - Acquired platelet dysfunction
- Alterations in leukocyte function

HEMATOLOGIC MANIFESTATIONS OF DISEASES RELATED TO VARIOUS ORGANS

Heart

Microangiopathic hemolysis occurs with prosthetic valves or synthetic patches utilized for correction of cardiac defects, particularly when there is failure of endothelialization ("Waring Blender" syndrome) or rarely after endoluminal closure of a patent ductus arteriosus. Microangiopathic hemolysis has the following characteristics:

- Hemolysis is secondary to fragmentation of the red cells as they are damaged against a distorted vascular surface.
- Hemolysis is intravascular and may be associated with hemoglobinemia and hemoglobinuria.
- Iron deficiency occurs secondary to the shedding of hemosiderin within renal tubular cells into the urine (hemosiderinuria).
- Thrombocytopenia secondary to platelet adhesion to abnormal surfaces.
- Autoimmune hemolytic anemia may occasionally occur after cardiac surgery with the placement of foreign material within the vascular system.

Cardiac Anomalies and Hyposplenism

Cardiac anomalies, particularly *situs inversus,* may be associated with hyposplenism and the blood smear may show Howell–Jolly bodies, Pappenheimer bodies, and elevated platelet counts.

Infective Endocarditis

Hematologic manifestations include anemia (due to immune hemolysis or chronic infection), leucopenia, or leukocytosis and rarely thrombocytopenia and pancytopenia.

Lanzkowsky's Manual of Pediatric Hematology and Oncology.
DOI: http://dx.doi.org/10.1016/B978-0-12-801368-7.00002-8

Coagulation Abnormalities

- A coagulopathy exists in some patients with cyanotic heart disease. The coagulation abnormalities correlate with the extent of the polycythemia. Hyperviscosity may lead to tissue hypoxemia, which could trigger disseminated intravascular coagulation (DIC).
- Marked derangements in coagulation such as DIC, thrombocytopenia, thrombosis, and fibrinolysis can accompany surgery involving cardiopulmonary bypass. Heparinization must be strictly monitored.

Platelet Abnormalities

Quantitative and qualitative platelet abnormalities are associated with cardiac disease:

- Thrombocytopenia occurs secondary to microangiopathic hemolysis associated with prosthetic valves.
- Cyanotic heart disease can produce polycythemia, thrombocytopenia, prolonged bleeding time, and abnormal platelet aggregation.
- Patients with chromosome 22q11.2 deletion (DiGeorge syndrome) can have platelet abnormalities including the Bernard−Soulier-like syndrome due to haploinsufficiency of the gene for GP1BB and thrombocytopenia due to autoimmunity.

Polycythemia

- The hypoxemia of cyanotic heart disease produces a compensatory elevation in erythropoietin and secondary polycythemia.
- Patients are at increased risk for cerebrovascular accidents secondary to hyperviscosity.
- Patients are also at risk for symptomatic hypoglycemia (especially in the neonatal period).
- The use of partial exchange transfusion has been suggested, although the long-term value of exchange has been challenged.

Gastrointestinal Tract

Esophagus

- Iron-deficiency anemia may occur as a manifestation of gastroesophageal reflux.
- Endoscopy may be required in unexplained iron deficiency.

Stomach

- The gastric mucosa is important in both vitamin B_{12} and iron absorption.
- Chronic atrophic gastritis produces iron deficiency. There may be an associated vitamin B_{12} malabsorption.
- Gastric resection may result in iron deficiency or in vitamin B_{12} deficiency due to lack of intrinsic factor.
- Zollinger−Ellison syndrome (increased parietal cell production of hydrochloric acid) may cause iron deficiency through mucosal ulceration.
- *Helicobacter pylori* infection, in addition to causing chronic gastritis, has been implicated in the initiation of iron-deficiency anemia, vitamin B_{12} deficiency, autoimmune thrombocytopenia, and platelet aggregation defects (ADP-like defect).

Small Bowel

- Celiac disease or tropical sprue may cause malabsorption of iron and folate. Table 2.1 lists the various hematologic manifestations of celiac disease.
- Inflammatory bowel disease (IBD) may cause anemia of chronic inflammation and iron deficiency from blood loss.
- Eosinophilic gastroenteritis can produce peripheral eosinophilia.
- Diarrheal illnesses of infancy can produce life-threatening methemoglobinemia.

Lower Gastrointestinal Tract

- Ulcerative colitis is often associated with iron-deficiency anemia.
- Peutz−Jeghers syndrome (intestinal polyposis and mucocutaneous pigmentation) predisposes to adenocarcinoma of the colon.
- Hereditary hemorrhagic telangiectasia (Osler−Weber−Rendu disease) may produce iron deficiency, platelet dysfunction, and hemostatic defects.

TABLE 2.1 Hematologic Manifestations of Celiac Disease

Problem	Frequency	Comments
Anemia: iron deficiency, folate deficiency, vitamin B_{12} deficiency, and other nutritional deficiencies	Common	The anemia is most commonly secondary to iron deficiency but may be multifactorial in etiology. Low serum levels of folate and vitamin B_{12} without anemia are frequently seen. Anemia due to other deficiencies appears to be rare
Thrombocytopenia	Rare	May be associated with other autoimmune phenomena
Thrombocytosis	Common	May be secondary to iron deficiency or hyposplenism
Thromboembolism	Uncommon	Etiology is unknown but may be related to elevated levels of homocysteine or other procoagulants
Leukopenia/neutropenia	Uncommon	Can be autoimmune or secondary to deficiencies of folate, vitamin B_{12}, or copper
Coagulopathy	Uncommon	Malabsorption of vitamin K
Hyposplenism	Common	Rarely associated with infections
IgA deficiency	Common	May be related to anaphylactic transfusion reactions
Lymphoma	Uncommon	The risk is highest for intestinal T-cell lymphomas

From: Halfdanarson et al. (2007), with permission.

Pancreas

- Hemorrhagic pancreatitis produces acute normocytic, normochromic anemia. It may also be associated with DIC.
- Shwachman—Diamond syndrome (SDS; see Chapter 13) is characterized by congenital exocrine pancreatic insufficiency, metaphyseal bone abnormalities, and neutropenia. There may also be some degree of anemia and thrombocytopenia.
- Cystic fibrosis produces malabsorption of fat-soluble vitamins (e.g., vitamin K) with impaired prothrombin production.
- Pearson syndrome is characterized by exocrine pancreatic insufficiency and severe sideroblastic anemia (see Chapter 13).

Liver

Anemia

Anemias of diverse etiologies occur in acute and chronic liver disease. Red cells are frequently macrocytic (mean corpuscular volume (MCV) of 100—110 fl). Target cells and acanthocytes (spur cells) are frequently seen. Some of the pathogenic mechanisms of anemia include:

- Shortened red cell survival and red cell fragmentation (spur cell anemia) in cirrhosis often occur in later-stage cirrhosis in the presence of dyslipidemia.
- Hypersplenism with splenic sequestration in the presence of secondary portal hypertension.
- Iron-deficiency anemia secondary to blood loss from esophageal varices in portal hypertension.
- Chronic hemolytic anemia in Wilson disease secondary to copper accumulation in red cells. Hemolytic anemia may be the presenting symptom in this disease.
- Aplastic anemia following acute hepatitis (usually seronegative) in certain immunologically predisposed hosts.
- Megaloblastic anemia secondary to folate deficiency in malnourished individuals.

Coagulation Abnormalities

The liver is involved in the synthesis of most of the coagulation factors. Liver dysfunction can be associated with either hyper- or hypocoagulable states because both procoagulant and natural anticoagulant synthesis are impaired. Table 2.2 lists the various coagulation abnormalities seen in liver disease and Table 2.3 lists the tests to differentiate between the coagulopathy of liver disease and other etiologies.

TABLE 2.2 Coagulation Abnormalities in Liver Disease

Hemorrhage	Thrombosis
(1) Thrombocytopenia/platelet dysfunction due to hypersplenism, altered TPO production	(1) Decreased anticoagulant—AT-III Protein C and S
(2) Decreased liver synthesis of procoagulant factors	(2) Portal hypertension-portal vein thrombosis
(3) Impaired carboxylation of vitamin K factors	
(4) Dysfibrinogenemia	
(5) Hyperfibrinolysis due to increased tPA and decreased PAI, α_2 antiplasmin	

TPO, thrombopoietin; tPA, tissue plasminogen activator; PAI, plasminogen activator inhibitor.

TABLE 2.3 Tests to Differentiate Coagulopathies of Different Etiologies

Procoagulant factors	Liver	Vitamin K	DIC
F V	Decreased (late)	Normal	Decreased
F VII	Decreased (early)	Decreased	Decreased
F VIII	Normal/increased	Normal	Decreased

Factor I (Fibrinogen)

Fibrinogen levels are generally normal in liver disease. Low levels may be seen in fulminant acute liver failure.

Factors II, VII, IX, and X (Vitamin K-Dependent Factors)

These factors are reduced in liver disease secondary to impaired synthesis. Factor VII is the most sensitive.

Factor V

Factor V does not require vitamin K for synthesis and is highly representative of actual liver function. Factor V levels at 36 h post liver injury have been used as a standalone marker for the possible need for transplant in patients with early liver failure.

Factor VIII

The procoagulant activity of Factor VIII is generally normal in liver disease. This makes Factor VIII an important factor to measure in distinguishing between DIC and severe liver disease in a patient with abnormal coagulation tests and thrombocytopenia. If there is associated DIC, factor VIII will be markedly depressed, whereas in severe liver disease Factor VIII remains close to or normal. Traditionally, the Factor VII and Factor VIII levels are measured along with the PT, PTT, and fibrinogen to distinguish liver disease from DIC.

Protein C, Protein S, and Antithrombin III

These natural anticoagulants are decreased in liver disease. Proteins C and S are most sensitive to vitamin K deficiency. In many cases this fall in the levels of natural anticoagulant creates a sensitive balance between loss of procoagulant activity and natural anticoagulant activity. Bleeding or thrombosis may appear quickly when additional illness (e.g., infection) may upset the balance.

Tissue plasminogen activator (TPA) and alpha-2-antiplasmin

Tissue plasminogen activator is cleared by the liver and as liver disease progresses TPA activity increases. Alpha-2-antiplasmin is also suppressed by liver disease, creating increased plasmin activity and ultimately the syndrome of hyperfibrinolysis with a tendency toward severe bleeding.

α_2-Macroglobulin and plasmin activator inhibitor

These opponents of plasma activity are still present in liver disease.

Kidneys

Renal disease may affect red cells, white cells, platelets, and coagulation.

Severe renal disease with renal insufficiency is frequently associated with chronic anemia (and sometimes pancytopenia). This type of anemia is characterized by:

- Hemoglobin as low as 4–5 g/dl
- Normochromic and normocytic red cell morphology unless there is associated microangiopathic hemolytic anemia (as in the hemolytic-uremic syndrome (HUS)), in which case schistocytes and thrombocytopenia are seen
- Reticulocyte count low
- Decreased erythroid precursors in bone marrow aspirate.

The following mechanisms are involved in the pathogenesis of this type of anemia:

- Erythropoietin deficiency is the most important factor (90% of erythropoietin synthesis occurs in the kidney)
- Shortened red cell survival is secondary to uremic toxins or in HUS secondary to microangiopathic hemolysis
- Renal failure itself inhibits erythropoiesis and in conjunction with decreased erythropoietin levels produces a hypoplastic marrow
- Increased blood loss from a hemorrhagic uremic state and into a hemodialysis circuit causes iron deficiency.
- Dialysis can lead to folic acid deficiency.

Treatment

- Recombinant human erythropoietin (rHuEPO):[1]
  - Determine the baseline serum erythropoietin and ferritin levels prior to starting rHuEPO therapy. If ferritin is less than 100 ng/ml, give ferrous sulfate 6 mg/kg/day aimed at maintaining a serum ferritin level above 100 ng/ml and a threshold transferrin saturation of 20%. With the advent of less immunoreactive forms of intravenous iron, prophylactic strategies utilizing intravenous iron infusion at the end of dialysis have been very successful. These are usually well tolerated compared to oral iron on a daily basis.
  - Start with rHuEPO treatment in a dose of 50–100 units/kg/day subcutaneously (SC) three times a week.
  - Monitor blood pressure closely (increased viscosity produces hypertension in 30% of cases) and perform complete blood count (CBC) weekly.
  - Titrate the dose:
    - If no response, increase rHuEPO up to 300 units/kg/day SC three times a week
    - If hematocrit (Hct) reaches 40%, stop rHuEPO until Hct is 36% and then restart at 75% dose
    - If Hct increases very rapidly (>4% in 2 weeks), reduce dose by 25%.

    Figure 2.1 shows a flow diagram, in greater detail, for the use of erythropoietin-stimulating agents in patients with chronic kidney disease.
- Folic acid 1 mg/day is recommended because folate is dialyzable
- Packed red cell transfusion is rarely required.

Endocrine Glands

Thyroid

Anemia is frequently present in hypothyroidism. It is usually normochromic and normocytic. The anemia is sometimes hypochromic because of associated iron deficiency and occasionally macrocytic because of vitamin B_{12} deficiency. The bone marrow is usually fatty and hypocellular and erythropoiesis is usually normocytic. The finding of a macrocytic anemia and megaloblastic marrow in children with hypothyroidism should raise the possibility of an autoimmune disease with antibodies against parietal cells as well as against the thyroid, leading to vitamin B_{12} deficiency (juvenile pernicious anemia with polyendocrinopathies).

---

[1]Thrombosis of vascular access occurs in 10% of cases treated with rHuEPO.

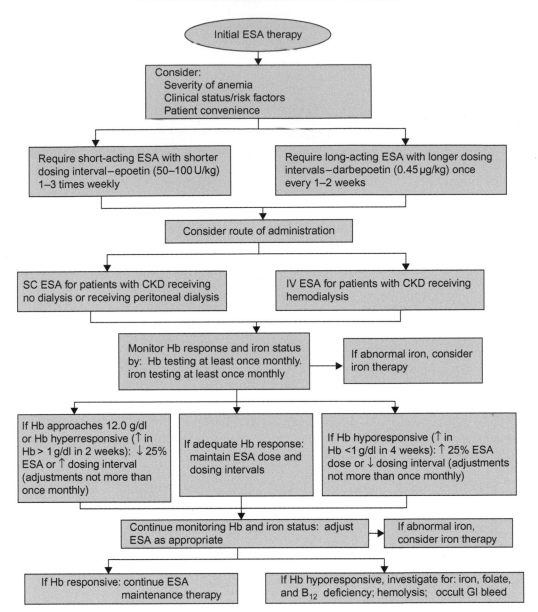

FIGURE 2.1 Recommended erythropoietin-stimulating agent (ESA) treatment in patients with chronic kidney disease. SC, subcutaneous; IV, intravenous; CKD, chronic kidney disease; ↑, increase; ↓, decrease. Source: *From Wish and Coyne (2007), with permission.*

Adrenal Glands

- Androgens stimulate erythropoiesis.
- Conditions of androgen excess, such as Cushing syndrome and congenital adrenal hyperplasia, can produce secondary polycythemia.
- In Addison disease, some degree of anemia is also present but may be masked by coexisting hemoconcentration. The association between Addison disease and megaloblastic anemia raises the possibility of an inherited autoimmune disease directed against multiple tissues, including parietal cells (juvenile pernicious anemia with polyendocrinopathies) (see Chapter 7).

Lungs

- Hypoxia secondary to pulmonary disease results in secondary polycythemia.
- Idiopathic pulmonary hemosiderosis is a chronic disease characterized by recurrent intra-alveolar microhemorrhages with pulmonary dysfunction, hemoptysis, and hemosiderin-laden macrophages, resulting

in iron-deficiency anemia. A precise diagnosis can be established by the presence of siderophages in the gastric aspirate. Apart from a primary idiopathic type, there is also a variant associated with hypersensitivity to cows' milk and one that occurs with a progressive glomerulonephritis (Goodpasture syndrome). Treatment is controversial and may involve:

- Corticosteroids
- Withdrawal of cow's milk
- Use of oral or IV iron when indicated
- Packed red cell transfusions when indicated.

Skin

Mast Cell Disease

Mast cell disease or mastocytosis is associated with an abnormal accumulation of mastocytes (more closely related to monocytes or macrophages rather than to basophils) in the dermis (cutaneous mastocytosis) or in an internal organ (systemic mastocytosis). The systemic form is rare in children. In children, this condition is more common under 2 years of age. It usually presents either as a solitary cutaneous mastocytoma or, more commonly, as urticaria pigmentosa. Involvement beyond the skin is unusual in children, but splenomegaly and bone lesions have been reported. No reports of bone marrow disease in either acquired or congenital mastocytosis have been reported.

Eczema and Psoriasis

Patients with extensive eczema and psoriasis commonly have anemia. The anemia is usually normochromic and normocytic (anemia of chronic disease) and mild in most cases, but severely affected individuals can have hemoglobin levels less than 9 g/dl.

Dermatitis Herpetiformis

- Macrocytic anemia secondary to malabsorption.
- Hyposplenism: Howell–Jolly bodies may be present on blood smear.

Dyskeratosis Congenita

This disease is characterized by ectodermal dysplasia and aplastic anemia (see Chapter 8). The aplastic anemia is associated with high MCV, thrombocytopenia, and elevated fetal hemoglobin. This may occur before the onset of skin manifestations.

Hereditary Hemorrhagic Telangiectasia

This autosomal dominant disorder is associated with a bleeding disorder. Easy bruisability, epistaxis, and respiratory and gastrointestinal bleeding may be caused by telangiectatic lesions.

Ehlers–Danlos Syndrome

This condition may be associated with platelet dysfunction: reduced aggregation with ADP, epinephrine, and collagen. An unusual sensitivity to aspirin is described in type IV Ehlers–Danlos syndrome (see Chapter 14).

CHRONIC ILLNESS

Chronic illnesses such as cancer, IBD, connective tissue disease, and chronic infection are associated with anemia. The anemia has the following characteristics:

- Normochromic, normocytic, occasionally microcytic
- Usually mild, characterized by decreased plasma iron and normal or increased reticuloendothelial iron
- Impaired flow of iron from reticuloendothelial cells to the bone marrow
- Decreased sideroblasts in the bone marrow.

The tests to differentiate the anemia of chronic illness from iron-deficiency anemia are listed in Table 2.4 and therapeutic options for the treatment of anemia in chronic disease are outlined in Table 2.5.

TABLE 2.4   Laboratory Tests to Differentiate Anemia of Chronic Disease from Iron-Deficiency Anemia[a]

Variable (serum levels)	Anemia of chronic disease	Iron-deficiency anemia	Both conditions[b]
Iron	Reduced	Reduced	Reduced
Transferrin	Reduced to normal	Increased	Reduced
Transferrin saturation	Normal to mildly reduced	Reduced	Reduced
Ferritin	Normal to increased	Reduced	Reduced to normal
Soluble transferrin receptor	Normal	Increased	Normal to increased
Cytokine levels	Increased	Normal	Increased

[a]Relative changes are given in relation to the respective normal values.
[b]Patients with both conditions include those with anemia of chronic disease and true iron deficiency.
Modified from: Weiss and Goodnough (2005), with permission.

TABLE 2.5   Therapeutic Options for the Treatment of Anemia of Chronic Disease

Treatment	Anemia of chronic disease	Anemia of chronic disease with true iron deficiency
Treatment of underlying disease	Yes	Yes
Transfusions[a]	Yes	Yes
Iron supplementation	No[b]	Yes[c]
Erythropoietin agents	Yes	Yes, in patients who do not have a response to iron therapy

[a]This treatment is for the short-term correction of severe or life-threatening anemia. Potentially adverse immunomodulatory effects of blood transfusions are controversial.
[b]Although iron therapy is indicated for the correction of anemia of chronic disease in association with absolute iron deficiency, no data from prospective studies are available on the effects of iron therapy on the course of underlying chronic disease.
[c]Overcorrection of anemia (hemoglobin >12 g/dl) may be potentially harmful to patients; the clinical significance of erythropoietin-receptor expression on certain tumor cells needs to be investigated.
From: Weiss and Goodnough (2005), with permission.

In inflammatory diseases, cytokines released by activated leukocytes and other cells exert multiple effects that contribute to the reduction in hemoglobin levels. The pathophysiology of anemia of chronic disease is shown in Figure 2.2:

1. Interleukins (IL), especially IL-6 along with endotoxin, induce hepcidin synthesis in the liver. Hepcidin in turn binds to Ferroportin located both in the GI tract and the reticuloendothelial system. Hepcidin binding induces the degradation of Ferroportin, sequestering oral intake iron from the GI tract and reticuloendothelial iron from storage sites. Hence in the classic patient with the anemia of chronic illness, the serum iron will be low, but there will also be a low level of transferrin iron-binding capacity secondary to a suppression of protein synthesis. This leads to a normal, or slightly diminished, iron saturation. The serum ferritin is then paradoxically elevated secondary to the hepcidin-induced sequester. The adult literature recognizes gray zone cases with normal or lower iron saturations and a low normal ferritin. In these situations laboratory testing for soluble transferrin receptor (which is elevated directly in response to iron deficiency) is used to direct iron treatment. These patients might also be simply given an intravenous iron challenge which will demonstrate improvement over 7–10 days if iron deficiency is present.
2. Inhibition of erythropoietin release from the kidney (especially by IL-1β and tumor necrosis factor-alpha (TNF-α)) leads to reduced erythropoietin-stimulated hematopoietic proliferation.
3. Direct inhibition of the proliferation of erythroid progenitors (especially by TNF-α, interferon-gamma (IFN-γ) and IL-1β).
4. Increased erythrophagocytosis by reticuloendothelial macrophages.

Treatment involves treating the underlying illness.

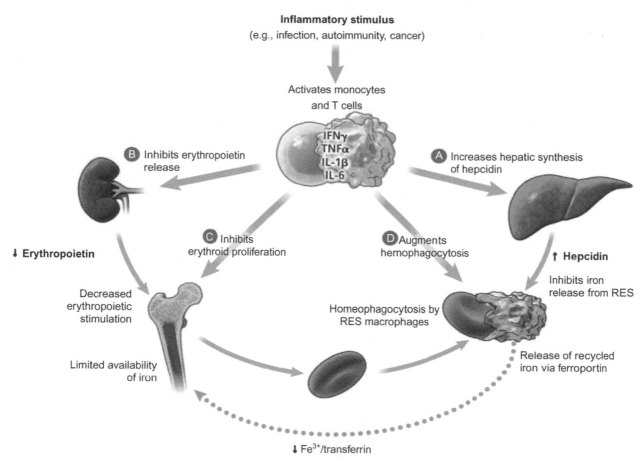

FIGURE 2.2   Pathophysiology of anemia of chronic disease. RES, Reticuloendothelial system. Source: *From Zarychanski and Houston (2008), with permission.*

Inflammatory Bowel Disease as a Model for Anemia of Chronic Illness

- In both Crohn's disease and ulcerative colitis the anemia of chronic illness is often seen—sometimes before gastrointestinal symptomatology manifests.
- It is often associated with concomitant iron deficiency due to bleeding from the involved bowel.
- The patient may present with mild normochromic anemia or severe microcytic anemia.

In an older child or adolescent presenting with iron deficiency a detailed history of gastrointestinal symptoms must be pursued and any suggestion of anemia of chronic illness may alter the type and route of iron medication. If a patient with IBD presents with anemia, iron saturation and ferritin should be assessed prior to the initiation of treatment. The anemia of chronic illness is not iron-deficient erythropoiesis. It is a balance between the effect of elevated hepcidin in sequestering iron and a direct effect of cytokines slowing down erythropoiesis—and hence diminishing the erythropoietic call for iron. If the patient has a very low iron saturation and an elevated ferritin, oral iron is likely to have less effect given the mucosal block to iron in the anemia of chronic illness. Intravenous iron preparations will bypass that block and also diminish the additional gastrointestinal toxicity of oral iron. The administration of iron alone may not ameliorate the situation as patients may have such a severe effect of inflammation on erythropoiesis that they may require simultaneous administration of erythropoietin in pharmacologic doses along with iron. These considerations are identical to those faced in other conditions with major ongoing inflammation (e.g., juvenile rheumatoid arthritis).

Connective Tissue Diseases

Rheumatoid Arthritis

- Anemia of chronic illness (normocytic, normochromic)
- High incidence of iron deficiency

- Leukocytosis and neutropenia common in exacerbations of juvenile rheumatoid arthritis (JRA)
- Thrombocytosis associated with a high level of IL-6 occurs in many patients, although there may be transient episodes of thrombocytopenia.

Felty Syndrome

- Triad of rheumatoid arthritis, splenomegaly, and neutropenia.
- Patients may be at risk for life-threatening bacteremia. Splenic dysfunction resulting in infection with encapsulated organisms has been observed.
- Treatment involves controlling the rheumatoid arthritis, which often leads to improvement in the anemia. Parenteral antibiotics, with coverage for encapsulated organisms, for febrile episodes is recommended.
- G-CSF may be used in urgent situations although but there are concerns about splenic rupture with the use of this G-CSF because of case reports of spontaneous rupture of the spleen in Felty syndrome.

Systemic Lupus Erythematous

- Two types of anemia are common: anemia of chronic illness (normocytic, normochromic) and acquired, autoimmune direct antiglobulin (DAT)-positive hemolytic anemia.
- Neutropenia is common as a result of decreased marrow production and immune-mediated destruction.
- Lymphopenia with abnormalities of T-cell function.
- Immune thrombocytopenia: Antiphospholipid antibodies may be present which prolong the aPTT but are associated with severe thrombosis (lupus anticoagulant).

Polyarteritis Nodosa

- Microangiopathic hemolytic anemia may be associated with renal disease or hypertensive crises.
- Prominent eosinophilia.

Wegener Granulomatosis

This autoimmune disorder is rare in children. Hematological features include:

- Anemia: normocytic; RBC fragmentation with microangiopathic hemolytic anemia
- Leukocytosis with neutrophilia
- Eosinophilia
- Thrombocytosis.

Kawasaki Syndrome

This syndrome is characterized by:

- Mild normochromic, normocytic anemia with reticulocytopenia
- Leukocytosis with neutrophilia and toxic granulation of neutrophils and vacuoles
- Decreased T-suppressor cells
- High C_3 levels
- Increased cytokines IL-1, IL-6, IL-8, interferon-α, and TNF
- Marked thrombocytosis (mean platelet count 700,000/mm^3)
- DIC.

Henoch–Schönlein Purpura

Henoch–Schönlein purpura (HSP), also called anaphylactoid purpura, is associated with systemic vasculitis characterized by unique palpable, erythema multiforma-like purpuric lesions, transient arthralgias or arthritis (especially affecting knees and ankles), colicky abdominal pain, and nephritis.

- Anemia occasionally occurs as a result of GI bleeding or decreased RBC production caused by renal failure.
- Transient decreased Factor XIII activity may occur, which may play a role in either gastrointestinal bleeding or HSP.
- Vitamin K deficiency from severe vasculitis-induced intestinal malabsorption has been reported.

Infections

Anemia

- Chronic infection is associated with the anemia of chronic illness.
- Acute infection, particularly viral infection, can produce transient bone marrow aplasia or selective transient erythrocytopenia.
- Parvovirus B19, with tropism for the developing red cell, infection in patients with an underlying hemolytic disorder (such as sickle cell disease, hereditary spherocytosis) can produce a rapid fall in hemoglobin and an erythroblastopenic crisis marked by anemia and reticulocytopenia. There may be an associated neutropenia and less commonly, thrombocytopenia.
- Many viral and bacterial illnesses may be associated with hemolysis.

White Cell Alterations

- Viral infections can produce leukopenia and neutropenia. Neutrophilia with an increased band count and left shift frequently results from bacterial infection.
- Neonates, particularly premature infants, may not develop an increase in white cell count in response to infection. Neonatal neutropenia may be serious and requires investigation and treatment. G-CSF has been used and found to be helpful in randomized clinical trials.
- Eosinophilia may develop in response to parasitic infections.

Clotting Abnormalities

Severe infections, for example Gram-negative sepsis, can produce DIC. Polymicrobial sepsis (including both aerobic and anaerobic organisms) in the head and neck region may cause thrombosis of major vessels. When this occurs in the jugular veins it leads to a constellation of findings called Lemierre's syndrome (suppurative thrombophlebitis with inflammation starting in the pharynx and spreading to the lateral parapharyngeal tissues in association with jugular vein thrombosis).

Thrombocytopenia

Infection can produce thrombocytopenia through decreased marrow production, immune destruction, or DIC.

Viral and Bacterial Illnesses Associated with Marked Hematologic Sequelae

Parvovirus

Parvovirus B19 has a peculiar predilection for red cell precursors in the bone marrow. It has preference for the red cell precursors because it uses P antigen as a receptor. This viral infection is associated with a transient erythroblastopenic crisis, particularly in individuals with an underlying hemolytic disorder. In addition, it can produce thrombocytopenia, neutropenia, and a hemophagocytic syndrome. In immunocompromised individuals, parvovirus B19 infection can produce prolonged aplasia. Bone marrow aspirate shows decreased or arrested maturation of erythroid precursors and the pathognomonic "giant pronormoblasts."

Epstein–Barr Virus

Epstein–Barr Virus (EBV) infection is associated with the following hematologic manifestations:

- Atypical lymphocytosis
- Acquired immune hemolytic anemia
- Agranulocytosis
- Aplastic anemia, rarely
- Lymphadenopathy and splenomegaly
- Immune thrombocytopenia.

EBV infection also has immunologic and oncologic associations (see Chapter 16). Some of the EBV-associated lymphoproliferative disorders are given in Table 2.6.

TABLE 2.6 EBV-Associated Lymphoproliferative Disorders

EBV-ASSOCIATED B-CELL LYMPHOPROLIFERATIVE DISORDERS

1. Classic Hodgkin lymphoma
2. Burkitt lymphoma
3. Posttransplantation lymphoproliferative disorders
4. HIV-associated lymphoproliferative disorders
   a. Primary CNS lymphoma
   b. Diffuse large B-cell lymphoma, immunoblastic
   c. HHV-8-positive primary effusion lymphoma
   d. Plasmablastic lymphoma

EBV-ASSOCIATED T/NK-CELL LYMPHOPROLIFERATIVE DISORDERS

1. Peripheral T-cell lymphoma, unspecified
2. Angioimmunoblastic T-cell lymphoma
3. Extranodal nasal T/NK-cell lymphoma

EBV, Epstein–Barr virus; HHV-8, human herpes virus-8; NK, natural killer.
Modified from: Carbone et al. (2008), with permission.

Human Immunodeficiency Virus

The main pathophysiology of human immunodeficiency virus (HIV) infection is a constant decline in CD4 lymphocytes, leading to immune failure and death. The other bone marrow cell lines also decline as HIV disease (acquired immunodeficiency syndrome (AIDS)) progresses.

HIV infection has the following hematologic manifestations.

Thrombocytopenia

Thrombocytopenia occurs in about 40% of patients with AIDS. Initially, the clinical findings resemble those of immune thrombocytopenic purpura (ITP). Some degree of splenomegaly is common and the platelet-associated antibodies are often in the form of immune complexes that may contain antibodies with anti-HIV specificity. Megakaryocytes are normal or increased and production of platelets is reduced in the bone marrow.

Thrombotic thrombocytopenic purpura (TTP) is also associated with HIV disease. This occurs in advanced AIDS.

Anemia and Neutropenia

HIV-infected individuals develop progressive cytopenia as immunosuppression advances. Anemia occurs in approximately 70–80% of patients and neutropenia in 50%. Cytopenias in advanced HIV disease are often of complex etiology and include the following:

- A production defect in the marrow appears to be most common.
- Antibody and immune complexes associated with red and white cell surfaces may contribute. Up to 40% have erythrocyte-associated antibodies. Specific antibodies against i and U antigens have occasionally been noted. About 70% of patients with AIDS have neutrophil-associated antibodies. The pathogenesis of the hematologic disorders includes:
  - Infections: Myelosuppression is frequently caused by involvement of the bone marrow by infecting organisms (e.g., mycobacteria, cytomegalovirus (CMV), parvovirus, fungi and, rarely, *Pneumocystis jiroveci*).
  - Neoplasms: Non-Hodgkin lymphoma (NHL) in AIDS patients is associated with infiltration of the bone marrow in up to 30% of cases. This is particularly prominent in the small non-cleaved histologic subtype of NHL.
  - Medications: Widely used antiviral agents in AIDS patients are myelotoxic, for example, zidovudine (AZT) causes anemia in approximately 29% of patients. Ganciclovir and trimethoprim/sulfamethoxazole or pyrimethamine/sulfadiazine may cause neutropenia. In general, bone marrow suppression is related to the drug dosage and to the stage of HIV disease. Importantly, the other nucleoside analogs of anti-HIV compounds (dideoxycytidine (ddC), dideoxyinosine (ddI), stavudine (d4T), or lamivudine (3TC)), are usually not associated with significant myelotoxicity.
  - Nutrition: Poor intake is common in advanced HIV disease and is occasionally accompanied by poor absorption. Vitamin B_{12} levels may be significantly decreased in HIV infection although vitamin B_{12} is not effective in treatment. The reduction in serum vitamin B_{12} levels is due to vitamin B_{12} malabsorption and abnormalities in vitamin B_{12}-binding proteins.

TABLE 2.7   AIDS-Related Neoplasms in Children

1. Classic Hodgkin lymphoma (lymphocyte depleted)
2. Non-Hodgkin lymphoma
    a. Burkitt lymphoma
    b. Central nervous system lymphoma
    c. Diffuse large B-cell lymphoma
    d. Mucosa-associated lymphoid tissue (MALT)-type lymphoma
3. Leiomyoma and leiomyosarcoma
4. Kaposi's sarcoma
5. Acute leukemias
6. Miscellaneous tumors—isolated cases of hepatoblastoma, fibrosarcoma of liver, embryonal rhabdomyosarcoma of biliary tree, Ewing's tumor of the bone

Modified from: Balarezo and Joshi (2002), with permission.

TABLE 2.8   Spectrum of Systemic Lymphoproliferative Lesions in Children with AIDS

1. Hyperplasia involving
    a. Lymph nodes
    b. Peyer's patches of ileum
    c. Lymphoid nodules in esophagus and colon
    d. Thymus
    e. Pulmonary lymphoid hyperplasia (PLH)
2. Lymphoplasmacytic infiltrates in
    a. Lungs (lymphoid interstitial pneumonitis (LIP))
    b. Salivary glands
    c. Liver
    d. Thymitis and multilocular thymic cyst
3. Polyclonal polymorphic B-cell lymphoproliferative disorder (PBLD) involving
    a. Lungs
    b. Liver, spleen, lymph nodes
    c. Kidneys
    d. Salivary glands
    e. Muscle, periadrenal fat
4. Myoepithelial sialadenitis
5. Myoepithelial sialadenitis with focal lymphoma
6. MALT lymphoma (involving nodal and extranodal sites)
7. Non-MALT lymphoma (involving nodal and extranodal sites)

Modified from: Balarezo and Joshi (2002), with permission.

Coagulation Abnormalities

The following abnormalities occur:

- Dysregulation of immunoglobulin production may affect the coagulation cascade through antibody-mediated effects. The dysregulation of immunoglobulin production may also occasionally result in beneficial effects, as in the resolution of anti-Factor VIII antibodies in HIV-infected patients with hemophilia.
- Lupus-like anticoagulant (antiphospholipid antibodies) or anticardiolipin antibodies occur in 82% of patients. The titers or specificities have not led to thrombosis in most patients.
- Thrombosis may occur secondary to protein S deficiency. Low levels of protein S occur in 73% of patients.

Role of Hematopoietic Growth Factors in Acquired Immunodeficiency Syndrome

- rHuEPO results in a significant improvement in Hct and reduces transfusion requirements while the patient is receiving AZT. rHuEPO therapy should be initiated if the erythropoietin level is less than 500 IU/l.
- G-CSF in a dose of 5 mg/kg/day SC is the most widely used growth factor in neutropenia.
- GM-CSF at doses starting at 250 μg/day is also effective but has more side effects than G-CSF and is used less often.

Cancers in Children with Human Immunodeficiency Virus Infection

Malignancies in children with HIV infection are not as common as in adults. Table 2.7 lists the AIDS-related neoplasms in children with HIV infection and Table 2.8 lists the spectrum of lymphoproliferative lesions in children with AIDS.

NHL is the most common malignancy secondary to HIV infection in children. It is usually of B-cell origin as in Burkitt's (small non-cleaved cell) or immunoblastic (large cell) NHL. The mean age at presentation of malignancy in congenitally transmitted disease is 35 months, with a range of 6–62 months. In transfusion-transmitted disease, the latency from the time of HIV seroconversion to the onset of lymphoma is 22–88 months. The CD4 lymphocyte count is less than $50/mm^3$ at the time of diagnosis of the malignancy.

The presenting manifestations include:

- Fever
- Weight loss
- Extranodal manifestations (e.g., hepatomegaly, jaundice, abdominal distention, bone marrow involvement, or central nervous system (CNS) symptoms). Some patients will already have had lymphoproliferative diseases such as lymphocytic interstitial pneumonitis or pulmonary lymphoid hyperplasia. These children usually have advanced (stage III or IV) disease at the time of presentation. Children with CNS lymphomas present with developmental delay or loss of developmental milestones or encephalopathy (dementia, cranial nerve palsies, seizures, or hemiparesis). Differential diagnosis includes infections such as toxoplasmosis, cryptococcosis, or tuberculosis. Contrast-enhanced computed tomography (CT) studies of the brain show hyperdense mass lesions that are usually multicentric or periventricular. CNS lymphomas in AIDS are fast-growing and often have central necrosis and a "rim of enhancement" as in an infectious lesion. A stereotactic biopsy will provide a definitive diagnosis.

Treatment of HIV Infection-Related Lymphomas Treatment consists of standard protocols as described in Chapter 22 on NHL. In addition, a concomitant approach to improving HIV viral load is critical in achieving positive survival outcomes in infected patients. Treatment of CNS lymphomas is more difficult. Intrathecal therapy is indicated even for those without evidence of meningeal or mass lesions at diagnosis of NHL. Radiation therapy may be a helpful adjunct for CNS involvement.

The following are more favorable prognostic features in NHL secondary to AIDS:

- CD4 lymphocyte count above $100/mm^3$
- Normal serum LDH level
- No prior AIDS-related symptoms
- Good Karnofsky score (80–100).

Proliferative Lesions of Mucosa-Associated Lymphoid Tissue Mucosa-associated lymphoid tissue (MALT) shows reactive lymphoid follicles with prominent marginal zones containing centrocyte-like cells, lymphocytic infiltration of the epithelium (lymphoepithelial lesion), and the presence of plasma cells under the surface epithelium. These lesions may be associated with the mucosa of the gastrointestinal tract, Waldeyer's ring, salivary glands, respiratory tract, thyroid, and thymus. Proliferative lesions of MALT can be benign or malignant (such as lymphomas). The proliferative lesions arising from MALT form a spectrum or a continuum extending from reactive to neoplastic lesions. The neoplastic lesions are usually low grade, but may progress into high-grade MALT lymphomas. MALT lymphomas characteristically remain localized, but if dissemination occurs, they are usually confined to the regional lymph nodes and other MALT sites. MALT lesions represent a category of pediatric HIV-associated disease that may arise from a combination of viral etiologies, including HIV, EBV, and CMV.

Treatment of Low-Grade MALT Lymphoma
1. α-Interferon: 1 million units/m^2 SC three times a week (continued until regression of disease or severe toxicity occurs).
2. Rituxan (monoclonal antibody-anti-CD20): 375 mg/m^2 IV weekly for 4 weeks (courses may be repeated as clinically indicated). Some patients may not require any treatment because of the indolent nature of the disease.

Leiomyosarcomas and Leiomyomas

Malignant or benign smooth muscle tumors, leiomyosarcomas (LS) and leiomyomas (LM), are the second most common type of tumor in children with HIV infection. The incidence in HIV patients is 4.8% (in non-HIV children, it is 2 per million). The most common sites of presentation are the lungs, spleen, and gastrointestinal tract. Patients with endobronchial LM or LS often have multiple nodules in the pulmonary parenchyma. Bloody diarrhea, abdominal pain, or signs of obstruction may signal intraluminal bowel lesions.

These tumors are clearly associated with EBV infection. *In situ* hybridization and quantitative polymerase chain reaction studies of LM and LS demonstrated that high copy numbers of EBV are present in every tumor cell. The EBV receptor (CD21/C3d) is present on tumor tissue at very high concentrations but it is present at lower concentrations in normal smooth muscle or control LM/LS that had no EBV DNA in them. In AIDS patients, the EBV receptor may be unregulated, allowing EBV to enter the muscle cells and cause their transformation.

Treatment:

- Complete surgical resection where possible
- Radiation therapy
- Decreasing viral load in patients if they present with an acceleration
- Possible use of immunomodulators or chemotherapy.

Kaposi Sarcoma

Kaposi Sarcoma (KS) is rare in children and constitutes the third most common malignancy in pediatric AIDS patients; it occurs in 25% of adults with AIDS. KS occurs only in those HIV-infected children who were born to mothers with HIV. The lymphadenopathic form of KS is seen mostly in Haitian and African children and may represent the epidemic form of KS unrelated to AIDS. The cutaneous form is a true indicator of the disease related to AIDS. Visceral involvement has not been pathologically documented in children with AIDS. The incidence is falling in patients with early HAART treatment as KS is an AIDS-defining cancer.

Leukemias

Almost all HIV-associated leukemia is of B-cell origin. They represent the fourth most common malignancy in children with AIDS. The clinical presentation and biologic features are similar to those found in non-HIV children. Treatment involves chemotherapy designed for B-cell leukemia and lymphomas, as well as lowering viral load where necessary.

Miscellaneous Tumors

There is no increase in Hodgkin disease in children with AIDS. Children with AIDS rarely develop hepatoblastoma, embryonal rhabdomyosarcoma, fibrosarcoma, and papillary carcinoma of the thyroid. The occurrence of these tumors is probably unrelated to the HIV infection.

Infections

Torchs

This is a group of congenital infections including toxoplasma, rubella, CMV, herpes simplex virus (HSV), and syphilis. They can all cause neonatal anemia, jaundice, thrombocytopenia, and hepatosplenomegaly. They have significant sequelae so prevention, early identification and treatment are required.

Salmonella typhi

Typhoid fever usually produces profound leukopenia and neutropenia in the initial stages of the illness and is often accompanied by thrombocytopenia. Bone marrow examination may show marrow suppression but also hematophagocytosis. Diminished absolute eosinophil counts may be a clue to the diagnosis.

Acute Infectious Lymphocytosis

Acute infectious lymphocytosis is caused by a *Coxsackie* virus and is a rare benign, self-limiting childhood condition. It is associated with a low-grade fever, diarrhea, and marked lymphocytosis (50,000/mm^3). Lymphocytes are mainly CD4 T-cells. The condition resolves in 2−3 weeks without treatment.

Bartonellosis

Bartonellosis is caused by a Gram-negative bacillus *Bartonella bacilliformis* confined to the mountain valleys of the Andes. The vector is a local sand fly. Infection from this organism causes a fatal syndrome of severe hemolytic anemia with fever (Oroya fever). Another species of Bartonella, *B. hensele* causes cat scratch fever. It is associated with a regional lymphadenitis following a scratch by a cat. Thrombocytopenia may occur in this condition.

Tuberculosis

Tuberculosis is caused by *Mycobacterium tuberculosis*. Hematologic manifestations include leukemoid reaction mimicking CML, monocytosis, and rarely pancytopenia from diffuse granulomatous marrow infiltration (often associated with leukoerythroblasosis).

Leptospirosis (Weil Disease)

This disease is caused by a leptospira, *L. icterohemorrhagiae*. A coagulopathy occurs which is complex and can be corrected with vitamin K administration. Thrombocytopenia commonly occurs and is the cause of bleeding when this occurs.

Parasitic Illnesses Associated with Marked Hematologic Sequelae

Malaria

The etiology of anemia in acute infections is multifactorial:

- Intracellular parasite metabolism alters negative charges on the RBC membrane, which causes altered permeability with increased osmotic fragility. Spleen removes the damaged RBC or the parasites are "pitted" during the passage from spleen which results in microspherocytes of RBC.
- Autoimmune hemolytic anemia may also occur. An IgG antibody is formed against the parasite and resulting immune complex attaches nonspecifically to RBC, complement is activated and cell destruction occurs. Positive Coombs test due to IgG is found in 50% of patients with *Plasmodium falcifarum* malariae.
- Thrombocytopenia without DIC is common. IgG antimalarial antibody bonds to the platelet-bound malaria antigen and the IgG platelet parasite complex is removed by the reticuloendothelial system.

Babesiosis

Babesiosis is caused by several species from the genus *Babesia* that colonize erythrocytes. It is a zoonotic disease transmitted by the *Ixodid* tick and has similar clinical features to malaria. The clinical features include fever, myalgia and arthralgia with hepatosplenomegaly and hemolysis. Blood film may reveal intraerythrocytic trophozoites arranged in the form of a "Maltese cross."

Leishmaniasis

The protozoal species *Leishmania* causes progressive splenomegaly and subsequent pancytopenia (anemia, neutropenia, and thrombocytopenia). The bone marrow is usually hypercellular with hemophagocytosis. Some children may show coagulopathy.

Hookworm

Worldwide hookworm is a major cause of iron-deficiency anemia. Two species infest humans:

- *Ancylostoma duodenale* is found in the Mediterranean region, North Africa, and the west coast of South America.
- *Necator americanus* is found in most of Africa, Southeast Asia, Pacific islands, and Australia.

Hookworms penetrate exposed skin, usually soles of bare feet and migrate through the circulation to the right side of the heart, then lungs (causing hypereosinophilic syndrome), through the airway down to the esophagus. They mature in the small intestine and attach their mouthparts to the mucosa. They suck blood, with each adult *A. duodenale* consuming about 0.2 ml/day. Heavily infested children may present with profound iron-deficiency anemia, hypoproteinemia, and marked eosinophilia.

Tapeworm

Diphyllobothrium latum is a fish tapeworm. It is acquired by eating uncooked freshwater fish. This worm infestation in the intestine results in vitamin B_{12} deficiency.

Trypanosomiasis

This disease may cause immune-mediated anemia and less often thrombocytopenia and neutropenia. More importantly, as making the diagnosis is so important for survival, the trypanosomes are more likely to be seen early in the illness on classic thick smears.

LEAD INTOXICATION

The most striking hematologic feature of lead intoxication is basophilic stippling (coarse basophilia) of red cells. It is caused by precipitation of denatured mitochondria secondary to inhibition of prymidine-5'-nucleotidase. Lead also produces ring sideroblasts in the marrow and it is associated with hypochromic microcytic anemia and markedly elevated free erythrocyte protoporphyrin levels.

NUTRITIONAL DISORDERS

Protein-Calorie Malnutrition

Protein deficiency in the presence of adequate carbohydrate caloric intake (kwashiorkor) is associated with mild normochromic, normocytic anemia secondary to reduced RBC production despite normal or increased erythropoietin levels as well as reduced red cell survival. Protein-calorie malnutrition is also associated with impaired leukocyte function.

Scurvy

Mild anemia is common. There is a bleeding tendency due to loss of vascular integrity which may result in petechiae, subperiosteal, orbital or subdural hemorrhages. Hematuria and melena may occur.

Anorexia Nervosa

Anorexia nervosa is associated with hematologic changes which may be helpful in diagnosis:

- Red cell morphology is striking for the unusual morphology of acanthocytosis. This occurs due to acquired hypobetalipoproteinemia secondary to nutritional failure
- Mild anemia (macrocytic), neutropenia, and thrombocytopenia
- Mild predisposition to infection associated with neutropenia
- Gelatinous changes of bone marrow which may become severely hypoplastic.

BONE MARROW INFILTRATION

The bone marrow may be infiltrated by non-neoplastic disease (storage disease), granulomata from infection, sarcoid or rheumatologic disease, or neoplastic disease. In storage disease, a diagnosis is made on the basis of the family history, clinical picture, enzyme assays of white cells or cultured fibroblasts, and bone marrow aspiration revealing the characteristic cells of the disorder. Mutation analysis, where available, is the standard for diagnosis. Differential diagnosis of granulomatous conditions begins with recognition of the granulomas in bone marrow morphology (with culture if suspicious at the time of bone marrow examination). Neoplastic disease may arise *de novo* in the marrow (leukemias) or invade the marrow as metastases from solid tumors (neuroblastoma or rhabdomyosarcoma). Table 2.9 lists the diseases that may infiltrate the marrow.

Gaucher Disease

Gaucher disease is the most common lysosomal storage disease, resulting from deficient activity of β-glucocerebrosidase. It is inherited in an autosomal-recessive manner. There are more than 200 mutations identified in the β-glucocerebrosidase gene located on 1q21, including point mutations, crossovers, and recombinations,

TABLE 2.9   Diseases Invading Bone Marrow

1. Non-neoplastic
   a. Storage diseases
      i. Gaucher disease
      ii. Niemann–Pick disease
      iii. Cystine storage disease
   b. Marble bone disease (osteopetrosis)
   c. Langerhans cell histiocytosis (see Chapter 20)
2. Neoplastic
   a. Primary
      i. Leukemia (see Chapters 18 and 19)
   b. Secondary
      i. Neuroblastoma (see Chapter 24)
      ii. Non-Hodgkin lymphoma (see Chapter 22)
      iii. Hodgkin lymphoma (see Chapter 21)
      iv. Wilms tumor (rarely) (see Chapter 25)
      v. Retinoblastoma (see Chapter 28)
      vi. Rhabdomyosarcoma (see Chapter 26)

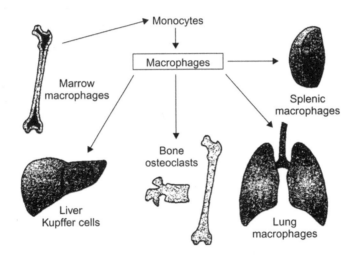

FIGURE 2.3   Diagram of the cellular pathophysiology of Gaucher disease. Monocytes are produced in the bone marrow and mature to macrophages in the marrow or in specific sites of distribution, such as liver Kupffer cells, bone osteoclasts, and lung and tissue macrophages. Once resident, they accumulate glucosylceramide by phagocytosis and become end-stage Gaucher cells. Source: *From Grabowski and Leslie (2000), with permission.*

yet prediction of clinical course can only be broadly ascribed on the basis of genotyping. Generally, the presence of the 1226G (N370S) mutation on one allele is synonymous with type-I disease (i.e., it is apparently protective against neurologic involvement), whereas homozygosity for the allele 1488C (L444P) is invariably correlated with neurological disease. The degree of clinical involvement differs greatly in individual patients, even those with the same genotype and those affected within the same family.

Pathogenesis

β-Glucocerebrosidase is necessary for the catabolism of glucocerebroside. Deficiency of β-glucocerebrosidase leads to accumulation of glucocerebroside in the lysosomes of macrophages in tissues of the reticuloendothelial system. Figure 2.3 shows a diagram of the cellular pathophysiology of Gaucher disease. Accumulation in splenic macrophages and in the Kupffer cells of the liver produces hepatosplenomegaly.

Hypersplenism produces anemia and thrombocytopenia. Glucocerebroside accumulation in the bone marrow results in osteopenia, lytic lesions, pathologic fractures, chronic bone pain, bone infarcts, osteonecrosis, and acute excruciating bone crises.

Gaucher disease is classified into three types based on the presence and degree of neuronal involvement. Table 2.10 outlines the clinical manifestations of the three types of Gaucher disease.

TABLE 2.10 Clinical Manifestations of Subtypes of Gaucher Disease

Characteristic	Type I Symptomatic	Type I Asymptomatic	Type II Infantile	Type II Neonatal	Type III IIIa	Type III IIIb	Type III IIIc
Most common genotype	1226G compound heterozygous	1226G (IN370S) homozygous	None	Two null mutations	None	1448C (L444P) homozygous	1342C (D409H) homozygous
Ethnic predilection	Ashkenazi Jews	Ashkenazi Jews	None	None	None	Norbottnians (Northern Sweden)	Palestinian Arab Japanese
Common presenting features	Hepatosplenomegaly Hypersplenism Bleeding Bone Pains	None	OMA Strabismus Opisthotonus Trismus	Hydrops fetalis Congenital icthyosis	OMA Myoclonic seizures	OMA Hepatospleno-megaly Growth retardation	Cardiac valve calcification
Central nervous system involvement	None	None	Severe	Lethal	Slow progressive neurological deterioration	OMA Slow cognitive deterioration	OMA
Bone involvement	Mild to severe	None	None	None	Mild	Moderate to severe	Small
Lung involvement	None to severe	None	Severe	Severe	Mild to moderate	Moderate to severe	Small
Enzyme replacement therapy	Indicated and efficient	Not indicated	Ethically problematic	Not relevant	Recommended for visceral features only		
Life expectancy	Normal	Normal	Death before age 2 years	Neonatal death	Death during childhood	Possible survival to adulthood	Survival to teenage

OMA, oculomotor apraxia.
From: Elstein et al. (2001), with permission.

Patients with type 1 Gaucher disease (non-neuropathic), which accounts for 90% of all cases of Gaucher disease, present with:

- Asymptomatic splenomegaly (rarely, portal hypertension develops). Splenic infarction is common and presents with pain, rigid abdomen, and fever. Splenic nuclide scanning is helpful in the presence of an acute abdomen.
- Pancytopenia secondary to hypersplenism (rarely from infiltration of the bone marrow with Gaucher cells).
- Skeletal manifestations include bone marrow infiltration with Erlenmeyer flask deformity from bone marrow expansion, generalized bone mineral loss, and infarction on radiographs. The resultant osteopenia and infarction can lead to pathologic fractures.
- Bone crises characterized by fever and excruciating local pain most frequently along femurs.
- Growth delay: 50% of the symptomatic children are at or below the third percentile for height and another 25% are shorter than expected based on their mid-parental height.
- Typical Gaucher cells in the bone marrow.[2]
- Decreased glucocerebrosidase activity of white cells.
- Characteristic mutations of the β-glucocerebrosidase gene on chromosome 1 on DNA analysis.

Diagnosis

β-Glucocerebrosidase assay on leukocytes or cultured skin fibroblasts is the most efficient method of diagnosis in the absence of molecular testing. The typical child with type 1 Gaucher disease will have enzyme activity that is 10−30% of normal.

[2]Large tissue cells of macrophage origin. Cytoplasm is pale gray-blue and they have eccentrically placed nuclei. Morphologically "Gaucher-like" cells are observed in chronic granulomatous disease, thalassemia, multiple myeloma, Hodgkin disease, AIDS, and acute lymphoblastic leukemia, but can be readily distinguished from true Gaucher cells.

Further Evaluation

1. DNA evaluation for β-glucocerebrosidase gene abnormalities in patient, parents, and siblings
2. CBC: Often, if a patient presents with no prior diagnosis, pancytopenia, splenomegaly, and leukoerythroblastosis (appearance of early white cell forms, nucleated red cells, and tear drop red cells) leads to a concern for leukemia. Of interest is the common appearance of the CBC that may make a diagnosis of Gaucher disease more likely
3. Serum chemistry with liver function tests
4. Acid phosphatase level
5. Angiotensin-converting enzyme
6. Chitotriosidase
7. Liver:spleen volume
8. MRI of femora
9. Bone density of the spine and hips (DXA)
10. Chest radiograph.

Treatment

Enzyme replacement therapy (ERT) is recommended for the treatment of symptomatic type 1 patients. Recombinant human macrophage-targeted human glucocerebrosidase (imiglucerase, Cerezyme)[3] is used for ERT. The initial dose is 30–60 units/kg IV every 2 weeks. The initial dose must be individualized for each patient based on disease severity and rate of progression. The maintenance dose is 15–60 units/kg IV every 2 weeks. Children who require treatment need to continue therapy indefinitely to maintain their clinical improvement. Prolonged periods without therapy are not appropriate.

Substrate reduction therapy (SRT) is available using miglustat (Zavesca; Actelion Pharmaceuticals, Allschwill, Switzerland). Miglustat is an inhibitor of glucosylceramide (GlcCer) synthase, the enzyme responsible for GlcCer synthesis and hence synthesis of all GlcCer-based glycolipids. Unlike Cerezyme, Zavesca is given orally and does cross the blood–brain barrier. It is important to note that Zavesca causes a number of side effects. Currently the role of substrate reduction is still evolving. See Table 2.10 for ERT in the other types of Gaucher disease.

Recommendations for monitoring of children with type 1 Gaucher disease receiving and not receiving enzyme replacement therapy are outlined in Table 2.11.

Iron therapy in Gaucher patients with anemia is not recommended, because Gaucher cells avidly take up iron, which leads to hemochromatosis and decreased iron availability for erythropoiesis.

Response to Therapy

The earliest response is an improvement in hematologic parameters. A progressive decrease in liver/spleen size is regarded as a positive response. Skeletal response occurs more slowly (after 2–4 years), along with a decrease in pain and bone crises.

Approximately 5% of patients develop hypersensitivity to ERT. These reactions respond to interruption of infusion and administration of antihistamine and glucocorticoids. Subsequent reactions can usually be prevented by reducing the initial rate of infusion, so that no more than 10 units/min are administered. These reactions commonly occur during the first 12 months of treatment. For this reason, the first year of treatment should be administered under the direct supervision of a physician. Following one year therapy can be administered at home by home nursing services. The non-neutralizing IgG antibodies that develop in up to 13% of patients are not clinically relevant.

Niemann–Pick Disease

Niemann–Pick disease types A and B result from deficient activity of acid sphingomyelinase, encoded by a gene on chromosome 11. The defect results in accumulation of sphingomyelin in the monocyte–macrophage

[3]Manufactured by Genzyme, Cambridge, MA.

TABLE 2.11 Recommendations for Monitoring Children with Type 1 Gaucher Disease (Minimal Evaluations Only)

	All patients, baseline	Patients not receiving enzyme therapy		Patients receiving enzyme therapy		
		Every 12 months	Every 12–24 months	Every 3 months[a]	Every 12 months[a]	At time of dose change
HEMATOLOGIC						
Hemoglobin	X	X		X		X
Platelet count	X	X		X		X
Acid phosphatase (total, non-prostatic), angiotensin converting enzyme, chitotriosidase[b]	X	X		X		X
VISCERAL[c]						
Spleen volume (volumetric MRI or CT)	X		X		X	X
Liver volume (volumetric MRI or CT)	X		X		X	X
SKELETAL[d]						
MRI (coronal; T1- and T2-weighted) of entire femora[e]	X		X		X	X
Radiograph: AP view of entire femora[e] and view of lateral spine	X		X		X	X
Bone density (DEXA): spine and hips	X		X		Every 12–24 months	
QUALITY OF LIFE[f]						
Patient reported functional health and well-being	X	X		X		

[a]For patients who have reached clinical goals and for whom there has been no change in dose, the frequency of monitoring can be decreased to every 12–24 months.
[b]One or more of these markers should be consistently monitored (at least once every 12 months) in conjunction with other clinical assessments of disease activity and response to treatment. Of the three currently recommended biochemical markers, chitotriosidase activity, when available as a validated procedure from an experienced laboratory, may be the most sensitive indicator of changing disease activity and is therefore preferred.
[c]Obtain contiguous transaxial 10-mm-thick sections for sum of region of interest.
[d]Additional skeletal assessments that are optional include bone age for patients ≤ 14 years old. Follow-up is recommended if baseline is abnormal.
[e]Optimally, obtain hips to below knees. As an alternative, obtain hips to distal femur.
[f]Ideally, quality of life should be assessed every 6 months using a standard and valid instrument.
DEXA, dual energy X-ray absorptiometry.
From: Charrow et al. (2004), with permission.

system. The progressive deposition of sphingomyelin in the CNS leads to type A and in non-neuronal tissues leads to type B. Type C is a neuronopathic form that results from the defective cholesterol transport.

Clinical Manifestations

Depending on the type, Niemann–Pick disease has classic signs, including:

- Hepatosplenomegaly.
- Cherry red spot in macula.
- Psychomotor deterioration.
- Reticular pulmonary infiltrates.
- Foamy cells in the bone marrow.

Diagnosis

Diagnosis involves examining leukocytes or cultured fibroblasts to determine sphingomyelinase activity.

Treatment

There is no specific treatment for Niemann–Pick disease. Miglustat (a GlcCer synthase inhibitor) may have some marginal value. Bone marrow transplant has helped a few specific subtypes of patients (type B).

Splenectomy in type B patients frequently causes progression of pulmonary disease and should be avoided if possible.

"Foam Cells" in Bone Marrow

Foam cells, with numerous uniform vacuoles often described as having a "honeycomb" appearance, are seen in the bone marrow in the following conditions:

- Neimann–Pick disease (types A, B, C, D)
- Gm1 gangliosidosis (type 1)
- Gm2 gangliosidosis (Sandhoff variant)
- Lactosyl ceramidosis
- Sialidosis I
- Sialidosis II, late infantile type
- Mucolipidosis II
- Mucolipidosis III
- Mucolipidosis IV
- Fucosidosis
- Mannosidosis
- Neuronal ceroid-lipofuscinosis
- Farber's disease
- Wolman's disease
- Cholesteryl ester storage disease
- Cerebrotendinous xanthomatosis
- Chronic hyperlipidemia
- Chronic corticosteroid therapy
- Hematologic malignancies (e.g., Hodgkin disease, leukemia, myeloma)
- Hematologic disease (e.g., aplastic anemia, ITP).

Careful history (including ethnic and family history), physical examination, examination of bone marrow using phase electron microscopy, and special stains and enzyme assays on white blood cells or cultured skin fibroblasts and liver biopsy for biochemical analysis can assist in making a specific diagnosis of these storage diseases.

Cystinosis

An autosomal-recessive defect, cystinosis is associated with generalized deposits of cystine in the tissues. Cystinosis occurs in the first year of life with the following manifestations:

- Thermal instability, polydipsia, polyuria
- Failure to thrive
- Recurrent episodes of vomiting and dehydration
- Dwarfism and rickets often prominent
- Early renal involvement with tubular dysfunction manifesting as a secondary Fanconi syndrome, leading to chronic renal failure.

Diagnosis

- Cystine crystals in the bone marrow
- Elevated cystine levels in leukocytes or fibroblasts.

Infantile Malignant Osteopetrosis (Marble Bone Disease)

Osteopetrosis is a hereditary disorder that may be present in either a severe or a mild form.

Severe Form (Autosomal Recessive)

The marrow space is progressively obliterated by excessive osseous growth. The difficulty in obtaining marrow by aspiration is a diagnostic clue. Radiologic changes are characteristic and diagnostic, consisting of generalized osteosclerosis. The cranial foramina progressively narrows resulting in blindness due to optic atrophy, deafness, and other cranial nerve lesions.

The hematologic characteristics include the following:

- Progressive pancytopenia due to encroachment on the hematopoietic marrow by the overgrowth of bone
- Compensatory extramedullary hematopoiesis with resultant leukoerythroblastic anemia (circulating normoblasts, teardrop-shaped poikilocytosis, and early myelocytes), hepatosplenomegaly, and lymphadenopathy
- Bone marrow hypoplasia
- Hemolysis due to splenic sequestration of red cells and perhaps general overactivity of the reticuloendothelial system.

Treatment

Allogeneic stem cell transplantation provides multipotent hematopoietic stem cells, which serve as a source of normal osteoclasts.

Mild Form (Autosomal Dominant)

Pathologic fractures occur in sclerotic bone. Nerve entrapment syndromes may also be present.

Neoplastic Disease

Neoplastic disease can be associated with the following hematologic alterations:

- Hemorrhage
- Nutritional deficiency states
- Dyserythropoietic anemias (including erythroid hypoplasia, sideroblastic anemia, and anemia similar to that seen in chronic inflammation)
- Defect in erythropoietin production
- Hemodilution
- Hemolysis
- Pancytopenia secondary to marrow invasion or to cytotoxic therapy
- Acquired von Willebrand disease as in Wilms tumor
- Hypercoagulable states as in NHL
- Coagulopathy as in acute promyelocytic leukemia
- Leukoerythroblastic anemia and marrow
- Infiltration
- Cytotoxic drug therapy.

Marrow infiltration is suspected when leukoerythroblastic anemia develops. This term signifies the presence of myelocytes, normoblasts, and teardrop-shaped red cells with anemia, thrombocytopenia, and neutropenia. This presentation is due to extramedullary erythropoiesis that occurs when the marrow is infiltrated, permitting the escape of early myeloid and erythroid cells into the circulation. Normal blood findings, however, do not exclude marrow infiltration.

Bone marrow examination frequently demonstrates infiltration with tumor cells in the presence of pancytopenia. Because metastatic bone marrow involvement from solid tumors may be patchy, a single aspiration is not diagnostic. At least two aspirates and two biopsies should be performed.

The hematologic alterations associated with malignancy should be managed supportively and resolve if the underlying neoplasms can be successfully treated.

Table 2.12 summarizes some of the peripheral blood manifestations of systemic illness.

TABLE 2.12 Peripheral Blood Manifestations of Systemic Illness

Condition	Red blood cells (RBC)	White blood cells (WBC)	Platelets	Comments
Hypersplenism	Spherocytes, schistocytes	Leucopenia	Thrombocytopenia	Splenectomy usually corrects the peripheral blood changes
Hyposplenism	Target cells, Howell–Jolly bodies		Thrombocytosis	
Leuco-erythroblastosis (marrow infiltration)	Teardrop-shaped cells, tailed RBCs, nucleated RBCs	Leucocytosis (increased immature granulocytes)	Thrombocytopenia	Triad of NRBC, teardrop-shaped cells and immature granulocytes
Megaloblastosis	Macrocytosis, fragmentation	Leucopenia, Hypersegmented granulocytes	Thrombocytopenia	Deficiency of B_{12}, folic acid
MALIGNANCY				
Acute lymphoblastic leukemia (ALL)	Anemia	Lymphoblasts, leucopenia, Hyperleucocytosis	Thrombocytopenia	Pancytopenia due to marrow infiltration
Acute myeloid leukemia (AML)		Myeloblasts, hyperleucocytosis, increased promyelocytes (M3), monocytes (M5), and eosinophils (M5eo)	Thrombocytopenia	Pancytopenia due to marrow infiltration
Hodgkin disease	Immune hemolytic anemia	Eosinophilia, neutrophilia	Thrombocytopenia	These paraneoplastic manifestations can precede the illness
INFECTIONS				
Sepsis (especially bacterial infections):	Agglutination, rouleaux formation, hemolytic anemia	Neutrophilic changes include toxic granulation, Döhle bodies, cytoplasmic vacuolation	Thrombocytosis, thrombocytopenia	Presence of immature myeloid precursors indicate leukemoid reaction
• AIH, warm antibody type	Nucleated RBCs, spherocytes, schistocytes	Leucocytosis, rarely leucopenia	Thrombocytopenia	Common causes include pneumococcal infection, typhoid fever, hepatitis C
• AIH, cold agglutinin type	Agglutination, rarely erythrophagocytosis	Reactive lymphocytes	Thrombocytopenia	Mycoplasma, parvovirus, legionella, Chlamydia, EBV, CMV, VZV, HIV
• AIH, cold IgG antibody type	Spherocytes. Intravascular hemolysis		Thrombocytopenia	Measles, mumps, influenza-A, adenovirus
Bacterial infections		Granulocytes may contain *Staphylococcus aureus*, *Streptococcus*, *Pneumococcus*, meningococci, *Clostridia* and *Bartonella*. Ehrlichiosis (morula within neutrophils/monocytes)		Relapsing fever (*Borellia recurrentis*), legionella, and *Klebsiella* can have extracellular organisms in the smear
Parasitic infections	Intracellular parasitic forms seen in malaria, babesiosis	Eosinophilia, trophozoite forms of *Toxoplasma* in neutrophils and monocytes		Extracellular forms include filariasis (microfilariae), trypanosomiasis (trypomastigote forms)
Fungal infections		*Candida*, *Histoplasma* and *Cryptococcus* in neutrophils and monocytes		*Candida* and *Cryptococcus* can also be found extracellularly
Viral infections	Rouleaux formation and acquired Pelger–Huet anomaly in HIV Agglutination in EBV infection	Reactive lymphocytosis (Downey type II) in EBV infection	Thrombocytopenia	Rarely plasma cells seen in Hepatitis B, C infections

NRBC, nucleated red blood cells; AIH, autoimmune hemolytic anemia.

Further Reading and References

Aird, W.C., 2003. The hematologic system as a marker of organ dysfunction in sepsis. Mayo Clin. Proc. 78, 869–881.

Balarezo, F.S., Joshi, V.V., 2002. Proliferative and neoplastic disorders in children with AIDS. Adv. Anat. Pathol. 9 (6), 360–370.

Carbone, A., Gloghini, A., Dotti, G., 2008. EBV associated lymphoproliferative disorders: classification and treatment. Oncologist 13 (5), 577–585.

Charrow, J., Anderson, H.C., Kaplan, P., et al., 2004. Enzyme replacement therapy and monitoring for children with type 1 Gaucher disease: consensus recommendations. J. Pediatr. 144, 112–120.

Cullis, J.O., 2011. Diagnosis and management of anaemia of chronic disease: current status. Br. J. Haematol. 154, 289–300.

Elstein, D., Abrahamov, A., Hadas-Halpern, I., Zimarn, A., 2001. Gaucher's disease. Lancet 358, 324–327.

Gasche, C., Dejaco, C., et al., 1997. Intravenous iron and erythropoietin for anemia associated with Crohn disease. A randomized, controlled trial. Ann. Intern. Med. 126, 782–787.

Goodhand, J.R., Kamperidis, N., et al., 2012. Prevalence and management of anemia in children, adolescents and adults with inflammatory bowel disease. Inflamm. Bowel Dis. 18, 513–519.

Grabowski, G.A., et al., 2015. Lysosomal storage diseases. In: Orkin, S., Fisher, D., Ginsburg, D., Thomas, L.A., Lux, S., Nathan, D.G. (Eds.), Nathan and Oski's Hematology and Oncology of Infancy and Childhood, eighth ed. Saunders Elsevier, Philadelphia.

Grabowski, G.A., Leslie, N., 2000. Lysosomal storage diseases: perspectives and principles. In: Hoffmann, R., Benz, E.J., Shattil, S.J., Furie, B., Cohen, H.J., Silberstein, L.E., McGlave, P. (Eds.), Hematology Basic Principles, third ed. Lippincott-Raven, Philadelphia.

Grace, R.F., 2015. Hematologic manifestations of system disease. In: Orkin, S., Fisher, D., Ginsburg, D., Thomas, L.A., Lux, S., Nathan, D.G. (Eds.), Nathan and Oski's Hematology of Infancy and Childhood, eighth ed. Saunders Elsevier, Philadelphia.

Granovsky, M.O., Mueller, B.U., Nicholson, H.S., Rosenberg, P.S., Rabkin, C.S., 1998. Cancer in human immunodeficiency virus-infected children: a case series from the Children's Cancer Group and the National Cancer Institute. J. Clin. Oncol. 16, 1729.

Halfdanarson, T.R., Litzow, M.R., Murray, J.A., 2007. Hematologic manifestations of celiac disease. Blood 109, 412–421.

McGovern, M.M., Desnick, R.J., 2007. Lipidoses. In: Kliegman, R.M., Behrman, R.E., Jenson, H.B., Stanton, B.F. (Eds.), Nelsons Textbook of Pediatrics, eighteenth ed. WB Saunders, Philadelphia.

Mueller, B.U., Pizza, P.A., 1995. Cancer in children with primary or secondary immunodeficiencies. J. Pediatr. 126, 1.

Thomas, A.S., Mehta, A.B., Hughes, D.A., 2013. Diagnosing Gaucher disease: an on-going need for increased awareness amongst haematologists. Blood Cells Mol. Dis. 50 (3), 212–217.

Volberding, P.A., Baker, K.R., Levine, A.M., 2003. Human immunodeficiency virus hematology. Hematology, 294–313.

Weiss, G., Goodnough, L.T., 2005. Anemia of chronic disease. N. Engl. J. Med. 352, 1011–1023.

Wish, J.B., Coyne, D.W., 2007. Use of erythropoiesis stimulating agents in patients with anemia of chronic kidney disease: overcoming the pharmacologic and pharmacoeconomic limitations of existing therapies. Mayo Clin. Proc. 82 (11), 1372–1380.

Zarychanski, R., Houston, D.S., 2008. Anemia of chronic disease. A harmful or beneficial response. CMAJ 179 (4), 333–337.

3

Classification and Diagnosis of Anemia in Children

Philip Lanzkowsky

CLASSIFICATION AND DIAGNOSIS

Anemia can be defined as a reduction in hemoglobin concentration, hematocrit, or number of red blood cells per cubic millimeter. The lower limit of the normal range is set at two standard deviations below the mean for age and sex for the normal population.[1]

This definition will result in 2.5% of the normal population being classified as anemic and by definition 2.5% of the population may receive unnecessary hematological investigations. Clinicians should be cognizant of this and temper investigations of the patient in the light of other clinical and laboratory findings.

The first step in diagnosis of anemia is to establish whether the abnormality is isolated to a single cell line (red blood cells only) or whether it is part of a multiple cell line abnormality (red cells, white cells, and platelets). Abnormalities of two- or three-cell lines usually indicate one of the following:

- bone marrow involvement (e.g., aplastic anemia, leukemia), or
- an immunologic disorder (e.g., connective tissue disease or immunoneutropenia, idiopathic thrombocytopenic purpura (ITP), or immune hemolytic anemia singly or in combination), or
- sequestration of cells (e.g., hypersplenism).

Table 3.1 presents an etiologic classification of anemia and the diagnostic features in each case.

Once the anemia has been established to be exclusively a red cell problem a useful approach to understanding the anemia is to separate it into two pathogenetic categories

1. Disorders of depressed red cell formation due to ineffectual erythropoeisis in which the marrow has many erythroblasts and there is a reticulocytosis or failure of erythropoiesis in which there is a paucity of erythroblasts in the marrow, an absolute erythroblastopenia and there is a reticulocytopenia (usually due to marrow failure diseases).
2. Disorders of erythrocyte destruction (hemolysis) or red cell loss (hemorrhage).

The *blood smear* is very helpful in the diagnosis of anemia. It establishes whether the anemia is hypochromic, microcytic, normocytic, macrocytic, or shows specific morphologic abnormalities suggestive of red cell membrane disorders (e.g., spherocytes, stomatocytosis, or elliptocytosis) or hemoglobinopathies (e.g., sickle cell disease, thalassemia). Table 3.2 lists the differential diagnoses based on the specific red cell morphological abnormalities.

The mean corpuscular volume (MCV) confirms the findings on the smear with reference to red cell size, for example, microcytic (<70 fl), macrocytic (>85 fl), or normocytic (72–79 fl). Figure 3.1 delineates diagnosis of anemia by examination of the smear and consideration of the MCV value. The mean corpuscular hemoglobin (MCH) and mean corpuscular hemoglobin concentration (MCHC) are calculated values and generally of less diagnostic value. The MCH usually parallels the MCV. The MCHC is a measure of cellular hydration status. A high value (>35 g/dl) is characteristic of spherocytosis and a low value is commonly associated with iron deficiency.

---

[1]Children with cyanotic congenital heart disease, chronic respiratory insufficiency, arteriovenous pulmonary shunts, or hemoglobinopathies that alter oxygen affinity can be functionally anemic with hemoglobin levels in the normal range.

TABLE 3.1   Etiologic Classification and Major Diagnostic Features of Anemia in Children

Etiologic classification	Diagnostic features
1. Impaired red cell formation	
a. Deficiency	
i. Decreased dietary intake (e.g., excessive cows' milk (iron-deficiency anemia), vegan (vitamin B_{12} deficiency))	
ii. Increased demand, for example, growth (iron) hemolysis (folic acid)	
iii. Decreased absorption	
• Specific: intrinsic factor lack (vitamin B_{12})	
• Generalized: malabsorption syndrome (e.g., folic acid, iron)	
iv. Impairment in red cell formation can result from one of the following deficiencies:	
i. Iron deficiency	Hypochromic, microcytic red cells; low MCV, low MCH, low MCHC, high RDW,[a] low serum ferritin, high FEP, guaiac positivity
ii. Folate deficiency	Macrocytic red cells, high MCV, high RDW, megaloblastic marrow, low serum, and red cell folate
iii. Vitamin B_{12} deficiency	Macrocytic red cells, high MCV, high RDW, megaloblastic marrow, low serum B_{12}, decreased gastric acidity; Schilling test positive
iv. Vitamin C deficiency	Clinical scurvy
v. Protein deficiency	Kwashiorkor
vi. Vitamin B_6 deficiency	Hypochromic red cells, sideroblastic bone marrow, high serum ferritin
vii. Thyroxine deficiency	Clinical hypothyroidism, low T_4, high TSH
b. Bone marrow failure	
i. Failure of a single cell line	
• Megakaryocytes[b]	
Amegakaryocytic thrombocytopenic purpura with absent radii (TAR)	Limb abnormalities, thrombocytopenic purpura absent megakaryocytes
• Red cell precursors	
Congenital red cell aplasia (Diamond–Blackfan anemia)	Absent red cell precursors
Acquired red cell aplasia (transient erythroblastopenia of childhood, TEC)	Absent red cell precursors
• White cell precursors[b]	
Congenital neutropenias	Neutropenia, recurrent infection
ii. Failure of all cell lines (characterized by pancytopenia and acellular or hypocellular marrow)	
• Congenital	
Fanconi anemia	Multiple congenital anomalies, chromosomal breakage
Familial without anomalies	Familial history, no congenital anomalies
Dyskeratosis congenita	Marked mucosal and cutaneous abnormalities
• Acquired	
Idiopathic	No identifiable cause
Secondary	History of exposure to drugs, radiation, household toxins, infections; (parvovirus B19, HIV) associated immunologic disease
iii. Infiltration	
• Benign (e.g., osteopetrosis, storage diseases)	
• Malignant primary (e.g., leukemia, myelofibrosis)	Bone marrow: morphology, cytochemistry, immunologic markers, cytogenetics, molecular features
• Secondary (e.g., neuroblastoma, lymphoma)	VMA, imaging studies, skeletal survey, bone marrow
iv. Dyshematopoietic anemias (decreased erythropoiesis, decreased iron utilization)	
• Anemia of chronic disease	Evidence of systemic illness
• Renal failure and hepatic disease	BUN and liver function tests
• Disseminated malignancy	Clinical evidence
• Connective tissue diseases	Rheumatoid arthritis
• Malnutrition	Clinical evidence
• Sideroblastic anemias	Hypochromic anemia, ring sideroblasts Overt or occult guaiac positive
1. Blood loss	
2. Hemolytic anemia	
a. Corpuscular	Splenomegaly, jaundice
i. Membrane defects (spherocytosis, elliptocytosis)	Morphology, osmotic fragility
ii. Enzymatic defects (pyruvate kinase, G6PD)	Autohemolysis, enzyme assays
iii. Hemoglobin defects	
• Heme	
• Globin	
Qualitative (e.g., sickle cell)	Hb electrophoresis
Quantitative (e.g., thalassemia)	Quantitative HbF, A_2 content

(Continued)

TABLE 3.1 *(Continued)*

Etiologic classification	Diagnostic features
b. Extracorpuscular	
i. Immune	Direct antiglobulin test (Coombs' test)
• Isoimmune	
• Autoimmune	
Idiopathic	Direct antiglobulin test, antibody identification
Secondary	
Immunologic disorder (e.g., lupus)	Decreased C_3, C_4, CH_{50}-positive ANA
One-cell line (e.g., red cells)	Anemia—direct antiglobulin test positive
Multiple cell line (e.g., white blood cells, platelets)	Neutropenia—immunotropenia, thrombocytopenia—ITP
ii. Nonimmune (idiopathic, secondary)	

[a]RDW 5 *coefficient of variation of the RBC distribution width (normal between 11.5% and 14.5%).*
[b]*Not associated with anemia.*
FEP, free erythrocyte protoporphyrin; G6PD, glucose-6-phosphate dehydrogenase; Hb, hemoglobin; ITP, idiopathic thrombocytopenic purpura, MCH, mean corpuscular hemoglobin; MCHC, mean corpuscular hemoglobin concentration; MCV, mean corpuscular volume; RBC, red blood cell; RDW, red cell distribution width (see definition); VMA, vanillylmandelic acid.

TABLE 3.2 Specific Red Cell Morphologic Abnormalities

1. **Target cells**
   Increased surface/volume ratio (generally does not affect red cell survival)
   i. Thalassemic syndromes
   ii. Hemoglobinopathies
       – Hb AC or CC
       – Hb SS, SC, S-Thal
       – HbE (heterozygote and homozygote)
       – HbD
   iii. Obstructive liver disease
   iv. Postsplenectomy or hyposplenic states
   v. Severe iron deficiency
   vi. LCAT deficiency: congenital disorder of lecithin/cholesterol acyltransferase deficiency (corneal opacifications, proteinuria, target cells, moderately severe anemia)
   vii. Abetalipoproteinemia
   viii. Hereditary xerocytosis
2. **Spherocytes**
   Decreased surface/volume ratio, hyperdense (>MCHC)
   i. Hereditary spherocytosis
   ii. ABO incompatibility: antibody-coated fragment of RBC membrane removed
   iii. Autoimmune hemolytic anemia: antibody-coated fragment of RBC membrane removed
   iv. G6PD deficiency
   v. Microangiopathic hemolytic anemia (MAHA): fragment of RBC lost after impact with abnormal surface
   vi. SS disease: fragment of RBC removed in reticuloendothelial system
   vii. Hypersplenism
   viii. Burns: fragment of damaged RBC removed by spleen
   ix. Posttransfusion
   x. Pyruvate kinase deficiency
   xi. Water-dilution hemolysis: fragment of damaged RBC removed by spleen
3. **Acanthocytes (spur cells)**[a]
   Cells with 5–10 spicules of varying length; spicules irregular in space and thickness, with wide bases; appear smaller than normal cells because they assume a spheroid shape
   i. Liver disease
   ii. Disseminated intravascular coagulation (and other MAHA)
   iii. Postsplenectomy or hyposplenic state
   iv. Vitamin E deficiency
   v. Hypothyroidism
   vi. Abetalipoproteinemia: rare congenital disorder; 50–100% of cells acanthocytes; associated abnormalities (fat malabsorption, retinitis pigmentosa, neurologic abnormalities)
   vii. Malabsorptive states

(Continued)

TABLE 3.2 (*Continued*)

4. **Echinocytes (burr cells)**[a]

10–30 spicules equal in size and evenly distributed over RBC surface; caused by alteration in extracellular or intracellular environment

 i. Artifact
 ii. Renal failure
 iii. Dehydration
 iv. Liver disease
 v. Pyruvate kinase deficiency
 vi. Peptic ulcer disease or gastric carcinoma
 vii. Immediately after red cell transfusion
 viii. Rare congenital anemias due to decreased intracellular potassium

5. **Pyknocytes**[a]

Distorted, hyperchromic, contracted RBC; can be similar to echinocytes and acanthocytes

6. **Schistocytes**

Helmet, triangular shapes, or small fragments. Caused by fragmentation upon impact with abnormal vascular surface (e.g., fibrin strand, vasculitis, artificial surface in circulation)

 i. Disseminated intravascular coagulation (DIC)
 ii. Severe hemolytic anemia (e.g., G6PD deficiency)
 iii. MAHA
 iv. Hemolytic uremic syndrome
 v. Prosthetic cardiac valve, abnormal cardiac valve, cardiac patch, coarctation of the aorta
 vi. Connective tissue disorder (e.g., SLE)
 vii. Kasabach–Merritt syndrome
 viii. Purpura fulminans
 ix. Renal vein thrombosis
 x. Burns (spheroschistocytes as a result of heat)
 xi. Thrombotic thrombocytopenia purpura
 xii. Homograft rejection
 xiii. Uremia, acute tubular necrosis, glomerulonephritis
 xiv. Malignant hypertension
 xv. Systemic amyloidosis
 xvi. Liver cirrhosis
 xvii. Disseminated carcinomatosis
 xviii. Chronic relapsing schistocytic hemolytic anemia

7. **Elliptocytes**

Elliptical cells, normochromic; seen normally in less than 1% of RBCs; larger numbers occasionally seen in a normal patient

 i. Hereditary elliptocytosis
 ii. Iron deficiency (increased with severity, hypochromic)
 iii. SS disease
 iv. Thalassemia major
 v. Severe bacterial infection
 vi. SA trait
 vii. Leukoerythroblastic reaction
 viii. Megaloblastic anemias
 ix. Any anemia may occasionally present with up to 10% elliptocytes
 x. Malaria

8. **Teardrop cells**

Shape of drop, usually microcytic, often also hypochromic

 i. Newborn
 ii. Thalassemia major
 iii. Leukoerythroblastic reaction
 iv. Myeloproliferative syndromes

9. **Stomatocytes**

Has a slit-like area of central pallor

 i. Normal (in small numbers)
 ii. Hereditary stomatocytosis
 iii. Artifact
 iv. Thalassemia
 v. Acute alcoholism
 vi. Rh null disease (absence of Rh complex)
 vii. Liver disease
 viii. Malignancies

(*Continued*)

TABLE 3.2 *(Continued)*

10. **Nucleated red blood cells**
    Not normal in the peripheral blood beyond the first week of life
    i. Newborn (first 3–4 days)
    ii. Intense bone marrow stimulation
        – Hypoxia (especially postcardiac arrest)
        – Acute bleeding
        – Severe hemolytic anemia (e.g., thalassemia, SS hemoglobinopathy)
    iii. Congenital infections (e.g., sepsis, congenital syphilis, CMV, rubella)
    iv. Postsplenectomy or hyposplenic states: spleen normally removes nucleated RBC
    v. Leukoerythroblastic reaction: seen with extramedullary hematopoiesis and bone marrow replacement; most commonly leukemia or solid tumor—fungal and mycobacterial infection may also do this; leukoerythroblastic reaction is also associated with teardrop red cells, 10,000–20,000 WBC with small to moderate numbers of metamyelocytes, myelocytes, and promyelocytes; thrombocytosis with large bizarre platelets
    vi. Megaloblastic anemia
    vii. Dyserythropoietic anemias
11. **Blister cells**
    i. Red cell area under membrane, free of hemoglobin, appearing like a blister
    ii. G6PD deficiency (during hemolytic episode)
    iii. SS disease
    iv. Pulmonary emboli
12. **Basophilic stippling**
    Coarse or fine punctate basophilic inclusions that represent aggregates of ribosomal RNA
    i. Hemolytic anemias (e.g., thalassemia trait)
    ii. Iron-deficiency anemia
    iii. Lead poisoning
13. **Howell–Jolly bodies**
    Small, well-defined, round, densely stained nuclear-remnant inclusions; 1 mm in diameter; centric in location
    i. Postsplenectomy or hyposplenia
    ii. Newborn
    iii. Megaloblastic anemias
    iv. Dyserythropoietic anemias
    v. A variety of types of anemias (rarely iron-deficiency anemia, hereditary spherocytosis)
14. **Cabot's Ring bodies**
    Nuclear remnant ring configuration inclusions
    i. Pernicious anemia
    ii. Lead toxicity
15. **Heinz bodies**
    Denatured aggregated hemoglobin
    i. Normal in newborn
    ii. Thalassemia
    iii. Asplenia
    iv. Chronic liver disease
    v. Heinz body hemolytic anemia

[a]May be morphologically indistinguishable.

MCV and red cell distribution width (RDW) indices, available from automated electronic blood-counting equipment, are extremely helpful in defining the morphology and the nature of the anemia and have led to a classification based on these indices (Table 3.3).

The MCV and reticulocyte count are helpful in the differential diagnosis of anemia (Figure 3.2). An elevated reticulocyte count suggests chronic blood loss or hemolysis; a normal or depressed count suggests impaired red cell formation. The reticulocyte count must be adjusted for the level of anemia to obtain the reticulocyte index,[2] a more accurate reflection of erythropoiesis. In patients with bleeding or hemolysis, the reticulocyte index should be at least 3%, whereas in patients with anemia due to decreased production of red cells, the reticulocyte index is <3% and frequently <1.5%.

[2]Reticulocyte index = reticulocyte count × patient's hematocrit/normal hematocrit. Example: reticulocyte count 6%, hematocrit 15%, reticulocyte index = 6 × 15/45 = 2%.

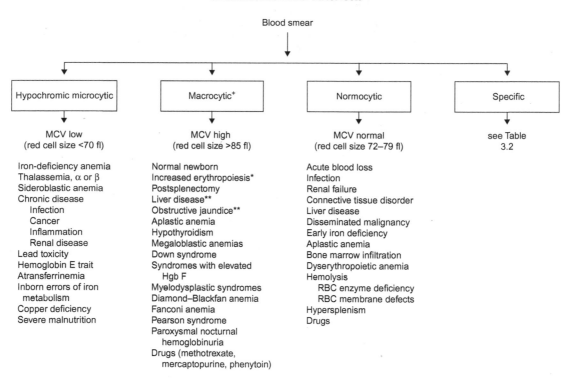

FIGURE 3.1 An approach to the diagnosis of anemia by examination of the blood smear. [+]Spurious macrocytosis (high MCV) may be caused by macroagglutinated red cells (e.g., *Mycoplasma pneumoniae* and autoimmune hemolytic anemia). *Increased number of reticulocytes. **On the basis of increased membrane resulting in an increased membrane/volume ratio. Increased membrane results from exchanges between red cell lipids and altered lipid balance in these conditions.

TABLE 3.3 Classification of Nature of the Anemia Based on MCV and RDW

	MCV low	MCV normal	MCV high
RDW normal	**Microcytic homogeneous**	**Normocytic homogeneous**	**Macrocytic homogeneous**
	Heterozygous thalassemia	Normal	Inherited bone marrow failure syndromes
	Chronic disease	Chronic disease	Preleukemia
		Chronic liver disease	
		Non-anemic hemoglobinopathy (e.g., AS, AC)	
		Chemotherapy	
		Chronic myelocytic leukemia	
		Hemorrhage	
		Hereditary spherocytosis	
RDW high	**Microcytic heterogeneous**	**Normocytic heterogeneous**	**Macrocytic heterogeneous**
	Iron deficiency	Early iron or folate deficiency	Folate deficiency
	S β-thalassemia	Mixed deficiencies	Vitamin B_{12} deficiency
	Hemoglobin H	Hemoglobinopathy (e.g., SS)	Immune hemolytic anemia
	Red cell	Myelofibrosis	Cold agglutinins
	Fragmentation disorders	Sideroblastic anemia	

MCV, mean corpuscular volume; RDW, red cell distribution width, which is coefficient of variation of RBC distribution width (normal, 11.5–14.5%).

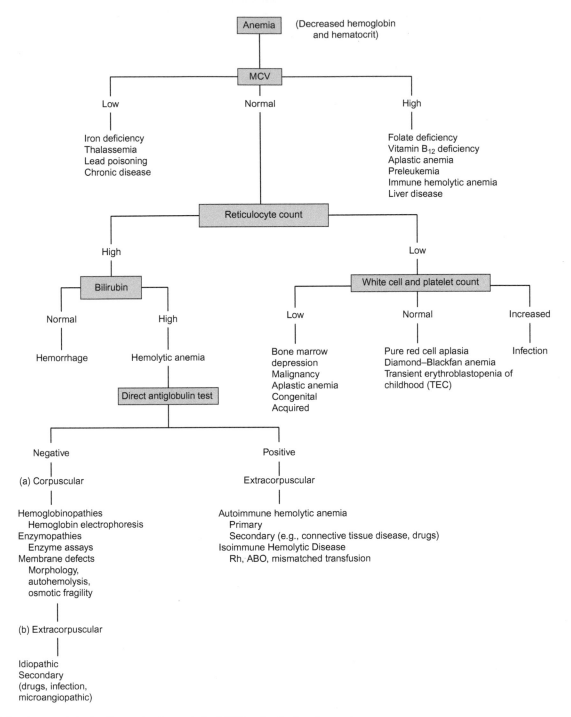

FIGURE 3.2 Approach to the diagnosis of anemia by MCV and reticulocyte count.

In more refractory cases of anemia, bone marrow examination may be indicated. A bone marrow smear should be stained for iron, where indicated, to estimate iron stores and to diagnose the presence of a sideroblastic anemia. Bone marrow examination may indicate a normoblastic, megaloblastic, or sideroblastic morphology. Figure 3.3 presents the causes of each of these findings.

Table 3.5 lists various laboratory studies helpful in the investigation of a patient with anemia.

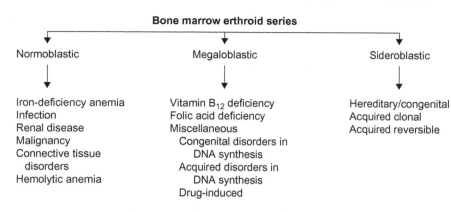

FIGURE 3.3 Causes of normoblastic, megaloblastic, and sideroblastic bone marrow morphology.

TABLE 3.4 The Clinical Features Suggestive of a Hemolytic Process

- Ethnic factors:—incidence of sickle gene carrier in the black population (8%), high incidence of thalassemia trait in people of Mediterranean ancestry and high incidence of glucose-6-phosphate dehydrogenase (G6PD) deficiency among Sephardic Jews
- Age factors:—anemia and jaundice in an Rh-positive infant born to a mother who is Rh negative or a group A or group B infant born to a group O mother (setting for a hemolytic anemia)
- History of anemia, jaundice, or gallstones in family
- Persistent or recurrent anemia associated with reticulocytosis
- Anemia unresponsive to hematinics
- Intermittent bouts or persistent indirect hyperbilirubinemia
- Splenomegaly
- Hemoglobinuria
- Presence of multiple gallstones
- Chronic leg ulcers
- Development of anemia or hemoglobinuria after exposure to certain drugs
- Cyanosis without cardiorespiratory distress
- Polycythemia: 2,3-diphosphoglycerate mutase deficiency
- Dark urine due to dipyrroluria: unstable hemoglobins, thalassemia, and ineffective erythropoiesis

The investigation of anemia entails the following steps:

- Detailed *history*. Table 3.1 lists the various causes of anemia with the associated diagnostic laboratory and clinical features.
- *Complete blood count*, to establish whether the anemia is only due to a one-cell line (red cell line) or part of a three-cell line abnormality (abnormality of red cell count, white blood cell count, and platelet count).
- Determination of the *morphologic characteristics* of the anemia based on blood smear (Table 3.2) and consideration of the *MCV* (Figures 3.1, 3.2 and Table 3.3) and *RDW* (Table 3.3) and morphologic consideration of *white blood cell and platelet morphology*.
- Reticulocyte count as a reflection of erythropoiesis (Figure 3.2).
- Determination of whether there is evidence of a hemolytic process by:
  - Consideration of the clinical features suggesting hemolytic disease (Table 3.4).
  - Laboratory demonstration of the presence of a hemolytic process (Table 3.5).
  - Determination of the precise cause of the hemolytic anemia by special hematologic investigations (Table 3.5).
- Bone marrow aspiration, if required, to examine erythroid, myeloid, and megakaryocytic morphology to determine whether there is normoblastic, megaloblastic, or sideroblastic erythropoiesis and to exclude marrow pathology (e.g., aplastic anemia, leukemia, and benign or malignant infiltration of the bone marrow) (Figure 3.3).
- Determination of underlying cause of anemia by additional tests (Table 3.5).
- Few symptoms occur until the hemoglobin level falls below 7—8 g/dl after which pallor becomes evident in the skin and mucus membranes, weakness tachycardia and tachypnea on exertion may occur. Physiological adjustments to anemia include:
  - Tachycardia and increased cardiac output.
  - Increased 2,3-DPG within red cells resulting in a "shift to the left" of the oxygen dissociation curve which reduces affinity of hemoglobin for oxygen resulting in more complete transfer of oxygen to tissues.

TABLE 3.5   Laboratory Studies in the Investigation of a Patient with Anemia

1. **Usual initial studies**
   a. Hemoglobin and hematocrit determination
   b. Erythrocyte count and red cell indices, including MCV and RDW
   c. Reticulocyte count
   d. Study of stained blood smear
   e. Leukocyte count and differential count
   f. Platelet count
2. **Suspected iron deficiency**
   a. Free erythrocyte protoporphyrin
   b. Serum ferritin levels
   c. Stool for occult blood
   d. ^{99m}Tc pertechnetate scan for Meckel's diverticulum—if indicated
   e. Endoscopy (upper and lower bowel)—if indicated
3. **Suspected vitamin B$_{12}$ or folic acid deficiency**
   a. Bone marrow
   b. Serum vitamin B$_{12}$ level
   c. Serum folate level
   d. Gastric analysis after histamine injection
   e. Vitamin B$_{12}$ absorption test (radioactive cobalt) (Schilling test)
4. **Suspected hemolytic anemia**
   a. Evidence of red cell breakdown
      i. Blood smear—red cell fragments (schistocytes), spherocytes, target cells
      ii. Increased unconjugated bilirubin-urinary and blood
      iii. Lower or absent serum haptoglobin (normal level 125 ± 25 mg/dl)
      iv. Raised plasma hemoglobin level (normal level <1 mg hemoglobin per dl of plasma, visibly red plasma when hemoglobin >50 mg per dl of plasma)
      v. Increased fecal and urinary urobilinogen
      vi. Hemoglobinuria (Figure 9.3 lists the causes of hemoglobinuria)
      vii. Hemosiderinuria (due to sloughing of iron-laden tubular cells into urine)
      viii. Raised plasma methemalbumin and methemoglobin
   b. Evidence of increased erythropoeisis (in response to hemoglobin reduction)
      i. Reticulocytosis—frequently up to 10–20%; rarely, as high as 80%
      ii. Increased MCV due to the presence of reticulocytosis and increased RDW as the hemoglobin level falls
      iii. Increased normoblasts in blood smear
      iv. Specific morphologic abnormalities—sickle cells, target cells, basophilic stippling, irregularly contracted cells (schistocytes), and spherocytes
      v. Erythroid hyperplasia of the bone marrow—erthroid/myeloid ratio in the marrow increasing from 1:5 to 1:1
      vi. Expansion of marrow space in chronic hemolysis resulting in:
         − prominence of frontal bones
         − broad cheekbones
         − widened intratrabecular spaces, hair-on-end appearance of skull radiographs
         − biconcave vertebrae with fish-mouth intervertebral spaces
   c. Evidence of type of corpuscular hemolytic anemia
      a. Membrane defects
         i. Blood smear: spherocytes, ovalocytes, pyknocytes, stomatocytes
         ii. Osmotic fragility test (fresh and incubated)
         iii. Autohemolysis test
      b. Hemoglobin defects
         i. Blood smear: sickle cells, target cells
         ii. Sickling test
         iii. Hemoglobin electrophoresis
         iv. Quantitative hemoglobin F determination
         v. Kleihauer–Betke smear
         vi. Heat-stability test for unstable hemoglobin
      c. Enzymes defects
         i. Heinz-body preparation
         ii. Autohemolysis test
         iii. Specific enzyme assay
   d. Evidence of type of extracorpuscular hemolytic anemia
      a. Immune
         i. Direct antiglobulin test: IgG (gamma), Cu3 (complement), broad-spectrum, both gamma and complement
         ii. Flow cytometric analysis of red cells with monoclonal antibodies to GP1-linked surface antigens for PNH
         iii. Donath–Landsteiner antibody
         iv. ANA

(Continued)

TABLE 3.5   (*Continued*)

5. **Suspected aplastic anemia or leukemia**
   a. Bone marrow (aspiration and biopsy)—cytochemistry, immunologic markers, chromosome analysis
   b. Skeletal radiographs
6. **Other tests often used especially to diagnose the primary disease**
   a. Viral serology, for example, HIV
   b. ANA, complement, CH_{50}
   c. Blood urea, creatinine, T_4, TSH
   d. Tissue biopsy (skin, lymph node, liver)

Further Reading and References

Bessman, J.D., Gilmer, P.R., Gardner, F.H., 1983. Improved classification of anemias by MCV and RDW. Am. J. Clin. Pathol. 80, 322.

Blanchette, V., Zipursky, A., 1984. Assessment of anemia in newborn infants. Clin. Perinatol. 11, 489.

Lanzkowsky, P., 1980. Diagnosis of anemia in the neonatal period and during childhood, p. 3. Pediatric Hematology-Oncology: A Treatise for the Clinician. McGraw-Hill, New York, NY.

4

Lymphadenopathy and Diseases of the Spleen

Philip Lanzkowsky

Lymphadenopathy and splenomegaly are common findings in children. Both benign and malignant processes can produce these findings and it is important to distinguish between the two so that appropriate management can be undertaken.

LYMPHADENOPATHY

Enlarged lymph nodes are commonly found in children. Lymphadenopathy might be caused by proliferation of cells intrinsic to the node, such as lymphocytes, plasma cells, monocytes, or histiocytes, or by infiltration of cells extrinsic to the node, such as neutrophils and malignant cells. In most instances, lymphadenopathy represents transient, self-limited proliferative responses to local or generalized infections.

Reactive hyperplasia, defined as a polyclonal proliferation of one or more cell types, is the most frequent diagnosis found in lymph node biopsies in children.

Lymphadenopathy, however, may be a presenting sign of malignancies such as leukemia, lymphoma, or neuroblastoma. It is important to be able to differentiate benign from malignant lymphadenopathy clinically. Lymphadenopathy in the head and neck region must be differentiated from several other nonlymphatic masses due to congenital malformations (Table 4.1).

Systematic palpation of the lymph nodes is important and should include examination of the occipital, posterior auricular, preauricular, tonsillar, submandibular, submental, upper anterior cervical, lower anterior cervical, posterior upper and lower cervical, supraclavicular, infraclavicular, axillary, epitrochlear, and popliteal lymph nodes. Many children have small palpable nodes in the cervical, axillary, and inguinal regions which are usually benign in nature.

When a child presents with lymphadenopathy, management is based on the following factors.

History

This involves the duration of the lymphadenopathy; presence of fever; recent upper respiratory tract infection; sore throat; skin lesions or abrasions, or other infections in the lymphatic region drained by the enlarged lymph nodes; immunizations; medications; previous cat scratches, rodent bites, or tick bites; arthralgia; sexual history; transfusion history; travel history; and consumption of unpasteurized milk. Significant weight loss, night sweats, or other systemic symptoms should also be recorded as part of the patient's history.

Age

Although in young children cervical lymphadenopathy, especially in the upper cervical region, is usually due to infection, more serious disorders may have to be considered.

In children younger than 6 years, the most common cancers of the head and neck are neuroblastoma, rhabdomyosarcoma, leukemia, and non-Hodgkin lymphoma. In children 7–13 years of age, non-Hodgkin lymphoma and Hodgkin lymphoma are equally common, followed by thyroid carcinoma and rhabdomyosarcoma; and for those older than 13 years Hodgkin lymphoma is the more common cancer encountered.

42

TABLE 4.1   Differential Diagnosis of Nonlymph Node Masses in Neck

Cystic hygroma
Branchial cleft anomalies, branchial cysts
Thyroglossal duct cysts
Epidermoid cysts
Neonatal torticollis
Lateral process of lower cervical vertebra may be misdiagnosed as supraclavicular node

TABLE 4.2   Differential Diagnosis of Lymphadenopathy

1. **Nonspecific reactive hyperplasia (polyclonal)**
2. **Infection**
   a. Bacterial: *Staphylococcus*, *Streptococcus*, anaerobes, tuberculosis, atypical mycobacteria, *Bartonella henselae* (cat scratch disease, brucellosis, *Salmonella typhi*, diphtheria, *Chlamydia trachomatis* lymphogranuloma venereum), *Calymmatobacterium granulomatis*, *Francisella tularensis*
   b. Viral: Epstein—Barr virus, cytomegalovirus, adenovirus, rhinovirus, coronavirus, respiratory syncytial virus, influenza, coxsackie virus, rubella, rubeola, varicella, HIV, herpes simplex virus, human herpes virus 6 (HHV-6)
   c. Protozoal: Toxoplasmosis, malaria, trypanosomiasis
   d. Spirochetal: Syphilis, *Rickettsia typhi* (murine typhus)
   e. Fungal: Coccidioidomycosis (valley fever), histoplasmosis, *Cryptococcus*, aspergillosis
   f. Postvaccination: Smallpox, live attenuated measles, DPT, Salk vaccine, typhoid fever
3. **Connective tissue disorders**
   a. Rheumatoid arthritis
   b. Systemic lupus erythematosus
4. **Hypersensitivity states**
   a. Serum sickness
   b. Drug reaction (e.g., Dilantin, mephenytoin, pyrimethamine, phenylbutazone, allopurinol, isoniazid, antileprosy, and antithyroid medications)
5. **Lymphoproliferative disorders** (Chapter 16)
   a. Angioimmunoblastic lymphadenopathy with dysproteinemia
   b. X-linked lymphoproliferative syndrome
   c. Lymphomatoid granulomatosis
   d. Sinus histiocytosis with massive lymphadenopathy (Rosai—Dorfman disease)
   e. Castleman disease benign (giant lymph node hyperplasia, angiofollicular lymph node hyperplasia)
   f. Autoimmune lymphoproliferative syndrome (ALPS) (Canale—Smith syndrome)
   g. Posttransplantation lymphoproliferative disorder (PTLD)
6. **Neoplastic diseases**
   a. Hodgkin and non-Hodgkin lymphomas
   b. Leukemia
   c. Metastatic disease from solid tumors: neuroblastoma, nasopharyngeal carcinoma, rhabdomyosarcoma, thyroid cancer
   d. Histiocytosis
      i. Langerhans cell histiocytosis
      ii. Familial hemophagocytic lymphohistiocytosis
      iii. Macrophage activation syndrome
      iv. Malignant histiocytosis
7. **Storage diseases**
   a. Niemann—Pick disease
   b. Gaucher disease
   c. Cystinosis
8. **Immunodeficiency states**
   a. Chronic granulomatous disease
   b. Leukocyte adhesion deficiency
   c. Primary dysgammaglobulinemia with lymphadenopathy
9. **Miscellaneous causes**
   a. Kawasaki disease (mucocutaneous lymph node syndrome)
   b. Kikuchi—Fujimoto disease (self-limiting histiocytic necrotizing lymphadenitis)
   c. Sarcoidosis
   d. Beryllium exposure
   e. Hyperthyroidism
   f. Periodic fever, aphthous stomatitis, pharyngitis and cervical adenitis syndrome (PFAPA syndrome)

Location

Enlargement of tonsillar and inguinal lymph nodes is most likely secondary to localized infection; enlargement of supraclavicular and axillary lymph nodes is more likely to be of a serious nature. Enlargement of the left supraclavicular node, in particular, should suggest a malignant disease (e.g., lymphoma or rhabdomyosarcoma) arising in the abdomen and spreading via the thoracic duct to the left supraclavicular area. Enlargement of the right supraclavicular node indicates intrathoracic lesions because this node drains the superior areas of the lungs and mediastinum. Palpable supraclavicular nodes are an indication for a thorough search for intrathoracic or intraabdominal pathology.

Localized or Generalized

Lymphadenopathy is either localized (one region affected) or generalized (two or more noncontiguous lymph node regions involved). Although localized lymphadenopathy is generally due to local infection in the region drained by the particular lymph nodes, it may also be due to malignant disease, such as Hodgkin lymphoma or neuroblastoma. Generalized lymphadenopathy is caused by many disease processes. Lymphadenopathy may initially be localized and subsequently become generalized.

Size

Nodes in excess of 2.5 cm should be regarded as pathologic. In addition, nodes that increase in size over time are significant.

Character

Malignant nodes are generally firm and rubbery. They are usually not tender or erythematous. Occasionally, a rapidly growing malignant node may be tender. Nodes due to infection or inflammation are generally warm, tender, and fluctuant. If infection is considered to be the cause of the adenopathy, it is reasonable to give a 2-week trial of antibiotics. If there is no reduction in the size of the lymph node within this period, careful observation of the lymph node is necessary. If the size, location, and character of the node suggest malignant disease, the node should be biopsied.

Diagnosis of Lymphadenopathy

Table 4.2 outlines the differential diagnosis of lymphadenopathy.

Figure 4.1 provides a diagnostic algorithm for evaluation of mononucleosis-like illness and Figure 4.2 for diagnostic evaluation of cervical lymphadenitis.

The following investigations should be carried out to elucidate the cause of either localized or generalized lymphadenopathy:

- Thorough history of infection, contact with rodents or cats, and systemic complaints.
- Careful examination of the lymphadenopathy including size, consistency, mobility, warmth, tenderness, erythema, fluctuation, and location. All the lymph-node-bearing areas as outlined above should be carefully examined.
- Physical examination for evidence of hematologic disease, such as hepatosplenomegaly and petechiae.
- Blood count and erythrocyte sedimentation rate (ESR).
- Skin testing for tuberculosis.
- Bacteriologic culture of regional lesions (e.g., throat).
- Specific serologic tests for Epstein–Barr virus (EBV), *Bartonella henselae* (IFA), syphilis (VDRL) toxoplasmosis, cytomegalovirus (CMV), human immunodeficiency virus (HIV), tularemia, brucellosis, histoplasmosis, coccidioidomycosis.
- Chest radiograph and CT scan (if necessary); abdominal sonogram and CT, if indicated.
- Ultrasonography is useful in an acute setting in assessing whether a swelling is nodal in origin, an infected cyst or other soft tissue mass. It may detect an abscess requiring drainage.
- EKG and echocardiogram if Kawasaki disease is suspected.
- Lymph node aspiration and culture; helpful in isolating the causative organism and deciding on an appropriate antibiotic when infection is the cause of the lymphadenopathy.
- Fine needle aspiration; may yield a definite or preliminary cytologic diagnosis and occasionally obviate the need for lymph node biopsy. It provides limited material in the event flow cytometry is required and negative results cannot rule out a malignancy because the sample may be inadequate.

- Bone marrow examination if leukemia or lymphoma is suspected.
- Lymph node biopsy is indicated if:
  - Initial physical examination and history suggest malignancy.
  - Lymph node size is greater than 2.5 cm in absence of signs of infection.
  - Lymph node persists or enlarges.
  - Appropriate antibiotics fail to shrink node within 2 weeks.
  - Supraclavicular adenopathy.

Close communication between surgeon, oncologist, and pathologist is critical to maximize results from lymph node biopsy. In addition, the following precautions should be observed:

- Upper cervical and inguinal areas should be avoided; lower cervical and axillary nodes are more likely to give reliable information.
- The largest node should be biopsied, not the most accessible one. The oncologist should select the node to be biopsied in consultation with the surgeon.
- The node should be removed intact with the capsule, not piecemeal.
- The lymph node should be immediately submitted to the pathologist fresh or in sufficient tissue culture medium to prevent the tissue from drying out. The node must not be left in strong light, where it will be subject to heat and it should not be wrapped in dry gauze, which may produce a drying artifact. Fresh and frozen samples should be set aside for additional studies, as noted below.

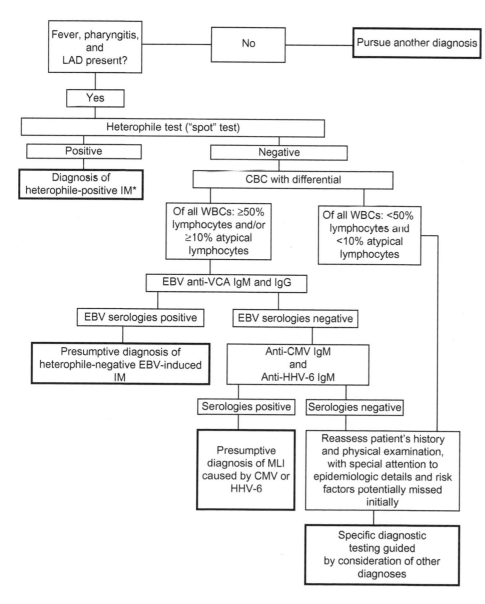

FIGURE 4.1 Diagnostic algorithm for evaluation of mononucleosis-like illness (MLI). CMV, cytomegalovirus; EBV, Epstein—Barr virus; HHV-6, human herpes virus 6; IM, infectious mononucleosis; LAD, lymphadenopathy; VCA, viral capsid antigen; WBC, white blood cell. *Consider possibility of false-positive heterophile test due to HIV-1 before finalizing diagnosis. *Source: Adapted from Hurt and Tammaro (2007), with permission.*

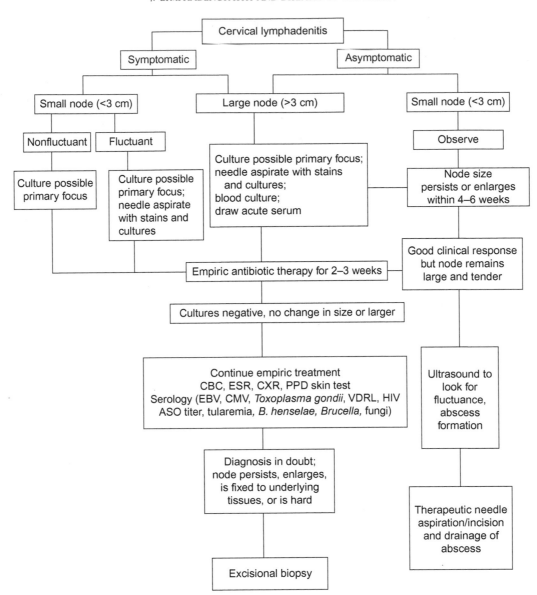

FIGURE 4.2 Diagnostic evaluation of cervical lymphadenitis. ASO, antistreptolysin titer; CXR, chest radiography; CBC, complete blood cell count; CMV, cytomegalovirus; EBV, Epstein–Barr virus; ESR, erythrocyte sedimentation rate; HIV, human immunodeficiency virus; PPD, purified protein derivative; VDRL, Venereal Disease Research Laboratories. *Source: Adapted from Gosche and Vick (2006), with permission.*

Intraoperative frozen section and cytologic smears should be performed. These findings, together with the clinical data, will determine which of the following additional studies may be required:

- Gram stain and culture (bacterial including mycobacterial, viral, and/or fungal) if clinically warranted or if intraoperative frozen section suggests an infection.
- Tissue in tissue culture medium for cytogenetic analysis in cases of suspected malignancy. Smears or touch preparations of the node on slides can be air dried for fluorescent *in situ* hybridization studies to confirm certain malignancies.
- Tissue frozen immediately for molecular studies.
- Immunohistochemical stains to help differentiate and classify tumor types.
- Flow cytometry for classifying and subtyping leukemias and lymphomas.
- Gene rearrangement studies for the T-cell receptor and the immunoglobulin gene may be required to determine monoclonality in leukemia or lymphoma. (These can be performed on fresh frozen tissue, or less optimally in formalin-fixed paraffin-embedded tissue.)
- Formalin fixation for light microscopic analysis.

Once the cause of the lymphadenopathy is ascertained, appropriate management can be undertaken.

DISEASES OF THE SPLEEN

The tip of the spleen is frequently palpable in otherwise normal infants and young children. It is usually palpable in premature infants and in about 30% of full-term infants. It may normally be felt in children up to 3 or 4 years of age. At an older age, the spleen tip is generally not palpable below the costal margin and a palpable spleen usually indicates splenic enlargement two to three times its normal size.

Asplenia

Congenital asplenia is found in the rare Ivemark syndrome—trilobed lungs, centralized liver, and cardiac defects as well as a risk of infection. Diagnosis of asplenia (congenital or postsurgical) is made by the presence of Howell–Jolly bodies and the presence of intracellular vesicles (appearing as pits or pocks) in erythrocytes and no uptake by Tc99 colloid sulfur radionuclide.

Congenital Polysplenia

This condition is characterized by the presence of several spleens of varying size and function, hepatobiliary abnormalities and cardiac anomalies.

Accessory Spleen

Accessory spleens occur in 15% of normal people and are usually present with no other abnormalities. The most frequent location is the splenic hilum or the tail of the pancreas or other locations in the abdomen or pelvis. Its identification is important when splenectomy is carried out for hematologic indications and the desired clinical effect is not obtained due to a functional accessory spleen.

Splenosis

Splenosis is autotransplantation of splenic tissue into the peritoneum or omentum and results from rupture of the spleen and spillage and subsequent implantation of splenocytes.

Sequestration of Spleen

Sequestration of the spleen refers to splenic enlargement when blood enters the spleen but is unable to exit properly, for example, sickle cell anemia in young infants and congenital spherocytosis, and is characterized by a sudden severe drop in hemoglobin, occasionally hypovolemic shock, and abdominal pain in the left upper quadrant.

Splenoptosis (Splenic Visceroptosis)

Splenoptosis occurs when the spleen is not fixed within the retroperitoneum, and a palpable spleen may be due to visceroptosis rather than true splenomegaly. This distinction is important to make so that extensive investigations for the cause of splenomegaly are not undertaken unnecessarily. Visceroptosis may result from congenital or acquired defects in the supporting mechanism responsible for maintaining the spleen in the correct position. The visceroptosed spleen may be felt anywhere from the upper abdomen to the pelvis and may undergo torsion. When the spleen is felt in the upper abdomen, it can easily be pushed under the left costal margin. This finding is helpful in diagnosing visceroptosis and in differentiating it from true splenomegaly.

In addition to this finding, an abdominal radiograph in the upright position may reveal intestinal gas bubbles between the left dome of the diaphragm and the spleen. This sign may be helpful in differentiating true splenomegaly from visceroptosis of the spleen.

SPLENOMEGALY

The significance of splenomegaly depends on the underlying disease. Splenomegaly can be caused by diseases that result in hyperplasia of the lymphoid and reticuloendothelial systems (e.g., infections, connective tissue disorders), infiltrative disorders (e.g., Gaucher disease, leukemia, lymphoma), hematologic disorders (e.g., thalassemia, hereditary spherocytosis), and conditions that cause distention of the sinusoids whenever there is increased pressure in the portal or splenic veins (portal hypertension). Table 4.3 lists the various causes of splenomegaly.

Diagnostic Approach to Splenomegaly

Detailed History

1. Fever or rigors indicative of infection (e.g., subacute bacterial endocarditis (SBE), infectious mononucleosis, malaria).
2. History of neonatal omphalitis, umbilical venous catheterization leading to inferior vena cava, or portal vein thrombosis resulting in portal hypertension.
3. Jaundice (evidence of liver disease) leading to portal hypertension.
4. Abnormal bleeding or bruising (hematologic malignancy).
5. Family history of hemolytic anemia (e.g., hereditary spherocytosis or thalassemia major).
6. Travel to endemic areas (e.g., malaria).
7. Trauma (splenic hematoma).

TABLE 4.3   Causes of Splenomegaly

1. **Infectious splenomegaly** (due to antigenic stimulation with hyperplasia of the reticuloendothelial and lymphoid systems)
   a. Bacterial: Acute and chronic systemic infection, subacute bacterial endocarditis, abscesses, typhoid fever, miliary tuberculosis, tularemia, plague
   b. Viral: Infectious mononucleosis (Epstein–Barr virus), cytomegalovirus, HIV, hepatitis A, B, C
   c. Spirochetal: Syphilis, Lyme disease, leptospirosis
   d. Rickettsial: Rocky Mountain spotted fever, Q fever, typhus
   e. Protozoal: Malaria, babesiosis, toxoplasmosis, *Toxocara canis*, *Toxocara catis*, leishmaniasis, schistosomiasis, trypanosomiasis
   f. Fungal: Disseminated candidiasis, histoplasmosis, coccidioidomycosis, South American blastomycosis
2. **Hematologic disorders**
   a. Hemolytic anemias, such as thalassemia, splenic sequestration crisis in sickle cell disease, hereditary spherocytosis
   b. Extramedullary hematopoiesis, as in osteopetrosis and myelofibrosis
   c. Myeloproliferative disorders (e.g., polycythemia vera, essential thrombocythemia)
3. **Infiltrative splenomegaly**
   a. Nonmalignant
      i. Langerhans cell histiocytosis
      ii. Storage diseases such as Gaucher disease, Niemann–Pick disease, GM-1 gangliosidosis, glycogen storage disease type IV, Tangier disease, Wolman disease, mucopolysaccharidoses, hyperchylomicronemia types I and IV, amyloidosis, and sarcoidosis
   b. Malignant
      i. Leukemia
      ii. Lymphoma: Hodgkin and non-Hodgkin
4. **Congestive splenomegaly**
   a. Intrahepatic (portal hypertension): Cirrhosis of the liver (e.g., neonatal hepatitis, α_1-antitrypsin deficiency, Wilson disease, cystic fibrosis)
   b. Prehepatosplenic or portal vein obstruction (e.g., thrombosis, vascular malformations)
5. **Immunologic diseases**
   a. Serum sickness, graft-versus-host disease (GVHD)
   b. Connective tissue disorders (e.g., systemic lupus erythematosus, rheumatoid arthritis—Felty syndrome, mixed connective tissue disorder, Sjogren syndrome, macrophage activation syndrome, systemic mastocytosis)
   c. Common variable immunodeficiency
   d. Autoimmune lymphoproliferative syndrome (ALPS) (Canale–Smith syndrome)
6. **Primary splenic disorders**
   a. Cysts
   b. Benign tumors (e.g., hemangioma, lymphangioma)
   c. Hemorrhage in spleen (e.g., subcapsular hematoma)
   d. Partial torsion of splenic pedicle leading to congestive splenomegaly, cyst, and abscess formation

Physical Examination

1. Size of spleen (measured in centimeters below costal margin); consistency, tenderness, audible rub. It is critical to differentiate splenoptosis from true enlargement of the spleen.
2. Hepatomegaly.
3. Lymphadenopathy.
4. Fever.
5. Ecchymoses, purpura, petechiae.
6. Stigmata of liver disease, such as jaundice, spider angiomata, or caput medusa.
7. Stigmata of rheumatoid arthritis or SLE.
8. Cardiac murmurs, Osler nodes, Janeway lesions, splinter hemorrhages, fundal hemorrhages as evidence of SBE.

Laboratory Investigations

The extent to which the following investigations are undertaken must be guided by clinical judgment. It is not necessary to perform all the evaluations. If the child appears well and the index of suspicion is low, it is reasonable to do no further investigations and reexamine the child in 1–2 weeks. If the splenomegaly persists, the following investigations should be done:

- *Blood count*: Red cell indices, reticulocyte count, platelet count, differential white blood cell count, and blood film (which may demonstrate evidence of hematologic malignancy, hemolytic disorders, viral and protozoal infections).
- *Evaluation for infection*: Blood culture and viral studies (CMV, EBV panel, HIV, toxoplasmosis, smear for malaria, tuberculin test).
- *Evaluation for evidence of hemolytic disease*: Blood count, reticulocyte count, blood smear, serum bilirubin, urinary urobilinogen, direct antiglobulin test (Coombs test), and red cell enzyme assays (if indicated).
- *Evaluation for liver disease*: Liver function tests, α_1-antitrypsin deficiency, serum copper, ceruloplasmin (to exclude Wilson disease), and liver biopsy (if indicated).
- *Evaluation for portal hypertension*: Ultrasound and Doppler of portal venous system and endoscopy (if indicated to exclude esophageal varices).
- *Evaluation for connective tissue disease*: ESR, C3, C4, CH_{50}, antinuclear antibody, rheumatoid factor, urinalysis, blood urea nitrogen, and serum creatinine.
- *Evaluation for infiltrative disease (benign and malignant)*:
  - Bone marrow aspiration and biopsy, looking for blasts, Langerhans cell histiocytes, or storage cells.
  - Enzyme assay for Gaucher and other storage diseases.
- *Lymph node biopsy*: If there is significant lymphadenopathy, lymph node biopsy may provide the diagnosis.
- *Imaging studies*:
  - Abdominal CT scan, if indicated.
  - Magnetic resonance imaging (MRI) if indicated.
  - Liver–spleen scans with ^{99m}Tc-sulfur colloid.
- *Splenectomy or partial splenectomy*: If less invasive studies have failed to provide the diagnosis, it may be necessary to perform a splenectomy or a partial splenectomy on rare occasions to establish a diagnosis. Splenic tissue must be processed for cultures and Gram stain, as well as for histology, flow cytometry, histochemical stains, electron microscopy, and gene rearrangement studies.

Once the etiology of the splenomegaly is ascertained, further management for the underlying disorder can be instituted.

Surgery Involving Spleen

Splenectomy is usually done laparoscopically and partial splenectomy has become a therapeutic alternative to total splenectomy. Partial splenectomy leaving at least 20% splenic tissue is sufficient to preserve immune competence and is suitable when splenectomy is performed for indications other than for immune-mediated hematologic disorders such as autoimmune hemolytic anemia or immune thrombocytopenic purpura.

The primary risk of splenectomy is overwhelming postsplenectomy infection (OPSI) and sepsis. Risk factors for OPSI are:

1. age of splenectomy—under 5 and especially infants under 2 years of age,
2. failure to receive presplenectomy immunization, and
3. noncompliance with prophylactic antibiotics

Reduction in incidence of OPSI can be achieved by:

1. presplenectomy immunization with protein conjugated vaccines against *Streptococcus pneumoniae* and *Haemophilus influenzae* type b.
2. postsplenectomy prophylactic antibiotics.
3. prompt, early, and effective medical treatment for fever.

Further Reading and References

Behrman, R., Kliegman, R.M., Jenson, H.B., 2004. Nelson Textbook of Pediatrics, seventeenth ed. Philadelphia, Saunders.

Gosche, J.R., Vick, L., 2006. Acute, subacute and chronic cervical lymphadenitis in children. Semin. Pediatr. Surg. 15 (2), 99−106.

Hoffman, R., Benz, E.J., Shattil, S.J., et al., 2005. Hematology: Basic Principles and Practice, fourth ed. Churchill Livingstone.

Hurt, C., Tammaro, D., 2007. Diagnostic evaluation of mononucleosis-like illnesses. Am. J. Med. 120 (10), 911.e1−911.e8.

Kim, D.S., 2006. Kawasaki disease. Yonsei Med. J. 47 (6), 759−772.

La Barge III, D.V., Salzmam, K.L., Harnsberger, H.R., et al., 2008. Sinus histiocytosis with massive lymphadenopathy (Rosai-Dorfman disease): imaging manifestations in the head and neck. Am. J. Roentgenol. 191 (6), W299−W306.

Leung, A.K.C., Robson, W.L.M., 2004. Childhood cervical lymphadenopathy. J. Pediatr. Health Care 18, 3−7.

Nathan, D., Orkin, S., 2003. Nathan and Oski's Hematology of Infancy and Childhood, sixth ed. Saunders, Philadelphia, PA.

Paradela, S., Lorenzo, J., Martinez-Gomez, W., et al., 2008. Interface dermatitis in skin lesions of Kikuchi-Fujimoto's disease: a histopathological marker of evolution into systemic lupus erythematosus? Lupus 17 (12), 1127−1135.

Tracy Jr, T.F., Muratore, C.S., 2007. Management of common head and neck masses. Semin. Pediatr. Surg. 16 (1), 3−13.

Twist, C.J., Link, M.P., 2002. Assessment of lymphadenopathy in children. Pediatr. Clin. N. Am. 49, 1009−1025.

5

Anemia During the Neonatal Period

Philip Lanzkowsky

Anemia during the neonatal period is caused by:

- *Hemorrhage*: acute or chronic.
- *Hemolysis*: congenital hemolytic anemias or due to isoimmunization.
- *Failure of red cell production*: inherited bone marrow failure syndromes, for example, Diamond−Blackfan anemia (pure red cell aplasia).

Table 5.1 lists the causes of anemia in the newborn.

HEMORRHAGE

Blood loss may occur during the prenatal, intranatal, or postnatal periods. Prenatal blood loss may be transplacental, intraplacental, or retroplacental or may be due to a twin-to-twin transfusion.

Prenatal Blood Loss

Transplacental Fetomaternal

In 50% of pregnancies fetal cells can be demonstrated in the maternal circulation, in 8% the transfer of blood is estimated to be between 0.5 and 40 ml and in 1% of cases exceeds 40 ml and is of sufficient magnitude to produce anemia in the infant. Transplacental blood loss may be acute or chronic. Table 5.2 lists the characteristics of acute and chronic blood loss in the newborn. It may be secondary to procedures such as diagnostic amniocentesis or external cephalic version. Fetomaternal hemorrhage is diagnosed by demonstrating fetal red cells by the acid-elution method of staining for fetal hemoglobin (Kleihauer−Betke technique) in the maternal circulation. Diagnosis of fetomaternal hemorrhage may be missed in situations in which red cells of the mother and infant have incompatible ABO blood groups. In such instances the infant's A- and B-cells are rapidly cleared from the maternal circulation by maternal anti-A or anti-B antibodies. In these cases an increase in maternal immune anti-A anti-B titers in the weeks after delivery may be helpful. The optimal timing for demonstrating fetal cells in maternal blood is within 2 h of delivery and no later than the first 24 h following delivery. This technique cannot be relied upon when maternal fetal hemoglobin is raised for other reasons such as:

1. Maternal thalassemia minor.
2. Maternal sickle cell anemia.
3. Hereditary persistence of fetal hemoglobin.
4. Some normal women have a pregnancy-induced increase in fetal hemoglobin.

In the presence of these conditions other techniques based on differential agglutination have to be employed.

Lanzkowsky's Manual of Pediatric Hematology and Oncology.
DOI: http://dx.doi.org/10.1016/B978-0-12-801368-7.00005-3

TABLE 5.1   Causes of Anemia in the Newborn

1. **Hemorrhage**
   a. Prenatal
      i. Transplacental fetomaternal (spontaneous, traumatic amniocentesis, external cephalic version)
      ii. Intraplacental
      iii. Retroplacental
      iv. Twin-to-twin transfusion
   b. Intranatal
      i. Umbilical cord abnormalities
         - Rupture of normal cord (unattended precipitous labor)
         - Rupture of varix or aneurysm of cord
         - Hematomas of cord or placenta
         - Rupture of anomalous aberrant vessels of cord (not protected by Wharton's jelly)
         - Vasa previa (umbilical cord is presenting part)
         - Inadequate cord tying
      ii. Placental abnormalities
         - Multilobular placenta (fragile communicating veins to main placenta)
         - Placenta previa—fetal blood loss predominantly
         - Abruptio placentae—maternal blood loss predominantly
         - Accidental incision of placenta during cesarean section
         - Traumatic amniocentesis
         - Placental chorioangioma
      iii. Hemorrhagic diathesis
         - Plasma factor deficiency
         - Thrombocytopenia
   c. Postnatal
      i. External
         - Bleeding from umbilicus
         - Bleeding from gut
         - Iatrogenic (diagnostic venipuncture, postexchange transfusion)
      ii. Internal
         - Cephal hematomata
         - Subgaleal (subaponeurotic) hemorrhage
         - Subdural or subarachnoid hemorrhage
         - Intracerebral hemorrhage
         - Intraventricular hemorrhage
         - Intra-abdominal hemorrhage
         - Retroperitoneal hemorrhage (may involve adrenals)
         - Subcapsular hematoma or rupture of liver
         - Ruptured spleen
         - Pulmonary hemorrhage
2. **Hemolytic anemia** (see Chapters 9–11)
   a. Congenital erythrocyte defects
      i. Membrane defects (with characteristic morphology)
         - Hereditary spherocytosis (p. 139)
         - Hereditary elliptocytosis (p. 143)
         - Hereditary stomatocytosis (p. 145)
         - Hereditary xerocytosis
         - Infantile pyknocytosis[a]
         - Pyropoikilocytosis
      ii. Hemoglobin defects[b]
         - α-Thalassemia syndromes[c]
           Single α-globin gene deletion (asymptomatic carrier state)
           Two α-globin gene deletion (α-thalassemia trait)
           Three α-globin gene deletion (hemoglobin Hβ_4 and hemoglobin Barts γ_4)
           Four α-globin gene deletion (death *in utero* or shortly after birth)
         - $\gamma\beta$-Thalassemia
         - $\epsilon\gamma$ $\delta\beta$-Thalassemia
         - β-Thalassemia[d]
         - Unstable hemoglobins (Hb Köln[c], Hg Zürich[c], Hb F Poole[e], Hb Hasharon[e]) (see Chapter 11)
      iii. Enzyme defects
         - Embden–Meyerhof glycolytic pathway
              Pyruvate kinase
              Other enzymes, for example, 5′-nucleotidase deficiency, glucose phosphate isomerase deficiency
         - Hexose-monophosphate shunt
              G6PD (Caucasian and Oriental) with or without drug exposure[c]
              Enzymes concerned with glutathione reduction or synthesis[c]

(Continued)

TABLE 5.1   *(Continued)*

  b. Acquired erythrocyte defects
      i. Immune
          • Maternal autoimmune hemolytic anemia
          • Isoimmune hemolytic anemia: Rh disease, ABO, minor blood groups (M, S, Kell, Duffy, Luther)
      ii. Nonimmune
          • Infections (cytomegalovirus, toxoplasmosis, herpes simplex, rubella, adenovirus, malaria, syphilis, bacterial sepsis, e.g., *Escherichia coli*)
          • Microangiopathic hemolytic anemia with or without disseminated intravascular coagulation: Disseminated herpes simplex, coxsackie B infections, Gram-negative septicemia, renal vein thrombosis
          • Toxic exposure (drugs, chemicals) ± G6PD ± prematurity:[c] Synthetic vitamin K analogs, maternal thiazide diuretics, antimalarial agents, sulfonamides, naphthalene, aniline-dye marking ink, penicillin
          • Vitamin E deficiency
          • Metabolic disease (galactosemia, osteopetrosis)
3. **Failure of red cell production**
  a. Congenital (Chapter 8)
      i. Diamond–Blackfan anemia (pure red cell aplasia)
      ii. Dyskeratosis congenita
      iii. Fanconi anemia
      iv. Aase syndrome
      v. Pearson syndrome
      vi. Sideroblastic anemia
      vii. Congenital dyserythropoietic anemia
  b. Acquired
      i. Viral infection (hepatitis, HIV, CMV, rubella, syphilis, parvovirus B_{19})
      ii. Malaria
      iii. Anemia of prematurity

[a]*Not permanent membrane defect but has characteristic morphology.*
[b]*β-Chain mutations (e.g., sickle cell) uncommonly produce clinical symptomatology in the newborn. In homozygous sickle cell disease, the HbS concentration at birth is usually about 20%.*
[c]*All these conditions can be associated with Heinz-body formation and in the past were grouped together as congenital Heinz-body anemia.*
[d]*β-Thalassemia syndromes only become clinically apparent after 2 or 3 months of age. The first sign is the presence of nucleated red cells on smear or continued high HgbF concentration.*
[e]*Hemolysis subsides after the first few months of life as fetal hemoglobin ($\alpha^2 \gamma^2$) is replaced by adult hemoglobin ($\alpha_2 \beta_2$).*

TABLE 5.2   The Characteristics of Acute and Chronic Blood Loss in the Newborn

Characteristic	Acute blood loss	Chronic blood loss
Clinical		
	Acute distress; pallor; shallow, rapid and often irregular respiration; tachycardia; weak or absent peripheral pulses; low or absent blood pressure; no hepatosplenomegaly	Marked pallor disproportionate to evidence of distress. On occasion signs of congestive heart failure may be present, including hepatomegaly
Venous pressure	Low	Normal or elevated
Laboratory		
Hemoglobin concentration	May be normal initially; then drops quickly during the first 24 h of life	Low at birth
Red cell morphology	Normochromic and macrocytic	Hypochromic and microcytic anisocytosis and poikilocytosis
Serum iron	Normal at birth	Low at birth
Course	Prompt treatment of anemia and shock necessary to prevent death	Generally uneventful
Treatment	Normal saline bolus or packed red blood cells. If indicated, iron therapy	Iron therapy. Packed red blood cells on occasion

From Oski and Naiman (1982), with permission.

Intraplacental and Retroplacental

Occasionally, fetal blood accumulates in the substance of the placenta (intraplacental) or retroplacentally and the infant is born anemic. Intraplacental blood loss from the fetus may occur when there is a tight umbilical cord around the neck or body or there is delayed cord clamping. Retroplacental bleeding from abruptio placenta is diagnosed by ultrasound or at surgery.

Twin-to-Twin Transfusion

Significant twin-to-twin transfusion occurs in at least 15% of monochorionic twins. Velamentous cord insertions are associated with increased risk of twin-to-twin transfusion. The hemoglobin level differs by 5 g/dl and the hematocrit by 15% or more between individual twins (by contrast, the maximal discrepancy in cord blood hemoglobin in dizogotic twins is 3.3 g/dl). The donor twin is anemic, pale and smaller, and may have evidence of oligohydramnios and show evidence of congestive heart failure and shock. The recipient is polycythemic and larger, with evidence of polyhydramnios and may show signs of hyperviscosity syndrome—hypoglycemia, central nervous system injury, hypocalcemia; disseminated intravascular coagulation, hyperbilirubinemia, and congestive heart failure (Chapter 12).

Intranatal Blood Loss

Hemorrhage may occur during the process of birth as a result of various obstetric accidents, malformations of the umbilical cord or the placenta, or a hemorrhagic diathesis (due to a plasma factor deficiency or thrombocytopenia; Table 5.1).

Postnatal Blood Loss

Postnatal hemorrhage may occur from a number of sites and may be internal (enclosed) or external. Hemorrhage may be due to:

- Traumatic deliveries (resulting in intracranial or intra-abdominal hemorrhage).
- Plasma factor deficiencies (see Chapter 15).
  - Congenital—hemophilia or other plasma factor deficiencies.
  - Acquired—vitamin K deficiency, disseminated intravascular coagulation.
- Thrombocytopenia (see Chapter 14)
  - Congenital—Wiskott–Aldrich syndrome, Fanconi anemia, thrombocytopenia absent radius syndrome.
  - Acquired—isoimmune thrombocytopenia, sepsis.
  - Rare causes—neonatal adenovirus infection, fetal cytomegalovirus infection, hemangiomas of the gastrointestinal tract, hemangioendotheliomas of the skin.

When the products of hemoglobin are absorbed from entrapped hemorrhage, hyperbilirubinemia may develop after several days.

Clinical and Laboratory Findings of Anemia due to Hemorrhage

The clinical manifestations of hemorrhage depend on the volume of the hemorrhage and the rapidity with which it occurs.

1. Anemia—pallor, tachycardia, and hypotension (if severe e.g., ≥ 20 ml/kg blood loss). Nonimmune hydrops can occur in severe anemia.
2. Liver and spleen not enlarged (except in chronic transplacental bleed).
3. Jaundice absent (except after several days in entrapped hemorrhage).
4. Laboratory findings:
   a. Reduced hemoglobin (as low as 2 g/dl has been observed).
   b. Increased reticulocyte count.
   c. Polychromatophilia.
   d. Nucleated RBCs raised.
   e. Fetal cells in maternal blood (in fetomaternal bleed).
   f. Direct antiglobulin test negative.

Treatment

1. Severely affected.
   a. Administer 10–20 ml/kg packed red blood cells (hematocrit usually 50–60%) via an umbilical vein catheter.
   b. Cross-match blood with the mother. If unavailable, use group O Rh-negative blood or saline boluses, temporarily for shock, while awaiting available blood.
   c. Use partial exchange transfusion with packed red cells for infants in incipient heart failure.
2. Mild anemia due to chronic blood loss.
   a. Ferrous sulfate (2 mg elemental iron/kg body weight three times a day) for 3 months.

HEMOLYTIC ANEMIA

Hemolytic anemia in the newborn is usually associated with an abnormally low hemoglobin level, an increase in the reticulocyte count, and with unconjugated hyperbilirubinemia. The hemolytic process is often first detected as a result of investigation for jaundice during the first week of life. The causes of hemolytic anemia in the newborn are listed in Table 5.1.

Congenital Erythrocyte Defects

Congenital erythrocyte defects involving the red cell membrane, hemoglobin, and enzymes are listed in Table 5.1 and discussed in Chapters 10 and 11. Any of these conditions may occur in the newborn and manifest clinically as follows:

- Hemolytic anemia (low hemoglobin, reticulocytosis, increased nucleated red cells, morphologic changes).
- Unconjugated hyperbilirubinemia.
- Direct antiglobulin test negative.

Infantile Pyknocytosis

The cause of this condition has not been clearly defined and should only be contemplated when other established causes of pyknocytes in the blood have been excluded such as glucose-6-phosphate dehydrogenase (G6PD) deficiency, pyruvate kinase deficiency, microangiopathic hemolytic anemia, neonatal hepatitis, vitamin E deficiency, neonatal infections, and hemolysis caused by drugs and toxic agents. Infantile pyknocytosis is characterized by:

- Hemolytic anemia—Direct antiglobulin test negative (nonimmune).
- Distortion of as many as 50% of red cells with several to many spiny projections (up to 6% of cells may be distorted in normal infants). Abnormal morphology is extracorpuscular in origin.
- Disappearance of pyknocytes and hemolysis by the age of 6 months. This is a self-limiting condition.
- Hepatosplenomegaly.

Anemia due to Hemoglobinopathies

Certain hereditary defects of hemoglobin occur in the newborn. Defects due to gamma chain defects resolve spontaneously whereas defects of beta chain disorders are clinically inapparent at birth and only occur at a later age. Defects of the alpha chain manifest clinically differently in the newborn when they are paired with the gamma rather than the beta chain of hemoglobin.

Gamma Chain Defects

These variants spontaneously resolve as gamma chain production diminishes in the newborn and generally produce no hematologic disturbance. The condition would be lethal if no gamma chains were produced *in utero*. If only one or two genes are involved slight hypochromia and mild anemia may occur and the condition is usually found during newborn screening programs.

Beta Chain Defects

Beta chain defects generally produce no clinical symptomatology in the newborn. However, homozygous hemoglobin S, the most common beta chain variant, may rarely present in the newborn as a hemolytic anemia. In beta thalassemia syndromes hematologic findings at birth are normal and only manifest later, typically in the first 6 months of life.

Alpha Chain Defects

A large spectrum of alpha-thalassemia syndromes occur in the newborn and they are discussed in the chapter on hemoglobinopathies (Chapter 11).

Acquired Erythrocyte Defects

Acquired erythrocyte defects may be due to immune (direct antiglobulin test-positive) or nonimmune (direct antiglobulin test-negative) causes. The immune causes are due to blood group incompatibility between the fetus and the mother, for example, Rh (D), ABO, or minor blood group incompatibilities such as anti-c, Kell, Duffy, Luther, anti-C, anti-Cw, anti-E, and anti-Jk(b) causing isoimmunization. Kell antigen is second to Rh (D) in its immunizing potential and occurs in about 9% of Whites and 2% of Blacks.

Immune Hemolytic Anemia

Rh Isoimmunization

Clinical Features

1. Anemia, mild to severe (if severe, may be associated with hydrops fetalis[1]).
2. Jaundice (unconjugated hyperbilirubinemia).
   a. Presents during first 24 h.
   b. Kernicterus may occur whenever the bilirubin level in full-term infants rises to, or exceeds, 20 mg/dl and is an indication for exchange transfusion.
      Certain factors predispose to the development of kernicterus at lower levels of bilirubin, such as prematurity, hypoproteinemia, metabolic acidosis, drugs (sulfonamides, caffeine, sodium benzoate), and hypoglycemia. When these conditions are present exchange transfusions should be performed even if the bilirubin level is below 20 mg/dl.
      See Table 5.3 for a list of various causes of unconjugated hyperbilirubinemia. Figure 5.1 outlines an approach to the diagnosis of both unconjugated and conjugated hyperbilirubinemia.
3. Hepatosplenomegaly; varies with severity.
4. Petechiae (only in severely affected infants). Hyporegenerative thrombocytopenia and neutropenia may occur during the first week.
5. Hydrops fetalis, stillbirth, or death *in utero* and delivery of a macerated fetus may occur with severe illness.
6. Late hyporegenerative anemia with absent reticulocytes. This occurs occasionally during the second to the fifth weeks and is due to a diminished population of erythroid progenitors (serum concentration of erythropoietin is low and the marrow concentrations of BFU-E and CFU-E are not elevated).

Laboratory Findings

1. Serologic abnormalities (incompatibility between blood group of infant and mother), with a direct antiglobulin test positive in infant. Mother's serum has the presence of immune antibodies detected by the indirect antiglobulin test.
2. Decreased hemoglobin level, elevated reticulocyte count, smear-increased nucleated red cells, marked polychromasia and anisocytosis.
3. Raised indirect bilirubin level.

---

[1] Infants have ascites, pleural and pericardial effusions and marked edema. Pathogenesis of hydrops may be due to heart failure, hypoalbuminemia, distortion and dysfunction of hepatic architecture and circulation due to islets of extramedullary erythropoiesis.

TABLE 5.3 Causes of Unconjugated Hyperbilirubinemia

1. **"Physiologic" jaundice:** Jaundice of hepatic immaturity
2. **Hemolytic anemia** (see Chapters 10 and 11 for more complete list of causes)
   a. Congenital erythrocyte defect
      i. Membrane defects: Hereditary spherocytosis, ovalocytosis, stomatocytosis, infantile pyknocytosis
      ii. Enzyme defects (nonspherocytic)
         • Embden–Meyerhof glycolytic pathway (energy potential): Pyruvate kinase, triose phosphate isomerase, etc.
         • Hexose-monophosphate shunt (reduction potential): G6PD
      iii. Hemoglobin defects
         • Sickle cell hemoglobinopathy[a]
   b. Acquired erythrocyte defect
      i. Immune: Alloimmunization (Rh, ABO, Kell, Duffy, Lutheran)
      ii. Nonimmune
         • Infection
            Bacterial: *E. coli*, streptococcal septicemia
            Viral: Cytomegalovirus, rubella, herpes simplex
            Protozoal: Toxoplasmosis
            Spirochetal: Syphilis
         • Drugs: Penicillin
         • Metabolic: Asphyxia, hypoxia, shock, acidosis, vitamin E deficiency in premature infants, hypoglycemia
3. **Polycythemia** (see Table 12.1 for more complete list of causes)
   a. Placental hypertransfusion
      i. Twin-to-twin transfusion
      ii. Maternal–fetal transfusion
      iii. Delayed cord clamping
   b. Placental insufficiency
      i. Small for gestational age
      ii. Postmaturity
      iii. Toxemia of pregnancy
      iv. Placenta previa
   c. Endocrinal
      i. Congenital adrenal hyperplasia
      ii. Neonatal thyrotoxicosis
      iii. Maternal diabetes mellitus
   d. Miscellaneous
      i. Down syndrome
      ii. Hyperplastic visceromegaly (Beckwith–Wiedemann syndrome), associated with hypoglycemia
4. **Hematoma**
   Cephal hematoma, subgaleal, subdural, intraventricular, intracerebral, subcapsular hematoma of liver; bleeding into gut
5. **Conjugation defects**
   a. Reduction in bilirubin glucuronyl transferase
      i. Severe (type I): Crigler–Najjar (autosomal-recessive)
      ii. Mild (type II): Crigler–Najjar (autosomal-dominant)
      iii. Gilbert disease
   b. Inhibitors of bilirubin glucuronyl transferase
      i. Drugs: Novobiocin
      ii. Breast milk: Pregnane-3α, 20β-diol
      iii. Familial: Transient familial hyperbilirubinemia
6. **Metabolic**
   Hypothyroidism, maternal diabetes mellitus, galactosemia
7. **Gut obstruction** (due to enterohepatic recirculation of bilirubin)
   (e.g., pyloric stenosis, annular pancreas, duodenal atresia)
8. **Maternal indirect hyperbilirubinemia**
   (e.g., homozygous sickle cell hemoglobinopathy)
9. **Idiopathic**

[a]*Not usually a cause of jaundice in the newborn because of the predominance of Hgb F (unless associated with concomitant G6PD deficiency).*

Severity of disease is predicted by:

• History indicating the severity of hemolytic disease of the newborn in previous infants.
• Maternal antibody titers.
• Amniotic fluid spectrophotometry.
• Fetal ultrasound.
• Percutaneous fetal blood sampling.

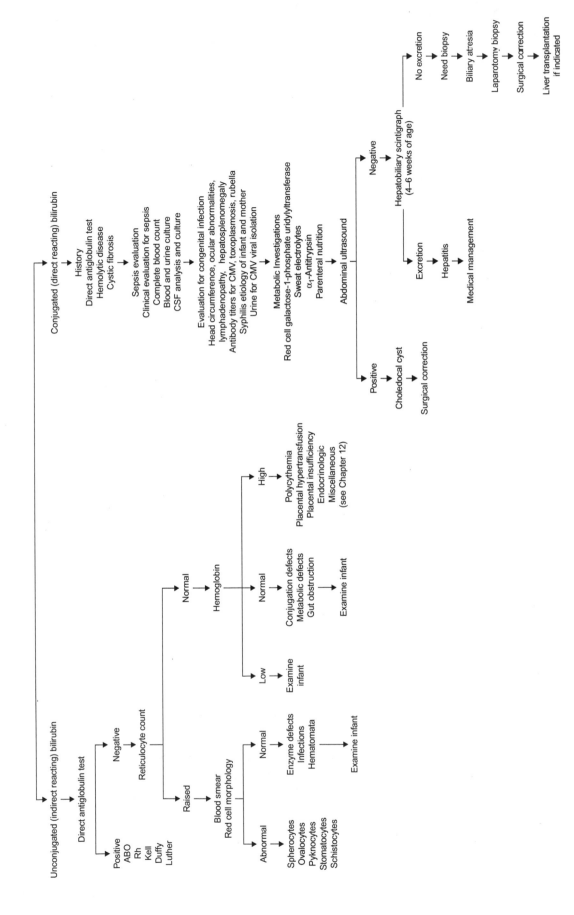

FIGURE 5.1 Approach to investigation of jaundice in the newborn.

Management

Antenatal   Patients should be screened at their first antenatal visit for Rh and non-Rh antibodies. Figure 5.2 shows a schema of the antenatal management of Rh disease.

If an immune antibody is detected in the mother's serum, proper management includes the following:

- Detailed past obstetric history and outcome of previous pregnancies including neonates who required exchange transfusion, hydrops fetalis, or stillbirth. History of prior blood transfusions.
- Blood group and indirect antiglobulin test (to determine the presence and titer of irregular antibodies). Most irregular antibodies can cause erythroblastosis fetalis, therefore screening of maternal serum is important. Titers should be determined at various weeks of gestation (Figure 5.2). The frequency depends on the initial or subsequent rise in titers. Theoretically, any blood group antigen (with the exception of Lewis and I, which are not present on fetal erythrocytes) may cause erythroblastosis fetalis. Anti-Le[a], Le[b], M, H, P, S, and I are IgM antibodies and rarely, if ever, cause erythroblastosis fetalis and need not cause concern.
- Zygosity of the father: If the mother is Rh negative and the father is Rh positive, the father's zygosity becomes critical. If he is homozygous, all his future children will be Rh positive. If the father is heterozygous, there is a 50% chance that the fetus will be Rh negative and unaffected. The Rh genotype can be accurately determined by the use of polymerase chain reaction (PCR) of chorionic villus tissue, amniotic cells, and fetal blood when the father is heterozygous or his zygosity is unknown. Mothers with fetuses found to be Rh D negative (dd) can be reassured and further serologic testing and invasive procedures can be avoided. Fetal zygosity can be determined by molecular genetic techniques. Fetal Rh D genotyping can be performed rapidly on maternal plasma in the second trimester of pregnancy without invading the fetomaternal circulation. This is performed by extracting DNA from maternal plasma and analyzing it for the Rh D gene with a fluorescent-based PCR test sensitive enough to detect the Rh D gene in a single cell. The advantage of this test is that neither the mother nor the fetus is exposed to the risks of amniocentesis or chorionic villus sampling.

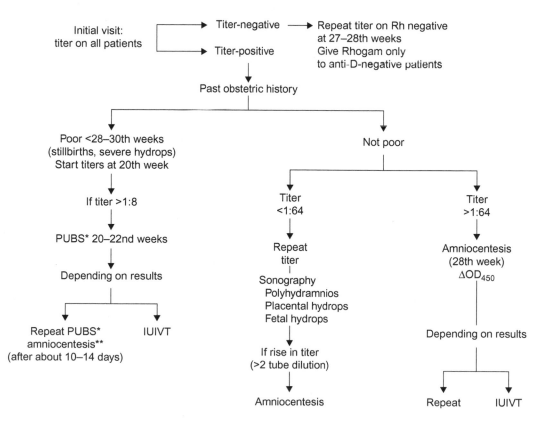

FIGURE 5.2   Schema of antenatal management of Rh disease.
*Percutaneous umbilical vein blood sampling.
**Amniotic fluid analysis is less reliable prior to the 26th week of gestation and PUBS is recommended.
IUIVT, intrauterine intravenous transfusion.

- Examination of the amniotic fluid for spectrophotometric analysis of bilirubin. Past obstetric history and antibody titer are indications for serial amniocentesis and spectrophotometric analyses of amniotic fluid to determine the condition of the fetus. Amniotic fluid analysis correlates well with the hemoglobin and hematocrit at birth ($r = 0.9$) but does not predict whether the fetus will require an exchange transfusion after birth.

The following are indications for amniocentesis:

1. History of previous Rh disease severe enough to require an exchange transfusion or to cause stillbirth.
2. Maternal titer of anti-D, anti-c, or anti-Kell (or other irregular antibodies) of 1:8 to 1:64 or greater by indirect antiglobulin test or albumin titration and depending on previous history.

An assessment of the optical density difference at 450 mm (ΔOD_{450}) at a given gestational age permits reasonable prediction of the fetal outcome (Figure 5.3). Determination of the appropriate treatment depends on the ΔOD_{450} of the amniotic fluid, the results of the fetal biophysical profile scoring and the assessment of the presence or absence of fetal hydrops (seen on ultrasound) and amniotic phospholipid determinations (lung profile).[2]

Features of lung profile	Immature fetus	Mature fetus
Lecithin/sphingomyelin ratio	<2.0	>2.0
Acetone-precipitable fraction	<45%	>50%
Phosphatidylinositol	Absent	Present (small amounts)
Phosphatidylglycerol	Absent	Present (prominent)

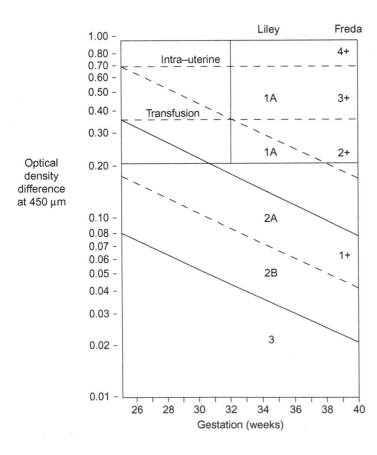

FIGURE 5.3 Assessment of fetal prognosis by the methods of Liley and Freda. Liley's method of prediction:

Zone 1A: Condition desperate, immediate delivery or intrauterine transfusion required, depending on gestational age.

Zone 1B: Hemoglobin less than 8 g/dl, delivery or intrauterine transfusion urgent, depending on gestational age.

Zone 2A: Hemoglobin 8–10 g/dl, delivery at 36–37 weeks.

Zone 2B: Hemoglobin 11.0–13.9 g/dl, delivery at 37–39 weeks.

Zone 3: Not anemic, deliver at term. Freda's method of prediction.

Zone 4+: Fetal death imminent, immediate delivery or intrauterine transfusion, depending on gestational age.

Zone 3+: Fetus in jeopardy, death within 3 weeks, delivery or intrauterine transfusion as soon as possible, depending on gestational age.

Zone 2+: Fetal survival for at least 7–10 days, repeat amniocentesis indicated, possible indication for intrauterine transfusion, depending on gestational age.

Zone 1+: Fetus in no immediate danger.

Source: From Robertson (1966), with permission.

[2]Ultrasound for the assessment of gestational age must be done early in pregnancy. The fetal biophysical profile scoring uses multiple variables: fetal breathing movements, gross body movements, fetal tone, reactive fetal heart rate, and quantitative amniotic fluid volume. This scoring system provides a good short-term assessment of fetal risk for death or damage *in utero*.

If the amniotic fluid optical density difference at 450 mm (ΔOD_{450}) indicates a severely affected fetus and phospholipid estimations indicate lung maturity, the infant should be delivered. If the ΔOD_{450} indicates a severely affected fetus and the phospholipid estimations indicate marked immaturity, maternal plasmapheresis and/or intrauterine intravascular transfusion (IUIVT) via cordocentesis should be carried out. IUIVT has many advantages over intraperitoneal fetal transfusions and is the procedure of choice. This decision is made in conjunction with the biophysical profile score.

Intensive maternal plasmapheresis antenatally using a continuous-flow cell separator can significantly reduce Rh antibody levels, reduce fetal hemolysis and improve fetal survival in those mothers carrying highly sensitized Rh-positive fetuses. This procedure, together with IUIVT, should be carried out when a high antibody titer exists early, before a time that the infant could be safely delivered.

If the risk of perinatal death resulting from complications of prematurity is high, then an IUIVT should be carried out. Percutaneously, the umbilical vein is used for blood sampling (PUBS) and venous access and permits a fetal transfusion via the intravascular route (IUIVT). With the availability of high-resolution ultrasound guidance, a fine (20 gauge) needle is inserted directly into the umbilical cord, either at the insertion site into the placenta or into a free loop of cord. This allows the same blood sampling as is available postnatally in the neonate. Temporary paralysis of the fetus with the use of pancuronium bromide facilitates the procedure, which may be applied to fetuses from 18 weeks' gestation until the gestational age when fetal lung maturity is confirmed. The interval between procedures ranges from 1 to 3 weeks.

Blood used for IUIVT should be cytomegalovirus-negative packed RBCs with a packed cell volume of 85−88%. Cells should be fresh, leukocyte-depleted, and irradiated to prevent the low risk of graft-versus-host disease (GVHD). The use of kell antigen-negative blood is optimal, if available.

Indications for IUIVT:

1. Bilirubin results from amniocentesis show the fetus to be moderately to severely affected by Rh sensitization and risk of delivery due to prematurity too great.
2. Ultrasound shows evidence of hydrops.
3. Fetal blood sampling shows that the fetus has severe anemia.

Risks of IUIVT:

- Fetal loss (2%).
- Premature labor and rupture of membranes.
- Chorioamnionitis.
- Fetal bradycardia.
- Cord hematoma or laceration.
- Fetomaternal hemorrhage.

The overall survival rate is 88%. Intraperitoneal transfusion can be performed in addition to IUIVT to increase the amount of blood transfused and to extend the interval between transfusions.

Modern neonatal care, including attention to metabolic, nutritional, and ventilatory needs and the use of artificial surfactant insufflation, makes successful earlier delivery possible. The need for IUIVT is rarely indicated.

Postnatal Hydropic infant at birth. In addition to phototherapy[3] the following measures are employed:

- Adequate ventilation must be established.
- Drainage of pleural effusions and ascites to improve ventilation.
- Use of resuscitation fluids and drugs, surfactant, and glucose infusions to counteract hyperinsulinemic hypoglycemia should be employed.
- Partial exchange transfusion may be necessary to correct severe anemia.
- Double-volume exchange transfusion may be required later.

Hyperbilirubinemia is the most frequent problem and can be managed by exchange transfusion. Phototherapy is an adjunct rather than the first line of therapy in hyperbilirubinemia due to erythroblastosis fetalis. Postnatal

---

[3]Intensive phototherapy implies the use of high levels of irradiance (430−490 nm, i.e., usually 30 mW/cm^2 per nm or higher).

management and criteria for exchange transfusion have changed over the years and still remain somewhat controversial. We currently use the following indications for exchange transfusion:

- A rapid increase in the bilirubin level of greater than 1.0 mg/h and/or a bilirubin level approaching 20 mg/dl at any time during the first few days of life in the full-term infant is an indication for exchange transfusion. In preterm or high-risk infants, exchange transfusion should be carried out at lower levels of bilirubin (e.g., 15 mg/dl).
- Clinical signs suggesting kernicterus at any time at any bilirubin level are an indication for exchange transfusion.

The blood for exchange transfusion should be ABO-compatible and for anti-D hemolytic disease of the newborn, Rh negative. If the mother is alloimmunized to an antigen other than D, the blood should not have that antigen. It should be crossmatched compatible with the mother's serum. Ideally, the blood should be leukocyte-depleted and be negative for Kell antigen (to avoid sensitizing the infant) and be hemoglobin S negative.[4] If the initial exchange transfusion is carried out using the group O blood, any further exchange transfusions should use O blood. Otherwise, brisk hemolysis and jaundice due to ABO incompatibility may become a further complication. Graft-versus-host disease (GVHD) occurs rarely after exchange transfusion, but blood should be irradiated, especially for premature infants.

Prevention of Rh Hemolytic Disease

Rh hemolytic disease can be prevented by the use of Rh immunoglobulin at a dose of 300 μg, which is indicated in the following circumstances:

- For all Rh-negative, Rh0 (D^u)-negative mothers who are unimmunized to the Rh factor. In these patients Rh immunoglobulin is given at 28 weeks' and 34 weeks' gestation and within 72 h of delivery. Antenatal administration of Rh immunoglobulin is safe for the mother and the fetus.
- For all unimmunized Rh-negative mothers who have undergone spontaneous (1.5–2% risk of sensitization) or induced abortion (5% risk of immunization). If surgical evacuation is done, 50 μg (250 IU) Rh immunoglobulin should be given. The D antigen is detectable on embryonic red cells by 38 days after conception and Rh immunoglobulin should be given beyond the seventh or eighth weeks of gestation in these circumstances. The precise risk of alloimmunization associated with these events is less clear compared to the risk after delivery.
- After ruptured tubal pregnancies in unimmunized Rh-negative mothers.
- Following any event during pregnancy that may lead to transplacental hemorrhage, such as external version, amniocentesis, or antepartum hemorrhage in unimmunized Rh-negative women.
- Following tubal ligation or hysterotomy after the birth of a Rh-positive child in unimmunized Rh-negative women.
- Following chorionic villus sampling at 10–12 weeks' gestation. In these patients 50 μg of Rh immunoglobulin should be given.

Acute drug-induced hypersensitivity and delayed transfusion-related reaction including anaphylaxis with deleterious effects on the mother and the fetus have been described.

ABO Isoimmunization

ABO incompatibility is milder than hemolytic disease of the newborn caused by other antibodies.

Clinical Features

1. Jaundice (indirect hyperbilirubinemia) usually within first 24 h; may be of sufficient severity to cause kernicterus.
2. Anemia at birth is usually absent or moderate and late anemia is rare.
3. Hepatosplenomegaly.

---

[4]In hypoxic infants sickle cell trait blood could cause an iatrogenic sickle cell crisis or death.

TABLE 5.4 Clinical and Laboratory Features of Isoimmune Hemolysis Caused by Rh and ABO Incompatibility

Feature	Rh disease	ABO incompatibility
CLINICAL EVALUATION		
Frequency	Unusual	Common
Occurrence in first born	5%	40–50%
Predictably severe in subsequent pregnancies	Usually	No
Stillbirth and/or hydrops	Occasional	Rare
Pallor	Marked	Minimal
Jaundice	Marked	Minimal (occasionally marked)
Hepatosplenomegaly	Marked	Minimal
Incidence of late anemia	Common	Uncommon
LABORATORY FINDINGS		
Blood type, mother	Rh-negative	O
Blood type, infant	Rh-positive	A or B or AB
Antibody type	Incomplete (7S)	Immune (7S)
Direct antiglobulin test	Positive	Usually positive
Indirect antiglobulin test	Positive	Usually positive
Hemoglobin level	Very low	Moderately low
Serum bilirubin	Markedly elevated	Variably elevated
Red cell morphology	Nucleated RBCs	Spherocytes
TREATMENT		
Need for antenatal management	Yes	No
Exchange transfusion		
Frequency	~2:3	~1:10
Donor blood type	Rh-negative group specific, when possible	Rh same as infant group O only

Table 5.4 lists the clinical and laboratory features of isoimmune hemolysis due to Rh and ABO incompatibility.

Diagnosis

1. Hemoglobin decreased.
2. Smear: Spherocytosis in 80% of infants, reticulocytosis, marked polychromasia.
3. Elevated indirect bilirubin level.[5]
4. Demonstration of incompatible blood group.
   a. Group O mother and an infant who is group A or B.
   b. Rarely, mother may be A and baby B or AB or mother may be B and baby A or AB.

---

[5]In the era of early discharge of newborns the use of the critical bilirubin level of 4 mg/dL at the 6th hour of life predicts significant hyperbilirubinemia and 6 mg/dL at the 6th hour will predict severe hemolytic disease of the newborn. The reticulocyte count, a positive antiglobulin test, and a sibling with neonatal jaundice are additional predictors of significant hyperbilirubinemia and reason for careful surveillance of the newborn.

5. Direct antiglobulin test on infant's red cells usually positive.
6. Demonstration of antibody in infant's serum.
   a. When free anti-A is present in a group A infant or anti-B is present in a group B infant, ABO hemolytic disease may be presumed. These antibodies can be demonstrated by the indirect antiglobulin test in the infant's serum using adult erythrocytes possessing the corresponding A or B antigen. This is proof that the antibody has crossed from the mother's to the baby's circulation.
   b. Antibody can be eluted from the infant's red cells and identified.
7. Demonstration of antibodies in maternal serum. When an infant has signs of hemolytic disease, the mother's serum may show the presence of immune agglutinins persisting after neutralization with A and B substance and hemolysins.

Treatment

In ABO hemolytic disease, unlike Rh disease, antenatal management or premature delivery is not required. After delivery, management of an infant with ABO hemolytic disease is directed toward controlling the hyperbilirubinemia by frequent determination of unconjugated bilirubin levels, with a view to the need for phototherapy or exchange transfusion. The principles and methods are the same as those described for Rh hemolytic disease. Group O blood of the same Rh type as that of the infant should be used. Whole blood is used to permit maximum bilirubin removal by albumin.

Late-Onset Anemia in Immune Hemolytic Anemia

Infants with hemolytic anemia of the newborn due to isoimmunization and particularly in those infants who have had severe isoimmunization treated with intrauterine transfusion and in those infants who do not require an exchange transfusion for hyperbilirubinemia following immune hemolytic anemia may develop severe anemia during the first 6 weeks of life. This is because of persistent maternal IgG antibodies hemolyzing the infant's red blood cells and reticulocytes associated with suppressed erythropoiesis and reticulocytopenia. For this reason, follow-up hemoglobin levels and reticulocyte count weekly for 4–6 weeks should be done in those infants.

Nonimmune Hemolytic Anemia

The causes of nonimmune hemolytic anemia are listed in Table 5.1.

Vitamin E Deficiency

Premature infants (<36 weeks' gestation or weighing less than 2000 g) are susceptible to vitamin E deficiency because of decreased absorption of the vitamin. With improvement in infant formulas recognizing this propensity this condition has virtually disappeared.

Clinical Findings

1. Hemolytic anemia and reticulocytosis. Hemolytic anemia develops under the following conditions: Diets high in PUFA supplemented with iron, which is a powerful oxidant; prematurity; oxygen administration.
2. Thrombocytosis.
3. Pyknocytes (acanthocytes), small number of spherocytes, and fragmented red cells.
4. Peripheral edema.
5. Neurologic signs:
   a. Cerebellar degeneration.
   b. Ataxia.
   c. Peripheral neuropathy.

Diagnosis

Peroxide hemolysis test: Red cells are incubated with small amounts of hydrogen peroxide and the amount of hemolysis is measured.

FAILURE OF RED CELL PRODUCTION

Congenital

The inherited bone marrow failure syndromes are discussed in Chapter 8.

Acquired

Viral Diseases

Viral interference (e.g., CMV, HIV) with fetal hematopoiesis may cause anemia, leukopenia, and thrombocytopenia in the newborn.

ANEMIA OF PREMATURITY

This anemia is due to impaired erythropoietin (EPO) production characterized by reduced bone marrow erythropoietic activity (hypoproliferative anemia) with low reticulocyte count and low serum EPO levels. It may be compounded by folic acid, vitamin E, and iron availability and frequent blood sampling.

The low hemoglobin concentration is due to:

- Preterm infants deprived of third-trimester hematopoiesis and iron transport.
- Decreased red cell production (premature infants have low EPO levels which reach a nadir between days 7 and 50, independently of weight at birth and are less responsive to EPO) associated with decreased marrow erythroid elements.
- Shorter red cell lifespan.
- Increased blood volume with growth.
- Marked blood loss from phlebotomy to monitor problems related to prematurity.

The nadir of the hemoglobin level is 4–8 weeks and is 8 g/dl in infants weighing less than 1500 g. However, small-for-gestational-age infants who have had intrauterine hypoxia exhibit increased erythropoiesis. The anemia of prematurity rarely occurs in association with cyanotic congenital heart disease or with respiratory insufficiency; indicating that higher oxygen-carrying capacities can be maintained in infants in the first few weeks of life if the need arises.

Clinical Features

Tachycardia, increased apnea and bradycardia, increased oxygen requirement, poor weight gain. The anemia is normocytic and normochromic.

Treatment

Delaying cord clamping for 30–60 s in infants who do not require immediate resuscitation may reduce the severity of anemia of prematurity. In addition, limiting blood loss by phlebotomy is important.

Recombinant human erythropoietin (rHuEPO) in a dose of 75–300 units/kg/week subcutaneously for 4 weeks starting at 3–4 weeks of age has been employed to increase reticulocyte counts and raise hemoglobin. It takes about 2 weeks to raise the hemoglobin to a biologically significant degree, which limits its usefulness when a prompt response is needed. Despite extensive studies, many of which have shown a reduction in need for transfusions in premature infants, particularly less than 1000 g who have significant phlebotomy losses, there is still no definite consensus as to whether rHuEPO minimizes the need for blood transfusion. A potential advantage of rHuEPO is the associated right shift in the oxyhemoglobin dissociation curve, most likely due to the increased erythrocyte 2,3-DPG content. The incidence of necrotizing enterocolitis has been shown to be lower in very-low-birth-weight infants treated with rHuEPO. However, the risk of severe retinopathy of prematurity (stage 3 or higher) is increased with early rHuEPO treatment compared with placebo.

Supplemental oral iron in a dose of at least 2 mg/kg/day or intravenous iron supplementation may also be required to prevent the development of iron deficiency. Adequate intake of folate, vitamin E, and protein are important to support erythropoiesis.

The criteria for transfusion of preterm infants vary considerably among different institutions. All transfusions should be provided from a single donor and be less than 7–10 days old and be leukodepleted. Packed red cells

TABLE 5.5   Indications for Small-Volume RBC Transfusions in Preterm Infants

Transfuse well infant at hematocrit ≤ 20% or ≤ 25% with low reticulocyte count and tachycardia, tachypnea, poor weight gain, poor suck, or apnea

Transfuse infants at hematocrit ≤ 30%
    If receiving <35% supplemental hood oxygen
    If on CPAP or mechanical ventilation with mean airway pressure <6 cmH₂O
    If significant apnea (>6/day) and bradycardia are noted while receiving therapeutic doses of methylxanthines
    If heart rate >180 beats/min or respiratory rate >80 breaths/min persists for 24 h
    If weight gain <10 g/day is observed over 4 days while receiving ≥ 100 kcal/kg/day
    If undergoing surgery

Transfuse for hematocrit ≤ 35%
    If receiving >35% supplemental hood oxygen
    If intubated on CPAP or mechanical ventilation with mean airway pressure >6 cmH₂O

Do not transfuse
    To replace blood removed for laboratory tests alone
    For low hematocrit value alone

CPAP, continuous positive airway pressure by nasal or endotracheal route.
Modified from Hume (1997), with permission.

should be adjusted to a hematocrit of 60−79% with normal saline or 5% albumin solution. Low-risk cytomegalovirus blood products (cytomegalovirus-negative or leukodepleted red cells) should be used only for neonates with birth weight less than 1200 g who are cytomegalovirus-negative or have unknown cytomegalovirus status. As a general rule, hemoglobin values should be maintained above 12 g/dl during the first 2 weeks of life. After that period indication for transfusion should not be based on hemoglobin concentration alone but on available tissue oxygen which is determined by:

- Hemoglobin concentration.
- Position of the oxyhemoglobin dissociation curve.
- Arterial oxygen saturation.
- Infant's clinical condition which includes:
    - Weight gain.
    - Fatigue during feeding.
    - Tachycardia.
    - Tachypnea.
    - Evidence of hypoxemia by an increase in blood lactic acid concentration.

Table 5.5 gives indications for small-volume red cell transfusions in preterm infants.

PHYSIOLOGIC ANEMIA

Physiologic anemia is a developmental response of the infant's erythropoietic system.

In utero the oxygen saturation of the fetus is 70% (hypoxic levels) and this stimulates EPO, produces a reticulocytosis (3−7%), and increases red cell production causing a high hemoglobin at birth. After birth the oxygen saturation is 95%, EPO is undetectable and red cell production by day 7 is 10% of the level *in utero*. As a result of this, the hemoglobin level falls to 11.4 + 0.9 g/dl at the nadir at 8−12 weeks (physiologic anemia). At this point oxygen delivery is impaired, EPO stimulated, and red cell production increases. Infants born prematurely experience a more marked decrease in hemoglobin concentration. Premature infants weighing less than 1500 g have a hemoglobin level of 8 g/dl at age 4−8 weeks.

DIAGNOSTIC APPROACH TO ANEMIA IN THE NEWBORN

Figure 5.4 is a flow diagram of the investigation of anemia in the newborn and stresses the importance of the direct antiglobulin test, the reticulocyte count, the mean corpuscular volume, and the blood smear as key investigative tools in elucidating the cause of the anemia and Table 5.6 lists the clinical and laboratory evaluations required in anemia in the newborn.

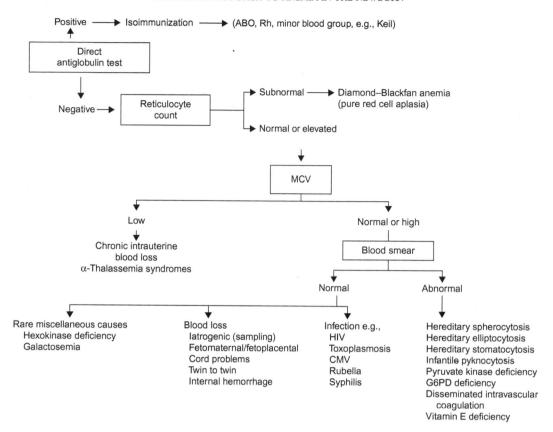

FIGURE 5.4 Approach to the diagnosis of anemia in the newborn.

TABLE 5.6 Clinical and Laboratory Evaluation in Anemia in the Newborn

HISTORY
Obstetric history
Family history
PHYSICAL EXAMINATION
LABORATORY TESTS
Complete blood count
Reticulocyte count
Blood smear
Antiglobulin test (direct and indirect)
Blood type of baby and mother
Bilirubin level
Kleihauer–Betke test on mother's blood (fetal red cells in maternal blood)
Studies for neonatal infection
Ultrasound of abdomen and head (if indicated)
Red cell enzyme assays (if clinically indicated)
Bone marrow (if clinically indicated)

Further Reading and References

Academy of Pediatrics, 1994. Provisional committee for quality improvement and subcommittee on hyperbilirubinemia. Pediatrics 94, 558–565.

Aher, S., Malwatkar, K., Kadam, S., 2008. Neonatal anemia. Semin. Fetal Neonatal Med. 13 (4), 239–247.

Bishara, N., Ohls, R.K., 2009. Current controversies in the management of the anemia of prematurity. Semin. Perinatol. 33 (1), 29–34.

Bowman, J., 1997. The management of hemolytic disease in the fetus and newborn. Semin. Perinatol. 21, 39–44.

Halperin, D.S., Wacker, P., LaCourt, G., et al., 1990. Effects of recombinant human erythropoietin in infants with the anemia of prematurity: a pilot study. J. Pediatr. 116, 779–796.

Hann, I.M., Gibson, B.E.S., Letsky, E.A., 1991. Fetal and Neonatal Haematology. Bailliere Tindall, London.

Hume, H., 1997. Red blood cell transfusions for preterm infants: the role of evidence-based medicine. Semin. Perinatol. 21, 8–19.

Maier, R.H., Obladen, M., Müller-Hansen, I., et al., 2002. Early treatment with erythropoietin [beta] ameliorates anemia and reduces transfusion requirements in infants with birth weights below 1000 g. J. Pediatr. 141, 8–15.

Messer, J., Haddad, J., Donato, L., et al., 1993. Early treatment of premature infants with recombinant human erythropoietin. Pediatrics 92, 519–523.

Murray, N.A., Roberts, I.A.G., 2007. Haemolytic disease of the newborn. Arch. Dis. Child. Fetal Neonatal Ed. 92 (2), F83–F88.

Oski, F.A., Naiman, J.L., 1982. Hematologic Problems in the Newborn, third ed. Saunders, Philadelphia.

Pilgrim, H., Lloyd-Jones, M., Rees, A., 2009. Routine antenatal anti-D prophylaxis for Rh-D-negative women: a systematic review and economic evaluation. Health Technol. Assess. 13 (10), 1–103.

Robertson, J.G., 1966. Evaluation of the reported methods of interpreting spectrophotometric tracings of amniotic fluid analysis in Rhesus isoimmunization. Am. J. Obstet. Gynecol. 95, 120.

Smits-Wintjens, V.E.H.G., Walther, F.J., Lopriore, E., 2008. Rhesus haemolytic disease of the newborn: postnatal management, associated morbidity and long-term outcome. Semin. Fetal Neonatal Med. 13 (4), 265–271.

Steiner, L.A., Gallagher, P.G., 2007. Erthrocyte disorders in the perinatal period. Semin. Perinatol. 31 (4), 254–261.

Urbaniak, S.J., Greiss, M.A., 2000. RhD haemolytic disease of the fetus and the newborn. Blood Rev. 14 (1), 44–61.

CHAPTER

6

Iron-Deficiency Anemia

Philip Lanzkowsky

Iron deficiency is the most common nutritional deficiency in children and is worldwide in distribution, affecting 2 billion people. The prevalence of iron deficiency worldwide is twice as high as iron-deficiency anemia.

PREVALENCE

The incidence of iron-deficiency anemia is high in infancy. It is estimated that 40–50% of children under 5 years of age in developing countries are iron-deficient. The prevalence is 5.5% in inner-city school children ranging in age from 5 to 8 years, 2.6% in pre-adolescent children and 25% in pregnant teenage girls. There is a higher prevalence of iron-deficiency anemia in African-American children than in Caucasian children. Although no socioeconomic group is spared, the prevalence of iron-deficiency anemia is inversely proportional to economic status.

Peak prevalence occurs during late infancy and early childhood when the following may occur:

- Rapid growth with exhaustion of gestational iron.
- Low levels of dietary iron.
- Blood loss due to internal or external bleeding.
- Complicating effect of cow's milk-induced exudative enteropathy due to whole cow's milk ingestion.

A second peak is seen during adolescence due to rapid growth and suboptimal iron intake. This is amplified in females due to menstrual blood loss.

Table 6.1 lists causes of iron deficiency and Table 6.2 lists infants at high risk for iron deficiency.

ETIOLOGIC FACTORS

Diet

1. In normal infants, 1 mg/kg/day to a maximum of 15 mg/day (assuming 10% absorption) is required.
2. In low-birth-weight infants, infants with low initial hemoglobin values, and those who have experienced significant blood loss, 2 mg/kg/day to a maximum of 15 mg/kg/day is required.

Food Iron Content

A newborn infant is fed predominantly on milk. Breast milk and cow's milk contain less than 1.5 mg iron per 1000 calories (0.5–1.5 mg/l). Although cow's milk and breast milk are equally poor in iron, breast-fed infants absorb 20–80% of the iron, in contrast to about 10% absorbed from cow's milk. The bioavailability of iron in breast milk is much greater than in cow's milk, for this reason the iron status of breast-fed infants at 6 months of age is better than infants fed cow's milk. However, after 6 months of age breast-feeding does not protect against iron deficiency and a supplemental source of dietary or medicinal iron is required for optimal iron nutrition.

Lanzkowsky's Manual of Pediatric Hematology and Oncology.
DOI: http://dx.doi.org/10.1016/B978-0-12-801368-7.00006-5

69

TABLE 6.1   Causes of Iron-Deficiency Anemia

1. **Deficient intake**
   Dietary (milk, 0.75 mg iron/l)
2. **Inadequate absorption**
   a. Poor bioavailability: absorption of heme $Fe > Fe^{2+} > Fe^{3+}$; breast milk iron > cow's milk
   b. Antacid therapy or high gastric pH (gastric acid assists in increasing solubility of inorganic iron). Bran, phytates, starch ingestion (contain organic polyphosphates which bind iron avidly), loss or dysfunction of absorptive enterocytes (inflammatory bowel disease, celiac disease)
   c. Cobalt, lead ingestion (share the iron absorption pathways)
3. **Increased demand**
   Growth (low birth weight, prematurity, twins or multiple births, adolescence, pregnancy), cyanotic congenital heart disease
4. **Blood loss**
   a. Perinatal (Chapter 5)
      i. Placental
         • Transplacental bleeding into maternal circulation
         • Retroplacental (e.g., premature placental separation)
         • Intraplacental
         • Fetal blood loss at or before birth (e.g., placenta previa)
         • Feto-fetal bleeding in monochorionic twins
         • Placental abnormalities (Table 5.1)
      ii. Umbilicus
         • Ruptured umbilical cord (e.g., vasa previa) and other umbilical cord abnormalities (Table 2.1)
         • Inadequate cord tying
         • Post exchange transfusion
   b. Postnatal
      i. Gastrointestinal tract
         • Primary iron-deficiency anemia resulting in gut alteration with blood loss aggravating existing iron deficiency: 50% of iron-deficient children have positive guaiac stools
         • Hypersensitivity to whole cow's milk? Due to heat-labile protein, resulting in blood loss and exudative enteropathy (leaky gut syndrome) (Table 3.4)
         • Anatomic gut lesions (e.g., esophageal varices, hiatus hernia, peptic ulcer disease, leiomyomata, Meckel's diverticulum, duplication of gut, hereditary hemorrhagic telangiectasia, arteriovenous malformation, polyps, hemorrhoids); exudative enteropathy caused by underlying bowel disease (e.g., allergic gastroenteropathy, intestinal lymphangiectasia); inflammatory bowel disease; substantial intestinal resection.
         • Gastritis from aspirin, adrenocortical steroids, indometacin, phenylbutazone
         • Intestinal parasites (e.g., hookworm (*Necator americanus* or *Ancylostoma duodenale*) and whipworm (*Trichuris Trichiura*))
         • Henoch–Schönlein purpura
      ii. Hepato-biliary system: hematobilia
      iii. Lung: idiopathic pulmonary hemosiderosis, Goodpasture syndrome, defective iron mobilization with IgA deficiency, tuberculosis, bronchiectasis
      iv. Nose: recurrent epistaxis
      v. Uterus: menstrual loss
      vi. Heart: intracardiac myxomata, valvular prostheses, or patches
      vii. Kidney:[a] infectious cystitis, microangiopathic hemolytic anemia, nephritic syndrome (urinary loss of transferrin), Berger disease, Goodpasture syndrome, chronic intravascular hemolysis (e.g., paroxysmal nocturnal hemoglobinuria, paroxysmal cold hemoglobinuria, march hemoglobinuria)
      viii. Extracorporeal: hemodialysis, trauma
5. **Impaired absorption**
   Malabsorption syndrome, celiac disease, severe prolonged diarrhea, postgastrectomy, inflammatory bowel disease, *Helicobacter pylori* infection-associated chronic gastritis
6. **Inadequate presentation to erythroid precursors**
   a. Atransferrinemia
   b. Anti-transferrin receptor antibodies
7. **Abnormal intracellular transport/utilization**
   a. Erythroid iron trafficking defects
   b. Defects of heme biosynthesis

---

[a]*Hematuria to the point of iron deficiency is extremely uncommon.*

Most environmental iron exists as insoluble salts and gastric acidity assists in converting it to an absorbable form. Any factors reducing gastric acidity (e.g., drugs—histamine-2 blockers, acid pump blockers; surgical procedures—vagotomy, gastrectomy) impair absorption from nonheme sources. The iron present in plant products is limited both by low solubility and the presence of powerful natural chelators, for example, phytates. Heme iron derived from animal sources is the most readily absorbed iron, is independent of gastric pH, and is increased in patients with high erythroid activity associated with reticulocytosis.

TABLE 6.2   Infants at High Risk for Iron Deficiency

1. Increased iron needs
   a. Low birth weight
   b. Prematurity
   c. Multiple gestation
   d. High growth rate
   e. Chronic hypoxia-high altitude, cyanotic heart disease
   f. Low hemoglobin level at birth
2. Blood loss
   a. Perinatal bleeding
3. Dietary factors
   a. Early cow's milk intake
   b. Early solid food intake
   c. Rate of weight gain greater than average
   d. Low-iron formula
   e. Frequent tea intake[a]
   f. Low vitamin C intake[b]
   g. Low meat intake
   h. Breast-feeding >6 months without iron supplements
   i. Low socioeconomic status (frequent infections)

[a]*Tea inhibits iron absorption.*
[b]*Vitamin C enhances iron absorption.*

TABLE 6.3   Iron Content of Infant Foods

Food	Iron, mg	Unit
Milk	0.5—1.5	Liter
Eggs	1.2	Each
Cereal, fortified	3.0—5.0	Ounce
VEGETABLES (STARCHED)		
Yellow	0.1—0.3	Ounce
Green	0.3—0.4	Ounce
MEATS (STRAINED)		
Beef, lamb, liver	0.4—2.0	Ounce
Pork, liver, bacon	6.6	Ounce
FRUITS (STRAINED)		
	0.2—0.4	Ounce

Table 6.3 lists the iron content of infant foods.

Growth

Growth is particularly rapid during infancy and during puberty. Blood volume and body iron are directly related to body weight throughout life. Iron-deficiency anemia can occur at any time when rapid growth outstrips the ability of diet and body stores to supply iron requirements. In the first year of life body weight triples and circulating hemoglobin mass doubles. Each kilogram gain in weight requires an increase of 35—45 mg body iron.

The amount of iron in the newborn is 75 mg/kg. If no iron is present in the diet or blood loss occurs the iron stores present at birth will be depleted by 6 months in a full-term infant and by 3—4 months in a premature infant.

The commonest cause of iron-deficiency anemia is inadequate intake during the rapidly growing years of infancy and childhood.

Blood Loss

Blood loss, an important cause of iron-deficiency anemia, may be due to prenatal, intranatal, or postnatal causes (see Chapter 5, Table 5.1). Hemorrhage occurring later in infancy and childhood may be either occult or

TABLE 6.4   Classification of Iron-Deficiency Anemia in Relationship to Gut Involvement

| | Primary iron deficiency (dietary, rapid growth) | | | | |
	Mild or severe	Severe[a]			
Gut changes	None	Leaky gut syndrome		Malabsorption syndrome	

Effect	No blood loss	Loss of:
		  Red cells only
			Loss of:
			  Red cells
			  Plasma protein
			  Albumin
			  Immune globulin
			  Copper
			  Calcium
				Impaired absorption of iron only	Impaired absorption of xylose, fat and vitamin A
					
					Duodenitis

Result	Iron-deficiency anemia (IDA)	IDA, guisac-positive	IDA, exudative enteropathy	IDA, refractory to oral iron	IDA, transient enteropathy
Treatment	Oral iron	Oral iron	Oral iron		IM iron-dextran complex

| | Secondary iron deficiency | |
	Mild or severe	Severe
Pathogenesis	Cow's milk-induced? Heat-labile protein	Anatomic lesion (e.g., Meckel's diverticulum, polyp, intestinal duplication, peptic ulcer)

Effect	Leaky gut syndrome
	
	Loss of:
	  Red cells
	  Plasma protein
	  Albumin
	  Immune globulin
	  Copper
	  Calcium
		Blood loss

Results	Recurrent IDA, exudative enteropathy	Recurrent IDA
Treatment	Discontinue whole cow's milk; soya milk formula; oral iron	Surgery, specific medical management, iron PO or IM iron dextran

[a]Can occur in severe chronic iron-deficiency anemia from any cause.

apparent (Table 6.1). A number of fecal occult blood tests should be performed in order to exclude intermittent occult gastrointestinal (GI) bleeding.

Iron deficiency by itself, irrespective of its cause, may result in occult blood loss from the gut. More than 50% of iron-deficient infants have guaiac-positive stools. This blood loss is due to the effects of iron deficiency on the mucosal lining (e.g., deficiency of iron-containing enzymes in the gut), leading to mucosal blood loss. This sets up a vicious cycle in which iron deficiency results in mucosal change, which leads to blood loss and further aggravates the anemia. The bleeding due to iron deficiency is corrected with iron treatment. In addition to iron deficiency per se causing blood loss it may also induce an enteropathy, or leaky gut syndrome. In this condition, a number of blood constituents, in addition to red cells, are lost in the gut (Table 6.4).

Cow's milk can result in an exudative enteropathy associated with chronic GI blood loss resulting in iron deficiency. Whole cow's milk should be considered the cause of iron-deficiency anemia under the following clinical circumstances:

- One quart or more of whole cow's milk consumed per day.
- Iron deficiency accompanied by hypoproteinemia (with or without edema) and hypocupremia (dietary iron-deficiency anemia not associated with exudative enteropathy is usually associated with an elevated serum copper level). It is also associated with hypocalcemia, hypotransferrinemia, and low serum immunoglobulins due to the leakage of these substances from the gut.
- Iron-deficiency anemia unexplained by low birth weight, poor iron intake, or excessively rapid growth.

- Iron-deficiency anemia recurring after a satisfactory hematologic response following iron therapy.
- Rapidly developing or severe iron-deficiency anemia.
- Suboptimal response to oral iron in iron-deficiency anemia.
- Consistently positive stool guaiac tests in the absence of gross bleeding and other evidence of organic lesions in the gut.
- Return of GI function and prompt correction of anemia on cessation of cow's milk and substitution by soybean or heat-treated cow's milk formula. Bleeding stops in 3—4 days of cessation of whole cow's milk.

Blood loss can thus occur as a result of gut involvement due to primary iron-deficiency anemia (Table 6.4) or secondary iron-deficiency anemia as a result of gut abnormalities induced by hypersensitivity to cow's milk, or as a result of demonstrable anatomic lesions of the bowel, for example, Meckel's diverticulum.

In postpubescent girls, menstrual blood loss and the increased iron requirements of pregnancy are important causes of iron deficiency. Menstrual losses average approximately 40 ml (20 mg iron) per period and in pregnancy there is an increased requirement of 1200—1500 ml blood (680 mg iron). Most of this iron requirement is in the third trimester when the daily iron requirements rise to 3—7.5 mg requiring supplemental iron to prevent iron deficiency.

Impaired Absorption

Impaired iron absorption due to a generalized malabsorption syndrome (e.g., celiac disease) is an uncommon cause of iron-deficiency anemia. Severe iron deficiency, because of its effect on the bowel mucosa, may induce a secondary malabsorption of iron as well as malabsorption of xylose, fat, and vitamin A (Table 6.4).

CLINICAL FEATURES

Iron-deficiency anemia is chronic and frequently asymptomatic and may go undiagnosed. Patients are characteristically between 6 months and 3 years of age or between 11 and 17 years of age because these are ages of rapid growth and expanding blood volume. In mild anemia there are usually no symptoms. In severe deficiency, pallor, irritability, anorexia, listlessness, fatigue, and pica (strange craving for nonfood items such as sand, dirt, ice, and clay) may occur. These symptoms are reversed in a few days on iron treatment before the anemia corrects itself. Restless leg syndrome, a syndrome of uncontrollable movement of legs, has been described in iron deficiency. It has been postulated that this is due to tissue iron deficiency of parts of the central nervous system concerned with movement control. The spleen in iron deficiency may be enlarged but of normal consistency.

Iron-Refractory Iron-Deficiency Anemia

This is a rare autosomal recessive disorder (OMIM number 206200). Iron-deficiency anemia is defined as "refractory" when there is absence of hematologic response (an increase of <1 g/dl, of hemoglobin) after 4—6 weeks of treatment with oral iron. Iron-refractory iron-deficiency anemia (IRIDA) is caused by a mutation of *TMPRSS6*, the gene encoding transmembrane protease, serine 6, also known as matriptase-2, which inhibits the signaling pathway which activates hepcidin. This type of anemia is variable, more severe in children, and unresponsive to treatment with oral iron. It is characterized by striking microcytosis, extremely low transferrin saturation, normal or borderline-low ferritin levels, and high hepcidin levels. The diagnosis is confirmed by sequencing of *TMPRSS6*. IRIDA occurs in less than 1% of cases of iron-deficiency anemia seen in medical practice. Most cases of iron resistance are due to disorders in the GI tract (see Table 6.1).

Non-hematological Manifestations

Iron deficiency is a systemic disorder involving multiple systems rather than exclusively a hematologic condition associated with anemia. Table 6.5 lists important iron-containing compounds in the body and their function and Table 6.6 lists the tissue effects of iron deficiency.

TABLE 6.5   Important Iron-Containing Compounds and their Function

Compound	Function
α-Glycerophosphate dehydrogenase	Work capacity
Catalase	RBC peroxide breakdown
Cytochromes	ATP production, protein synthesis, drug metabolism, electron transport
Ferritin	Iron storage
Hemoglobin	Oxygen delivery
Hemosiderin	Iron storage
Mitochondrial dehydrogenase	Electron transport
Monoamine oxidase	Catecholamine metabolism
Myoglobin	Oxygen storage for muscle contraction
Peroxidase	Bacterial killing
Ribonucleotide reductase	Lymphocyte DNA synthesis, tissue growth
Transferrin	Iron transport
Xanthine oxidase	Uric acid metabolism

DIAGNOSIS

1. *Hemoglobin*: Hemoglobin is below the acceptable level for age (Appendix 1).
2. *Red cell indices*: Lower than normal mean corpuscular volume (MCV), mean corpuscular hemoglobin (MCH), and mean corpuscular hemoglobin concentration (MCHC) for age. Widened red cell distribution width (RDW) in association with a low MCV is one of the best screening tests for iron deficiency. In general, the decrease in indices parallels the decrease in hemoglobin.
3. *Blood smear*: Red cells are hypochromic and microcytic with anisocytosis and poikilocytosis, generally occurring only when hemoglobin level falls below 10 g/dl. Basophilic stippling can also be present but not as frequently as is present in thalassemia trait. The RDW is high (>14.5%) in iron deficiency and normal in thalassemia (<13%).
4. *Reticulocyte count*: The relative number of reticulocytes is often increased but when corrected for anemia the reticulocyte count is usually normal. In severe iron-deficiency anemia associated with bleeding, a reticulocyte count of 3–4% may occur.
5. *Platelet count*: The platelet count varies from thrombocytopenia to thrombocytosis. Thrombocytopenia is more common in severe iron-deficiency anemia. Thrombocytosis is present when there is associated bleeding.
6. *Bone marrow*: This is not indicated to diagnose iron deficiency. If performed it shows hypercellularity of red cell precursors and distortion of normoblast nuclei may occur. Little or no iron is shown in normoblast and reticulum cells by Prussian blue staining.
7. *Free erythrocyte protoporphyrin (FEP)*: The incorporation of iron into protoporphyrin represents the ultimate stage in the biosynthetic pathway of heme. Failure of iron supply will result in an accumulation of free protoporphyrin not incorporated into heme synthesis in the normoblast and the release of erythrocytes into the circulation with high FEP levels.
    a. The normal FEP level is 15.5 ± 8.3 mg/dl. The upper limit of normal is 40 mg/dl. Table 6.7 gives the causes of elevated levels of FEP and its advantages over transferrin saturation levels as a diagnostic tool.
    b. In both iron deficiency and lead poisoning, the FEP level is elevated. It is much higher in lead poisoning than in iron deficiency. The FEP is normal in α- and β-thalassemia minor. FEP elevation occurs as soon as the body stores of iron are depleted, before microcytic anemia develops. An elevated FEP level is therefore an indication for iron therapy even when anemia and microcytosis have not yet developed.

TABLE 6.6   Tissue Effects of Iron Deficiency

1. **Gastrointestinal tract**
   a. Anorexia: common and an early symptom
      i. Increased incidence of low-weight percentiles
      ii. Depression of growth
   b. Pica: pagophagia (ice) geophagia (sand)
   c. Atrophic glossitis with flattened, atrophic, lingual papillae which makes the tongue smooth and shiny
   d. Dysphagia
   e. Esophageal webs (Kelly–Paterson syndrome)
   f. Reduced gastric acidity
   g. Leaky gut syndrome
      i. Guaiac-positive stools: isolated
      ii. Exudative enteropathy: gastrointestinal loss of protein, albumin, immunoglobulins, copper, calcium, red cells
   h. Malabsorption syndrome
      i. Iron only
      ii. Generalized malabsorption: xylose, fat, vitamin A, duodenojejunal mucosal atrophy
   i. Beeturia
   j. Decreased cytochrome oxidase activity and succinic dehydrogenase
   k. Decreased disaccharidases, especially lactase with abnormal lactose tolerance tests
   l. Increased absorption of cadmium and lead (iron-deficient children have increased lead absorption)
   m. Increased intestinal permeability index
2. **Central nervous system**
   a. Irritability
   b. Fatigue and decreased activity
   c. Conduct disorders
   d. Lower mental and motor developmental test scores on the Bayley scale which may be long-lasting
   e. Decreased attentiveness, shorter attention span
   f. Significantly lower scholastic performance
   g. Reduced cognitive performance
   h. Breath-holding spells
   i. Papilledema
3. **Cardiovascular system**
   a. Increase in exercise and recovery heart rate and cardiac output
   b. Cardiac hypertrophy
   c. Increase in plasma volume
   d. Increased minute ventilation values
   e. Increased tolerance to digitalis
4. **Musculoskeletal system**
   a. Deficiency of myoglobin and cytochrome C
   b. Impaired performance of a brief intense exercise task
   c. Decreased physical performance in prolonged endurance work
   d. Rapid development of tissue lactic acidosis on exercise and a decrease in mitochondrial alpha-glycerophosphate oxidase activity
   e. Radiographic changes in bone-widening of diploic spaces
   f. Adverse effect on fracture healing
5. **Immunologic system**
   There is conflicting information as to the effect on the immunologic system of iron-deficiency anemia.
   a. Evidence of increased propensity for infection
      i. Clinical
         * Reduction of acute illness and improved rate of recovery in iron-replete compared to iron-deficient children
         * Increased frequency of respiratory infection in iron deficiency
      ii. Laboratory
         * Impaired leukocyte transformation
         * Impaired granulocyte killing and nitroblue tetrazolium reduction by granulocytes
         * Decreased myeloperoxidase in leukocytes and small intestine
         * Decreased cutaneous hypersensitivity
         * Increased susceptibility to infection in iron-deficient animals
   b. Evidence of decreased propensity for infection
      i. Clinical
         * Lower frequency of bacterial infection
         * Increased frequency of infection in iron-overload conditions
      ii. Laboratory
         * Transferrin inhibition of bacterial growth by binding iron so that no free iron is available for growth of microorganisms
         * Enhancement of growth of nonpathogenic bacteria by iron

(Continued)

TABLE 6.6   (Continued)

6. **Cellular changes**
   a. Red cells
      i. Ineffective erythropoiesis
      ii. Decreased red cell survival (normal when injected into asplenic subjects)
      iii. Increased autohemolysis
      iv. Increased red cell rigidity
      v. Increased susceptibility to sulfhydryl inhibitors
      vi. Decreased heme production
      vii. Decreased globin and α-chain synthesis
      viii. Precipitation of α-globin monomers to cell membrane
      ix. Decreased glutathione peroxidase and catalase activity
         • Inefficient H_2O_2 detoxification
         • Greater susceptibility to H_2O_2 hemolysis
         • Oxidative damage to cell membrane
         • Increased cellular rigidity
      x. Increased rate of glycolysis-glucose 6-phosphate dehydrogenase, 6-phosphogluconate dehydrogenase, 2,3-diphosphoglycerate (2,3-DPG), and glutathione
      xi. Increase in NADH-methemoglobin reductase
      xii. Increase in erythrocyte glutamic oxaloacetic transaminase
      xiii. Increase in free erythrocyte protoporphyrin
      xiv. Impairment of DNA and RNA synthesis in bone marrow cells
   b. Other tissues
      i. Reduction in heme-containing enzymes (cytochrome C, cytochrome oxidase)
      ii. Reduction in iron-dependent enzymes (succinic dehydrogenase, aconitase)
      iii. Reduction in monoamine oxidase
      iv. Increased excretion of urinary norepinephrine
      v. Reduction in tyrosine hydroxylase (enzyme converting tyrosine to dihyroxyphenylalanine)
      vi. Alterations in cellular growth, DNA, RNA, and protein synthesis in animals
      vii. Persistent deficiency of brain iron following short-term deprivation
      viii. Reduction in plasma zinc

TABLE 6.7   Causes of Elevated Levels of Free Erythrocyte Protoporphyrin (FEP) and Advantages of FEP Compared to Transferrin Saturation as a Diagnostic Tool

Causes of raised levels of FEP
   1. Iron-deficiency anemia
   2. Conditions with high reticulocyte count[a]
   3. Lead poisoning (very high levels)
   4. Chronic infection
   5. Erythropoietic protoporphyria
   6. Acute myelogenous leukemia
   7. Rare cases of dyserythropoietic and sideroblastic anemias
Advantages of FEP compared with transferrin saturation
   1. FEP is not subject to daily fluctuations
   2. FEP remains elevated during iron treatment (returns to normal after cells with excess FEP are replaced)[b]
   3. FEP is not elevated in α- and β-thalassemia

[a]Reticulocytes have a slightly higher concentration of FEP. It occurs in hemolytic anemias (e.g., hemoglobin SS disease).
[b]Useful to know whether a patient who is in the process of receiving iron treatment was iron-deficient before commencement of iron therapy.

8. *Serum ferritin*: The level of serum ferritin reflects the level of body iron stores; it is quantitative, reproducible, specific, and sensitive; and requires only a small blood sample. A concentration of less than 12 ng/ml is considered diagnostic of iron deficiency. Normal ferritin levels, however, can exist in iron deficiency when bacterial or parasitic infection, malignancy, or chronic inflammatory conditions coexist because ferritin is an acute-phase reactant and its synthesis increases in acute or chronic infection or inflammation. Figure 6.1 depicts the normal range of serum ferritin concentrations at different ages.

9. *Serum iron and iron saturation percentage*: Serum iron estimation as a measure of iron deficiency has serious limitations. It reflects the balance between several factors, including iron absorbed, iron used for hemoglobin synthesis, iron released by red cell destruction, and the size of iron stores. The serum iron concentration represents an equilibrium between the iron entering and leaving the circulation. Serum iron has a wide range

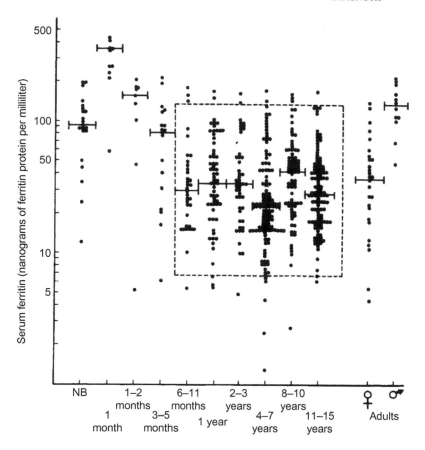

FIGURE 6.1 Serum ferritin concentrations during development in the healthy non-anemic newborn, in infants and in children of various age groups, together with adult male and female values. The median value in each age group is indicated by a horizontal line. The dashed line encloses a square, which includes the 95% confidence levels of the values between the ages of 6 months and 15 years. Note: Normal ferritin levels can occur in iron deficiency in the presence of bacterial or parasitic infection, malignancy, or chronic inflammatory conditions because ferritin is an acute-phase reactant. *From: Siimes et al. (1974), with permission.*

of normal, varies significantly with age (see Appendix 1) and is subject to marked circadian changes (as much as 100 mg/dl during the day). The author has abandoned the use of serum iron for the routine diagnosis of iron deficiency (in favor of MCV, RDW, FEP, and serum ferritin) because of the following limitations:

a. Wide normal variations (age, sex, laboratory methodology).
b. Unnecessary expense.
c. Subject to error from iron ingestion.
d. Diurnal variation.
e. Falls in mild or transient infection.

10. *Therapeutic trial*: The most reliable criterion of iron-deficiency anemia is the hemoglobin response to an adequate therapeutic trial of oral iron. Ferrous sulfate, in a dose of 3 mg/kg per day is given for 1 month. A reticulocytosis with a peak occurring between the fifth and tenth days followed by a significant rise in hemoglobin level occurs (a hemoglobin rise of more than 1 g/dl in 1 month). The absence of these changes implies that iron deficiency is not the cause of the anemia. Iron therapy should then be discontinued and further diagnostic studies implemented. Table 6.8 summarizes the diagnostic tests in the investigation of iron-deficiency anemia.

11. Other tests for iron deficiency not in common clinical usage include: Serum transferrin receptor levels (STfR), STfR/log ferritin ratio, and red blood cell zinc protoporphyrin/heme ratio.

a. *STfR levels*: This is a sensitive measure of iron deficiency and correlates with hemoglobin and other laboratory parameters of iron status. It may not be better than serum ferritin in diagnosing iron deficiency. The STfR is increased in instances of hyperplasia of erythroid precursors such as iron-deficiency anemia and thalassemia. It is unaffected by infection and inflammation, unlike serum ferritin which is raised. It is therefore of great value in distinguishing iron deficiency from the anemia of chronic disease and in identifying iron deficiency in the presence of chronic inflammation or infection. With erythroid hypoplasia or aplasia, for example, aplastic anemia, red cell aplasia, or chronic renal failure, the STfR concentration is decreased. It can be measured by a sensitive enzyme-linked immunosorbent assay technique

  b. *STfR/log ferritin ratio*: Calculating the ratio of serum transferring receptor concentration to the logarithm of the serum ferritin concentration provides the highest sensitivity and specificity in the presence of chronic inflammation or infection. Values of less than 2.2 mg/l exclude iron deficiency and values of more than 2.9 mg/l confirm iron deficiency.

  c. *Red blood cell zinc protoporphyrin/heme ratio*: This is increased whenever there is disruption of normal heme production, when available bone marrow iron is insufficient to support heme synthesis. It is not specific as to cause and is raised in iron deficiency and lead poisoning and markedly raised in protoporphyria and congenital erythropoietic porphyria. In these conditions zinc substitutes for iron in protoporphyrin IX and the concentration of zinc protoporphyrin relative to heme increases. False-positive results may occur in hyperbilirubinemia and falsely low results if the specimen is not protected from light. This is more sensitive than plasma ferritin levels, is inexpensive and simple. It is useful in identifying iron deficiency before anemia occurs.

12. Serum hepcidin levels are not available for clinical use. It could help to identify patients in whom a response to oral iron is probable (those with low hepcidin levels) and those in whom it is not probable (those with normal or elevated hepcidin levels).

STAGES OF IRON DEPLETION

1. Iron depletion: This occurs when tissue stores are decreased without a change in hematocrit or serum iron levels. This stage may be detected by low serum ferritin measurements.
2. Iron-deficient erythropoiesis: This occurs when reticuloendothelial macrophage iron stores are completely depleted. The serum iron level drops and the total iron-binding capacity increases without a change in the hematocrit. Erythropoiesis begins to be limited by a lack of available iron and STfR levels increase.
3. Iron-deficiency anemia: This is associated with erythrocyte microcytosis, hypochromia, increased RDW, and elevated levels of FEP. It is detected when iron deficiency has persisted long enough that a large proportion of circulating erythrocytes were produced after iron became limiting.

TABLE 6.8 Diagnostic Tests for Iron-Deficiency Anemia

1. Blood smear
   a. Hypochromic microcytic red cells, confirmed by RBC indices
      i. MCV less than acceptable normal for age (see Appendix 1)
      ii. MCH less than 27.0 pg
      iii. MCHC less than 30%
   b. Wide red cell distribution width greater than 14.5%
2. Free erythrocyte protoporphyrin: elevated
3. Serum ferritin: decreased
4. Serum iron and iron-binding capacity
   a. Decreased serum iron
   b. Increased iron-binding capacity
   c. Decreased iron saturation (16% or less)
5. Therapeutic responses to oral iron
   a. Reticulocytosis with peak 5–10 days after institution of therapy
   b. Following peak reticulocytosis hemoglobin level rises on average by 0.25–0.4 g/dl/day or hematocrit rises 1%/day
6. Serum transferrin receptor level[a]
7. Red blood cell zinc protoporphyrin/heme ratio[a]
8. Bone marrow[b]
   a. Delayed cytoplasmic maturation
   b. Decreased or absent stainable iron

[a]*Rarely required or readily available.*
[b]*Used only if difficulty is experienced in elucidating cause of anemia.*
MCV, mean corpuscular volume; MCH, mean corpuscular hemoglobin; MCHC, mean corpuscular hemoglobin concentration.

DIFFERENTIAL DIAGNOSIS

Iron-deficiency anemia must be differentiated from thalassemia minor and the anemia of chronic disease, which are common in infants. The RDW is normal in patients with thalassemia and anemia of chronic disease but high in those with iron deficiency. The MCV is decreased in iron-deficiency anemia and in thalassemia minor and normal or decreased in chronic disease. The FEP is increased in iron-deficiency anemia and in chronic disease and normal in thalassemia. The serum ferritin is decreased in iron deficiency, normal in thalassemia, and increased in chronic infection. The plasma in iron deficiency is clear and in thalassemia it is straw-colored. Sometimes these conditions coexist. It may be necessary to use a trial of iron for 1 month to sort out the diagnosis and determine the contribution of iron-deficiency anemia to the clinical picture. It should be noted that since hemoglobin A2 is decreased in iron deficiency a diagnosis of β-thalassemia minor (in which hemoglobin A2 is increased) may not be able to be made until after iron therapy and the iron deficiency is corrected.

Although hypochromic anemia in children is usually due to iron deficiency, it is not necessarily attributable to this condition. A list of the causes of hypochromia is given in Table 6.9. In some of these cases, there is an inability to synthesize hemoglobin normally in spite of adequate iron (e.g., thalassemia, lead poisoning). In unusual or obscure cases of hypochromic anemia, it is necessary to do additional investigations, such as determination of serum ferritin, STfR levels, hemoglobin electrophoresis, and examination of the bone marrow for stained iron, in order to establish the cause of the hypochromia.

Table 6.10 lists the investigations employed in the differential diagnosis of microcytic anemias and Figure 6.2 depicts a flow chart for the diagnosis of microcytic anemia using MCV and RDW.

In addition to making a diagnosis of iron-deficiency anemia, its pathogenesis must be established. The history should include conditions resulting in low iron stores at birth, dietary history and consideration of all factors leading to blood loss. The most common site of bleeding is into the bowel and the most important investigation is examination of the stool for occult blood. If occult blood is found, its cause should be established by examination of stools for ova, rectal examination, barium enema, upper GI series, ^{99m}Tc-pertechnetate scan for detection of a Meckel's diverticulum, upper endoscopy, and colonoscopy.

Negative guaiac tests for occult bleeding may occur if bleeding is intermittent. For this reason, occult bleeding should be tested for on at least five occasions when GI bleeding is suspected. The guaiac test is only sensitive enough to pick up more than 5 ml occult blood. Excessive uterine bleeding, epistaxis, renal blood loss (hematuria or hemoglobinuria from paroxysmal nocturnal hemoglobinuria) and, on rare occasions, bleeding into the lung (idiopathic pulmonary hemosiderosis and Goodpasture syndrome) may all be causes of iron-deficiency anemia. Bleeding into these areas requires specific investigations designed to detect the cause of bleeding.

TABLE 6.9 Disorders Associated with Hypochromia

1. Iron deficiency
2. Hemoglobinopathies
   a. Thalassemia (α and β)
   b. Hemoglobin Köln
   c. Hemoglobin Lepore
   d. Hemoglobin H
   e. Hemoglobin E
3. Disorders of heme synthesis caused by a chemical
   a. Lead
   b. Pyrazinamide
   c. Isoniazid
4. Sideroblastic anemias
5. Chronic infections or other inflammatory states
6. Malignancy
7. Hereditary orotic aciduria
8. Hypo- or atransferrinemia
   a. Congenital
   b. Acquired (e.g., hepatic disorders); malignant disease, protein malnutrition (decreased transferrin synthesis), nephrotic syndrome (urinary transferrin loss)
9. Copper deficiency
10. Inborn error of iron metabolism
    a. Congenital defect of iron transport to red cells

TABLE 6.10   Summary of Laboratory Studies in Microcytic Anemias

	Ethnic origin	Hb	MCV	MCV in parents	RDW	FEP	Ferritin	Serum iron	TIBC	Bone marrow iron status	Hb electrophoresis	Other features
Iron deficiency	Any	↓	↓	N	↑	↑	↓	↓	↑	↓	Normal	Dietary deficiency
β-Thalassemia β⁺ trait (heterozygous)	Mediterranean	Slight ↓	↓	One parent ↓	N	N	N or ↑	N	N	N	A₂ raised F normal or ↑	Normal examination
β⁰ (homozygous)	Mediterranean	↓	↓	Both parents ↓	N	↑	↑	↑	↑	↑	F raised (60–90%)	Hepatosplenomegaly transfusion-dependent
α-Thalassemia silent carrier (α-thal-2)	Asians, Blacks, Mediterranean	N	N	N	N	N	N	N	N	N	Normal	No hematologic abnormalities
Trait (α-thal-1)	Asians, Blacks, Mediterranean	N or slightly ↓	↓	One parent ↓	N	N	N or ↑	N	N	N	Normal	
Hemoglobin H disease		↓	↓		↑	N	N or ↑	N or ↑	N	↑	Hemoglobin H (2–40%)	Hemolytic anemia of variable severity inclusion bodies in RBCs
Anemia of chronic infection	Any	↓	N	N	N	↑	N or ↑	↓	N or ↑	N or ↑	Normal	
Sideroblastic	Any	↓	N	N	↑	N or ↑	N or ↑	N or ↑	N or ↓	↑	Normal	

FEP, free erythrocyte protoporphyrin; Hb, hemoglobin; MCV, mean corpuscular volume; RDW, red cell distribution width; TIBC, total iron-binding capacity, ↑, abnormally high; ↓, abnormally low; N, normal.

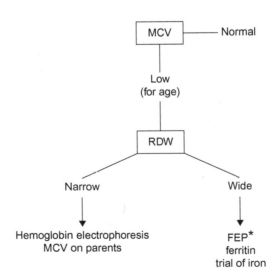

FIGURE 6.2   Flow chart depicting the diagnosis of microcytic anemia using mean corpuscular volume (MCV) and red cell distribution width (RDW). *Also elevated in lead poisoning. Do serum lead level (if clinically investigated).

TREATMENT

Nutritional Counseling

1. Maintain breast-feeding for at least 6 months, if possible.
2. Use an iron-fortified (6–12 mg/l) infant formula until 1 year of age (formula is preferred to whole cow's milk). Restrict milk to 1 pint/day. Avoid cow's milk until after the first year of age because of the poor bioavailability of iron in cow's milk and because the protein in cow's milk can cause occult GI bleeding.
3. Use iron-fortified cereal from 6 months to 1 year.
4. Evaporated milk or soy-based formula should be used when iron deficiency is due to hypersensitivity to cow's milk.
5. Provide supplemental iron for low-birth-weight infants:
   a. Infants 1.5–2.0 kg: 2 mg/kg/day supplemental iron.
   b. Infants 1.0–1.5 kg: 3 mg/kg/day supplemental iron.
   c. Infants <1 kg: 4 mg/kg/day supplemental iron.
6. Facilitators of iron absorption such as vitamin C-rich foods (citrus, tomatoes, and potatoes), meat, fish, and poultry should be included in the diet and inhibitors of iron absorption such as tea, phosphate, and phytates common in vegetarian diets should be eliminated.

Oral Iron Medication

The goal of therapy for iron deficiency is both correction of the hemoglobin level and replenishment of body iron stores.

1. *Product*: Ferrous iron (e.g., ferrous sulfate, ferrous gluconate, ferrous ascorbate, ferrous lactate, ferrous succinate, ferrous fumarate, or ferrous glycine sulfate) is effective. Ferric irons and heavily chelated iron should not be used as they are poorly and inefficiently absorbed.
2. *Dose*: 1.5–2.0 mg/kg elemental iron three times daily. Older children: ferrous sulfate (0.2 g) or ferrous gluconate (0.3 g) given three times daily, to provide 100–200 mg elemental iron. In children with GI side effects, iron once every other day may be better tolerated with good effect.
3. *Duration*: 10–12 weeks in order to replete the iron stores and restore the ferritin levels to normal.
4. *Response*: The lower the hemoglobin to start the higher the reticulocyte response and rise in hemoglobin.
   a. Peak reticulocyte count on days 5–10 following initiation of iron therapy.
   b. Following peak reticulocyte level, hemoglobin rises on average by 0.25–0.4 g/dl/day or hematocrit rises 1%/day during first 7–10 days.
   c. Thereafter, hemoglobin rises slower: 0.1–0.15 g/dl/day.
5. *Failure to respond to oral iron*: The following reasons should be considered:
   a. Poor compliance: Failure or irregular administration of oral iron; administration can be verified by change in stool color to gray-black or by testing stool for iron.
   b. Inadequate iron dose.
   c. Ineffective iron preparation.
   d. Insufficient duration.
   e. Persistent or unrecognized blood loss.
   f. Incorrect diagnosis: Thalassemia, sideroblastic anemia.
   g. Coexistent disease that interferes with absorption or utilization of iron (e.g., chronic inflammation, inflammatory bowel disease, malignant disease, hepatic or renal disease, concomitant deficiencies (vitamin B_{12}, folic acid, thyroid, associated lead poisoning)).
   h. Impaired GI absorption due to high gastric pH (e.g., antacids, histamine-2 blockers, gastric acid pump inhibitors). *Helicobacter pylori*, estimated to affect half the world's population, decreases iron absorption because the organism competes with its human host for available iron and should be eradicated in iron-resistant iron-deficiency anemia. Gluten sensitivity, which may be asymptomatic, interferes in iron absorption. Five percent of iron-refractory iron-deficient individuals have gluten sensitivity.
   i. Congenital IRIDA (see Section 'Iron-Refractory Iron-Deficiency Anemia' in this chapter).
6. *Side effects*: Constipation, diarrhea, abdominal cramps, nausea, and metallic taste. Although stools may be dark it does not produce false-positive results on tests for occult blood.

Parenteral Therapy

Indications

1. Non compliance, poor tolerance of oral iron or failure of oral iron therapy.
2. Severe bowel disease (e.g., inflammatory bowel disease) where use of oral iron might aggravate the underlying disease of the bowel or iron absorption is compromised, after gastrectomy or duodenal bypass surgery, atrophic gastritis, celiac disease.
3. Chronic hemorrhage (e.g., congenital coagulation disorders, hereditary telangiectasia, menorrhagia, chronic hemoglobinuria from prosthetic heart valves). Losing blood at a rate too rapid for oral intake to compensate for the loss.
4. Acute diarrheal disorder in underprivileged populations with iron-deficiency anemia.
5. Severe iron deficiency requiring rapid replacement of iron stores.
6. Patients anemic after receiving erythropoietin therapy (e.g., renal dialysis and in patients receiving chemotherapy) to ensure an ample and steady supply of iron.
7. Genetically induced IRIDA.
8. Substitution for blood transfusion when not accepted by patient for religious reasons.
9. Iron deficiency in heart failure.

Intramuscular

Iron-dextran, a parenteral form of elemental iron, is available for intramuscular use. It is safe, effective, and well-tolerated even in infants with a variety of acute illnesses, including acute diarrheal disorders.

An increased risk for clinical attacks of malaria and other infections has been demonstrated in malarious regions, particularly with parenteral or high-dose oral iron supplementation.

Dose

For intramuscular iron-dextran the following formula is used to raise the hemoglobin level to normal and to replenish iron stores:

$$\frac{\text{Normal hemoglobin} - \text{initial hemoglobin}}{100} \times \text{blood volume (ml)} \times 3.4 \times 1.5$$

1. Normal hemoglobin (see Appendix 1).
2. Blood volume: 80 ml/kg or 40 ml/lb body weight.
3. Multiplication by 3.4 converts grams of hemoglobin into milligrams of iron.
4. Factor 1.5 provides extra iron to replace depleted tissue stores.

Iron—dextran complex provides 50 mg elemental iron/ml.

Intravenous

Sodium ferric gluconate ((Ferrlecit) contains 12.5 mg/ml or polynuclear iron hydroxide sucrose complex (Venofer) contains 20 mg/ml elemental iron) for intravenous use is effective and relatively safe. Iron dextran should not be used intravenously because it is less safe and requires a test dose prior to use. They are especially useful in anemia associated with chronic renal failure and hemodialysis. Dosage ranges from 1 to 4 mg/kg per week.

Side Effects

Flushing, headache, muscle and joint pain, nausea, dizziness, rashes, fever, and chills may occur with parenteral iron. A very small number of patients experience anaphylaxis requiring emergency treatment. With intramuscular injection staining at the site may occur, especially in cases in which the solution is accidentally administered into the superficial tissues. Staining is of a transient type, disappearing after a few weeks or months. A "Z-track" injection into the muscle minimizes the chance of a subcutaneous leak. The local inflammatory reaction is slight. Because of the painful nature and the skin discoloration that occurs with intramuscular injection the preferred route for parenteral iron administration is intravenous.

Contraindications to Parenteral Iron Therapy

1. Anemias not due to iron deficiency.
2. Iron overload.
3. History of hypersensitivity to parenteral iron preparations.
4. History of severe allergy or anaphylactic reactions.
5. Clinical or biochemical evidence of liver damage.
6. Acute or chronic infection.
7. Neonates.

Blood Transfusion

A packed red cell transfusion should be given in severe anemia requiring correction more rapidly than is possible with oral iron or parenteral iron or because of the presence of certain complicating factors. This should be reserved for debilitated children with infection, especially when signs of cardiac dysfunction are present and the hemoglobin level is 4 g/dl or less.

Partial Exchange Transfusion

A partial exchange transfusion has been recommended in the management of a severely anemic child under two circumstances:

- In a surgical emergency, when a final hemoglobin of 9–10 g/dl should be attained to permit safe anesthesia.
- When anemia is associated with congestive heart failure, in which case it is sufficient to raise the hemoglobin to 4–5 g/dl to correct the immediate anoxia.

Further Reading and References

Ballin, A., Berar, M., Rubinstein, U., et al., 1992. Iron state in female adolescents. Am. J. Dis. Child. 146, 803–805.

Carmaschella, C., 2015. Iron deficiency anemia. N. Engl. J. Med. 372, 1832–1843.

Chaparro, C.M., 2008. Setting the stage for child health and development: prevention of iron deficiency in early infancy. J. Nutr. 138 (12), 2529–2533.

Clark, S.F., 2008. Iron deficiency anemia. Nutr. Clin. Pract. 23 (2), 128–141.

Committee on Nutrition, 1976. Iron supplementation for infants. Pediatrics 58, 765–768.

Dallman, P.R., 1987. Iron deficiency and related nutritional anemias. In: Nathan, D.G., Oski, F.A. (Eds.), Hematology of Infancy and Childhood. WB Saunders, Philadelphia.

Dallman, P.R., 1990. Progress in the prevention of iron deficiency in infants. Acta Paediatr. Scand. 365 (Suppl.), 28–37.

Dallman, P.R., Reeves, J.D., 1984. Laboratory diagnosis of iron deficiency. In: Steckel, A. (Ed.), Iron Nutrition in Infancy and Childhood. Raven Press, New York, p. 11.

Grant, C.C., Wall, C.R., Brewster, D., et al., 2007. Policy statement on iron deficiency in pre-school-aged children. J. Paediatr. Child Health 43 (7–8), 513–521.

Lanzkowsky, P., 1978. Iron metabolism and iron deficiency anemia. In: Miller, D.R., Pearson, M.A., Baehner, R.L., McMillan, C.W. (Eds.), Blood Diseases of Infancy and Childhood, fourth ed. CV Mosby, Saint Louis.

Lanzkowsky, P., 1980. Iron-Deficiency Anemia. Pediatric Hematology-Oncology: A Treatise for the Clinician. McGraw-Hill, New York.

Lanzkowsky, P., Iron deficiency anemias: A systemic disease. Transactions of the College of Medicine of South Africa, July–December, 1982, pp. 67–113.

Lozoff, B., 2007. Iron deficiency and child development. Food Nutr. Bull. 28 (4 Suppl.), S560–S571.

Lukens, J.N., 1984. Iron metabolism and iron deficiency anemia. In: Miller, D.R., Baehner, R.L., McMillan, C.W. (Eds.), Blood Diseases of Infancy and Childhood. CV Mosby, St. Louis.

Lynch, S., Stoltzfus, R., Rawat, R., 2007. Critical review of strategies to prevent and control iron deficiency in children. Food Nutr. Bull. 28 (4 Suppl.), S610–S620.

Oski, F.A., 1993. Iron deficiency in infancy and childhood. N. Engl. J. Med. 199s (329), 190–193.

Siimes, M.A., Addrego, J.E., Dallman, P.R., 1974. Ferritin in serum: diagnosis of iron deficiency and iron overload in infants and children. Blood 43, 581.

7

Megaloblastic Anemia

Philip Lanzkowsky

Megaloblastic anemias are characterized by the presence of megaloblasts in the bone marrow and macrocytes in the blood. In more than 95% of cases, megaloblastic anemia is a result of folate and vitamin B_{12} deficiency. Megaloblastic anemia may also result from rare inborn errors of metabolism of folate or vitamin B_{12}. In addition, deficiencies of ascorbic acid, tocopherol, and thiamine may be related to megaloblastic anemia. The causes of megaloblastosis are listed in Table 7.1. The clinical features, diagnosis, and treatment of cobalamin and folate deficiency are discussed later in this chapter (see section 'General clinical features of cobalamin and folate deficiency').

VITAMIN B_{12} (COBALAMIN) DEFICIENCY

Absorption and Metabolism

Dietary vitamin B_{12} (Cbl),[1] acquired mostly from animal sources, including meat and milk, is absorbed in a series of steps that include:

- Proteolytic release of Cbl from its associated proteins and Cbl binds to haptocorrin, a cobalamin-binding protein, produced by salivary and esophageal glands.
- In the duodenum, after exposure to pancreatic proteases, Cbl is released from haptocorrin.
- In the proximal ileum Cbl binds to intrinsic factor (IF), a gastric secretory protein, to form the IF–Cbl complex.
- Recognition of the IF–Cbl complex by specific receptors on ileal mucosal cells, which is taken into lysosomes where the IF–Cbl complex is released and IF is degraded.
- Transport across ileal cells in the presence of calcium ions.
- Release into the portal circulation bound to transcobalamin II (TC II)—the serum protein that carries newly absorbed Cbl throughout the body.

Figure 7.1 shows the pathway of cobalamin absorption, transport, and cellular uptake.

Cobalamin is converted into the two required coenzyme forms, adenosylcobalamin (AdoCbl) and methylcobalamin (MeCbl). The cellular metabolism by which the coenzymes are formed involves the following:

- Receptor-mediated binding of the TC II–Cbl complex to the cell surface.
- Adsorptive endocytosis of the complex.
- Intralysosomal degradation of the TC II.
- Release of Cbl into cytoplasm.
- Enzyme-mediated reduction of the central cobalt atom and cytosolic methylation to form MeCbl or mitochondrial adenosylation to form AdoCbl.

Causes of Vitamin B_{12} Deficiency

- The causes of vitamin B_{12} deficiency are listed in Table 7.2.

---

[1]For the purposes of this chapter, vitamin B_{12}, cobalamin, and cbl are used interchangeably. Vitamin B_{12} contains a metal ion in the form of cobalt and therefore is also known as cobalamin.

TABLE 7.1 Causes of Megaloblastosis

1. **Vitamin B₁₂ (cobalamin) deficiency** (see Table 7.2)
2. **Folate deficiency** (see Table 7.6)
3. **Miscellaneous**
   a. Congenital disorders in DNA synthesis
      i. Orotic aciduria (uridine-responsive) pyrimidine biosynthesis is interrupted
      ii. Thiamine-responsive megaloblastic anemia[a]
      iii. Congenital familial megaloblastic anemia requiring massive doses of vitamin B₁₂ and folate
      iv. Associated with congenital dyserythropoietic anemia (Tables 8.15 and 8.16)
      v. Lesch−Nyhan syndrome (adenine-responsive) purine nucleotide regeneration is blocked
   b. Acquired defects in DNA synthesis
      i. Liver disease
      ii. Sideroblastic anemias (Table 8.18)
      iii. Leukemia, especially acute myeloid leukemia (M6) (Chapter 19)
      iv. Aplastic anemia (congenital or acquired)
      v. Refractory megaloblastic anemia
   c. Drug-induced megaloblastosis
      i. Purine analogs (e.g., 6-mercaptopurine, azathioprine and thioguanine)
      ii. Pyrimidine analogs (5-fluorouracil, 6-azauridine)
      iii. Inhibitors of ribonucleotide reductase (cytosine arabinoside, hydroxyurea)

[a]*Associated in some cases with diabetes and sensorineural hearing impairment and in others with the DIDMOAD syndrome. There is wide clinical heterogeneity of this rare disorder. Only the anemia is responsive to high doses of thiamine.*

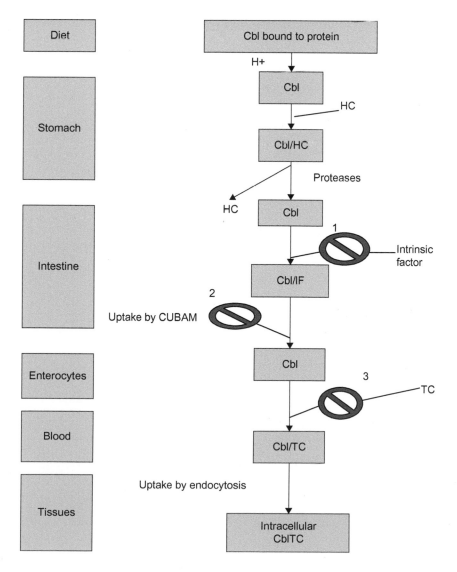

FIGURE 7.1 Summary of cobalamin absorption, transport, and cellular uptake. Cbl, cobalamin; Cbl/HC, cobalamin haptocorrin complex; Cbl/IF, cobalamin intrinsic factor complex; Cbl/TC, cobalamin transcobalamin complex; CUBAM, ileal receptors made up of cubilin and amnionless proteins; HC, haptocorrin; TC, transcobalamin; 1, intrinsic factor deficiency; 2, Imerslund−Gräsbeck syndrome; 3, transcobalamin deficiency. *Source: Adapted from Morel and Rosenblatt (2006), with permission.*

TABLE 7.2   Causes of Vitamin B$_{12}$ Deficiency

1. **Inadequate vitamin B$_{12}$ intake**
   a. Dietary (<2 mg/day): food fads, lacto-ovo vegetarianism, low animal-source food intake, veganism, malnutrition, poorly controlled phenylketonuria diet
   b. Maternal deficiency leading to B$_{12}$ deficiency in breast milk
2. **Defective vitamin B$_{12}$ absorption** (Table 7.3)
   a. Failure to secrete intrinsic factor
      i. Congenital intrinsic factor deficiency (gastric mucosa normal) (OMIM 261000)
         * Quantitative
         * Qualitative (biologically inert)[a]
      ii. Juvenile pernicious anemia (autoimmune) (gastric atrophy)[b]
      iii. Juvenile pernicious anemia (gastric autoantibodies) with autoimmune polyendocrinopathies (OMIM 240300)
      iv. Juvenile pernicious anemia with IgA deficiency
      v. Gastric mucosal disease
         * Chronic gastritis, gastric atrophy (elevated serum gastrin and/or low serum pepsinogen 1 concentrations) often caused by *Helicobacter pylori*
         * Corrosives
         * Gastrectomy (partial/total)
   b. Failure of absorption in small intestine
      i. Specific vitamin B$_{12}$ malabsorption
         * Abnormal intrinsic factor[a]
         * Defective cobalamin transport by enterocytes—abnormal ileal uptake (Imerslund–Gräsbeck syndrome) (OMIM 261100)
         * Ingestion of chelating agents (phytates, EDTA) (binds calcium and interferes with vitamin B$_{12}$ absorption)
      ii. Intestinal disease causing generalized malabsorption, including vitamin B$_{12}$ malabsorption:
         * Intestinal resection (e.g., congenital stenosis, volvulus, trauma)
         * Crohn's disease
         * Tuberculosis of terminal ileum
         * Lymphosarcoma of terminal ileum
         * Pancreatic insufficiency[c]
         * Zollinger–Ellison syndrome (caused by gastrinoma in duodenum or pancreas)
         * Celiac disease (gluten enteropathy), tropical sprue
         * Other less specific malabsorption syndromes
         * HIV infection
         * Long-standing medication that decreases gastric acidity (H$_2$-receptor antagonists and proton pump inhibitors)
         * Parasites (*Giardia, Lamblia, Diphyllobothrium latum*)
         * Neonatal necrotizing enterocolitis
      iii. Competition for vitamin B$_{12}$
         * Small-bowel bacterial overgrowth (e.g., small-bowel diverticulosis, anastomoses and fistulas, blind loops and pouches, multiple strictures, scleroderma, achlorhydria, gastric trichobezoar)
         * *Diphyllobothrium latum*, the fish tapeworm (takes up free B$_{12}$ and B$_{12}$-intrinsic factor complex), *Giardia lamblia, Plasmodium falciparum, Strongyloides stercoralis*
3. **Defective vitamin B$_{12}$ transport**
   a. Congenital TC II deficiency (OMIM 275350)
   b. Transient deficiency of TC II
   c. Partial deficiency of TC I (haptocorrin deficiency) (OMIM 193090)
4. **Disorders of vitamin B$_{12}$ metabolism**
   a. Congenital
      i. Adenosylcobalamin deficiency *CblA* (OMIM 251100) and *CblB* diseases (OMIM 251110)
      ii. Deficiency of methylmalonyl-CoA mutase (mut, mut^2)
      iii. Methylcobalamin deficiency *CblE* (OMIM 236270) and *CblG* diseases (OMIM 250940)
      iv. Combined adenosylcobalamin and methylcobalamin deficiencies: *CblC* (OMIM 277400), *CblD* (OMIM 277410), and *CblF* diseases (OMIM 277380)
   b. Acquired
      i. Liver disease
      ii. Protein malnutrition (kwashiorkor, marasmus)
      iii. Drugs associated with impaired absorption and/or utilization of vitamin B$_{12}$ (e.g., *p*-aminosalicylic acid, colchicine, neomycin, ethanol, oral contraceptive agents, Metformin)

[a]*Same condition.*
[b]*Pernicious anemia is the final stage of an autoimmune disorder in which autoantibodies against H$^+$K$^+$-adenosine triphosphatase destroy parietal cells in the stomach.*
[c]*Because of lack of the enzymes needed to liberate B$_{12}$ from haptocorrin, the protein that initially binds ingested vitamin B$_{12}$.*
OMIM, Online Mendelian Inheritance in Man (http://www.ncbi.nlm.nih.gov/omim/).

Nutritional Deficiency

The recommended dietary allowance of vitamin B$_{12}$ for children is 0.9—2.4 µg/day. The most common cause of Cbl deficiency in infants is dietary deficiency in the mother. Mothers following vegetarian, vegan, macrobiotic, and other special diets are at particular risk. Cbl in breast milk parallels that in serum and is deficient when the mother is a vegan or has unrecognized pernicious anemia, has had previous gastric bypass surgery or short gut syndrome.

Defective Absorption

Table 7.3 lists the features of congenital and acquired defects of vitamin B$_{12}$ absorption, Table 7.4 lists the main features of genetic defects in processing of vitamin B$_{12}$.

TABLE 7.3 Features of Congenital and Acquired Defects of Vitamin B$_{12}$ Absorption

Condition	Stomach			Schilling Test		Serum Antibodies		
	Histology	Intrinsic factor[a]	Hydrochloric acid (HCl)	Without IF	With IF	Intrinsic factor	Parietal cell	Associated features
Congenital pernicious anemia	Normal	Absent	Normal	Decreased	Normal	Absent	Absent	None; relative of patient may exhibit defective vitamin B$_{12}$ malabsorption
Juvenile pernicious anemia (autoimmune)	Atrophy	Absent	Achlorhydria	Decreased	Normal	Present (90%)	Present (10%)	Occasional lupus erythematosus, IgA deficiency, moniliasis, endocrinopathy in siblings
Juvenile pernicious anemia with polyendocrino-pathies or selective IgA deficiency	Atrophy	Absent	Achlorhydria	Decreased	Normal	Present	Present	Hypothyroidism (chronic auto-immune thyroiditis—Hashimoto's thyroiditis) insulin-dependent diabetes mellitus, primary ovarian failure, myasthenia gravis, hypoparathyroidism, Addison disease, moniliasis, or selective IgA deficiency
Enterocyte vitamin B$_{12}$ malabsorption (Imerslund-Gräsbeck)	Normal	Present	Normal	Decreased	Decreased	Absent	Absent	Benign proteinuria, amino-aciduria, no generalized malabsorption[b]
Generalized malabsorption	Normal	Present	Normal	Decreased	Decreased	Absent	Absent	Malabsorption; syndrome; history of ileal resection, Crohn's disease, lymphoma

[a]*Either absent secretion of immunologically recognizable IF or secretes immunologically reactive protein that is inactive physiologically. The latter group includes patients whose IF has reduced affinity for the ileal IF receptor, reduced affinity for cobalamin or increased susceptibility for proteolysis.*
[b]*Rare cases have been described of this syndrome associated with generalized malabsorption reversed by vitamin B$_{12}$ administration and rare cases have been described without proteinuria or aminoaciduria.*
IF, intrinsic factor.

TABLE 7.4 Main Features of Genetic Defects in Processing of Vitamin B$_{12}$

Defect	Serum B$_{12}$	Clinical/biochemical
Food cobalamin malabsorption	Low	N.A. ± Anemia, mild ↑ MMA/tHcy
Intrinsic factor deficiency	Low	Anemia, delayed development, mild ↑ MMA/tHcy
Enterocyte cobalamin malabsorption (Imerslund—Gräsbeck)	Low	Anemia, proteinuria, delayed development, mild ↑ MMA/tHcy
Transcobalamin I (R-Binder) deficiency	Low	No abnormality, No ↑ MMA/tHcy
Transcobalamin II deficiency	Normal	N.A. ± Anemia, failure to thrive, mild ↑ MMA/tHCy
Intracellular defects of cobalamin	Normal	Severe disease, ↑ MMA/tHcy

N.A., neurologic abnormalities; MMA, methylmalonic acid; tHcy, total homocysteine; ↑, increased.

Gastric Acidity and Peptic Activity Deficiency

Acid pH and peptic activity are required to release cobalamin from the protein-bound state in which it occurs in food. Impaired absorption occurs when there is impaired gastric function, for example atrophic gastritis, partial gastrectomy resulting in a low serum cobalamin, mild increase in methylmalonic acid and homocysteine, and a normal Schilling test.

Intrinsic Factor Deficiency

Patients with absent or defective IF (also known as S-binder) have low serum B_{12} and megaloblastic anemia. This autosomal recessive disorder usually appears early in the second year of life, but may be delayed until adolescence or adulthood. The abnormal absorption of cobalamin is corrected by mixing the vitamin with a source of normal IF. Some patients have no detectable IF, whereas others have IF that can be detected immunologically but lacks function. The gene for human IF (GIF gene) has been cloned and localized to chromosome 11. Mutations have been identified in the GIF gene, together with a polymorphism ($68A \rightarrow G$) which may be a marker for this inheritance. Homozygous GIF mutations result in complete loss of IF function.

Defective Cobalamin Transport by Ileal Enterocyte Receptors for the Intrinsic Factor-Cobalamin Complex (Imerslund-Gräsbeck Syndrome)

Most of the known patients reside in Norway, Finland, Saudi Arabia, and among Sephardic Jews in Israel. It is due to a selective defect in cobalamin absorption in the ileum that is not corrected by treatment with IF. In these patients, gastric IF level is normal, they do not have antibodies to IF, and the ileal intestinal morphology is normal. In some cases the ileal receptor for IF—cobalamin complex is absent, whereas in other patients it is present. It is an autosomal recessive disorder associated with a low serum B_{12}.

The locus for Imerslund—Gräsbeck syndrome has been assigned to chromosome 10. Imerslund—Gräsbeck-causing mutations are found in either of two genes encoding the epithelium proteins: cubilin (CUBN) and amnionless (AMN). The gene receptor, cubilin P1297L (OMIM 602997)[2] is a 640-kDa protein which recognizes IF—cobalamin and various other proteins to be endocytosed in the intestine and kidney. The exact function of AMN is unknown but mutations affecting either of the two proteins may cause Imerslund—Gräsbeck syndrome.

The Imerslund—Gräsbeck syndrome usually presents with pallor, weakness, anorexia, failure to thrive, delayed development, recurrent infections, and gastrointestinal symptoms within the first 2 years of life but has been reported up to 15 years of age. In many patients, proteinuria of the tubular type is found that is not corrected by systemic cobalamin.

There has been a decrease in the number of new cases, suggesting that dietary or other factors may influence the expression of this disease.

Defective Transport

Table 7.5 lists clinical manifestations, laboratory finding and treatment of inborn errors of cobalamin transport and metabolism and Figure 7.2 shows the pathways of vitamin B_{12} metabolism and sites of inborn errors of vitamin B_{12} metabolism.

Transcobalamin II Deficiency (OMIM 275350)

TC II is the principal transport carrier protein system of cobalamin. The TC II gene is located on chromosome 22 and is inherited as an autosomal recessive trait. In the absence of TC II, a serious and potentially fatal condition occurs. It presents clinically at 3—5 weeks of age with

- Failure to thrive, weakness.
- Vomiting and diarrhea.
- Neurologic disease that appears 3—6 months after onset of symptoms.

---

[2]The six-digit number is the entry number for the disorder in Online Mendelian Inheritance in Man (OMIM) a continuously updated electronic catalog of human genes and genetic disorders. The online version is accessible through the world wide web (http://www.ncbi.nlm.nih.gov/omim/).

TABLE 7.5 Clinical Manifestations, Laboratory Findings, and Treatment of the Autosomal Recessive Inborn Errors of Cobalamin Transport and Metabolism

Condition (OMIM no.)	Defect	Typical clinical manifestations	Typical onset	Laboratory findings	Treatment and response
TC II deficiency (OMIM 275350)	Defective/absent TC II	Failure to thrive, megaloblastic anemia, later neurologic features and immunodeficiency	Early infancy 3–5 weeks	Usually normal serum Cbl; elevated serum MMA, homocysteine; absent/defective TC II	High doses of Cbl by injection; good response to treatment if begun early
TC I (R-binder) deficiency (OMIM 193090)	Deficiency/absence of TCI in plasma, saliva, leukocytes	Neurologic symptoms (myelopathy) reported, but unclear if these are related to condition	Unclear if observed symptoms are related to condition	Low serum Cbl, normal TC II-Cbl levels. No increase in MMA or homocysteine	Cbl therapy does not appear to be of benefit
Defective synthesis of AdoCbl: cblA (OMIM 251100) cblB (OMIM 251110)	Defective synthesis of AdoCbl	Lethargy, failure to thrive, recurrent vomiting, dehydration, hypotonia, keto acidosis hypoglycemia	First weeks or months of life	Normal serum Cbl, homocysteine and methionine; elevated MMA, ketones, glycine, ammonia; leukopenia, thrombocytopenia, anemia	Pharmacologic doses of Cbl, dietary protein restriction, oral antibiotics. Treatment response for cblA better than for cblB
Defective synthesis of MeCbl: cblE (OMIM 236270) cblG (OMIM 250940)	Defective synthesis of MeCbl	Vomiting, poor feeding, lethargy, severe neurologic dysfunction, megaloblastic anemia	Most in first 2 years of life	Normal serum Cbl and folate; homocystinuria, hypomethioninemia	Pharmacologic doses of Cbl, betaine; good treatment response in some patients treated early
Defective synthesis of AdoCbl and MeCbl: cblC (OMIM 277400) cblD (OMIM 277410) cblF (OMIM 277380)	Impaired synthesis of both AdoCbl and MeCbl	Failure to thrive, developmental delay, neurologic dysfunction, megaloblastic anemia, some cases with retinal findings	Variable from neonatal period to adolescence majority with neonatal onset	Normal serum Cbl, TC II; methylmalonic aciduria, homcystinuria, hypomethioninemia	Pharmacologic doses of hydroxocobalamin, moderate protein restriction, betaine treatment. Response often not optimum

TC II, transcobalamin II; OMIM, Online Mendelian Inheritance in Man; Cbl, cobalamin; MMA, methylmalonic acid; TCI, transcobalamin I; AdoCbl, adenosylcobalamin; MeCbl, methylcobalamin.

Modified from Rasmussen et al. (2001), with permission.

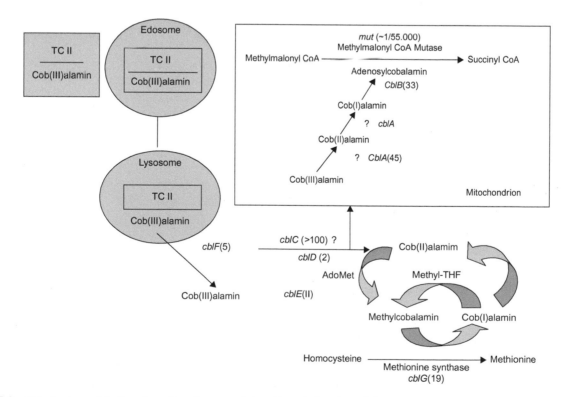

FIGURE 7.2 Cobalamin metabolism in cultured mammalian cells and the sites of the known inborn errors of cobalamin metabolism. AdoMet, S-adenosylmethionine, cob(III)alamin, cob(II)alamin, cob(II) represent cobalamin with its cobalt in the 3+, 2+, or 1+ oxidation state, methyl-THF is 5-methyltetrahydrofolate. The incidence or minimum numbers of patients with a given diseases are shown in parentheses. *Source: Adapted from Rosenblatt and Whitehead (1999), with permission.*

- Severe megaloblastic anemia. Some present with progressive pancytopenia or isolated erythroid hypoplasia. Defective granulocyte function may occur. Serum cobalamin levels are normal (most of the cobalamin in serum is bound to transcobalamin I (TC I)).
- Cellular and humoral immunologic deficiency.
- Hyperhomocysteinemia, homocystinuria, and methylmalonic aciduria occurs.

Diagnosis

Diagnosis requires demonstration of the absence of protein capable of binding radiolabeled cobalamin and migrating with TC II on chromatography or gel electrophoresis, or by immunologic techniques. TC II is synthesized by amniocytes, permitting prenatal diagnosis.

Treatment

One thousand μg vitamin B_{12} intramuscularly 1−2 times weekly. Serum cobalamin levels must be kept very high (1,000−10,000 pg/ml) in order to treat TC II deficiency patients successfully.

Partial Deficiency of Transcobalamin I (Haptocorrin Deficiency) (OMIM 193090)

In partial deficiency of TC I (also known as haptocorrin or R-binder) serum vitamin B_{12} concentrations are low but patients show no signs of vitamin B_{12} deficiency (no megaloblastic hematologic features, normal values for hemocysteine and methylmolonic acid) because their TC II-cobalamin levels are normal and patients are not clinically deficient in vitamin B_{12}. TC I concentrations range from 25 to 54% of the mean normal concentration.

Clinically this syndrome is characterized by a myelopathy, not attributable to other causes and the etiology of these symptoms remains unclear.

Disorders of Metabolism: Congenital

The conversion of a vitamin to its active coenzyme and subsequent binding to an apoenzyme producing active holoenzyme are fundamental biochemical processes. Therefore deficient activity of an enzyme can result not only from a defect of the enzyme protein itself, which may involve interaction of a coenzyme with an apo-enzyme, but also from a defect in the conversion of the vitamin to a coenzyme.

Once vitamin B_{12} has been taken up into cells, it must be converted to an active coenzyme in order to act as a cocatalyst with vitamin B_{12}-dependent apoenzymes. Two enzymes known to depend for activity on vitamin B_{12} derivatives are:

- Methylmalonyl coenzyme A (CoA) mutase, which requires AdoCbl. Methylmalonyl CoA mutase catalyzes the conversion of methylmalonyl CoA to succinyl CoA. A decreased activity of methylmalonyl CoA mutase is reflected by the excretion of elevated amounts of methylmalonic acid.
- N^5-methyltetrahydrofolate homocysteine methyltransferase, which requires MeCbl. Lack of MeCbl leads to deficient activity of N^5-methyltetrahydrofolate homocysteine methyltransferase, with reduced ability to methylate homocysteine, resulting in hyperhomocysteinemia and homocysteinuria.

Patients with inborn errors of cobalamin utilization present with methylmalonic acidemia and hyperhomocysteinemia, either alone or in combination. Methylmalonic acidemia occurs as a result of a functional defect in the mitochondrial methylmalonyl CoA mutase or its cofactor AdoCbl, which catalyzes the conversion of L-methylmalonyl CoA to succinyl CoA. Hyperhomocysteinemia occurs as a result of a functional defect in the cytoplasmic methionine synthase or its cofactor MeCbl. Those disorders causing methylmalonic aciduria are characterized by severe metabolic acidosis, with the accumulation of large amounts of methylmalonic acid in blood, urine, and cerebrospinal fluid.

The incidence is estimated at 1:61,000. All the disorders of Cbl metabolism are inherited as autosomal recessive traits and prenatal diagnosis is possible. Classification has relied on somatic cell complementation studies in cultured fibroblasts. Prenatal detection of fetuses with defects in the complementation groups *cblA, cblB, cblC, cblE,* and *cblF* has been accomplished using cultured amniotic cells and chemical determinations on amniotic fluid or maternal urine. In several cases, in utero cbl therapy has been attempted with apparent success.

Adenosylcobalamin Deficiency CblA (OMIM 251100) and CblB (OMIM 251110) Diseases

Deficiency of AdoCbl synthesis leads to impaired methylmalonyl CoA mutase activity and results in methylmalonic acidemia. Cobalamin-responsive methylmalonic aciduria characterizes both *CblA* and *CblB* diseases. Intact cells from both *CblA* and *CblB* patients fail to synthesize AdoCbl. However, cell extracts from *CblA* patients can synthesize AdoCbl when provided with an appropriate reducing system, whereas extracts from *CblB* patients cannot. The defect in *CblA* may be related to a deficiency of a mitochondrial nicotinamide adenine dinucleotide phosphate (NADPH)-linked aquacobalamin reductase. The defect in *CblB* affects adenosyltransferase, which is involved in the final step in AdoCbl synthesis.

This group of patients presents with:

- Life-threatening or fatal ketoacidosis in the first few weeks or months of life.
- Hypoglycemia and hyperglycinemia.
- Failure to thrive or developmental retardation (may be a consequence of the acidosis and reversed by relief of the ketoacidosis).
- Serum cobalamin concentrations are normal.
- Both *CblA* and *CblB* are autosomal recessive diseases.

Studies of these patients have shown that intact cells fail to oxidize propionate normally. Methylmalonyl CoA arises chiefly through the carboxylation of propionate, which in turn derives largely from degradation of valine, isoleucine, methionine, and threonine.

Treatment

Ninety percent of *CblA* patients respond to therapy with systemic hydroxocobalamin or cyanocobalamin, whereas only 40% of *CblB* patients respond to this therapy. Only 30% have long-term survival.

Methylmalonyl-CoA Mutase (mut, mut^2) Deficiency

Defects in methylmalonyl CoA mutase apoenzyme formation can occur and result in methylmalonic aciduria. This is an autosomal recessive disease, and prenatal diagnosis is possible. Culture of patients' fibroblasts shows two classes of mutase deficiency: those having no detectable enzyme activity are designated mut, whereas those with residual activity, which can be stimulated by high levels of cobalamin, are called mut^2.

Clinical Findings

Infants are well at birth but become rapidly symptomatic on protein feeding and develop lethargy, failure to thrive, muscular hypotonia, respiratory distress, and recurrent vomiting and dehydration. Basal ganglia infarcts, tubulointerstitial nephritis, acute pancreatitis, and cardiomyopathy may occur as complications despite treatment. Patients excrete greater than 100 mg to several grams of methylmalonic acid daily (normal excretion is less than 15–20 micrograms per gram of creatinine), develop life-threatening or fatal ketoacidosis, elevated levels of ketones, glycine, and ammonia in the blood and urine. Hypoglycemia, leukopenia, and thrombocytopenia may occur.

Treatment

Patients do not respond to vitamin B$_{12}$ therapy. Protein restriction using a formula deficient in valine, isoleucine, methionine, and threonine (with the goal of limiting amino acids that use the propionate pathway) is employed. Lincomycin and metronidazole may reduce enteric propionate production by anaerobic bacteria. Therapy with carnitine is employed for those patients who are carnitine deficient.

Methylcobalamin Synthesis Deficiency: CblE (OMIM 236270) and CblG (OMIM 250940) Diseases

Abnormalities in MeCbl synthesis result in reduced N^5-methyltetrahydrofolate: homocysteine methyltransferase and consequently lead to homocysteinuria with hypomethioninemia. Thus homocysteinuria and hypomethioninemia, usually without methylmalonic aciduria, characterize functional methionine synthase deficiency (*CblE, CblG*), although one *CblE* patient has been described showing transient methylmalonic aciduria. Fibroblasts from *CblE* and *CblG* patients show a decreased accumulation of MeCbl with a normal accumulation of AdoCbl after incubation with cyanocobalamin. Their fibroblasts show decreased incorporation of labeled methyltetrahydrofolate as well. Cyanocobalamin uptake and binding to both cobalamin-dependent enzymes is normal in *CblE* fibroblasts and in most *CblG* fibroblasts.

Clinical Findings

1. Most patients become ill with failure to thrive during the first 2 years of life, but a number have been diagnosed in adulthood.
2. Megaloblastic anemia.
3. Various neurological deficits including developmental delay, cerebral atrophy, EEG abnormalities, nystagmus, hypotonia, hypertonia, seizures, blindness, and ataxia.

Treatment

Hydroxocobalamin is administered systemically, daily at first, then once or twice weekly. Usually this corrects the anemia and the metabolic abnormalities. Betaine supplementation may be helpful to reduce the homocysteine further. The neurological findings are more difficult to reverse once established, particularly in *CblG* disease. There has been successful prenatal diagnosis of *CblE* disease in amniocytes and the mother with an affected fetus can be treated with twice-weekly hydroxocobalamin after the second trimester.

Combined Adenosylcobalamin and Methylcobalamin Deficiencies CblC (OMIM 277400), CblD (OMIM 277410), and CblF (OMIM 277380) Diseases

These disorders result in failure of cells to synthesize both MeCbl (resulting in homocysteinuria and hypomethioninemia) and AdoCbl (resulting in methylmalonic aciduria) and accordingly, deficient activity of methylmalonyl CoA mutase (leading to homocystinuria and hypomethioninemia with methylmalonic aciduria) and N^5-methyltetrahydrofolate: homocysteine methyltransferase. Fibroblasts from *CblC* and *CblD* patients accumulate virtually no AdoCbl or MeCbl when incubated with labeled cyanocobalamin. In contrast, fibroblasts from *CblF* patients accumulate excess cobalamin, but it is all unmetabolized cyanocobalamin, nonprotein-bound and localized to lysosomes. In *CblC* and *CblD*, the defect is believed to involve cob(III)alamin[3] reductase or reductases, whereas in *CblF*, the defect involves the exit of cobalamin from the lysosome. Partial deficiencies of cyanocobalamin beta-ligand transferase and microsomal cob(III)alamin reductase have been described in *CblC* and *CblD* fibroblasts as well.

Clinical Findings

These patients present in the first month or before the end of the first year of life with:

- Poor feeding, failure to thrive and lethargy.
- Developmental retardation.
- Spasticity, delirium, and psychosis (in older children and adolescence).
- Hydrocephalus, cor pulmonale, and hepatic failure have been described, as well as a pigmentary retinopathy with perimacular degeneration.
- Methylmalonic acid levels are less than in methylmalonyl CoA mutase deficiency but greater than in defects of cobalamin transport.
- Macrocytosis, hypersegmented neutrophils, thrombocytopenia, and megaloblastic anemia.
- Many patients with the onset of symptoms in the first month of life die, whereas those with a later onset have a better prognosis.

CblC, *CblD*, and *CblF* diseases can be differentiated using cultured fibroblasts. Failure of uptake of labeled cyanocobalamin distinguishes *CblC* and *CblD* from all other cbl mutations. There is reduced incorporation of propionate and methyltetrahydrofolate into macromolecules in all three disorders and reduced synthesis of AdoCbl and MeCbl. Complementation analysis between an unknown cell line and previously defined groups establishes the specific diagnosis. Prenatal diagnosis has been successfully accomplished in *CblC* disease using chorionic villus biopsy material and cells.

Treatment

The treatment of *CblC* disease can be difficult. Daily therapy with oral betaine and twice-weekly injections of hydroxocobalamin improve lethargy, irritability, and failure to thrive, reduce methylmalonic aciduria and return serum methionine and homocysteine concentrations to normal. There has been incomplete reversal of the

---

[3]In this form of cobalamin the cobalt atom is trivalent (cob (III)) and must be reduced before it can bind to the respective enzyme.

neurologic and retinal findings. Surviving patients usually have moderate to severe developmental delay, even with good metabolic control.

Disorders of Metabolism: Acquired

In protein malnutrition (kwashiorkor, marasmus) and liver disease, impaired utilization of vitamin B_{12} has been reported. Certain drugs are associated with impaired absorption or utilization of vitamin B_{12} (see Table 7.2).

FOLIC ACID DEFICIENCY

Absorption and Metabolism

Food folate occurs in the polyglutamate form, which must be hydrolyzed by conjugase in the brush border of the intestine to folate monoglutamates. These are absorbed in the duodenum and upper small intestine and transported to the liver becoming 5-methyltetrahydrofolate, the principal circulating folate form.

Folate binds to and acts as a coenzyme for enzymes that mediate single-carbon transfer reactions. They accept and donate single-carbon atoms at different states of oxidation. 5,10-Methylenetetrahydrofolate is used unchanged for the synthesis of thymidylate, reduced to 5-methyltetrahydrofolate for the synthesis of methionine, or oxidized to 10-formyl tetrahydrofolate for the synthesis of purines.

The recommended dietary allowance of folate increases from 150 to 400 μg/day from age 1 year to 18 years.

The causes of folic acid deficiency are listed in Table 7.6.

Acquired Folate Deficiency

Folate deficiency, next to iron deficiency, is one of the commonest micronutrient deficiencies worldwide.

Folate sufficiency for the fetus is necessary to prevent neural tube defects. It is best achieved by administering folate (and cobalamin) to mothers during the periconceptional period. Low daily folate intake is associated with a two fold increased risk for preterm delivery and low infant birth-weight.

Folate requirements are increased in the following conditions:

- Rapid growth in the first few weeks of life. Folate supplement is necessary at this time, in doses of 0.05−0.2 mg daily, particularly in premature infants.
- Pregnant women (due to the requirements of the growing fetus) and lactating women.
- Goat's milk diet (goat's milk contains almost no folate).
- Medication ingestion, for example, antiepileptic medication, oral contraceptive pills.
- Diseases of the small intestine causing malabsorption.
- Hemolytic anemia with rapid red cell turnover, for example, sickle cell anemia, thalassemia.
- Chronic infections, for example, diarrhea, hepatitis (may disturb folate stores), HIV infection.

Inborn Errors of Folate Transport and Metabolism

These include hereditary folate malabsorption, methylenetetrahydrofolate reductase (MTHFR) deficiency, and glutamate formiminotransferase deficiency. In addition to these rare severe deficiencies, polymorphisms in the MTHFR gene have been implicated with neural defects and vascular thrombosis. Table 7.7 lists the clinical and biochemical features of inherited defects of folate metabolism.

Hereditary Folate Malabsorption (OMIM 229050)

Hereditary folate malabsorption (congenital malabsorption of folate) is due to a rare autosomal recessive trait.

It is characterized by an abnormality in the absorption of oral folic acid or of reduced folates (5-methyltetrahydrofolate or 5-formyltetrahydrofolate (folinic acid)) associated with elevated excretion of formiminoglutamate (FIGLU) and of orotic acid. This disease is due to a specific transport system defect for folates across both the intestine and the choroid plexus. It is caused by loss-of-function mutations in the protein-coupled folate transporter gene (pcft/SLC46A1). Even when blood folate levels are increased sufficiently to correct the anemia, folate levels in the cerebrospinal fluid may remain low. These patients are unable to achieve the normal 3:1 CSF:serum

TABLE 7.6   Causes of Folic Acid Deficiency

1. **Inadequate intake**
   a. Poverty, ignorance, faddism
   b. Method of cooking (sustained boiling loses 40% folate)
   c. Goat's-milk feeding (6 mg folate/l)
   d. Malnutrition (marasmus, kwashiorkor)
   e. Special diets for phenylketonuria or maple syrup urine disease
   f. Prematurity
   g. Post bone marrow transplantation (heat-sterilized food)
2. **Defective absorption**
   a. Congenital, isolated defect of folate malabsorption[a]
   b. Acquired
      i. Idiopathic steatorrhea
      ii. Tropical sprue
      iii. Partial or total gastrectomy
      iv. Multiple diverticula of small intestine
      v. Jejunal resection
      vi. Regional ileitis
      vii. Whipple's disease
      viii. Intestinal lymphoma
      ix. Broad-spectrum antibiotics
      x. Drugs associated with impaired absorption and/or utilization of folic acid, for example, methotrexate, diphenylhydantoin (Dilantin), primidone, barbiturates, oral contraceptive agents, cycloserine, metformin, ethanol, dietary amino acids (glycine, methionine), sulfasalazine, and pyrimethamine
      xi. Post bone marrow transplantation (total body irradiation, drugs, intestinal GVH disease)
3. **Increased requirements**
   a. Rapid growth (e.g., prematurity, pregnancy)
   b. Chronic hemolytic anemia, especially with ineffective erythropoiesis (e.g., thalassemia major)
   c. Dyserythropoietic anemias
   d. Malignant disease (e.g., lymphoma, leukemia)
   e. Hypermetabolic states (e.g., infection, hyperthyroidism)
   f. Extensive skin disease (e.g., dermatitis herpetiformis, psoriasis, exfoliative dermatitis)
   g. Cirrhosis
   h. Post bone marrow transplantation (bone marrow and epithelial cell regeneration)
4. **Disorders of folic acid metabolism**
   a. Congenital[b]
      i. Methylenetetrahydrofolate reductase deficiency (OMIM 236250)
      ii. Glutamate formiminotransferase deficiency (OMIM 229100)
      iii. Functional N^5-methyltetrahydrofolate: homocysteine methyltransferase deficiency caused by *cblE* (OMIM 236270) or *cblG* (OMIM 250940) disease
      iv. Dihydrofolate reductase deficiency (less well established)
      v. Methenyl-tetrahydrofolate cyclohydrolase (less well established)
      vi. Primary methyltetrahydrofolate: homocysteine methyltransferase deficiency (less well established)
   b. Acquired
      i. Impaired utilization of folate
         • Folate antagonists (drugs that are dihydrofolate reductase inhibitors, e.g., methotrexate, pyrimethamine, trimethoprim, pentamidine)
         • Vitamin B_{12} deficiency
         • Alcoholism
         • Liver disease (acute and chronic)
         • Other drugs (IIB10 above)
5. **Increased excretion** (e.g., chronic dialysis, vitamin B_{12} deficiency, liver disease, heart disease)

[a]*Rare disorder. Isolated disorder of folate transport resulting in low CSF folate and mental retardation. The ability to absorb all other nutrients is normal. The defect is overcome by pharmacologic oral doses of folic acid or intramuscular folic acid (Lanzkowsky et al., 1969).*
[b]*These disorders are associated with megaloblastic anemia, mental retardation, disorders in gait, and both peripheral and central nervous system disease.*
OMIM, Online Mendelian Inheritance in Man (http://www.ncbi.nlm.nih.gov/omim/).

folate ratio indicative of failure to transport folates across the choroid plexus. The uptake of folate into other cells is probably not defective and the uptake of folate into cultured cells is not abnormal.

Clinical Presentation

This disorder manifests in the first few months of life and is characterized by megaloblastic anemia associated with low serum, red cell, and cerebrospinal fluid folate levels as well as chronic or recurrent diarrhea, mouth ulcers, failure to thrive, and usually loss of developmental milestones, seizures, and progressive neurological deterioration.

TABLE 7.7 Clinical and Biochemical Features of Inherited Defects of Folate Metabolism

| | Hereditary folate malabsorption | Methylene-H$_4$ folate reductase deficiency | Glutamate formimino-transferase deficiency | Functional methionine synthase deficiency | |
				CblE	CblG
CLINICAL SIGNS					
Prevalence	13 cases	>30 cases	13 cases	8 cases	12 cases
Megaloblastic anemia	A	N	N[a]	A	A
Developmental delay	A	A	N[a]	A	A
Seizures	A	A	N[a]	A	A
Speech abnormalities	N	N	A[a]	N	N
Gait abnormalities	N	A	N[a]	N	A[a]
Peripheral neuropathy	N[a]	A	N[a]	N	A[a]
Apnea	N	A	N[a]	N[a]	N
BIOCHEMICAL FINDINGS					
Homocystinuria/homocysteinemia	N	A	N	A	A
Hypomethioninemia	N	A	N	A	A
Formininoglutamic aciduria	A[a]	N	A	N	N[a]
Folate absorption	A	N	N	N	N
Serum Cbl	N	N	N[a]	N	N
Serum folate	A	A	N[a]	N	N
Red blood cell folate	A	A[a]	N[a]	N	N
DEFECTS DETECTABLE IN CULTURED FIBROBLASTS					
Whole cells					
CH$_3$H$_4$ folate uptake	N	N	N	A	A
CH$_3$H$_4$ folate content	N	A	N	N	N
CH$_3$B$_{12}$ content	N	N[a]	N	A	A
Extracts					
Activity of holoenzyme of methionine synthase	N	N[a]	N	N[b]	A
Glutamate formiminotransferase				Activity undetectable in cultured cells Abnormal in liver and erythrocytes	
Methylene-H$_4$ folate reductase	N	A	N	N	N
TREATMENT					
	Folic acid or reduced folates in pharmacologic doses	Folates, betaine methionine	Folates	OH-Cbl, folinic acid, betaine	

[a]*Exceptions described in some cases.*
[b]*Abnormal activity with low concentrations of reducing agent in assay.*
N, normal; A, abnormal; (i.e., clinical or laboratory findings).
From Rosenblatt (1995), with permission.

Treatment

It is essential to maintain normal levels of folate in the blood and in the cerebrospinal fluid. Oral folic acid in doses of 5–40 mg daily and lower parenteral doses correct the hematologic abnormality, but cerebrospinal fluid folate levels may remain low. Oral doses of folates may have to be increased to 100 mg or more daily if necessary. Oral methyltetrahydrofolate and folinic acid can increase cerebrospinal fluid folate levels, but only slightly. If oral therapy is not effective, systemic therapy with reduced folates should be tried. It may be necessary to give intrathecal reduced folates if cerebrospinal fluid levels cannot be normalized.

Methylenetetrahydrofolate Reductase Deficiency (OMIM 236250)

This condition is a rare autosomal recessive disorder. The gene for MTHFR is located on chromosome 1p36.3 and there are more than 30 mutations. MTHFR deficiency results in elevated plasma homocysteine and homocystinuria and decreased plasma methionine levels, because levels of methyltetrahydrofolate serve as one of three methyl donors for the conversion of homocysteine to methionine.

Clinical Findings

This condition can present severely in early infancy (first month of life) or much more mildly as late as 16 years of age. Clinical symptoms vary and consist of developmental delay which is the most common clinical manifestation, hypotonia, motor and gait abnormalities, recurrent strokes, seizures, mental retardation, psychiatric manifestations, and microcephaly. Pathologic findings include internal hydrocephalus, microgyria, perivascular changes, demyelination, macrophage infiltration, gliosis, astrocytosis, and subacute combined degeneration of the spinal cord. Methionine deficiency may cause demyelination by interfering with methylation. Vascular changes include thrombosis of both cerebral arteries and veins. Megaloblastic anemia is uncommon in patients with this disease because reduced folates are still available for purine and pyrimidine synthesis.

Diagnosis

MTHFR deficiency can be diagnosed by measuring enzyme activity in liver, white blood cells, and cultured fibroblasts.

Prognosis

Prognosis is poor in early-onset severe MTHFR deficiency.

Treatment

MTHFR deficiency is resistant to treatment. Regimens have included folic acid, methyltetrahydrofolate, methionine, pyridoxine, various cobalamins, carnitine, and betaine. Betaine therapy after prenatal diagnosis has resulted in the best outcome to date since it has the theoretical advantage of both lowering homocysteine levels and supplementing methionine levels. Prenatal diagnosis is possible by enzyme assay in amniocytes, chorionic villus biopsy samples, or cultured chorionic villus cells. The phenotypic heterogenicity in MTHFR deficiency is reflected by genotypic heterogeneity.

Glutamate Formiminotransferase Deficiency (OMIM 229100)

Glutamate formiminotransferase and formiminotetrahydrofolate cyclodeaminase are involved in the transfer of a formimino group to tetrahydrofolate followed by the release of ammonia and the formation of 5,10-methyltetrahydrofolate. These activities are found only in the liver and kidneys and are performed by a single octameric enzyme. It is not clear that glutamate formiminotransferase deficiency is associated with disease, even though formiminoglutamic acid (FIGLU) excretion is the one constant finding. There have been 20 patients described, with ages ranging from 3 months to 42 years at diagnosis. Some have been asymptomatic and several patients have macrocytosis and hypersegmentation of neutrophils.

Mild and severe phenotypes have been described. Patients with the severe form show mental and physical retardation, abnormal EEG activity and dilatation of the cerebral ventricles with cortical atrophy. In the mild form, there is no mental retardation but massive excretion of FIGLU.

Liver-specific activity ranges from 14 to 54% of control values. It is not possible to confirm the diagnosis using cultured cells because the enzyme is not expressed. There is dispute as to whether the enzyme is expressed in red cells.

Patients may have elevated to normal serum folate levels and elevated FIGLU levels in the blood and urine after a histidine load. Plasma amino acid levels are usually normal, but hyperhistidinemia, hyperhistidinuria, and hypomethioninemia have been found. The excretion of hydantoin-5-propionate, the stable oxidation product of the FIGLU precursor, 4-imidazolone-5-propionate and 4-amino-5-imidazolecarboxamide, an intermediate in purine synthesis, has been seen in some patients.

Autosomal recessive inheritance is the probable means of transmission because there have been affected individuals of both sexes with unaffected parents.

Functional Methionine Synthase Deficiency (OMIM 250940)

Functional methionine synthase deficiency due to the *cblE* and *cblG* mutations is characterized by homocystinuria and defective biosynthesis of methionine. Most patients present in the first few months of life with megaloblastic anemia and developmental delay. The distribution of cobalamin derivatives is altered in cultured cells, with decreased levels of MeCbl as compared with normal fibroblasts. The *cblE* mutation is associated with low methionine synthase activity when the assay is performed with low levels of thiol, whereas the *cblG* mutation is associated with low activity under all assay conditions. *cblE* and *cblG* represent distinct complementation classes. Both diseases respond to treatment with hydroxycobalamin (OH-cbl).

Other Megaloblastic Anemias

1. *Thiamine-responsive anemia in DIDMOAD (Wolfram) syndrome*: This is a rare autosomal recessive disorder of thiamine transport, possibly deficient thiamine pyrophosphokinase activity, due to mutations in a gene on chromosome 1q23. Megaloblastic anemia and sideroblastic anemia with ringed sideroblasts may be present. Neutropenia and thrombocytopenia are present. It is accompanied by diabetes insipidus (DI), diabetes mellitus (DM), optic atrophy (OA) and deafness (D). *Treatment*: Anemia responds to 100 mg thiamine daily but megaloblastic changes persist. Insulin requirements decrease.
2. *Hereditary orotic aciduria*: Rare autosomal recessive defect of pyrimidine synthesis with failure to convert orotic acid to uridine and excretion of large amounts of orotic acid in the urine, sometimes with crystals. It is associated with severe megaloblastic anemia, neutropenia, failure to thrive, and physical and mental retardation are frequently present. *Treatment*: Oral uridine in a dose of 100–200 mg/kg/day. The anemia is refractory to vitamin B_{12} and folic acid.
3. *Lesch–Nyhan syndrome*: Mental retardation, self-mutilation, and choreoathetosis result from impaired synthesis of purines due to lack of hypoxanthine phosphoribosyltransferase. Some patients have megaloblastic anemia. *Treatment*: Megaloblastic anemia responds to adenine therapy (1.5 g daily).
4. *Miscellaneous conditions*: Chemotherapeutic agents that interfere in nucleoprotein synthesis such as hydroxyurea and 5-fluorouracil cause megaloblastic anemia. Myelodysplastic syndromes, erythroleukenia, and some cases of acute myeloblastic leukemia may show megaloblastic changes.

CLINICAL FEATURES OF COBALAMIN AND FOLATE DEFICIENCY

1. *Insidious onset*: Pallor, lethargy, fatigability and anorexia, sore red tongue and glossitis, diarrhea—episodic or continuous.
2. *History*: Poor diet, maternal vitamin B_{12} or folate deficiency. Maternal folate deficiency results in neural tube defects, prematurity, fetal growth retardation, and fetal loss. History of similarly affected sibling because inborn errors of metabolism of cobalamin and folate result in failure to thrive, neurologic disorders, unexplained anemias, or cytopenias.
3. *Vitamin B_{12} deficiency*: Infants may show signs of developmental delay, apathy, weakness, irritability, or evidence of neurodevelopmental delay, loss of developmental milestones, particularly motor achievements (head control, sitting, and turning). Athetoid movements, hypotonia, and loss of reflexes occur.

    In older children *subacute combined degeneration of the spinal cord* occurs, resulting in signs of degeneration of the posterior and lateral columns often with associated peripheral nerve loss. Loss of vibration and position sense with an ataxic gait and positive Romberg's sign are features of posterior column and peripheral nerve loss. Spastic paresis may occur, with knee and ankle reflexes increased because of lateral tract loss, but flaccid weakness may also occur when these reflexes are lost (secondary to peripheral nerve loss) but the Babinski sign remains extensor. Paresthesia in the hands or feet and difficulty in walking and use of the hands may occur due to peripheral neuropathy. Long-term cognitive and developmental retardation are irreversible following B_{12} treatment.

    MRI findings include increased signals on T2-weighted images of the spinal cord, brain atrophy, and retarded myelination.
4. Increased risk of vascular thrombosis due to hyperhomocysteinemia may occur.
5. Elevation of methylmalonic acid and homocysteine levels reflect a functional lack of cobalamin and/or folate by tissues even when plasma vitamin levels are at the lower level of normal.

TABLE 7.8　Disorders Giving Rise to Megaloblastic Anemia in Early Life and Their Likely Age at Presentation

Disease	Likely age at presentation (months)		
	2–6	7–24	>24
FOLATE DEFICIENCY			
Inadequate supply			
Prematurity	+		
Dietary (e.g., goat's milk)	+		
Chronic hemolysis			+
Defective absorption			
Celiac disease/sprue			+
Anticonvulsant drugs			+
Congenital	+		
COBALAMIN DEFICIENCY			
Inadequate supply			
Maternal cobalamin deficiency		+	
Nutritional			+
Defective absorption			
Juvenile pernicious anemia			+
Congenital malabsorption		+	±
Congenital absence of intrinsic factor		+	±
Defective metabolism			
Transcobalamin II deficiency	+		
Inborn errors of cobalamin utilization	+		
OTHER			
Thiamine responsive			+
Orotic aciduria	+		
Lesch–Nyhan syndrome			+

DIAGNOSIS OF COBALAMIN AND FOLATE DEFICIENCY

The age of presentation may help to focus on the most likely diagnosis. Table 7.8 lists disorders giving rise to megaloblastic anemia in early life and their likely age at presentation. The following are the laboratory findings:

1. Red cell changes.
   a. Hemoglobin: usually reduced, may be marked.
   b. Red cell indices: MCV increased for age and may be raised to levels of 110–140 fl; MCHC normal. Patients with associated iron deficiency or thalassemia may have a normal or low MCV.
   c. Red cell distribution width (RDW): increased.
   d. Blood smear: many macrocytes[4] and macro-ovalocytes; marked anisocytosis and poikilocytosis with tear drop cells, presence of Cabot rings, Howell–Jolly bodies, and punctuate basophilia. These misshapen cells have a shorter survival time.
   e. Absolute reticulocyte counts fall slightly.

[4]Macrocytosis can be masked by associated iron deficiency and thalassemia.

2. White blood cells: count reduced to 1500–4000/mm^3; neutrophils show hypersegmentation, that is, nuclei with more than five lobes and at least 4–5% of neutrophils have five lobes. Hypersegmentation also occurs as a benign hereditary condition, in patients receiving chemotherapy or steroids and rarely in myelofibrosis and chronic myelogenous leukemia.

3. Platelet count: moderately reduced to 50,000–180,000/mm^3. Platelet function is sometimes impaired.

4. Bone marrow: megaloblastic appearance. The marrow is hyperplastic due to increased levels of erythropoietin acting on erythroid progenitor cells.

   a. The cells are large and the nucleus has an open, stippled, or lacy appearance due to its retarded condensation. The cytoplasm is comparatively more mature than the nucleus and this dissociation (nuclear–cytoplasmic dissociation) is best seen in the more mature erythroid red cell precursors. Orthochromatic cells may be present with nuclei that are still not fully condensed.

   b. Mitoses are frequent and sometimes abnormal;[5] nuclear remnants, Howell–Jolly bodies, bi- and trinucleated cells and dying cells are evidence of gross dyserythropoiesis.

   c. The metamyelocytes are abnormally large (giant) and have a large horseshoe-shaped nucleus.

   d. Hypersegmented polymorphs may be seen and the megakaryocytes show an increase in nuclear lobes.

5. The functional pathophysiology of megaloblastic anemia is ineffective erythropoiesis in all cell lines (manifest by increased levels of lactate dehydrogenase, indirect bilirubin, ferritin, serum iron and transferrin saturation, and low serum haptoglobin levels) due to programmed cell death or apoptosis of megaloblastic cells during maturation rather than when they are mature, resulting in a predominance of younger erythroid cells in the bone marrow.

6. Serum vitamin B_{12} level: Levels <80 pg/ml are almost always indicative of vitamin B_{12} deficiency (normal values 200–800 pg/ml).

7. Serum and red cell folate levels: There is wide variation in normal range; serum levels less than 3 ng/ml indicate low level, 3–5 ng/ml borderline level, and greater than 5–6 ng/ml normal level. Red cell folate levels below 160 ng/ml are considered to be low. The reference range varies widely.

8. Homocysteine rises in both folate and vitamin B_{12} deficiency and its non-specificity limits its diagnostic value but a normal result helps to exclude clinical deficiency. Methylmalonic acid increases in the serum and urine in cobalamin deficiency but not in folate deficiency and is therefore a more specific test for cobalamin deficiency. Most laboratories accept methylmalonic acid levels above 280 nmol/l as elevated, but published cutoff points have varied between 210 and 480 nmol/l.

9. Urinary excretion of orotic acid to exclude orotic aciduria.

10. Deoxyuridine suppression test: This test can discriminate between folate and cobalamin deficiencies.

If vitamin B_{12} is suspected as a cause of the megaloblastic anemia then the following investigations should be carried out:

- Detailed dietary history, history of previous surgery.
- Gastric acidity after histamine stimulation, IF content in gastric juice, serum antibodies to intrinsic factor and parietal cells and gastric biopsy establishes a gastric cause.
- Schilling urinary excretion test:[6] This test measures both IF availability and intestinal absorption of vitamin B_{12}.
  If the Schilling test is abnormal, it should be repeated with commercial IF. If absorption occurs, the abnormality is due to lack of IF. If no absorption occurs then the abnormality is due to specific ileal vitamin B_{12} malabsorption (Imerslund–Gräsbeck) or TC II deficiency. The Schilling test should be repeated after treatment with tetracycline and will often revert to normal when bacterial competition (blind-loop syndrome) is the cause of vitamin B_{12} deficiency. Persistent proteinuria is a feature of specific ileal vitamin B_{12} malabsorption.
- Ileal disease should be investigated by barium studies and small-bowel biopsy
- Measurement of serum holo-TC II (cobalamin bound to TC II). In patients with vitamin B_{12} deficiency holo-TC II falls below the normal range before total serum cobalamin does.

---

[5]Megaloblastic cells exhibit increased frequency of chromosomal abnormalities, especially random breaks, gaps, and centromere spreading. A rare case of non random, transient 7q has been described in acquired megaloblastic anemia.

[6]The test is performed by administering 0.5–2.0 mg radioactive ^{57}cobalt-labeled vitamin B_{12} PO. This is followed in 2 hours by an intramuscular injection of 1000 mg non radioactive vitamin B_{12} to saturate the B_{12}-binding proteins and allow the subsequently absorbed oral radioactive vitamin B_{12} to be excreted in the urine. All urine is collected for 24 h and may be collected for a second 24 h, especially if there is renal disease. Normal subjects excrete 10–35% of the administered dose; those with severe malabsorption of vitamin B_{12}, because of lack of intrinsic factor or intestinal malabsorption, excrete less than 3%.

- Disorders of vitamin B_{12} metabolism should be excluded by serum and urinary levels of excessive methylmalonic acid and of total homocysteine, as well as by other sophisticated enzymatic assays. In folate deficiency, serum methylmalonic acid is normal whereas homocysteine is increased. Therefore, evaluation of both methylmalonic acid and total homocysteine is helpful in distinguishing between folate and vitamin B_{12} deficiency.
- Persistent proteinuria is a feature of specific ileal vitamin B_{12} malabsorption.

If folic acid is suspected as a cause of the megaloblastic anemia then the following investigations should be carried out:

- Detailed dietary and drug history (e.g., antibiotics, anticonvulsants), gastroenterologic symptoms (e.g., malabsorption, diarrhea, dietary history).
- Tests for gastrointestinal diseases with associated malabsorption: gluten autoantibodies, 24-h stool fat, upper gastrointestinal barium study and follow-through, upper-gut endoscopy and jejunal biopsy.
- Oral doses of 5 mg pteroylglutamic acid should yield a plasma level in excess of 100 ng/ml in 1 h. If there is no rise in plasma level congenital folate malabsorption should be considered.
- Sophisticated enzyme assays to diagnose congenital disorders of folate metabolism.

TREATMENT

Vitamin B_{12} Deficiency

Prevention

In conditions in which there is a risk of developing vitamin B_{12} deficiency (e.g., total gastrectomy, ileal resection), prophylactic vitamin B_{12} should be prescribed.

Active Treatment

Once the diagnosis has been accurately determined, several daily doses of 25–100 mg cyanocobalamin or OH-cbl may be used to initiate therapy as well as potassium supplements. Alternatively, in view of the ability of the body to store vitamin B_{12} for long periods, maintenance therapy can be started with monthly intramuscular injections in doses between 200 and 1000 mg cyanocobalamin or OH-cbl. Depending on the cause vitamin B_{12} deficiency may require life-long treatment.

Patients with defects of intestinal absorption of vitamin B_{12} (abnormalities of IF or of ileal uptake) will respond to 100 mg of B_{12} injected subcutaneously monthly.

Patients with complete TC II deficiency respond only to large amounts of vitamin B_{12} and the serum cobalamin level must be kept very high. Doses of 1000 mg IM two or three times weekly are required to maintain adequate levels.

Patients with methylmalonic aciduria with defects in the synthesis of cobalamin coenzymes are likely to benefit from massive doses of vitamin B_{12}. These children may require 1–2 mg vitamin B_{12} parenterally daily. However, not all patients in this group are benefited by vitamin B_{12}. It may be possible to treat vitamin B_{12}-responsive patients in utero. Congenital methylmalonic aciduria has been diagnosed in utero by measurements of methylmalonate in amniotic fluid or maternal urine.

Response to Vitamin B_{12} Treatment

In vitamin B_{12}-responsive megaloblastic anemia, the reticulocytes begin to increase on the third or fourth days, rise to a maximum on the sixth to eighth days, and fall gradually to normal about the twentieth day. The height of the reticulocyte count is inversely proportional to the degree of anemia. Beginning bone marrow reversal from megaloblastic to normoblastic cells occurs within 6 h and is complete in 72 h. The level of alertness and responsiveness improves within 48 h and developmental delays may catch up in several months in young infants. However, permanent neurologic sequelae often occur.

The use of oral folic acid is *contraindicated in vitamin B_{12} deficiency* since it has no effect on neurologic manifestations and may precipitate or accelerate their development despite the fact that prompt hematologic response may occur with folic acid.

Folic Acid Deficiency

Optimal response occurs with 100—200 mg folic acid per day. Before folic acid is given vitamin B_{12} deficiency must be excluded.

It is often possible to correct the cause of folate deficiency, for example, improved diet, a gluten-free diet in celiac disease, or treatment of an inflammatory disease such as Crohn's disease. In these cases, there is no need to continue folic acid for life. In other situations, folic acid is required life-long, for example, chronic hemolytic anemia such as thalassemia or in patients with malabsorption who do not respond to a gluten-free diet.

Response to Folic Acid Treatment

The clinical and hematologic response to folic acid is prompt. Within 1—2 days, the appetite improves and a sense of well-being returns. There is a fall in serum iron (often to low levels as accelerated hematopoiesis occurs with folic acid treatment) in 24—48 h and a rise in reticulocytes in 2—4 days, reaching a peak at 4—7 days. This is followed by a return of hemoglobin levels to normal in 2—6 weeks. Leukocytes and platelets increase and megaloblastic changes in the marrow diminish within 24—48 h. However, large myelocytes, metamyelocytes, and band forms may be present for several days. Folic acid is usually administered for several months until a new population of red cells has been formed.

Folinic acid is reserved for treating the toxic effects of dihydrofolate reductase inhibitors (e.g., methotrexate, pyrimethamine).

Cases of hereditary dihydrofolate reductase deficiency respond to N-5-formyl tetrahydrofolic acid and not to folic acid.

Further Reading and References

Allen, L.H., 2008. Causes of vitamin B_{12} and folate deficiency. Food Nutr. Bull. 29 (2 Suppl.), S20—34.

Dror, D.K., Allen, L.H., 2008. Effects of vitamin B_{12} deficiency on neurodevelopment in infants: current knowledge and possible mechanisms. Nutr. Rev. 66 (5), 250—255.

Gordon, N., 2009. Cerebral folate deficiency. Dev. Med. Child Neurol. 51 (3), 180—182.

Hvas, A.M., Nexo, E., 2006. Diagnosis and treatment of vitamin B_{12} deficiency — an update. Haematologica 91 (11), 1506—1512.

Lanzkowsky, P., 1987. The megaloblastic anemias: vitamin B_{12} cobalamin deficiency and other congenital and acquired disorders. Clinical, pathogenetic and diagnostic considerations of vitamin B_{12} (cobalamin) deficiency and other congenital and acquired disorders. In: Nathan, D.G., Oski, F.A. (Eds.), Hematology of Infancy and Childhood. WB Saunders, Philadelphia, PA.

Lanzkowsky, P., 1987. The megaloblastic anemias: folate deficiency II. Clinical, pathogenetic and diagnostic considerations in folate deficiency. In: Nathan, D.G., Oski, F.A. (Eds.), Hematology of Infancy and Childhood. WB Saunders, Philadelphia, PA.

Lanzkowsky, P., Erlandson, M.E., Bezan, A.I., 1969. Isolated defect of folic acid absorption associated with mental retardation and cerebral calcifications. Blood 34, 452—465, Amer J Med 48, 580—583, 1970.

Rasmussen, S.A., Fernhoff, P.M., Scanlon, K.S., 2001. Vitamin B_{12} deficiency in children and adolescents. J. Pediatr. 138, 110.

Rosenblatt, D.S., 1995. Inherited disorders of folate transport and metabolism. In: Scriver, C.R., Beaudet, A., Sly, W.S., Valle, D. (Eds.), The Metabolic and Molecular Bases of Inherited Disease, seventh ed. McGraw-Hill, New York, NY.

Rosenblatt, D.S., Whitehead, V.M., 1999. Cobalamin and folate deficiency: acquired and hereditary disorders in children. Semin. Hematol. 36, 19.

Whitehead, V.M., 2006. Acquired and inherited disorders of cobalamin and folate in children. Br. J. Haematol. 134 (2), 125—136.

8

Bone Marrow Failure

Adrianna Vlachos and Jeffrey M. Lipton

Bone marrow failure may manifest as a single cytopenia (e.g., erythroid, myeloid, or megakaryocytic), or as pancytopenia. It may present with a hypoplastic or aplastic marrow or result from invasion of the bone marrow by non-neoplastic (e.g., storage cells) or neoplastic cells.

Bone marrow failure may be congenital or acquired (Table 8.1). Table 8.2 lists the inherited bone marrow failure syndromes (IBMFS) with their causative genes. IBMFS can manifest with pancytopenia (e.g., Fanconi anemia (FA) and dyskeratosis congenita (DC)) or single cytopenias (e.g., Diamond–Blackfan anemia (DBA), Shwachman–Diamond syndrome (SDS), severe congenital neutropenia (SCN), Kostmann syndrome (KS), cyclic neutropenia, amegakaryocytic thrombocytopenia (AMT), and thrombocytopenia absent radii (TAR) syndrome). The "single cell line cytopenias" may develop abnormalities in other hematopoietic cell lines (Table 8.1).

Congenital dyserythropoietic anemias (CDAs) result in moderate erythroid failure due to ineffective erythropoiesis with characteristic morphological abnormalities of erythroblasts.

Mitochondrial diseases may present with bone marrow failure (Pearson syndrome, Wolfram syndrome, and various types of sideroblastic anemia).

Figure 8.1 shows the differential diagnosis of pancytopenia, and Table 8.3 lists the investigations to be carried out in a patient with pancytopenia.

APLASTIC ANEMIA

Aplastic anemia is characterized by a marked decrease or absence of blood-forming elements with resulting pancytopenia and can be inherited or acquired. Various degrees of lymphopenia may be present. Splenomegaly, hepatomegaly, and lymphadenopathy do not generally occur in this condition.

ACQUIRED APLASTIC ANEMIA

Definition

Severe aplastic anemia (SAA) is defined by:

a. Bone marrow cellularity of less than 25%.
b. At least two of the following cytopenias: granulocyte count $<500/mm^3$ ($<200\ mm^3$ defines very SAA); platelet count $<20,000/mm^3$; and/or reticulocyte count $<20,000/mm^3$.

Non-SAA occurs when the above criteria are not met. There is little consensus on distinguishing between mild and moderate aplastic anemia.

Pathophysiology

Aplastic anemia results from an immunologically mediated, tissue-specific, organ-destructive mechanism. It is postulated that after exposure to an inciting antigen, cells and cytokines of the immune system destroy stem cells in the marrow resulting in pancytopenia. Treatment with immunosuppression leads to marrow recovery.

102

TABLE 8.1 Causes of Single Cytopenias and Trilineage Bone Marrow Failure

SINGLE CYTOPENIAS

1. Red cells
   a. Inherited
      i. Diamond–Blackfan anemia (pure red cell aplasia)
      ii. Congenital dyserythropoietic anemia
   b. Congenital (can be inherited)
      i. Pearson syndrome
   c. Acquired
      i. Idiopathic
         * Transient erythroblastopenia of childhood (TEC)
      ii. Secondary
         * Drugs or toxins
         * Infection—parvovirus B19 infection in immunodeficiency patients (chronic bone marrow failure)
         * Malnutrition
         * Thymoma
         * Chronic hemolytic anemia with associated parvovirus B19 infection (transient bone marrow failure)
         * Connective tissue disease and autoimmune disease associated
         * Malignancy associated
2. White blood cells
   a. Inherited
      i. Shwachman–Diamond syndrome (SDS)
      ii. Severe congenital neutropenia
      iii. Reticular dysgenesis
      iv. Other rare genetic disorders
   b. Acquired
      i. Primary
         * Benign neutropenia of childhood
      ii. Secondary
         * Drugs
         * Autoimmune neutropenia of infancy
         * Other autoimmune associated (systemic lupus erythematosis, etc.)
3. Platelets
   a. Inherited
      i. Amegakaryocytic thrombocytopenia
      ii. Thrombocytopenia absent radii syndrome
   b. Acquired
      i. Idiopathic/immune thrombocytopenia purpura
      ii. Other autoimmune associated (systemic lupus erythematosis, Crohn's disease, etc.)

TRILINEAGE BONE MARROW FAILURE (GENERALIZED PANCYTOPENIA)

1. Inherited
   a. Fanconi anemia (associated with chromosomal breakages induced by clastogens, e.g., diepoxybutane (DEB) or mitomycin C (MMC))
   b. Dyskeratosis congenita (associated with short telomeres)
   c. Shwachman–Diamond syndrome (predominantly neutropenia)[a]
   d. Congenital amegakaryocytic thrombocytopenia (predominantly thrombocytopenia)[a]
   e. Diamond–Blackfan anemia (predominantly anemia)[a]
   f. Aplastic anemia with constitutional chromosomal abnormalities
      i. Dubowitz syndrome (congenital abnormalities, mental retardation, aplastic anemia)
2. Acquired
   a. Idiopathic (more than 70% of cases)
   b. Secondary
      i. Drugs[b]
         * Predictable, dose-dependent, rapidly reversible (affects rapidly dividing maturing hematopoietic cells rather than pluripotent stem cells)
            i. 6-Mercaptopurine
            ii. Methotrexate
            iii. Cyclophosphamide
            iv. Busulfan
            v. Chloramphenicol
         * Unpredictable to normal doses (defect or damage to pluripotent stem cells)
            i. Antibiotics: chloramphenicol, sulfonamides
            ii. Anticonvulsants: mephenytoin (Mesantoin), hydantoin
            iii. Antirheumatics: phenylbutazone, gold
            iv. Antidiabetics: tolbutamide, chlorpropamide
            v. Antimalarial: quinacrine

(Continued)

TABLE 8.1 *(Continued)*

ii. Chemicals: insecticides (e.g., DDT pesticide, Parathion, Chlordane)
iii. Toxins (e.g., benzene, carbon tetrachloride, glue, toluene)
iv. Radiation
v. Infections
   - Viral hepatitis (hepatitis A, B, and C and serotype-negative hepatitis)
   - Human immunodeficiency virus infection (HIV/AIDS)
   - Infectious mononucleosis (Epstein−Barr virus)
   - Rubella[c]
   - Influenza[c]
   - Parainfluenza[c]
   - Measles[c]
   - Mumps[c]
   - Venezuelan equine encephalitis
   - Rocky Mountain spotted fever[c]
   - Cytomegalovirus (in newborn)
   - Herpes virus (in newborn)
   - Chronic parvovirus
vi. Immunologic disorders
   - Graft-versus-host reaction in transfused immunologically incompetent subjects
   - X-linked lymphoproliferative syndrome (see Chapter 16)
   - Eosinophilic fasciitis
   - Hypogammaglobinemia
vii. Aplastic anemia preceding acute leukemia (hypoplastic preleukemia)
viii. Myelodysplastic syndromes (see Chapter 17)
ix. Thymoma
x. Paroxysmal nocturnal hemoglobinuria (see Chapter 9)
xi. Malnutrition
   - Kwashiorkor
   - Marasmus[c]
   - Anorexia nervosa[c]
   - Pregnancy

[a]*Can have reduction in other cell lines.*
[b]*Partial listing.*
[c]*Pancytopenia with temporary marrow hypoplasia.*

Gamma-Interferon (γ-IFN) plays a central role in the pathophysiology of aplastic anemia. *In vitro* studies show that the T-cells from aplastic anemia patients secrete γ-IFN and tumor necrosis factor (TNF). Long-term bone marrow cultures have shown that γ-IFN and TNF are potent inhibitors of both early and late hematopoietic progenitor cells. Both of these cytokines suppress hematopoiesis by their effects on the mitotic cycle and, more importantly, by the mechanism of cell killing. The mechanism of cell killing involves the pathway of apoptosis (i.e., γ-IFN and TNF upregulate each other's cellular receptors, as well as the Fas receptors in hematopoietic stem cells). Cytotoxic T-cells also secrete interleukin-2 (IL-2), which causes polyclonal expansion of the T-cells. Activation of the Fas receptor on the hematopoietic stem cell by the Fas ligand present on the lymphocytes leads to apoptosis of the targeted hematopoietic progenitor cells. Additionally, γ-IFN mediates its hematopoietic suppressive activity through IFN regulatory factor 1 (IRF-1), which inhibits the transcription of cellular genes and their entry into the cell cycle. γ-IFN also induces the production of nitric oxide, diffusion of which causes additional toxic effects on the hematopoietic progenitor cells. Direct cell−cell interactions between effective lymphocytes and targeted hematopoietic cells probably also occur. The oligoclonal expansion of CD4+ and CD8+ T-cells that fluctuate with disease activity further supports an immune etiology.

Table 8.1 lists the various causes of acquired aplastic anemia.

Clinical Manifestations

Acquired aplastic anemia may be idiopathic or secondary. At least 70% of cases are idiopathic. The incidence is approximately two cases per million per year in the West and higher in parts of Asia (~4−7.5 cases per million

TABLE 8.2 Inherited Bone Marrow Failure Syndrome Genes, Known, and Presumed

Disorder	Gene	Number of cases	Locus	Genetics	Gene product
Fanconi anemia	*FANCA*	60–70%	16q24.3	Autosomal recessive	FANCA
	FANCB	~2%	Xp22.31	X-linked recessive	FANCB
	FANCC	14%	9q22.3	Autosomal recessive	FANCC
	FANCD1	~3%	13q12.3	Autosomal recessive	FANCD1/BRCA2
	FANCD2	~3%	3p25.3	Autosomal recessive	FANCD2
	FANCE	~3%	6p21.3	Autosomal recessive	FANCE
	FANCF	~2–10%	11p15	Autosomal recessive	FANCF
	FANCG	~1%	9p13	Autosomal recessive	FANCG/XRCC9
	FANCI	~2%	15q25-26	Autosomal recessive	FANCI/KIAA1794
	FANCJ	<1%	17q22.3	Autosomal recessive	FANCJ/BRIP1
	FANCL	<1%	2p16.1	Autosomal recessive	FANCL/PHF9/POG
	FANCM[a]	<1%	14q21.3	Autosomal recessive	FANCM
	FANCN	<1%	16p12.1	Autosomal recessive	FANCN/PALB2
	FANCO	<1%	17q25.1	Autosomal recessive	FANCO/RAD51C
	FANCP	<1%	16p13.3	Autosomal recessive	FANCP/SLX4
	FANCQ	<1%	16p13.12	Autosomal recessive	FANCQ/XPF/ERCC4
	UBE2T		1q32.1	Autosomal recessive	FANCT
Dyskeratosis congenita	*DKC1*	17–36%	Xq28	X-linked recessive	Dyskerin
	TERC	6–10%	3q26.2	Autosomal dominant	TERC/telomerase RNA component
	TERT	1–7%	5p15.33	Autosomal dominant and autosomal recessive	TERT/telomerase reverse transcriptase
	TINF2	11–24%	14q12	Autosomal dominant	TIN2
	NOP10	<1%	15q14	Autosomal recessive	NOP10
	NHP2	<1%	5q35.3	Autosomal recessive	NHP2
	WRAP53	3%	17p13.1	Autosomal recessive	WRAP53
	CTC1	1–3%	17p13.1	Autosomal recessive	CTC1
Diamond–Blackfan anemia	*RPS19*	25%	19q13.2	Autosomal dominant	RPS19
	RPS17	1%	15q25.2	Autosomal dominant	RPS17
	RPS24	2%	10q22-23	Autosomal dominant	RPS24
	RPL5	7%	1p22.1	Autosomal dominant	RPL5
	RPL11	5–10%	1p36.11	Autosomal dominant	RPL11
	RPL35A	2–4%	3q29	Autosomal dominant	RPL35a
	RPS7	1%	2p25.3	Autosomal dominant	RPS7
	RPS10	2–6%	6p21.31	Autosomal dominant	RPS10
	RPS26	2–6%	12q13.2	Autosomal dominant	RPS26
	RPS29	<1%	14q21.3	Autosomal dominant	RPS29
	RPL26	<1%	17p13.1	Autosomal dominant	RPL26
	RPL15	<1%	3p24.2	Autosomal dominant	RPL15
	GATA1	<1%	Xp11.23	X-linked recessive	GATA1

(Continued)

TABLE 8.2   (*Continued*)

Disorder	Gene	Number of cases	Locus	Genetics	Gene product
Shwachman–Diamond syndrome (SDS)	*SBDS*	~90%	7q11.21	Autosomal recessive	SBDS
Severe congenital neutropenia (SCN)	*ELANE*	60%	19p13.3	Autosomal dominant	Neutrophil elastase
	HAX1 (Kostmann syndrome)	Rare	1q21.3	Autosomal recessive	HAX1
	G6PC3	Rare	17q21.31	Autosomal recessive	G6PC3
	GFI1	Rare	1p22	Autosomal dominant	GFI1
	WAS	Rare	Xp11.4-p11.21	X-linked recessive	WASP
	JAGN1	Rare	3p25.2	Autosomal recessive	JAGN1
	GATA2 (MonoMac syndrome)	Rare	3q21.3	Autosomal dominant	GATA2
	CXCR4 (WHIM syndrome)	Rare	2q22.1	Autosomal dominant	CXCR4
Amegakaryocytic thrombocytopenia	*MPL*	100%	1p34	Autosomal recessive	Thrombopoietin receptor/TPOR
Thrombocytopenia absent radii (TAR) syndrome	*RBM8A*	100%	1q21.1	Autosomal recessive	RBM8A

*a*FANCM is a member of the "core complex" but homozygosity not yet identified in patients with Fanconi anemia.

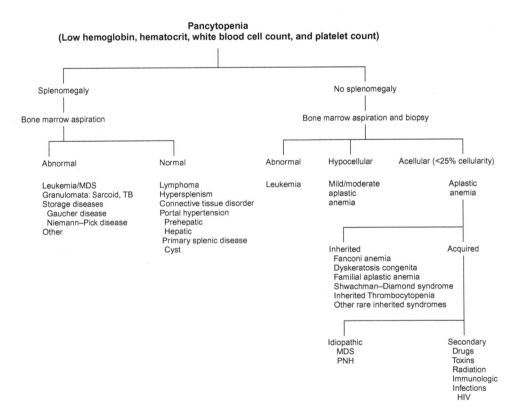

FIGURE 8.1 Approach to the differential diagnosis of pancytopenia. MDS, myelodysplastic syndrome; PNH, paroxysmal nocturnal hemoglobinuria.

TABLE 8.3   Investigations in Patients with Pancytopenia

1. Detailed past history, medication history, toxin and radiation exposure, thorough physical examination for congenital anomalies, detailed family history of aplastic anemia, MDS or leukemia, and congenital anomalies in other family members
2. Blood count: Hb, Hct, MCV, absolute reticulocyte count, WBC, absolute neutrophil count, platelet count
3. ANA and anti-dsDNA titers, direct antiglobulin test (DAT), rheumatoid factor, liver function tests, and tuberculin test
4. Viral serology: hepatitis A, B, C, EBV, parvovirus, varicella, CMV, HIV. PCR for virus when indicated
5. Serum vitamin B12, red cell and serum folate levels
6. Bone marrow aspirate and trephine biopsy
7. Cytogenetic studies on bone marrow along with fluorescence *in situ* hybridization for chromosomes 5, 7, 8, 22 and perhaps others to evaluate for MDS
8. Chromosome breakage assay on blood lymphocytes or skin fibroblasts using clastogen stimulation (e.g., diepoxybutane or mitomycin C) to diagnose Fanconi anemia
9. Skeletal radiographs; renal, cardiac, abdominal ultrasounds; chest radiograph to determine congenital anomalies
10. Telomere length determination for dyskeratosis congenita
11. Flow cytometric immunophenotypic analysis of erythrocytes for deficiency of GPI-linked surface protein (e.g., CD59) to exclude paroxysmal nocturnal hemoglobinuria
12. Skeletal radiograph, chest radiograph, pancreatic ultrasound, or CT scan, 72-h fecal fat, serum trypsinogen and isoamylase, fecal elastase for Shwachman–Diamond syndrome (SDS)
13. Mutation analysis for specific inherited bone marrow failure syndromes when suspected

Hb, hemoglobin; Hct, hematocrit; MCV, mean corpuscular volume; WBC, white blood cell; CMV, cytomegalovirus; HIV, human immunodeficiency virus; EBV, Epstein–Barr virus; PCR, polymerase chain reaction.

per year) and the male:female ratio is 1:1. The onset of acquired aplastic anemia is usually gradual, and the symptoms are related to the pancytopenia:

- Anemia results in pallor, easy fatigability, weakness, and loss of appetite.
- Thrombocytopenia leads to petechiae, easy bruising, severe nosebleeds and gastrointestinal bleeding and hematuria.
- Leukopenia leads to increased susceptibility to infections and oral ulcerations and gingivitis that respond poorly to antibiotic therapy.
- Hepatosplenomegaly and lymphadenopathy do not occur and their presence suggests an underlying malignant process.

Laboratory Investigations

- *Anemia*: normocytic or macrocytic, normochromic.
- *Reticulocytopenia*: absolute count more reliable.
- *Leukopenia*: granulocytopenia often less than 1500/mm^3.
- *Thrombocytopenia*: platelets often less than 30,000/mm^3.
- *Fetal hemoglobin*: may be slightly to moderately elevated.
- *Bone marrow*:
  - Marked depression or absence of hematopoietic cells and replacement by fatty tissue containing reticulum cells, lymphocytes, plasma cells, and usually tissue mast cells.
  - Megaloblastic changes and other features indicative of dyserythropoiesis frequently seen in the erythroid precursors.
  - Bone marrow biopsy essential to assess cellularity for diagnosis and to exclude the possibility of poor aspiration technique or poor bone marrow sampling; additionally it will help to rule out granulomas, myelofibrosis, or leukemia.
  - Chromosomal analysis is normal and assists in excluding myelodysplastic syndromes (MDSs).
  - Bone marrow cultures, antigen-based evaluation, and molecular testing for infectious agents and/or DNA; when indicated.
- *Chromosome breakage assay*: performed on peripheral blood to rule out FA.
- *Flow cytometry (CD59)*: to exclude paroxysmal nocturnal hemoglobinuria.
- *Telomere length*: to screen for DC.
- *Liver function chemistries*: to exclude hepatitis.
- *Renal function chemistries*: to exclude renal disease.

TABLE 8.4   Recommendations for Treatment of Children with Aplastic Anemia

1. Moderate aplastic anemia:
   Observe with close follow-up and supportive care
   If the patient develops:
   - Severe aplastic anemia, and/or
   - Severe thrombocytopenia with significant bleeding, and/or
   - Chronic anemia requiring transfusion treatments, and/or
   - Serious infections

   Then treat as severe aplastic anemia

2. Severe aplastic anemia:
   Allogeneic bone marrow transplantation when HLA-matched sibling donor available.
   In the absence of an HLA-matched sibling marrow donor:
   - Treat the patient with antithymocyte globulin (ATG) and cyclosporine A (CSA), with methylprednisolone for serum sickness prophylaxis
   - The use of growth factors such as G-CSF or GM-CCF is no longer routinely done (see Table 8.4)
   - Hematopoietic stem cell transplant clinical trials are ongoing and experience suggests that, particularly in light of late clonal evolution following immunosuppressive therapy, matched unrelated donor (MUD) transplants are likely to become a treatment of choice for newly diagnosed SAA.
   If no response or waning of response and recurrence of severe aplastic anemia a second course of immunosuppressive therapy is controversial. The following is recommended by most experts:
   - HLA-matched unrelated bone marrow, peripheral blood or umbilical cord blood transplant if a suitable donor is available. If not available a second course of immunosuppression is warranted
   - High-dose cyclophosphamide without stem cell transplant is controversial and not commonly considered
   - Newer thrombopoietin-mimetic drugs such as eltrombopag are under investigation for patients with refractory severe aplastic anemia

Partial response: absence of infections and transfusion dependency and sustained increase in all cell counts as follows: reticulocyte count $\geq 20,000/mm^3$; platelet count $\geq 20,000/mm^3$; absolute neutrophil count $\geq 500/mm^3$. Complete response: Normal counts. Partial response and complete response are considered as responses for the evaluation of the success of immunosuppressive therapy.

- *Viral serology testing*: hepatitis A, B, and C antibody panel; Epstein–Barr virus (EBV) antibody panel; parvovirus B19 IgG and IgM antibodies; varicella antibody titer; cytomegalovirus (CMV) antibody titer; human immunodeficiency virus antibody test.
- *Quantitative immunoglobulins (Igs)* to rule out immunodeficiency.
- *Autoimmune disease evaluation*: antinuclear antibody (ANA), total hemolytic complement (CH50), C3, C4, direct antiglobulin test (DAT).
- *HLA typing*: patient and family done at the diagnosis of SAA to identify a suitable donor and ensure a timely transplant.

Physical examination, appropriate laboratory screening assays and imaging studies, and, if warranted, mutation analysis should be performed to rule out other IBMFS (DC, DBA, SDS, AMT).

Table 8.4 shows the recommendations for the treatment of moderate and SAA.

Treatment

Supportive Care

- Transfusion of red cells and platelets should be minimized, but should not be withheld if clearly indicated. The risk of serious bleeding and symptomatic anemia must be balanced against the risk of transfusion sensitization and iron overload.
  - Prior to any transfusion, perform complete blood group typing to minimize the risk of sensitization to minor blood group antigens and to permit identification of antibodies should they subsequently develop.
  - Transfusions should be restricted, if possible, to unrelated donors to decrease the likelihood of sensitization to donor antigens.
  - In all patients, blood products should be leukocyte-depleted to reduce the risk of sensitization and CMV infection. CMV-negative blood products are equivalent to CMV-safe blood products, even for CMV-seronegative patients who may require future transplant.
  - Patients receiving chronic red cell transfusion should be followed for evidence of iron overload and receive appropriate chelation.
  - The use of single donor platelets, when available, is recommended.

- Menses should be suppressed by the use of contraceptives.
- Drugs that impair platelet function, such as aspirin, should be avoided.
- Intramuscular injections should be given carefully, followed by ice pack application to injection sites.
- The antifibrinolytic agent, ε-aminocaproic acid (Amicar) (100 mg/kg/dose every 6 h, daily maximum 24 grams) can be used to reduce mucosal bleeding in thrombocytopenic patients with good hepatic and renal function. Hematuria is a contraindication to its use. Teeth should be brushed with a cloth or soft toothbrush to avoid gum bleeding.
- Avoid infection. Keep patients out of the hospital as much possible. Good dental care is important. Rectal temperatures should not be taken, and the rectal areas should be kept clean and free of fissures. If a patient is febrile:
  - Culture possible sources, including blood, sputum, urine, stool, skin, and sometimes spinal fluid and bone marrow, for aerobes, anaerobes, fungi, and tubercle bacilli.
  - Patients with fever and neutropenia should be treated with broad-spectrum antibiotic coverage (Chapter 33). The specific therapy depends upon the clinical status of the patient, the presence of indwelling vascular access devices and knowledge of the local flora pending specific culture results and antibiotic sensitivities. Patients who remain febrile for 4−7 days, even with broad antibacterial coverage, should be started on antifungal therapy empirically. Therapy should be continued until the patient is afebrile and cultures are negative or a specific organism is identified.
- Patients previously treated with immunosuppressive therapy (IST) should receive irradiated cellular blood products to prevent transfusion-acquired graft-versus-host disease (GVHD) (Chapter 36). Patients receiving IST should also receive *Pneumocystis jiroveci* prophylaxis with trimethoprim/sulfamethoxazole (Bactrim/Septra) or pentamidine.

Patients with mild to moderate aplastic anemia should be observed for spontaneous improvement or complete resolution. Hematopoietic stem cell transplantation (HSCT) is the treatment of choice for SAA for patients who have an HLA-matched related donor.

Specific Therapy

Hematopoietic Stem Cell Transplantation

HLA typing should be performed as soon as the diagnosis of SAA is suspected in children. Patients with related histocompatible donors should have an HSCT (FA, DC, paroxysmal nocturnal hemoglobinuria (PNH) or other IBMFSs should be ruled out prior to HSCT). Rapidly treating with HSCT is critical as prolonged neutropenia and multiple transfusions increase the risk of transplant-related morbidity and mortality. See Chapter 34 for preparatory regimens employed pre-transplant.

Immunosuppressive Therapy

Patients unable to undergo matched related HSCT (because no suitable donor is present) should receive IST consisting of antithymocyte globulin (ATG) and cyclosporine (CSA) (Table 8.5). Methylprednisolone or prednisone should be used to prevent serum sickness. The response rate using this regimen is 85%.

TABLE 8.5 Immunosuppressive Therapy for Severe Aplastic Anemia

1. Antithymocyte globulin: Atgam® (lymphocyte immune globulin, antithymocyte globulin (equine) sterile solution) (Pharmacia & Upjohn Co, Pfizer, Inc.) is preferred at 40 mg/kg/d IV once daily, or thymoglobulin (antithymocyte globulin (rabbit)) (Genzyme Corp) 5 mg/kg/d IV once daily on days 1−4 infused over 6−8 h. Premedication with diphenhydramine and acetaminophen is recommended
2. Methylprednisolone, 2 mg/kg/d IV on days 1−4. Divide into 0.5 mg/kg/dose IV every 6 h
3. Prednisone oral taper following the 4-day course of IV methylprednisolone. On days 5 through 14, start prednisone, 1.5−2 mg/kg/d PO to be divided into two equal daily doses. After day 14 institute a slow taper (e.g., on days 15 and 16, prednisone, 1 mg/kg/d PO to be divided into two equal daily doses). On days 17 and 18, prednisone, 0.5 mg/kg/d PO to be divided into two equal daily doses. On day 19, prednisone, 0.25 mg/kg/d PO to be given in one dose
4. CSA (Sandimmune®, Gengraf®, Neoral®) 10 mg/kg/d PO initially starting on day 1, divided into two equal daily doses. Serum drug levels should be monitored as needed with the first level at 72 h post initiation of therapy. CSA dose to be adjusted to keep serum trough levels between 200 and 400 ng/ml. CSA should be continued for one year until and after a trilineage response is achieved. Decrease the dose by 2.0 mg/kg every 2 weeks once the taper is begun, watching for signs of recurrence of aplastic anemia
5. G-CSF is no longer routinely used in most centers. Occasionally G-CSF is utilized at a dose of 5 μg/kg SC once daily to improve neutrophil counts in selected patients who are actively infected and persistently severely neutropenic ($<200/mm^3$), with reassessment and discontinuation after no more than a few days or weeks if there is no significant response

CSA, cyclosporine (formerly cyclosporine A); G-CSF, granulocyte colony-stimulating factor.

Contraindications to the use of immunosuppressive drugs include:

- Serum creatinine, >2 mg/dl
- Concurrent pregnancy.
- Concurrent hepatic, renal, cardiac, or metabolic problems of such severity that death is likely to occur within 7−10 days.
- Moribund patients.

Antithymocyte Globulin

Test dose: To identify those at greatest risk of systemic anaphylaxis, skin testing is strongly recommended by the manufacturer prior to ATG treatment. An intradermal ATG test dose consisting of 0.02 ml of a 1:1000 dilution in 0.9% sodium chloride solution for injection (5 μg equine IgG) is used with a saline control injection administered on the contralateral side. The results are read at 10 min: a wheal at the ATG site 3 mm or larger in diameter than that at the saline control site suggests clinical sensitivity and increased possibility of a systemic allergic reaction should the drug be dosed intravenously. Many centers have eliminated the test dose in favor of premedication.

The dose of ATG is shown in Table 8.5.

Usual adverse reactions to ATG:

- Thrombocytopenia: Patients may require daily platelet transfusions to maintain a platelet count of more than 20,000/mm^3 (during administration of ATG). Only irradiated and leukocyte-filtered cellular blood products should be used.
- Headache.
- Myalgia.
- Arthralgia.
- Chills and fever: Treatment with an antipyretic, an antihistamine, and corticosteroid is indicated as premedication.
- Chemical phlebitis: A central line (high flow vein) for infusion of ATG should be used and peripheral veins should be avoided.
- Itching and erythema: Treatment with an antihistamine with or without corticosteroids is indicated.
- Leukopenia.
- Serum sickness may occur approximately 7−10 days following ATG administration. This should be treated by increasing the daily dose of Solumedrol until the symptoms abate.

Uncommon adverse reactions to ATG: Dyspnea, chest, back and flank pain, diarrhea, nausea, vomiting, hypertension, herpes simplex infection, stomatitis, laryngospasm, anaphylaxis, tachycardia, edema, localized infection, malaise, seizures, gastrointestinal bleeding/perforation, thrombophlebitis, lymphadenopathy, hepatosplenomegaly, renal function impairment, liver function abnormalities, myocarditis, and congestive heart failure.

Cyclosporine (CSA) Preparations:

Cyclosporine modified:
- Capsules, soft gelatin 50 mg.

Gengraf:
- Capsules 25 mg (as cyclosporine modified).
- Capsules 100 mg (as cyclosporine modified).
- Solution, oral 100 mg/ml (as cyclosporine modified).

Neoral:
- Capsules, soft gelatin 25 mg (as cyclosporine modified).
- Capsules, soft gelatin 100 mg (as cyclosporine modified).
- Solution, oral 100 mg/ml (as cyclosporine modified).

Sandimmune:
- Capsules, soft gelatin 25 mg.
- Capsules, soft gelatin 100 mg.
- Solution, oral 100 mg/ml.

Administer capsules and oral solution on a consistent schedule with respect to time of day and meals.

Gengraf and Neoral oral solution may be diluted with orange juice or apple juice to make the solution more palatable. Sandimmune oral solution may be diluted with milk, chocolate milk, or orange juice, preferably at room temperature, to make it more palatable. Stir mixture well and administer immediately after mixing. Do not allow mixture to stand before administering.

Sandimmune is not bioequivalent to Neoral or Gengraf. Conversion using a 1:1 ratio may result in lower blood concentrations.

The starting dose of cyclosporine is 10 mg/kg per day. CSA levels should be performed once a week for the first 2 weeks and then once every 2 weeks for the remainder of the treatment or as necessary to maintain a whole-blood CSA level between 200 and 400 ng/ml. An elevated serum creatinine level is the principal criterion for dose change. An increase in creatinine level of more than 30% above baseline warrants a reduction in the dose of CSA by 2 mg/kg/day each week until the creatinine level has returned to normal. A serum CSA level of less than 100 ng/ml may be evidence of inadequate absorption and/or non compliance; a CSA level above 500 ng/ml is considered an excessive dose and CSA should be discontinued. Levels should be repeated daily or every other day. When the level returns to 200 ng/ml or less, CSA should be resumed at a 20% reduced dose. In responders, CSA should be tapered very slowly, beginning at 6 months to a year from initiation of therapy although there is little to no evidence available for guidance regarding the target levels and tapering schedule.

Principal side effects of CSA: Renal dysfunction, tremor, hirsutism, hypertension, and gingival hyperplasia.

Uncommon side effects of CSA: Significant hyperkalemia, hyperuricemia, hypomagnesemia, hepatotoxicity, lipemia, central nervous system toxicity (including seizures), and gynecomastia. An increase of more than 100% in the bilirubin level or of liver enzymes is treated in the same way as an increase of more than 30% in creatinine and warrants a reduction in the dose of CSA by 2 mg/kg/day each week until the bilirubin and/or liver enzymes return to the normal range.

Contraindications to the use of CSA: Hypersensitivity to CSA.

Pharmacokinetic interactions with CSA:

- Carbamazepine, phenobarbital, phenytoin, rifampin—decreases half-life and blood levels of CSA.
- Sulfamethoxazole/trimethoprim IV—decreases serum levels of CSA.
- Erythromycin, fluconazole, ketoconazole, nifedipine—increases blood levels of CSA.
- Imipenem-cilastatin—increases blood levels of CSA and central nervous system toxicity.
- Methylprednisolone (high dose), prednisolone—increases plasma levels of CSA.
- Metoclopramide (Reglan)—increases absorption and increases plasma levels of CSA.

Pharmacologic interactions with CSA:

- Aminoglycosides, amphotericin B, nonsteroidal anti-inflammatory drugs, trimethoprim/sulfamethoxazole—nephrotoxicity.
- Melphalan, quinolones—nephrotoxicity.
- Methylprednisolone—seizures.
- Azathioprine, corticosteroids, cyclophosphamide—increases immunosuppression, infections, malignancy.
- Verapamil—increases immunosuppression.
- Digoxin—elevates digoxin level with toxicity.
- Non-depolarizing muscle relaxants—prolongs neuromuscular blockade.

Hematopoietic Growth Factors

Granulocyte colony-stimulating factor, G-CSF (Neupogen), had been used to achieve a more rapid increment in the granulocyte count and theoretically to improve protection from infectious complications by stimulating granulopoiesis. G-CSF added to standard ATG and CSA reduces the rate of early infectious episodes and days of hospitalization in very SAA patients but has no effect on overall survival, event-free survival, remission, relapse rates or mortality.

Treatment Choices and Long-Term Follow-Up

Although the short-term outcome with IST is comparable to that obtained with HLA-matched related HSCT, the decision to choose HSCT for younger patients with a histocompatible donor is based on the result of long-term follow-up. There are low rates of late mortality (due to chronic GVHD and therapy-related cancer) in patients undergoing HSCT, and the survival curves are relatively flat. Improved GVHD prophylaxis and safer preparative regimens should further improve these results. In contrast, there is a high risk of clonal hematopoietic disorders (MDS, AML, and PNH) in patients treated with IST compared to HSCT. Those undergoing IST must be closely followed for the development of clonal disorders.

Salvage Therapy

For patients who fail sibling donor HSCT, or have a partial response (ANC $\geq 500/mm^3$, but are red cell and platelet transfusion dependent) or relapse following IST, management choices include alternative donor HSCT or further IST. HSCT is preferred to IST if a suitable donor is available. Children and teenagers for whom a fully HLA-matched unrelated donor exists (as determined by high-resolution typing) are excellent candidates for an alternative donor HSCT. For patients without a good alternative donor, a second course of ATG/CSA is warranted.

Long-Term Sequelae and Outcomes for SAA

Outcomes for both IST and HSCT have improved considerably in recent years. The results of multiple cohorts report a slightly different response rate and incidence of the clonal evolution of SAA to MDS, AML or PNH. These data vary based on length of follow-up, age of patient, as well as institution/consortium.

- Complete or partial response rates in the range of 60–70%, largely from studies in adults, have been reported with IST. Horse ATG appears to be superior to rabbit ATG in these studies. Although the outcomes in children for IST are generally superior to those described for adults, disease-free survival for matched related HSCT is ~95%.
- IST improves hematopoiesis and achieves transfusion independence in the majority of patients, but the time to response is long. Hematopoietic response may be partial and relapses are relatively common.
- The incidence of clonal hematopoietic disorders including PNH, myelodysplasia, and leukemia in patients with SAA treated with IST ranges from 10–40%. The European Bone Marrow Transplantation Working Party compared the rate of secondary malignancies following HSCT and IST. Forty-two malignancies developed in 860 patients receiving IST, compared to nine in 748 patients who underwent HSCT. In this study, acute leukemia and myelodysplasia were seen exclusively in IST-treated patients while the incidence of solid tumors was similar in the two groups of patients.
- From the aggregate data there are a number of conclusions:
  - Matched sibling donor HSCT is always superior as primary therapy in young patients (<20 years of age) at any neutrophil count.
  - Immunosuppression, due to transplant-related morbidity and mortality in older patients, is superior to HSCT in older patients (41–50 years).
  - For the 21–40-year-old age group, the differences are less clear.
  - In all age groups there are a higher percentage of late failures and clonal evolution in the immunosuppression-treated patients.
  - When considering the response rate (partial and complete) for IST, the low rate of transplant failure with MUD transplant, the incidence of GVHD and the evolution of clonal disease after IST, the difference in survival between patients treated with MUD donor HSCT and IST increases with time. Thus, MUD transplant (preferably with controlled clinical trials) should be considered as primary therapy for SAA.

Treatment of Moderate Aplastic Anemia

The natural history of moderate aplastic anemia is uncertain and clinical experience varies widely. For this reason, it is generally thought that these patients should be treated initially with supportive therapy with very close follow-up. The majority of patients progress to SAA or develop significant and severe thrombocytopenia and bleeding, serious infections, or a chronic red blood transfusion requirement. These patients should be treated with the same treatment options as described for SAA.

INHERITED BONE MARROW FAILURE SYNDROMES

The key shared clinical manifestations of IBMFSs are:

- Bone marrow failure.
- Congenital anomalies.
- Cancer predisposition.
- Occasional presentation in adulthood.

The common pathophysiology is low apoptotic threshold of mutant cells.

FANCONI ANEMIA

Fanconi Anemia (FA) is rare, with a heterozygote frequency in the general population of 1/181 in North America; 1/93 in Israel and less than 1/100 in Ashkenazi Jews (*FANCC, BRACA2/FANCD1*), South African Afrikaners (*FANCA*), Northern Europeans (*FANCC*), sub-Saharan Blacks (*FANCG*) and Spanish Gypsies (*FANCA*) due to the "founder effect." It is an IBMFS associated with multiple congenital anomalies and a predisposition to cancer.

The details of guidelines for the diagnosis and management of FA as reviewed in the text and tables are beyond the scope of this chapter but are available from the Fanconi Anemia Research Fund. Patients with FA should be registered with the International Fanconi Anemia Registry (IFAR). This registry collects and maintains long-term outcome data as well as providing resources for physicians, patients, and families.

Pathophysiology

Somatic cell hybridization studies have thus far defined 16 FA complementation groups. All 16 FA genes have been cloned (Table 8.2). Additional genes, including *FANCT*, are being identified. Complementation groups FANCA, C and G represent ~90% of the cases. The gene products of these genes have been shown to cooperate in a common pathway. After FANCM and the FA-associated protein FAAP24 detect the DNA damage, eight of the FA proteins (FANCA, B, C, E, F, G, L, and M) assemble to form the FA core complex that is required to monoubiquinate and activate FANCD2 and FANCI. Ubiquinated FANCD2 and FANCI form a dimer which stabilizes the stalled replication fork and then in turn interacts in nuclear repair foci with the downstream FA gene products (FANCO, D1, N, and J) in the FA/BRCA DNA damage repair pathway. Damage repair is then achieved by the late FA proteins in cooperation with proteins from other DNA repair pathways. Of note FANCD1 has been identified as BRCA2. Despite the identification of this pathway, the manner in which disruption in this cascade of events results in a faulty DNA damage response and genomic instability leading to hematopoietic failure, birth defects and cancer predisposition is only incompletely understood.

FA cells are characterized by hypersensitivity to chromosomal breakage as well as hypersensitivity to G2/M cell cycle arrest induced by DNA crosslinking agents. In addition there is sensitivity to oxygen free radicals and to ionizing radiation.

Clinical Manifestations

- FA is inherited as an autosomal recessive disorder (>99%) and rarely as an X-linked recessive (FANCB, <1%) and is the most frequently inherited aplastic anemia. FANCA is the most common complementation group, representing about 70% of cases. FANCC and FANCG are the next most common, representing 10% of cases each. The remaining complementation groups are quite rare, representing the remainder of cases (Table 8.2).
- Genotype—phenotype correlations are complex and are emerging and relate to the complementation group as well as the specific allelic mutation (i.e., null versus hypomorphic gene product). In particular, certain associations relating genotype to specific congenital anomalies, early-onset aplastic anemia, leukemia, as well as Wilms tumor and medulloblastoma, have been confirmed.
- All racial and ethnic groups are affected.
- Pancytopenia is the usual finding.
  - The median age at hematologic presentation of patients with aplastic anemia is approximately 8—10 years. Leukemia tends to appear later in the teenage years and solid tumors appear in young adulthood and continue to occur as patients age.
  - Hematologic dysfunction usually presents with macrocytosis, followed by thrombocytopenia, often leading to progressive pancytopenia and SAA. FA frequently terminates in MDS and/or acute myeloid leukemia (AML).
  - The diagnosis of FA should always be considered in any child with an isolated cytopenia even when the classical somatic anomalies are absent as a significant number of these cases are physically normal.

- FA cells are hypersensitive to chromosomal breaks induced by DNA crosslinking agents. This observation is the basis for the commonly used chromosome breakage test for FA. The clastogens diepoxybutane (DEB) and mitomycin C (MMC) are the agents most frequently used *in vitro* to induce chromosome breaks, gaps, rearrangements, quadriradii, and other structural abnormalities. Clastogens also induce cell cycle arrest in G2/M. The hypersensitivity of FA lymphocytes to G2/M arrest, detected using cell cycle analysis by flow cytometry either de novo or clastogen-induced, is being used by some as a screening tool for FA.
- Bone marrow examination reveals hypocellularity and fatty replacement consistent with the degree of peripheral pancytopenia. Residual hematopoiesis may reveal dysplastic erythroid (megaloblastoid changes, multinuclearity) and myeloid (abnormal granulation) precursors and abnormal megakaryocytes.
- Congenital anomalies include increased pigmentation of the skin along with café-au-lait and hypopigmented areas, short stature (impaired growth hormone secretion), skeletal anomalies (especially involving the thumb, radius, and long bones), male hypogenitalism, microcephaly, abnormalities of the eyes (microphthalmia, strabismus, ptosis, nystagmus) and ears (deafness), hyperreflexia, developmental delay, and renal and cardiac anomalies. However up to 40% of patients lack obvious physical abnormalities. There is great clinical heterogeneity even within a genotype (affected siblings may be phenotypically different). Table 8.6 lists the anomalies and frequency in FA.
- There is a nearly 800-fold increased relative risk of developing AML and perhaps an even greater relative risk of non-hematologic tumors (e.g., squamous cell carcinoma of head and neck, cancer of the breast, kidney, lung, colon, bone, retinoblastoma, and female gynecologic) in patients with FA. In general these occur at much younger ages than those seen in the general population. A relatively large number of patients only become aware that they have FA when they are diagnosed with cancer. Androgen-related, usually benign liver neoplasia may also occur. The risk of solid tumors may become even higher as death from aplastic anemia is reduced, and as post-HSCT patients survive longer. These data must be considered in the context of HSCT, in particular when the risk of non-hematologic malignancy is likely to increase as a result of HSCT conditioning regimens and chronic GVHD. Treatment for cancer is generally ineffective.
- Prenatal diagnosis is possible by amniotic fluid cell cultures and chorionic villus biopsy.

Diagnosis

Table 8.7 lists the indications for FA screening studies.
Table 8.8 lists the laboratory studies required to make the diagnosis of FA.
Table 8.9 lists the initial and follow-up investigations to be performed in a patient with an established diagnosis of FA.

Complications

Table 8.10 describes the complications of malignancy and liver disease associated with FA.

TABLE 8.6 Congenital Anomalies and Frequency in Fanconi Anemia

Anomaly	Approximate frequency (%)
Skin	55
Skeletal	51
Reproductive organs	35
Small head or eyes	26
Renal	21
Low birth weight	11
Cardiopulmonary	6
Gastrointestinal	5

Modified from Alter B and Lipton J. Anemia, Fanconi. EMedicine [serial online]. 2009. Available at http://www.emedicine.com/ped/topic3022.htm.

TABLE 8.7 Indications for Fanconi Anemia Screening Studies

All children with aplastic anemia or unexplained cytopenias
All children with MDS or AML
Patients with classic birth defects suggestive of FA
  VATER/VACTRL (vertebral anomalies, anal atresia, cardiac anomalies, tracheoesophageal fistula, renal anomalies and limb anomalies)
  Structural anomalies of the upper extremity and/or genitourinary system
Patients with:
  Excessive café-au-lait spots, hypo- or hyper pigmentation of skin (especially if increasing with age)
  Microcephaly
  Microphthalmia
  Growth failure
Development of FA-associated cancers at a young age (e.g., squamous cell carcinoma in esophagus, head, and neck <50 years of age, vulvar cancer <40 years of age and uterine cervical cancer <30 years of age or liver tumors)
  Patient with leukemia or solid tumor with unusual sensitivity to chemotherapy
  Karyotype with spontaneous chromosome breaks
  Patients with unexplained macrocytosis and an elevated HbF
  Patients with non immune thrombocytopenia
  Males with unexplained infertility
  Siblings of known FA patients

TABLE 8.8 Laboratory Studies for Diagnosis of Fanconi Anemia

1. Screening tests:
  * Demonstration of the presence of increased chromosomal breakage in T-lymphocytes cultured in the presence of DNA crosslinking agents such as diepoxybutane (DEB) or mitomycin C (MMC). DEB test is used more widely. Chromosome fragility includes breaks, gaps, rearrangements, radials, exchanges, and endoreduplication. Fibroblasts should be studied in patients for whom mosaicism is suspected[a]
  * A flow cytometric technique for the analysis of alkylating-agent-treated cells can determine the percentage of cells arrested in G2/M because a characteristic distribution clearly distinguishes FA cells from normal cells[b]
  * Western blot for D2-L (long protein formed by ubiquitination of FANCD2[b])
2. Definitive tests:
  * Complementation group analysis[b]
  * Targeted mutation analysis, copy number variant analysis[b] (see Table 8.2 for cloned FANC genes)
3. Prenatal diagnosis of FA:
  * DEB test can be used in either chorionic villus or amniocentesis derived samples
4. Detection of carrier state:
  * In a FA family, if proband has been identified to have a defect in one of the cloned genes, molecular testing is available for the extended family members. Population-based screening is only done in at-risk populations

[a]Some patients with FA may have two populations of cells exhibiting either a normal or an FA phenotype. Such mosaicism may result in a false-negative chromosome breakage study if the percentage of normal cells is high. The study of fibroblasts is useful in this circumstance.
[b]Done only in specialized laboratories.

Differential Diagnosis

* The differential diagnosis of FA generally includes acquired aplastic anemia, AMT, TAR syndrome as well as VATER/VACTRL (vertebral anomalies, anal atresia, cardiac anomalies, tracheoesophageal fistula, renal anomalies, limb anomalies) syndromes. FA is easily distinguished from TAR syndrome. There is an intercalary defect in TAR consisting of absent radii with normal thumbs, whereas in FA the defect is terminal, an abnormal radius always being associated with anomalies of the thumb. Table 8.11 lists the features differentiating FA from TAR syndrome.
* FA testing is warranted in any child who presents with hematologic cytopenias, unexplained macrocytosis, aplastic anemia, or AML, as well as representative congenital abnormalities or solid tumors typical of FA such as head and neck, esophageal or gynecologic tumors presenting at an early age (Table 8.8).
* The critical investigations are aspiration and biopsy of the bone marrow and demonstration in peripheral blood of increased chromosomal fragility or G2/M arrest induced by clastogens (e.g., DEB, MMC). Complementation group analysis and/or mutation analysis are helpful after the demonstration of a positive screening test and should be obtained if possible.
* FA somatic mosaics with DEB-positive and DEB-negative (double population) cells belong to distinct groups based upon the degree of mosaicism and may present diagnostic problems. Mosaicism leading to a "normal" T-cell that is resistant to the less dose-intense HSCT conditioning, used for FA, may result in graft rejection. In cases where mosaicism is suspected cultured skin fibroblasts should be studied.

TABLE 8.9   Initial and Follow-Up Investigations to be Performed in a Patient with an Established Diagnosis of Fanconi Anemia

1. Endocrine studies for
    1. Short stature (growth hormone deficiency)
    2. Glucose intolerance (diabetes mellitus)
    3. Hypothyroidism
    4. Pubertal delay
    5. Evaluation of undescended testes
    6. Reduced fertility
2. Imaging studies[a] and evaluation of
    1. Cardiac anomalies
    2. Orthopedic anomalies
    3. Genitourinary abnormalities
3. Hepatic ultrasound every 6 months if taking androgens, serum chemistries for
    1. Liver function
    2. Kidney function

Hearing test

Monitoring for iron overload for patients on red cell transfusion therapy
    1. Ferritin
    2. Liver enzymes
    3. Liver biopsy
    4. MRI T2[a] for cardiac and hepatic liver iron quantification

Survey of family members:
    1. Exclude diagnosis of FA in any other family members
    2. Type family members for the potential availability of an HLA-matched sibling for future consideration of stem cell transplant
    3. Provide genetic counseling to parents and patient

Prospective counseling and screening:
    1. Avoid exposure to potential mutagens or carcinogens (e.g., insecticides, organic solvents, hair dye, papilloma virus)
    2. Cancer surveillance:
        a. Examine bone marrow yearly with histologic and cytogenetic studies for evidence of myelodysplasia or leukemia
        b. Yearly head and neck examination over age 7 years
        c. Yearly gynecologic examination beginning at age 16 years
        d. Breast self-examination beginning at age 16 years
        e. Periodic oral cancer and dermatologic screening

Mutation analysis:
    These studies are performed in specialized laboratories only. Mutation analysis may help predict the phenotype as more data become available

[a]Limit exposure to radiation by using appropriate restraint and non-radiologic imaging studies.

TABLE 8.10   Malignancy and Liver Disease in Fanconi Anemia

	Total number[a]	Male	Female	Median age, years (range)
Leukemia	175	86	71	13 (0.1−49)
Myelodysplastic syndrome	110	56	51	14 (2−49)
Solid tumors	124	42	76	23 (0.2−56)
Liver—hepatoma	30	20	10	14 (5−50)
Liver—adenoma	16	7	9	11 (8−48)

[a]Number of cancers. Total number of patients without transplant with any cancer = 320. Twenty-five patients had multiple solid tumors (22 patients had two and three patients had three). Fourteen with solid tumors also had leukemia and six had liver tumors; six with leukemia also had liver tumors.
Adapted from Shimamura and Alter (2010).

Management

Serial assessments of the bone marrow should be performed to provide evidence of progression and the development or evolution of cytogenetic abnormalities:

- Bone marrow aspiration should be performed for cytology, cytogenetics with FISH analysis for cytogenetic abnormalities that may be predictive of leukemia (e.g., 3q26q29 amplification and 7q deletion) approximately yearly and more often if indicated by the emergence of specific clonal or morphological abnormalities.
- Bone marrow biopsy should be done for cellularity.

TABLE 8.11 Features Differentiating Fanconi Anemia from Thrombocytopenia Absent Radii Syndrome (TAR Syndrome)

Feature	Fanconi anemia	TAR
Age of onset of aplastic anemia symptoms	Median of 8–10 years	Birth to infancy (first year of life)
Low birth weight	~10%	~10%
Stature	Short	Short
Skeletal deformities	66%	100%
Absent radii with fingers and thumbs present	0%	100%
Other hand deformities	~40%	~40%
Lower extremity deformities	~40%	<10%
Cardiovascular anomalies	5–10%	5–10%
Anomalous pigmentation of skin	77%	0%
Hemangiomas	0%	~10%
Mental retardation	17%	7%
Peripheral blood	Pancytopenia Macrocytosis (high MCV)	Thrombocytopenia, eosinophilia, Leukemoid reactions, anemia
Bone marrow	Aplastic	Absent or abnormal megakaryocytes, normal myeloid and erythroid precursors
Marrow CFU-GM, CFU-E	Decreased	Normal (decreased CFU-megakaryocytes)
HbF	Increased	Normal
Hexokinase in blood cells	Decreased in some	Not defined in TAR
Chromosomal breaks in leukocytes	Present	Absent
Malignancy	Common	Rare (leukemia only)
Sex ratio (male/female)	~1:1	~1:1
Inheritance pattern	Autosomal recessive	Autosomal recessive
Associated leukemia	Yes	Rare
Prognosis	Poor	Good if patient survives first year when platelet count Improves

Modified from the classic review, Hall et al. (1969), with permission.

The patient's complete blood counts should be monitored. The degree of cytopenia guides management as follows:

	Mild	Moderate	Severe
Hemoglobin level (g/dl)	≥8.0	<8.0	<8.0
ANC (/mm³)	<1500	<1000	<500
Platelet count (/mm³)	50,000–150,000	<50,000	<30,000

When cytopenias are in the mild to moderate range and in the absence of cytogenetic abnormalities, counts should be monitored every 3–4 months and bone marrow aspiration should be performed yearly. Monitoring of blood counts and bone marrow should be increased to every 1–2 months and every 1–6 months, respectively, for cytopenia in the presence of cytogenetic abnormalities or more significant dysplasia without frank MDS. With falling (or in some cases rising) counts surveillance must be increased.

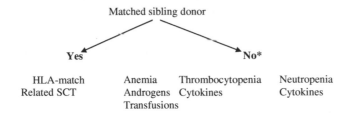

Androgens — oxymetholone: 2–5 mg/kg/day (may be able to taper)
Cytokines — G-CSF 5 µg/kg every day or every other day

*The use of matched unrelated and other alternative donor transplants is evolving.
A center with extensive transplant experience should be consulted for all FA patients with significant bone marrow failure.

FIGURE 8.2   Treatment of Fanconi anemia.

Treatment

Androgen therapy: Oxymetholone 2–5 mg/kg/day and tapered to the lowest effective dose is effective in approximately 50% of patients. Recent anecdotal reports and small studies suggest that Danazol may be an effective synthetic androgen with less virilizing effects. Doses in the range of ~5 mg/kg have been effective.

Cytokines: G-CSF at a starting dose of 5 µg/kg/day, tapered to the lowest effective dose may be administered when moderate to severe neutropenia is present.

Transfusions: Treatment with packed red blood cells and platelets should be minimized and reserved for patients who fail androgen therapy. Blood products should be irradiated, leukocyte-depleted and of single donor origin, when possible. Blood relatives should not be used as blood donors until a matched allogeneic related donor transplant is ruled out. Iron status should be monitored at regular intervals to determine the degree of iron overload and the institution of chelation treatment in chronically transfused patients.

HSCT: HLA typing should be done at diagnosis to facilitate therapeutic planning. If an HLA-matched related donor is available, stem cell transplantation should be carried out. Indications for alternative donor HSCT are rapidly evolving and every patient, regardless of the availability of a matched related donor should be considered for transplant as part of a clinical trial. Patients should be evaluated for transplant at one of only a few centers, worldwide, that have FA transplant experience. Evidence of true MDS (as opposed to benign clonal abnormalities) or evolution to leukemia is a clear indication for transplant. The sensitivity of FA patients to traditional transplant conditioning regimens requires the use of lower dosages of chemotherapy and radiation therapy (Chapter 31). Before a family member is used as a donor, the donor should be evaluated to exclude a diagnosis of FA.

HPV vaccination: Vaccination is recommended in patients with FA.

Growth hormone therapy: The majority of patients with FA have short stature. Up to 50% have deficient growth hormone. Given a theoretical association of growth hormone and leukemia, growth hormone should be used with that understanding in patients with FA.

Gene therapy: This approach is experimental and will only be performed in approved clinical trials in the future.

The treatment of FA is shown in Figure 8.2.

Prognosis

Current results of matched sibling transplantation prior to development of overt leukemia show a long-term disease-free survival of 80–90%. However, the long-term risks of late sequelae from HSCT include an increase in cancer risk. Alternative donor transplant had historically been reserved for androgen-refractory patients and those with MDS or leukemia. However, this recommendation is rapidly evolving as improvements in HLA typing, conditioning regimens and overall care as well as HSCT experience have improved outcomes considerably. Thus, every patient should be evaluated for HSCT.

DYSKERATOSIS CONGENITA

- DC is characterized by the classic triad of ectodermal dysplasia consisting of:
  - Abnormal skin pigmentation of the upper chest and neck.
  - Dysplastic nails.
  - Leukoplakia of oral mucous membranes.
- Predisposition to bone marrow failure.
- Predisposition to cancer—hematologic (leukemia, myelodysplasia) and epithelial cancers.
- Somatic findings in DC include: epiphora (tearing due to obstructed tear ducts), blepharitis, developmental delay, pulmonary disease (fibrosis), short stature, esophageal webs, liver fibrosis, dental carries, tooth loss, premature gray hair and hair loss; ocular, dental, skeletal, cutaneous, genitourinary, gastrointestinal, neurologic abnormalities and immunodeficiency have been reported.
- The classical clinical diagnosis of DC requires two of the three elements of the classic diagnostic triad and any other associated abnormality in patients with a known mutation or very short telomeres. The presence of short telomeres in a member of a pedigree with definitive DC is sufficient for the diagnosis.
- The median age at diagnosis is approximately 15 years. The median age for the onset of mucocutaneous abnormalities is 6–8 years. Nail changes occur first but hematologic abnormalities may precede mucocutaneous changes. The median age for the onset of pancytopenia is 10 years. Approximately 50% of patients develop SAA by age 50 years and greater than 90% develop at least a single cytopenia by 40 years of age. The anemia is associated with a high mean corpuscular volume (MCV) and elevated fetal hemoglobin. As with FA it is the non-hematologic manifestations of DC that are of particular concern, especially when HSCT for bone marrow failure is considered.

Pathophysiology

The DC phenotype results from deficient telomerase activity. Telomerase adds DNA sequence back to the ends of chromosomes that are eroded with each DNA replication. Telomerase activity is found in tissues with rapid turnover such as the basal layer of the epidermis, squamous epithelium of the oral cavity, hematopoietic stem cells and progenitors and in other tissues affected in DC. The lack of telomerase activity may also give rise to chromosome instability resulting in the high rate of premature cancer. Table 8.12 shows the cells in various organs expressing telomerase and the defects that occur in telomerase failure. Epithelial malignancies develop at or beyond the third decade of life. Type 2 alveolar epithelial cells express telomerase, and as a consequence, about one in five patients develop progressive pulmonary disease characterized by fibrosis, resulting in diminished diffusion capacity and/or restrictive lung disease. It is likely that more pulmonary disease would be evident if patients did not succumb earlier to the complications of SAA and cancer.

TABLE 8.12 Cells Expressing Telomerase and Defects Occurring in Telomerase Failure

Organ system	Cells expressing telomerase	Defect
Hair	Hair follicle	Alopecia
Oral cavity	Squamous epithelium	Leukoplakia
Skin	Epidermis, basal layer	Abnormal pigmentation and dyskeratotic nails
Lungs	Type II alveolar cells	Pulmonary fibrosis
Liver	Unknown cell type	Cirrhosis
Intestines	Crypt cells	Enteropathy
Testes	Spermatogonia	Hypogonadism
Bone marrow	Progenitors	Bone marrow failure

Genetics

Mutations in eight genes in the telomerase maintenance pathway have been associated with DC. DC is most commonly inherited as an X-linked recessive but may also be autosomal dominant or recessive. The gene responsible for the X-linked form was mapped to Xq28 and subsequently identified as DKC1. DKC1 codes for dyskerin, a nucleolar protein associated with nucleolar RNAs. Dyskerin is associated with the telomerase complex. This latter function appears to be the one involved in the pathophysiology of DC, as all the genes found to date to be mutated in DC (Table 8.2) are involved in telomere biology. There are many features in common to all three genetic subtypes, however the clinical phenotypes may vary widely in severity even within different mutations of the same allele. Affected members within the same family may exhibit wide variability in clinical presentation suggesting the influence of modifying genes and environmental factors. In addition different allelic mutations in these genes may result in bone marrow failure in the absence of any physical anomalies. Thus, in addition to FA, DC must be ruled out in all cases of aplastic anemia.

Clinical Manifestations

The clinical manifestations are quite variable.

Bone marrow failure: The incidence of bone marrow failure is 50% at 50 years of age. The majority of deaths (67%) are a result of bone marrow failure, followed by cancer and lung disease (pulmonary fibrosis) with or without HSCT. Overall median survival has improved to 49 years from 34 years in the past decade.

Malignancy: The causes of death are similar to those reported for FA with the exception of pulmonary fibrosis which is unique to DC. MDS and AML are less frequent than in FA, as are solid tumors including head and neck squamous cell carcinoma followed by anorectal, esophagus/stomach, brain, renal, and others. All of these cancers occur at younger ages than these cancers occur in the population at large, but older than in patients with FA.

Neurological: Patients with the severe form of DC known as Hoyeraal–Hreidarsson (HH) syndrome have symptomatic cerebellar hypoplasia, microcephaly, and developmental delay. Revesz syndrome (RS) is associated with CNS calcification, occasionally cerebellar hypoplasia and exudative retinopathy. Multiple DC genes have been implicated in HH and TINF2 gene has been implicated in RS.

Immunodeficiency: Significant progressive immunodeficiency occurs in DC. Although DC is predominantly a cellular immune defect, humoral immunodeficiency as well as neutropenia probably play a significant role in the infectious morbidity and mortality in DC.

Outcome: The median survival is approximately 40–45 years for patients with DC. In HH it is approximately 5 years of age and in RS the median has not yet been defined.

Treatment

Supportive care: Blood products, antibiotics, and antifibrinolytic agents are similar to those used for idiopathic aplastic anemia.

HSCT should be considered for those patients with an HLA-matched related donor or an acceptable alternative donor and no DC-related contraindications. The results of HSCT have been poor predominantly due to hepatic and pulmonary fibrosis. All DC patients are at a high risk of interstitial pulmonary disease when undergoing HSCT. There have been too few transplant survivors to determine whether an increase in the prevalence of cancer will follow as a consequence of HSCT. An immunoablative rather than a myeloablative approach is currently being used to potentially reduce the incremental risk of pulmonary toxicity as well as the non-hematologic cancer risk.

Others: Although responses to androgens, G-CSF as well as erythropoietin and rarely splenectomy have been documented, they have been transient. IST is ineffective.

CONGENITAL APLASTIC ANEMIAS OF UNKNOWN INHERITANCE

Rare cases of aplastic anemia have been associated with Down syndrome; congenital trisomy-8 mosaicism; familial Robertsonian translocation (13;14); non-familial translocation in a male with t(1;20); (p22;q13.3); cerebellar ataxia with bone marrow monosomy-7 manifesting prior to pancytopenia (familial ataxia–pancytopenia syndrome); and increased spontaneous chromosomal breakage without further increase in breakage with MMC as well as other very rare cases with familial associations. Many of these cases were reported before the discovery of the genes associated with the IBMFSs and can now be categorized. However, a large number of cases are yet to be genetically diagnosed (20–40% in some series) and await the identification of new mutated genes while some represent known syndromes with atypical presentations that are missed with targeted mutation analysis.

DIAMOND–BLACKFAN ANEMIA (CONGENITAL PURE RED CELL APLASIA)

Pathophysiology

Diamond–Blackfan anemia (DBA) is a rare pure red cell aplasia predominantly, but not exclusively, of infancy and childhood resulting from defective ribosome biosynthesis. As a consequence erythroid progenitors and precursors are highly sensitive to death by apoptosis. All of the genetically known cases are the consequence of either small or large subunit-associated ribosomal protein (RP) haploinsufficiency. Patients in North America with DBA should be registered with the Diamond–Blackfan Anemia Registry (DBAR) or other national registries. These registries collects and maintains long-term outcome data as well as providing resources for physicians, patients, and families.

Genetics

- Dominant inheritance
  - The first "DBA gene" was cloned in 1997 and identified as *RPS19*, a gene that codes for an RP, located at chromosome 19q13.2. *RPS19* mutations account for 20–25% of both sporadic and familial cases. Since that time an additional 11 genes have been identified (Table 8.2) comprising approximately 50–70% of DBA cases analyzed. Mutations leading to RP haploinsufficiency, both of the small and large subunit-associated proteins, account for the majority of mutations. The functions of the RP are not fully understood.
  - Laboratory studies used for identification of dominant inheritance in family members of a proband with DBA include: hemoglobin level, MCV, erythrocyte adenosine deaminase (eADA) activity (the absence of these markers clearly does not exclude dominant inheritance) and mutation analysis when available. By carefully evaluating families it appears that at least 40–50% of cases of DBA may be dominantly inherited.
- X-linked recessive inheritance
  - One gene implicated in DBA is inherited as X-linked recessive. *GATA1* encodes an erythroid transcription regulator.
- Recessive inheritance
  - New preliminary studies strongly support evidence of the autosomal recessive inheritance of DBA.

To provide genetic counseling, it is important to perform the previously mentioned laboratory studies to reduce the possibility of missing dominant inheritance in presumed recessive or sporadic cases. It is also important to perform these laboratory studies in potential family stem cell donors to increase the likelihood of detection of a silent phenotype. When there is a known mutation the parents should be evaluated, as well as extended family members as indicated.

Clinical Manifestations[1]

- The median age at presentation of anemia is 2 months and the median age at diagnosis of DBA is 3–4 months. Over 90% of the patients present during the first year of life. A small percentage of affected infants may be anemic at birth.
- Platelet and white cell counts are usually normal; thrombocytosis occurs rarely, neutropenia and/or thrombocytopenia may occur. Instances of significant cytopenias including aplastic anemia have also been noted.
- Physical anomalies, excluding short stature, are found in 47% of the patients. Of these, 50% are of the face and head (microcephaly, eye anomalies), 38% upper limb and hand (thumb deformity, triphalangeal thumb, duplication of thumb and bifid thumb), 39% genitourinary, and 30% cardiac. Twenty-one percent of the patients have more than one anomaly.
- Low birth weight occurs in approximately 10% of all affected patients, with about half of this group being small for gestational age. Over 60% are below the 25th percentile for height. There appears to be a slight

---

[1]Includes findings of the DBA Registry (DBAR) of North America.

increase in the incidence of miscarriages, stillbirths, and complications of pregnancy among the mothers who have given birth to infants with this syndrome.

- Karyotype is generally normal.
- Hepatosplenomegaly is not a common feature.
- Malignant potential: DBA has been recognized as a cancer predisposition syndrome, with an incidence of cancer of 22% by age 46 years. Colorectal and other GI malignancies are the most common, with osteogenic sarcoma being the next most common. Cases of breast cancer and other solid tumors have been reported, all occurring at a younger age than expected for these malignancies. As patients age MDS and AML become increasingly more likely.

Diagnosis

A number of *RP* genes mutations have been identified in DBA and a number of genetically defined individuals have been identified who lack some or all of the classical clinical criteria.

The following laboratory findings occur in DBA:

- Macrocytosis associated with reticulocytopenia. The white cell count and platelet counts are usually normal at presentation but neutropenia and thrombocytopenia are being more frequently recognized and trilineage marrow failure may become evident with increasing age.
- An elevated eADA activity is found in approximately 80−85% of patients.
- Elevated fetal hemoglobin.
- Bone marrow with virtual absence of normoblasts, in some cases with relative increase in proerythroblasts or normal number of proerythroblasts with a maturation arrest; normal myeloid and megakaryocytic series.
- Importantly, these laboratory findings (macrocytosis, elevated fetal hemoglobin, and eADA activity) may be useful in avoiding potential matched related HSCT donors with genotypic DBA, and have been helpful in distinguishing DBA from transient erythroblastopenia of childhood (TEC).

Table 8.13 lists the diagnostic criteria for DBA.

Differential Diagnosis

This condition must be differentiated from:

- TEC. Table 8.14 lists the differentiating features of TEC from DBA.
- Congenital hypoplastic anemia due to transplacental infection with parvovirus B19 can be differentiated from DBA by performing reverse transcriptase polymerase chain reaction for parvovirus B19 on a bone marrow sample. Parvovirus may result in a transient red cell failure in a patient with underlying hemolytic anemia or chronic red cell failure in a patient with underlying immune deficiency.
- Late hyporegenerative anemia due to severe Rh or ABO hemolytic disease of the newborn. This may rarely last for a few months and should be considered in the differential diagnosis of DBA.
- Pearson syndrome, which is characterized by refractory aregenerative macrocytic sideroblastic anemia, neutropenia, vacuolization of bone marrow precursors with sideroblasts (usually ring sideroblasts), exocrine pancreatic dysfunction, and metabolic acidosis. The anemia presents at 1 month of age in 25% and at 6 months of age in 70% of affected individuals. A deletion in mitochondrial DNA has been found in Pearson syndrome. In many instances the anemia may resolve with age. However, many patients will develop neurodegenerative disease (Kearns−Sayre syndrome) later in childhood or adulthood. The natural history of Pearson syndrome is not well characterized.
- Other IBMFSs, as well as acquired marrow dysfunction due to viral infections or medications.

Treatment

Prednisone at a dose of 2 mg/kg/day in single or divided dosages is used to initiate therapy. Reticulocytosis usually occurs in 1−2 weeks but may take slightly longer. When the hemoglobin level reaches ~10 g/dl, the prednisone dose should be tapered slowly to the minimum dose necessary to maintain a reasonable hemoglobin level on an alternate-day schedule. A dose equivalent of 1 mg/kg/every other day (0.5 mg/kg/day) is generally safe but the corticosteroid dose must be individualized. Any patient who experiences significant steroid-related side effects, including growth failure, should have steroid medication temporarily discontinued and should be

TABLE 8.13 Diagnostic Criteria for Diamond—Blackfan Anemia[a]

Diagnostic criteria:

- **Classical**
  - Normochromic, usually macrocytic anemia, relative to patient's age and occasionally normocytic anemia developing in early childhood with no other significant cytopenias
  - Reticulocytopenia
  - Normocellular marrow with selective paucity of erythroid precursors
  - Age less than 1 year

Supporting criteria:

- **Definitive but not essential**
  - Presence of mutation described in classical DBA
- **Major**
  - Positive family history
- **Minor**
  - Congenital abnormalities described in classical DBA
  - Macrocytosis
  - Elevated fetal hemoglobin
  - Elevated erythrocyte adenosine deaminase activity

[a]*These criteria are under constant analysis and may be modified as new DBA genes are identified. The diagnosis becomes less certain when there are fewer diagnostic criteria and the patient does not have a positive family history or a known mutation.*

TABLE 8.14 Differentiating Transient Erythroblastopenia from Diamond—Blackfan Anemia

Feature	Transient erythroblastopenia	Diamond—Blackfan anemia
Frequency	Common	Rare (5—10 per 10^6 live births)
Etiology	Acquired (viral, idiopathic)	Genetic
Age at diagnosis	6 months—4 years, occasionally older	90%, by 1 year; 25%, at birth or within first 2 months
Familial	No	Yes (in at least 10—20% of cases)
Antecedent history	Viral illness	None
Congenital abnormalities	Absent	Present ~50% cases (heart, kidneys, musculoskeletal system)
Course	Spontaneous recovery in weeks to months	Prolonged, 20% actuarial probability of remission
Transfusion dependence	Not dependent	Transfusion or steroid dependent
MCV increased		
At diagnosis	20%	80%
During recovery	90%	100%
In remission	0%	100%
Hemoglobin F increased		
At diagnosis	25%	100%
During recovery	100%	100%
In remission	0%	85%
i antigen	Usually normal	Elevated
Erythrocyte adenosine deaminase activity	Not elevated	Elevated (~85% of cases)
Treatment	Packed cell transfusion, if required	Packed red cell transfusion until 1 year of age
		Prednisone 2 mg/kg/day and taper to lowest effective dose
		Stem cell transplantation

placed on a red cell transfusion regimen. Patients with DBA on low-dose alternate-day therapy of long duration, starting in early infancy, may manifest significant steroid toxicity. Forty percent manifest cushingoid features, 12% pathologic fractures, and 7% have cataracts.

- *Packed red cell transfusion*: When the blood product supply is safe and venous access is available packed red cell transfusion should be used to support patients until they are 1 year of age to avoid significant steroid-related problems encountered in infants and to allow for adequate growth and safe and effective immunization. Leukocyte-depleted packed red cell transfusion should be used to reduce the incidence of non hemolytic, febrile transfusion reactions, as well as the risk of transmission of CMV and the risk of HLA alloimmunization. Patients who are receiving or who have recently been treated with immunosuppressive drugs should receive irradiated blood products. Patients in whom stem cell transplantation is contemplated should receive CMV-safe blood products. Effective iron chelation must accompany a transfusion protocol.
- *HSCT*: HLA-matched sibling donor transplantation should be considered for any patient with DBA, particularly those who are transfusion dependent. Consideration should be given to the fact that 20% of all patients attain spontaneous remission, balanced by the risk of hematologic malignancy, myelodysplasia, or SAA. A family marrow donor must be tested for the presence of a "silent phenotype." Matched unrelated or incompletely matched related donor transplants have proven to be more risky and should be reserved for patients with leukemia, MDS, SAA or patients with clinically significant neutropenia or thrombocytopenia. Recently, the results for alternative donor transplants have improved considerably and these recommendations may change for selected patients. The risk of developing non-hematologic malignancy is increased following preparative regimens used in HSCT.
- *Alternative therapy*: A number of treatments, including erythropoietin, immunoglobulin, megadose corticosteroids, and androgens have been utilized in DBA patients with little success. Cyclosporine, IL-3, and metoclopramide have resulted in anecdotal responses in DBA and have been largely abandoned as therapeutic options. The toxicity of cyclosporine and the lack of availability of IL-3 preclude their use for most patients. Based upon anecdotal data a clinical trial with leucine is ongoing to determine efficacy in DBA.

Prognosis

- Approximately 80% of DBA patients initially respond to corticosteroid therapy, however only approximately half of the responders can remain on steroids at a dose required for an adequate red cell response due to significant side effects. The remaining patients will require transfusion therapy or stem cell transplantation.
- The actuarial remission rate in DBA is approximately 20% by age 25, irrespective of their pattern of response to treatment, with the majority remitting during the first decade of life.
- The major complication of transfusion is iron overload, the consequences of which include diabetes mellitus, cardiac and hepatic dysfunction, growth failure, as well as endocrine dysfunction. Iron chelation with either deferoxamine or deferasirox is therefore an essential component of a transfusion program. The oral chelator Deferiprone (L1) has caused significant neutropenia in DBA and should only be considered with extreme caution in specific cases. New oral iron chelators are in development. Many patients find nearly daily subcutaneous and even oral chelation therapy onerous and compliance is poor.
- In summary both chronic corticosteroid therapy and chronic transfusion therapy may lead to a number of significant short-term and long-term complications, supporting a role for HSCT. Improvements in matched unrelated donor transplantation suggest that patients should be evaluated on an individual basis to determine the utility of that approach in a given patient. Survival of patients into adulthood in remission or sustainable on steroids is 85–100%. Only about 60% of transfusion-dependent patients currently survive to middle age. The overall actuarial survival for DBA at 40 years of age is $75.1 \pm 4.8\%$.
- HLA-matched sibling stem cell transplant patients have long-term survival of over 90% if performed at age 9 years or younger. Well matched unrelated donor transplants done recently have resulted in a survival rate of 80%. Favorable transplantation outcomes are most likely if the patient is in good health at the time of HSCT without iron overload or allosensitization. Improvements in supportive care, GVHD prophylaxis, and infection

control have resulted in a marked decrease in HLA-matched related HSCT transplant-related morbidity and mortality. Sibling HSCT is recommended for young DBA patients, prior to development of significant allosensitization or iron overload, when there is an available HLA-matched unaffected related donor.

- Death in DBA is primarily due to treatment-related causes (iron overload, infection, complications of stem cell transplant) in 67% of cases as opposed to disease-related causes (solid tumors, leukemia and MDS, SAA). This will likely change as management of iron overload and HSCT improve.
- Patients with DBA who become pregnant may develop either an increased requirement for steroid therapy or red cell transfusions due to worsening anemia and should be considered high risk. This appears to be a hormonally induced problem because oral contraceptives have the same effect.
- Fetal hydrops secondary to fetal DBA has been reported.

TRANSIENT ERYTHROBLASTOPENIA OF CHILDHOOD

TEC is much more common than DBA and must be differentiated from DBA (Table 8.14), in order to avoid unnecessary corticosteroid use. TEC has the following features:

- *Pathophysiology*: The following clinical and laboratory observations have shed light on the basic mechanisms of the pathogenesis of TEC:
  - Viral: There is usually a history of a preceding non-specific viral illness 1−2 months prior to TEC.
  - Erythropoietin levels: Serum erythropoietin levels are high, in keeping with the degree of anemia.
  - CFU-E and BFU-E: Both are decreased in 30−50% of patients, suggesting that the defect might be at the CFU-E and BFU-E levels.
  - Serum inhibitors of erythropoiesis: IgG inhibitors of normal progenitor cells have been found in 60−80% of patients with TEC.
  - Cellular inhibitors of erythropoiesis: Inhibitory mononuclear cells have been observed in approximately 25% of patients with TEC.

  On the basis of these observations, it has been speculated that a non-specific virus is cleared as the host develops IgG antibody. This IgG antibody probably recognizes shared viral and erythroid progenitor epitopes.
- *Age*: Usually between 6 months and 4 years of age. With more children attending daycare programs younger patients with TEC are being identified.
- *Sex*: Equal frequency in boys and girls.
- *Hematologic values*:
  - Hemoglobin falls to levels ranging from 3 to 8 g/dl.
  - Reticulocyte count is 0%.
  - White blood cell and platelet count are usually normal.

  Approximately 10% of patients may have significant neutropenia (absolute neutrophil count (ANC), $<1000/mm^3$) and 5% have thrombocytopenia (platelet count, $<100,000/mm^3$) (Table 8.14 lists the hematologic characteristics). An analysis of 50 patients presenting with TEC at our institution revealed a high incidence of neutropenia (64% with an ANC of less than $1500/mm^3$).
- *Bone marrow*: Absence of red cell precursors, except when the diagnostic bone marrow is performed during early recovery (prior to a reticulocytosis) when variable degrees of erythroid maturation may be observed.
- *Prognosis*: Spontaneous recovery occurs within weeks to months with the vast majority of patients recovering within 1 month. TEC rarely recurs.
- *Treatment*: Transfusion of packed red blood cells if there is impending cardiovascular compromise. As recovery is usually prompt restraint should be exercised with regard to red cell transfusions.

Other instances of transient red cell failure may occur secondary to:

- *Drugs*—chloramphenicol, penicillin, phenobarbital, and diphenylhydantoin.
- *Infections*—viral infections (e.g., mumps, EBV, parvovirus B19, atypical pneumonia) and bacterial sepsis.
- *Malnutrition*—kwashiorkor and other disorders.
- *Chronic hemolytic anemia*—hereditary spherocytosis, sickle cell anemia, ß-thalassemia, and other congenital or acquired hemolytic anemias. The etiologic agent is often human parvovirus B19.

CONGENITAL DYSERYTHROPOIETIC ANEMIA

The CDAs are a group of conditions characterized by ineffective erythropoiesis (intramedullary red cell death, anemia with reticulocytopenia, and marrow erythroid hyperplasia) and by specific morphologic abnormalities in the bone marrow consisting of increased numbers of morphologically abnormal red cell precursors. There are three major types of CDA (I−III) for which the mutated genes have been identified (I-*CDAN1*, II-*SEC23B*, III-*KIF23*) and other variants (IV−VIII plus others) that have been described, with genes identified in two (*KLF1* and *GATA1*).

Clinical Manifestations

CDA has the following diagnostic/clinical manifestations:

- Chronic mild congenital anemia (red cells have non-specific abnormalities; basophilic stippling, occasional normoblasts) usually presenting in childhood.
- Reticulocyte response insufficient for the degree of anemia in the context of erythroid hyperplasia in marrow.
- Normal granulopoiesis and thrombopoiesis.
- Chronic or intermittent mild jaundice.
- Splenomegaly.
- High plasma iron turnover rate and low iron utilization by erythrocyte (ineffective erythropoiesis) resulting in progressive iron overload (hemosiderosis).
- Red cell survival time shortened.
- Marrow with abnormal erythroid morphology that can usually distinguish the three types of CDA.

Other manifestations of CDA include the following:

- CDA associated with atypical hereditary ovalocytosis.
- CDA of neonatal onset (with severe anemia at birth, hepatosplenomegaly, jaundice, syndactyly, and small for gestational age).
- CDA associated with hydrops fetalis and hypoproteinemia.

Table 8.15 lists the clinical and laboratory features of CDA, types I−III. Cases with clinical manifestations that do not fit the classical categories of CDA have been described as CDA variants. Table 8.16 represents Wickramasinghe's attempt at the classification of CDA variants. These types share the common features of a congenital, perhaps hereditary, anemia with an inappropriately low reticulocyte count for the degree of anemia, and ineffective marrow dyserythropoiesis. Mutations in two transcription factors, GATA1 and KLF1, have been identified in patients with CDA variants and should be considered in patients lacking mutations described in CDA I−III lacking a known CDA-mutated gene.

Thalassemia and other metabolic abnormalities must be excluded. Table 8.17 lists the diagnostic tests necessary when CDA is suspected.

Differential Diagnosis

The diagnosis of CDA can only be made after the exclusion of other causes of congenital hemolytic anemias associated with ineffective erythropoiesis, such as thalassemia syndromes and hereditary sideroblastic anemias.

Treatment

- Splenectomy performed in severely affected patients with CDA type I having the poorest response results in moderate to marked improvement.
- Transfusion program with the use of deferoxamine, deferasirox, or deferiprone to ameliorate the effects of iron overload may be required to maintain an acceptable hemoglobin level.
- Folic acid 1 mg per week should be administered. Iron therapy is contraindicated.
- Vitamin E has been used in the treatment of CDA type II, with an apparent improvement in red cell survival and a reduction in serum bilirubin and reticulocyte count.
- Recombinant α-IFN 2a has been used in CDA type I, resulting in increased hemoglobin level, decrease in MCV and red cell distribution width, reduction in serum bilirubin and lactic dehydrogenase levels, improvement in morphology of erythroblasts, and reduction in ineffective erythropoiesis.
- Successful stem cell transplant has been performed in CDA types I and II.

TABLE 8.15 Clinical and Laboratory Features of Congenital Dyserythropoietic Anemia, Types I–III

Feature	Type I	Type II (HEMPAS)	Type III (familial); Type III (sporadic)
Inheritance	Autosomal recessive	Autosomal recessive	Autosomal dominant; variable
Clinical	Hepatosplenomegaly	Hepatosplenomegaly	Hepatosplenomegaly
	Jaundice	Variable jaundice	Hair-on-end appearance on skull radiograph
		Gallstones	
		Hemochromatosis	Increased prevalence of lymphoproliferative disorders
Gene	*CDAN1*	*SEC23B*	*KIF23*; Unknown
Gene locus (in some cases)	15q15.1–15.3	20q11.2	15q22
Red cell size	Macrocytic	Normo- or macrocytic	Macrocytic; macrocytic
Anemia	Mild to moderate (usually presenting in neonatal period)	Moderate	Mild to moderate; mild to moderate
	Hemoglobin 8–12 g/dl	Hemoglobin 6–7 g/dl	Hemoglobin 7–8.5 g/dl; hemoglobin 7–8.5 g/dl
Reticulocytes	1.5%	±2%	2–4%; 2–4%
Smear	Macrocytic:	Normocytic:	Macrocytic: macrocytic
	Marked anisocytosis and poikilocytosis; basophilic stippling	Anisocytosis and poikilocytosis; basophilic stippling; "tear drop" cells; irregular contracted cells; occasionally, normoblasts	Anisocytosis; anisocytosis and poikilocytosis; poikilocytosis: basophilic stippling;
Marrow normoblasts	Megaloblastoid: Binucleated, 2–5%; internuclear chromatin bridges, 1–2%	Normoblastic: Bi- and multi-nucleated 10–50% binuclearity predominates	Megaloblastic: multinuclearity (up to 12 nuclei gigantoblasts), 10–50%; megaloblastic: multinuclearity (up to 12 nuclei gigantoblasts) 10–50%
Marrow iron	Scant increase	Increased	Increased; increased
Serum bilirubin and urine urobilinogen	Elevated	Elevated	Elevated; elevated
Treatment	Some patients respond to α interferon 2a treatment or undergo HSCT	Splenectomy; HSCT	HSCT possibly; HSCT possibly

Modified from Alter (2003). and Iolascon et al. (2013).

SIDEROBLASTIC ANEMIAS (MITOCHONDRIAL DISEASES WITH BONE MARROW FAILURE SYNDROMES)

The sideroblastic anemias are a heterogeneous group of disorders characterized by iron deposition in erythroblast mitochondria. They are often secondary to defects in the enzymes of the heme biosynthetic pathway (δ-aminolevulinic acid synthase (ALA-S) deficiency). Impaired production of heme resulting from defects in these enzymes results in mitochondrial iron accumulation, damage to the mitochondrial machinery, and formation of ring sideroblasts. Porphyrias, however, do not display sideroblastic anemia because they are characterized by defects in the cytoplasmic steps of heme synthesis.

Laboratory Findings

- Anemia that may be normocytic, normochromic or microcytic, and hypochromic except in Pearson syndrome, which is characterized by macrocytic anemia probably due to fetal-like erythropoiesis.

- Reticulocytopenia.
- Ineffective erythropoiesis (i.e., erythroid hyperplasia in bone marrow despite anemia).
- Presence of iron-loaded normoblasts demonstrated as ring sideroblasts (greater than 10% of erythroid precursor) by Pearls Prussian blue stain (this stain serves as a surrogate technique for electron microscopy or energy dispersive x-ray analysis used for the demonstration of iron-loaded mitochondria in normoblasts).
- Mild to moderate hemolysis due to peripheral red blood cell destruction of unknown etiology.

The sideroblastic anemias can arise from the primary or secondary defects of mitochondria.

TABLE 8.16   Clinical and Laboratory Features of Congenital Dyserythropoetic Anemia, Types IV–VI

	Type IV	Type V	Type VI
Clinical	Mild to moderate splenomegaly	Spleen palpable in few cases Unconjugated hyperbilirubinemia due to intramedullary destruction of morphologically normal, but functionally abnormal erythroblasts/marrow reticulocytes	Spleen not palpable
Hemoglobin	Very low, transfusion dependent	Normal or near normal	Normal or near normal
MCV	Normal or mildly elevated	Normal or mildly elevated	Very high (119–125) without vitamin B12, folic acid, or other causes of megaloblastic anemia
Erythropoiesis	Normoblastic or mildly to moderately megaloblastic	Normoblastic	Grossly megaloblastic
Non-specific erythroblast dysplasia	Present	Absent or little	Present

From: Wickramasinghe (1997).

TABLE 8.17   Diagnostic Tests for Congenital Dyserythropoietic Anemia[a]

Complete blood count, including MCV, red cell distribution width (RDW), blood smear examination

Absolute reticulocyte count

Quantitative light and, if needed, electron microscope analysis of the bone marrow

Serum vitamin B12 and red cell folate measurements

Parvovirus B19

Serum bilirubin levels

Hemoglobin (Hb) electrophoresis: Hb A2, Hb F assays

Red cell enzyme assays (pyruvate kinase, glucose-6-phosphate dehydrogenase [G6PD])

Sodium dodecyl sulfate polyacrylamide gel electrophoresis of red cell membranes

Test for urinary hemosiderin

Cytogenetic studies of bone marrow cells

Mutation analysis for known CDA genes

Studies of globin chain synthesis

Studies of globin gene analysis

[a]This list is not exhaustive nor is it required in all patients. These tests may be required when there is a need to rule out ß-thalassemia, thiamine-responsive sideroblastic anemia, megaloblastic (B12, folate) anemia, iron deficiency and other causes of ineffective erythropoiesis.

TABLE 8.18   Classification of the Sideroblastic Anemias (SAs) Hereditary/Congenital SA

Isolated heritable
    X-linked (XLSA)
    Glutaredoxin 5 deficiency
    Associated with erythropoietic protoporphyria
    Presumed autosomal
    Suggested maternal
    Sporadic congenital
Associated with genetic syndromes
    X-linked with ataxia (XLSA/A)
    Thiamine-responsive megaloblastic anemia (TRMA)
    Myopathy, lactic acidosis and sideroblastic anemia (MLASA)
    Mitochondrial cytopathy (Pearson syndrome)
Acquired clonal SA
    Refractory anemia with ring sideroblasts (RARS)/pure SA (PSA)
    Refractory anemia with ring sideroblasts and thrombocytosis (RARS-T)
    Refractory cytopenia with multilineage dysplasia and ring sideroblasts (RCMD-RS)
Acquired reversible SA
    These are associated with:
    Alcoholism
    Certain drugs (isoniazid, chloramphenicol)
    Copper deficiency (idiopathic, zinc ingestion, copper chelation, nutritional, malabsorption)
    Hypothermia

Bottomly (2009), with permission.

In congenital sideroblastic anemias, iron rings are predominantly seen in late normoblasts (i.e., orthochromatic and polychromatophilic normoblasts), whereas they are seen in earlier erythroid cells (i.e., basophilic normoblasts) in the acquired form.

Table 8.18 shows a classification of the sideroblastic anemias.

Pathophysiology

Heme biosynthesis involves eight enzymes, four of which are cytoplasmic and four that are localized in the mitochondria.

δ-ALA-S: There are two distinct types of ALA-S. ALA-S1 (housekeeping form) occurs in non-erythroid cells and its gene maps on the autosome, and ALA-S2 (erythroid-specific form) occurs in erythroid cells and its gene maps on the X chromosome. Distinct aspects of heme synthesis regulation in non-erythroid and erythroid cells are related to the differences between these two ALA-S enzymes. In non-erythroid cells, the synthesis and activity of ALA-S1 is subject to feedback inhibition by heme, thus making ALA-S1 the rate-limiting enzyme for the heme pathway. In erythroid cells, heme does not inhibit either the activity or the synthesis of ALA-S2 but it does inhibit cellular iron uptake from transferrin without affecting its utilization for heme synthesis.

There are distinct features of iron and heme metabolism in erythroid and non-erythroid cells. These differences explain the large amount of heme production by erythroid cells compared to the low amount produced by non-erythroid cells. They also explain the mitochondrial deposition of iron in iron-loaded erythroid precursors.

Sideroblastic anemias result from injury to the mitochondria. Defects attributed to the mitochondrial pathways of heme synthesis result in sideroblastic anemias.

Mitochondrial injury results from:

- Defective heme synthesis and the accumulation of iron, especially in erythroid precursors. This iron accumulation causes oxidative damage to the mitochondrial machinery through a Fenton reaction (i.e., the formation of a hydroxyl radical catalyzed by iron and reactive oxygen species damaging mitochondrial DNA by crosslinking DNA strands or by promoting the formation of DNA protein crosslinks).
- Congenital deletions of mitochondrial DNA.
- As a result of mitochondrial damage, there is increased deposition of iron in heme-containing cells (e.g., erythroid cells). Additionally, there is decreased oxidative phosphorylation and decreased adenosine triphosphate (ATP) synthesis in many organs as observed in Pearson syndrome. Figure 8.3 shows a simplified view of the pathophysiologic relationship of various mitochondrial diseases in the context of sideroblastic anemias, bone marrow failure, and/or mitochondrial cytopathies.

SEVERE CONGENITAL NEUTROPENIA AND
KOSTMANN SYNDROME

A
Erythroid cell normoblast

Damaged mitochondria, e.g., as a
result of mitochondrial DNA
damage, ALAS deficiency,
antibiotic toxins

↓

Increased accumulation of
non-utilized iron

↓

Formation of hydroxyl radical
through Fenton reaction

↓

Cross-linking of OH radical
to DNA, proteins, and lipids

↓

Damage to cellular organelles
and cell membranes

B
Non-erythroid cell

Damaged mitochondria due to
mitochondrial DNA mutation

↓

Decreased ATP production

↓

Damage to cellular organelles
and cell membranes

FIGURE 8.3   Simplified view of pathophysiologic consequences of mitochondrial diseases.

Treatment

- Oral pyridoxine is used in some patients with either congenital or acquired sideroblastic anemia with partial response.
- Removal of the toxin/drug responsible for causing sideroblastic anemia may be effective.
- Stem cell transplantation is employed for the treatment of sideroblastic anemia secondary to MDS.

Treatment of *Pearson syndrome* is largely palliative and consists of the following:

- Correction of the metabolic acidosis (e.g., avoidance of fasting, administration of thiamine, riboflavin, carnitine, and coenzyme Q to bypass deleted respiratory enzymes).
- Removal of reactive oxygen radical by the use of ascorbate, vitamin E, or lipoic acid. The efficacy of these therapies is not clear at this time.
- Anemia is treated with red cell transfusions. G-CSF may be used to support clinically significant neutropenia. If patients do not succumb to metabolic acidosis and organ failure the majority will improve within the first decade of life.
- HSCT has been performed and although engraftment occurred the patient succumbed to non-hematopoietic manifestations of the disease.

SEVERE CONGENITAL NEUTROPENIA AND KOSTMANN SYNDROME

Epidemiology

Severe congenital neutropenia (SCN) includes a heterogeneous group of disorders with different patterns of inheritance. Kostmann syndrome (KS) follows an autosomal recessive pattern of inheritance. Its underlying genetic defect is due to homozygous mutations in the *HAX1* gene on chromosome 1. Other SCN may follow autosomal dominant or sporadic patterns of inheritance. In this group of patients, ~60% have diverse mutations in the neutrophil elastase gene (*ELANE*). These mutations affect only one allele. The majority of patients present with a sporadic pattern, since autosomal dominant inheritance is relatively more lethal. In some patients, there may be germline mosaicism.

Incidence

The incidence of SCN is two per million population.

Pathogenesis and Genetics

In vitro bone marrow studies show a reduced number of granulocyte-macrophage colonies in SCN patients. There is also a reduced number of CD34 + /Kit + /G-CSFR + myeloid progenitor cells in the bone marrow.

Neutrophil elastase gene (*ELANE*) mutations: It has been hypothesized that mutations of ELANE in SCN result in a high rate of premature apoptosis in neutrophil precursors, which results in decreased myelopoiesis. Neutrophil elastase is a serine protease localized in the granules of neutrophils and monocytes. A mutant enzyme has a dominant negative effect on the normal wild-type elastase. This explains the defective proteolysis in the SCN neutrophils even though half of the normal amount of the elastase is present in the neutrophils of these patients.

The gene responsible for the autosomal recessive form of SCN in the original cases described by Kostmann has been identified as mutations in the *HAX1* gene located on chromosome 1. HAX1 is an inhibitor of apoptosis. In its absence, apoptosis proceeds unchecked. G-CSF can overcome this deficiency by activating other antiapoptotic pathways.

The mutated genes for other rare SCN syndromes are listed in Table 8.2. Some of these syndrome have unique features:

- Glucose 6 phosphatase, catalytic subunit (G6PC3) syndrome:
  - Cardiac and urogenital malformations, neurologic findings.
- *GFI1* mutation:
  - May be milder but clinically resembles the *ELANE* mutation as GFI1 is a repressor of ELANE.
- *WAS* mutation:
  - X-linked.
- *JAGN1* mutation:
  - Autosomal recessive.
- MonoMac syndrome (*GATA2* mutation):
  - Is inherited as an autosomal dominant with severe persistent monocytopenia.
- WHIM syndrome (*CXCR4* mutation):
  - Warts, Hypogammaglobulinemia, Infections, Myelokathexis with extensive human papilloma virus (HPV) infection is inherited as an autosomal dominant.

Clinical Manifestations and Laboratory Investigations

During the first year of life omphalitis, otitis media, upper respiratory tract infections, pneumonitis, skin abscesses, and liver abscesses occur commonly with positive cultures for staphylococci, streptococci, *Pseudomonas*, *Peptostreptococcus*, and fungi. Splenomegaly may be present. Other manifestations include the following:

- *Blood counts* reveal a normal WBC with an ANC less than $200/mm^3$ and a compensatory eosinophilia and monocytosis. Mild anemia and thrombocytosis may be present.
- *Bone marrow examination* shows a maturation arrest of myelopoiesis at the promyelocyte or myelocyte stage with marked paucity of mature neutrophils. There is an increase in monocytes, eosinophils, macrophages, and plasma cells.

Treatment

- *G-CSF*: The initial dose of G-CSF employed is 5 µg/kg/day. Response occurs 7–10 days from the start of treatment. The goal of therapy is to achieve an ANC of approximately $1000-1500/mm^3$ and maintain the patient free of infections. More than 95% of patients with SCN will respond to G-CSF. After beginning G-CSF therapy, the dose should be adjusted up or down at 1- to 2-week intervals until the lowest effective dose is reached.
  - Complications associated with the use of G-CSF include: bone pain, splenomegaly, hepatomegaly, thrombocytopenia, osteopenia/osteoporosis, Henoch–Schonlein purpura type of immune-complex-induced vasculitis of the skin, and/or glomerulonephritis.
  - Baseline bone marrow cytogenetics should be obtained prior to G-CSF therapy. Initial cytogenetic studies at diagnosis are usually normal. However, during the course of the disease, clonal abnormalities may emerge, 50% of which are monosomy-7. Since 12% of patients with SCN develop MDS and/or acute myelogenous

leukemia (AML), it is important to perform periodic bone marrow examinations for morphology and cytogenetic studies in the follow-up of these patients. Patients who require higher doses of G-CSF (more than 8 µg/kg/d) are at higher risk to develop MDS/AML than those who are more G-CSF responsive (40 vs 11% after 10 years of therapy).
- The G-CSF receptor is normal in patients with SCN. However, patients with SCN are predisposed to develop acquired (somatic) mutations of the cytoplasmic domain of the G-CSF receptor. There is a good correlation between the development of leukemia/MDS and the acquisition of G-CSF receptor mutations in these patients. The time interval between these two events varies considerably.
- *Hematopoietic Stem Cell Transplantation (HSCT)*
  The following are the indications for HSCT:
  - Patients who require greater than ~8 µg/kg/day of G-CSF are statistically much more likely to succumb to infection or leukemia.
  - Refractoriness to G-CSF treatment.
  - Emergence of MDS/AML.

HSCT can be considered as treatment for all patients with SCN who have a HLA-matched sibling donor available with matched unrelated transplant reserved for high-risk patients.

Prognosis

- The leading causes of death in SCN are infection and MDS/leukemia.
- Patients with severe chronic neutropenia should be registered with the Severe Chronic Neutropenia International Registry. This registry collects and maintains long-term outcome data as well as providing resources for physicians, patients, and families.

RETICULAR DYSGENESIS

Reticular dysgenesis is a disorder of stem cells in which maturation of both myeloid and lymphoid lineages is defective. Platelet and red cell production are normal. Affected individuals have severe neutropenia and moderate to severe lymphopenia. In addition, there is absence of peripheral lymphoid tissues, Peyer's patches, tonsils, and splenic follicles. The mortality rate is high from infection at an early age. Treatment: HSCT can be curative.

Cyclin neutropenia and SDS are discussed in Chapter 13.

Figure 13.1 shows an approach to the diagnosis of neutropenia.

Further Reading and References

Albers, C.A., Paul, D.S., Schulze, H., et al., 2012. Compound inheritance of a low-frequency regulatory SNP and a rare null mutation in exon-junction complex subunit RBM8A causes TAR syndrome. Nat. Genet. 44, 435–439, S1–2.

Alter, BP., 2003. Inherited bone marrow failure syndromes. In: Nathan, DG, Orkin, SH, Ginsburg, D, Look, AT (Eds.), Nathan and Oski's Hematology of Infancy and Childhood, sixth ed. Saunders, Philadelphia, PA.

Bottomley, S.S., Fleming, M.D., 2014. Sideroblastic anemia: diagnosis and management. Hematol. Oncol. Clin. North Am. 28, 653–670.

Bottomly, S.S., 2009. Sideroblastic Anemia. In: Wintrobe's Clinical Hematology, twelfth ed. Lippincott Williams & Wilkins, Philadelphia, PA.

Chirnomas, S.D., Kupfer, G.M., 2013. The inherited bone marrow failure syndromes. Pediatr. Clin. North Am. 60, 1291–1310.

Eiler, M., Frohnmayer, D. (Eds.), 2009. Fanconi Anemia: Guidelines for Diagnosis and Management. third ed. Fanconi Anemia Research Fund, Inc., Eugene, OR.

Fernández García, M.S., Teruya-Feldstein, J., 2014. The diagnosis and treatment of dyskeratosis congenita: a review. J. Blood Med. 5, 157–167.

Hall, J.G., Levin, J., Kuhn, J.P., et al., 1969. Thrombocytopenia with absent radius (TAR). Medicine (Baltimore) 48, 411.

Iolascon, A., Heimpel, H., Wahlin, A., Tamary, H., 2013. Congenital dyserythropoietic anemias: molecular insights and diagnostic approach. Blood 122, 2162–2166.

Kee, Y., D'Andrea, A.D., 2012. Molecular pathogenesis and clinical management of Fanconi anemia. J. Clin. Invest. 122, 3799–3806.

Lipton, J.M., Federman, N., Khabbaze, Y., Schwartz, C.L., Hilliard, L.M., Clark, J.I., et al., 2001. Osteogenic sarcoma associated with Diamond–Blackfan anemia: a report from the Diamond–Blackfan Anemia Registry. J. Pediatr. Hematol. Oncol. 23, 39–44.

Locasciulli, A., Bacigalupo, A., Bruno, B., Montante, B., Marsh, J., Tichelli, A., Severe Aplastic Anemia Working Party of the European Blood and Marrow Transplant Group, et al., 2010. Hepatitis-associated aplastic anaemia: epidemiology and treatment results obtained in Europe. A report of The EBMT aplastic anaemia working party. Br. J. Haematol. 149, 890–895.

Locasciulli, A., Oneto, R., Bacigalupo, A., et al., 2007. Outcome of patients with acquired aplastic anemia given first line bone marrow transplantation or immunosuppressive treatment in the last decade: a report from the European Group for Blood and Marrow Transplantation (EBMT). Haematologica 92, 11−18.

Rosenberg, P.S., Greene, M.H., Alter, B.P., 2003. Cancer incidence in persons with Fanconi's anemia. Blood 101, 822−826.

Saracco, P., Quarello, P., Iori, A.P., Zecca, M., Longoni, D., Svahn, J., et al., 2008. Bone Marrow Failure Study Group of the AIEOP (Italian Association of Paediatric Haematology Oncology). Cyclosporin A response and dependence in children with acquired aplastic anaemia: a multicentre retrospective study with long-term observation follow-up. Br. J. Haematol. 140, 197−205.

Shimamura, A., Alter, B.P., 2010. Pathophysiology and management of inherited bone marrow failure syndromes. Blood Rev. 24 (3), 101−122.

Socie, G., Rosenfeld, S., Frickhofen, N., et al., 2000. Late clonal diseases of treated aplastic anemia. Semin. Hematol. 37, 91−101.

Vlachos, A., Muir, E., 2010. How I treat Diamond−Blackfan anemia. Blood 116, 3715−3723.

Vlachos, A., Ball, S., Dahl, N., Alter, B.P., Sheth, S., Ramenghi, U., et al., on behalf of the participants of the Sixth Annual Daniella Maria Arturi International Consensus Conference 2008. Diagnosing and treating Diamond−Blackfan anemia: results of an International Clinical Consensus Conference. Br. J. Haematol. 142, 859−876.

Vulliamy, T., Dokal, I., 2006. Dyskeratosis congenita. Semin. Hematol. 43, 157−166.

Vulliamy, T., Marron, A., Goldman, F., Dearlove, A., Bessler, M., Mason, P.J., et al., 2001. The RNA component of telomerase is mutated in autosomal dominant dyskeratosis congenita. Nature 413, 432−435.

Wickramasinghe, S.N., 1997. Dyserythropoiesis and congenital dyserythropoietic anemias. Br. J. Haematol. 98, 785−797.

Williams, D.A., Bennett, C., Bertuch, A., Bessler, M., Coates, T., Corey, S., et al., 2014. Diagnosis and treatment of pediatric acquired aplastic anemia (AAA): an initial survey of the North American Pediatric Aplastic Anemia Consortium (NAPAAC). Pediatr. Blood Cancer, 869−874.

9

General Considerations of Hemolytic Diseases, Red Cell Membrane, and Enzyme Defects

Lionel Blanc and Lawrence C. Wolfe

Hemolysis is a reduction in the normal red cell survival of 120 days. It may result from corpuscular abnormalities such as membrane, cytoskeleton, enzyme, or hemoglobin defects; or from extracorpuscular abnormalities involving immune or non immune mechanisms (see Chapter 10).

CLINICAL FEATURES OF HEMOLYTIC DISEASE

The following clinical features suggest a hemolytic process in a child with anemia:

1. History of anemia, jaundice, or gallstones in family.
2. Persistent or recurrent anemia associated with reticulocytosis.
3. Anemia unresponsive to hematinics.
4. Intermittent bouts or persistent indirect hyperbilirubinemia/jaundice.
5. Splenomegaly.
6. Hemoglobinuria.
7. Presence of multiple gallstones.
8. Chronic leg ulcers.
9. Development of anemia or hemoglobinuria after exposure to certain drugs.
10. Cyanosis without cardiorespiratory distress.
11. Polycythemia (2,3-diphosphoglycerate mutase deficiency).
12. Dark urine due to dipyroluria (unstable hemoglobins, thalassemia, and ineffective erythropoiesis).
13. Ethnic factors:
    a. Incidence of sickle gene carrier in the African-American population (8%).
    b. High incidence of thalassemia trait in people of Mediterranean ancestry.
    c. High incidence of glucose-6-phosphate dehydrogenase (G6PD) deficiency among those with ethnic origins arising in territories near the Mediterranean Sea, Africa, and Southeast Asia.
    d. High incidence of hereditary ovalocytosis in Southeast Asian populations.
14. Age factors:
    Anemia and jaundice in a Rh-positive infant born to a mother who is Rh negative or a group A or group B infant born to a group O mother (setting for a hemolytic anemia).

LABORATORY FINDINGS

Laboratory findings of hemolytic anemia consist of:

* Evidence of accelerated hemoglobin catabolism due to reduced red cell survival.
* Evidence of increased erythropoiesis.

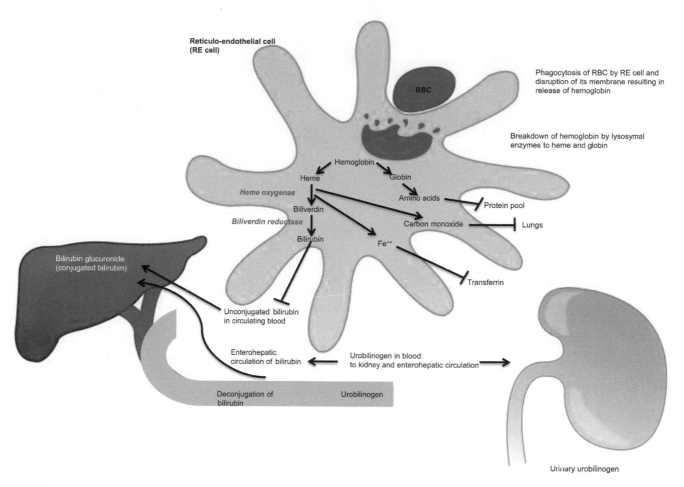

FIGURE 9.1 Extravascular hemoglobin catabolism following extravascular hemolysis.

Accelerated Hemoglobin Catabolism

Accelerated hemoglobin catabolism varies with the type of hemolysis as follows:

- Extravascular hemoglobin catabolism (see Figure 9.1).
- Intravascular hemoglobin catabolism (see Figure 9.2).

The two may not be easily distinguished if the cause for hemolysis is not obvious, hence the long lists of markers of testing indicated below. The presence of hemoglobinuria and hemosiderinuria and the absence of haptoglobin are the major markers of intravascular hemolysis in practice.

Markers of Extravascular Hemolysis

1. Increased unconjugated bilirubin.
2. Increased lactic acid dehydrogenase in serum.
3. Decreased plasma haptoglobin (normal level, 36–195 mg/dl).
4. Increased fecal and urinary urobilinogen.
5. Increased rate of carbon monoxide production.

Markers of Intravascular Hemolysis

1. Increased unconjugated bilirubin (although often less than extravascular hemolysis as urinary losses leave less hemoglobin to be scavenged and processed to bilirubin).
2. Increased lactic acid dehydrogenase in serum.
3. Hemoglobinuria (Figure 9.3 lists the causes of hemoglobinuria).
4. Low or absent plasma haptoglobin.

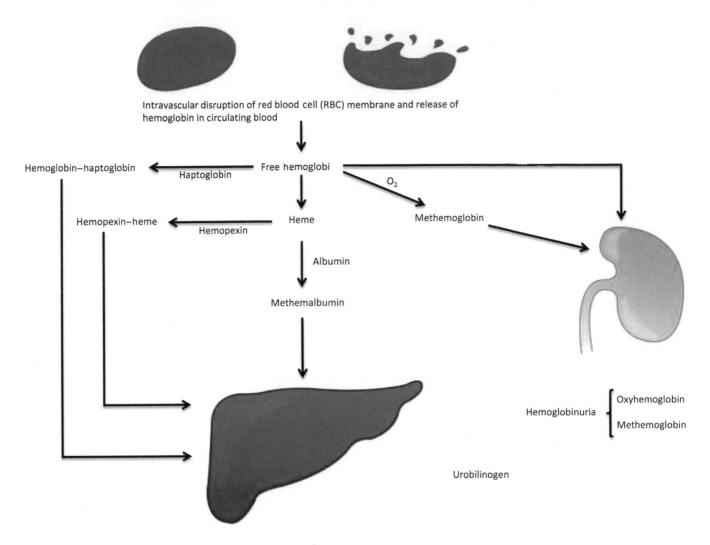

FIGURE 9.2 Intravascular hemoglobin catabolism following intravascular hemolysis. Hemoglobin—haptoglobin, hemopexin—heme, and methemalbumin are cleared by hepatocytes. Heme is converted to iron and bilirubin. The common pathway for both extravascular and intravascular hemolysis is the conjugation of bilirubin (bilirubin glucuronide) by the hepatocytes, its excretion in bile and ultimately formation of urobilinogen by the bacteria in the gut. Part of urobilinogen enters the enterohepatic circulation and part is excreted by the kidney in urine and the remainder of urobilinogen is excreted in stool.

5. Hemosiderinuria (due to sloughing of iron-laden tubular cells into urine).
6. Raised plasma hemoglobin level (normal value <1 mg hemoglobin/dl plasma, visibly red plasma contains greater than 50 mg hemoglobin/dl plasma).
7. Raised plasma methemalbumin (albumin bound to heme; unlike haptoglobin, albumin does not bind intact hemoglobin).
8. Raised plasma methemoglobin (oxidized free plasma hemoglobin) and raised levels of hemopexin—heme complex in plasma.

Increased Erythropoiesis

Erythropoiesis increases in response to a reduction in hemoglobin and is manifested by:

1. *Reticulocytosis*: Frequently up to 10—20%; rarely, as high as 80%.
2. Increased mean corpuscular volume (MCV) due to the presence of reticulocytosis.
3. An increased red cell distribution width (RDW) as the hemoglobin level falls.
4. Normoblasts in the peripheral blood.

I. Acute

A. Mismatched blood transfusions

B. Warm antibody-induced autoimmune hemolytic anemia

C. Drugs and chemicals

  C.1. Regularly causing hemolytic anemia

  C.1.1. Drugs: phenylhydrazine, sulfones (dapsone), phenacetin, acetanilid (large doses)

  C.1.2. Chemicals: nitrobenzene, lead, inadvertent infusion of water

  C.1.3. Toxins: snake and spider bites

  C.2. Occasionally causing hemolytic anemia

  C.2.1. Associated with G6PD deficiency: antimalarials (primaquine, chloroquine), antipyretics (aspirin, phenacetin), sulfonamides (Gantrisin, lederkyn), nitrofurans (Furadantin, Furacin), miscellaneous (naphthalene, vitamin K, British antilewisite [BAL], favism)

  C.2.2. Associated with Hb Zürich: sulfonamides

  C.2.3. Hypersensitivity: quinine, quinidine, *para*-aminosalicylic acid (PAS), phenacetin

D. Infections

  D.1. Bacterial:

  Clostridium perfringens, Bartonella bacilliformis (Oroya fever)

  D.2. Parasitic:

  Malaria

E. Burns

F. Mechanical

(*e.g.*, Prosthetic valves)

II. Chronic

A. Paroxysmal cold hemoglobinuria; syphilis; idiopathic

B. Paroxysmal nocturnal hemoglobinuria

C. March hemoglobinuria

D. Cold agglutinin hemolysis

FIGURE 9.3   Causes of hemoglobinuria.

5. *Specific morphologic abnormalities*: Sickle cells, target cells, basophilic stippling, irregularly contracted cells or fragments (schistocytes), elliptocytes, acanthocytes, and spherocytes.
6. *Erythroid hyperplasia of the bone marrow*: Erythroid:myeloid ratio in the marrow increasing from 1:5 to 1:1.
7. Expansion of marrow space in chronic hemolysis resulting in:
   a. Prominence of the frontal bones and broadened cheekbones.
   b. Widened intratrabecular spaces with hair-on-end appearance of skull radiographs.
   c. Biconcave vertebrae with fish-mouth intervertebral spaces.
8. Decreased red cell survival demonstrated by 51Cr red cell labeling.
9. Red cell creatine levels increased.

Figure 9.4 lists the tests used to establish the cause of the hemolytic anemia.

MEMBRANE DEFECTS

Figure 9.5 lists causes of hemolytic anemia due to corpuscular defects.

Structure of the Red Cell Membrane

Spectrin, the major red cell membrane protein, is largely responsible for maintaining the normal red cell shape and overall morphology. It is composed of two large subunits, α- and β-spectrin, which are encoded by separate genes and are structurally distinct. Spectrin is integrated vertically into the lipid bilayer of the red cell membrane

I. Corpuscular defects

A. Membrane defects

A.1. Blood smear [a]

spherocytes, ovalocytes, pyknocytes, stomatocytes

A.2. Osmotic fragility [a] (fresh and incubated)

A.3. Eosin-5-maleimide dye staining with flow cytometry [b]

A.4. Ektacytometry

A.5. Autohemolysis [a]

A.6. Cation permeability studies

A.7. Membrane phospholipid composition

A.8. Scanning electron microscopy

B. Hemoglobin defects

B.1. Blood smear: sickle cells, target cells (HbC) [a]

B.2. Sickling test [a]

B.3. Hemoglobin electrophoresis [a]

B.4. Quantitative fetal hemoglobin determination [a]

B.5. Kleihauer-Betke smear [a]

B.6. Heat stability test for unstable hemoglobin

B.7. Oxygen dissociation curves

B.8. Rates of synthesis of polypeptide chain production

B.9. Fingerprinting of hemoglobin

C. Enzyme defects

C.1. Heinz-body preparation [a]

C.2. Osmotic fragility [a]

C.3. Autohemolysis test [a]

C.4. Screening test for enzyme deficiencies [a]

C.5. Specific enzyme assays [a]

II. Extracorpuscular defects

Direct antiglobulin test: IgG-γ, C3 (complement), broad-spectrum (both IgG-γ and C3) [a]

III. Serological testing for unusual immune defects

IgA-induced hemolysis, Direct Antiglobulin Test (DAT) negative hemolytic anemia

Donath-Landsteiner test [a]

Flow cytometric analysis of red cells with monoclonal antibodies to GP1-linked surface antigens (for PNH) [a]

FIGURE 9.4   Tests used to establish a specific cause of hemolytic anemia. [a]Tests commonly employed and most useful in establishing a diagnosis; [b]Test available in reference laboratories. Test of choice for hereditary spherocytosis. DAT, direct antiglobulin test.

through the intercession of smaller proteins (ankyrin, 4.2) to integral membrane-spanning proteins (band 3, Rh antigen and glycophorin A). These vertical interactions maintain red cell membrane cohesion. Spectrin associates with itself head to head, while the tail associates with actin and other members of the junctional complex (4.1R, adducin). These horizontal interactions maintain membrane stability. Figure 9.6 summarizes the structure of the normal red cell membrane, highlighting the vertical and horizontal interactions.

The red cell membrane is semipermeable and must maintain its volume in order for the erythrocyte to negotiate the narrower spaces in the circulatory system. Red cell volume is maintained by a number of passive, gradient-driven cation and anion channels as well as active transporters.

Red Cell Membrane Disorders

Hereditary spherocytosis, elliptocytosis, stomatocytosis, acanthocytosis, xerocytosis, and pyropoikilocytosis can be diagnosed on the basis of their characteristic morphologic abnormalities.

Alterations in the quality and/or quantity of the proteins involved in the maintenance of the unique properties of the red cell membrane (deformability and stability) lead to the red cell membrane disorders:

- Hereditary spherocytosis (HS): perturbations in the vertical linkage.
- Hereditary elliptocytosis (HE): perturbations in the horizontal.
- Stomatocytosis: perturbations in the function of the ion transporters.

Note: Enzyme defects and many hemoglobinopathies have non-specific morphologic abnormalities related to secondary effects on red cell membrane proteins and pumps (e.g., ATP depletion).

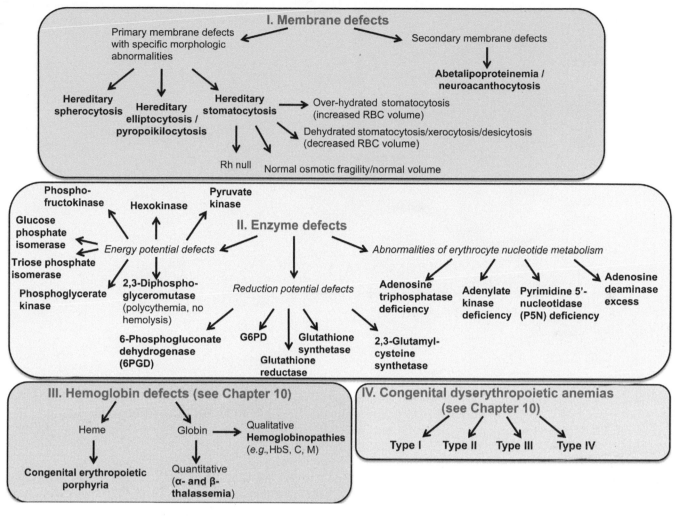

FIGURE 9.5 Causes of hemolytic anemia due to corpuscular defects. World Health Organization (WHO) classification of G6PD variant: Class I variant: Chronic hemolysis due to severe G6PD deficiency, e.g., G6PD deficiency Harilaou. Class II variant: Intermittent hemolysis in spite of severe G6PD deficiency, e.g., G6PD Mediterranean. Class III variant: Intermittent hemolysis associated usually with drugs/infections and moderate G6PD deficiency, e.g., G6PDA variant. Class IV variant: No hemolysis, no G6PD deficiency, e.g., normal G6PD (B variant). ATPase, adenosine triphosphatase; G6PD, glucose-6-phosphate dehydrogenase.

Hereditary Spherocytosis

Genetics

- Autosomal dominant inheritance (75% of cases). The severity of anemia and the degree of spherocytosis may not be uniform within an affected family.
- No family history in 25% of cases. Some show minor laboratory abnormalities suggesting a carrier (recessive) state. Others are due to a *de novo* mutation.
- Most common in people of northern European heritage, with an incidence of 1 in 5000.

Pathogenesis

The primary defect is membrane instability due to dysfunction or deficiency of a red cell skeletal or membrane protein, including:

- *Ankyrin mutations*: Account for 50—67% of HS. In many patients, both spectrin and ankyrin proteins are deficient. Mutations of ankyrin occur in both dominant and recessive forms of HS. The clinical course varies from mild to severe. Red cells are typically spherocytes.

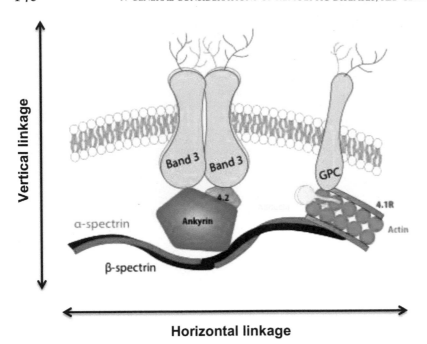

FIGURE 9.6 Schematic representation of the normal red cell membrane. The vertical and horizontal interactions leading respectively to HS and HE are highlighted.

- α-Spectrin mutations occur in recessive HS and account for less than 5% of HS. Clinical course is severe. Contracted cells, poikilocytes, and spherocytes are seen.
- β-Spectrin mutations occur in dominant HS and account for 15—20% of HS. Clinical course is mild to moderate. Acanthocytes, spherocytic elliptocytes, and spherocytes are seen.
- Protein 4.2 mutations occur in the recessive form of HS and account for less than 5% of HS. Clinical course is mild to moderate. Spherocytes, acanthocytes, and ovalocytes are seen.
- Band 3 mutations occur in the dominant form of HS and account for 15—20% of HS. Clinical course is mild to moderate. Spherocytes are occasionally mushroom-shaped or pincered cells.

  Deficiency of these proteins in HS results in a vertical defect, which causes progressive loss of membrane lipid and surface area. The loss of surface area results in characteristic microspherocytic morphology of HS red cells. The sequelae are as follows:

- Depletion of membrane lipid.
- Decrease in membrane surface area relative to volume, resulting in a decrease in surface area-to-volume ratio.
- Tendency to spherocytosis.
- Influx and efflux of sodium increased; cell dehydration.
- Sequestration of red cells in the spleen due to reduced erythrocyte deformability.
- Rapid adenosine triphosphate (ATP) utilization and increased glycolysis leading to increased loss of surface area under ATP-depleted conditions. This leads to the observation of splenic conditioning where the changes in glucose utilization, as well as cell volume control, are dramatically exacerbated with each circulatory passage through the spleen.
- Premature red cell destruction.

Hematologic Features

1. *Anemia*: Mild to moderate when there is a compensated hemolytic anemia. In erythroblastopenic (aplastic or hypoplastic) crisis, hemoglobin may drop to 2—3 g/dl.
2. MCV usually decreased; mean corpuscular hemoglobin concentration (MCHC) raised and RDW elevated.
   - The MCHC is raised in HS, hereditary xerocytosis, hereditary pyropoikilocytosis (HPP), pyruvate kinase (PK) deficiency (which has acquired xerocytosis), and cold agglutinin disease. The presence of elevated RDW and MCHC (performed by aperture impedance instruments, e.g., Coulter) makes the likelihood of HS very high, because these two tests used together are very specific for HS.
3. Reticulocytosis (3—15%).

4. *Blood film*: Spherocytes, microspherocytes (vary in number); hyperdense cells with or without polychromasia. The percentage of microspherocytes is the best indicator of the severity of the disease but not a good discriminator of the HS genotype. Hyperdense cells are seen in HbSC disease, HbCC disease, and xerocytosis. In HS, hyperdense cells are a poor indicator of disease severity but an effective discriminating feature of the HS phenotype.
5. Direct antiglobulin test (DAT) negative.
6. Increased red cell osmotic fragility (spherocytes lyse in higher concentrations of saline than normal red cells) occasionally only demonstrated after incubation of blood sample at 37 °C for 24 h (therefore, always do this test incubated). In spite of normal osmotic fragility, increased MCHC or an increase in hyperdense red cells is highly suggestive of HS.
7. Autohemolysis at 24 and 48 h increased, corrected by the addition of glucose.
8. Survival of ^{51}Cr-labeled cells reduced with increased splenic sequestration.
9. Marrow: Normoblastic hyperplasia; increased iron.
10. Eosin-5-maleimide dye staining of red cells and analysis by flow cytometry is the test of choice to diagnose HS but is only available in special reference laboratories.
11. Genetic analysis for the alpha and beta spectrin, ankyrin, and band 3 mutations is available, but rarely necessary to be performed for diagnosis.

Biochemical Features

1. Raised bilirubin, mainly indirect.
2. Obstructive jaundice with increased direct bilirubin; may develop due to gallstones, a consequence of increased pigment excretion.

Clinical Features

1. Anemia and jaundice: Severity depends on rate of hemolysis, degree of compensation of anemia by reticulocytosis and ability of liver to conjugate and excrete indirect hyperbilirubinemia.
2. Splenomegaly.
3. Presents in newborn period in 50% of cases with hyperbilirubinemia, reticulocytosis, normoblastosis, spherocytosis, negative DAT, and splenomegaly. Patients may present with a requirement for transfusion in the first 8 weeks of life that may not be reflective of their ultimate clinical severity.
4. Presents before puberty in most patients.
5. Diagnosis sometimes made much later in life, often after the birth of an infant with neonatal jaundice caused by HS.
6. Coinheritance of HS with hemoglobin S-C disease may increase the risk of splenic sequestration crisis.
7. Coinheritance of α- or β-thalassemia trait and HS has been reported to have variable effects on hemolysis.
8. Iron deficiency may correct the laboratory values (artificially reducing the MCHC, etc.) but not the red cell lifespan in HS patients.
9. HS with other system involvement:
   * Interstitial deletion of chromosome 8p11.1–8p21.1 causes ankyrin deficiency, psychomotor retardation, and hypogonadism.
   * HS may be associated with neurologic abnormalities such as cerebellar disturbances, muscle atrophy, and a tabes-like syndrome.
   * In patients presenting with common bile duct obstruction associated with gallstones, the increased cholesterol and triglyceride load from the induced dyslipidemia can correct the membrane defect and the resulting spherocyte morphology, MCHC, and osmotic fragility results, hence masking the diagnostic features of the disease. Removal of bile duct obstruction leads to a reappearance of the disease phenotype.

Classification

Table 9.1 lists a classification of HS in accordance with clinical severity and indications for splenectomy.

Diagnosis

* Clinical features and family history.
* Hematologic features.

TABLE 9.1 Classification of Spherocytosis and Indications for Splenectomy

Classification	Trait	Mild spherocytosis	Moderate spherocytosis	Severe spherocytosis[a]
Hemoglobin (g/dl)	Normal	11−15	8−12	6−8
Reticulocyte count (%)	≤3	3.1−6	≥6	≥10
Bilirubin (mg/dl)	≤1.0	1.0−2.0	≥2.0	≥3.0
Reticulocyte production index	<1.8	1.8−3	>3	
Spectrin per erythrocyte[b] (percentage of normal)	100	80−100	50−80	40−60
OSMOTIC FRAGILITY				
Fresh blood	Normal	Normal to slightly increased	Distinctly increased	Distinctly increased
Incubated blood	Slightly increased	Distinctly increased	Distinctly increased	Distinctly increased
AUTOHEMOLYSIS				
Without glucose (%)	>60	>60	0−80	50
With glucose (%)	<10	≥10	≥10	≥10
Splenectomy	Not necessary	Usually not necessary during childhood and adolescence	Necessary during school age before puberty	Necessary, not before 5 years of age
Symptoms	None	None	Pallor, erythroblastopenic crises, splenomegaly, gallstones	Pallor, erythroblastopenic crises, splenomegaly, gallstones

[a]*Value before transfusion.*
[b]*Normal (mean ± SD): $226 ± 54 × 10^3$ molecules per cell.*
From: Eber et al., 1990.

Complications

1. *Hemolytic crisis*: With more pronounced jaundice due to accelerated hemolysis (may be precipitated by viral infection).
2. *Erythroblastopenic crisis (hypoplastic crisis)*: Dramatic fall in hemoglobin level (and reticulocyte count); usually due to maturation arrest and often associated with giant pronormoblasts in the recovery phase; often associated with parvovirus B19 infection.

   Parvovirus B19 infects developing normoblasts, causing a transient cessation of production. The virus specifically infects CFU-E and prevents their maturation. Giant pronormoblasts are seen in bone marrow. Diagnosis is made by increased IgM antibody titer against parvovirus and polymerase chain reaction for parvovirus on bone marrow.
3. *Folate deficiency*: Caused by increased red cell turnover; may lead to superimposed megaloblastic anemia. Megaloblastic anemia may mask HS morphology as well as its diagnosis by osmotic fragility.
4. *Gallstones*: In approximately one-half of untreated patients; increased incidence with age, can occur as early as 4−5 years of age. Occasionally, HS may be masked or improved in obstructive jaundice due to increase in surface area of red cells and the coinheritance of Gilbert syndrome markedly increases the incidence of gallstones.
5. *Complications of chronic anemia*. Patients with more severe HS (see Table 9.1) may suffer growth retardation, anemic heart failure and failure to thrive, necessitating intermittent or chronic transfusion.
6. *Hemochromatosis:* Rarely. This may occur more frequently when a restricted or partial splenectomy is carried out (see below).
7. Splenic rupture: The risk of *splenic rupture* in HS is similar to that of the normal population. Nonetheless, a patient with a large spleen below the costal margin should be cautioned against contact sports or other activities known to lead to blunt trauma to the abdomen.

Treatment

1. Folic acid supplement (1 mg/day).
2. Leukocyte-depleted packed red cell transfusion for severe erythroblastopenic crisis.

3. Splenectomy for moderate to severe cases (see Table 9.1). Most patients with less than 80% of normal spectrin content require splenectomy. Splenectomy should be carried out early in severe cases but not before 5 years of age, if possible. The management of the splenectomized patient is detailed in Chapter 4. Although spherocytosis persists post splenectomy, the red cell lifespan normalizes and complications are prevented, especially transient erythroblastopenia and hyperbilirubinemia. There may, however, be an increased risk of arterial and venous thrombosis in later life as well as an increased risk for idiopathic pulmonary hypertension. Patients are at risk of sepsis after splenectomy, especially for those under 5 years of age. In partial splenectomy, up to 90% of the splenic mass is removed leaving enough splenic tissue to protect against infection. The technique is not widely utilized and its use should be primarily in transfusion-dependent patients who are under 5 years of age. There may be an increased risk for iron loading in patients with HS who have not undergone splenectomy.

Since patients in this situation are unlikely to tolerate phlebotomy, iron overload may make the decision for splenectomy or even partial splenectomy even more complex. Laparoscopic splenectomy is safe in children. Although it requires more operative time than open splenectomy, it is superior with regard to postoperative analgesia, smaller abdominal wall scars, duration of hospital stay and more rapid return to a regular diet and daily activities. It is not known whether accessory spleens are readily identified with the laparoscope although the magnification afforded by the laparoscope might be advantageous in some cases.

4. Ultrasound should be carried out before splenectomy to exclude the presence of gallstones. If present, cholecystectomy is also indicated.

Hereditary Elliptocytosis

HE is clinically and genetically a heterogeneous disorder.

Genetics

HE is characterized by an autosomal dominant or codominant mode of inheritance with variable penetrance, affecting about 1 in 25,000 of the population. The prevalence of HE is much higher in regions where malaria is endemic. This could be explained by the resistance of elliptocytes to malarial invasion.

Occasionally, patients who are severely affected appear to be the offspring of a family with only a single affected parent. In this case a "silent carrier"-like mutation in an α-spectrin gene of the unaffected parent may be the cause.

Pathogenesis

HE is due to various defects in the skeletal proteins, spectrin, and protein 4.1, but also in the integral protein glycophorin C. The basic membrane defects consist of:

- Defects of spectrin self-association involving the α-chains (65%).
- Defects of spectrin self-association involving the β-chains (30%).
- Deficiency of protein 4.1.
- Deficiency of glycophorin C.
- "Silent carrier" effect: α-spectrin mutant genes which produce less α-spectrin when paired with an α-spectrin structural mutant. They lead to more severe disease (see below).

The mechanically unstable membrane of HE leads to shape change from discocyte to elliptocyte as the membrane is buffeted by sheer stress in the circulation.

Patients who are heterozygotes for these defects have milder disease while double heterozygotes and homozygotes for these mutants have progressively more severe syndromes.

Hematologic Features

- Blood smear: 25–90% of cells have elongated oval elliptocytes.
- Osmotic fragility is normal or increased.
- Autohemolysis is usually normal but may be increased and usually corrected by the addition of glucose or ATP.

Clinical Features

- Varies from patients who are symptom-free to severe anemia requiring blood transfusions. The percentage of microcytes best reflects the severity of the disease.
- About 12% have symptoms indistinguishable from HS.
- The percentage of elliptocytes varies from 50 to 90%. No correlation has been established between the degree of elliptocytosis and the severity of the anemia.

Classification

HE has been classified into the following clinical subtypes:

- Common HE, which is divided into several groups:
  a. Silent carrier state.
  b. Mild HE.
  c. HE with infantile pyknocytosis.
- Common HE with chronic hemolysis, which is divided into two groups:
  a. HE with dyserythropoiesis.
  b. Homozygous common HE, which is clinically indistinguishable from HPP (see later discussion).
- Spherocytic HE, which clinically resembles HS; however, a family member usually has evidence of HE.
- Southeast Asian ovalocytosis, in which the majority of cells are oval. Some red cells contain either a longitudinal or transverse ridge.
- Infantile hemolytic elliptocytosis of infancy: These patients present with hemolytic elliptocytosis (mimicking HPP) which changes over the first 2 years of life to a clinical picture of mild HE as fetal hemoglobin changes to adult hemoglobin. Usually there is a single affected parent with HE.

Treatment

The indications and considerations for transfusion, splenectomy, and prophylactic folic acid are the same as for HS.

Hereditary Pyropoikilocytosis

Genetics

Homozygous or doubly heterozygous for spectrin chain mutants (e.g., Sp-a$^{1/74}$ and Sp-a$^{1/76}$). The spectrin chain defects found in HPP are similar to those found in HE.

Pathogenesis

HPP is a congenital hemolytic anemia associated with *in vivo* red cell fragmentation and marked *in vitro* fragmentation of red cells at 45 °C. Because of the similarities in the membrane defect in this condition and HE, it is viewed as a subtype of HE.

Biochemical and Biophysical Features

1. Increased ratio of cholesterol to membrane protein.
2. Decreased cell deformability.

Clinical Features

1. Anemia characterized by extreme anisocytosis and poikilocytosis:
   a. Red cell fragments, spherocytes, and budding red cells (the red cells are exquisitely sensitive to temperature and fragment after 10 min of incubation time at 45−46°C *in vitro*; heating for 6 h at 37°C explains *in vivo* formation of fragmented red cells and chronic hemolysis).
   b. Hemoglobin level reduced to 7−9 g/dl.
   c. Marked reduction in MCV and elevated MCHC.
2. Jaundice.
3. Splenomegaly.
4. Osmotic fragility and autohemolysis increased.
5. Mild HE present in a parent or sibling.

Differential Diagnosis

Similar cells are seen in microangiopathic hemolytic anemias, after severe burns or oxidant stress and in PK deficiency.

Treatment

In infancy these patients require intermittent transfusion for hypoplastic crises. Patients respond well to splenectomy with a rise in hemoglobin to 12 g/dl. Following splenectomy, hemolysis is decreased but not totally eliminated.

Hereditary Stomatocytosis

Definition and Genetics

The stomatocyte has a linear slit-like area of central pallor rather than a circular area. When suspended in plasma, the cells assume a bowl-shaped form. This hereditary hemolytic anemia of variable severity is characterized by an autosomal dominant mode of inheritance. There are two forms of this inherited disorder related to failure to maintain normal red cell volume:

1. Overhydrated stomatocytosis (previously referred to as "hereditary stomatocytosis").
2. Dehydrated stomatocytosis (previously referred to as "hereditary desicytosis or xerocytosis").
   This is characterized by a relative paucity of stomatocytes with cells that appear very hyperchromic.

Etiology

The cells contain high Na^+ and low K^+ concentrations. The disorder is probably due to a membrane and protein defect. Although both forms share the relative increase in red cell sodium, overhydrated stomatocytosis is associated with an increase in red cell volume as the total cation content increases from unbridled sodium entry while dehydrated stomatocytosis has a reduced red cell volume as the potassium cation loss is not matched by sodium accumulation. The cells are abnormally rigid and poorly deformable, contributing to their rapid rate of destruction. There are many biochemical variants. The properties of the stomatocytosis syndromes are listed in Table 9.2.

Red cells from most patients with overhydrated stomatocytosis lack the membrane protein Stomatin (Band 7.2b). Mutations in PIEZO1 (a mechanotransduction protein) lead to dehydrated stomatocytosis.

TABLE 9.2 Properties of the Stomatocytosis Syndromes

	Severe stomatocytosis	Mild stomatocytosis	Cryohydrocytosis	Xerocytosis
Hemolysis	Severe	Mild−moderate	Mild−moderate	Moderate
Smear	Stomatocytes	Stomatocytes	Stomatocytes	Target cells
MCV fl.	110−150	95−130	90−105	85−125
MCHC %	24−30	26−29	34−38	34−38
Osmotic fragility	Very increased	Increased	Normal/slightly increased	Very decreased
RBC Na^+	60−100	30−60	6−25 at 20°C	10−30
RBC K^+	20−55	40−85	55−90 at 20°C	60−90
Cation leak[a]	10−50	~3−10	2−10 at 20°C	2−4
Cold lysis	No	No	Yes	No
Pseudohyperkalemia	? Yes	? Yes	Yes	Occasionally
Perinatal ascites	No	No	No	Occasionally
Genetics	AD	AD	AD	AD

[a]Times normal value.
AD, autosomal dominant.
Table provided by Dr. Samuel Lux, personal communication, 2009.

Clinical Features

Overhydrated Stomatocytosis

1. Jaundice at birth.
2. Pallor: marked variability depending on severity of anemia.
3. Splenomegaly.
4. Hematology
   a. Variable degrees of anemia
   b. Smear, 10–50% stomatocytes
   c. Reticulocytosis
   d. Increased MCV
   e. Decreased MCHC
   f. Increased osmotic fragility and autohemolysis.

Dehydrated Stomatocytosis

1. Mild anemia.
2. Variable neonatal presentation.
3. Splenomegaly and gallstones.
4. Mild increase of MCV.
5. Increased MCHC.
6. Decreased osmotic fragility (i.e., increased osmotic resistance).
7. Increased heat stability (46 and 49°C for 60 min).

Differential Diagnosis

Stomatocytosis morphology may occur with thalassemia, some red cell enzyme defects (glutathione peroxidase deficiency, glucose phosphate isomerase deficiency), Rh_{null} red cells, viral infections, lead poisoning, some drugs (e.g., quinidine and chlorpromazine), some malignancies, liver disease, and alcoholism. Dehydrated stomatocytosis syndrome resembles PK deficiency and infantile pyknocytosis on blood smear.

Treatment

Most patients have mild to moderate hemolysis that occasionally requires transfusion. Splenectomy should be avoided in these syndromes as there seems to be a consistent finding of significant venous thromboembolic complications postsplenectomy in these disorders.

Hereditary Acanthocytosis

Definition

Acanthocytes have thorn-like projections that vary in length and width and are irregularly distributed over the surface of red cells. There are apparently a number of genetic syndromes associated with acanthocytosis and their molecular basis is not yet well defined.

Genetics

The mode of inheritance is autosomal recessive.

Clinical Features

1. *Steatorrhea*: In cases when acanthocytosis is associated with severe fat malabsorption.
2. *Neurologic symptoms*: Weakness, ataxia and nystagmus, atypical retinitis pigmentosa with macular atrophy, blindness.
3. *Anemia*: Mild hemolytic anemia; 10–80% acanthocytes; slight reticulocytosis.

Diagnosis

1. Inherited acanthocytosis is associated with the following clinical syndromes:
   • Abetalipoproteinemia (absent beta-lipoprotein in blood).
   • Chorea-acanthocytosis.

- Huntington-like disease 2.
- Pantothenate kinase-associated neurodegeneration.
- HARP syndrome (hypo-betaliproteinemia, acanthocytosis, retinitis pigmentosa, and pallidal degeneration).
- McLeod syndrome (X-linked anomaly of Kell blood group syndrome).

2. Acquired acanthocytosis is associated with the following clinical conditions:
- Anorexia nervosa.
- Renal failure.
- Microangiopathic hemolytic anemia.
- Subgroup of HS.
- Thyroid disease.
- Liver disease: When associated with liver disease, the acanthocytosis is due to an imbalanced loading of cholesterol and phospholipid on to the red cell membrane. Hemolysis may be more brisk in this situation.

Differential Diagnosis

During the neonatal period, hereditary acanthocytosis may have to be distinguished from the benign non-hereditary disorder of infantile pyknocytosis. Later, acquired causes of acanthocytosis must be considered.

PAROXYSMAL NOCTURNAL HEMOGLOBINURIA

Paroxysmal nocturnal hemoglobinuria (PNH) is characterized by a non-malignant clonal expansion of hematopoietic stem cells that are mutated at *PIGA*. *PIGA* encodes the glycosylphosphatidylinositol (GPI) anchor, the mutation of which results in a deficiency of GPI-anchor proteins. Many of these are complement regulatory surface proteins, a deficiency of which results in hemolytic anemia by increasing sensitivity to complement-induced hemolysis.

Pathogenesis

Patients with PNH have a somatic mutation in the PIG-A gene (phosphatidylinositol glycan complementation group A).

This mutation occurs in primitive hematopoietic stem cells.

A protein product (probably α-1, 6N-acetylglucosamine transferase) of the PIG-A gene is normally responsible for the transfer of N-acetylglucosamine to phosphatidylinositol. In patients with PNH, there is a mutation in the PIG-A gene, which results in a decrease in its protein product and leads to a metabolic block in the biosynthesis of the glycolipid (i.e., GPI) anchor. This anchoring molecule is required for several surface proteins of the hematopoietic cells.

Table 9.3 lists the surface proteins missing on PNH blood cells as a result of a deficiency in the GPI anchor. Thus, the primary defect in PNH resides in the deficient assembly of the GPI anchor and, as a result, all GPI-linked antigens are absent on the surface of PNH cells.

Mechanism of Hemolysis and Hemoglobinuria in PNH

The absence of the surface complement-regulatory proteins CD55 and CD59 allows deposition of complement factors and C3 convertase complexes. This leads to chronic complement-mediated intravascular hemolysis, resulting in hemoglobinuria.

Mechanism of Hypercoagulable State

The mechanism of a hypercoagulable state in PNH is not well understood. A theory is that complement deposition on platelets results in vesiculations of their plasma membranes, which leads to increased procoagulant activity of the platelets. The monocytes and granulocytes of PNH cells lack the receptor for the GPI-linked urokinase plasminogen activator and this deficiency may lead to impaired fibrinolysis.

The antithrombin (AT), protein C, and protein S levels are normal in PNH patients.

TABLE 9.3   Surface Proteins Missing on Paroxysmal Nocturnal Hemoglobinuria Blood Cells

Protein	Expression pattern
ENZYMES	
Acetylcholinesterase (AChE)	Red blood cells
5′-ectonucleotidase (CD73)	Some B and T lymphocytes
Leukocyte alkaline phosphatase	Neutrophils
ADHESION MOLECULES	
Blast-1/CD48	Lymphocytes
Lymphocyte function-associated antigen-3 (LFA-3 or CD58)	All blood cells[a]
COMPLEMENT REGULATING SURFACE PROTEINS	
Decay accelerating factor (DAF or CD55)	All blood cells[b]
Homologous restriction factor (HRF or C8bp)	All blood cells[c]
Membrane inhibitor of reactive lysis (MIRL or CD59)	All blood cells
RECEPTORS	
Fcγ receptor III (Fcγ III or CD16)	Neutrophils, NK cells[d], macrophages[d], some T lymphocytes[d]
Endotoxin binding protein (CD14)	Monocytes, macrophages, granulocytes
Urokinase-type plasminogen activator receptor (CD87)	Monocytes, granulocytes
BLOOD GROUP ANTIGENS	
Comer antigens (DAF)	Red blood cells
Yt antigens (AChE)	Red blood cells
Holley Gregory antigen	Red blood cells
John Milton Hagen (JMH) bearing protein (CD108)	Red blood cells, lymphocytes
Dombrock residue	Red blood cells
NEUTROPHIL ANTIGENS	
NA1/NA2 (CD16)	Neutrophils
NB1/NB2	Neutrophils
Other surface proteins	Various
CD52 (CAMPATH)	CD109
CD24	CD157
CD48	GP500
CD66c	GP175
CD67	Folate receptor
CD90	

[a]On lymphocytes expressed in GPI-linked and transmembrane form.
[b]Level of expression on T lymphocytes varies.
[c]Expression of C8bp on human blood cells is controversial (personal communication, Taroh Kinoshita).
[d]Expressed in a transmembrane form.
From: Young et al. (1996); with permission; and Ware, (2014), with permission.

Mechanism of Defective Hematopoiesis

The mechanism of defective hematopoiesis (macrocytosis with bone marrow erythroid dysplasia) evolving to severe aplastic anemia (AA) in some patients is not well understood. Possible explanations include:

- The initial step is the development of the PIG-A mutation. This is followed by a bone marrow insult.
- Resistance of PNH clones to injury compared with the susceptibility of normal hematopoietic stem cells.
- Intrinsic proliferation advantage of PNH stem cells compared with normal hematopoietic stem cells.
- Suppression of normal hematopoietic stem cells by PNH cells and evolution to MDS or AML.

In the preceding explanation, it is assumed that two populations of stem cells normally reside in bone marrow: (i) a large population of normal stem cells and (ii) a minor population of PNH stem cells.

Clinical Manifestations

The three main clinical features of PNH are:

- Paroxysmal intravascular hemolysis more frequent at night associated with hemoglobinuria and abdominal and back pain. In most cases hemolytic episodes occur every few weeks although some patients have chronic unrelenting hemolysis with severe anemia.
- Bone marrow failure (macrocytosis, pancytopenia to severe AA).
- Tendency to venous thrombosis. PNH can present as a primary "classic" intravascular hemolysis or it may arise during the course of AA as AA-PNH syndrome. The nature of the pathogenetic link between the two conditions remains unknown. They may be differentiated from each other by the clinical findings shown in Table 9.4.

Many patients have an overlap of the aforementioned findings and do not fit precisely into one of these two groups.

Course of the Disease

The onset of PNH is insidious. There is no familial tendency. Venous thrombosis is more often responsible for death than bone marrow failure in patients with PNH. Spontaneous long-term remission or leukemia transformation or AA may occur in some patients. Anemia is the most common finding and AA is found in approximately 10% of patients.

Patients with classic PNH may have cytopenia of one or all blood cell lineages and the degree of bone marrow failure may vary from mild to severe. About 15% of patients with AA develop overt PNH; however, 35–50% of AA patients may have flow cytometric evidence of deficiency of GPI-linked molecules at some stage of their disease as evidence of subclinical PNH.

TABLE 9.4 Clinical Findings in Classic PNH Syndrome and in Aplastic Anemia—PNH Syndrome

Findings	Classic PNH syndrome	Aplastic anemia—PNH syndrome
Hemolysis	Chronic with acute exacerbation	Hemolysis clinically subtle
Thrombotic complications	More often present	Occurs less frequently
	Acute hemolysis may be preceded by abdominal pain, thought to be due to temporary occlusion of the gastrointestinal veins. Thrombosis of larger abdominal veins may be present	Bone marrow failure predominant clinical finding
Abnormal erythrocyte or granulocyte CD55/CD59	Positive from the time of diagnosis	Positive in 20–50% of patients with SAA. May evolve post immunosuppressive therapy

SAA, Severe aplastic anemia.

Complications

Intravascular Hemolysis (DAT Negative)

- Hemoglobinuria (dark urine).
- Iron deficiency.
- Acute renal failure.

Venous Thrombosis

- Peripheral veins.
- Superior and inferior vena cava.
- Hepatic veins (Budd–Chiari syndrome).
- Mesenteric veins.
- Sagittal sinus.
- Splenic vein.
- Abdominal wall veins.
- Intrathoracic veins.

Defective Hematopoiesis

- AA.
- Macrocytosis.
- Evolution to myelodysplastic or AML.

Infectious

- Sinopulmonary.
- Bloodborne.

Other

- Dysphagia.

Table 9.5 lists the laboratory findings in PNH.

TABLE 9.5   Laboratory Findings in Paroxysmal Nocturnal Hemoglobinuria

Non-specific findings	Cytopenia involving one or more cell lineages
	Macrocytosis, anisocytosis, polychromasia
	Reticulocytosis
	Decreased neutrophil alkaline phosphatase
	Increased level of lactate dehydrogenase
	Decreased haptoglobin
	Hemoglobinuria, hemosiderinuria
	Iron deficiency, folate deficiency
Bone marrow findings	Varies from hyperplastic with predominant erythropoiesis to hypoplastic with little or patchy hematopoiesis
	Hypoplasia or aplasia of one or more hematopoietic lineages
	Increased number of mast cells
Cytogenesis	Usually normal
Specific test for PNH	Flow cytometric analysis for glycosylphosphatidylinositol (GPI)-linked cell surface proteins (e.g., CD59) on peripheral blood or bone marrow cells

Adapted from: Young et al., 1996; with permission.

Diagnosis

PNH is definitively diagnosed through flow cytometric analysis of blood cells with the use of monoclonal antibodies to GPI-linked surface antigens.

All blood cell lineages (i.e., red blood cells, lymphocytes, monocytes, granulocytes) can be analyzed by the flow cytometric technique, and heterogeneous patterns of the phenotypic expressions of various blood cells can be identified. For example, red blood cell phenotypes can be identified by their CD59 expression:

PNH type I: Normal expression of CD59.
PNH type II: Partially deficient or residual expression of CD59.
PNH type III: Complete absence of expression of CD59.

The proportion of the three different phenotypes may vary from patient to patient.

Because other blood cell lineages can be analyzed, the transfusion of red blood cells to a patient does not interfere with the diagnosis of PNH.

The percentage of granulocytes with a PNH phenotype is usually higher than the percentage of red cells lacking CD59. Thus, flow cytometric analysis of the granulocytes increases sensitivity in the diagnosis of PNH.

Management

Hematopoietic Stem Cell Transplantation

Hematopoietic stem cell transplantation (HSCT) is the only curative treatment for PNH. If a fully matched family donor is available, then HSCT is the treatment of choice, especially for patients who develop bone marrow failure. In the absence of a matched unrelated donor, alternative donor transplantations can be considered based on the quality of the available alternative donor and the severity of the PNH.

Ecluzumab

Ecluzumab, a humanized monoclonal antibody that blocks complement activation at C5 preventing the formation of C5a is the standard of care for PNH. It reduces hemolysis and thromboembolism, and dramatically improves the patient's quality of life. Due to the importance of complement in immunity against *Neisseria meningitidis* patients receiving ecluzumab must be vaccinated.

Immunosuppressive Therapy

Therapy with cyclosporine and ATG is indicated in the setting of PNH-associated AA. This treatment may lead to improvement in AA but not in the hemolysis of PNH.

Prednisone 1–2 mg/kg daily can ameliorate hemolysis and can be used for 24–72 h around the time of a hemolytic episode.

Use of Hematopoietic Growth Factor

The use of G-CSF may be attempted in the setting of a pertinent cytopenia.

Supportive Therapy

- Long-term anticoagulant therapy (e.g., with warfarin or low-molecular weight heparin) is indicated for patients with venous thrombosis. Women with PNH should be discouraged from using birth control pills.
- Iron and folate supplements are indicated due to chronic hemoglobinuria accompanied by iron loss and chronic hemolysis with increased erythroid marrow activity requiring supplementation of additional folate.
- Sidenafil may be effective in treating dysphagia and intestinal spasm and impotence, which are the consequence of decreased nitric oxide secondary to consumption by plasma free hemoglobin.
- Red blood cell transfusion as needed for symptomatic anemic patients.

ENZYME DEFECTS

There are two major biochemical pathways in the red cell: the Embden–Meyerhof anaerobic pathway (energy potential of the cell) and the hexose monophosphate shunt (reduction potential of the cell). Figure 9.7 illustrates the enzyme reactions in the red cell.

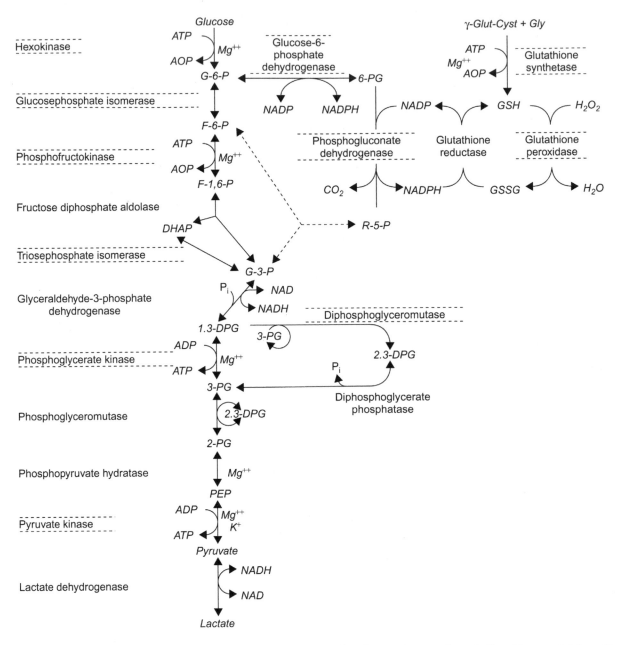

FIGURE 9.7 Enzyme reactions of Embden–Meyerhof and hexose monophosphate pathways of metabolism. Documented hereditary deficiency diseases are indicated by enclosing dotted lines.

PK Deficiency

PK is an enzyme active in the penultimate conversion in the Embden–Meyerhof pathway. Although deficiency is rare, it is the most common enzyme abnormality in the Embden–Meyerhof pathway.

Genetics

- Autosomal recessive inheritance.
- Significant hemolysis seen in homozygotes.
- Found predominantly in people of northern European origin.
- Deficiency not simply quantitative; probably often reflects the production of PK variants with abnormal characteristics.

Pathogenesis

- Defective red cell glycolysis with reduced ATP formation.
- Red cells rigid, deformed and metabolically and physically vulnerable (reticulocytes less vulnerable because of ability to generate ATP by oxidative phosphorylation).

Hematology

- Features of non-spherocytic hemolytic anemia: macrocytes, oval forms, polychromatophilia, anisocytosis, occasional spherocytes, contracted red cells with multiple projecting spicules, rather like echinocytes or pyknocytes.
- Erythrocyte PK activity decreased to 5–20% of normal; 2,3-diphosphoglycerate (2,3-DPG) and other glycolytic intermediary metabolites increased (because of two- to threefold increase in 2,3-DPG, there is a shift to the right in P_{50}).[1]
- Autohemolysis markedly increased, showing marked correction with ATP but not with glucose.

Clinical Features

- Variable severity; can cause moderately severe anemia (not drug-induced). Patients may tolerate their anemia better because of the increase in 2,3-DPG shifting the hemoglobin oxygen dissociation curve to the right. This leads to superior off-loading of oxygen to the tissues and may mitigate the anemia.
- Usually presents with neonatal jaundice.
- Splenomegaly common but not invariable.
- Late: gallstones, bone changes of chronic hemolytic anemia, cardiomegaly secondary to severe anemia.
- Erythroblastopenic crisis due to parvovirus B19 infection.
- Hemochromatosis. These patients seem to have a risk of hemochromatosis beyond the number of transfusions they received. Careful attention should be paid to their iron loading.

Treatment

- Folic acid supplementation.
- Transfusions as required.
- Splenectomy (if transfusion requirements increase); splenectomy does not arrest hemolysis, but decreases transfusion requirements. Note that there is a paradoxical increase in reticulocytosis after splenectomy even as transfusion requirement and hemolytic rate abate.
- Surveillance for iron overload.

Other Enzyme Deficiencies

1. Hexokinase deficiency, with many variants.
2. Glucose phosphate isomerase deficiency.
3. Phosphofructokinase deficiency, with variants.
4. Aldolase.
5. Triosephosphate isomerase deficiency.
6. Phosphoglycerate kinase deficiency.
7. 2,3-DPG deficiency due to deficiency of diphosphoglycerate mutase.
8. Adenosine triphosphatase deficiency.
9. Enolase deficiency.
10. Pyrimidine 5'-nucleotidase deficiency.
11. Adenosine deaminase overexpression.
12. Adenylate kinase deficiency.

[1]Because of the right shift of P50, patients do not exhibit fatigue and exercise intolerance proportionate to the degree of anemia.

These enzyme deficiencies have the following features:

1. General hematologic features:
   a. Autosomal recessive disorders except phosphoglycerate kinase deficiency, which is sex linked and enolase deficiency which presents as an autosomal dominant.
   b. Chronic non-spherocytic hemolytic anemias (CNSHAs) of variable severity.
   c. Osmotic fragility and autohemolysis normal or increased.
   d. Improvement in anemia after splenectomy.
   e. Diagnosed by specific red cell assays.
2. Specific non-hematologic features:
   a. Phosphofructokinase deficiency associated with type VII glycogen storage disease. Hematologic symptoms are mild compared to the significant myopathy.
   b. Triosephosphate isomerase deficiency associated with progressive debilitating neuromuscular disease with generalized spasticity and recurrent infections (some patients have died of sudden cardiac arrest).
   c. Phosphoglycerate kinase deficiency associated with mental retardation, myopathy, and a behavioral disorder.

Note the three exceptions to the general hematologic features listed above:

- Adenosine deaminase excess (i.e., not an enzyme deficiency) is an autosomal dominant disorder.
- Pyrimidine 5′-nucleotidase deficiency is characterized by marked basophilic stippling, although the other CNSHAs lack any specific morphologic abnormalities.
- Deficiency of diphosphoglycerate mutase results in polycythemia.

Glucose-6-Phosphate Dehydrogenase Deficiency

Glucose-6-phosphate dehydrogenase (G6PD) is the first enzyme in the pentose phosphate pathway of glucose metabolism. Deficiency diminishes the reductive energy of the red cell and may result in hemolysis, the severity of which depends on the quantity and type of G6PD and the nature of the hemolytic agent (usually an oxidation mediator that can oxidize NADPH, generated in the pentose phosphate pathway in red cells).

Genetics
- Sex-linked recessive mode of inheritance by a gene located on the X chromosome (similar to hemophilia).
- Disease is fully expressed in hemizygous males and homozygous females.
- Variable intermediate expression is shown by heterozygous females (due to random deletion of X chromosome, according to Lyon hypothesis).
- As much as 3% of the world's population is affected; most frequent among African-American, Asian, and Mediterranean peoples.
      The molecular basis of G6PD deficiency and its clinical implications follow:
   - Deletions of G6PD genes are incompatible with life because it is a housekeeping gene and complete absence of G6PD activity, called hydeletions, will result in death of the embryo.
   - Point mutations are responsible for G6PD deficiencies. They result in:
- *Sporadic mutations*: They are not specific to any geographic areas. The same mutation may be encountered in different parts of the world that have no causal (e.g., encountering G6PD Guadalajara in Belfast) relationship with malarial selection. These patients manifest with CNSHA WHO Class I.
- *Polymorphic mutations*: These mutations have resulted from malaria selection; hence, they correlate with specific geographic areas. They are usually WHO Class II or III and not Class I.

The World Health Organization (WHO) classifies G6PD variants on the basis of magnitude of the enzyme deficiency and the severity of hemolysis (Table 9.6).

Pathogenesis
- Red cell G6PD activity falls rapidly and prematurely as red cells age.
- Decreased glucose metabolism.
- Diminished NADPH/NADP and GSH/GSSG ratios.
- Impaired elimination of oxidants (e.g., H_2O_2).

TABLE 9.6 WHO Classification of G6PD Variants

WHO class	Variant	Magnitude of enzyme deficiency	Severity of hemolysis
I	Harilaou, Tokyo, Guadalajara, Stonybrook, Minnesota	2% of normal activity	Chronic non-spherocytic hemolytic anemia
II	Mediterranean	3% of normal activity	Intermittent hemolysis
III	A^2	10–60% of normal activity	Intermittent hemolysis usually associated with infections or drugs
IV	B (Normal)	100% normal activity	No hemolysis

- Oxidation of hemoglobin and of sulfhydryl groups in the membrane.
- Red cell integrity impaired, especially on exposure to oxidant drugs, oxidant response to infection and chemicals.
- Oxidized hemoglobin precipitates to form Heinz bodies which are plucked out of the red cell leading to hemolysis and "bite cell" and "blister cell" morphology.

Clinical Features

Episodes of hemolysis may be produced by:

- Drugs. Table 9.7 lists the agents capable of inducing hemolysis in G6PD-deficient subjects.
- Fava bean (broad bean, *Vicia fava*): ingestion or exposure to pollen from the bean's flower (hence favism).
- Infection.
  1. Drug-induced hemolysis
     a. Typically in African-Americans but also in Mediterranean and Canton types.
     b. List of drugs (see Table 9.7); occasionally need additional stress of infection or the neonatal state.
     c. Acute self-limiting hemolytic anemia with hemoglobinuria.
     d. Heinz bodies in circulating red cells.
     e. Blister cells, fragmented cells, and spherocytes.
     f. Reticulocytosis.
     g. Hemoglobin normal between episodes.
  2. Favism
     a. Acute life-threatening hemolysis, often leading to acute renal failure, caused by ingestion of fava beans.
     b. Associated with Mediterranean and Canton varieties.
     c. Blood transfusion required.
  3. Neonatal jaundice
     a. Usually associated with Mediterranean and Canton varieties but can occur with all variants.
     b. Infants may present with pallor, jaundice (can be severe and produce kernicterus), and dark urine.
        i. The excessive jaundice resulting in kernicterus is not only due to hemolysis but may be due to reduced glucuronidation of bilirubin caused by defective G6PD activity in the hepatocytes.
     Often no exposure to drugs, occasionally exposure to naphthalene (mothballs), aniline dye, marking ink, or a drug. In a majority of neonates, the jaundice is not hemolytic but hepatic in origin.
  4. CNSHA
     a. Occurs mainly with sporadic inheritance.
     b. Clinical picture:
        i. Chronic non-spherocytic anemia variable but can be severe with transfusion dependence.
        ii. Reticulocytosis.
        iii. Intense neonatal presentation.
        iv. Shortened red cell survival.
        v. Increased autohemolysis with only partial correction by glucose.
        vi. Slight jaundice.
        vii. Mild splenomegaly.

TABLE 9.7   Agents Capable of Inducing Hemolysis in G6PD-deficient Subjects[a]

Clinically significant hemolysis	Usually not clinically significant hemolysis
ANALGESICS AND ANTIPYRETICS	
Acetanilid	Acetophenetidin (phenacetin)
	Acetylsalicylic acid (large doses)
	Antipyrine[a,b]
	Aminopyrine[b]
	p-Aminosalicylic acid
ANTIMALARIAL AGENTS	
Pentaquine	Quinacrine (Atabrine)
Pamaquine	Quinine[b]
Primaquine	Chloroquine[c]
Quinocide	Pyrimethamine (Daraprim)
	Plasmoquine
FLOUROQUINONES	
Ciprofloxacin	
SULFONAMIDES	
Sulfanilamide	Sulfadiazine
N-Acetylsulfanilamide	Sulfamerazine
Sulfapyridine	Sulfisoxazole (Gantrisin)[c]
Sulfamethoxypyridazine (Kynex)	Sulfathiazole
Salicylazosulfapyridine (Azulfidine)	Sulfacetamide
NITROFURANS	
Nitrofurazone (Furacin)	
Nitrofurantoin (Furadantin)	
Furaltadone (Altafur)	
Furazolidone (Furoxone)	
SULFONES	
Thiazolsulfone (Promizole)	
Diaminodiphenylsulfone (DDS, dapsone)	Sulfoxone sodium (Diasone)
MISCELLANEOUS	
Naphthalene	
Phenylhydrazine	Menadione
Acetylphenylhydrazine	Dimercaprol (BAL)
Toluidine blue	Methylene blue
Nalidixic acid (NegGram)	Chloramphenicol[b]
Neoarsphenamine (Neosalvarsan)	Probenecid (Benemid)
Infections	Quinidine[b]
Diabetic ketoacidosis	Fava beans[b]
Doxorubicin	
Urate oxidase (Rasburicase)	
Foods	

[a]Many other compounds have been tested but are free of hemolytic activity. Penicillin, the tetracyclines, and erythromycin, for example, will not cause hemolysis and the incidence of allergic reactions in G6PD-deficient persons is not any greater than that observed in others.
[b]Hemolysis in Whites only.
[c]Mild hemolysis in African-Americans, if given in large doses.
Note: Drugs associated with hemolysis in any WHO class are listed as clinically significant.

Treatment

1. Avoidance of agents that are deleterious in G6PD deficiency. (For a consistent, up-to-date list of drug susceptibilities visit: www.favism.org.)
2. Education of families and patients in recognition of food prohibition (fava beans), drug avoidance, heightened vigilance during infection and the symptoms and signs of hemolytic crisis (orange/dark urine, lethargy, fatigue, jaundice).
3. Indication for transfusion of packed red blood cell in children presenting with acute hemolytic anemia:
   a. Hemoglobin (Hb) level below 7 g/dl.
   b. Persistent hemoglobinuria and Hb below 9 g/dl.
4. CNSHA:
   a. In patients with severe chronic anemia: transfuse red blood cells to maintain Hb level 8–10 g/dl and iron chelation, when needed.
   b. Splenectomy has only occasionally ameliorated severe anemia in this disease. Indications for splenectomy are as follows:
      i. Hypersplenism.
      ii. Severe chronic anemia.
      iii. Splenomegaly causing physical impediment.
   c. Genetic counseling and prenatal diagnosis for severe CNSHA if the mother is a heterozygote.

Other Defects of Glutathione Metabolism

Glutathione Reductase

In this autosomal dominant disorder, hemolytic anemia is precipitated by drugs having an oxidant action. Thrombocytopenia has occasionally been reported. Neurologic symptoms occur in some patients. The disease is mimicked by riboflavin deficiency.

Glutamyl Cysteine Synthetase

In this autosomal recessive disorder, there is a well-compensated hemolytic anemia. This very rare disease has been associated with spinocerebellar degeneration in one patient.

Glutathione Synthetase

In this autosomal recessive disorder, there is a well-compensated hemolytic anemia, exacerbated by drugs having an oxidant action. This is the most common disorder of the group and can also present as a systemic metabolic disorder with acidosis, hemolysis, and susceptibility to infection.

Glutathione Peroxidase

In this autosomal recessive disorder, acute hemolytic episodes occur after exposure to drugs having an oxidant action.

Further Reading and References

An, X., Mohandas, N., 2008. Disorders of red cell membrane. Br. J. Haematol. 141, 367–375.

Becker, P., Lux, S., 1993. Disorders of the red cell membrane. In: Nathan, D., Oski, F. (Eds.), Hematology of Infancy and Childhood. WB Saunders, Philadelphia.

Bolton-Maggs, P.H., Stevens, R.F., et al., on behalf of the General Haematology Task Force of the British Committee for the Standards in Haematology 2004. Guidelines for the diagnosis and management of hereditary spherocytosis. Br. J. Haematol. 126, 455–474.

Da Costa, L., Gallimand, J., Fenneteau, O., Mohandas, N., 2013. Hereditary spherocytosis, elliptocytosis, and other red cell membrane disorders. Blood Rev. 27 (4), 167–178.

Dacie, J., 1992. The Haemolytic Anaemias 3. The Auto-Immune Haemolytic Anaemias, third ed. Churchill Livingstone, Edinburgh.

Eber, S.W., Armburst, R., Schröter, W.J., 1990. Variable clinical severity of Hereditary Spherocytosis: relation to erythrocytic spectrin concentration, osmotic fragility, and autohemolysis. Pediat 117, 409.

Gallagher, P.G., 2013. Abnormalities of the erythrocyte membrane. Pediatr. Clin. North Am. 60 (6), 1349–1362.

Hillmen, P., Lewis, S.M., Bressler, M., et al., 1995. Natural history of paroxysmal nocturnal hemoglobinuria. N. Engl. J. Med. 333, 1253–1258.

Inoue, N., Izui-Sarumaru, T., et al., 2006. Molecular basis of clonal expansion of hematopoiesis in 2 patients with paroxysmal nocturnal hemoglobinuria (PNH). Blood 108 (13), 4232–4236.

King, M.-J., Behrens, J., et al., 2000. Rapid flow cytometric test for the diagnosis of membrane cytoskeleton-associated haemolytic anemia. Br. J. Haematol. 111, 924–933.

Lanzkowsky, P., 1980. Hemolytic anemia. Pediatric Hematology Oncology: A Treatise for the Clinician. McGraw-Hill, New York.

Schilling, R.F., Gangnon, R.E., et al., 2007. Delayed diverse vascular events after splenectomy in hereditary spherocytosis. J. Thromb. Haemost. 6, 1289–1295.

Tracy, E., Rice, H., 2008. Partial splenectomy for hereditary spherocytosis. Ped. Clin. North Am. 55, 503–519.

Ware, R.E., 2014. Autoimmune hemolytic anemia. In: Orkin, S.H., Fisher, D.E., Ginsburg, D., Look, T.A., Lux, S.E., Nathan, D.G. (Eds.), Nathan and Oski's Hematology of Infancy and Childhood, eighth ed. Elsevier Saunders, Philadelphia.

Wolfe, L., Manley, P., 2006. Disorders of erythrocyte metabolism including porphyria. In: Arceci, R., Hann, I., Smith, O. (Eds.), Pediatric Hematology. Blackwell Publishing Ltd., Boston.

Young, N.S., Bressler, M., Casper, J.T., Liu, J., 1996. Biology and therapy of aplastic anemia. In: Schacter, G.P., McArthur, T.R. (Eds.), Hematology. American Society of Hematology, Washington, D.C.

10

Extracorpuscular Hemolytic Anemia

Lawrence C. Wolfe

The causes of hemolytic anemia due to extracorpuscular defects are listed in Table 10.1; they may be immune or nonimmune.

IMMUNE HEMOLYTIC ANEMIA

Immune hemolytic anemia can be either isoimmune or autoimmune. Isoimmune hemolytic anemia results from a mismatched blood transfusion or from hemolytic disease of the newborn. In autoimmune hemolytic anemia (AIHA), shortened red cell survival is caused by the action of immunoglobulins with or without the participation of complement on the red cell membrane. The red cell autoantibodies may be of the warm type, the cold type, mixed, with both types present, or the cold Donath—Landsteiner type. The presence of antibody or complement on the red cell surface is measured by the direct antiglobulin test (DAT; previously known as the Coombs test).

Complement participation is usually confined to the immunoglobulin M (IgM) type of antibody; only rarely is it associated with IgG (see Paroxysmal Cold Hemoglobinuria, below). AIHA may be idiopathic or secondary to a number of conditions listed in Table 10.1.

Warm AIHA

Antibodies of the IgG class are most commonly responsible for AIHA in children. The antigen to which the IgG antibody is directed is one of the Rh erythrocyte antigens in more than 70% of cases. This antibody usually has its maximal activity at 37°C and the resultant hemolysis is called warm antibody-induced hemolytic anemia.

Rarely, warm reacting IgA and IgM antibodies may be responsible for hemolytic anemia. As in all patients with AIHA, erythrocyte survival is generally proportional to the amount of antibody on the erythrocyte surface, although rarely hemolysis can occur in patients with too few antibodies on the surface of the red cell to cause a positive DAT (DAT-negative hemolytic anemia).

Clinical Features
- Severe, life-threatening condition.
- Sudden onset of pallor, jaundice, dark urine.
- Splenomegaly.

Laboratory Findings
- Hemoglobin level: very low in fulminant disease or normal in indolent disease.
- Reticulocytosis: common although often the reticulocytes are destroyed by the antibody as well and reticulocytopenia may occur.
- Mean corpuscular hemoglobin concentration may be elevated.
- Smear: prominent spherocytes, polychromasia, macrocytes, autoagglutination (IgM), nucleated red blood cells, erythrophagocytosis.

Lanzkowsky's Manual of Pediatric Hematology and Oncology.
DOI: http://dx.doi.org/10.1016/B978-0-12-801368-7.00010-7

TABLE 10.1   Causes of Hemolytic Anemia Due to Extracorpuscular Defects

1. **Immune**
   a. Isoimmune
      i. Hemolytic disease of the newborn
      ii. Incompatible blood transfusion
   b. Autoimmune: IgG only; complement only; mixed IgG and complement, other antibody mediated mechanisms
      i. Idiopathic
         * Warm antibody
         * Cold antibody
         * Cold−warm hemolysis (Donath−Landsteiner antibody)
      ii. Secondary
         * Infection, viral: infectious mononucleosis: Epstein−Barr virus (EBV), cytomegalovirus (CMV), hepatitis, herpes simplex, measles, varicella, influenza A, coxsackie virus B, human immunodeficiency virus (HIV); bacterial: streptococcal, typhoid fever, *Escherichia coli* septicemia, *Mycoplasma pneumonia* (atypical pneumonia)
         * Drugs and chemicals: quinine, quinidine, phenacetin, *p*-aminosalicylic acid, sodium cephalothin (Keflin), ceftriaxone, penicillin, tetracycline, rifampin, sulfonamides, chlorpromazine, pyradone, dipyrone, insulin; lead
         * Hematologic disorders: leukemias, lymphomas, lymphoproliferative syndrome, paroxysmal cold hemoglobinuria, paroxysmal nocturnal hemoglobinuria
         * Immunopathic disorders: systemic lupus erythematosus, periarteritis nodosa, scleroderma, dermatomyositis, rheumatoid arthritis, ulcerative colitis, agammaglobulinemia, Wiskott−Aldrich syndrome, dysgammaglobulinemia, IgA deficiency, thyroid disorders, giant cell hepatitis, Evans syndrome (immune-mediated anemia associated with immune thrombocytopenia), autoimmune lymphoproliferative syndrome (ALPS), common variable immune deficiency
         * Tumors: ovarian teratomata, dermoids, thymoma, carcinoma, lymphomas
2. **Nonimmune**
   a. Idiopathic
   b. Secondary
      i. Infection, viral: infectious mononucleosis, viral hepatitis; bacterial: streptococcal, *E. coli* septicemia, *Clostridium perfringens*, *Bartonella bacilliformis*; parasites: malaria, histoplasmosis
      ii. Drugs and chemicals: phenylhydrazine, vitamin K, benzene, nitrobenzene, sulfones, phenacetin, acetinalimide; lead
      iii. Hematologic disorders: leukemia, aplastic anemia, megaloblastic anemia, hypersplenism, pyknocytosis
      iv. Microangiopathic hemolytic anemia: thrombotic thrombocytopenic purpura, hemolytic-uremic syndrome, chronic relapsing schistocytic hemolytic anemia, burns, post cardiac surgery, march hemoglobinuria
      v. Miscellaneous: Wilson disease, erythropoietic porphyria, osteopetrosis, hypersplenism

- Neutropenia and thrombocytopenia (occasionally).
- Increased osmotic fragility and autohemolysis proportional to spherocytes.
- DAT positivity establishes the diagnosis of AIHA.
- Hyperbilirubinemia and increased serum lactate dehydrogenase.
- Haptoglobin level is markedly decreased.
- Hemoglobinuria usually only at first presentation, increased urinary urobilinogen.

Management

Because this is potentially a life-threatening condition, the following must be monitored carefully:

- Hemoglobin level (every 4 h).
- Reticulocyte count (daily).
- Splenic size (daily).
- Hemoglobinuria (daily).
- Haptoglobin level (weekly).
- DAT (weekly).

Close attention should always be paid to supportive care issues such as folic acid supplementation, hydration status, urine output, and cardiac status.

Treatment

Blood Transfusion

Transfusion should be avoided, when possible, because there will be no truly compatible blood available and the survival of transfused cells in this situation is quite limited and may fail to elevate the hemoglobin level significantly. Nonetheless, using the "least incompatible" blood may be required in properly selected situations in order to avoid cardiopulmonary compromise. The guidelines listed below should be followed:

- If a specific antibody is identified, a compatible donor may be selected. The antibody usually behaves as a panagglutinin and no totally compatible blood can be found.
- Washed packed red cells should be used from donors whose erythrocytes show the least agglutination in the patient's serum.
- The volume of transfused blood should only be of sufficient quantity to relieve any cardiopulmonary embarrassment from the anemia. Aliquots of 5 ml/kg are taken from a single unit and transfused at a rate of 2 ml/kg/h.
- The use of such incompletely matched blood is made relatively safe by biologic cross-matching, transfusing of relatively small volumes of blood at any given time and concomitant use of high-dose corticosteroid therapy.

Corticosteroid Therapy

- Prednisone 2–10 mg/kg/day orally or methylprednisone 4–8 mg/kg/day IV (both given over four doses each day) for 3 days followed by oral prednisone.
- High-dose corticosteroid therapy should be maintained for several days. Thereafter, corticosteroid therapy in the form of prednisone should be slowly tapered over a 3- to 4-week period.

The dose of prednisone should be tailored to maintain the hemoglobin at a reasonable level; when the hemoglobin stabilizes, the corticosteroids should be discontinued. The presence of a continued positive DAT does not preclude continuing to taper steroids as long as the hemoglobin is stable or rising and reticulocytosis continues to decrease or remain normal.

About 50% of patients respond within 4–7 days to corticosteroid therapy, but there are a number of patients who continue with profound hemolysis for the first week after initiation of therapy. For these patients and patients who appear dependent on steroids alternative treatment needs to be considered.

Intravenous Gammaglobulin

Doses in the range of 1–5 g/kg have been employed but response in children is poor. It should be considered in patients with severe hemolysis who are requiring transfusion and are having poor responses to transfusion.

Rituximab

In patients with severe disease not responding early on or in patients exhibiting steroid dependence, rituximab (a manufactured antibody targeting CD20) should be used in doses of 375 mg/m^2 once a week for 4 weeks. It has a very high rate of remission induction in AIHA in children. The short-term side effects include:

- Itching.
- Hives.
- Hypotension.
- Chest pain.

These can be prevented through premedication with acetaminophen 15 mg/kg, diphenhydramine 1 mg/kg and, if necessary, corticosteroids. Patients should be monitored carefully during each infusion. Although rituximab eliminates the B-cell compartment, there have not been increased rates of infection, and intravenous gammaglobulin has been administered to offset the loss of B-cell function.

Plasmapheresis

Plasmapheresis has been successful in slowing the rate of hemolysis in patients with severe IgG-induced immune hemolytic anemia. The effect is short-lived if antibody production is ongoing and success is limited. The limited efficacy of plasmapheresis is likely because:

- More than half of the IgG is extravascular.
- Most of the antibody is on the red cell surface with little remaining free in the plasma.

In IgG warm immune hemolytic anemia, plasmapheresis should always be combined with moderate immunosuppression (e.g., rituximab). This insures that both antibody production and antibody titer reduction are employed concomitantly.

Immunomodulating Agents

1. *Mycophenolate mofetil (MMF)*. This drug is showing promise in the treatment of a number of autoimmune diseases, including AIHA. It is also effective in Evans syndrome. MMF, as well as the antimetabolites listed below, often require 4–12 weeks for their effects to begin and are usually started as steroids are being weaned. Tacrolimus and sirolimus have also been used in these settings although there are only a few case reports available.
2. *Cyclosporine*. This drug has been frequently used in immune cytopenias, and in Evans syndrome for patients who are poorly responsive to steroids. Rituximab and MMF are preferred to cyclosporine due to the side-effect profile.
3. *Danazol*. There has been some success with Danazol (synthetic androgen). Danazol's early effect appears to be due to decreased expression of macrophage Fc-receptor activity. Its use is less desirable as it has a virilizing effect.

Antimetabolites

Azathioprine and 6-mercaptopurine: As with the immunomodulators they may take 4–12 weeks to provide a steroid-sparing effect.

Alkylating Agents

Cyclophosphamide: This drug is used in patients with severe disease that is unresponsive to steroids, rituximab, or immunomodulators. It is rarely used.

Mitotic Inhibitors

Vincristine and vinblastine: These drugs are rarely used, but when given, are used as a bridge to suppress hemolysis while waiting for an immunomodulator or cytotoxic agent to take effect.

Splenectomy

Splenectomy is indicated if the hemolytic process is brisk despite the use of high-dose corticosteroid therapy, rituximab, and transfusions and the patient cannot maintain a reasonable hemoglobin level safely, or if chronic hemolysis develops.

Splenectomy is beneficial in 60–75% of patients. Whenever possible, children should be older than 5 years of age and the disease should be present for at least 6–12 months with no significant response to medical treatment prior to undertaking splenectomy. Presplenectomy immunization should be instituted.

Giant Cell Hepatitis and DAT-Positive AIHA

This is a specific rare entity of unknown etiology, although an autoimmune component has been suggested because of the association of DAT-positive AIHA and response to immunosuppression.

Clinical Findings

- Age: 6–24 months, occasionally older age.
- Fever.
- Pallor.
- Jaundice (progressing to cirrhosis and liver failure).
- Firm hepatomegaly and splenomegaly.
- Associated convulsions.

Prognosis

Poor.

Laboratory Findings

- DAT: mixed (IgG and complement); no evidence of other autoimmunity.
- Hemolytic anemia.
- Liver function abnormality: high direct bilirubin, transaminase, and serum globulin values; prolonged prothrombin time.
- Liver histology: marked lobular fibrosis, extensive necrosis with central-portal bridging and giant cell transformation.

Treatment

Corticosteroids in combination with immunosuppressive agents are the primary therapy as steroids alone cannot typically control the hemolysis. Cytoxan, rituximab, cyclosporine, splenectomy, and azathioprine have all been used. Survival has been achieved with more intensive immunosuppression.

Cold AIHA

IgM antibody-associated hemolysis occurs less often in children than in adults. Most IgM autoantibodies causing immune hemolytic anemia are cold agglutinins, and cold hemagglutinin disease is almost always caused by an IgM antibody. The destruction of red blood cells is triggered by cold exposure.

Cold hemagglutinin disease typically occurs during *Mycoplasma pneumoniae* infection, although it may also occur with infectious mononucleosis, cytomegalovirus, mumps, and rarely other infections. Cold hemagglutinin disease or IgM-induced hemolysis is usually due to the production of antibodies targeting the I/i system (red cell surface antigens). Anti-I is characteristic of *M. pneumoniae*-associated hemolysis, and anti-I cold agglutinins are found in infectious mononucleosis. *Mycoplasma pneumoniae* adherence to the red cell membrane appears to be mediated by sialic-acid-containing receptors, associated with terminal galactose residues of the I antigen. The association of the infecting organism with the red blood cell may alter the antigenic structure of red blood cell membrane antigen, rendering it immunogenic. In children, the IgM antibody is usually polyclonal and immunologically heterogeneous.

Clinical Features

This disease may be idiopathic but is more frequently seen in conjunction with infections such as *M. pneumoniae* (atypical pneumonia) and less commonly with lymphoproliferative disorders. The features are:

- Hemoglobin is usually normal or mildly decreased and the reticulocyte count may be elevated.
- The blood smear may show agglutination and polychromatophilia.
- Spherocytosis is usually absent.
- The DAT is positive for complement (polyspecific and anti C3-agents) only and is negative for anti-IgG.
- Most blood banks do cold agglutinin testing only when the DAT is positive for complement.

Treatment

1. Control of the underlying disorder.
2. Transfusions may be necessary for patients with significant hemolysis who may be symptomatic. Identification of compatible blood may prove difficult and the blood bank may have to release a "least incompatible" unit of blood. Warming the blood to 37°C during administration by means of a heating coil or water bath is indicated to avoid further temperature activation of the antibody. Efficient in-line blood warmers (McGaw Water Bath; Fenwall Dry Heat Warmer) are designed to deliver blood at 37°C to the patient. Unmonitored or uncontrolled heating of blood is extremely dangerous and should not be attempted. Red cells heated too long are rapidly destroyed *in vivo* and can be lethal to the patient.
3. Warm the patient's room. Keeping a patient warm will help diminish hemolysis and peripheral agglutination.
4. Plasmapheresis is very efficient for the treatment of IgM disease as IgM is largely intravascular. Patients with severe hemolysis should undergo plasmapheresis.
5. Autotransfusion of red blood cells can be performed if the blood is obtained at 37°C, with the patient's arm warmed by hot pads. The warm unit can be separated quickly by centrifugation and the red cells returned to the patient through an efficient in-line blood warmer.
6. Drug therapy. If the anemia is severe, a drug trial is appropriate. Rituximab and cyclophosphamide have been used with plasmapheresis. Steroids are of marginal value in cold agglutinin disease.

Paroxysmal Cold Hemoglobinuria Due to Donath—Landsteiner Cold Hemolysin

This is an unusual IgG antibody with anti-P specificity and a cold thermal amplitude, originally described in cases of syphilis. This antibody, although uncommon, is most frequently found in young children with viral infections. Hemolysis is most commonly intravascular as a result of the unusual complement-activating efficiency of this IgG antibody.

Clinical Features

The most common clinical finding is a sudden bout of hemolysis with a drop in hemoglobin and hemoglobinuria. The hemoglobin drop is often serious enough to require transfusion (and sudden death from this disease has been reported). Children usually have a short-lived, explosive illness where the antibody is only produced for a short time. Although blood for transfusion will appear compatible, all red cells carry the P blood group specificity against which the antibody is directed.

Laboratory Findings

A positive complement test is present on antiglobulin testing and this should lead to testing for the Donath–Landsteiner antibody (the IgG cold binding antibody) in the absence of an obvious IgM cold agglutinin.

Treatment

Keeping a patient warm is the mainstay of treatment and warming blood in a blood warmer prior to transfusion is important. Patients with ongoing severe hemolysis may benefit from plasmapheresis despite the fact that the disease is caused by an IgG antibody. Steroid use should be reserved for refractory patients. Rituximab may be effective.

NONIMMUNE HEMOLYTIC ANEMIA

This group of conditions is due to extracorpuscular causes of hemolytic anemia in which the DAT is negative. The various causes are listed in Table 10.1. Conditions caused by various infections, drugs, and underlying hematologic disease respond to treatment of the underlying condition, as well as the necessary acute supportive care including red cell transfusions as needed.

Microangiopathic Hemolytic Anemia

Microangiopathic hemolytic anemia (MAHA) is a result of diverse causes that have in common a relatively uniform hematologic picture and in general a common pathogenesis. Table 10.2 lists the various causes of MAHA.

Diagnosis

The blood smear is characterized by the presence of burr erythrocytes, schistocytes, helmet cells, and microspherocytes. This occurs in association with evidence of hemolysis and there may be associated thrombocytopenia depending on the initial etiology of the microangiopathy (e.g., DIC and hemolytic-uremic syndrome, see Table 10.2). The severity of anemia varies and is often dependent on the nature of the original pathophysiology. Intravascular hemolysis occurs in all forms; plasma hemoglobin levels may be elevated, haptoglobin absent, hemosiderinuria present and urinary iron excretion increased in the more chronic forms. Treatment involves improving the underlying condition and transfusion as needed. Patients can become iron-deficient due to the ongoing hemoglobinuria.

Hypersplenism

Shortened red cell survival with excessive sequestration can be demonstrated in many patients with clinical splenomegaly. This can occur whether splenic enlargement is caused by infection or is secondary to such diseases as thalassemia, portal hypertension, or storage diseases. Typically, hypersplenism is accompanied by moderate neutropenia and thrombocytopenia with active erythropoiesis, myelopoiesis, and thrombopoiesis in the marrow. There may also be mild spherocytosis. The blood values return to normal following splenectomy.

Wilson Disease

Wilson disease is a rare inherited disease of copper metabolism that leads to copper deposition most prominently in the liver and central nervous system (CNS). It has an autosomal recessive inheritance pattern and usually presents with liver or CNS symptoms. Wilson disease rarely presents with anemia which is normochromic and normocytic without an intense reticulocytosis or indirect hyperbilirubinemia. Because of the lethal nature of the disease without treatment and the potential successful treatment if the disease is detected early, Wilson disease should be considered in any patient with unexplained hemolytic anemia that has no abnormal morphology and a negative DAT.

TABLE 10.2   Causes of Microangiopathic Hemolytic Anemia

Renal disease

    Hemolytic-uremic syndrome

    Renal vein thrombosis

    Renal transplantation rejection

    Radiation nephritis

    Chronic renal failure

Cardiac conditions

    Malignant hypertension

    Coarctation of aorta

    Severe valvular heart disease

    Subacute bacterial endocarditis of aortic valve

    Intracardiac prosthesis

Liver disease

    Severe hepatocellular disease

Infections

    Disseminated herpes infection

    Meningococcal septicemia

    Cerebral falciparum malaria

Hematologic

    Thrombotic thrombocytopenic purpura (hereditary or secondary) (see Chapter 12)

Miscellaneous

    Severe burns

    Giant hemangioma (Kasabach—Merritt syndrome)

    Disseminated intravascular coagulation of any causation; sometimes accompanied by consumption of circulating coagulation factors (consumption coagulopathy)

Further Reading and References

Dacie, J., 1992. The Haemolytic Anaemias: The Auto-Immune Haemolytic Anaemias, third ed. Churchill Livingstone, Edinburgh.

Gertz, M., 2007. Management of cold haemolytic syndrome. Br. J. Haematol. 138, 422–429.

Giulino, L., Bussel, J.B., Neufeld, E., 2007. Treatment with rituximab in benign and malignant hematologic disorders in children. J. Pediatr. 150, 338–344.

Glader, B., 2006. Autoimmune hemolytic anemia. In: Arceci, R., Hann, I., Smith, O. (Eds.), Pediatric Hematology. Blackwell Publishing Ltd, Boston.

Acquired haemolytic anemias. In: Lewis, S.M., Bain, B.J., Bates, I. (Eds.), Dacie and Lewis Practical Hematology. Elsevier, Ltd, Philadelphia.

Hoffman, P.C., 2009. Immune hemolytic anemia-selected topics. ASH Education Book, 80–86. http://dx.doi.org/10.1182/asheducation-2009.1.80.

Koppel, A., Lim, S., Osby, M., Garratty, G., Goldfinger, D., 2007. Rituximab as successful therapy in a patient with refractory paroxysmal cold hemoglobinuria. Transfusion 47 (10), 1902–1904.

Maggiore, G., Sciveres, M., Fabre, M., Gori, L., Pacifico, L., Resti, M., et al., 2011. Giant cell hepatitis with autoimmune hemolytic anemia in early childhood: long-term outcome in 16 children. J. Pediatr. 159 (1), 127–132.e1, http://dx.doi.org/10.1016/j.jpeds.2010.12.050. Epub February 24, 2011.

Sève, P., Philippe, P., Dufour, J.F., Broussolle, C., Michel, M., 2008. Autoimmune hemolytic anemia: classification and therapeutic approaches. Expert Rev. Hematol. 1 (2), 189–204.

Zanella, A., Barcellini, W., 2014. Treatment of autoimmune hemolytic anemias. Haematologica 99 (10), 1547–1554.

11

Hemoglobinopathies

Janet L. Kwiatkowski

SICKLE CELL DISEASE

Incidence

Sickle hemoglobin is the most common abnormal hemoglobin found in the United States (approximately 8% of the African-American population has sickle cell trait). The incidence of sickle cell disease (SCD) at birth is approximately one in 600 African-Americans. Fifty percent of the world's patients with SCD reside in Nigeria, India, and the Democratic Republic of the Congo.

Genetics

1. SCD is transmitted as an autosomal codominant trait.
2. Homozygotes (two abnormal genes, SS) do not synthesize hemoglobin A (HbA); beyond infancy, red cells contain >75% HbS.
3. Compound heterozygotes for HbS and HbC have a form of SCD.
4. Compound heterozygotes for HbS and beta thalassemia trait (beta^0 or beta$^+$) have a form of SCD (S-beta^0 thalassemia or S-beta + thalassemia, respectively).
5. Heterozygotes (AS), sickle cell trait, have red cells containing 35–45% HbS.
6. Sickle cell trait provides selective advantage against *Plasmodium falciparum* malaria (balanced polymorphism).
7. ∝-Thalassemia (frequency of one or two ∝ gene deletions is 35% in African-Americans) may be coinherited with sickle cell trait or disease. Individuals who have both ∝-thalassemia and SCD-SS tend to be less anemic than those who have SCD-SS alone. The coinheritance of SCD and ∝-thalassemia trait is associated with a reduction in risk of some complications, such as stroke, but with no effect on the frequency or severity of vaso-occlusive pain.

Pathophysiology

HbS arises as a result of a point mutation (A—T) in the sixth codon of the β-globin gene on chromosome 11, which causes a single amino acid substitution (glutamic acid to valine at position 6 of the β-globin chain). HbS is more positively charged than HbA and hence has a different electrophoretic mobility. Deoxygenated HbS polymerizes, leading to cellular alterations that distort the red cell into a rigid, sickled form. Vaso-occlusion with ischemia—reperfusion injury is the central event, but the underlying pathophysiology is complex, involving a number of factors including hemolysis-associated reduction in nitric oxide bioavailability, chronic inflammation, oxidative stress, altered red cell adhesive properties, activated white blood cells and platelets, altered hemostasis including platelet activation, thrombin activation, lowered levels of anticoagulants, impaired fibrinolysis, and increased viscosity. HbF affects HbS by decreasing polymer content in cells. The effect of HbF on HbS may have direct and indirect effects on other RBC characteristics (i.e., percentage of HbF affects the RBC adhesive properties in patients with SCD). Elevated HbF concentration is associated with a reduction in certain complications of SCD.

166

TABLE 11.1 Common Location of Vaso-Occlusive Pain

Site	Manifestations
Hands/feet (dactylitis)	Most common in children younger than 3 years old. Painful swelling of the hands and/or feet. Often can be managed with acetaminophen or nonsteroidal anti-inflammatory medication. Unusual in older children because as the child ages, the sites of hematopoiesis move from peripheral locations such as the fingers and toes to more central locations such as arms, legs, ribs, and sternum
Bone	More common after age 3 years. Often involves long bones, sternum, ribs, spine, and pelvis. May involve more than one site during a single episode. Swelling and erythema may be present. May be difficult to differentiate from osteomyelitis because clinical symptoms, laboratory studies, and radiological imaging may be similar. Features that may aid in distinguishing these two diagnoses are shown in Table 11.2
Abdomen	Caused by microvascular occlusion of mesenteric blood supply and infarction in the liver, spleen, or lymph nodes that results in capsular stretching. Symptoms of abdominal pain and distension mimic acute abdomen

TABLE 11.2 Differentiation Between Bone Infarction and Osteomyelitis

Features	Favoring osteomyelitis	Favoring vaso-occlusion
History	No previous history	Preceding painful crisis
Pain, tenderness, erythema, swelling	Single site	Multiple sites
Fever	Present	Present
Leukocytosis	Elevated band count ($>1000/mm^3$)	Present
Erythrocyte sedimentation rate	Elevated	Normal to low
Magnetic resonance imaging	Abnormal	Abnormal
Bone scan[a]	Abnormal ^{99m}Tc-diphosphonate	Abnormal ^{99m}Tc-diphosphonate
	Normal ^{99m}Tc-colloid marrow uptake	Decreased ^{99m}Tc-colloid marrow uptake
Blood culture	Positive (*Salmonella*, *Staphylococcus*)	Negative
Recovery	Only with appropriate antibiotic therapy	Spontaneous

[a]*Obtained within 3 days of symptom onset.*

Clinical Features

Hematology

1. Anemia—moderate to severe in SS and S-β^0-thalassemia, milder with SC or Sβ^+-thalassemia.
2. Mean corpuscular volume (MCV) is normal with SS; microcytic with concomitant $\propto$-thalassemia or with S-β-thalassemia.
3. Reticulocytosis.
4. Neutrophilia common.
5. Platelet count often increased.
6. Blood smear—sickle cells (not infants or others with high HbF) increased polychromasia, nucleated red cells, and target cells (Howell–Jolly bodies may indicate hyposplenism).
7. Erythrocyte sedimentation rate—usually low despite inflammation (sickle cells fail to form rouleaux).
8. Hemoglobin electrophoresis—HbS migrates slower than hemoglobin A. Newborn screening shows FS, FSC, or FSA pattern depending on genotype.

Acute Complications

1. *Vaso-occlusive pain event (VOE)*
   a. Episodic microvascular occlusion at one or more sites resulting in pain and inflammation. Common locations and manifestations of VOE are shown in Table 11.1. Symptoms of fever, erythema, swelling, and focal bone pain may accompany VOE, making it difficult to distinguish from osteomyelitis. Unfortunately, no test clearly distinguishes these two entities but Table 11.2 describes clinical, laboratory, and radiographic features that may be helpful in differentiating bone infarction from osteomyelitis.

b. The average rate of VOE prompting medical evaluation in SCD-SS is 0.8 events/year. Approximately 40% of patients never seek medical attention for pain, while about 5% of patients account for a third of all VOE. These numbers underestimate the true incidence of VOE because many episodes are managed at home.

c. Risk factors for pain include high baseline hemoglobin level, low hemoglobin F levels, nocturnal hypoxemia, and asthma.

d. Evidence-based guidelines for the management of VOE are lacking. The typical approach to pain management involves a stepwise progression, beginning with a nonsteroidal anti-inflammatory pain medication for mild to moderate pain and adding an opioid pain medication rapidly if pain is not resolving or for moderate to severe pain. The management of vaso-occlusive pain is shown in Table 11.3, and a guideline for dosing of commonly utilized pain medications is provided in Table 11.4. Higher opioid dosing will be required for patients who have developed tolerance. Rapid alleviation of VOE pain is a basic tenet of SCD management.

e. Other drug therapies are under investigation for treatment of acute VOE, but are not yet clinically available. A number of treatments have been tested or are currently being tested in early Phase 1 or 2 trials for patients with SCD and acute VOE. These include:

   i. Drugs and supplements that cause vasodilatation, including arginine and nitrous oxide

   ii. Drugs that inhibit platelet aggregation

   iii. Drugs that lower blood viscosity

   iv. Drugs that affect red blood cell adhesion

f. Prevention of pain

   i. Hydroxyurea (HU)

   ii. Prophylactic red blood cell transfusions

2. *Acute chest syndrome (ACS)*

   a. ACS is the most common cause of death and the second most common cause of hospitalization in children with SCD. It is generally defined as the development of a new pulmonary infiltrate accompanied by symptoms including fever, chest pain, tachypnea, cough, hypoxemia, and wheezing.

TABLE 11.3   Management of Vaso-Occlusive Pain Episodes

At home
    Ibuprofen and/or acetaminophen
    If continued pain, add oral opiod
        Mild pain—codeine
        Moderate pain—oxycodone, hydrocodone, morphine
    Supportive measures
        Heating pad
        Fluids
        Stool softeners and/or laxative if taking opiods for more than 1–2 days
        If pain persists or worsens, patient should be evaluated and treated in an acute care setting
In Emergency Department/Acute Care Unit
    Rapid triage and administration of pain medication
    If no pain medications were taken prior to arrival and pain not severe, may use ibuprofen and oral opioid
    If prior pain medications were taken or pain is severe
        Ketorolac tromethamine (NSAID)
        IV opioid
    Fluids to maintain euvolemia. IV normal saline bolus should only be used if evidence of decreased oral intake/dehydration
Inpatient
    Continue nonsteroidal anti-inflammatory agent
    Continue IV opioids. Should be given as scheduled medication rather than "as needed"
    Consider patient-controlled analgesia pump if pain not adequately controlled
    Ongoing evaluation of adequacy of pain control is essential—utilize pain scales
    Supportive care
        Fluids (oral +IV) to maintain euvolemia
        Incentive spirometry
        Heating pad—must be used carefully to avoid burns
        Bowel regimen to prevent/treat constipation secondary to opioid use
            Stool softeners (e.g. docusate)
            Laxative (e.g., senna)
        Antihistamines (e.g., diphenhydramine, hydroxyzine) for pruritis
    Transition to oral nonsteroidal and oral opiod as pain level improves. Addition of long-acting opioid (e.g., sustained-release morphine)

TABLE 11.4 Dosages of Common Analgesics for Management of Sickle Cell Vaso-Occlusive Pain

Medication	Usual dose	Maximum dose	Route	Interval (h)
NONSTEROIDAL ANTI-INFLAMMATORY MEDICATIONS				
Ibuprofen	10 mg/kg	800 mg	PO	Q 6–8
Naproxen	5–7 mg/kg	500 mg	PO	Q 12
Ketorolac	0.5 mg/kg	30 mg	IV, IM	Q 6–8[a]
OPIOID PAIN MEDICATIONS				
Codeine	0.5–1 mg/kg	60 mg	PO	Q 4–6
Oxycodone				
<6 years	0.05–0.15 mg/kg	2.5 mg	PO	Q 4–6
6–12 years	0.05–0.2 mg/kg	5 mg	PO	Q 4–6
>12 years	0.05–0.2 mg/kg	10 mg	PO	Q 4–6
Hydromorphone				
<50 kg	0.03–0.08 mg/kg		PO	Q 3–4
≥50 kg	1–4 mg/dose	8 mg	PO	Q 3–4
<50 kg	0.01 mg/kg		IV, IM, SQ	Q 3–4
≥50 kg	0.01 mg/kg	4 mg	IV, IM, SQ	Q 3–4
Morphine (immediate release)				
<6 months	0.1–0.3 mg/kg		PO	Q 3–4
6 months-18 years	0.2–0.5 mg/kg		PO	Q 3–4
Adults	10–30 mg/dose		PO	Q 3–4
Morphine (controlled release)				
>30 kg	0.3–0.6 mg/kg	60 mg	PO	Q 8–12
Morphine				
<6 months	0.05–0.1 mg/kg		IV, IM, SQ	Q 3–4
≥6 months	0.1–0.2 mg/kg	15 mg	IV, IM, SQ	Q 3–4

[a]*Duration of therapy should not exceed 5 days.*
PO, orally; IV, intravenously; IM, intramuscularly; SQ, subcutaneously.

b. ACS is caused by infection, infarction, and/or fat embolization and iatrogenically by overhydration. About 50% of ACS events are associated with infections, including viruses, atypical bacteria including *Mycoplasma* and *Chlamydia*, and less frequently with *Streptococcus pneumoniae*. Parvovirus B19 infection can also result in ACS. In about half of cases, ACS develops during hospitalization, often for vaso-occlusive pain, where fat embolization, hypoventilation, and iatrogenic overhydration contribute to the pathophysiology.

c. The incidence of ACS in SCD-SS is about 24 events per 100 patients in children. The incidence in other sickle cell genotypes is lower (SS > Sβ^0-thalassemia > SC > Sβ^+-thalassemia), and concomitant α-thalassemia does not appear to affect ACS rates.

d. The risk of ACS is directly proportional to the hemoglobin level and white blood cell count; increased levels of cytokines and/or white cell adhesion to the endothelium may play a role. Rates of ACS are also higher in children with asthma. Higher hemoglobin F levels appear to be protective.

e. Laboratory findings:
   i. White blood cell count is often elevated.
   ii. Hemoglobin level often falls to 1.5 g/dl below baseline values.
   iii. Thrombocytosis may be present, and often follows an episode of ACS.

f. The management of ACS is described in Table 11.5.

TABLE 11.5   Management of Acute Chest Syndrome in Children

Evaluations
    Chest radiograph
    Complete blood count and reticulocyte count
    Blood type and screen
    Blood culture
    Viral studies
    Pulse oximetry
    Consider arterial blood gas in room air
Treatment
    Antibiotics: Broad-spectrum intravenous antibiotic such as cefuroxime plus an oral macrolide (erythromycin or azithromycin) to cover atypical bacteria
    Supplemental oxygen if hypoxemic
    Fluids: Intravenous + oral fluids should be kept at maintenance. Avoid overhydration.
    Pain control: Must be carefully monitored. Goal is to relieve pain to reduce splinting/poor aeration but avoid oversedation with hypoventilation
    Transfusion:
        Simple transfusion (10–15 cc/kg)—do not exceed post transfusion hemoglobin level of ~10 g/dl
        Exchange transfusion—if no improvement with simple transfusion or with severe hypoxemia/respiratory distress
    Bronchodilators—particularly if history of reactive airways disease or if wheezing present
    Steroids may be beneficial for severe acute chest syndrome or if reactive airways disease component. There is a risk of rebound pain with discontinuation of the steroids
    Incentive spirometry to reduce atelectasis
    Mechanical ventilation as needed
    Consider thoracentesis if significant pleural effusion

    g. Prevention of ACS:
        Patients with a history of recurrent ACS are candidates for preventative/curative therapies including;
        i. Hydroxyurea (HU).
        ii. Prophylactic red cell transfusions. Optimal target HbS level is not known, but usually a goal of 30–50% is used.
        iii. Stem cell transplantation.
3. *Overt stroke*
    a. Acute symptomatic stroke is usually infarctive in children, although hemorrhagic stroke may occur, particularly in older children and adults.
    b. The most common underlying lesion is intracranial arterial stenosis or occlusion, usually involving the large arteries of the circle of Willis, particularly the distal internal carotid artery (ICA) and the middle (MCA) and anterior cerebral arteries (ACAs).
    c. Chronic injury to the endothelium of vessels by sickled red blood cells results in changes in the intima with proliferation of fibroblasts and smooth muscle; the lumen is narrowed or completely obliterated. Small friable collateral blood vessels known as moyamoya may develop. Infarction of brain tissue occurs acutely as a result of *in situ* occlusion of the damaged vessel or distal embolization of a thrombus. Perfusional and/or oxygen delivery deficits related to changes in blood pressure or other factors also may contribute to infarction, particularly in watershed zones.
    d. Stroke is most common in homozygous SCD-SS. Prior to transcranial Doppler (TCD) ultrasound screening with transfusions for high-risk children, stroke prevalence in children with SCD-SS was estimated at 11%, with the highest incidence rates occurring in the first decade of life (1.02 per 100 patient-years in 2–5 year olds and 0.79 per 100 patient-years in 6–9 year olds).
    e. A number of clinical, laboratory, and radiological risk factors for stroke have been identified (Table 11.6).
    f. Symptoms of stroke include:
        i. Focal motor deficits (e.g., hemiparesis, gait dysfunction).
        ii. Speech defects.
        iii. Altered mental status.
        iv. Seizures.
        v. Headache.
    g. Gross neurological recovery occurs in approximately two-thirds of children, but neurocognitive deficits are common.

TABLE 11.6   Factors Associated with Increased Risk of Overt Infarctive Stroke in Sickle Cell Disease

Clinical
  History of transient ischemic attacks
  History of bacterial meningitis
  Sibling with SCD-SS and stroke
  Recent episode of acute chest syndrome (within 2 weeks)
  Frequent acute chest syndrome
  Systolic hypertension
  Nocturnal hypoxemia
Laboratory
  Low steady-state hemoglobin level
  No alpha gene deletion
  Certain HLA haplotypes
Radiological
  Abnormal transcranial Doppler ultrasound
  Silent infarct

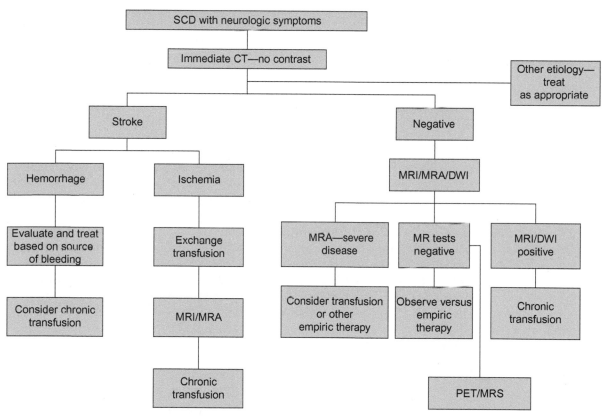

FIGURE 11.1   Management of the child with sickle cell disease and neurological symptoms. PET, positron emission tomography; MRA, magnetic resonance arterial angiography; DWI, diffusion-weighted imaging; MRS, magnetic resonance spectroscopy. *From The Management of Sickle Cell Disease, 4th Ed, 2002, National Heart, Lung and Blood Institute of the National Institutes of Health and The US Department of Health and Human Services, NIH Publication No. 02-2117, with permission.*

   h. In untreated patients, about 70% of patients experience a recurrence within 3 years. Outcome after recurrent stroke is worse.
   i. Any child with SCD who develops acute neurological symptoms requires immediate medical evaluation. A guideline for management is presented in Figure 11.1. The acute management involves prompt diagnosis and treatment.
   • *Diagnosis*:
     • Physical examination with detailed neurological examination. Treatment in the face of clinical suspicion must not await imaging confirmation. Head CT scan is useful for detecting intracranial hemorrhage and often more readily available than magnetic resonance imaging (MRI). May not be positive for acute infarction within the first 6 h.

- Brain MRI with diffusion-weighted imaging is more sensitive to early ischemic changes and may be abnormal within 1 h. Should be performed as soon as possible in a child with SCD presenting with acute neurological symptoms, but should not delay empiric treatment.
- Magnetic resonance arterial angiography (MRA)—demonstrates large-vessel disease.
- *Treatment*:
  - Transfusion. Exchange transfusion, either automated or manual, should be performed as soon as possible. The goal is to reduce the amount of HbS to less than 30% and to raise the hemoglobin level to approximately 10 g/dl. If exchange transfusion is not readily available, a simple transfusion to raise the hemoglobin level to no greater than 10 g/dl, although suboptimal, may be used. Exchange transfusion may be associated with a decreased risk of stroke recurrence.
  - Supportive therapy including avoiding hypotension and maintaining adequate oxygenation and euthermia should be initiated as adjunctive therapy.

j. Long-term management
  i. Prevention of recurrent stroke
   - A chronic red cell transfusion program should be instituted, with the goal of maintaining the pretransfusion HbS level at less than 30%. Transfusions must be continued indefinitely, due to the high risk of stroke recurrence after discontinuation of therapy. After a period of 3–4 years after the initial stroke, it may be possible to allow the pretransfusion HbS level to rise to less than 50% in low-risk patients, without increased risk of stroke recurrence. This approach is associated with decreased transfusional iron loading. The high erythroid drive in SCD permits a transfusion approach in which chronic exchange transfusion to a hemoglobin that does not suppress erythropoiesis allows for erythropoiesis that utilizes iron and still permits a HbS <30 resulting in iron balance rather than iron loading.
   - Hematopoietic stem cell transplantation (HSCT).
   - The uses of revascularization procedures such as encephalodurosynangiosis are controversial and believed by some to be beneficial in children with significant vasculopathy, particularly if symptomatic (transient ischemic attacks, recurrent stroke) although published data on use in SCD are limited.
   - Treatment with HU with a several month overlap period with transfusions (until the hematological effects of HU were evident), was associated with a stroke recurrence rate of 3.6 events per 100 patient-years, which is lower than the risk of stroke recurrence without treatment. Furthermore, the use of phlebotomy during treatment with HU successfully reduced iron stores. However, multicenter phase 3 trial of HU and phlebotomy compared with transfusions and chelation therapy was terminated due to a higher rate of recurrent stroke (10%) in the HU arm compared with continued transfusions (0%) without any difference in iron reduction between the two treatment arms.
   - Prophylactic aspirin may also be useful in children with progressive vasculopathy, but the risks of hemorrhage must be weighed against the potential benefit.
  ii. Rehabilitation
   - Physical and occupational therapy as needed.
   - Neuropsychological testing should be performed with educational interventions if indicated.

k. Primary stroke prevention
  i. Screening
   - TCD ultrasonography is a noninvasive study used to measure the blood flow velocity in the large intracranial vessels of the circle of Willis.
   - The highest time-averaged mean velocity (TAMMvel) in the ICA, its bifurcation, and the MCA are used to categorize studies into risk groups:
     - Normal (velocity <170 cm/sec), low risk.
     - Conditional (170–199 cm/sec), moderate risk.
     - Abnormal (≥200 cm/sec), high risk.
     - Inadequate—unable to obtain velocity in the ICA or MCA on either side, in the absence of a clearly abnormal value in another vessel. Inadequate TCD may be due to technique, skull thickness, or severely stenosed vessel.
   - Very low velocity (ICA/MCA velocity <70 cm/s) may indicate vessel stenosis and increased risk of stroke.
   - Elevated velocity in the ACA is associated with increased stroke risk. Treatment of children with isolated high ACA velocities has not been established. Brain MRI/A should be obtained. Chronic transfusion should be instituted for children with ACA velocity ≥200 cm/s, and any child if silent infarcts and/or cerebral blood vessel stenosis are present on MRI/A.

TABLE 11.7 Transcranial Doppler Ultrasonography Screening Protocol

Last TCD result (TAMMvel in ICA/MCA)	Screening interval
Normal (<170 cm/s)	Annual
Low conditional (170–184 cm/s)	3–6 months[a]
High conditional (185–199 cm/s)	6 weeks–3 months[a]
Abnormal (200–219 cm/s)	1 week
High abnormal (220 cm/s or higher)	No confirmation needed—recommend treatment

[a]*Use the shorter time interval for children <10 years old.*

- TCD screening is recommended for children with SCD-SS or SCD-Sβ⁰-thalassemia ages 2–16 years. Screening is performed annually, but more frequently if the prior study was not normal. An approach to screening is shown in Table 11.7. In addition, more frequent screening should also be considered if other known stroke risk factors are present (such as sibling with SCD-SS and stroke or abnormal TCD).
- Brain MRI/MRA should be obtained in children with abnormal TCD and should be considered for children with conditional TCD.
- Brain MRA is helpful to evaluate cerebral vasculature in children with repeatedly inadequate TCD or with very low velocity.

ii. Treatment
- Chronic transfusion to maintain the HbS level <30% reduces the risk of stroke by >90% in children with abnormal TCD.
- Discontinuation of transfusion therapy after at least 30 months of transfusion with normalization of TCD results is associated with a high risk of reversion to abnormal TCD and stroke. Thus, transfusions are continued indefinitely.
- Stem cell transplantation with an HLA-identical sibling donor should be considered.
- HU therapy is associated with a lowering of TCD velocities and is currently under study for primary stroke prevention.

4. *Priapism*
   a. Priapism is a sustained, painful erection of the penis. Priapism may be prolonged (lasts more than 3 h), or stuttering (lasts less than 3 h). Stuttering episodes often recur or may develop into a prolonged episode.
   b. Occurs in 30–45% of patients with SCD, most commonly in the SS type. The prevalence is likely underestimated due to underreporting by patients.
   c. Mean age at the first episode of priapism in patients with SCD is about 12–15 years; 75% have their first episode before age 20 years.
   d. Priapism often occurs in the early morning hours, when normal erections occur, and is probably related to nocturnal acidosis and dehydration. The normal slow blood flow pattern in the penis is similar to the blood flow in the spleen and renal medulla. Failure of detumescence is due to venous outflow obstruction or to prolonged smooth muscle relaxation, either singly or in combination.
   e. A history of priapism in childhood is associated with later sexual dysfunction, with 10–50% of adults with SCD and a history of priapism reporting impotence.
   f. Treatment
      i. At home, patients may try warm baths, oral analgesics, increased oral hydration, and pseudoephedrine.
      ii. Patients should be evaluated in an emergency room for episodes lasting over 2 h.
      iii. Initial treatment includes intravenous hydration and parenteral analgesia.
      iv. Episodes lasting ≥4 h are associated with an increased risk of irreversible ischemic injury and thus warrant more aggressive management. Urological consultation should be obtained. Treatment involves aspiration of the corpus cavernosum followed by irrigation with or without intracavernous administration of a dilute (1:1,000,000) epinephrine solution. Although published data in SCD are lacking, a dilute solution of phenylephrine, an alpha adrenergic agent, rather than epinephrine, has also been utilized in some centers.

    v. The role of transfusion for the management of priapism is controversial and the clinical response is variable. Furthermore, exchange transfusion for acute priapism has been associated with the development of acute neurological events.

    vi. Surgical shunting procedures (cavernosa spongiosum or cavernosaphenous vein) may be considered if the above treatments fail although shunt occlusion is a common complication.

  g. Prevention of priapism

    i. Pseudoephedrine, 30–60 mg orally at bedtime.

    ii. HU therapy has been employed, although this treatment has not been studied for this indication.

    iii. Leuprolide injections, a gonadotropin-releasing hormone analog that suppresses the hypothalamic–pituitary access, reducing testosterone production.

    iv. Phosphodiesterase type 5 inhibitors (excluding Sildenafil) may have some benefit, though further research is needed.

    v. Transfusion protocol for 6–12 months following an episode of priapism requiring irrigation and injection.

5. *Splenic sequestration*

  a. Highest prevalence between 5 and 24 months of age in SCD-SS (may occur at older ages in other sickling syndromes).

  b. May occur in association with fever or infection.

  c. Splenomegaly due to pooling of large amounts of blood in the spleen.

  d. Rapid onset of pallor and fatigue. Abdominal pain is often present.

  e. Hemoglobin level may drop precipitously, followed by hypovolemic shock and death.

  f. Reticulocytosis and nucleated red blood cells often present.

  g. Platelet and white blood cell count also usually fall from baseline.

  h. Treatment of splenic sequestration is shown in Table 11.8

6. *Transient pure red cell aplasia*

  a. Cessation of red cell production that may persist for 7–14 days with profound drop in hemoglobin (as low as 1 g/dl).

  b. Reticulocyte count and the number of nucleated red cells in the marrow sharply decrease; platelet and white blood cell counts are generally unaffected.

  c. May occur in several members of a family and can occur at any age.

  d. Almost invariably associated with parvovirus B19 infection.

  e. Terminates spontaneously usually after about 10 days (recovery occurs with reticulocytosis and nucleated red cells in the blood).

  f. Vaso-occlusive pain and/or splenic sequestration may occur in association with parvovirus B19/transient pure red cell aplasia.

  g. Treatment

    i. Close monitoring of complete blood count (CBC) and reticulocyte count.

    ii. Red cell transfusion to raise hemoglobin level to no greater than 9–10 g/dl.

    iii. Monitor siblings with SCD closely (CBC, reticulocyte count, parvovirus polymerase chain reaction (PCR), and/or titers).

TABLE 11.8   Management of Splenic Sequestration

Treatment of acute splenic sequestration episode
    Monitor cardiovascular status, spleen size, and hemoglobin level closely
    Normal saline bolus of 10–20 cc/kg
    Red cell transfusion. Administer in small aliquots because transfusion often results in reduction in spleen size with "autotransfusion" of previously trapped red cells. Rapid infusion used for cardiovascular instability
    Pain management
Prevention of recurrent splenic sequestration
    Splenectomy if history of one major or two minor acute splenic sequestration episodes
    For children <2 years old, chronic transfusion therapy can be considered to postpone splenectomy

Chronic Complications and End-Organ Damage

1. *Central nervous system*
   a. Silent stroke
      i. Defined as one or more focal T2-weighted signal hyperintensities demonstrated on brain MRI, in the absence of a focal neurological deficit corresponding to the anatomical distribution of the brain lesion.
      ii. Present in approximately 35% of children with SCD-SS and occurs less commonly in other sickle cell genotypes.
      iii. Associated with neuropsychologic deficits and impaired school performance.
      iv. Silent infarcts may progress in size and number over time and are associated with an increased risk of overt stroke.
      a. Treatment
         i. Management of children with silent infarcts includes neuropsychological testing and monitoring of academic performance.
         ii. Chronic transfusion therapy to maintain the hemoglobin level above 9 g/dl and the HbS below 30% for children with silent cerebral infarcts is associated with a reduction in infarct recurrence. New or enlarged silent infarcts or overt stroke occurred in 6% of children receiving transfusions compared with 14% of children in the observation group.
         iii. HU has not been studied for this indication.
2. *Cardiovascular system*
   a. Abnormal cardiac findings are present in most patients as a result of chronic anemia and the compensatory increased cardiac output.
   b. Cardiomegaly is found in most patients and left ventricular hypertrophy occurs in about 50%.
   c. Prolonged QTc >440 msec occurs in 9–38% of children with SCD, most commonly the SS type. Prolonged QTc is associated with increased risk of mortality in adults with SCD.
   d. A moderate-intensity systolic flow murmur is often present.
   e. Echocardiogram may show left and right ventricular dilatation, increased stroke volume, and abnormal septal motion.
   f. Pulmonary hypertension
      i. Defined as a resting pulmonary artery systolic pressure equal to or greater than 25 mmHg. Right heart catheterization is required to make a definitive diagnosis. Noninvasive echocardiography often is used to screen for the possible presence of pulmonary hypertension. Tricuspid regurgitant jet velocity (TRV) of at least 2.5 m/s is an indicator of possible pulmonary hypertension.
      ii. Prevalence of pulmonary hypertension documented by right heart catheterization in adults is estimated at 6–11%, with 10% of these adults having moderate to severe pulmonary hypertension (pressure above 45 mmHg). An elevated TRV is found in approximately 30% of adults with SCD. The prevalence of pulmonary hypertension in children appears to be about 11% and is most common with the SS genotype. Diagnosis of pulmonary hypertension by TRV alone has been questioned. Children with elevated TRV should be managed along with a cardiologist.
      iii. In adults, pulmonary hypertension by right heart catheterization, elevated TRV, and increased serum N-terminal pro-brain natriuretic peptide are independent risk factors for mortality; the significance of these findings in children is unclear.
      iv. A central role for hemolysis and altered NO bioavailability has been postulated.
      v. The optimal treatment is unknown, but HU or red cell transfusions have been used. Treatment with sildenafil, an agent used to treat pulmonary hypertension in other patient groups, is associated with an increased risk of vaso-occlusive pain episodes.
3. *Pulmonary*
   a. Reduced PaO$_2$.
   b. Reduced PaO$_2$ saturation. Pulse oximetry may not correlate with PaO$_2$ in steady state. Changes in pulse oximetry are useful for monitoring children with ACS. Daytime and/or nocturnal hypoxemia may be present.
   c. Pulmonary fibrosis—Chronic lung disease: Early identification of progressive lung disease using pulmonary function testing is imperative. Aggressive treatment has little benefit in end-stage lung disease and this should be avoided by prophylactic transfusions.
   d. Asthma—Prevalence appears to be higher than in the general population in children with SCD. Asthma is associated with complications of SCD including pain, ACS, stroke, and pulmonary hypertension. Aggressive management is warranted.

4. *Kidney*
   a. Increased renal flow.
   b. Increased glomerular filtration rate.
   c. Enlargement of kidneys; distortion of collecting system.
   d. Hyposthenuria (urine concentration defect): Hyposthenuria is the first manifestation of sickle cell-induced obliteration of the vasa recta of the renal medulla. Edema in the medullary vasculature is followed by focal scarring, interstitial fibrosis, and destruction of the countercurrent mechanism. Hyposthenuria results in a concentration capacity of more than 400–450 mOsmol/kg and an obligatory urinary output as high as 2000 ml/m^2/day, causing the patient to be particularly susceptible to dehydration. The increased urine output is associated with nocturia, often manifesting as enuresis.

      Treatment of nocturnal enuresis includes behavioral modifications such as waking to void and the use of a bedwetting alarm. 1-deamino-8-D-arginine vasopressin at bedtime has also been used. The intranasal form is no longer recommended for childhood nocturnal enuresis. An oral dose of 0.2 mg has been used in children over age 6 years.
   e. Hematuria: Papillary necrosis is usually the underlying anatomic defect. Treatment of papillary necrosis is IV hydration and rest. Frank hematuria usually resolves, although bleeding can be prolonged. Antifibrinolytic agents such as epsilon-aminocaproic acid have been used for recalcitrant bleeding with variable success. However, great caution must be taken when using this drug because of the risk of thrombosis and urinary obstruction. Evaluation for other causes of hematuria (e.g., renal medullary carcinoma) is indicated for the first episode of hematuria.
   f. Renal tubular acidification defect.
   g. Increased urinary sodium loss (may result in hyponatremia).
   h. Hyporeninemic hypoaldosteronism and impaired potassium excretion are results of renal vasodilating prostaglandin increase in patients with SCD.
   i. Proteinuria: Persistent increasing proteinuria is an indication of glomerular insufficiency, perihilar focal segmental sclerosis, and renal failure. Intraglomerular hypertension with sustained elevations of pressure and flow is the prime etiology of the hemodynamic changes and subsequent proteinuria. If proteinuria persists for more than 4–8 weeks, angiotensin-converting enzyme inhibitors (i.e., enalapril) are recommended.
   j. Nephrotic syndrome: A 24-h urine protein of more than 2 g/day, edema, hypoalbuminemia, and hyperlipidemia may indicate progressive renal insufficiency. The efficacy of steroid therapy in the management of nephrotic syndrome in SCD is not clear. Carefully monitored use of diuretics is indicated to control edema.
   k. Chronic renal failure: Uremia. Renal failure can be managed with peritoneal dialysis, hemodialysis, and transplantation.
5. *Liver and biliary system*
   a. Chronic hepatomegaly.
   b. Liver function tests: Increased serum aspartate transaminase and serum alanine transaminase.
   c. Cholelithiasis
      i. Chronic hemolysis with increased bilirubin turnover causes pigmented stones.
      ii. Occurs as early as 2 years old and affects at least 30% by age 18 years.
      iii. Sonographic examinations of the gallbladder should be performed in children with symptoms.
      iv. The treatment for symptomatic cholelithiasis is laparoscopic cholecystectomy. The role of screening and treatment of asymptomatic patients is unclear.
   d. Transfusion-related hepatitis. Hepatitis C is more common in older patients who received red cell transfusions prior to the availability of screening of blood products.
   e. Intrahepatic crisis: Intrahepatic sickling can result in massive hyperbilirubinemia, elevated liver enzyme values, and a painful syndrome mimicking acute cholecystitis or viral hepatitis. Progression to multiorgan system failure may occur. Early exchange transfusion is indicated.
   f. Hepatic necrosis, portal fibrosis, regenerative nodules, and cirrhosis are common postmortem findings that may be a consequence of recurrent vascular obstruction and repair.
   g. Transfusional iron overload, secondary to repeated intermittent or chronic transfusions may cause hepatic fibrosis.

6. *Bones*

Skeletal changes in SCD are common because of expansion of the marrow cavity, bone infarcts, or both.

a. Avascular necrosis (AVN): The most common cause of AVN of the femoral head is SCD. The incidence is much higher with coexistent α-thalassemia in patients who have frequent painful events and in those with the highest hematocrits. The pathophysiology is sludging in marrow sinusoids, marrow necrosis, healing with increased intramedullary pressure, bone resorption, and eventually collapse. About 50% of patients are asymptomatic. Symptomatic patients have significant chronic pain and limited joint mobility. The diagnosis is made radiographically and shows:

  i. Subepiphyseal lucency and widened joint space.

  ii. Flattening or fragmentation and scarring of the epiphysis.

  iii. On MRI, AVN of femoral head can be detected before deformities are apparent on radiograph.

- Treatment:

Therapy for AVN is largely supportive, with bed rest, nonsteroidal anti-inflammatory drugs (NSAIDs), and limitation of movement during the acute painful episode. Transfusion therapy and HU do not seem to delay progression of AVN. Physical therapy is helpful and may reduce the risk of progression. Core decompression of the affected hip has been reported to reduce pain and stop progression of the disease. In this procedure, avascularized bone is removed to decompress the area with the potential for subsequent new bone formation. This procedure seems to be beneficial only in the early stages of AVN and before loss of the integrity of the femoral head. AVN of the hip may have its onset in childhood, so thorough musculoskeletal examination with concentration on the hips should be performed at least yearly in children with SCD. This ensures that AVN is detected early when it is in its most treatable form. Total hip replacement may be the only option for severely compromised patients; 30% of replaced hips require surgical revision within 4.5 years, and more than 60% of patients continue to have pain and limited mobility postoperatively. AVN of the humeral head is less common. Patients are less symptomatic, and arthroplasty is exceedingly rare.

b. Widening of medullary cavity and cortical thinning: Hair-on-end appearance of skull on radiograph.

c. Fish-mouth vertebra sign on radiograph.

7. *Eyes*

a. Retinopathy: Sickle retinopathy is common in all forms of SCD, but particularly in those patients with SCD, type SC.

  i. Nonproliferative: Occlusion of small blood vessels of the eye detected on dilated ophthalmological exam and usually not associated with defects in visual acuity.

  ii. Proliferative: Occlusion of small blood vessels in the peripheral retina may be followed by enlargement of existing capillaries or development of new vessels. Clusters of neovascular tissue "sea fans" grow into vitreous and along the surface of the retina. Sea fans may cause vitreous hemorrhage, which results in transient or prolonged loss of vision. Small hemorrhages resorb, but repeated leaks cause formation of fibrous strands. Shrinkage of these strands can cause retinal detachment.

- Treatment:
  - Nonproliferative: Treatment not usually needed.
  - Proliferative: Neovascularization may not progress or may even regress spontaneously. Indications for treatment include bilateral proliferative disease, rapid growth of neovascularization, and large elevated neovascular fronds. Laser photocoagulation and other methods are used to induce regression of neovascularization. With proper screening and new methods such as laser surgery, most of the complications of retinopathy can be avoided. Annual ophthalmologic examinations including inspection of the retina are indicated for children from the age of 5 years for children with SCD-SC and 8 years for children with SCD-SS.

b. Angioid streaks: These are pigmented striae in the fundus caused by abnormalities in the Baruch membrane due to iron or calcium deposits or both. They usually produce no problems for the patient, but occasionally they can lead to neovascularization that can bleed into the macula and decrease vision.

c. Hyphema: Blood in the anterior chamber (hyphema) rarely occurs secondary to sickling in the aqueous humor, because of its low pH and pO_2. Traumatic hyphema may occur as in any individual. Anterior chamber paracentesis should be performed if pressure is increased.

    d. Conjunctivae: Comma-shaped blood vessels, seemingly disconnected from other vasculature, can be seen in the bulbar conjunctiva of patients with SCD and variants (SS > SC > Sβ-thalassemia). These produce no clinical disability. Their frequency may be related to the number of irreversibly sickled cells in the blood. This abnormality can be identified by using the +40 lens of an ophthalmoscope.

8. *Ears*

    Up to 12% of patients have high-frequency sensorineural hearing loss. The pathophysiology may involve sickling in the cochlear vasculature with destruction of hair cells.

9. *Adenotonsillar hypertrophy*

    Adenotonsillar hypertrophy giving rise to upper airway obstruction can become a problem from the age of 18 months. The marked hypertrophy is postulated to be compensation for the loss of lymphoid tissue in the spleen. It occurs in at least 18% of patients. In severe cases, this can cause hypoxemia at night with consequent sickling. Early tonsillectomy and adenoidectomy may be indicated in these patients.

10. *Skin*

    Cutaneous ulcers of the legs occur over the external or internal malleoli. Leg ulcers occur less commonly in children, and rarely before age 10 years. Ulcers are most common in homozygous SCD. Ulceration may result from increased venous pressure in the legs caused by the expanded blood volume in the hypertrophied bone marrow.

    Treatment:

    a. Rest; elevation of the leg.
    b. Protection of the ulcer by the application of a soft sponge–rubber doughnut.
    c. Debridement and scrupulous hygiene.
    d. Low-pressure elastic bandage and above-the-knee elastic stockings to improve venous circulation.
    e. Transfusion therapy for 3–6 months course if ulcers persist despite optimal care.
    f. Antibiotic therapy if acutely infected (typical organisms are *Staphylococcus*, *Streptococcus*, and *Pseudomonas* species).
    g. Oral administration of zinc sulfate (220 mg three times a day) may promote healing of leg ulcers.
    h. Split-thickness skin grafts.

11. *Growth and development*

    a. Birth weight is normal. However, by 2–6 years of age, the height and weight are significantly delayed. The weight is more affected than the height, and patients with SCD-SS and Sβ^0-thalassemia experience more delay in growth than patients with SCD-SC and Sβ$^+$-thalassemia. In general, by the end of adolescence, patients with SCD have caught up with controls in height but not weight. The poor weight gain is likely to represent increased caloric requirements in anemic patients with increased bone marrow activity and cardiovascular compensation. Zinc deficiency may be a cause of poor growth. In these patients, zinc supplementation (dose of 220 mg three times a day) at about 10 years of age should be administered. Growth hormone levels and growth hormone stimulation studies appear to be normal in most children who have impaired growth.
    b. Delayed sexual maturation: Tanner 5 is not achieved until the median ages of 17.3 and 17.6 years for girls and boys, respectively. In males, decreased fertility with abnormal sperm motility, morphology, and numbers is prominent. Zinc sulfate 220 mg three times a day may be effective for sexual maturity in these patients; females are more responsive than males.

12. *Functional hyposplenism*

    a. By 6 months of age, mild splenomegaly may be apparent and persists during early childhood, after which the spleen undergoes progressive fibrosis (autosplenectomy).
    b. Functional reduction of splenic activity occurs in early life. This is the consequence of altered intrasplenic circulation caused by intrasplenic sickling. It can be temporarily reversed by transfusion of normal red cells. Children with functional hyposplenia are 300–600 times more likely to develop overwhelming pneumococcal and *Haemophilus influenzae* sepsis and meningitis than are normal children; other organisms involved are Gram-negative enteric organisms and *Salmonella*. The period of greatest risk of death from severe infection occurs during the first 5 years of life.
    c. Functional hyposplenism may be demonstrated by the following:
      i. Presence of Howell–Jolly bodies on blood smear.
      ii. ^{99m}Tc-gelatin sulfur colloid spleen scan—no uptake of the radioactive colloid by enlarged spleen.
      iii. Pitted red blood cell count >3.5%.

Diagnosis

1. **In utero:** SCD can be diagnosed accurately in utero by mutation analysis of DNA prepared from chorionic villus biopsy or fetal fibroblasts (obtained by amniocentesis). With the advent of PCR amplification of specific DNA sequences, sufficient DNA can be obtained from a very small number of fetal cells, thereby eliminating the necessity of culturing fetal fibroblasts from amniotic fluid. These techniques should be employed before 10 weeks' gestation.

2. **During the newborn period:** The diagnosis of SCD can be established by electrophoresis using:
   a. Isoelectric focusing (most commonly used in screening programs).
   b. High-performance liquid chromatography.
   c. Citrate agar with a pH of 6.2, a system that provides distinct separation of HbS, HbA, and HbF.
   d. DNA-based mutation analysis.

   These tests are commonly performed on a dried blood specimen blotted on filter paper (Guthrie cards) used in newborn screening programs.

3. **In older children:** Table 11.9 lists the diagnosis and differential diagnosis of various sickle cell syndromes.

TABLE 11.9 Differential Diagnosis in Sickle Cell Syndromes

Syndrome[a]	Clinical severity	Splenomegaly	Mean hemoglobin (g/dl)	Mean hematocrit (%)	Mean corpuscular volume (fl)	Reticulocytes (%)	Red cell morphology	Electrophoresis
AS	Asymptomatic	(−)	Normal	Normal	Normal	Normal	Few target cells	35−45% S; 55−60% A; F[b]
SS	Severe	YC(+) OC(−)	7.5	22	85	5−30	Many target cells, ISCs (4+) and NRBCs	80−96% S; 2−20% F[b]
SC	Mild/ moderate	(+)	11	33	80	2−6	Many target cells, few ISCs (1+)	50−55% S; 45−50% C; F[b]
S/β-thalassemia	Moderate/ severe	(+)	8.5	28	65	3−20	Marked hypochromia and microcytosis; many target cells, ISCs (3+) and NRBCs	50−85% S; 2−30% F[b]; >3.5% A2
S/β+-thalassemia	Mild/ moderate	(+)	10	32	72	2−6	Mild microcytosis and hypochromia; many target cells few ISCs (1+)	50−80% S; 10−30% A; 0−20% F[b]; <3.5% A2
SS/α-thalassemia-1	Mild/ moderate	(+)	10	27	70	5−10	Mild hypochromia and microcytosis; few ISCs (2+)	80−100% S; 0−20% F[b]
S/HPFH	Asymptomatic	(−)	14	40	85	1−3	Occasional target cells, no ICSs	60−80% S; 15−35% F[c]

[a]All syndromes have positive sickle preparations.
[b]Hemoglobin F distribution; heterogeneous.
[c]Hemoglobin F distribution; homogeneous.
HPFH, high persistent fetal hemoglobin; ISC, irreversible sickle cell; NRBC, nucleated red blood cell; OC, older child; YC, young child. (−) absent; (+) present.

Prognosis

The survival time is unpredictable and is related in part to the severity of the disease and its complications. Data from 2005 showed the median age of death was 38 years for men and 42 years for women.

Causes of death include:

- Infection (sepsis, meningitis) with a peak incidence between 1 and 3 years of age,
- ACS/respiratory failure,
- Stroke (especially hemorrhagic), and
- Organ failure including heart, liver, and renal failure.

Management

1. Comprehensive care: Prevention of complications is as important as treatment.

   Optimal care is best provided in a comprehensive setting. Recommended screening studies are shown in Table 11.10.

2. Infection: Because of a marked incidence of bacterial sepsis and meningitis and fatal outcome under 5 years of age, the following management is recommended:

   a. All children with SCD should receive oral penicillin prophylaxis starting by 3—4 months of age:

      i. 125 mg bid (<3 years)
      ii. 250 mg bid (3 years and older).

   b. In patients allergic to penicillin erythromycin ethyl succinate 10 mg/kg orally twice a day should be prescribed.

Penicillin prophylaxis should be continued at least through age 5 years. Because the incidence of invasive bacterial infections declines with age, it may be reasonable to discontinue penicillin in older children. However, given that the rate of infection remains higher than the rate in individuals with spleens, some centers advocate continuing penicillin indefinitely.

TABLE 11.10   Routine Health Maintenance Related Laboratory and Special Studies in Patients with Sickle Cell Disease

Laboratory studies	Starting age	Frequency
Complete blood count/reticulocyte count	At diagnosis	Quarterly to yearly with differential monthly if receiving HU
Hemoglobin quantitation	At diagnosis	Yearly
Red cell antigen typing	At diagnosis	—
Liver and renal functions	At diagnosis	Yearly
Urinalysis	1 year	Yearly
HIV, hepatitis B, C		Yearly if receiving transfusions
Special studies		
Pulse oximetry	At diagnosis	Quarterly to yearly
Pulmonary function	5 years	Every 3 years
Sleep study		If symptoms present
Eye examinations	5 years for SCD-SC	Yearly
	8 years for SCD-SS	Yearly
Transcranial Doppler	2 years	Based on prior results
Brain MRI/A		If school difficulties, abnormal or repeatedly conditional TCD, neurological symptoms
Abdominal ultrasound		If symptoms of cholelithiasis
Hip radiograph/MRI		If symptoms of AVN
Echocardiogram	10 years	Every 3 years or more frequent if abnormal

AVN, avascular necrosis; TCD, transcranial Doppler; HU, hydroxyurea.

All children with SCD should receive routine childhood immunizations including conjugate *H. influenzae*. The 23-valent pneumococcal vaccine (PPV-23) should be administered at 2 years of age with a booster administered 5 years later. The conjugate 13-valent pneumococcal vaccine (PCV-13) and hepatitis B should be administered according to the routine childhood schedule. Children aged 6—18 years who have not previously received PCV-13 should receive a single dose of the vaccine. Adults who have received PPV-23 should receive a single dose of PCV-13 ≥1 year after receiving PPV-23. Meningococcal vaccination should also be administered. Influenza virus vaccine should be given yearly, each Fall.

Early diagnosis of infections requires:

Education of the family to identify a child with fever: Families should be instructed to call their physician immediately if their child develops a single temperature greater than 38.5°C (by mouth) or three elevations between 38°C and 38.5°C. The child should be seen immediately by a physician.

The patient should be investigated to determine the etiology of the fever, which should include a CBC with differential WBC and reticulocyte count and blood culture in all children. Chest radiograph is obtained in children under 3 years of age and in older children with respiratory symptoms. Urinalysis and culture are indicated in children <3 years or in older children with symptoms. Lumbar puncture is performed in young infants (<2—3 months) and in older infants and children with symptoms of meningitis. Other studies such as viral studies, stool cultures, and sputum cultures are performed based on symptoms.

Prompt antibiotic treatment with a broad-spectrum intravenous antibiotic that covers encapsulated organisms, such as a third-generation cephalosporin should be given.

Many centers recommend inpatient hospitalization for all children younger than 5 years because this group is at highest risk of infection. In addition, all children, regardless of age, with the following high-risk features should be admitted:

- Ill appearance.
- High fever (>39.5°C).
- ACS.
- Meningeal signs.
- Enlarging spleen
- Elevated leukocyte count (>30,000/mm3).
- Falling blood counts or low reticulocyte count.

A subset of lower-risk children, over age 12 months and without the above high-risk features, may be considered for discharge after a shorter period of observation (4—18 h) after having received a long-acting antibiotic such as ceftriaxone. This option should only be considered if the family can be contacted readily, follow-up is ensured, and continuous blood culture monitoring is available.

3. Treatment of specific complications of SCD are provided in the acute and chronic complication sections above.
4. Transfusion therapy increasingly is used to manage acute and chronic complications of SCD. Indications for transfusions in SCD are shown in Table 11.11. Risks of transfusion include infection (hepatitis B virus, hepatitis C virus, HIV, bacterial), alloimmunization, and iron overload.

TABLE 11.11   Generally Accepted Indications for Transfusions in Sickle Cell Disease

Episodic transfusion
Overt stroke
Transient pure red cell aplastic episode
Splenic sequestration
Acute chest syndrome
Preoperatively for surgical procedure with general anesthesia[a]
Acute multiorgan failure
Retinal artery occlusion
Chronic transfusion
Stroke
Abnormal transcranial Doppler ultrasound
Silent cerebral infarcts
Recurrent acute chest syndrome
Pulmonary hypertension
Recurrent severe pain

[a]*Moderate to high-risk surgical procedures. Controversial for low-risk procedures.*

The incidence of alloimmunization is 17.6%: mostly Kell (26%) and Rh (E (24%) and C (16%), respectively) antibodies. Other antibodies also occur in the following order of frequency: Jk^b (10%), Fy^a (6%), M (4%), Le^a (4%), S (3%), Fy^b (3%), e (2%), and Jk^a (2%).

All children with SCD should have a red cell phenotype when available identified at diagnosis. This allows determination of the child's red cell antigen phenotype before any transfusion. The patients should receive blood that is phenotypically matched to the patient for the Rh and Kell antigens. However, a high rate of Rh alloimmunization may still be seen with such an approach, likely due to the high prevalence of Rh variants that are not detected by routine phenotyping in the African-American population. Whether RH genotyping of patients and donors may further reduce the rate of alloimmunization due to variant RH alleles needs to be studied. In addition, blood should be leukoreduced and sickle negative blood should be administered to children receiving chronic transfusion therapy to allow accurate monitoring of HbS levels.

Chronic red cell transfusion therapy or repeated intermittent transfusions lead to iron overload. Complications of iron overload include hepatic fibrosis, endocrinopathies, and cardiac disease, and are best defined for thalassemia. The prevalence of certain complications such as heart disease is lower in SCD than in thalassemia. The treatment is similar to the approach used for thalassemia described later in this chapter. In addition, in SCD, exchange transfusion limits or prevents iron loading and should be utilized when possible for chronic transfusion therapy.

5. Induction of fetal hemoglobin (HbF)

Sustained elevations in HbF ($\geq$20%) are associated with reduced clinical severity in SCD. HU is the only approved drug for HbF modulatory therapy. HU results in the upregulation of HbF. HbF, within the red cell, interferes with polymerization of HbS, and therefore decreases the propensity of the red cell to sickle. Other effects of HU include increased red cell hydration and decreased expression of red cell adhesion molecules, increased NO production, and lowering of white blood cell count, reticulocytes, and platelets. Numerous studies in adults and children have shown the beneficial effects of HU in SCD-SS and Sβ^0-thalassemia:

a. Reduces number of VOEs.
b. Reduces incidence of ACS.
c. Reduces mortality.

Similarly, in infants ages 9−18 months, treatment with HU significantly reduces the number of episodes of dactylitis, pain, and ACS. Early use of HU is not associated with a reduction in the development of splenic dysfunction or glomerular hyperfiltration in these children. HU is not yet approved in the United States by the FDA for use in children with SCD.

There are no large studies on the use of HU for patients with SCD-SC and S-beta- + -thalassemia.

Dose

The starting dose of HU is 15−20 mg/kg/day. It is increased every 8 weeks by 5 mg/kg/day until a total dose of 35 mg/kg/day is reached or until a favorable response is obtained or until signs of toxicity appear. Evidence of toxicity includes:

- Neutrophil count <1000/mm^3.
- Platelet count <80,000/mm^3.
- Hemoglobin drop of 2 g/dl.
- Absolute reticulocyte count <80,000/mm^3.

Response is indicated by clinical improvement (reduction in VOE, ACS, etc.) and by laboratory response including rise in HbF (typically 10−20%), a rise in hemoglobin level of 1−2 g/dl, and increased MCV.

Follow-up

When HU is started, the patient should be monitored with a CBC every 2−4 weeks and HbF level at least quarterly. Once a stable and maximum tolerated dose is obtained, the patient can be monitored with CBCs monthly.

Indications

Given the clinical benefits of HU, treatment with this drug should be discussed with the families of all children, 9 months of age and older, with SCD-SS or S-beta-0-thalassemia. Frequent VOE and/or ACS are indications for treatment.

Side effects

- Myelosuppression.
- Rarely hair loss; skin and nail pigment changes.
- Headache.
- Gastrointestinal (GI) disturbance.

- Potential birth defects (female patients on HU should not become pregnant or should be on birth control because of potential for birth defects).
- Reduced sperm count and motility.

6. Hematopoietic stem cell transplantation

Currently HSCT (including umbilical cord blood (UCB)) is the only curative therapy.

The results of transplantation are best when performed in children with a sibling donor who is HLA-identical.

Eligibility criteria for HSCT for SCD-SS or SCD-Sβ^0 thalassemia include one or more of the following complications:

- Stroke or CNS event lasting longer than 24 h.
- Impaired neuropsychological performance with abnormal brain MRI.
- Recurrent ACS (at least two episodes in the last 2 years) or stage 1 or 2 sickle lung disease[1].
- Recurrent severe, debilitating VOE (three or more severe pain events per year for the past 2 years).
- Recurrent priapism.
- Sickle nephropathy.
- Bilateral proliferative retinopathy with major visual impairment in at least one eye.
- AVN of multiple joints
- Significant red cell alloimmunization during long-term transfusion therapy

With HLA-matched sibling donor HSCT, the survival rate is more than 90%. Over 85% survive free from SCD after HLA-matched sibling HSCT. Patients who have stable engraftment of donor cells experience no subsequent sickle-cell-related events and stabilization of preexisting organ damage. The majority of patients have stabilization or improvement of cerebrovascular disease after transplantation. Similarly, other organ toxicity (such as lung disease) related to SCD tends to stabilize post-transplantation. Linear growth is normal or accelerated after transplantation in the majority of patients. About 5% of the patients develop clinical grade III acute or extensive graft-versus-host disease (GVHD) (see Chapter 31). The risk of secondary cancers is estimated to be less than 5%.

Only about 15% of patients with SCD are likely to have an HLA-identical sibling donor. Unrelated donor stem cell transplantation, haplo-identical stem cell transplantation, and reduced-intensity conditioning protocols are under investigation and should be considered only in the context of a clinical trial in an experienced center.

Recommendations:

- Children with SCD who experience significant sickle cell complications should be considered for HSCT.
- HLA typing should be performed on all siblings.
- Families should be counseled about the collection of UCB from prospective siblings/donors.
- For severely affected children who have HLA-identical sibling donors, families should be informed about the benefits, risks, and treatment alternatives regarding HSCT.

7. Psychological support.

As for any chronic disease, patients require psychological support. Major problems that occur are:

- Coping with chronic pain.
- Inability to keep up with peers.
- Fears of premature death.
- Delayed sexual maturity.
- Increased doubts about self-worth.

Sickle Cell Trait (Heterozygous Form, AS)

The concentration of HbS in red cells is low, and sickling does not occur under normal conditions.

---

[1]The staging system for chronic lung disease is based on clinical, physiological, and roentgenographical criteria. Stage I is characterized by a mild reduction in lung volumes (vital capacity and total lung capacity) and FEV_1/FVC ratio (defines airflow obstruction). Stage 3 is where hypoxemia is first observed during stable periods and a severe reduction in lung volumes and flows are seen with associated borderline pulmonary hypertension and fibrosis on chest radiograph. Stage 4 is characterized by severe pulmonary fibrosis and pulmonary hypertension. Patients progress from one stage to the next every 2–3 years. Chronic lung disease is a prime contributor to mortality in young adults with sickle cell anemia.

Hematology

1. Indices—usually normal.
2. Blood smears—normal with few target cells.
3. Sickle cell preparation—reducing agents (e.g., sodium metabisulfite) to induce sickling *in vitro*.
4. Hemoglobin electrophoresis—AS pattern (HbA 55–60%; HbS, 35–45%).

Clinical Features

1. Usually asymptomatic.
2. Hematuria rarely.
3. Increased propensity for renal medullary cancer.
4. Exertional rhabdomyolysis/exercise-related sudden death. Ensure adequate hydration with sports activities.
5. Complicated hyphema—with secondary hemorrhage, increased intraocular pressure, central retinal artery occlusion. This requires evaluation/treatment by an ophthalmologist.
6. Infarction rare, occurring during flights in unpressurized aircraft.

Significance

The genetic implications mandate counseling. Table 11.9 lists the differential diagnosis of sickle cell syndromes.

Hemoglobin C

Basic Features and Pathology

1. Carrier state—2% in African-Americans.
2. Amino acid substitution (the same codon in the β-chain as in HbS)—lysine for glutamic acid.
3. HbC tendency to form rhomboidal crystals with increases in osmolality—red cell deformability impaired and splenic sequestration increased.

HbC Disease (Homozygous CC)

Hematology

1. Anemia—usually mild, hemolytic.
2. Blood smear—numerous target cells, as well as some spherocytes (the result of membrane loss in the spleen); a bar of crystalline hemoglobin across cell due to alteration in intracellular hemoglobin is a frequent finding.
3. Hemoglobin electrophoresis—CC pattern.

Clinical Features

1. Less severe than hemoglobin SS.
2. Splenomegaly.
3. Dehydration, leading to marked hemolysis and microcirculatory problems.

HbC Trait (Heterozygous Form, AC)

Asymptomatic with only genetic significance.

Hemoglobin SCD

HbS and HbC compound heterozygote.

Hematology

1. Anemia—if present, usually mild, hemolytic.
2. Blood smear—many target cells; sickle cells occasionally seen.
3. Sickle cell preparations—positive.
4. Hemoglobin electrophoresis—SC pattern (HbS ± 50%; HbC, ± 50%).

Clinical Features

1. Similar to, but usually less severe than, SCD-SS.
2. Severe infarctions on occasion (e.g., during pregnancy or the puerperium); may prove fatal.

HbS/β-Thalassemia

1. HbS and β-thalassemia trait compound heterozygote.
2. Hematology and clinical features vary; severity depends on the amount of normal adult hemoglobin synthesized (0–30%).
3. With no HbA (S-β⁰-thalassemia), disease comparable to SCD-SS.

Hemoglobin E

1. Mutation in β-globin gene that creates an alternate splice site which leads to decreased production of an abnormal globin chain.
2. Heterozygotes (HbE trait) and homozygotes (HbE disease) are asymptomatic. The MCV is reduced and target cells are seen on peripheral blood smear. Mild anemia is seen with HbE disease and less commonly with HbE trait. Important to distinguish HbE disease from HbE/β-thalassemia as the latter is clinically significant.
3. HbE/β-thalassemia — causes a thalassemia intermedia or thalassemia major phenotype (see later in chapter).

Unstable Hemoglobins

Unlike the amino acid substitutions in HbS and HbC, which affect the polarity of the external surface of the hemoglobin molecule, resulting in polymerization (HbS) or crystallization (HbC), the substitutions in unstable hemoglobins occur within the heme cavity or pocket of the α- or β-polypeptide chain. Substitution in the region of heme attachment causes gross molecular instability.

Changes in the oxygen affinity have also been found in some of the unstable hemoglobins and some of the M hemoglobins. An increase in oxygen affinity results in greater tissue anoxia and greater erythropoietin stimulation for a given level of anemia. In at least one hemoglobinopathy, hemoglobin Chesapeake, the only clinical manifestation is mild polycythemia.

Table 11.12 lists the various clinical manifestations that suggest unstable hemoglobinopathies. Table 11.13 presents laboratory data that suggest unstable hemoglobinopathies.

The hereditary methemoglobinopathies are closely related to the unstable hemoglobins. The substitution in these cases is also in the region of heme attachment, but it results in increased susceptibility to oxidation of heme Fe^{2+} to Fe^{3+} with consequent methemoglobin accumulation and cyanosis rather than hemolysis. There is some overlap between these two disorders, insofar as there is an increase in methemoglobin formation in most types of unstable hemoglobinopathies.

TABLE 11.12 Clinical Manifestations of Unstable Hemoglobins

Chronic nonspherocytic hemolytic anemia, varying from mild to severe

Intraerythrocyte inclusions (Heinz bodies) demonstrable by incubation of the cells with brilliant cresyl blue or methyl violet

Urinary dipyrrolic pigment excretion

Drug-induced hemolytic anemia

Methemoglobinemia

Cyanosis

Polycythemia

Chronic hemolytic anemia with normal hemoglobin electrophoresis

Variable response of hemolytic anemia to splenectomy

TABLE 11.13 Laboratory Data in Unstable Hemoglobinopathies

Chronic hemolytic anemia with normal red cell morphology, red cell enzymes, and hemoglobin electrophoresis

Abnormal heat stability test; tendency to precipitate on heating at 50°C

Presence of Heinz bodies

Raised methemoglobin levels

THALASSEMIAS

Basic Features

Thalassemia syndromes are characterized by varying degrees of ineffective hematopoiesis and increased hemolysis. Clinical syndromes are divided into α- and β-thalassemias, each with varying numbers of their respective globin genes mutated. There is a wide array of genetic defects and a corresponding diversity of clinical syndromes. Most β-thalassemias are due to point mutations, usually in both of the two β-globin genes (chromosome 11), which can affect every step in the pathway of β-globin expression from initiation of transcription to messenger RNA synthesis to translation and post-translation modification. A mutation in a single β-globin gene inherited along with triplicated alpha genes also may cause a β-thalassemia syndrome. Autosomal dominant forms of β-thalassemia also occur rarely. Figure 11.2 shows the organization of the genes (i.e., ε and γ, which are active in embryonic and fetal life, respectively) and activation of the genes in the locus control region, which promote transcription of the β-globin gene.

There are four genes for α-globin synthesis (two on each chromosome 16). Most α-thalassemia syndromes are due to deletions of the α-globin genes rather than to point mutations.

Mutations of β-globin genes occur predominantly in children of Mediterranean, African, and Southeast Asian ancestry. Deletions of α-globin genes are most common in those of Southeast Asian and African ancestry.

The main genetic variants include:

β-Thalassemia

1. β^0-Thalassemia: No detectable β-chain synthesis due to absent β-chain messenger RNA (mRNA).
2. β^+-Thalassemia: Reduced β-chain synthesis due to reduced or non-functional β-chain mRNA.
3. δβ-Thalassemia: δ- and β-chain genes deleted.
4. Eβ-Thalassemia: HbE (lysine → glutamic acid at 26) combined with β-thalassemia mutation. May be β^0 or β^+.
5. Hb Lepore: A fusion globin due to unequal crossover of the β- and δ-globin genes (the globin is produced at a low level because it is under δ-globin regulation).

α-Thalassemia

1. Silent carrier α-thalassemia: Deletion of one α-globin gene.
2. α-Thalassemia trait: Deletion of two α-globin genes.
3. Hb Constant Spring: Abnormal α-chain variant produced in very small amounts, thereby mimicking deficiency of the gene.
4. HbH disease: Deletion of three α-globin genes resulting in significant reduction of α-chain synthesis.
5. Hydrops fetalis: Deletion of all four α-globin genes; no normal adult or HbF production.

In many populations, α- and β-thalassemia and structural hemoglobin variants (hemoglobinopathies) exist together, resulting in a wide spectrum of clinical disorders.

Tables 11.14 and 11.15 list some features of the heterozygous and homozygous states of β-thalassemia and its variants. Table 11.16 lists the α-thalassemia syndromes.

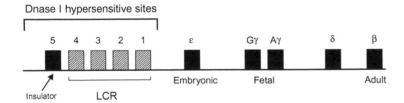

FIGURE 11.2 The structure of the human β-globin locus in chromosome 11. Beta globin gene transcription is regulated by activation of the genes of the locus control region and repression of the early genes. *From Nathan D and Orkin S (1998) with permission.*

TABLE 11.14 Heterozygous States of β-Thalassemia and Variants

Type	HbA2	HbF
β^+-Thalassemia	Increased	Normal to slightly increased
β^0-Thalassemia	Increased	Normal to slightly increased
δβ-Thalassemia	Normal	Increased (5–15%)
HPFH	Normal	Increased (15–30%)

TABLE 11.15   Homozygous or Doubly Heterozygous States of β-Thalassemia and Variants

Type	Anemia	δ-Globin chain	β-Globin chain	β-Globin mRNA	β-Globin gene mutation
β + -Thalassemia	Severe	Present	Decreased	Decreased	Point mutations or deletions
β⁰-Thalassemia	Severe	Present	Absent	Absent or abnormal	Point mutations or deletions
δβ-Thalassemia	Mild	Absent	Absent	Absent	Deletion mutation
HPFH	None	Absent	Absent	Absent	Point mutations or deletions

TABLE 11.16   α-Thalassemia Syndromes

Syndrome	Genetics	Number of α-genes deleted	Newborn Hb Barts (δ4) (%)	α/β synthesis ratio	Comments
Silent carrier of α-thalassemia	Heterozygous silent carrier	1	1–2	0.8–0.9	No anemia; no microcytosis; detectable by genetic interaction (i.e., two silent carriers can produce a child with α-thalassemia trait; a silent carrier and a person with α-thalassemia trait can produce a child with HbH disease); also detectable by molecular studies
α-Thalassemia trait	Heterozygous α-thalassemia trait OR Homozygous silent carrier OR Homozygous Hb Constant Spring	2	3–10	0.7–0.8	Microcytosis; hypochromia; mild anemia
Hemoglobin H disease	Heterozygous α-thalassemia trait/ silent carrier OR α-thalassemia trait/ constant spring	3	25	0.3–0.6	Hemolytic anemia of variable severity; relatively little ineffective erythropoiesis; no transfusion requirement; intermittent transfusions may be needed (especially with illness or stressors); HbH (β4) present
Hydrops fetalis	Homozygous α-thalassemia trait Hb Barts (δ4)	4	80–100	0	Death in utero or shortly after birth

β-Thalassemia: Homozygous or Doubly Heterozygous Forms (Major and Intermedia)

Pathogenesis

1. Variable reduction of β-chain synthesis (β⁰, β + , and variants).
2. Relative α-globin chain excess resulting in intracellular precipitation of insoluble α-chains.
3. Increased but ineffective erythropoiesis with many red cell precursors prematurely destroyed; related to α-chain excess.
4. Shortened red cell life span; variable splenic trapping.

Sequelae

1. Hyperplastic marrow (bone marrow expansion with cortical thinning and bony abnormalities).
2. Increased iron absorption and iron overload (especially with repeated blood transfusion), resulting in
   a. Fibrosis/cirrhosis of the liver.
   b. Endocrine disturbances (e.g., diabetes mellitus, hypothyroidism, hypogonadism, hypoparathyroidism, hypopituitarism).

c. Skin hyperpigmentation.
d. Cardiac hemochromatosis causing arrhythmias and cardiac failure.
3. Hypersplenism
  a. Shortened red cell life (of autologous and donor cells).
  b. Leukopenia.
  c. Thrombocytopenia.

Hematology

1. Anemia—hypochromic, microcytic.
2. Reticulocytosis.
3. Leukopenia and thrombocytopenia (may develop with hypersplenism).
4. Blood smear—target cells and nucleated red cells, extreme anisocytosis, contracted red cells, polychromasia, punctate basophilia, circulating normoblasts.
5. Hemoglobin F raised; HbA2 increased.
6. Bone marrow—erythroid hyperplasia, may be megaloblastic (due to folate depletion).

Biochemistry

1. Raised bilirubin (chiefly indirect).
2. Evidence of liver dysfunction (late, as cirrhosis develops).
3. Evidence of endocrine abnormalities (e.g., diabetes (typically late), hypogonadism (low estrogen and testosterone), hypothyroidism (elevated thyroid stimulating hormone)).
4. Elevated transferrin saturation and ferritin levels.

Clinical Features

Because of the variability in the severity of the fundamental defect, there is a spectrum of clinical severity (intermedia to major), which considerably influences management. β-Thalassemia intermedia is defined as homozygous or doubly heterogeneous thalassemia, which is not transfusion-dependent. β-Thalassemia major is defined as homozygous or doubly heterogeneous thalassemia (β^0 or β^+), which requires regular transfusions to manage clinical complications.

If untreated, 80% of patients die in the first decade of life. With current management, the life expectancy has dramatically increased. Patients now reach the sixth decade of life and are expected to live even longer.

Complications

Complications develop as a result of:

1. Ineffective erythropoiesis and hemolysis (in patients who are undertransfused or in untransfused thalassemia intermedia patients). These complications include:
  a. Anemia.
  b. Failure to thrive in early childhood.
  c. Growth retardation, delayed puberty, primary amenorrhea in females, and other endocrine disturbances secondary to chronic anemia and iron overload.
  d. Jaundice, usually mild; cholelithiasis.
  e. Hepatosplenomegaly, which may be massive; hypersplenism.
  f. Bone abnormalities
    i. Abnormal facies, prominence of malar eminences, frontal bossing, depression of bridge of the nose, and exposure of upper central teeth.
    ii. Skull radiographs showing hair-on-end appearance due to widening of diploic spaces.
    iii. Fractures due to marrow expansion and abnormal bone structure.
    iv. Osteopenia and osteoporosis are common and the risk is directly proportional to age (the prevalence of osteoporosis is about 60% in patients 20 years and older). The causes of this include medullary expansion, deficiency of estrogen and testosterone, nutritional deficiency (including calcium, vitamin D, and zinc), and chelator toxicity. Genetic factors likely also contribute.
  g. Pulmonary hypertension (suggested by TRV of at least 2.5 cm/s) occurs in both β-thalassemia major and intermedia. Splenectomy may exacerbate this risk, particularly in patients who are not regularly transfused.
  h. Thrombosis, which is exacerbated by splenectomy.
  i. Leg ulcers.

2. Iron overload—Due to repeated red cell transfusions in β-thalassemia major. In patients not treated with chelation therapy, cardiac disease from iron loading typically develops in late teens and early 20s. Iron overload also develops in β-thalassemia intermedia due to increased absorption of dietary iron. Complications of iron overload include:
   a. Endocrine disturbances (e.g., growth retardation, pituitary failure with impaired gonadotropins, hypogonadism, insulin-dependent diabetes mellitus, adrenal insufficiency, hypothyroidism, hypoparathyroidism, osteopenia, and osteoporosis).
   b. Cirrhosis of the liver and liver failure (exacerbated if concomitant hepatitis B or C infection).
   c. Cardiac failure and arrhythmias due to myocardial iron overload.
   d. Skin bronzing

Causes of Death

1. Congestive heart failure.
2. Arrhythmia.
3. Sepsis secondary to increased susceptibility to infection postsplenectomy.
4. Multiple organ failure due to hemochromatosis.

Management

Transfusion Therapy

Indications for initiation of regular red cell transfusions include:

1. Hemoglobin level <7 g/dl (on at least two measurements).
2. Poor growth.
3. Facial bone changes.
4. Fractures.
5. Development of other complications (pulmonary hypertension, extramedullary hematopoiesis, etc.).

The goal of transfusions is to maintain the pretransfusion hemoglobin greater than 9– 9.5 g/dl. Typical programs involve transfusion of 10–15 cc/kg of packed leukodepleted red cells every 3–5 weeks. Blood should be matched for ABO, C, E, and Kell antigens to reduce the risk of alloimmunization (some centers perform extended red cell antigen matching).

Transfusions result in:

1. Maximizing growth and development.
2. Minimizing extramedullary hematopoiesis and decreasing facial and skeletal abnormalities.
3. Reducing excessive iron absorption from gut.
4. Retarding the development of splenomegaly and hypersplenism by reducing the number of red cells containing α-chain precipitates that reach the spleen.
5. Reducing and/or delaying the onset of complications (e.g., cardiac).

Iron overload results from:

1. Ongoing transfusion therapy.
2. Increased GI absorption of iron (more important in β-thalassemia intermedia).

Monitoring Iron Overload

A number of tests are available to monitor iron loading, including:

1. Serum ferritin—particularly useful to follow trends but a measurement of ferritin in a chronically transfused patient may not accurately reflect the iron burden. Value may be altered by infection, inflammation, and vitamin C deficiency.
2. Liver iron concentration (LIC) may be measured by different techniques. LIC ≥ 15 mg/g dry weight of liver is associated with an increased risk of cardiac disease and death. Methods to measure LIC include:
   a. MRI: Using R2 or R2* methodologies. MRI is noninvasive and has become the most frequently utilized modality to assess LIC.

   b. Superconducting quantum interference device (SQUID): highly specialized equipment is required and is available in only a few centers worldwide.

   c. Liver biopsy: The gold standard, but invasive. This is the method of choice if histopathological examination is needed.

3. Cardiac iron measurement by T2* MRI. Cardiac iron may be high even if the LIC is low, particularly in patients with a history of poorly controlled iron overload in the past with recent intensification of chelation.

   a. T2* $\geq$ 20 ms indicates minimal cardiac iron loading.

   b. T2* of 10–19 ms indicates some cardiac iron loading. This result should prompt a discussion with patient/family about adherence with chelation. Intensification of chelation may be warranted.

   c. T2* <10 ms is associated with a high risk of cardiac disease (arrhythmias, congestive heart failure). Improved adherence and/or intensification of chelation therapy is indicated.

Chelation Therapy

The objectives of chelation therapy are:

1. To bind and detoxify free (non-transferrin bound) extracellular iron.

2. To remove excess intracellular iron.

3. To maintain a safe level of body iron burden

   a. Reduce previous iron loading.

   b. Reverse organ dysfunction.

   c. Prevent new iron loading.

Chelation therapy typically is not used in children younger than 2 years old and is generally initiated after 1–2 years of regular transfusions. Indications for chelation therapy in patients receiving chronic transfusions include:

1. Cumulative transfusion load of 120 ml/kg or greater.

2. Serum ferritin level persistently >1000 ng/ml.

3. LIC >5–7 mg/g dry weight.

Transfusion requirements and iron burden should be monitored closely and doses of chelation adjusted to maintain LIC at 2–7 mg/g dry weight. If LIC determination is not available a serum ferritin level between 500 and 1500 ng/ml is a reasonable goal. Some centers advocate more aggressive chelation, often using deferiprone, to maintain "normal" body iron burden, which has been associated with reversal of some endocrine complications. The safety of this approach has not been studied in children.

Currently available options for chelation therapy in the United States include deferoxamine, deferasirox, and deferiprone (approved as a second-line agent). The properties of the common chelators are summarized in Table 11.17.

Deferoxamine was the first available chelator, in clinical use for about 40 years. Due to its poor oral bioavailability, this drug must be administered parenterally, usually as a subcutaneous infusion over 8–24 h. Potential complications of deferoxamine are listed in Table 11.17. Audiological and ophthalmological toxicities are more common when the iron burden is low relative to the chelator dose. Similarly, bone changes, including metaphyseal dysplasia, are more common in young children with lower iron burden. Thus, it is important to avoid "over-chelation" in all patients, and lower doses of chelation therapy should be used in young children to avoid toxicity.

Nightly subcutaneous administration of desferrioxamine is time-consuming and painful and interferes in many ways with the lifestyle of the patient. For this reason, treatment adherence is often suboptimal and patients develop iron overload. The availability of oral chelation may help improve adherence to therapy. Two oral chelators are currently available in the United States:

1. Deferasirox (ICL-670, Exjade). The drug is supplied as orally dispersible tablets, which are dissolved in a glass of water or apple juice and administered $^{1}/_{2}$ h before meals. Studies have shown efficacy similar to that of deferoxamine. GI disturbances including abdominal pain, nausea, vomiting, and diarrhea are common, and may improve with continued administration of the drug. The GI effects may be related to lactose intolerance as lactose is present in the drug preparation. Elevations in hepatic transaminases to more than five times above normal can occur, and fulminant hepatic failure has been reported rarely. Liver function tests should be measured every 2 weeks for the first month after starting the medication, and tested monthly thereafter. Elevations in serum creatinine are also common, although renal insufficiency is rare. Renal function should be monitored monthly.

TABLE 11.17 Properties of Common Chelators

Property	Deferoxamine	Deferiprone	Deferasirox
Chelator:iron binding	1:1	3:1	2:1
Route of administration	Subcutaneous or intravenous	Oral	Oral
Usual dosage	25–50 mg/kg/day	75–99 mg/kg/day	20–40 mg/kg/day
Schedule	Administered over 8–24 h, 5–7 days/week	Divided three times a day	Daily
Route of excretion	Urine/feces	Urine	Feces
Adverse effects	Local reactions—swelling, rash	Gastrointestinal disturbances	Gastrointestinal disturbances
	Ophthalmologic—cataracts, reduction of visual fields and visual acuity and night vision	Transaminase elevations	Transaminase elevations
	Hearing impairment	Agranuloctyosis/neutropenia	Hepatic failure
	Bone abnormalities	Arthralgia	Gastrointestinal bleeding
	Pulmonary		Rise in serum creatinine
	Neurologic		Proteinuria
	Allergic reactions		Rash
Special monitoring considerations		Weekly complete blood count with differential	Monthly blood urea nitrogen, creatinine, hepatic transaminases (also obtain 2 weeks after starting the medication), and urinalysis

Adapted from Kwiatkowski, 2013 with permission.

2. Deferiprone (L1) is a second-line agent for patients with transfusion-dependent thalassemia. The most significant side effect is agranulocytosis; milder forms of neutropenia also may occur. Therefore, weekly monitoring of the CBC with differential is required. The drug should be held during all febrile illnesses and a CBC with differential should be checked. Other adverse effects include arthropathy and elevated hepatic transaminases. Deferiprone appears to be particularly useful in reducing cardiac iron overload either as a single agent, or in combination with deferoxamine.

Splenectomy

1. Splenectomy reduces the transfusion requirements in patients with hypersplenism. It is used in patients with severe leukopenia and/or thrombocytopenia due to hypersplenism and for patients with very high annual packed red blood cell requirements (>250 ml/kg/year) and uncontrolled iron overload. More recently, splenectomy is utilized less frequently due to the increased risk of pulmonary hypertension, thromboembolism, and infection after splenectomy.
2. At least 2 weeks prior to splenectomy, a polyvalent pneumococcal and meningococcal vaccine should be given. If the patient has not received a *Haemophilus influenzae* vaccine, this should also be given. Following splenectomy, prophylactic penicillin 250 mg bid is given to reduce the risk of overwhelming postsplenectomy infection. Management of the febrile splenectomized patient is detailed in Chapter 4.

Supportive Care

1. Folic acid, 1 mg daily orally, is given to patients who are not receiving regular red cell transfusions.
2. Hepatitis A and B vaccination should be given to all patients.
3. Appropriate inotropic, antihypertensive, and antiarrhythmic drugs should be administered when indicated for cardiac dysfunction.

4. Endocrine intervention (i.e., thyroxine, growth hormone, estrogen, testosterone) should be implemented when indicated.
5. Cholecystectomy should be performed if symptomatic gallstones are present.
6. Referral to gastroenterology for management of chronic hepatitis C infection.
7. HIV-positive patients should be treated with the appropriate antiviral medications.
8. Genetic counseling and antenatal diagnosis (when indicated) should be carried out using chorionic villus sampling or amniocentesis.
9. Management of osteoporosis includes:
   a. Periodic screening and prevention through early hormonal replacement.
   b. Yearly bone densitometry and gonadal hormone evaluation should be performed starting by age 10 years.
   c. Calcium and vitamin D intake should be monitored and supplements administered if poor intake or low vitamin D levels.
   d. Zinc supplementation can be considered. In a pilot study of patients aged 10–30 years with thalassemia major and low bone mass, supplementation with zinc, 25 mg daily, for 18 months resulted in a greater rise in whole-body bone mineral content compared to placebo.
   e. Hormonal replacement therapy (estrogen/progesterone; testosterone) should be administered to those with gonadal insufficiency.
   f. Encourage physical activity. Discourage smoking.
   g. Bisphosphonates, which inhibit osteoclast-mediated bone resorption have been used to treat osteoporosis in thalassemia, with some efficacy.

Follow-up of patients with thalassemia major includes:

Baseline	Alpha and beta globin genotype
	Transplant evaluation/HLA typing
	Red blood cell antigen profile
Monthly	Complete blood count
	Complete blood chemistry (including liver function tests, BUN, creatinine) if taking deferasirox
	Record transfusion volume
Every 3 months	Measure height and weight
	Measure ferritin (trends in ferritin used to adjust chelation); perform complete blood chemistry, including liver function tests
Every 6 months	Complete physical examination including Tanner staging, monitor growth and development, dental examination
Every year	Cardiac function—echocardiogram, ECG, Holter monitor (as indicated)
	Endocrine function (TFTs, PTH, FSH/LH), vitamin D levels starting at age 6 years; fasting glucose, testosterone/estradiol, FSH, LH, IGF-1, starting at age 10 years
	Vitamin C level
	Ophthalmological exam and auditory acuity
	Viral serologies (HAV, HBV panel, HCV (or if HCV +, quantitative HCV RNA PCR), HIV)
	Bone densitometry (from age 10 years)
	Ongoing psychosocial support
Every 1–2 years	Evaluation of tissue iron burden:
	Liver iron measurement—MRI, SQUID, or biopsy T2* MRI measurement of cardiac iron (beginning at age 10 years; earlier if chelation history unknown such as with international adoption, or if there is a history of elevated liver iron >15 mg/g dw)

Pharmacologic Enhancement of HbF Synthesis

High levels of HbF ameliorate the symptoms of β-thalassemia by increasing the hemoglobin concentration of the thalassemic red cells and decreasing the accumulation of unmatched α-chains, which cause ineffective erythropoiesis. HU has been demonstrated to increase HbF production and mean hemoglobin levels in patients with thalassemia intermedia or Eβ-thalassemia, decreasing or eliminating the need for transfusion. Additionally, there are reports of a few β-thalassemia major patients who became transfusion-independent using HU. However, neutropenia may limit adequate dose escalation. Decitabine is another HbF-inducing agent that is currently being studied in thalassemia. Butyric acid analogs and erythropoietin, as well as further testing with HU, are avenues of further investigation. Side effects of these agents include neutropenia, increased susceptibility to infection, and possible oncogenicity. Additional agents that increase hemoglobin levels through novel mechanisms also are under study.

Hematopoietic Stem Cell transplantation

1. Stem cell transplantation is a curative mode of therapy.
2. Outcome is best for children <17 years with an HLA-identical sibling donor. Overall survival is greater than 90%.
3. The presence of hepatomegaly, liver fibrosis, and/or history of poor adherence with chelation therapy has been associated with worse outcome; however, with the use of modified conditioning regimens for those with two or more of these risk factors, outcome is similar.
4. Although limited data are available, the outcome for matched unrelated donor transplantation with high-resolution molecular testing at HLA Class 1 and 2 loci appears to be comparable to matched sibling donor transplantation. Chronic GVHD is seen in 18%.
5. Reduced intensity conditioning regimens are under study.

Gene Therapy

Research is underway on methods of inserting a normal β-globin gene into mammalian cells. Ultimately, the aim is to insert the gene into stem cells and utilize these for stem cell transplant.

MANAGEMENT OF THE ACUTELY ILL THALASSEMIC PATIENT

Acute illness requiring urgent treatment occurs secondary to:

- Sepsis, usually with encapsulated organisms. Iron overload and chelation with deferoxamine also increase the risk of infection with *Yersinia enterocolitica*.
- Cardiomyopathy secondary to myocardial iron overload.
- Endocrine crises such as diabetic ketoacidosis.

Prevention of these complications should be the primary treatment. Preventive measures include:

- Management of the splenectomized patient as outlined in Chapter 4.
- Adequate chelation to prevent secondary hemochromatosis.
- Routine monitoring of cardiac and endocrine function.

If a patient presents with signs of shock, the following measures should be instituted:

1. Determine hemoglobin, electrolyte, calcium, and glucose levels; perform urinalysis.
2. Obtain blood cultures.
3. Distinguish between cardiogenic shock and septic shock because the management of each differs. To distinguish between the two, obtain:
   - ECG.
   - Echocardiograph, looking at left ventricular contractility.
   - Central venous pressure (CVP).
4. If the patient is in cardiogenic shock, management includes:
   - Continuous electrocardiographic and hemodynamic monitoring.
   - Immediately initiate intensive chelation, with deferoxamine as a continuous intravenous infusion at a dose of 50–60 mg/kg/day administered over 24 h; add deferiprone, 75–99 mg/kg/day divided tid as soon as possible.
   - Diuretics—use carefully because baseline preload is high due to chronic anemia and overdiuresis can precipitate acute renal failure.
   - Presumptive administration of hydrocortisone given the high rate of adrenal insufficiency.
   - Meticulous glucose control.
   - Obtain cardiac T2* as soon as practical.

5. If the patient is in septic shock, management consists of:
   - Blood cultures, at least two peripheral sites.
   - Broad-spectrum antibiotics IV (e.g., third-generation cephalosporin and an aminoglycoside).
   - Hold chelation until infection is under control.
   - Fluid boluses of 10 cc/kg normal saline to restore blood pressure.
   - Pressors such as dopamine, as indicated.
   - Presumptive administration of hydrocortisone given the high rate of adrenal insufficiency.
   - Coagulation studies to evaluate for disseminated intravascular coagulation.
   - CVP monitoring to guide fluid management.
   - Arterial blood gas and chest radiograph.
6. If the patient is in diabetic ketoacidosis, manage the ketoacidosis in the usual manner with careful monitoring of cardiac function when the patient is being vigorously hydrated.

β-Thalassemia Intermedia

Although patients are homozygous or doubly heterozygous, the resultant anemia is milder than in thalassemia major.

Clinical Features

1. Patients generally do not require transfusions and maintain a hemoglobin level between 7 and 10 g/dl.
2. Medullary expansion may result in nerve compression, extramedullary hematopoiesis, hepatosplenomegaly, growth retardation, and facial anomalies
3. Pulmonary hypertension and increased risk of thrombosis, particularly in splenectomized patients.
4. Patients are most healthy if management is as vigorous as that for thalassemia major.

Management

1. Folic acid 1 mg/day PO should be administered.
2. Iron-fortified foods should be avoided. A cup of tea with every meal will reduce the absorption of nonheme iron.
3. Chelation therapy is required at an older age than in thalassemia major because patients have received fewer transfusions. Ferritin levels may not correlate well with total iron burden (usually lower than expected for the degree of iron loading). Indications for chelation include elevated transferrin saturation of 70% or ferritin >1000 ng/ml. Liver iron quantitation may also be used to guide treatment.
4. Transfusions generally are not required except during periods of erythroblastopenia (aplastic crises) or during acute infection. If hemoglobin falls below 7 g/dl, growth is poor, or other complications develop, chronic transfusion therapy should be initiated. Children should be monitored for facial bone changes, which can be prevented, but not reversed, by chronic transfusions.
5. Splenectomy may improve hemoglobin level. However, the risk of infection with encapsulated organisms, pulmonary hypertension, and hypercoagulability are increased following splenectomy; therefore, splenectomy is often avoided.
6. Cardiac (including evaluation for pulmonary hypertension) and endocrine evaluation and bone densitometry should be performed as in thalassemia major.

β-Thalassemia Minor or Trait (Heterozygous β⁰ or β+)

Clinical Features

1. Asymptomatic (physical examination is normal)
   a. Discovered on routine blood test—slightly reduced hemoglobin, basophilic stippling, low MCV, normal RDW.
   b. Discovered in family investigation or family history of heterozygous or homozygous β-thalassemia.
   c. Confirmed with hemoglobin electrophoresis, demonstrating slightly decreased HbA (90−95% typically) increased HbA2 (>3.5%); hemoglobin F mildly elevated in 50% of cases.
2. Thalassemia trait of unusual severity. There are cases of β-thalassemia trait of unusual severity secondary to the coinheritance of α-gene duplication with increased α-globin synthesis, thereby increasing α- and β-chain imbalance, causing a β-thalassemia intermedia phenotype.

α-Thalassemias

The syndromes resulting from decreased α-chain synthesis are listed in Table 11.16. α-Thalassemia may present as a silent carrier, thalassemia trait, HbH disease, or hydrops fetalis. HbH disease is clinically milder than homozygous β-thalassemia and usually does not require regular red cell transfusions. HbH Constant Spring generally results in a more severe phenotype, with more severe anemia. Acute worsening of anemia with infections that requires treatment with acute red cell transfusion may occur in HbH Constant Spring, and less commonly with HbH disease. Hydrops fetalis is not compatible with life and presents with intrauterine or neonatal death, though some babies have survived with fetal packed red blood cell transfusions when antenatal diagnosis was made. These patients should continue on hypertransfusion regimens and be treated like β-thalassemia major, or treated with allogeneic stem cell transplant.

Further Reading and References

Adams, R.J., Brambilla, D., 2005. Discontinuing prophylactic transfusions used to prevent stroke in sickle cell disease. N. Engl. J. Med. 353, 2769—2778.

Adams, R.J., McKie, V.C., Hsu, L., Files, B., Vichinsky, E., Pegelow, C., et al., 1998. Prevention of a first stroke by transfusions in children with sickle cell anemia and abnormal results on transcranial Doppler ultrasonography. N. Engl. J. Med. 339, 5—11.

Aessopos, A., Kati, M., Tsironi, M., 2008. Congestive heart failure and treatment in thalassemia major. Hemoglobin 32, 63—73.

Bhatia, M., Walters, M.C., 2008. Hematopoietic cell transplantation for thalassemia and sickle cell disease: past, present and future. Bone Marrow Transplant. 41, 109—117.

Borgna-Pignatti, C., 2007. Modern treatment of thalassaemia intermedia. Br. J. Haematol. 138, 291—304.

Chou, S.T., Jackson, T., Vege, S., Smith-Whitley, K., Friedman, D.F., Westhoff, C.M., 2013. High prevalence of red blood cell alloimmunization in sickle cell disease despite transfusion from Rh-matched minority donors. Blood 122, 1062—1071.

Cunningham, M.J., Macklin, E.A., Neufeld, E.J., Cohen, A.R., 2004. Complications of beta-thalassemia major in North America. Blood 104, 34—39.

DeBaun, M.R., Gordon, M., McKinstry, R.C., Noetzel, M.J., White, D.A., Sarnaik, E.R., et al., 2014. Controlled trial of transfusions for silent cerebral infarcts in sickle cell anemia. N. Eng. J. Med. 371, 699—710.

Fung, E.B., Harmatz, P., Milet, M., Ballas, S.K., De Castro, L., Hagar, W., et al., 2007. Morbidity and mortality in chronically transfused subjects with thalassemia and sickle cell disease: a report from the multi-center study of iron overload. Am. J. Hematol. 82, 255—265.

Fung, E.B., Kwiatkowski, J.L., Huang, J.N., Gildengorin, G., King, J.C., Vichinsky, E.P., 2013. Zinc zupplementation improves bone density in patients with thalassemia: a double-blind, randomized, placebo-controlled trial. Am. J. Clin. Nutr. 98, 960—971.

Howard, J., Llwelyn, C., Choo, L., Hodge, R., Johnson, T., Purohit, S., et al., 2013. The Transfusion Alternatives Preoperatively in Sickle Cell Disease (TAPS) study: a randomised, controlled, multicentre clinical trial. Lancet 381, 930—938.

Koshy, M., Weiner, S.J., Miller, S.T., Sleeper, L.A., Vichinsky, E., Brown, A.K., et al., 1995. Surgery and anesthesia in sickle cell disease. Cooperative study of sickle cell diseases. Blood 86, 3676—3684.

Kwiatkowski, J.L., 2013. Evaluation and treatment of transfusional iron overload in children. Pediatr. Clin. North Am. 60 (6), 1393—1406.

Lal, A., Goldrich, M.L., Haines, D.A., Azimi, M., Singer, S.T., Vichinsky, E.P., 2011. Heterogeneity of hemoglobin H disease in childhood. N. Eng. J. Med. 364, 710—718.

Lanzkron, S., Carroll Jr., C.P., Haywood Jr., C., 2013. Mortality rates and age at death from sickle cell disease: US, 1979—2005. Public Health Rep. 128, 110—116.

Locatelli, F., Kabbara, N., Ruggeri, A., Ghavamzadeh, A., Roberts, I., Li, C.K., et al., Eurocord and European Blood and Marraow Transplantation Group 2013. Outcome of patients with hemoglobinopathies gven either cord blood or bone marrow transplantation from an HLA-identical sibling. Blood 122, 1072—1078.

Nathan, D., Orkin, S., 1998. Nathan and Oski's Hematology of Infancy and Childhood, fifth ed. Saunders, Philadelphia, p. 817.

Ohene-Frempong, K., Weiner, S.J., Sleeper, L.A., Miller, S.T., Embury, S., Moohr, J.W., et al., 1998. Cerebrovascular accidents in sickle cell disease: rates and risk factors. Blood 91, 288—294.

Pegelow, C.H., Macklin, E.A., Moser, F.G., Wang, W.C., Bello, J.A., Miller, S.T., et al., 2002. Longitudinal changes in brain magnetic resonance imaging findings in children with sickle cell disease. Blood 99, 3014—3018.

Pennell, D.J., 2005. T2* magnetic resonance and myocardial iron in thalassemia. Ann. N. Y. Acad. Sci. USA 1054, 373—378.

Pennell, D.J., Udelson, J.E., Arai, A.E., Bozkurt, B., Cohen, A.R., Galaello, R., et al., The American Heart Association Committee on Heart Failure and Transplantation of the Council on Clinical Cardiology and Council on Cardiovascular Radiology and Imaging 2013. Cardiovascular function and treatment in β-thalassemia major: a consensus statement from the American Heart Association. Circulation 128, 281—308.

Platt, O.S., Thorington, B.D., Brambilla, D.J., Milner, P.F., Rosse, W.F., Vichinsky, E., et al., 1991. Pain in sickle cell disease. Rates and risk factors. N. Engl. J. Med. 325, 11—16.

Rosse, W.F., Gallagher, D., Kinney, T.R., Castro, O., Dosik, H., Moohr, J., et al., 1990. Transfusion and alloimmunization in sickle cell disease. The Cooperative Study of Sickle Cell Disease. Blood 76, 1431—1437.

Rund, D., Rachmilewitz, E., 2005. Beta-thalassemia. N. Engl. J. Med. 353, 1135—1146.

Smith-Whitley, K., Zhao, H., Hodinka, R.L., Kwiatkowski, J., Cecil, R., Cecil, T., et al., 2004. Epidemiology of human parvovirus B19 in children with sickle cell disease. Blood 103, 422—427.

St Pierre, T.G., Clark, P.R., Chua-anusorn, W., Fleming, A.J., Jeffrey, G.P., Olynyk, J.K., et al., 2005. Noninvasive measurement and imaging of liver iron concentrations using proton magnetic resonance. Blood 105, 855—861.

Vichinsky, E., 2007. Hemoglobin E syndromes. Hematology Am. Soc. Hematol. Educ. Program, 2007, 79–83.

Vichinsky, E.P., Haberkern, C.M., Neumayr, L., Earles, A.N., Black, D., Koshy, M., et al., 1995. A comparison of conservative and aggressive transfusion regimens in the perioperative management of sickle cell disease. The Preoperative Transfusion in Sickle Cell Disease Study Group. N. Engl. J. Med. 333, 206–213.

Vichinsky, E.P., Styles, L.A., Colangelo, L.H., Wright, E.C., Castro, O., Nickerson, B., 1997. Acute chest syndrome in sickle cell disease: clinical presentation and course. Cooperative Study of Sickle Cell Disease. Blood 89, 1787–1792.

Voskaridou, E., Terpos, E., 2008. Pathogenesis and management of osteoporosis in thalassemia. Pediatr. Endocrinol. Rev. 6 (Suppl. 1), 86–93.

Wang, W.C., Miller, Ware, R.E., Miller, S.T., Iyer, R.V., Casella, J.F., Minniti, C.P., et al., BABY HUG Investigators 2011. Hydroxycarbamide in very young children with sickle-cell aneamia: a multicentre, randomised, controlled trial (BABY HUG). Lancet 377, 1663–1672.

Ware, R.E., Helms, R.W., for the SWiTCH Investigators, 2012. Stroke with transfusions changing to hydroxyurea (SWiTCH). Blood 119, 3925–3932.

Ware, R.E., Zimmerman, S.A., Sylvestre, P.B., Mortier, N.A., Davis, J.S., Treem, W.R., et al., 2004. Prevention of a secondary stroke and resolution of transfusional iron overload in children with sickle cell anemia using hydroxyurea and phlebotomy. J. Pediatr. 145, 346–352.

12

Polycythemia

Tsewang Tashi and Josef T. Prchal

The term polycythemia, particularly as it pertains to newborns and children, should be more accurately termed erythrocytosis because it generally refers to conditions in which only erythrocytes are increased in number. It is usually a response to tissue hypoxia such as the presence of high-oxygen-affinity hemoglobins or arterial hypoxia, increased production of erythropoietin (EPO) or other circulating erythropoietic stimulating factors, or mutations making erythroid progenitors intrinsically hyperproliferative. As there is no consensus for using the term polycythemia or erythrocytosis, the term polycythemia is used in this chapter, as this is the term in predominant use.

Polycythemia can be primary or secondary. Primary polycythemias can further be congenital (germline mutations) or acquired (somatic mutations). Congenital primary polycythemias, such as primary familial and congenital polycythemia (PFCP), are associated with polyclonal hematopoiesis, while acquired primary polycythemias such as polycythemia vera (PV) have clonal hematopoiesis due to mutations in hematopoietic/erythroid progenitors that make them hypersensitive to, or even independent of, EPO.

Secondary polycythemias, on the other hand, have normally responsive erythroid progenitors but excessive production of erythropoietic stimulating factors such as EPO. From a functional perspective, the increased red cell mass in secondary polycythemias can be physiologically *appropriate* or *inappropriate*. *Appropriate* responses are due to tissue hypoxia that can be either from congenital causes (high-oxygen-affinity hemoglobins) or acquired causes (cyanotic congenital heart disease, exposure to high altitude, lung disease). *Inappropriate* responses can also be congenital (mutations of normal hypoxia-sensing pathways) or acquired causes (EPO-producing tumors, postrenal transplant erythrocytosis).

Elevated hematocrit can be associated with normal red cell mass and decreased plasma volume, so-called *spurious*, *stress*, or *relative* polycythemia.

POLYCYTHEMIA (ERYTHROCYTOSIS) IN THE NEWBORN

Hypoxia is the major regulator and determinant of red cell mass and EPO transcription. Neonatal polycythemia is an appropriate physiological response to intrauterine hypoxia and is also contributed to by fetal hemoglobin, which has increased oxygen affinity. Humans have the highest hematocrit at birth, which dramatically decreases during the first 2 weeks of life. This decrease is consistent with preferential destruction of young red blood cells, a process called *neocytolysis*. A venous hematocrit reading of more than 65% or venous hemoglobin concentration in excess of 22 g/dl at any time during the first week of life should be considered evidence of polycythemia. Capillary blood samples should not be relied on for the diagnosis of polycythemia because their measured hemoglobin and hematocrit are significantly higher than venous hemoglobin or venous hematocrit and these measurements also vary with the temperature of the extremity from which the sample is obtained. Hematocrit values determined on a microcentrifuge include a small amount of trapped plasma and have a higher value than hematocrit values determined from automated analyzers.

The incidence of neonatal polycythemia is 0.4–4.0% of all births and is higher at high altitudes than at sea level. However, the normal range of hematocrit progressively increases with altitude, and appropriate adjustment needs to be made. The causes of neonatal polycythemia are listed in Table 12.1.

Lanzkowsky's Manual of Pediatric Hematology and Oncology.
DOI: http://dx.doi.org/10.1016/B978-0-12-801368-7.00012-0

TABLE 12.1   Causes of Neonatal Polycythemia

1. **Intrauterine hypoxia**
    a. Placental insufficiency
        i. Small-for-gestational-age (intrauterine growth factor)
        ii. Dysmaturity
        iii. Postmaturity
        iv. Placenta previa
        v. Maternal hypertension syndromes (toxemia of pregnancy)
    b. Severe maternal cyanotic heart disease
    c. Maternal smoking
2. **Hypertransfusion**
    a. Twin-to-twin transfusion
    b. Maternal-to-fetal transfusion
    c. Placental cord transfusion (delayed cord clamping, cord stripping, third stage of labor underwater at body temperature, holding baby below mother with cord attached)
3. **Endocrine causes**
    a. Congenital adrenal hyperplasia
    b. Neonatal thyrotoxicosis
    c. Congenital hypothyroidism
    d. Maternal diabetes mellitus
4. **Miscellaneous**
    a. Chromosomal abnormalities
        i. Trisomy 13
        ii. Trisomy 18
        iii. Trisomy 21 (Down syndrome)
    b. Beckwith–Wiedemann syndrome (hyperplastic visceromegaly)
    c. Oligohydramnios
    d. Maternal use of propranolol
    e. High-altitude conditions
    f. High-oxygen-affinity hemoglobinopathies (Table 12.4)

TABLE 12.2   Factors Increasing Viscosity

1. Hematocrit >60%
2. Larger mean cell volume
3. Decreased deformability of fetal erythrocytes
4. Plasma protein levels, especially high fibrinogen
5. Decreased flow rate—vessel diameter and endothelial integrity, for example, increased levels of EPO, in addition to inducing erythrocytosis, may induce other effects such as:
    a. an hematocrit-independent, vasoconstriction-dependent hypertension
    b. upregulation of tissue renin
    c. increased endothelin production
    d. stimulation of endothelial and vascular smooth muscle proliferation
    e. change in vascular tissue prostaglandin production
    f. stimulation of angiogenesis

Symptoms

Symptoms are due mainly to an increase in blood viscosity. A hematocrit level up to 65% has a linear correlation with viscosity and beyond 65% has an exponential relationship. Viscosity depends on a number of factors, as listed in Table 12.2. However, optimal oxygen delivery to the tissues is also markedly influenced by total blood volume. In secondary polycythemias, both hematocrit and total blood volume may be increased for optimal tissue oxygenation. This is a normal physiologic response and decreasing the hematocrit may be detrimental. Therefore, it is important to take the total blood volume into consideration while assessing the symptoms. Dehydration should be considered in neonates if polycythemia persists beyond the first 48 h of life.

Table 12.3 lists the clinical and laboratory findings and complications in neonatal polycythemia. Some of the symptoms may result from an underlying cause, such as intrauterine hypoxia, maternal diabetes, or placental insufficiency.

TABLE 12.3 Clinical and Laboratory Findings and Complications in Neonatal Polycythemia

CLINICAL FINDINGS

Feeding problems (20%)	Hypotonia (7%)	Hepatomegaly
Plethora (20%)	Tremulousness (7%)	Vomiting
Cyanosis (15%)	Difficult to arouse	Tachycardia
Lethargy (15%)	Weak suck	Cardiomegaly
Respiratory distress (9%)	Easily startled	Jaundice

LABORATORY FINDINGS

Venous hemoglobin >22 g/dl	Unconjugated hyperbilirubinemia (22%)	Chest radiograph • Increased vascularity • Pleural fluid • Hyperaeration • Alveolar infiltrates • Cardiomegaly
Venous hematocrit >65%		
Thrombocytopenia	Hypoglycemia (12–40%)	
Reticulocytosis	Hypocalcemia (1–11%)	
Normoblastemia	Hypomagnesemia	
Increased blood viscosity (normal 12.1 cP ± 3.9)		

COMPLICATIONS

Transient tachypnea of newborn	Intracranial hemorrhage (<1.0%)	Acute renal failure
Respiratory distress	Peripheral gangrene	Testicular infarction
	Priapism	Disseminated intravascular coagulation
Congestive heart failure	Necrotizing enterocolitis	
Convulsions	Ileus	

% indicates the frequency of the symptoms.

Laboratory Findings

When polycythemia is due to maternofetal transfusion, the following laboratory findings may be present:

- Increased quantities of immunoglobulin (IgA and IgM) in the infant's serum.
- Reduction in fetal hemoglobin to less than 60%.
- The presence of red cells bearing maternal blood group antigens in the infant's circulation.
- Presence of cells of maternal origin (bearing XX chromosomes) in the infant's circulation if the infant is male.

If polycythemia is due to intrauterine hypoxia, it is usually accompanied by an increase in nucleated red blood cells (nRBC) in peripheral blood during the early neonatal period. The mean value of nRBC in the first few hours of life in a healthy full-term neonate is 500 nRBC/mm^3 or 0–12 nRBC/100 white blood cells (WBC). A value of greater than 1000 nRBC/mm^3 or 10–20 nRBC/100 WBC is considered abnormal. Other hematologic indices of fetal hypoxia include higher absolute lymphocyte count and lower platelet count in comparison with normal full-term neonates without hypoxia during fetal life.

Treatment

Because instruments to measure viscosity are not widely available, neonatal hyperviscosity is diagnosed by a combination of symptoms and an abnormally high hematocrit.

Treatment should be reserved for infants who have a venous hematocrit of >65%, with respiratory, cardiac, or central nervous system symptoms, or an asymptomatic infant with a venous hematocrit >70%. All polycythemic infants, however, should be carefully monitored for evidence of hypoglycemia, hypocalcemia, and hyperbilirubinemia. Treatment should be designed to reduce the venous hematocrit to approximately 50–55%. This can be accomplished by a partial exchange transfusion using 5% human albumin, Ringer's lactate, or normal saline. It is better to avoid the use of fresh frozen plasma because it may potentially transmit infectious agents. Normal saline or Ringer's lactate solutions have the advantage that they are easily available and equally effective. Serum sodium level and renal function should be carefully monitored during the exchange transfusion procedure to avoid sodium overload.

The following formula is employed to approximate the volume of exchange required to reduce the hematocrit reading to the desired level:

$$\text{Volume of exchange (ml)} = [(\text{observed hct} - \text{desired hct}) \times \text{blood volume (ml)}]/\text{observed hct}$$

Partial exchange transfusion has been shown to increase capillary perfusion, cerebral blood flow, and cardiac function and reduces the risk of tissue ischemia in various organs resulting from severe microcirculation slowing due to high hematocrit and low shear rates. However, there is little evidence that the long-term outcome of infants is improved by this procedure. A few cases of necrotizing enterocolitis after partial exchange transfusions have been reported; however, a causative association has not been conclusively established.

POLYCYTHEMIA IN CHILDHOOD

The term polycythemia applies to an increase in circulating red cell mass to above the normal upper limits of 30 ml/kg body weight (excluding hemoconcentration due to dehydration). A hemoglobin level greater than the 99th percentile of method-specific reference range for age and sex, and adjusted for normal range at the altitude of residence should be applied. For practical purposes, this means a hemoglobin level higher than 17 g/dl or a hematocrit level of 50% or more during childhood. See Table 12.4 for various causes and classification of polycythemia.

TABLE 12.4 Classification of Polycythemia

1. **Relative polycythemia** (hemoconcentration, dehydration)
2. **Primary polycythemia** (results from somatic or germline mutations of erythroid progenitor cells that make them exquisitely sensitive to EPO or other cytokines)
   Congenital: primary familial congenital polycythemia (*EPOR* mutation resulting from germline mutation)
   Acquired: Polycythemia Vera (PV) (results from somatic mutation)
3. **Secondary polycythemia**
   a. Insufficient oxygen delivery (also known as *appropriate* polycythemia)
      i. Physiologic
         * Fetal life
         * Low environmental O_2 (high altitude)
      ii. Pathologic
         * Impaired ventilation: cardiopulmonary disease, obesity
         * Pulmonary arteriovenous fistula
         * Congenital heart disease with left-to-right shunt (e.g., tetralogy of Fallot, Eisenmenger syndrome)
         * Abnormal hemoglobins (reduced P_{50} in whole blood)
            Methemoglobinemia (congenital and acquired)
            Carboxyhemoglobin
            Sulfhemoglobinemia
            High-oxygen-affinity hemoglobinopathies (hemoglobin Chesapeake, Ranier, Yakima, Osler, Tsurumai, Kempsey, and Ypsilanti)
            2,3-Bisphosphoglycerate (BPG) deficiency
   b. Increase in EPO (also known as *inappropriate* polycythemia since it results from an aberrant production of EPO or other growth factors)
      i. Endogenous
         * Renal: Wilms' tumor, hypernephroma, renal ischemia, e.g., renal vascular disorder, congenital polycystic kidney, benign renal lesions (cysts, hydronephrosis), renal cell carcinoma
            Postrenal transplantation erythrocytosis (occurs in 10–15% of renal graft recipients). Contributing factors include persistence of EPO secretion from the recipient's diseased and ischemic kidney and secretion of angiotensin II androgen and insulin-like growth factor
         * Endocrine: pheochromocytoma, Cushing's syndrome, congenital adrenal hyperplasia, adrenal adenoma with primary aldosteronism
         * Liver: hepatoma, focal nodular hyperplasia, hepatocellular carcinoma, hepatic hemangioma, Budd–Chiari syndrome (some of these patients may have overt or latent myeloproliferative disorder)
         * Cerebellum: hemangioblastoma, hemangioma, meningioma
         * Uterus: leiomyoma, leiomyosarcoma
         * Ovaries: dermoid cysts
      ii. Exogenous
         * Administration of testosterone and related steroids
         * Administration of growth hormone
   c. Polycythemia with characteristics of both primary and secondary polycythemias
      * Chuvash Polycythemia (CP) and other *VHL* gene mutations
      * *HIF2α* gene (*EPAS1*) and *PHD2* gene (*EGLN1*) mutations

POLYCYTHEMIA VERA

Polycythemia Vera (PV) is a clonal disorder arising from a pluripotent hematopoietic stem cell manifesting by excess production of erythrocytes with low EPO levels and variable overproduction of leukocytes and platelets. It is one of the *Philadelphia chromosome negative myeloproliferative disorders* and can be differentiated from other myeloproliferative disorders by the predominance of erythrocyte production (other myeloproliferative syndromes are discussed in Chapter 17). This is a well-characterized disease in middle- to older-age adults, but it is extremely rare in childhood and adolescence, and thus literature on clinical presentation, treatment, and long-term prognosis in children is very limited. The overall clinical course, disease biology, and management do not differ significantly from adults and much of the information available is extrapolated from adult literature.

Pathophysiology

The biology of PV is characterized by clonality and EPO independence. In PV, a single clonal population of erythrocytes, granulocytes, platelets, and variable clonal B-cells arises when a hematopoietic stem cell gains a proliferative advantage over other nonmutated stem cells.

Genome-wide scanning, which compared clonal PV and nonclonal cells from the same individuals, revealed a loss of heterozygosity in chromosome 9p, in approximately 30% of patients. This is not a classical chromosomal deletion, but, rather, duplication of a portion of the chromosome and loss of the corresponding parental region, a process referred to as uniparental disomy. The 9p region contains a gene encoding for JAK2 tyrosine kinase, which transmits an activating signal in the EPO receptor-signaling pathway. A point mutation involving valine-to-phenylalanine substitution at codon 617 in the pseudokinase JAK2 domain on exon 14, known as $JAK2^{V617F}$, leads to constitutive gain-of-function of the kinase, which at least partly explains EPO hypersensitivity/independence. Over 95% of adult patients with PV carry the $JAK2^{V617F}$ mutation, as well as approximately 50% of adults with essential thrombocythemia and idiopathic myelofibrosis. In children, the frequency of the $JAK2^{V617F}$ mutation is reported in the range of about 39%. This underestimation is very likely due to the fact that many of these children did not actually have PV and had some other, possibly inherited, polycythemic disorder.

In about 2% of $JAK2^{V617F}$-negative adult PV patients, other JAK2 mutations have been found in exon 12 and these mutations are heterogeneous, consisting of insertions, deletions, or stop codons. These patients may have marked polycythemia without other affected cell lines. However, the risk of thrombosis and transformation to myelofibrosis is similar to $JAK2^{V617F}$-positive PV patients.

There is compelling evidence against $JAK2^{V617F}$ and JAK2 exon 12 mutations being disease-initiating mutations, but rather that these mutations play a major role in the behavior of the PV clone. Most leukemic transformation, however, arises from $JAK2^{V617F}$-negative PV progenitor cells.

Clinical Features

PV in children is extremely rare. The incidence in adults is approximately 10–20 per 100,000, of which 1% is present before 25 years of age and 0.1% present before the age of 20. Patients usually present with elevated hemoglobin and hematocrit found on routine testing. Some patients may initially present with isolated elevated platelet count, thus often initially diagnosed as essential thrombocytosis, but later develop erythrocytosis, transforming to PV. Some patients are asymptomatic, while others may have had various nonspecific symptoms recognized retrospectively to be consistent with PV. Overall, children tend to be less symptomatic than adults.

In adults, about one-third of patients present with thrombosis or hemorrhage. Thrombosis is about equally distributed between arterial and venous thrombosis. Less frequent, but more specific for PV, is Budd–Chiari syndrome (hepatic vein thrombosis). In younger adults, about 20–30% may present with Budd–Chiari syndrome. The presence of leukocytosis at presentation has been shown to be an independent risk factor for thrombosis. The rate of thrombosis is much lower, about 5%, in children and the thrombosis invariably occurs in the setting of leukocytosis associated with infections. Children may have much better vascular integrity than adults, which may negate some prothrombotic factors associated with PV.

Less than 5% of patients will have erythromelalgia, that is, erythema and warmth of the distal extremities, especially the hands and feet, with a painful burning sensation that can progress to digital ischemia. Erythromelalgia is associated with augmented platelet aggregation and frequently responds within hours to low- or regular-dose aspirin therapy. Less commonly, PV may present with elevated uric acid, with associated gout, due to increased cell turnover. Hemorrhagic presentations are usually mild, with gum bleeding and easy bruising, although serious gastrointestinal

TABLE 12.5 Revised WHO Criteria

Diagnosis requires the presence of both major criteria and one minor criterion *or* the presence of the first major criterion together with two minor criteria

MAJOR CRITERIA

1. Hemoglobin >18.5 g/dl in men, 16.5 g/dl in women *or* other evidence of increased red cell volume (hemoglobin or hematocrit >99th percentile of method-specific reference range for age, sex, altitude of residence *or* hemoglobin >17 g/dl in men, 15 g/dl in women if associated with a documented and sustained increase of at least 2 g/dl from the individual's baseline value that cannot be attributed to correction of iron deficiency, *or* elevated red cell mass >25% above mean normal value)
2. Presence of *JAK2*^{V617F} or other functionally similar mutation such as *JAK2* exon 12 mutation

MINOR CRITERIA

1. Bone marrow biopsy showing hypercellularity for age with trilineage growth (panmyelosis) with prominent erythroid, granulocytic, and megakaryocytic proliferation (not validated in prospective studies)
2. Serum EPO level below the reference range for normal
3. Endogenous erythroid colony formation *in vitro*

hemorrhage can occur, typically in essential thrombocytosis when the platelet count is more than 1 million/mm^3, associated with acquired von Willebrand disease. About 40% of adult patients present with pruritis, which typically gets worse after a warm bath or shower, known as aquagenic pruritis. This has been attributed to increased numbers of mast cells and elevated histamine levels, and these patients may have plethora and ruddiness of the face.

Diagnosis

The World Health Organization (WHO) criteria, listed in Table 12.5, are used for diagnosis. While the presence of EPO-independent erythroid colonies is specific for PV, this test is difficult and not widely available. Due to a lack of data on the frequency of *JAK2*^{V617F} and *JAK2* exon 12 mutations in children, the WHO diagnostic criteria may not be wholly applicable in children.

Treatment

Thromboembolism is the major cause of morbidity for patients with PV and the goal of the treatment is primarily directed at reducing vascular events by restraining monoclonal proliferation with cytoreductive therapy. However, the rate of thrombosis in children is significantly lower than adults, and a conservative initial approach may be justified.

1. *Phlebotomy* is performed to maintain hematocrits less than 45%. As more blood is removed and the patient becomes iron-deficient, the hematocrit becomes easier to control and the phlebotomy schedule should be adjusted accordingly. However, in some patients, iron deficiency can become symptomatic and can cause neurocognitive impairment and decreased exercise tolerance. Although phlebotomy is effective for controlling erythrocytosis, it does not affect variable leukocytosis, thrombocytosis, risk of thromboembolic events, or the overall natural course of the disease. In fact, there have been reports that the risk of thrombosis is slightly higher during the immediate postphlebotomy period.
2. *Low-dose aspirin* is employed to reduce the risk of thromboembolic events and results in a decreased risk of cardiovascular death, nonfatal myocardial infarction, nonfatal stroke, pulmonary embolism, and major venous thrombosis without a significant increase in rates of hemorrhage.
3. *Chemotherapeutic cytoreductive therapy*: Cytoreductive therapy can be achieved either with hydroxyurea or interferon-alpha. Indications are as follows:
   a. Prior history of thrombosis or transient ischemic attacks.
   b. A platelet count greater than 1.5 million/mm^3. Platelet counts at this level are a risk factor for bleeding.

The following cytoreductive therapy is used:

a. Hydroxyurea. Initial dose of 20–30 mg/kg daily. This dose is adjusted depending on the hematological response or signs of toxicity. Hydroxyurea reduces the risk of thrombosis compared to phlebotomy or phlebotomy and aspirin. The safety and efficacy are unclear in pediatric patients. Although there is a small theoretical risk of leukemogenesis, such an association with hydroxyurea has not been conclusively proven.

b. Interferon-alpha (IFN-α) achieves a complete hematological response in a high percentage of cases, and some patients achieve a complete molecular response of the $JAK2^{V617F}$ mutant allele that is durable; however, many patients do not tolerate IFN-α well because of a high rate of side effects and inconvenience of frequent intravenous administration. Newer pegylated formulations are much better tolerated, with less side effects and better efficacy, and some consider pegylated-IFNα as the first-line therapy in children. However, safety and efficacy are unclear in patients younger than 18 years old.

c. Anagrelide is useful to decrease platelet counts in patients presenting with thrombocytosis. The induction dose of anagrelide in children is 0.5 mg twice daily, followed by a maintenance dose of 0.5–1.0 mg twice a day, adjusted to the lowest effective dosage required to maintain platelet counts below 600,000/mm^3 and ideally to maintain it in the normal range.

PRIMARY FAMILIAL AND CONGENITAL POLYCYTHEMIA

PFCP is a primary polycythemia that is an autosomal dominant condition where the defect exists in erythroid progenitor cells. In contrast to PV, PFCP does not present with leukocytosis, thrombocytosis, or splenomegaly and does not progress to myelofibrosis or leukemia. Although PFCP is a rare disease, it is frequently misdiagnosed as PV. To date, 12 erythropoietin receptor (EPOR) mutations associated with PFCP have been described. These EPOR mutations lead to a hyperfunctional EPO receptor (a gain-of-function mutation) involving deletion of the cytoplasmic negative regulatory subunit of EPOR. These patients invariably have low EPO levels because the progenitor cells are extremely sensitive to EPO. They are generally asymptomatic, besides high hematocrit levels; however, there may be a predisposition in these families to cardiovascular disease and other thrombotic complications. Phlebotomy should only be used in those patients with hyperviscosity symptoms.

CONGENITAL POLYCYTHEMIA DUE TO ALTERED HYPOXIA SENSING WITH NORMAL P$_{50}$

Chuvash Polycythemia and Other von Hippel Lindau Mutations

Chuvash polycythemia (CP) is an endemic polycythemia found with high frequency on the west bank of the Volga River in the Chuvash Autonomous Republic in western Russia, the Italian island of Ischia, and sporadically worldwide in all ethnic and racial groups. It is an autosomal recessive disorder characterized by a loss-of-function mutation of the von Hippel Lindau (VHL) gene (VHL^{R200W}), thus impairing degradation of alpha subunits of hypoxia-inducible factors (HIFs) and resulting in their accumulation, leading to upregulation of transcription in a number of target genes, including EPO and vascular endothelial growth factor. Because EPO can be high-normal or increased, CP can be grouped with the secondary polycythemias. However, because the erythroid progenitors are hypersensitive to EPO for reasons not fully explained, CP also has features of primary polycythemia.

Clinical Manifestations

Patients with CP have normal arterial blood gases and normal P$_{50}$. (P$_{50}$ is defined as the partial pressure of oxygen at which hemoglobin is 50% saturated with oxygen, and reference range is 22–28 mmHg.) They often have a relatively low blood pressure, varicose veins, benign vascular abnormalities, and increased risk of pulmonary hypertension. There is increased risk for arterial and venous thrombotic and hemorrhagic complications and strokes, but no predisposition for developing malignancies typical of VHL.

Other congenital VHL mutations have been described in which there are compound heterozygous and even homozygous genotypes. One such example includes VHL^{H191D}, which is mainly found in Croatia. These patients may present with isolated erythrocytosis, elevated EPO level, and a normal P$_{50}$.

Table 12.6 compares the clinical manifestations of PV, PFCP, and CP.

HIF2α Mutations

There are three isoforms of HIFα: HIF-1α, HIF-2α, and HIF-3α. HIF2α is the main regulator of EPO transcription encoded by the EPAS1 gene. Several missense gain-of-function mutations of the EPAS1 gene associated with

TABLE 12.6   Clinical Manifestations of PV, PFCP, and CP

Clinical entities	PV	PFCP	CP
Frequency	Rare	Unknown	Unknown
Inheritance	None	Dominant	Recessive
Underlying cause	Acquired somatic mutation	EPO receptor mutation is found only in 12%	Functional deficiency of VHL
Symptoms of polycythemia (e.g., headache, dizziness lethargy, blurred vision)	Present	Usually diagnosed on routine blood count	Present
Signs	Plethora, splenomegaly	Plethora, no splenomegaly	Plethora, no splenomegaly Varicosities of peripheral veins
EPO level	Undetectable	Normal or low	Increased but high or normal in sporadic non-CP
Course	Thrombosis or hemorrhage	Benign	Thrombosis or hemorrhage
Diagnosis	$JAK2^{V617F}$ or $JAK2$ exon 12 mutation, endogenous erythroid colonies	Molecular analysis for truncation of cytosolic portion of ER and *in vitro* hypersensitivity to EPO	Molecular analysis of VHL, EPAS1, and EGLN1 mutations
Treatment	Phlebotomy, peg-IFN-α, ASA, hydroxyurea, anagrelide	Phlebotomy	Phlebotomy

VHL, von Hippel Lindau; VEGF, vascular endothelial growth factor; PAI-1, plasminogen activator inhibitor; IFN-α, interferon-alpha; ASA, aspirin; EPAS1, endothelial PAS domain-containing protein 1; EGLN1, Egl-9 family hypoxia-inducible factor 1; peg-IFN, pegylated IFN; EPO, erythropoietin; PV, polycythemia vera; PFCP, primary familial and congenital polycythemia; CP, Chuvash polycythemia.

isolated polycythemia and elevated EPO levels have been reported. A unique syndrome of congenital polycythemia and mosaicism of the *EPAS1* gene associated with recurrent paraganglioma/pheochromocytoma and somatostatinoma in later life have been reported.

Prolyl Hydroxylase Domain-2 (PHD2) Mutations

Prolyl hydroxylase domain (PHD)-containing enzymes hydroxylate alpha subunits of HIF, which facilitates binding to VHL, thus leading to ubiquitin-mediated proteasome degradation of HIFα subunits. Several mutations of the *EGLN1* gene, which encodes for PHD2, have been reported. These mutations lead to loss-of-function of PHD2, thus decreasing hydroxylation of HIFα subunits and subsequent proteasomal degradation, leading to increased HIF stability and levels. These patients present with polycythemia but generally normal EPO levels.

Phlebotomy in these patients does not alter the risk of thrombotic complications or the natural course of the disease. Similar to other congenital disorders of hypoxia sensing, several HIF inhibitors have been identified; however, no data from preclinical trials yet exist.

CONGENITAL POLYCYTHEMIA DUE TO ALTERED HYPOXIA SENSING WITH DECREASED P$_{50}$

High-Affinity Hemoglobinopathies

High-affinity hemoglobinopathies are autosomal dominant conditions. Most of these mutations occur within the β-globin chain where α1 and β2 chains come in contact. This change impairs intramolecular rotation or 2,3-bisphosphoglycerate (BPG) binding, making hemoglobin unable to transition from high-oxygen-affinity to low-oxygen-affinity states, thus causing tissue hypoxia and compensatory polycythemia. Hemoglobin electrophoresis is insufficient to identify hemoglobin structural defects, and some hemoglobin mutants will be missed.

The only reliable screening test is P_{50} measurement, either by Hemox Analyzer or calculated by using pH, PaO_2 and O_2% saturation obtained from venous blood gas (see the formula at the end of the chapter). Identification of these mutations can be only determined by sequencing of globin genes.

Since the polycythemia in patients with rare high-affinity hemoglobinopathies (e.g., unstable and hemoglobin M, see Chapter 11, as well as Badalona, Bethesda, Columbia-Missouri, Johnstown, La Coruna, Malmo, Olympia, San Diego, Strasbourg, and Syracuse) is a compensatory mechanism for tissue hypoxia, phlebotomy in these patients is not beneficial, but, in fact, detrimental as it only further augments tissue hypoxia characterized by a low P_{50} value with a normal 2,3-BPG level.

2,3-Bisphosphoglycerate Deficiency

2,3-Bisphosphoglycerate (BPG), also known as 2,3-Disphosphoglycerate (2,3-DPG), promotes hemoglobin transition from a high-oxygen-affinity state to a low-oxygen-affinity state. The 2,3-BPG binds to the central compartment of the hemoglobin tetramer, changing its conformation and shifting the oxygen dissociation curve to the right. The deficiency is created by ineffective bisphosphoglyceratemutase (BPGM), a red cell enzyme of the early glycolytic pathway that converts 1,3-BPG to 2,3-BPG. Mutations of BPGM are extremely rare and are typically autosomal recessive. Diagnosis is confirmed by establishing a decreased P_{50} and excluding hemoglobin mutants, and by establishing a decreased 2,3-BPG level and BPGM enzyme activity.

Methemoglobinemia

Methemoglobinemia is usually considered when the patient is cyanotic with low oxygen saturation by pulse oximetry, but with normal PaO_2 level. Methemoglobin levels are often included in blood gas measurements. Methemoglobin is generated when oxygen-carrying ferrous iron (Fe^{2+}) has been oxidized to ferric iron (Fe^{3+}) which is unable to bind oxygen. In normal physiological conditions, methemoglobin is converted to hemoglobin by enzyme cytochrome b5 reductase (also known as methemoglobin reductase or b5R). Congenital methemoglobinemia, an autosomal recessive disorder, is most commonly due to a cytochrome b5 reductase (b5R) deficiency. Methemoglobinemia can also be caused by various mutations of globin genes, known as hemoglobins M, inherited as an autosomal dominant phenotype.

Acquired methemoglobinemia is usually caused by exposure to oxidizing substances or drugs, including nitrates and sulfa-containing antibiotics. Acute methemoglobinemia is a medical emergency and early recognition is critical because it can be life-threatening. It usually does not cause polycythemia unless it has been chronic for weeks, allowing a compensatory increase of erythropoiesis responding to tissue hypoxia to take place. Patients with chronic methemoglobinemia can be asymptomatic to minimally symptomatic.

OTHER CAUSES OF POLYCYTHEMIA

Physiologically inappropriate polycythemia is often due to exogenous sources of EPO. Several malignancies, for example, hepatocellular carcinoma, renal cell carcinoma, uterine myomas, and cerebellar hemangiomas have been shown to produce EPO. Large, bulky tumors produce erythrocytosis by mechanical interference with blood supply to the kidneys, resulting in false sensing of hypoxia and EPO production. Renal polycythemia is due to EPO produced by renal cysts, polycystic kidney disease, or hydronephrosis.

Endocrine disorders such as pheochromocytomas, aldosterone-producing adenomas, Barter syndrome, and dermoid cysts of the ovary, can result in inappropriate EPO production through mechanical interference with renal blood supply or hypertensive damage to renal parenchyma resulting in a false sensing of hypoxia by the kidneys.

Postrenal transplantation erythrocytosis occurs in some patients following kidney transplantation and is associated with dysregulation of angiotensin receptor. These patients respond to drugs that cause inactivation of the renin—angiotensin system, such as angiotensin-converting enzyme inhibitors (ACE inhibitors). Patients unable to tolerate ACE inhibitors can be treated with an angiotensin II AT1 receptor antagonist.

Excess androgens (exogenous use or endogenous production) can also cause an increase in hematocrit by two mechanisms: stimulation of EPO production or an independent hyperproliferative effect on erythrocyte precursors.

Severe lung diseases affecting gas exchange and causing hypoxia, such as cystic fibrosis and α-1 antitrypsin deficiency-related lung diseases, may cause secondary compensatory polycythemia. However, quite often, polycythemia may not be clinically apparent because of anemia due to chronic inflammation, as well as concomitant increase in plasma volume. Smoking causes carboxyhemoglobinemia (HbCO), which can be measured with standard arterial blood gas measurements. Carbon monoxide has a 200-fold higher affinity for hemoglobin than oxygen and HbCO is fairly stable. Long-term smokers have compensatory polycythemia, mainly due to chronic hypoxia caused by both carboxyhemoglobinemia and lung disease, as well as due to decreased plasma volume caused by smoking.

In Eisenmenger complex, patients have right-to-left shunting of blood, resulting in increased pulmonary vascular resistance. Because of the admixture of desaturated venous blood with arterial circulation, they develop compensatory polycythemia due to tissue hypoxia. Clinically, they may manifest symptoms of hyperviscosity. Phlebotomy may alleviate symptoms of hyperviscosity; however, excessive phlebotomies may cause iron deficiency, leading to an increase in HIF, which may further complicate pulmonary vasoconstriction.

High-altitude polycythemia is a physiologic response to hypoxia. As altitude increases, a decrease in atmospheric pressure reduces the partial pressure of inspired oxygen, even though the percentage of oxygen in the air is constant. At an altitude of 5000 m, the atmospheric pressure and partial pressure of oxygen decrease by about 50% compared to sea level. This fall in pressure reduces the driving force for gas exchange in the lungs, leading to arterial hypoxemia and tissue hypoxia, which causes secondary polycythemia as a compensatory response.

Acute and chronic exposure to high-altitude hypoxia leads to many physiologic changes that have overall detrimental effects, such as acute mountain sickness and chronic mountain sickness. However, in native high-altitude dwellers such as Tibetans highlanders, distinct evolutionary adaptations have been observed which protect them from polycythemia and many of the detrimental effects of chronic exposure to hypoxia, unlike the Andean and Ethiopian highlanders in whom polycythemia is more common. With recent advances in genomics, we are beginning to understand more about the molecular and genetic mechanisms underlying these high-altitude human adaptations.

DIAGNOSTIC APPROACH

Figure 12.1 shows a diagnostic algorithm for diagnosis of polycythemia in children. The initial step is to apply the appropriate age-specific reference range for confirmation and then repeat laboratory studies, as hemoglobin concentration may reflect a transient decrease in plasma volume due to dehydration, causing hemoconcentration (relative polycythemia). Next, a determination has to be made as to whether the increased hemoglobin is acquired or congenital and, if congenital, whether it is familial; obtaining a thorough family history is crucial. If the hemoglobin is persistently elevated, hypoxia should be considered as the most common cause. An arterial oxygen saturation level of less than 92% suggests cardiac or pulmonary etiologies. A complete blood count (CBC), serum EPO level, arterial blood gas, and P_{50} value (either by Hemox Analyzer or calculated) should be done, as well as red cell and plasma volume studies to rule out spurious polycythemia due to chronic contraction of plasma volume. Arterial blood gas studies are helpful in determining the presence of arterial hypoxemia due to inadequate oxygenation, carboxyhemoglobinemia due to smoking or carbon monoxide poisoning, or methemoglobinemia.

The CBC may reveal increased leukocytes, platelets, and erythrocytes, which may often coexist in PV. $JAK2^{V617F}$ mutation testing is now a widely available test, and 98% of adult PV patients are positive for the mutation. Testing for endogenous erythroid colony formation can be considered for $JAK2^{V617}$-negative patients who harbor a strong suspicion for PV. However, this test is not standardized or widely available. Bone marrow biopsy may be helpful in determining marrow hyperproliferation.

PFCP is another primary polycythemia due to gain-of-function mutations in the EPO receptor that presents with isolated elevated erythrocytes without leukocytosis or thrombocythemia; it has an autosomal dominant pattern of inheritance (although *de novo* cases have been known to occur).

EPO levels will be low in primary polycythemias, while in secondary polycythemias, the EPO level will be inappropriately elevated or high normal, that is inappropriate to the high hemoglobin level.

Disorders resulting from high hemoglobin oxygen affinity such as high-affinity hemoglobin mutants or low 2,3-BPG concentrations are diagnosed with a decreased P_{50} from a Hemox Analyzer, an instrument which

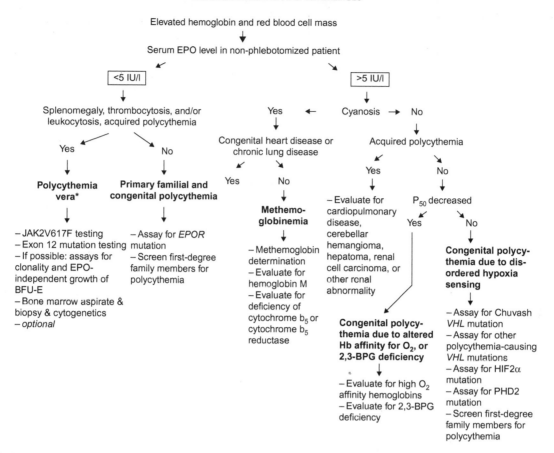

FIGURE 12.1 Diagnostic algorithm for polycythemia. *Some patients may have a normal EPO level. EPO, erythropoietin; EPOR, erythropoietin receptor; VHL, von Hippel Lindau; 2,3-BPG, 2,3-bisphosphoglycerate.

records blood oxygen equilibrium curves. If a Hemox Analyzer is not available, the P_{50} value can be calculated from freshly obtained venous blood gases by applying the formula:

$$P_{50} \text{ std} = \text{antilog} \frac{\log\left(\frac{1}{k}\right)}{n};$$

where

$$\frac{1}{k} = \left[\text{antilog}\left(n \log PO_{2(7.4)}\right)\right] \cdot \frac{100 - SO_2}{SO_2};$$

n = Hill's constant. The PO_2 in venous blood at 37°C can be converted to PO_2 at pH 7.4 with the formula:

$$\log PO_{2(7.4)} = \log PO_2 - \left[0.5(7.40 - pH)\right]$$

where pH is measured from the antecubital venous blood.

P_{50} can be calculated using this formula in an Excel spreadsheet which can be found on the Internet at http://www.medsci.org/v04/p0232/ijmsv04p0232s1.xls.

Further Reading and References

Alsafadi, T.R., Hashmi, S.M., Youssef, H.A., Suliman, A.K., Abbas, H.M., Albaloushi, M.H., 2014. Polycythemia in neonatal intensive care unit, risk factors, symptoms, pattern, and management controversy. J. Clin. Neonatal. 3 (2), 93—98.

Cario, H., McMullin, M.F., Bento, C., Pospisilova, D., Percy, M.J., Hussein, K., et al., 2013. Erythrocytosis in children and adolescents-classification, characterization, and consensus recommendations for the diagnostic approach. Pediatr. Blood Cancer 60 (11), 1734—1738.

Carobbio, A., Finazzi, G., Antonioli, E., Guglielmelli, P., Vannucchi, A.M., Delaini, F., et al., 2008. Thrombocytosis and leukocytosis interaction in vascular complications of essential thrombocythemia. Blood 112, 3135.

DiNisio, M., Barbui, T., Di Gennaro, L., Borrelli, G., Finazzi, G., Landolfi, R., et al., 2007. The haematocrit and platelet target in polycythemia vera. Br. J. Haematol. 136, 249.

Goina, F., Teofili, L., Moleti, M.L., Martini, M., Palumbo, G., Amendola, A., et al., 2012. Thrombocythemia and polycythemia in patients younger than 20 years at diagnosis: clinical and biologic features, treatment, and long-term outcome. Blood 119 (10), 2219—2227.

Gordeuk, V.R., Stockton, D.W., Prchal, J.T., 2005. Congenital polycythemias/erythrocytoses. Haematologica 90, 110.

Ismael, O., Shimada, A., Hama, A., Sakaguchi, H., Doisaki, S., Muramatsu, H., et al., 2012. Mutations profile of polycythemia vera and essential thrombocythemia among Japanese children. Pediatr. Blood Cancer 59 (3), 530—535.

Kiladjian, J., Cassinat, B., Chevret, S., Turlure, P., Cambier, N., Roussel, M., et al., 2008. Pegylated interferon-alfa-2a induces complete hematologic and molecular responses with low toxicity in polycythemia vera. Blood 112, 3065.

Lorenzo, F.R., Yang, C., Lanikova, L., Butros, L., Zhuang, Z., Prchal, J.T., 2013. Novel compound VHL heterozygosity (VHL T124A/L188V) associated with congenital polycythaemia. Br. J. Haematol. 162 (6), 851—853.

Patnaik, M.M., Tefferi, A., 2009. The complete evaluation of erythrocytosis: congenital and acquired. Leukemia 23, 834.

Sarangi, S., Lanikova, L., Kapralova, K., Acharya, S., Swierczek, S., Lipton, J.M., et al., 2014. The homozygous VHL^{D126N} missense mutation is associated with dramatically elevated erythropoietin levels, consequent polycythemia, and early onset severe pulmonary hypertension. Pediatr. Blood Cancer 61, 2104—2106.

Teofili, L., Giona, F., Martini, M., Cenci, T., Guidi, F., Torti, L., et al., 2007. Markers of myeloproliferative diseases in childhood polycythemia vera and essential thrombocythemia. J. Clin. Oncol. 25, 1048.

Uslu, S., Ozdemir, H., Bulbul, A., Comert, S., Can, E., Nuhoglu, A., 2011. The evaluation of polycythemic newborns: efficacy of partial exchange transfusion. J. Matern. Fetal Neonatal Med. 24 (12), 1492—1497.

Zhuang, Z., Yang, C., Lorenzo, F., Merino, M., Fojo, T., Kebebew, E., et al., 2012. Somatic HIF2A gain-of-function mutations in paraganglioma with polycythemia. N. Engl. J. Med. 367 (10), 922—930.

CHAPTER

13

Disorders of White Blood Cells

Mary Ann Bonilla and Jill S. Menell

White blood cells (WBCs) are cells of the immune system that defend the body against infectious disease and foreign material. The total WBC and the differential counts are valuable guides in the diagnosis, treatment, and prognosis of various childhood illnesses. The normal leukocyte counts and the absolute counts of different classes of leukocytes vary with age in children and their ranges are listed in Appendix 1.

It is important to calculate the absolute count of each class of WBC rather than relative percentage count for the purposes of quantitative interpretation. If nucleated red blood cells (NRBCs) are present, the total WBC count includes the total nucleated cell count (TNCC). Under these circumstances, the true total WBC count is calculated by subtracting the absolute NRBC count from the TNCC. This correction is generally required in the hemolytic anemias and in newborns when NRBCs tend to be raised.

Blood smear examination of the white cell morphology is important in the diagnosis of various causes of leukocytosis. For example, in severe infections or other toxic states, the neutrophils may contain fine deeply basophilic granules (toxic granulations) or larger basophilic cytoplasmic masses (Döhle bodies). Vacuolization of neutrophils may also occur. Döhle bodies are also found on examination of the blood smear in pregnancy, burns, cancer, leukemia blasts, May Hegglin anomaly, and many other conditions.

QUANTITATIVE DISORDERS OF LEUKOCYTES

Leukocytosis

Leukocytosis is an increase in the total WBC count more than two standard deviations above the mean for age. It is most commonly due to an increase in the absolute number of mature neutrophils (neutrophilia), but can be due to an increase in the absolute numbers of lymphocytes, monocytes, eosinophils, or basophils. Leukocytosis may be acute or chronic. Table 13.1 lists the causes of leukocytosis and Table 13.2 lists the causes of neutrophilia.

In infants and children, there is a tendency to release immature granulocytes into the circulation, and the WBC count may reach very high levels ($>50,000/mm^3$). This is called a *leukemoid reaction* and is often associated with bacterial infections. The term *left shift* is a $>5\%$ increase in the percentage of immature precursors (primarily bands) due to rapid release of the bone marrow reserve. The shift to the left may be so marked as to suggest myeloid leukemia. Table 13.3 lists the distinguishing features of leukemoid reaction and true leukemia.

Monocytosis is defined as an absolute monocyte count of $>500-800/mm^3$. Table 13.4 lists the causes of monocytosis and monocytopenia. The causes of basophilia, defined as an absolute basophil count $>200/mm^3$ are listed in Table 13.5. Eosinophils and lymphocytes are discussed later in this chapter.

Leukopenias

Leukopenia exists when the total WBC count is less than $4000/mm^3$. Leukopenia may result from a decrease in one or more specific classes of leukocyte. The causes of neutropenia are listed in Table 13.6 and lymphopenia in Table 13.7. Leukopenia can result from a number of conditions. However, isolated leukopenia resulting from a decrease in all classes of leukocytes is observed uncommonly.

TABLE 13.1   Causes of Leukocytosis

Physiologic
  Newborn (maximal 38,000/mm³)
  Strenuous exercise
  Emotional disorders; fear, agitation
  Ovulation, labor, pregnancy
Acute infections
  Bacterial, viral, fungal, protozoal, spirochetal
Metabolic causes
  Diabetic coma
  Acidosis
  Anoxia
  Azotemia
  Thyroid storm
  Acute gout
  Burns
  Seizures
Drugs
  Steroids
  Epinephrine
  Endotoxin
  Lithium
  Ranitidine
  Serotonin
  Histamine
  Heparin
  Acetylcholine

Poisoning
  Lead
  Mercury
  Camphor
Acute hemorrhage
Malignant neoplasms
  Carcinoma
  Sarcoma
  Lymphoma
Connective tissue diseases
  Rheumatic fever
  Rheumatoid arthritis
  Inflammatory bowel disease
Hematologic diseases
  Splenectomy, functional asplenia
  Leukemia and myeloproliferative disorders
  Hemolytic anemia
  Transfusion reaction
  Infectious mononucleosis
  Megaloblastic anemia during therapy
  Postneutropenia rebound

TABLE 13.2   Causes of Neutrophilia

Increased production
  Clonal disease
    Myeloproliferative disorders
      Chronic myelogenous leukemia
      Chronic neutrophilic leukemia
      Juvenile myelomonocytic leukemia
      Transient myeloproliferative disorder of Down syndrome
  Hereditary
    Autosomal dominant form of hereditary neutrophilia
    Familial cold urticaria
  Reactive
    Chronic infection
    Chronic inflammation
      Juvenile idiopathic arthritis
      Inflammatory bowel disease
      Kawasaki disease
    Hodgkin disease
    Drugs: Lithium, G-CSF, GM-CSF, chronic use of corticosteroids
    Leukemoid reaction
    Chronic idiopathic neutrophilia
Increased mobilization from marrow storage pool
  Drugs: Corticosteroids, G-CSF
  Stress
    Acute infection
    Hypoxia
Decreased margination
  Exercise
  Epinephrine
Decreased egress from circulation
  Leukocyte adhesion deficiency (LAD)
    LAD type I: deficiency of CD11/CD18 integrins on leukocytes
    LAD type II: absence of neutrophil sialyl Lewis X structures
    LAD type III: mutations of Kindlin-3
Asplenia

Modified from Dinauer (1998), with permission.

TABLE 13.3 Features of Leukemoid Reaction and Leukemia

Feature	Leukemoid reaction	Leukemia
Clinical	Evidence of infection	Hepatosplenomegaly
		Lymphadenopathy
Hematologic	No anemia	Anemia
	No thrombocytopenia	Thrombocytopenia
Bone marrow	Normal, hypercellular	Blasts
		Decreased megakaryocytes
		Decreased erythroid precursors
Leukocyte alkaline phosphatase	High	Absent

TABLE 13.4 Causes of Monocytosis and Monocytopenia

Monocytosis
  Hematologic disorders
    Leukemia
    Acute myelogenous leukemia
    Chronic myelogenous leukemia
    Lymphoma (Hodgkin and non-Hodgkin)
    Chronic neutropenia
    Histiocytic medullary reticulosis
    Recovery from myelosuppressive chemotherapy
  Connective tissue disorders
    Systemic lupus erythematosus
    Rheumatoid arthritis
    Myositis
    Granulomatous diseases
    Inflammatory bowel disease
    Sarcoidosis
  Infections
    Subacute bacterial endocarditis
    Tuberculosis
    Syphilis
    Rocky Mountain spotted fever
    Kala-azar
  Malignant disease
    Leukemia
    Acute myelogenous leukemia
    Chronic myelogenous leukemia
    Lymphoma (Hodgkin and non-Hodgkin)
  Miscellaneous disorders
    Postsplenectomy state
    Tetrachlorethane poisoning
    Lipoidoses (e.g., Niemann—Pick disease)
Monocytopenia
  Glucocorticoid administration
  Infections associated with endotoxemia

Neutropenia

Neutropenia is defined as a decrease in the absolute neutrophil count (ANC). The ANC is calculated by multiplying the total WBC count by the percentage of segmented neutrophils and bands: ANC = WBC (cells/mm^3) × percent (segmented + bands). In Caucasians, neutropenia is defined as an ANC of less than 1000/mm^3 in infants between 2 weeks and 1 year of age and less than 1500/mm^3 beyond 1 year of age. African Americans may have lower counts, with ANC levels 200—600/mm^3 less than in Caucasians. Neutropenia can be transient or chronic.

TABLE 13.5   Causes of Basophilia

Hypersensitivity reactions
    Drug and food hypersensitivity
    Urticaria
Inflammation and infection
    Ulcerative colitis
    Rheumatoid arthritis
    Influenza
    Chickenpox
    Smallpox
    Tuberculosis
Myeloproliferative diseases
    Chronic myeloid leukemia
    Myeloid metaplasia
    Polycythemia vera
    Essential thrombocythemia

Neutropenia is considered "chronic" when it persists beyond 3 months and is due to reduced production or increased destruction of neutrophils. It may be an inherited, intrinsic disorder or an acquired, extrinsic defect.

In pseudoneutropenia, a normal neutrophil population may be shifted toward the marginating compartment, leaving fewer cells in the circulating compartment. The WBC count measures only the circulating cells and not the marginating pool; therefore, this represents a pseudoneutropenia. The bone marrow is normal in appearance. The neutrophils function normally and the leukocyte changes are usually found incidentally on blood count. Marginating neutrophils may be uncovered by the injection of epinephrine.

The severity and duration of neutropenia correlate with susceptibility to develop various types of bacterial infections. The severity of neutropenia is graded according to ANC as follows:

- Mild neutropenia: ANC 1000–1500/mm^3
- Moderate neutropenia: ANC 500–1000/mm^3
- Severe neutropenia: ANC less than 500/mm^3

Clinical Manifestations

Patients with severe neutropenia are at an increased risk of infection from their own endogenous bacterial flora that reside in the mouth, oropharynx, gastrointestinal tract, and skin. For this reason, the frequency of Gram-negative bacterial infections and *Staphylococcus aureus* infections is high in these patients. The risk of infection is inversely proportional to the ANC. With mild neutropenia, stomatitis, gingivitis, and cellulitis may develop. More severe infections occur when the ANC is below 500/mm^3, with perirectal abscesses, colitis, pneumonia, and sepsis being common. Neutropenia alone does not predispose to parasitic, viral, or fungal infections unless other parts of the immune system are compromised, such as postchemotherapy.

Figure 13.1 shows an approach to the diagnosis of neutropenia.

DECREASED PRODUCTION OR INTRINSIC DEFECTS OF NEUTROPHILS

The inherited congenital neutropenias are primary bone marrow failure disorders involving the myeloid series. These are discussed in more detail in Chapter 8.

Table 13.6 lists the congenital neutropenias.

Benign Ethnic Neutropenia

Benign ethnic neutropenia is observed in a variety of populations, including Africans, West Indians, Yemenite and Ethiopian Jews, Bedouin Arabs, and Jordanians. In these groups, an ANC as low as 1000/mm^3 may be considered as normal. There is no increased infection risk in these individuals.

TABLE 13.6 Causes of Neutropenia

1. **Decreased production or intrinisc defects**
   a. Neutropenia in various ethnic groups[a]
   b. Severe congenital neutropenia: sporadic (most common) or autosomal dominant or Kostmann disease—autosomal recessive
   c. Familial benign chronic neutropenia—autosomal dominant
   d. Cyclic neutropenia
   e. Reticular dysgenesis
   f. Pancreatic insufficiency syndromes (Shwachman–Diamond syndrome and Pearson syndrome)
   g. Neutropenia associated with metabolic disease
      i. Glycogen storage disease (type IB)
      ii. Barth syndrome
      iii. Idiopathic hyperglycinemia
      iv. Isovaleric acidemia
      v. Methylmalonic acidemia
      vi. Propionic acidemia
      vii. Thiamine-responsive anemia in DIDMOAD syndrome (Diabetes Insipidus, Diabetes Mellitus, Optic Atrophy, and Deafness)
   h. Neutropenia in. Bone marrow failure syndromes (Chapter 10)
      i. Fanconi anemia
      ii. Familial congenital aplastic anemia without anomalies
      iii. Dyskeratosis congenital
2. **Increased destruction or extrinsic defects**
   a. *Congenital*
      i. Neutropenia associated with immunodeficiency disorders
         * XLA and dysgammaglobulinemia
         * Abnormal cellular immunity in cartilage hair hypoplasia
         * Common variable immune deficiencies
         * Hyperimmunoglobulin M syndrome
         * IgA deficiency
         * Dubowitz syndrome
         * Myelokathexis and WHIM syndrome
   b. *Acquired*
      i. Drug-induced
         * Idiosyncratic—antibiotics (sulfonamide, penicillin) antithyroid, antipsychotics
         * Toxic suppression—cytotoxic drugs, sulfasalazine, phenothiazines
         * Drug-hapten—penicillin, propylthiouracil
      ii. Infection
         * Viral infection (e.g., HIV, EBV, hepatitis A and B, respiratory syncytial virus, measles, rubella, varicella, influenza)
         * Bacterial infection (e.g., typhoid, paratyphoid, tuberculosis, brucellosis)
         * Rickettsial infection (e.g., ehrlichiosis)
      iii. Bone marrow aplasia
      iv. Chronic idiopathic neutropenia
      v. Secondary: chemicals, irradiation, immune reaction, malnutrition, copper deficiency, vitamin B_{12} deficiency, folate deficiency
      vi. Bone marrow infiltration, neoplastic
         * Primary: leukemia
         * Secondary: neuroblastoma, lymphoma, rhabdomyosarcoma
      vii. Bone marrow infiltration: non-neoplastic
         * Osteopetrosis, Cystinosis, Gaucher disease, Niemann–Pick disease
      viii. Immune
         * Drug-induced (e.g., anticonvulsants)
         * Alloimmune (isoimmune)
         * Maternofetal
         * Primary: autoimmune neutropenia
         * Secondary: autoimmune: systemic lupus erythematosus, lymphoma, leukemia, rheumatoid arthritis, HIV infection (in 20–44% of AIDS patients), infectious mononucleosis, associated with autoimmune thrombocytopenia and/or AIHA
      ix. Autoimmune lymphoproliferative syndrome
      x. Hypersplenism

[a]*Chronic, mild with a benign course.*

Neutropenia Associated with X-linked Agammaglobulinemia

This primary humoral immunodeficiency syndrome is inherited as an X-linked recessive trait. It is caused by mutations in the gene encoding a tyrosine kinase known as Bruton's or B-cell tyrosine kinase (BTK). BTK is expressed in all stages of B-cells lineage development, and in myeloid maturation.

TABLE 13.7 Causes of Lymphocytosis and Lymphopenia

1. Lymphocytosis
   a. Physiologic: 4 months—4 years
   b. Infections
      i. Acute
         - Moderate lymphocytosis: measles, rubella, varicella, mumps, roseola infantum, brucellosis, typhoid, paratyphoid, autoimmune diseases, granulomatous diseases, postimmunization states, drug reactions, graft rejection
         - Marked lymphocytosis: acute infectious lymphocytosis,[a] infectious mononucleosis, X-linked lymphoproliferative disorder, cytomegalovirus infection, toxoplasmosis, pertussis
      ii. Chronic
         - Tuberculosis
         - Syphilis
   c. Leukemia: acute lymphoblastic leukemia
2. Lymphopenia
   a. X-linked agammaglobulinemia
   b. Reticular dysgenesis
   c. Severe combined immunodeficiency

[a] *A case associated with Coxsackievirus B2 has been described.*

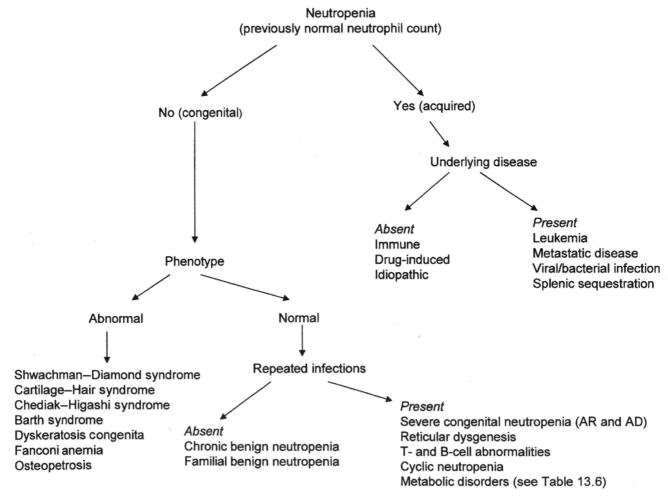

FIGURE 13.1 Approach to diagnosis of neutropenia. *Source: From Roskos and Boxer (1991).*

Clinical Manifestations

- Severe hypogammaglobulinemia, antibody deficiencies, and increased susceptibility to infection.
- Most patients present by 5 years of age.
- 15–25% of patients may develop neutropenia, which can be severe, during the stress of active infections when rapid production of these cells is required. Those with neutropenia are more prone to develop fungal or *Pneumocystis jirovecii* infections.
- Laboratory findings include severe decrease in serum IgG, IgA, and IgM levels and marked decrease in B-lymphocytes, but normal T-cell function.

Treatment

- Immunoglobulin therapy and aggressive antibiotic regimens for infections.
- Short-term use of Granulocyte Colony Stimulating Factor (G-CSF) has been found to be effective in some reported cases, but no evidence-based data are available.

Neutropenia Associated with Autosomal Recessive Agammaglobulinemia

There are at least six forms of autosomal recessive agammaglobulinemia. Mutations of the gene encoding for the μ heavy chain, surrogate light chain, and Ig-alpha are found in 85–90% of patients affected by this disease. Similar clinically to X-linked agammaglobulinemia (XLA), there are absent B-cells and associated neutropenia.

Treatment

Immunoglobulin replacement therapy, antibiotics, and a short course of G-CSF may be used, although no evidence-based data are available.

Neutropenia Associated with Abnormal Cellular Immunity in Cartilage Hair Hypoplasia

Cartilage hair hypoplasia is a rare autosomal recessive disorder found more commonly in Amish and Finnish populations. It is due to mutations in the ribonuclease mitochondrial RNA-processing gene that result in clinical heterogeneity.

Clinical Manifestations

- Short-limbed dwarfism, fine hair, moderate to severe neutropenia, and impaired cell-mediated immunity. There is a greater susceptibility to infection in children, including sino pulmonary, and a higher incidence of leukemia/lymphoma among adults.
- Hematologic findings: Lymphopenia, macrocytic anemia, neutropenia, immune-mediated thrombocytopenia (ITP), autoimmune hemolytic anemia (AIHA).

Treatment

- Allogeneic stem cell transplantation corrects both the immunologic defects and neutropenia. G-CSF therapy has been effective in one case report.

Neutropenia Associated with Common Variable Immunodeficiency

Common variable immunodeficiency (CVID) is a heterogeneous group of disorders involving immune dysfunction of B- and T-lymphocytes.

Clinical Manifestations

- Defined as an age-specific reduction in serum IgG. IgM and IgA levels vary.
- Presence of B-cells with poor or absent response to immunizations, and absence of other immunodeficiency states.
- Usual onset is between 20–30 years of age and only occasionally in childhood. Most patients manifest recurrent infections of the sinopulmonary and GI tracts. They are also prone to develop noncaseating granulomas of the lung, skin, gut, and other organs.
- These patients are predisposed to develop autoimmune disorders such as ITP, systemic lupus erythematosus (SLE), AIHA, rheumatoid arthritis, and thyroiditis.

Treatment

Neutropenia in CVID is explained on an autoimmune basis. Patients respond to G-CSF treatment, IV immuno-globulin, and antibiotics.

Myelokathexis and WHIM Syndromes

WHIM is a rare primary immunodeficiency disorder associated with warts, hypogammaglobulinemia, recurring bacterial infections, and retention of neutrophils in the bone marrow. If warts are absent, the condition is called "myelokathexis."

Genetics

The majority of patients have an autosomal dominant mutation of *CXCR4* that prevents the normal release of mature neutrophils from the marrow into the blood.

Clinical Manifestations

- Moderate to severe neutropenia.
- WBC count is usually <1000/mm^3 with severe neutropenia and lymphocytopenia. Neutrophils and eosinophils contain vacuoles, prominent granules, and nuclear hypersegmentation with pyknotic nuclei connected to each other by thin filaments. Normal morphology of lymphocytes, monocytes, and basophils.
- Neutrophil function is usually normal.
- Bone marrow shows granulocytic hyperplasia and neutrophils with similar degenerative changes as in the blood.

Treatment

- Immune globulin replacement therapy can reduce bacterial infections.
- G-CSF or granulocyte—macrophage colony-stimulating factor (GM-CSF) results in increased neutrophils.
- Preliminary evidence suggests that long-term treatment with the CXCR4 antagonist, plerixafor, is effective in increasing circulating leukocytes and decreasing infection frequency.

Selective IgA Deficiency and Neutropenia

Selective IgA deficiency is the most common form of immunodeficiency. Infections most commonly affect the respiratory and GI tracts. Neutropenia may be autoimmune in nature. There are no reports of the efficacy of G-CSF in this condition.

Dubowitz Syndrome

Dubowitz syndrome is an autosomal recessive disorder characterized by dysmorphic facies, mental retardation, microcephaly, growth retardation, eczema, and recurrent neutropenia.

Laboratory findings include low IgG and IgA with elevated IgM levels.

Neutropenia Associated with Hyperimmunoglobulin M Syndrome

The hyperimmunoglobulin M syndromes are rare heterogeneous disorders in which defective immunoglobulin (Ig) class switch recombination leads to normal or increased levels of serum IgM associated with deficiency of IgG, IgA, IgE and poor antibody function. These can be genetic or acquired secondary to congenital rubella syndrome, phenytoin use, or T-cell malignancy.

Genetics

- The most common form is an X-linked recessive trait, caused by mutations in the gene for CD40 ligand (*TNFSF5*). It has associated abnormal cellular immunity and patients develop infections by opportunistic organisms including *P. jirovecii*, histoplasmosis, *Cryptosporidium*, and toxoplasmosis.
- Two rare hyper-IgM syndromes with autosomal recessive inheritance and normal T-cell function are caused by defects in the gene for enzymes called activation-induced cytidine deaminase and uracil nucleoside glycosylate. These are humoral immunodeficiencies.

Clinical Manifestations

- Presents in infancy with severe recurrent bacterial infections due to encapsulated organisms or opportunistic organisms including *P. jirovecii*, histoplasmosis, *Cryptosporidium*, and *Toxoplasma*.
- Neutropenia is common. It may be transient, cyclic (10%), or chronic (50% of cases). Mechanism is not known but antineutrophil antibodies are not detected.
- DAT (direct antiglobulin test—Coombs)-positive AIHA and thrombocytopenia may be present.
- Lymphoid hyperplasia and predisposition to lymphoid malignancy.
- Low serum IgA, IgE and IgG; elevated IgM.
- Number of circulating B-cells is normal or increased and T-cells are normal.

Treatment

- Immunoglobulin replacement therapy reduces bacterial infections.
- Neutropenia responds to G-CSF.
- Hematopoietic stem cell transplantation is reported to be curative in patients with CD40L mutations.

Neutropenia Associated with Metabolic Diseases

The presenting clinical features in neutropenia associated with metabolic diseases (Table 13.6) are lethargy, vomiting, ketosis, and dehydration during the neonatal period, failure to thrive, and growth retardation. The marrow is hypoplastic in these conditions with decreased numbers of myeloid precursors. Idiopathic hyperglycinemia and methylmalonic acidemia also have associated thrombocytopenia. Individuals with isovaleric acidemia have a characteristic odor of "smelly feet."

Glycogen Storage Disease Type 1b

- Glycogen storage disease type Ib (GSD-1b) is a rare autosomal recessive disorder characterized by hypoglycemia, hepatosplenomegaly, seizures, and failure to thrive in infants.
- It results from a deficiency of glucose-6-phosphate-translocase enzyme and is characterized by impaired glucose homeostasis.

Clinical Manifestations

- Patients with GSD-1b frequently have neutropenia and/or neutrophil dysfunction and are susceptible to recurrent bacterial infections involving the perirectal area, ears, skin, and urinary tract. Life-threatening infections occur less frequently.
- Mechanism of neutropenia in GSD-1b may be caused by both increased levels of apoptosis and egress of neutrophils from the blood to the tissues.

Treatment

- G-CSF corrects neutropenia, defects in neutrophil chemotaxis, and intracellular bacterial killing, reducing the incidence of infections.

Barth Syndrome

- Rare X-linked recessive disorder characterized by cardiomyopathy, mild neutropenia, proximal skeletal myopathy, growth retardation, mitochondrial abnormalities, neutropenia, and increased urinary excretion of 3-methylglutaconic acid.
- Due to mutations in the *TAZ* gene (G4.5) resulting in an inborn error of lipid metabolism.
- Neutropenia is variable (absent to severe; persistent, intermittent, or cyclical). Despite the mild neutropenia, their cardiac defect places them at increased risk for morbidity when they develop infections.

Treatment

- G-CSF is used concomitant with appropriate prophylactic antibiotics if clinically indicated. Long-term G-CSF may be used based on severity.

Bone Marrow Disease

Bone marrow failure from any cause may present with neutropenia as a component of the pancytopenia that occurs in these disorders. These include:

- Congenital (e.g., Fanconi anemia, dyskeratosis congenita) aplastic anemias.
- Acquired (idiopathic or secondary) aplastic anemias.
- Bone marrow infiltrative disorders (non-neoplastic storage diseases such as Gaucher disease, Niemann–Pick disease).
- Neoplastic (leukemia, neuroblastoma).
- Osteopetrosis.

Bone marrow failure syndromes are discussed in detail in Chapter 8.

INCREASED DESTRUCTION OR EXTRINSIC DEFECTS OF NEUTROPHILS

Drug-Induced Neutropenia

Drug-induced neutropenia is defined as an idiosyncratic reaction to an offending drug. Onset may be days to months after exposure and may be acute or insidious. It may be due to:

- Idiosyncratic suppression of myeloid production affecting a few exposed persons—for example, antibiotics (novobiocin, methicillin), sulfonamides, antidiabetics (tolbutamide, chlorpropamide), antithyroids (propylthiouracil, methimazole), antihistamines, and antihypertensives (chlorothiazides, Aldomet).
- Regularly occurring dose-dependent myeloid suppression from cytotoxic drugs or antimetabolites—for example, 6-mercaptopurine, methotrexate, and nitrogen mustard.
- Toxic suppression of myeloid production in the marrow due to differences in individual ability to metabolize a drug—for example, phenothiazine and thiouracil.
- Drug-hapten disease, in which antibodies to the drug–leukocyte complex are produced. Clinically presents with rash, fever, lymphadenopathy, hepatitis, or pneumonia—for example, amidopyrine-related drugs (dipyrone, phenylbutazone), sulfapyridine, phenobarbital, mercurial diuretics, and chlorpropamide.

Treatment
- Immediate withdrawal of the offending drug is indicated.
- G-CSF may be indicated for persistent neutropenia or if severe infections develop.

IMMUNE NEUTROPENIA

Neonatal Immune Neutropenia—Alloimmune and Autoimmune

Immune neutropenias may be alloimmune (isoimmune), due to alloantibodies directed against epitopes (most commonly involving HNA-1a, HNA-1b, and HNA-1c) inherited from the father, analogous to Rh isoimmunization. In infants born to mothers with autoimmune neutropenia, antibodies are due to transplacental transfer of IgG antibodies against both maternal and fetal cells (e.g., a mother who has systemic lupus erythematosus).

Clinical Manifestations

Neutropenia may be moderate to severe. Infants may be asymptomatic or they may develop sepsis, cellulitis, omphalitis, or pneumonia.

- Neutropenia usually resolves by 2 months of age but may last up to 6 months.
- Bone marrow is hypercellular with an increase in neutrophil precursors and a paucity of mature neutrophils.

Treatment
- Antibiotics for the treatment and prevention of infections.
- Immunoglobulin or G-CSF therapy have been effective for more serious infections.

Autoimmune Neutropenia

Autoimmune neutropenia can be primary or secondary.

Primary Autoimmune Neutropenia

Chronic autoimmune neutropenia of childhood is a common disorder resulting from antibodies against leukocytes. It is analogous to AIHA or autoimmune thrombocytopenia. Neutrophil antibodies may adversely affect the function of neutrophils, producing qualitative defects in the neutrophils and amplifying the risk of infection associated with neutropenia. Neutrophil autoantibodies may also affect myeloid precursor cells. When this occurs, it can produce profound neutropenia. The disease is characterized by the following:

- Age of onset: 3–30 months; median age, 8 months.
- Neutropenia may be mild to severe. During infections, the ANC may transiently increase to normal levels.
- Physical examination is normal with mild splenomegaly noted occasionally.
- Benign infections of skin and upper respiratory tract, as well as otitis media, are seen in 90% of cases. These are not usually life-threatening and tend to be responsive to standard antibiotics.

Diagnosis

- Neutrophil counts range from 0 to 1000 cells/mm^3. Monocytosis is common.
- The bone marrow may be normal or may show evidence of myeloid hyperplasia with marked reduction in segmented neutrophils due to their destruction by antibodies.
- Epinephrine or hydrocortisone administration results in a rise in neutrophil count.
- Antineutrophil antibodies are not always detectable and screening has to be repeated for antibody detection from a reliable reference laboratory. Immunoassay is more sensitive than leukoagglutination testing for diagnosing immune neutropenia. In most patients, the autoantibody is auto-anti-HNA1 and some have auto-anti-HNA2.

The different types of antineutrophil assays are:

- *Granulocyte immunofluorescence test (GIFT)* detects neutrophil-bound IgG by binding of glutaraldehyde-fixed patient neutrophils to fluorescent-labeled antihuman IgG.
- *Granulocyte indirect immunofluorescence test (GIIFT)* uses the patient's serum with normal neutrophils or previously typed neutrophils and subsequent incubation with fluorescent-labeled antihuman IgG.
- *Granulocyte agglutination test (GAT)* uses the patient's serum incubated with normal neutrophils followed by microscopic evaluation for leukoagglutination.
- *Enzyme-linked immunoassay (ELISA)* uses microtiter plates with bound glutaraldehyde fixed normal neutrophils to detect antineutrophil antibodies in patient's sera. An antihuman IgG conjugated to a reporter enzyme is used for detection.
- *Monoclonal antibody-specific immobilization of granulocyte antigens (MAIGA)* is the most specific. It involves incubation of type-specific neutrophils with both patient's serum and mouse monoclonal antibodies directed against another neutrophil surface antigen. The mixture is passed over an affinity column containing antimouse IgG antibodies and then is assayed for the presence of human IgG. This allows for the detection of antibody and knowing its specificity in one assay.

GIFT and GAT are used most commonly to diagnose autoimmune neutropenia.

Prognosis

Spontaneous recovery usually occurs within a few months to a few years. The median age at recovery is 30 months (range, 7–73 months). In 95% of cases, recovery occurs by 4 years of age.

Treatment

Infections are not related to the degree of neutropenia and respond to standard antibiotics. Usually infections are not life-threatening but in patients with severe neutropenia who develop severe or recurrent infections, the following management is recommended:

- Appropriate antibiotics for acute bacterial infections or prophylactic antibiotics such as trimethoprim/sulfamethoxazole.

- Mouth care with oral rinses and good dental hygiene for mouth ulcers or gingivitis.
- In patients with more serious infections, G-CSF can be used at low doses of $1-2\,\mu g/kg$ to improve $ANC > 1000/mm^3$. The dose can be further decreased or given on alternate days or less often.
- Immunoglobulin therapy has been used with variable response and duration.

Secondary Autoimmune Neutropenia

Secondary neutropenia is seen in other acquired disorders of the immune system. It occurs more commonly in adolescents and adults than in children. In these patients, chronic immune neutropenia warrants screening for associated immunodeficiency or autoimmune disorders.

Autoantibody specificity: Pan-FcR γ IIIb.

The following diseases are associated with secondary autoimmune neutropenia:

- Evans syndrome.
- AIHA.
- Autoimmune thrombocytopenia.
- Thyroiditis.
- Insulin-dependent diabetes mellitus.
- Common variable immune deficiency.
- Systemic lupus erythematosus.
- Rheumatoid arthritis.

Treatment
- Treat the underlying condition.
- For patients with severe infection, low-dose G-CSF can be used.

Autoimmune Lymphoproliferative Syndrome

Autoimmune lymphoproliferative syndrome (ALPS) is a disorder of defects in lymphocyte apoptosis. ALPS is reviewed in detail in Chapter 16.

NONIMMUNE NEUTROPENIA

Chronic Idiopathic Neutropenia

This is a group of neutropenia disorders due to decreased or ineffective production or excessive apoptosis of neutrophils due to unknown causes. It occurs in late childhood or adulthood with a female predominance.

Clinical Manifestations
- The neutropenia may be variable ($<500-1000$ cells/mm^3) with normal to hypocellular bone marrow and normal cytogenetics.
- Myeloid maturation arrest at the myelocyte and band stage occurs.
- Gingivitis and mouth ulcers are common. Patients with severe neutropenia may develop infections of the skin, mucous membranes, and lungs.

Treatment
- Low-dose G-CSF is effective.

TABLE 13.8   Investigations of Patients with Neutropenia[a]

1. History of drug ingestion, toxin exposure, infectious history
2. Physical examination—nature of infectious lesions, growth and development, presence of anomalies, presence of enlarged lymph nodes or hepatosplenomegaly
3. Familial: absolute neutrophil count in family members
4. Complete Blood Count (CBC)—with differential and platelet count, absolute neutrophil count and reticulocyte count; CBC and differential three times per week for 6–8 weeks (to exclude cyclic neutropenia)
5. Bone marrow
   a. Maturation characteristics of myeloid series; there is a reduction in mature neutrophils
   b. Maturation and number of megakaryocytes and erythroid precursors
   c. Karyotype (to identify myelodysplasia or acute myelocytic leukemia) and FISH studies for chromosome 7 and 5q
   d. Electron microscopy (subcellular morphology, congenital dysgranulopoiesis)
6. Detection of antineutrophil antibodies (see text for details)
   a. Granulocyte immunofluorescence test (GIFT)
   b. Granulocyte indirect immunofluorescence test (GIIFT)
   c. Granulocyte agglutination test (GAT)
   d. Enzyme-linked immunoassay (ELISA)
   e. Monoclonal antibody-specific immobilization of granulocyte antigens (MAIGA)
7. Immunologic tests
   a. Immune globulins (IgA, IgG, IgM, IgE)
   b. Cellular immunity (skin-test activity, purified protein derivative (PPD), lymphocyte subsets; suppressor T-cell assay)
   c. Antinuclear antibodies, C_3, C_4, CH_{50}
8. Evidence of metabolic disease
   a. Plasma and urine amino acid screening
   b. Serum vitamin B_{12}, folic acid, and copper
9. Evidence of pancreatic disease
   a. Exocrine pancreatic function: stool fat, pancreatic enzyme assays, CT scan of pancreas for pancreatic lipomatosis, serum levels of trypsinogen and isoamylase
10. Chromosomal breakage analysis (Fanconi anemia)
11. Radiographic bone survey (cartilage hair hypoplasia, Shwachman–Diamond syndrome, Fanconi anemia)
12. Serum muramidase (ineffective myelopoiesis)
13. Flow cytometry for CD59 (or other GPI linked protein) (paroxysmal nocturnal hemoglobinuria)
14. Bone density studies (14% of patients with chronic neutropenia show nonclinical osteoporosis or osteopenia)
15. Many gene mutation analyses are commercially available including: neutrophil elastase (*ELANE*) (SCN and cyclic neutropenia), *GFI-1* (SCN), *WAS* (X-linked neutropenia), *SBDS* (Shwachman–Diamond syndrome), *HAX 1, TAZ* (Barth syndrome), Fanconi family of genes, *CHS1* (Chediak Higashi syndrome) and others that are continually being discovered.

[a]*Absolute neutrophil count less than 1500/mm³.*

Infections

Viral infections and certain bacterial infections, such as typhoid fever, paratyphoid fever, and rickettsial disease, may be associated with neutropenia. Staphylococcal or Pneumococcal infections associated with neutropenia indicate a grave prognosis.

Hypersplenism

Hypersplenism causes peripheral sequestration not only of red cells and platelets, but of granulocytes as well. The marrow in such cases shows myeloid hyperplasia with normal maturation to the polymorph stage. Splenomegaly from any cause (e.g., thalassemia, storage diseases, portal hypertension, inflammation, or neoplasia) may produce hypersplenism. Usually treating the underlying cause will improve the neutropenia. Occasionally, splenomegaly and neutropenia of unknown cause (primary splenic neutropenia) may require splenectomy.

Investigations in Neutropenia

Table 13.8 lists the investigations to be considered when evaluating patients with neutropenia.

Management of Neutropenia

The therapy of neutropenia differs between the cause and severity of neutropenia as noted previously. Table 13.9 lists the management required in the care of neutropenic patients.

TABLE 13.9   Management of the Neutropenic Patient[a]

1. Fever	Patients with poor marrow reserve (i.e., congenital neutropenias) require aggressive approach to fever. Obtain appropriate cultures (blood, throat, urine, infected area) and sensitivity and administer broad-spectrum antibiotics. If an organism is isolated, 10−14 days intravenous treatment is required. If no organism is isolated, antibiotic is continued until afebrile and clinically stable.
2. Periodontal disease	Good dental hygiene and regular cleaning are essential. Treat mouth ulcerations and gingivitis with appropriate systemic antibiotics if secondary bacterial infection is found. Use of periodontal washes may alleviate symptoms and aid recovery.
3. Hematopoietic growth factors	G-CSF is effective in increasing the neutrophil count in various neutropenic disorders. The starting dose differs for each diagnosis (ranging from 1−5 μg/kg with dose modification according to the patient's ANC).
4. Hematopoietic stem cell transplantation	Used for severe congenital neutropenia refractory to G-CSF, increased risk of MDS/AML, and those where matched sibling transplant is available. As better reduced conditioning regimens and unrelated donors are used, increased use for other neutropenia disorders may be seen.

[a]Neutrophil count less than 500 cells/mm^3.

EOSINOPHILS

Eosinophils are effector cells of myeloid descent. They are increased in a wide variety of disorders. They are produced in the bone marrow and have a half-life in the blood similar to that of the neutrophil (6−10 h). The majority of eosinophils in the body are localized to areas exposed to the external environment, for example the tracheobronchial tree, the gastrointestinal tract, mammary glands, vagina, cervix, and the connective tissue beneath the epithelium. The estimated blood-to-tissue ratio is 1:300−500. The lifespan in tissues is not known but is thought to be several weeks.

Figure 13.2 illustrates mechanisms of eosinophil production in bone marrow, release in circulation and migration in the tissues. Eosinophil production is stimulated by a number of cytokines including IL-5, GM-CSF, and IL-3. Following activation, the eosinophil expresses its effector function, which includes the release of highly toxic granule proteins and other mediators of inflammation. The granule contents are highly toxic to helminths and other parasitic organisms. They also can cause significant damage to any tissue, but particularly to the skin, heart, lung, and brain.

Eosinophilia

Normal mean eosinophil count in the circulating blood is 350−500/mm^3. The absolute eosinophil count (AEC) is calculated in the same manner as the ANC: AEC = WBC count (cells/mm^3) × percent eosinophils. Under normal circumstances, their number and activation increase as a response to antigens, especially when these antigens are deposited in the above tissues. A response is characterized by an immediate hypersensitivity reaction, mediated by IgE, or delayed hypersensitivity reaction mediated by T-lymphocytes.

The severity of eosinophilia is graded according to the presence of their absolute number in the circulating blood as follows:

- Mild eosinophilia: 500−1500/mm^3
- Moderate eosinophilia: 1500−5000/mm^3
- Severe eosinophilia: greater than 5000/mm^3.

Since the total eosinophil count does not always predict the degree of organ damage, a thorough history and physical to exclude this possibility should be done promptly. The history and physical exam should address: fever, weight loss, fatigue, rash, pruritis, allergic symptoms, wheezing, cough, gastrointestinal symptoms (diarrhea), myalgias, central nervous system (CNS) symptoms, lymphadenopathy, hepatosplenomegaly, signs and symptoms of cardiac dysfunction, exposure to drugs, foods, or travel history.

The initial work-up of eosinophilia should include a CBC with differential and review of the peripheral smear, comprehensive metabolic panel to assess liver and renal function, troponin, and chest X-ray. Additional work-up is guided by the history. Travel to endemic areas would indicate stool for ova and parasites or specific serologic testing such as *Strongyloides*, *Toxocara*, *Trichinella*, and others. If the blood

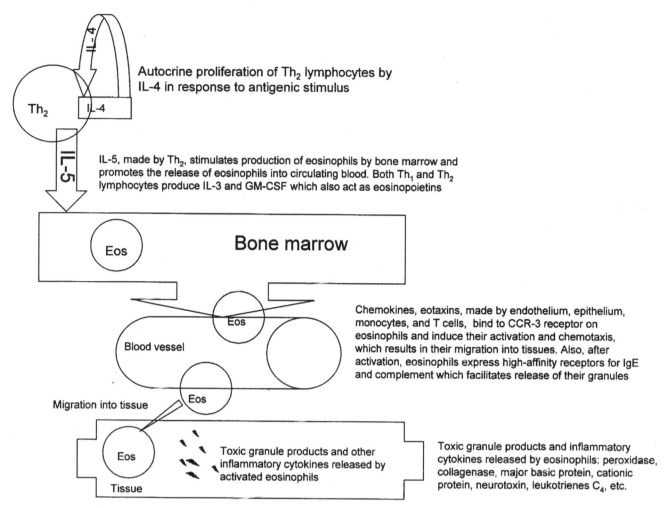

FIGURE 13.2 Mechanism of eosinophil production in bone marrow, release in circulation and migration in tissue. *Abbreviations*: IL, interleukin; Eos, eosinophil; CCR, chemokine receptor; Th, helper T-lymphocytes; GM-CSF, granulocyte–macrophage colony-stimulating factor.

smear shows immature cells or abnormalities on physical examination, such as lymphadenopathy or hepatosplenomegaly, suggestive of a malignant disorder, bone marrow flow cytometry and cytogenetic studies (leukemia or lymphoma) or CT scan and lymph node biopsy (lymphoma) are indicated. If frequent infections are noted, an immunologic work-up such as quantitative immunoglobulins and lymphocyte subsets may be indicated to exclude an underlying immunodeficiency. Figure 13.3 is a guide for the diagnostic work-up for eosinophilia.

Classification of Eosinophilia

Eosinophils can be increased as a reactive process, as part of a malignant clone, or in response to cytokines produced by malignant cells. In the United States, the most common cause of eosinophilia in children is allergy. Outside the United States, parasitic infections are the most common causes. Table 13.10 lists the causes of reactive eosinophilia.

Based on the consensus from the 2011 Working Conference on Eosinophilic Disorders and Syndromes, eosinophilia can be classified as transient, episodic, or persistent (chronic). Hypereosinophilia (HE) is defined by an eosinophil count $\geq 1500/mm^3$ on at least two occasions with a minimal interval of 4 weeks. HE can be documented on tissue specimens as well. Figure 13.4 shows a schema for the classification of the eosinophilias.

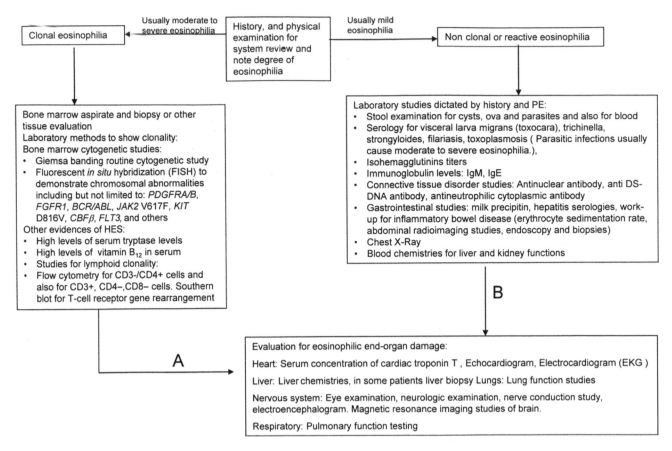

FIGURE 13.3   Diagnostic studies for evaluation of eosinophilia. (A) In all patients. (B) In patients with prolonged eosinophilia.

TABLE 13.10   Causes of Reactive (Nonclonal) Eosinophilia

Allergic disorders
  Asthma, hayfever, urticaria, drug hypersensitivity
Immunologic disorders
  Omenn syndrome (SCID and eosinophilia), hyper IgE syndrome
Skin disorders
  Eczema, scabies, mites, erythema toxicum, dermatitis herpetiformis, angioneurotic edema, pemphigus
Parasitic infestations
  Helminthic: *Ascaris lumbricoides*,[a] trichinosis, echinococcosis, visceral larva migrans,[a,b] hookworm,[a] strongyloidiasis,[a] filariasis[a], schistosomiasis, toxoplasmosis, toxocara
  Protozoal: malaria
Fungal infections
  Aspergillus, pneumocystis, coccidiomycosis, basidiomycosis, paracoccidiomycosis, disseminated histoplasmosis, *Cryptococcus*
Viral infections
  HTLVII, HIV1 (may be related to drugs), EBV
Hematologic disorders
  Hodgkin disease, postsplenectomy state, eosinophilic leukemoid reaction, Fanconi anemia, thrombocytopenia with absent radii, congenital neutropenia, familial reticuloendotheliosis
Familial eosinophilia
Irradiation
Pulmonary disorders
  Eosinophilic pneumonitis (Loeffler syndrome), pulmonary eosinophilia with asthma, tropical eosinophilia
Gastrointestinal disorders
  Eosinophilic gastroenteritis, milk precipitin disease, ulcerative colitis, protein-losing enteropathy, regional enteritis, allergic granulomatosis
Miscellaneous
  Idiopathic hypereosinophilic syndrome,[b] periarteritis nodosa, metastatic neoplasm, cirrhosis, peritoneal dialysis, chronic renal disease, Goodpasture syndrome, sarcoidosis, thymic disorders, hypoxia, adrenal insufficiency, connective tissue disorders

[a]*Helminth infestations associated with eosinophilia and pulmonary infiltrates.*
[b]*Conditions associated with striking eosinophilia. Leukocyte counts of 30,000–100,000/mm³ are characteristic, with 50–90% of leukocytes being eosinophils. In all other conditions, the WBC count is normal or only slightly elevated and eosinophils make up 10–40% of the leukocyte count.*

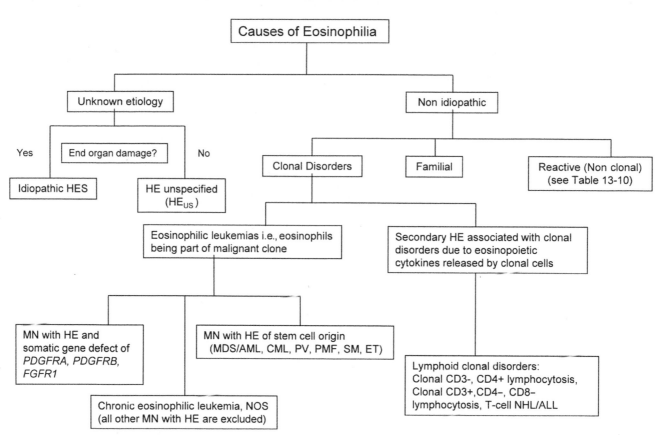

FIGURE 13.4 Etiologic classification of eosinophilia. *Abbreviations*: NHL, non-Hodgkin lymphoma; ALL, acute lymphoblastic leukemia; MDS, myelodysplastic syndrome; AML, acute myelocytic leukemia; MN, myeloid neoplasm; CML, chronic myelogenous leukemia; PV, polycythemia vera; PMF, primary myelofibrosis; SM, systemic mastocytosis; ET, essential thrombocythemia; PDGFRA/B: platelet-derived growth factor receptor alpha/beta; FGFR1, fibroblast growth factor receptor 1.

(Idiopathic) Hypereosinophilic Syndrome

Definition

The term hypereosinophilic syndrome (HES) is defined by:

- A persistent eosinophilia consistent with HE as described above.
- Absence of evidence of known causes of eosinophilia despite a comprehensive work-up for such causes.
- Signs and symptoms of organ involvement, directly attributable to eosinophilia.

It is important to recognize that even moderate eosinophilia can be associated with significant organ damage, while some patients with significant eosinophilia can remain completely asymptomatic. HE without any evidence of an underlying reactive disorder or molecular or clonal abnormality *and* without organ damage has been designated HE unspecified or HE$_{US}$. These patients should be monitored carefully as the cause of the eosinophilia may become clear over time and/or chronic eosinophilia may eventually lead to end organ damage and HES.

Clinical Presentation

The disease generally has a gradual onset. The chief complaints include anorexia, fatigue, weight loss, recurrent abdominal pain, fever, night sweats, persistent nonproductive cough, chest pain, pruritus, skin rash, and congestive heart failure. They occasionally can present with a more aggressive disease process.

Clinical Manifestations

Any organ can be damaged due to sustained exposure to increased eosinophils.

- Skin is the most commonly affected tissue. The most common lesions include pruritic papules and nodules, urticarial plaques, and angioedema. Vesiculobullous lesions, generalized erythroderma, and aquagenic pruritus occur in some patients. Digital necrosis may result from vasculitis and microthrombi.
- The next most common organ is the lung. Pulmonary complications include nocturnal cough, fever, and diaphoresis that can occur due to accumulation of eosinophils in the lungs. Pulmonary fibrosis can also occur.
- Gastrointestinal manifestations include hepatomegaly due to eosinophilic infiltration of the liver with resulting liver function abnormalities, enteropathy due to blunting of the villi and cellular infiltration in the lamina propria resulting in diarrhea and fat malabsorption, and eosinophilic infiltration of the colon results in colitis.
- Eosinophilia can cause a hypercoagulable state, the etiology of which is unclear. Eosinophil major basic proteins inactivate thrombomodulin resulting in unavailability of activated protein C. Intracardiac thrombus, deep venous thrombosis, dural sinovenous thrombosis and/or arterial thrombosis can occur.
- Cardiac involvement is less common but can be the most debilitating. Cardiac disease involves both ventricles and can cause incompetence of mitral and tricuspid valves. HES-associated heart disease evolves through three stages:
  - The early acute phase is associated with degranulating eosinophils in the heart muscle (5–6 weeks into eosinophilia).
  - The subacute thrombotic stage occurs 10 months into eosinophilia.
  - The chronic stage of fibrosis occurs 24 months into eosinophilia.
- CNS manifestations of HES include: encephalopathy, thrombotic strokes, peripheral neuropathies including mononeuritis multiplex, symmetrical sensory–motor neuropathy and radiculopathy, and retinal hemorrhages.

Treatment

Table 13.11 lists the treatments for idiopathic HES.

TABLE 13.11   Treatment of Eosinophilia

Treatment of reactive or non-clonal eosinophilia: Treat the underlying cause e.g. treatment of parasitic infections with appropriate anti-parasitic drugs

Treatment of clonal disease with HE: Treat with appropriate therapy for the underlying malignancy. Refer to chapters on AML (Chapter 19), ALL (Chapter 18) and NHL (Chapter 22).

Treatment of PDGFRA/B rearranged MN with associated HE: Imatinib mesylate, adult dose: 100–400 mg/day. Pediatric dose: not established. Use of steroids in the first 7–10 days is recommended for patients with elevated troponin or other signs of cardiac compromise until the eosinophil count is in the normal range.

Treatment of FGFR1 associated HE: These disorders tend to be aggressive and AML-like therapy followed by stem cell transplant is required to induce a durable remission.

Treatment of idiopathic HES:
- Glucocorticoids, hydroxyurea, alpha-interferon, vincristine, cyclophosphamide, or etoposide. Case reports of 2-chlorodeoxy adenosine, cytarabine and cyclosporine A have been published.
- Steroids and hydroxyurea are considered first line therapy either alone or in combination. During acute life threatening presentation of HES, high dose 10–20 mg/kg of methylprednisolone may be required, but usually 1–2 mg/kg of prednisone is sufficient.
- Use the other agents sequentially.
- If the response is unsatisfactory, then treat with imatinib at doses of ≥ 400 mg/day have been used with some success.
- Use of antibody therapy against IL-5 and CD52 are under investigation in clinical trials for patients that are steroid refractory or steroid dependent. Treatment with allogeneic hematopoietic stem cell transplantation is reserved for patients with HES refractory to above mentioned therapies.

Treatment of patients with HE_US without organ involvement: None. Treatment is not necessary, but continuous periodic monitoring for organ involvement and emergence of clonality is warranted. Also, continue search for rare reactive causes of eosinophilia.

The following eosinophilic disorders with single organ involvement may progress into HES:
- Eosinophilic gastroenteritis
- Gleich syndrome (episodic eosinophilia with angioedema)
- Loeffler syndrome
- Schulman syndrome (eosinophilic fascitis)
- Well syndrome (eosinophilic cellulitis)
- Parasitic infections with eosinophilia

- Glucocorticoids are the primary therapy, but if the disease process is not well controlled other therapies including hydroxyurea, alpha-interferon, vincristine, cyclophosphamide, or etoposide may be used sequentially.
- Imatinib may be helpful in idiopathic HES but generally not if IL-5 levels are elevated.
- Monoclonal antibody therapies including mepolizumab (anti IL-5) and alemtuzumab (anti CD52) are in clinical trials for patients with steroid-refractory or steroid-dependent disease.

Nonidiopathic Eosinophilia

Eosinophilia, for which a cause is ascertained, can be clonal or reactive. Patients with reactive eosinophilia should be treated with therapy directed at the underlying disease process, for example, antiparasitic therapy (see Table 13.10). Clonal disorders can be subdivided into primary clonal diseases, secondary clonal involvement such as in myeloproliferative disorders, or eosinophilia due to release of cytokines by lymphoid or myeloid neoplastic disease (see Figure 13.4).

Primary Clonal Eosinophilic Disorders

In 2008, the WHO developed definitions for HE associated with myeloid, lymphoid, and hematopoietic stem cell neoplasms based upon molecular and chromosomal defects. They proposed separate categories for eosinophilic neoplasms with molecular rearrangements of platelet-derived growth factor receptor alpha and beta (*PDGFRA* and *PDGFRB*) and fibroblast growth factor receptor 1 (*FGFR1*).

PDGFRA, PDGFRB, FGFR1 Rearranged Neoplasms

A myeloid neoplasm (MN) resulting from a fusion gene *FIP1L1-PDGFRA* occurring as a result of interstitial deletion on chromosome 4q12 has been described and accounts for 10–20% of patients with clonal HE. The fusion gene makes an activated tyrosine kinase, which results in a myeloproliferative variant of HES. Rarely, other partners of *PDGFRA* have been identified that lead to a similar phenotype. It is characterized by increased levels of tryptase, increased atypical mast cells in bone marrow, and tissue fibrosis (myelofibrosis, endomyocardial fibrosis, pulmonary fibrosis).

They respond well to imatinib mesylate treatment, which targets the fusion tyrosine kinase, at a dose of 100–400 mg/day. Hematologic remissions are nearly universal within 1 month of therapy and molecular remissions are achieved in 3 months to a year. Some patients with *FIP1L1-PDGFRA* fusion gene may not have the classic characteristics of a myeloproliferative variant of HES and also respond well to imatinib. Some patients with HES may not have *FIP1L1-PDGFRA* fusion gene and still respond to imatinib.

Imatinib treatment for these patients is likely indefinite, similar to chronic myelogenous leukemia (CML), as discontinuation of therapy has been associated with recurrence of the clone. Occasional patients have had durable remissions after stopping therapy. The majority of patients who have recurred responded to reinitiating imatinib therapy. Rare resistant clonal disease after years of treatment with imatinib has been reported.

Clonal rearrangements of *PDGFRB* and *FGFR1* have been less commonly associated with HE. These gene defects are easily identified by fluorescent *in situ* hybridization (FISH) looking for rearrangements of *PDGFRA*, *PDGFRB*, *FGFR1*. The majority of *FIP1L1-PDGFRA* patients are male, whereas, the other chromosomal defects do not demonstrate gender bias.

Secondary Clonal Eosinophilic Disorders

Secondary clonal involvement of eosinophil lineage can occur in myeloproliferative disorders of stem cell origin, for example Ph[1]-chromosome (*BCR/ABL*)-positive CML. Eosinophilia is also observed in *JAK2* V617F myeloproliferative neoplasms, (polycythemia vera and essential thrombocythemia), *KIT* D816V systemic mastocytosis, core binding factor beta-fusion related acute myelocytic leukemia (AML) (formerly FAB M4Eo), primary myelofibrosis, and MDS-Eo. The following cytogenetic abnormalities associated with acute myeloid leukemia with eosinophilia have been reported: inv (16) (p13:q22), t(16:16) (p13:q22), t(5;16) (q33:q22), and monosomy 7.

Secondary Eosinophilic Clonal Disorders

In secondary eosinophilic clonal disorders, clonal cells release eosinopoietic cytokines and, thus, are associated with eosinophilia. Lymphoid clonal disorders associated with eosinophilia include dermatologic patients with abnormal clones of T-cells producing interleukin-5, patients with acute lymphoblastic leukemia, and T lymphoblastic lymphoma. Patients who present with T-cell lymphoblastic lymphoma and eosinophilia may be predisposed to developing secondary AML.

Eosinophilia in Newborn Period

A mild eosinophilia, with eosinophil count greater than $700/mm^3$, is observed in 75% of growing preterm infants. It is present in the second or third weeks of life and persists for several days or sometimes for weeks. Eosinophilia of prematurity is considered to be benign, although it could be associated with a higher incidence of sepsis, especially with Gram-negative bacteria.

A complete absence of eosinophils is observed in neonates who fare poorly and subsequently die.

Familial Eosinophilia

Familial eosinophilia is an autosomal dominant disorder. A genome-wide search showed evidence of linkage on chromosome 5q31-33 between markers D55642 and D55816. Some of the affected members are found to have high WBC counts, lower red blood cell counts, intermittent thrombocytopenia, cellular infiltration with mast cells in the liver and bone marrow, or involvement of the heart and nervous system. The levels of IL-3, IL-5, and GM-CSF are normal.

Figure 13.3 shows a list of diagnostic studies for evaluation of eosinophilia.

Table 13.11 outlines the treatment of some of the conditions that are associated with eosinophilia.

LYMPHOCYTES

Lymphocytes are the effector cells for adaptive immunity. They include B-cells, T-cells, and natural killer (NK) cells. The breadth of their function and disease processes is beyond the scope of this chapter. Readers are directed to textbooks or review articles on immunology and immunodeficiency states. This section will briefly address quantitative disorders of lymphocytes.

Normal age-related lymphocyte counts are listed in Appendix 1. After the neutrophilia in the newborn period, lymphocytes are the predominant WBC for the first 2 years of life. The ratio of lymphocytes:neutrophils gradually reverses to the adult neutrophil predominance by 5 years of age.

Table 13.7 lists the causes of lymphocytosis and lymphopenia.

Atypical Lymphocytosis

Infectious Mononucleosis

Infectious mononucleosis (IM) is an acute infectious disease caused by Epstein–Barr virus (EBV). This common disease occurs in epidemic form in children of all ages, but is rare in infants under 6 months of age. IM is usually a benign, self-limited disease; occasionally, it is associated with severe and fatal complications. In young children, the symptoms are often mild and infection may not be apparent. They often exhibit rashes, neutropenia, and may have pneumonia. "Typical" IM is more frequently diagnosed when the primary exposure occurs in adolescents and associated with a higher socioeconomic status in industrialized countries.

Patients with IM are often referred to a hematologist for a suspected malignant disorder given the presentation of a high WBC count with atypical lymphocytosis, lymphadenopathy (sometimes massive), and hepatosplenomegaly. X-linked lymphoproliferative disorder is a fulminant form of EBV infection associated with mutations in the *SH2D1A* gene. EBV infections are confirmed by serologic testing for EBV-specific IgM and IgG antibodies and can also be detected using polymerase chain reaction (PCR) techniques.

TABLE 13.12   Causes of Atypical Lymphocytosis

1. **Less than 20%**
   a. Infections
      i. Bacterial: brucellosis, tuberculosis
      ii. Viral: mumps, varicella, rubeola, rubella, atypical pneumonia, herpes simplex, herpes zoster, roseola infantum, HIV
      iii. Protozoal: toxoplasmosis
      iv. Rickettsial: rickettsialpox
      v. Spirochetal: congenital syphilis, tertiary syphilis
   b. Radiation
   c. Hematologic: Langerhans cell histiocytosis, leukemia, lymphoma, agranulocytosis
   d. Other: lead intoxication, stress
2. **More than 20%**
   a. Infectious mononucleosis
   b. Infectious hepatitis
   c. Posttransfusion syndrome
   d. Cytomegalovirus syndrome
   e. Drug hypersensitivity: *p*-aminosalicylic acid (PAS), phenytoin (Dilantin), mephenytoin (Mesantoin), organic arsenicals

Differential Diagnosis

Mononucleosis-like syndromes are common in infants and children and are generally due to EBV infection. The EBV antibody-negative mononucleosis-like syndrome may be due to the following agents:

- Cytomegalovirus.
- *Toxoplasma gondii*.
- Drugs (para-aminosalicylic acid (PAS), Dilantin, sulfone).
- Other agents (adenovirus, herpes simplex, rubella).

Leukemia and lymphomas must always be considered and can be excluded by a bone marrow examination if sufficient doubt exists about the diagnosis. See Table 13.12 for a list of the causes of atypical lymphocytosis.

Lymphopenia

Decreased or absent lymphocytes in the differential on a CBC should alert one to a diagnosis of severe combined immunodeficiency (SCID) or other immunodeficiency state. Reticular dysgenesis is a severe form of SCID in which neutropenia and lymphopenia are present. Lack of a thymic shadow on chest X-ray, frequent and severe infections, especially with *P. jirovecii* or other opportunistic organisms, and mucocutaneous *Candida* infections can be clues to these disorders. Prompt evaluation by an immunologist to characterize the disorder is warranted.

DISORDERS OF LEUKOCYTE FUNCTION

Patients with defects in neutrophil function typically present in infancy or childhood with recurrent infections. These disorders comprise 20% of immune deficiencies. Normal neutrophil functions against microbes require egress from the marrow, migration from circulating blood to tissues, and effective pathogen killing. Upon activation, fucosylated proteins on neutrophils engage selectins on the vascular endothelium resulting in *tethering* and *rolling* along the vessel. *Firm adhesion*, mediated by the family of adhesion molecules called β2 integrins, controls movement between endothelial cells (*diapedesis*) and *chemotaxis* toward the inflammation in the tissues. Following *phagocytosis*, neutrophil granules release proteases, enzymes, and antibacterial proteins resulting in *oxidative (respiratory burst) killing*. The following will detail the disorders of neutrophil function based on one or more defects.

Table 13.13 classifies the diseases of leukocyte dysfunction and lists the investigations to be carried out in patients with suspected leukocyte dysfunction.

TABLE 13.13 Classification and Investigation of Diseases of Leukocyte Dysfunction

Function	Disease	Genetic defect[a]	Investigations
Adhesion	Leukocyte adhesion deficiency	LAD I- *ITGB2* encodes CD18	Flow cytometry for CD11b/CD18 (Mac-1)
		LAD 2-*SLC35C1* encodes GDP-fucose transporter I	Flow cytometry for CD15s (SLeX) Bombay (hh) red cell phenotype
		LAD-3 FERMT3 endodes kindling-3	Functional assays for neutrophil and platelet adhesion
Chemotaxis	Hyper IgE	*STAT3* *DOCK8* *TYK2*	Serum IgE > 2500 IU/ml Rebuck skin window
Opsonization	Complement deficiency Specific bacterial antibody deficiency		Serum complement levels Immunoglobulin levels Specific bacterial antibody titers
Degranulation	Chédiak–Higashi (CH) syndrome	*CHS1*	Morphologic tests for CH giant granules
	Specific granule deficiency	*C/EBPε*	Morphology—bilobed nuclei and assay of granule-specific protein (lactoferrin)
Oxidative metabolism	Myeloperoxidase deficiency	*MPO* gene (chromosome 17)	Peroxidase stain
	Chronic granulomatous disease (CGD) Leukocyte glutathione synthetase deficiency Glucose-6-phosphate dehydrogenase (G6PD) deficiency	Phagocyte oxidase (see Table 13.14)	Bacterial test Killing of catalase-positive (*S. aureus*) and catalase-negative (streptococcal) bacteria Metabolic tests Nitroblue tetrazolium (NBT) reduction test[b] Flow cytometry for DHR_{123} to R_{123}[c] Glucose $1-^{14}C$ oxidation with phagocytosis Oxygen consumption during phagocytosis H_2O_2-dependent ^{14}C-formate oxidation with phagocytosis Iodine-125 fixation during phagocytosis[d]

[a]*Gene testing is available for many of the diseases listed (see text).*
[b]*Impaired in CGD due to NADPH oxidase deficiency, deficiency of glutathione reductase of G6PD. Normal in myeloperoxidase deficiency.*
[c]*Flow cytometric analysis to detect reduction of dihydroxyrhodamine 123 to rhodamine 123.*
[d]*Impaired in CGD and in myeloperoxidase deficiency.*

DISORDERS OF ADHESION AND CHEMOTAXIS

Leukocyte Adhesion Disorders

Leukocyte Adhesion Deficiency Type I

This rare autosomal recessive disorder is characterized by genetic defects of CD18, common chain of β2 integrins, required for CD11a/CD18 (LFA-1, lymphocyte function antigen), CD11b/CD18 (MAC-1), and CD11c/CD18. Mutations are heterogeneous, with varying severity of disease. Because of the mutation and the absence of Mac-1, neutrophils in Leukocyte adhesion disorders (LAD) type I are not able to attach to the endothelium and undergo transendothelial migration. Their ability for chemotaxis, phagocytosis, degranulation, and respiratory burst activity is also impaired. Survival of neutrophils is prolonged in LAD type I.

Clinical Manifestations
- Neonatal presentation with delayed umbilical cord separation, complicated by omphalitis.
- Persistent neutrophilia and lack of pus formation.
- Frequent skin and periodontal infections.
- Perirectal abscess.
- Sepsis.
- Necrotizing enterocolitis.
- Pneumonia.
- Sinusitis.
- Infecting organisms: *S. aureus, Pseudomonas aeruginosa, Proteus, Escherichia coli, Klebsiella* sp., *Candida albicans, Aspergillus* sp.

Diagnosis

Flow cytometry for assessment of expression of CD11b or CD18 on neutrophil cell surfaces with the use of specific monoclonal antibodies.

Treatment

- For mild and moderate disease: Oral hygiene with antimicrobial mouthwash (e.g., chlorhexidine gluconate), prophylactic use of trimethoprim/sulfamethoxazole (cotrimoxazole), and aggressive treatment of infections.
- For severe cases: Hematopoietic stem cell transplantation.

Leukocyte Adhesion Deficiency Type II

LAD type II is an extremely rare autosomal recessive disorder caused by mutations in GDP-fucose transporter 1, which results in generalized loss of fucosylated proteins including neutrophil Sialyl Lewis X (CD15s).

Clinical Manifestations

- Compared with LAD type I, these patients suffer from less serious types of infections of the skin, lung, and periodontal tissue. They are unable to form pus in spite of extreme leukocytosis of $30,000-150,000/mm^3$.
- Patients have rare Bombay (hh) red cell phenotype due to absence of fucosylated proteins on red blood cells.
- Evidence of lymphocyte dysfunction with reduced delayed-type hypersensitivity reactions.
- Severe mental retardation, short stature.

Diagnosis

- Flow cytometry for leukocyte CD15s.

Treatment

- Antibiotics for active infections and prophylaxis with trimethoprim-sulfamethoxazole.
- A trial of fucose supplementation is recommended.

Leukocyte Adhesion Deficiency Type III

LAD type III is a rare autosomal recessive disorder due to mutations in the gene *FERMT3* encoding for Kindlin-3, a protein that acts as a switch to activate integrins.

Clinical Manifestations

- Severe recurrent infections similar to LAD I.
- There are defects in integrin-mediated platelet aggregation and patients may have a severe bleeding tendency including intracranial hemorrhage and mucosal purpura.
- NK cell activity is also impaired in LAD III.

Diagnosis

- Normal expression of β2 integrins with defective activation.
- Abnormal platelet aggregation.

Treatment

- Hematopoietic stem cell transplantation.

Rac2 GTPase Mutation

Dominant-negative mutation in the Rac2 GTPase binding site causing impaired neutrophil adhesion and motility, decreased nicotinamide adenine dinucleotide phosphate (NADPH) oxidase activity and degranulation has been reported in one infant. Mutation of *Rac2* (located at 22q13.1) results in clinical symptoms of recurrent skin and mucous membrane infections similar to LAD 1.

Treatment

- Hematopoietic stem cell transplantation.

Hyperimmunoglobulin E Syndrome (Job Syndrome)

Genetics

Hyperimmunoglobulin E (hyper-IgE or HIES, Job syndrome) is most commonly an inherited or sporadic auto-somal dominant mutation in *STAT3*, a *JAK* activation transcription factor. Rare forms include autosomal recessive mutations in *DOCK8* or *TYK2* which encode proteins regulating leukocyte signaling.

Clinical Manifestations

- Chronic eczema, delay in shedding primary teeth, hyperextensible joints, scoliosis, osteopenia and tendency to fractures, growth retardation, and coarse facies.
- Recurrent severe Staphylococcal abscesses.
- Recurrent cutaneous, pulmonary, and joint abscesses.
- Chronic candidiasis of mucosa and nails.

Laboratory Findings

- Very high serum IgE level (greater than 2500 IU/ml).
- Defect in T-lymphocytes that results in reduced production of IFN-γ and tumor necrosis factor.
- Molecular basis of hyper-IgE syndrome is unknown.
- Striking defect in neutrophil granulocyte chemotactic responsiveness; neutrophil migration, phagocytosis and bactericidal activity are normal.

Treatment

- Prophylactic antibiotics: Trimethoprim/sulfamethoxazole.

Localized Aggressive Periodontitis in Children (Formerly Known as Localized Juvenile Periodontitis)

Clinical Manifestations

- This disease is characterized by severe alveolar bone loss localized to the first molars and incisors, absence of other infections, and normal neutrophil counts.
- It is frequently associated with the presence of the bacterium *Actinobacillus actinomycetemcomitans*.
- Age of onset: Usually at puberty.

Genetics

- Evidence suggests homozygous mutation of the *Cathepsin C (CTSC)* gene on chromosome 11q14.

Laboratory Findings

- Defective neutrophil chemotaxis *in vitro*. This defect is attributed to neutrophil chemotactic inhibitors. Phagocytosis is also abnormal in many of these patients.
- Unstimulated neutrophils from these patients show reduced Lewis X, sialyl Lewis X, and L-selectin expression.

Treatment

- Deep subgingival scaling and debridement.
- Adjunctive systemic use of antibiotics: Simultaneous use of metronidazole and amoxicillin or azithromycin.

DISORDERS OF NEUTROPHIL GRANULES

Chédiak—Higashi Syndrome

Genetics

- Autosomal recessive syndrome caused by a defect in the *CHS1* gene—a lysosomal trafficking protein leading to abnormal granule formation.
- Most mutations are nonsense or null mutations, resulting in an absent CHS1/LYST protein.

Clinical Manifestations

- Most patients are identified in infancy or early childhood with partial oculocutaneous albinism and repeated pyogenic infections involving the skin, respiratory tract, and mucous membranes.
- The most common causative organisms are *S. aureus*, *Streptococcus pyogenes*, and *Pneumococcus*.
- Most cases are fatal with the mean age at death 6 years.
- Altered granules are seen in all cells including melanocytes, resulting in photophobia, pale optic fundi, nystagmus, partial oculocutaneous albinism, and excessive sweating.
- Neurologic features: ataxia, muscle weakness, decreased deep tendon reflexes, sensory loss, and abnormal electrical evaluations on electroencephalograms, visual- and auditory-evoked potentials.

Hematologic Features

- Anemia, neutropenia, and thrombocytopenia.
- Moderate neutropenia (500−2000 cells/mm^3) due to intramedullary destruction.
- Giant refractile peroxidase-positive granules (1−4 mm) are present in the neutrophils, eosinophils, basophils, and platelets. They stain greenish gray. Ring-shaped lysosomes are present in monocytes.
- Granules fail to discharge lysosomal enzymes into the phagocytic vacuole, leading to increased susceptibility to infection.
- A defect in chemotaxis along with the neutropenia contributes to a further increase in the risk of bacterial infection.
- Platelets are deficient in dense bodies resulting in a storage pool deficiency and mild bleeding diathesis with easy bruising and epistaxis.
- An accelerated phase occurs in 85% of cases. It is characterized by lymphoproliferative infiltration of the CNS and peripheral nerves, liver, spleen, and other organs by histiocytes and atypical lymphocytes. Patients develop fever, jaundice, hepatosplenomegaly, lymphadenopathy, bleeding tendency, and pancytopenia. This inherited hemophagocytic lymphohistiocytosis syndrome (HLH) results from the granule defects in NK cells and lymphocytes and may be triggered by viral infections including Ebstein−Barr virus. Patients with an absence of cytotoxic T-lymphocyte function have a particularly high risk for developing HLH.

Diagnosis

- Examination of the peripheral blood and/or bone marrow reveals giant azurophilic cytoplasmic granules in leukocytes and platelets.

Treatment

- High-dose ascorbic acid (20 mg/kg/day) may normalize the chemotactic defect and bactericidal function, however, efficacy not established.
- Antibiotics:
  − Prophylactic antibiotics: Trimethoprim/sulfamethoxazole.
  − Therapeutic use of antibiotics to treat infections.
- G-CSF has been used to improve neutropenia.
- Treatment of HLH: Current treatments include corticosteroids and etoposide-based regimens followed by hematopoietic stem cell transplantation. For Chédiak−Higashi syndrome (CHS) and EBV-associated HLH, the addition of rituximab is effective.
- Allogeneic hematopoietic stem cell transplantation is potentially curative for hematologic and immunologic defects but does not affect progressive neurologic deterioration.

Neutrophil-Specific Granule Deficiency

Neutrophil-specific granule deficiency (SGD, previously called lactoferrin deficiency) is an autosomal recessive rare disorder due to mutation in the *C/EBPε* gene. It is characterized by the absence of specific or secondary granules, including lactoferrin and defensins, during late myeloid development. Neutrophils exhibit severe chemotactic defects.

Clinical Manifestations

- Recurrent and refractory infections, lung abscesses, and mastoiditis.
- Major pathogens include *S. aureus*, *P. aeruginosa*, other enteric Gram-negative bacteria, and *C. albicans*.

Diagnosis

- Blood smear examination reveals neutrophils that lack specific granules and have bilobed nuclei, resembling the Pelger–Huet anomaly.
- Assay or staining of granule-specific proteins (lactoferrin or gelatinase).

Treatment

- Early diagnosis of infections, antimicrobial prophylaxis, and aggressive management of infectious complications are critical.
- Successful hematopoietic cell transplantation was reported in one patient.

DISORDERS OF OXIDATIVE METABOLISM

Chronic Granulomatous Disease

Pathogenesis

In chronic granulomatous disease (CGD), neutrophils show normal phagocytosis but defective killing of micro-organisms as a result of markedly deficient or absent superoxide production due to inherited mutations of polypeptides of reduced NADPH oxidase (also known as respiratory burst oxidase). Superoxide is a precursor of microbicidal oxidants such as hydrogen peroxide and hypochlorous acid. Figure 13.5 lists the reactions of the respiratory burst pathway.

Organisms that make their own catalase are more often responsible for severe infections in these patients. These organisms are able to convert their own hydrogen peroxide to water, thus making it unavailable to the phagocyte for the microbicidal purpose. As a consequence, the ingested bacteria or fungi remain viable in the phagocytes and are protected from host humoral immunity and from antibiotics, which fail to penetrate the cell.

The mobility of the phagocytes leads to generalized seeding of the reticuloendothelial system with live microorganisms. This results in the formation of generalized chronic granulomatous lesions, characterized by recurrent suppurative infection with bacteria of low virulence (e.g., *S. aureus, S. epidermidis, Aerobacter aerogenes, Serratia marcescens, Burkholderia (Pseudomonas) cepacia*, and *Salmonella*) or infection with mycotic organisms (e.g., *Aspergillus*). *Staphylococcus aureus* is the most frequently isolated organism. However, the most common causes of death are pneumonia and/or sepsis due to *Aspergillus* or *B. cepacia*.

Genetics

CGD results from mutations in any of the four genes encoding essential subunits of the NADPH oxidase. Table 13.14 shows the genetic classification of CGD according to the component of NADPH oxidase affected.

1. Assembly of respiratory oxidase, also known as NADPH oxidase, a multisubunit enzyme complex consisting of four essential phagocyte oxidase (phox) polypeptides: gp91^{phox}, p22^{phox}, p47^{phox}, and p67^{phox}. NADPH oxidase catalyzes the transfer of an electron from NADPH to molecular oxygen as a result of which superoxide ($O_2^{\cdot-}$) is formed.
   $$NADPH + 2O_2 \xrightarrow{\text{NADPH oxidase}} NADP + H^+ + 2O_2^{\cdot-}$$
2. Conversion of superoxide to hydrogen peroxide (H_2O_2) by superoxide dismutase or spontaneously and also, formation of hydroxyl radical ($OH^{\cdot}$),
3. Hypochlorous acid (HOCL) formation: $H_2O_2 \xrightarrow{\text{Myeloperoxidase}} HOCl$
4. Conversion of hydrogen peroxide to water by glutathione peroxidase:
   GSH (reduced glutathione) + H_2O_2 $\xrightarrow{\text{glutathone peroxidase}}$ GSSG (oxidized glutathione) + H_2O
5. Hydrogen peroxide is also converted to water by catalase
6. Restoration of GSH by conversion of GSSG by glutathione reductase:
   NADPH + GSSG $\xrightarrow{\text{glutathione reductase}}$ GSH + NADP
7. Generation of NADPH through G6PD (glucose-6-phosphate dehydrogenase) reaction.

FIGURE 13.5   Reactions of the respiratory burst pathway in a neutrophil (activated for phagocytosis).

TABLE 13.14 Genetic Classification of CGD

Component affected	Gene	Gene locus	Inheritance	Frequency (% of Cases)
gp91^{phox}	CYBB	Xp21.1	X-linked	65
p22^{phox}	CYBA	16q24	AR	7
p47^{phox}	NCF1	7q11.23	AR	23
p67^{phox}	NCF2	1q25	AR	5
P40phox *	NCF4	22q12.3	AR	<1

Abbreviations: phox, phagocyte oxidase; AR, autosomal recessive.

Clinical Manifestations

- Incidence, between one in 200,000 and one in 250,000 live births.
- First symptoms may manifest in infancy or childhood.
- Lymphadenopathy (nodes suppurate and drain pus) and hepatosplenomegaly.
- Recurrent suppurative infections—pneumonitis, subcutaneous abscesses, impetiginous rashes, and osteomyelitis (often small bones of the hands and feet).
- Urologic problems (e.g., granulomatous ureteral or urethral strictures, bladder granulomas, and urinary tract infections) in 38% of cases.
- Gastrointestinal problems: colitis, enteritis, granulomatous obstruction of gastric outlet.
- Hematologic features include:
  - Appropriate neutrophil leukocytosis.
  - Anemia due to infection.
  - McLeod syndrome (i.e., mild hemolytic anemia, acanthocytosis, decreased expression of Kell antigen due to defect in K_x antigen on red cells): This occurs in rare patients with large deletions of *Xp 21.1* gene.
  - Hypergammaglobulinemia.
- Noninfectious complications: Noninfectious granulomas resulting in colitis, granulomatous cystitis and urethritis, cutaneous granulomata, pericarditis, and recurrent gastrointestinal strictures occur in patients with CGD, as a result of a failure of phagocytes to clear both exogenous and endogenous debris.

Diagnosis

Testing for CGD is performed through flow cytometric analysis to detect reduction of dihydroxy-rhodamine$_{123}$ to rhodamine$_{123}$, a fluorescent compound. This can also be used to detect carrier states in X-linked forms. As a quantitative assay, it can also distinguish between forms with no activity, such as gp91^{phox}, versus those with low activity, such as p47^{phox}.

Treatment

- The use of prophylactic antibiotics and antifungal agents:
  - Antibacterial: Trimethoprim/sulfamethoxazole (cotrimoxazole) at a dose of 5 mg/kg/day, once a day orally (based on trimethoprim component). For those patients allergic to sulfa, dicloxacillin or a cephalosporin may be an alternative.
  - Antifungal antibiotic: Itraconazole: 3–5 mg/kg/day once a day orally.
- Aggressive infection management:
  - Meticulous wound care—prompt cleansing of wounds and abrasions with hydrogen peroxide.
  - Prompt treatment with broad-spectrum antibiotics to cover *B. cepacia* and *Staphylococcus* species when serious infection is suspected.
  - Aggressive surgical intervention to drain abscesses when not responding to antibiotics appropriately.
- The use of rhIFN-γ (interferon-γ) in a dose of 50 μg/m^2 subcutaneously three times a week: Beneficial effect of interferon-γ is probably related to increased synthesis of nitric oxide (NO) through nitric oxide synthase pathway. NO causes nitration of bacteria. Interferon-γ may also be stimulating other nonoxidative microbicidal pathways. It also increases superoxide production in phagocytes. Interferon-γ as prophylaxis has been recommended only in patients with significant infections despite appropriate oral agents, or as an adjunct to the treatment of deep-seated infections. Oral antibiotic prophylaxis appears to be adequate in most patients.

- Granulocyte transfusions may provide short-term relief at times of crisis.
- Granulomatous disease in the gastrointestinal or urinary tracts is managed with a combination of antibiotics and steroids. However, steroids should be used cautiously, even at low doses.
- Allogeneic stem cell transplantation can be successful but is reserved for patients with severe disease. Current improved outcomes with decreased complications, even with refractory chronic infections, are reported with matched unrelated donors and reduced-intensity conditioning regimens.
- Gene therapy is also a consideration for the future but no therapies are currently available.

Prognosis

With the use of prophylactic antibiotics and recombinant human interferon-γ (rhIFN-γ) the prognosis of CGD has improved remarkably.

Myeloperoxidase Deficiency

Myeloperoxidase (MPO) deficiency is an autosomal recessive inherited disorder due to mutations of the *MPO* gene localized to 17q23. It is the most common inherited disorder of phagocytes with a lack of clinical symptoms in most patients. Acquired or secondary MPO deficiency can occur in patients with AML, MDS and CML, iron deficiency, or pregnancy.

MPO is found in the azurophilic granules and catalyzes the formation of hypochlorous acid from chloride and hydrogen peroxide. Although neutrophils show marked defect in the killing of *Candida in vitro*, MPO deficiency is rarely associated with an increased incidence of infection in patients except for patients with diabetes mellitus.

Diagnosis

- Measurement of peroxidase activity and histochemical staining of neutrophils. This is available in commercial laboratories.

Treatment

- No treatment or prophylaxis is usually required.

Patient with MPO deficiency and diabetes can develop disseminated *Candida* and require an aggressive approach to prevent and treat fungal infections.

Glutathione Pathway Disorders

The reduced form of glutathione (GSH) protects cells from the effects of reactive oxygen species. Glutathione reductase (on 8p21.1) and glutathione synthetase (GSS) (on 20q11.2) are involved in maintaining adequate intracellular levels of GSH and critical for preserving (NADPH) oxidase activity in neutrophils.

Glutathione Synthetase Deficiency

- Rare autosomal recessive disorder.
- Associated with hemolysis after increased oxidant stress.
- Severely affected patients have hemolytic anemia, metabolic acidosis with increased 5-oxoproline levels, and progressive CNS dysfunction.
- Recurrent bacterial infections, and impairments in phagocytosis and intracellular bacterial killing.

Diagnosis

- Detection of elevated concentrations of 5-oxoproline in urine and low GSS activity in erythrocytes or cultured skin fibroblasts and by mutational analysis.

Treatment

- Avoid exposure to drugs and chemicals with oxidant.
- Daily vitamin E 400 units can improve neutrophil function and decrease infection.
- Aggressive antibiotic therapy for infections.

Glutathione Reductase Deficiency

- Rare autosomal recessive disorder.
- Accumulation of H_2O_2 in neutrophils.
- Hemolysis induced by oxidant stress.
- Frequency of infections *not* increased.

Glucose-6-Phosphate Dehydrogenase Deficiency in Leukocytes

Glucose-6-phosphate dehydrogenase (G6PD) deficiency, the most common enzyme defect in humans, is inherited in an X-linked pattern (Xq28). In rare cases of G6PD deficiency, the enzyme is severely depressed (less than 5% of normal) in the neutrophil. As a result the conversion of NADP to NADPH is decreased, leading to decreased respiratory burst. There is persistent and eventually fatal bacterial infection because of the inability of the leukocytes to generate H_2O_2. Clinical manifestations are similar to CGD, with predominantly catalase-positive bacteria. Hemolytic anemia due to low erythrocyte G6PD level is the most common symptom.

Treatment

- Similar to CGD except rIFN-γ not proven to be effective.
- Aggressive treatment of infections.
- Avoidance of oxidant drugs and foods.
- Red cell transfusions for anemia.

NEUTROPHIL PRODUCTION AND DESTRUCTION IN NEWBORN INFANTS

Neutrophil Production

Using soluble Fc receptor III as a surrogate marker for estimation of total neutrophil mass, it has been shown that newborn infants born before 32 weeks of gestation have 20% of the adult neutrophil mass. Normal levels of neutrophil mass are attained at 4 weeks of age in premature neonates. However, full-term neonates have neutrophil stores within the normal adult range.

NEUTROPHIL FUNCTION IN THE NEWBORN

Chemotaxis

Chemotaxis is decreased in newborn infants. Normal chemotactic ability is attained at 2 weeks of age in term and preterm neonates.

Vascular Rolling of Neutrophils

Vascular rolling is also decreased due to decreased expression of L-selectin on the neutrophil cell membrane.

Vascular Endothelial Adhesion

Impaired endothelial adhesion due to decreased expression of β_2 integrin Mac-1 on the neutrophil cell membrane in neonates.

Dynamics of Change of Shape

Newborn neutrophils are rigid because of impaired ability to form polymers of actin (P-actin) and reduced formation of microtubules.

Phagocytosis

Neutrophils of term neonates have a normal ability for phagocytosis of Gram-positive and Gram-negative bacteria. However, their ability for phagocytosis of *Candida* is abnormal up to 2 weeks of age. Preterm neonates have abnormal bacterial phagocytosis. However, when treated with therapeutic doses of intravenous immunoglobulin the neutrophils are able to ingest bacteria normally.

Respiratory Burst

Respiratory burst in term neonates is normal under normal conditions. In contrast, it is less active under the conditions of stress and sepsis. For this reason, neonates are more susceptible to overwhelming infections with group B Streptococci, *Staphylococcus epidermis*, *Staphylococcus aureus*, and *E. coli*. Respiratory burst performance of neutrophils in preterm neonates can remain abnormal for more than 2 months of age. In term neonates, the generation of superoxide and hydrogen peroxide is increased. However, since the levels of lactoferrin and myeloperoxidase are low there is truncation of the later respiratory burst activity, resulting in abnormal bacterial killing.

Therapeutic Implications

G-CSF or GM-CSF may be helpful in term and preterm neonates during sepsis. The prophylactic use of G-CSF is more effective in preterm infants.

Neonatal Preeclampsia-Associated Neutropenia

Neutropenia occurs in low-birth-weight neonates with a maternal history of pregnancy-induced hypertension.

Treatment

- Prophylactic use of G-CSF.

Further Reading and References

Antachopouloss, C., Walsh, T., 2007. Roilides. Fungal infections in primary immunodeficiencies. Eur. J. Pediatr. 166, 1099—1117.

Amulic, F., Cazalet, C., Metzler, K.D., Zychinsky, A., 2012. Neutrophil function from mechanism to disease. Annu. Rev. Immunol. 30, 459—489.

Berliner N., 2004. Acquired Neutropenia. American Society of Hematology (ASH) Education Book. 69 p. San Diego, CA.

Bux, J., 2002. Molecular nature of antigens implicated in immune neutropenias. Int. J. Hematol. 76 (Suppl. 1), 399—403.

Bux, J., Mueller-Eckhardt, C., 1992. Autoimmune neutropenia. Semin. Hematol. 29, 45—53.

Carr, R., 2000. Neutrophil production and function in newborn infants. Br. J. Haematol. 110, 18—28.

Cartron, J., Tchernia, G., Cleton, J., et al., 1991. Alloimmune neonatal neutropenia. Am. J. Pediatr. Hematol. Oncol. 13, 21—25.

Cogan, E., Roufosse, F., 2012. Clinical management of the hypereosinophilic syndromes. Exp. Rev. Hematol. 5 (3), 275—289, quiz 290. http://dx.doi.org/10.1586/ehm.12.14.

Dale, D.C. (Ed.), 2002. Guest Editor: severe chronic neutropenia. Semin. Hematol. Saunders, Philadelphia, Vol. 39, pp. 73—133.

Dale, D.C., Cottle, T.E., Fier, C.J., et al., 2003. Severe chronic neutropenia: treatment and follow-up of the patients in the Severe Chronic Neutropenia International registry. Am. J. Hematol. 72, 82.

Dale, D.C., Link, D.C., 2009. The many causes of severe congenital neutropenia. N. Engl. J. Med. 360, 3.

Dinauer, M.C., 1998. The phagocyte system and disorders of granulopoiesis and granulocyte function. In: Nathan, D.G., Orkin, S.H., Ginsburg, D., Look, A.T. (Eds.), Nathan and Oski's Hematology of Infancy and Childhood, fifth ed. Saunders, Philadelphia.

Dinauer, M.C., 2014. Disorders of neutrophil function: an overview. Methods Mol. Biol. 1124, 501—515.

Gotlib, J., 2014. World Health Organization-defined eosinophilic disorders: 2014 update on diagnosis, risk stratification, and management. Am. J. Hematol. 89 (3), 325—337.

Hager, M., Cowland, J.B., Boregaard, N., 2010. Neutrophil granules in health and disease. J. Int. Med. 268, 25—34.

Johannsen, E.C., Schooley, R.T., Kaye, K.M., 2005. Epstein—Barr Virus (Infectious Mononucleosis). In: Mandell, G.L., Bennett, J.E., Dolin, R. (Eds.), Principles and Practice of Infectious Diseases, sixth ed. Churchill Livingstone, Philadelphia.

Karlsson, S., 1991. Treatment of genetic defects in hematopoietic cell function by gene transfer. Blood 78, 2481—2492.

Lalezari, P., Khorshidi, M., Petrosova, M., 1986. Autoimmune neutropenia of infancy. J. Pediatr. 109, 764.

Newburger, P.E., Dale, D.C., 2013. Evaluation and management of patients with isolated neutropenia. Semin. Hematol. 50, 198—206.

Roskos, R.R., Boxer, L.A., 1991. Clinical disorders of neutropenia. Pediatr. Rev. 12, 202—218.

Shastri, K.A., Logue, G.L., 1993. Autoimmune neutropenia. Blood 81, 1984—1995.

Sullivan, J.L., 1987. Hematologic consequences of Epstein—Barr virus infection. Hematol. Oncol. Clin. N. Am. 1, 397.

Valent, P., Klion, A., et al., 2012. Contemporary consensus proposal on criteria and classification of eosinophilic disorders and related syndromes. J. Allergy Clin. Immunol. 130, 607—612.

Vickery, J.D., Michael, C.F., Lew, D.B., 2013. Evaluation of B lymphocyte deficiencies. Cardiovasc. Hematol. Disord. Drug Targets 13, 133—143.

Yang, K., Hill, H., 1991. Neutrophil function disorders: pathophysiology, prevention and therapy. J. Pediatr. 119, 343—354.

14

Disorders of Platelets

Catherine McGuinn and James B. Bussel

Platelets are an important component in primary hemostasis. Defects in platelet number or function may lead to bleeding and bruising. Bleeding due to platelet disorders usually involves skin and mucous membranes, presenting as petechiae, purpura, ecchymosis, epistaxis, hematuria, menorrhagia, as well as gastrointestinal and even intracranial hemorrhage (ICH).

Platelet characteristics include:

- Size: 1–4 μm (younger platelets are larger). Table 14.1 lists the causes of thrombocytopenia based on platelet size.
- Mean platelet volume (MPV): 8.9 ± 1.5 fl.
- Distribution: one-third in the spleen, two-thirds in circulation.
- Average lifespan: 9–10 days.

Table 14.2 lists the causes of thrombocytopenia according to pathophysiology and Table 14.3 lists the common and uncommon causes of thrombocytopenia in the neonate and child.

THROMBOCYTOPENIA IN THE NEWBORN

Neonatal thrombocytopenia is relatively common, occurring in 1–3% of healthy term infants and in 20–30% in the neonatal intensive care unit population. Thrombocytopenia in sick neonates is often secondary to an underlying pathology such as sepsis, disseminated intravascular coagulation (DIC), or respiratory distress syndrome or secondary to maternal factors such as pregnancy-induced hypertension, gestational diabetes, and intrauterine growth retardation (IUGR).

Table 14.4 lists the causes of neonatal thrombocytopenia.

Neonatal Alloimmune Thrombocytopenia

Neonatal alloimmune thrombocytopenia (NAIT) is the most common cause of severe thrombocytopenia in the newborn, with an overall incidence of approximately 1 in 1000 births but is perhaps as frequent as 1:3–5000 births when defined as a platelet count <50,000/mm³ NAIT typically resolves in 2–4 weeks. First-born infants are 25–50% of those affected and subsequent affected pregnancies have increasingly severe presentation and require antenatal treatment.

Pathophysiology

NAIT can be thought of as a platelet analog of Rh incompatibility (i.e., hemolytic disease of the fetus and newborn). It differs from Rh incompatibility because many cases are first-born infants, suggesting the antigenic exposure occurs early in pregnancy unlike in Rh, which occurs primarily at the time of delivery. NAIT occurs when fetal platelets that express platelet-specific antigens inherited from the father, are the target of maternal alloantibodies. Mothers who lack the paternally inherited platelet-specific surface antigen, and who possess the immunologic predisposition to make antibodies to it, can become sensitized when they are expressed on fetal platelets or

TABLE 14.1   Platelet Diseases Based on Platelet Size

MACROTHROMBOCYTES (MPV RAISED)
ITP or any condition with increased platelet turnover (e.g., DIC)
Bernard—Soulier syndrome
May—Hegglin anomaly and other MYH9-related diseases
Montreal platelet syndrome
Gray platelet syndrome
Various mucopolysaccharidoses

NORMAL SIZE (MPV NORMAL)
Conditions in which marrow is hypocellular or infiltrated with malignant disease

MICROTHROMBOCYTES (MPV DECREASED)
Wiskott—Aldrich syndrome
TAR syndrome
Some storage pool diseases
Cytomegalovirus infection

MPV, mean platelet volume (as determined by automated electronic counters); normal, $8.9 \pm 1.5\ \mu m^3$; ITP, idiopathic thrombocytopenic purpura; DIC, disseminated intravascular coagulation; MYH-9, non-muscle myosin heavy chain 9 gene; TAR, thrombocytopenic absent radii. *Inherited thrombocytopenias are not well characterized and may present with normal/large platelet size.*

possibly also from the A_V/B_{III} receptor on the trophoblast. The most common antigen involved is HPA-1a, which accounts for approximately 75% of cases. A further 10—20% of cases are due to maternal sensitization to HPA-5b. More than 20 other antigens are known to be involved in NAIT. HPA-4 is important in Asian populations which do not have the HPA-1A/B polymorphism. Mothers who possess the HLA-DR type DRB30101 represent >90% of cases of sensitization to HPA-1a. These IgG antibodies cross the placenta and attach to the surface of fetal platelets, causing platelet destruction and perhaps inhibition of platelet production.

Clinical Features

- Typically infants are otherwise healthy full-term babies, who manifest symptomatic thrombocytopenia with generalized petechiae, ecchymosis, cephalohematomata, umbilical bleeding, oozing from skin puncture sites, and/or gastrointestinal or renal tract bleeding.
- Affected neonates have rates of ICH up to 10—20% and, when present, the ICH tends to be severe and intraparenchymal. ICH is now thought to frequently occur in utero and may be detected on ultrasonography during apparently uncomplicated pregnancies. Death in utero may occur.
- Platelet count is very low at birth, usually <50,000/mm^3. Cases of HPA-5b incompatibility are milder.

Diagnosis

NAIT should be considered in all newborns with thrombocytopenia.

Ninety percent of cases of HPA-1a incompatibility NAIT have a platelet count of <50,000/mm^3, making it a reasonable screening tool for identifying NAIT. Two other reasons to suspect NAIT, even if the neonatal count is >50,000/mm^3, include:

- No clinically apparent etiology of thrombocytopenia.
- Family history of transient neonatal thrombocytopenia.

Response to random platelet transfusion or lack thereof is NOT diagnostically useful.

It is important to investigate and establish the diagnosis because of the impact on subsequent pregnancies and hence their management. Laboratory evaluation should include screening for HPA-1, 3, and 5 antibodies, as well as HPA-4 antibodies in those of Asian descent. HPA-9 and 15 are the next most common antigen incompatibilities. To confirm the diagnosis of NAIT, ideally testing must show both platelet antigen incompatibility and antibodies to the discordant antigen.

TABLE 14.2 Pathophysiological Classification of Thrombocytopenic States

1. Increased platelet destruction (normal or increased megakaryocytes in the marrow—megakaryocytic thrombocytopenia)
   a. Immune thrombocytopenias (also with decreased production)
      i. Idiopathic
         - Immune (idiopathic) thrombocytopenic purpura
      ii. Secondary
         - Infection induced (e.g., viral—HIV, CMV, EBV, varicella, rubella, rubeola, mumps, measles, pertussis, hepatitis, parvovirus B19; bacterial—tuberculosis, typhoid)
         - Drug-induced (see Table 14.4)
         - Posttransfusion purpura
         - Autoimmune hemolytic anemia (Evans syndrome)
         - Systemic lupus erythematosus
         - Hyperthyroidism
         - Lymphoproliferative disorders
      iii. Neonatal immune thrombocytopenias
         - Neonatal autoimmune thrombocytopenia
         - Neonatal alloimmune thrombocytopenia
         - Erythroblastosis fetalis—Rh incompatibility
   b. Nonimmune thrombocytopenias
      i. Due to platelet consumption
         - Microangiopathic hemolytic anemia: hemolytic-uremic syndrome (HUS), thrombotic thrombocytopenic purpura (TTP), hematopoietic stem cell transplantation (HSCT) associated microangiopathy
         - Disseminated intravascular coagulation
         - Virus-associated hemophagocytic syndrome
         - Kasabach–Merritt syndrome (giant hemangioma)
         - Cyanotic heart disease
      ii. Due to platelet destruction
         - Drugs (e.g., ristocetin, protamine sulfate, bleomycin)
         - Left ventricular outflow obstruction
         - Infections
         - Cardiac (e.g., prosthetic heart valves, repair of intracardiac defects, left ventricular outflow obstruction)
         - Malignant hypertension
2. Disorders of platelet distribution or pooling
   a. Hypersplenism (e.g., portal hypertension, Gaucher disease, cyanotic congenital heart disease, neoplastic, infectious)
   b. Hypothermia
3. Decreased platelet production—deficient thrombopoiesis (decreased or absent megakaryocytes in the marrow—amegakaryocytic thrombocytopenia)
   a. Hypoplasia or suppression of megakaryocytes[a]
      i. Drugs (e.g., chlorthiazides, estrogenic hormones, ethanol, tolbutamide)
      ii. Constitutional
         - Thrombocytopenia absent radii (TAR) syndrome
         - Congenital amegakaryocytic thrombocytopenia
         - Amegakaryocytic thrombocytopenia with radio-ulnar synostosis
         - Thrombocytopenia agenesis of corpus callosum syndrome
         - Paris–Troussea syndrome
         - Rubella syndrome
         - Trisomies 13, 18
      iii. Ineffective thrombopoiesis
         - Megaloblastic anemias (folate and vitamin B12 deficiencies)
         - Severe iron-deficiency anemia
         - Certain familial thrombocytopenias
         - Paroxysmal nocturnal hemoglobinuria
      iv. Disorders of control mechanism
         - Thrombopoietin deficiency
         - Tidal platelet dysgenesis
         - Cyclic thrombocytopenias
      v. Metabolic disorders
         - Methylmalonic acidemia
         - Ketotic glycinemia
         - Holocarboxylase synthetase deficiency
         - Isovaleric acidemia
         - Idiopathic hyperglycinemia
         - Infants born to hypothyroid mothers

(Continued)

TABLE 14.2   (*Continued*)

      vi. Hereditary platelet disorders[b]
- Bernard—Soulier syndrome
- May—Hegglin anomaly and other MYH-9 gene-related disorders (Table 14.9)
- Wiskott—Aldrich syndrome
- Pure sex-linked thrombocytopenia
- Mediterranean thrombocytopenia

      vii. Acquired aplastic disorders
- Idiopathic
- Drug-induced (e.g., dose-related: antineoplastic agents; benzene, organic and in organic arsenicals, Mesantoin, Tridione, antithyroids, antidiabetics, antihistamines, phenylbutazone, insecticides, gold compounds; idiosyncrasy: chloramphenicol)
- Radiation-induced
- Viral infections (e.g., hepatitis, HIV, EBV)

  b. Marrow infiltrative processes
    i. Benign
- Osteopetrosis
- Storage diseases

    ii. Malignant
- *De novo*—leukemias, myelofibrosis, Langerhans cell histiocytosis, histiocytic medullary reticulosis
- Secondary—lymphomas, neuroblastoma, other solid tumor metastases

4. Pseudothrombocytopenia:
  a. Platelet activation during blood collection
  b. Undercounting of megathrombocytes
  c. *In vitro* agglutination of platelets due to EDTA
  d. Monoclonal antibodies that bind to platelet glycoprotein receptors such as abciximab, eptifibatide, tirobifan

[a]*A bone marrow biopsy, in addition to marrow aspiration, should always be carried out to avoid sampling errors and to establish the presence of a decreased number of megakaryocytes in the marrow.*
[b]*These conditions are associated with normal or increased bone marrow megakaryocytes.*
MPV, mean platelet volume; MYH-9, non-muscle myosin heavy chain 9 gene; EBV, Epstein—Barr virus; CMV, cytomegalovirus.

Additional useful clinical criteria for the diagnosis in addition to severe congenital thrombocytopenia ($<50,000/mm^3$) include:

- Normal non-pregnant maternal platelet count and negative history of maternal immune thrombocytopenic purpura (ITP).
- Exclusion of alternate diagnoses.
- Recovery of normal platelet count within 2—3 weeks.
- History of NAIT in a prior pregnancy.
- Increased megakaryocytes in bone marrow examination (if performed).

Treatment

The following treatment is recommended:

1. Platelet transfusion 10—20 ml/kg body weight. Maternal or matched platelets are infrequently necessary for effective therapy since there is often good efficacy with transfusion of unmatched platelets.
2. Matched donor platelets or concentrated maternal platelets may be used, if available, especially if unrelated donor platelet transfusions are ineffective.
3. Intravenous immunoglobulin (IVIG) 1 g/kg/day for 1—3 days, especially in combination with random platelet transfusion, depending on response with the goal of platelet count being above 30—50,000/ mm^3.
4. Methylprednisolone (1 mg IV) every 8 h with IVIG until the IVIG is stopped (no tapering is necessary).
5. Head ultrasound is mandatory for the thrombocytopenic neonate. If there are any abnormal neurological findings, a CT or MRI should also be done. If ICH is present in NAIT on ultrasound, the target platelet count should be greater than $100,000/mm^3$ and a head CT or MRI should be performed to better define the hemorrhage. Imaging should be repeated to document stabilization/improvement and then monthly for 3 months to identify early hydrocephalus, along with head circumference measurements.
6. With or without treatment, follow-up until the platelet count is within the normal range will avoid missing inherited causes of thrombocytopenia.

TABLE 14.3 Classification of Thrombocytopenic Purpura by Age and Frequency

	Common	Uncommon
Neonate	Sepsis	Cardiac (prosthetic heart valves, repair of intracardiac defects, left ventricular outflow obstruction)
	Asphyxia	
	Alloimmune thrombocytopenia	Maternal hypertension
	Necrotizing enterocolitis	Infections (rubella, CMV, HIV, hepatitis B, syphilis)
	Maternal ITP	Amegakaryocytic thrombocytopenias
		Congenital amegakaryocytic thrombocytopenia without anomalies (CAMT)
		Congenital amegakaryocytic thrombocytopenia with bilateral absence of radii (TAR syndrome)
		Amegakaryocytic thrombocytopenia with radio-ulnar synostosis (ATRUS)
		Wiskott–Aldrich and X-linked thrombocytopenia
		Bernard–Soulier syndrome
		MYH9 disorders
		Montreal platelet syndrome
		Quebec syndrome
		Gray platelet syndrome
		Inborn errors of metabolism
		Congenital leukemia (trisomies 13, 18, 21)
Child	ITP	Type II von Willebrand disease
	Drug-induced	Autoimmune diseases such as SLE, JIA
	Immunodeficiency	Infections
		Fanconi anemia
		Leukemia and other malignancies with bone marrow involvement
		Autoimmune hemolytic anemia (Evans syndrome)
		TTP/HUS
		Hyperthyroidism
		Megaloblastic anemias (folate and vitamin B12 deficiencies)
		Severe iron-deficiency anemia
		Cyclic thrombocytopenias
		Lymphoproliferative disorders
		Aplastic anemia
		Drug-induced
		Radiation-induced

CMV, cytomegalovirus; SLE, systemic lupus erythematosus; ITP, idiopathic thrombocytopenic purpura; JIA, Juvenile Idiopathic Arthritis; HUS, hemolytic-uremic syndrome; TTP, thrombotic thrombocytopenic purpura.

Management of Subsequent Pregnancies

Identification of a family at risk for NAIT is critical to help stratify the antenatal management of future pregnancies with the goal of preventing fetal and neonatal ICH. Previous stratification included the sampling of fetal blood to determine antigen expression and platelet count, but given the invasiveness of this procedure and the risks of serious complications, current approaches are based on noninvasive information. Management of subsequent pregnancies should be undertaken by specialists in maternal fetal medicine, when available.

TABLE 14.4   Causes of Neonatal Thrombocytopenia

1. **Normal or increased megakaryocytes in the marrow (consumptive thrombocytopenia)**
    a. Immune disorders
        i. Autoimmune (passive transfer of platelet antibody) (NITP)
            - Maternal ITP
            - Maternal drug-induced thrombocytopenia
            - Maternal SLE
        ii. Alloimmune (NAIT)[a]
            - Isolated platelet antigen incompatibility
    b. Infection
        i. Bacterial: Gram-negative and Gram-positive septicemia, listeriosis
        ii. Viral: cytomegalovirus, rubella, herpes simplex, coxsackievirus, HIV[a]
        iii. Protozoal: toxoplasmosis
        iv. Spirochetal: syphilis
    c. Drugs
        i. Immune: drug-hapten disease (e.g., quinine, quinidine, sedormid)
        ii. Nonimmune: thiazide, tolbutamide (given to mother)
        iii. Prolonged antibiotics or ganciclovir
        iv. Chemotherapy
    d. Disseminated intravascular coagulation
        i. Antenatal causes
            - Preeclampsia and eclampsia
            - Abruptio placentae
            - Dead twin fetus
            - Amniotic fluid embolism
        ii. Intranatal causes
            - Breech delivery
            - Fetal distress
        iii. Postnatal causes
            - Infections
            - Hypoxia and acidosis
            - Respiratory distress syndrome
            - Renal vein thrombosis
            - Indwelling catheters
            - Giant hemangioma (usually in the months after birth)
    e. Inherited thrombocytopenia
        i. Sex-linked
            - Gata-1
            - Wiskott–Aldrich syndrome[a]
        ii. Autosomal
            - Bernard–Soulier syndrome
            - MYH9-RD (was called May–Hegglin and other rarer types)
2. **Decreased or absent megakaryocytes in the marrow (amegakaryocytic thrombocytopenia)**
    a. Isolated megakaryocytic hypoplasia
        i. Congenital amegakaryocytic thrombocytopenia associated with bilateral absent radii (TAR syndrome)[a]
        ii. Congenital amegakaryocytic thrombocytopenia associated with radio-ulnar synostosis (ATRUS)
        iii. Congenital megakaryocytic hypoplasia without anomalies (CAMT)[a]
        iv. Congenital hypoplastic thrombocytopenia with microcephaly
        v. Rubella syndrome
        vi. Congenital hypoplastic thrombocytopenia associated with trisomy syndromes
        vii. Thrombocytopenia agenesis of corpus callosum
        viii. Fanconi anemia
        ix. Hoyeraal–Hreidarsson syndrome
    b. Generalized bone marrow disorders
        i. Bone marrow aplasia
3. Fanconi anemia
4. Pancytopenia without congenital anomalies
5. Osteopetrosis
    i. Bone marrow infiltration
6. Congenital leukemia
7. Langerhans cell histiocytosis
8. Congenital neuroblastoma
    a. Metabolic causes
        i. Associated with acidosis and ketosis
9. Hyperglycinemia
10. Methylmalonic acidemia
11. Isovaleric acidemia
12. Propionic acidemia
    i. Other
13. Maternal hyperthyroidism

[a]*Characterised by severe thrombocytopenia at presentation.*
ITP, immune thrombocytopenic purpura; SLE, systemic lupus erythematosus; NAIT, neonatal alloimmune thrombocytopenia; MYH-9, non-muscle myosin heavy chain 9 gene.

In cases where there is identification of paternal incompatibility, but not active alloimmunization, the recommendation is for heightened screening. These are potential cases of NAIT and crossmatching paternal platelets and maternal serum for anti-HPA antibodies can be serially performed with initiation of therapy if anti-HPA antibodies (rarely) are detected. With certain incompatibilities (i.e., HPA-3a and -b and HPA-9b) antibodies can be very difficult to detect. Clinical judgment in conjunction with the expertise of the laboratory needs to be combined in these cases.

If a previous affected sibling was serologically diagnosed with severe NAIT, the likelihood of the next fetus being affected depends on the father's platelet typing. If the father is homozygous for the antigen responsible (as is the case in 75% of men with HPA-1A) then essentially all later fetuses will be affected. If the father is heterozygous, or if typing is unavailable or uncertain, PCR testing via amniocentesis or maternal blood can determine whether the fetus is at risk. If the fetus is affected, stratified antenatal therapy should be initiated.

For mothers with previous NAIT pregnancy without ICH, recommended treatment starts later (20 weeks) and is less intensive than for those in which a previous sibling had an ICH (12 weeks). Even if there was no history of ICH, IVIG 1 g/kg/week by itself may not be sufficient treatment.

Neonatal Autoimmune Thrombocytopenia

Neonatal autoimmune thrombocytopenia is due to a passive transfer of autoantibodies from mothers with ITP to their fetus. It may also be seen in association with other conditions such as maternal systemic lupus erythematosus (SLE) and lymphoproliferative states. Neonates born to mothers who have autoimmune thrombocytopenia are typically well after an uncomplicated delivery. Maternal history and platelet counts can help to distinguish auto- from alloimmune thrombocytopenia, but such a distinction can be confusing in the presence of gestational thrombocytopenia (GTP). However, a history of maternal thrombocytopenia during the pregnancy is not diagnostic of maternal ITP. GTP occurs in 5–10% of pregnancies. By definition it is almost always mild ($70-100,000/mm^3$), is not associated with neonatal thrombocytopenia, and the maternal platelet count normalizes after delivery.

Neonatal thrombocytopenia in infants born to mothers with autoimmune thrombocytopenia is usually less severe than that seen in NAIT. Only 10–15% of these newborns have a platelet count less than $50,000/mm^3$. There is a lower risk of bleeding and only rare reports of ICH. The platelet count may be near normal at delivery (e.g., $90,000/mm^3$), but then fall to a clinically significant nadir over the next 1–3 days.

Table 14.5 lists the pathogenesis and clinical differences between NAIT and autoimmune thrombocytopenia.

TABLE 14.5 Pathogenesis and Clinical Differences Between Alloimmune Thrombocytopenia (NAIT) and Autoimmune Thrombocytopenia

	Alloimmune thrombocytopenia	Autoimmune thrombocytopenia
Platelet antigens	Antigens found on fetal platelets not present on maternal platelets (usually HPA-1A or HPA-5b)	Antigens common to both maternal and fetal platelets (usually GPIIB/IIIA and GPIb/IX complexes)
Platelet count	Often $<20,000/mm^3$	Birth counts often $>50,000/mm^3$
Time of presentation	Birth	Platelet count can be near normal at birth, and then fall
Maternal history	Normal platelet count, no history of ITP, SLE, or hypothyroidism, may have GTP (unrelated)	Low platelet counts (unless mother is splenectomized) History of ITP, SLE, hypothyroidism
Intracranial hemorrhage	10–20%	<1–2%
Treatment	Random donor platelets IVIG +/− Methylprednisolone Matched platelets	IVIG +/− Methylprednisolone Random platelets (if hemorrhage)
Resolution of thrombocytopenia	Usually in 2–4 weeks	Usually 3–12 weeks

ITP, immune thrombocytopenic purpura; SLE, systemic lupus erythematosus; GTP, gestational thrombocytopenic purpura; IVIG, intravenous immunoglobulin.

Pathophysiology

Neonatal ITP occurs as a result of passive transfer of maternal antibodies across the placenta, as is seen in neonatal alloimmune thrombocytopenia. However, the target of the antibodies is an antigen present on maternal platelets that is also on fetal platelets (in contrast to alloimmune thrombocytopenia, where the antigen is not present on maternal platelets). The most frequently targeted antigens are the GPIIB/IIIA or GPIb/IX complexes. Differences in glycosylation of fetal and maternal platelets may explain mothers with normal or near normal counts having thrombocytopenic neonates. A careful history must be obtained to identify mothers who have previously undergone splenectomy for ITP and may still pass antibodies transplacentally.

Diagnosis

Pregnant women with the following conditions may give birth to a neonate with autoimmune thrombocytopenia:

- History of previously affected infant.
- Mother who was previously splenectomized for ITP (mother may have platelet antibodies without being thrombocytopenic).
- Mother with thrombocytopenia ($<100,000/mm^3$) in current pregnancy especially if the platelet count is $<50,000/mm^3$.
- Mother with SLE, hypoythyroidism, preeclampsia—HELPP syndrome (hemolytic anemia, elevated liver enzymes, and low platelets).
- Maternal drug ingestion (e.g., thiazide).

Treatment

Follow the platelet count (as the initial count may be near normal) until 3—7 days of life to capture nadir and monitor until stable or a rising count to $>150,000/mm^3$ without intervention. All infants with severe thrombocytopenia should have a head ultrasound to exclude ICH.

Treatment is required when the infant's platelet count falls below $30,000/mm^3$ or if significant bleeding is present. The regimen is similar to that of NAIT, utilizing IVIG and IV methylprednisolone but unrelated donor platelet transfusion is used only if indicated by severe bleeding symptoms or a platelet count that remains $<30,000/mm^3$ despite maximal care.

The duration of neonatal thrombocytopenia is usually about 3 weeks. If persistent in an infant of a breastfeeding mother, a trial of discontinuation of breastfeeding may be considered. Unlike NAIT, there is no need for maternal treatment during pregnancy except under extraordinary circumstances.

General Diagnostic Approach to a Newborn with Thrombocytopenia

If the maternal platelet count is low, this suggests maternal ITP or another etiology for the maternal thrombocytopenia such as inherited thrombocytopenias. If the maternal platelet count is normal but the mother has had a splenectomy for ITP, the neonatal ITP can still be seen from circulating antiplatelet antibodies. Gestational thrombocytopenia, which complicates up to 10% of all normal deliveries, may have a platelet count as low as $50-70,000/mm^3$ but hardly ever results in neonatal thrombocytopenia.

If the platelet count is $<50,000/mm^3$ on the first day of life in an otherwise normal neonate, NAIT should be considered and appropriate laboratory testing performed; imaging is required based on the platelet count and clinical findings. Platelet transfusion and possibly IVIG should be given. The next most common cause of early severe neonatal thrombocytopenia is a TORCH infection (toxoplasmosis, rubella, cytomegalovirus or herpes virus) which presents as a variably sick newborn with associated fever, microcephaly, IUGR, conjunctivitis, hearing loss, hepatosplenomegaly, or occasionally blueberry muffin rash. If the baby is sick and the platelet count is $>50,000/mm^3$, it is likely that the underlying illness (e.g., respiratory distress syndrome, sepsis, DIC) is the cause of the low platelet count and the underlying condition should be treated with the expectation that with the resolution of the underlying illness resolution of the thrombocytopenia will occur.

If there is persistent severe thrombocytopenia, congenital amegakaryocytic thrombocytopenia (CAMT) and other inherited thrombocytopenias should be considered (see Chapter 8).

Thrombocytopenia Associated with Hemolytic Disease of Fetus and Neonate

Newborns with severe erythroblastosis frequently develop petechiae and purpura in the first few hours after birth. Thrombocytopenia with red cell alloimmunization may be due to an alloimmune mechanism, decreased production and overlap with known risk factors for thrombocytopenia such as SGA, lower birth weight, and maternal factors (e.g., hypertension/hypoxic-ischemic encephalopathy).

Exchange transfusion, the emergent treatment for hemolytic disease of the newborn, is an independent risk factor for transient thrombocytopenia because of the absence of platelets in stored, reconstituted blood, although this may be balanced somewhat by procoagulant alterations in endogenous thrombin generation from adult blood used for the exchange transfusion.

Thrombocytopenia Secondary to Chronic Fetal Hypoxia, Maternal Diabetes, Pregnancy-Induced Hypertension, or IUGR

Neonatal thrombocytopenia may be caused by chronic intrauterine hypoxia resulting in placental insufficiency in association with pregnancy-induced hypertension, preeclampsia, HELLP (*h*emolysis, *e*levate *l*iver enzymes, *l*ow *p*latelet count) syndrome, maternal diabetes mellitus, and IUGR. These may be due in part to increased platelet destruction, but typically there is impaired megakaryopoiesis and an elevated thrombopoietin level. Neonates can increase the number of megakaryocytes, but not their size, therefore potentially limiting their platelet-producing capability. Thrombocytopenia is usually not severe and is self-limited. The nadir tends to occur around days 3–4 with recovery by days 7–10. Often no treatment is required, but it may be appropriate to treat if there is sufficient asphyxia and thrombocytopenia raising the risk of ICH, in which case platelet transfusions are indicated.

Thrombocytopenia Secondary to Congenital Infections

Perinatal viral, bacterial and fungal infection may present with early- or late-onset thrombocytopenia. Often the presentation may be in an ill neonate with accompanying low birth weight, microcephaly, hepatosplenomegaly, chorioretinitis, and impaired hearing. Infections to consider include: toxoplasmosis, rubella, cytomegalovirus, or herpes simplex (TORCH), group B *Streptococcus*, *Listeria monocytogenes*, *Escherichia coli*, or HIV. Of the TORCH infections, cytomegalovirus (CMV) infection most commonly causes severe thrombocytopenia in 50–77% of affected infants. Thrombocytopenia in early-onset neonatal sepsis can occur because of:

- Platelet consumption associated with disseminated intravascular coagulation.
- Bacteria, viruses, or immune complexes that adhere to platelets and are cleared by the mononuclear phagocyte system.
- Sequestration secondary to hepatosplenomegaly.
- Impaired thrombopoiesis—often there is insufficient compensation for platelet destruction or increased platelet clearance. In these infants, thrombocytopenia resolves with effective treatment of the underlying infection.

Late-Onset Thrombocytopenia Secondary to Late-Onset Infections, Necrotizing Enterocolitis, or Thrombosis

Thrombocytopenia occurring at greater than 72 h after birth is more likely to be related to late-onset sepsis, necrotizing enterocolitis (NEC), thrombosis, or liver disease. NEC presents with feeding intolerance, abdominal distension, bloody stools, and pneumatosis. Severe thrombocytopenia, when present, has been associated with poor outcomes.

The causes of this type of thrombocytopenia are varied and often multifactorial including:

- Consumption related to infection (as in NEC).
- Deficient platelet production.
- DIC.
- Consumption secondary to thrombosis (e.g., renal vein or hepatic vein thrombosis).

Thrombocytopenia due to Aneuploidy

Thrombocytopenia is common in Down syndrome, occurring in up to 85% of cases. Some of these cases may represent a transient myeloproliferative disorder (TMD). TMD occurs in 10% of trisomy 21 cases and is associated with mutations in the second exon of GATA-1 transcription factor. TMD may present in the first month of life with thrombocytopenia, leukocytosis, hepatosplenomegaly, and hepatic fibrosis from megakaryotic infiltration. The majority of TMD will resolve with observation, but ~30% will progress, often later, to AMKL (see Chapter 19). Thrombocytopenia may also be seen in trisomies 13 and 18 and in Turner syndrome and triploidy, in which case other congenital anomalies suggest the diagnosis. In these cases bleeding complications are not frequent and thrombocytopenia may not become obvious for 3–4 days. Other medical problems in these patients far outweigh thrombocytopenia in their importance.

Rare Bone Marrow Disease or Inborn Errors of Metabolism

The following marrow diseases, many relatively rare, may be associated with thrombocytopenia in the newborn:

- Osteopetrosis—generalized hyperostosis of bone and obliteration of bone marrow cavity resulting in extramedullary hematopoiesis and pancytopenia.
- Metastatic neuroblastoma.
- Gaucher disease, often including hypersplenism and other coagulopathies.
- Niemann–Pick disease.
- Hemophagocytic lymphohistiocytosis (HLH), autosomal recessive, patients present with hepatosplenomegaly, fever, cytopenia, jaundice, tachypnea, hyperferritinemia, hypofibrinogenemia, hypertriglyceridemia, elevated soluble IL-2 receptor and absent NK cell activity.
- Congenital leukemia—more commonly with AML. It may be associated with skin infiltrates, hepatosplenomegaly, and megakaryoblasts in the peripheral blood smear. Congenital leukemia carries a poor prognosis unless associated with Down syndrome. Classic "blueberry muffin" rash, which is not actually petechial or purpuric in nature, but rather represents sites of extramedullary hematopoiesis in the skin. These skin lesions may also be seen in other etiologies of bone marrow infiltration and replacement, TORCH infections, or severe hemolysis.

Metabolic Causes

Hyperglycinemia with ketosis and the closely related metabolic disorder of *methylmalonic acidemia* may cause periodic thrombocytopenia, as well as neutropenia, during infancy. Infants with these metabolic disorders present with lethargy, vomiting, and ketosis during the neonatal period. A similar disorder, *isovaleric acidemia*, is associated with a "sweaty feet odor" but may present with a generalized marrow hypoplasia causing thrombocytopenia and neutropenia. Thrombocytopenia has been reported in rare infants born with neonatal Graves disease, resulting from maternal transfer of thyroid-stimulating antibodies with associated symptoms of tachycardia, cardiac dysfunction, liver disease, and hepatosplenomegaly.

Vascular Anomalies

Vascular anomalies are heterogeneous disorders involving the skin, subcutaneous tissues, and visceral organs. Current classification divides lesions into proliferative neoplasms and malformations. Included within this classification of proliferative neoplasms are infantile hemangioma, congenital hemangioma, and kaposiform hemagioendothelioma/tufted angioma (previously known as giant capillary hemangioma). Kaposiform hemagioendothelioma, a rapidly proliferating neoplasm, usually presents as a >5 cm red-purple indurated plaque with a pebbled texture and ill-defined margins typically occurring on the lateral neck, axilla, groin, extremities, or trunk. Kaposiform hemagioendothelioma can be associated with high morbidity and mortality secondary to the association with Kasabach–Merritt phenomenon. Infantile hemangiomas, the most common tumor of infancy, are not associated with the complication of Kasabach–Merritt phenomenon.

Kasabach–Merritt phenomenon is a form of localized intravascular microangiopathic anemia presenting with severe thrombocytopenia, hypofibrinogenemia, and elevated fibrin split products or D-dimers in the setting of this proliferative vascular anomaly. Platelet trapping and fibrinogen consumption has been demonstrated through immunohistochemistry and radiolabeling and leads to the consumption of coagulation factors and increased risk for hemorrhage and high-output cardiac failure. MRI imaging of the lesion is used to determine the extent of dermal, lymphatic and muscular involvement as well identifying the arterial supply and venous drainage which may help to direct therapy. CT, x-ray, and ultrasound are of limited utility in determining the extent of the lesion and directing appropriate therapy. Since these vascular anomalies often require time to grow after birth, they may not present until after the first month of life.

Treatment

1. Supportive care with transfusions of platelets and other coagulation factors (e.g., fibrinogen concentrates, fresh frozen plasma, cryoprecipitate, and antifibrinolytic drugs) have only transient effects. These modalities may help with acute management and stabilization in cases of intralesional or systemic bleeding. Prophylactic platelet transfusions are not recommended as they may lead to increased trapping and pain with tumor enlargement. Antiplatelet agents (aspirin, ticlopidine, and clopidogrel) have been used with mixed results. Low-dose heparinization may be useful.
2. Surgical resection, the gold standard for cure, is often difficult given the extent of tumor infiltration and coagulopathy at presentation.
3. Embolization may be an adjunctive therapy to minimize bleeding but is often a temporizing measure that may improve the thrombocytopenia rather than be a curative modality.
4. Radiation therapy is not recommended.
5. Pharmacologic therapy:
   a. Steroids: Corticosteroids are the most commonly used first-line therapy at high oral doses (2–3 mg/kg/d of oral prednisone) or equivalent intravenous doses. There is limited evidence supporting their effectiveness as a single agent therapy.
   b. Chemotherapy: Vincristine, a vinca alkaloid and inhibitor of endothelial proliferation has shown successful response in tumor regression. Doses of vincristine 0.05 mg/kg weekly have been used. Barriers to its use include the need for IV access and risk of neurologic side effects. Other reports have looked at cyclophosphamide and actinomycin-D in refractory disease, with less clear data for efficacy, safety, and dosing.
   c. Interferon α-2a has been shown to be effective. It inhibits angiogenesis, in part by inhibiting the proliferation of endothelial cells, smooth muscle cells, and fibroblasts that have been stimulated by fibroblast growth factor, decreasing collagen production and increasing endothelial prostacyclin production. Despite its efficacy, use of this therapy has been limited due to concerns over its association with spastic diplegia in infants <8 months of age.
   d. New agents:
   e. Propranolol, a non-selective beta-blocker, has been used successfully in the management of infantile hemangioma. The proposed mechanism is multifactorial with vasoconstriction, antiangiogenesis, and apoptosis of capillary endothelial cells contributing to the clinical outcome. Isolated case reports have demonstrated successful patient outcomes.
   f. Sirolimus, an inhibitor of m-TOR, targets antiangiogenesis pathways and results in apoptosis as the proposed mechanism of action. Due to multiple case reports with good outcomes, Sirolimus is currently in an ongoing prospective trial.

INHERITED THROMBOCYTOPENIAS

Thrombocytopenias due to inherited causes are usually distinguished by characteristic clinical features including family history, platelet morphology, long-term stable low counts, and lack of response to therapies of immune thrombocytopenia, for example, IVIG. Most of the syndromes are associated with "giant" platelets (Table 14.1). In general, the thrombocytopenia is not severe and often there are characteristic features that define each syndrome other than the thrombocytopenia. In certain cases (i.e., Wiskott–Aldrich), bleeding symptoms can be more severe than would be expected by the platelet count alone, due to platelet dysfunction.

Table 14.6 lists the characteristics of congenital thrombocytopenic syndromes.

TABLE 14.6   Characteristics of Congenital Thrombocytopenic Syndromes

Clinical features	TACC	TAR	ATRUS	FA	AMT	HH (DC)	WAS	BS	MH	Trisomies 13, 18, 21
Agenesis of corpus callosum	1	2	2	2	2	2	2	2	2	2
Hypoplasia of cerebellar vermis	1	2	2	2	2	1	2	2	2	2
Low birth weight	1	1	2	1	2	1	2	2	2	1
Growth delay	1	1	2	1	1	1	2	2	2	1
Dysmorphic face	1	1	2	1	2	1	2	2	2	1
Developmental delay	1	2	2	6	1	1	2	2	2	1
Thrombocytopenia	1	1	1	6	1	1	1	1	1	6
Platelet size	N	N	N	N	N	N	↓	↑	↑	N
Pancytopenia	2	2	1	1	6	1	2	2	2	2
Immunodeficiency	2	2	2	2	2	1	1	2	2	2
Megakaryocytes, bone marrow	N/↓	↓	↓	↓	↓	↓	N/↑	N/↑	N/↑	N/↑
Skeletal deformities										
Radial x-ray	2	1	2	6	2	2	2	2	2	2
Clinodactyly/syndactyly	6	2	1	6	2	2	2	2	2	6
Enamel hypoplasia	6	2	2	2	2	2	2	2	2	2
Cardiac defects	6	6	2	6	6	2	2	2	2	6
Renal malformations	6	2	2	6	6	2	2	2	2	6
Cutaneous abnormalities	2	1	2	1	2	1[a]	1	2	2	2
Karyotypic abnormalities	2	2	2	2	2	2	2	2	2	1
Chromosome breaks	2	2	2	1	2	2	2	2	2	2

[a]Present at a mean age of 9 years.

TACC, thrombocytopenia agenesis of corpus callosum; TAR, thrombocytopenia absent radii; ATRUS, amegakaryocytic thrombocytopenia with radio-ulnar synostosis; FA, Fanconi anemia; AMT, amegakaryocytic thrombocytopenia; HH, Hoyeraal–Hreidarsson syndrome; DC, dyskeratosis congenital; WAS, Wiskott–Aldrich syndrome; BS, Bernard–Soulier syndrome; MH, May–Hegglin anomaly; N, normal; ↑, increased; ↓, decreased; 1, present; 2, absent; 6, present occasionally.

Bernard—Soulier Syndrome

Bernard—Soulier syndrome is a relatively rare autosomal recessive disorder. It is characterized by:

- Moderate thrombocytopenia (automated counting usually underestimates the true platelet count because of undercounting of very large platelets).
- Prolonged bleeding time.
- Characteristic platelet morphology.

Platelets are very large, equaling or exceeding the size of a red cell. These platelets contain two to four times the normal protein content and three times the usual number of dense granules. These dense granules can gather and give the appearance of a pseudonucleus. The findings are more severe in the rare homozygote (or compound heterozygote) than in the far more common heterozygote. Morphologically, the megakaryocytes have an abnormality in the demarcation membrane system, likely explaining the large platelets and the thrombocytopenia.

There can be complete absence of GPIb glycoprotein complex or a point mutation in the GPIb-alpha subunit known as the Bolzano variant; all of which leads to inability of von Willebrand factor (vWF) binding. Platelets fail to agglutinate in response to ristocetin, despite normal aggregation and secretion in response to adenosine diphosphate (ADP), epinephrine, thrombin, and collagen. Because vWF acts as a bridge in adhesion of platelets to exposed subendothelium, the absence of the vWF receptors inhibits normal platelet adhesion and causes a significant degree of bleeding (even in cases of mild thrombocytopenia), with mucocutaneous bleeding that can begin in early infancy, especially severe epistaxis.

TABLE 14.7   Clinical and Morphological Features of the MYH9 Macrothrombocytopenias

Disorder	Clinical feature				
	Thrombocytopenia	Leukocyte inclusions	Nephritis	High-tone sensorineural deafness	Cataracts
May–Hegglin anomaly	1	1	2	2	2
Sebastian syndrome	1	1	2	2	2
Epstein syndrome	1	2	1	2	2
Fechtner syndrome	1	1	1	1	1
Alport syndrome	2	2	1	1	1

1, Present; 2, Absent.

Treatment

- Antifibrinolytic therapy is the mainstay of treatment.
- Platelet transfusion is the only reliable therapy, but is frequently reserved for managing severe bleeding episodes. There can be antibody formation with repeated platelet transfusions, because these patients lack the GPIb/IX complex. The alloantibody can inhibit normal platelet adhesion and obviate the benefit of transfusion.
- Activated Factor VIIa can be used in an emergency.

MYH9 Disorders

A mutation in the *MYH9* gene encoding the non-muscle myosin-heavy chain expressed only in neutrophils and platelets (myosin IIa) has been reported in the May–Hegglin anomaly, Sebastian, Fechtner, Epstein, and Alport syndromes. These disorders are referred to as MYH9-related diseases. Currently, no clear genotype–phenotype correlation has been established for this group of disorders which may have other manifestations like high-tone hearing loss and glomerulonephritis. Primary hemostatic bleeding, as in Bernard–Soulier syndrome, can be seen and local control measures are required. If bleeding is uncontrolled despite antifibrinolytic agents such as desmopressin acetate and aminocaproic acid (Amicar), platelet transfusions may be necessary. The newly licensed thrombopoietic agents are being explored.

Table 14.7 lists the clinical and morphological features of the MYH 9 macrothrombocytopenias.

- *May–Hegglin anomaly*: May–Hegglin anomaly is a MYH9-related syndrome associated with giant platelets and variable thrombocytopenia, although frequently of moderate degree. It has an autosomal dominant inheritance and its distinguishing features are the large bluish cytoplasmic inclusions in granulocytes and monocytes known as Döhle bodies. Bleeding is usually minor and is thought to be due to the sheer size of the platelets and the difficulty in shape change necessary for adhesion to damaged endothelium with collagen exposure in small vessels. Bleeding is mainly manifest with trauma, surgery, or dental extractions.
- *Sebastian platelet syndrome*: Sebastian platelet syndrome is very similar to May–Hegglin, but even more rare. The Döhle body-like inclusions are smaller, and are composed of ribosomes and dispersed filaments, and lack an enclosing membrane when examined under the microscope, compared with those of May–Hegglin. It is inherited in an autosomal dominant manner and bleeding tendency is considered to be mild to moderate.
- *Fechtner syndrome*: Fechtner syndrome is a variant of Sebastian syndrome, but also has specific additional features that are associated with Alport syndrome, including high-tone sensorineural deafness, cataracts, and renal failure. White cell inclusions are also seen and resemble toxic Döhle bodies and May–Hegglin granulocyte inclusions. There is a mild to moderate bleeding tendency with significant hemorrhage following trauma, dental extraction, and surgery. Patients bleed more as renal failure progresses.
- *Epstein syndrome*: Epstein syndrome, like Fechtner syndrome, is another variant of Alport syndrome and is associated with sensorineural deafness and renal failure (but not cataracts). In addition to thrombocytopenia, it is characterized by bleeding secondary to defective platelet aggregation and secretion in response to ADP and collagen. Alport syndrome itself is not associated with thrombocytopenia.

Wiskott—Aldrich Syndrome (WAS)

This syndrome has X-linked inheritance and has the classic features of thrombocytopenia, eczema, recurrent bacterial and viral infections secondary to abnormalities in T-cell function, and a propensity to develop autoimmune disorders.

Pathophysiology

The molecular defect in this syndrome is the abnormal Wiskott—Aldrich syndrome protein (WASP) resulting from a mutation of the Wiskott—Aldrich syndrome (*WASP*) gene located on band Xp11-12. WASP is known to be involved in signal transduction, and it regulates actin filament assembly which explains the abnormalities in platelet and lymphocyte cytoskeleton and signaling.

Clinical Manifestations

- Infants present with thrombocytopenia in the first few months of life.
- Bleeding is frequently ushered in by melena during the neonatal period, later followed by purpura. Bleeding is often out of proportion to thrombocytopenia.
- The clinical course is punctuated by recurrent pyogenic infections, including otitis media, pneumonia, and skin infections. There is also lowered resistance to non-bacterial infections, including herpes simplex and *Pneumocystis jiroveci* (formerly *carinii*) pneumonia.
- The disease is often fatal by the early teens due to infection, lymphoproliferative malignancy, or bleeding.

Hematologic Findings

- Thrombocytopenia (platelet count $10,000-100,000/m^3$); microthrombocytes; low MPV. This is not obvious in the newborn and the MPV is unreliable when the platelet count is low.
- Platelets have abnormal aggregation in response to agonists such as ADP, epinephrine, and collagen.
- Platelets have reduced survival to half of normal.
- Ineffective megakaryocytopoiesis reflected by a platelet turnover 25% that of normal megakaryocyte mass.
- Anemia (due to blood loss).
- Leukocytosis (due to infection).
- Normal or increased megakaryocytes.
- Absent isohemagglutinins, reduced IgM, and normal or elevated IgG and IgA.
- Defective cell-mediated immunity in some cases.

Treatment

Allogeneic stem cell transplantation is the treatment of choice when there is a fully matched donor available. If no matched donor is available the patient should be managed as follows:

- Aggressive treatment of infections.
- Platelet transfusions for hemorrhagic episodes.
- Steroid cream for eczema.
- Splenectomy: reserved only for severe cases because of risk of overwhelming postsplenectomy infection. Usual pre- and postsplenectomy precautions pertain.
- Thrombopoietic agents are being explored.

X-Linked Thrombocytopenia

A variant of WAS, known as inherited X-linked thrombocytopenia (XLT), has thrombocytopenia which is less severe than WAS, and does not have the associated features of eczema or immunodeficiency. Like WAS, there is an abnormality in the WASP. However the mutations in XLT occur more likely in exons 1 and 2 of *WASP*, which give rise to this milder phenotype, because only the megakaryocyte binding domain is affected. The distinction of the XLT form from full-blown WAS is not always clinically clear. Patients with XLT may be mistaken for chronic "refractory" ITP but have a higher incidence of postsplenectomy sepsis than patients with ITP.

Anemia and Thrombocytopenia with *GATA1* Mutation

Several families have been identified with variable anemia and thrombocytopenia with *GATA1* mutations recognized as the underlying cause. GATA1 is a transcription factor important in erythrocyte and megakaryocyte development. Bone marrow shows many large megakaryocytes with nuclei pushed to the side and with unorganized granular content.

Thrombocytopenia with Absent Radii Syndrome

- Thrombocytopenia with absent radii (TAR) syndrome is a rare disorder with autosomal recessive inheritance.
- Most cases are diagnosed either in utero or in the first day of life.
- There is bilateral absence of the radii manifesting as a shortening of the forearms and flexion at the elbows.
- Both thumbs are present, which helps to distinguish TAR from Fanconi anemia. Other defects of phalanges, humeri, and lower limbs can be present, as well as cardiac anomalies.
- White blood cell count elevation is frequently seen, and anemia can be part of the clinical picture.
- Significant bleeding episodes, such as gastrointestinal and even intracerebral hemorrhage, occur in the first 6 months of life.
- Death can occur in the first year of life due to hemorrhage, but this can be mitigated by the use of prophylactic single-donor platelet transfusions when counts are very low.
- Typically, the thrombocytopenia improves with time for reasons which are not understood, and the platelet count can be normal beyond the first year of life.
- There is a subgroup of patients who continue to have low counts and bleeding through adulthood.
- Compound inheritance of a rare null allele and one of two low-frequency single nucleotide polymorphisms in the regulatory regions of RBM8A, encoding the Y14 subunit of the exon-junction complex (EJC) causes TAR. The EJC performs essential RNA processing tasks. TAR is the first human disorder caused by deficiency in one of the four EJC subunits.
- There have been several case reports of acute myeloid leukemia in early childhood in these patients, but there is no associated chromosomal fragility as seen in Fanconi anemia.

Treatment

- Transfusion of red cells for anemia and single-donor platelet transfusions for severe bleeding may be necessary.
- Allogeneic stem cell transplantation may be required for symptomatic patients.
- Corticosteroids and splenectomy are of no longlasting benefit.

Amegakaryocytic Thrombocytopenia (AMT)

- CAMT is inherited as an autosomal recessive.
- Patients may present with isolated thrombocytopenia in the neonatal period. However the most common age at presentation of thrombocytopenia is within the first month, because of petechiae and other bleeding symptoms.
- The thrombocytopenia is severe and, unlike NAIT, does not resolve.
- The diagnosis of CAMT is not usually made until the infant is several weeks or months old when the bone marrow is examined.
- Bone marrow evaluation reveals absent or greatly reduced numbers of megakaryocytes with normal granulopoietic and erythroid elements.
- Thrombopoietin levels are very high as a result of the markedly decreased megakaryocytes and their progenitors.
- Physical anomalies that may be seen in some of the patients are orthopedic, renal, or cardiac, but most have no physically distinguishable features.
- CAMT is caused by mutations in the c-mpl gene (thrombopoietin receptor). Both frameshift and nonsense mutations have been described, which result in loss of c-mpl function. The type of mutation, such as frameshift versus missense mutation, may determine the degree of clinical severity, and whether megakaryocytes are present in small numbers or are altogether absent.
- Due to c-mpl function in preventing stem cell apoptosis, many patients progress to total aplastic anemia which is thought to result from stem cell depletion.

- Recombinant thrombopoietin would likely play no role in therapy for these patients, as its receptor is unable to signal normally.
- Current treatment for CAMT is supportive, using platelet transfusion.
- The only curative treatment is allogeneic stem cell transplantation.
- Gene therapy is being developed.

Amegakaryocytic Thrombocytopenia with Radio-Ulnar Synostosis

- Patients present at birth with severe thrombocytopenia (similar to CAMT), but they have the following characteristic physical examination findings that can identify this specific syndrome:
  - Proximal radio-ulnar synostosis (fusion of the radius and ulnar at the elbow).
  - Clinodactyly (minor).
  - Shallow acetabulae (minor).
- An *HOXA11* mutation has been identified in two kindreds but not in others. The mutation appears to inhibit megakaryocytic differentiation.
- The disease is associated with aplastic anemia and possible leukemia.

Familial Platelet Syndrome with Predisposition to Acute Myelogenous Leukemia (FPS/AML)

- FPS/AML involves a mild thrombocytopenia (usually a platelet count around 100,000/mm^3) with dysfunctional platelets.
- There is an associated approximately 50% chance of malignancy, with two-thirds developing AML and one-third developing a solid tumor.
- The genetic defect has been traced to the transcription factor RUNX1 (also known as AML1).

Paris—Trousseau Syndrome

This syndrome consists of mild thrombocytopenia with a subpopulation of platelets with giant α granules. There is an expansion of immature megakaryocyte progenitors in the bone marrow with normal erythroid and granulocytic maturation. A subset of these patients will have Jacobsen syndrome, which includes the same platelet defects as Paris—Trousseau with additional abnormalities such as trigonocephaly, facial dysmorphism, cardiac defects, syndactyly, and psychomotor retardation.

Congenital Hypoplastic Thrombocytopenia with Microcephaly

Three infants have been reported with congenital hypoplastic thrombocytopenia with microcephaly and the persistence of thrombocytopenia beyond 1 year of life. Rubella syndrome, which may present in a similar manner, should be excluded.

Thrombocytopenia Agenesis of Corpus Callosum Syndrome

Three female patients have been reported with thrombocytopenia, agenesis of the corpus callosum, low birth weight, growth delay, and dysmorphic facial features. Megakaryocytes are absent and dysmorphic.

IMMUNE THROMBOCYTOPENIC PURPURA (ITP)

Immune thrombocytopenia is a disorder caused by antiplatelet antibodies leading to accelerated destruction of platelets and inhibition of the production of platelets. ITP is the most common cause of thrombocytopenia in children of any age but with a peak occurrence between 2 and 5 years of age. In most children the disease is self-limited, with resolution in 60—80% of patients within 6—12 months from diagnosis. North American studies report an incidence of 7.2—9.5/100,000 children between 1 and 14 years of age. There is a seasonal pattern to ITP with a peak in late winter and early springtime, presumably mimicking the pattern of viral illnesses. There does not appear to be a race or sex predilection.

Pathophysiology

ITP is a heterogeneous disease with a complex pathogenesis. An acute infection often is the initial trigger, but may only potentiate an already-established autoimmune disturbance.
Mechanisms:

1. Antibody-mediated destruction:
   a. Most identified autoantibodies are directed against GPIIb-GPIIIa, GPIb-GPIX, and GPIa-IIa; follicular B-cells in the spleen are heavily involved in synthesis.
   b. Antibody-coated platelets are destroyed by activated Fc receptors on reticuloendothelial cells (mostly splenic) via recognition of the IgG Fc region of the antiplatelet antibody.
   c. Antiplatelet antibodies rarely have a significant effect on platelet function.
2. Impaired megakaryopoiesis
   a. Antibody, cellular cytotoxicity, and/or immune-cell-derived cytokines have been implicated in impairment of megakaryocytes.
   b. Platelet kinetic studies have shown that autologous platelets labeled with indium have a 2–3-day half-life (longer than expected for platelet count) suggestive of reduced platelet production.
   c. Absolute platelet reticulocyte counts are reduced, which is also suggestive of reduced platelet production.
   d. Thrombopoietin levels are minimally elevated.
3. T-cell activity
   a. Glycoprotein-specific antibodies are absent in 20–40% of cases of ITP.
   b. There is an upregulation of genes involved in cell-mediated toxicity (e.g., granzyme b, perforin) in CD3 + CD8 + T-cells in ITP patients.
   c. CD4 + T-helper cells stimulate antiplatelet-antibody-secreting B-cells.
   d. Th1-associated cytokines predominate in ITP.

Infections in ITP

Acute infections have been implicated in the initiation of ITP, yet also may cause an acute decrease (or paradoxically an acute increase) in the platelet count in patients with ongoing ITP.

1. Virus-specific antibodies that cross-react with platelets have been demonstrated in several children with varicella.
2. HIV
   a. A clear relationship between platelet count and viral load has been demonstrated; suppressing the virus increases the platelet count in more than 80–90% of cases.
   b. The pathogenesis of HIV-related ITP may be different from non-HIV-related ITP. First, severe T-cell depletion and immune dysregulation occurs but second, antiplatelet antibodies may cause intravascular platelet lysis more than FcR-mediated clearance.
   c. Megakaryocytes express receptors for HIV including CD4, CXCR4, and CCR5, suggesting that direct viral infection may play a role.
3. *H. pylori*—studies indicate that low platelet counts are often increased with *H. pylori* treatment in countries with high background incidence of *H. pylori*, for example, Italy and Japan (but not the United States).
4. Hepatitis C and its treatment with interferon are each associated with thrombocytopenia, which may prevent successful eradication of the virus. Thrombopoietic agents may raise the platelet counts in these cases (e.g., Eltrombopag 50–100 mg orally daily). The effect of eradication of hepatitis C on thrombocytopenia is not clear. Newer hepatitis C treatments which do not cause thrombocytopenia have made low platelet counts associated with hepatitis C much less of an issue.
5. ITP refractory to therapy may be caused or exacerbated by CMV infection (reactivation or *de novo*).

Clinical Manifestations

Typically, patients are otherwise well and present with petechiae, purpura, and non-palpable ecchymoses 1–3 weeks after almost any viral infection. ITP may also occur after rubella, rubeola, chickenpox, EBV, or live virus vaccination. Occasionally patients may present with mucosal bleeding (hematuria, hematochezia, menometrorrhagia, or epistaxis). Most often, bleeding symptoms are mild, but rarely patients may develop severe bleeding including ICH, protracted epistaxis, hematuria, hemoptysis, menometrorrhagia, and gastrointestinal bleeding.

Table 14.8 lists the clinical manifestations of ICH in ITP.

With the exception of hemorrhagic manifestations, the physical examination is normal. Pallor is usually absent unless there has been significant bleeding. Spleen tip is palpable in fewer than 10% of patients. The finding of splenomegaly suggests the probability of leukemia, SLE, infectious mononucleosis, or hypersplenism. Cervical lymphadenopathy is not present unless the precipitating factor is a viral illness. There are no congenital anomalies suggestive of an inherited bone marrow failure syndrome.

Table 14.9 lists the features of newly diagnosed, persistent and chronic ITP.

TABLE 14.8 Characteristics of ICH in ITP

Incidence	0.2−0.8%
Age	13 months−16 years
Platelet count	<20,000 in 90% of cases
	<10,000 in 75% of cases
Interval between diagnosis of ITP and ICH	At presentation (25% of cases)
	<1 week (45% of cases)
	1 week−6 months (25% of cases)
	>6 months (30%)

Identifiable risk factors for ICH include:

- Head injuries (33%) (vs 1% in ITP without ICH)
- Hematuria (22%) (vs 0% in ITP without ICH)
- Hemorrhage more than petechiae and bruises (63%) (vs 44% in ITP without ICH)
- Arteriovenous malformation
- Aspirin treatment

Site of ICH

- Intracerebral (77% of cases)—87% supratentorial; 13% posterior fossa
- Subdural hematoma (23% of cases)

Prior treatment

- 70% had prior treatment

Survival

- 75% survive, but 1/3 have neurologic sequelae

ITP, immune thrombocytopenic purpura; ICH, intracranial hemorrhage.
Source: Butros (2003), Medeiros and Buchanan (1996), Psaila et al. (2009).

TABLE 14.9 Features of Newly Diagnosed and Chronic Immune Thrombocytopenic Purpura

Feature	Newly diagnosed/persistent	Chronic
Age	Children 2−6 years old	Adults
Sex distribution	Equal	Female:male = 2:1
Preceding infection	~80%	Unusual
Seasonal predilection	Springtime	None
Associated autoimmunity	Uncommon	More common
Onset	Acute	Insidious
Platelet count	<20,000/mm^3	<20,000−80,000/mm^3
Eosinophilia-lymphocytosis	Not uncommon	Rare
IgA/IgG levels	Normal	Infrequently low
Duration	2−8 weeks/3−12 months	1 to many years
Prognosis	Spontaneous remission in 70−80% of cases	Ongoing thrombocytopenia with occasional remission

Diagnosis

The diagnosis of ITP remains almost entirely one of exclusion. Demonstrating antiplatelet antibodies has not been shown to be of diagnostic or prognostic importance since antiplatelet antibodies are only present in approximately 60—80% of cases and specificity may be poor as well. There are three diagnostic criteria:

1. Isolated thrombocytopenia with otherwise completely normal CBC and blood smear (particularly red cell and white cell morphology).
   * Smear often has large platelets (too many megathrombocytes suggest other disorder, e.g., Bernard—Soulier, MYH9 disorders) (Table 14.1).
   * Artifactual low platelet count ("pseudothrombocytopenia") must be excluded.
   * Other possibilities with secondary ITP such as Evans syndrome and also TTP (anemia, high reticulocyte count, schistocytes) must be excluded.
2. Absence of hepatosplenomegaly, lymphadenopathy, and congenital anomalies such as radial ray anomaly (oligodactyly, aplasia or hypoplasia of the thumb or radius, radio-ulnar synostosis).
3. Platelet response to ITP therapy (especially IVIG or anti-D, possibly steroids) is the only finding that helps to positively support the diagnosis of ITP (including secondary ITP).

Coagulation screening tests—PT, PTT, fibrinogen are all normal (unless there is an incidental finding or antiphospholipid syndrome).

Other causes of thrombocytopenia must be excluded:

1. Inherited thrombocytopenia by:
   * Family history, duration, response to therapy.
   * Presence of congenital defects (skeletal, cardiac, renal, neurologic).
   * Usually large platelet size on smear (MPV is unreliable in severe thrombocytopenia).
2. Pregnancy.
3. Chronic infection with HIV, hepatitis C, *H. pylori*, and CMV.
4. Immunodeficiency (e.g., CVID (hypogammaglobulinemia), Wiskott—Aldrich).
5. Lymphoproliferative disease (benign or malignant) (see Chapter 16).
6. Type II von Willebrand disease (see Chapter 15).
7. Medications, especially quinine, valproate, heparin, estrogen in any form, diphenylhydantoin and dietary supplements. Many drugs have been proven or suspected to induce drug-dependent antibody-mediated immune thrombocytopenia. Reference http://www.ouhsc.edu/platelets/DITP.html, for a current reference source.

Additional laboratory analyses may be required for non-responsive, persistent or chronic cases or for those cases which have *specific clinical indications*:

1. Autoimmune screening—ANA and antidouble-stranded DNA antibodies, RA, C3, C4, etc.
2. Thyroid screening (TSH, free T4, thyroid antibodies).
3. Immune globulin measurements (IgG, IgA, and IgM) and possibly specific antibody levels, for example, to multiple pneumococcal serotypes.
4. Liver function tests.
5. EBV, CMV, parvovirus, hepatitis C, and HIV testing by PCR.
6. *H. pylori* testing (in countries with a high background incidence).
7. Bone marrow aspiration and biopsy.
   Patients with isolated thrombocytopenia, who fit the diagnostic criteria above, generally do not require bone marrow sampling to rule out leukemia. Failure to respond (not even a transient increase in platelets) to ITP therapy (e.g., anti-D and/or IVIG) strongly suggests the performance of a bone marrow aspirate and biopsy.
8. Antiphospholipid antibodies (present in up to 15% of cases) and lupus anticoagulant in cases of prolonged PTT, persistent headache, and/or thrombosis.

Treatment

The goal of therapy in ITP is to increase the platelet count enough to prevent serious hemorrhage and possibly also to alleviate fatigue or difficulty with activities of daily living. Treatment decisions should be based on the potential for bleeding including the physical activity level, patient's history of bleeding, current platelet count, signs and symptoms and other factors as follows:

- Patients who have a platelet count greater than $30,000/mm^3$ with no signs of bleeding generally do not need treatment, although certain risk factors (Table 14.8) and surgery may necessitate treatment.
- Patients with a platelet count less than $20,000/mm^3$ and moderate bleeding and those with counts less than $10,000/mm^3$ should be treated.
- Quality of life can be a major concern in ITP (in particular fatigue) and this may be a reason to treat even in the absence of moderate to severe bleeding.

Competitive contact sports should be avoided when the platelet count is known to be less than $30,000/mm^3$. Depo-Provera or any other long-acting or short-term progesterone is useful to suspend menstruation for several months in order to prevent excessive menorrhagia in menstruating females. Aspirin, nonsteroidal anti-inflammatory agents, fish oil, and any other drugs that interfere with platelet function should be avoided.

The following drugs are employed in the treatment of ITP.

Corticosteroids

Prednisone 1–2 mg/kg/day in divided doses orally for 2–4 weeks with tapering after there is a platelet response. The initial response rate is 50–90% (children tend to be at the higher end of the response range). Another approach is to use 4 mg/kg/day of prednisone for a short period of time with rapid taper (total dosing period of 7–14 days). As with any initial therapy of newly diagnosed ITP, many children go into sustained remission after a single course.

Dexamethasone in a daily dose of 40 mg/kg/day (24 mg/m²) orally for 4 days given every 14 days for three cycles has been shown to have an initial response rate of 85% in adults and children but durable response is not nearly as good.

Prolonged use of steroids in ITP is undesirable. Large doses or prolonged usage may perpetuate the thrombocytopenia and depress platelet production. It also leads to side effects including gastritis, ulcers, weight gain, cushingoid facies, fluid retention, acne, hyperglycemia, hypertension, mood swings, pseudotumor cerebri, cataracts, growth retardation, and avascular necrosis.

Mechanism of action of steroids:

- Inhibit the phagocytosis of antibody-coated platelets.
- Inhibit platelet antibody production.
- Suppress activation of T-cells driving the autoimmune response.

Immune Globulin (IVIG)

Immune Globulin (IVIG) can be administered in a dose of 0.4–1 g/kg/day for 1–5 days for initial therapy or for relapsed disease. IVIG is preferred over steroids in children less than 2 years of age because they tend to have lower response rate to steroids and more challenging behavioral risk factors for bleeding.

Meta-analysis of randomized controlled trials has shown a more rapid response to IVIG in children compared to corticosteroids. In addition, a large, retrospective study has suggested that there may be a lower rate of chronic disease in patients initially treated with IVIG compared to those treated with prednisone. However, IVIG as an alternative therapy to corticosteroid therapy is much more expensive and has significant side effects (see below).

Mechanism of action of IVIG
  Early studies suggested that IVIG inhibits clearance of Ig-coated platelets. Recent studies of mouse models of ITP suggested the hypothesis that IVIG upregulates FcγRIIb, the inhibitor of FcγR, on phagocytes.
Adverse effects of IVIG
  - Post-infusion headache in >50% of patients. It is transient but occasionally severe (in severe cases, administer IV steroids, e.g., dexamethasone 0.15–0.3 mg/kg IV). Severe headache in ITP may suggest the presence of intracranial hemorrhage and, if clinically indicated, may require a CT scan, although most post-IVIG headaches occur with good platelet counts. Amelioration of this adverse effect with acetaminophen and prednisone is thus important.

- Fever and chills in 1−3% of patients. Prophylactic acetaminophen (10−15 mg/kg, 4-hourly, as required) and diphenhydramine (1 mg/kg, 6−8 hourly, as required) may be useful to reduce the incidence and severity.
- DAT-positive hemolytic anemia especially in patients with blood group A because of the presence of blood group antibodies (anti-A, anti-B, and sometimes anti-D) present in IVIG.
- Anaphylaxis in IgA-deficient patients because of pre-existing IgA antibodies that react with small amounts of IgA present in commercially available gammaglobulin.
- Aseptic meningitis (rare).
- Acute renal failure (very uncommon in children and much less overall since elimination of higher osmolality IVIG preparations).
- Pulmonary insufficiency.
- Thrombosis (recently demonstrated in certain cases to be caused by FXIa in IVIG).
- Viral transmission (hepatitis C in the past, but not currently. No cases of HIV in the United States).

Anti-D Therapy

IV anti-D is used in a dose of 50−75 µg/kg for initial therapy or for recurrent disease. Approximately 70% of patients have an initial response to 75 µg/kg of anti-D therapy within 1 day (comparable to high-dose IVIG). The effect is more pronounced after 48−72 h. Anti-D is plasma-derived, immune globulin with high titers against the Rhesus D antigen. It can be used only in Rh+ (and in DAT-negative and non-anemic) patients for the treatment of ITP. Hemolysis is expected when anti-D is used, and hemoglobin levels usually decrease by 0.5−2 g/dl.

Mechanism of action
Anti-D works by binding to Rhesus D antigen expressed on red blood cells, which leads to their recognition by Fc receptors on cells of the reticuloendothelial system. The coated red cells slow clearance of antiplatelet antibody-coated platelets.

Adverse effects (largely preventable by premedication with high-dose steroids)
- Fever and chills.
- Intravascular hemolysis.
- Headache, vomiting.
- Anaphylaxis (rare).

Splenectomy

- Splenectomy is rarely indicated because of the increasing number of effective medical therapies that have been developed combined with the favorable natural history of ITP in children.
- Splenectomy is indicated for severe ITP with acute life-threatening bleeding that is non-responsive to medical treatment or in patients with chronic ITP with bleeding and/or limitation of a patient's activities, for example, contact sports (because of potential danger of ICH), and non-responsive to medical treatment. It is rarely performed within 1 year of diagnosis and in children less than 5 years of age. Splenectomy can restore platelet counts in at least two-thirds of patients. Other modalities should be tried before splenectomy in pediatric patients because of:
  - Perioperative complications, for example, thrombosis or infection or bleeding.
  - Long-term risk of infection with encapsulated organisms postsplenectomy (splenectomy should be proceeded by appropriate immunizations).
  - Unpredictable response to splenectomy.
  - Unknown late side effects of splenectomy.

Rituximab

Rituximab is a chimeric human−mouse monoclonal antibody directed against the transmembrane CD20 antigen present on B-cells, licensed for the treatment of B-cell non-Hodgkin lymphoma and also rheumatoid arthritis. Rituximab has shown a substantial initial response rate (40−50%) in children with chronic ITP after a four-dose course, but long-term response rates in children after 2 years are less than 25%. A recent randomized controlled study in adults suggested very limited benefit although there was a substantial reduction of steroid use in the first year after treatment and a longer time until "relapse."

1. Dosage

The standard dosage of Rituximab is 375 mg/m^2 IV weekly for 4 weeks. Very limited studies of lower doses have suggested comparable efficacy but shorter duration of response and thus the optimal dose has not yet been determined.

  a. *Mechanism of action*

    Rituximab works by attaching to CD20 antigen on B-cells causing:

    i. Cell lysis via induction of apoptosis, antibody-dependent cellular cytotoxicity and complement deposition on antibody-coated B-cells.

    ii. Antibody-mediated opsonization resulting in phagocytosis.

  b. *Adverse effects*

    Fever and chills—common with first infusion.

    Serum sickness—5–10% in children with persistent and chronic ITP.

    Headache, nausea, and vomiting.

    Hypotension (rare).

    Tachycardia.

    Mucocutaneous reactions including hives during first infusion, and rarely Stevens–Johnson syndrome, lichenoid dermatitis, vesiculobulbar dermatitis, toxic epidermal necrolysis.

    Profound and prolonged peripheral B-cell depletion (hepatitis B reactivation has been seen so hepatitis B carriers should not receive rituximab).

    Progressive multifocal leukoencephalopathy: one case has been reported in an adult with ITP more than 3 years after treatment.

Thrombopoietic Agents

These agents are very effective in adults with ITP with efficacy in adults of more than 60–70%. Experience with thrombopoietic agents in children with ITP is increasing. One study of 22 children treated with romiplostim given subcutaneously weekly at doses of 1–10 µg/kg showed responses in 15/17 children with chronic ITP. A retrospective study found similar results in >20 patients and results are pending from a large randomized study which has finished enrolling. Dosing of eltrombopag from two large randomized studies encompassing >150 children suggested that adult doses ranging from 25–75 mg orally daily are also needed in children, even very small children. A suspension is being developed. Single responses in both studies were 81% over 6 months of treatment and more durable responses 50–70%. Toxicity is important to consider in adults, although thromboembolism and development of malignancy in patients with apparent ITP in the context of myelodysplasia are apparently very rare; none have been reported to date. One report described 24 bone marrows in children with only one grade 2 reticulin fibrosis in the marrow; this issue remains to be settled in children and in adults as longer courses of treatment are pursued. Rebound thrombocytopenia may occur when treatment is discontinued. Eltrombopag infrequently results in transient abnormal liver tests but in both pediatric studies, these have been the cause of the small number (<5%) of withdrawals from study, emphasizing the need for monitoring of liver tests.

Plasmapheresis

Plasmapheresis may only rarely be useful to accelerate the effect of other therapies by the removal of previously synthesized platelet antibodies.

Platelet Transfusions

Platelet transfusions may be transiently effective and required for rare emergency situations including ICH, internal bleeding, and emergency surgery. Some short-term hemostatic activity is usually obtained.

Drugs that May Be Effective but Are Infrequently to Rarely Used in Pediatric ITP

- Danazol—if given before puberty, may shorten final height; in adolescent females will abrogate menses.
- Dapsone—used more commonly in underdeveloped countries because of cost considerations.
- Azathioprine (or 6MP)—used in ITP for >40 years, can be combined with Danazol.
- Cyclophosphamide—seemingly more effective IV than PO, often last resort because of toxicity.
- Vinca alkaloids (Vincristine, Vinblastine)—efficacy usually short-term.

- Cyclosporine A—relatively effective but may not be curative; levels need to be monitored and targeted to 100–200 (below transplant levels to minimize toxicity).
- Mycophenolate mofetil—in class with azathioprine and cyclosporine; seemingly relatively effective in Evans syndrome.

Emergency Therapy

Patients with profound mucosal bleeding or internal bleeding require immediate therapy. Combination therapy is optimal:

- IV methylprednisolone 30 mg/kg/day for 1–3 days.
- IVIG 1 g/kg/day for 2–3 days with or without:
  - Anti-D 75 μg/kg (one dose).
  - Platelet transfusion (bolus followed by continuous infusion if required).
  - Recombinant human factor VIIa (rhuVIIa).
  - Emergency splenectomy in urgent, life-threatening bleeding is not usually recommended but can be pursued.

Chronic ITP

Chronic ITP is currently defined as thrombocytopenia that persists beyond 12 months from diagnosis. Patients with ITP of 3–12 months' duration are defined as persistent disease. It is more common in older children and those with altered immunity.

OTHER CAUSES OF THROMBOCYTOPENIA

HIV-Associated Thrombocytopenia

Thrombocytopenia is common in patients with HIV, and can be an early finding in the disease process. HIV-associated thrombocytopenia should be considered in a child with known HIV or a patient who presents with thrombocytopenia in conjunction with a compatible family or transfusion history and on physical examination has axillary or inguinal lymphadenopathy. The thrombocytopenia in this group of patients is multifactorial, but includes immune-mediated thrombocytopenia, sequestration, and ineffective thrombopoiesis. The antibodies are most commonly to GPIIb/IIIa, as in ITP without HIV, but have been shown to be directed at a specific peptide of GPIIIa (amino acids 49–66) leading to platelet intravascular lysis. The virus may also invade megakaryocytes and precursors, leading to reduced thrombopoiesis.

Aggressive treatment of the underlying HIV infection with highly active antiretroviral therapy has almost eliminated HIV-related thrombocytopenia as a clinical problem; especially since a formulation containing multiple agents in a single pill has resulted in the need to take a single pill only once or twice daily. Treatments used in non-HIV ITP are often effective. Steroids, however, should be avoided, as they may contribute to the development of secondary infections.

Drug-Induced Thrombocytopenia

Many drugs have been proven or suspected to induce drug-dependent antibody-mediated immune thrombocytopenia. http://www.ouhsc.edu/platelets/DITP.html, is a current reference source containing a list of drugs which have been implicated in causing thrombocytopenia. Almost any drug can, but very few frequently do, cause thrombocytopenia. Selective serotonin reuptake inhibitor drugs used as antidepressants may affect platelet function, thereby leading to bleeding symptoms without thrombocytopenia.

Heparin-Induced Thrombocytopenia

Heparin-induced thrombocytopenia (HIT) is a potentially life-threatening cause of thrombocytopenia, more common in adults than children. HIT is an immune complication of exposure to unfractionated heparin or, less frequently, low-molecular-weight heparin. HIT is caused by antibodies directed against complexes of heparin and platelet factor 4. The antibodies attached to the complex bind to and activate FcγRIIa on the platelet surface and thus cause platelet activation. This in turn creates an increased risk for both venous and arterial thromboses. HIT is a challenging clinical diagnosis because of the combination of the frequency of heparin use and thrombocytopenia in hospitalized patients and the low incidence in pediatrics. HIT occurs at a rate of 1–2% in pediatric patients who are in a pediatric ICU or who have recently undergone cardiac surgery.

Features of HIT include (4 T's)

1. Thrombocytopenia: Platelet count drops below 150,000/mm^3 or 30–50% drop from baseline. Moderate thrombocytopenia with mean platelet counts 50–70,000/mm^3.
2. Timing: Onset of thrombocytopenia usually 5–10 days following first heparin exposure; may be 3–5 days, for example, if sensitization occurred previously.
3. Thrombosis: Venous or arterial thrombosis. Less commonly skin necrosis at subcutaneous heparin injection sites.
4. OTher: Absence of alternative explanation for thrombocytopenia.

    To confirm diagnosis, in the setting of high clinical probability, heparin-dependent antibodies are screened for by immunologic testing for antibodies against heparin-PF4 by ELISA. This is sensitive but not always specific so, if required, confirmation can be obtained by functional antibody testing with the serotonin release assay demonstrating platelet activation.

When a child is receiving heparin, no matter how small the dose (e.g., to keep a catheter patent or in total parenteral nutrition), a watchful eye should be kept on the daily platelet count. When HIT is suspected to be the most likely cause of thrombocytopenia, heparin must be discontinued immediately and alternative (not Coumadin) anticoagulation initiated (see Chapter 15 for details of management).

A separate entity called nonimmune heparin-associated thrombocytopenia can also occur. Its exact pathophysiology is not known but may be mediated by platelet clumping. It is mild and occurs in the first day of heparin administration.

Thrombotic Microangiopathies

Thrombotic microangiopathies (TMA) are an infrequent, related group of syndromes with overlapping clinical features of including microangiopathic anemia, thrombocytopenia, and organ injury. Pathologic features of TM include vascular endothelial damage with microthrombi of the arterial and capillary vessels, leading to a final common pathway of end-organ damage and tissue ischemia.

Thrombotic Thrombocytopenic Purpura

Thrombotic thrombocytopenic purpura (TTP) is a rare disease characterized by microangiopathic hemolytic anemia, thrombocytopenia, neurologic symptoms, renal impairment, fever, and elevated LDH. Both the acquired and congenital forms are a consequence of ADAMTS13 (*a d*isintegrin *a*nd *m*etalloprotease with *t*hrombospondin type 1 repeats) deficiency-mediated thrombotic microangiopathy. ADAMTS13 is a protein responsible for cleaving unusually large multimers of vWF (UL-vWF) into biologically less active multimers. Absence of this protein, either due to antibody binding (acquired) or decreased production (congenital), results in an increase in ultra-large multimers, which cause "spontaneous" platelet adhesion and aggregation.

Clinical Features
The spectrum of systemic involvement is heterogeneous in severity and varies from patient to patient and from time to time and may be acute, chronic or relapsing. Clinical features include: fever, bleeding, pallor, jaundice, malaise, nausea, vomiting, abdominal pain, chest pain, seizures, fluctuating neurologic signs and symptoms, progressive renal failure (may occur in 25% of chronic patients).

Laboratory Features
- Microangiopathic hemolytic anemia (blood smear reveals polychromasia, basophilic stippling, schistocytes, microspherocytes, and nucleated red blood cells).
- DAT negative.
- Thrombocytopenia (often more severe than the degree of hemorrhage).
- DIC (prolonged PT/PTT, elevated D-dimer and hypofibrinogenemia).
- Haptoglobin level reduced.
- Hemoglobulinuria and hemosiderinuria usually present.
- Unconjugated bilirubin increased.
- Lactate dehydrogenase levels very increased.
- Elevated BUN/creatinine.

Acquired TTP

In the majority of cases of acquired TTP, there is an immune-mediated inhibition of ADAMTS13 activity. Incidence is 4–11 cases per million with additional risk factors that include: female sex, African-American descent, history of autoimmune disease, pregnancy, or infection. One-third of patients successfully treated relapse, and the mortality is approximately 10–20%. The disease is far more prevalent in adults compared to children who are typically adolescents.

Routine diagnostic evaluation includes: CBC, reticulocyte count, peripheral smear, BUN/Cr, bilirubin, LDH, and direct antiglobuluin test (DAT; and direct Coombs). A diagnosis is suspected in the setting of microangiopathic anemia (based on smear review) and thrombocytopenia and confirmatory testing with an ADAMTS13 assay. The diagnosis is made clinically and treatment should not await the results of an ADAMTS13 assay which involves identification of severe ADAMTS13 activity <10% and documentation of the presence of inhibitory anti-ADAMTS13 IgG autoantibodies. There are additional patients who present with a clinical picture of acquired TTP, who do not demonstrate ADAMTS13 deficiency who may nonetheless benefit from TTP directed therapy.

Treatment for acquired TTP is exchange plasmapheresis combined with high-dose steroids. Plasmaphereis provides a 50–80% remission rate by removing antibodies to ADAMTS13 and donor plasma replaces the missing enzyme. Given the mortality and possibility of stroke, there is urgency to begin plasma exchange and steroids prior to confirmation of diagnosis. Platelet transfusion is not generally recommended in TTP as it has the potential to worsen the consumptive coagulopathy but can be used in emergent situations. The majority of patients respond to serial plasma exchange but refractory or relapsing patents may also require the addition of second-line therapies. Additional adjuvant therapies include rituximab and other immunosuppressive agents. Rituximab targets antibody product through B-cell-directed therapy to decrease the inhibitory ADAMTS13 antibody production. There has been benefit in administration to refractory patients and there may be benefit to early administration to reduce the number of plasmaphereses required, morbidity, and potentially rates of relapse. Rituximab is administered at 375 mg/m^2 weekly for 2–8 weeks, although the schedule will vary with plasma exchange schedule and clinical response. Ideally it is not given within 48 h of plasmapheresis. Immunosuppressive agents like cyclosporine, cyclophosphamide, vincristine, mycophenalate mofetil, and azathioprine have been used, but no systematic evaluation of these agents has been performed.

Congenital TTP (Upshaw–Shulman Syndrome)

Congenital TTP occurs secondary to a genetic mutation in the gene for ADAMTS13 on chromosome 9q34. This is a rare disorder and occurs much less frequently than the acquired form. Patients typically present with neonatal jaundice and thrombocytopenia, although some patients will not have episodes of overt TTP until late childhood or adulthood following an environmental trigger (i.e., pregnancy or infection). Patients with congenital TTP have low levels of ADAMTS13 activity, but no antibodies.

Because there are no circulating antibodies to ADAMTS13 and a small amount of the enzyme is needed for function; plasma infusion is the mainstay of therapy. Many patients require plasma therapy only with symptomatic episodes, while others are maintained on a prophylactic schedule administered at 2–3-week intervals. An additional therapeutic option for patients with severe reactions to plasma is administration of plasma-derived FVIII concentrate that contains ADAMTS13.

Hemolytic-Uremic Syndrome

Hemolyic-uremic syndrome (HUS) is the clinical trial of microangiopathic hemolytic anemia, thrombocytopenia, and renal failure. HUS is often associated (>90%) with the production of Shiga toxin from *E. coli* 0157:H7 or *Shigella dysenteriae* 1. These Shiga-toxin-producing bacteria are common in cattle and outbreaks often occur secondary to contaminated beef, water, or vegetables. The bacteria first bind to the gut endothelium and penetrate the lining so that Shiga toxin can enter the bloodstream. The Shiga toxin then binds to CD77 (globsyl ceramide 3) on endothelial cells and renal mesangial cells resulting in apoptosis and a proinflammatory state, enhancing the secretion of vWF and creating a thrombogenic microenvironment.

Clinical presentation initially includes abdominal pain and classically bloody diarrhea followed by several days of apparent improvement and then development of symptomatic anemia, thrombocytopenia, and renal failure.

HUS treatment consists of supportive care (e.g., dialysis and antihypertensive medication). Red cell transfusions are commonly needed, but platelet transfusions should be used only when necessary, for example, for placement of a catheter for peritoneal dialysis. HUS is a common cause of renal failure in children and approximately 50% of children who develop the disorder will require dialysis, but it is rare that a patient does not recover and develops end-stage renal disease (<5%).

Atypical Hemolytic-Uremic Syndrome

Atypical hemolytic-uremic syndrome (aHUS) is thrombotic microangiopathy mediated by dysregulation of the alternative complement pathway. The alternative complement pathway is a constitutively active part of the innate immune system and with unregulated activation from a mutation in a critical component, there is deposition of C3b in tissue leading to increased formation of the C5b-9 membrane attack complex. Ineffective, or inhibition of, regulatory proteins of this alternative pathway through genetic mutations or antibody inhibition have been identified in 60–70% of aHUS patients. Genetic mutations have been identified in complement factor H, membrane cofactor protein, complement factor I, C3, complement factor B, thrombomodulin, and CFH-related proteins. The clinical presentation of aHUS is similar to that other TMA syndromes described with microangiopathic anemia, thrombocytopenia, and predominance of acute kidney injury and hypertension. Suspicion of aHUS, other than by laboratory findings indicated below, is based on either recurrence or the known occurrence of HUS in family members. CNS manifestations may occur in 10% of patients with seizures, diplopia, irritability, cortical blindness, hemiparesis, or hemiplegia. Similar to HUS, diarrheal symptoms may proceed clinical diagnosis in 24% of cases. Additional triggers may include respiratory illness or pregnancy. Given the limited ability to test the complement system inhibition at the time of presentation, this is often a diagnosis of exclusion with identification of normal ADAMTS13 > 5–10%, negative stool studies for Shiga-producing organisms in the setting of elevated serum creatinine and TMA triad (microangiopathic anemia, thrombocytopenia, and elevated LDH). Complement genetic testing can be performed for the known mutations, but measurement of C3, C4, and complements H, I, and B at normal levels does not exclude this diagnosis. Recognition of aHUS is important as there is targeted anticomplement therapy available. Eculizumab is a recombinant humanized monoclonal immunoglobulin targeting C5 to prevent the cleavage of C5 to C5a, thereby preventing the downstream formation of the membrane attack complex. This therapy was developed for management of paroxysmal nocturnal hemoglobinuria, but has been used in some patients with TTP and aHUS. A side effect of this therapy is increased risk for infection with encapsulated organisms, particularly *Neisseria meningitis*. Given the difficulty of diagnosis and the expense of eculizmab, the role as adjuvant to plasma exchange therapy or a first-line therapy in children is still debated. A microangiopathic picture following hematopoietic stem cell transplant is very likely aHUS.

Disseminated Intravascular Coagulation

Thrombocytopenia is seen in several syndromes associated with DIC, including purpura fulminans, overwhelming sepsis, and Kasabach–Merrit syndrome. In an unusual case of thrombocytopenia, PT, PTT, fibrinogen, and fibrin split products or D-dimers should be determined to exclude DIC as the cause of the thrombocytopenia.

Autoimmune Disorders

Thrombocytopenia is often associated with a variety of autoimmune disorders.

1. SLE: Thrombocytopenia occurs in 15–25% of patients with SLE. The initial treatment is similar to that of ITP, including steroids, IVIG, and immunosuppressive agents. Thrombopoietic agents may be effective but should be avoided as they may lead to a higher incidence of thrombosis in SLE.
2. Autoimmune lymphoproliferative syndrome (ALPS): This syndrome is characterized by lymphadenopathy, hepatosplenomegaly, hypergammaglobulinemia, and autoimmune cytopenias (i.e., Evans syndrome) and ITP. It is discussed in detail in Chapter 16.
3. Antiphospholipid antibody syndrome: Antiphospholipid antibodies enhance platelet activation. These patients have recurrent arterial or venous thrombi. The treatment is anticoagulation; steroids and/or immunosuppressive agents may be used in severe presentations. Thrombocytopenia is consumptive.
4. Evans syndrome: Evans syndrome is a combination of at least two of autoimmune hemolytic anemia, thrombocytopenia, and/or neutropenia. Patients typically have poor responses to steroids, IVIG, or

splenectomy. A combination of these therapies is often required and rituximab and MMF seem to be particularly effective as described above under ALPS and is often associated with ALPS.

5. Other autoimmune processes: Hodgkin lymphoma, non-Hodgkin lymphoma, juvenile rheumatoid arthritis, dermatomyositis, Graves disease, Hashimoto thyroiditis, myasthenia gravis, inflammatory bowel disease, sarcoidosis, and protein-losing enteropathy may be associated with autoimmune thrombocytopenia. Treatment of the underlying autoimmune disease may or may not improve the secondary ITP. Hypothyroidism is by far the most common association.

Cyanotic Congenital Heart Disease

Thrombocytopenia frequently occurs in children with severe cyanotic congenital heart disease, usually when the hematocrit levels are more than 65% and when the arterial oxygen saturation is less than 65%. This may be due to margination of platelets in the small blood vessels, which may occur in the presence of a high hematocrit level, although it could be partly artifactual. Cyanotic congenital heart disease may be associated with impairment of platelet aggregation by ADP, norepinephrine, and collagen. This impairment is correlated with the severity of hypoxia. Affected children usually experience little bleeding during corrective surgery, which, if successful, results in an improved platelet count.

Hypersplenism

A variety of conditions characterized by splenomegaly are associated with thrombocytopenia, presumably resulting from the sequestration or destruction of platelets by the enlarged spleen. This is usually associated with neutropenia and anemia. Megakaryocytes are plentiful in the marrow. Hypersplenism occurs in patients who have splenomegaly, irrespective of cause. Treatment is extremely difficult and platelet transfusion must be timed exactly to the invasive intervention required. Bleeding may be associated with other issues related to liver disease if severe.

THROMBOCYTOSIS

Thrombocytosis is defined as a platelet count greater than two standard deviations above the mean or greater than $450,000/mm^3$. Severity of thrombocytosis can be classified into the following groups:

- Mild ($450-700,000/mm^3$).
- Moderate ($700-900,000/mm^3$).
- Severe ($900,000-1$ million$/mm^3$).
- Extreme (>1 million$/mm^3$).

Thrombocytosis is relatively common in young children, but is usually a transient, benign finding occurring secondary to an underlying infectious or inflammatory process. Platelets are an acute-phase reactant and thrombocytosis can be part of the inflammatory response with upregulation of thrombopoietin (TPO receptors) and interleukin-6 (IL-6). Reactive thrombocytosis may also occur secondary to iron deficiency, major trauma, surgery, and postsplenectomy. Table 14.10 lists the conditions associated with thrombocytosis in infants and children.

Primary Thrombocytosis

Primary thrombocytosis is not well characterized in the pediatric population, due to the limited number of identified genetic mutations and small case series of long-term clinical information reported.

Essential Thrombocythemia

Essential thrombocythemia (ET) is a well-characterized myeloproliferative neoplasma (MPN) in adults associated with a mutation in *JAK2V617F* or relevant JAK-STAT and TPO proliferation pathway genes (cMPL PRV-1 and CALR). It is characterized by a sustained thrombocytosis, hyperplasia of megakaryocytes in the bone marrow in the presence of a pathogenic mutation, and absence of evidence for reactive thrombocytosis. ET diagnosis in

TABLE 14.10   Conditions Associated with Thrombocytosis in Infants and Children

Hereditary	Autoimmune diseases or chronic inflammation
Asplenia	Kawasaki disease
Myeloproliferative disorder in Down syndrome	Inflammatory bowel disease
	Rheumatoid arthritis
	Henoch–Schönleïn purpura
	Polyarteritis nodosa
Nutritional	Myelodysplastic states
Iron deficiency (chronic blood loss)	5q-syndrome
Vitamin E deficiency	Sideroblastic anemia
Megaloblastic anemia	
Metabolic	Traumatic
Hyperadrenalism	Surgery
Immune	Fractures
Graft-versus-host reaction	Hemorrhage
Nephrotic syndrome	Burns
Infectious	Miscellaneous
Viral (e.g., CMV)	Splenectomy
Bacterial	Caffey disease
Mycobacterial	Pulmonary embolism
Fungal	Thrombophlebitis
Drug-induced	Cerebrovascular accident
Vinca alkaloids	Sarcoidosis
Citrovorum factor	Acute blood loss
Corticosteroid therapy	Hemolytic anemia
Epinephrine	Exercise
	Ankylosing spondylitis
	Spurious
	Idiopathic
Neoplastic	
Chronic myeloid leukemia	
Polycythemia vera	
Essential thrombocythemia	
Histiocytosis	
Lymphoma (Hodgkin and non-Hodgkin)	
Carcinoma of colon, lung	
Hepatoblastoma	
Wilms tumor	
Neuroblastoma	
Leukemia	

children is rare, occurring in approximately 1/1,000,000 children with a decreased percentage of patients with identified JAK/STAT mutations.

In adult populations there is a well-described risk stratification system predicated on the risk of the development of vascular events (arterial and venous thrombosis) and the progression to MDS/AML that stratifies treatment and interventions in ET and MPNs. While the molecular pathogenesis is less well characterized in the pediatric population, the clinical presentation of primary thrombocytosis in children also has limited characterization. In children there appears to be a more benign course compared to adults, with a decreased risk of thrombosis.

Diagnosis of ET requires the following criteria (World Health Organization (WHO) 2008 Criteria):

- Persistent thrombocytosis ($>450,000/mm^3$).
- Bone marrow biopsy with megakaryocyte proliferation without increased neutrophil granulopoiesis or erythropoiesis.
- Absence of criteria for PV, PMF, CML, MDS, or myeloid neoplasm.
- Presence of JAK2V617F or clonal marker.
- No known cause for reactive thrombocytosis.

Genetic testing may confirm the diagnosis, but the absence of mutations does not rule out primary thrombocytosis in a pediatric patients. There are lower reported rates of JAK2V617F mutation, PRV-1 expression, and alternatively an increase in MPL mutations identified.

Hereditary Thrombocytosis

Familial or hereditary thrombocytosis is reported in families with causative mutations in the TPO (THPO) or TPO-receptor (cMPL) gene leading to constitutive activation or upstream regulation leading to increased TPO expression. Mutations in the *TPO* gene leads to increased translation and subsequent thrombocytosis, whereas the *mpl mutation* (S505N) leads to constitutive activation of the protein. The course of the disease seems to be largely benign with rare thrombosis.

Treatment of Pediatric Primary Thrombocytosis

Adults are stratified for treatment using risk factors or age, thrombosis history and cardiovascular risk factor into low-, intermediate-, and high-risk groups. In pediatrics, given the lower risk of thrombosis, there are no clear consensus guidelines for management but the general approach is adding observation for asymptomatic patients, adding antiplatelet agents for lower risk patients with additional thrombophilia risk factors and escalating to cytoreductive therapy in high-risk patients with extreme thrombocytosis or symptoms of thrombosis/bleeding. In these patients the risk of bleeding can be associated with acquired VWD and testing for ristocetin cofactor activity can be important prior to starting therapy.

Antiplatelet Agents

1. Low-dose acetylsalicylic acid (ASA)
   Risk for Reye syndrome should be considered in pediatric patients.
2. Additional antiplatelet agents: Clopidigrel, prasugrel, Dipyridamole, or Ibuprofen can be considered.

Cytoreductive Therapy (Platelet-Lowering Drugs)

1. Hydroxyurea can be used as a daily oral medication to lower platelet count to between 300,00 and 600,000/mm^3. Continued maintenance is necessary to sustain the effect. The use of HU in children with sickle cell disease suggests the risk of induction of malignancy is very low.
2. Anegrelide hydrochloride has also been shown to be effective in adults as a daily oral medication to lower platelet count between 300,000 and 600,000/mm^3 with maintenance therapy necessary to sustain response.
3. Additional and future considerations: Alpha interferon, low-dose busulfan, and especially Ruxolitinib (targeted JAK1/2 inhibitor).

QUALITATIVE PLATELET DISORDERS

Qualitative platelet disorders result from a variety of congenital or acquired conditions, but all demonstrate a bleeding tendency including petechiae, purpura, and mucosal bleeding as well as bleeding following surgery or trauma. Various tests may demonstrate abnormalities depending upon the particular condition and the platelet count.

Table 14.11 provides a classification by laboratory finding of congenital platelet function disorders, Table 14.12 lists the laboratory findings in inherited platelet function disorders and Table 14.13 lists the genetic transmission and recommended treatment for commonly known hereditary disorders of platelet function. In some of these disorders hemostasis will be improved by administration of DDAVP, as is used in management of mild hemophilia and von Willebrand disease. Platelet transfusion is often necessary for hemostasis in those disorders not responsive to DDAVP. Leukodepleted single-donor platelets, HLA matched if available, are preferred in these patients to reduce the risk of platelet allo-sensitization. Iso-antibodies against the platelet proteins absent from the patient are common in these disorders and may render patients refractory to platelet transfusion. Administration of rFVIIa 90 µg/kg can effect hemostasis in some patients not responsive to platelet transfusion. For non-responders, higher doses may be hemostatic. Antifibrinolytic agents, topical treatments, and hormonal therapy are useful adjuncts in patients with these disorders.

Defects in Platelet Receptor−Agonist interactions

Selective Impairments in Platelet Responsiveness to Epinephrine

The interaction of platelets with epinephrine (adrenaline), *in vitro* in aggregation tests, is mediated by α_2-adrenargic receptors and results in several responses, including the exposure of fibrinogen receptors, an increase

TABLE 14.11   Classification of Congenital Platelet Function Disorders

Defects in platelet–agonist interaction (receptor defects)
    Selective epinephrine defect
    Selective collagen defect
    Selective thromboxane A_2 defect
    Selective ADP defect
Defects in platelet–vessel wall interaction (disorders of adhesion)
    von Willebrand's disease (deficiency or defect in plasma von Willebrand factor)
    Platelet-type von Willebrand disease
    Bernard–Soulier syndrome (deficiency or defect in GP1b)
Defects in platelet–platelet interaction (disorders of aggregation)
    Congenital afibrinogenemia
    Glanzmann's thrombasthenia (deficiency or defect in GP11b/111a)
Disorders of platelet secretion
    Storage pool deficiency
        δ-storage pool deficiency
            Hermansky–Pudlak syndrome
            Chediak–Higashi syndrome
        α-Storage pool deficiency (gray platelet syndrome)
    Abnormalities in arachidonic acid pathway
        Impaired liberation of arachidonic acid
        Cyclooxygenase deficiency
        Thromboxane synthetase deficiency
        Altered nucleotide metabolism
        Glycogen storage disease
        Fructose 1,6, bisphosphatase deficiency
    Primary secretion defect with normal granule stores and normal thromboxane synthesis
        Defects in calcium mobilization
        Defects in phosphatidylinositol metabolism
        Defects in myosin phosphorylation
Disorders of platelet–coagulant protein interaction
    Defect in factor V^a–X^a interaction on platelets
Vascular or connective tissue defect
    Ehlers–Danlos syndrome
    Pseudoxanthoma elasticum
    Marfan syndrome
    Osteogenesis imperfecta
    Hereditary hemorrhagic telangiectasia (Osler–Weber–Rendu disease)

in intracellular ionized calcium, the inhibition of adenylate cyclase activity, and platelet aggregation. Some patients have impaired aggregation and secretion only to epinephrine, associated with a decrease in the number of platelet α_2-adrenargic receptors, with a history of easy bruising with minimally prolonged bleeding times. On the other hand, platelets in 10% of apparently normal people may fail to aggregate in response to epinephrine. Therefore the clinical significance of this finding is unknown.

Selective Impairment in Platelet Responsiveness to Collagen

Collagen is a substrate for platelet adhesion, binding site for vWF and agonist for platelet secretion and aggregation. There are two major collagen receptors, GPVI and α2β1, that mediate these hemostatic interactions. Defects in GPVI structure and signaling have been identified in case reports in congenital deficiency and autoantibody development leading to mild-severe mucocutaneous bleeding symptoms. α2β1 integrin allow for a firm platelet adhesion to collagen and in the absence of this interaction impaired aggregation and adherence to collagen. Very few patients have been shown to have bleeding due to abnormalities in α2β1.

Defects in Thromboxane A_2

Defects in thromboxane A_2 (TXA$_2$) formation and TXA$_2$ receptor function have been identified as causes of mild but lifelong bleeding symptoms, similar to the effects of chronic aspirin treatment.

TABLE 14.12 Laboratory Findings in Inherited Platelet Function Disorders

	Glanzmann's thrombasthenia	Storage pool deficiency[a]	Collagen receptor defect	Release defect[b]	Bernard–Soulier syndrome	von Willebrand disease
Platelet count	Normal	Normal	Normal	Normal	Decreased	Normal
Platelet size	Normal	Microplatelets	Normal	Normal	Giant platelets	Normal[c]
Bleeding time	Prolonged	Variable	Variable	Variable	Prolonged	Prolonged
PLATELET AGGREGATION						
ADP	Absent	No second wave	Normal	No second wave	Normal	Normal
Arachidonic acid	Absent	Variable	Absent	Decreased	Normal	Normal
Collagen	Absent	Decreased	Normal	Absent	Normal	Normal
Ristocetin (1.5 mg/ml)	Normal	Normal	Normal	Normal	Absent	Decreased[d]
Ristocetin (0.5 mg/ml)	Absent	variable	Absent	Absent	Normal	Decreased[d]
Storage nucleotide pool	Normal	Decreased	Normal	Normal	Normal	Normal
Others	HPA-1[a] absent; GP IIb and III deficient	Dense bodies reduced; ATP/ADP ratio increased	GP Ia and IIa deficient	Cyclo-oxygenese; thromboxane synthetase or TXA$_2$ receptor deficient	GP Ib deficiency	Decreased FVIII, vWF antigen and ristocetin cofactor

[a]In Hermansky–Pudlak, Chediak–Higashi, and Wiskott–Aldrich syndromes.
[b]Classically occurs with acetylsalicylic acid, omega-3 ingestion, and other drugs affecting arachidonic acid and prostaglandin pathway.
[c]Platelet in type IIB will often be clumped together.
[d]Types I and II, decreased; type IIb increased; type III absent.
vWF, von Willebrand factor; ATP, adenosine triphosphate; ADP, adenosine diphosphate.

TABLE 14.13 Genetic Transmission and Recommended Treatment for Commonly Known Hereditary Disorders of Platelet Function

Disorder	Transmission/ frequency	Defect	Primary treatment for platelets	Alternative treatment
DISORDERS OF RECEPTORS				
Glanzmann's thrombasthenia	Autosomal recessive, rare	GPIIb/IIIa complex	Platelet transfusion	rFVIIa in instances of platelet alloimmunization
Bernard–Soulier syndrome	Autosomal recessive, Rare	GP Ib/V/IX complex	Platelet transfusion	rFVIIa in instances of platelet alloimmunization
DEFECTS IN GRANULE CONTENT, STORAGE POOL DEFICIENCY				
Gray platelet syndrome	Rare	Absent alpha granules (platelets look gray)	DDAVP (treatment rarely required)	Platelet transfusion for non-responders
Chediak–Higashi syndrome	Autosomal recessive, rare	Abnormal granules	DDAVP	Platelet transfusion for non-responders
Wiskott–Aldrich syndrome	X-linked recessive	WAS protein. Primarily a quantitative defect, storage pool deficiency may also be present	Platelet transfusion	Splenectomy TPO-R agonists HSCT
Hermansky–Pudlak	Autosomal recessive	Absent dense granules	DDAVP	Platelet transfusion
Storage pool release defects	Variable	Impaired secondary wave of aggregation	DDAVP	Platelet transfusion

TPO-R, thrombopoietin receptor; HSCT, hematopoietic stem cell transplantation.

Selective Impairment in Platelet Response to ADP

Patients have been identified with defects in P2Y12 (the ADP receptor responsible for macroscopic platelet aggregation), with associated bleeding symptoms. This should be suspected when ADP at high >10 μM concentrations fail to produce full aggregation. This receptor is targeted by clopidogrel and prasugrel.

Defects in Platelet Vessel—Wall Interaction

Bernard—Soulier Syndrome

Bernard—Soulier syndrome is a relatively rare disorder inherited as either an autosomal dominant or recessive. It is characterized by:

- Moderate thrombocytopenia (automated counting often underestimates the true platelet count because of undercounting of very large platelets).
- Prolonged bleeding time.
- Characteristic platelet morphology.

Platelets are very large, equaling or exceeding the size of a red cell. These platelets contain two to four times the normal protein content, and three times the usual number of dense granules. These dense granules can gather and give the appearance of a pseudonucleus. The findings are more severe in cases of rare homozygote (or compound heterozygote) than in the more common heterozygote inheritance. Morphologically, the megakaryocytes have an abnormality in the demarcation membrane system, likely explaining the large platelets and the thrombocytopenia.

There can be complete absence of GPIb glycoprotein complex or a point mutation in the GPIb-alpha subunit known as the Bolzano variant; all of which leads to inability of vWF binding. Platelets fail to agglutinate in response to ristocetin, despite normal aggregation and secretion in response to ADP, epinephrine, thrombin, and collagen. Because vWF acts as a bridge in adhesion of platelets to exposed subendothelium especially in high shear states, the absence of the vWF receptors prevents normal platelet adhesion and causes a significant degree of bleeding (even in cases of mild thrombocytopenia) with mucocutaneous bleeding that can begin in early infancy, especially severe epistaxis.

Confirmation of the diagnosis is based on GP1b-IX-V deficiency by flow cytometry or immunoblotting or genetic diagnosis especially for the Bolzano variant.

Type 2B von Willebrand Disease and Platelet-Type (Pseudo-von Willebrand) Disease

Type 2B von Willebrand and platelet-type von Willebrand both cause mucocutaneous bleeding out of proportion to any thrombocytopenia that may be present. Stress-induced worsening of thrombocytopenia is common. Type 2B von Willebrand disease is approximately 10 times more common than platelet-type von Willebrand disease. The management of these conditions is described in detail in Chapter 15.

Defects in Platelet—Platelet Interaction

Glanzmann Thrombasthenia

Glanzmann thrombasthenia (GT) is an autosomal recessive bleeding disorder characterized by the following:

- Normal platelet count and morphology.
- Prolonged bleeding time.
- Mild to severe bleeding symptoms.
- Reduced clot formation.
- Defective platelet aggregation with all agonists except ristocetin.

GT represents a family of disorders of the platelet GPIIb/IIIa receptor complex. As this complex functions in platelet aggregation in low shear states via fibrinogen, vWF, and fibronectin, disruption of this pathway leads to impaired platelet aggregation. GT platelets attach normally to damaged subendothelium but fail to spread normally to form platelet aggregates. Platelets from patients with GT fail to aggregate in response to most physiologic agonists like ADP, thrombin, and epinephrine; they do agglutinate in response to ristocetin, however.

The treatment of GT is platelet transfusions; DDAVP is ineffective. Patients may become refractory to transfusions, and in such cases activated recombinant human factor VIIa (rVIIa) may be required.

Disorders of Platelet Secretion

The category of storage pool disease includes patients with deficiencies of dense granules (δ-storage pool deficiency (δ-SPD)), α-granules (α-SPD), or both types of granules (αδ-SPD).

δ-Storage Pool Deficiency

The total adenosine triphosphate (ATP) and ADP platelet granule content in patients with δ-SPD is decreased as are other dense granule contents, including calcium, pyrophosphate, and serotonin. Patients tend to have mild to moderate bleeding symptoms. Platelet aggregation studies show an absent second wave of aggregation with ADP and epinephrine. The aggregation response to collagen is impaired or absent. Response to arachidonic acid is variable. Disorders of δ-granules are seen in Hermansky–Pudlak syndrome (HPS), Chédiak–Higashi syndrome, WAS, and TAR syndrome.

Hermansky–Pudlak Syndrome

HPS is an autosomal recessive disorder. The syndrome includes oculocutaneous albinism photophobia, rotatory nystagmus, loss of visual acuity, and ceroid-like material accumulation in reticuloendothelial cells. Platelets have platelet dense-body granule storage pool deficiency and the bleeding tendency is usually mild (related to storage pool defect and not thrombocytopenia, which is not a feature of the syndrome). Patients generally present in childhood with mucosal bleeding symptoms, or prolonged bleeding after surgical procedures or dental extraction. HPS is a group of eight autosomal recessive disorders (HPS 1–8). The most common and severe form is HPS 1, which is a defect on chromosome 10q23. While rare worldwide, it is more common in northwest Puerto Rico, where 1 in 1800 individuals are affected. In this disorder, platelets contain very low levels of serotonin, adenine nucleotides, and calcium. The most serious complication and most common cause of death is pulmonary fibrosis, which may be due to ceroid-like lipfuchsin deposits in the lungs. Granulomatous colitis is also seen in this variant; conversely pulmonary fibrosis does not occur in HPS-III even though granulomatous colitis does. HPS has been reported to be associated with HLH. Treatment is DDAVP for bleeding and/or platelet transfusion if necessary.

Chédiak–Higashi Syndrome

This rare autosomal recessive syndrome includes partial oculocutaneous albinism, recurrent infections, mild coagulation defects, and progressive neurologic dysfunction. Mutations in lysosomal trafficking regulator on chromosome 1 are the cause of this disease. Leukocytes, lymphocytes, monocytes, and platelets have large peroxidase-positive intracytoplasmic granules. Severe immunologic deficiency with abnormal chemotaxis and NK function leads to HLH and death.

Patients also develop lymphoproliferative infiltration of bone marrow and reticuloendothelium in >85% of affected individuals. DDAVP is used for bleeding and platelet transfusions are indicated for non-responders. Stem cell transplantation may cure patients of immune and hemostatic defects, but does not improve the neurologic dysfunction.

α-Granule SPD (Gray Platelet Syndrome)

The term "gray platelet" describes the morphological appearance of platelets in Romanovsky-stained peripheral blood smears prepared from patients with a deficiency of α-granules. Electron microscopy reveals the virtual absence of α-granules and platelets appear gray because they are devoid of "purplish" cytoplasmic granulation on blood smear. Platelets from these patients contain absent or markedly reduced α-granule proteins including PF4, vWF, fibronectin, and factor V. Affected patients have mild thrombocytopenia with a platelet count of around 100,000/mm^3, prolonged bleeding times, and a lifelong bleeding diathesis. These patients tend to develop myelofibrosis due to an inability of megakaryocytes to store newly synthesized, platelet-derived granule proteins such as TGF-beta and PDGF-alpha, which "leak out" in the marrow. The most consistent laboratory abnormality has been impairment in thrombin-mediated aggregation and secretion. Aggregation responses to collagen and ADP are variable.

Autosomal recessive, autosomal dominant and often X-linked (GATA-1 mutation) inheritance has been observed with the GPS gene mapping to chromosome 3p21. Additional observations have noted these patients often have mild β-thalassemia in the X-linked form.

Treatment is rarely required for this condition but DDAVP is used if bleeding occurs and platelet transfusions are indicated for non-responders.

Quebec Platelet Disorder (Platelet Factor V Quebec)

Quebec syndrome is as an autosomal dominant inherited disorder with mild thrombocytopenia. However, it is associated with a significant degree of bleeding, including mucosal and joint bleeding often with a delayed presentation 12–24 h from time of injury. In this disorder, there is an overproduction of the proteolytic enzyme urokinase-type plasminogen activator (uPA). This enzyme overexpression occurs in megakaryocytes and leads to degradation of several proteins, including factor V. There is a platelet aggregation deficiency, in the setting of epinephrine exposure in particular, for unclear reasons. The disorder is diagnosed by examination of platelet uPA levels and α-granule fibrinogen degradation products. Genetic analysis has linked random duplication of the urokinase plasminogen activator gene. The treatment of choice is antifibrinolytic agents. Patients do not respond adequately to platelet transfusions.

Arthogryposis–Renal Dysfunction–Cholestasis

Arthogryposis–renal dysfunction–cholestasis (ARC) syndrome is a multisystem disorder that includes platelet dysfunction and low granule content.

May–Hegglin Anomaly

This is so-called MYH-RD (Myosin Heavy Chain 9 Related Disorder). Infrequent patients with this disorder may have platelet dysfunction.

Miscellaneous

Montreal Platelet Syndrome

Montreal platelet syndrome is a rare autosomal dominant disorder characterized by thrombocytopenia that tends to be severe, with large platelets seen on smear. Bleeding is typically delayed for several days. There may be spontaneous platelet aggregation and, unlike Bernard–Soulier syndrome, aggregation testing demonstrates normal response to ristocetin. There is a reduced response to thrombin-induced aggregation. These features also accompany deficient activity of platelet calpain activity, which is thought to underlie this disorder. Calpain is a calcium-dependent protease that is active on the platelet cytoskeleton.

Platelet function abnormalities have been reported in WAS, TAR syndrome, hexokinase deficiency, and glucose-6-phosphate deficiency.

Impaired Liberation of Arachidonic Acid Pathways

A major response of platelets during activation is the release of arachidonic acid from membrane-bound phospholipids and its subsequent oxygenation to TXA_2. TXA_2 forms an important positive feedback that enhances platelet activation. Ingestion of omega-3 fatty acids in sufficient quantity over a several-week period can reduce arachidonic release by replacing the natural omega-4 fatty acids in the platelet membrane with omega-3.

Cyclooxygenase and Thromboxane Synthetase Deficiency

Defects in TXA_2 due to deficiencies of cyclooxygenase and thromboxane synthetase have been reported.

Platelet Intracellular Signaling Defects

Kindlin-3 (Leukocyte Adhesion Defect III)

Kindlin-3 is a combination of mild leukocyte adhesion deficiency and platelet dysfunction. The intracellular signaling molecules, kindlin 2 and 3, are important in the activation of integrin αIIBβ3. All patients identified with this disease have been shown to have mutations in the cytoskeleton linking protein kindlin-3 (FERMTS3) gene. The presentation is mucocutaneous bleeding symptoms and predisposition to infections without pus formation, and with delayed wound healing, delayed umbilical stump separation, and variable osteopetrosis. Cure can be achieved by stem cell transplantation.

Dysregulated Calcium Signaling

Calcium is an essential signaling molecule involved in platelet activation; several patients have been reported with defective calcium mobilization.

Deficiency of Platelet Procoagulant Activity

Scott Syndrome

Scott syndrome is a rare autosomal recessive disorder involving mucosal and post surgical bleeding. The disorder arises from impaired translocation of phosphatidylserine from the inner to the outer leaflet of the plasma membrane, resulting in impaired binding of coagulation proteins and subsequent decrease in thrombin generation and fibrin formation. In these patients, bleeding time, platelet aggregation, and secretion studies are normal. Mutations in TMEM16F, a protein associated with membrane scramblase activity, have recently been identified in two patients with Scott syndrome.

Isolated Defect in Membrane Vesiculation

Patients in four families have been reported with lifelong bleeding disorders who have impaired microvesicle secretion.

ACQUIRED PLATELET DISORDERS

Table 14.14 lists the acquired disorders causing defective platelet function.

Medications

ASA (or aspirin) impairs platelet aggregation usually attributed to inhibition of cyclooxygenase-1 (COX-1) and subsequent reduction in TXA_2 production, although other mechanisms have been implicated. A dose of 81 mg of

TABLE 14.14 Acquired Disorders Causing Defective Platelet Function

1. Vascular or connective tissue defects
    a. Scurvy
    b. Amyloidosis
2. Adhesion defects
    a. Acquired von Willebrand disease, e.g., high-flow cardiac lesion
    b. Renal failure
    c. Drugs: dipyridamole
3. Platelet aggregation defects
    a. Fibrin or fibrinogen split products: DIC, liver disease
    b. Macromolecules: paraproteins, dextran
    c. Drugs: penicillin, semisynthetic penicillins, cephalosporins
4. Release reaction defects
    a. Storage pool deficiency
        i. ∝-Granules: cardiopulmonary bypass
        ii. Dense granules: ITP, SLE
        iii. Drugs: reserpine, tricyclic antidepressants, phenothiazines
    b. Defective release
        i. Platelet dyspoiesis: myelodysplastic syndromes, acute leukemias, myeloproliferative syndromes
        ii. Drugs: aspirin, other nonsteroidal anti-inflammatory agents, furosemide, nitrofurodantin
        iii. Ethanol
    c. Altered nucleotide metabolism
        i. Drugs: phosphodiesterase inhibitors or stimulators of adenylcyclase
5. Other defects
    a. Drugs: heparin, sympathetic blockers, clofibrate, antihistamines
    b. Infection: viral
    c. Hypothyroidism

ITP, idiopathic thrombocytopenic purpura; DIC, disseminated intravascular coagulation; SLE, systemic lupus erythematosus.

aspirin daily can permanently affect platelets in circulation, such that the length of drug effect is related to platelet lifespan, not drug half-life. Other nonsteroidal anti-inflammatory drugs (NSAIDs) impair TXA_2 production as well but the effects are related to drug half-life.

Platelet dysfunction caused by aspirin is important in several clinical settings:

- Excessive post-tonsillectomy bleeding after ASA ingestion.
- Development of purpura in children after ingesting aspirin.
- Prolonged bleeding time in children with suspected hemostatic defect.
- Misdiagnosing children with mild von Willebrand disease because of ASA/NSAID ingestion prior to testing.

Platelet aggregation may be impaired in the newborn following maternal drug therapy. For this reason a history of drug ingestion is an essential part of the investigation of hemorrhagic states in the newborn period.

Renal Failure

A generalized hemorrhagic state is known to occur in advanced renal failure. Thrombocytopenia is present in a minority of patients, and reduced platelet adhesiveness to glass and defective platelet factor 3 occurs with platelet dysfunction being the primary hemostatic problem. Treatment with estrogen and DDAVP can be useful but the former may take a week to have its effect and reversal of renal failure with dialysis may be required to control bleeding.

Liver Disease

In addition to a deficiency of coagulation factors of the prothrombin complex and others in liver disease, an abnormality of platelet function has been described. Platelet aggregation by ADP and thrombin is significantly impaired in patients with cirrhosis and prolonged bleeding time. This is due to known inhibition of platelet function by fibrinogen degradation products, resulting from excessive fibrinolysis seen in advanced liver disease.

Management of Defects in Platelet Function

The management of defects of platelet function may include:

1. Removing any exogenous cause of platelet dysfunction (e.g., drugs).
2. Treating the underlying disorder.
3. Using platelet transfusions for hemorrhagic episodes or surgery.
4. DDAVP or cryoprecipitate—this may shorten bleeding time in some patients with:
   a. Renal failure.
   b. Inherited or acquired defects in release reaction.
   c. Inherited or acquired von Willebrand disease.
5. Antifibrinolytic therapy—EACA (ε-aminocaproic acid) in a dose of 50–100 mg/kg orally or IV every 6 h—this may have some benefit in mucosal hemorrhage; there is no direct platelet effect.
6. Activated recombinant factor VIIa (rVIIa).

Inherited Vascular and Connective Tissue Disorders

Disorders of the vascular system and connective tissues do not generally lead to clotting or platelet abnormalities; bleeding is instead caused by the fragility of the connective tissues of the skin, subcutaneous tissues, and vessel wall.

Ehlers–Danlos Syndrome

Ehlers–Danlos syndrome (EDS) is a heterogeneous group of disorders of connective tissue characterized by skin hyperextensibility, delayed wound healing, joint hypermobility, bleeding tendency, and connective tissue fragility. Patients may present with easy bruising, bleeding from the gums after dental extraction, and prolonged menstruation. This condition may be associated with reduced aggregation with ADP, epinephrine, and collagen. Epistaxis, petechiae, hematuria, hemoptysis, and hemarthrosis are usually not seen. There are many subtypes of EDS. Vascular type (type IV) is an autosomal dominant disorder with a defect in type III collagen. Molecular testing identifies a mutation in the COL3A1 gene with a generalized vascular fragility that may manifest as arterial rupture and sudden death. Type IV EDS carries the worst prognosis due to the vascular complications.

Pseudoxanthoma Elasticum

Pseudoxanthoma elasticum is a rare disorder resulting in mineralization of elastic tissues in the skin, eyes, and blood vessels, and less frequently in other areas such as the digestive tract. It is inherited in an autosomal dominant fashion with mutations in ABCC6. Clinical presentation includes yellowish bumps called papules on the necks, underarms, and other areas of skin that touch when a joint bends. There is a low reported risk of bleeding, mostly gastrointestinal.

A pseudoxanthoma-elasticum-like condition has been associated with an increased risk of bleeding. There is a similar clinical presentation but the mutation is the GGCX gene encoding a vitamin-K-dependent carboxylase necessary for activation of the vitamin-K-dependent clotting factors (II, VII, IX, and X). These patients present with bleeding including epistaxis, gingival bleeding hematemesis, and vaginal/postpartum bleeding.

Marfan's Syndrome

Marfan's syndrome is characterized by skeletal abnormalities (scoliosis, pectus excavatum, and arachnodactyly), cardiovascular abnormalities, and dislocation of the lens. The syndrome is inherited in an autosomal dominant fashion with mutations in fibrillin-1. Affected individuals demonstrate easy bruising and may bleed excessively during surgery. While there is no vessel fragility in this syndrome there may be a predisposition for bruising due to less protective cushioning from lack of subcutaneous fat and increased incidence of trauma given the joint laxity and diminished visual acuity.

Osteogenesis Imperfecta

Osteogenesis imperfecta is an autosomal dominant disorder characterized by variable bone fragility, short stature, and blue sclera and hearing impairment. Patients may present with a bleeding disorder characterized by bruising, epistaxis, hemoptysis, and ICH. The abnormality occurs as a result of mutations in COLA1 and COLA2, which code for type I collagen.

Hereditary Hemorrhagic Telangiectasia

Hereditary hemorrhagic telangiectasia (HHT) (Osler–Weber–Rendu disease) is the most common, yet underdiagnosed, of the inherited vascular disorders. There are five underlying genetic mutations with ENG and ACVRL/ALK1 implicated in 85% of cases. HHT is a heterogeneous autosomal dominant disease characterized by multiple arteriovenous malformations (AVM) of the skin, liver, lung, GI tract, and brain. Telangiectasias are most prominent on the lips, tongue, face, and buccal and GI mucosa and are prone to rupture and bleeding. Epistaxis is usually the most common symptomatology. Lesions consist of dilated arterioles and capillaries lined by a thin endothelial layer. They are typical in appearance (1–3 mm in diameter, flat, round, red or violet in color) and they blanch on pressure. Histology of the abnormal vessels shows a deficiency of supporting elastic fibers, making them prone to rupture. HHT is a progressive disease with age-related penetrance with presenting symptoms from bleeding or shunting for blood through visceral AVM with transient ischemic attacks, embolic stroke, migraines, hypoxemia, or high-output cardiac failure. Treatment is generally symptomatic and expectant management with observational screening and supportive care.

Laboratory Evaluation of Platelets and Platelet Function

The following tests can be helpful in determining platelet number and function.

Examination of the Blood Smear

The blood smear should be evaluated in all cases where a platelet disorder is suspected. A review of the smear can quickly rule out pseudothrombocytopenia (the low platelet count by automatic electronic cell counter compared to an abundance of usually clumped platelets on blood smear). A disproportionate number of large platelets suggests increased platelet turnover. The large platelets represent a younger population. If the MPV using automatic electronic counters is raised (normal, 8.9 ± 1.5 fl) it indicates peripheral platelet destruction rather than impairment of platelet formation. Table 14.1 lists the various platelet diseases based on platelet size. If a true platelet abnormality is present, however, the morphology of the platelets, including size and color, may suggest one or more specific diagnoses.

Bleeding Time

The bleeding time test is an *in vivo* screening examination for the interaction between platelets and the blood vessel wall. Several techniques exist for measuring bleeding time. The test has not been standardized for young children or infants (who have been shown to have shorter bleeding times). In general, bleeding time for children has been shown to be less than 9 min, and is prolonged in cases where the platelet count is less than $100,000/mm^3$ or when disorders of platelet adhesion or aggregation are present (e.g., von Willebrand disease, acquired or congenital platelet function abnormalities).

This test is now rarely used because of difficulties with standardization.

Closure Time

The platelet function analyzer-100 (PFA-100; Dade-Behring, Deerfield, IL) is a rapid *in vitro* screening test for platelet function at high shear rates. Citrated whole blood is aspirated through a 150-μm diameter aperture in a membrane coated with collagen and either epinephrine or adenosine 5′-diphosphate (ADP). The machine measures closure time, which is the time it takes for a platelet plug to form and occlude the aperture. Closure time is useful in identifying most cases of von Willebrand disease, aspirin effect, and for some cases of platelet dysfunction. There is a high negative predictive value for these disorders. Closure time can be affected by hematocrit and platelet count. Optimal results require a hematocrit between 25% and 50% and a platelet count of at least $50,000/mm^3$. Closure time is useful in attempting to identify a suspected bleeding disorder, but is a poor screening test in the general population (i.e., in preoperative screening for bleeding disorders).

Platelet Aggregation in Platelet-Rich Plasma

Platelet-rich plasma (PRP) aggregometry is considered the gold standard for examining platelet function (at very low shear rates). Platelet aggregation is measured by a photo-optical instrument, which compares platelet-rich plasma to platelet-poor plasma. As aggregating agents like ADP, thrombin, and epinephrine are added, platelet aggregation is stimulated and platelet clumps form, causing the turbid PRP to become clear. The photo-optical device measures the increased light transmittance through the sample. The light received through the sample is converted through the electronic signals, amplified and recorded on chart paper. The following are normal wave responses to the addition of aggregating agents:

1. A biphasic response (i.e., a primary and secondary wave of aggregation) is seen in response to:
   a. ADP (low concentration).
   b. Epinephrine.
   c. Thrombin.
   d. Ristocetin: aggregation with ristocetin may be reduced or absent in von Willebrand disease (ristocetin may be used in the quantitative assay of von Willebrand disease).
2. A single wave of aggregation is seen with collagen.
3. Arachidonic acid causes rapid secondary aggregation. NSAIDs, such as indometacin and aspirin, are potent inhibitors of aggregation induced by arachidonic acid.
4. A platelet count in the PRP used in the actual aggregation testing must exceed $100,000/mm^3$ for accurate results.

Aspirin and other NSAIDs inhibit platelet aggregation by interfering with the release reaction of platelets, thereby reducing or eliminating the secondary wave of aggregation.

Several weaknesses are inherent in PRP aggregometry. Specifically, the centrifugation required can injure or activate platelets and may remove giant platelets. Centrifugation also removes erythrocytes and leukocytes, which are known to affect platelet function.

Platelet Aggregation in Whole Blood

Platelet aggregation may be measured with an impedance aggregometer (Lumi aggregometer). In this test, whole blood is placed in a cuvette containing two electrodes. An electric current is applied to the electrodes and agonist-stimulated platelets aggregate on the surface of the electrodes and impede the current flow in proportion to the degree of platelet aggregation. Because this method does not require centrifugation, it avoids some of the problems encountered in PRP aggregometry.

Either method of testing platelet aggregation may identify platelet disorders like Bernard–Soulier syndrome, Glanzmann's thrombasthenia, cyclooxygenase deficiency, storage pool disease, and von Willebrand disease.

NONTHROMBOCYTOPENIC PURPURA

Anaphylactoid purpura

Henoch—Schönlein is a vasculitis that leads to nonthrombocytopenic, nonthrombopathic purpura. The purpuric eruptions are different to those seen in thrombocytopenia in that the lesions are maculopapular, initially resembling urticaria because of edema and perivascular infiltration and later becoming erythematous, with central areas of hemorrhage that finally fade to brown because of denaturation of the extravasated hemoglobin. The rash appears on the buttocks and on the extensor surfaces of the lower extremities. Accompanying the rash are joint or gastrointestinal symptoms, localized areas of edema, and renal damage. Initial hematuria occurs in about one-third of patients. The platelet count and tests for hemostasis are normal. Treatment is symptomatic, and steroids may be beneficial for severe abdominal pain. Follow-up for the patients with suspected renal disease to prevent irreversible renal damage is important.

Infections

Infections may present with nonthrombocytopenic purpura, including acute bacterial endocarditis, meningococcal septicemia, coxsackievirus and echovirus infections, rubella, and atypical measles.

Drugs

Diffuse, benign, and self-limited purpura has been described after exposure to certain drugs, particularly sulfonamides.

Purpura Factitia

Although rare, purpura factitia may present a diagnostic dilemma. It is self-inflicted purpura, usually linear and always on accessible parts of the body. It is more common in females than in males and frequently indicates deep-seated psychopathology. All hematologic tests are normal.

Gardner–Diamond Syndrome

Gardner—Diamond syndrome is a poorly characterized syndrome, also known as psychogenic purpura, which presents with bruising in the context of physical or psychological stress. It occurs most commonly in adolescent girls. The etiology remains unknown. Often there is prodrome of warmth and pain at preceding bruising with associated nausea and headache, and the lack of evidence for hematological, vascular, immunological or infectious abnormalities. The "diagnostic test" involves injection of the patient's own red cells into the patient's skin and demonstrating an allergic reaction at the site, which supports a theory of erythrocyte autosensitization as a contributing etiology. Although in practice, this is often a diagnosis of exclusion.

Scurvy

Scurvy is a vascular disease occurring with profound deficiency of vitamin C (ascorbic acid) secondary to a diet lacking in fresh fruits and vegetables. Vitamin C is necessary for the formation of collagen and bleeding results from this defective synthesis in blood vessels. Platelet number and function are normal. It presents with bruising and oozing from the gums with poor wound healing.

Further Reading and References

Althaus, K., Greinacher, A., 2009. MYH9-related platelet disorders. Semin. Thromb. Hemost. 35, 213—223.

Audia, S., Rossato, M., Santegoets, K., Spijkers, S., Wichers, C., Bekker, C., et al., 2014. Splenic TFH expansion participates in B-cell differentiation and antiplatelet-antibody production during immune thrombocytopenia. Blood 124 (18), 2858—2866.

Ballow, M., Cates, K.L., Rowe, J.C., Goetz, C., Desbonnet, C., 1986. Development of the immune system in very low birth weight (less than 1500 g) premature infants: concentrations of plasma immunoglobulins and patterns of infections. Pediatr. Res. 20 (9), 899—904.

Barsam, S.J., Psaila, B., Forestier, M., Page, L.K., Sloane, P.A., Geyer, J.T., et al., 2011. Platelet production and platelet destruction: assessing mechanisms of treatment effect in immune thrombocytopenia. Blood 117 (21), 5723—5732.

Blatt, J., McLean, T.W., Castellino, S.M., Burkhart, C.N., 2013. A review of contemporary options for medical management of hemangiomas, other vascular tumors, and vascular malformations. Pharmacol Ther. 139 (3), 327—333.

Bouw, M.C., Dors, N., van Ommen, H., et al., 2009. Thrombotic Thrombocytopenic Purpura in Childhood. Pediatr. Blood Cancer 53, 537—542.

Bussel, J.B., Zacharoulis, S., Kramer, K., et al., 2005. Clinical and diagnostic comparison of neonatal alloimmune thrombocytopenia to non-immune cases of thrombocytopenia. Pedaitr. Blood Cancer 45, 176—183.

Butros, L.J., Bussel, J.B., 2003. Intracranial hemorrhage in immune thrombocytopenic purpura: a retrospective analysis. J. Pediatr. Hematol. Oncol. 25 (8), 660–664.

Cines, D.B., Bussel, J.B., Leibman, A.L., et al., 2009. The ITP syndrome: pathologic and clinical diversity. Blood 113, 6511–6521.

Dame, C., Sutor, H.A., 2005. Primary and secondary thrombocytosis in childhood. Br. J. Haematol., 165–177.

De Marco, L., Mazzucato, M., Fabris, F., De Roia, D., Coser, P., Girolami, A., et al., 1990. Variant Bernard–Soulier syndrome type bolzano. A congenital bleeding disorder due to a structural and functional abnormality of the platelet glycoprotein Ib-IX complex. J. Clin. Invest. 86 (1), 25–31.

Drolet, B.A., Trenor III, C.C., Brandão, L.R., Chiu, Y.E., Chun, R.H., Dasgupta, et al., 2013. Consensus-derived practice standards plan for complicated Kaposiform hemangioendothelioma. J Pediatr. 163 (1), 285–291.

George, J.N., Platelets on the Web. University of Oklahoma Health Sciences Center. <http://www.ouhsc.edu/platelets/DITP.html>. Accessed 12.01.14.

George, J.N., Nester, C.M., 2014. Syndromes of thrombotic microangiopathy. N. Engl. J. Med. 371 (7), 654–666.

Ghanima, W., Khelif, A., Waage, A., Michel, M., et al., 2015. Rituximab as second line treatment for adult immune thrombocytopenia (the RITP study)—A multicentre, randomized, double blind, placebo-controlled study. Lancet 385 (9978), 1653–1661.

Immuno polymorphism database, 2014. <http://www.ebi.ac.uk/ipd/hpa/table1.html>. Accessed 12.01.14.

Isreals, S.J., 2009. Diagnostic evaluation of platelet function disorders in neonates and children: an update. Semin. Thromb. Hemost. 35, 181–188.

Kucine, N., Chastain, K.M., Mahler, M.B., Bussel, J.B., 2014. Primary thrombocytosis in children. Haematologica. 99 (4), 620–628.

Lee, G.M., Arepally, G.M., 2013. Heparin-induced thrombocytopenia. Hematol. Am. Soc. Hematol. Educ. Program. 2013, 668–674.

Lewis, K.A., Engle, W., Hainline, B.E., Johnson, N., Corkins, M., Eugster, E.A., 2011. Neonatal Graves' disease associated with severe metabolic abnormalities. Pediatrics 128 (1), e232–e236.

Malfait, F., De Paepe, A., 2009. Bleeding in the heritable connective tissue disorders: mechanisms, diagnosis and treatment. Blood Rev. 23 (5), 191–197.

McDonald, J., Bayrak-Toydemir, P., Pyeritz, R.E., 2011. Hereditary hemorrhagic telangiectasia: an overview of diagnosis, management, and pathogenesis. Genet. Med. 13 (7), 607–616.

Medeiros, D., Buchanan, G.R., 1996. Current controversies in the management of idiopathic thrombocytopenic purpura during childhood. Pediatr. Clin. North Am. 43 (3), 757–772.

Millikan, P.D., Balamohan, S.M., Raskind, W.H., Kacena, M.A., 2011. Inherited thrombocytopenia due to GATA-1 mutations. Semin. Thromb. Hemost. 37 (6), 682–689.

Nardi, M.A., Gor, Y., Feinmark, S.J., Xu, F., Karpatkin, S., 2007. Platelet particle formation by anti GPIIIa49-66 Ab, Ca2+ ionophore A23187, and phorbol myristate acetate is induced by reactive oxygen species and inhibited by dexamethasone blockade of platelet phospholipase A2, 12-lipoxygenase, and NADPH oxidase. Blood 110 (6), 1989–1996, Epub June 1, 2007.

Pacheco, L.D., Berkowitz, R.L., Moise Jr., K.J., Bussel, J.B., McFarland, J.G., Saade, G.R., 2011. Fetal and neonatal alloimmune thrombocytopenia: a management algorithm based on risk stratification. Obstet Gynecol. 118 (5), 1157–1163.

Podda, G., Femia, E.A., Pugliano, M., Cattaneo, M., 2012. Congenital defects of platelet function. Platelets. 23 (7), 552–563.

Psaila, B., Petrovic, A., Page, L.K., Menell, J., Schonholz, M., Bussel, J.B., 2009. Intracranial hemorrhage (ICH) in children with immune thrombocytopenia (ITP): study of 40 cases. Blood 114 (23), 4777–4783.

Provan D., Stasi R., Newland A. et al., 2010. International consensus report on the investigation and management of primary immune thrombocytopenia. Blood 115 (2),168–186.

Ramaswamy, K., Hsieh, L., Leven, E., Thompson, M.V., Nugent, D., Bussel, J.B., 2014. Thrombopoietic agents for the treatment of persistent and chronic immune thrombocytopenia in children. J Pediatr. 165 (3), 600–5.e4.

Rao, A.K., 2013. Inherited platelet function disorders: overview and disorders of granules, secretion, and signal transduction. Hematol. Oncol. Clin. North Am. 27 (3), 585–611.

Rodeghiero, F., Statsi, R., Gernsheimer, T., et al., 2009. Standardization of terminology, definitions and outcome criteria in immune thrombocytopenic purpura in adults and children: report from an international working group. Blood 113, 2386–2393.

Salles, I.I., Feys, H.B., Iserbyt, B.F., et al., 2008. Inherited traits affecting platelet function. Blood Rev. 22, 155–172.

Sola-Visner, M., Sallmon, H., Brown, R., 2009. New insights into the mechanisms of nonimmune thrombocytopenia in neonates. Semin. Perinatol. 33 (1), 43–51.

Wei, H.W., Schoenwaelder, S.M., Andrews, R.K., et al., 2009. New insights into the haemostatic function of platelets. Br. J. Haematol. 147, 415–30.

15

Disorders of Coagulation

Suchitra S. Acharya and Susmita N. Sarangi

HEMOSTATIC DISORDERS

Physiology of Hemostasis

As a result of injury to the blood vessel endothelium, three events take place concurrently:

1. Vasoconstriction (vascular phase).
2. Platelet plug formation (primary hemostatic mechanism—platelet phase).
3. Fibrin thrombus formation (initiation, amplification, and propagation phases).

Relevant Components of Hemostasis

There are three integral components to hemostasis:

1. Endothelial cells.
2. Platelets.
3. Plasma coagulation factors.

Endothelial cells secrete substances that:

- Repel platelets (prostaglandin I2 (PGI2), adenosine diphosphate (ADP), and nitric oxide).
- Initiate coagulation (collagen, fibronectin).
- Promote platelet adhesion (von Willebrand factor (vWF)) and fibrin dissolution (tissue plasminogen activator, t-PA).
- Catalyze the inhibition of thrombin (heparin and thrombomodulin).
- Inhibit the initiation of fibrin dissolution (t-PA inhibitor).

Participation of *platelets* in hemostasis is a fundamental component of the physiologic process of coagulation. Platelet interactions in coagulation are initiated by adhesion to areas of vascular injury. Subsequent activation of platelets results in release of ADP, serotonin, and calcium from "dense bodies" and fibrinogen, vWF, factor V (FV), high-molecular-weight (HMW) kininogen, fibronectin, α-1-antitrypsin, α-thromboglobulin, platelet factor 4 (PF4), and platelet-derived growth factor from α granules. Platelets provide surfaces for the assembly of coagulation factors (e.g. VIIIa/Ca^{2+}/IXa and Va/Ca^{2+}/Xa complexes). The platelets aggregate and increase the mass of the hemostatic plug. They also mediate blood vessel constriction (by releasing serotonin) and neutralize heparin.

All of the *plasma coagulation factors* are produced in the liver; factor VIII is also produced by endothelial cells. Table 15.1 lists the half-life and plasma levels of the coagulation factors. Factors II, VII, IX, and X are vitamin-K-dependent and require vitamin K in order to undergo post-translational gamma carboxylation. These vitamin-K-dependent factors circulate in zymogen form, are activated on platelet phospholipid surfaces and, upon activation, have serine protease activity. The plasma coagulation factors work in an interdependent manner to generate thrombin (factor IIa) from prothrombin (factor II); thrombin then converts fibrinogen to form fibrin monomers. Fibrin monomers polymerize and establish a network. Thrombin activates factor XIII which in turn crosslinks the fibrin network. By incorporating into the hemostatic plug, thrombin becomes inactivated.

TABLE 15.1　Half-Life and Plasma Levels of Coagulation Factors

Factors	Common name	Biologic half-life (h)	Plasma concentration (nM)	Plasma levels (units/dl)
I	Fibrinogen	56–82	8800	200–400[a]
II	Prothrombin[b]	45–60	1400	50–150
III	Tissue thromboplastin	N/A	—	0
V	Proaccelerin, labile factor	36	20	50–150
VII	Proconvertin[b], stable factor	5	10	50–150
VIII	Antihemophilic factor	8–12	0.7	50–150
IX	Christmas factor[b]	12–24	90	50–150
X	Stuart factor[b]	24–60	170	50–150
XI	Plasma thromboplastin antecedent	48	30	50–150
XII	Hageman factor	48–52	375	50–150
XIII	Fibrin-stabilizing factor	168–240	70	50–150
HMW kininogen	Fitzgerald factor	136	6,000	—
Prekallikrein	Flecher factor	N/A	450	—

[a]In mg/dl.
[b]Vitamin-K-dependent.

Thrombin plays a central bioregulatory role, promoting platelet aggregation and release reactions and generating a biofeedback-positive loop to form more thrombin at a faster rate. Thrombin and thrombin complexed to thrombomodulin also activate thrombin-activatable fibrinolysis inhibitor (TAFI), a procarboxypeptidase found in plasma, which attenuates fibrinolysis of the clot. The three components of hemostasis (blood vessels, platelets, and plasma coagulation factors) do not function independently but are integrated and interdependent.

Primary Hemostatic Mechanism (Platelet Phase)

This mechanism leads to the formation of a reversible aggregate of platelets: a temporary hemostatic plug. Endothelial injury exposes vWF and collagen from the subendothelial matrix to flowing blood and shear forces. Plasma vWF binds to the exposed collagen, uncoils its structure and, in synergy with collagen supports the adhesion of platelets. Initially the vWF interacts with the GPIb platelet receptor, tethering the platelets. As the platelet collagen receptors GPVI and α2β1 bind to collagen, the platelets adhere and become activated with a resulting release of platelet alpha and dense granule contents. Platelet activation results in a conformational change in the αIIbβ3 receptor, activating it and enhancing its avidity for vWF, for vessel wall ligands and for fibrinogen. The enhanced avidity for vWF and fibrinogen mediates platelet-to-platelet interactions which eventually lead to platelet plug formation.

Fibrin Thrombus Formation

The fibrin thrombus formation component of hemostasis occurs in three overlapping phases (Figure 15.1):

1. Initiation.
2. Amplification.
3. Propagation.

The *initiation phase* begins with cell-based expression of tissue factor (TF) at the site of endothelial injury. Factor VII binds to the exposed TF and is rapidly activated. The factor VIIa/TF complex in turn generates factor Xa (FXa) and factor IXa (FIXa). FXa can activate FV which complexes with FXa and generates small amounts of thrombin. During the *amplification phase* the procoagulant stimulus is transferred to the surface of platelets at the site of injury. The small amounts of thrombin enhance platelet adhesion, fully activate the platelets and activate factors V, VIII, and XI. In the *propagation phase* the "tenase" complex of FIXa–FVIIIa is assembled on the platelet surface and efficiently generates FXa. Similarly the "prothrombinase" complex of FXa–FVa is assembled on the platelet surface and efficiently generates thrombin. Unlike FXa generated from TF–FVIIa interactions, FXa complexed to FV is protected from inactivation by tissue factor pathway inhibitor (TFPI), assuring adequate thrombin generation.

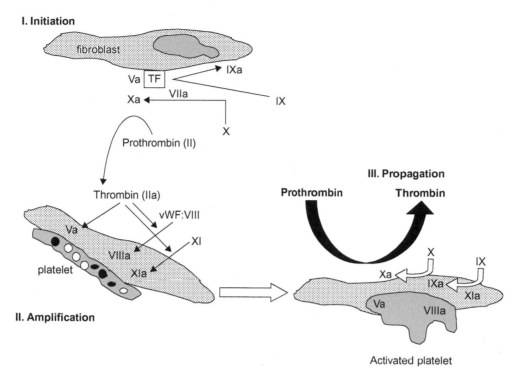

I. Initiation

fibroblast

Va | TF → IXa

VIIa

Xa ← IX

X

Prothrombin (II)

Thrombin (IIa)

vWF:VIII

Va

platelet

VIIIa   XI

XIa

II. Amplification

III. Propagation

Prothrombin   Thrombin

X   IX

Xa   IXa

XIa

Va   VIIIa

Activated platelet

FIGURE 15.1   A Cell based Model of Coagulation. The three phases of coagulation occur on different cell surfaces: initiation on the tissue-factor bearing cell, amplification on the platelet as it becomes activated; and propogation on the activated platelet surface. *Adapted from: Hoffman et al. (2001). Used with permission.*

The resulting procoagulant, thrombin, activates factor XIII and cleaves fibrinopeptides (FPs) A and B from fibrinogen. The residual peptide chains aggregate by means of loose hydrogen bonds to form fibrin monomers. Under the influence of FXIIIa, fibrin monomers are converted into fibrin polymers, forming a stable fibrin clot. In the presence of thrombin, the mass of loosely aggregated intact platelets is transformed into a densely packed mass that is bound together by strands of fibrin to form a definitive hemostatic barrier against the loss of blood.

Platelet Vessel Interaction

The vasculature forms a circuit which maintains blood in a fluid state, free of leaks. With vascular injury, platelets and the coagulation system temporarily close the rent and repair the leak. Blood vessel wall characteristics exhibit properties that contribute to hemostasis or stop hemorrhage as well as prevent thrombosis. The media and adventitia of the vessel wall enable vessels to dilate or constrict. The subendothelial basement membranes contain adhesive proteins which provide binding sites for platelet and leukocytes. Remodeling which occurs after injury is enhanced by extracellular matrix metalloproteinases.

Fibrinolysis

The fibrinolytic system provides a mechanism for removal of physiologically deposited fibrin. Clot lysis is brought about by the action of plasmin on fibrin. Fibrinolytic events are shown in Figure 15.2. Plasminogen from circulating plasma is laid down with fibrin during the formation of thrombin. Plasminogen is primarily synthesized in the liver and circulates in two forms, one with a NH2-terminal glutamic acid residue (glu-plasminogen) and a second form with a NH2-terminal lysine, valine, or methionine residue (lys-plasminogen). Glu-plasminogen can be converted to lys-plasminogen by limited proteolytic degradation. Lys-plasminogen has a higher affinity for fibrin and cellular receptors; it is also more readily activated to plasmin than glu-plasminogen. Both forms of plasminogen bind to fibrin through specific lysine-binding sites. These lysine-binding sites also mediate the interaction of plasminogen with its inhibitor, α_2-antiplasmin (α_2AP). TAFI-mediated removal of C-terminal lysine and arginine residues will prevent high-affinity plasminogen binding and will attenuate fibrinolysis. Plasminogen is converted to its enzymatically active form, plasmin, by several activators. These activators are widely distributed in body tissues and fluids. t-PA is the principal intravascular activator of plasminogen. t-PA is a serine protease that binds to fibrin through lysine-binding sites. When t-PA is bound to fibrin its plasmin generation efficiency increases

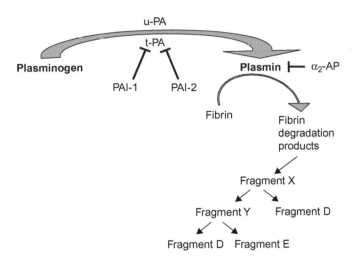

FIGURE 15.2   The Fibrinolytic Pathway. Plasminogen is converted enzymatically to plasmin by t-PA or by u-PA. Plasmin cleaves fibrin and fibrinogen into fibrin degradation products. Major inhibitors of the fibrinolytic pathway are depicted. PAI-1 and PAI-2 inhibit t-PA. Plasmin activity is inhibited by α_2-AP. t-PA: Tissue Plasminogen activator, u-PA: urokinase, PAI-1: plasminogen activator inhibitor 1, PAI-2: plasminogen activator inhibitor 2, α_2-AP: alpha 2 anti-plasmin. T : inhibition.

markedly. Urokinase (UK)-type plasminogen activator (u-PA), a second physiological activator of plasminogen, is present in urine and activates plasminogen to plasmin independent of the presence of fibrin. Plasmin splits fibrin and fibrinogen into fibrin-degradation products (FDPs): fragments X, Y, D, and E.

Properties attributed to the various fibrin split products include heparin-like effects, inhibition of platelet adhesion and aggregation, potentiation of the hypotensive effect of bradykinin and chemotactic properties (monocytes and neutrophils). Increased fibrinolysis is usually a reaction to intravascular coagulation (secondary fibrinolysis) rather than the initial event (primary fibrinolysis).

The action of plasmin is negatively regulated by several inhibitors (shown in Figure 15.2) these include α_2AP and α_2-macroglobulin. PA is in turn regulated by two inhibitors, plasminogen activator inhibitor-1 (PAI-1) and plasminogen activator inhibitor-2 (PAI-2). PAI-1 is the more physiologically important of these inhibitors.

Natural Inhibitors of Coagulation

In addition to the physiologic role of fibrinolysis, other inhibitors play critical roles in the control of hemostasis. Table 15.2 lists the plasma fibrinolytic components and hemostatic inhibitors and their principal substrates. All members of this group have overlapping roles in the control of coagulation and fibrinolysis. Major antiproteases of this group of inhibitors include antithrombin (AT), α_2AP, α_2-macroglobulin, the inhibitor of the activated first component of complement (C1 inhibitor), and α_1-antitrypsin.

AT neutralizes the procoagulants thrombin, FIXa, Xa, and XIa (Figure 15.3). When bound to circulating heparin or heparin sulfate on endothelial cells, AT undergoes a conformational change with a dramatic increase in this activity. TFPI is responsible for inactivation of the FXa/FVIIa/TF complex.

The vitamin-K-dependent zymogen, protein C, and its cofactor protein S, which is also a vitamin K protein, play an important role in the control of hemostasis by inhibiting activated factors V and VIII (Figure 15.3). Binding of thrombin to thrombomodulin on endothelial cells of small blood vessels neutralizes the procoagulant activities of thrombin and activates protein C. Protein C (PC) binds to a specific receptor and the binding augments the activation of protein C by thrombin. Activated protein C inactivates factors Va and VIIIa in a reaction that is greatly accelerated by the presence of free protein S and phospholipids, thereby inhibiting the generation of thrombin. Free protein S itself has anticoagulant effects: it inhibits the prothrombinase complex (FXa, factor Va, and phospholipid), which converts prothrombin to thrombin and inhibits the complex of FIXa, factor VIIIa, and phospholipid, which converts factor X to FXa.

HEMOSTASIS IN THE NEWBORN

Table 15.3 lists the hemostatic values in healthy preterm and term infants. In comparison with hemostatic mechanisms in older children and adults, those of newborn infants are not uniformly developed. The following physiological differences are present in the normal newborn infant:

TABLE 15.2 Plasma Fibrinolytic Components and Hemostatic Inhibitors

	Biologic half-life	Proteases inhibited	Concentration in plasma (mg/dl)
FIBRINOLYTIC COMPONENTS			
Plasminogen[a]	48 h	—	10−15
PAs			
Tissue	3−4 min	—	—
Urokinase	9−16 min	—	—
PAI[a]	—	PA, XII[a]	60−200[b]
INHIBITORS			
AT[a] (heparin cofactor)	17−76 h	XII[a], XI[a], IX[a], X[a], thrombin, kallikrein, plasmin	10−14 104−121[b]
α_2-Plasmin inhibitor[a] (AP)	30 h	XII[a], XI[a], kallikrein, plasmin, thrombin	6−8 80−120[b]
α_2-Macroglobulin[a]	—	XII[a], XI[a], thrombin, kallikrein, plasmin	190−310
C1 inhibitor[a]	—	XII[a], kallikrein	20−25
α_1-Antitrypsin[a]	—	Thrombin, XI[a], kallikrein	245−325
TAFI			20−400[c]
TFPI	—	Factor VII[a]/TF complex	Endothelial bound
Protein C[a]	6 h	V[a], VIII[a], PAI	0.4−0.6 71−109[b]
Protein S	60 h	V[a], VIII[a]	95−125[b]
Protein C inhibitor[a]	—	Protein C[a]	0.5

[a]*Enzymatic activity: serine protease.*
[b]*Activity in plasma (%).*
[c]*Activity expressed as nM/l.*

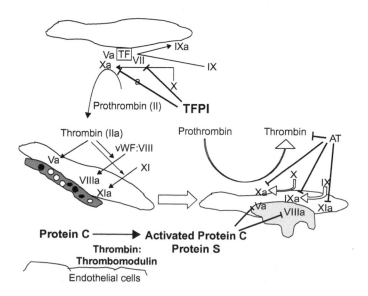

FIGURE 15.3 Major Inhibitory Proteins of Coagulation. TFPI, AT, protein C and protein S are depicted with their target coagulation factor substrates. TFPI: tissue factor pathway inhibitor, AT: antithrombin. T: inhibition.

TABLE 15.3 Hemostatic Values in Healthy Preterm and Term Infants

	Normal adults/children	Preterm infant (28–32 weeks)	Preterm infant (33–36 weeks)	Term infant
PT (s)	10.8–13.9	14.6–16.9	10.6–16.2	10.1–15.9
APTT (s)	26.6–40.3	80–168	27.5–79.4	31.3–54.3
Fibrinogen (mg/dl)	95–425	160–346	150–310	150–280
II (%)	100[a]	16–46	20–47	30–60
V (%)	100[a]	45–118	50–120	56–138
VII (%)	100[a]	24–50	26–55	40–73
VIII (%)	100[a]	75–105	130–150	154–180
vWF Ag (%)	100[a]	82–224	147–224	67–178
vWF (%)	100[a]	83–223	78–210	50–200
IX (%)	100[a]	17–27	10–30	20–38
X (%)	100[a]	20–56	24–60	30–54
XI (%)	100[a]	12–28	20–36	20–64
XII (%)	100[a]	9–35	10–36	16–72
XIII (%)	100[a]	—	35–127	30–122
PK (%)	100[a]	14–38	20–46	16–56
HMW-K (%)	100[a]	20–36	40–62	50–78

[a]*Expressed as a percentage of activity in pooled control plasma.*
PK, prekallikrein; HMW-K, high-molecular-weight kininogen.

Plasma Factors:

In newborns, plasminogen levels are only 50% of adult values and α_2AP levels are 80% of adult values, whereas PAI-1 and t-PA levels are significantly increased over adult values. The increased plasma levels of t-PA and PAI-1 in newborns on day 1 of life are in marked contrast to values from cord blood, in which concentrations of these two proteins are significantly lower than in adults. Newborns also have decreased activity of anticoagulant factors, especially AT, protein C, and protein S.

Blood vessels:

- Capillary fragility is increased.
- Prostacyclin production is increased.

Platelets:

1. Platelet adhesion is increased due to increased vWF and increased HMW vWF multimers.
2. Platelet aggregation abnormalities
   a. Epinephrine-induced aggregation is decreased due to decreased platelet receptors for epinephrine.
   b. Ristocetin-induced aggregation is increased due to increased vWF and increased HMW vWF multimers.
3. Platelet activation is increased: as evidenced by elevated levels of thromboxane A2, B thromboglobulin, and PF4.

DETECTION OF HEMOSTATIC DEFECTS

Evaluation of a patient for a hemostatic defect generally entails the following:

1. Detailed history (see Table 15.4 for initial features suggestive of pathological bleeding in children)
   a. Symptoms: Epistaxis, gingival bleeding, easy bruising, menorrhagia, hematuria, neonatal bleeding (heel stick, umbilicus), gastrointestinal (GI) bleeding, hemarthrosis, prolonged bleeding after lacerations.
   b. Response to hemostatic challenge: Circumcision, surgery, phlebotomy, immunization/intramuscular injection, suture placement/removal, dental procedures.

TABLE 15.4 Initial Features Suggestive of Pathologic Bleeding in Children

---

Age-related

Bleeding (e.g., umbilical stump bleeding, Intracranial hemorrhage, excessive and prolonged bleeding postcircumcision or after heel stick or intramuscular injection) during neonatal period

Palpable and multiple bruises in infants and older children who are not independently mobile

Persistent palpable bruising in an older mobile child

Spontaneous bleeding in the absence of anatomic causes

Personal history of

Recurrent (especially excessive and spontaneous) mucocutaneous bleeding

Atypical bleeds (e.g., hemarthroses, retroperitoneal bleeding), whether spontaneous or provoked

Excessive or prolonged bleeding after hemostatic challenges (i.e., trauma, dental procedures, or surgery)

Menorrhagia in adolescent girls: menstrual bleeding for >7 days or 80-ml blood loss per menstrual cycle (as evidenced by soaking through a pad or tampon within 1 h or change of pads or tampons every hour or passage of large (1.1-inch diameter) clots)

Traumatic bleeding that is out of proportion to or inconsistent with reported injury (consider nonaccidental trauma)

Family history of

Recurrent bleeding symptoms

Excessive or prolonged bleeding after trauma or invasive procedures

Known or suspected bleeding diatheses

Physical findings

Multiple bleeding stigmata

Physical findings suggestive of specific underlying causes (e.g., petechiae in platelet disorders, jaundice in liver disease, hypermobility, vascular malformations, musculoskeletal abnormalities)

Pallor/anemia

Reproduced with permission from Pediatrics, *p. 883, Copyright © 2013 by the AAP.*

   c. Underlying medical conditions: Known associations with hemostatic defects (liver disease, renal failure, vitamin K deficiency).

   d. Medications: Antiplatelet drugs (nonsteroidal anti-inflammatory drugs), anticoagulants (warfarin, heparin, low-molecular-weight heparin (LMWH)), antimetabolites (L-asparaginase), prolonged use of antibiotics causing vitamin K deficiency, long-term use of iron suggestive of ongoing blood loss causing iron-deficiency anemia.

   e. Family history: Symptoms, response to hemostatic challenge (siblings, parents, aunts, uncles, grandparents), transfusions after surgeries, iron deficiency after surgery or menorrhagia.

2. Complete physical examination

   a. Signs consistent with past coagulopathy: Petechiae, ecchymosis, hematomas, synovitis/joint effusion, arthropathy, muscle atrophy, evidence of joint laxity or hyperextensibility which can exacerbate the bleeding phenotype. In very young patients parental joint mobility should be assessed.

3. Laboratory evaluation (Tables 15.1, 15.2, 15.3, and 15.5):

Initial screening tests

1. Complete blood count (CBC): Quantitative assessment of platelets and review of blood smear to assess platelet morphology.

2. Assessments of platelet function:

   a. Platelet Function Analyzer (PFA-100): Assesses flow through a membrane, membrane closure time is measured in response to ADP and to epinephrine. The closure time is often prolonged with impaired platelet function (see Chapter 14).

   b. Bleeding time no longer used where PFA-100 is available because of difficulties in validation.

3. Coagulation factor screening tests:

   a. From a laboratory perspective the coagulation system is divided into the intrinsic pathway, the extrinsic pathway and the common pathway. Such an artificial division is not physiologically based but is useful for conceptualizing *in vitro* laboratory testing (Figure 15.4).

4. Prothrombin time (PT) assay (assesses the extrinsic system): This test utilizes tissue thromboplastin and calcium chloride, to initiate the formation of thrombin via the extrinsic pathway. The international normalized ratio (INR) is used to correct for differences between thromboplastin potency across laboratories.

5. Activated partial thromboplastin time (aPTT) assay (assesses the intrinsic system): This test utilizes a phospholipid reagent, a particulate activator (e.g., ellagic acid, kaolin, silica, soy extract) and calcium chloride to start the enzyme reaction that leads to the formation of thrombin via the intrinsic pathway.

TABLE 15.5   Coagulation Tests and Normal Values

Test	Normal value	Clinical application
PLATELET FUNCTION		
Template bleeding time (min)	<9	Crude, lack of reproducibility
Platelet aggregation		(see Chapter 14)
Platelet Factor3 availability		Screens for platelet procoagulant activity
Clot retraction	Starts at hour 1; completes at hour 24	Measures platelet interaction with fibrin
INTRINSIC SYSTEM		
Activated partial thromboplastin time(s)	25–35	
EXTRINSIC SYSTEM		
Prothrombin time(s)	10–12	
Factor assays	See Tables 15.1 and 15.3	
Thrombin time(s)	<24	Prolonged in hypofibrinogenemia, dysfibrinogenemia, hypoalbuminemia, liver disease, neonates
Reptilase time(s)	<25	Modification of thrombin time, unaffected by presence of heparin
ANTIPHOSPHOLIPID ASSAYS		
Dilute Russell's viper venom time(s)	29–42	
Kaolin clotting time(s)		Sensitive test even in presence of heparin
FIBRINOLYTIC SYSTEM		
Euglobulin clot lysis time (minutes)	90–240	Prolonged with hypofibrinolysis

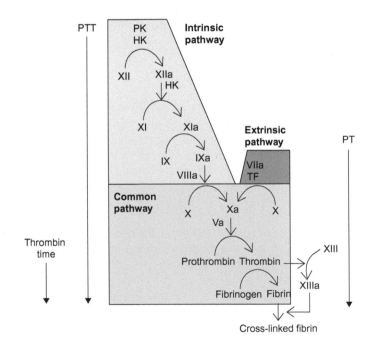

FIGURE 15.4   A conceptualization of commonly used screening tests of coagulation and the coagulation parameters they measure. PTT: partial thromboplastin time, PT: prothrombin time.

Common Confirmatory Coagulation Assays

- Fibrinogen: Quantitative measurement of fibrinogen, useful when both the PT and the aPTT are prolonged.
- Thrombin time: Prolonged when fibrinogen is reduced or abnormal, in the presence of inhibitors (FDPs, D-dimers) and in the presence of thrombin-inhibiting drugs and hypoalbuminemia. This is a useful test to diagnose dysfibrinogenemia (qualitatively abnormal fibrinogen). Also useful when both the PT and aPTT are prolonged. The reptilase time is a modification of the TT using the purified enzyme reptilase instead of thrombin and is unaffected by heparin and heparin-like anticoagulants.
- Mixing studies (performed to evaluate a prolonged PT or aPTT): The respective assay is performed following addition of normal pooled plasma to patient plasma. Normalization indicates a clotting factor deficiency that was corrected by addition of normal pooled plasma. Continued prolongation indicates presence of a coagulation inhibitor. Such inhibitors may be physiologically relevant or only detected *in vitro*.
- Clotting factor activity assays: Performed to identify clotting factor deficiencies if mixing studies normalize. FXII, FXI, FIX, and FVIII assays are useful if the aPTT normalizes in mixing studies. The FVII assay is useful if the PT normalizes in mixing studies. FX, FV, FII, and fibrinogen assays are useful if both PT and aPTT normalize in mixing studies. (*Note*: Factor XIII deficiency does not result in prolongation of the PT or aPTT.)
- von Willebrand antigen: Quantitative assay for vWF, useful when PFA-100 closure time is prolonged or when von Willebrand disease (vWD) is suspected.
- vWF (ristocetin cofactor activity): Functional/qualitative assay for vWF, useful when the PFA-100 closure time is prolonged or when vWD is suspected.
- vWF multimers are useful when discrepancies between vWF antigen (normal to low) and ristocetin cofactor activity (extremely low; ratio of ristocetin cofactor activity to vWF antigen <0.6).
- Platelet aggregation studies: A qualitative assessment of platelet function, useful when platelet function disorders are suspected or PFA-100 closure time is prolonged.
- Urea clot lysis assay: Useful screening for FXIII deficiency outside the newborn period although not sensitive enough to make an accurate diagnosis. Fetal fibrinogen interferes with this assay giving a false-positive result in the newborn period. In the absence of fibrin crosslinkage by FXIII, a clot will degrade with incubation in 5M urea. As this assay detects only severe factor XIII deficiency, functional FXIII assays should be conducted if mild deficiency is suspected. FXIII activity, antigen, and FXIII A and B subunit sequencing are essential to make an accurate diagnosis as per recommendations of the International Society of Thrombosis and Hemostasis.

Global Hemostatic Tests

Hemostasis is a complex interplay of simultaneously occurring events, with current assays only reflecting snapshots of this dynamic process. Global hemostatic tests can provide detailed information on thrombin generation and processes downstream including fibrin polymerization and fibrin dissolution.

- Thrombin generation assay: The Calibrated Automated Thrombogram (CAT) system uses a fluorogenic substrate to continuously measure the generated thrombin. The endogenous thrombin potential (ETP), which can be measured by calculating the area under the curve from the thrombogram, has shown correlation with the bleeding phenotypes in hemophilia, in hemophilia patients with inhibitors, and factor XI deficiency.
- Thromboelastography (TEG): TEG is performed on whole blood, assessing the viscoelastic property of clot formation under low shear condition after the addition of specific coagulation activators. Improvement in the methodology is widely expanding the use of TEG from preclinical research settings to point of care testing in intensive care units and operating rooms. The device has a metal pin suspended by a torsion wire immersed into a cup which holds the whole blood. Once clotting starts, fibrin strands formed increase the torque between the pin and the cup which is measured electronically. TEG provides various data relating to clot formation and fibrinolysis (the lag time before the clot starts to form, the rate at which clotting occurs, the maximal amplitude of the trace or clot strength, and the extent and rate of amplitude).

Preoperative Evaluation of Hemostasis

1. History

The history is perhaps the most important element of the evaluation. In an effort to standardize bleeding histories, a number of quantitative bleeding assessment tools (BATs) have been developed and validated in individuals with known vWD (Vicenza-based BAT, the condensed MCMDM-1 VWD, ISTH, and the more

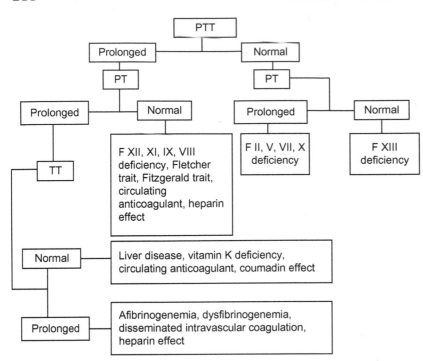

pediatric-specific PBQ), although whether these tools can directly predict future bleeding episodes needs further study.

  a. If negative: No coagulation tests are indicated; only a CBC.

  b. If positive or unreliable: The following tests should be performed: CBC, PFAs, PT, aPTT, and fibrinogen, TT, von Willebrand panel.

2. Abnormal tests require further investigation (see Figure 15.5).

3. In a patient with a significant bleeding history, if all screening tests and von Willebrand panel are normal consider FXIII, PAI-1 activity, α_2AP, and platelet aggregation studies, vitamin C levels and rule out hyperextensibility on clinical examination.

ACQUIRED COAGULATION FACTOR DISORDERS

Vitamin K Deficiency

The normal full-term infant is born with levels of factors II, VII, IX, and X that are low by adult standards (Table 15.3). The coagulation factors fall even lower over the first few days of life, reaching their nadir on about the third day. This is due to the low body stores of vitamin K at birth. As little as 25 μg vitamin K can prevent this fall in activity of the vitamin-K-dependent clotting factors. The vitamin K content of cow's milk is only about 6 μg/dl and that of breast milk 1.5 μg/dl. Moreover, breastfed infants are colonized by lactobacilli that do not synthesize gut vitamin K. It is a combination of low initial stores and subsequent poor intake of vitamin K that occasionally produces an aggravation of the coagulation defect causing primary hemorrhagic disease of the newborn. Vitamin K deficiency results in hemorrhagic disease between the second and fourth days of life and is manifested by GI hemorrhage, hemorrhage from the umbilicus, or internal hemorrhage (classic hemorrhagic disease of the newborn). Bleeding attributable to this cause is responsive to parenteral vitamin K therapy; for this reason parenteral vitamin K is routinely administered to newborns. Serious occurrences of vitamin K deficiency bleeding continue to occur due to parental refusal of vitamin K prophylaxis after birth and increased prevalence of home deliveries with the aid of midwives who may not be licensed to administer intramuscular vitamin K after birth.

In premature infants of low birth weight, both the vitamin K stores and the level of coagulation factors are even lower than in term infants. The response to vitamin K is slow and inconsistent, suggesting that the immature liver has reduced synthetic capability.

TABLE 15.6 Laboratory Findings in Vitamin K Deficiency, Liver Disease, and Disseminated Intravascular Coagulation

Component	Vitamin K deficiency	Liver disease	DIC
Red cell morphology	Normal	Target cells	Fragmented cells, burr cells, helmet cells, schistocytes
aPTT	Prolonged	Prolonged	Prolonged
PT	Prolonged	Prolonged	Prolonged
Fibrin split products	Normal	Normal or slightly increased	Markedly increased
Platelets	Normal	Normal	Reduced
Factors decreased	II,VII, IX, X	I, II, V, VII, IX, X	Assays are of limited utility

DIC, disseminated intravascular coagulation; PT, prothrombin time; PTT, partial thromboplastin time.

TABLE 15.7 Treatment of Disseminated Intravascular Coagulation and Purpura Fulminans

Treatment of the underlying disorder
    Treatment of infections with appropriate anti-infectives (antibiotics, antiviral drugs, antifungal drugs)
    Correction of electrolyte imbalances, acidosis, and shock
    Appropriate antineoplastic therapy
    Removal of triggering stimulus
Replacement therapy as indicated
    Platelet concentrates (1 unit/10 kg)
    Cryoprecipitate (50–100 mg/kg fibrinogen)[a]
    FFP (10–15 ml/kg, initially; may need 5 ml/kg q6h)
Intravenous heparinization[b]
Intravenous direct thrombin inhibitors
Antiplatelet drugs
Anti-thrombin concentrate
Activated protein C concentrate

[a]One bag of cryoprecipitate contains about 200 mg fibrinogen.
[b]See Heparin Therapy.

Maternal ingestion of certain drugs may result in neonatal hypoprothrombinemia and reduction in factors VII, IX, and X, resulting in early hemorrhagic disease of the newborn. These drugs include oral anticoagulants and anticonvulsants (phenytoin, primidone, and phenobarbital). This can be prevented by administration of vitamin K to the mother 2–4 weeks prior to delivery.

Late hemorrhagic disease of the newborn occurs between day 7 and 6 months of life in the absence of adequate vitamin K prophylaxis at birth. It may be idiopathic or exacerbated by malabsorption or liver disease. Consider investigating for biliary atresia, alpha 1 antitrypsin deficiency, malabsorption conditions like celiac disease, cystic fibrosis if hemorrhagic disease of the newborn occurs after adequate vitamin K prophylaxis at birth.

Table 15.6 lists the laboratory findings in vitamin K deficiency in relationship to the findings in liver disease and disseminated intravascular coagulation (DIC).

Hepatic Dysfunction

Any transient inability of the newborn liver to synthesize necessary coagulation factors, even in the presence of vitamin K, can result in hemorrhagic disease that is nonresponsive to vitamin K therapy. Hepatic dysfunction as a result of immaturity, infection, hypoxia, or underperfusion of the liver can all result in transient inability of the liver to synthesize coagulation factors. This is more prominent in small premature infants. The sites of bleeding in these cases are usually pulmonary and intracerebral, with a high mortality. Other causes of hepatocellular dysfunction affecting patients of all ages, include hepatitis, cirrhosis, Wilson disease, and Reye syndrome. In liver disease vitamin-K-dependent factors, FV, and fibrinogen are usually decreased, fibrin split products may be elevated due to impaired clearance (Table 15.6). In contrast, factor VIII levels are usually normal. There is no response to vitamin K. There is usually a clinical response to clotting factor replacement therapy, using fresh frozen plasma (FFP) and cryoprecipitate (replacement guidelines are the same as those outlined in Table 15.7).

Disseminated Intravascular Coagulation

DIC is characterized by the intravascular consumption of platelets and plasma clotting factors. Widespread coagulation within the vasculature results in the deposition of fibrin thrombi and the production of a hemorrhagic state when the rapid utilization of platelets and clotting factors results in levels inadequate to maintain hemostasis. The accumulation of fibrin in the microcirculation leads to mechanical injury to the red cells, resulting in erythrocyte fragmentation and microangiopathic hemolytic anemia.

Widespread activation of the coagulation cascade rapidly results in the depletion of many clotting factors as fibrinogen is converted to fibrin throughout the body as follows:

1. The generation of thrombin results in intravascular coagulation, rapidly falling platelet count, fibrinogen, and FV, FVIII, and FXIII levels. Paradoxically, *in vitro* bioassays for these factors may be elevated owing to generalized activation of the coagulation system.
2. Concurrently, plasminogen is converted to its enzymatic form (plasmin) by t-PA. Plasmin digests fibrinogen and fibrin (secondary fibrinolysis) into fibrin split products (FSPs), resulting in clot lysis.

Diagnosis of DIC relies on the presence of a well-defined clinical situation associated with a thrombo-hemorrhagic disorder. Table 15.6 lists the typical diagnostic findings in DIC. Disease states associated with DIC and low-grade DIC are listed in Table 15.8. Generally available treatment options for treatment of DIC are shown in Table 15.7.

Low-grade DIC has the potential to accelerate into fulminant DIC. Careful monitoring in terms of detecting the presence of fragmented red blood cells, mild decrease in platelets and low fibrinogen levels usually indicate high fibrinolysis and increased risk of bleeding. Treatment of low-grade DIC involves treating the underlying disease triggering it.

INHERITED COAGULATION FACTOR DISORDERS

Table 15.9 lists the genetics, prevalence, coagulation studies, and symptoms of inherited coagulation factor disorders.

Hemophilia A and B

The most common coagulation disorders after vWD are hemophilia A and B. Hemophilia A is an X-linked recessive bleeding disorder attributable to decreased blood levels of functional procoagulant factor VIII (FVIII, VIII: C, antihemophilic factor). Hemophilia B is also an X-linked recessive disorder and is indistinguishable from hemophilia A with respect to its clinical manifestations. In hemophilia B, the defect is a decreased level of functional procoagulant factor IX (FIX, IX: C, plasma thromboplastin component or Christmas factor). The incidence of hemophilia is probably 1 per 5,000 live male births with hemophilia B being one-fifth as common as hemophilia A. Thus factor VIII deficiency accounts for 80–85% of cases of hemophilia, with factor IX deficiency accounting for the remainder. Given X-linked recessive inheritance, both types occur with similar incidence among all races and in all parts of the world.

Hemophilia A Carrier Detection

Excessive lyonization may result in reduced FVIII levels in female carriers of hemophilia; hence a reduced FVIII level can have utility in diagnosing the carrier state. However, a normal FVIII level does not rule out carrier status, which is most accurately diagnosed by genetic testing.

Direct gene mutation analysis: The FVIII common intron 22 inversion, resulting from an intrachromosomal recombination, is identifiable in 45% of severe hemophilia A patients. For the remaining 55% of patients with severe hemophilia A, as well as all those with mild and moderate hemophilia A, the molecular defects can usually be detected by efficient screening of all 26 FVIII exons and splice junctions. Therefore, targeted mutation analysis is the most accurate test for carrier detection and prenatal diagnosis for severe hemophilia A. For rare patients in whom a precise mutation cannot be identified or gene sequencing is not an option, intragenetic and extragenetic linkage analysis of DNA polymorphisms can be useful with up to 99.9% precision (when an affected male patient and his related family members are available). Preimplantation genetic diagnosis is a reproductive option available to carrier females.

TABLE 15.8 Disease States Associated with Disseminated Intravascular Coagulation

Causative factors	Clinical situation
Tissue injury	Trauma/crush injuries
	Head injury
	Major surgery
	Heat stroke
	Burns
	Venoms
	Malignancy
	Obstetrical accidents
	Amniotic fluid embolism
	Placental abruption
	Stillborn fetus
	Abortion
	Fat embolism
Endothelial cell injury	Infection (bacterial, viral, protozoal)
AND/OR	Immune complexes
Abnormal vascular surfaces	Eclampsia
	Postpartum renal failure
	Oral contraceptives
	Cardiopulmonary bypass
	Giant hemangioma
	Vascular aneurysm
	Cirrhosis
	Malignancy
	Respiratory distress syndrome
Platelet, leukocyte, or red cell injury	Incompatible blood transfusion
	Infection
	Allograft rejection
	Hemolytic syndromes
	Drug hypersensitivity
	Malignancy
Localized intravascular coagulopathy	Kasabach–Merritt syndrome
	Chronic inflammatory disorders
	Arteriovenous fistulae
	Vascular prosthesis
	Glomerulonephritis

When definitive diagnosis of the carrier state cannot be made, determination of the FVIII/vWF:Ag ratio (<1.0) can be used to detect 80% of hemophilia A carriers with 95% accuracy. Use of this methodology requires careful standardization of the laboratory performing the testing.

TABLE 15.9 Genetics, Prevalence, Coagulation Studies, and Symptoms of Inherited Coagulation Factor Deficiencies

Factor deficiency	Genetics	Estimated prevalence	Prevalence of ICH (upper limits)	APTT	PT	Associated with bleeding episodes
Afibrinogenemia	AR	1:500,000	10%	P	P	++
Dysfibrinogenemia	AR	1:1 million	Single case	N/P	P	+/− thrombosis
II	AR	1:1 million	11%	P	P	++
V (parahemophilia)	AR	1:1 million	8% of homozygotes	P	P	++
VII	AR	1:300,000	4−6.5%	N	P	+
VIII (hemophilia A)	XLR	1:5000 males	5−12%	P	N	+++
von Willebrand's disease		1:1000	Extremely rare			
Type 1	AD			N/P	N	+
Type 2	AD			N/P	N	++
Type 3	AR			P	N	++
IX (hemophilia B)	XLR	1;25,000 males	5−12%	P	N	+++
X	AR	1:1 million	21%	P	N	++
XI (hemophilia C)	AV	1:100,000	Extremely rare	P	N	+
XII	AD			P	N	−
XIII	AR	1:2 million	33%	N	N	+[a]
Prekallikrein (Fletcher trait)	AD			P[b]	N	−
HMW kininogen (Fitzgerald trait)	AR			P	N	−
Passovoy (?)	AR			P	N	+/−

[a]Umbilical stump bleeding; need FXIII activity and gene sequencing for FXIII A and B subunits for accurate diagnosis. The 5M urea lysis solubility test is not a sensitive test for diagnosis.

[b]Shortened with prolonged exposure to kaolin.

ICH, intracranial hemorrhage; AD, autosomal dominant; aPTT, activated partial thromboplastin time; AR, autosomal recessive; AV, autosomal variable; N, normal; P, prolonged; PT, prothrombin time; XLR, X-linked recessive.

Bleeding episodes occur in most individuals homozygous for the disorder. However, with FXI and FVII deficiency there is no correlation between factor levels and bleeding phenotype. Most studies consider factor levels < 10% to be associated with bleeding symptoms, recent data from EN-RBD study (Peyvandi et al., 2012).

Hemophilia B Carrier Detection

Hemophilia B carriers have a wide range of FIX levels but, in a subset of cases, can be detected by the measurement of reduced plasma factor IX activity (in 60−70% of cases).

Targeted mutation analysis: The factor IX gene is located centromeric to the factor VIII gene in the terminus of the long arm of the X chromosome. There is no linkage between the FVIII and FIX genes. The 34 kb FIX coding sequence comprises eight exons and encodes a 461-amino-acid precursor protein that is approximately one-third the size of the factor VIII cDNA. Because of the smaller gene size, FIX mutations can be identified in nearly all patients. Direct FIX mutation testing is available through DNA diagnostic laboratories, with linkage analysis used in those cases where the responsible mutation cannot be identified.

Prenatal Diagnosis

Prenatal diagnosis of hemophilia can be performed by either chorionic villus sampling (CVS) at 10−12 weeks gestation or by amniocentesis after 15 weeks gestation. If DNA analysis is not available or if a woman's carrier status cannot be determined, fetal blood sampling can be performed at 18−20 weeks gestation for direct fetal factor VIII plasma activity level measurement. The normal fetus at 18−20 weeks gestation has a very low FIX level, which an expert laboratory can distinguish from the virtual absence of FIX in a fetus with severe hemophilia B.

TABLE 15.10 Relationship of Factor Levels to Severity of Clinical Manifestations of Hemophilia A and B

Type	Percentage factor VIII/IX	Type of hemorrhage
Severe	<1	Spontaneous; hemarthroses and deep soft tissue hemorrhages
Moderate	1–5	Gross bleeding following mild to moderate trauma; some hemarthrosis; seldom spontaneous hemorrhage
Mild	5–40	Severe hemorrhage only following moderate to severe trauma or surgery
High-risk carrier females	Variable	Gynecologic and obstetric hemorrhage common, other symptoms depend on plasma factor level.

TABLE 15.11 Common Sites of Hemorrhage in Hemophilia

Hemarthrosis
Intramuscular hematoma
Hematuria
Mucous membrane hemorrhage
    Mouth
    Dental
    Epistaxis
    Gastrointestinal
High-risk hemorrhage
    Central nervous system
        Intracranial
        Intraspinal
    Retropharyngeal
    Retroperitoneal
    Hemorrhage causing compartment syndrome/nerve compression
        Femoral (iliopsoas muscle)
        Sciatic (buttock)
        Tibial (calf muscle)
        Perineal (anterior compartment of leg)
        Median and ulnar nerve (flexor muscles of forearm)

Maternal–fetal combined complication rates for amniocentesis and CVS are 0.5–1.0% and 1.0–2.0%, respectively. Fetal blood sampling is less available; the fetal loss rate for these procedures ranges from 1% to 6%.

Clinical Course of Hemophilia

Hemophilia should be suspected when unusual bleeding is encountered in a male patient. Clinical presentations of hemophilia A and hemophilia B are indistinguishable. The frequency and severity of bleeding in hemophilia are usually related to the plasma levels of factor VIII or IX (Table 15.10), although some genetic modifiers of hemophilia severity have been identified. The median age for first joint bleed is 10 months, corresponding to the age at which the infant becomes mobile. Table 15.11 shows the common sites of hemorrhage in hemophilia. The incidence of severity and clinical manifestations of hemophilia are listed in Table 15.12.

For hemophilia B, the Leyden phenotype (severe hemophilia as a child that becomes mild after puberty) has been described in families with defects in the androgen-sensitive promoter region of the gene.

Treatment (Factor Replacement Therapy)

Factor replacement therapy is the mainstay of hemophilia treatment. The degree of factor correction required to achieve hemostasis is largely determined by the site and nature of the particular bleeding episode. Commercially available products for replacement therapy are listed in Table 15.13. Commercially available FVIII products include high-purity recombinant preparations, highly purified plasma-derived concentrates (monoclonal/immunoaffinity purified) and intermediate-purity plasma-derived preparations. Available FIX products include recombinant

TABLE 15.12 Incidence of Severity and Clinical Manifestations of Hemophilia

Severity	Severe	Moderate	Mild
INCIDENCE			
Hemophilia A	70%	15%	15%
Hemophilia B	50%	30%	20%
BLEEDING MANIFESTATIONS			
Age of onset	≤1 year	1—2 years	2 years—adult
Neonatal hemorrhages			
Following circumcision	Common	Common	None
Intracranial	Occasionally	Rare	Rare
Post Neonatal period			
Muscle/joint hemorrhage	Spontaneous	Following minor trauma	Following major trauma
CNS hemorrhage	High risk	Moderate risk	Rare[a]
Postsurgical hemorrhage	Common	Common	Rare[a]
Oral hemorrhage[b]	Common	Common	Rare[a]

[a]FVIII, >25; FIX, >15.
[b]Following trauma or tooth extraction.

FIX and plasma-derived high-purity FIX concentrate (coagulation FIX concentrate). Recently recombinant products with the FVIII or FIX product fused to the Fc portion of immunoglobulin 1 (Ig1) have become available which prolong the half-life of the factor in circulation. This enables prophylaxis dosing up to 3—5 days for FVIII-deficient patients and up to weekly dosing for FIX-deficient patients. New products in the pipeline include PEGylation of the factor molecule, which by increasing the molecular mass, reduces glomerular filtration, proteolytic degradation, and clearance. The combined effect is to increase the half-life of the PEGylated protein. Several other strategies being investigated include single-chain rFVIII which more effectively binds to vWF, bioengineered antibodies, and short interfering RNAs which slow down the endogenous AT synthesis. Gene therapy offers the promise of curing hemophilia by inserting the deficient gene into the patient's tissue which can then permanently restore circulating factor levels. Moreover, the gene for FIX is small and easy to insert into many vectors.

In 2011 Nathwani and colleagues successfully transduced six hemophilia B patients with all of them experiencing stable FIX expression at 1—7% of normal levels. The same group has shown long-term therapeutic factor IX expression in 10 patients associated with clinical improvement and no late toxic effects when assessed for a period of up to 3 years.

Prothrombin complex concentrates (PCCs), formerly a mainstay of hemophilia B treatment, are not utilized because of the risk of thrombotic complications associated with intensive treatment. Source plasma for all plasma-derived factor concentrates undergoes donor screening and nucleic acid testing for a variety of viral pathogens. In addition all plasma-derived and many recombinant factor concentrates undergo a viral inactivation treatment, typically with solvent detergent, wet or dry heat treatment, pasteurization or nanofiltration. Factor concentrates are preferred over FFP or cryoprecipitate because the plasma-derived concentrates undergo viral inactivation treatment in addition to donor screening. Recombinant factor concentrates are widely accepted as the treatment of choice for previously untreated patients, minimally treated patients, and patients who have not had transfusion-associated infections.

Strategies for hemophilia care include on-demand treatment of acute bleeding episodes or, for severe hemophilia patients, prophylactic administration of clotting factor concentrate to maintain trough factor levels >1% augmented with on-demand treatment of breakthrough bleeding episodes. The latter strategy pharmacologically converts the severe hemophilia phenotype to a moderate phenotype with an attendant reduction in frequency of bleeding episodes. A randomized multicenter US national study suggested a markedly reduced incidence of hemophilic arthropathy when prophylaxis was instituted prior to onset of recurrent joint bleeds. This benefit must be balanced with the need for frequent prophylactic infusions (3—4 times/week for FVIII, 2 times/week for FIX), venous access considerations, potential requirements for central venous access devices, increased cost of treatment, and the occasional patients who have a very mild clinical course. Table 15.14 provides generally

TABLE 15.13 Commercially Available Coagulation Factor Concentrates

Class	Product	Purity	Procedure	Primary use
FVIII Plasma Derived	Koate-DVI (Bayer Corp)	Intermediate purity	SD/Dry Heat	Hemophilia A
	Humate P (Aventis Behring)	Intermediate purity	P	Hemophilia A/vWD
	Alphanate SD-HT (Grifols Corp)	High purity	SD/Dry Heat	Hemophilia A /vWD
	Hemofil M (Baxter Bioscience)	Ultra high purity	SD	Hemophilia A
	Monoclate P (Aventis Behring)	Ultra high purity	P	Hemophilia A
FVIII Recombinant	1st generation - Recombinate (Baxter Bioscience)	Ultra high purity	None	Hemophilia A
	2nd generation - Helixate (Aventis Behring)	Ultra high purity	SD	Hemophilia A
	2nd generation - ReFacto (Wyeth)	Ultra high purity	SD	Hemophilia A
	2nd generation - Kogenate FS (Bayer Corp)	Ultra high purity	SD	Hemophilia A
	3rd generation - Advate (Baxter Bioscience)	Ultra high purity	SD	Hemophilia A
	3rd generation - Xyntha (Wyeth)	Ultra high purity	SD	Hemophilia A
FVIII Fc Fusion Long Acting	Eloctate (Biogen)	High	SD/ Nanofiltration	Hemophilia A
FIX Plasma Derived	Alphanine SD-VF (Grifols Corp)	High	SD/ nanofiltration	Hemophilia B
	Mononine (Aventis Behring)	High	SD/ nanofiltration	Hemophilia B
FIX Recombinant	BeneFIX (Wyeth)	High	Nanofiltration	Hemophilia B
FIX Fc Fusion Long Acting	Alprolix (Biogen)	High	Nanofiltration	Hemophilia B
Activated PCC	FEIBA VH (Basxter Bioscience)	N/A	Vapor heat	Inhibitor bypass therapy
PCC	Profilnine SD (Grifols Corp)	Intermediate	SD	Rare bleeding disorders
	Bebulin VH (Baxter Bioscience)	Intermediate	Vapor heat	Rare bleeding disorders
	Proplex T (Baxter Bioscience)	Intermediate	Dry heat	FVII deficiency in places where recombinant product not available.
Bypassing agent:	NovoSeven (NovoNordisk)	N/A	None	Inhibitor bypass therapy, FVII deficiency
Specialty items	Kcentra (CSL Behring)		Ion exchange	Anticoagulant reversal

Notes: PCC: Prothrombin Complex Concentrate. Purity (international units/mg protein before albumin is added); intermediate <100, high >100, ultra high >1,000; SD, solvent detergent; P, pasteurization; N/A, not applicable.

accepted guidelines for treatment of most types of hemophilic bleeding. When a bleeding episode is suspected hemostatic treatment should be rendered first then diagnostic evaluation(s) may be performed.

Ancillary Therapy

1-Deamino-8-ᴅ-Arginine Vasopressin

In hemophilia A patients 1-deamino-8-ᴅ-arginine vasopressin (DDAVP) increases plasma FVIII levels two- to fivefold. It is commonly used to treat selected hemorrhagic episodes in mild hemophilia A patients. When used intravenously the dose is 0.3 μg/kg administered in 25—50 ml normal saline over 15—20 min

TABLE 15.14   Treatment of Bleeding Episodes

Type of hemorrhage	Hemostatic factor level	Hemophilia A	Hemophilia B	Comment/adjuncts
Hemarthrosis	30–50% minimum	FVIII 20–40 U/kg q12–24h as needed; if joint still painful after 24 h, treat for further 2 days	FIX 30–40 U/kg q24h as needed; if joint still painful after 24 h, treat for further 2 days	Rest, immobilization, cold compress, elevation (RICE)
Muscle	40–50% minimum, for iliopsoas or compartment syndrome 100% then 50–100% × 2–4 days	20–40 U/kg q12–24h as needed For iliopsoas or compartment syndrome initial dose is 50 U/kg	40–60 U/kg q24 h as needed For iliopsoas or compartment syndrome initial dose is 60–80 U/kg	Calf/forearm bleeds can be limb-threatening. Significant blood loss can occur with femoral-retroperitoneal bleed.
Oral mucosa	Initially 50%, then EACA at 50 mg/kg q6h × 7 days usually suffices	25 U/kg	50 U/kg	Antifibrinolytic therapy is critical. Do not use with PCC or APCC
Epistaxis	Initially 30–40%, use of EACA 50 mg/kg q6h until healing occurs may be helpful	15–20 U/kg	30–40 U/kg	Local measures: pressure, packing
Gastrointestinal	Initially 100% then 50% until healing occurs	FVIII 50 U/kg, then 25 U/kg q12h	FIX 100 U/kg, then 50 U/kg q day	Lesion is usually found, endoscopy is recommended, antifibrinolytic may be helpful
Hematuria	Painless hematuria can be treated with complete bed rest and vigorous hydration for 48 h. For pain or persistent hematuria 100%	FVIII 50 units/kg, if not resolved 30–40 U/kg q day until resolved	FIX 80–100 units/kg, if not resolved then 30–40 U/kg q day until resolved	Evaluate for stones or urinary tract infection. Lesion may not be found. Prednisone 1–2 mg/kg/day × 5–7 days may be helpful. Avoid antifibrinolytics
Central nervous system	Initially 100% then 50–100% for 14 days	50 U/kg; then 25 U/kg q12 h	80–100 U/kg; then 50 U/kg q24 h	Treat presumptively before evaluating, hospitalize. Lumbar puncture requires prophylactic factor coverage
Retroperitoneal or retropharyngeal	Initially 80–100% then 50–100% until complete resolution	FVIII 50 units/kg; then 25 U/kg q12 h until resolved	FIX 100 U/kg; then 50 U/kg q24 h until resolved	Hospitalize
Trauma or surgery	Initially 100%; then 50% until would healing is complete	50 U/kg; then 25 U/kg q12 h	100 U/kg; then 50 U/kg q24 h	Evaluate for inhibitor prior to elective surgery

Antifibrinolytic = EACA, epsilon-aminocaproic acid (Amicar); syrup, 250 mg/5 ml; tablet, 500 mg, 1,000 mg; PCC, prothrombin complex concentrates; aPCC, activated PCC.

(maximum dose 25 μg). Its peak effect is observed in 30–60 min. Subcutaneous DDAVP is as effective as intravenous DDAVP, facilitating treatment of very young patients with limited venous access.

Concentrated intranasal DDAVP (Stimate), available as a 1.5 mg/ml preparation, has approximately two-thirds the effect of intravenous DDAVP. Care should be exercised to avoid inadvertent dispensing of the dilute intranasal DDAVP commonly used for treatment of diabetes insipidus. The peak effect of intranasal Stimate is observed 60–90 min after administration.

Recommended dosage for use of intranasal Stimate:

Body weight <50 kg: 150 μg (one metered dose).
Body weight >50 kg: 300 μg (two metered doses).

Recommendations during the administration of DDAVP are:

1. Mild fluid restriction to two-thirds maintenance fluids and drinking to thirst (for 24 hours after DDAVP dose), only electrolyte-containing fluids; avoidance of free water.
2. Monitoring urinary output and daily weights may be useful to track fluid retention.

Responses to DDAVP vary from patient to patient, but in a given patient are reasonably consistent over time. Therefore a test dose of DDAVP should be administered at the time of diagnosis or in advance of an invasive procedure to assess the magnitude of the patient's response. DDAVP administration may be repeated at 24-h intervals according to the severity and nature of the bleeding. Administration of DDAVP at shorter intervals results in a progressive tachyphylaxis over a period of 4–5 days. Depending on the response after a trial, DDAVP could be recommended for most minor procedures. For major procedures requiring maintenance of factor activity >80%, factor concentrates may be needed even in mild hemophilia A patients.

Side effects of DDAVP:

Asymptomatic facial flushing.

Thrombosis (a rarely reported complication).

Hyponatremia, more common in very young patients, in patients receiving repeated doses of DDAVP or large volumes of oral or intravenous fluid; hyponatremic seizures have been reported in children under 2 years of age. DDAVP is contraindicated in children less than 2 years of age.

Antifibrinolytic Therapy

Antifibrinolytic drugs inhibit fibrinolysis by preventing activation of the proenzyme plasminogen to plasmin, mostly by activating TAFI. This intervention is useful for preventing clot degradation in areas rich in fibrinolytic activity including the oral cavity, the nasal cavity, and the female reproductive tract. Approved antifibrinolytic drugs are:

Epsilon aminocaproic acid (EACA; Amicar): Orally; dose is 50–100 mg/kg every 6 h (maximum, 24 g total dose per day). GI symptoms may occur at higher doses, therefore the preferred starting dose is 50 mg/kg. The drug is available as 500 mg, 1000 mg tabs or as a flavored syrup (250 mg/ml).

Tranexamic acid (Cyklokapron, Lysteda): 20–25 mg/kg (maximum, 1.5 g) orally or 10 mg/kg (maximum, 1.0 g) intravenously every 8 h. This is approved for use in women with bleeding disorders with menorrhagia.

Antifibrinolytic therapy should not be utilized in patients with urinary tract bleeding because of the potential of intrarenal clot formation.

To treat spontaneous oral hemorrhage or to prevent bleeding from dental procedures in pediatric patients with hemophilia, either drug is begun in conjunction with DDAVP or factor replacement therapy and continued for up to 7 days or until mucosal healing is complete. Antifibrinolytic drugs also have efficacy as an adjunct treatment for epistaxis and for menorrhagia. Antifibrinolytic drugs are safe to use in hemophilia B patients receiving coagulation factor IX concentrates but should not be used contemporaneously with PCCs because of the thrombotic potential of these concentrates. Initiation of oral antifibrinolytic drug therapy 4 h after the last dose of PCCs appears to be well tolerated.

Management of Inhibitors in Hemophilia

Approximately 30% of patients with hemophilia A develop neutralizing alloantibodies (inhibitors) directed against factor VIII. Inhibitors are a major cause of morbidity and mortality in hemophilia. Risk factors for inhibitors include early age at exposure, presence of the common inversion mutation, large deletions of the FVIII gene, African-American ethnicity and a sibling with hemophilia and an inhibitor. Other factors not proven in randomized studies include recombinant factor treatment and continuous infusion of factor concentrates for treatment. Inhibitors are quantified using the Bethesda assay. Low responder inhibitors have titers ≤5 Bethesda Units (BU) and do not exhibit anamnesis upon repeated exposure to FVIII. One Bethesda Unit neutralizes 50% of factor VIII/IX activity. Approximately half of patients with inhibitors will be low responders and, of these, approximately half will have transient inhibitors. Hemophilia A patients with low-responder inhibitors can generally be treated with factor VIII concentrate, albeit at an increased dosing intensity because of reduced *in vivo* recovery and a shortened half-life of the FVIII. High-responder inhibitors have titers >5 BU and, although the titer may decay in the absence of FVIII exposure, these patients will display an anamnestic rise in titer upon rechallenge with FVIII.

The clinical approach is different for high and low responders (Table 15.15).

TABLE 15.15 Recommendations for Replacement Therapy for Treatment of Bleeding in Patients with Factor VIII Inhibitors

Type of patient	Type of bleed	Recommended treatment
Low responder[a] (<5 BU)	Minor or major bleed	Factor VIII infusions using adequate amounts of factor VIII to achieve a circulating hemostatic level
High responder[b] with low inhibitor level (<5 BU)	Minor/major bleed	aPCC, or rVIIa infusions
	Life-threatening bleed	Factor VIII infusions until anamnestic response occurs, then aPCC or rVIIa infusions
High responder with high inhibitor level (>5 BU)	Minor bleed	aPCC, or rVIIa infusions
	Major bleed	aPCC, or rVIIa infusions

[a]Rise of inhibitor titer is slow to factor VIII challenge.
[b]Rise of inhibitor titer is rapid to factor VIII challenge.
BU, Bethesda units; PCC, prothrombin complex concentrates. Activated PCC, for example, Autoplex (Hyland) and FEIBA (factor VIII inhibitor bypassing activity).

Low Responders

For serious limb- or life-threatening bleeding, a bolus infusion of 100 units/kg factor VIII is administered, repeat doses of 100 units/kg are administered at 12-h intervals or, alternatively, the level is maintained with a continuous infusion rate based on the inhibitor titer and recovery and survival studies which estimate half-life of the factor. A factor VIII assay should be obtained 15 min after the bolus infusion and trough or steady-state FVIII levels should be followed at least daily thereafter. As an example, if a child has a 3 BU inhibitor titer, for treatment of an intracranial bleed a 100% FVIII level is needed; he needs to receive a bolus of 125 U/kg which will raise his level to 250% of which 150% will be neutralized given his 3 BU titer leaving him with 100% factor activity level. This should be followed by a continuous infusion of at least 12 U/kg/h if his half-life is 4 h since 4 U/kg/h gives a 100% level in an individual with a 12-h half-life. Of course all these doses need to be adjusted based on FVIII activity levels drawn at least once a day on continuous infusions. Normal saline at 20 cc/h should be piggy-backed to the infusion distally to dilute the factor in order to reduce thrombophlebitis caused by infusion of concentrated FVIII protein.

During prolonged treatment in FVIII *in vivo* recovery and half-life may transiently improve as inhibitor antibody is adsorbed by the FVIII.

High Responders

Treatment of an acute bleed is achieved by use of bypassing agents (aPCCs and rVIIa) and inhibitor eradication is achieved by achieving immune tolerance. In high-responder inhibitor patients with limb- or life-threatening bleeding, if the inhibitor titer is <20 BU, high-dose continuous infusion of factor VIII may saturate the antibody permitting a therapeutic FVIII level. The dose can be based on recovery and survival studies.

If a factor VIII level is not attainable or the antibody level is greater than 20 BU, then bypassing agents to initiate hemostasis independent of FVIII should be employed, using aPCC or rFVIIa.

Activated Prothrombin Complex Concentrate (Autoplex and FEIBA)

These products have increased amounts of activated factor VII (VIIa), factor X (Xa), and thrombin and are effective in patients even with high-titer inhibitors (>50 BU). The initial dose of 75–100 units/kg can be repeated in 8–12 h. Generally for a joint bleed, 4–5 doses at 12-h intervals are employed after which the risk of thrombogenicity is increased. Approximately 75% of patients with inhibitors respond to aPCC infusions. For some patients trace amounts of FVIII in aPCC products may cause anamnesis of the inhibitor titer. If multiple doses are administered, the patient should be monitored for the development of DIC or even myocardial infarction. The simultaneous use of antifibrinolytic therapy (e.g., Amicar) should be avoided. Oral antifibrinolytic drug therapy 4 h after the last dose of a PCC, however, appears to be well tolerated.

Recombinant Factor VIIa (NovoSeven)

Recombinant activated factor VII (rFVIIa) concentrate can be administered to achieve hemostasis in patients with high-titer inhibitors. The usual dose is 90 μg/kg rFVIIa repeated every 2 h for 2–3 infusions. Higher doses up to 200 μg/kg for 2–3 doses have been used in pediatric patients in life-threatening situations to achieve better hemostasis. Recently, a single high dose of 270 µg/kg has been shown to be effective in the outpatient setting for the treatment of an acute bleed instead of using a q 3–4 h dosing schedule. The subsequent frequency of infusion and duration of therapy must then be individualized, based on the clinical response and severity of bleeding. Early initiation of hemostatic treatment (within 8 h of a bleed) with rFVIIa produces response rates on the order of 90%. Treatment failures with conventional doses of rFVIIa may respond to higher doses. The incidence of thrombotic complications with this product has been low and anamnesis of the inhibitor does not occur.

Plasmapheresis with Immunoadsorption

When bleeding persists despite active treatment, extracorporeal plasmapheresis (4 liter exchange) over staphylococcal A columns may rapidly reduce the inhibitor titer (up to 40%) by adsorbing offending inhibitory IgG antibodies. This approach is cumbersome and may produce significant fever and hypotension due to the release of staphylococcal A protein from the solid phase matrix of the chromatographic column. However, it can be life-saving in desperate situations and its efficacy can be enhanced by concomitant replacement therapy with FVIII containing concentrates. Staphylococcal protein A columns for immunoadsorption are not readily available.

Immune-Tolerance Induction

This intervention is instituted for inhibitor eradication in high-responder inhibitors and involves frequent administration of FVIII concentrate to induce immune tolerance to exogenous FVIII. The objectives of immune-tolerance induction (ITI) are an undetectable inhibitor titer, restoring the ability to treat bleeds with FVIII concentrates and restoration of normal *in vivo* FVIII recovery and half-life. A variety of regimens have been used for ITI (Table 15.16), including daily high-dose FVIII (100 U/kg twice daily or 200 U/kg daily regimen) with or without immunomodulatory therapy, daily intermediate dose FVIII (50–100 U/kg/day) and alternate-day low-dose FVIII (25 IU/kg). Immune tolerance is eventually achieved in 60–75% of patients. In the recently concluded International Immune Tolerance Study (Hay & DiMichele, 2012) low dose (50 IU/kg 3 times/week) versus high-dose (200 IU/kg/day) regimens were compared in 115 pediatric hemophilia A patients with high-titer inhibitor. Both regimens reported equally successful tolerization rates (41% vs 39%) and similar overall time to achieve tolerance with the high-dose arm achieving negative titers and normal recovery significantly faster than the low-dose arm. However, children in the low-dose arm showed a significantly increased number of bleeding episodes at all stages of ITI including prophylaxis after ITI termination but specifically in the ITI induction phase when inhibitors were still detectable. Until tolerance is successfully attained episodic bleeds require treatment with bypassing agents. A low historical peak inhibitor titer, a low inhibitor titer at initiation of ITI (<10 BU), and a low maximum inhibitor titer during ITI all favor success. Success rates may also be higher in young patients and in patients treated on higher-dose regimens. Data from the Italian ITI study indicate that the relationship between FVIII mutations and rate of inhibitor development is likely to correlate with ITI outcome. Large FVIII deletions known to be associated with the highest risk of inhibitor development also showed highest ITI failure rates. Cost and venous access are added obstacles to successfully completing immune tolerance. The role of using thrombin generation assays to individualize ITI therapy is being explored in the Predict TGA Study.

In the subset of children who fail ITI, Anti-CD20-antibody (Rituximab) has been used recently with and without ITI with varying success rates. Clinically significant responses were observed with concurrent ITI in 47% of patients with only 14% patients achieving durable responses. Larger studies are needed to determine efficacy and treatment toxicity.

Treatment of Factor IX Inhibitors

The frequency of inhibitory antibodies to factor IX is much lower (3–6%) than the frequency seen with factor VIII. Bypassing agents such as aPCC and rFVIIa have hemostatic efficacy in hemophilia B patients with inhibitors. Their use and dosages are the same as for the treatment of hemophilia A patients with inhibitors. Immune tolerance induction has had limited success in hemophilia B patients with inhibitors. However many hemophilia B patients, typically those with a history of anaphylaxis to FIX concentrates, develop nephrotic syndrome in response to the frequent, high-dose, infusions of coagulation FIX concentrates, necessitating termination of the ITI effort. More recently, rituximab (anti-CD20 antibody) in combination with mycophenolate mofetil, dexamethasone, and immunoglobulin has been shown to be successful along with ITI in anecdotal cases.

Spontaneously Acquired Inhibitory Antibodies to Coagulation Factors

Spontaneous autoantibodies to factors VIII and IX and other coagulation factors may arise in nonhemophilia patients in a variety of clinical conditions.

TABLE 15.16   Selected Immune Tolerance Induction Regimens for Hemophilia-Associated Inhibitors

Protocol	FVIII dose (IU/kg)	Other agents	% Success rate (# patients in trial)	Inhibitor elimination time (range in months)	Predictors of success (P value)
HIGH DOSE					
Bonn[a]	100−150 twice daily	APCC as required	100 (60)	1−2	
Malm[b]	Maintain FVIII >0.40 U/ml	Cyclophosphamide; 12−15 mg/kg IV daily for 2 days followed by 2−3 mg/kg orally given daily for 8−10 days IV Ig; dose is 0.4 g/kg daily for 5 days. Immunoadsorption	63 (16)		
INTERMEDIATE DOSE					
Kasper/ Ewing[c]	50−100 daily	Oral prednisone PRN	79 (12)	1−10	
LOW DOSE					
Dutch[d]	25 alternate days		87 (24)	2−28	
REGISTRY					
International, IITR[e]	>200 daily:32%*	Steroids	51%	10.5 (time to success)	Age at ITI start (0.008)
	100−199 daily:20%				Pre-ITI inhibitor titer (0.04)
	50−100 daily:23%				Historical peak titer (0.04)
	<50 daily:25%				FVIII dose (higher,0.03)
North American NAITR[f]	>200 daily:14%	Immunomodulators	63%	16.3 (time to success)	Pre-ITI inhibitor titer (0.005)
	100−199 daily:33%				Historical peak titer (0.04)
	50−100 daily:28%				Peak titer on ITI (0.0001)
	<50 daily:25%				FVIII dose (lower, 0.01)
German GITR[g]	200−300 daily		76%	15.5 (time to success)	Historical peak titer (0.0012)

[a]Oldenburg et al. (1999).
[b]Freiburghaus et al. (1999).
[c]Ewing et al. (1988).
[d]van Leeuwen et al. (1986).
[e]Mariani et al. (2001).
[f]DiMichele et al. (2002).
[g]Lenk (2000).
*Percentages indicate number of patients in each group.

Acquired Hemophilia A

Autoantibodies (acquired inhibitors) may occur in "nonhemophiliacs," resulting in acquired hemophilia A. These develop mainly in adults, more frequently in elderly subjects and in postpartum patients. Associated settings include pregnancy, immune diseases, and hematologic neoplasia (CLL, lymphoma). Symptoms are hemorrhage, typically into soft tissues and mucous membranes with occasional cerebral hemorrhage. Diagnosis is based upon history (hemorrhage in the absence of prior patient or family history), prolonged aPTT with failure to correct on mixing studies, reduced FVIII level, and the presence of an anti-FVIII inhibitor on Bethesda assay. Affinity of acquired inhibitors for FVIII differs from that of inhibitors (alloantibodies) associated with hemophilia A, resulting in complex kinetics and often incomplete inactivation of FVIII. Treatment options for minor bleeding episodes in patients with low-titer autoantibodies include augmentation of endogenous FVIII using DDAVP or FVIII concentrates. For more severe bleeding associated with high-titer autoantibodies hemostatic treatment options include bypassing agents as used for hemophilia A inhibitors (see above and Table 15.16). Immunosuppression-immunomodulation has been shown to be effective in eliminating autoantibodies. Treatment modalities alone or in combination include high-dose IVIG infusion (400 mg/kg/day for 5 days), prednisone (1 mg/kg/day as a single agent), cyclophosphamide (2 mg/kg/day) or other agents (vincristine, azothioprine, etc.). Cyclosporine A and rituximab are other agents reported to be effective in elimination of autoantibodies. Responses to any individual regimen are variable, requiring individualization of therapy. A thorough evaluation for autoimmune disease is warranted.

Acquired Antibodies to Other Coagulation Factors

Acquired autoantibodies (inhibitors) to FV are very rare. Some patients may have major hemorrhage (patients with antibodies that also bind to Platelet factor5). Others may not bleed, even at the time of major surgery. Available treatments for acute hemostasis include platelet transfusion (the FV contained within the platelet α granule may be locally protected from the inhibitor) and the use of bypassing agents.

Autoantibodies (inhibitors) to other factors (including prothrombin or factor XI) may occur, most commonly in the setting of systemic lupus erythematosus (SLE) (either isolated or in addition to the lupus anticoagulant (LA)) or hematologic malignancy. Some patients, usually those with a profound decrease in prothrombin concentration, may have hemorrhagic symptoms—LA hypoprothrombinemia syndrome. Acute hemostatic treatment strategies include infusion of FFP or PCCs. Inhibitor elimination strategies, as for other acquired inhibitors, include the use of plasmapheresis, glucocorticoids, and/or immunosuppression form the mainstay of therapy for inhibitor eradication for these complications.

Lupus Anticoagulant

See the section on Thrombotic Disorders.

VON WILLEBRAND DISEASE

Von Willebrand disease (vWD) is an autosomally inherited congenital bleeding disorder caused by a deficiency (type 1), dysfunction (type 2) or complete absence (type 3) of vWF. vWF has two functions:

- It plays an integral role in mediating adherence of platelets at sites of endothelial damage, promoting formation of the platelet plug.
- It binds and transports FVIII, protecting it from degradation by plasma proteases.

vWF is a large multimeric glycoprotein that is synthesized in megakaryocytes and endothelial cells as pre-pro-vWF. Sequential cleavage releases mature vWF which undergoes multimerization and is stored in specific cellular storage granules such as the Weibel−Palade body in endothelial cells and the α granule in platelets. It is present in normal amounts in plasma and levels can be significantly increased by administering drugs, such as desmopressin (DDAVP), that induce the release of vWF from storage sites into plasma. Deficiency of vWF results in mucocutaneous bleeding and prolonged oozing following trauma or surgery. vWD is the most common hereditary bleeding disorder, with biochemical evidence present in 1−2% of the population and biochemical evidence combined with a bleeding history in 0.1% of the population. Differences between vWD and hemophilia are shown in Table 15.17. Table 15.18 shows the current classification of vWD, Figure 15.6 shows the structure of vWF and indicates the location of mutations giving rise to variants of vWD.

Diagnosis and Treatment of vWD

The recommended diagnostic profile and treatment of vWD are indicated in Table 15.19.

TABLE 15.17 Differences Between vWD and Hemophilia A

	vWD	Hemophilia A
Symptoms	Bruising and epistaxis	Joint bleeding
		Muscle bleeding
	Menorrhagia or mucosal bleeding	
Sexual distribution	Males = females	Males
Frequency	1:200 to 1:500	1:6,000 males
Abnormal protein	vWF	Factor VIII
Molecular weight	$0.6-20 \times 10^6$ Da	280 kDa
Function	Platelet adhesion	Clotting cofactor
Site of synthesis	Endothelial cell or megakaryocytes	Liver and other tissues
Chromosome	Chromosome 12	X chromosome
Inhibitor frequency	Rare	14–25% of patients
LABORATORY TESTS		
aPTT	Normal or prolonged	Prolonged
Factor VIII	Borderline or decreased	# or absent
vWF Ag	Decreased or absent	Normal or increased
vWF R:Co	Decreased or absent	Normal or increased
vWF multimers	Normal or abnormal	Normal

From Montgomery et al. (1998), with permission.

TABLE 15.18 Classification of vWD

	Type 1	Type 2A	Type 2B	Type 2N	Type 2M	Type 3	Platelet type pseudo-vWD
Genetic transmission	AD	AD	AD	AR	AD	AR	AD
Frequency (%)	70–80	10–12	3–5	1–2	1–2	1–3	0–3
Platelet count	N	N	N/decreased	N	N	N	Decreased
FVIII	Decreased	N	N	Markedly decreased	N	Absent	N
VWF:Ag	Decreased	Decreased	N/decreased	Decreased	N	Absent	N/decreased
R:Co (VWF activity)	Decreased	Decreased	N/decreased	Decreased	Decreased	Absent	N/decreased
RIPA	N/decreased	Decreased	N	N	Decreased	Absent	N
RIPA-low dose	Absent	Absent	Increased	Absent	Absent	Absent	N
Multimeric structure	N	Absence of large multimers from plasma and platelets	Reduced large multimers from plasma	N	N	Absent	Reduced large multimers
RESPONSE TO DDAVP							
FVIII, vWFAg, R:Co	Increases	Increases	Increases	Increases	Increases	NR	

AD, autosomal dominant; N, normal; VWF:Ag, von Willebrand's factor antigen (FVIII-related antigen); RIPA, ristocetin-induced platelet aggregation; RIPA-LD, low-dose RIPA; NR, no response.

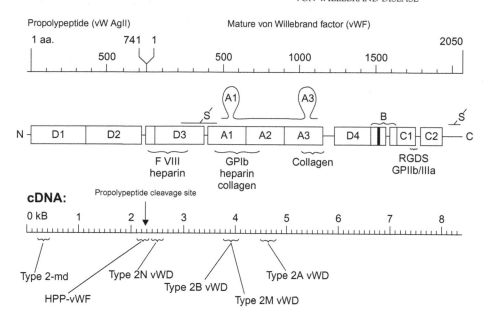

FIGURE 15.6 Protein structure of vWF and its large propolypeptide, von Willebrand's antigen II (vW Ag II). Domains of the vWF molecule have high degrees of similarity and various protein interactions have been localized to specific regions such as the interaction with platelet glycoprotein Ib or glycoprotein IIb/IIIa. This latter site may be related to an Arg-Gly-Asp-Ser (RGDS) sequence located in the C1 domain. The lower portion of this figure represents the clusters of complementary DNA mutations that cause some of the variants of von Willebrand's disease, including types 2A, 2B, 2M and 2N, as well as the less common variants that prevent propolypeptide cleavage (hereditary persistence of pro-vWF (HPP-vWF) or a variant that prevents N-terminal multimerization (type 2-md). *Adapted from: Montgomery et al. (1998). Used with permission.*

TABLE 15.19   Diagnostic Profile and Treatment of vWD

	Type 1	Type 2A	Type 2B	Type 2N	Type 2M	Type 3	Platelet type pseudo-vWD
Clinical phenotype	Use quantifiable Bleeding Assessment Tools (BATs) like ISTH-BAT, PBQ						
Laboratory phenotype	vWF antigen, vWF activity, Factor VIII level, vWF multimer profile, Ristocetin-induced platelet aggregation (RIPA), Blood type						
Genotype	Not in routine use	Not usually needed	Helpful in confirmation and ruling platelet type	Needed to differentiate from hemophilia A	Sometimes helpful	Very helpful for genetic counseling	Helpful in confirming platelet type
TREATMENT							
Severe hemorrhages; major surgical procedures	DDAVP	FVIII:vWF concentrates	FVIII:vWF concentrates	FVIII:vWF concentrates	FVIII:vWF concentrates	FVIII:vWF concentrates	Platelet transfusion
Mild hemorrhages; minor surgical procedures	DDAVP	FVIII:vWF concentrates (may respond to DDAVP)	FVIII:vWF concentrates (DDAVP contraindicated)	FVIII:vWF concentrates (may respond to DDAVP)	FVIII:vWF concentrates (may respond to DDAVP)	FVIII:vWF concentrates	Platelet transfusion
Oral surgical procedures	DDAVP and EACA	FVIII:vWF concentrates and EACA	FVIII:vWF concentrates and EACA	FVIII:vWF concentrates and EACA	FVIII:vWF concentrates and EACA	FVIII:vWF concentrates and EACA	Platelet transfusion and EACA

DDAVP, desmopressin; EACA, Amicar. Factor VIII:vWF concentrated (see Table 15.13).

Type 1 vWD

This is the most common form of the disorder and is characterized by a mild to moderate decrease in the plasma levels of vWF. Plasma vWF from these individuals has a normal structure. Levels of ristocetin cofactor activity (R:Co or vWF activity) and vWF antigen tend to be decreased in parallel. DDAVP can be used to manage most hemostatic problems in patients with type I VWD. DDAVP treatment generally normalizes FVIII and vWF levels. The standard dose may be repeated daily as necessary. Response to DDAVP should be assessed for each individual patient. Tachyphylaxis with repeated doses of DDAVP needs to be considered when prescribing

DDAVP for a major surgical procedure when prolonged high levels of vWF may be needed to achieve hemostasis and promote healing. As described previously in the hemophilia treatment section a two- to fivefold rise above baseline levels for vWF antigen and ristocetin cofactor activity can be expected but not guaranteed, necessitating a DDAVP trial in an individual before any surgical procedure to document a response.

Type 2A vWD

This variant is associated with decreased platelet-dependent vWF function and a lack of large HMW multimers in plasma and platelets. In some patients with type 2A vWD, abnormalities have been localized to the A2 domain of vWF (Figure 15.6) and specific gene defects have been described. There is either abnormal synthesis or increased proteolysis of vWF due to the genetic defect. A much lower level of ristocetin cofactor activity as compared to the vWF antigen and FVIII level should raise suspicion for type 2 vWD (RCo:Ag ratio <0.6). DDAVP may suffice to control some types of bleeding in patients with this variant. All such patients should undergo a therapeutic trial with DDAVP with measurements of vWF:Ag, vWF:RCo, and FVIII:C for at least 4 h after administration to evaluate the possibility of accelerated clearance of the newly released protein. A three- to fivefold rise above baseline levels for vWF antigen and ristocetin cofactor activity can be expected. Other patients will require treatment with clotting factor concentrates containing both FVIII and vWF. Use of cryoprecipitate for vWF replacement therapy should not be utilized because this product does not undergo viral inactivation treatment.

Type 2B vWD

This variant is a rare form of vWD characterized by a highly penetrant dominant gain in function mutations in the A1 domain of the vWF gene. The HMW multimers bind spontaneously to platelets and are continuously removed from circulation which is aggravated in conditions of acute stress and pregnancy. There may be mild thrombocytopenia due to spontaneous agglutination of the platelets. *In vitro* this variant is characterized by a reduced threshold for aggregation of platelets upon exposure to the antibiotic ristocetin. A low-dose ristocetin induced platelet aggregation (low-dose RIPA) in which platelet-rich plasma from patients with type 2B vWD will show increased aggregation but not in type 2A. DDAVP is contraindicated in type 2B vWD because of transient thrombocytopenia resulting from release and clearance of abnormal vWF. Clotting factor concentrates containing both FVIII and vWF are the mainstay of treatment for this vWD variant.

Type 2N vWD

Type 2N vWD is a result of an abnormal vWF molecule that does not bind factor VIII due to a genetic defect in the FVIII-binding domain of the vWF molecule (Figure 15.6). The unbound factor VIII is rapidly cleared from circulation resulting in a disproportional reduction of plasma factor VIII as compared with vWF levels. It is transmitted as an autosomal recessive trait with a variety of homozygous and compound heterozygous genotypes leading to FVIII levels between 5% and 40%. Some of these patients may be misdiagnosed as having mild hemophilia A since FVIII levels tend to be in the 8–10% range. It appears that the phenotype is expressed only in the presence of a second allele that carries type 1 vWD (compound heterozygous) or in the rare situation of recessive inheritance. vWD FVIII levels increase following DDAVP infusion but the released FVIII circulates only for a short time because of impaired binding to vWF. For the same reason the half-life of infused high-purity factor VIII is markedly shortened. For major bleeding or surgery, the recommended treatment is a factor VIII concentrate containing high levels of vWF.

Type 2M vWD

Type 2M vWD results from an abnormal binding site on vWF for platelet GP Ib, resulting in reduced ristocetin cofactor (R:Co) activity (functional defect). In this variant multimers of all sizes are present. The vWF protein is released by DDAVP so that patients heterozygous for this variant may respond clinically to standard doses of DDAVP. For homozygotes the vWF released by DDAVP is defective and a poor clinical response occurs. Major bleeding episodes or surgery should be managed with vWF replacement therapy.

Type 3 vWD

This vWD variant occurs in patients with homozygous or doubly heterozygous null mutations, deletions or missense mutations that prevent vWF synthesis and secretion. The clinical phenotype is a severe bleeding disorder with major deficits in both primary and secondary hemostasis. The plasma level of FVIII and vWF is virtually undetectable. Patients are unresponsive to DDAVP (no releasable vWF stores) and require episodic treatment with vWF containing FVIII concentrates. Alloantibodies that inactivate vWF develop in 10–15% of patients with type 3 disease who have received multiple transfusions. Administration of FVIII concentrates containing vWF to these patients is contraindicated since life-threatening anaphylactic reactions may result. In this setting, administration of recombinant FVIII, which is devoid of vWF, can raise FVIII levels to hemostatic levels. In the absence of vWF the half-life of the infused FVIII will be short (1–2 h) and administration of high doses, at short intervals or by continuous infusion, will be required. Administration of rFVIIa at a dose of 90 µg/kg every 3 h or the use of a PCC has produced hemostasis for some of these patients.

Platelet-Type Pseudo-vWD

Platelet-type pseudo-vWD is due to a gain-of-function mutation defect in the platelet GP Ib receptor on platelets. This platelet disorder has a phenotype similar to type 2B vWD. Excessive binding of vWF to platelet GP Ib receptor causes platelet activation and vWF removal from the circulation, plasma concentrations of vWF are reduced and platelet aggregation is increased. Bleeding in this disorder may be treated with platelet transfusions or low-dose vWF-containing concentrates.

Acquired vWD

Acquired vWD may present as a marked reduction in levels of von Willebrand antigen in a person who does not have a lifelong bleeding disorder. The onset has been associated with a variety of conditions including Wilms tumor, other neoplasms, hypothyroidism, autoimmune diseases (e.g., SLE), myeloproliferative disease, lymphoproliferative disorders, use of various drugs, as well as in individuals with angiodysplastic lesions. Proposed mechanisms include specific autoantibodies, adsorption onto malignant cell clones and depletion in conditions of high vascular shear force. Therapeutic interventions are directed at controlling acute bleeding episodes, treating the underlying disorder in the hope of correcting the abnormalities of vWF and effecting antibody elimination. DDAVP infusion is the initial treatment of choice for achieving hemostasis, plasma-derived FVIII/vWF concentrates are the second choice. In either case clearance of the vWF will be accelerated and levels must be monitored. IVIG, plasmapheresis, and/or immunosuppressive drugs may be useful for eliminating antibody.

RARE COAGULATION FACTOR DISORDERS (FII, V, VII, X, XI, XIII, FIBRINOGEN DEFICIENCIES)

The exact prevalence of these disorders is unclear due to lack of epidemiological data; estimates range from 1:500,000 to 1:2,000,000. Registries such as the EN-RBDD (www.rbdd.org) and the Rare Coagulation Disorder Resource Room (www.rarecoagulationdisorders.org) have improved our understanding of these disorders in recent times. They seem to occur in a relatively low frequency and most of them, except for dysfibrinogenemia (autosomal dominant), are autosomal recessive in inheritance. They comprise 3–5% of inherited bleeding disorders. The natural history and spectrum of clinical features is not well established. Severe deficiencies are seen in populations with a high rate of consanguineous marriages (Iran, India; three- to sevenfold higher). FXI deficiency is especially prevalent among Ashkenazi Jews; 1 in 450 affected and 8% heterozygotes. Clinical manifestations are usually seen in homozygous or compound heterozygous individuals and rarely in the heterozygous state. They generally present with mild–moderate bleeding tendency with a potential for severe bleeds. Bleeding manifestations tend to be less severe than hemophilia with less frequent GI, genitourinary, central nervous system (CNS), and musculoskeletal (MSK) bleeds. Severe bleeding and MSK bleeds occur with fibrinogen disorders, FX, FII, and FXIII deficiencies. GI and CNS bleeds are commonly seen with FX deficiency while umbilical cord bleeds are seen with fibrinogen FII, V, XIII, and fibrinogen deficiencies. Mucocutaneous bleeds are the most common bleeding sites, with menorrhagia occurring in 50% women with rare bleeding disorders. These women are also at increased risk of developing hemorrhagic ovarian cysts, which are a less common but perhaps more specific manifestation of an underlying bleeding disorder than menorrhagia. They may be diagnosed after prolonged postoperative bleeding for the first time in adulthood.

In children being evaluated for suspected nonaccidental trauma, it is sometimes necessary to rule out underlying bleeding disorders. It is important to remember that most bleeding disorders are rare and history and clinical evaluation are vital in guiding further testing. Tests should be chosen based on the prevalence of the condition, patient and family history, ease of testing and the probability of a bleeding disorder causing the bleeding event (such data are available for intracranial hemorrhage). The initial testing panel should include PT/aPTT, VWF antigen/activity, factor VIII/IX levels, and a CBC with platelet counts.

TABLE 15.20   Treatment of Rare Coagulation Factor Deficiencies[a]

Factor deficiency	Half-life (h)	Therapeutic option		Increase in plasma concentration after 1 unit/kg IV (%)	Percentage required for hemostasis	Percentage required for minor trauma	Percentage required for surgery and major trauma
		First line	Second line				
Afibrinogenemia	56–82	RiaSTAP[b] (pd FI)	CPP	1–1.5[c]	80[c]	150[c]	200[c]
II	45–60	PCCs	FFP	1	20–30	30	50
V	36	FFP		1.5	10–15	10–15	25
VII	5	NovoSeven[d] (rVIIa)		1	10–15	10–15	20
X	24–60	PCCs		1	10–15	10–15	25
XI	48	FFP, FXI concentrate		1	30	30	>45
XIII	168–240	Corifact[e] (pd FXIII)	CPP	1	2–3	15	25

[a]See Table 15.14 for factor VIII and factor IX deficiency treatment guidelines.
[b]RiaSTAP®; CSL Behring GmbH, Marburg, Germany.
[c]In mg/dl.
[d]NovoSeven® RT (Novo Nordisk A/S, Bagsvaerd, Denmark).
[e]Corifact®; CSL Behring GmbH, Marburg, Germany.
pd, plasma derived; PCCs, prothrombin complex concentrates; CPP, cryoprecipitate; FFP, fresh frozen plasma.

Diagnostically the screening of PT/aPTT is abnormal, except in FXIII deficiency. Specific assays are required to measure activity and immunoassays for antigen to differentiate type 1 (hypoproteinemias) and type 2 (dysproteinemias) deficiencies. Genetic mutations have been identified in most deficiencies. These mutations are unique to each kindred. Deletions, null, nonsense, missense mutations have been identified, but in 10–20% no mutation has been identified. A genotype–phenotype correlation is not well established. The homozygous patients present with clinically significant bleeding while the heterozygotes generally have no bleeding except with FXI and FVII deficiencies wherein severe bleeding may be observed in the heterozygous state. Unlike hemophilia, there is no significant association between bleeding phenotype and coagulant factor activity for FVII, XI, and V deficiencies.

For treatment purposes purified concentrates are the treatment of choice. Dosing is based on minimal hemostatic level and half-life of the particular factor. FFP (FV deficiency, fibrinolytic factor deficiency) can be used for most disorders when diagnosis is not clearly established at first encounter in an emergency), PCCs (used for FII and X deficiencies) cryoprecipitate (FXIII and fibrinogen disorders when concentrates are not readily available), purified concentrates (fibrinogen and FXIII) and in some cases recombinant products (FVII and FXIII deficiency) are generally used.

Antifibrinolytics such as epsilon aminocaproic acid and tranexamic acid may be used as adjuvant therapy for skin and mucocutaneous bleeds. Caution needs to be exercised when used concurrently with PCCs. Oral contraceptive pills are used in women for menorrhagia in these bleeding disorders to control bleeding.

Table 15.20 lists the treatment of rare coagulation deficiencies.

THROMBOTIC DISORDERS

Mechanisms of Thrombosis in Inherited Thrombophilia

In inherited thrombophilias, impaired neutralization of thrombin or a failure to control the generation of thrombin causes thrombosis. There is a malfunction in a system of natural anticoagulants that maintain the fluidity of the blood. Decreases in AT III activity impair the neutralization of thrombin and reduced activity of protein C or protein S impairs or increases thrombin generation. Both of these mechanisms increase susceptibility to thrombosis (Figure 15.7).

Table 15.21 lists the clinical manifestations of a hypercoagulable state and laboratory findings in hypercoagulable states are listed in Table 15.22.

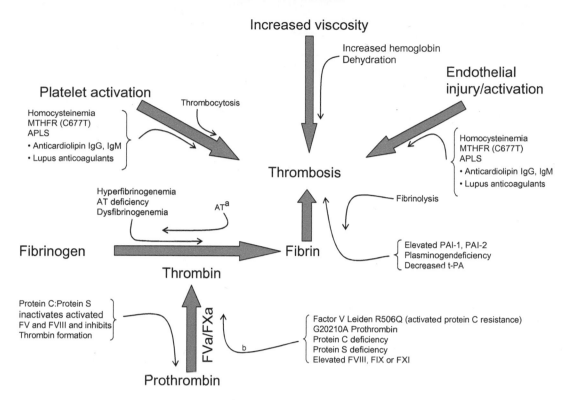

FIGURE 15.7   Natural inhibitors of coagulation and the effect(s) of pro-thrombotic states. [a]Leads to neutralization of thrombin; [b]Decreased neutralization or increased generation of thrombin.

TABLE 15.21   Clinical Manifestations of Hypercoagulable States

Family history of thrombosis
Recurrent spontaneous thromboses
Thrombosis in unusual sites
Thrombosis at an early age—myocardial infarcts, strokes, DVTs
Resistance to anticoagulant therapy
Warfarin-induced skin necrosis syndrome
Recurrent spontaneous abortions
Spinal surgeries associated with prolonged immobilization and thrombosis
Thrombosis associated with long airplane flights
Northern European ancestry
Thrombosis during pregnancy/postpartum/use of oral contraceptives
Migratory superficial thrombophlebitis
Antiphospholipid syndrome
Autoimmune disorders
    Inflammatory bowel disease
    Lupus erythematosus (SLE)
Malignancy
Nephrotic syndrome
Infections
    Varicella
    HIV
    Osteomyelitis or serious infection requiring hospitalization

TABLE 15.22   Laboratory Findings in Hypercoagulable States

PRIMARY HEMOSTASIS
Thrombocytosis
Platelet aggregation abnormalities
Increased platelet adhesiveness
Elevated levels of β-thromboglobulin, vWF, and PF4
Short platelet lifespan

SECONDARY HEMOSTASIS
Shortened PT/aPTT
Elevated coagulation factors (i.e., I, II, V, VII, VIII, IX, X, XI), fibrinogen
Reduced AT, heparin cofactor II, plasminogen, t-PA, protein C, protein S, FXII, prekallikrein
FV Leiden/APC resistance
Prothrombin gene mutation (G20210A)
Elevated α_2AP, α-2-macroglobulin, PAI-1
Increased Lp (a)
Thrombomodulin deficiency
Presence of fibrinopeptide A (FPA)
Increased FDPs (some measured as the D-dimer)
Short fibrinogen lifespan
Dysfibrinogenemia—diagnosed by prolonged thrombin and reptilase times, normal fibrinogen antigen with reduced activity
Homocysteinemia
Presence of circulating LAs, anticardiolipin, and anti-β-2-GPI antibodies

VENOUS THROMBOSIS

Venous thrombotic events (VTEs) develop under conditions of slow blood flow. They may occur by way of activation of the coagulation system with or without vascular damage. Venous thrombi are composed of large amounts of fibrin containing numerous erythrocytes, platelets, and leukocytes (red thrombus). The incidence of a VTE in children is estimated to be between 0.7 and 1.9 per 100,000 children. However there has been a recent three- to tenfold increase in diagnosis of VTE in hospitalized children. Venous thrombosis usually produces significant obstruction to blood flow. The most serious consequence is embolization from a deep vein thrombosis resulting in pulmonary embolism. Table 15.23 lists the predisposing causes of venous thrombosis. The detection of venous thrombosis and diagnosis of pulmonary emboli are given in Table 15.24 and Table 15.25, respectively. The treatment of VTEs is given in Table 15.26

Inherited causes of thrombophilia are discussed later and some specific risk factors for the development of venous thrombosis are discussed below.

Specific Risk Factors for Venous Thrombosis

Homocysteine

Homocysteine is a sulfur-containing amino acid formed during the metabolism of methionine, an amino acid derived from dietary protein. When excess methionine is present, homocysteine is linked with serine to form cystathionine in a reaction catalyzed by *cystathionine β-synthetase*, a vitamin-B6-dependent enzyme. Otherwise homocysteine acquires a methyl group from N^5-methylhydrofolate in a reaction catalyzed by the vitamin-B12-dependent enzyme *methionine synthetase*. Hyperhomocysteinemia, a risk factor for venous thrombosis, can be caused by genetic disorders affecting the trans-sulfuration or remethylation pathways of homocysteine

TABLE 15.23 Predisposing Factors for Venous Thrombosis

ACQUIRED[a]

Age
Trauma
Venipuncture
Indwelling venous access devices (intravenous catheters)
Surgery
Thermal injury
Immune complexes
Infections (HIV, varicella, supportive thrombophlebitis)
Severe dehydration
Shock
Prolonged immobilization
Cancer
L-Asparaginase therapy, prednisone
Oral contraceptives or hormone replacement therapy
Cyanotic heart disease
Pregnancy and the puerperium
Nephrotic syndrome
Liver disease
Neonatal asphyxia
Infant of a diabetic mother
Vitamin B12/folic acid deficiency (due to associated hyperhomocysteinemia)
Antiphospholipid syndrome
    Primary
    Secondary
        Systemic lupus erythematosus
        Rheumatoid arthritis
Resistance to activated protein C not due to FV Leiden
Systemic lupus erythematosus
    Anticardiolipin antibodies
    Lupus anticoagulants
Hyperviscosity syndromes
    Myeloproliferative disorders
    Polycythemia vera
    Chronic myelogenous leukemia
    Essential thrombocythemia
    Paroxysmal nocturnal hemoglobinuria
Hyperleukocytosis (acute leukemias)
Sickle cell disease, thalassemia intermedia not on chronic transfusion program
Infusion of concentrated vitamin-K-dependent (II, VII, IX and X) factors (aPCC), FFP frequently without close monitoring

---

CONGENITAL/INHERITED[b]

R506Q mutation in the FV gene ((FV Leiden))—activated protein C resistance)
G20210A mutation in the prothrombin (factor II) gene
AT deficiency
Protein C deficiency
Protein S deficiency
Plasminogen deficiency
Reduced levels of tissue thromboplastin activator (t-PA)
Dysfibrinogenemia
FXII (Hageman factor) deficiency
Prekallikrein deficiency
Heparin cofactor II (HC-II) deficiency
Hyperhomocysteinemia (due to C677T MTHFR or folate or vitamin B12 deficiency)
Increased levels of factor VIII, factor IX, factor XI, or fibrinogen
Increased TAFI

ANATOMIC RISK FACTORS

Paget–Schroetter syndrome
May–Thurner syndrome
    Atretic inferior vena cava

---

[a]In inherited thrombophilia these acquired factors increase the risk of venous thrombosis.
[b]The first thrombotic event may occur at an early age in patients who have more than one thrombophilia or who are homozygous for FV Leiden or the G20210A prothrombin gene mutation or asymptomatic heterozygotes who are relatives of patients with inherited thrombophilia who have had venous thrombosis. The combination of hyperhomocysteinemia with either FV Leiden or G20210A prothrombin gene mutation significantly increase the risk of venous thrombosis. Recurrent thrombosis is more common in those with more than one thrombophilia.

TABLE 15.24 Detection of Venous Thrombosis

Clinical assessment:
    symptoms and signs of deep vein thrombosis
    presence or absence of an alternative diagnosis
    presence and number of predisposing factors for VTE (see Table 15.23)
Venous ultrasound
D-dimer blood testing:
    high sensitivity and negative predictive value for pulmonary embolism more than deep vein thrombosis
Venogram
MRI/magnetic resonance venography (MRV)/CT venography especially for sinus venous thrombosis where CT scan may not be accurate

TABLE 15.25 Diagnosis of Pulmonary Embolism

Clinical assessment—typical signs and symptoms
High-resolution helical CT scan with good sensitivity and specificity
Ventilation-perfusion lung scanning
Pulmonary angiography
    The gold standard for the diagnosis of pulmonary emboli
Risk factors
    Recent surgery
    Immobilization (>3 days bed rest)
    Previous deep vein thrombosis or pulmonary emboli
    Lower extremity plaster cast
    Lower extremity paralysis
    Strong family history for deep vein thrombosis/pulmonary emboli
    Cancer
    Postpartum

TABLE 15.26 Treatment of VTEs and Thromboembolism

Treatment	Indications
Anticoagulant therapy	Majority of patients with a VTE (see Anticoagulant therapy)
Thrombolytic therapy	Massive pulmonary embolism with hemodynamic compromise; patients with extensive iliofemoral thrombosis (see Thrombolytic therapy)
Intracaval filter	Pulmonary embolism despite adequate anticoagulant therapy
Pulmonary embolectomy	Massive pulmonary embolism despite thrombolytic therapy
Surgical thrombectomy	Acute thrombosis of Blalock–Taussing shunts, life-threatening intracardiac thrombosis, prosthetic valve thrombosis, septic thrombosis, peripheral arterial thrombosis related to vascular access in neonates

metabolism, or by folic acid deficiency, vitamin B12 deficiency, vitamin B6 deficiency, renal failure, hypothyroidism, increasing age, and smoking. A rare example of excessive hyperhomocysteinemia is homozygous homocystinuria due to cystathionine β-synthetase deficiency; 50% of affected patients present with venous or arterial thrombosis by the age of 29 years. Homozygosity for the C677T mutation in the methylenetetrahydrofolate reductase (MTHFR) gene is a cause of mild hyperhomocysteinemia.

It has been shown that excess homocysteine:

- Has a toxic effect on the endothelium.
- Promotes thrombosis by platelet activation.
- Causes oxidation of LDL cholesterol.
- Increases levels of vWF and thrombomodulin.
- Increases smooth muscle proliferation.

The mean total homocysteine level in the newborn in various studies is 5.84 μmol/l, 7.4 μmol/l, and 7.8 μmol/l. Daily folic acid (1 mg), vitamin B6 (50 mg) and vitamin B12 (1 mg) lower plasma homocysteine levels. In countries with folate fortification of various food items, hyperhomocystinemia is rarely found.

Lipoprotein (a)—Lp (a)

Lp (a) is a distinct serum lipoprotein composed of a low-density lipid particle with disulfide links along a polypeptide chain (*apolipoprotein a*). Lp (a) regulates fibrinolysis by competing with plasminogen. An elevated level of Lp (a) is associated with coronary artery disease and is an independent risk factor for spontaneous stroke and venous thrombosis in childhood. Measurement is ideally performed in a fasting state and interpretation of Lp (a) levels needs to be performed in a reference laboratory as normal values vary with ethnicity.

Fibrin Degradation Products (D-Dimers)

The breakdown of crosslinked fibrin yields a number of degradation products, most notably D-dimers. Elevated levels of D-dimers are associated with risk of recurrence of thrombosis. Low or normal levels of D-dimers are useful for excluding a diagnosis of pulmonary embolus and acute thrombosis. D-dimers are a sensitive marker for fibrin turnover that allow recognition of covert coagulation. Alternately, it may indicate preclinical atherosclerosis, because levels are elevated for years before arterial thrombosis occurs.

Risk Factors for Recurrence

Recurrence is uncommon as long as the acquired risk factor (e.g., central venous line) is removed. The presence of any of three kinds of antiphospholipid antibodies increases the risk of recurrence. Persistent elevation of factor VIII and or D-dimer has been shown as a predictor of poor outcome including lack of thrombus resolution, recurrent thrombosis, or post-thrombotic syndrome. The first thrombotic events in heterozygotes with inherited thrombophilia in the presence of a triggering factor (e.g., infection or inflammation) may not warrant life-long anticoagulation. In these cases careful monitoring of serological markers of inflammation (ESR, CRP) and recurrence for thrombosis (FVIII, fibrinogen, D-dimer, LA, ACA, anti-beta-2 glycoprotein-1 antibodies) is required before a decision is made to discontinue anticoagulation therapy.

ARTERIAL THROMBOSIS

Arterial thrombosis initially occurs under conditions of rapid blood flow and often is the result of a process that damages the vessel wall. The thrombus is composed of tightly coherent platelets that contain small amounts of fibrin and few erythrocytes and leukocytes (white thrombus). The most serious consequence of arterial thrombosis is vascular occlusion. Arterial thrombotic events occur in a number of congenital and acquired diseases (Table 15.27).

Arterial Catheterization

The most common cause of an arterial thrombotic event in children is the use of arterial vascular catheters. Without prophylactic anticoagulation, the incidence of an arterial thrombotic event from femoral artery catherization is about 40%. Anticoagulation with heparin (100−150 U/kg) reduces the incidence of arterial thrombotic events to 8%. The incidence of an arterial thrombotic event from an umbilical arterial catheter is markedly reduced by a low-dose continuous heparin infusion (3−5 U/kg/h).

Cardiac Procedures

Blalock−Taussig Shunts

The short length and very high flow in these shunts may result in arterial thrombosis. The incidence varies from 1% to 17%. For anticoagulant management see later.

Fontan Operation

In this procedure the incidence of an arterial thrombotic event ranges from 3% to 19%. Arterial thrombotic events may occur anytime following the surgery and it is the major cause of early and late morbidity and mortality. These children may need to be on life-long anticoagulation.

TABLE 15.27 Predisposing Causes of Arterial Thrombosis

ACQUIRED

Catherization
  Cardiac
  Umbilical artery
  Renal artery—kidney transplantation
  Hepatic artery—liver transplantation
Vascular
  Injury
  Infections
  Periarteritis nodosa
  Systemic lupus erythematosus
  Kawasaki disease
  Hemolytic-uremic syndrome (HUS)
  Thrombotic thrombocytopenic purpura (TTP)
Cardiac
  Blalock—Taussig shunts
  Fontan operation
  Endovascular stents
  Cyanotic congenital heart disease
  Primary endocardial fibroelastosis
  Enlarged left atrium with arterial fibrillation
  Hypertension
  Myocarditis
Hematologic/hemostatic
  Elevated levels of LP (a)
  Elevated PA1-1
  Reduced t-PA
  Elevated level of fibrinogen
  Antiphospholipid syndrome (APLS)
  Chronic disseminated intravascular coagulation (DIC)
  Elevated levels of FDPs (D-dimers)
  Hyperhomocysteinemia
  Hypereosinophilic syndrome
  Myeloproliferative disorders (MDS)
  Polycythemia vera (PV)
  Chronic myelogenous leukemia (CML)
  Essential thrombocythemia (ET)
  Paroxysmal nocturnal hemoglobinuria (PNH)
  Hyperleukocytosis (acute leukemia)
  Sickle cell disease. Hemoglobin SC disease, thalassemia intermedia not on chronic transfusions
  Thrombotic thrombocytopenic purpura (TTP)
  Vitamin B12/folic acid deficiency (due to associated hyperhomocysteinemia)
  Activated prothrombin complex concentrate (aPCC) administration for prolonged periods without close monitoring
Other
  Shock
  Nephrotic syndrome
  Diabetes mellitus
  Hyperlipidemia
  Hypercholesterolemia
  Cigarette smoking
  Elevated CRP
  Malignancy
  Obesity
  Physical inactivity

CONGENITAL/INHERITED

R506Q mutation in the FV gene ((FV Leiden)—activated protein C resistance)
G20210A mutation in the prothrombin (factor II) gene
Hyperhomocystinemia
AT deficiency
PC deficiency
Protein S deficiency
Marfan syndrome
Familial hypercholesterolemia
Mitral valve prolapse

Lp (a), lipoprotein a; PAI-1, plasminogen activator inhibitor-1; t-PA, tissue plasminogen activator; CRP, C-reactive protein.

Endovascular Stents

These stents are used increasingly to manage some patients with congenital heart disease (e.g., pulmonary artery stenosis, pulmonary vein stenosis, coarctation of aorta). Therapeutic doses of heparin are given at the time of stent insertion, followed by aspirin therapy (5 mg/kg/day).

Umbilical Artery Catherization

The incidence of an arterial thrombotic event from an umbilical arterial catheter is markedly reduced by a low-dose continuous heparin infusion (3–5 U/h).

Renal Artery Thrombosis

Renal arterial thrombosis is commonly associated with kidney transplant. The incidence is about 0.2–3.5% in children. Prophylactic administration of LMWH 0.5 mg/kg twice daily for 21 days is effective anticoagulant therapy.

Hepatic Artery Thrombosis

Hepatic artery thrombosis is a very serious complication of liver transplantation. It usually occurs within 2 weeks. The reported incidence in children can be as high as 42%. The methods of detection of a hepatic artery thrombosis include Doppler, angiography, and CT of the liver. The prophylactic use of LMWH and aspirin is controversial. The mortality from hepatic artery thrombosis may be as high as 70% in children. Anticoagulant therapy alone is not effective in hepatic artery thrombosis. Surgical intervention and retransplantation usually are necessary. Thrombolytic therapy has been successful in about 20% of children.

Kawasaki Disease

The acute phase of Kawasaki disease may be associated with arteritis, arterial aneurysms, valvulitis, and myocarditis. The incidence of coronary artery aneurysm, stenosis, or thrombosis is about 25% without initial treatment. There is strong evidence that early use of IV gamma-globulin and aspirin can reduce the coronary artery involvement of Kawasaki disease.

ANTIPHOSPHOLIPID ANTIBODY SYNDROME

Antiphospholipid syndrome (APLS) may result in an arterial thrombotic event including stroke in children. APLS is defined by the presence of the following two criteria:

1. The presence in the plasma of at least one type of autoantibody known as an antiphospholipid antibody.
2. The occurrence of at least one of the following clinical manifestations: venous or arterial thromboses, or pregnancy-related morbidity.

Lupus Anticoagulant (LA) is detected by prolonged aPTT with no/partial correction on mixing studies. A greater than 3 sec difference in aPTT value between control and patient after mixing studies is indicative of an LA. Antiphospholipid antibodies (APLAs) are directed against plasma proteins bound to anionic phospholipids and include:

LA
Anticardiolipin antibodies (ACLAs) detected by immunoassays
    Anticardiolipin antibody IgA[1]
    Anticardiolipin antibody IgG
    Anticardiolipin antibody IgM
Subgroups of APLAs detected by immunoassays
    Anti-beta-2-glycoprotein-I (β-2-GPI) (IgA, IgG, IgM)[1]

[1]These APLAs are not included in the laboratory diagnosis of APLS (see Table 15.28).

Antiphosphatidylserine (IgG IgA, IgM)[1]
Antiphosphatidylethenolamine (IgG, IgA, IgM)[1]
Antiphosphatidylcholine (IgG, IgA, IgM)[1]
Antiphosphatidylinositol (IgG, IgA, IgM)[1]
Antiannexin-V[1]

Both ACLAs and β-2-GPI can have LA activity, the detection of which needs functional clotting tests.

Antiphospholipid antibodies (APLA) are found in 1—5% of young healthy subjects and the prevalence increases with advancing age. Among patients with SLE the prevalence is 23—47% for ACLA, 15—34% for LA, and 20% for β-2-GPI. APLA's have also been detected in patients with infections, sickle cell disease, malignancies, and with certain medications (procainamide, amoxicillin, oral contraceptive pills, propranolol, etc.). Many patients have laboratory evidence of APLA without clinical consequences.

The APLAs may promote thrombosis by:

- Antibody-induced activation of endothelial cells and the secretion of cytokines.
- Platelet activation.
- Impaired activation of protein C, annexin V, and TFs.
- Inhibition of β2-GPI (thought to act as a natural anticoagulant).
- Impaired fibrinolytic activity.
- Antibody against endothelial cells.

APLS is the most common acquired condition associated with venous and/or arterial thrombotic events. The thrombotic events associated with APLS include deep vein thrombosis, pulmonary embolus, coronary artery thrombosis, stroke, transient ischemic attacks, retinal vascular thrombosis, and placental vascular thrombosis (leading to recurrent-miscarriage syndrome). Virtually any organ can be involved and the range of disorders observed within any one organ spans a diverse spectrum. The effect of APLS depends on the nature and size of the vessel affected and the acuteness or chronicity of the thrombotic process.

Venous thrombosis, especially deep venous thrombosis of the legs, is the most common manifestation of the APLS. Up to half these patients have pulmonary emboli. Arterial thromboses are less common than venous thromboses and most frequently manifest with features consistent with ischemia or infarction. The severity of presentation relates to the acuteness and extent of occlusion. The brain is the most common site, with strokes and transient ischemic attacks accounting for almost 50% of arterial occlusions. Coronary occlusions account for an additional 23%; the remaining 27% involve diverse vessels, including the subclavian, renal, retinal, and pedal arteries.

In patients with APLS there is an increased rate of heart valve abnormalities, mainly valve masses or thickening affecting primarily the mitral valve. Up to 63% of patients with APLS have at least one valvular abnormality on echocardiography. Emboli, especially from mitral or aortic valve vegetation, can lead to cerebral events. Not all arterial episodes of ischemia or infarction are thrombotic in origin.

Acute involvement at the level of the capillaries, arterioles, or venules may result in a clinical picture resembling hemolytic-uremic syndrome and thrombotic thrombocytopenia purpura as well as other thrombotic microangiopathies. Thrombotic microangiopathy may also occur as a more chronic process, resulting in slow, progressive loss of organ function, the underlying reason for which may only be determined by biopsy. Thus, organ involvement in patients with the APLS can present as a spectrum from rapidly progressive to clinically silent and indolent. Depending on the size of the vessels affected, organ failure has two predominant causes, thrombotic microangiopathy and ischemia secondary to thromboembolic events. A small subset of patients with APLS has widespread thrombotic disease with multiorgan failure, which is called "catastrophic APLS" which leads to death in 50% of cases. This is associated with laboratory features such as elevated FDPs, depressed fibrinogen levels, or elevated D-dimer concentrations. Pregnant women with APLS can suffer from morbidities including fetal death after 10 weeks' gestation, premature birth due to severe preeclampsia or placental insufficiency or multiple embryonic losses (<10 weeks' gestation).

Other prominent manifestations of the antiphospholipid syndrome include thrombocytopenia, hemolytic anemia, and livedo reticularis. Although renal manifestations are a very common feature of systemic lupus erythematosus, they were only recently recognized as part of the antiphospholipid syndrome.

Histopathologic features of APLS consist of:

- Thrombotic microangiopathy.
- Ischemia secondary to arterial thromboses or emboli.
- Peripheral embolization from venous, arterial or intracardiac sources.

Table 15.28 shows the criteria for diagnosis of APLS.

TABLE 15.28 Criteria for Diagnosis of antiphospholipid Antibody Syndrome (APS)[a]. APS is Present if at Least One of the Clinical Criteria and One of the Laboratory Criteria are Present

CLINICAL CRITERIA

1. **Vascular thrombosis**
One or more clinical episodes of arterial, venous, or small vessel thrombosis,[b] in any tissue or organ. Thrombosis must be confirmed by objective validated criteria (i.e. unequivocal findings of appropriate imaging studies or histopathology). For histopathologic confirmation, thrombosis should be present without significant evidence of inflammation in the vessel wall.

2. **Pregnancy morbidity**
   a. One or more unexplained deaths of a morphologically normal fetus at or beyond the 10th week of gestation, with normal fetal morphology documented by ultrasound or by direct examination of the fetus, or
   b. One or more premature births of a morphologically normal neonate before the 34th week of gestation because of:
      i. eclampsia or severe preeclampsia defined according to standard definitions, or
      ii. recognized features of placental insufficiency[c]
   c. Three or more unexplained consecutive spontaneous abortions before the 10th week of gestation, with maternal anatomic or hormonal abnormalities and paternal and maternal chromosomal causes excluded.
   In studies of populations of patients who have more than one type of pregnancy morbidity, investigators are strongly encouraged to stratify groups of subjects according to a, b, or c above.

LABORATORY CRITERIA[d]

1. Lupus anticoagulant (LA) present in plasma, on two or more occasions at least 12 weeks apart, detected according to the guidelines of the International Society on Thrombosis and Haemostasis (Scientific Subcommittee on LAs/phospholipid-dependent antibodies)
2. Anticardiolipin (aCL) antibody of IgG and/or IgM isotype in serum or plasma, present in medium or high titer (i.e. >40 GPL or MPL, or >the 99th percentile), on two or more occasions, at least 12 weeks apart, measured by a standardized ELISA [100,129,130]. 3. Anti-b2 glycoprotein-I antibody of IgG and/or IgM isotype in serum or plasma (in titer >the 99th percentile), present on two or more occasions, at least 12 weeks apart, measured by a standardized ELISA.
3. Anti-β2 glycoprotein-I antibody of IgG and/or IgM isotype in serum or plasma (in titer greater than the 99th percentile), present on two or more occasions, at least 12 weeks apart, measured by a standardized ELISA, according to recommended procedures.

[a]Classification of APS should be avoided if less than 12 weeks or more than 5 years separate the positive aPL test and the clinical manifestation. Coexisting inherited or acquired factors for thrombosis are not reasons for excluding patients from APS trials. However, two subgroups of APS patients should be recognized, according to: (a) the presence, and (b) the absence of additional risk factors for thrombosis such as age (>55 in men and >65 in women), and the presence of any of the established risk factors for cardiovascular disease (hypertension, diabetes mellitus, elevated LDL or low HDL cholesterol, cigarette smoking, family history of premature cardiovascular disease, body mass index >30 kg /m², microalbuminuria, estimated GFR <60 ml/min/1.73 m²) inherited thrombophilias, oral contraceptives, nephrotic syndrome, malignancy, immobilization, and surgery. Thus, patients who fulfil criteria should be stratified according to contributing causes of thrombosis. A thrombotic episode in the past could be considered as a clinical criterion, provided that thrombosis is proved by appropriate diagnostic means and that no alternative diagnosis or cause of thrombosis is found.
[b]Superficial venous thrombosis is not included in the clinical criteria.
[c]Generally accepted features of placental insufficiency include: (i) abnormal or non-reassuring fetal surveillance test(s), e.g. a non-reactive non-stress test, suggestive of fetal hypoxemia, (ii) abnormal Doppler flow velocimetry waveform analysis suggestive of fetal hypoxemia, e.g. absent end-diastolic flow in the umbilical artery, (iii) oligohydramnios, e.g. an amniotic fluid index of 5 cm or less, or (iv) a postnatal birth weight less than the 10th percentile for the gestational age.
[d]Investigators are strongly advised to classify APS patients in studies into one of the following categories. I, more than one laboratory criteria present (any combination); IIa, LA present alone; IIb, aCL antibody present alone; IIc, anti-b2 glycoprotein-I antibody present alone.
Adapted from: Miyakis et al., 2006, with permission.

LA and Thrombosis

The so-called "lupus anticoagulant antibody" (LA) was first described in patients with SLE who presented with prolonged aPTT. The term is actually a misnomer because only few patients with SLE have the LA and the inhibitor is not an anticoagulant. Despite the name, LA antibodies are associated with thromboembolic events rather than clinical bleeding. Approximately 10% of patients with SLE have LA. LA is occasionally found in patients with immune thrombocytopenic purpura in whom SLE subsequently develops. Human leukocyte antigen (HLA)-DR antigens (DR7 and DRW35) may play a role in the development of LA in some individuals.

Thromboembolism occurs in about 10% of patients with SLE; however, in patients with SLE and LA the occurrence of thromboembolism may be as high as 50%. The LA is estimated to account for approximately 6–8% of thrombosis in otherwise healthy individuals. Primary LA thrombosis syndrome consists of patients with LA and thrombosis but who have no secondary underlying disease such as SLE or other autoimmune disorders, malignancy, infection; inflammation or ingestion of drugs inducing the LA. It is fivefold more common than the secondary type associated with an underlying disease.

The LA is an acquired circulating immunoglobulin (usually IgG, sometimes IgM or IgA), which binds to negatively charged phospholipid components of factors Xa and Va, phospholipid, and calcium. This circulating anticoagulant is detected by a prolongation of phospholipid-dependent clotting tests, such as APTT, that does not correct to normal when mixed with normal plasma. More specific tests include a dilute tissue thromboplastin inhibition test, platelet neutralization procedure, dilute Russell viper venom test (DRVVT), or APLAs by enzyme-linked immunosorbent assay. To exclude the presence of LA, two or more assays that are sensitive to these antibodies must be negative.

Because this antibody does not usually cause a bleeding disorder, the primary reason to diagnose it is to avoid intensive workup and therapy prior to surgical procedures. If an antibody is identified in a young child during an acute viral illness, testing should be repeated in several months in order to document its disappearance.

Despite the frequent concordance between LA and either anticardiolipin or anti-β-2-GPI antibodies, these antibodies are not identical. In general, LAs are more specific for APLS whereas anticardiolipin antibodies (ACLAs) are more sensitive.

Anticardiolipin Antibodies

ACLAs are generally IgG, IgA, and IgM anticardiolipin idiotypes. These antibodies are found in both SLE and non-SLE patients. The ACLAs are associated with venous and arterial thrombotic events. Most individuals with ACLAs do not have an LA, and most with the LA do not have ACLAs.

Treatment

Nonobstetric thrombotic events in APLS are treated with standard anticoagulation with heparin, LMWH, warfarin, and with some role for antiplatelet agents. Many patients with coexisting SLE are also treated with hydroxychloroquine. The frequency of recurrent thrombotic events in APLS ranges between 11% and 29% per year and provides a strong argument for lifelong anticoagulation in these patients. The role of primary prophylaxis for patients with APLA without thrombotic events is less clearly defined. Pregnant women with APLS are usually treated with a combination of low-dose aspirin with or without LMWH. The management of incidentally detected APLAs in asymptomatic pregnant women is more of a challenge with some advocating for the use of low-dose aspirin.

HEREDITARY THROMBOTIC DISORDERS

Inherited thrombophilia[2] should be suspected when the patient has:

- Recurrent or life-threatening venous thromboembolism.
- A family history of venous thrombosis.
- An age younger than 45 years.
- No apparent risk factors for thrombosis.
- A history of multiple abortions, stillbirths, or both.

Factor V Leiden (Activated Protein C Resistance)

Activated protein C (APC) is a natural anticoagulant that acts by cleaving and inactivating FV and factor VIII. In Caucasians, FV Leiden is the single most common inherited disorder predisposing to thrombosis. It results from a single G to A point mutation at nucleotide 1691 within the FV gene, arginine is replaced by glutamine at position (R506Q), rendering activated FV relatively resistant to inactivation by protein C. Approximately 95% of patients with activated protein C resistance test positively for FV Leiden mutation. The remaining 5% of cases are attributable to oral contraceptive usage, presence of a LA, or other rare mutations in the FV gene.

Three to eight percent of Caucasians carry the FV Leiden mutation and approximately 0.1% are homozygous. In contrast this mutation is relatively uncommon in African and Asian populations with a prevalence of <1.0% (heterozygous). Children with homozygous and heterozygous FV Leiden usually have their first thrombotic event following puberty with an estimated annual incidence of 0.28%. Heterozygous FV Leiden increases the risk of thrombosis five- to tenfold, whereas homozygous individuals have an 80-fold increased risk. APC resistance even in heterozygotes is a significant risk factor for thrombosis because >20% of patients presenting with a

---

[2]In thrombophilia multiple gene defects often coexist with the clinical penetrance of the syndrome being the end result of the number of gene defects present in an individual. This has been shown for patients with inherited deficiencies of antithrombin III, protein C or protein S whose risk of developing thrombotic manifestations is enhanced when there is coexistence of factor V Leiden.

thrombotic event exhibit APC resistance. Forty-two percent of patients who sustain a first VTE before 25 years of age and 21% of those between 54 and 70 years of age have APC resistance.

Both heterozygous and homozygous cases of FV Leiden mutation have increased risk for either venous or arterial thrombosis throughout life but usually are asymptomatic in youth unless associated with other acquired or genetic prothrombotic conditions, including central venous catheters, trauma, surgery, cancer, pregnancy, oral contraceptives, deficient protein C, deficient protein S, or homocysteinemia.

FV Leiden mutation has been associated with cerebral infarction and with venous thrombosis in children.

Diagnosis

An aPTT-based assay can be used to screen for APC resistance. Patients positive for APC resistance in the clotting assay should undergo genetic testing for FV Leiden mutation by analyzing genomic DNA in peripheral blood mononuclear cells.

Prothrombin G20210A Mutation (FII G20210A)

The genetic defect is at nucleotide position 20210 in the prothrombin gene. The mutation results in abnormally high prothrombin levels and contributes to thrombotic risk by promoting increased thrombin generation. Prothrombin gene mutation is the second most common inherited thrombotic defect. The frequency of this mutation is about 2–3% in Caucasian populations and 4–5% in Mediterranean populations. The homozygous state is extremely rare. Homozygotes for the prothrombin G200210A mutation have less severe clinical presentations than homozygotes for AT, protein C, and protein S deficiencies. The individual with this genetic defect usually presents with thrombotic episodes in adulthood.

Anti-thrombin Deficiency

Anti-thrombin (AT) functions by forming a complex with the activated clotting factors thrombin, Xa, IXa, and XIa. The relatively slow formation of this complex is greatly accelerated in the presence of heparin or cell-surface heparin sulfate. The incidence of AT deficiency is about 0.2–0.5%. Congenital AT deficiency is a heterozygous disorder; homozygous deficiency, probably incompatible with life, has not been reported.

Two major types of AT deficiency have been described. Type I AT deficiency (low activity and low antigen level) is a quantitative defect caused by a mutation resulting in both decreased synthesis and functional activity of AT, whereas type II AT deficiency is a qualitative defect characterized by decreased AT activity and normal antigenic levels.

Thrombotic complications of AT deficiency usually occur in the second decade of life. In affected individuals AT levels are about 40–60% (normal range, 80–120%). The risk of developing thrombotic complications depends on the particular subtype of AT deficiency and on other coexisting inherited or acquired risk factors. In children with heterozygous AT deficiency the relative risk of developing thrombotic episodes is increased 10-fold.

Treatment

AT deficiency is usually treated with oral warfarin-type anticoagulants (Coumadin) to decrease the level of vitamin-K-dependent procoagulants so that they are in balance with the level of AT. Heparin requires binding with AT to anticoagulate blood and, for this reason, heparin administration is usually ineffective in AT deficiency states. Because the administration of warfarin may take several days to decrease the vitamin-K-dependent factors, acute thrombotic episodes are usually managed with AT replacement therapy, using either FFP or recombinant AT concentrates. Patients are initially treated with approximately 50 units/kg AT, which will increase the baseline level of AT by approximately 50%. AT levels should be maintained at 80% or higher. In patients with severe recurrent thrombosis who cannot be managed with oral anticoagulants, therapeutic or prophylactic replacement therapy can be monitored with AT concentrates.

5,10-Methylenetetrahydrofolate Reductase Mutation

In this condition there is a cytosine to thymine mutation at nucleotide 677 (C677T) of MTHFR gene. This mutation may be a risk factor for stroke in children, venous thrombosis in the young, and coronary artery disease in adults. In children, recent studies have not shown the MTHFR mutation to be a big risk factor for thrombosis. MTHFR is essential for the remethylation of homocysteine to methionine. Homozygosity for mutation of MTHFR is associated with hyperhomocysteinemia. Hyperhomocysteinemia is a common risk factor for deep-vein thrombosis and increases the risk for deep-vein thrombosis in patients with FV Leiden. In general, testing for this genetic defect is not recommended. Testing for a fasting homocysteine level seems to be more relevant since not all individuals with this defect have an elevated homocysteine level.

Protein C Deficiency

PC is a vitamin-K-dependent plasma glycoprotein which when activated functions as an anticoagulant by inactivating factors Va and VIIIa. PC activity is enhanced by another vitamin-K-dependent inhibitory cofactor, protein S.

PC deficiency is inherited in an autosomal-dominant manner and is subdivided into two types. Type I PC deficiency (low activity and low antigen level) is a quantitative deficiency with decreased plasma concentration and functional activity to approximately 50% of normal. Type II PC deficiency (low activity and normal antigen level) is less common and is characterized by a qualitative decrease in functional activity, despite normal levels of PC antigen. Plasma levels of PC below 50% (normal range, 70–110%) are associated with the risk of thrombotic complications.

The population prevalence of heterozygous PC deficiency is estimated at 0.2%. The clinical manifestation of heterozygous PC deficiency is primarily venous thrombotic episodes during the second decade of life or young adulthood. Heterozygous deficiency of PC is not a significant problem during the neonatal period, but homozygous deficiency or compound heterozygous individuals with protein C deficiency may cause a fatal thrombotic disorder. It may present with neonatal purpura fulminans, DIC, progressive skin necrosis with microvascular thrombosis, and thrombosis of the renal veins, mesenteric veins, and dural venous sinuses, and rarely stroke in the newborn period. Several of the neonates with homozygous deficiency have been subsequently found to be blind.

Treatment

Initially, affected neonates are treated for several months with replacement therapy with FFP (10–20 ml/kg q12h); later, they can usually be managed with long-term oral warfarin or protein C replacement (FFP or PCC). A purified plasma-derived protein concentrate has been approved for use in children with protein C deficiency.

Heterozygous PC deficiency is also one of the major causes of warfarin-induced skin necrosis. Because the half-life of PC is extremely short (2–8 h), warfarin decreases PC more rapidly than factors IX, X, and prothrombin, resulting in a hemostatic imbalance with resultant microvascular thrombosis. Warfarin-induced skin necrosis is most likely to occur in these patients when loading doses of warfarin are used. This complication can usually be avoided by overlapping heparin and warfarin usage until the desired INR is attained and only then discontinuing the heparin.

Protein S Deficiency

Protein S deficiency is inherited as an autosomal-dominant manner and has a prevalence of 1:33,000. Protein S (PS) is also a vitamin-K-dependent anticoagulant that circulates in the plasma in two forms: a free active form (40%) and an inactive form bound to C4b-free binding protein (60%). PS functions as a cofactor to PC by enhancing its activity against factors Va and VIIIa. Free PS levels (normal range, 44–92%) correlate clinically with thrombotic episodes. PS deficiency is classified into three subtypes with total and free PS reduced as follows:

	Total PS	Free PS	PS function
Type 1	↓	↓	↓
Type II a	N	↓	↓
Type II b	N	N	↓

The clinical presentation of protein S deficiency is indistinguishable from that of protein C deficiency. Heterozygous PS deficiency manifests in adulthood as either venous or arterial thrombotic events. Only a few cases with heterozygous protein S deficiency have been reported with thrombotic episodes during childhood. A small number of newborns with either homozygous or compound heterozygous PS deficiency have been reported. These infants, like those with homozygous PC deficiency, may present with purpura fulminans.

Treatment

The acute episodes are usually treated with standard anticoagulation therapy with heparin, followed by oral warfarin administration for 3–6 months. Recurrent thrombosis or a life-threatening thromboembolic event in a patient with PS deficiency is usually managed with long-term oral anticoagulation. Asymptomatic patients are not usually treated but need prophylaxis for high-risk procedures such as surgery. Pregnancy may be accompanied with a reduction in free PS. Oral contraceptives will also result in a reduction of PS and may precipitate thrombosis in a patient with heterozygous PS deficiency.

Usually first thrombotic events in heterozygous thrombophilia may not warrant life-long anticoagulation in the presence of a triggering factor such as infection or inflammation. Careful monitoring of serological markers of inflammation and recurrence for thrombosis (ESR, CRP, FVIII, fibrinogen, D-dimer, LA, ACA, anti-β-2-GPI antibodies) may be justified in individual cases especially before a decision to discontinue anticoagulation is made (refer to Monagle et al., 2012)

Dysfibrinogenemia

Approximately 45 different dysfibrinogenemias that predispose to thrombotic events have been described, with the majority due to single-point mutations. Dysfibrinogenemia, a very rare condition estimated to occur in 0.8% of adults presenting with thrombotic episodes, is usually inherited as an autosomal-dominant condition. Impaired binding of thrombin to an abnormal fibrin and defective fibrinolysis occurs. t-PA and plasminogen activation on the abnormal fibrin have been implicated in the development of thrombosis. Congenital dysfibrinogenemia may also be associated with abnormal fibrin interactions with platelets and defective calcium binding. The clinical presentation is variable, consisting of venous thrombosis, pulmonary embolism, and arterial occlusion in 20% of cases. Most cases end up with bleeding rather than thrombosis. Some cases of dysfibrinogenemia may have an associated bleeding diathesis. Although it is very rare, intracranial hemorrhage is the most common cause of death in these patients. Homozygosity for dysfibrinogenemia is very rare and associated with juvenile arterial stroke, thrombotic abdominal aortic occlusions, and postoperative thrombotic episodes. Fibrinogen infusions may precipitate the thrombotic events. A normal fibrinogen antigen with low functional fibrinogen along with a prolonged thrombin time (TT) or reptilase time is indicative of dysfibrinogenemia.

Heparin Cofactor II (HC-II) Deficiency

HC-II deficiency is inherited as an autosomal dominant trait. Clinical manifestations of HC-II deficiency are arterial and VTEs. HC-II deficiency seems to be a rare cause of unexplained thrombotic events.

Hereditary Defects of Fibrinolytic System

Type I Dysplasminogenemia

Homozygous deficiency of this condition clinically manifests as pseudomembranous conjunctivitis, hydrocephalus, obstructive airway disorder, and abnormal wound healing secondary to failure of removal of fibrin deposits in various organs. Replacement therapy with plasminogen corrects these defects by allowing the lysis of fibrin deposits. Infants with homozygous deficiency do not seem to have a higher incidence of thrombotic events.

Type II Dysplasminogenemia

Type II is inherited as a mutation at various loci in the plasminogen molecule leading to functional abnormalities and failure of plasminogen activation. The type II defect is also associated with an increased incidence of thrombotic events.

Tissue Plasminogen Activator Deficiency

There are several families reported with this defect in which there is a failure to release fibrinolytic activity following VTEs.

Plasminogen Activator Inhibitor Deficiency

These polymorphic variations in the human PAI-1 gene have been reported where specific alleles may be associated with decreased PAI-1 levels. PAI-1 genotype abnormality may represent a risk factor for VTEs in the setting of PS deficiency.

Prophylaxis in Relatives of Patients with Thrombophilia

First-degree relatives of index patients with thrombophilias who are asymptomatic should be advised of the risk of venous thrombosis. Primary prophylaxis in these persons includes the administration of LMWH in high-risk conditions, such as during surgery, trauma, immobilization, and 6 weeks postpartum. These individuals should maintain a normal weight and homocysteine levels and careful consideration must be observed with regard to the risk–benefit ratio of contraceptives and hormone-replacement therapy. Women with AT deficiency, combined thrombophilia, or homozygosity for FV Leiden or the G20210A mutation in the prothrombin gene should be treated throughout pregnancy and for 6 weeks postpartum with LMWH.

THROMBOTIC DISORDERS IN NEWBORNS

Newborns are at a higher risk for a VTE because of decreased activity of anticoagulant factors, specifically AT, protein C, and protein S. Fibrinolytic activity in the newborn period is also decreased by lower serum plasminogen levels. Renal, caval, portal and hepatic venous system thrombosis are well-known complications of peripartum asphyxia, sepsis, dehydration, and maternal diabetes. The estimated annual incidence of VTEs is about 0.5 per 10,000 newborns. However, the majority of VTEs within the first year of life are associated with central venous access devices.

Congenital

Newborns comprise the largest group of children developing thrombotic events. Childhood stroke appears to be a diverse condition with many potential risk factors. Inherited thrombophilia is a multigene disorder in which the likelihood of developing thrombotic episodes increases with the number of genetic risk factors present in a subject. The "thrombophilia burden"[3] is found to be increased in children with stroke.

Acquired

Systemic Venous Thromboembolic Disorders

The incidence of symptomatic VTEs in newborns (exclusive of CNS) is 0.24 per 10,000. The incidence of neonatal thrombotic events, including CNS events, is 0.51 per 10,000 births with about 70% of cases being venous events.

The incidence of *sinovenous thrombosis* is 2.6 per 100,000 children per year. The following factors predispose a newborn to sinovenous thrombosis:

- During birth, the normal molding and overlapping of cranial sutures may damage the cerebral sinus structures and provoke sinovenous thrombosis.
- Asphyxia.
- Dehydration.
- Sepsis and meningitis.

The noncontrast CT scan alone could miss a diagnosis of sinovenous thrombosis and in suspected cases a CT venogram should be considered. MRI scanning with venography is an alternative test for diagnosis of sinus venous thrombosis.

Central-Venous-Catheter-Related Thrombosis

More than 80% of VTEs are secondary to central venous catheters. Thrombotic events result from damage to vessel walls, disrupted blood flow, infusion of substances in total parenteral nutrition (TPN) which damage endothelial cells and thrombogenic catheter materials. Right atrial thrombosis commonly occurs in newborns with central venous catheters. Right atrial thrombosis may result in cardiac failure, persistent sepsis, and fatal pulmonary emboli.

---

[3]Concomitant presence or any combination of APLA, factor V Leiden, FIIG20210A, AT deficiency, PC deficiency, PS deficiency, Lp (a). A full thrombophilia work-up should be carried out in newborn infants with unexplained prenatal or neonatal cerebral infarction especially if there is a family history of venous thrombosis, early stroke, or heart disease.

Umbilical Venous Catheters

Major clinical symptoms of umbilical arterial catheter-related thrombotic events occur in about 3% of infants. Loss of patency due to umbilical arterial catheter-related thrombotic events occurs in 13–73% without unfractionated heparin being administered and in 0–13% with the administration of unfractionated heparin.

Central Venous Catheters

Central venous catheters are the most common risk factor for VTEs in newborns. Right atrial thrombosis commonly occurs in newborns with central venous catheters. Right atrial thrombosis may result in cardiac failure, persistent sepsis, and fatal pulmonary emboli.

Renal Vein Thrombosis

Renal vein thrombosis is the most common noncatheter-related VTE in newborns presenting with the classical triad of hematuria, flank mass, and thrombocytopenia. The incidence is about 10% of all VTEs. Almost 80% present within the first week of life. Renal vein thrombosis may occur bilaterally in 24% of newborns with almost half extending into the IVC. The risks of renal vein thrombosis include perinatal asphyxia, shock, polycythemia, cyanotic congenital heart disease, maternal diabetes, and sepsis, which result in reduced renal flow, hyperviscosity, hyperosmolality, and hypercoagulability. Diagnostic ultrasonography is the radiographic test of choice. The long-term consequences include irreversible renal atrophy, hypertension, and chronic renal failure.

Systemic Arterial Thromboembolic Disorders

Arterial Ischemic Stroke

The incidence of arterial ischemic stroke in newborns is 93 per 100,000 live births. A primary risk factor is identifiable in about 70% of affected newborns. Systemic risk factors include:

- Iatrogenic causes secondary to indwelling arterial catheters. This is usually related to catheter material, duration of placement, diameter, length, solution infused, and arterial site.
- Cardiac disease.
- Perinatal complications (trauma, hypoxia-ischemia, maternal cocaine abuse).
- Dehydration.
- Congenital thrombophilia (e.g., PC and PS deficiencies, AT deficiency, FII G20210A, MTHR C 677T, FV Leiden, elevated LP (a), maternal APLAs).

Diagnostic radiographic studies include CT scan, MRI, MRA of the brain and neck, and less frequently conventional angiogram. MRA can identify carotid artery dissection and congenital vascular anomalies in intracranial vessels. Cranial ultrasound has a limited role in arterial ischemic stroke.

Use of anticoagulants should be considered in specific situations associated with sinus venous thrombosis or when a documented cardioembolic source is identified. Thrombolytic therapy is rarely indicated. Diagnosis and treatment of thrombotic events are shown in Table 15.29.

The risk factors for perinatal stroke are listed in Table 15.30. More than one risk factor is identified in many cases.

ANTITHROMBOTIC AGENTS

Heparin Therapy

Heparin mediates its activities through catalysis of the natural inhibitor AT. The activities of heparin are considered anticoagulant and antithrombotic. The antithrombotic activities of heparin are influenced by plasma concentration of AT. In the presence of heparin, thrombin generation is decreased in young children compared to adults. Clearance of heparin is also faster in the young. For this reason the optimal dosing of heparin is different in children from adults. In pediatric patients the aPTT value correctly predicts heparin dosing levels 70% of the time. The following dosage schedule is utilized for heparinization:

- Loading dose: 75 units/kg IV over 10 min.
- Initial maintenance dose:
    ≤1 year of age: 28 units/kg/h.
    >1 year of age: 20 units/kg/h.

TABLE 15.29 Diagnosis and Treatment of Neonatal Thromboembolism

Thrombosis	Diagnosis	Treatment
Systemic VTEs	US	LMWH/UFH/TT(*)
Pulmonary emboli	CT Angio	LMWH/UFH
Central venous catheter-related thromboembolic event	US	LMWH/UFH/TT(*)
Right atrial thrombosis	ECHO	LMWH/UFH
Cerebral sinovenous thrombosis	Contrast-enhanced MRV	LMWH/UFH if no hemorrhage present
Umbilical venous catheter	US	LMWH/UFH
Renal vein thrombosis	US	LMWH/UFH indicated if thrombosis extends to IVC or renal failure
Umbilical arterial catheter	US	LMWH/UFH
Peripheral arterial catheter	US	LMWH/UFH + thrombolysis or surgical thrombectomy

US, ultrasonogram; LMWH, low-molecular-weight heparin; UFH, unfractionated heparin; TT, thrombolytic therapy; *recommended only if potential loss of life, organ, or limb; MRV, magnetic resonance venography; ECHO, echocardiogram; IVC, inferior vena cava.

TABLE 15.30 Risk Factors for Perinatal Stroke

Cardiac disorders
  Congenital heart disease
  Patent ductus arteriosus
  Pulmonary valve atresia
Hematologic disorders
  Polycythemia
  Disseminated intravascular coagulopathy
  Factor-V Leiden mutation
  Protein S deficiency
  Protein C deficiency
  Prothrombin G20210A mutation
  High homocysteine level
  High Lp (a) level
  MTHFR C677T mutation
Infectious disorders
  CNS infection
  Systemic infection
Maternal disorders
  Autoimmune disorders
  Coagulation disorders
  Anticardiolipin antibodies
  Twin-to-twin transfusion syndrome
  In utero cocaine exposure
  Smoking during pregnancy
  Infection
  Diabetes mellitus—type 1/2 or gestational
Placental disorders
  Placental thrombosis
  Placental abruption
  Placental infection (chorioamnionitis)
  Fetomaternal hemorrhage
Vasculopathy
  Vascular maldevelopment
Trauma and catheterization
  Central venous catheters, umbilical venous catheters
Birth asphyxia
Dehydration
Extracorporeal membrane oxygenation

- Repeat the aPTT 4 h after administration of the heparin loading dose and monitor heparin dose to maintain the aPTT at $2^1/_2$ times normal for the reference laboratory as follows:

APTT times of normal	Bolus (units/kg)	Dose change (units/kg/h) (increase or decrease of bolus dose)	Intervals of testing APTT (h)
<2	50	20% increase	4
2	–	10% increase	4
>2	–	No change	4
$2^1/_2$	–	No change	24
>$2^1/_2$[a]	–	10% decrease	4
>3[b]	–	20% decrease	4

[a]Hold the dose 30 min.
[b]Hold the dose 1 h.
Heparin solution at concentrations of 80 units/ml for children 10 kg or less or 40 units/ml for children greater than 10 kg.

- When the patient achieves a therapeutic aPTT level repeat the aPTT and CBC with platelet count daily.
- The anti-Xa level should be monitored after a therapeutic aPTT is achieved in order to ensure a level between 0.35 and 0.70 units/ml.
- When heparin is interrupted for more than 1 h, re-establish the heparin maintenance infusion at the previous rate until the aPTT result is available. After that, administer heparin in accordance with the aPTT results.
- When prophylactic dosing of unfractionated heparin is desired, a dose of 10 units/kg/h has been commonly used although the efficacy of this dosing has not been proved.
- When the platelet count is 100,000/mm^3 or less, consider discontinuing heparin therapy and instituting alternative therapy, because the risk of heparin-induced thrombocytopenia (HIT) is greater after 5 days of therapy.

Duration of Heparin Therapy

- Deep-vein thrombosis: A minimum of 5–7 days; maintenance warfarin therapy can be instituted on day 1 or 2 of heparin therapy.
- Pulmonary embolus: 7–14 days; start warfarin therapy on day 5.
- Neonates may be treated for 10–14 days without warfarin.

Avoid acetylsalicylic acid (ASA, aspirin) or other antiplatelet drugs, arterial sticks, intramuscular injections during heparin therapy when possible.

Heparin Antidote

- If heparin needs to be discontinued, termination of the heparin infusion will be sufficient (because of the rapid clearance of heparin).
- If an immediate effect is required, protamine sulfate administration may be indicated.
- Following administration of IV protamine sulfate, neutralization occurs within 5 min.

The dose of protamine sulfate required to neutralize heparin is as follows:

Last dose of heparin	Protamine dose[a] (per 100 mg heparin)
<30 min	1 mg
30 min	0.5 mg
1	0.75 mg
>1 h	0.375 mg
>2 h	0.25 mg

[a]Maximum dose of protamine sulfate, 50 mg, to be administered in a concentration of 10 mg/ml at a rate not to exceed 5 mg/min. If administered faster, it may cause cardiovascular collapse; patients with known hypersensitivity reactions to fish and those who have received protamine-containing insulin or previous protamine therapy may be at risk for hypersensitivity reactions to protamine sulfate. Repeat aPTT 15 min after the administration of protamine sulfate.

Low-Molecular-Weight Heparin

The anticoagulant activities of LMWH are also mediated by catalysis of the natural inhibitor AT. LMWHs preferentially inhibit FXa over thrombin due to the decreased capacity to bind both AT and thrombin simultaneously for thrombin inhibition. Preparations used for children are enoxaparin (Lovenox, Rhone-Poulenc), reviparin (Clivarine, Knoll Pharma), and dalteparin (Pfizer, Inc.). Young infants have an accelerated clearance of LMWH compared to older children. The advantages of LMWH include reduced need for monitoring, reduced risk of HIT and lack of interference by other drugs or diet.

Indications for LMWH Therapy

- Neonates.
- Patients requiring anticoagulation and deemed to be at increased risk for hemorrhage.
- Patients in whom venous access for administration and monitoring of standard heparin therapy is difficult.

Dose (Enoxaparin) (Lovenox, Aventis)

Age (months)	Treatment[a] (dose)	Prophylactic (dose)
<2	1.5 mg/kg every 12 h SC	0.75 mg/kg every12 h SC
>2	1.0 mg/kg every12 h SC	0.5 mg/kg every12 h SC

[a]Maximum dose, 2.0 mg/kg every 12 h.
SC, subcutaneously.

Monitoring of LMWH Therapy

- Prior to the initiation of LMWH therapy, a CBC, including platelet count, PT, and aPTT, should be determined.
- Aspirin or other antiplatelet drugs should be avoided during the therapy.
- Intramuscular injections and arterial punctures should be avoided during the therapy.
- If the platelet count drops to $100,000/mm^3$ or less, HIT must be ruled out; it rarely occurs with LMWH therapy.
- The post-treatment anti-FXa level should be determined after three doses; thereafter, weekly monitoring should be sufficient 4–6 h after the SC administration of LMWH—recent ACCP guidelines recommend this. The therapeutic anti-FXa level is 0.5–1.0 units/ml and the prophylactic anti-FXa level is 0.1–0.3 units/ml.
- For long-term LMWH therapy, bone densitometry studies should be performed at baseline and then at 6-month intervals to assess for possible treatment-related osteoporosis.

Duration of LMWH Therapy

- LMWH is usually administered up to 3 months without warfarin.
- For extensive thrombus or pulmonary embolus administer LMWH for 7–14 days and warfarin should be started on day 5.
- Newborns may be treated for 10–14 days with LMWH alone.

Adjusting LMWH Dose

The dose of LMWH can be adjusted according to the anti-FXa level achieved:

Anti-FXa level (units/ml)[a]	Dose
<0.35	25% increase
<0.5	10% increase
0.5–1.0	No change
>1.0	20% decrease
>1.5	30% decrease
>2.0	Hold for 24 h

[a]Repeat anti-FXa level 4 h post next dose until 0.5–1.0 units/ml, then once weekly at 4 h postdose.

Antidote for LMWH

- Termination of LMWH is usually sufficient.
- When an immediate effect is required, 1 mg protamine per 100 units (1 mg) of LMWH may be given. It is generally not as effective as when used for unfractionated heparin.
- The protamine should be administered intravenously over a 10-min period. A rapid infusion of protamine may cause hypotension.

Heparin-Induced Thrombocytopenia

HIT, although an uncommon complication of heparin therapy, can cause significant morbidity and mortality. The incidence is about 3% in adult patients receiving unfractionated heparin. The incidence in pediatric patients is unknown. It usually begins 5—15 days after commencing heparin therapy (median 10 days) but can occur earlier in patients with prior exposure to heparin. An abrupt decrease of platelet count or a decrease of platelet count by half in 1—2 days should raise suspicion of HIT. Any type of heparin including LMWH or unfractionated heparin used to flush lines should be discontinued. Discontinuation of heparin will result in the platelet count returning to normal. In the setting of HIT associated with thrombotic events heparin must be discontinued and alternative anticoagulation therapy must be initiated until the antibody complex causing HIT is cleared. Alternative intravenous anticoagulation therapy may include the following.

Fondaparinux

- Indirect Xa inhibitor (high-affinity binding to AT which induces a conformational change and increases the ability of AT to inactivate FXa).
- Synthesized compound that is almost free of risk of contamination by animal proteins.
- Has no effect on bone metabolism.
- Long half-life allows for once-daily injections.
- Recently concluded FondaKIDS study determined a dose on 0.1 mg/kg once daily resulted in similar PK concentrations known to be efficacious in adults and with acceptable safety data.

Argatroban

- No loading dose.
- Direct thrombin inhibitor.
- Maintenance dose: Begin at 2 μg/kg/min continuous IV infusion (maximum 10 μg/kg/min).
- Monitor with aPTT beginning 2 h after the start of therapy; target aPTT to 1.5- to 3-fold greater than the pretreatment value and <100 s.
- When switching to oral anticoagulation with warfarin, do not give loading dose of warfarin due to additive effect on INR.
- Excretion: hepatic, reduce dose in patients with hepatic insufficiency.
- Clinical data for use in patients under age 18 is extremely limited.

Orgaran (Heparin Sulfate) (Danaproid Sodium)

- Loading dose: 30 units/kg.
- Maintenance dose: 1.2—4 units/kg/h.
- Anti-FXa activity can be monitored immediately following the bolus dose and every 4 h until a steady state is reached and then daily to maintain a therapeutic range of 0.4—0.8 units/ml.
- Orgaran is contraindicated in patients with severely impaired renal function.

Dabigatran (Pradaxa)

- Orally active direct thrombin inhibitor.
- Adult dose: 150 mg twice daily.
- Routine monitoring of coagulation tests not necessary. aPTT values >2.5 times over control may indicate excessive anticoagulation.

There are several newer anticoagulants with improved pharmacological properties that have been approved in adults and are awaiting study completion in pediatrics. These include bivalirudin, fondaparinux, and rivaroxaban. Direct thrombin inhibitors (argobatran, bivalirudin) have several potential advantages including more predictable pharmacokinetics, not subject to fluctuating levels of AT and not causing HIT. The most significant drawback is the lack of a suitable antidote, although preliminary studies show that rVIIa and aPCC can probably reverse the anticoagulant effects. The direct FXa inhibitor, rivaroxaban, is currently undergoing pharmacokinetic studies in pediatric patients. These agents cannot be recommended until they are approved for pediatric use.

Warfarin Therapy

Warfarin (4-hydroxycumarin) (Coumadin (Bristol-Myers, Squibb)) competitively inhibits vitamin K, an essential cofactor for the post-translational carboxylation of gamma glutamic acid residues on factors II, VII, IX, and X. Coumadin is the trade name for the most commonly available warfarin preparation.

- A loading dose of 0.2 mg/kg by mouth as a single daily dose (maximum initial dose, 10 mg) should be employed when the INR is less than 1.3. When it is greater than 1.3 reduce the loading dose to 0.1 mg/kg. For a patient having undergone a Fontan procedure or liver dysfunction, the daily dose should be reduced by 50%.
- The subsequent dose is age-dependent (infants having the highest—0.32 mg/kg and teenagers the lowest—0.09 mg/kg) and based on the INR response (see the following table). Before the INR was adopted, the PT test was used for many years to control oral anticoagulant therapy. The variability of different thromboplastin reagents and instruments used in the performance of the PT test made the transferability of results among laboratories difficult and impeded the development of universal guidelines for patient therapy.
- Generally coumadin is started after the patient is on a regular diet. To facilitate absorption the patient is instructed to take it on an empty stomach or 2 h after a meal, without any other medications. Usually green leafy vegetables which contain large amounts of vitamin K are discouraged but moderate and consistent intake will help prevent fluctuations in INR values and the need for frequent monitoring.

Warfarin Daily Loading Doses (Approximately 3–5 Days)

INR[a]	Warfarin loading doses
1.1–1.3	Repeat initial loading dose
1.1–1.9	50% of initial loading dose
1.1–3.0	50% of initial loading dose
1.1–1.5	25% of initial loading dose
>3.5	Hold until INR < 3.5, then restart at 50% less than previous dose

[a]*The international normalized ratio (INR) provides a standardized scale for monitoring patients who are receiving oral anticoagulant therapy. The INR is effectively the PT ratio upon which the patient would have been measured had the test been made using the primary World Health Organization international reference preparation (IRP). The INR is calculated by use of the international sensitivity index (ISI), which is established by the manufacturer for each lot of thromboplastin reagent using a specific instrument: $INR = (Patient\ PT/Normal\ PT)^{ISI}$.*

Warfarin Maintenance Doses for Long-Term Therapy

INR	Warfarin dose
1.1–1.4	Increase dose by 20%
1.5–1.9	Increase dose by 10%
2.0–3.0	No change
3.1–3.5	Decrease dose by 10%
>3.5	Hold until INR <3.5, then restart at 20% less than previous dose

- The warfarin loading period is approximately 3–5 days for most patients before a stable maintenance phase is achieved.
- Warfarin should be started on day 1 or day 2 of heparin therapy. Heparin should be continued for a minimum of 5 days' duration. When the target INR is greater than 2.0 for 2 consecutive days and at least 5 days of heparin are completed, heparin can be discontinued.
- For extensive deep vein thrombosis with or without pulmonary emboli warfarin should be started on day 5 of heparin therapy.
- The INR should be maintained between 2.0 and 3.0 for the vast majority of patients. Children with mechanical heart valves require an INR between 2.5 and 3.5.
- Once the patient has two INRs between 2.0 and 3.0 (or 2.5–3.5 for mechanical valves) obtained weekly, the INR determinations could be carried out every 2 weeks. If the INR remains stable the INR could then be determined once monthly.
- The INR should be monitored a minimum of once monthly.
- If the patient is receiving TPN, vitamin K should be removed from the amino acid solution before warfarin therapy begins.
- If the patient is receiving other medications (i.e., antibiotics which can affect warfarin), the loading dose may require adjustment.
- The INR should be obtained 5–7 days after initiating a new dose. The maintenance guidelines should be employed for making changes in dosage.
- Children with mechanical heart valves or repeated thromboembolic complications should receive warfarin indefinitely.
- Children with uncomplicated deep vein thrombosis and pulmonary emboli should receive warfarin for a minimum of 3 months.
- Children with a thrombotic event and a persistent, significant, underlying predisposing factor (e.g., continued presence of a central venous catheter, persistence of an antiphospholipid antibody) may be placed on low-dose warfarin (0.1 mg/kg) target INR—1.5–2.0 following 3 months of treatment with full-dose warfarin until the predisposing factor is no longer present.
- Vitamin K is the antidote for warfarin. The patient may also be given FFP or PCC infusions to reverse the effects of warfarin:
  If there is no active bleeding, vitamin K_1 0.5–2 mg SC can be given.
  If there is active bleeding, vitamin K_1 0.5–2 mg SC (not IM) plus FFP 20 cc/kg IV can be given.
  If bleeding is significant and life-threatening, vitamin K_1 5 mg IV (by slow infusion over 10–20 min because of the risk of anaphylactic shock) and PCCs 50 units/kg IV can be given. The FDA recently approved the PCC Kcentra for the urgent reversal of warfarin therapy in adult patients with an acute major bleed and for patients on warfarin needing urgent surgery or invasive procedures.
- If low-risk procedures have to be carried out the INR should be reduced to 1.5 or less prior to the procedures. Warfarin should be discontinued 72 h prior to any high-risk surgery.
- Where surgical procedures have to be carried out and where the risk of thrombosis is high and anticoagulant therapy cannot be reversed for even a short period of time, the following may be considered:
  Discontinue warfarin 72 h prior to surgery.
  Initiate heparin infusion therapy without a bolus at appropriate dose for age.
  If the INR is more than 1.5 at 12 h prior to surgery, low-dose vitamin K_1 0.5 mg SC should be given and the INR should be determined approximately 6 h later.
  Intravenous heparin infusion should be discontinued approximately 6 h prior to surgery. Preoperative PT and a PTT should be within normal limits.
  Heparin should be resumed 8 h following surgery.
  If the patient develops any signs of bleeding, heparin infusion should be immediately discontinued.
  Oral warfarin should be resumed on the second day postsurgery.
  Heparin infusion should be discontinued when a therapeutic INR is reached.
- Where the risk of thrombosis is low and there is no thrombotic event for several weeks:
  Warfarin should be discontinued 72 h prior to surgery.
  Warfarin maintenance should be initiated on the day following surgery.
- Both diet and co-administration of certain drugs can have a marked effect on the magnitude of warfarin action (see Table 15.31).

TABLE 15.31   Effect of Drugs on Warfarin Response

MEDICATIONS THAT POTENTIATE THE EFFECT OF WARFARIN

Acetaminophen	Isoniazid
Acetohexamide	Mefenamic acid
Allopurinol	Methimazole
Androgenic and anabolic steroids	Methotrexate
α-Methyldopa	Methylphenidate
Antibiotics that disrupt intestinal flora (tetracyclines, streptomycin erythromycin, kanamycin, nalidixic acid, neomycin)	Nalidixic acid
	Nortriptyline
	Oxyphenbutazone
Cephaloridine	p-Aminosalicyclic acid
Chloramphenicol	Paromomycin
Chlorpromazine	Phenylbutazone
Chlorpropamide	Phenytoin
Chloral hydrate	Phenyramidol
Cimetidine	Propylthiouracil
Clofibrate	Quinidine
Diazoxide	Salicylate
Disulfiram	Sulfinpyrazone
Ethacrynic acid	Sulfonamides
Glucagon	Thyroid hormone
Guanethidine indomethacin	Tolbutamide

MEDICATIONS THAT REDUCE THE EFFECT OF WARFARIN

Antipyrine	Glutethimide
Barbiturates	Griseofulvin
Carbamazepine	Haloperidol
Chlorthalidone	Oral contraceptives
Cholestyramine	Phenobarbital
Digitalis	Prednisone
Ethanol	All vitamin preparations containing vitamin K
Ethchlorvynol	

Antiplatelet Therapy

The two most commonly used antiplatelet agents for children are aspirin and dipyridamole. Aspirin acetylates the enzyme cyclo-oxygenase and thereby interferes with the production of thromboxane A2 and platelet aggregation. Dipyridamole interferes with platelet function by increasing the cellular concentration of adenosine 3,5-monophosphate (cyclic AMP). This latter effect is mediated by inhibition of cyclic nucleotide phosphodiesterase and/or by blockade of uptake of adenosine, which acts at A2, receptors for adenosine to stimulate platelet adenylcyclase.

Antiplatelet agents are used in the following conditions:

- Cardiac disorders:
    mechanical prosthetic heart valves.
    Blalock—Tausig shunts.
    endovascular shunts.
- Cardiovascular events.
- Kawasaki disease.

Dosage:

Aspirin: 1—5 mg/kg/day.
Dipyridamole (2—5 mg/kg/day).

Thrombolytic Therapy

In contrast to the anticoagulants heparin and warfarin, which function to prevent fibrin clot formation, the thrombolytic agents act to dissolve established thrombi by converting endogenous plasminogen to plasmin, which can lyse an existing thrombus. Thrombolytic therapy should be considered for arterial thrombi using t-PA, UK, or streptokinase (SK).

Tissue Plasminogen Activator

t-PA infusion should be given at a rate of 0.5 mg/kg/h IV for 6 h (standard dose t-PA).

- Heparin should be given (20 units/kg/h) during t-PA infusion if the patient is not already on heparin.
- Following 6 h of t-PA infusion and if there is no response to treatment, the plasminogen level should be determined. If the plasminogen level is low, FFP 20 cc/kg IV q8h should be administered. A repeat infusion of t-PA may be considered.
- Given the risk of bleeding with standard dose t-PA low-dose t-PA (0.03—0.06 mg/kg/h up to 96 h has been used with equivalent efficacy and lower bleeding risk has been utilized more recently. Continue heparin at 20 U/kg/h along with t-PA.
- No arterial sticks or intramuscular injections during t-PA administration.

Urokinase

UK should be administered in a loading dose of 4000 units/kg over 10 min, followed by 4000 units/kg/h for 6 h.

- Heparin should be given (20 units/kg/h) during UK infusion if the patient is not already on heparin.
- Following 6 h of UK infusion and if there is no response to treatment, the plasminogen level should be determined. If the plasminogen level is low, FPP 20 cc/kg IV q8h should be administered. A repeat infusion of UK may be considered.
- SK can be administered when t-PA and UK are not available.

Streptokinase

SK should not be administered if it was used previously or if the plasminogen level is low (i.e., newborns). The loading dose consists of 4000 units/kg (maximum 250,000 units) over 10 min, followed by 2000 units/kg/h for 6 h.

- Heparin should be given (20 units/kg/h) during SK infusion if the patient is not already on heparin.
- If there is no response to treatment, the plasminogen level should be determined. If the plasminogen level is low, FPP 20 cc/kg IV q8h should be given. UK or t-PA should be utilized (not SK) for further thrombolytic therapy.
- Patients must be premedicated with Tylenol and Benadryl before SK (repeat every 4—6 h).
- Patients should not receive more than one course of SK because of the potential for allergic reactions. An anaphylactic reaction can occur in 1—2% of patients receiving SK. In the event of an anaphylactic reaction:
    Discontinue SK immediately.
    Administer epinephrine, steroids, and antihistamines.

Monitoring Response of Thrombolytic Therapy

- Obtain the PT and aPTT every 6 h.
- Determine the fibrinogen level and/or FDPs or D-dimer, every 6 h.
- Determine the plasminogen level at the end of the 6-h infusion if there is no response or prior to proceeding to another course of therapy.
- The fibrinogen concentration may decrease by at least 20–50%; and the fibrinogen concentration must be maintained at approximately 100 mg/dl by cryoprecipitate infusions (1 unit/5 kg).
- When the fibrinogen concentration is less than 100 mg/dl and the patient is still receiving infusion of SK, UK, or t-PA, the dose of the thrombolytic agent should be decreased by 25%.
- The platelet count should be maintained at $100,000/m^3$ or higher. Dipstick every void and test every stool for occult blood.
- Six hours following thrombolytic therapy, heparin therapy may be sufficient for 24 h before reinstituting thrombolytic therapy. There may be ongoing thrombolysis even in the absence of continued administration of the thrombolytic agent.

Complications of Thrombolytic Therapy

Minor bleeding may occur in up to 50% of patients (i.e., oozing from a wound or puncture site). Supportive care and application of local pressure may be sufficient.

- If severe bleeding occurs:
  The infusion of the thrombolytic agent should be terminated.
  Cryoprecipitate should be administered (usual dose of 1 unit/5 kg).
- When life-threatening bleeding occurs:
  The infusion of the thrombolytic agent should be terminated.
  The fibrinolytic process can be reversed by infusing Amicar 100 mg/kg (maximum, 5 g) bolus, then 30 mg/kg/h (maximum, 1.25 g/h) until bleeding stops (maximum, $18 g/m^2/day$).
  Protamine sulfate may be required to reverse the heparin effect.

The table below lists the management of blocked central venous catheters (CVC) using t-PA.

Weight	Single-lumen CVC	Double-lumen CVC	Subcutaneous CVC (Mediport)
<10 kg	0.5 mg t-PA diluted in 0.9% NaCl to volume required to fill line	0.5 mg t-PA diluted in 0.9% NaCl per lumen to fill volume of line. Treat one lumen at a time	0.5 mg t-PA diluted with 0.9% NaCl to 3 ml
>10 kg	1.0 mg t-PA in 1.0 ml 0.9% NaCl. Use amount required to fill volume of line, to maximum of 2 ml (2 mg t-PA)	1 mg t-PA in 1.0 ml 0.9% NAC1. Use amount required to fill volume of line, to a maximum of 2 ml (2 mg t-PA per lumen). Treat one lumen at a time	2.0 mg t-PA diluted with 0.9% NaCl to 3 ml

ANTITHROMBOTIC THERAPY IN SPECIAL CONDITIONS

Central Venous Access Devices/Umbilical Venous Catheters

- Start anticoagulation with either LMWH or heparin followed by LMWH.
- Remove line after 3–5 days of therapeutic anticoagulation.
- Duration of anticoagulation between 6 weeks and 3 months.
- If venous access is critical and it is decided to leave the line in place, after therapeutic anticoagulation duration, consider prophylactic anticoagulation till the line can be removed.

Valve Replacement

Mechanical Valve Replacement

Heparin:

- Heparinization starting at 48 h postoperatively.
- For dosage see Heparin normogram.
- Heparin should be stopped for 2 h before intracardiac lines are removed.
- Chest tube removal or insertion is not a contraindication for heparin.

Warfarin:

- Warfarin begins with oral intake and is continued indefinitely.
- See warfarin dose adjustment schedule.
- Aim for an INR of 2.5–3.5.

Aortic Valve Replacement

- Low-dose ASA begins with oral intake (3–5 mg/kg/day) and is continued for 3 months.

Mitral and Tricuspid Valve Replacement

- Atrial fibrillation or has proven intra-atrial thrombus, treatment as per mechanical valve replacement (INR of 2.5–3.5).
- If normal sinus rhythm, anticoagulation will be with warfarin for 3 months (INR of 2.0–3.0), and low-dose ASA (3–5 mg/kg/day), indefinitely.

Fontan Procedure

Heparin:

- No heparin loading dose.
- Low-dose heparin infusion of 10 units/kg/h starting 48 h postoperatively.
- Heparin is stopped for 2 h before intracardiac lines are removed.
- Chest tube removal or insertion is not a contraindication for heparin.

Warfarin:

- Warfarin begins with oral intake and is continued for 3 months.
- Patients after a Fontan procedure are very sensitive to oral anticoagulants. Loading doses of warfarin should be decreased.
- Warfarin dosing should be adjusted according to the warfarin protocol (INR of dose 2.0–3.0).

Aspirin:

- Low-dose ASA (3–5 mg/kg/day) starts with oral intake and continues indefinitely.

Blalock–Taussig Shunts

Heparin:

- Loading dose of heparin: 75 units/kg over 10 min starts intraoperatively.
- Maintenance dose: 28 units/kg/h (≤1 year old): 20 units/kg/h (>1 year old).

Aspirin:

- Low-dose ASA (3–5 mg/kg/day) starts with oral intake and continues indefinitely.

Acute Arterial Infarct with Evidence of Dissection in Cerebral or Carotid Arteries

Heparin:

- Loading dose of heparin followed by maintenance dose (see heparin normogram). Can switch to LMWH once stable to be continued for minimum of 6 weeks.

  Warfarin:

- Warfarin therapy for a minimum of 6 weeks (see adjustment schedule).

Idiopathic Arterial Infarct Without Evidence of Dissection

- Start with UFH, LMWH, or aspirin as initial therapy till dissection and embolic causes have been excluded.
- Daily aspirin prophylaxis (3–5 mg/kg/day) for 2 years.
- Consider switching to clopidogrel or use of anticoagulants like warfarin or LMWH in cases of recurrent stroke or transient ischemic attacks.

Arterial Infarct with Moyamoya Syndrome

- Low-dose ASA (3–5 mg/kg/day) starts at diagnosis and continues indefinitely.
- Patient should be referred to a center equipped to perform revascularization procedure which leads to improvement in symptoms in over 90% of patients.

Transient Ischemic Attacks

- Anticoagulation with heparin or LMWH for 5 days.
- Low-dose ASA (3.5 mg/kg/day) starts at diagnosis and continues indefinitely.

Acute Stroke Without Hemorrhage

- Anticoagulation with heparin or LMWH for 5 days.
- Subsequent anticoagulation will be according to the underlying etiology, depending on the presence or absence of dissection.

Cerebral Sinovenous Thrombosis

- Anticoagulation with heparin or LMWH for total duration of 6 weeks to 3 months in newborns. In infants or older children heparin or LMWH for 5 days followed by warfarin for 3 months (for dosage see heparin nomogram, LMWH dose adjustment schedule, and warfarin dose adjustment schedule).
- Thrombolysis, thrombectomy, or surgical decompression should be restricted to patients with severe cerebral venous sinus thrombosis with no improvement on anticoagulation.

Kawasaki Disease

- High-dose aspirin 80–100 mg/kg/day for 14 days during acute phase followed by 1–5 mg/kg/day for 6–8 weeks.
- IVIG 2 g/kg within 10 days of onset of symptoms.
- With moderate to giant coronary aneurysms as sequalae of Kawasaki disease, anticoagulation with warfarin in addition to low-dose aspirin.
- Thrombolysis or surgical thrombectomy for acute coronary thrombosis.

Further Reading and References

Acharya, S.S., 2013. Rare bleeding disorders in children: identification and primary care management. Pediatrics 132, 882–892.

Acharya, S.S., DiMichele, D.M., 2006. Management of FVIII inhibitors, Francis, C.W., Blumberg, N., Heale, J. (Eds.), Best Practice & Research. Clin. Haemat 19 (1), 51–66.

Andrew, M., Vegh, P., Johnston, M., Bowker, J., Ofosu, F., Mitchell, L., 1992. Maturation of the hemostatic system during childhood. Blood 80, 1998–2005.

Baker Jr, W.F., 2003. Thrombolytic therapy clinical application. Hematol. Oncol. Clin. N. Am. 17, 283–311.

Barnard, T., Goldenberg, N.A., 2008. Pediatric arterial ischemic stroke in children. Pedaitr. Clin. North Am. 55, 323–338.

Bick, R., 1995. Disseminated intravascular coagulation: objective criteria for clinical and laboratory diagnosis and assessment of therapeutic response. Clin. Appl. Thrombosis/Hemostasis 1, 3–23.

Bick, R.L., 2003. Antiphospholipid thrombosis syndrome. Hematol. Oncol. Clin. N. Am. 17, 115–147.

Bick, R.L., 2003. Prothrombin G 20210A mutation, heparin Cofactor II, protein C and protein S defects. Hematol. Oncol. Clin. N. Am. 17, 9–36.

Carpenter, S.L., et al., 2013. Evaluating for suspected child abuse conditions that predispose to bleeding. Pediatrics 131, e1357.

Crowther, M.A., Kelton, J.G., 2003. Congenital thrombophilic states associated with venous thrombosis. A qualitative overview and proposed classification system. Ann. Intern. Med. 138, 128–134.

DiMichele, D.M., Kroner, B.L., North American Immune Tolerance Study Group, 2002. The North American Immune Tolerance Registry: practices, outcomes, outcome predictors. Thromb. Haemost. 87, 52–57.

Ewing, N.P., Sanders, N.L., Dietrich, S.L., et al., 1988. Induction of immune tolerance to factor VIII in hemophilia A patients with inhibitors. JAMA 259, 65–68.

Freiburghaus, C., Berntorp, E., Ekman, M., et al., 1999. Tolerance induction using the Malmo treatment model 1982–1995. Haemophilia 5, 32–39.

Hay, C.R.M., DiMichele, D.M., 2012. The principal results of the International Immune Tolerance Study: a randomized dose comparison. Blood 119, 1335–1344.

Hoffman, M., Monroe III, D.M., 2001. A cell-based model of hemostasis. Thromb. Haemost. 85, 958–965.

Kearon, C., Kahn, S.R., Agnelli, G., Goldhaber, S., Raskob, G.E., Comerota, A.J., 2008. Antithrombotic therapy for venous thrombo-embolic disease. Chest 133, 454S–545S.

Keeling, D., Tait, C., Makris, M., 2008. Guideline on the selection and use of therapeutic products to treat haemophilia and other hereditary bleeding disorders: A UKHCDO guideline. Haemophilia 14, 671–684.

Kwann, H.C., Nabhan, C., 2003. Hereditary and acquired defects in the fibrinolytic system associated with thrombosis. Hematol. Oncol. Clin. N. Am. 17, 103–114.

Lenk, H., 2000. The German registry of immune tolerance treatment in hemophilia – 1999 update. Haematologica 85, 45–47.

Miyakis, S., Lockshin, M.D., Atsumi, T., et al., 2006. International consensus statement on an update of the classification criteria for definite antiphospholipid syndrome (APS). J. Thromb. Haemost. 4 (2), 295–306.

Van Leeuwen, E.F., Mauser-Bunschoten, E.P., van Dijken, P.J., 1986. Disappearance of factor VIII:C antibodies in patients with haemophilia A upon frequent administration of factor VIII in intermediate or low dose. Br. J. Haematol. 64, 291–297.

Levine, J.S., Branch, W., Rauch, J., 2002. Antiphospholipid syndrome. New Engl. J. Med. 346, 752–763.

Manco-Johnson, M.J., et al., 2007. Prophylaxis versus episodic treatment to prevent joint disease in boys with severe hemophilia. N. Engl. J. Med. 357 (6), 535–544.

Mannucci, P., 2004. Treatment of von Willebrand Disease. N. Eng. J. Med. 351, 683–694.

Mannucci, P.M., 1988. Desmopressin: a nontransfusional form of treatment for congenital and acquired bleeding disorders. Blood 72, 1449–1455.

Mariani, G., Kroner, B., for the Immune Tolerance Study Group, 2001. Immune tolerance in hemophilia with inhibitors: predictors of success. Haematologica 86, 1186–1193.

Meinardi, J.R., Middeldorp, S., De Kam, P.J., et al., 2002. The incidence of recurrent venous thromboembolism in carriers of factor V Leiden is related to concomitant thrombophilic disorders. Brit. J. Hematol. 116, 625–633.

Monagle, P., Goldenberg, N., Journeycake, J.M., 2012. Antithrombotic therapy in Neonates and Children. Chest 141 (2), e737S–e801S.

Montgomery, R.R., Gill, J.C., Scott, J.P., 1998. Hemophilia and von Willebrand disease. In: Nathan, D.G., Orkin, S.H. (Eds.), Nathan and Oski's Hematology of Infancy and Childhood, fifth ed. Saunders, Philadelphia, PA, pp. 1631–1675.

Nathwani, A.C., et al., 2011. Adenovirus Associated virus vector-mediated gene transfer in Hemophilia B. N. Engl. J. Med. 365 (25), 2358–2365.

Nichols, W.L., et al., 2008. von Willebrand disease (VWD): evidence-based diagnosis and management guidelines, the National Heart, Lung, and Blood Institute (NHLBI) Expert Panel report (USA) 1. Haemophilia 14 (2), 171–232.

Oldenburg, J., Schwaab, R., Brackmann, H.H., 1999. Induction of immune tolerance in haemophilia A inhibitor patients by the "Bonn Protocol": predictive parameter for therapy duration and outcome. Vox Sanguinis 77 (Suppl. 1), 49–54.

Peyvandi, F., Palla, R., Mengatti, M., et al., 2012. Coagulation factor activity and clinical bleeding severity in rare bleeding disorders: results from the Europeon Network of Rare Bleeding Disorders. J. Thromb. Haemostat. 10 (4), 615–621.

Roach, E.S., Golomb, M.R., Adama, R., Biller, J., et al., 2008. Management of stroke in infants and children. Stroke 39, 2644–2691.

Selighsohn, U., Lubertsky, A., 2001. Genetic susceptibility to venous thrombosis. N. Eng. J. Med. 344, 1222–1231.

The diagnosis and management of factor VIII and IX inhibitors: a guideline from the UK Haemophilia Centre Doctor's Organization (UKHCDO). Br. J. Haematol. 2000, 111, 78–90.

Wang, M., et al., 2003. Low-dose tissue plasminogen activator thrombolysis in children. J. Pediatr. Hematolol. Oncolol. 25 (5), 379–386.

16

Lymphoproliferative Disorders

David T. Teachey

LYMPHOPROLIFERATIVE DISORDERS

Lymphoproliferative disorders (LPDs) manifest with uncontrolled hyperplasia of lymphoid tissues (lymph nodes, spleen, bone marrow, liver). They are a heterogeneous group of diseases that range from reactive polyclonal hyperplasia (immunologic disorders) to true monoclonal (malignant) diseases.

Angioimmunoblastic Lymphadenopathy with Dysproteinemia

Manifestations of this clinicopathological syndrome include:

- Generalized lymphadenopathy (80%).
- Hepatosplenomegaly (70%).
- Fever (70%), malaise, weight loss, polyarthralgia.
- Quantitative changes in serum proteins (polyclonal hypergammaglobulinemia, 70%); hypocomplementemia.
- Autoantibodies; circulating immune complexes, antismooth muscle antibody.
- Rashes.
- Pulmonary infiltrates, pleural effusions.
- Thrombocytopenia.
- Hemolytic anemia (often direct antiglobulin test (Coombs') positive).

Diagnosis

The lymph node shows architectural effacement, absence of germinal center, arborization of postcapillary venules, and a polymorphous infiltrate that includes immunoblasts and plasma cells. Immunoblasts are CD4-positive. Lymph node cytogenetic studies have shown non-random abnormalities, including +3, 14q +, and del (8) (p21). Angioimmunoblastic lymphadenopathy with dysproteinemia (AILD) was previously considered to be a benign LPD with the potential to transform into lymphoma. Since it is a monoclonal disorder, it is now recognized as a type of angioimmunoblastic peripheral T-cell lymphoma.

Prognosis

The prognosis is poor. Death usually results from overwhelming infection. The median overall survival is 1.5 years.

Treatment of AILD-Type Lymphoma

Chemotherapy: Adriamycin-containing regimens such as CHOP (see Table 16.7) have been used. Various combinations including fludarabine, lenalidomide, and bortezomib are often used in salvage regimens with variable success.

334

Small Lymphocytic Infiltrates of the Orbit and Conjunctiva (Ocular Adnexal Lymphoid Proliferation, Pseudolymphoma, Benign Lymphoma, Atypical Lymphocytic Infiltrates)

Lymphocytic infiltrates of the orbit and conjunctiva may be divided into three histologic groups:

- Monomorphous infiltrates of clearly atypical lymphocytes.
- Infiltrates composed of small lymphocytes with minimal or no cytologic atypia.
- Benign inflammatory pseudotumor or reactive follicular hyperplasia.

On the basis of immunophenotypic criteria, they can be divided into two classes:

- Infiltrates with monotypic immunoglobulin expression.
- Infiltrates with polytypic immunoglobulin expression.

For the localized small lymphocytic infiltrates, monotypic (monoclonal) immunoglobulin expression confers a 50% risk of dissemination. The initial immunophenotypic (monoclonal or polyclonal) and molecular studies of various histologic groups fail to correlate with the eventual outcome of these cases because the initial polyclonal tumors may become monoclonal.

All patients presenting with small lymphocyte infiltrates of the orbit and conjunctiva should have a systemic evaluation with serum chemistries, blood counts, and appropriate imaging studies at initial diagnosis and every 6 months for 5 years thereafter.

For localized disease, local radiotherapy is commonly used, regardless of histologic grading. DNA from *Chlamydia psittaci* has been isolated in a large percentage of biopsies and treatment with doxycyline may be efficacious.

Angiocentric Immunolymphoproliferative Disorders

These are a collection of entities classified as peripheral T-cell disorders and include lymphomatoid granulomatosis, midline granuloma, and post-malignancy angiocentric immunolymphoproliferative (AIL) lymphoma.

There are three grades of AIL disorders:

- *Grade I*: Polymorphic infiltrates with minimal necrosis, few large atypical lymphoid cells, and small lymphocytes lacking nuclear irregularities.
- *Grade II*: Cytologic atypia of small lymphocytes, scattered large atypical lymphoid cells, and intermediate amount of necrosis.
- *Grade III*: Lymphoma, either diffuse, mixed, large cell, or immunoblastic, with prominent necrosis.

The cellular origin of AIL lymphoma remains uncertain because of the following findings:

- Immunophenotype of T-cells (CD2 + , CD3 + , CD4 ± , CD5 ± , CD7 ±).
- Absence of clonal rearrangements of T-cell receptors (TCRs).
- Expression of natural killer (NK) cell antigens (CD16 + , CD56 + , CD57 +).

AIL lymphoma has been postulated to be a clonal process induced by Epstein—Barr virus (EBV) infection of T lymphocytes.

Clinical Features

Lymphomatoid Granulomatosis

This is a systemic disease in which the lungs are typically involved. Patients may present with cough, dyspnea, and chest pain, or they may be asymptomatic and discovered incidentally on a chest radiograph. Chest radiographs may show bilateral nodules, consolidation, diffuse bilateral reticulonodular infiltrates, lymphadenopathy (mediastinal and/or hilar), and/or pleural effusion. Purplish skin nodules occur and may undergo spontaneous central necrosis and ulceration. The kidneys, central nervous system (CNS), skeletal muscles, nasopharynx, or peripheral nerves may also be involved.

Midline Lethal Granuloma

This presents with a progressive necrotizing and destructive process involving the upper airways. There may or may not be a tumor mass. The most common sites of the disease are the nasal fossa, nasal septum, nasopharynx, palate, and adjacent soft tissue or bony structures. There is often a history of longstanding sinusitis with purulent and foul-smelling nasal discharge.

TABLE 16.1 Relationship Between Histologic Types and Clinical Features in Castleman Disease

Histologic type (frequency)	Disease sites	Symptoms
Hyaline vascular type (80%)	Solitary lymph node 1.5–16 cm or chain of lymph nodes; two-thirds in mediastinum; sometimes other sites such as peripheral lymph nodes, abdomen and pelvis	>90% Asymptomatic; pressure effects referable to the location of the mass may be present; 5–10% patients have constitutional symptoms
Plasma cell variety (20%)	• Two types: 　• Unicentric (single node or chain) 　• Multicentric (multiple nodal groups) splenomegaly may be present	• Associated with systemic symptoms, such as fever, sweats, arthralgia, rashes, growth retardation, peripheral neuropathy, nephrotic syndrome • Laboratory data: 　• Hypergammaglobulinemia, microcytic anemia, raised ESR, amyloidosis; plasma cell proliferation usually polyclonal; increased interleukin-6 production

ESR, erythrocyte sedimentation rate.

Postmalignancy Angiocentric Immunolymphoproliferative Lymphoma

This rarely occurs in children and most commonly occurs after a history of acute lymphoblastic leukemia (ALL). The interval between the remission of ALL to the diagnosis of AIL lymphoma ranges from 1 month to 4–5 years. The prognosis is poor.

Treatment

There is no standard treatment; the majority of patients are treated with chemotherapy based on adult non-Hodgkin lymphoma protocols and rituximab. The most commonly used agents are cyclophosphamide, prednisone, adriamycin, and vincristine.

Castleman Disease (Angiofollicular Lymph Node Hyperplasia, Benign Giant Lymph Node Hyperplasia, Angiomatous Lymphoid Hamartoma)

Castleman disease is characterized by an accumulation of nonmalignant lymphoid tissue interspersed with plasma cells and blood vessels.

Vascular hyperplasia has been attributed to a humoral vasoproliferative factor.

Table 16.1 shows the relationship between histologic types and clinical features in Castleman disease.

The production of interleukin-6 (IL-6) by B-cells in the germinal centers of hyperplastic lymph nodes in Castleman disease plays a central role in inducing the variety of symptoms in this disease. In localized disease, human herpes virus 8 (HHV-8) DNA sequences have been detected in CD19 B-cells. In multicentric disease, HHV-8 sequences have been detected in CD19 B-cells and CD2 T-cells.

Clinical Features

Unicentric Hyaline Vascular Variant

- Single lymph node or chain: cervical and mediastinal nodes most common.
- Five to ten percent of patients have constitutional symptoms, but most are asymptomatic.
- May have reactive diffuse shotty nonpathologic lymphadenopathy.
- Associated with thrombotic thrombocytopenic purpura.

Unicentric Plasma Cell Variant

- Single lymph node or chain: abdominal nodes most common.
- Associated with increased IL-6.
- May have reactive diffuse shotty nonpathologic lymphadenopathy and splenomegaly.
- Frequently have constitutional symptoms, hematologic abnormalities (anemia, thrombocytopenia, lymphocytosis), hypoalbuminemia, and hypergammaglobulinemia.

Multicentric Plasma Cell Variant

- Similar clinical presentation as unicentric plasma cell variant; however, it involves multiple lymph nodes and chains. Hepatosplenomegaly is more common.
- May also have neurologic manifestations, peripheral edema, ascites, and pleural effusions.
- Associated with HIV and HHV-8 infections.

Prognosis

Localized disease: Excellent.

Multicentric disease: Poor. A minority with multicentric disease progress rapidly and develop hemolytic anemia and fatal intercurrent infection. Some develop non-Hodgkin lymphoma or Kaposi sarcoma.

Treatment

Localized disease: Surgical resection is curative. If complete resection is not possible, prednisone may be used to reduce size to enable complete surgical resection. Local radiation may be used for nonresectable disease.

Multicentric disease:

- Glucocorticoids alone or with vincristine.
- Monoclonal antibodies: anti-IL-6, anti-IL-6 receptor, anti-IL-1 receptor, and anti-CD20 antibodies have been used successfully.
- Multiagent chemotherapy (see Table 16.7) with rituximab is considered the standard of care.

EBV-Associated LPDs in Immunocompromised Individuals

EBV is a gamma herpes virus, a subfamily distinguished by its limited tissue tropism and latency. EBV enters via the oropharyngeal route and infects resting B lymphocytes through the interaction between the EBV viral envelope glycoprotein (gp350/220) and the C3d complement receptor (CD21). Infected B-cells induce immunologic response of both virus-specific and −non-specific T-cells during primary infection which leads to regression of the majority of infected B-cells. However, the virus persists in its latent state lifelong, by its continued presence in a small number of B-cells, which express latent membrane protein 2A and small EBV-encoded RNA 1 and 2 (EBERs 1 and 2). Table 16.2 lists the functions of EBV latent antigens.

EBV Antigens Associated with the Lytic Cycle

- Early antigen.
- Viral capsid antigen (VCA).

During acute infectious mononucleosis (AIM) humoral responses are directed against both lytic and latent proteins. Immunoglobulin M (IgM) antibodies to cellular proteins and heterophile antigens are also present during AIM.

Detectable levels of IgG antibodies to VCA and EBNA1 persist through lifetime.

In conjunction with humoral response, the CD4 and CD8 T-cell-mediated responses also play an important role in controlling the EBV infection by identifying and destroying latently infected cells.

TABLE 16.2 Functions of Epstein−Barr Virus Latent Antigens

EBV antigen	Function
LMP1	Prevention of apoptosis (through NFκB and MAPK, induction of antiapoptotic proteins AP-1, bcl-2) and immortalization of B-cells, increase in tumor invasiveness (by induction of matrix metalloproteinase)
LMP 2A	Inhibits lysis of B-cells
Lytic antigens	vIL-10 and BHRF1 facilitate lysis of cells and induce survival and proliferation of nearby, latently infected B-cells
EBNA 1	Maintenance and replication of the EBV episome in latency
EBNA 2	Has an immortalizing function

MAPK, mitogen activated protein kinase; NFκB, nuclear factor κ B; vIL-10, viral homolog of cellular interleukin-10; BHRF1, viral homolog of cellular bcl-2; bcl-2, B-cells lymphoma protein; EBV, Epstein−Barr virus.

Cellular Responses in the Control of EBV Infection

The lymphocytosis during AIM consists of activated CD4 and CD8 T lymphocytes. They are specific for EBV lytic and latent proteins.

Cell	Reactivity (function)
CD8 + and CD4 + T-cells	Directed toward latent phase proteins: EBNA3, EBNA4, LMP1, EBNA6
Natural killer cells	Lysis of virus-containing cells during lytic phase

Reasons for Persistence of Latency in EBV Infection

- Paucity of production of neutralizing antibodies to the gp 350/220 component of the viral glycoprotein, a ligand for interaction with C3d on B-cells.
- Absence of cytotoxic T lymphocytes (CTL) response to EBNA1 which is primarily responsible for replication of and maintenance of viral genome during host cell division.

A state of balance between the control of EBV infection in latent state and host's immune response is achieved in a healthy host after the control of AIM phase. However, this state of balance is offset if the host is immunocompromised. Under this situation, the EBV can proliferate and can cause chronic active EBV infection, hemophagocytic lymphohistiocytosis (HLH), LPD, lymphomas, and other rare complications.

The following immunodeficiencies are associated with the development of LPDs and often require hematopoietic stem cell transplant for cure.

- Inherited non-EBV specific immunodeficiencies.
    Patients with the following immunodeficiencies have difficulty fighting EBV, as well as other infections and viruses:
- Ataxia telangiectasia.
- Wiscott−Aldrich syndrome.
- Common variable immunodeficiency (CVID).
- Severe combined immunodeficiency.
- Antibody deficiency syndromes such as hypergammaglobulinemia, hyper IgM syndrome, X-linked agammaglobulinemia, IgA and IgG subclass deficiency.
- Bloom syndrome.
- Chédiak−Higashi syndrome.
- Inherited EBV-specific immunodeficiencies.
- X-linked lymphoproliferative (XLP).
- XMEN: X-linked magnesium deficiency with EBV infection and neoplasia. This is a rare immune deficiency caused by low CD4 T cells and uncontrolled chronic EBV infection. Caused by mutations in magnesium transporter 1 (MagT1).
- ITK deficiency: A rare inherited autosomal recessive disorder caused by mutations in ITK, leading to severe immune dysregulation following EBV infection.
- X-linked inhibitor of apoptosis protein (XIAP) deficiency: A rare disorder caused by mutations in BIRC4 and originally thought to be a form of XLP in patients without SH2D1A mutations because the initial reported cohorts only included patients with EBV HLH. However, it is now recognized to include a wider phenotype that can manifest with splenomegaly, uveitis, inflammatory bowel disease, EBV LPD, antibody deficiency, skin infections, and/or EBV HLH.

Biological Factors of Significance Involved in the Pathogenesis of LPDs in this Population

- EBV, which causes B-cell proliferation.
- Imbalanced production of cytokines (e.g., predominance of IL-4 and IL-6 activates B-cells, i.e., humoral arm of immune system).
- Genetic defects resulting in ineffective or aberrant rearrangement of TCR and immunoglobulin genes.

TABLE 16.3   Classification of Post-Transplantation Lymphoproliferative Disorders

1. Plasmacytic hyperplasia
   a. Commonly arise in the oropharynx or lymph nodes
   b. Nearly always polyclonal
   c. Usually contains multiple clones of EBV
   d. No oncogenes and tumor suppressor gene alterations
2. Polymorphic B-cell hyperplasia and polymorphic B-cell lymphoma
   a. Can involve lymph nodes or extranodal sites
   b. Nearly always monoclonal
   c. Usually contains a single clone of EBV
   d. No oncogenes or tumor suppressor gene alterations
3. Immunoblastic lymphoma or multiple myeloma
   a. Present with widely disseminated disease
   b. Always monoclonal
   c. Contains a single clone of EBV
   d. Contains alterations of one or more oncogenes or tumor suppressor genes (N-*ras*, p53, c-*myc*)

Magrath et al. (1997), with permission.

Iatrogenically Induced Immunodeficiencies

- *Organ transplantation recipients*: Long-term treatment with immunosuppressive drugs, including the calcineurin inhibitors cyclosporine and tacrolimus, increases the risk of infections with EBV. EBV status of the organ transplantation recipient plays a significant role. If the patient is EBV-seronegative and the donor is seropositive, then the risk of developing clinical infectious mononucleosis followed by LPD increases remarkably. The incidence is higher in children than adults and it is more likely to present with clinical features of fulminating infectious mononucleosis. Many children have not been previously exposed to EBV and, thus, are more likely to develop a primary EBV infection. Post-transplantation LPDs (PT-LPDs) after SOT are of B-cell origin and are associated with EBV infection in most patients. A few cases of T-cell PT-LPDs have been reported.

- *Hematopoietic stem cell transplantation (HSCT) recipients*: HSCT differs in the following aspects: The patient receives a new immune system from a donor who is either a full histocompatibility locus antigen (HLA) match or a partial HLA match. The patient's immune system is ablated prior to HSCT. HSCT may require T-cell depletion of the donor's bone marrow.

  The complete recovery of the donor's immune system is delayed by the use of cyclosporine, methotrexate, prednisone, and calcineurin inhibitors, which are used to prevent rejection of graft and graft-versus-host disease (GVHD). It takes 6 months to 2 years for immunologic recovery to occur. LPD after HSCT is also associated with EBV-infected B-cells. The incidence of LPD in standard matched HSCT is low, but it is 6–12% in HLA-mismatched HSCT due to an increased use of T-cell depletion of the donor's marrow. Also, the use of antithymocyte globulin (ATG) or alemtuzumab (Campath) in a preparative regimen increases the risk of PT-LPD. Depletion of both T- and B-cells from the graft is associated with a lower incidence of PT-LPD, since, by removing B-cells, it depletes the graft of latent virus creating a balance between latent virus and cellular T-cell immunity. PT-LPD has also been reported after autologous transplantation after CD34 + selection, resulting in complete depletion of lymphocytes.

Table 16.3 shows a classification of PT-LPDs.

The frequency of organ involvement with PT-LPDs varies with the type of transplantation. For example, liver involvement is 50% in HSCT recipients; lung involvement is 54% in heart transplantation recipients and 38% in HSCT recipients; CNS involvement is 24% in renal transplantation recipients; and kidney involvement is 24% in HSCT recipients and 33% in liver transplantation recipients.

Table 16.4 lists the World Health Organization (WHO) classification of PT-LPDs.

Diagnosis of PT-LPDs

1. Physical examination: Table 16.5 shows the common sites of involvement in B-cell PT-LPDs.
2. Biopsy of the appropriate presenting site for histologic studies of immune phenotyping, cytogenetics, molecular analysis, and the examination of EBV genomes in the lymphoid cells (i.e., *in situ* hybridization, Southern blot, polymerase chain reaction, terminal repeat probes for clonality). Also, immunostaining of tissues for EBERs and EBV antigens (proteins) is required for a diagnosis of EBV LPD.

3. Complete blood count.
4. Kidney and liver chemistries, lactic dehydrogenase.
5. EBV serology: PT-LPD patients have very high anti-VCA titers but lack anti-EBNA antibodies; however, this is not always a consistent finding.
6. A real-time quantitative polymerase chain reaction (qPCR) for EBV DNA in serum or plasma.
7. A real-time qPCR for EBV DNA copies in serum or plasma can be used for preemptive diagnosis of post-hematopoietic stem cell post-transplant lymphoproliferative disease (PTLD) and post-solid organ transplantation PTLD.

Table 16.6 shows the comparison of characteristics of lymphoproliferative diseases in post-solid organ transplantation patients and post HSCT patients.

TABLE 16.4   WHO Classification of Post-Transplantation Lymphoproliferative Disorders (PT-LPDS)

- Early lesions
  - Reactive plasmacytic hyperplasia
  - Infectious mononucleosis type
- PT-LPD, polymorphic
  - Polyclonal (rare)
  - Monoclonal
- PT-LPD, monomorphic (classify according to lymphoma classification)
  - B-cell lymphomas
    - Diffuse large-cell lymphoma (immunoblastic, centroblastic, anaplastic)
    - Burkitt/Burkitt-like lymphoma
    - Plasma cell myeloma
  - T-cell lymphomas
    - Peripheral T-cell lymphoma, not otherwise categorized
    - Other types (hepatosplenic, gamma/delta, T/NK)
- Other types, rare
  - Hodgkin disease-like lesions (associated with methotrexate therapy)
  - Plasmacytoma-like lesions

T/NK, T-cell natural killer.
Harris et al. (1999).

TABLE 16.5   Common Sites of Involvement in B-Cell Post-Transplantation Lymphoproliferative Disorders

Site	Percentage of patients
Lymph nodes	59
Liver	31
Lung	29
Kidney	25
Bone marrow	25
Small intestine	22
Spleen	21
Central nervous system	19
Large intestine	14
Tonsils	10
Adrenals	9
Skin/soft tissue	7
Blood	6
Heart	5
Salivary glands	4

Magrath et al. (1997), with permission.

TABLE 16.6 Comparison of Lymphoproliferative Disorders: Post-Solid Organ Transplantation and Post-Hematopoietic Stem Cell Transplantation

Characteristic	LPD post-solid organ transplantation	LPD post-hematopoietic stem cell transplantation
Immune system	Patient's own, i.e., organ recipient's own	Replaced by donor
Immunosuppressive therapy	Lifelong	About 6 months post-transplantation or as long as graft-versus-host disease (GVHD) persists
Onset of LPD	Highest incidence: 6–12 months after transplantation	Highest incidence: 6–12 months after transplantation
EBV association	Early-onset LPD: always associated with EBV; late onset LPD: multifactorial, many patients being EBV-negative	Almost always EBV-associated
Cellular origin of LPD	LPD arises in patient's, i.e., organ recipient's, own lymphocytes	LPD arises in HSCT donor's lymphocytes; however, sometimes it may arise from blood transfusions

Incidence	• EBV status of recipient
	  • Seronegative 15.8%
	  • Seropositive 5.4%
	• Organ transplantation
	  • Kidney 1.2–9%
	  • Liver 6.8–13%
	  • Thoracic organ 3.8–11.7%
	  • Heart 5–15%
	  • Lung 10–20%
	  • Intestinal/multivisceral 31%
	• Type of immunosuppression
	  • Antithymocyte globulin 11.4%
		• Cumulative incidence 1% at 10 years
		• Mismatched 1%
		  • Mismatched and T-cell depletion 1–8%
		  • Unrelated 1.5%
		  • Unrelated and T-cell depletion 5–29%
		  • Unrelated and Campath depletion 1.3%

Treatment	• Preemptive therapy
	  • Intravenous immunoglobulin infusion
	  • Acyclovir or ganciclovir
	  • Infusion of patient's (organ recipient's) own EBV-specific cytotoxic T-cells (CTL)
	  • Infusion of anti-CD20 antibody (Rituximab)
	• Therapeutic
	  • Same as preemptive plus chemotherapy
		• Preemptive therapy
		  • Intravenous immunoglobulin infusion
		  • Acyclovir or ganciclovir
		  • Infusion of donor's EBV-specific cytotoxic T-cells
		  • Infusion of anti-CD20 antibody (Rituximab)
		• Therapeutic
		  • Same as preemptive plus chemotherapy

HSCT, hematopoietic stem cell transplantation; LPD, lymphoproliferative disorder; EBV, Epstein–Barr virus.

Treatment

The indications for treatment of LPD with chemotherapy are:

- The patient in whom all other measures to control LPD have failed and the patient has widespread LPD.
- The patient who has developed monoclonal LPD.

Treatment of B-Cell Lymphoproliferative Disease in Immunosuppressed Patients

General Treatment

The following treatment strategy is often employed:

- Reduction or withdrawal of immunosuppressive therapy (including prednisone and other drugs such as cyclosporine A, azathioprine, and tacrolimus) is attempted as first-line treatment. However, this can aggravate GVHD in HSCT patients and cause graft rejection in organ transplantation patients. In addition, transition of immunosuppressive therapy to mammalian target of rapamycin (mTOR) inhibitors may be effective in treating PT-LPD.
- Antiviral agents, acyclovir (500 mg/m^2 every 8 h with adjustment for impaired renal function) or ganciclovir (5 mg/kg intravenous (IV) every 12 h with adjustment for impaired renal function) without or without IV IgG (500 mg/kg/day once weekly for 2–4 weeks) may be effective as a preemptive measure in some patients. These measures are largely not effective in HSCT patients because the regenerating donor's immune system cannot provide enough immunity to eradicate EBV-infected B-cells.
- In solid organ transplantation (SOT) patients, regression of LPD has been observed in small studies using antiviral therapy.

For localized LPD, local radiation may be helpful.

Specific Treatment

1. Inherited immunodeficiency syndromes: Often these patients need HSCT for cure. Radiation and certain chemotherapeutics should be avoided in patients with chromosomal instability/DNA repair defects.
   a. *Polyclonal or monoclonal with no lymphoma-specific genetic abnormalities*
      i. Localized LPD: Radiation (low dose) or surgery are often employed; however, recent data suggest that a combination of chemotherapy with anti-B-cells monoclonal antibodies (rituximab in particular) may be more effective (Table 16.7).
      ii. Generalized LPD: Combination of chemotherapy with anti-B-cell monoclonal antibodies (Table 16.7).
2. Post-solid organ transplantation.
   a. *Polyclonal or monoclonal with no lymphoma-specific genetic abnormalities.*
      Reduce dose of immunosuppressive drugs if possible.
      i. Localized LPD: Radiation, surgery, or chemotherapy (Table 16.7).
      ii. Generalized LPD:
      - Interferon (rarely used): Interferon-α (IFN-α). Dose: 3 million units/m^2 once a day subcutaneously. IFN-α stimulates NK cells. It also inhibits proliferation of EBV-infected B-cells.
      - Anti-B-cell monoclonal antibodies:
        – Humanized anti-CD20 monoclonal antibody (Rituximab) can be administered as monotherapy. It is administered at a dose of 375 mg/m^2 once a week for 4 weeks after reducing the dose of immunosuppressive drugs. Complete remission rates of 50–65% have been reported after SOT. However, it should be used cautiously since it may increase the risk of graft rejection, especially when a reduction in immunosuppressive therapy is attempted at the same time. A second course of anti-CD20 may be given for the relapsed patients, although it should be given with chemotherapy (Table 16.7).
        – Preemptive therapy with 375 mg/m^2 of Rituximab can be used in patients who develop asymptomatic reactivation of EBV (1000 or more genome-equivalent per milliliter).
      - Adoptive immunotherapy: Infusion of autologous EBV-specific cytotoxic T-cells (CTL) has been tried in a number of patients with high success rates.
      - Other therapies:
        – Anti- IL-6 antibody treatment: IL-6 induces proliferation and maturation of EBV-infected B-cells. Anti-IL-6 antibody has been tried in a few patients with some success.
        – Vaccination of EBV-seronegative recipient before undergoing transplantation.
        – Chemotherapy: If there is no response or it is rapidly progressive, or recurrent disease chemotherapy should be utilized (Table 16.7).
3. Post-HSCT
   a. Anti-B-cell monoclonal antibodies.
   b. Donor leukocyte infusions:
      i. Use of unmanipulated donor's T-cells.
      - Without transduction of donor's T-cells with a suicide gene that can be turned on in case of worsening of GVHD.
      - Transduced donor's T-cells with a suicide gene, herpes simplex virus thymidine kinase, which can be activated by ganciclovir therapy in the event of worsening of GVHD.
      ii. Use of selectively ex vivo expanded EBV-specific donor's cytotoxic T-cells (CTL).

Chemotherapy (Table 16.7) should be employed if no response or recurrence occurs.

TABLE 16.7 Treatment of Post-Transplantation Lymphoproliferative Disorder with Cyclophosphamide, Doxorubicin, Vincristine, Prednisone (CHOP) OR Cyclophosphamide, Prednisone, Rituximab (CPR)

Cyclophosphamide	600 mg/m^2 or 750 mg/m^2 intravenous on day 1 of each cycle
Doxorubicin	50 mg/m^2 intravenous on day 1 of each cycle
Vincristine	1.5 mg/m^2 intravenous (with maximum dose 2 mg) on day 1 of each cycle
Prednisone	1 mg/kg twice orally per day on days 1 through 5
Rituximab	375 mg/m^2 intravenous on days 1, 8, and 15 of cycles 1 and 2

CHOP chemotherapy is often used in adult centers to treat post-transplant lymphoproliferative disease (PTLD), as well as most forms of non-hodgkin lymphoma (NHL). In children, CPR is frequently used to treat PTLD based on data from COG trial ANHL0221. Either regimen can also be utilized for treatment of angioimmunoblastic lymphadenopathy with dysproteinemia-type lymphoma and Castleman disease.

Monoclonal with a Lymphoma-Specific Genetic Abnormality

See chapters on the treatment of Hodgkin disease and non-Hodgkin lymphoma. Patients with PTLD that transforms into lymphoma should be treated with antilymphoma chemotherapy regimens with or without rituximab.

XLP Syndrome

XLP syndrome is a rare disorder characterized by dysregulation of T-cell-mediated immune response, which is induced by EBV.

Clinical manifestations of XLP such as dysgammaglobulinemia, aplastic anemia, and lymphoproliferative disease have been described in the absence of EBV infection.

Pathophysiology

Familial XLP results from mutations in *SH2D1A* gene that makes SAP (signaling lymphocyte activation molecule family (SLAM)-associated protein) protein. Normally, in activated T-cells and NK cells SAP levels increase. SAP interacts with SLAM, a molecule expressed on the T-cell surface, B-cell surface, and dendritic cell surface. In T-cells, SAP regulates TCR-induced IFN-γ. In NK cells, SAP binds to 2B4 and NTB-A (which also belong to SLAM family receptors) and activates NK-cell-induced cytotoxicity.

In XLP, various types of *SAP (SH2D1A)* gene mutations have been described including deletion, nonsense, missense, and splice site mutations. As a result of this, SAP protein may be absent or truncated, or may contain altered amino acid residues at highly conserved sites.

No correlation has been found between genotypes and phenotypes and outcomes of these patients. It has been postulated that the mutation of SAP protein in XLP causes defective helper and cytotoxic T-cell function. In some patients, XLP is related to a mutation in XIAP. In contrast to familial XLP, no mutations of SAP occur in sporadic XLP.

Clinical Manifestations

1. Fulminant infectious mononucleosis (frequency 58%, survival 4%): This is characterized by infiltration of various organs with polyclonal B and T-cells, production of inflammatory cytokines and necrosis of liver, bone marrow, lymph nodes, and spleen caused by the invading cytotoxic T-cell and uncontrolled killer cell activity. Death is generally attributable to liver failure with hepatic encephalopathy or bone marrow failure with fatal hemorrhage in the lungs, brain, or gastrointestinal tract and occurs within 1 month of onset of symptoms.
2. Secondary dysgammaglobulinemia (frequency 30%, survival 55%).
3. B-cell lymphoproliferative disease including malignant lymphoma (extranodal non-Hodgkin lymphoma) (frequency 25%, survival 35%).
4. Aplastic anemia.
5. Virus-associated hemophagocytic syndrome. Patients usually die within 1 month of onset of symptoms.
6. Vasculitis and pulmonary lymphomatoid granulomatosis.
7. In the same patient, several sequential phenotypes of the disease may manifest over time. This phenotypic variation most often includes dysgammaglobulinemia, malignant lymphoma, and marrow aplasia.

Laboratory Manifestations

Patients with the XLP syndrome reveal many humoral and cellular immunologic defects, which include the following:

- Selective impaired immunity to EBV but normal immune responses to other herpes viruses.
- Uncontrolled T_H1 responses (with high levels of IFN-γ in some patients).
- Inverted CD4/CD8 ratio (due to increase in CD8 + cells).
- Dysgammaglobulinemia: low IgG, high IgM, and high IgA.
- Defective NK cell activity.
- Decreased T-cell regression assay.
- Failure to switch from IgM to IgG class response after secondary challenge with OX174 and diminished mitogen-induced transformation of lymphocytes.
- Unchanged lymphocyte-mediated antibody-dependent cellular cytotoxicity.

Diagnosis is established by sequencing analysis of *SH2D1A* and *XIAP* genes and SAP protein quantification.

Treatment

1. Prevention of EBV infection: Prophylactic use of intravenous immunoglobulin (IVIG) has been used without much success.
2. Treatment of acute EBV infection in XLP patients: High-dose IVIG and/or acyclovir are ineffective; however, rituximab may be helpful.
3. Treatment of virus-associated hemophagocytosis: Corticosteroids, etoposide, and cyclosporine A are effective agents. Etoposide decreases macrophage activity and cyclosporine A decreases T-cell activity. The HLH (2004) protocol of the Histiocytic Society may be used (see Chapter 20). Rituximab is often used in combination with multi-agent HLH-directed therapy.
4. Treatment of aplastic anemia: ATG and cyclosporine A can be used.
5. Treatment of lymphoma: Standard therapy for lymphoma is used, that is, chemotherapy and radiation therapy where indicated.

Allogeneic HSCT is the only curative therapy for XLP syndrome.

Prognosis

Seventy percent of patients with XLP die before 10 years of age.

Autoimmune Lymphoproliferative Syndrome (ALPS) (Canale—Smith Syndrome)

Autoimmune lymphoproliferative syndrome (ALPS) is a disorder of disrupted lymphocyte homeostasis caused by defective fas-mediated apoptosis. Normally, as part of the downregulation of the immune response, activated B and T lymphocytes upregulate fas expression and activated T lymphocytes upregulate expression of fas ligand. Fas and fas ligand interact triggering the caspase cascade, leading to cellular apoptosis. Patients with ALPS have a defect in this apoptotic pathway leading to chronic lymphoproliferation, autoimmunity, and secondary malignancies.

Clinical Manifestations

1. Chronic (>6 months) nonmalignant lymphoproliferation manifesting as lymphadenopathy and/or hepatomegaly and/or splenomegaly. Lymphoproliferation may be massive and tends to wax and wane. The majority of patients present with lymphoproliferation prior to 2 years of age; however, lymphoproliferation may not occur until adult years in rare cases.
2. Autoimmune disease. Greater than 80% of patients develop autoimmune disease. The most common manifestation is autoimmune destruction of blood cells (autoimmune cytopenias) including immune thrombocytopenia, autoimmune hemolytic anemia, and autoimmune neutropenia. Patients may also develop autoimmune disease of almost any organ system, including autoimmune hepatitis, nephritis, gastritis, and colitis. Patients may develop urticaria, alopecia, and bronchiolitis obliterans.
3. Secondary malignancy. Occurs in greater than 10% of ALPS patients, most commonly lymphoma.
4. CVID. A subset of ALPS patients develops secondary CVID.

Laboratory Manifestations

1. Elevated double-negative T-cells (DNTs); cell phenotype CD3 + /TCRα/β + /CD4 − /CD8 −). DNTs are a rare population, representing <1% of circulating T lymphocytes in normal individuals. When markedly elevated, DNTs are virtually pathognomonic for ALPS. Mild elevations can be found in systemic lupus erythematosus and other autoimmune diseases.
2. *In vitro* evidence of defective fas-mediated apoptosis. Only performed in a few specialized laboratories.
3. Elevated gamma/delta DNTs, CD57 + T-cells, CD5 + B-cells, CD8 + T-cells and HLA-DR + T-cells.
4. Elevated serum IL-10.
5. Increased soluble Fas-ligand.
6. Increased vitamin B_{12}.
7. Hypergammaglobulinemia (or hypogammaglobulinemia).
8. Autoantibodies (DAT, antiplatelet, antineutrophil, ANA, Rf, antiphospholipid).
9. Eosinophilia.

Pathophysiology

ALPS has been attributed to defective apoptosis (programmed cell death) of lymphocytes, most often arising as a result of mutations in the gene encoding the lymphocyte apoptosis receptor FAS/APO-1/CD95. Because of the failure of the affected lymphocytes to die after their response to antigen has been completed, there is an accumulation and buildup of an excessive number of polyclonal lymphocytes, which leads to hepatosplenomegaly and lymphadenopathy.

ALPS is classified into subtypes based on genotype. In 2009, an international consensus conference was held at the NIH to develop a gene-based nomenclature mirroring the WHO classification.

- *ALPS-FAS*: ALPS-FAS is the most common type of ALPS, affecting 60–70% of patients. It is caused by germline mutations or deletions in *FAS (TNFRSF-6, tumor necrosis factor receptor super family member 6)*. These can be homozygous mutations (formerly called ALPS-0) or heterozygous mutations (formerly called ALPS 1a). Homozygous mutations are rare and lead to a complete absence of functional Fas protein. Heterozygous mutant FAS can exert a transdominant effect on wild-type FAS, leading to absent to near-absent functional Fas protein or can lead to haploinsufficiency and a partial defect in Fas protein function. Studies have shown there is not a phenotype–genotype correlation based on *FAS* mutation type. The one exception appears to be that patients with dominant-negative mutations in the death domain of *FAS* (exon 9) are predisposed to secondary malignancies and patients with other mutations may not be predisposed.

- *ALPS sFAS*: ALPS-sFAS is the second most common type of ALPS (~10% of patients) with an identifiable genetic mutation. ALPS-sFAS (formerly ALPS-1s) is caused by somatic mutations in *FAS* restricted to the DNT compartment. Occasionally, ALPS-sFAS can occur as a consequence of somatic mutations in multiple lymphocyte subsets. Patients with ALPS-sFAS present with similar clinical features as ALPS-FAS with the exception that these patients tend to develop disease at an older age. The average age of presentation for ALPS-FAS is ~18 months. The average age of presentation for ALPS-sFAS is over 5 years. DNTs do not survive in routine culture medium. Thus, the former gold standard test to diagnosis ALPS, the Fas-mediated apoptosis assay (described below) is normal (false negative). Diagnosis is often made by genetic analysis of sorted DNTs.

- *ALPS-FASL*: ALPS-FASL (formerly ALPS-1b) is caused by mutations in *FASL* (Fas ligand). Clinically, it manifests with features of systemic lupus erythematosus. However, it often lacks the classical features of ALPS, that is, expansion of DNT-cells and splenomegaly. It is quite rare, with only a handful of reported cases. Nevertheless, as the phenotype of ALPS-FASL mirrors SLE it may be more common as a subset of patients diagnosed with SLE who may in fact have ALPS-FASL.

  ALPS-CASP10: ALPS-CASP10 (formerly ALPS-II) is caused by caspase 10 deficiency. It is also very rare and represents ~2% of ALPS patients. These patients present clinically similar to ALPS-FAS; however, they may have less prominent lymphoproliferation. They also may have comorbid CVID.

  Caspase 8 deficiency was formerly considered a subset of ALPS; however, it is now considered a separate disease as these patients have defective apoptosis of B, T, and NK cells and have difficulty with mucocutaneous herpes virus infections. CASP 8 mutations are extremely rare with only two reported cases in one family with a history of consanguinity.

- *ALPS-U* (formerly ALPS-III): The molecular defects to account for this subtype of ALPS are unknown; 20–30% of ALPS patients are type III and clinically they are indistinguishable from other ALPS variants.

Patients with somatic mutations of *NRAS* and *KRAS* in lymphocytes used to be classified as having ALPS IV. These patients, however, are considered to have a distinct disease termed Ras-associated autoimmune leukoproliferative disease. These rare patients have a mixed phenotype with features similar to ALPS and juvenile myelomonocytic leukemia (JMML).

Diagnostic Criteria

The following changes to the diagnostic criteria for ALPS were made in 2010:

- Required criteria:
  - Chronic nonmalignant lymphoproliferation.
  - Elevated peripheral blood DNTs.
- Primary accessory criteria:
  - Defective *in vitro* Fas-mediated apoptosis (verified in two separate assays).
  - Somatic or germline mutation in ALPS causative gene (*FAS, FASL, CASP10*).

- Secondary accessory criteria:
  - Elevated biomarkers (any of the following)
    - Plasma sFASL >200 pg/ml.
    - Plasma IL-10 >20 pg/ml.
    - Plasma or serum vitamin B12 >1500 ng/l.
    - Plasma IL-18 >500 pg/ml.
  - Family history of ALPS or nonmalignant lymphoproliferation.
  - Immunohistochemical findings consistent with ALPS as determined by experienced hematopathologist.
  - Autoimmune cytopenias and polyclonal hypergammaglobulinemia.

In order to diagnose ALPS, a patient must have the two required criteria plus one primary accessory criteria (definitive diagnosis) or the two required criteria plus one secondary accessory criteria (probable diagnosis). A definitive and probable diagnosis should be treated the same and patients and families counseled that the patient has ALPS. The reason to distinguish between a definitive and probable diagnosis is a useful differentiation for the purposes of the medical literature and clinical trials.

Treatment

Lymphoproliferation rarely requires treatment unless patients develop hypersplenism and/or organ compression, in which case it should be treated with immunosuppressive agents similarly to autoimmune disease. *Splenectomy should be avoided because ALPS patients have a high risk of post-splenectomy sepsis even with vaccination and antimicrobial prophylaxis.* Splenectomy is only indicated in patients with severe clinically significant hypersplenism who do not respond to immunosuppression. Other treatments include:

- Corticosteroids. Most commonly used and is very effective. Many patients however, have chronic problems and develop steroid side-effects. Corticosteroids should be used short-term (days to weeks) only.
- Second-line therapies: Mycophenolate mofetil (MMF) and sirolimus. Patients who are refractory to or intolerant of corticosteroids may be treated with alternative immunosuppressants. MMF has the longest published track record; however, many patients are refractory to MMF, relapse or have partial responses. MMF is not effective for lymphoproliferation and does not affect DNTs. Sirolimus has been demonstrated to result in complete responses in MMF and steroid-refractory patients. Sirolimus is effective against autoimmune disease and lymphoproliferation. Sirolimus also directly targets DNTs as DNTs have dysregulation of the PI3K/Akt/mTOR signaling pathway. Other agents that may be effective include azathioprine, tacrolimus, mercaptopurine, methotrexate, vincristine, and cyclosporine. Rituximab may be effective; however, a percentage of ALPS patients treated with rituximab have developed CVID and it should be avoided if possible. Allogeneic stem cell transplantation is occasionally used but rarely necessary.

Prognosis

ALPS may improve with age.

The severity of ALPS varies from mild to severe within the same family. This may be because other pathways of apoptosis are compensating for the FAS pathway.

The following malignancies have been reported in ALPS families:

- Burkitt lymphoma, T-cell-rich B-cell lymphoma, and atypical lymphoma.
- Nodular lymphocyte predominant Hodgkin lymphoma.
- Breast cancer, lung cancer, basal cell carcinoma of the skin, squamous cell carcinoma of the tongue, and colon cancer.

Dianzani Autoimmune Lymphoproliferative Disease

This ALPS-like syndrome is characterized by:

- Defective function of the Fas receptor.
- Autoimmune conditions, predominantly involving blood cells; immune thrombocytopenic purpura (ITP), autoimmune hemolytic anemia, and autoimmune neutropenia.
- Polyclonal accumulation of lymphocytes in the spleen and lymph nodes.

Dianzani autoimmune lymphoproliferative disease (DALD) differs from ALPS in that it lacks expansion of DNT-cells. DALD patients have a high level of osteopontin (OPN) due to an increase in the frequency of B and C haplotypes of OPN. These haplotypes are responsible for the increased levels of OPN. *In vitro*, observations show that high levels of OPN decrease activation-induced T-cell apoptosis. For this reason, high levels of OPN have been implicated in the apoptotic defect of DALD.

Lymphomatoid Papulosis in Children

Lymphomatoid papulosis (LyP) is a clinically benign skin disorder characterized by chronic and recurrent self-healing papulonodular lesions. This disorder occurs rarely in children. The sites of lesions include the limbs and/or trunk. The eruption recurs episodically, often ulcerates and heals spontaneously in 3–8 weeks, occasionally leaving an atrophic scar. LyP results from a clonal T-cell proliferation, which may explain its evolution or coexistence with Hodgkin disease, mycosis fungoides, or anaplastic large-cell lymphoma.

Histology

LyP is characterized by superficial and deep infiltrates consisting of atypical lymphocytes with a hyperchromatic, pleomorphic nucleus that can obscure the dermoepidermal junction. Similar cellular infiltrates can also be found in the epidermis along with spongiosis, parakeratosis, and neutrophils. Two types of activated T-cell infiltrates are found:

- LyP type A lesions: LyP lesions of this cell type predominantly contain Ki-1 + (CD30 +), Reed–Sternberg-type cells with a pale-staining, convoluted nucleus, prominent nucleoli, and moderately basophilic cytoplasm.
- LyP type B lesions: LyP lesions of this cell type contain the Sezary-type cell with a cerebriform, hyperchromatic nucleus and scant cytoplasm. Only some of these cells are CD30 + .

The atypical cells of LyP have been shown to be of T-cell origin, in most cases expressing CD2, CD3, and CD4. These cells lack the expression of some of the usual pan-T-cell antigens, such as CD5, CD7, or both. Clonal rearrangement of the T-cell antigen receptor (TCR- and/or clone genes) has been demonstrated in most of the cases studied. Thus, LyP serves as an example of a benign LPD in spite of the evidence of clonal TCR gene rearrangement.

A lymphoma evolving from LyP can be recognized by enlarging or persistent skin lesions, peripheral lymphadenopathy, or circulating atypical lymphocytes. A biopsy of suspicious skin lesions or enlarged lymph nodes for histologic, immunologic, cytogenetic, or gene rearrangement studies is indicated. It has also been suggested that a thorough physical examination every 6 months in children with LyP with attention to growth and development (as a tool in assessing occult malignancies in children), skin lesions, and lymph nodes should be carried out. Additional studies such as bone marrow examination and radioimaging studies should be performed as needed.

Treatment

Treatment involves oral antibiotics, systemic corticosteroids, low-dose methotrexate, psoralen with ultraviolet radiation, topical steroids, or ultraviolet beam.

Prognosis

The duration of LyP ranges from 1 to 40 years. Approximately 23 children have been described and evolution to lymphomas has occurred in two of the 23 cases.

Further Reading and References

Dispenzieri, A., Armitage, J., et al., 2012. The clinical spectrum of Castleman's disease. Am. J. Hematol. 87 (11), 997–1002.

Gross, T.G., Savolodo, B., 2010. Punnet A Posttransplant lymphoproliferative diseases. Pediatr. Clin. North Am. 57 (2), 481–503.

Harris, N.L., Jaffe, E.S., Diebold, J., et al., 1999. World Health Organization classification of neoplastic disease of the hematopoietic and lymphoid tissues: Report of the Clinical Advisory Committee Meeting-Airlie House, Virginia, November 1997. J. Clin. Oncol. 17, 3835–3849.

Magrath, I.T., Shad, A.T., Sandlund, J.T., 1997. Lymphoproliferative disorders in immunocompromised individuals. In: Magrath, I.T. (Ed.), The Non-Hodgkin Lymphoma, second ed. Arnold, London, pp. 955–974.

Teachey, D.T., 2012. New advances in the diagnosis and treatment of autoimmune lymphoproliferative syndrome. Curr. Opin. Pediatr. 24 (1), 1–8.

Veillete, A., Perez-Quintero, L., Latour, S., 2013. X-linked lymphoproliferative syndromes and related autosomal recessive disorders. Curr. Opin. Allergy Clin. Immunol. 13 (6), 614–622.

Myelodysplastic Syndromes and Myeloproliferative Disorders

Inga Hofmann and Tarek M. Elghetany

MYELODYSPLASTIC SYNDROMES

Myelodysplastic syndromes (MDS) are clonal hematopoietic stem cell disorders characterized by varying degrees of cytopenias secondary to ineffective and dysplastic hematopoiesis and increased propensity to evolve into acute myeloid leukemia (AML). In contrast to adult MDS patients, who usually present with a hypercellular bone marrow (BM), the majority of pediatric MDS patients present with a hypocellular BM, making them difficult to distinguish from acquired and inherited BM failure disorders (IBMFS) (see Chapter 8).

Diagnostic Criteria and Classification

Table 17.1 provides diagnostic criteria for pediatric MDS.

The classification of MDS has been updated over the years to accommodate for newer diagnostic findings. Table 17.2 provides the historical evolution of the MDS classification and compares the French—American—British (FAB) classification with the 2001 and 2008 World Health Organization (WHO) classifications and diagnostic criteria. The 2008 WHO classification introduced a pediatric MDS classification for the first time. Refractory cytopenia of childhood (RCC) was proposed as the most common subtype of pediatric MDS, accounting for about 50% of cases. The majority of RCC patients have normal cytogenetics and present with a hypocellular BM with varying degrees of dysplasia resembling BMF. Higher-grade pediatric MDS include refractory anemia with excess blasts (RAEB) and refractory anemia with excess blasts in transformation (RAEB-T).

Juvenile myelomonocytic leukemia (JMML) and MDS associated with Down syndrome (DS), previously grouped under MDS, are now recognized as distinct entities and are therefore classified and discussed separately. Table 17.3 provides the diagnostic categories of MDS and myeloproliferative neoplasms in children.

Epidemiology

Incidence: 1.8 per million children per year in age group 0—14 years.

Constitutes 4% of all hematological malignancies.

Table 17.4 provides constitutional and acquired abnormalities associated with secondary pediatric MDS. Primary MDS occurs *de novo* without an apparent underlying cause.

Familial MDS has been observed in 10% of children with MDS, but is likely more frequent than previously considered. It is commonly associated with partial or complete loss of chromosome 7 (7q- or monosomy 7), particularly in association with *GATA2* mutation.

TABLE 17.1 Minimal Diagnostic Criteria for MDS

At least two of the following

- Sustained unexplained cytopenia (neutropenia, thrombocytopenia, or anemia)
- At least bilineage morphologic myelodysplasia
- Acquired clonal cytogenetic abnormality in hematopoietic cell
- Increased blasts (5%)

Hasle et al. (2003) with permission.

TABLE 17.2 Historic and Current 2008 WHO Classifications of MDS

FAB (1982)	WHO (2001)	WHO (2008)	WHO (2008) diagnostic criteria
RA	RA	RCUD (includes RA, RN, RT)	
RA	RA	RA	Anemia (Hb <10 g/dl); ± neutropenia or thrombocytopenia; <1% circulating blasts; <5% medullary blasts; unequivocal dyserythropoiesis in ≥10% erythroid precursors; dysgranulopoiesis and dysmegakaryopoiesis, if present, in <10% nucleated cells; <15% RS; no Auer rods
n/a	n/a	RN	Neutropenia (absolute neutrophil count <1.8 × 10^9/l); ± anemia or thrombocytopenia; <1% circulating blasts; <5% medullary blasts; ≥10% dysplastic neutrophils; <10% dyserythropoiesis and dysmegakaryopoiesis; <15% RS; no Auer rods
n/a	n/a	RT	Thrombocytopenia (platelet count <100 × 10^9/l); ± anemia or neutropenia; <1% circulating blasts; <5% medullary blasts; ≥10% dysplastic megakaryocytes of ≥30 megakaryocytes; <10% dyserythropoiesis and dysgranulopoiesis; <15% RS; no Auer rods
RARS	RARS	RARS	Anemia; no circulating blasts; <5% medullary blasts; dyserythropoiesis only, with RS among >15% of 100 erythroid precursors; no Auer rods
n/a	RCMD	RCMD	Cytopenia(s); <1 × 10^9/l circulating monocytes; <1% circulating blasts; <5% medullary blasts; dysplasia among >10% cells of ≥2 lineages; no Auer rods
	and RCMD with RS	n/a	n/a
RAEB	RAEB-1	RAEB-1	Cytopenia(s); <1 × 10^9/l circulating monocytes ; <5% circulating blasts; 5–9% medullary blasts; dysplasia involving ≥1 lineage(s); no Auer rods
RAEB	RAEB-2	RAEB-2	Cytopenia(s); <1 × 10^9/l circulating monocytes; 5–19% circulating blasts; 10–19% medullary blasts; dysplasia involving ≥1 lineage(s); ± Auer rods[a]
n/a	n/a	RAEB-F	Similar to RAEB-1 or RAEB-2, with at least bilineage dysplasia and with diffuse coarse reticulin fibrosis, with or without collagenous fibrosis
RAEB-T (blasts 20–29%)	(AML)	(AML)	n/a
n/a	MDS, U	MDS, U	RCUD with 1% circulating blasts OR pancytopenia with unilineage dysplasia OR <1 × 10^9/l circulating monocytes, ≤1% circulating blasts, <5% circulating blasts, dysplasia in <10% cells of ≥1 lineage(s) and demonstration of MDS-associated chromosomal abnormality(ies), exclusive of +8, del(20q), and loss of chromosome Y (-Y)
	MDS, isolated del(5q)	MDS, isolated del(5q)	Anemia; platelet count may be normal or increased; <1% circulating blasts; <5% medullary blasts; megakaryocytes with characteristic nuclear hypolobulation; isolated del(5q) cytogenetic abnormality involving bands q31–q33
CMML	(MDS/MPN)	(MDS/MPN)	n/a
n/a	n/a	RCC	Thrombocytopenia, anemia, and/or neutropenia; <2% circulating blasts; <5% medullary blasts; unequivocal dysplasia in ≥2 lineages, or in >10% cells of one lineage; no RS

[a]*In the FAB Classification, the presence of Auer rods is one of the criteria for RAEB-T regardless of the number of blasts.*
AML, acute myeloid leukemia; RA, refractory anemia; RCUD, refractory cytopenia with unilineage dysplasia; CMML, chronic myelomonocytic leukemia; MDS, U, myelodysplastic syndrome, unclassifiable; MPN, myeloproliferative neoplasm; n/a, not applicable; RAEB-F, refractory anemia with excess blasts with fibrosis; RAEB-T, refractory anemia with excess blasts in transformation; RCC, refractory cytopenia of childhood; RN, refractory neutropenia; RS, ring sideroblasts; RT, refractory thrombocytopenia; RARS, refractory anemia with ringed sideroblasts; RCMD, refractory cytopenia with multilineage dysplasia.
Modified from Nguyen (2009) with permission.

TABLE 17.3　Diagnostic Categories of Myelodysplastic and Myeloproliferative Diseases in Children

I. Myelodysplastic/myeloproliferative disease
  • Juvenile myelomonocytic leukemia (JMML)
  • Chronic myelomonocytic leukemia (CMML) (secondary only)
  • BCR/ABL negative chronic myeloid leukemia
II. Myeloid proliferations related to Down syndrome (DS)
  • Transient abnormal myelopoiesis
  • Myeloid leukemia of DS
III. Myelodysplastic syndrome
  • Refractory cytopenia (peripheral blood blasts <2% and bone marrow blasts <5%)
  • Refractory anemia with excess blasts (peripheral blood blasts 2−19% and bone marrow blasts 5−19%)
  • Refractory anemia with excess blasts in transformation (peripheral blood and/or bone marrow blasts 20−29%)

Modified from Hasle et al. (2003) with permission and the 2008 WHO Classification of Tumors of Haematopoietic and Lymphoid Tissues.

TABLE 17.4　Inherited and Acquired Conditions Associated with Pediatric Myelodysplastic Syndrome (MDS) Leading to Secondary MDS

Conditions associated with MDS

INHERITED CONDITIONS

Inherited bone marrow failure syndromes (IBMFS)
　Fanconi anemia
　Shwachman−Diamond syndrome (SDS)
　Severe congenital neutropenia
　Dyskeratosis congenita
　Diamond−Blackfan anemia
GATA2 haploinsufficiency (MonoMac syndrome, Emberger syndrome, familial MDS/AML)
Familial nonsyndromic MDS due to mutations in *ETV6, RUNX1/AML1,* or *CEBPA*
Other familial MDS (at least one first degree relative with MDS/AML) without identified genetic cause[a]
Trisomy 8 mosaicism

ACQUIRED CONDITIONS

Prior chemotherapy
Prior radiation therapy
Acquired aplastic anemia[b]

[a]*Familial cases of MDS not due to GATA2, ETV6, RUNX1/AML1, CEBPA.*
[b]*A subset of cases currently classified as acquired aplastic anemia might be due to in underlying genetic predisposition and may in the future be considered as an inherited condition.*

Therapy-Related Myeloid Neoplasms

In the 2008 WHO Classification, this category includes therapy-related acute myeloid leukemia (t-AML) and therapy-related MDS (t-MDS). These disorders occur secondary to treatment with alkylating agents, topoisomerase II inhibitors, other chemotherapeutic agents, and radiation therapy.

• Alkylating agents- and ionizing irradiation-induced MDS are characterized by deletions or loss of whole chromosome or complex cytogenetics. Latency period: 5−10 years.
• Topoisomerase II inhibitor-induced MDS is characterized by balanced translocations, commonly involving chromosome band 11q23, which harbors the MLL gene. Latency period: 1−3 years.

Incidence of therapy-related MDS: represents 5% of all childhood MDS and occurs in 13% of children treated for malignancies.

Children with t-MDS or t-AML compared to children with *de novo* AML or MDS, have the following characteristics:

1. Older at presentation, have lower white blood cell (WBC) counts, and are less likely to have hepatomegaly or splenomegaly or hepatosplenomegaly.
2. More likely to have trisomy 8 and less likely to have classic AML translocations.
3. Less likely to attain remission after induction therapy (50% vs 72%) and less likely to have a longer overall survival (OS) (26% vs 47%), and event-free survival (21% vs 39%).

Their disease-free survival (DFS) after attaining remission is similar to children with *de novo* AML or MDS (45% vs 53%).

Pathophysiology

MDS is a heterogeneous disease with different pathophysiologic mechanisms playing roles in its initiation and progression. Initially, apoptosis dominates the process and is responsible for the characteristic ineffective hematopoiesis. With time, as more genetic abnormalities accumulate in the MDS cells, arrest of maturation and proliferation occurs, resulting in transformation to AML. Recent evidence suggests that an underlying inherited predisposition (germline mutations, e.g., *GATA2*) is the disease-initiating event in a subset of patients with pediatric MDS. In those cases secondary acquired somatic mutations may lead to disease progression.

Clinical Features

The clinical presentation can be variable. A third of the patients are asymptomatic and come to medical attention because of an incidental finding of cytopenia. If symptoms occur they are usually related to cytopenias, for example, pallor, bruises, petechiae, infections. In contrast to adults that frequently present with anemia, children typically present with thrombocytopenia, neutropenia, macrocytosis, and sometimes show an elevated fetal hemoglobin levels. Lymphadenopathy or hepatosplenomegaly are uncommon and are associated with advanced disease.

Cytogenetics

- The presence of a clonal cytogenetic marker can confirm the diagnosis; however, about 61–67% of patients with RCC have normal cytogenetics.
- Monosomy 7 is the most common cytogenetic abnormality in childhood MDS followed by trisomy 8.
- Aberrations in chromosome 5, in particular the 5q- syndrome commonly seen in adults, is rare in children.
- Monosomy 7 and complex karyotype (≥three abnormalities) have been associated with increased risk for leukemic transformation and poor prognosis.

The concept of monosomy 7 as a distinct syndrome has been abandoned. A subset of these cases ultimately fit the diagnosis of JMML. Another subset of cases initially reported as "monosomy 7 syndrome" were cases of familial MDS/AML and were likely due to GATA2 haploinsufficiency in a significant portion of the cases.

Molecular Genetics

In adult MDS, somatic mutations in splicing factors (*SF3B1, U2AF1, ZRSR2*) and epigenetic regulators (*TET2, ASXL1, EZH2, DNMT3A, IDH1/IDH2*) are present in about 75% of cases, followed by isolated *TP53* mutations and mutations in a variety of other genes including transcription factors (*RUNX1, ETV6, GATA2, PHF6*), kinase signaling (*NRAS, KRAS, JAK2, CBL*), and cohesion complex genes. Interestingly, mutations commonly found in adult MDS, particularly genes controlling the RNA splicing machinery and epigenetics, are only rarely present in pediatric MDS, suggesting different pathogenic mechanisms between the two groups. This evidence and the recent discovery of germline mutations in *RUNX1/AML1, GATA2*, and *ETV6* suggest that pediatric MDS is more frequently due to an inherited genetic predisposition.

Recent studies suggest a role for epigenetics in both adult and pediatric MDS, resulting in gene silencing through methylation or histone deacetylation.

Differential Diagnosis

1. It is to be noted that dysplastic features may be seen in non-MDS diseases, such as immunologic, rheumatologic, metabolic, mitochondrial (e.g., Pearson syndrome) and nutritional disorders, viral infections, and drug or toxin exposure.
2. Some degree of dyspoiesis/dysplasia can be seen in IBMFS. Given the different treatment strategies and clinical implications these need to be ruled out by molecular testing in all patients with suspected MDS.
3. It is difficult to distinguish hypocellular RCC from acquired aplastic anemia. The 2008 WHO Classification has established criteria to discriminate between both entities (summarized in Table 17.5). It is important to note that RCC currently remains a provisional entity for which the clinical and prognostic implications remain under investigation.

TABLE 17.5 Histopathologic Criteria of Hypocellular Refractory Cytopenia of Childhood (RCC) and Severe Aplastic Anemia (SAA) as Outlined in the WHO Classification

Lineage characteristics	RCC	SAA
Erythroid	Patchy, left-shifted erythropoiesis with increased mitoses	Lacking foci or left-shifted erythroid cells or only showing single small focus of <10 cells of erythroid cells with maturation
Myeloid	Markedly decreased, left-shifted myelopoiesis	Lacking or markedly decreased myelopoiesis with very few small foci of granulopoiesis with maturation
Megakaryocytes	Markedly decreased megakaryopoiesis	Lacking or only very few megakaryocytes present
	Dysplastic changes (micromegakaryocytes)[a]	No dysplastic changes or micromegakaryocytes[a]
Lymphoid	Lymphocytes, PC, MC may be focally increased or dispersed	Lymphocytes, PC, MC may be focally increased or dispersed
CD34+ cells	Not increased	Not increased

[a]Immunohistochemistry with CD61 staining is required for the detection of micromegakaryocytes. PC, plasma cells; MC, mast cells.
Adapted from the 2008 WHO classification.

4. The majority of children who develop MDS following a diagnosis of acquired aplastic anemia present with MDS within the first 3 years from the diagnosis of aplastic anemia.
5. Patients with mild to moderate aplastic anemia may be more likely to develop a clonal disease than a patient with severe aplastic anemia. Repeated evaluation for both conditions including BM examinations may become necessary to reach a diagnosis.
6. If ringed sideroblasts are observed in a pediatric BM with concern for MDS or BMF, a search for other etiologies, particularly nutritional deficiencies, drug toxicity, congenital sideroblastic anemias, especially those related to mitochondrial cytopathies, including Pearson marrow-pancreas syndrome should be considered as refractory anemia with ringed sideroblasts (RARS), is rare in children.
7. Clinical course and the response to therapy for *de novo* AML are different from advanced MDS with excess blasts (RAEB). For this reason, it is important that these conditions are diagnosed accurately.
8. The MDS subtype RAEB-T has been abandoned in adults. However, that term is still in use for pediatric patients, particularly for those patients with a slowly progressive disease rather that the classic abrupt course of AML.

Table 17.6 highlights some investigations and other diagnostic possibilities in patients suspected of having MDS.

Prognosis

- Monosomy 7 and complex karyotype (≥3 abnormalities) are known to be associated with increased risk for disease progression to leukemia.
- The International Prognostic Scoring System (IPSS) and the recently revised IPSS (IPSS-R) that are frequently used in adult MDS have less value in children and their applicability in pediatric MDS has not been yet verified.
- An analysis by the European Working Group of Childhood MDS (EWOG-MDS) suggested a poor prognosis in patients with two- to three-lineage cytopenia and a blast count >5% in BM.

Patients with low-grade pediatric MDS, in particular hypocellular RCC can have, relatively stable disease for months to years.

Studies in children undergoing myeloablative hematopoietic stem cell transplantation (HSCT) for MDS demonstrated highly variable outcomes and contain a heterogeneous group of patients and treatments. Three-year OS ranges between 18% and 74% depending on stage (RCC, 74%; RAEB, 68%; RAEB-T, 18%).

HSCT results for 3-year OS for children with MDS:

De novo MDS patients (all grades)	
HLA-matched family donor	50%
Matched unrelated donor	35%
Secondary MDS	20−30%

TABLE 17.6 Investigation for the Diagnosis of Myelodysplastic Syndrome (MDS) and Pertinent Differential Diagnosis (D/D)

Blood

Hemoglobin level and red cell indices:

Macrocytosis D/D—drugs, folate, or B12 deficiency (including abnormalities in their metabolic pathways), IBMFS, MDS, juvenile myelomonocytic leukemia (JMML)

Microcytic and ring sideroblasts: unlikely to be MDS, exclude mitochondrial diseases, copper deficiency.

Normocytic D/D—anemia of chronic diseases

White cell count, differential count, platelet count

Blood film for morphologic review

Fetal hemoglobin concentration[a] D/D—MDS, JMML, IBMFS

Immunodeficiencies: Occasionally, immunodeficiency may be associated with MDS or vice versa

Pertinent tests to be performed on peripheral blood in the context of above D/D:

Fetal hemoglobin

Cytogenetics (mitomycin C or di-epoxybutane study for excessive chromosomal breakage)[b]

Erythrocyte adenosine deaminase level

Vitamin B12 and folate levels

Quantitative immunoglobulin levels and T- and B-lymphocyte quantitation

Flow cytometry studies for paroxysmal nocturnal hemoglobinuria using standard panel for granulocytes, monocytes, and red blood cells (particularly for hypocellular marrow). Recommended panels include a combination of fluorescent acrolysin, CD14, CD16, CD24, CD59, CD33, CD15 for granulocytes and monocytes

Bone marrow

Aspirate and trephine biopsy[c], including appropriate immunohistochemical studies, such as CD34 and CD61

Iron stain

Cytogenetics: conventional and fluorescent *in situ* hybridization for chromosomes 7, 8, 20, and BCR/ABL (Ph chromosome)

Molecular analysis: RT-PCR for BCR/ABL and FLT3 ITD.

Additional specialized tests (not routinely performed)

Neutrophil function

Platelet function

Colony-forming unit assay for various lineages on bone marrow cells

[a]*Draw blood for hemoglobin fractionation studies, paroxysmal nocturnal hemoglobinuria panel, and adenosine deaminase (if macrocytosis), before transfusing patient with red blood cells.*

[b]*Patients with Fanconi anemia may present with MDS. The significance of this observation is that all patients with MDS should have chromosomal breakage analysis performed to exclude Fanconi anemia because Fanconi anemia is a recessive disorder and warrants genetic counseling. Additionally, preparative regimen for hematopoietic stem cell transplantation is different for patients with Fanconi anemia.*

[c]*Bone marrow trephine biopsy: Biopsy may help differentiate refractory cytopenia of childhood from acquired aplastic anemia. Two additional types of MDS have been recognized recently. (A) Hypoplastic MDS and (B) MDS with myelofibrosis. Also, in some reports emphasis is placed on recognizing abnormal location of immature precursor cells, that is, presence of blasts in intertrabecular areas in bone marrow biopsy specimens, since it may have prognostic significance. However, no significance has been found thus far, in the major reports of pediatric MDS series.*

If bone marrow is hypocellular aplastic anemia and/or paroxysmal nocturnal hemoglobinuria should be considered. Hepatomegaly or splenomegaly or hepatosplenomegaly favors diagnosis of JMML or acute myeloid leukemia.

Results of long-term (8-year) DFS after *unrelated* HSCT:

RCC	51%
RAEB	35%
RAEB-T	29%

Probability of relapse after *unrelated* HSCT (8 years)

RCC	4%
RAEB	23%
RAEB-T	29%

Recurrence rate was similar between primary and secondary MDS.
Adapted from Woodard et al. (2011).

A recent study of children with advanced primary MDS (RAEB, RAEB-T, and myeloid dysplasia-related AML) showed improved of 5-year OS of 63% after HSCT (related and unrelated donors).

Treatment

HSCT remains the only curative therapy for pediatric MDS. Chemotherapy alone is ineffective. Therefore, high-resolution HLA typing should be performed at diagnosis to expedite a matched related donor (MRD) or unrelated donor search. Given the clinical heterogeneity and lack of precise prognostic factors, guidelines on the optimal timing for HSCT for pediatric MDS and the need for upfront chemotherapy in advanced MDS do not currently exist.

1. RCC: Children with RCC, in particular those without:
   * High-risk cytogenetic abnormalities, such as monosomy 7 or complex abnormalities.
   * Need for transfusions.
   * Severe neutropenia (absolute neutrophil count remains above 1000/mm^3) have a long and stable clinical course without treatment for months to years. For this reason some investigators recommend a careful "watch-and-wait" strategy for this particular low-risk cohort of patients that fulfills the histologic criteria of *hypocellular* RCC.

   Recent reports have suggested that immunosuppressive therapy with antithymocyte globulin (ATG), corticosteroids, and cyclosporine may be effective in a subset of patients with RCC, for which a MRD is not available for HSCT. Nevertheless, patients with RCC remain at risk for progression to leukemia of ~30% over 5 years, with a median time to RAEB of 47 months. Therefore, for the majority of RCC patients, HSCT with the best available donor is recommended. BM is the preferred stem cell source. Pre-HSCT chemotherapy is usually not indicated. Myeloablative-conditioning regimens are used for the majority of patients, although reduced intensity-conditioning regimens have been utilized in a subset of patients with low-risk hypocellular RCC.

2. RAEB and RAEB-T: Debate exists as to whether patients with RAEB should receive cytoreduction prior to HSCT. Historically many investigators recommend AML-like induction chemotherapy before HSCT in children with RAEB, although there are no clear data that this improves outcomes. Therefore, other investigators recommend proceeding directly to HSCT with the best available donor in patients with RAEB. Upfront chemotherapy continues to be recommended for patients with blasts over 20–30% (RAEB-T). A suitable MRD is preferred for HSCT over an unrelated or alternate donor.

Biological agents: Hypomethylating agents (azacitidine, decitabine) have been shown to improve OS in high-risk adult MDS patients, but have limitations as they are not curative and the time to treatment response can be several months. Other newer agents including thalidomide analogs (used in 5q- syndrome in adults), multikinase inhibitors, and thrombopoietin receptor agonists are being used more frequently in adult MDS. None of these agents has been extensively studied in pediatric MDS and should be limited to clinical trials and compassionate use in children.

Myeloid Proliferations in Children with DS

Incidence

It has been suggested that ~1 in 150 children with DS develop MDS or AML by the age of 3 years. DS children are at 10–100 times risk of developing leukemia compared with non-DS children. Myeloid leukemia in children with DS is distinct from the disease in non-DS children. Transient abnormal myelopoiesis (TAM) often precedes myeloid leukemia associated with DS. The most common form of AML in DS children is acute megakaryoblastic leukemia (FAB M7).

Ten percent of DS infants develop TAM in the first 3 months of life. It is characterized by accumulation of immature megakaryoblasts in blood and liver, and to a much lesser degree in the BM. The majority of patients attain spontaneous remission in 3 months. However, 30% of TAM patient develop myeloid leukemia associated with DS by 3 years of age.

Biology

Acquired mutations in *GATA1* (a hematopoietic transcription factor gene) are associated with M7 AML and TAM in DS. A subset of patients with myeloid leukemia associated with DS has mutations in *JAK3* and *FLT3* genes.

Treatment

DS patients with MDS or AML have improved survival rates (DFS at 4 years is 88%) compared to their non-DS counterparts. Nevertheless, DS patients do not tolerate (nor need) as intensive therapy as non-DS patients with AML or MDS since their blasts are very sensitive to chemotherapeutic agents, particularly Ara-C. DS patients are treated with similar agents as non-DS patients but with a less intensive schedule (see Chapter 19).

Treatment of TAM in DS

Supportive treatment is given.

In case of severe organ dysfunction exchange transfusion, leukapheresis, and chemotherapy with cytarabine are used.

Uncommon complications of TAM: hydrops fetalis, renal failure, organ infiltration, pleural effusion, respiratory failure, hepatic fibrosis, disseminated intravascular coagulopathy.

JUVENILE MYELOMONOCYTIC LEUKEMIA

JMML is a rare and aggressive myeloid neoplasm of early childhood classified as an overlap myelodysplastic syndrome/myeloproliferative neoplasm (MDS/MPN) in the 2008 WHO Classification. JMML is characterized by hyperproliferation of monocytic and granulocytic cells with infiltration of spleen, liver, lungs, and the gastrointestinal tract. The majority of patients carry a mutation in *PTPN11, KRAS, NRAS, CBL*, or neurofibromatosis type 1 (*NF-1*) leading to hyperactive RAS signaling. Allogeneic HSCT remains the only curative therapy with a cure rate of about 50–60%.

Epidemiology

Incidence: 1.2 per million children per year comprises about 2% of all pediatric hematologic malignancies.
Median age at diagnosis: 1.8 years; 35% below 1 year of age, and only 4% above 5 years of age.
Male: Female ratio of 2:1.
Association with inherited syndromes due to germline mutations in *PTPN11, NF1* or *CBL*
Increased risk of developing JMML in trisomy 8 mosaicism.
Children with NF-1 have a 200- to 500-fold increased risk of JMML.
Children with Noonan syndrome (NS) show multiple developmental defects and can develop a transient MDS/MPN with a JMML-like presentation due to germline mutations in *PTPN11*. NS patients are also at higher increased risk of developing JMML.

Clinical Features

Age: usually presents before 2 years of age.
Physical findings

- Constitutional/general symptoms: fever, failure to thrive, poor weight gain, infections.
- Skin: skin rash, xanthoma, café-au-lait spots, petechiae, bruising.
- Hematologic: splenomegaly, hepatomegaly, lymphadenopathy, enlarged tonsils, bleeding, pallor.
- Respiratory symptoms: tachypnea, cough, wheezing, respiratory distress.
- Gastrointestinal: abdominal pain, distended abdomen, diarrhea (sometimes bloody).

Laboratory Features

Blood smear

- Blood smear shows anemia, thrombocytopenia, and leukocytosis with immature myeloid forms and often striking monocytosis with dysplasia and nucleated red blood cells.
- The majority of patients have a WBC count >25,000/mm^3 (median 33,000/mm^3). WBC <10,000/mm^3 may be seen in patients with monosomy 7.
- Absolute monocytosis of >1000/mm^3 is required for the diagnosis. Monocytosis may precede overt symptoms of JMML.
- Less than 20% blasts.
- Increased fetal hemoglobin (HbF) for age.

TABLE 17.7   Diagnostic Criteria for Juvenile Myelomonocytic Leukemia (JMML)

Category 1 Mandatory clinical and hematologic features	Category 2 Molecular diagnostics	Category 3 Additional supportive hematologic features
All of the following: • Peripheral blood monocyte count >1 × 10^9/l • Blasts in peripheral blood and bone marrow <20% • Splenomegaly[b] • Absence of the t(9;22) BCR/ABL fusion gene	1 of the following: • Somatic mutation in *PTPN11*[c], *KRAS*[c], or *NRAS*[d] • Clinical diagnosis of NF-1 or germline *NF1* mutation • Germline *CBL* mutation and loss of heterozygosity of *CBL*	*At least 2 of the following must be fulfilled if none of Category 2 criteria are met:*[a] • Monosomy 7 or any other clonal cytogenetic abnormality[e] • Increased fetal hemoglobin (HbF) for age • Circulating myeloid precursors on peripheral blood smear • White blood cell (WBC) >10 × 10^9/l[f] • Granulocyte-macrophage colony stimulating factor hypersensitivity in colony assay • Hyperphosphorylation of STAT5[g]

[a]*10% of patients with clinical JMML lack a pathogenic mutation outlined in Category 2. For those patients all of Category 1 criteria and at least 2 of Category 3 criteria need to be met to confirm the diagnosis of JMML.*
[b]*7–10% of patients do not have splenomegaly at initial presentation. For those patients all remaining Category 1 and one parameter in Category 2, or at least 2 parameters in Category 3 must be fulfilled.*
[c]*Germline mutations in PTPN11 (indicating Noonan syndrome and KRAS) need to be excluded.*
[d]*Clinical diagnosis of JMML with spontaneous regression of myeloproliferation has been noted in rare cases with NRAS mutation and normal HbF.*
[e]*Monosomy 7 has been categorized under Category 2 criteria in Chan et al., but moved to Category 3 in a recently updated version by Locatelli et al.*
[f]*WBC >10 × 10^9/l is outlined in Chan et al., but has been removed by recent updated criteria by Locatelli et al.*
[g]*Hyperphosphorylation of STAT5 is a new parameter that was added by Locatelli et al.*
Modified from Chan et al. (2009) and updated criteria from Locatelli et al. (2015).

Bone marrow

• Increased cellularity, increased myeloid series, increased monocytes, <20% blasts.
• Hypersensitivity of myeloid progenitors to granulocyte-macrophage colony stimulating factor (GM-CSF).

Cytogenetics

• Most patients have a normal karyotype (65%).
• Monosomy 7 is the most frequent abnormality in JMML (25%) and is associated with a lower WBC count, higher percentage of monocytes (but similar absolute monocyte count) and macrocytosis. The BM shows relative erythroid hyperplasia. HbF is normal to moderately elevated in association with monosomy 7.
• Other cytogenetic abnormalities occur in 10%.

Genetics
Eighty-five to ninety percent of patients carry mutations in genes involved in *PTPN11*, *NRAS*, *KRAS*, *NF-1*, or *CBL*. Table 17.7 provides diagnostic guidelines for JMML.

Differential Diagnosis

The clinical presentation of JMML can be nonspecific and thus it can be challenging to distinguish JMML from other more common conditions. Viral infections, such as human herpes virus-6, Epstein–Barr virus, or cytomegalovirus, can mimic JMML in young infants and need to be excluded. A number of other conditions share clinical and laboratory features similar to those of JMML, including Wiskott–Aldrich syndrome, leukocyte-adhesion deficiency, hemophagocytic lymphohistiocytosis (HLH), and infantile malignant osteopetrosis (IMO). Finally, AML and chronic myeloid leukemia (CML) can present with leukocytosis, monocytosis, and splenomegaly and need to be ruled out.

Our understanding of the genetic underpinnings in JMML has greatly improved and driver mutations in the RAS signaling pathway can now be found in ~90% of patients, which has facilitated a molecular diagnosis.

If the diagnosis remains uncertain the following studies can provide clues to the diagnosis:

1. GM-CSF hypersensitivity on BM.
2. Cytogenetics: presence of t(9;22) would exclude JMML.
3. Hemophagocytosis on BM aspirate may favor viral illness or HLH.
4. Radiographs, serum alkaline phosphatase, and calcium to differentiate from IMO.

Biology

JMML is a clonal disorder that arises from a pluripotent hematopoietic stem cell. The disease leads to clonal proliferation of myeloid, erythroid, and possibly lymphoid cells originating from the leukemic progenitor cell. It is a heterogeneous disease in the context of monoclonal involvement of different lineages.

Disturbed signal transduction through the RAS signaling pathway plays a central role in the pathogenesis of JMML and leads to the distinct phenomenon of GM-CSF hypersensitivity. JMML mononuclear cells yield excessive numbers of colony-forming units granulocyte-macrophage when cultured in a semisolid system (methylcellulose). This is due to endogenous production of interleukin-1 (IL-1), GM-CSF, and tumor necrosis factor-α (TNF-α) by monocytes.

TNF-α inhibits normal hematopoiesis causing BM suppression and results in anemia and thrombocytopenia. It also induces proliferation of the JMML clone-derived monocyte-macrophage elements. IL-1 stimulates accessory cells to produce more GM-CSF.

Splenomegaly also contributes to the development of anemia and thrombocytopenia.

Molecular Genetic Events

The discovery of RAS signaling as the key molecular pathway in JMML enabled the identification of genetic mutations in *NF-1*, *NRAS*, *KRAS*, *PTPN11* and *CBL*. These are mutually exclusive from each other and account for ~90% of the genetic lesions in JMML. These mutations all lead to hyperactive RAS signaling. In JMML, there may be mutations of the *RAS* gene or there may be defective regulation of the *RAS* gene, which results in the aberrant transmission of proliferative signals from GM-CSF to the nucleus.

NF-1 was the first mutation found to be associated with JMML and occurs in 10–15% of cases.

Germline mutations in *PTPN11* were first identified in patients with NS that can develop a JMML-like transient MPN that will usually self-resolve within several months, but may last up to 1–2 years. Subsequently, somatic mutations were identified in 35% of *de novo* nonsyndromic cases of JMML. Somatic *PTPN11* mutations appear to encode stronger gain-of-function compared to germline mutations found in NS, likely accounting for the phenotypic differences and spontaneous resolution of the transient MPN in patients with NS.

Somatic mutations in *NRAS* and *KRAS* occur in 25% of JMML cases. Rare cases of germline mutations in *NRAS* and *KRAS* have been described and can lead to improvement of the MPN over time. Those patients appeared to have a normal HbF and a higher platelet count and might benefit from a watch-and-wait strategy.

Most recently, homozygous *CBL* mutations have been identified in 15% of JMML cases. These patients initially have germline mutations in *CBL* (inherited autosomal dominant or occur spontaneously) with risk of developing loss of heterozygosity for the *CBL* locus in the hematopoietic stem cells. Most of these patients experience a spontaneous resolution of their JMML phenotype. They are, however, at high risk for the development of vasculopathy in the second decade of life, which can potentially be prevented by HSCT.

Genetic mutations, such as *JAK2*, *TET2*, and *RUNX1*, commonly seen in adult MPN are usually not found in children with JMML. *ASXL1*, *FLT3*, and *SRSF2* mutations have been reported in rare cases of JMML. Secondary mutations in *SETBP1* and *JAK3* have been identified in 17% of JMML patients and are associated with a poor prognosis.

Figure 17.1 shows the biology of JMML and its correlation with hematologic findings in JMML.

Natural History

- The clinical course of JMML can be variable. A third of patients have aggressive disease with rapid progression while others have a more indolent course. The median survival without HSCT is <1 year.
- Most often patients die of respiratory failure, due to leukemic infiltrates and/or infections.
- Spontaneous remission has been reported in a very few cases associated with KRAS and NRAS mutations and can occur in patients with *CBL* mutations. Resolution of MPN is expected in patients with NS (germline *PTPN11* mutations).
- Blastic transformation occurs in 15% of patients and is associated with additional cytogenetic abnormalities in some patients.
- Occasionally, B-cell acute lymphoblastic leukemia (ALL) develops.

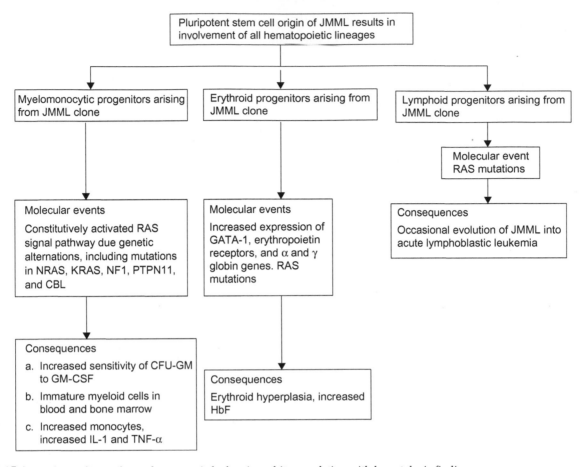

FIGURE 17.1   Biology of juvenile myelomonocytic leukemia and its correlation with hematologic findings.

Prognosis

Despite significant improvements in understanding the molecular genetics of JMML the outcome of the disease is still poor, with a 5-year OS of 52—64% after HSCT. JMML patients presenting in blast crisis have a dismal prognosis. The following adverse prognostic factors have been described:

- Age at diagnosis: 2 years and older.
- High fetal hemoglobin level at diagnosis (>10%).
- Platelet count below 33,000/mm^3 at diagnosis (considered to be the strongest indicator of prognosis).
- Presence of PTPN11 mutation.
- An AML-like gene expression signature (vs non-AML like)
- Monosomy 7 as an adverse risk factor for poor prognosis has been controversial in JMML.
- Increased hypermethylation.

Treatment

JMML responds poorly to standard chemotherapy and HSCT remains the only curative treatment option.

Hematopoietic Stem Cell Transplantation

Early HLA typing to identify a suitable family donor or best-unrelated donor is critical. Umbilical cord blood (UCB) is an alternative if a suitable donor is not available. The EWOG-MDS published the largest series of HSCT in JMML with a myeloablative-conditioning regimen consistent of busulfan, cyclophosphamide, and melphalan (Bu-Cy-Mel) and reported a DFS of 52%. Relapse is the most common treatment failure (35%) followed by transplant-related mortality (TRM). Bu-Cy-Mel is the most commonly used preparative regimen for JMML.

Serotherapy with ATG should be added in unrelated or UCB donors.

There is sufficient evidence to suggest that a graft-versus-leukemia (GVL) effect is important in JMML. Therefore an early and rapid taper of graft-versus-host disease (GVHD) prophylaxis is generally recommended.

Pre-HSCT Approaches

Pre-HSCT splenectomy has been used in JMML with the rationale to decrease disease burden and recurrence and promote engraftment. Available data on the role of pre-HSCT splenectomy are controversial. Therefore splenectomy is not generally recommended, but may be considered in selected cases.

Similarly, pre-HSCT chemotherapy for JMML remains controversial. There is no clear evidence that standard chemotherapy improves survival or relapse after HSCT and is not recommended. For asymptomatic patients a "watch-and-wait" strategy can be considered. In an effort to decrease disease burden and symptomatology (secondary to organomegaly and pulmonary infiltrates) the following agents have been used as single agents or in combination as a bridge to HSCT and can be given over an extended period of time:

- Oral 6-mercaptopurine ($50 \, mg/m^2/day$).
- Cis-retinoic acid ($100 \, mg/m^2/day$).
- Low-dose cytarabine ($40 \, mg/m^2/day \times 5 days$).

In children with aggressive disease and life-threatening pulmonary infiltration the following regimen has been used:

- Fludarabine ($30 \, mg/m^2/day \times 5$ days) *plus* high-dose cytarabine ($2 \, g/m^2/day \times 5$ days).

Note: Patients should be monitored for pulmonary rebound phenomena after recovery from any chemotherapy.

Targeted/Investigational Therapies

Despite rapid advances in the molecular genetics, this knowledge has not yet been translated into successful targeted therapies for children with JMML.

Inhibition of the RAS/MAPK pathway has been challenging. Earlier trials of the farnesyl transferase inhibitor tipifarnib given prior to HSCT did not lead to improved survival. MEK inhibitors show promising results in murine models and are in development for clinical trials. The DNA hypomethylating agent 5-azacytidine has been reported in a single anecdotal case of JMML with *KRAS* mutation and JMML. A clinical trial for newly diagnosed or relapsed high-grade MDS and JMML patients is now offered in Europe.

Treatment Options for Patients with Relapse After Allogeneic HSCT

Early detection of molecular relapse by dropping donor chimerism or detection of the genetic clone allows for early intervention. Rapid withdrawal of immunosuppression and/or donor lymphocyte infusion (DLI) may prevent clinically overt relapse, but is usually not curative. Second allogeneic HSCT can salvage about 50% of relapsed patients and should be performed as soon as possible.

MYELOPROLIFERATIVE NEOPLASMS

Table 17.8 provides the 2008 WHO Classification of myeloproliferative neoplasms.

TABLE 17.8 WHO Classification of Myeloproliferative Neoplasms

Chronic myelogenous leukemia (Ph chromosome, t(9;22) (q34;q11), BCR/ABL1 positive)
Chronic neutrophilic leukemia
Chronic eosinophilic leukemia (and the hypereosinophilic syndrome)
Polycythemia vera
Primary myelofibrosis
Essential thrombocythemia
Mastocytosis
Myeloproliferative neoplasms, unclassifiable.

Swerdlow et al. (2008).

CML, BCR−ABL1 POSITIVE

CML is a clonal myeloproliferative disorder of the pluripotent hematopoietic stem cell or granulocyte-macrophage progenitor cell and is characterized by the presence of the Philadelphia (Ph[1]) chromosome. This abnormal chromosome results from reciprocal translocation involving the long arms of chromosomes 9 and 22, t(9;22)(q34;q11).

Incidence

The overall incidence worldwide is 1−2 cases per 100,000. The incidence is approximately one per one million for persons younger than 20 years and 1−3% of all childhood leukemia. There are about 100 cases of pediatric CML diagnosed in the United States per year.

Clinical Phases

Chronic Phase

- Blood or BM contain <10% leukemic blasts.
- Clinically stable for several years.

Signs and Symptoms

- Nonspecific complaints: fever, night sweats, abdominal pain, bone pain.
- Symptoms resulting from hyperviscosity:
  - Neurologic dysfunction: headache, strokes.
  - Visual disturbances: retinal hemorrhages, papilledema.
  - Priapism.
- Hepatomegaly, splenomegaly.
- Pallor.

Laboratory Findings

Hematologic findings:

- Mild normocytic, normochromic anemia.
- Leukocytosis with all stages of neutrophilic maturation with peaks in the myelocytic and segmented neutrophilic stages.
- Monocytes usually <3% of WBCs.
- Increased absolute eosinophil and basophil counts.
- Thrombocytosis.
- Decreased leukocyte alkaline phosphatase score.
- Progressive deterioration of neutrophil function.
- BM examination: hypercellularity, granulocytic hyperplasia with foci of left-shifted maturation, increase eosinophilic and basophilic series, increased megakaryocytes. There may be myelofibrosis. Gaucher-like cells or sea-blue histiocytes may be present.
- Cytogenetics/genetics: presence of Ph[1] chromosome and/or *BCR−ABL1* fusion gene detected by FISH, PCR, or Southern blot analysis.

Blood chemistry:

Elevation of lactic dehydrogenase, uric acid, vitamin B12 associated with haptocorrins (formerly a form of transcobalamins).

Accelerated Phase

- Poorly defined intermediate phase characterized by increasing blood and BM leukemic blasts (10−19%), persistent cytopenias or persistent thrombocytosis or leukocytosis.
- Dyspoiesis, rising basophil count, progressive splenomegaly, progressive myelofibrosis, or refractoriness to therapy.
- Clonal karyotypic evolution.
- Signs and symptoms: fever, night sweats, weight loss.

Blast Crisis

- Blasts >20% in blood or BM; or extramedullary or intramedullary clusters of blasts.
- *Myeloid blast crisis*: most common type of blast crisis (80%), may be myeloblastic or myelomonocytic. Associated karyotypic evolution: duplication of Ph[1] chromosome, trisomies of 8, 19, or 21 chromosomes, i(17q), t(7;11), AML-specific rearrangements, for example, t(15;17).
- *Lymphoid blast crisis*: less common type of blast crisis (15–20%), more often B lineage than T. Associated karyotypic evolution: duplication of Ph[1] chromosome, trisomy 21, inv(7), t(14;14).
- *Mixed phenotype blast crisis*
  - Blasts may co-express antigens of different lineage or two distinct populations of blasts may be present, such as B/myeloid and T/myeloid.
- *Erythrocytic blast crisis*: rare in pure form.
- *Megakaryoblastic blast crisis*: rare in pure form, associated with inv(3)(q21;q26) or t(3)(q21;q26).
- *Mast cell crisis*: extremely rare.
- Signs and symptoms: pallor, easy bruisability, pruritis, urticaria, bone pain

Biology of CML

Molecularly targeted therapies have revolutionized cancer therapeutics and are much more selective in their actions than traditional chemotherapy agents. CML is a disease that exemplifies how the discoveries in molecular biology have helped design molecularly targeted therapies.

Figure 17.2 depicts the Ph[1] chromosome, the molecular genetic studies of which have revealed that the ABL (Abelson) segment from chromosome 9 is translocated to chromosome 22 at its major breakpoint cluster region (BCR)

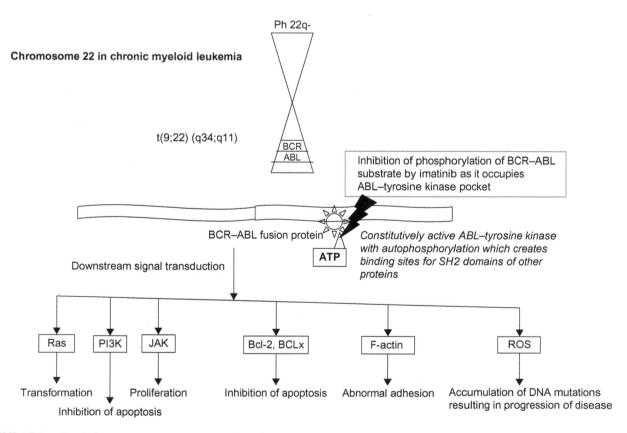

FIGURE 17.2 t(9;22), fusion gene BCR–ABL, cellular effects of BCR–ABL and inhibition of BCR–ABL by imatinib in chronic myeloid leukemia.

and thus forming a novel fusion gene termed the BCR–ABL. The BCR–ABL fusion gene usually makes an oncoprotein of 210 kDa molecular weight. The following are some of the functional domains of BCR–ABL oncoprotein.

Domain	Location	Function
Oligomerization domain	BCR	Activation of ABL kinase
Y^{177}	BCR	Y^{177} regulates the Ras and P13K pathways through recruitment of GRB2-SOS and GAB2-P13K. These pathways are important for transformation and proliferation
Serine–threonine kinase	BCR	Activation of signal transduction proteins
Y kinase (SH1)	ABL	Constitutive tyrosine kinase activity, which is responsible for phosphorylation of signal and adaptor proteins and plays a central role in leukemogenesis
Actin-binding domain	ABL	Interference with adhesion

The BCR/ABL oncoprotein activates directly or indirectly many downstream signal transduction pathways, for example, Ras, P13K, JAK-STAT, Bcl-2, BCLx, and ROS, which induce transformation, proliferation, inhibition of apoptosis, and oxidative damage to DNA. Additionally, cytoskeletal proteins are tyrosine phosphorylated and this results in alteration of cytoskeletal function. CML cells are less adherent and egress from the marrow into circulating blood prematurely. Figure 13.2 also depicts various cellular effects of BCR/ABL oncoprotein through downstream transduction pathways.

Several mechanisms play roles in disease progression, which lead eventually to blast crisis of CML. They include:

- Drug resistance:
    Mechanisms
    1. BCR/ABL gene amplification.
    2. Overexpression of BCR/ABL transcripts.
    3. Mutations in the ABL tyrosine kinase domain at adenosine triphosphate (ATP) binding site.
    4. Clonal evolution.
    5. Drug efflux.
- Genomic instability.
- Impaired DNA repair.
- Tumor suppressor gene inactivation.
- Differentiation block.
- BCR/ABL-independent activation of more oncogenes and thus, participation of more oncoproteins resulting in redundant pathways for proliferation, decreased apoptosis, and differentiation block.

Treatment

Frontline use of a BCR–ABL targeting tyrosine kinase inhibitor (TKI) is now the standard of care for most patients with CML.

Treatment of Chronic Phase of CML

Firstline: Imatinib (see dosage below).
Definitions of response criteria

1. Complete hematologic response (CHR): platelet count <450,000/mm^3; WBC <10,000/mm^3; differential without immature granulocytes and with <5% basophils; nonpalpable spleen.
2. Major cytogentic response (MCR): <35% Ph+ cells in BM.
3. Complete cytogenetic response (CCR): No detectable Ph+ cells in BM.
4. Major molecular response (MMR) : BCR–ABL ratio <0.1% by reverse transcriptase polymerase chain reaction (RT-PCR) (3 log reduction in BCR–ABL transcript).
5. Complete molecular response (CMR): No detectable BCR–ABL by RT-PCR.

Frequency of monitoring response to imatinib:

1. Hematologic: monitor every 2 weeks until CHR then every 3 months.
2. Cytogenetic: monitor every 6 months until CCR then annually.
3. Molecular: monitor every 3 months.

TABLE 17.9 Response Criteria for Treatment with Imatinib

Category	Suboptimal response	Treatment failure
Hematologic	No complete hematologic response after 3 months	No response after 3 months
Cytogenetic	No major cytogenic response (MCR) after 6 months; No complete cytogenic response (CCR) after 12 months	No response after 6 months; no MCR after 12 months; no CCR after 18 months
Molecular	No MMR after 18 months	

Any loss of a hematologic or cytogenetic complete response that is confirmed on repeat is a treatment failure. Any loss of a major molecular response (MMR) is considered a suboptimal response but not a treatment failure. See text for definitions.

Table 17.9 lists the definitions of a suboptimal response and treatment failure.

- If there is failure or resistance to imatinib, test for highly resistant mutations. If mutation is found and mildly resistant (IC50 <fivefold unmutated BCR—ABL), increase imatinib dose.
- If mutation is highly resistant (IC50 ≥ fivefold unmutated BCR—ABL) change to second-generation TKI (dasatinib or nilotinib).
- If fails (or intolerant to) second-generation TKI, consider experimental TKIs, HSCT, or hydroxyurea.
- If imatinib suboptimal response, increase dose.

Metabolic problems

Tumor lysis syndrome: hydration, alkalinization, and allopurinol are employed (see Chapter 33). Tumor lysis is rare in chronic phase of CML and usually occurs in blast crisis or advanced accelerated phase.

Hyperleukocytosis: consider hydroxyurea and/or leukapheresis. Leukapheresis is rarely indicated. Some investigators recommend leukapheresis when WBC >600,000/mm^3.

Mechanism of Action of Imatinib

Imatinib inhibits ABL-tyrosine kinase by occupying the ATP-binding site in the kinase domain of the ABL component of the BCR/ABL oncoprotein (see Figure 17.2). As a result phosphorylation of ABL-tyrosine kinase substrates fails to occur, and activation of downstream leukemogenic signal transduction is prevented.

Dose

- In adults: chronic phase of CML 400 mg/day orally.
  Accelerated phase of CML 600 mg/day orally.
  Blastic transformation of CML 600 mg/day orally.
- In children: chronic phase of CML 340 mg/m^2/day orally.

Side Effects of Imatinib

1. Myelosuppression.
2. Gastrointestinal toxicity: nausea, vomiting, diarrhea.
3. Edema and fluid retention: use of diuretics may be necessary.
4. Muscle cramps, bone pain, and arthralgia: use of calcium and magnesium supplements is effective in controlling muscle cramps, in spite of normal serum levels of ionized calcium and magnesium.
5. Skin rashes: use of antihistamines and/or topical steroids may control rashes. Occasionally, Stevens—Johnson syndrome may develop. In that case, imatinib should be discontinued and systemic steroids started.
6. Hepatotoxicity: transaminitis.
7. Drug interactions: imatinib is metabolized in the liver by the CYP3A4/5 cytochrome P450 enzyme system.

Treatment of Advanced Phases of CML

Treatment of CML in Accelerated Phase

Patients with accelerated phase are treated similarly to patients with chronic phase; however, they are more likely to fail TKIs and more likely to require HSCT.

Treatment of CML in Blastic Phase (Blast Crisis)

Patients with blastic phase who are newly diagnosed may response to TKIs and these agents should be tried first. A majority of patients will need HSCT and cytoreductive chemotherapy resembling AML therapy for patients with myeloid blasts or ALL therapy for lymphoid blasts can be used (see Chapters 18 and 19).

The remission rates in blast cell phase of CML are low and duration of remission is usually short.

Allogeneic Stem Cell Transplantation for CML

For patients who fail TKIs, HSCT may be indicated. HSCT is very successful in patients with CML, partially because of a strong GVL effect. GVL results from genetic disparity between donor and recipient. Therefore, greatest GVL effect occurs when the donor is unrelated to the recipient, because of a greater number of incompatible minor histocompatibility antigens, and also, disparities between more major histocompatibility antigens.

Strongest GVL occurs when the transplantation is performed during the chronic phase in the presence of minimal residual disease, detectable by molecular testing. The weakest GVL effect occurs when the transplant is performed in a blastic phase and/or T-cell depletion is used for prevention of GVHD.

Specific antigens implicated in the GVL response in CML are:

Antigen	Protein	Peptide
Minor histocompatibility	–	HA-1
antigens	–	HA-2
	SMYC	HY-1
Tissue-restricted	Proteinase 3	PR-1
		PR-7
Leukemia-specific	BCR/ABL	KQSSKALQR

The immune response also generates CD4+ Th_1 and CD4+ Th_2 cells. Th_1 cells make IL-2, interferon-γ (IFN-γ), and TNF, which induce proliferation of T cells (induced by IL-2) and apoptosis of CML cells (induced by IFN-γ and TNF). BCR/ABL-specific Th_1 cells exist in circulating blood, but they do not recognize the CML cells, indicating involvement of antigens other than BCR/ABL in GVL effect induced by Th_1 cells. In this regard, Th_1 cells exhibit leukemia-specific cytotoxicity through tissue-restricted antigens. Th_2 cells make IL-4, IL-5, and IL-10, and are implicated in humoral responses.

Treatment of post-transplant relapse:

- Donor lymphocyte infusion (DLI).
- TKIs.
- DLI plus TKIs.
- Re-transplant.
- If the patient relapses with accelerated or blast phase, intensive chemotherapy followed by a stem cell transplant may be performed.

ESSENTIAL THROMBOCYTHEMIA AND POLYCYTHEMIA VERA

See Chapter 14 for Essential Thrombocythemia, and Chapter 12 for Polycythemia Vera.

PRIMARY MYELOFIBROSIS

Primary myelofibrosis (PMF) is a clonal myeloproliferative disorder characterized by leukoerythroblastosis, proliferation of predominantly megakaryocytes and granulocytes, varying degrees of reactive myelofibrosis, and extramedullary hematopoiesis. It is a rare disease in children and they have a better outlook than adults. Its etiology is unknown.

Clinical Features

Malaise, night sweats, weight loss, and discomfort from splenomegaly.

Hematologic Findings

- Blood smear—teardrop cells, leukoerythroblastosis, increased platelet count, and leukocytosis in the initial phase followed by cytopenias.
- BM is difficult to aspirate in the fibrotic phase. BM biopsies show fibrosis with abnormal megakaryocytes with increased hyperchromatic and naked nuclei. Osteosclerosis may occur in the terminal phases.
- Cytogenetic abnormalities may be present in BM cells.

Differential Diagnosis

Myelofibrosis can occur secondary to metastatic, autoimmune, immunologic, acute leukemic diseases, and vitamin D deficiency. These disorders are more common in children with PMF.

Complications

- Thrombocytosis, bleeding.
- Splenomegaly: hypersplenism, refractory thrombocytopenia, hemolytic anemia, some children may not have hepatosplenomegaly.
- Portal hypertension.
- Transformation into acute leukemia.

Genetic Mechanisms

In adults ~50–60% of cases are due to mutation in the *JAK2* gene (V617F) and 5–10% are related to MPL gene mutation. Recently, calreticulin gene mutations or deletions were reported in the majority of cases without *JAK2* or *MPL* mutation, which amount to 30% of all PMF cases. With the exception of older children and young adults, these mutations are usually not present in young children. Familial cases of PMF occur suggesting an underlying genetic cause, but the underlying genetic mechanism, particularly in infants and young children with PMF, remains unknown.

Recently mutations in *VPS45* were described in children with a distinct syndrome of congenital neutrophil defect syndrome, neutropenia, immunodeficiency, nephromegaly, and marrow fibrosis.

Treatment

- Some adult patients with PMF will be observed. A number of agents have been used in adults with PMF. These include hydroxyurea, IFN-α, corticosteroids, androgens, danazol, lenalidomide, thalidomide (in adults with 5q- and PMF).
- Recently, JAK2 inhibitors have become the first line of therapy in most patients resulting in a markedly improved outcome.
- None of these agents are routinely used in children and only limited data are available on their efficacy since the disease is exceedingly rare.
- Splenectomy has occasionally been used for hypersplenism and/or severe discomfort in adults.

HSCT is the only curative therapy and should be considered for most pediatric patients for whom a diagnosis of PMF has been established. Most of the reported pediatric PMF cases in the literature were successfully treated with HSCT.

Further Reading and References

Andolina, J.R., Neudorf, S.M., Corey, S.J., 2012. How I treat childhood CML. Blood 119, 1821–1830.

Baumann, I., Führer, M., Behrendt, S., et al., 2012. Morphologic differentiation of severe aplastic anemia from hypocellular refractory cytopenia of childhood: reproducibility of histopathologic diagnostic criteria. Histopathology 61, 10–17.

Chan, R.J., Cooper, T., Kratz, C.P., et al., 2009. Juvenile myelomonocytic leukemia: a report from the 2nd international JMML symposium. Leuk Research 33, 355–362.

Chang, T.Y., Dvorak, C.C., Loh, M.L., 2014. Bedside to bench in juvenile myelomonocytic leukemia: insights into leukemogenesis from a rare pediatric leukemia. Blood 124, 2487–2497.

Dvorak, C.C., Loh, M.L., 2014. Juvenile myelomonocytic leukemia: molecular pathogenesis informs current approaches to therapy and hematopoietic cell transplantation. Front. Pediatr. 2, 25.

Elghetany, M.T., 2007. Myelodysplastic syndromes in children: a critical review of issues in the diagnosis and classification of 887 cases from 13 published series. Arch. Pathol. Lab. Med. 131, 1110–1116.

Gowin, K., Emanuel, R., Geyer, H., et al., 2013. The new landscape of therapy for myelofibrosis. Curr. Hematol. Malig. Rep. 8, 325–332.

Greenberg, P.L., Tuechler, H., Schanz, J., et al., 2012. Revised International Prognostic Scoring System for myelodysplastic syndromes. Blood 120, 2454–2465.

Hasle, H, et al., 2003. A pediatric approach to the WHO classification of myelodysplastic and myeloproliferative diseases. Leukemia 17, 277–282.

Hasle, H., Baumann, I., Bergsträsser, E., et al., 2004. The International Prognostic Scoring System (IPSS) for childhood myelodysplastic syndrome (MDS) and juvenile myelomonocytic leukemia (JMML). Leukemia 18, 2008–2014.

Hasle, H., Niemeyer, C.M., 2011. Advances in the prognostication and management of advanced MDS in children. Br. J. Haematol. 154, 185–195.

Jabbour, E., Kantarjian, H., 2014. Chronic myeloid leukemia: 2014 update on diagnosis, monitoring, and management. Am. J. Hematol. 89, 548–556.

Klampfl, T., Gisslinger, H., Harutyunyan, A.S., et al., 2013. Somatic mutations of calreticulin in myeloproliferative neoplasms. N. Engl. J. Med. 369, 2379–2390.

Locatelli, F., Niemeyer, C.M., 2015. How I treat juvenile myelomonocytic leukemia. Blood 125, 1083–1090.

Nguyen, P., 2009. The myelodysplastic syndromes. Hematol. Oncol. Clin. North Am. 23, 680–681.

Quintás-Cardama, A., Cortes, J., 2009. Molecular biology of bcr–abl1-positive chronic myeloid leukemia. Blood 113, 1619–1630.

Swerdlow, S.H., Campo, E., Harris, N.L., Jaffe, E.S., Pileri, S.A., Stein, H., et al., 2008. WHO Classification of Tumors of Haematopoietic and Lymphoid Tissues. IARC, Lyon.

Vilboux, T., Lev, A., Malicdan, M.C., et al., 2013. A congenital neutrophil defect syndrome associated with mutations in VPS45. N. Engl. J. Med. 369, 54–65.

Woodard, P., Carpenter, P.A., Davies, S.M., et al., 2011. Unrelated donor bone marrow transplantation for myelodysplastic syndrome in children. Biol. Blood Marrow Transplant. 17 (5), 723–728.

18

Acute Lymphoblastic Leukemia

William L. Carroll and Teena Bhatla

Acute leukemias represent a clonal expansion and arrest at a specific stage of normal myeloid or lymphoid hematopoiesis. They constitute 97% of all childhood leukemias and consist of the following types:

- Acute lymphoblastic leukemia (ALL)—75%: Amongst ALL, the nomenclature of the subtypes has been changed from "precursor B-lymphoblastic leukemia/lymphoma" and "precursor T-lymphoblastic leukemia/ lymphoma" to "B or T-lymphoblastic leukemia/lymphoma" respectively.
- Acute myeloblastic leukemia (AML), also known as acute non-lymphocytic leukemia—20%.
- Acute undifferentiated leukemia— < 0.5%.
- Acute mixed-lineage leukemia.

Chronic myeloid leukemias constitute 3% of all childhood leukemias and consist of

- Philadelphia chromosome-positive (Ph[1] positive) myeloid leukemia.
- Juvenile myelomonocytic leukemia.

INCIDENCE OF ALL

1. In the United States, 3–4 cases per 100,000 white children and 2500–3000 children diagnosed per year.
2. Peak incidence between 2 and 5 years of age.
3. Accounts for 25–30% of all childhood cancers.

ETIOLOGY

The etiology of acute leukemia is unknown. The following factors are important in the pathogenesis of leukemia:

- Ionizing radiation.
- Chemicals (e.g., benzene in AML)
- Drugs (e.g., use of alkylating agents either alone or in combination with radiation therapy increases the risk of AML).
- Genetic considerations:
  - Identical twins—If one twin develops leukemia during the first 5 years of life the risk of the second twin developing leukemia is 20%.
  - Incidence of leukemia in siblings of leukemia patient is four times greater than that of the general population.
  - Chromosomal abnormalities:

Group	Risk	Time interval
Trisomy 21 (Down syndrome)	1 in 95	<10 years of age
Bloom syndrome	1 in 8	<30 years of age
Fanconi anemia	1 in 12	<16 years of age

Lanzkowsky's Manual of Pediatric Hematology and Oncology.
DOI: http://dx.doi.org/10.1016/B978-0-12-801368-7.00018-1

- Increased incidence with the following genetically determined conditions:
  - Congenital agammaglobulinemia.
  - Poland syndrome.
  - Shwachman—Diamond syndrome.
  - Ataxia telangiectasia.
  - Li—Fraumeni syndrome (germline p53 mutation)—the familial syndrome of multiple cancers in which acute leukemia is a component malignancy.
  - Neurofibromatosis.
  - Diamond—Blackfan anemia.
  - Kostmann disease.
  - Bloom syndrome.

Most cases of leukemia do not stem from an inherited genetic predisposition but from somatic genetic alterations. However, recent studies indicate a possible genetic linkage with inherited polymorphisms in *ARID5B* and *IKZF1* genes for childhood ALL.

CLINICAL FEATURES OF ALL

Table 18.1 shows the common clinical and laboratory presenting features in ALL.

TABLE 18.1   Clinical and Laboratory Presenting Features in Acute Lymphoblastic Leukemia

Clinical and laboratory presenting features	Percentage of patients
SYMPTOMS AND PHYSICAL FINDINGS	
Fever	61
Bleeding (e.g., petechia or purpura)	48
Bone pain	23
Lymphadenopathy	50
Splenomegaly	63
Hepatosplenomegaly	68
LABORATORY FEATURES	
Leukocyte count (mm³)	
<10,000	53
10,000—49,000	30
>50,000	17
Hemoglobin (g/dl)	
<7.0	43
7.0—11.0	45
>11.0	12
Platelet count (mm³)	
<20,000	28
20,000—99,000	47
>100,000	25
Lymphoblast morphology	
L1	84
L2	15
L3	1

From: Pizzo and Poplack (2006), with permission.

General Systemic Effects

1. Fever (60%).
2. Lassitude (50%).
3. Pallor (40%).

Hematologic Effects Arising from Bone Marrow Invasion

1. Anemia—causing pallor, fatigability, tachycardia, dyspnea, and sometimes congestive heart failure.
2. Neutropenia—causing fever, ulceration of buccal mucosa, and infection.
3. Thrombocytopenia—causing petechia, purpura, easy bruisability, bleeding from mucous membrane, and sometimes internal bleeding (e.g., intracranial hemorrhage).
4. One to 2% of patients present initially with pancytopenia and may be erroneously diagnosed as having aplastic anemia or bone marrow failure (represents 5% of acquired aplastic anemia) and ultimately develop acute leukemia. In these cases the illness is characterized by:
   * Pancytopenia or single cytopenia.
   * Hypocellular bone marrow.
   * No hepatosplenomegaly.
   * Diagnosis of leukemia 1–9 months after onset of symptoms.

The treatment consists of supportive transfusions initially and specific antileukemic chemotherapy, when leukemia is diagnosed.

Clinical Manifestations Arising from Lymphoid System Infiltration

1. Lymphadenopathy—sometimes presents with bulky mediastinal lymphadenopathy causing superior vena cava syndrome. This is more common in T-cell leukemia in adolescents.
2. Splenomegaly.
3. Hepatomegaly.

Clinical Manifestations of Extramedullary Invasion

Central Nervous System Involvement

This occurs in less than 5% of children with ALL at initial diagnosis. It may present with the following:

* Signs and symptoms of raised intracranial pressure (e.g., headache, morning vomiting, papilledema, bilateral sixth-nerve palsy).
* Signs and symptoms of parenchymal involvement (e.g., focal neurologic signs such as hemiparesis, cranial nerve palsies, convulsions, cerebellar involvement—ataxia, dysmetria, hypotonia, hyperflexia).
* Hypothalamic syndrome (polyphagia with excessive weight gain, hirsutism, and behavioral disturbances).
* Diabetes insipidus (posterior pituitary involvement).
* Chloromas of the spinal cord (very infrequent in ALL)—may present with back pain, leg pain, numbness, weakness, Brown–Séquard syndrome, and bladder and bowel sphincter problems.
* Central nervous system (CNS) hemorrhage—complication that occurs more frequently in patients with AML than in ALL. It is caused by:
   * Leukostasis in cerebral blood vessels, leading to leukothrombi, infarcts, and hemorrhage.
   * Thrombocytopenia and coagulopathy, contributing to CNS hemorrhage.

Genitourinary Tract Involvement

Testicular Involvement

1. Usually presents with painless enlargement of the testis.
2. Occurs in less than 2% of boys at diagnosis. Isolated testicular relapse accounts for less than 10% of all ALL relapses.

Ovarian Involvement

Occurs very rarely.

Priapism

Occurs rarely. It is due to involvement of sacral nerve roots or mechanical obstruction of the corpora cavernosa and dorsal veins by leukemic infiltrates.

Renal Involvement

1. Occasionally may present with hematuria, hypertension, and renal failure.
2. Evaluated by ultrasonography; renal involvement is more common in T-cell ALL or mature B-cell ALL (Burkitt).

Gastrointestinal Involvement

1. The gastrointestinal (GI) tract may be involved in ALL. The most common manifestation is bleeding.
2. Leukemic infiltrates in the GI tract are usually clinically silent until terminal stages when necrotizing enteropathy might occur. The most common site for this is the cecum, giving rise to a syndrome known as typhlitis.

Bone and Joint Involvement

Bone pain is one of the initial symptoms in 25% of patients. It may result from direct leukemic infiltration of the periosteum, bone infarction, or expansion of marrow cavity by leukemic cells. Radiologic changes occasionally seen include:

- Osteolytic lesions involving medullary cavity and cortex.
- Transverse metaphyseal radiolucent bands.
- Transverse metaphyseal lines of increased density (growth arrest lines).
- Subperiosteal new bone formation.

Skin Involvement

Skin involvement occurs occasionally in neonatal leukemia or AML.

Cardiac Involvement

One-half to two-thirds of patients have demonstrated cardiac involvement at autopsy, although symptomatic heart disease occurs in less than 5% of cases due to late toxicity of treatment. Pathologic findings may include leukemic infiltrates and hemorrhage of the myocardium or the pericardium.

Lung Involvement

Lung involvement is uncommon and may be due to leukemic infiltrates or hemorrhage in patients with very high white blood cell (WBC) counts (leukostasis).

Diagnosis

Laboratory Studies

1. *Blood count*
   a. *Hemoglobin*: Moderate to marked reduction. Normocytic; normochromic red cell morphology. Low hemoglobin indicates longer duration of leukemia; higher hemoglobin may indicate a more rapidly proliferating leukemia.
   b. *WBC count*: Low, normal, or increased.
   c. *Blood smear*: Blasts are present on blood smear. Very few to none (in patients with leukopenia). When the WBC count is greater than 10,000/mm^3, blasts are usually abundant. Eosinophilia is seen uncommonly in children with ALL.
   d. *Thrombocytopenia*: 92% of patients have platelet counts below normal. Serious hemorrhage (GI or intracranial) occurs at platelet counts less than 20,000/mm^3.

2. *Bone marrow*: Bone marrow is usually replaced by 80–100% blasts. Megakaryocytes are usually absent. Leukemia must be suspected when the bone marrow contains more than 5% blasts. The hallmark of the diagnosis of acute leukemia is the blast cell, a relatively undifferentiated cell with diffusely distributed nuclear chromatin, one or more nucleoli (more prominent in AML) and scant cytoplasm (more abundant in AML). Special bone marrow studies, which help in detailed cell classification, include the following:
   a. Immunophenotyping using flow cytometry
   b. Cytogenetics—karyotype and molecular studies.
3. *Chest radiograph*: Mediastinal mass in T-cell leukemia.
4. *Blood chemistry*: Electrolytes, blood urea, uric acid, LDH, liver function tests, immunoglobulin levels.
5. *Cerebrospinal fluid* (CSF): Chemistry, cell count, and cytospin for pathological examination. Cerebrospinal fluid findings for the diagnosis of CNS leukemia require:
   a. Presence of >5 WBCs/mm^3.
   b. Identification of blast cells on cytocentrifuge examination. CNS involvement in leukemia is classified as follows:
      i. CNS1 <5 WBCs/mm^3, no blasts on cytocentrifuge slide.
      ii. CNS2 <5 WBCs/mm^3, blasts on cytocentrifuge slide.
      iii. CNS3 >5 WBCs/mm^3, blasts on cytocentrifuge slide.
   c. TdT stain for suspicious cells.
   d. If a lumbar puncture is traumatic in a patient with peripheral blasts the following formula can be helpful in defining the presence of CNS leukemia. CNS disease is present if:
      i. CSF WBC/CSF RBC >2 × Blood WBC/Blood RBC.
6. *Coagulation profile*: PT, PTT, and fibrinogen.
7. *Cardiac function*: Electrocardiogram and echocardiogram.
8. *Infectious disease profile*: Varicella antibody titer, cytomegalovirus antibody titer, herpes simplex antibody, hepatitis antibody screening.
9. *Immunologic screening*: Serum for immunoglobulin levels, C3 and C4.

Classification

Acute leukemia can be classified based on morphologic characteristics (Table 18.2), cytochemical features (Table 18.3), immunologic characteristics (Figure 18.1), and cytogenetic and molecular characteristics (Figure 18.2). The World Health Organization has developed a new classification of ALL based on cytogenetic and molecular characteristics (Table 18.4).

TABLE 18.2 Cytologic Features of the Morphologic Types of Acute Lymphoblastic Leukemias According to French–American–British Classification

Cytologic features[a]	L1	L2	L3[b]
Cell size	Small cells predominate	Large, heterogeneous in size	Large and heterogeneous
Nuclear chromatin	Homogeneous	Variable, heterogeneous	Finely stippled and homogeneous
Nuclear shape	Regular, occasional clefting or indentation	Irregular, clefting and indentation common	Regular, oval to round
Nucleoli	Not visible, or small and inconspicuous	One or more present, often large	Prominent, one or more vesicular
Amount of cytoplasm	Scanty	Variable, often moderately abundant	Moderately abundant
Basophilia of cytoplasm	Slight or moderate, rarely intense	Variable, deep in some	Very deep
Cytoplasmic vacuolation	Variable	Variable	Often prominent

[a]For each of the features considered, up to 10% of the cells may depart from the characteristic of the type.
[b]The only immunologically pure type of ALL that can be consistently recognized morphologically and invariably carries IgM surface receptor on its membrane.
From: Bennett et al. (1976), with permission.

TABLE 18.3 Cytochemical Characteristics of the Various Types of Acute Leukemias

			Acute nonlymphoblastic leukemia		
Staining reaction	ALL	AML	Acute myelomonocytic leukemia	Erythroleukemia	Megakaryoblastic leukemia
NONENZYMATIC					
PAS	Present as coarse granules or blocks in a variable number of cells	Negative or diffusely positive	Negative or fine granulation	Strongly positive granular	Positive or negative
Sudan black	Negative	Positive	Positive	Positive	Negative
ENZYMATIC					
Peroxidase	Negative	Positive[a]	Usually negative	Positive	Negative
Alkaline phosphatase	Normal	Low	High	Normal or high	—
ESTERASES					
Naphthol ASD Chloroacetate	Negative	Positive	Negative	Negative	
Naphthol ASD acetate	Negative or weakly positive	Positive (not inhibited by fluoride)	Strongly positive (inhibited by fluoride)	Weakly positive	Positive or negative
α-Naphthyl acetate	Negative	Negative	Strongly positive	Strongly positive	Positive or negative
Acid phosphatase	Positive in T-ALL	Negative	Negative	Negative	Positive (localized pattern)

[a]The peroxidase reaction, when positive, is considered to indicate the presence of myeloblastic rather than lymphoblastic elements. Unfortunately, myeloblasts that contain no specific granules will be peroxidase-negative. It is these cells that cause the greatest difficulty in morphologic classification. Generally, when morphologic classification is difficult, these histochemical tests are of little help. However, demonstration of MPO (myeloperoxidase) by immunologic techniques or expression of MPO by molecular methods can be performed in specialized research laboratories.

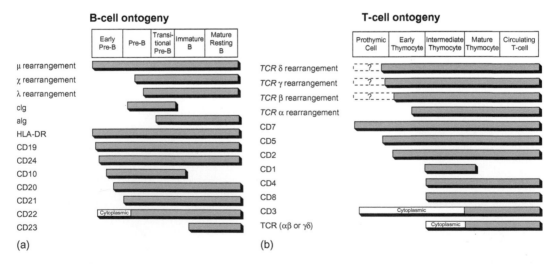

FIGURE 18.1 Schematic representation of human lymphoid differentiation. (a) Hypothetical schema of marker expression and gene rearrangement during normal B-cell ontogeny. (b) Hypothetical schema of marker expression and gene rearrangement during normal T-cell ontogeny. *From: Pui et al. (1993), with permission.*

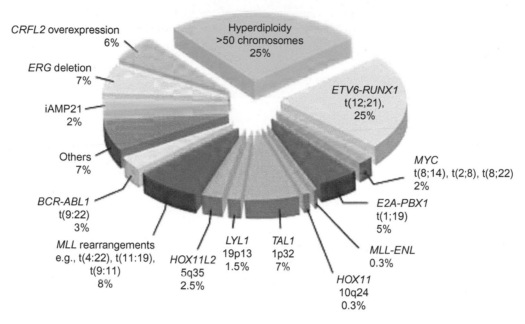

FIGURE 18.2 Estimated frequencies of specific genotypes of ALL in childhood leukemias. *From: Pui et al. (2011), with permission.*

TABLE 18.4 World Health Organization Classification of Acute Lymphoblastic Leukemia

B-lymphoblastic leukemia/lymphoma
  B-lymphoblastic leukemia/lymphoma, not otherwise specified (NOS)
  B-lymphoblastic leukemia/lymphoma with recurrent genetic abnormalities
    B-lymphoblastic leukemia/lymphoma with t(9;22)(q34;q11.2); *BCR−ABL 1*
    B-lymphoblastic leukemia/lymphoma with t(v;11q23); *MLL* rearranged
    B-lymphoblastic leukemia/lymphoma with t(12;21)(p13;q22) *TEL-AML1 (ETV6-RUNX1)*
    B-lymphoblastic leukemia/lymphoma with hyperdiploidy
    B-lymphoblastic leukemia/lymphoma with hypodiploidy
    B-lymphoblastic leukemia/lymphoma with t(5;14)(q31;q32) *IL3-IGH*
    B-lymphoblastic leukemia/lymphoma with t(1;19)(q23;p13.3); *TCF3-PBX1*
T-lymphoblastic leukemia/lymphoma

Light microscopy, cytochemistry, immunophenotyping, and cytogenetics are necessary studies to characterize leukemic subtypes. Current risk classification schema and treatment regimens however primarily utilize immunophenotypic and cytogenetic features only.

Morphology

Light Microscopy

Cytochemical criteria have been established to differentiate lymphoblasts from myeloblasts (Table 18.3). ALL can be further subclassified according to the French−American−British classification as L1, L2, and L3 morphologic types (Table 18.2).

Immunology

The putative immunologic classification and cellular characteristics of B-lineage ALL and T-cell ALL are shown in Figure 18.1.

A panel of antibodies is used to establish the diagnosis of leukemia and to distinguish among the immunologic subclones. The panel should include at least one marker that is highly lineage-specific, for example, CD19 for B-lineage, cytoplasmic CD3 for T lineage and myeloperoxidase or markers of monocytic differentiation such as non-specific esterase, CD11c, CD14, CD64, lysozyme for myeloid lineage malignancies. In addition, the use of cytoplasmic CD79a, cytoplasmic CD22, CD10 for B-lineage, surface CD3, CD7, and CD5 for T lineage and CD13 and CD33 for myeloid cells can be helpful in differentiating unclear immunophenotypes.

Immunophenotype Distribution of ALL

B-precursor cell accounts for 80% of ALL cases.
T-cell accounts for 15—20% of ALL cases. This subtype is associated with

- Older age at presentation.
- High initial white cell count.
- Presence of extramedullary disease, for example, mediastinal mass.
- Traditionally poor prognosis but treatment on high-risk intensive therapies has improved the outcome.

Mature B-cell accounts for 1—2% of ALL cases. These are surface immunoglobulin-positive and are treated as Burkitt's lymphoma. The prognosis is similar to other subtypes of high-risk ALL.

Acute Leukemia of Ambiguous Lineage

Please see discussion in Chapter 19.

Cytogenetics and Molecular Characteristics

The cytogenetic abnormalities observed in leukemia have biologic and prognostic significance. The realization of the biologic significance of leukemia cytogenetics has resulted in the appreciation of distinct biological subsets associated with prognosis (Table 18.5) and treatment stratification.

Molecular Genetics of ALL

Figure 18.2 shows the distribution of molecular rearrangements in childhood ALL.
Seventy-five percent of childhood ALL cases have evidence of chromosomal gains/losses and/or translocations:

- Hyperdiploid ALL is characterized by whole chromosomal gains and is observed in 30% of patients. The gains are non-random (trisomy 21 is the most common). Specific trisomies (4, 10 and 17 and 18 in some studies) are associated with a particularly good outcome.
- Hypodiploid ALL is found in approximately 6% of patients, characterized by fewer than 46 chromosomes in the leukemic blasts. Patients with either a karyotype with <44 chromosomes or a DNA index of <0.81 have a worse outcome.
- ETV6-RUNX1 fusion gene t(12;21) (p13q22). t(12;21), previously referred to as TEL-AML1 is detected by standard cytogenetics in less than one in 1000 cases; whereas, using molecular techniques such as fluorescent *in situ* hybridization (FISH), it is detected in approximately 25% of B-ALL cases. This translocation is associated with an excellent prognosis.
- BCR—ABL fusion gene t(9;22) (q34q11). t(9;22) occurs in only 3% of pediatric ALL cases. This is in contrast to adult ALL where this translocation is present in 25% of cases and 95% of chronic myelogenous leukemia (CML)

TABLE 18.5 Prognostic Significance of Chromosomal Abnormalities in Acute Lymphoblastic Leukemia

Chromosomal abnormalities	EFS
Trisomies of chromosomes 4, 10, and 17	89% 8 years
ETV6-RUNX1 Fusion positive	86% 5 years
Hypodiploidy <45 chromosomes	39% 8 years
MLL-REARRANGED	
Infants (0—12 months)	37% 4 years
Non infants (more than 12 months)	50% 8 years
t(9;22)	39% 8 years[a]

[a]*Studies have shown a 3-year EFS of 80% using intensive chemotherapy with Imatinib.*
EFS, event-free survival.

patients. In pediatric BCR–ABL-positive ALL, the BCR breakpoint characteristically produces a 190-kDa protein (p190) in contrast to CML where a different protein (p210) is usually produced. The t(9;22) in pediatric ALL is usually associated with older age, higher WBC count, and frequent CNS involvement at diagnosis.

- TCF3-PBX1 fusion gene t(1;19) (q23p13.3). t(1;19) is frequently associated with an elevated WBC at diagnosis and occurs in approximately 5% of newly diagnosed ALL. While earlier studies have indicated it as a relatively poor prognostic marker, the contemporary intensive treatment on current ALL regimens has shown no difference in the outcomes.
- On the other hand, TCF3-HLF fusion gene t(17:19) occurs in approximately 1% of newly diagnosed patients and has been associated with hypercalcemia, DIC and poor prognosis. This translocation will be classified under "very high category" in the upcoming COG classification schema, and patients will receive intensive upfront treatment including the option of stem cell transplant in first remission.
- MLL gene rearrangement at chromosome band 11q23 affects 80% of ALL cases in infants, 3% of ALL cases in older children, and 85% of secondary AML involving the use of topoisomerase II inhibitors. This translocation carries a very poor prognosis despite intensive therapy.
- Intrachromosomal amplification of chromosome 21 (iAMP21), being defined as at least four copies of RUNX1 in a single chromosome, is found in approximately 2% of pediatric B-ALL patients. iAMP21 is associated with inferior outcome, particularly for standard-risk patients, however, outcomes are better when these patients are treated with intensive chemotherapy regimens.
- Mature B-cell ALL translocations involve MYC genes on chromosome 8q24. Eighty percent of mature B-ALL cases contain t(8;14) (q24q32); the remaining cases have t(2;8) (p12q24) or t(8;22) (q24q11). All of these translocations deregulate MYC expression through association with highly active immunoglobulin genes and need to be treated with short and intensive chemotherapy comprising the same agents used to treat Burkitt lymphoma.
- More than 50% of the cases of T-cell ALL have activating mutations that involve *NOTCH1*, a gene encoding a transmembrane receptor that regulates normal T-cell development. An additional 20% of cases have mutations in *FBXW7* which degrades intracellular NOTCH. When *NOTCH* receptors become activated there is a cascade of proteolytic cleavages, which causes intracellular *NOTCH1* to translocate into the nucleus, which then regulates the transcription of a set of oncogenes including *MYC*. It is believed that aberrant *NOTCH* signaling causes constitutive expression of *MYC* and activation of *RAS*. Interference with *NOTCH* signaling by small-molecule inhibition has the potential for induction of remission in T-cell ALL.

Future Directions in ALL Classifications

1. Gene expression profiling using DNA microarray technology and next-generation sequencing technology can define biologically and prognostically distinctive ALL subsets, identify genes which may be responsible for leukemogenesis and may identify genes for which targeted therapy could be developed. The discovery of "Ph-like" ALL in up to 15% of patients (older age, inferior prognosis) that is associated with novel tyrosine kinase fusions and *JAK* mutations, provides an opportunity for targeted therapy with tyrosine kinase and JAK inhibitors.
2. Host pharmacogenomics: Polymorphisms for genes that encode drug-metabolizing enzymes can influence the efficacy and toxicity of chemotherapy. Gene polymorphisms for thiopurine methyltransferase is a gene that catalyzes the inactivation of mercaptopurine. Ten percent of the population carries at least one variant allele. This results in high levels of active metabolites of mercaptopurine. These patients, especially homozygous, have an increased risk of side effects and require marked reduction of 6-mercaptopurine doses.
3. Identification of high-risk genetic lesions in newly diagnosed ALL: Investigators have identified DNA copy number abnormalities in genes that encode regulators of B-cell development, such as *CDKN2A/B*, *PAX5*, *IKZF1*, and *EBF1*. The alteration of *IKZF1*, a gene that encodes the lymphoid transcription factor *IKAROS*, has been associated with poor outcome. A deletion or a mutation of *IKZF1* is strongly associated with elevated levels of MRD. In addition, *JAK2* mutations, somatic alterations in RAS pathway genes (*NRAS*, *KRAS*, *PTPN11*, and *NF1*) and *CRLF2* alterations have been identified in increased frequency in high-risk patients. The Janus genes *JAK1* and *JAK2* have been shown to be mutated in Down syndrome patients with B-lymphoblastic ALL and T-lymphoblastic ALL as well. Current high-risk Children's Oncology Group (COG) ALL trials are utilizing these markers to determine their frequency and prognostic significance, with a goal to further refine the risk stratification schema.

PROGNOSTIC FACTORS

Table 18.6 lists the prognostic factors in childhood ALL.

Age and WBC count are two of the most important predictors of outcome. Patients who are between the ages of 1–9 years with an initial WBC of less than <50,000/mm^3 (standard risk), which includes two-thirds of B-lymphoblastic leukemia patients, have a 4-year event-free survival (EFS) of over 80%. The remaining patients (high risk) have a 4-year event-free survival of 75%. Factors that are included in risk classification are:

- Age—Patients under 1 year of age and greater than 10 years of age have a worse prognosis than children more than 1 year and less than 10 years of age. Infants under 1 year of age have the worst prognosis.
- White cell count—Children with the higher WBC tend to have a poor prognosis.
- Immunophenotype—B-lymphoblastic ALL has the best prognosis. T-lymphoblastic ALL has a worse survival in part due to its association with older age and with higher WBC at diagnosis. Mature B-cell ALL previously had a poor prognosis with early relapses and CNS involvement but recent aggressive therapies have improved prognosis.
- Cytogenetics—The combinations of trisomies of chromosomes 4, 10, and 17 have been associated with good outcome. Likewise, translocations involving ETV6-RUNX1 are also associated with excellent prognosis. Translocations involving the MLL rearrangement on 11q23 (not MLL deletion) have been associated with an inferior prognosis. The Philadelphia chromosome t(9;22)(q34;q11) ALL has been traditionally associated with a bad prognosis; however, the incorporation of tyrosine kinase inhibitors to the intensive chemotherapy backbone has dramatically improved the outcomes.
- DNA index—Modal chromosome number can also be assessed by the DNA index and a value greater than 1.16 (chromosome number >50) is associated with good outcome. This is thought to be due to decreased apoptosis threshold and increased sensitivity to chemotherapeutic agents. Hypodiplody (less than 44 chromosomes and/or DNA index <0.81) on the other hand are associated with poor outcome.
- CNS disease—The presence of CNS disease at diagnosis is an adverse prognostic factor despite intensification of therapy with CNS irradiation and additional intrathecal therapy. The presence of blasts on cytospin without an increased WBC (CNS2 status) is also associated with an inferior outcome.
- Early response to induction therapy. Patients who are not in remission at the end of induction therapy (<2–5% of all patients) have a very poor prognosis. Laboratory methods such as polymerase chain reaction of antigen receptor genes or flow cytometry can be used to detect minimal residual disease (MRD) in a background of normal cells. Among those patients who are in remission the presence of detectable MRD levels >0.01% has a worse prognosis and requires treatment intensification. In contrast, those patients whose peripheral blood MRD clears by day 8 (and have no detectable bone marrow MRD at day 29) have an excellent prognosis. Table 18.7 lists the classification of bone marrow remission status in ALL. Table 18.8 shows the levels of detection of MRD in leukemia by various methods.

TABLE 18.6   Prognostic Factors in Childhood Acute Lymphoblastic Leukemia

Factor	Favorable	Less favorable
Age (years)	1–9	<1 or >10
White blood cell count (×10^9/l)	<50	>50
Immunophenotype	B-precursor cell	T-cell
Sex	Girls	Boys
Genetics	Hyperdiploidy >50 chromosomes or DNA index >1.16	Hypodiploid <44 chromosomes or DNA index <0.81
	Trisomies 4,10, and 17	MLL rearrangement
	t(12;21)/ETV6-RUNX1	t(9;22)/BCR−ABL1 iAMP21
CNS status	Absent	Present
MRD	End of induction day 29 MRD <0.01%	End of Induction day 29 MRD >0.01% Positive MRD at end consolidation

MRD, minimal residual disease.

TABLE 18.7 Classification of Bone Marrow Remission Status in Acute Lymphoblastic Leukemia

Classification	% Blasts in bone marrow
M_1 bone marrow	≤5
M_2 bone marrow	5–25
M_3 bone marrow	>25

Based on 200-cell count, true remission requires M_1 marrow status with normal marrow cellularity and presence of all cell lines.

TABLE 18.8 Levels of Detection of Minimal Residual Disease in Leukemia

Method	Lowest levels of detection[a]
Morphology	5 per 100
Conventional cytogenetics	2 per 100
Fluorescent in situ hybridization	1 per 1000
Multicolor flow cytometry	1 per 10,000
Polymerase chain reaction	1 per 1,000,000

[a]Number of leukemic cells per number of normal bone marrow cells.

TREATMENT

The general supportive care for a child newly diagnosed with leukemia requires special attention to the prevention and management of tumor lysis syndrome and febrile neutropenia (see Chapter 33) as well as paying detailed attention to the psychosocial aspects of the care of the patient and the family (see Chapter 35).

Treatment of Newly Diagnosed ALL

There are many successful treatment regimens for ALL. All ALL regimens include certain treatment elements: remission induction, consolidation (intensification of remission), prevention of CNS leukemia, and maintenance therapy. Remission induction is achieved with three or four drugs (i.e., vincristine, prednisone, asparaginase, with or without an anthracycline depending on the risk classification) and intrathecal chemotherapy. This regimen produces a 95–98% remission induction rate.

The aim of therapy in acute leukemia is to cure the patient and includes the following:

- To induce a clinical and hematologic remission.
- To further consolidate remission with rounds of non-cross-resistant and more intensive periods of treatment early in treatment course (e.g., consolidation, interim maintenance, and delayed intensification (DI)).
- To prevent the emergence of disease in CNS sanctuary sites by using prophylactic CNS therapy (usually intrathecal chemotherapy).
- To further eradicate residual low-level disease as measured through MRD assessment through prolonged maintenance therapy.
- To prevent and treat the complications of therapy and of the disease.

A complete remission usually achieved after 1 month of treatment is defined as:

- No symptoms attributable to the disease (e.g., fever and bone pain).
- No physical findings related to the disease such as hepatosplenomegaly, lymphadenopathy, or other clinical evidence of residual leukemic tissue infiltration.
- A normal blood count, with minimal levels of $500/mm^3$ granulocytes, $75,000/mm^3$ platelets and 12 g/dl hemoglobin with no blast cells seen on the blood smear.

- A moderately cellular bone marrow with a moderate number of normal granulocytic and erythroid precursors, together with adequate megakaryocytes and less than 5% blast cells, none of which possess frank leukemic features.
- A normal CSF examination (including cytology).

Treatment of B-Lineage ALL

ALL trials of the COG have determined the following therapeutic principles:

- Treatment is stratified based on the following prognostic factors: age, WBC, CNS status at diagnosis, immunophenotype, blast genotype, and response to therapy as measured by peripheral blood MRD on day 8 (B-ALL) and bone marrow MRD at day 29 (B- and T-ALL).
- DI improves outcome for standard- and high-risk patients but there is no additional benefit of second DI.
- Standard-risk patients who receive DI only require dexamethasone, vincristine, and PEG-asparaginase and intrathecal methotrexate for induction.
- Improved systemic therapy eliminates cranial irradiation for almost all patients except those who are CNS3 at diagnosis.
- Substitution of dexamethasone for prednisone improves outcome for standard-risk patients, although its use in induction chemotherapy for patients >10 years of age increases the rate of osteonecrosis.
- Standard-risk patients who receive the Capizzi methotrexate regimen during interim maintenance have improved survival.
- High-dose methotrexate given during interim maintenance improves outcome in high-risk patients.

Using prognostic variables obtained at diagnosis and during the first month of therapy allows patient stratification into four groups at varying risks for relapse:

1. Low risk (15% of patients, predicted survival >95%).
2. Average risk (35% of patients, predicted survival 90–95%).
3. High risk (25% of patients, predicted survival 85–90%).
4. Very high risk (25% of patients, predicted survival <80%) ALL.

Details of risk classification are enumerated in Table 18.9.

Table 18.10 lists the protocol used for standard-risk B ALL and the event-free survival for this group of patients at 5 years is over 85%.

Table 18.11 lists a protocol for the treatment of high-risk and very high risk B ALL while Table 18.12 shows the treatment of low-risk ALL. Treatment duration for low-risk patients is about 2.5 years. This group has 5-year EFS of $97 \pm 2\%$ and overall survival (OS) of 98.8%.

Treatment of Mature B-Cell Lymphoma/Leukemia

Mature B-cell ALL is treated with short intensive chemotherapy courses (no maintenance cycles) and has an EFS of about 85% at 5 years. The COG has demonstrated the safety and tolerability of adding rituximab (an anti-CD20 monoclonal antibody) to the standard ALL chemotherapy backbone (detailed in Table 18.10) in children and adolescents, with a 3-year EFS of 90% in patients with advanced disease (with bone marrow and CNS involvement). A phase 3 randomized study is currently underway to investigate whether incorporation of rituximab improves the event-free survival in patients with advanced stage B-cell lymphoma/leukemia.

Treatment of T-Cell ALL

Table 18.13 lists the protocol for treatment of T-cell ALL.
T-cell ALL has an EFS of approximately 75% at 5 years using this protocol.

TABLE 18.9 Classification of Newly Diagnosed Acute Lymphoblastic Leukemia in Current Children's Oncology Group Protocols

Low risk	NCI risk group—Standard risk (age 1–9.99 years, WBC <50,000/μl)
	Presence of favorable cytogenetic features—Trisomies four and 10 or ETV6-RUNX1 fusion
	Absence of unfavorable genetic features
	Day 8 peripheral blood MRD <0.01%
	Day 29 bone marrow MRD <0.01%
	Absence of CNS and testicular disease
Average risk	NCI risk group—Standard risk (age 1–9.99 years, WBC <50,000/μl)
	Absence of unfavorable cytogenetic features
	Day 8 peripheral blood MRD ≥0.01% but <1%
	Day 29 bone marrow MRD <0.01%
	Patient can be CNS1 or CNS2
	Absence of CNS3 or testicular disease
High risk	Age 1–13 years[a]
	Absence of unfavorable cytogenetic features
	Absence of CNS3 disease
	Day 29 bone marrow MRD <0.01%[a]
	NCI standard- and high-risk patients with testicular disease
Very high risk	Age >13 years at diagnosis, regardless of other prognostic indicators
	Presence of unfavorable cytogenetic features (DNA ploidy <0.81 and/or <44 chromosomes, iAMP21, MLL rearrangement, BCR–ABL1[b])
	Presence of CNS3 disease
	Day 29 bone marrow MRD >0.01%
	Induction failure (M3 marrow at day 29)

[a]NCI standard-risk patients who have day 8 peripheral blood MRD ≥1% with no favorable cytogenetic risk features or those who have day 29 bone marrow MRD >0.01% with the presence of favorable cytogenetic features are also classified as high risk.
[b]Treatment of BCR–ABL1 fusion positive (Philadelphia-chromosome-positive) are treated with tyrosine kinase inhibitors on an intensive chemotherapy backbone.

INFANT LEUKEMIA

Infant ALL accounts for 2–5% of childhood leukemias. Infants under 12 months of age with ALL have an extremely poor prognosis, worse than any other age group. Patients go into remission but have a high incidence of early bone marrow or extramedullary relapse. These patients have a high incidence of the following poor prognostic features:

- High initial WBC.
- Massive organomegaly.
- Thrombocytopenia.
- CNS leukemia.
- Failure to achieve complete remission by day 14.

Infant ALL is biologically unique. The leukemia arises from a very early stage of commitment to B-cell differentiation and has the following characteristics:

- The leukemic cells are usually CD10-negative.
- A chromosomal abnormality on chromosome 11, particularly band 11q23 where the MLL/ALL1 gene is located, is commonly found in infant ALL and is associated with a poor prognosis.
- The cells frequently express myeloid antigens.
- Blasts have fetal characteristics and have a greater resistance to therapy.

TABLE 18.10　Therapy for Standard/Average-Risk Acute Lymphoblastic Leukemia

Induction (4 weeks)	Oral dexamethasone for 28 days (6 mg/m^2/day in three divided doses) IV vincristine (1.5 mg/m^2 on days 0, 7, 14 and 21), IV Pegylated L-asparaginase (2500 units/m^2, on day 4), Age-adjusted intrathecal cytarabine (age 1 to less than 2 years 30 mg; age 2 to less than 3 years 50 mg; age 3 years and older 70 mg) on Day 1 Age-adjusted intrathecal methotrexate (age 1 to less than 2 years, 8 mg; age 2 to less than 3 years, 10 mg; older than 3–8.99 years, 12 mg; older than 9 years, 15 mg on day 8 and 29)
Consolidation (4 weeks)	Oral 6-mercaptopurine (75 mg/m^2/d on days 1–28 of consolidation) IV vincristine (1.5 mg/m^2 on day 1) Age-adjusted (see above) intrathecal methotrexate on days 1, 8, and 15 for patients without CNS disease at diagnosis
Interim maintenance 1 (8 weeks)	IV vincristine. 1.5 mg/m^2 (max dose 2 mg) on days 1, 11, 21, 31 and 41 IV methotrexate starting dose of 100 mg/m^2/dose on day 1 thereafter escalate by 50 mg/m^2/dose on days 11, 21, 31, and 41 (discontinue escalation and resume at 80% of last dose if there is a delay because of myelosuppression or mucositis) Age-adjusted intrathecal methotrexate (see Induction) on day 31
Delayed intensification (8 weeks)	Oral dexamethasone (10 mg/m^2/d on days 1–7 and 15–21 days) IV vincristine (1.5 mg/m^2 on days 1, 8, and 15) IV pegylated L-asparaginase (2500 u/m^2 on day 4) Doxorubicin (25 mg/m^2, IV push, on days 1, 8, and 15), IV cyclophosphamide (1000 mg/m^2 over 30 min on day 29) Oral 6-thioguanine (60 mg/m^2/day on days 29–42), IV Cytarabine (75 mg/m^2/day, on days 29–32 and 36–39) Age-adjusted intrathecal methotrexate (see Induction) on day 1 and 29
Interim Maintenance 2 (8 weeks)	IV vincristine. 1.5 mg/m^2 (max dose 2 mg) on days 1, 11, 21, 31 and 41 IV methotrexate starting dose is two-thirds of the maximum tolerated dose attained in interim maintenance 1 on day 1 thereafter escalate by 50 mg/m^2/dose on days 11, 21, 31, and 41 (discontinue escalation and resume at 80% of last dose if there is a delay because of myelosuppression or mucositis). Age-adjusted intrathecal methotrexate (see Induction) on day 1 and 31
Maintenance (12-week cycles and is repeated until 2 years for girls and 3 years for boys from the start of interim maintenance 1)	Oral dexamethasone 3 mg/m^2/dose BID on Days 1–5, 29–33, and 57–61 IV Vincristine 1.5 mg/m^2 on day 1, 29, and 57 Oral mercaptopurine 75 mg/m^2/dose on days 1–84 Oral methotrexate 20 mg/m^2/dose weekly (omit on the days when receive IT methotrexate) IT Methotrexate (age adjusted) on day 1

Independent adverse prognostic features in this group are:

- Age less than 3 months.
- High WBC.
- Slow response to induction.
- Presence of translocation at 11q23.

Infant ALL requires very intensive therapy with advanced supportive care. Many centers advocate allogeneic stem cell transplantation for infant ALL with an 11q23 cytogenetic abnormality or a molecular MLL abnormality, although there are only very limited data to indicate that it is superior to chemotherapy alone. The COG has studied an intensive therapy and reported 5-year EFS rate of approximately 48% in these patients regardless of the use of stem cell transplant, suggesting that the routine use of stem cell transplantation for infants with MLL-rearranged ALL is not indicated.

PHILADELPHIA-POSITIVE ALL

Philadelphia-positive ALL is present in only 3% of children with ALL. Previously less than 40% of these patients were cured with intensive chemotherapy. The use of imatinib (340 mg/m^2/day) added to an intensive chemotherapy regimen has improved the outcome in this population at 3 years to an EFS of 80%. These patients are no longer candidates for bone marrow transplantation routinely. Currently, a phase 2 multicenter study of the second-generation tyrosine kinase inhibitor, dasatinib, is being tested in addition to standard chemotherapy in pediatric patients with newly diagnosed Philadelphia-chromosome-positive ALL.

TABLE 18.11 High-Risk/Very-High-Risk B-Cell Acute Lymphoblastic Leukemia Protocol

Phase	Treatment	Dose
Induction	Prednisone	60 mg/m^2/day PO for 28 days (dexamethasone 10 mg/m^2/day is used for children <10 years of age for 14 days)
	Vincristine	1.5 mg/m^2/week IV, days 1, 8, 15, 22
	Daunomycin	25 mg/m^2/week IV, days 1, 8, 15, 22
	PEG-asparaginase	2500 units/m^2/day IV, day 4
	Cytarabine[a]	Age-adjusted IT, day 0
	Methotrexate[a]	Age-adjusted IT, day 8
Consolidation (9 weeks)	Cyclophosphamide	1000 mg/m^2/day IV, days 1 and 29
	Cytarabine	75 mg/m^2/day IV, days 1–4, 8–11, 29–32, 36–39
	Mercaptopurine	60 mg/m^2/day PO, days 1–14 and 29–42
	Vincristine	1.5 mg/m^2/day IV, days 15, 22, 43, 50
	PEG-asparaginase	2500 units/m^2 IV days 15, 43
	Methotrexate[a]	Age-adjusted IT, days 1, 8, 15, 22
Interim maintenance 1 (63 days)	Vincristine	1.5 mg/m^2 per day IV days 1, 15, 29 and 43
	High-dose methotrexate	5000 mg/m^2 IV over 24 h on days 1, 15, 29, and 43
	Leucovorin	15 mg/m^2/dose starting at hour 42 after the start of high-dose methotrexate infusion
	Methotrexate	Age-adjusted IT days 1 and 29
	6-Mercaptopurine	25 mg/m^2/dose by mouth from days 1–56
Delayed Intensification (8 weeks)		
Reinduction (4 weeks)	Dexamethasone	10 mg/m^2/day PO, days 1–7, 15–21
	Vincristine	1.5 mg/m^2/day IV, days 1, 8, 15
	Doxorubicin	25 mg/m^2/day IV, days 1, 8, and 15
	PEG-asparaginase	2500 units/m^2/day IM, day 4,
	Methotrexate	Age-adjusted IT day 1
Reconsolidation (4 weeks)	Cyclophosphamide	1000 mg/m^2/day IV day 29
	Thioguanine	60 mg/m^2/day PO days 29–42
	Cytarabine	75 mg/m^2/day SC or IV days 29–32 and 36–39
	Methotrexate	Age-adjusted IT days 29 and 36
	Vincristine	1.5 mg/m^2 IV days 43 and 50
	PEG-asparaginase	2500 units/m^2 IM day 43
Interim maintenance II (56 days), given for very-high-risk patients only	Vincristine	1.5 mg/m^2 per day IV days 1, 11, 21, 31, and 41
	Capizzi style Methotrexate	Starting dose is 100 mg/m^2, then escalate by 50 mg/m^2/dose on days 1, 11, 21, 31, and 41
	PEG-asparaginase	2500 IU/m^2/dose on days 2 and 22
	Methotrexate	Age-adjusted IT days 1 and 31
Maintenance (12 weeks)	Vincristine	1.5 mg/m^2/day IV days 1, 29, and 57
	Prednisone	40 mg/m^2/day PO days 1–5, 29–33, and 57–61
	Mercaptopurine	75 mg/m^2/day PO days 1–84
	Methotrexate	20 mg/m^2/day PO days 8, 15, 22, 29 (hold cycles 1–2 when receiving IT methotrexate), 36, 43, 50, 57, 64, 71, and 78
	Methotrexate	Age-adjusted IT day 1 also day 29 of cycles 1 and 2 for patients who did not receive CNS radiation

[a]The doses are age-adjusted: IT methotrexate age 1–1.9 years 8 mg; age 2–2.9 years 10 mg; age >3–8.99 years 12 mg; age >9 years 15 mg, IT cytarabine age 1–1.9 years 30 mg, age 2–2.9 years 50 mg, age >3 years 70 mg. The cycles of maintenance are repeated until the duration of therapy beginning with the first interim maintenance period reaches 2 years for girls and 3 years for boys. Patients with CNS3 disease at diagnosis (classified as very high risk) should receive cranial irradiation during the first 4 weeks of maintenance chemotherapy.
IV, intravenously; PO, orally; IT, intrathecally; SQ, subcutaneously; IM, intramuscularly.
Modified from: Siebel et al. (2008).

TABLE 18.12   Treatment of Low-Risk B-Lineage Acute Lymphoblastic Leukemia

Induction (4 weeks) same as standard/average-risk ALL

Consolidation (19 weeks)

Methotrexate IV 1 g/m^2 as 24-h infusion on days 8, 29, 50, 71, 92, and 113 with delayed leucovorin rescue (10 mg/m^2) orally or IV every 6 h for five doses beginning 42 h after start of methotrexate infusion

6-Mercaptopurine 50 mg/m^2 orally daily on weeks 1–133

Intrathecal methotrexate (age-adjusted as above) on days 8, 29, 50, 71, 92, and 113

Vincristine 1.5 mg/m^2 IV on days 15, 22, 78, and 85

Dexamethasone 3 mg/m^2/dose BID on days 15–21 and 78–84

Maintenance (16-week cycle)

Maintenance lasts for total of 2.5 years timed from the date of diagnosis. It includes vincristine and dexamethasone pulses every 16 weeks and PO methotrexate weekly. Age-adjusted intrathecal methotrexate is given every 12 weeks.

IV, intravenously; IM, intramuscularly. This regimen is based on the excellent outcomes with pediatric oncology group's P9904 therapy.

TABLE 18.13   Therapy for T-Cell Acute Lymphoblastic Leukemia

Induction (4 weeks)	IV vincristine 1.5 mg/m^2 weekly on days 1, 8, 15, and 22
	Oral prednisone 30 mg/m^2/dose BID for 28 days
	IV PEG-asparaginase 2500 IU/m^2 on day 4
	IV daunorubicin 25 mg/m^2 weekly on days 1, 8, 15, and 22
	IT cytarabine (age adjusted) at the time of diagnostic lumbar puncture or day 1
	IT methotrexate (age adjusted) on days 8 and 29
Consolidation (8 weeks)	IV vincristine 1.5 mg/m^2 on days 15, 22, 43, and 50
	IV or SubQ cytarabine 75 mg/m^2 days 1–4, 8–11, 29–32, and 36–39
	IV PEG-asparaginase 2500 IU/m^2 on days 15 and 43
	IV cyclophosphamide 1000 mg/m^2 on days 1 and 29
	Oral 6-mercaptopurine 60 mg/m^2/dose on days 1–14 and 29–42
	IT methotrexate (age adjusted) on days 1, 8, 15, and 22
Interim maintenance (Capizzi methotrexate) (8 weeks)	IV vincristine 1.5 mg/m^2 on day 1, 11, 21, 31, and 41
	IV methotrexate starting at 100 mg/m^2/dose on day 1 then escalate by 50 mg/m^2/dose on days 11, 21, 31, and 41
	IV PEG-asparaginase 2500 IU/m^2 on days 2 and 22
	IT methotrexate (age adjusted) on days 1 and 31
Delayed intensification (8 weeks)	IV vincristine 1.5 mg/m^2 on days 1, 8, 15, 43, and 50
	IV or SubQ cytarabine 75 mg/m^2 on days 29–32 and 36–39
	IV PEG-asparaginase 2500 IU/m^2 on days 4 and 43
	Oral dexamethasone 5 mg/m^2/dose BID on days 1–7 and 15–21
	IV doxorubicin 25 mg/m^2 on days 1, 8, and 15
	IV cyclophosphamide 1000 mg/m^2 on day 29
	Oral thioguanine 60 mg/m^2/dose on days 29–42
	IT methotrexate (age adjusted) on days 1, 29, and 36
Maintenance (12-week cycles and is repeated until 2 years for girls and 3 years for boys from the start of interim maintenance)	IV vincristine 1.5 mg/m^2 on days 1, 29, and 57
	Oral prednisone 20 mg/m^2/dose BID on days 1–5, 29–33, and 57–61
	Oral mercaptopurine 75 mg/m^2/dose on days 1–84
	Oral methotrexate 20 mg/m^2/dose weekly (dose needs to be skipped on the days of IT methotrexate)
	IV doxorubicin 25 mg/m^2 on days 1, 8, and 15
	IV cyclophosphamide 1000 mg/m^2 on day 29
	Oral thioguanine 60 mg/m^2/dose on days 29–42
	IT methotrexate (age adjusted) on days 1, 29, and 36

Intermediate- and high-risk patients receive prophylactic cranial radiation therapy (1200 cGy) during delayed intensification. All T-ALL patients who are CNS 3 at diagnosis receive 1800 cGy during delayed intensification. T-ALL patients with testicular disease at diagnosis that does not resolve by the end of induction receive testicular irradiation (2400 cGy) during consolidation. Current COG T-ALL protocol is evaluating the safety and efficacy of the addition of nelarabine (prodrug of araG 9-B-arabinofuranosylguanine) to the standard augmented BFM chemotherapy backbone and of high-dose methotrexate (5 g/m^2) in comparison to Capizzi methotrexate/PEG-asparaginase delivered during interim maintenance in a 2 × 2 randomized fashion.
Chemotherapy backbone modified from: Siebel et al. (2008).

DOWN SYNDROME AND ALL

Children with Down syndrome have a 20-fold higher risk of ALL and these children have a unique biological subtype characterized by absence of T-ALL and favorable subtypes like hyperdiploid and *EVT6/RUNX1* B-ALL. Moreover a third of samples have a JAK2 mutation and a unique fusion between *P2RY8* and *CRLF2* leading to overexpression of CRLF2 protein on the surface of blasts. Importantly these patients can have many side effects of treatment, particularly in response to higher doses of methotrexate. Given a high risk of infection most protocols include special supportive care guidelines for these patients including that they:

1. Remain hospitalized through induction (a high-risk period) until they show signs of bone marrow recovery.
2. Have IgG levels periodically checked, and intravenous immunoglobulin administered if they are low.
3. Have a blood culture drawn and parenteral antibiotics administered for any change in clinical status, even if the criteria for febrile neutropenia are not specifically met.

RELAPSE IN CHILDREN WITH ALL

Leukemia relapse is defined as:

- *Isolated bone marrow relapse*: More than 25% blasts (M3 marrow) at any point after achieving remission in a single bone marrow aspirate or biopsy without involvement of the CNS and/or testicles.
- *Isolated CNS relapse*: Positive cytomorphology and WBC >5/µl and/or clinical signs of CNS leukemia such as facial nerve palsy, brain/eye involvement, or hypothalamic syndrome. If any CSF evaluation shows positive cytomorphology with CSF 0–4/µl, a second CSF evaluation is required within 2–4 weeks. Identification of the leukemic clone in CSF by flow cytometry (CD19, CD10, TdT, etc.) or FISH for diagnostic karyotype abnormality is also useful.
- *Isolated testicular relapse*: Leukemic infiltration of testicles, confirmed by testicular biopsy.
- *Combined relapse*: M2 (5–25% blasts in the bone marrow) or M3 bone marrow at any point after achieving remission with concomitant CNS and/or testicular relapse. Despite current intensive frontline treatments, 10–20% of children with ALL experience bone marrow relapse. Relapse may be an isolated event in the bone marrow or may be combined with relapse in other sites. The prognosis depends on the timing of the relapse, the site of relapse, and the immunophenotype (T-ALL patients do very poorly). Patients with isolated bone marrow relapse have a worse prognosis than those who have combined bone marrow and CNS relapse while those patients with isolated extramedullary relapse (e.g., CNS relapse) fare better.

Reinduction remission rates for patients with first relapse range from 71 to 93%, depending on the timing and site of relapse. Survival of patients experiencing relapse can be predicted by the site of relapse and the length of first complete remission. Children who have a bone marrow relapse during therapy have a very poor long-term survival (<36 months from initial diagnosis, "early relapse"). Allogeneic stem cell transplantation should be considered for these patients.

Children whose bone marrow relapse occurs after completion of therapy (>6 months, "late relapse") have a significantly longer second remission. Although clinical remission can be achieved in most relapses, long-term survival rates range from 40 to 50%.

Reinduction of patients with relapsed ALL commonly includes conventional agents largely identical to those used at initial diagnosis.

Studies have shown the impact of MRD on outcome for patients with relapsed ALL. Patients who are MRD-negative at the end of the first block of chemotherapy have improved survival compared with those who are MRD-positive. MRD positivity also correlates strongly with the duration of initial remission. Patients experiencing relapse less than 18 months from initial diagnosis have the highest proportion of MRD positivity.

Many reinduction protocols have been piloted for bone marrow relapse and not one has proved to be superior to others with the exception of the ALL R3 regimen listed in Table 18.14. COG has used a triple reinduction platform (Table 18.15), which is suitable for testing in combination with novel agents emerging into the relapsed ALL clinical trials.

TABLE 18.14   Reinduction Regimen for Children with First Relapse of Acute Lymphoblastic Leukemia with Mitoxantrone (Acute Lymphoblastic Leukemia R3 Trial)

Drug and dose	Days
PHASE 1—INDUCTION (WEEKS 1–4)	
Dexamethasone oral (20 mg/m^2/day)	1–5, 15–19
Mitoxantrone IV (10 mg/m^2)	1, 2
Vincristine IV (1.5 mg/m^2)	3, 10, 17, 24
PEG-asparaginase IV (1000 mg/m^2)	3, 18
Age-adjusted intrathecal methotrexate	1, 8
PHASE 2—CONSOLIDATION (WEEKS 5–8)	
Dexamethasone oral (6 mg/m^2/day)	1–5
Vincristine IV (1.5 mg/m^2)	3
Methotrexate IV (1000 mg/m^2)	8
PEG-asparaginase IV (1000 mg/m^2)	9
Cyclophosphamide IV (440 mg/m^2)	15–19
Etoposide (100 mg/m^2)	15–19
Age-adjusted intrathecal methotrexate	8
PHASE 3—INTENSIFICATION (WEEKS 9–12)	
Dexamethasone oral (6 mg/m^2/day)	1–5
Vincristine IV (1.5 mg/m^2)	3
Cytarabine IV (3000 mg/m^2 every 12 h)	1, 2, 8, 9
Erwinase IM (20,000 U/m^2)	2, 4, 9, 11, 23
Methotrexate IV (1000 mg/m^2)	22
Intrathecal methotrexate	1, 22

All patients receive three consecutive blocks of therapy as described above, after which patients can further receive chemotherapy or allogeneic stem cell transplantation depending on the risk group and minimal residual disease.
Adapted from: Parker et al. (2010).

Many new drugs have been tested for patients with relapsed disease often on a backbone of conventional therapy.

- Clofarabine is a second-generation purine nucleoside analog, which inhibits DNA synthesis and repair through inhibition of both DNA polymerase and ribonucleotide reductase and directly induces apoptosis. It has been tested in multiply relapsed refractory lymphoblastic or myeloblastic leukemia in phase 1 and 2 settings as a single agent and in combination with cyclophosphamide and etoposide. When used in combination, one regimen of administration is to give each drug daily for 4 days as follows:
  - Clofarabine 20 mg/m^2/dose IV over 2 h.
  - Etoposide 100 mg/m^2/dose IV over 1–2 h.
  - Cyclophosphamide 440 mg/m^2 IV over 15–30 min.
    The three-drug combination showed a superior response rate when compared to the historical observed data. Based on these encouraging results, clofarabine is being tested in very-high-risk pediatric leukemia patients in the upfront newly diagnosed setting in the context of a randomized clinical trial through the COG.
- Bortezomib is a proteasome inhibitor and a recently completed COG study has evaluated efficacy in T- and B-ALL. Encouraging results have led to its inclusion in a randomized phase III trial for T-ALL.
- Temsirolimus belongs to a class of drugs that are inhibitors of the MTOR pathway and its impact in relapsed disease is currently being investigated.
- Epratuzumab or anti-CD22 has been used in patients with relapsed disease although a recent study did not indicate its addition was superior to chemotherapy alone.

TABLE 18.15 Triple Reinduction Regimen for Recurrent Acute Lymphoblastic Leukemia

Drug and dosage	Days
BLOCK 1	
Vincristine, 1.5 mg/m² IV	1, 8, 15, and 22
Prednisone, 40 mg/m²/d PO	1–29
PEG-asparaginase, 2500 U/m² IM	2, 9, 16, and 23
Doxorubicin, 60 mg/m² IV	1
Intrathecal cytarabine	1
Intrathecal methotrexate	8 and 29 (CNS negative)
Triple intrathecal therapy	8, 15, 22, and 29 (CNS1)
BLOCK 2	
Cyclophosphamide, 440 mg/m² IV	1–5
Etoposide, 100 mg/m² IV	1–5
Methotrexate, 5 g/m² IV	22 (pending blood count recovery)
Intrathecal methotrexate	1 and 22 (CNS negative)
Triple intrathecal therapy	1 and 22 (CNS1)
G-CSF, 5 mg/kg SQ	6 until ANC > 1500/ml 32 days
BLOCK 3	
Cytarabine, 3 g/m² IV every 12 h	1, 2, 8, and 9
L-asparaginase, 6000 U/m² IM	2 and 9 at hour 42 after cytarabine
G-CSF, 5 mg/kg SQ	10 until ANC > 1500/ml 32 days

IV, intravenous; PO, oral; IM, intramuscular; SQ, subcutaneous; Ph, Philadelphia chromosome; G-CSF, granulocyte colony-stimulating factor; LP, lumbar puncture; CNS, central nervous system; ANC, Absolute Neutrophil Count.
Triple intrathecal therapy (methotrexate, cytarabine, and hydrocortisone) is continued weekly beyond four doses until two successive lumbar puncture CSF are free of blasts. All intrathecal medications are dosed based on age.
Modified from: Raetz et al. (2008).

Central Nervous System Relapse

CNS relapse occurs in about 3–8% of patients, although the use of effective CNS prophylactic regimens has resulted in a significant decrease in the incidence of CNS relapse. Risk factors predicting CNS relapse are:

- T-cell immunophenotype.
- High WBC at diagnosis.
- Presence of leukemic cells in the CSF at diagnosis.
- Unfavorable genetic abnormalities.

Submicroscopic involvement of the bone marrow is a frequent finding in patients with isolated CNS relapse.

Treatment

1. Intrathecal chemotherapy alone fails to cure CNS leukemia. However, temporary remission can be achieved with intrathecal chemotherapy alone.
2. Bone marrow relapse occurs in more than 50% of patients achieving CNS remission, regardless of the use of intensified chemotherapy at the time of CNS relapse.
3. Systemic administration of dexamethasone, PEG-asparaginase, high-dose methotrexate, and high-dose cytarabine has been shown to be efficacious in treating CNS leukemia.

One of the recommended regimens in children relapsing only in the CNS, who have not previously received cranial irradiation, is listed in Table 18.16. This protocol should be employed in CNS relapses occurring after

TABLE 18.16   Treatment Schema for Acute Lymphoblastic Leukemia Patients with Isolated Central Nervous System Relapse

Induction (weeks 1–4)
    DEX: 10 mg/m^2 orally daily for 28 weeks
    VCR: 1.5 mg/m^2 IV weekly for 4 weeks
    DNR: 25 mg/m^2 weekly for 3 weeks
    TIT: MTX/HC/Ara-C (age-adjusted dose)[a] IT weekly for 4 weeks
Consolidation (weeks 5–10)
    Ara-C: 3 g/m^2 IV every 12 h for 4 doses on each of weeks 5 and 8 (a total of 8 doses)
    L-asp: 10,000 IU/m^2 IM weekly on weeks 5, 6, 8, 9
    TIT: MTX/HC/Ara-C (age-adjusted dose) IT week 10
Intensification 1 (weeks 11–22)
    MTX: 1000 mg/m^2 IV over 24 h, followed by
    MP: 1000 mg/m^2 IV over 8 h weeks 11, 14, 17, and 20
    VP-16: 300 mg/m^2 IV, followed by
    CYC: 500 mg/m^2 IV weeks 12, 15, 18, and 21
    TIT: MTX/HC/Ara-C (age-adjusted dose)[a] IT weeks 13, 16, 19, and 22
Reinduction (weeks 23–26)
    DEX: 10 mg/m^2 orally daily for 28 days
    VCR: 1.5 mg/m^2 IV weekly for 4 weeks
    DNR: 25 mg/m^2 IV weekly for 3 weeks
Intensification 2 (weeks 27–50)
    Ara-C: 3 g/m^2 IV every 12 h for 4 doses in weeks 27, 33, 39, and 45
    L-asp: 10,000 IU/m^2 IM on weeks 27, 28, 33, 34, 39, 40, 45, and 46
    TIT: MTX/HC/Ara-C (age-adjusted dose)[a] IT weekly on weeks 30, 36, 42, and 48
    MTX: 1000 mg/m^2 IV over 24 h on weeks 31, 37, 43, and 49
    6-MP: 100 mg/m^2 IV over 8 h on weeks 31, 37, 43, and 49
    VP-16: 300 mg/m^2 IV on weeks 32, 38, 44, and 50
    CYC: 500 mg/m^2 IV on weeks 32, 38, 44, and 50
Irradiation (weeks 51–54)
CR1 < 18 months Craniospinal: 24 Gy cranial and 15 Gy spinal
CR1 ≥ 18 months Cranial 18 Gy ONLY (not spinal)
    Dex 10 mg/m^2/ orally daily for 21 days
    VCR 1.5 mg/m^2 IV weekly for 3 weeks 51, 52, and 53
    L-asp: 10,000 IU/m^2 IM × 9 on weeks 51–54
Maintenance (weeks 55–104)
    MP: 75 mg/m^2 orally daily ×42
    MTX: 20 mg/m^2 IM weekly ×6
    Alternate with
    VCR: 1.5 mg/m^2 IV weekly ×4
    CYC: 300 mg/m^2 IV weekly ×4

[a]*See Table 18.10 for age-adjusted doses for intrathecal therapy.*
DEX, dexamethasone; VCR, vincristine; DNR, daunomycin; TIT, triple intrathecal therapy; HC, hydrocortisone; Ara-C, Cytarabine; L-asp, L-asparaginase; MP, 6-mercaptopurine; VP-16, Etoposide; CYC, cyclophosphamide; IV, intravenous, IT, intrathecally; IM, intramuscularly; CR1, first complete remission.
The authors have substituted 6-mercaptopurine 50 mg/m^2/day orally for 7 days instead of IV 6-mercaptopurine.
Adapted from: Barredo et al. (2006b).

18 months of first remission. However, early CNS relapses (duration of first remission <18 months) should be treated with chemotherapy followed by allogeneic stem cell transplantation. In patients who had previously received cranial irradiation and then develop a CNS relapse the author favors a reinduction regimen followed by allogeneic bone marrow transplantation.

Toxicity of CNS Treatment

1. Significant decline in mean values on global IQ scale.
2. Possibility of increased risk of leukoencephalopathy.

For these reasons recent protocols have further delayed radiation therapy to permit more extensive chemotherapy to be delivered prior to radiation therapy and have decreased cranial radiation therapy to 1800 cGy.

The 4-year EFS after CNS relapse is 70%, for patients with a first complete remission ≥18 months it is 78%, while for those with a first complete remission <18 months it is 52%.

Testicular Relapse

As frontline therapies for newly diagnosed ALL continues to improve, isolated testicular relapse has become less common, an incidence of approximately 2% was reported in recent studies.

The treatment for isolated testicular relapse includes:

- Local radiotherapy to both testes.
- Reinduction and continuation of systemic chemotherapy and CNS chemoprophylaxis (Table 18.16). However, trials have been using the same protocol as for CNS relapse with the elimination of cranial irradiation.

A recent COG study demonstrated an excellent 5-year OS rate in children with first isolated testicular relapse by using intensive chemotherapy including high-dose methotrexate and limiting testicular radiation.

FUTURE DRUGS IN ALL THERAPY

Various new therapies are under investigation for relapsed and refractory childhood ALL. Table 18.17 describes the mechanism of action of new drugs being studies in pediatric ALL.

TABLE 18.17 Selected Antileukemic Drugs Being Tested in Clinical Trials

Drug	Mechanism of action	Subtype of leukemia targeted
Clofarabine	Inhibits DNA polymerase and ribonucleotide reductase; disrupts mitochondria membrane	All
Nelarabine	Inhibits ribonucleotide reductase and DNA synthesis	T-cell
Rituximab	Anti-CD20 chimeric murine-human monoclonal antibody	CD20-positive
Blinatumomab	Bispecific T-cell engager monoclonal antibody against CD19	CD19-positive
Epratuzumab	Anti-CD22 humanized monoclonal antibody	CD22-positive
Imatinib mesilate	ABL kinase inhibitor	BCR−ABL-positive
Nilotinib	ABL kinase inhibition	BCR−ABL-positive
Dasatinib	BCR−ABL kinase inhibition	BCR−ABL-positive
MLN8237	Aurora kinase inhibition	BCR−ABL-positive
Ruxolitinib	JAK2 inhibitor	All
Lestaurtinib	FMS-like tyrosine kinase 3 inhibition	MLL-rearranged
Tipifarnib	Farnesyltransferase inhibition	All
Azacytidine, decitabine	DNA methyltransferase inhibition	All
Panobinostat, vorinostat	Histone deacetylase inhibition	All
Sirolimus, Temsirolimus	Mammalian target of rapamycin inhibition	All
Sorafenib	Multikinase inhibitor	All
Bortezomib	Inhibition of ubiquitin proteasome pathway	All
Flavopiridol	Serine−threonine−cyclin-dependent kinase inhibition	All
Obatoclax	Pan antiapoptotic BCL-2 family small-molecule inhibitor	All
17-AAG	Heat shock protein-90 inhibition	BCR−ABL-positive
		ZAP-70-positive

Novel therapeutic strategies include molecularly targeted therapy, global epigenetic approaches, and immuno-therapeutic agents.

1. Amongst the molecularly targeted therapies, the success of imatinib in the treatment of CML and Ph-positive ALL has been phenomenal and has set the stage for second- and third-generation tyrosine kinase inhibitors, such as dasatinib and ponatinib, respectively. Dasatinib is currently under investigation in frontline therapy for pediatric Ph-positive ALL. The identification of a Ph-like subset that lacks the classic BCR—ABL fusion protein but has translocations involving other tyrosine kinases, particularly PDGF beta and ABL, offers opportunities for treatment with imatinib or other tyrosine kinase inhibitors. Other patients within this subset have blasts with JAK mutations and therefore are candidates for treatment with JAK inhibitors.

2. Preclinical studies have shown the feasibility of incorporation of epigenetic therapy, histone deacetylase inhibitors, and DNA methyltransferase inhibitors, into the backbone chemotherapy regimen for relapsed and refractory leukemias and a clinical trial to test the tolerability and efficacy is currently underway. These agents reverse the epigenetic changes of hypermethylation and histone deacetylation and allow for the re-expression of genes silenced by these mechanisms.

3. Immunological approaches offer great promise. Rituximab, an anti-CD20 monoclonal antibody, is currently being investigated in a phase 3 setting in children with advanced mature B-cell malignancies, after the earlier published results of its safety and tolerability in combination with multiagent chemotherapy backbone. Recently, blinatumomab, a bispecific T-cell engager anti-CD19 antibody, which directs the patient's cytotoxic CD3-positive T-cells to CD19-positive leukemic cells, thus activating the T-cells to destroy the leukemic blasts, has entered into a clinical trial for relapsed and refractory ALL. Another novel promising approach is the use of autologous or allogenic genetically engineered T-cells to express chimeric antigen receptor targeted against the tumor cells (e.g., CD19 in B-lineage ALL). Likewise, cytotoxic abilities of natural killer cells are also being utilized for the treatment of ALL.

Further Reading and References

Barredo, J.C., Devidas, M., Lauer, S.J., et al., 2006a. Isolated CNS relapse of acute lymphoblastic leukemia treated with intensive systemic chemotherapy and delayed CNS radiation: a pediatric oncology group study. J. Clin. Oncol. 19, 3142–3149.

Barredo, J.C., Devidas, M., Lauer, S.J., et al., 2006b. Isolated CNS relapse of acute lymphoblastic leukemia treated with intensive systemic chemotherapy and delayed CNS radiation: a pediatric oncology study. J. Clin. Oncol. 24, 3142–3149.

Bennett, J.M., Catovsky, D., Daniel, M.T., et al., 1976. Proposals for the classification of the acute leukemias. Br. J. Haematol. 33, 451.

Bhatla, T., Jones, C.L., Meyer, J.A., et al., 2014. The biology of relapsed acute lymphoblastic leukemia: opportunities for therapeutic interventions. J. Pediatr. Hematol. Oncol. 36 (6), 413–418.

Bhojwani, D., Pui, C.H., 2013. Relapsed childhood acute lymphoblastic leukemia. Lancet Oncol. 14 (6), e205–e217.

Bhojwani, D., Yang, J.J., Pui, C.H., 2015. Biology of childhood acute lymphoblastic leukemia. Pediatr. Clin. North Am. 168 (6), 845–853.

Borowitz, M.J., Devidas, M., Hunger, S.P., et al., 2008. Clinical significance of minimal residual disease in childhood acute lymphoblastic leukemia and its relationship to other prognostic factors: a Children's Oncology Group study. Blood 111, 5477–5485.

Carroll, W.L., Raetz, E.A., 2012. Clinical and laboratory biology of childhood acute lymphoblastic leukemia. J. Pediatr. 160 (1), 10–18.

deBoton, S., Coiteux, V., Chevret, S., et al., 2004. Outcome of childhood acute promyelocytic leukemia with all-trans-retinoic acid and chemotherapy. J. Clin. Oncol. 22 (8), 1404–1412.

Goldberg, J.M., Silverman, L.B., Levy, D.E., et al., 2003. Childhood T-cell acute lymphoblastic leukemia: the Dana Farber Cancer Institute. Acute lymphoblastic leukemia consortium experience. J. Clin. Oncol. 21, 3616–3636.

Gregory, J., Kim, H., Alonzo, T., et al., 2009. Treatment of children with promyelocytic leukemia: results of the first North American Intergroup trial INT0129. Pediatr. Blood Cancer 53, 1005–1010.

Hunger, S.P., Lu, X., Devidas, M., et al., 2012. Improved survival for children and adolescents with acute lymphoblastic leukemia between 1990 and 2005: a report from the Children's Oncology Group. J. Clin. Oncol. 30 (14), 1663–1669.

Ko, R.H., Ji, L., Barnette, P., et al., 2009. Outcome of patients treated for relapsed or refractory acute lymphoblastic leukemia: a therapeutic advances in childhood leukemia consortium study. J. Clin. Oncol. 22, 2950.

Meshinchi, S., Arceci, R.J., 2007. Prognostic factors and risk-based therapy in pediatric acute myeloid leukemia. Oncologist 12, 341–355.

Nguyen, K., Devidas, M., Cheng, S.C., et al., 2008. Factors influencing survival after relapse from acute lymphoblastic leukemia: a Children's Oncology Group study. Leukemia 22, 2142–2150.

Parker, C., Waters, R., Leighton, C., 2010. Effect of mitoxantone on outcome of children with first relapse of acute lymphoblastic leukemia (ALL R3): an open label randomized trial. Lancet 376 (9757), 2009–2017.

Pieters, R., Schrappe, M., Delorenzo, P., et al., 2007. A treatment protocol for infants younger than 1 year with acute lymphoblastic leukemia (interfant-99): an observational study and multicentre randomized trial. Lancet 370, 240–250.

Pizzo, P.A., Poplack, D.G., 2006. Principles and Practice of Pediatric Oncology, sixth ed. Lippincott-Williams and Wilkins, Philadelphia.

Pui, C.H., Howard, S., 2008. Current management and challenges of malignant disease in the CNS in paediatric leukemia. Lancet Oncol. 9, 257–268.

Pui, C.H., Behm, F.G., et al., 1993. Clinical and biologic relevance of immunologic marker studies in childhood acute lymphoblastic leukemia. Blood 82, 343.

Pui, C.H., Pei, D., Campana, D., et al., 2011. Improved prognosis for older adolescents with acute lymphoblastic leukemia. J. Clin. Oncol. 29 (4), 386–391.

Pui, C.H., Carroll, W.L., Meshinchi, S., Arceci, R.J., 2011. Biology, risk straitification, and therapy of pediatric acute leukemis: an update. J. Clin Oncol. 29 (5), 551–565.

Rabin, K.R., Whitlock, J.A., 2009. Malignancy in children with trisomy 21. Oncologist 14, 164–173.

Raetz, E.A., Borowitz, M.J., Devidas, M., et al., 2008. Reinduction platform for children with first marrow relapse of acute lymphoblastic leukemia: a Children's Oncology Group study. J. Clin. Oncol. 26, 3971–3978.

Schultz, K.H., Bowman, P., Aledo, A., et al., 2009. Improved early event-free survival with imatinib in Philadelphia chromosome-positive acute lymphoblastic leukemia: a Children's Oncology Group study. J. Clin. Oncol. 27, 5175–5181.

Siebel, N.L., Steinherz, P.G., Harland, H.N., et al., 2008. Early post induction intensification therapy improves survival for children and adolescents with high risk acute lymphoblastic leukemia: a report from the Children's Oncology Group. Blood 111, 2548–2555.

Vardiman, J.W., Thiele, J., Arber, D.A., et al., 2009. The 2008 revision of the World Health Organization (WHO) classification of myeloid and acute leukemia: rationale and important changes. Blood 114, 937–951.

19

Acute Myeloid Leukemia

Arlene Redner and Rachel Kessel

Acute myelogenous leukemia (AML) is characterized by the abnormal proliferation and differentiation of myeloid precursors in the bone marrow. While the etiology of primary AML is unknown, certain predisposing factors can lead to secondary AML as discussed below. While AML is the less common of the two acute leukemias of childhood, it is responsible for most acute leukemia deaths.

INCIDENCE AND EPIDEMIOLOGY

- Leukemia accounts for 25–30% of all childhood cancers.
- AML accounts for 20% of childhood leukemia.
- Incidence: 500 new cases of childhood AML in the United States per year.
- Age distribution: Peaks in neonatal period and during adolescence.

ETIOLOGY AND PREDISPOSING CONDITIONS

Development of AML is thought to follow a multihit hypothesis. An initial oncogenic mutation creates a "preleukemic" cell that eventually develops into a leukemic cell via a second promotional mutation. Genetic and molecular mutations resulting in AML are classified as Class I and Class II mutations as listed in Table 19.1.

Table 19.2 lists the predisposing conditions associated with an increased risk of developing AML.

- Fanconi anemia: Patients have a 50% risk of developing AML. They have increased toxicity from chemotherapy, necessitating protocol modifications.
- Severe congenital neutropenia: Patients have a 21% risk of developing AML, most often preceded by a mutation in the granulocyte colony-stimulating factor receptor gene. Partial or total loss of chromosome 7 occurs in half of the patients.
- Shwachman–Diamond syndrome (SDS): Patients have a 30% risk of developing AML associated with abnormalities of chromosome 7.

Therapy-related myelodysplastic syndrome and therapy-related AML (t-AML) typically present within 3–5 years after treatment but cases up to 10 years or more have been described. Patients with t-AML tend to have a worse prognosis than patients with *de novo* disease with identical cytogenetics. t-AML related to topoisomerase II inhibitors have a short latency (6–36 months).

- Epipodophyllotoxins: Exposure typically results in French–American–British (FAB) M4 or M5 AML and often involves a mixed lineage leukemia (*MLL*) gene rearrangement (11q.23).
- Anthracyclines: Often involves an *MLL* gene rearrangement.
- Alkylating agents: Often results in AML with poor-risk cytogenetics.

Lanzkowsky's Manual of Pediatric Hematology and Oncology.
DOI: http://dx.doi.org/10.1016/B978-0-12-801368-7.00019-3

TABLE 19.1 Categories of Genetic and Molecular Mutations Resulting in Acute Myeloid Leukemia

	Mechanism	Examples
Class I mutations	Proliferative and/or survival advantage to cells, without altering cellular differentiation	*RAS, FLT-3, KIT, CBL*
Class II mutations	Impair differentiation and dysregulate apoptosis	*RUNX1, MLL, PML/RARA*

MLL, mixed lineage leukemia.

TABLE 19.2 Conditions Predisposing to Acute Myeloid Leukemia

Inherited conditions	Acquired conditions	Environmental exposures
• Down syndrome • Fanconi anemia • Severe congenital neutropenia • Diamond–Blackfan anemia • Dyskeratosis congenital • Neurofibromatosis-1 • Bloom syndrome • Shwachman–Diamond syndrome • Li–Fraumeni syndrome • Ataxia telangiectasia • Twinning	• Aplastic anemia • Myelodysplastic syndrome • Myeloproliferative syndrome • Acquired amegakaryocytic thrombocytopenia • Paroxysmal nocturnal hemoglobinuria	• Ionizing radiation • Alkylating chemotherapeutic agents • Epidophyllotoxin chemotherapeutic agents

Twin concordance

- Concordance studies on identical twins show that an identical twin is twice as likely as the general population to develop leukemia if his/her twin developed leukemia before age 7.
- If the affected identical twin was diagnosed as an infant, the concordance rate is nearly 100%.
- Analysis of unique genomic fusion gene sequences of the leukemias suggests a common clonal origin.

CLINICAL FEATURES

See Chapter 18 on ALL for some of the common clinical features that may occur in all cases of acute leukemia.

Children with AML may present with a wide range of signs and symptoms, ranging from fever, anemia, or thrombocytopenia to life-threatening coagulopathy or complications from extramedullary disease (EMD) resulting in organ dysfunction.

EMD consists of a collection of myeloblasts or immature myeloid cells outside of the bone marrow. It is seen in approximately 10–20% of patients with AML. It can occur as a myeloid sarcoma (MS), previously referred to as a chloroma, or as the presence of leukemic cells in the cerebrospinal fluid (CSF). The most common presentations are gingival hypertrophy, lymphadenopathy, and leukemia cutis. MS is also seen within the central nervous system (CNS) and the orbit, periorbital areas, and paraspinal areas. EMD typically occurs concurrently with AML but occasionally may present as the first manifestation, even before bone marrow involvement. Even in the absence of bone marrow involvement EMD should be treated using an AML protocol. EMD has been associated with t(8;21), inv(16), and 11q23 *MLL* rearrangements. CNS disease is associated with a high white blood cell (WBC) and is more often seen with M4 and M5 AML.

DIAGNOSIS

Laboratory Studies

Blood count and bone marrow

- Leukocytosis: Median WBC count at diagnosis is 20,000/mm^3.
- Approximately 20% of patients present with WBC count above 100,000/mm^3.
- Auer rods, needle-shaped intracytoplasmic azurophilic inclusion bodies are often, but not always seen in AML, particularly in M2 or M3 AML.
- Anemia: Hemoglobin <9 g/dl in 50% of patients.
- Thrombocytopenia: Platelets <100,000/mm^3 in 75% of patients.

- The WHO classification of AML defines that 20% blasts are required for the diagnosis of AML, patients with clonal cytogenetic abnormalities including t(8;21) (q22;q22), inv(16) (p13;q22) or t(16;16) (p13;q22) and t(15;17) (q22;q12) are considered to have AML regardless of the blast percentage.
- Special bone marrow studies, which help in detailed cell classification, include: histochemistry, immunophenotying, and cytogenetics.

The morphologic features of myeloblasts and the cytochemical features of AML are shown in Tables 19.3 and 18.3 (see previous chapter), respectively.

Cerebrospinal fluid (CSF): The diagnosis of CSF leukemia is the same as described for ALL (see Chapter 18).

CNS involvement at diagnosis and at relapse is seen in 5–10% of pediatric AML patients. See Table 19.4 for factors associated with CNS leukemia. Unlike acute lymphoblastic leukemia (ALL), CNS disease in AML is not a factor within the AML risk group stratification because it does not affect overall survival (OS), although those with CNS disease have an increased incidence of isolated CNS relapse. Intrathecal (IT) chemotherapy is given to all patients, including those without any detectable CNS involvement. Patients with CNS involvement at diagnosis receive additional intensified IT chemotherapy consisting of weekly IT chemotherapy until blasts clear from the CSF and monthly thereafter until the end of therapy.

Coagulation profile: Decreased coagulation factors that frequently occur are hypofibrinogenemia, decreased levels of factors V, IX, and X.

Monitor for tumor lysis syndrome: The extent of electrolyte disturbances and degree of tumor lysis varies depending on the leukemic burden and rate of cell turnover. Tumor lysis syndrome occurs less frequently in AML than in ALL, and is more often seen in FAB M4 or M5 AML than other subtypes. Complete metabolic panel, including lactic dehydrogenase and uric acid should be sent at diagnosis (see Chapter 32).

Cardiac function assessment: Electrocardiogram and echocardiogram should be performed at baseline and with each cycle of chemotherapy to monitor for cardiotoxicity.

Infectious disease evaluation: Patients often present with fever, thought to be due to pyrogens released by leukemic cells and as an inflammatory response. Blood cultures should be drawn and antibiotics initiated (see Chapter 33 for details). Viral studies should be sent including: varicella antibody titer, cytomegalovirus antibody titer, herpes simplex antibody, and hepatitis antibody screening at baseline.

TABLE 19.3   Morphologic Characteristics of Lymphoblasts and Myeloblasts

Characteristic	Lymphoblasts	Myeloblasts
Size	10–20 mm	14–20 mm
NUCLEUS		
Shape	Round or oval	Round or oval
Chromatin	Smooth, homogeneous	Spongy, loose, finely developed meshwork
Nucleoli	0–2 and indistinct	2–5 and distinct "punched-out"
Nuclear membrane	Smooth, round	Irregular
Nuclear–cytoplasmic ratio	High	Low
CYTOPLASM		
Color	Blue	Blue-gray
Amount	Thin rim	More abundant
Granules	Absent	Present
Auer rods	Absent	Present

TABLE 19.4   Factors Associated with Central Nervous System Disease

Hyperleukocytosis

Monocytic leukemia (FAB M4 or M5, including M4eo with inv(16))

MLL rearrangement

Younger age (<2 years)

FAB, French–American–British; *MLL*, mixed lineage leukemia.

CLASSIFICATION OF AML

Acute leukemia can be classified based on morphologic characteristics, cytochemical features, immunologic characteristics, and cytogenetic and molecular characteristics. Table 19.5 lists the WHO classification of acute myeloid leukemia and related neoplasms.

Table 19.6 lists the FAB classification of AML.

TABLE 19.5   WHO Classification of Myeloid Neoplasms and Acute Leukemia

---

ACUTE MYELOID LEUKEMIA AND RELATED NEOPLASMS

---

1. Acute myeloid leukemia with recurrent genetic abnormalities
   a. AML with t(8;21)(q22;q22); *RUNX1-RUNX1T1*
   b. AML with inv(16)(p13.1q22) or t(16;16)(p13.1;q22); *CBFB-MYH11*
   c. APL with t(15;17)(q22;q12); *PML-RARA*
   d. AML with t(9;11)(p22;q23); *MLLT3-MLL*
   e. AML with t(6;9)(p23;q34); *DEK-NUP214*
   f. AML with inv(3)(q21q26.2) or t(3;3)(q21;q26.2); *RPN1-EVI1*
   g. AML (megakaryoblastic) with t(1;22)(p13;q13); *RBM15-MKL1*
   h. *Provisional entity: AML with mutated NPM1*
   i. *Provisional entity: AML with mutated CEBPa*
2. Acute myeloid leukemia with myelodysplasia-related changes
3. Therapy-related myeloid neoplasms
4. Acute myeloid leukemia, not otherwise specified
   a. AML with minimal differentiation
   b. AML without maturation
   c. AML with maturation
   d. Acute myelomonocytic leukemia
   e. Acute monoblastic/monocytic leukemia
   f. Acute erythroid leukemia
      i. Pure erythroid leukemia
      ii. Erythroleukemia, erythroid/myeloid
   g. Acute megakaryoblastic leukemia
   h. Acute basophilic leukemia
   i. Acute panmyelosis with myelofibrosis
5. Myeloid sarcoma
6. Myeloid proliferations related to Down syndrome
   a. Transient abnormal myelopoiesis
   b. Myeloid leukemia associated with Down syndrome
7. Blastic plasmacytoid dendritic cell neoplasm

Adapted from Vardiman et al. (2009).

TABLE 19.6   French−American−British Classification of Acute Myeloid Leukemia

---

1. *Type M0*—acute undifferentiated leukemia
2. *Type M1*—myeloblastic leukemia without maturation; morphologically indistinguishable from L2 morphology
3. *Type M2*—myeloblastic leukemia with differentiation
4. *Type M3*—acute promyelocytic leukemia; most cells abnormal hypergranular promyelocytes; cytoplasm contains multiple auer rods
5. *Type M3V*—microgranular variant of acute promyelocytic leukemia; cells with deeply notched nucleus; typical hypergranular promyelocytes less frequent
6. *Type M4*—both myelocytic and monocytic differentiation present in varying proportions
7. *Type M4EOS*—associated with prominent proliferation of eosinophils
8. *Type M5*—monocytic leukemia containing poorly differentiated and/or well-differentiated monocytoid cells (the M4 and M5 subtypes are particularly common in children under 2 years of age)
9. *Type M6*—ertholeukemia (Di Guglielmo disease)
10. *Type M7*—megakaryoblastic leukemia; associated with myelofibrosis; frequently observed in children with trisomy 21. M7 leukemia has the following characteristics:
    a. The blast morphology is heterogeneous in appearance, resembling L1 or L2 cells with or without granules and having one to three nucleoli; the cytoplasm has blebs
    b. Immunophenotype is CD41-, CD42-, CD61-positive—in addition to CD13 and CD33 positivity
    c. Electron microscopy demonstrates positive platelet peroxidase reaction localized exclusively on the nuclear membrane and the endoplasmic reticulum

---

TABLE 19.7   2008 WHO Classification: Acute Leukemias of Ambiguous Lineage

Lineage	Marker

Myeloid	• Myeloperoxidase OR
	• Monocytic differentiation (at least two of the following):
	  ◦ NSE
	  ◦ CD11c
	  ◦ CD14
	  ◦ CD64
	  ◦ Lysozyme

T lineage	• Cytoplasmic CD3 OR • Surface CD3

B lineage	• Strong CD19 AND
	• At least two of the following with strong expression of
	  ◦ CD79a
	  ◦ Cytoplasmic CD22
	  ◦ CD10 OR
	• Weak CD19 AND
	• At least two of the following with strong expression of
	  ◦ CD79a
	  ◦ Cytoplasmic CD22 OR
	  ◦ CD10

NSE, Non-specific esterase.

TABLE 19.8   Quantitative Bone Marrow Criteria for the Diagnosis of Acute Myeloblastic Leukemia Subtypes[a]

Bone marrow cells	M1 (%)	M2 (%)	M4 (%)	M5 (%)	M6 (%)
BLASTS					
All nucleated cells	–	>30	>30	–	<30 or >30
Nonerythroid cells	90	>30	>30	>80[b]	>30
Erythroblasts < all nucleated cells	–	<50	<50	–	>50
Granulocytic component[c] < nonerythroid cells	<10	>10	>20[d]	<20	Variable
Monocytic component[e] < nonerythroid cells	<10	<20	>20	>80[b]	Variable

[a]*Lysozyme estimations and cytochemical tests are required if the peripheral blood monocyte count is 5 × 10⁹/l or more, but the marrow suggests M2 and marrow suggests M4, but the peripheral blood monocyte count is less than 5 × 10⁹/l.*
[b]*Monoblasts in M5a; in M5b, the predominant cells are promonocytes and monocytes.*
[c]*Promyelocytes, myelocytes, metamyelocytes, and neutrophils.*
[d]*May include myeloblasts.*
[e]*Promyelocytes and monocytes.*
From: Bennett et al. (1985), with permission.

FAB M5 and M7 are more common in early childhood, while older children are more likely to have FAB M0, M1, M2, and M3. AML in patients with Down syndrome (DS) is frequently associated with FAB M7 (megakaryoblastic leukemia). Table 19.7 shows the WHO classification of acute leukemias of ambiguous lineage which are discussed below. The quantitative bone marrow criteria for the diagnosis of acute myeloblastic leukemia are summarized in Table 19.8.

Immunophenotype of AML

Antibodies to cell surface proteins are useful in the diagnosis of AML and can be correlated with the FAB subtypes. Table 19.9 lists the relationship among immunologic surface markers with FAB subtypes of AML.

Molecular Genetics of AML

• Several significant genetic anomalies are seen in patients with AML. Fifteen percent of AML patients have a t(8;21) (q22;q22), which is associated with FAB M2. The translocation creates an *AML1-ETO* fusion gene. The translocation seems to interfere with the expression of a myeloid-specific gene. AML-ETO is associated with a high rate of long-term remission.

- A related myeloid transcription factor is also altered by the cytogenetic inv(16) and t(16;16), which occurs in 15% of AML cases. These cases are associated with myelomonocytic differentiation with abnormal bone marrow eosinophils and a favorable prognosis. They result in a chimeric protein (CBFB-MYH11). CBFB-MYH11 is associated with a high rate of long-term remission.
- Acute promyelocytic leukemia (APML) is associated with a balanced translocation of the retinoic acid receptor-alpha (*RARA* gene at 17q21) and the *PML* gene at 15q21. RARA is a transcription factor that binds retinoids and interacts directly with DNA. APML can be treated with trans- retinoic acid due to its binding to the RARA receptor.
- FMS-like tyrosine kinase mutations internal tandem duplications (*FLT3-ITD*) or point mutations have been identified in 15—30% of pediatric AML patients. ITDs in the juxtamembrane domain result in ligand-independent constitutive activation of the kinase and uncontrolled proliferation of the leukemic blasts. *FLT3-ITD* mutation is associated with high WBC at diagnosis and increased blast percentage as well as cytogenetically normal AML. Patients with *FLT 3-ITD* mutations have a poorer prognosis.

Table 19.10 lists the common cytogenetic abnormalities in pediatric AML, the FAB subtype associations, the affected genes, and frequency of clinical presentations of these subtypes. Figure 19.1 identifies the distribution of translocations in pediatric AML.

TABLE 19.9 Relationship Among Immunologic Surface Markers with FAB Subtypes of Acute Myeloblastic Leukemia

FAB subtype of AML	Immunologic surface marker										
	HLA-DR	CD11b	CD13	CD14	CD15	CD33	CD34	Glycophorin	CD41	CD42	CD61
M1/M2	+				+	+	+				
M3/M3V		+	+		+	+	+				
M4/M5	+	+	+	+	+	+	+				
M6	+		+			+	+	+			
M7	+		+			+	+		+	+	+
M0			+			+	+				

FAB, French—American—British; AML, acute myeloid leukemia.

TABLE 19.10 Cytogenetic Abnormalities in Childhood Acute Myelogenous Leukemia[a]

Chromosome abnormality	AML FAB type	Affected genes	Frequency	Comments
t(8;21)(q22;q22)	M1, M2	ETO-AML1	5—15%	Auer rods common; chloromas
t(15;17)(q22;q12)	M3, M3v	PML-RARA	6—15%	Coagulopathy; ATRA responsiveness
t(11;17)(q23;q12)	M3	PLZF-RARA	Rare	Coagulopathy; ATRA unresponsiveness
inv(16)(p13q22); t(16;16)	M4Eo	MYH11-CBFB	2—11%	CNS leukemia; eosinophilia with basophilic granules
t(8;16)	M5b	MOZ-CBP	1%	Infants or young adults; high WBC; chloromas; erythrophagocytosis; secondary leukemia after epipodophyllotoxins
t(9;11)(p22;q23)	M4, M5a	AF9-MLL	5—13%	Infants or young adults; high WBC; chloromas; erythrophagocytosis; secondary leukemia after epipodophyllotoxins
t(10;11)(p12;q23)	M5	AF10-MLL	Rare	Infants or young adults; high WBC; chloromas; erythrophagocytosis; secondary leukemia after epipodophyllotoxins

(Continued)

TABLE 19.10 *(Continued)*

Chromosome abnormality	AML FAB type	Affected genes	Frequency	Comments
t(11;17)(q23;q21)	M5	*MLL-AF17*	Rare	Infants or young adults; high WBC; chloromas; erythrophagocytosis; secondary leukemia after epipodophyllotoxins
t(11q23)[b]	M4, M5	*MLL*, other partners	2–10%	Infants; high WBC, CNS and skin involvement; poor prognosis often associated with these, especially t(4;11)
t(1;22)	M7	*RBM15-MKL1*	2–3%	M7 AML in infants with DS; myelofibrosis
t(6;9)	M2, M4, MDS	*DEK-CAN*	1%	Basophilia nuclear protein
inv(3)(q21;q26) t(3;3)(q21;q26)	M2, M4, MDS	*EV11*	1%	Prior MDS; thrombocytosis and abnormal platelets
−7/del(7)(q22-q36)	All subtypes, MDS		2–7%	Toxic exposure; prior MDS; more common in other adults, bacterial infections common
−5/del(5)(q11-q35)	All subtypes, MDS		Rare	Toxic exposure; prior MDS; more common in older adults
+8	All subtypes		5–13%	Prior MDS; older patients

[a]*This table does not contain a complete list of all chromosomal abnormalities in childhood AML.*
[b]*11923 translocation involves* MLL *gene have been shown to have many fusion partners.*
AML, acute myeloid leukemia; FAB, French–American–British; *MLL*, mixed lineage leukemia; CNS, central nervous system; WBC, white blood cell; DS, Down syndrome; ATRA, all-trans retinoic acid; MDS, myelodysplastic syndrome.
Adapted from Pizzo and Poplack (2001).

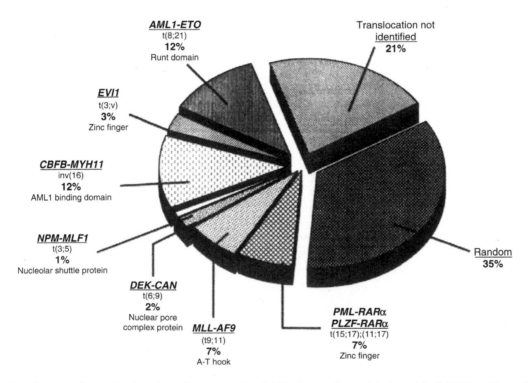

FIGURE 19.1 Distributions of translocations in pediatric acute myeloid leukemia. *From: Rubnitz and Look (1998), with permission.*

An increasing number of genetic abnormalities have been identified in patients with AML. The key will be to determine prognostic relevance and risk stratify based on this information. Table 19.11 categorizes AML cytogenetics as favorable and unfavorable.

TABLE 19.11 Favorable- and Unfavorable-Risk Cytogenetics and Molecular Studies

Favorable-risk molecular studies	Unfavorable-risk molecular studies
• CEBP α • NPM1 • GATA1s[a]	• FLT3/ITD • c-KIT[a]

Favorable-risk cytogenetics	Unfavorable-risk cytogenetics

• RUNX1-ETO
  • t(8; 21)(q22;q22)
• inv(16)
  • inv(16)(p13.1q22), t(16;16)(p13.1;q22), del(16)(q22)
• PML-RARA
  • t(15;17)(q22;q12)
• MLL
  • t(1;11)(q21;q23)
	• Preexisting MDS, particularly affecting chromosomes 5 and 7
	  • monosomy 5
	  • monosomy 7
	  • del(5q)
	  • del(7q)
	• MLL
	  • t(6;11)(q27;q23), t(10;11)(p12;q23)
	• MISC
	  • t(7;12)(q36;p13) t(6;9)(p23;q34), t(5;11)(q35;p15.5)

[a]Not used prognostically.
MDS, myelodysplastic syndrome; MLL, mixed lineage leukemia.

Gene expression profiling by microassay technology has been able to define distinct expression profiles of leukemia based on genetic mutations. These profiles may in the future be able to uncover linkages between molecular subclass and clinical outcome, which cannot be identified by standard cytogenetic analysis and clinical variables at present.

TREATMENT

General Care

More detailed information on supportive medical care is provided in Chapter 33.

Initial and ongoing evaluation should include complete blood count, electrolytes, creatinine, liver enzymes, serum protein, uric acid, lactate dehydrogenase, and coagulation studies, as well as blood cultures every 24 h when febrile.

Bleeding, leukostasis, and complications from infections are the main causes of death in children with AML. Death within the first 2 weeks of diagnosis is typically related to complications of AML itself (leukostasis and hemorrhage), while bacterial infections are the most common cause later in the course of therapy.

Management of Infectious Complications

Due to the use of highly myelotoxic chemotherapy in the treatment of AML, up to 70% of children have a bacterial infection during each course of AML therapy. The cumulative incidence of bacteremia with *Streptococcus viridans* is >40% and is particularly associated with use of high-dose cytarabine and mucositis. Invasive fungal infections occur in approximately 20% of patients. There is a significant reduction in toxic mortality and improvement in OS when the following stringent supportive care guidelines are implemented:

- Mandatory hospitalization during the first course of chemotherapy and following each course until the absolute neutrophil count rises for 2 successive days and the patient is afebrile and clinically stable.
- Empiric antibiotic coverage for first fever of 38.4°C.
- Empiric antifungal coverage with prolonged fever.

Given the very high risk of bacteremia associate with AML, our institution has been providing infection prophylaxis with:

- Cefepime and vancomycin at therapeutic doses empirically upon completion of chemotherapy until blood count recovery.
- *Pneumocystis jiroveci* pneumonia prophylaxis (typically bactrim or pentamidine if bactrim is contraindicated) to start as soon as possible and continue for at least 3 months following therapy completion.
- Fungal prophylaxis with fluconazole is started. Once febrile for >2 days, micafungin or voriconazole is used to broaden antifungal coverage.

Studies are underway to determine the effect that bathing with chlorhexadine, a broad-spectrum antiseptic may have on reducing central-line-associated bloodstream infection rates and multidrug-resistant organisms. In addition, a Children's Oncology Group study is investigating the use of caspofungin versus fluconazole to prevent invasive fungal infections.

See Chapter 33 for the management of the febrile neutropenic patient.

Transfusions

Supportive care including the use of packed red cells and platelet transfusions is required fairly frequently, especially when aggressive multiagent chemotherapy is employed. Platelet transfusions should be administered to patients with overt bleeding or when the platelet count is below 10,000/mm³. Packed red blood cell transfusions (3−5 cc/kg raises hemoglobin 1 g) should be administered to patients with symptomatic anemia, typically once hemoglobin is <8 gm/dl. G-CSF in children with AML, is generally not recommended. While it decreases the duration of neutropenia in induction therapy, it does not influence the incidence of febrile neutropenia, documented infections, or affect infection-associated mortality. Its use has been restricted to neutropenic patients with serious infections.

Management of Tumor Lysis Syndrome

The management of tumor lysis syndrome is reviewed in Chapter 32.

Management of Hyperleukocytosis

Due to the large size of AML blasts and their adhesive nature, patients with elevated WBC are at increased risk of complications from leukostasis. Clumping and sludging of blasts in small blood vessels may result in serious neurologic or respiratory symptoms, such as confusion, headache, and coma as well as pulmonary edema, respiratory failure, and pulmonary hemorrhage, respectively. In cases involving hyperleukocytosis (WBC >100,000/mm³), leukapheresis can be performed to rapidly reduce the level of circulating blasts within 2−4 h. Leukapheresis, however, does not correct the metabolic problems (e.g., hyperphosphatemia, hypocalcemia, and hyperuricemia) or correct the anemia. See Chapter 33 for management of metabolic disorders. In patients who weigh less than 10 kg partial exchange transfusion may have to be performed.

Cardioprotection

Anthracyclines, the backbone in AML therapy, are associated with acute and late cardiotoxicity. These effects are carefully monitored via echocardiography at the onset of each cycle.

Risk factors for anthracycline-associated cardiotoxicity include:

- Female sex.
- Younger age at exposure.
- Higher cumulative anthracycline dose.

    Time from Exposure to Anthracyclines
    Dexrazoxane cardioprotection:

- Anthracyclines produce iron-dependent oxygen free radicals that lead to intracellular damage and death of cardiac myocytes.

Dexrazoxane, a topoisomerase II inhibitor that also chelates intracellular free iron and iron bound to anthracyclines, results in the reduction of iron-dependent oxygen free radicals.

Cardioprotection with dexrazoxane is not universally utilized at this time, although current data support dexrazoxane's safety.

Dose: 10:1 ratio of dexrazoxane to doxorubicin dose intravenous (IV) push 30 min prior to the administration of an anthracycline.

Minimal Residual Disease and Its Implication in the Management of AML

Response to therapy has traditionally been determined on the basis of morphologic findings alone. However, nearly half of patients who achieved a complete remission (CR), defined as <5% blasts following induction, subsequently relapse. Methods with increased sensitivities are utilized for the detection of occult disease below the level of morphologic detection, referred to as minimal residual disease (MRD). Methods used to detect MRD include polymerase chain reaction of AML-associated genetic abnormalities or multi-dimensional flow cytometry profiling of AML-associated immunophenotypes. MRD is now being used in AML to identify patients at higher risk of relapse and is being incorporated as part of risk-stratification in clinical trials and treatment decisions.

Application of these techniques for MRD may have a profound influence on the management of leukemia for the following reasons:

- A more precise definition of remission can be obtained based on morphologic, cytogenetic, immunologic, and molecular findings.
- Treatments may be modified, depending on the level of MRD.
- Patients who have a molecular or immunologic remission (defined as leukemic cells less than 0.01% of nucleated bone marrow cells) are predicted to have a lower risk of relapse, and therefore a better outcome. Patients with >0.01% leukemic cells after the end of induction have a worse prognosis and may require more intensive therapy.

Table 18.8 shows the levels of detection of MRD in leukemia by various methods.

Treatment of Newly Diagnosed AML

While the prognosis for children with AML has improved over the years, the OS rate is still approximately 65%. Treatment failure is mainly due to relapse and treatment-related mortality (TRM). Improvements in survival rates are due to intensification of regimens, aggressive supportive care, use of risk stratification based on cytogenetic and molecular markers, and use of better salvage options including transplant, as discussed below. Table 19.12 lists therapies for newly diagnosed AML.

TABLE 19.12   Therapy for Newly Diagnosed Acute Myeloid Leukemia Protocol (Modified from MRC 15)

Course 1	ADE	Daunorubicin 50 mg/m² IV days 1, 3, 5
		Cytarabine 100 mg/m² IV bolus every 12 h days 1–10 (20 doses)
		Etoposide 100 mg/m² IV 1 h infusion days 1–5
		Intrathecal cytarabine age adjusted doses at time of diagnostic LP
Course 2	ADE	Daunorubicin 50 mg/m² IV daily on days 1, 3, 5
		Cytarabine 100 mg/m² IV bolus every 12 h on days 1–8 (16 doses)
		Etoposide 100 mg/m² IV daily (1 h infusion) days 1–5
		Intrathecal cytarabine age adjusted dosing on day 1
Course 3[a]	MACE	Amsacrine[b] 100 mg/m² IV daily (1 h infusion) days 1–5
		Cytarabine 200 mg/m²/d IV (continuous infusion) days 1–5
		Etoposide 100 mg/m² IV daily (1 h infusion) days 1–5
		Intrathecal cytarabine age adjusted dosing on day 1
Course 4	Mid-AC	Mitoxantrone 10 mg/m² IV daily (short infusion) days 1–5
		Cytarabine 1.0 gram/m² 12-hourly IV (2 h infusion) days 1–3 (6 doses)
		Intrathecal cytarabine age adjusted dosing on day 1

[a]Course 3 is modified by the Children's Oncology Group. Since amascarine is not available in the United States, COG uses high-dose cytarabine 1000 mg/m² q12h or 33 mg/kg/dose q12h if BSA <0.6 m².
[b]Not available in the United States.
All doses are reduced by 25% for children less than 1 year of age. Age-adjusted intrathecal chemotherapy with cytarabine: Age 0–1 year 20 mg; 1–2 years 30 mg; 2–3 years 50 mg; 3 year or older 60 mg. For patients with CNS disease at diagnosis IT therapy with cyatarabine is given twice per week until CSF is clear with two additional doses after clearing of CSF with a minimum of 4 doses of intrathecal therapy.
LP, lumbar puncture; IV, intravenous.
Adapted from Gibson et al. (2005).

Hematopoietic Stem Cell Transplant for AML

The role of hematopoietic stem cell transplant (HSCT) for particular subgroups of AML patients remains controversial. Patients with favorable risk disease based on cytogenetics or molecular status do not benefit from transplant as a first-line treatment. Patients with disease positive for the *FLT3-ITD* and those with monosomy 7 may benefit from HSCT in the first remission. Disease status at transplant is a significant predictor of recurrence and OS. The 5-year probabilities of OS are 47%, 28%, and 17% for patients undergoing transplant in CR, relapse, and primary induction failure, respectively. While gross morphologic disease has generally been considered a contraindication to transplant, HSCT for MRD-positive patients may still allow a significant chance of cure. HSCT is discussed in more detail in Chapter 31.

Prognosis of Newly Diagnosed AML

Data collected from the Surveillance, Epidemiology, and End Results Program collected between 1975 and 2008 show that 5-year survival rates for children with AML younger than 20 years have improved from <20% to >60%. Approximately 60–70% of patients become long-term survivors. Despite advances, refractory disease exists in nearly 20% of patients and disease recurs in as many as 30–40% of children who achieve remission. Two subsets with significantly better outcomes are APML and DS-ML. Studies report APML survival of >95% and survival for DS-ML approximates 80%.

Although up to 85% of AML patients achieve a first remission, relapse rates range between 20% and 40%. While MRD provides benefit in detecting disease beyond morphologic methods, attaining low MRD does not exclude subsequent relapse. Nearly one quarter of patients who achieve MRD-negative status still relapse. The adverse prognostic factors among MRD-negative patients remain unknown. In the AML02 study, St. Jude identified prognostic factors among this challenging group. Among MRD-negative patients after either Induction I or II, patients with certain 11q23 abnormalities, such as t(6;11) and t(10;11), acute megakaryoblastic leukemia (AMKL) without t(1;22), and age ≥0 years had inferior outcomes. Patients with rearrangement of CBF genes had superior outcomes. In addition, FLT3 mutations are associated with a poorer prognosis.

Relapsed and Refractory AML

Failure to achieve a CR after first induction is associated with a poor outcome in *de novo* and relapsed AML. Achievement of second CR in relapsed AML is the most important prognostic factor of OS. Duration of CR1 is an important prognostic factor in relapsed AML, with a relapse at <1 year from diagnosis considered early and >1 year considered late. Five-year survival is 13% if disease recurs <12 months after diagnosis and 36% if recurrence occurs >1 years after diagnosis.

Treatment of relapsed or refractory disease is difficult. However, induction of remission may be attempted with the FLAG-IDA Protocol as shown in Table 19.13. Fludarabine, a purine analog, inhibits DNA and RNA synthesis, induces apoptosis, and potentiates the activity of cytarabine. Cytarabine, an anti-metabolite which inhibits DNA synthesis, is an integral part of AML therapy. The addition of idarubicin, an anthracycline, has been shown to significantly increase the antileukemic effect of the regimen. Idarubicin use is limited by cardiotoxicity, particularly in heavily pretreated patients. Infection predominantly with pulmonary involvement is the most common regimen-related toxicity of this regimen.

TABLE 19.13 Therapy for Relapsed/Refractory AML: FLAG-IDA Protocol

Granulocyte colony-stimulating factor: 5 µg/kg/day from day 0 until absolute neutrophil count >1.0×10^3/mm^3

Fludarabine: 30 mg/m^2/day by a 30 min IV infusion daily (days 1–4)

Cytarabine: 2000 mg/m^2/day by a 30 min IV infusion (days 1–4) starting 4 h after beginning of fludarabine

Idarubicin[a] 12 mg/m^2/day by 1 h IV infusion daily (days 2–4) starting 1 h prior to cytarabine infusion

[a]*Heavily pretreated patients who have reached the limit for anthracycline exposure receive FLAG without Idarubicin.*
IV, intravenous; AML, acute myeloid leukemia.

TABLE 19.14   Novel Therapeutic Approaches

Antibody therapy

Tyrosine kinase inhibitors

Proteasome inhibitors

Epigenetic/demethylating agents

Chimeric antigen receptor T-cell immunotherapy

Killer immunoglobulin receptor-mismatched natural killer cells

NOVEL THERAPEUTIC APPROACHES

Despite improvements, survival curves have plateaued. Future protocols will likely incorporate stratification based on molecular characterization of AML, use of novel agents in addition to tailoring of therapies to account for host factors regarding toxicity and response to chemotherapy. See Table 19.14 for a list of novel approaches.

Monoclonal Antibodies

Monoclonal antibodies target surface markers on multiple cell types and can provide precise killing of malignant clones. Gemtuzumab ozogamicin is a recombinant humanized anti-CD33 monoclonal antibody conjugated to calicheamicin, an antitumor antibiotic. CD33 is an adhesion protein expressed on the surface of myeloid blasts in approximately 90% of patients with AML. The Children's Oncology Group investigated the use of gemtuzumab added in Induction Course 1 and Intensification Course 2 at a dose of $3\,mg/m^2$. Event-free survival (EFS) was improved but OS was not due to an increase in TRM in several subgroups. Gemtuzumab was withdrawn from the US market in 2010 because of concerns for increased induction mortality and lack of efficacy. It may be available via compassionate use.

Tyrosine Kinase Inhibitors

Sorafenib, a small-molecule tyrosine kinase inhibitor, is currently under investigation in a COG trial in which FLT3-ITD-positive patients receive sorafenib in addition to a high-risk chemotherapy regimen.

Proteasome Inhibitors

Proteasome inhibitors selectively deplete leukemia-initiating cells and augment effects of cytotoxic chemotherapeutics. Bortezomib, a proteasome inhibitor, is currently under investigation.

Other Novel Agents

Novel agents are currently in early phases of clinical trials for treatment of patients with relapsed/refractory AML. Agents utilize methods such as cytotoxic effects, epigenetic strategies, and antibody-directed therapies and immune modulation, among others.

CPX-351, a liposomal formulation of a synergistic 5:1 molar ratio of cytarabine and daunorubicin followed by FLAG is being investigated for use in relapsed/refractory patients. This agent has shown superior pharmacology compared with conventional chemotherapy and completed Phase 2 clinical trials in adults, with Phase 3 trials ongoing. The liposomal formulation delivers a 5:1 molar ratio for >24 h, is given on days 1, 3, and 5 and accumulates and persists in the bone marrow and may mitigate cardiotoxicity.

Selinexor (KPT-330), an oral selective inhibitor of nuclear export, is being investigated for use in relapsed/refractory leukemia patients. This agent results in accumulation of tumor suppressor proteins in the nucleus leading to G2 arrest and apoptosis of cancer cells. Phase I pediatric studies are underway.

Epigenetics

Genes that regulate DNA methylation and demethylation are often mutated in adult patients with AML. While the frequency of these mutations is lower in pediatric patients, there may be a role for demethylating agents in a subset of pediatric AML patients. Decitabine, a demethylating agent with multiple mechanisms of action including induction of terminal differentiation of myeloid blasts has shown activity in several adult studies. Azacitidine, another demethylating agent is being evaluated in combination with FLAG therapy (Table 19.13). Studies looking at decitabine + ADE and azacitabine + FLAG have been recently completed. Vorinostat, a histone deacetylase (HDAC) inhibitor, with multiple functions, including induction of cell cycle arrest and apoptosis in CD33-positive cells from AML patients, has shown activity in adult studies. Suberoylanilide hydroxamic acid, another HDAC inhibitor, is under investigation. The combination of a hypomethylating followed by a HDAC inhibitor is also being considered. EPZ-5676, a DOT1 inhibitor, is in early trials for patients with MLL rearrangements, a poor prognostic indicator, which causes aberrant methylation and leukemogenesis.

Chimeric Antigen Receptor T-cell Immunotherapy in AML

Cancer immunotherapy uses human T-cells genetically engineered with chimeric antigen receptors to target tumor antigens such as CD33 or CD34. *In vitro* studies have demonstrated cytotoxicity against AML cell lines and some *in vivo* results have been promising; however, depletion of normal CD34 + hematopoietic cells has been reported, highlighting the myeloablative nature of this technology. This technique has shown early-phase promise in pediatric ALL and will now be studied in pediatric AML.

Use of Killer Immunoglobulin Receptors-mismatched Natural Killer Cells in HSCT

With appropriate activation, natural killer (NK) cells release cytokines and induce apoptosis even against targets to which they have not previously been exposed. Their cytotoxic nature relies on signals including those from killer immunoglobulin receptors (KIRs), inhibitory regulators of NK cells that recognize human leukocyte class I alleles. Often, cancerous cells downregulate MHC class I surface expression, thereby avoiding eradication of T-cells, but convey *de novo* stimulatory NK cell signals, making them good targets for KIR-mediated detection and destruction. A few studies in adults have been completed and a small study demonstrated safety and feasibility in pediatric patients in bone marrow transplants.

ACUTE PROMYELOCYTIC LEUKEMIA

Background

APML comprises about 5−10% of pediatric AML cases. *PML- RARA* t(15;17 translocation) fuses a promyelocytic leukemia gene to a retinoic acid receptor gene, causing maturation arrest in the promyelocytic stage. APML cells are highly sensitive to differentiation therapy with all-trans retinoic acid (ATRA), except in <5% of cases where RARA is fused to an alternative gene, causing variable sensitivity to ATRA. Patients are considered standard risk if WBC <10,000/mm^3 at diagnosis and high risk if WBC >10,000/mm^3 at diagnosis. Untreated APML carries a high risk of life-threatening complications, particularly hemorrhage.

Treatment of APML

ATRA should be started immediately when APML is suspected. ATRA drives leukemic cells into terminal differentiation and eliminates malignant proliferation. Over the past decade APML has had a significant improvement in survival, due to the use of ATRA and anthracyclines, with CR rates of ≥90% and EFS of 70−80%. Arsenic trioxide (ATO) has been incorporated into treatment regimens, as it specifically binds the PML moiety of the PML-RARA oncoprotein, leading to its degradation and resulting in partial differentiation and induction of apoptosis of leukemic promyelocytes. Synergy of ATRA and arsenic has been demonstrated at biologic and clinical levels. A large European prospective, randomized, multicenter, phase 3 non-inferiority study demonstrated a CR for all evaluable patients in the ATRA-arsenic group. Two-year EFS was 97% in the ATRA-arsenic group vs 86% in ATRA-chemotherapy group, and OS was better with ATRA-arsenic compared with ATRA-chemotherapy. This study provided prednisone 0.5 mg/kg throughout induction as prophylaxis for differentiation syndrome. At the earliest sign of differentiation syndrome, ATRA and arsenic are temporarily stopped and dexamethasone 10 mg q12h was administered until disappearance of signs and symptoms for a minimum of 3 days. See Table 19.15 for a recommended approach to treat standard-risk patients with ATRA and ATO alone. High-risk patients receive idarubicin in addition.

Supportive Care for Patients with APML

Perform surveillance coagulation studies periodically and treat as follows:

- *Disseminated intravascular coagulation*: Treat with platelet and fresh frozen plasma transfusion to maintain a platelet count of 50,000/mm^2 or more and a fibrinogen level of at least 100 mg/dl. Heparin is indicated for patients with marked or persistent elevation of fibrin degradation products.
- *Fibrinolysis*: The use of epsilon-aminocaproic acid is reserved for patients with life-threatening hemorrhages.

Differentiation Syndrome

Formerly referred to as retinoic acid syndrome, differentiation syndrome is a common and potentially life-threatening complication seen in patients with APML treated with ATRA and/or ATO therapy. The clinical features of differentiation syndrome are also observed with other complications (i.e., sepsis, fluid overload, pulmonary hemorrhage, pneumonia, renal failure, and congestive heart failure). These conditions should be considered in the differential diagnosis and treated if suspected. See Table 19.16 for characteristics of differentiation syndrome and see Table 19.17 for suspected pathogenesis of differentiation syndrome.

There is no consensus on a prophylactic strategy to prevent differentiation syndrome. Corticosteroids (i.e., dexamethasone 2.5 mg/m^2/12 h × 15 days), should be administered for patients with WBC >5–10,000/mm^3.

TABLE 19.15 Therapy for Standard-Risk Acute Promyelocytic Leukemia

	Induction[a]	Consolidation
Arsenic Trioxide (ATO) (IV)	0.15 mg/kg/day	0.15 mg/kg/day
		5 days per week
		(4 weeks on/4 weeks off for a total of four courses)
ATRA (oral)	45 mg/m^2/day in 2 equally divided doses	45 mg/m^2 day
		(2 weeks on/2 weeks off for a total of seven courses)

[a]*Induction continues until hematologic complete remission or for a maximum of 60 days.*

TABLE 19.16 Characteristics of Differentiation Syndrome

Unexplained fever
Weight gain
Peripheral edema
Dyspnea with pulmonary infiltrates
Pleuropericardial effusion
Hypotension
Acute renal failure

TABLE 19.17 Suspected Pathogenesis of Differentiation Syndrome

Systemic inflammatory response syndrome
Endothelium damage with capillary leak syndrome
Occlusion of microcirculation
Tissue infiltration

Once the condition is suspected, treatment (dexamethasone 10 mg/m^2/12 h) should be started immediately. Additional supportive measures may be required, including diuretics, blood products, dialysis, or mechanical ventilation. Temporary discontinuation of ATRA or arsenic is indicated only for critically ill patients, with severe renal or pulmonary dysfunction.

AML SPECIAL SUBGROUPS

Infant AML

The biology of AML in children <2 years differs from that of older children and generally consists of high-risk features. *MLL* rearrangements occur in approximately 50% of patients with some partner chromosomes associated with the *MLL* translocation, unique to this age group. Immaturity of organs and differences in pharmacokinetic and pharmacodynamic profiles of drugs increases the susceptibility of this age group to toxicities (especially anthracyclines), therefore drug dosage in infants is generally calculated by body weight, not body surface area. Toxicities, particularly cardiotoxicity, are more severe and occur more often in infants compared with older children.

Myeloid Leukemia of DS

Children with Down syndrome have up to a 20-fold increased risk of developing leukemia, (the myeloid leukemia of Down syndrome (DS-ML)). Patients usually present at a younger age than children without DS; the greatest risk of myeloid leukemia is in children with DS younger than 5 years, with only a modest increase in risk continuing into young adulthood.

Approximately 4−10% of DS and DS mosaic infants will be diagnosed with transient myeloproliferative disorder (TMD), a precursor to AMKL. TMD is characterized by megakaryoblasts in the peripheral blood and/or bone marrow with the same morphology and surface antigen expression as megakaryoblasts in AMKL. Patients will typically present with pancytopenia and hepatosplenomegaly. TMD spontaneously remits in a majority of patients within 4−10 weeks. However, some have severe life-threatening symptoms due to pancytopenia or organ infiltration. In these situations, intervention with exchange transfusion and/or low-dose cytarabine is utilized.

Approximately 1−2% of children with DS will develop DS-ML, up to age 4 years. Approximately 20% of DS infants with a history of TMD will subsequently develop AMKL after the clinical resolution of TMD. Children with DS-ML have a markedly superior outcome compared to children with non-DS AML. DS-ML patients usually present with a lower initial WBC, no CNS involvement, and fewer cytogenetic abnormalities. The remission rates are approximately 90%, with EFS approximating 70−80%. This may be partially explained by the increased sensitivity of DS myeloblasts to cytarabine due to a mutation in *GATA1*, a hematopoetic transcription factor.

DS-ML patients have successfully been treated with less-intensive chemotherapy regimens such as standard timing DCTER therapy (Woods et al., 1996). The COG designed the COG A2971 clinical trial with the objective of reducing acute morbidity and mortality in children with DS-ML while maintaining or improving efficacy. The trial eliminated etoposide and dexamethasone from the standard timing DCTER regimen and removed 3 months of systemic maintenance chemotherapy, leaving only three IT doses of cytarabine as maintenance therapy. The 5-year EFS was comparable, at approximately 79% or 77%, respectively. While most patients did very well, once stratified by age, it was noted that patients >4 years had a significantly worse EFS of approximately 33%. A possible explanation is that patients >4 years old with DS-ML often lack *GATA1* mutations, with the risk of relapse more similar to sporadic AML, as demonstrated in several trials.

While the majority of DS-ML patients respond to current therapy, a subset of 10−15% do not, suggesting the need for risk stratification for these patients. A future study will likely stratify based on age (<4 years old) and MRD status. The goal will be to decrease treatment intensity and therefore adverse effects, while improving survival for standard-risk patients, while intensifying therapy for high-risk patients to increase their survival.

For most patients with DS-ML, maintenance therapy and stem cell transplantation are typically not required. However, if transplantation is needed, reduced intensity conditioning for bone marrow transplantation (BMT) has been utilized. Supportive care is paramount, as children with DS are at an increased susceptibility to infections and other treatment-related side effects.

Acute Mixed Lineage Leukemia (Acute Leukemia of Ambiguous Lineage)

Acute MLLs are a group of rare pediatric leukemias that have characteristics of both the lymphoid and myeloid lineages. They represent 3—5% of pediatric leukemias. The leukemia can be:

- Biphenotypic (have both lineages on all of the cells).
- Bilineage (have a mixed population of lymphoid and myeloid blasts).
- Undifferentiated.
- Leukemia that switches lineage during therapy.

In 2008, the WHO proposed an algorithm to classify mixed phenotype acute leukemia. This classification system (Table 19.7) relies on lineage-specific markers.

The majority of the mixed lineage pediatric leukemias have blasts cells that simultaneously express T lineage and myeloid markers or B lineage plus myeloid markers. The OSs for the T/myeloid or B/myeloid leukemias are not significantly different and are similar to acute myeloid leukemia survival but are significantly inferior to the survival of ALL.

The initial treatment of patients with acute MLL utilizes myeloid induction therapy. If patients fail to go into remission, lymphoid induction therapy with prednisone, vincristine, and asparaginase should be instituted, followed by consolidation with intensive ALL therapy. Stem cell transplantation should be reserved for patients with >1% blasts by flow cytometry at the end of induction or persistent MRD.

Further Reading and References

Abildagaard, L., Ellebaek, E., Gustafsson, G., et al., 2006. Optimal treatment intensity in children with Down syndrome and myeloid leukemia: data from 56 children treated on NOPHO-AML protocols and a review of the literature. Ann. Hematol. 85, 275—280.

Bennett, J.M., Catovsky, D., Daniel, M.T., et al., 1985. Proposed revised criteria for the classification of acute myeloid leukemia. Ann. Intern. Med. 103 (4), 620—625.

Burnett, A.K., Goldstone, A., Hills, R.K., et al., 2013. Curability of patients with acute myeloid leukemia who did not undergo transplantation in first remission. J. Clin. Oncol. 31 (10), 1293—1301.

Burnett, A.K., Russell, N.H., Hills, R.K., et al., 2013. Optimization of chemotherapy for younger patients with acute myeloid leukemia: results of the Medical Research Council AML15 trial. J. Clin. Oncol. 31 (27), 3360—3368.

Creutzig, U., van den Heuvel-Eibrink, M.M., Gibson, B., et al., 2012. Diagnosis and management of acute myeloid leukemia in children and adolescents: recommendations from an international expert panel. Blood 120 (16), 3187—3205.

Davila, J., Slotkin, E., Renaud, T., 2014. Relapsed and refractory pediatric acute myeloid leukemia: current and emerging treatments. Pediatr. Drugs 16, 151—168.

de la Rubia, J., Regadera, A.I., Martin, G., et al., 2002. FLAG-IDA regimen (fludarabine, cytarabine, idarubicin and G-CSF) in the treatment of patients with high-risk myeloid malignancies. Leuk. Res. 26, 725—730.

Faulk, K., Gore, L., Cooper, T., 2014. Overview of therapy and strategies for optimizing outcomes in de novo pediatric acute myeloid leukemia. Pediatr. Drugs 16 (3), 213—227.

Gamis, A.S., 2005. Acute myeloid leukemia and Down syndrome evolution of modern therapy—state of the art review. Pediatr. Blood Cancer 44, 13—20.

Gamis, A.S., Alonzo, T.A., Gerbing, R.B., et al., 2011. Natural history of transient myeloproliferative disorder clinically diagnosed in Down syndrome neonates: a report from the Children's Oncology Group Study A2971. Blood 118, 6752—6759.

Gamis, A.S., Alonzo, T.A., Meshinchi, S., et al., 2014. Gemtuzumab ozogamicin in children and adolescents with de novo acute myeloid leukemia improves event-free survival by reducing relapse risk: results from the randomized phase III Children's Oncology Group trial AAML0531. J. Clin. Oncol. 32 (27), 3021—3032.

Gibson, B.E.S., Wheatley, K., Hann, I.M., et al., 2005. Treatment strategy and long-term results in paediatric patients treated in consecutive UK AML trials. Leukemia 19, 2130—2138.

Gregory, J., Kim, H., Alonzo, T., et al., 2009. Treatment of children with promyelocytic leukemia: results of the first North American Intergroup trial INT0129. Pediatr. Blood Cancer 53, 1005—1010.

Hasle, H., Abrahamsson, J., Arola, M., et al., 2008. Myeloid leukemia in children 4 years or older with Down syndrome often lacks GATA1 mutation and cytogenetics and risk of relapse are more akin to sporadic AML. Leukemia 22, 1428—1474.

Horton, T.M., Perentesis, J.P., Gamis, A.S., et al., 2014. A phase 2 study of bortezomib combined with either idarubicin/cytarabine or cytarabine/etoposide in children with relapsed, refractory or secondary acute myeloid leukemia: a report from the Children's Oncology Group. Pediatr. Blood Cancer 61 (10), 1754—1760.

Johnston, D.L., Alonzo, T.A., Gerbing, R.B., et al., 2010. The presence of central nervous system disease at diagnosis in pediatric acute myeloid leukemia does not affect survival: a Children's Oncology Group study. Pediatr. Blood Cancer 55, 414—420.

Karol, S.E., Coustan-Smith, E., Cao, X., et al., 2015. Prognostic factors in children with acute myeloid leukaemia and excellent response to remission induction therapy. Br. J. Haematol. 168 (1), 94—101.

Lancet, J.E., 2014. Postremission therapy in acute promyelocytic leukemia: room for improvement? J. Clin. Oncol. 32 (33), 3692—3696.

Lo-Coco, F., Avvisati, G., Vignetti, M., et al., 2013. Retinoic acid and aresinic trioxide for acute promyelocytic leukemia. N. Engl. J. Med. 369 (2), 111—121.

Loken, M.R., Alonzo, T.A., Pardo, L., et al., 2012. Residual disease detected by multidimensional flow cytometry signifies high relapse risk in patients with de novo acute myeloid leukemia: a report from Children's Oncology Group. Blood 120, 1581–1588.

Manola, K.N., Panitsas, F., Polychronopoulous, S., et al., 2013. Cytogenetic abnormalities and monosomal karyotypes in children and adolescents with acute myeloid leukemia: correlations with clinical characteristics and outcome. Cancer Genet. 206 (3), 63–72.

Paietta, E., 2012. Minimal residual disease in acute myeloid leukemia: coming of age. Hematol. Am. Soc. Hematol. Educ. Program 2012, 35–42.

Pizzo, P.A., Poplack, D.G., 2001. Principles and Practice of Pediatric Oncology, fourth ed. Lippincott Williams & Wilkins, Philadelphia.

Rubnitz, J.E., Look, A.T., 1998. Molecular genetics of childhood leukemia. J. Pediatr. Hematol. Oncol. 20, 1–11.

Sanz, M.A., Montesinos, P., 2014. How we prevent and treat differentiation syndrome in patients with acute promyelocytic leukemia. Blood 123 (18), 2777–2782.

Seif, A.E., Walker, D.M., Li, Y., et al., 2015. Dexrazoxane exposure and risk of secondary acute myeloid leukemia in pediatric oncology patients. Pediatr. Blood Cancer 62, 704–709.

Sorrell, A.D., Alonzo, T.A., Hilden, J.M., et al., 2012. Favorable survival maintained in children who have myeloid leukemia associated with Down syndrome using reduced-dose chemotherapy on Children's Oncology Group trial A2971. Cancer 118 (19), 4806–4814.

Vardiman, J.W., Thiele, J., Arber, D.A., et al., 2009. The 2008 revision of the World Health Organization (WHO) classification of myeloid and acute leukemia: rationale and important changes. Blood 114, 937–951.

Woods, W.G., Kobrinsky, N., Buckley, J.D., 1996. Timed-sequential induction therapy improves postremission outcome in acute myeloid leukemia: a report from the Children's Cancer Group. Blood. 87 (12), 4979–4989.

20

Histiocytosis Syndromes

Robert J. Arceci[†]

The term histiocytosis syndrome identifies a group of disorders that have in common the proliferation of cells of the mononuclear phagocyte system and/or the dendritic cell system.

Macrophages function predominantly as antigen-processing cells and dendritic cells function as accessory cells or antigen-presenting cells. Although the principal pathological cells of these syndromes are histiocytes, the term histiocytosis syndrome is a very selective term in the sense that it does not include storage diseases, hyperlipidemic xanthomatosis, or granulomatous reaction in chronic infections such as tuberculosis or foreign body granuloma.

The histiocytoses thus represent a diverse group of diseases characterized by the excess proliferation and accumulation of dendritic cells, monocytes, and macrophages in addition to other immune effector cells such as eosinophils and lymphocytes depending upon the specific type of disorder (Table 20.1).

LANGERHANS CELL HISTIOCYTOSIS

Incidence

The prevalence has been estimated to be between 2 and 10 cases per million children under age 15 years annually. The male to female ratio is between 1.3 and 1.9:1. The peak incidence is between the ages of 1 and 4 years. Isolated pulmonary Langerhans cell histiocytosis (LCH) occurs primarily in adults who smoke cigarettes (>90% of cases). LCH has also been shown to be commonly concordant in identical twins and less frequently in nonidentical twin siblings and relatives.

Pathology

The characteristic histopathology required for a "presumptive diagnosis" usually shows a granulomatous-like lesion with immature dendritic-appearing cells that have characteristic bean-shaped, folded nuclei and pale cytoplasm. Often, multinucleated giant cells are present. A "definitive diagnosis" of LCH requires the immunohistochemical identification of the presence of Langerhans cell antigen expression of cell surface CD1a, CD207 (langerin) or by the presence of cells with Birbeck granules by electron microscopy (Table 20.2). Early in the course of the disease the lesions are usually proliferative and locally destructive. In later or healing stages, they can become more fibrotic.

Pathogenesis

LCH can most accurately be considered a clonal neoplasm that arises from an immature dendritic cell or in some cases an earlier hematopoietic precursor cell. Multiple reports have furthermore now documented the

[†]Dr Robert Arceci died tragically in an accident on June 8, 2015. We pay tribute to his contribution to pediatric hematology and oncology as a researcher, teacher of future generations of leaders in the field, and in recognition of an individual of sterling character, an outstanding clinician with abundant energy, infectious enthusiasm, and exceptional compassion.

Lanzkowsky's Manual of Pediatric Hematology and Oncology.
DOI: http://dx.doi.org/10.1016/B978-0-12-801368-7.00020-X

TABLE 20.1 Classification of Histiocytoses

Type I. Dendritic cell and dermal dendritic cell disorders
- LCH, including eosinophilic granuloma, multisystem with bone involvement, DI, and exophthalmos (formerly called Hand−Schüller−Christian syndrome) and disseminated involvement with risk organ dysfunction (formerly called Abt−Letterer−Siwe syndrome)
- Juvenile xanthogranulomatosis
- Erdheim−Chester disease

Type II. Macrophage-related disorders
- Hemophagocytic lymphohistiocytosis
- Familial erythrophagocytic lymphohistiocytosis
- Infection-associated hemophagocytic syndrome
- Malignancy-associated hemophagocytic syndrome
- Rosai−Dorfman disease

Type III. Malignant histiocytic disorders
- Dendritic cell-related histiocytic sarcoma (localized or disseminated)
- Macrophage-related histiocytic sarcoma
- Monocyte-related (monocytic leukemia or sarcoma)

Adapted from WHO committee Classification Working Group of Histiocytic Society.

TABLE 20.2 Histological, Histochemical, and Electron Microscopic Diagnosis of LCH

1. Presumptive diagnosis: light morphologic characteristics
2. Designated diagnosis
    a. Light morphologic features plus
    b. Two or more supplemental positive stains for
        i. Adenosine triphosphatase
        ii. S-100 protein
        iii. ∝-D-mannosidase
        iv. Peanut lectin
3. Definitive diagnosis
    a. Light morphologic characteristics plus
    b. Birbeck granules in the lesional cell with electron microscopy or CD207 expression by immunohistochemistry and/or
    c. Immunohistochemical staining for CD1a antigen on the lesional cell.

Adapted from Writing Group of the Histiocyte Society (Chu et al. 1987), with permission.

presence of BRAF V600E mutations in 50−60% of cases. This mutation is nearly always monoallelic and thus appears to act like a dominant driving oncogene. There also appears to be no predilection for this BRAF V600E mutation in terms of the extent of organ involvement or age (Badalian-Very et al., 2013). Of particular interest, however, is that the cases which appear not to have BRAF V600E mutations do have evidence for activation of the RAS-RAF-MEK-ERK signaling pathway. This suggested that alternative activating mutations were present, which has been confirmed with the subsequent identification of mutations involving genes encoding CSF-1 receptor, RAS, and MAP2K1. Evidence for circulating CD34-positive hematopoietic progenitors harboring V600E mutations, usually in the more severe forms of LCH, along with RNA expression patterns placing the pathologic LCH LC in a category of an activated myeloid dendritic cell rather than a epidermal Langerhans cell, have contributed to LCH now being considered an oncogene-driven neoplasm of the myeloid lineage.

Clinical Features

Clinical manifestations depend on the site of lesions, number of involved areas, and extent to which the function of key involved organs is compromised. Although replaced by more prognostic categories, the classic designations, eosinophilic granuloma, Hand−Schüller−Christian disease, and Abt−Letterer−Siwe disease, are still present in the literature and are useful descriptions of the various clinical manifestations of LCH.

Eosinophilic granuloma, solitary (SEG) or multifocal (MEG), are found predominantly in older children, as well as in young adults, usually within the first three decades of life with the incidence peaking between 5 and 10 years of age. SEG and MEG represent approximately 60−80% of all instances of LCH. Patients with systemic involvement frequently have similar bone lesions in addition to other manifestations of disease.

Hand−Schüller−Christian disease (multisystem disease) was historically described as the clinical triad of lytic lesions of bone, exophthalmos, and diabetes insipidus (DI), although this form of LCH is now usually considered to be more typical multisystem LCH without key organ dysfunction. It most commonly occurs in younger children, 2–5 years of age, and represents 15–40% of such patients; this type of involvement can be observed, however, in all ages. Signs and symptoms include bone lesions with exophthalmos due to tumor mass in the orbital cavity. This usually occurs from involvement of the roof and lateral wall of the orbital bones. Orbital involvement may sometimes result in vision loss or strabismus due to optic nerve or orbital muscle involvement, respectively. The most frequent sites of skeletal involvement include the flat bones of the skull, ribs, pelvis, and scapulae. There may be extensive involvement of the skull, with irregularly shaped, lytic lesions. Less frequently, long bones and lumbosacral vertebrae, usually the anterior portion of the vertebral body, are involved. Oral involvement commonly affects the gums and/or palate, resulting in the characteristic floating tooth seen on dental radiographs. Large sections of the mandible may be involved, with loss of bone leading to diminished height of the mandibular rami. Chronic otitis media, due to involvement of the mastoid and petrous portion of the temporal bone, and otitis externa are common.

Abt–Letterer–Siwe disease occurs in about 10% of cases and represents the most severe manifestation of LCH. Typically, patients are less than 2 years of age and present with a scaly seborrheic, eczematoid, sometimes purpuric rash that involves the scalp, ear canals, abdomen, and intertriginous areas of the neck and face. The rash may be maculopapular or nodulopapular. Ulceration may result, especially in intertriginous areas. Draining ears, lymphadenopathy, hepatosplenomegaly, and, in severe cases, hepatic dysfunction with hypoproteinemia and diminished synthesis of clotting factors can occur. Gastrointestinal involvement may give rise to diarrhea, malabsorption, and bleeding. Anorexia, irritability, failure to thrive, and significant pulmonary symptoms such as cough, tachypnea, and pneumothorax may occur as well. Involvement of the hematopoietic system, with bone marrow infiltration, can result in pancytopenia.

Other presentations of LCH are commonly seen. LCH can have a strictly *nodal presentation*, not to be confused with sinus histiocytosis with massive lymphadenopathy (SHML) (Rosai–Dorfman disease). This presentation is characterized by significant enlargement of multiple lymph node groups, with little or no other signs of disease. Pulmonary disease, usually seen in young adults in their third or fourth decade (occasionally in adolescents) and associated with cigarette smoking, may follow a severe and often chronic, debilitating course; patients may present with pneuomthorax. Pulmonary disease may also occur as part of multisystem disease. *Cutaneous disease* with no evidence of dissemination has been described in infants, children, and adults.

Involvement by Site of Disease

Skeleton

Painful bone lesions affecting hematopoietically active bones are common. Radiographically, the lesions are lytic and sometimes may have sclerotic edges or bone islands within the lytic area. They occur commonly in the skull as punched-out lytic lesions. Bone involvement of the mandible and maxilla and soft-tissue involvement of the gingivae may result in loss of teeth. Involvement of vertebrae can result in vertebral collapse (vertebra plana) and lesions of long bones can result in fractures. There is often an inability to bear weight and tender, sometimes warm, swelling due to soft-tissue infiltration overlying the bone lesions occur. Radionuclide bone scan (^{99m}Tc-polyphosphate) or PET/CT may show localized increased uptake at the site of involvement. The differential diagnoses include osteomyelitis, malignant bone tumors, and bony cysts. Only rarely are bones involving the wrists, hands, and feet observed. Table 20.3 lists the distribution of the sites of bone lesions.

Skin

Cutaneous eruptions consist of:

1. Diffuse papular scaling lesions, resembling seborrheic eczema (most common).
2. Petechiae and purpura.
3. Granulomatous ulcerative lesions.
4. Xanthomatous lesions.

Lungs

Lung involvement may result in pulmonary dysfunction with tachypnea and/or dyspnea, cyanosis, cough, pneumothorax, or pleural effusion. Radiographic infiltrates consisting of diffuse cystic changes, nodular infiltrations, or extensive fibrosis can occur. The radiographic appearance may resemble miliary tuberculosis.

TABLE 20.3   Distribution of Bone Lesions in LCH

Site	Incidence (%)
Skull	49
Innominate bone	23
Femur	17
Orbit	11
Ribs	8
Humerus	7
Mandible	7
Tibia	7
Vertebra	7
Clavicle	5
Scapula	3
Fibula	2
Sternum	1
Radius	1
Metacarpal	1

Liver

The liver may be enlarged with dysfunction that results in hypoproteinemia (total protein less than 5.5 g/dl and/or albumin less than 2.5 g/dl), edema, ascites, and/or hyperbilirubinemia (bilirubin level greater than 1.5 mg/dl, not attributable to hemolysis). Pretreatment liver biopsy more often reveals portal triditis and less often fibrohistiocytic infiltrates or bile duct proliferation. Sclerosing cholangitis, fibrosis, and liver failure may lead to the need for liver transplantation.

Hematopoietic System

The pathophysiology of hematopoietic dysfunction can be due to hypersplenism as well as direct involvement of the bone marrow by Langerhans cells and/or reactive macrophages.

Hematopoietic system dysfunction may consist of the following:

Anemia (hemoglobin level less than 10 g/dl, not due to iron deficiency or superimposed infection), leukopenia (neutrophils less than 1500/mm^3), or thrombocytopenia (less than 100,000/mm^3). Excessive number of histiocytes in the marrow aspirate is not considered evidence of dysfunction.

Lymph Nodes

Occasionally, there is massive lymph node enlargement of cervical nodes occurs without other evidence of histiocytosis.

Endocrine System

Short stature has been found in up to 40% of children with systemic LCH. Chronic illness and steroid therapy play an important role in its causation. However, short stature may also be a consequence of anterior pituitary involvement and growth hormone deficiency, which may occur in up to about half of the patients with initial anterior pituitary dysfunction. Posterior pituitary involvement with DI is characteristic of systemic LCH. Other endocrine manifestations include hyperprolactinemia and hypogonadism due to hypothalamic infiltration. Pancreatic and thyroid involvement have also been reported.

Gastrointestinal System

Gastrointestinal tract disease can occur and 2–13% of patients have biopsy-proven gastrointestinal involvement and/or digestive tract symptoms, although this may be an underestimate. Diarrhea, malabsorption, and hematochezia are common manifestations.

Central Nervous System

Four groups of patients can be clinically distinguished:

1. Patients who present with hypothalamic pituitary system involvement.
2. Patients who present with site-dependent symptoms of space-occupying lesions leading to headache and seizures.
3. Patients who exhibit neurologic dysfunction characterized by a neurodegenerative picture of reflex abnormalities, ataxia, intellectual impairment, sometimes hydrocephalus, tremor, and dysarthria with variable progression to severe CNS deterioration.
4. Patients who present with an overlap of the aforementioned symptoms.

Patients who develop CNS disease are more likely to have multisystem disease and skull lesions. Table 20.4 shows the clinical characteristics of patients with LCH who developed CNS disease compared to those who did not develop CNS disease. It reveals that patients who developed CNS disease are more likely to have multisystem disease with skull and temporal bone lesions, orbital involvement, DI, and endocrinopathies. Table 20.5 shows the classification of CNS lesions according to MRI morphology.

Histopathology of CNS Lesions

There are likely four stages of LCH involvement of the CNS as follows:

1. Hyperplastic proliferative.
2. Granulomatous.
3. Xanthomatous.
4. Fibrosis.

TABLE 20.4 Extent of LCH and Organs Involved in Diagnosis of Patients Who Develop CNS Disease Compared to Those Who Do Not Develop CNS Disease

	Percentage in 38 CNS patients	Percentage in 275 LCH patients
Multisystem disease	72	40
Single-system bone disease	18	53
Single-system skin, lymph node	0	7
Primary CNS disease	10	0
Bone	84	79
Skull	74	40
Temporal bone	34	8
Skin	58	25
Diabetes insipidus	31	6
Orbits	24	2
Endocrinopathies	18	3
Lungs	16	6
Gastrointestinal tract	10	5
Liver	10	11
Spleen	10	9

From Grois et al. (1998), with permission.

TABLE 20.5   Classification of Central Nervous System Lesions According to Magnetic Resonance Imaging Morphology[a]

Type		Number	Percentage
Ia	White-matter lesions without enhancement	21	55
Ib	White-matter lesions with enhancement	9	24
IIa	Gray-matter lesions without enhancement	19	50
IIb	Gray-matter lesions with enhancement	3	8
IIIa	Extraparenchymal dural based	12	32
IIIb	Extraparenchymal arachnoidal based	2	5
IIIc	Extraparenchymal choroid plexus based	3	8
IVa	Infundibular thickening	8	21
IVb	(Partial) empty sella	14	37
IVc	Hypothalamic mass lesions	4	10
Va	Atrophy: diffuse	10	26
Vb	Atrophy: localized	6	16
VI	Therapy-related with enhancement	6	15

[a]*LCH-CNS Study*, n = 38.
From Grois et al. (1998), with permission.

Lesions in the hyperplastic proliferative stage are most likely to contain the diagnostic LCH cells. In the cerebellum and cerebrum, white matter may show demyelination and there may also be destruction of Purkinje cells in the absence of histiocytes. Gliosis with plasma cell infiltrates may also be found.

Hypothalamic Pituitary Involvement

Signs and symptoms: Disturbances in social behavior, appetite, temperature regulation, and sleep patterns are common.
Posterior pituitary involvement: DI, polyuria, polydipsia.
Anterior pituitary involvement: Growth failure, precocious or delayed puberty, amenorrhea, hypothyroidism.

Of these, DI is the most common manifestation. The incidence of this complication ranges from 5% to 35% depending upon the extent and location of disease. Most present within 4 years of diagnosis. DI is due to infiltration by the disease into the hypothalamus with or without involvement of the posterior pituitary gland. Local tissue damage may be a consequence of IL-1 and prostaglandin E2 production. Polydipsia and polyuria may develop at presentation, during active disease (even when there is improvement in other areas), or after therapy is discontinued and there is no other apparent active disease.

Laboratory Studies for the Diagnosis of DI

1. Water deprivation test with at least 3 h of no intake and with serum and urine electrolytes and osmolalities before and after deprivation plus arginine vasopressin levels is a definitive approach to diagnosis. Correction with a dose of desmopressin (DDAVP) further confirms the diagnosis. An early morning fasting urinalysis for specific gravity is also helpful as a screen.
2. Gadolinium-enhanced MRI studies show thickening of the hypothalamic pituitary stalk (>2.5 mm) and absence of a posterior pituitary bright signal in T1 weighted images. These lesions are caused by pathologic infiltrates. There is no convincing evidence that established DI can be reversed by any treatment modality, but new-onset DI is considered to represent active disease and rapid initiation of treatment, usually LCH-directed chemotherapy, is recommended, although there are few data about the true response rate.

Replacement therapy with DDAVP is recommended for patients with DI. The rapid institution of effective systemic chemotherapy for disseminated disease may prevent the occurrence of DI and might be responsible for the low frequency of DI, although this has not been definitively proven. A small pituitary or empty sella indicates combined anterior and posterior pituitary insufficiency. This may be a result of disease or may be observed following cranial radiotherapy for DI.

Patients presenting with isolated idiopathic DI and morphologic changes in the suprasellar area should be closely observed. Stereotactic biopsy performed because of an enlarged pituitary stalk can distinguish a variety of conditions such as sarcoidosis, granulomatosis, tuberculosis, nonspecific lymphocytic hypophysitis, and LCH. A biopsy is not always possible. Cerebrospinal fluid (CSF) analysis for markers of germ cell tumors can be an important differentiator. Patients without a definitive diagnosis and with DI are often followed with serial contrast MRIs. If the CNS lesion enlarges and no definitive diagnosis has been made, then a biopsy may be indicated.

Space-Occupying Central Nervous System Lesions

These lesions most often arise from adjacent bone lesions, brain meninges, or choroid plexus. They usually give rise to signs and symptoms of increased intracranial pressure. They are also site-specific and size-dependent. Symptoms include headaches, vomiting, papilledema, optic atrophy, seizures, and other focal symptoms. Even diffuse meningitis-like manifestations can occur. These lesions may occur without any other evidence of LCH. Mass lesions respond well to treatment, leaving minimal or no residual defects.

Neurodegenerative Disease

The cerebellum is the second most common site of LCH CNS involvement, the first being the hypothalamic—hypophyseal axis. Neurologic symptoms occasionally predate the diagnosis of LCH. Symptoms mainly follow the pontine—cerebellar pattern, beginning as a discrete reflex abnormality or gait disturbance, and/or nystagmus. Sometimes patients may also present with hydrocephalus. They can progress to disabling ataxia. Pontine symptoms include dysarthria, dysphagia, and other cranial nerve deficits, ultimately leading to fatal neurodegeneration. On MRI, enhancing lesions involving the pons, basal ganglion, and cerebellar peduncles are observed. MRI of the cerebral hemispheres may show white-matter lesions in the periventricular area.

Biopsy shows a primarily inflammatory, lymphocyte response associated with gliosis, demyelination, and neuronal cell death. The etiology of this neurodegenerative process is unknown, but is believed to be an immune-mediated paraneoplastic response. In some cases, the onset of CNS symptoms occurs many years after the initial diagnosis of LCH.

Clinical and Laboratory Evaluation of LCH

Diagnostic Evaluation

Physical Findings: Patients should have a thorough physical examination including temperature, height, weight and head circumference, pubertal status, skin and scalp for rash, pallor or jaundice, external and middle ear, face and orbits, oropharynx, dentition, chest and lungs, abdomen for organomegaly, extremity and spine. There are few evidence-based guidelines, but the ones listed represent general best practice recommendations (Tables 20.6 and 20.7).

Laboratory Testing

Routine Blood and Serum Tests

As part of an initial diagnostic evaluation, a complete blood count and white blood cell differential should be completed. In addition, liver function tests, including bilirubin, aspartate aminotransferase, alanine aminotransferase, and alkaline phosphatase, are done. Serum electrolytes and creatinine along with prothrombin time and partial thromboplastin time for patients with suspected liver disease should be done. Albumen and prealbumen are helpful in patients with suspected liver disease and failure to thrive. For patients with severe gastrointestinal involvement and diarrhea, serum immunoglobulin levels may also be useful to obtain.

Urine Testing

A routine urinalysis to assess particularly the specific gravity is useful and if low, suggesting DI, and the history is also suggestive, then a water deprivation of usually several hours (first morning void sample for instance) should be obtained. For infants and young children with suspected DI, a water deprivation test should be done under medical supervision.

TABLE 20.6 Laboratory and Radiographic Evaluation of Newly Diagnosed Patients with LCH

Test	Follow-up test interval when organ system is:		
	Involved	Not involved	Single-bone lesion
Hemoglobin and/or hematocrit	Monthly	6 months	None
White blood cell count and differential count	Monthly	6 months	None
Platelet count	Monthly	6 months	None
Ferritin, iron, transferrin ESR	Monthly	6 months	None
Liver function tests (SGOT, SGPT, alkaline phosphatase, bilirubin, total proteins, albumin)	Monthly	6 months	None
Coagulation studies (PT, PTT, fibrinogen)	Monthly	6 months	None
Chest radiograph (PA and lateral)	Monthly	6 months	None
Skeletal radiograph survey[a]	6 months	None	Once, at 6 months
Urine osmolality measurement after overnight water deprivation	6 months	6 months	None
Bone marrow aspirate and biopsy[b]			
HLA-typing[c]			

[a]Radionuclide bone scan is not as sensitive as the skeletal radiograph survey in most patients. It may be performed optionally but should not replace the skeletal survey. If suspicion of lesion exists (e.g., pain) and both radiographic and radionuclide tests are negative, an MRI should be performed.
[b]From patients with multisystem disease.
[c]From patients with high-risk disease (multisystem with major organ dysfunction).
Note: SGOT, serum glutamic-oxaloacetic transaminase; SGPT, serum glutamic pyruvate transaminase; PT, prothrombin time; PTT, partial thromboplastin time; PA, posteroanterior.
Modified from Broadbent et al. (1989).

TABLE 20.7 Recommended Evaluations Based on Specific Clinical Indications in Patients with LCH

Clinical scenario and recommended additional testing

1. History of polyuria or polydipsia
   a. Early morning urine specific gravity and osmolality
   b. Blood electrolytes
   c. Water deprivation test if possible
   d. MRI of the head
2. Bicytopenia, pancytopenia, or persistent unexplained single cytopenia
   a. Other cause of anemia or thrombocytopenia has to be ruled out according to standard medical practice. If no other causes are found, the cytopenia is considered LCH-related
   b. Bone marrow aspirate and trephine biopsy to exclude causes other than LCH[a] as exposant
   c. Evaluation for features of macrophage activation and hemophagocytic syndrome[b] (triglycerides and ferritin in addition to coagulation studies)
3. Liver dysfunction
   a. If frank liver dysfunction (liver enzymes fivefold upper limit of normal/bilirubin > fivefold upper limit of normal): consult a hepatologist and consider liver MRI which is preferable to retrograde cholangiography
   b. Liver biopsy is only recommended if there is clinically significant liver involvement and the result will alter treatment (i.e., to differentiate between active LCH and sclerosing cholangitis)
4. Lung involvement (further testing is only needed in case of abnormal chest X-ray or symptoms/signs suggestive of lung involvement, or pulmonary findings not characteristic of LCH or suspicion of an atypical infection)
   a. Lung high-resolution computed tomography (HR-CT) or preferably low-dose multidetector HR-CT if available. Note that cysts and nodules are the only images typical of LCH; all other lesions are not diagnostic. In children already diagnosed with MS-LCH (see section "Clinical Classification") low-dose CT is sufficient in order to assess extent of pulmonary involvement, and reduce the radiation exposure
   b. Lung function tests (if age-appropriate)
   c. Bronchoalveolar lavage (BAL): >5% CD1a + cells in BAL fluid may be diagnostic in a nonsmoker
   d. Lung biopsy (if BAL is not diagnostic)
5. Suspected craniofacial bone lesions including maxilla and mandible
   a. MRI of head including the brain, hypothalamus—pituitary axis, and all craniofacial bones. If MRI not available, CT of the involved bone and the skull base in recommended

(Continued)

TABLE 20.7   *(Continued)*

Clinical scenario and recommended additional testing

6. Aural discharge or suspected hearing impairment/mastoid involvement
   a. Formal hearing assessment
   b. MRI of head or HR-CT of temporal bone
7. Vertebral lesions (even if only suspected)
   a. MRI of spine to assess for soft-tissue masses and to exclude spinal cord compression
8. Visual or neurological abnormalities
   a. MRI of head
   b. Neurological assessment
   c. Neuropsychometric assessment
9. Suspected other endocrine abnormality (i.e., short stature, growth failure, hypothalamic syndromes, precocious, or delayed puberty)
   a. Endocrine assessment (including dynamic tests of the anterior pituitary and thyroid)
   b. MRI of head
10. Unexplained chronic diarrhea, failure to thrive, or evidence of malabsorption
    a. Endoscopy
    b. Biopsy

[a]The clinical significance of CD1a positivity in the bone marrow remains to be proven. An isolated finding of histiocytic infiltration on the bone marrow with no cytopenia is not a criterion for diagnosis or reactivation.
[b]Hemophagocytic syndrome with macrophage activation is a common finding in patients with hematological dysfunction.
Note: HR-CT, high-resolution computed tomography; MRI, magnetic resonance imaging.
From Haupt et al. (2013).

Diagnostic Biopsy

The optimal biopsy location should be the one associated with the least morbidity, while providing sufficient material to make a definitive biopsy. The most common sites include skin or lytic bone lesions as well as enlarged lymph nodes. For patients with hepatomegaly and liver dysfunction associated with liver dysfunction and hypoalbuminemia, but in the absence of significant gastrointestinal involvement, a liver biopsy may be indicated to assess for LCH involvement and/or for evidence of sclerosing cholangitis. For patients with isolated pulmonary involvement, a lung biopsy may be necessary to secure a diagnosis unless characteristic pathologic cells can be obtained from a bronchopulmonary lavage. Evidence for radiographic lung disease when a definitive diagnosis can be made from biopsy of another side does not usually require a lung biopsy. Endoscopy or colonoscopy may be necessary to document gastrointestinal involvement. Routine histology along with immunohistochemical staining for the expression of CD1a and CD207 should be done. With the evidence for greater than 50% of cases having BRAF V600E mutations, plus the potential for open clinical trials testing new agents targeting this gene mutation, determining whether this gene mutation is present by immunohistochemistry and/or molecular diagnostic testing, should be considered. A bone marrow aspirate and biopsy are indicated for patients who have otherwise unexplained cytopenias to assess involvement.

Radiographic Studies

The initial diagnostic work-up should include a chest X-ray, skeletal survey with skull series, as well as a bone scan, which can be complementary to the skeletal survey. Computed tomography (CT) scan of the head may provide more detail of cranial lesions, such as whether involvement is suspected of the orbit and mastoid anatomic sites. Similarly, if evidence of lung disease is suggested by chest X-ray or by signs/symptoms, then a chest CT scan plays an important role in defining this involvement along with assessing response to therapy.

18F-FDG PET scan is a particularly useful modality to assess the extent of disease in patients with LCH. It is often combined with CT scanning and can likely replace the need for a technetium bone scan.

MRI scanning is particularly useful in terms of evaluating CNS disease, including evaluation of pituitary/hypothalamic, parenchymal tumors as well as evaluation of dural and meningeal involvement from calvarial regions. The last issue can be important in determining whether a lesion is considered a "CNS-risk" lesion. Of note, nearly 30% of patients have been identified as having meningeal and/or choroid plexus involvement. For patients with signs/symptoms suggesting neurodegenerative disease, MRI is essential in evaluating for the presence of characteristic radiographic findings, such as bilateral enhancing involvement of the cerebellar peduncles, brain stem, and basal ganglion. MR-cholangiogram may be useful in assessing the diagnosis of sclerosing cholangitis.

Ultrasonography can play a role in particularly evaluating hepatosplenomegaly and response to therapy as well as assessment of sclerosing cholangitis.

At Diagnosis

A chest radiograph and skeletal survey should usually be performed. A PET/CT is also becoming a commonly used approach to evaluating the extent of disease in patients with newly diagnosed LCH.

Follow-Up Radiograph

A chest radiograph should be considered monthly if the lungs are involved and every 6 months if the lungs are not involved. A follow-up chest film is not required when a monostotic lesion is found at presentation; however, a one-time skeletal survey should be obtained at 6 months in this situation. When multiple bones are affected, skeletal surveys are obtained every 6 months. No follow-up skeletal survey is required if bones are not involved at presentation.

Special Situations

Presence of malabsorption, unexplained chronic diarrhea, or failure to thrive—endoscopic biopsy, upper gastrointestinal study with small-bowel follow-up; 72-h stool fat.
Patients with hormonal, visual, or neurologic abnormalities—MRI scan or contrast-enhanced CT scan of the brain and hypothalamic—pituitary axis.
Patients with oral involvement—panoramic dental radiographs of the mandible and maxilla every 6 months.
Patients with suspected spinal cord compression—MRI of the spine.
Patients with pulmonary symptoms or significant mediastinal widening of chest film—high-resolution CT of the lungs, pulmonary function tests.
Superior vena cava syndrome—CT with contrast.
Significant cervical lymphadenopathy—CT or MRI of the neck.
Ear involvement—CT of temporal bones.
Hepatosplenomegaly—ultrasound of the abdomen.
Soft-tissue tumors—MRI of involved tissue.

Technetium-99m labeled with methylene diphosphonate scintography for bone lesions and routine radiographic skeletal examinations are complementary to each other, because radiography is more likely to detect older and quiescent lesions and scintigraphy may detect early aggressive lesions. The ability of scintigraphy to determine the activity of a lesion can be useful in the evaluation of a persistent radiographic abnormality. Further, PET/CT has been reported be a sensitive approach to detecting and assessing both soft-tissue and skeletal lesions.

Treatment of LCH

A generally accepted standard for the initial treatment of patients with LCH is to use an appropriate amount of the least toxic therapy to treat the disease. In patients with potentially life-threatening disease at presentation, or in those developing life-threatening disease during the course of treatment, alternative and sometimes more aggressive treatment should be considered, including hematopoietic stem cell transplantation. Although recurrence rates have been shown to be significantly reduced in patients with multisystem low-risk disease when patients are treated for 12 months instead of 6 months with vinblastine and prednisone, the role of more intensive or prolonged treatment to prevent DI, central nervous system degeneration, and sclerosing cholangitis is less clear.

Specific Site

Solitary Bone Lesions

Curettage (radical surgery is not indicated).
Intralesional steroids (e.g., methylprednisolone acetate 40 mg/ml, 1—4 ml, depending on size of lesion).

Radiotherapy (600–900 cGy for most lesions although some lesions, especially in adults, may require 1500 cGy) is reserved for isolated lesions inaccessible to intralesional steroid treatment or lesions that have potential to compromise vital structures (e.g., optic nerve, spinal cord, and occasionally other sites, such as mastoids).

Systemic therapy for disease involving multiple bones or organ involvement.

Localized Skin Involvement

Topical steroid application.
Systemic steroids for patients with skin and multisystem disease.
Thalidomide has also been used.

Severe or Refractory Skin Disease

Local application of 20% nitrogen mustard only to involved skin, avoiding the surrounding normal skin.
Psoralen and ultraviolet A irradiation.
Topical tacrolimus.
Electron beam radiotherapy.
Systemic therapy with chemotherapy (vinblastine/prednisone standard in children, but other regimens have also been used).

Solitary Lymph Node

Excisional biopsy should be performed and observation of the patient.

Regional Lymph Node Involvement

Systemic therapy, usually with vinblastine plus prednisone should be administered.

Meningeal Disease and Brain Parenchyma Disease

Systemic therapy should be administered.

Multisystem Disease

For appropriate risk-adapted therapy, patients with multisystem disease are grouped as shown in Table 20.8.

Certain principles of treatment were derived from the Histiocyte Society-sponsored randomized studies, LCH-I, II, and III.

- LCH-I showed that the rapidity of the initial response correlates with prognosis.
- Results from the LCH-II study suggest that there is no significant advantage from adding etoposide in terms of survival or the frequency of disease recurrence, although a retrospective analysis of a subset of patients on this trial has suggested an increased response in patients with multisystem, risk organ involved disease when treated with vinblastine, etoposide, and prednisone. This analysis, however, was not felt to provide sufficient evidence to include etoposide in the standard treatment regimen recommendation.
- In an attempt to improve the response rate and possibly reduce the frequency of recurring disease, the LCH-III trial addressed whether the initial response rate is improved by the addition of intermediate-dose methotrexate to prednisone and vinblastine and whether the overall outcome is improved using 6 or 12

TABLE 20.8 Risk Groups According to the Histiocyte Society LCH-III Trial

Group 1—Multisystem "risk" patients
Multisystem patients with involvement of one or more risk organs (i.e., hematopoietic system, liver, spleen, or lungs)
Group 2—Multisystem "low-risk" patients
Multisystem patients with multiple organs involved but without involvement of risk organs
Group 3—Single-system "multifocal bone disease" or localized "special site" involvement
Patients with multifocal bone disease, that is, lesions in two or more different bones or patients with localized special site involvement, such as lesions with intracranial soft-tissue extension or vertebral lesions with intraspinal soft-tissue extension

TABLE 20.9 Systemic Therapy for LCH[a]

	Initial treatment	Continuation treatment	Duration
Group 1	Prednisone[b] Vinblastine[c] ± Methotrexate[d]	6-MP[e] Prednisone[f] Vinblastine[g] ± Methotrexate[h]	12 months
Group 2	Same as Group 1	Prednisone[f] Vinblastine[g]	6 vs 12 months
Group 3	Same as Group 1	Same as Group 2	6 months only

[a]Patients on systemic chemotherapy should receive standard supportive care including sulfamethoxazole/trimethoprim (5 mg/kg/day of trimethoprim) in two divided doses per day for 3 days per week. Sulfamethoxazole/trimethoprim should not be administered during methotrexate administration.

[b]Oral prednisone, 40 mg/m² daily in three divided doses as a 4-week course, followed by a tapering doses over a period of 2 weeks. Poor responders should receive a further 6-week course of prednisone 40 mg/m² daily on days 1–3 of each week for an additional 6 weeks.

[c]Vinblastine 6 mg/m² i.v. bolus on day 1 of weeks 1–6. This 6-week course of vinblastine is repeated for poor responders.

[d]Methotrexate 500 mg/m² 24-hinfusion with folinic acid (leucovorin) rescue day 1 of weeks 1, 3, and 5. Ten percent of the dose is given as i.v. bolus over 30 min, followed by 90% of the dose as a 23.5-h infusion with 2000 ml/m² hydration. Folinic acid 12 mg/m² is given 24 and 30 h after methotrexate infusion is completed (at 48 and 54 h after methotrexate therapy is started).

[e]Oral 6-MP 50 mg/m² daily until the end of the 12th month from commencement of therapy.

[f]Pulses of oral prednisone 40 mg/m² daily in three doses on days 1–5 every 3 weeks, starting on day 1 of week 7 in patients who have no active disease after course 1 or on day 1 of week 13 in patients who have no active disease or the active disease is better after course 2, continued until the end of month 12.

[g]Vinblastine 6 mg/m²/day i.v. bolus once weekly for 3 weeks, starting on day 1 of week 7 in patients with no active disease after course 1 or on day 1 of week 13 in patients with no active disease or the active disease is better after course 2, continued until the end of month 12.

[h]Methrotrexate 20 mg/m² orally, once weekly, until the end of month 12. It has not been established whether the addition of methotrexate improves results.

months of continuation therapy. This study has demonstrated no advantage to adding methotrexate but a significantly lower recurrence rate for patients treated for 12 months compared to 6 months. In addition, no change in the incidence of DI was observed. Table 20.9 lists systemic therapy for patients with LCH and summarizes the LCH-III study.

Recurrent or Refractory Disease

For patients with recurrent and/or refractory disease, alternative treatment has not been standardized. Patients with recurrent disease, that is, disease that reappears after a period of remission, often respond well to the drugs with which they were initially treated. Several studies have demonstrated significant activity to cladrabine (2-chlorodeoxyadenosine (2-CdA)) in recurrent and refractory LCH. In addition, the combination of 2-CdA and high-dose cystosine arabinoside (AraC) (cytarabine) has been used in highly refractory patients. Patients with relapsed, refractory disease can receive two courses of cladrabine (5 mg/m²/day in 50 ml normal saline over 2 h for 5 consecutive days every 3–4 weeks). If there is a good response, then patients receive 2–4 additional courses of 2-CdA for not more than six total courses. If there is a poor initial response after two courses of cladrabine or in some patients with extensive disease, the combination of cladrabine (9 mg/m²/day as a 2 h IV infusion given daily for 5 days. Cladrabine is started on the second day of the course and is given at hours 23, 47, 71, 95, and 119) and AraC (500 mg/kg in 250 ml/m² twice a day, i.e., every 12 h, for 5 days as a 2-h IV infusion) can be used. This latter regimen is that of the stem cell transplantation for high-risk or refractory patients and warrants further evaluation. A regimen from the Netherlands has used vincristine, low-dose cytosine arabinoside, and prednisone for both patients with newly diagnosed and recurrent disease. The use of single-agent low-dose cytarabine has also been reported, and may be better tolerated than vinblastine/prednisone or cladrabine regimens, especially in adults. There is anecdotal experience using TNF-inhibitors in recurrent or refractory disease as well as bisphosphonates for skeletal lesions. The use of *BRAF V600E* inhibitors has been reported in a small number of patients. Clinical trials using *BRAF V600E* and *MEK* (*MAPK1*) inhibitors are ongoing.

Prognosis

Historic prognostic factors used to stratify therapy include:

1. Response to initial therapy.
2. Age at diagnosis (<24 months, 55–60% mortality) was considered to be an important factor in early studies, although response to the initial 6–12 weeks of therapy has now been shown to outweigh this factor.

3. Number of organs involved at diagnosis:

Number of organs	Mortality (%)
1–2	0
3–4	35
5–6	60
7–8	100

4. Organ dysfunction (e.g., lung, liver, bone marrow) at diagnosis:

Organ dysfunction	Mortality (%)
Present	66
Absent	4

Of note, in multivariate analyses, the Histiocyte Society trials have demonstrated that lung involvement is not an independent adverse prognostic factor.

5. Natural history on treatment:

Group	Description	Mortality (%)
A	No disease progression over 6–12 months	0
B	Progressive disease without organ dysfunction	20
C	Development of organ dysfunction during course of disease	100

6. Congenital self-healing histiocytosis: condition that manifests in neonates with skin lesions, pulmonary nodules with or without bone lesions (these lesions regress spontaneously and require no treatment); close follow-up is required as resolution is estimated to occur in about 50% of cases.

Sequelae and Complications

The risk factors for developing residual adverse sequelae include:

1. Generalized disease with skeletal disease and especially lesions of the orbit, sphenoid, mastoid, or temporal bones with associated dural involvement.
2. Smoldering and/or recurrent disease.

Long-Term Complications

Pulmonary: Progressive fibrosis, pulmonary cyst formation and chronic pneumothoraces. There is no effective therapy for these complications and progression to cor pulmonale and respiratory failure commonly occurs.

Hepatic: Sclerosing cholangitis has been reported and may lead to secondary biliary cirrhosis, portal hypertension, and liver failure. The etiology is not understood. The only successful treatment has been for liver transplantation.

Neuropsychiatric: CNS manifestations can occur without any relationship to radiotherapy or other treatments. This may manifest with learning disability, ataxia, pyramidal signs, and behavioral changes. MRI studies with gadolinium contrast are helpful in localizing structural changes in the brain.

Endocrine: DI and growth retardation are the most frequent complications. They result from histiocytic infiltration of the pituitary and hypothalamus. Such lesions that initially present with DI will result in panhypopituitarism in approximately 50–60% of patients. A hyperphagic syndrome associated with extreme weight gain may also occur as a result of hypothalamic lesions.

Orthopedic: Deformities of the spine can result in long-term disabilities.

Dental: Loss of teeth and jaw abnormalities may occur.

Hearing: Patients with mastoid and middle ear involvement may develop permanent hearing loss.

Malignancies: Second primary malignancies associated with radiation therapy include astrocytoma, medulloblastoma, meningioma, hepatoma, osteosarcoma of the skull, and thyroid carcinoma. Secondary AML has been associated with the use of chemotherapy, and, in particular, cladrabine.

OTHER HISTIOCYTIC DISORDERS

Secondary Dendritic Cell Processes

The accumulation of dendritic cells and Langerhans cells occurs in the lymph nodes in Hodgkin disease, lymphoma, and other tumors, such as breast, lung, colon, and thyroid. The secondary dendritic cell processes involute with control of the primary disease. This is a pathological finding of no clinical significance in terms of the diagnosis of a primary histiocytic disorder.

Dermal Dendrocyte Disorders

Juvenile Xanthogranuloma

Clinically, juvenile xanthogranuloma is characterized by multiple cutaneous nodules consisting of dermal dendrocytes. These lesions sometimes involute spontaneously and no treatment is required. Occasionally there is systemic involvement. When this occurs therapy similar to that for LCH is used. When the disease is disseminated, it is referred to as xanthoma disseminatum.

Erdheim—Chester Disease

Erdheim—Chester Disease (ECD) occurs primarily in adults but can occur rarely in young individuals. ECD is characterized by accumulations of xanthomatous macrophages, particularly in the retroperitoneum, which often can lead to renal failure. In addition, ECD commonly affects the lungs and heart, along with bilateral long bone involvement leading to severe and chronic pain. In addition, ECD can lead to DI and other CNS signs and symptoms similar to LCH. Treatment for patients with ECD usually involves α-interferon or treatments similar to those used to treat patients with LCH.

Lesions from patients with ECD have been reported to show *BRAF V600E* mutations in 60—100% of cases. Other mutations that involve *RAS* and *MEK* (*MAPK1*) have also been reported. Responses to inhibitors of BRAF V600E have been reported in a small number of cases.

Solitary Histiocytomas with Dendritic Cell Phenotypes

These tumors are composed of dendritic cells without malignant features. They have variable phenotypes identified by various immunological and cytochemical markers, for example, indeterminate cell or interdigitating dendritic cell phenotype. They occur in the cutaneous tissue and less often in the central nervous system. Surgical resection when possible is a reasonable therapeutic approach; on occasion, radiation therapy or chemotherapy such as that used in patients with LCH or high-risk lymphoma regimens are used.

MACROPHAGE MEDIATED DISORDERS

Sinus Histiocytosis with Massive Lymphadenopathy (Rosai—Dorfman Disease)

Characteristics and manifestations of SHML include the following:

Worldwide occurrence with higher incidence among Blacks.
Onset usually within the first two decades of life.
Massive painless bilateral cervical lymphadenopathy with involvement of other groups of lymph nodes.
Snoring, when there is involvement of retropharyngeal lymphoid tissue; possibility of sleep apnea.
Extranodal infiltration in approximately 25% of patients (skin, orbit, eyelid, liver, spleen, testes, CNS, salivary glands, bone, respiratory tract).

Immunologic abnormalities with manifestations of autoimmune disorders (e.g., hematologic antibodies, glomerulonephritis, amyloidosis, or joint disease) in 10% of patients.
Fever.
Leukocytosis with neutropenia, mild anemia, elevated erythrocyte sedimentation rate (ESR).
Polyclonal hypergammaglobulinemia.

Complications

Retropharyngeal involvement causing respiratory compromise.
Epidural involvement causing spinal cord compression.
Visual loss due to optic nerve compression or corneal involvement.

Diagnosis

Lymph node shows marked dilatation of sinuses by proliferation of benign histiocytes with prominent phagocytosis of lymphocytes, plasma cells, and erythrocytes by sinus histiocytes. Plasma cell infiltrates in the medullary cords and capsular fibrosis may also be evident.

Prognosis

Twenty percent of patients have spontaneous resolution or improvement within 3–9 months. The majority of patients have stable but persistent disease lasting up to several years. Seven percent of patients have a fatal outcome, especially if immunologic abnormalities and extranodal involvement are present.

Treatment

No treatment is generally warranted because this it is a self-limited disease.
Prednisone 2 mg/kg PO may be administered for life-threatening complications. If prednisone fails, the combination of vinblastine plus steroid can be used with anecdotal information suggesting dexamethasone as the more effective steroid; 6-mercaptopurine (6-MP) 50–75 mg/m^2/day PO and methotrexate 10–20 mg/m^2/week PO have also been used with evidence of response. Intermediate- and high-dose methotrexate with leukovorin rescue, cladrabine, or clofarabine have been reported in small numbers of patients to result in good, partial responses. α-Interferon has also been reported to be effective in some cases.

Hemophagocytic Lymphohistiocytosis (Hemophagocytic Syndromes)

Hemophagocytic lymphohistiocytosis (HLH) falls into two categories: The HLH disorders may be primary (inherited/familial) or secondary such as in response to specific types of infections (infection-associated hemophagocytic syndrome or IAHS), malignancies (malignancy-associated hemophagocytic syndrome or MAHS), rheumatologic disorders or immune system modifying treatments such as chemotherapy and immunosuppressive agents.

Inherited disorders that can lead to HLH include Chédiak–Higashi syndrome, Griscelli syndrome, X-linked lymphoproliferative disease (XLP), Hermansky–Pudlak syndrome, and lysinuric protein intolerance.

These disorders are clinically characterized by excessive, systemic production of inflammatory cytokines leading to macrophage activation, hemophagocytosis, pancytopenia, hepatosplenomegaly, lymphadenopathy, fever, seizures, or central nervous system complications, capillary leak with pulmonary insufficiency, hypotension, and renal failure.

Table 20.10 lists the diagnostic guidelines for HLH. It should be noted that there is no specific diagnostic feature for familial or primary HLH (FHLH). For this reason, when the index of suspicion is strong for primary HLH, treatment may be started before extensive disease activity causes irreversible organ damage and the likelihood of a response to therapy decreases.

Familial or Primary Hemophagocytic Lymphohistiocytosis

Pathophysiology, Immunology, and Genetics

In the absence of perforin activity due to production, protein function, transport, or exocytosis, the resulting inability to kill infected target cells results in sustained NK and cytolytic T-cell (CTL) activity. This, in turn, results in the overexpression of inflammatory cytokines (soluble IL-2 receptor, IL-6, TNF-α, IL-10, and IL-12) leading to excessive macrophage activation, dissemination, and organ infiltration and the signs, symptoms, and laboratory abnormalities that characterize HLH.

Impaired NK-cell activity is key to the diagnosis. FHLH is usually inherited as an autosomal recessive disorder. The absence of intracytoplasmic perforin may be used as a reliable marker in the 20–40% of patients with familial HLH type 2 associated with the 10q21-22 mutations in the gene at chromosome 10q22 resulting in perforin gene (PRF1) mutations. Perforin functions by perforating the cytolytic target cell membrane allowing for the entry of cytolytic granules that in turn initiate the apoptotic cell death pathway. The pathways leading to the synthesis of perforin, subcellular compartmentalization, as well as directional targeting and release of cytolytic granules, all represent potential points that could be mutated and contribute to different genetic causes of inherited HLH syndromes. A list of mutated genes and syndromes associated with HLH is given in Table 20.11.

TABLE 20.10 Diagnostic Guidelines for HLH

The diagnosis HLH can be established if one of either 1 or 2 below is fulfilled

1. A molecular diagnosis consistent with HLH
2. Diagnostic criteria for HLH fulfilled (five out of the eight criteria below)
   a. Initial diagnostic criteria (*to be evaluated in all patients with HLH*)
      i. Fever
      ii. Splenomegaly
      iii. Cytopenias (affecting ≥2 of 3 lineages in the peripheral blood):
         • Hemoglobin <90 g/l (in infants <4 weeks: hemoglobin <100 g/l)
         • Platelets <100 × 10^9/l
         • Neutrophils <1.0 × 10^9/l
      iv. Hypertriglyceridemia and/or hypofibrinogenemia:
         • Fasting triglycerides ≥3.0 mmol/l (i.e., ≥265 mg/dl)
         • Fibrinogen ≤1.5 g/l
      v. Hemophagocytosis in bone marrow or spleen or lymph nodes
      vi. No evidence of malignancy
   b. New diagnostic criteria
      i. Low or absent NK-cell activity (according to local laboratory reference)
      ii. Ferritin ≥500 µg/l
      iii. Soluble CD25 (i.e., soluble IL-2 receptor) ≥2400 U/ml

Comments:
1. If hemophagocytic activity is not proven at the time of presentation, further search for hemophagocytic activity is encouraged. If the bone marrow specimen is not conclusive, material may be obtained from other organs. Serial marrow aspirates over time may also be helpful.
2. The following findings may provide strong supportive evidence for the diagnosis: (a) spinal fluid pleocytosis (mononuclear cells) and/or elevated spinal fluid protein, (b) histological picture in the liver resembling chronic persistent hepatitis (biopsy).
3. Other abnormal clinical and laboratory findings consistent with the diagnosis are: cerebromeningeal symptoms, lymph node enlargement, jaundice, edema, skin rash. Hepatic enzyme abnormalities, hypoproteinemia, hyponatremia, VLDL ↑, HDL ↓.
From Henter et al. (2007), with permission.

TABLE 20.11 Genes Mutated in Inherited HLH Disorders

Gene	Syndrome	Inheritance	Protein
PRF1	FHL2	AR	Perforin
UNC13D	FHL3	AR	Munc13-4
STX11	FHL4	AR	Syntaxin 11
STXBP2	FHL5	AR	Munc18-2
RAB27A	GS2	AR	RAB27A
LYST	CHS1	AR	LYST
SH2D1A	XLP1	XL	SAP
XIAP	XLP2	XL	XIAP

Note: FHL, familial hemophagocytic lymphohistiocytosis; GS, Griscelli syndrome (associated with albinism); CHS, Chédiak–Higashi syndrome (associated with albinism); XLP, X-linked lymphoproliferative disease; AR, autosomal recessive inheritance; XL, X-linked inheritance.

Those patients with absent perforin expression should undergo PRF1 and the MUNC 13-4 (FHLH type 3) mutation analysis. FHLH-4 is characterized by mutations in the *STX11* gene encoding syntaxin 11 on chromosome 6(q24). Syntaxin is believed to also be involved in the intracellular movement of cytolytic granules. FHLH type 5 is characterized by mutations in the MUNC 18-2 that appear to be involved in a multiprotein complex with syntaxin 11. FHLH type 1, linked to the 9q21.3 locus, represents approximately 10% of the cases and can be recognized by impaired NK-cell function not associated with the absence of perforin expression.

Patients with infection-associated HLH (IAHLH) may have transiently impaired NK activity, thus mutation analysis will distinguish these cases from familial HLH. In the absence of a positive mutation analysis, re-evaluation of NK function after successful treatment should be undertaken.

In summary, a wide array of immune dysfunction may result in defective NK and CTL target cell killing resulting in the failure to eliminate infected cells permitting a sustained inflammatory response complete with excessive cytokine production and sustained systemic macrophage activation that characterizes a final pathway of HLH.

Clinical Features

1. The age of onset is less than 1 year of age in 70% of cases. There is no known upper age limit for the onset of disease.
2. Signs and symptoms of FHLH include:
   a. Fever (91%), splenomegaly (98%), and hepatomegaly (94%) are the most common early findings.
   b. Lymph node enlargement (17%), skin rash (6%), and neurologic abnormalities (20%) may also occur. Neurologic findings include irritability, bulging fontanel, neck stiffness, hypotonia, hypertonia, convulsions, cranial nerve palsies, ataxia, hemiplegia, blindness, and unconsciousness.
   c. Multisystem involvement includes lungs, bone marrow, and leptomeninges.
      Occasionally, ocular, heart, skeletal muscle, and kidney involvement have been noted.

Treatment

Patients should be treated preferably on clinical trials. If not on a clinical trial, most patients will benefit by treatment as per established protocols such as HLH 2004 of the Histiocytic Society[1]. Nonfamilial disease is initially treated in a similar manner, although some patients, such as those with secondary HLH due to rheumatologic disorders, may benefit from initial treatment with intravenous immunoglobulin and steroids.

The following treatment regimen, based on Histiocyte Society clinical trials, includes:

1. Dexamethasone, 10 mg/m^2/day for 2 weeks followed by a decrease every 2 weeks to 5 mg/m^2, 2.5 mg/m^2, and 1.25 mg/m^2 for a total of 6 weeks.
2. Etoposide IV, 150 mg/m^2 IV 2-h infusions daily, twice-weekly for 2 weeks, then weekly.
3. Cyclosporine A, 3−5 mg/kg/day by continuous IV infusion starting week 8 to reach a blood trough level of 150−200 ng/ml and switching to oral administration of 6−10 mg/kg/day in two divided doses.
4. Intrathecal methotrexate (IT MTX), age-adjusted doses of IT MTX weekly for 3−6 weeks as follows if there are progressive neurological symptoms or if abnormal cells persist in the CSF:

Age	IT MTX dose (mg)
<1 year	6
1−2 years	8
2−3 years	10
>3 years	12

5. Allogeneic stem cell transplantation (BMT) after cytotoxic chemotherapy for patients with familial disease or those with persistent or recurrent nonfamilial disease is indicated.

Without treatment, FHLH is usually rapidly fatal, with a median survival of about 2 months. Chemotherapy and immunosuppressive therapy may prolong survival in FHLH but only stem cell transplantation may be curative.

[1]For details, contact the local chapter of the Histiocyte Society in various countries or the Histiocyte Society, 302 North Broadway, Pitman, NJ 08071, USA.

Patients with known familial disease or severe or persistent acquired disease should then receive hematopoietic stem cell transplantation. The 3-year actuarial survival in familial HLH with this approach has been reported as approximately 50–55% overall but 64% following HSCT.

Nonfamilial HLH

A number of inherited disorders associated with impaired cytotoxic T- or NK-cell function result in HLH. HLH may also be associated with a number of infectious agents and malignancies.

Infection-Associated HLH

The findings in children with IAHLH are similar to those in FHLH. However, decreased or absent NK cells are found more often in FHLH. NK-cell activity in IAHLH patients is reconstituted as soon as the infection is cleared.

Table 20.12 lists the triggering organisms and clinical outcomes for IAHLH. Viruses include human herpes virus-6, cytomegalovirus (most common of the viruses), adenovirus, parvovirus, varicella zoster, herpes simplex virus, Q-fever virus, and measles.

Treatment

Epstein–Barr virus (EBV)-related IAHLH: In addition to HLH-directed treatment, the addition of rituximab, a monoclonal antibody against the B-cell antigen, CD20, has been used.

Other infections: antibiotics for bacterial infections, antiviral drugs for viruses, in addition to corticosteroids and/or etoposide. Patients with persistent HLH may require FHLH treatment and hematopoietic stem cell transplantation. Patients with resolved disease may discontinue therapy at 8 weeks. If they recur therapy should be restarted and HSCT should be employed. Table 20.13 shows IAHLH in children by age and clinical outcome prior to the use of effective protocols.

TABLE 20.12　IAHS in Children: Associated Organisms and Clinical Outcome

Organism	Number of patients	Clinical outcome		
		Dead	Alive	No data
EBV	121	72	27	22
Other viruses	28	11	13	4
Bacteria	11	2	9	0
Fungi	2	1	1	0
Protozoae	1	0	1	0
No organism	56	13	34	11

From Janka et al. (1998), with permission.

TABLE 20.13　IAHS in Children by Age and Clinical Outcome

Age	Number of patients	Clinical outcome		
		Dead	Alive	No data
<3 years	77	40	26	11
>3 years	82	29	47	6
Children[a]	60	34	22	4
Totals	219	103/198	95/198	

[a]Age unknown from records reviewed.
From Janka et al. (1998).

TABLE 20.14 Malignancies Associated with the Development of Hemophagocytic Syndromes

1. Development of a hemophagocytic syndrome before and/or during the treatment for malignancy, such as:
   Acute lymphoblastic leukemia
   Acute myeloid leukemia
   Multiple myeloma
   Germ cell tumor
   Thymoma
   Carcinoma
2. Development of a hemophagocytic syndrome with a masked hematolymphoid malignancy in the background, such as:
   T/NK-cell leukemia
   Lymphomas
   Large cell anaplastic lymphoma
   Adult B-cell lymphoma

From Janka et al. (1998).

Malignancy-Associated Hemophagocytic Syndrome

Table 20.14 shows various malignancies associated with the development of hemophagocytic syndromes.

Treatment

If MAHS occurs in an immunocompromised host before treatment, therapy of malignancy and infection is suggested. If it develops in association with an infection during chemotherapy, cessation of chemotherapy may be considered if the malignancy is under control. Rapidly progressive MAHS may also need to be treated according to standard HLH regimens.

Macrophage Activation Syndrome in Systemic Juvenile Rheumatoid Arthritis and Other Chronic Conditions (Reactive HLH)

Macrophage activation syndrome (MAS) is caused by an excessive activation and proliferation of mature macrophages. It is observed in a number of conditions, including infections, neoplasms, and rheumatologic diseases. Triggers for MAS in juvenile rheumatoid arthritis (JRA) include gold therapy, aspirin, other nonsteroidal anti-inflammatory drugs, and viral infections. Typically, patients with such chronic conditions present with the following features of acute illness:

Persistent fever.
Hepatosplenomegaly.
Pancytopenia.
Low ESR.
Elevated liver enzymes.

In patients with systemic JRA, the acute deterioration is commonly preceded by either a viral infection or major changes in therapy (e.g., administration of gold therapy or nonsteroidal anti-inflammatory drugs).

MAS as a complication of JRA is associated with considerable morbidity and death. Table 20.15 shows the clinical differentiation of MAS associated with JRA from a typical exacerbation of systemic JRA.

Prolonged prothrombin time, prolonged partial thromboplastin time, hypofibrinogenemia, and low levels of vitamin-K-dependent clotting factors, easy bruisability and mucosal bleeding; fibrin degradation products and serum liver enzyme values may be elevated. Numerous, well-differentiated hemophagocytic histiocytes are a pathognomonic feature of this condition; hemophagocytic histiocytic cells are found in various organs.

The presence of hemophagocytic histiocytes in the bone marrow and lymph nodes strongly supports the diagnosis.

Treatment

Because of its seriousness, this syndrome should be recognized promptly on the basis of the previously mentioned findings. Early treatment, consisting of the use of corticosteroids, after performing a bone marrow examination, frequently along with intravenous gamma globulin, should be considered. This treatment usually results in a rapid resolution of symptoms. The dose of steroids should be tapered slowly to prevent relapse. If this type

TABLE 20.15　Features That May be Helpful to Differentiate Acute Exacerbation of JRA from MAS Complicating JRA

	Acute exacerbation of JRA	MAS in JRA[a]
	One spike or two spikes daily	Persistent
Generalized lymphadenopathy	Present	Not present
Hepatosplenomegaly	Present	Present
LABORATORY FINDINGS		
Blood count	Marked polymorphonuclear leukocytosis and thrombocytosis	Pancytopenia
Clotting studies	Hyperfibrinogenemia	Hypofibrinogenemia, prolonged PT, prolonged PTT, increased D-dimers
ESR	Increased ESR	Decreased ESR
Liver enzymes	Mildly elevated	Moderately elevated

[a]*Must be clearly differentiated from malignant histiocytic conditions.*

of treatment fails to rapidly result in a significantly improved condition, then treatment with HLH-directed regimens should be initiated.

Malignant Histiocytic Disorders in Children

These disorders usually represent high-grade cancers that arise within the dendritic cell lineage, monocytic/macrophage lineage, or their hematopoietic precursors. Thus, some of these malignancies express dendritic cell markers (called dendritic cell sarcomas) while others express primarily monocyte/macrophage markers (called histiocytic sarcomas). These malignancies are also commonly associated with malignant lymphomas and lymphoid malignancies.

Pathologically, they usually have large malignant cells with pleomorphic nuclei, prominent nucleoli, and eosinophilic cytoplasm. Giant cells may be present and demonstrate hemophagocytosis.

Immunophenotypic antigens typical of histiocytic sarcomas usually include CD68, lysozyme, CD4, and CD163, but not usually myeloperoxidase or CD33, lymphoid, or dendritic-cell-specific markers. When there are dendritic cell markers such as CD1a, Cd21, and CD35, along with often more variable expression of CD45, CD6,8 and S-100, the term follicular dendritic cell sarcoma is used.

Clinical manifestations are variable and depend on the site and extent of disease involvement. Commonly involved organs include lymph nodes, bone marrow, skin, liver, liver/spleen, lung, skeleton, gastrointestinal tract, and the central nervous system. The skin involvement is usually nodular but can be maculopapular. Fever and night sweats are common with disseminated disease.

A complete clinical staging is important and will often include a PET/CT and MRI of the brain and, when clinically indicated, spinal column. Complete blood counts, liver function tests, and metabolic panels along with coagulation studies should usually be done. Bone marrow aspiration and biopsy should be done in cases of disseminated disease.

Treatment/Outcome

Although these malignancies are usually referred to as "sarcomas," they should be treated as hematopoietic cell disorders and not with typical regimens used for soft-tissue or bone sarcomas such as rhabdomyosarcoma, Ewing sarcoma, or osteosarcoma. Most clinical treatment and response data are anecdotal, as no large, prospective clinical trials have been reported.

Localized Disease

When possible, complete resection can be curative, for localized disease, such as an isolated lesion or involved lymph node. Nevertheless, many clinicians will also recommend adjuvant radiation therapy at doses similar to

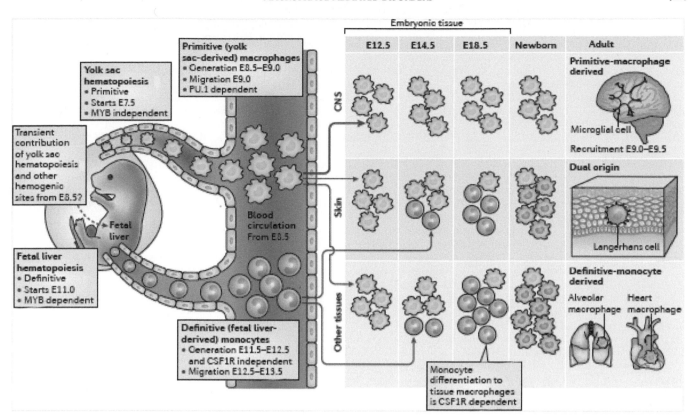

FIGURE 20.1 Much of the basis for this scheme of ontogeny derives from *in vitro* work in developing systems for the production of dendritic cells for immunotherapy. Cytokines (GM-CSF, CSF-1, α-TNF, IL-4, etc.) play important roles in regulating differentiation, expansion, and function of these cell types. Langerhans cells, indeterminate cells of the dermis, and interdigitating dendritic cells are considered to be in a cytological continuum. Indeterminate cells are precursors to Langerhans cells. Several recent studies have demonstrated that fetal yolk sac progenitors (macrophages) are the cell of origin for epidermal Langerhans cells, which have the ability to self-renew independently of bone-marrow-derived precursors. However, fetal liver and subsequently bone marrow, precursor cells generate blood monocytes, tissue macrophages, and dermal dendritic cells. Under certain physiological stresses, such as inflammation, epidermal Langerhans cells may be replaced by bone-marrow-derived circulating progenitors (Malissen et al., 2014; Ginhoux and Jung, 2014). *Adapted from Ginhoux and Jung (2014).*

those used in lymphomas (25–50 Gy). For localized disease, adjuvant chemotherapy is usually not indicated and prognosis is good although no large, outcome studies have been reported.

Disseminated/Multisystem Disease

The most effective regimens appear to be those designed to treat patients with high-risk lymphomas and lymphoid leukemias such as BFM-NHL/ALCL regimens, CHOP, and M-BACOD. For patients without marrow involvement, consolidation with autologous transplantation has been reported. For patients with disseminated disease involving the bone marrow, allogeneic transplant with the best available donor should be considered to consolidate initial responses. Overall survival nevertheless remains poor and in the 30–50% range.

Recurrent/Refractory Disease

Treatment regimens for patients with refractory dendritic or histiocytic sarcomas that have shown responses often include cladrabine or clofarabine alone or usually in combination with cytarabine. Other regimens include ICE or high-dose methotrexate. Outcomes are usually poor but if a good response can be achieved, then strong consideration for early, allogeneic transplant should be considered.

Potential cellular lineage relationships of the histiocytoses are shown in Figure 20.1.

Further Reading and References

Allen, C.E., Kelly, K.M., Bollard, C.M., 2015. Pediatric lymphomas and histiocytic disorders of childhood. Pediatr. Clin. North Am. 62 (1), 139–165.

Alston, R.D., Tatevossian, R.G., McNally, R.J., Kelsey, A., Birch, J.M., Eden, T.O., 2007. Incidence and survival of childhood Langerhans cell histiocytosis in Northwest England from 1954 to 1998. Pediatr. Blood Cancer 48 (5), 555–560.

Arceci, R.J., 2014. Dendritic cell disorders: matters of lineage and clinical drug testing in rare diseases. J. Clin. Oncol. 33, 383–385.

Arico, M., Nichols, K., Whitlock, J.A., Arceci, R., Haupt, R., Mittler, U., et al., 1999. Familial clustering of Langerhans cell histiocytosis. Br. J. Haematol. 107 (4), 883–888.

Badalian-Very, G., Vergilio, J.A., Degar, B.A., MacConaill, L.E., Brandner, B., Calicchio, M.L., et al., 2010. Recurrent BRAF mutations in Langerhans cell histiocytosis. Blood 116 (11), 1919–1923.

Badalian-Very, G., Vergilio, J.A., Fleming, M., Rollins, B.J., 2013. Pathogenesis of Langerhans cell histiocytosis. Annu. Rev. Pathol. 8, 1–20.

Berres, M.L., Lim, K.P., Peters, T., Price, J., Takizawa, H., Salmon, H., et al., 2014. BRAF-V600E expression in precursor versus differentiated dendritic cells defines clinically distinct LCH risk groups. J. Exp. Med. 211 (4), 669–683.

Broadbent, V., Gadner, H., Komp, D., Ladish, S., 1989. Histiocytosis in children II: approach to the clinical and laboratory evaluation of children with Langerhans cell histiocytosis. Med. Pediatr. Oncol. 17, 492.

Brown, N.A., Furtado, L.V., Betz, B.L., Kiel, M.J., Weigelin, H.C., Lim, M.S., et al., 2014. High prevalence of somatic MAP2K1 mutations in BRAF V600E-negative Langerhans cell histiocytosis. Blood 124 (10), 1655–1658.

Cangi, M.G., Biavasco, R., Cavalli, G., Grassini, G., Dal-Cin, E., Campochiaro, C., et al., 2014. BRAFV600E-mutation is invariably present and associated to oncogene-induced senescence in Erdheim-Chester disease. Ann. Rheum. Dis. 211, 669–683.

Chakraborty, R., Hampton, O.A., Shen, X., Simko, S.J., Shih, A., Abhyankar, H., et al., 2014. Mutually exclusive recurrent somatic mutations in MAP2K1 and BRAF support a central role for ERK activation in LCH pathogenesis. Blood 124 (19), 3007–3015.

Chu, T., D'Angio, G.J., Favara, B., Ladisch, S., Nebit, M., Pritchard, J., 1987. Histiocytosis syndromes in children. Lancet 1, 208–209.

Diamond, E.L., Dagna, L., Hyman, D.M., Cavalli, G., Janku, F., Estrada-Veras, J., et al., 2014. Consensus guidelines for the diagnosis and clinical management of Erdheim-Chester disease. Blood 124 (4), 483–492.

Emile, J.F., Diamond, E.L., Helias-Rodzewicz, Z., Cohen-Aubart, F., Charlotte, F., Hyman, D.M., et al., 2014. Recurrent RAS and PIK3CA mutations in Erdheim-Chester disease. Blood 124 (19), 3016–3019.

Ginhoux, F., Jung, S., 2014. Monocytes and macrophages: developmental pathways and tissue homeostasis. Nat. Rev. Immunol. 14 (6), 392–404.

Girschikofsky, M., Arico, M., Castillo, D., Chu, A., Doberauer, C., Fichter, J., et al., 2013. Management of adult patients with Langerhans cell histiocytosis: recommendations from an expert panel on behalf of Euro-Histio-Net. Orphanet J. Rare Dis. 8, 72.

Grois, N.G., Favara, B.E., Mostbeck, G.H., Prayer, D., 1998. Central nervous system disease in Langerhans cell histiocytosis. Hematol. Oncol. Clin. N. Am. 12, 287–305.

Grois, N., Fahrner, B., Arceci, R.J., Henter, J.I., McClain, K., Lassmann, H., et al., 2010. Central nervous system disease in Langerhans cell histiocytosis. J. Pediatr. 156 (6), 873–881, 881 e871.

Haroche, J., Arnaud, L., Cohen-Aubart, F., Hervier, B., Charlotte, F., Emile, J.F., et al., 2014a. Erdheim-Chester disease. Curr. Rheumatol. Rep. 16 (4), 412.

Haroche, J., Cohen-Aubart, F., Emile, J.F., Maksud, P., Drier, A., Toledano, D., et al., 2014b. Reproducible and sustained efficacy of targeted therapy with vemurafenib in patients with BRAFV600E-Mutated Erdheim-Chester disease. J. Clin. Oncol..

Haroche, J., Cohen-Aubart, F., Emile, J.-F., Maksud, P., Drier, A., Tledano, D., et al., Reproducible and sustained efficacy of targeted therapy with Vemurafenib in Eight patients with BRAF V600E mutated Erdheim-Chester disease. J. Clin. Oncol., in press.

Haupt, R., Minkov, M., Astigarraga, I., Schafer, E., Nanduri, V., Jubran, R., et al., 2013. Langerhans cell histiocytosis (LCH): guidelines for diagnosis, clinical work-up, and treatment for patients till the age of 18 years. Pediatr. Blood Cancer 60 (2), 175–184.

Henter, J.I., Horne, A., Arico, M., Egeler, R.M., Filipovich, A.H., Imashuku, S., et al., 2007. HLH-2004: diagnostic and therapeutic guidelines for hemophagocytic lymphohistiocytosis. Pediatr. Blood Cancer 48 (2), 124–131.

Janka, G., Imashuku, S., Elinder, G., Schneider, M., Henter, J.I., 1998. Infection and malignancy associated hemophagocytic syndromes. Hematol. Oncol. Clin. N. Am. 12, 435–443.

Kansal, R., Quintanilla-Martinez, L., Datta, V., Lopategui, J., Garshfield, G., Nathwani, B.N., 2013. Identification of the V600D mutation in Exon 15 of the BRAF oncogene in congenital, benign langerhans cell histiocytosis. Genes Chromosomes Cancer 52 (1), 99–106.

Malissen, B., Tamoutounour, S., Henri, S., 2014. The origins and functions of dendritic cells and macrophages in the skin. Nat. Rev. Immunol. 14, 417–428.

Nelson, D.S., Quispel, W., Badalian-Very, G., van Halteren, A.G., van den Bos, C., Bovee, J.V., et al., 2014. Somatic activating ARAF mutations in Langerhans cell histiocytosis. Blood 123 (20), 3152–3155.

Simko, S.J., Tran, H.D., Jones, J., Bilgi, M., Beaupin, L.K., Coulter, D., et al., 2014. Clofarabine salvage therapy in refractory multifocal histiocytic disorders, including Langerhans cell histiocytosis, juvenile xanthogranuloma and Rosai-Dorfman disease. Pediatr. Blood Cancer 61 (3), 479–487.

Vaiselbuh, S.R., Bryceson, Y.T., Allen, C.E., Whitlock, J.A., Abla, O., 2014. Updates on histiocytic disorders. Pediatr. Blood Cancer 61 (7), 1329–1335.

Weitzman, S., Jaffe, R., 2005. Uncommon histiocytic disorders: the non-Langernhans cell histiocytoses. Pediatr. Blood Cancer 45 (3), 256–264.

Weitzman, S., Braier, J., Donadieu, J., Egeler, R.M., Grois, N., Ladisch, S., et al., 2009. 2′-Chlorodeoxyadenosine (2-CdA) as salvage therapy for Langerhans cell histiocytosis (LCH). results of the LCH-S-98 protocol of the Histiocyte Society. Pediatr. Blood Cancer 53 (7), 1271–1276.

Wilejto, M., Abla, O., 2012. Langerhans cell histiocytosis and Erdheim-Chester disease. Curr. Opin. Rheumatol. 24 (1), 90–96.

Yousem, S.A., Dacic, S., Nikiforov, Y.E., Nikiforova, M., 2013. Pulmonary Langerhans cell histiocytosis: profiling of multifocal tumors using next-generation sequencing identifies concordant occurrence of BRAF V600E mutations. Chest 143 (6), 1679–1684.

21

Hodgkin Lymphoma

Debra L. Friedman

Hodgkin lymphoma (HL) is characterized by progressive enlargement of lymph nodes. It is considered unicentric in origin and has a predictable pattern of spread by extension to contiguous nodes.

ETIOLOGY AND EPIDEMIOLOGY

1. Specific etiology is unknown.
2. Overall there is a slight female predominance when considering all children less than 20 years (M:F = 0.9).
3. The Caucasian:African American ratio is 1.3:1.
4. Comprises 8.8% of all childhood cancers under the age of 20, but 17.7% of cancers in adolescents between the ages of 15 and 19 years.
5. Overall annual incidence rate in the United States is 12.1 per million for children under 20 years.
6. Incidence increases to 32 per million for adolescents 15–19 years.
7. Bimodal age—incidence curve with one peak at 15–35 years of age and the other above 50 years of age (incidence is highest among 15–19 year olds).
8. Association with Epstein–Barr virus (EBV) is known.
9. Incidence increased among consanguineous family members and among siblings of patients with HL.

RISK FACTORS

There are several factors that are known to increase the risk of HL, which include family history of HL, EBV infections, socioeconomic status, and social contacts.

Familial Hodgkin Lymphoma

Familial HL represents 4.5% of all HL cases. For adolescents and young adults there is a 99-fold increased risk among monozygotic twins and a sevenfold increased risk among siblings.

EBV-Associated Hodgkin Lymphoma

EBV incorporated into the tumor genome has been most commonly reported with the mixed cellularity histologic subtype. This subtype is most common in children from underdeveloped countries, in males under age 10 years, and in those with other immunodeficiencies. Conversely, in young adult HL, incorporation of EBV in the tumor genome is unusual but a history of infectious mononucleosis and high-titer antibodies to EBV are associated.

Tumor necrosis factor receptor-associated factor 1 (TRAF 1) is overexpressed in EBV-transformed lymphoid cells and Reed–Sternberg (RS) cells and is associated with activation of NF-κB and protection of lymphoid cells from antigen-induced apoptosis. Activation of NF-κB, in turn, leads to expression of TRAF1, thereby establishing

Lanzkowsky's Manual of Pediatric Hematology and Oncology.
DOI: http://dx.doi.org/10.1016/B978-0-12-801368-7.00021-1

a positive feedback loop that maximizes NF-κB-dependent gene expression. EBV latent membrane protein 1 (LMP1) interacts with TRAF1, and tumors with TRAF1-LMP1 aggregates exhibit high NF-κB activity. LMP1 activates NF-κB by promoting IκBa turnover. RS cells express CD30 and CD30 ligation promotes proliferation of HL-derived cells with constitutive activation of NF-κB.

EBV genome fragments can be found in approximately 30—50% of HL specimens, and may play a role in the rescue and repair of RS cells, further aiding in their evasion of apoptosis and enhanced survival. Three latent viral antigens are expressed in EBV-positive HL in RS cells: Epstein—Barr nuclear antigen-1, required for viral episome maintenance, LMP1 with transforming properties, and LMP2, which is nontransforming.

Socioeconomic Status and Hodgkin Lymphoma

There is an association between HL and socioeconomic status. In children less than 10 years of age and in underdeveloped nations, HL is associated with lower socioeconomic status and in households with more children. However, in young adult patients and in developed nations, HL incidence increases with higher socioeconomic status and with smaller households with fewer children. These findings may be related to an association with infections; increased infections in early childhood may decrease the risk of HL in young adults.

BIOLOGY

The hallmark of classical HL is the RS cell. Sequence analyses of RS cell clones reveal rearrangements of immunoglobulin variable-region genes resulting in deficient immunoglobulin production. RS cells then evade the apoptotic pathway, leading to the genesis of HL. The B lymphoid cells from which RS arise have high levels of constitutive nuclear NF-κB, a transcription factor known to mediate gene expression related to inflammatory and immune responses, and deregulation of NF-κB has been postulated as a mechanism by which RS cells evade apoptosis. NF-κB dimers are held in an inactive cytoplasmic complex with inhibitory proteins, the IκBs. B-cell stimulation by diverse signals results in rapid activation of the IκB kinase (IKK). The IKK complex phosphorylates two critical serine residues of IκBs, thereby targeting them for rapid ubiquitin-mediated proteasomal degradation. Active NF-κB dimers are then released and translocated to the nucleus, where they activate gene transcription. Activation of NF-κB appears to be a final common effect of costimulatory interactions, genetic aberrations, or viral proteins that operate in HL. RS cell survival is dependent on several downstream pathways. RS cells express CD40 and CD40 ligand (CD40L) is expressed on inflammatory T and dendritic cells that surround them. CD40/CD40L interactions normally provide a second signal from activated helper T-cells to normal B-cells, resulting in activation of NF-κB. NF-κB in turn causes proliferation and induces expression of BCL-x_L, which protects B-cells from apoptosis.

PATHOLOGY

Macroscopic Features

The spread of HL occurs most commonly by contiguity from one chain of lymph nodes to another. Involvement of the left supraclavicular nodes often follows abdominal para-aortic node involvement with spread up the thoracic duct, whereas involvement of the right supraclavicular nodes tends to be associated with mediastinal adenopathy. Para-aortic node involvement commonly occurs in association with involvement of the spleen, which in turn is commonly followed by liver or bone marrow involvement, or both. Nodular sclerosing classical HL shows the greatest propensity to spread by contiguity, whereas noncontiguous dissemination, when it occurs, is more than twice as frequent in the mixed cellularity and lymphocyte-depleted classical HL histologic types.

The RS cell is the hallmark of classical HL and is characterized by a binucleated or a multinucleated giant cell that is often characterized by a bilobed nucleus, with two large nucleoli, giving the classically described "owl's eye" appearance. The RS cells are embedded within a benign-appearing reactive infiltrate of lymphocytes, macrophages, granulocytes, and eosinophils. Important exceptions to this histologic picture apply to the categories of nodular sclerosis (NS) type and nodular variant lymphocyte predominance type in which peculiar variant forms of the tumor giant cells can be used to establish the diagnosis (lacunar cell variants).

TABLE 21.1 Histologic Variants of Hodgkin Lymphoma

CLASSICAL HODGKIN LYMPHOMA (CHL)

Nodular Sclerosis Classical HL (NSCHL)	Characterized by collagen bands and lacunar variants of RS cells; the presence of one or more sclerotic bands is the defining feature. These bands usually radiate from a thickened lymph node capsule, often following the course of a penetrating artery, and are composed of mature, laminated, relatively acellular collagen. The sclerotic bands are birefringent in polarized light. In most cases, several broad collagenous bands can be identified, or fibrosis can be so extensive that isolated nodules of lymphoid tissue remain. The collagenous bands of nodular sclerosis enclose nodules of lymphoid tissue containing variable numbers of HCs and reactive infiltrates. Lacunar cells are a common type of RS cell present and may be found in large numbers or in sheets. They tend to aggregate at the center of nodules, sometimes forming a rim around central areas of necrosis. Diagnostic RS cells are present in variable numbers and may be difficult to identify in small biopsy specimens. Eosinophils, histiocytes, and sometimes even neutrophils are often numerous; plasma cells are usually less conspicuous.
Mixed Cellularity Classical HL	This intermediate subtype falls between lymphocyte-rich classical HL and lymphocyte-depleted classical HL. The capsule is usually intact and of normal thickness. A vague nodularity may be present at low magnification, but the presence of any definite fibrous bands would warrant classification as nodular sclerosis rather than mixed cellularity. At high magnification, a heterogeneous mixture of HCs, small lymphocytes, eosinophils, neutrophils, epithelioid and nonepithelioid, histiocytes, plasma cells, and fibroblasts are present. Diagnostic RS cells and mononuclear variants are usually easy to find. Small foci of necrosis may be present, but the extent is much less than that seen in nodular sclerosis.
Lymphocyte-Depleted Classical HL	Lymphocyte-depleted HL encompasses two variants: Diffuse fibrosis, and reticular. The most characteristic features are a marked degree of reticulin fibrosis surrounding single cells along with lymphocyte depletion. In contrast to nodular sclerosis, this subtype is not characterized by the presence of thick fibrous bands and the fibrosis envelops individual cells, not nodules of cells. HCs are usually easily identified, but increased numbers of HCs are not essential to the diagnosis. In the reticular variant, sheets of HCs, often showing pleomorphic features, are found.
Lymphocyte-Rich Classical HL	Many cases of lymphocyte-rich classical HL have a resemblance to mixed cellularity HL, with vaguely nodular and less often diffuse pattern at low magnification. Hodgkin and RS cells are relatively rare and the background is dominated by small mature lymphocytes. Eosinophils and neutrophils are usually absent and if present are scanty and usually within the interfollicular areas. RS cells and variants are not easy to find but when encountered have identical features to the HCs of mixed cellularity. Some cases of lymphocyte-rich HCs may show a distinctly nodular appearance that may closely mimic nodular lymphocyte predominance HL and often contain relatively small germinal centers, with Hodgkin and RS cells present in and near the mantle zone, a pattern that has been called follicular HL.

UNCLASSIFIED CASES (UC)[a]

Nodular Lymphocyte-Predominant HL	This is a B-cell neoplasm with a nodular or nodular and diffuse proliferation of scattered large neoplastic cells termed "popcorn" cells (formerly called L&H cells, for lymphocytic and/or histiocytic RS cell variants). These large cells resemble centroblasts but are larger and have folded or multilobulated nuclei and multiple small basophilic nucleoli are often present adjacent to the nuclear membrane. The cytoplasm is broad and only slightly basophilic. These large cells are present within spherical nodules with numerous dendritic cells, histiocytes, and small lymphocytes. Ultrastructural studies demonstrate that popcorn cells have the appearance of centroblasts of germinal center. Epithelioid histiocytes are preferentially found in the outer rim of nodules. They are arranged in small groups or clusters and well-formed granulomas may be present in rare cases. Eosinophils and neutrophils are rare. Plasma cells are not common and are seen only between follicles. In diffuse areas, the popcorn cells are still often arranged in a vaguely nodular pattern. Classic Hodgkin and RS cells are completely lacking or are few in number. In some cases, popcorn cells may resemble lacunar cells because both cell types show irregularly shaped or lobulated nuclei, small nucleoli and broad pale to slight basophilic cytoplasm. The popcorn cells are often surrounded by rosettes of CD31, CD571 T-lymphocytes.

---

[a]Any histopathology that does not fit into a definite or provisional category (they may be T-cell, B-cell, or undefined; they may be borderline between HL and NHL).
HC, Hodgkin cells; HL, Hodgkin lymphoma; RS, Reed-Sternberg.

Histology

Table 21.1 describes the histologic variants in HL which consist of two disease groups:

1. Classical HL includes the NS classical HL, mixed cellularity HL (MCHL), lymphocyte-depleted and lymphocyte-rich classical HL subtypes. Adolescent and young adults are most likely to have disease of the nodular sclerosing subtype, which accounts for 74% of cases in those 15—19 years of age. Under the age of 20 years, the mixed cellularity subtype accounts for 16% of cases, but under the age of 10 years, 32% of cases and across the pediatric age group, it is more common in males.
2. Nodular lymphocyte-predominant HL (NLPHL) is nonclassical HL with expression of different immunophenotypic features.

TABLE 21.2   Immunophenotypic Markers in Classical Hodgkin Lymphoma and Nodular Lymphocyte-Predominant HL

	CD15	CD30	CD45	CD20	CD79a	Pax5
Classical HL	+/−	+	−	−/+	−/+	+
NLPHL	−	−	+	+	+	+

+, all cases are positive; +/−, majority of cases positive; −/+, minority of cases positive; −, all cases are negative.

Immunophenotypic Features

RS cells in classic HL do not express B-cell antigens such as CD45, CD19, and CD79A, but virtually all express CD30 and approximately 70% express CD15, with only 20–30% expressing CD20. In comparison, the tumor cells of NLPHL always express B-cell antigens such as CD20, CD79a, and are almost always negative for CD30 and CD15. The majority of RS cells and popcorn cells express B-cell-specific activator protein PSX-5. Table 21.2 shows the common immunophenotypic markers in HL.

CLINICAL PRESENTATION

Constitutional B Symptoms

Approximately 20% of patients have associated B symptoms defined as:

1. Unexplained weight loss of >10% of body weight in the 6 months preceding the diagnosis.
2. Unexplained fever with temperatures >38°C for more than 3 days.
3. Drenching night sweats.

Mild, moderate, or severe pruritus in the absence of rash can also be seen with HL but is not considered a B symptom.

Peripheral Lymphadenopathy

1. Painless swelling of one or more groups of superficial lymph nodes.
2. Cervical nodes involved in the majority of cases.
3. Other commonly affected peripheral nodal regions include supraclavicular, axillary, and inguinal.
4. The bulk of palpable lymph nodes is defined by product of the perpendicular diameters using the single largest dimension (in centimeters) of the lymph node or conglomerate and that perpendicular to the same in each region of involvement. A node or nodal mass of >6 cm is generally defined as *bulky*.

Mediastinal Adenopathy

1. Approximately 20% of patients have *bulky mediastinal disease* defined as a large mediastinal mass (>6–10 cm in maximum dimension) or which, on a posterior–anterior chest radiograph (CXR), has a maximum width equal to or greater than one-third of the internal transverse diameter of the thorax at the level of the T5–6 interspace (mass to thoracic ratio 0.33).
2. Adolescents and young adults present with a mediastinal mass in 75% of cases, as opposed to children less than 10 years of age where mediastinal disease is present in only 35% of cases. This is thought to be due to these younger patients having either mixed cellularity or lymphocyte-predominant histology, where peripheral adenopathy is more common.
3. With mediastinal involvement there may be persistent nonproductive cough; however, this site is often asymptomatic. Despite large mediastinal masses with compression on the airway, superior *vena cava* syndrome (enlargement of the vessels of the neck, hoarseness, dyspnea, and dysphagia) is uncommon.

Pulmonary

1. Lung parenchymal lesions may occur from direct extension from mediastinal or hilar adenopathy or may be discrete lesions.
2. Uncommonly lesions may be cavitating and must be differentiated from infectious etiologies. Such infections (e.g., fungal infections or tuberculosis) can exist alone or in combination with HL.
3. Pleural effusion can accompany hilar, mediastinal, or pulmonary disease.

Spleen

1. The spleen is commonly enlarged on physical examination or by imaging study.
2. Size is not indicative of splenic involvement with HL.
3. In splenic involvement, discrete hypodense lesions may be seen on computed tomography (CT) scan. Fluorodeoxyglucose positron emission tomography (FDG-PET) scan has high sensitivity and specificity for splenic lesions and is now used to identify splenic involvement from HL, although clinical trials may require CT-based lesions for definition as well. In 13% of cases, the spleen is the only site of subdiaphragmatic disease.

Bone

1. Osseous HL typically presents with bone pain and the majority of patients have concurrent nonosseous lesions detected at staging.
2. FDG-PET is an excellent imaging modality for osseous involvement and has generally replaced the need for technetium bone scans.

Hematology and Bone Marrow

1. Blood count abnormalities can include a normocytic normochromic anemia, thrombocytopenia, neutrophilia or neutropenia, eosinophilia, and lymphopenia.
2. Neutropenia, anemia, or thrombocytopenia may be related to marrow or splenic involvement or may be autoimmune in nature and can in those circumstances, precede the diagnosis of HL. Eosinophilia may be cytokine-mediated due to IL-5 production by RS cells.
3. A positive direct antiglobulin test may or may not be associated with overt hemolysis.
4. To confirm bone marrow involvement, multiple biopsies are indicated because HL tends to involve the marrow in a focal fashion. However, bone marrow disease is well characterized by FDG-PET and this may obviate the need for bone marrow biopsies in the future.

Liver

1. Mild hepatomegaly and abnormal liver function tests do not correlate with actual histologic involvement of the liver.
2. Discrete lesions should be noted on CT scan and liver biopsy is the most accurate method for confirmation of liver involvement.

Kidney

1. Renal involvement may be unilateral or bilateral and may be present as diffuse involvement, discrete nodules, or microscopic disease.
2. Renal involvement with HL may result from ureteral obstruction and patients with HL may also have renal dysfunction related to renal vein thrombosis, hypercalcemia, and hyperuricemia.

TABLE 21.3   Diagnostic Investigations for Hodgkin Lymphoma

Surgical	Excisional or core lymph node biopsy
	Bilateral bone marrow biopsies
Imaging studies	CT scan of neck, chest, abdomen and pelvis
	FDG-PET
Laboratory studies	Complete blood count
	Blood chemistries for renal and hepatic function
	Erythrocyte sedimentation rate
	Ferritin

CT, computed tomography; FDG-PET, fluorodeoxyglucose positron emission tomography.
Adapted from: Friedman and Schwartz (2008), with permission.

Nervous System

1. Neurologic dysfunction is usually a late manifestation and extremely rare.
2. There can be spread from paravertebral lymph nodes or hematogenous spread.
3. Symptoms are related to area of neurologic involvement.

DIAGNOSTIC EVALUATION AND STAGING

Diagnostic Evaluation

Table 21.3 summarizes the diagnostic evaluation. A detailed history is required to elucidate B symptoms, which becomes important in risk stratification. A thorough physical evaluation should be performed, documenting the location and size of adenopathy, presence of splenomegaly, and any evidence of organ dysfunction. CT scans of the neck, thorax, abdomen, and pelvis and FDG-PET imaging are required investigations in HL. Response to treatment is evaluated by FDG-PET alone or in combination with CT. Areas of disease can be evaluated simultaneously with both modalities in an overlapping fashion. Technetium-99 bone scintigraphy can be considered in patients with bone pain or elevated alkaline phosphatase, although FDG-PET may obviate the need for this modality. An upright CXR with posteroanterior (PA) and lateral views has been traditionally required for documentation of a large mediastinal mass (bulk mediastinal disease) for clinical trials, defined as tumor diameter greater than one-third of the thoracic diameter (measured transversely at the level of the dome of the diaphragm on a 6-foot upright PA CXR) although the ongoing value of this remains unclear.

Biopsy of an involved nodal site is required for definitive diagnosis by either excisional or core biopsy. Fine-needle aspiration should be avoided as it often yields insufficient tissue for definitive diagnosis. Bone marrow biopsy is still generally recommended for all stage III and IV patients or patients with B symptoms. There are less consistent recommendations for lower-stage patients without B symptoms, but bone marrow biopsies are often still required for staging on clinical trials. With ongoing use of FDG-PET, such imaging may obviate the need for bone marrow. However, this may result in stage shifting, as FDG-PET will identify bone marrow disease not noted on bone marrow biopsy. Laboratory studies include a complete blood count, blood chemistries to evaluate hepatic and renal function, and may include acute-phase reactants such as ferritin, erythrocyte sedimentation rate, and serum copper, which may be seen as nonspecific markers of tumor activity, but may correlate with prognosis or response.

Staging

The staging classification currently used for HL is the Ann Arbor Classification as shown in Table 21.4.

1. Approximately 80–85% of children and adolescents with HL have involvement limited to lymph nodes or direct extension from the lymph nodes and/or the spleen (stages I to III), whereas 15–20% of patients are stage IV with involvement of the lung, bone marrow, bone, or liver. Although staging definitions by nodal region, B symptoms, and definition of extranodal involvement are well-defined, bulk disease and substaging have not been consistent across studies.

TABLE 21.4   Clinical and Staging Criteria for Hodgkin Lymphoma

A. STAGE GROUPING

Stage I: Involvement of single lymph node region (I) or localized involvement of a single extralymphatic organ or site (IE)

Stage II: Involvement of two or more lymph node regions on the same side of the diaphragm (II) or localized contiguous involvement of a single extralymphatic organ or site and its regional lymph node(s) with involvement of one or more lymph node regions on the same side of the diaphragm (IIE)

Stage III: Involvement of lymph node regions on both sides of the diaphragm (III), which may also be accompanied by localized contiguous involvement of an extralymphatic organ or site (IIIE), by involvement of the spleen (IIIS), or both (IIIE + S)

Stage IV: Disseminated (multifocal) involvement of one or more extralymphatic organs or tissues, with or without associated lymph node involvement, or isolated extralymphatic organ involvement with distant (nonregional) nodal involvement

B. SYMPTOMS AND PRESENTATIONS

"A" Symptoms: Lack of "B" symptoms.

"B" Symptoms: At least one of the following:

> Unexplained weight loss >10% in the preceding 6 months

> Unexplained recurrent fever >38°C

> Drenching night sweats

> X Bulk disease (see C below)

> E Involvement of a single extranodal site that is contiguous or proximal to the known nodal site.

C. BULK DISEASE

One or both of the following presentations are considered "bulk" disease:

- *Large mediastinal mass:* tumor diameter >1/3 the thoracic diameter (measured transversely at the level of the dome of the diaphragm on a 6-foot upright PA CXR). In the presence of hilar nodal disease the maximal mediastinal tumor measurement may be taken at the level of the hilus. This should be measured as the maximum mediastinal width (at a level containing the tumor and any normal mediastinal structures at the level) over the maximum thoracic ratio
- *Large extramediastinal nodal aggregate:* A continuous aggregate of nodal tissue that measures >6 cm[a] in the longest transverse diameter in any nodal area

[a]Some studies use 10 cm for definition of extramediastinal bulk disease.
PA, posteroanterior; CXR, chest radiograph; E, Extra nodal disease.

PROGNOSTIC FACTORS

Adverse prognostic markers often form the basis for risk stratification and subsequent modification of therapeutic algorithms. In sequential trials, as treatment is then risk-based, these adverse factors are abrogated by the changes in therapy.

1. Early response to treatment, which may be a correlate for biology, may also be an important prognostic factor, allowing titration of therapy to the individual.
2. Histologic subtype is not consistently associated with prognosis but NLPHL appear to have an overall better prognosis than those with CHL.
3. Pretreatment factors that have been shown to be associated with adverse outcome include:
   a. Advanced stage (III, IV).
   b. B symptoms.
   c. Bulk disease.
   d. Extranodal extension.
   e. Male sex.
   f. Elevated erythrocyte sedimentation rate.
   g. Hemoglobin <10 g/dl.
   h. White blood cell count >11,500/mm^3.
   i. Age 5−10 years.
   j. Increased numbers of sites of disease.

4. Serum markers that may confer adverse prognostic risk include:
   a. Soluble vascular adhesion molecule-1.
   b. Tumor necrosis factor.
   c. Soluble CD30.
   d. Beta-2 microglobulin.
   e. Transferrin.
   f. Serum IL-10.
   g. Serum CD 8 antigen.
   h. High RS cell levels of caspase 3.

TREATMENT

In general, the use of chemotherapy with or without low-dose involved field radiation therapy is considered standard of care. Table 21.5 gives some of the commonly accepted standard strategies for newly diagnosed children and adolescent patients with HL.

Treatment is risk-adapted to presenting features as well as to response to initial treatment.

Surgery

The role of surgery is generally limited to a diagnostic biopsy. Resection of all disease is not indicated. The exception is for NLPHL, where observation following complete surgical resection of a single nonbulky peripheral node in patients with Stage IA disease is indicated.

Chemotherapy

Chemotherapy regimens commonly include the use of vincristine, vinblastine, doxorubicin, bleomycin, cyclophosphamide, procarbazine, dacarbazine, etoposide, prednisone, and methotrexate in varying combinations.

Radiation

HL is a radiosensitive disease. Treatment for HL in children and adolescents employs radiation therapy in the context of multimodality therapy. In general, doses of 15–25 Gy are used with modification based on patient's

TABLE 21.5 Common Standard Therapy Protocols for Pediatric Classical Hodgkin Lymphoma

Stage and presentation	Commonly accepted chemotherapy regimens
IA, IIA <4 nodal regions without B symptoms, bulk disease, or extranodal extension	VAMP × 4 COPP/ABV hybrid × 4 ABVE × 4 OEPA or OPPA × 2 CHOP × 3–4 ± Rituximab ABVD × 2–4
IA, IIA with bulk disease, ≥3 nodal regions, or extranodal extension IIB*, IIIA, IVA	COPP/ABV × 6 ABVE-PC × 3–5 OPPA/OEPA × 2 + COPP × 2 OEPA-COPAC ABVD × 4–6
IIB*, IIIB, IVB	ABVE-PC × 3–5 BEACOPP × 8 (or BEACOPP × 4 + ABVD × 2 or + COPP/ABV × 4) OPPA/OEPA × 2 + COPP × 4 OEPA-COPAC Stanford V ABVD × 6–8

All regimens should consider low-dose involved field radiotherapy; 15–25 Gy, but there are variations in recommendations based on risk and early and complete response to chemotherapy.

age, the presence of bulk disease, normal tissue concerns and potential acute and long-term effects. Photon-based radiotherapy has remained the standard of care for HL, but there are growing data on the use of proton beam radiotherapy, which may confer less toxicity due to decreased exposure to normal tissue. Volume considerations for treatment are also in evolution with the standard having been the involved field and now with regimens utilized an involved node approach. There are also regimens for which radiotherapy fields may be determined based on response to chemotherapy. Radiotherapy fields must be designed with the goal of delivering the optimum volume of radiotherapy for disease control, while avoiding normal tissue damage. Customized shielding blocks should be utilized as appropriate to protect normal tissue. Blocking the genitalia is of specific importance when pelvic fields are included. In females, oophoropexy generally results in preservation of fertility. For males, a frog-leg position and an individually fitted shield will provide the greatest shielding to the testes, reducing scatter.

Treatment for Refractory or Recurrent Disease

HL may be cured even if it has failed to respond to initial treatment or has recurred after initial treatment. Potentially curative options include:

1. Conventionally dosed combined modality protocols.
2. Radiotherapy alone for those relapsing in a limited nodal pattern.
3. Reinduction chemotherapy followed by autologous hematopoietic stem cell transplantation (HSCT), full- and reduced-intensity allogeneic transplants.
4. Evolving targeted antibodies.

Some common retrieval regimens are shown in Table 21.6 and the choice of regimen is dependent on previous treatment and patterns of relapse. There is no single standard of care.

Two ifosfamide-based regimens have been used with good response and survival rates, ifosfamide, carboplatin, and etoposide and ifosfamide and vinorelbine (IV). Gemcitabine has also been used in combination with vinorelbine with excellent overall response and survival rates. Autologous HSCT is most commonly used for recurrent or refractory HL, particularly when used following dose-intensive regimens or for high-risk disease. However, an allogeneic effect has been observed, and full- and reduced-intensity allogeneic approaches, as well as immune modulation and induction of autologous graft-versus-host disease, have been explored to enhance the allogeneic effect.

The most promising new agent in the treatment for HL is brentuximab vedotin, which is also being utilized in the earlier relapse period. There are several recently reported as well as ongoing trials, although pediatric experience is still limited. These data led to accelerated approval of brentuximab vedotin for the treatment of patients with HL after failure of autologous stem cell transplantation (ASCT) or after failure of at least two prior multiagent chemotherapy regimens in patients who are not ASCT candidates. Studies are now ongoing evaluating brentuximab vedotin as part of multiagent protocols in the upfront and recurrent setting. Other new agents showing promise in recurrent or refractory disease and under study in clinical trials are bendamustine and PD-1 inhibitors such as pembrolizumab and nivolumab. The use of radiation therapy in the recurrent and refractory setting is based on the prior radiation therapy exposure and the pattern of disease.

TABLE 21.6 Common Treatment Options for Recurrent Hodgkin Lymphoma

ICE (ifosfamide, carboplatin, and etoposide)
DECA (dexamethasone, etoposide, cisplatin, cytarabine)
IV (ifosfamide and vinorelbine)
GV (gemcitabine and vinorelbine)
IEP-ABVD-COPP (ifosfamide, etoposide, prednisone-doxorubicin, bleomycin, vinblastine, dacarbazine-cyclophosphamide, vincristine, procarbazine, prednisone)
APE (cytosine arabinoside, cisplatin, etoposide)
SGN-35 (brentuximab vedotin)-antibody-drug conjugate targeting CD30

These therapies can be used as standalone therapy or as reinduction prior to planned stem cell transplant. Radiotherapy can be considered together with this therapy.

LONG-TERM COMPLICATIONS

A significant consideration in deciding therapy is the risk of adverse long-term effects, which include organ dysfunction and second malignancies. The common potential long-term effects of radiotherapy and chemotherapy for HL are summarized together with general monitoring recommendations in Table 21.7. See Chapter 34 for more information on long-term complications.

Secondary Malignancies

The incidence of secondary malignancies in survivors is 7–18 times higher than the general population. Secondary hematological malignancies (most commonly acute myeloid leukemia and myeloid dysplasia) are related to the use of alkylating agents, anthracyclines, and etoposide. The risk of leukemia appears to plateau at 10–15 years post therapy while the risk of second solid malignancies can occur at any time post therapy.

Female survivors of HL have a particularly increased risk for breast cancer, directly related to the dose of radiation therapy received.

Cardiac Toxicity

HL survivors exposed to doxorubicin and/or thoracic radiation therapy are at an increased risk for long-term cardiotoxicity.

TABLE 21.7 General Guidelines for Risk and Surveillance for Adverse Long-Term Outcomes

System	Therapeutic exposure	Potential effects	Monitoring recommendations[a]
Cardiac	Thoracic RT Doxorubicin	Cardiomyopathy Pericarditis Coronary artery disease Valvular disease	Electrocardiogram and echocardiogram
Pulmonary	Thoracic RT Bleomycin	Pulmonary fibrosis Restrictive lung disease	Pulmonary function tests (including DLCO and spirometry)
Thyroid	Neck RT	Overt or compensated hypothyroidism Thyroid nodules or cancer Hyperthyroidism	Free T4; TSH
Gonadal (female)	Pelvic RT Alkylating agents	Delayed/arrested puberty Early menopause Ovarian failure	FSH, LH, estradiol
Gonadal (male)	Pelvic RT Alkylating agents	Germ cell failure Infertility/azoospermia Leydig cell dysfunction Hypogonadism Delayed/arrested puberty	FSH, LH, testosterone Semen analysis
Bone	Corticosteroids	Decreased bone mineral density	Bone density evaluation (DEXA or quantitative CT)
Second malignancies	Radiotherapy Doxorubicin Etoposide Mechlorethamine Cyclophosphamide	Sarcomas CNS tumors Breast cancer Melanoma Nonmelanoma skin cancer Thyroid cancer Other solid tumors Therapy-related myelodysplasia and acute leukemia	Routine cancer screening per general population guidelines Mammography to screen for female breast cancer at age 25- or 10-year post RT exposure, whichever is later Annual CBC with differential for 10-y post exposure

[a]*Monitoring recommendations all include a risk-adapted health history and physical examination. The frequency with which diagnostic studies should be performed is dependent on many factors including radiotherapy dose, chemotherapy exposures, age at exposure, and other clinical parameters. See Children's Oncology Group guidelines for more details (www.survivorshipguidelines.org).*
DEXA, Dual-energy X-ray absorptiometry; CBC, Complete blood count; CNS, Central nervous system; DLCO, Diffusing capacity; FSH, follicle-stimulating hormone; LH, luteinizing hormone; RT, radiation therapy; TSH, thyroid-stimulating hormone; CT, computed tomography.

Cardiac toxicities include:

- Pericarditis.
- Pancarditis.
- Cardiomyopathy and congestive heart failure.
- Coronary artery disease.
- Functional valve injury.
- Conduction injuries.

Risk factors include:

- Female sex (for doxorubicin exposure).
- Higher cumulative doxorubicin doses.
- Concomitant thoracic radiation therapy exposure, with a dose-effect.
- Younger age at time of exposure.
- Increased time from exposure.

Nontherapeutic risk factors for coronary heart disease including family history, obesity, smoking, hypertension, diabetes, and hypercholesterolemia are also likely to impact the frequency of disease.

Pulmonary Dysfunction

Pulmonary fibrosis is a late complication of HL associated with bleomycin and radiation therapy. Bleomycin-associated pulmonary fibrosis with decreased diffusion capacity is most commonly seen following doses greater than $200-400$ units/m^2.

Acute pneumonitis as evidenced by fever, congestion, cough, and dyspnea can follow radiation therapy alone at doses greater than 40 Gy to focal lung volumes or after lower doses (15–20 Gy) when combined with chemotherapy and including generous or whole lung volumes.

Thyroid Dysfunction

Primary hypothyroidism is the most common thyroid dysfunction noted following radiation therapy to the neck for HL, though hyperthyroidism, goiter, or nodules have also been noted. The incidence of thyroid dysfunction varies with the dose of radiation and the length of follow-up. The risk for both hypo- and hyperthyroidism increases in the first 3–5 years after diagnosis. The incidence of nodules increases 10 years from the time of diagnosis.

Gonadal Dysfunction and Infertility

Male Gonadal Toxicity

Infertility caused by azoospermia is the most common manifestation of gonadal toxicity. Alkylating agents and radiation therapy are common agents associated with male infertility. Males who receive less than 4 g/m^2 of cyclophosphamide without testicular radiation or any other alkylating agents are likely to retain their fertility. Cumulative doses of cyclophosphamide greater than 9 g/m^2 are likely to result in infertility. Radiation doses less than 30 Gy are unlikely to affect endocrine function and boys usually progress through puberty normally. There can be temporary oligospermia following low doses of radiation though permanent azoospermia may occur.

Female Gonadal Toxicity

The risk of female gonadal toxicity increases with age at treatment. Menstrual irregularities including amenorrhea and premature ovarian failure occur more commonly in older women treated with cyclophosphamide and other alkylating agents than in adolescent and prepubertal females. However, the incidence of early menopause in young female survivors of HL is increased over that seen in the general population.

Psychosocial and Neurocognitive Impairment

HL survivors are at risk for potential psychosocial and neurocognitive impairment, related to the disease, treatment, and nontreatment-related risk factors.

FOLLOW-UP EVALUATIONS

During Therapy

Monitoring during therapy is focused on response to therapy as well as toxicity associated with therapy. Frequency and precise monitoring is highly risk-adapted and dependent on the stage and extent of disease and type of therapy. In general, patients are followed closely with careful interim history and physical examination, imaging, and laboratory studies. These studies may include blood counts, chemistries, erythrocyte sedimentation rates, CT, and FDG-PET scans. Interval evaluation with diagnostic studies is often also performed to assess organ toxicity conferred by therapy, such as echocardiogram and/or ECG for cardiac function and pulmonary function tests for pulmonary assessment.

After Completion of Therapy

Following the completion of therapy, monitoring is required for disease recurrence as well as for long-term sequelae of therapy with the same modalities utilized during treatment. The frequency of evaluation is not completely standardized, and there are data to suggest that imaging studies no longer need to be conducted past 18 months–2 years post completion of therapy.

Monitoring for Long-Term Outcomes

Specific long-term follow-up guidelines after treatment of childhood cancer are available at www.survivorshipguidelines.org.

Monitoring for long-term outcomes is highly dependent on demographic, disease, and treatment factors, as well as nontreatment exposures and comorbidities which may alter treatment-related outcome risk. Patients should be followed lifelong with history and physical and routine medical and dental care.

Further Reading and References

Appel, B.E., Chen, L., Buxton, A., Wolden, S.L., Hodgson, D.C., Nachman, J.B., 2012. Impact of low-dose involved-field radiation therapy on pediatric patients with lymphocyte-predominant Hodgkin lymphoma treated with chemotherapy: a report from the Children's Oncology Group. Pediatr. Blood Cancer 59 (7), 1284–1289.

Castellino, S.M., Geiger, A.M., Mertens, A.C., et al., 2011. Morbidity and mortality in long-term survivors of Hodgkin lymphoma: a report from the childhood cancer survivor study. Blood 117, 1806–1816.

Cheson, B.D., Fisher, R.I., Barrington, S.F., Cavalli, F., Schwartz, L.H., Lister, T.A., 2014. Recommendations for initial evaluation, staging, and response assessment of Hodgkin and non-Hodgkin lymphoma: the Lugano classification. J. Clin. Oncol. 32 (27), 3059–3068.

Claviez, A., Canals, C., Dierickx, D., et al., 2009. Allogeneic hematopoietic stem cell transplantation in children and adolescents with recurrent and refractory Hodgkin lymphoma: an analysis of the European group for blood and marrow transplantation. Blood 114, 2060–2067.

Cole, P.D., Schwartz, C.L., Drachtman, R.A., et al., 2009. Phase II study of weekly gemcitabine and vinorelbine for children with recurrent or refractory Hodgkin's disease: a children's oncology group report. J. Clin. Oncol. 27, 1456–1461.

Dharmarajan, K.V., Friedman, D.L., Schwartz, C.L., Chen, L., FitzGerald, T.J., McCarten, K.M., et al., 2014. Patterns of relapse from a Phase III study of response-based therapy for intermediate-risk hodgkin lymphoma (AHOD0031): a report from the Children's Oncology Group. Int. J. Radiat. Oncol. Biol. Phys 92 (1), 60–66.

Donaldson, S.S., Link, M.P., Weinstein, H.J., et al., 2007. Final results of a prospective clinical trial with VAMP and low-dose involved-field radiation for children with low-risk Hodgkin's disease. J. Clin. Oncol. 25, 332–337.

Friedman, D., Schwartz, C., 2010. Hodgkin lymphoma in children. In: Magrath, I. (Ed.), The Lymphoid Neoplasms, third ed. CRC Press, Taylor and Francis, Boca Raton, FL, USA.

Friedman, D.L., Constine, L.S., 2006. Late effects of treatment for Hodgkin lymphoma. J. Natl. Compr. Canc. Netw. 4, 249–257.

Friedman, D.L., Whitton, J., Leisenring, W., et al., 2010. Subsequent neoplasms in 5-year survivors of childhood cancer: the childhood cancer survivor study. J. Natl. Cancer Inst. 102, 1083–1095.

Friedman, D.L., Chen, L., Wolden, S., Buxton, A., McCarten, K., FitzGerald, T.J., et al., 2014. Dose-intensive response-based chemotherapy and radiation therapy for children and adolescents with newly diagnosed intermediate-risk hodgkin lymphoma: a report from the Children's Oncology Group Study AHOD0031. J. Clin. Oncol. 32 (32), 3651–3658.

Kelly, K.M., Sposto, R., Hutchinson, R., et al., 2011. BEACOPP chemotherapy is a highly effective regimen in children and adolescents with high-risk Hodgkin lymphoma: a report from the Children's Oncology Group. Blood 117, 2596–2603.

Kelly, K.M., Hodgson, D., Appel, B., Chen, L., Cole, P.D., Horton, T., et al., 2013. Children's Oncology Group's 2013 blueprint for research: Hodgkin lymphoma. Pediatr. Blood Cancer 60 (6), 972–978.

Mauz-Korholz, C., Hasenclever, D., Dorffel, W., et al., 2010. Procarbazine-free OEPA-COPDAC chemotherapy in boys and standard OPPA-COPP in girls have comparable effectiveness in pediatric Hodgkin's lymphoma: the GPOH-HD-2002 study. J. Clin. Oncol. 28, 3680–3686.

Nachman, J.B., Sposto, R., Herzog, P., et al., 2002. Randomized comparison of low-dose involved-field radiotherapy and no radiotherapy for children with Hodgkin's disease who achieve a complete response to chemotherapy. J. Clin. Oncol. 20, 3765–3771.

Schwartz, C., Constine, L.S., Villaluna, D., et al., 2009. Risk-adapted, response-based approach using ABVE-PC for children and adolescents with intermediate- and high-risk Hodgkin lymphoma: the results of P9425. Blood 114, 2051–2059.

Tebbi, C.K., Mendenhall, N.P., London, W.B., et al., 2012. Response-dependent and reduced treatment in lower risk Hodgkin lymphoma in children and adolescents, results of P9426: a report from the children's oncology group. Pediatr. Blood Cancer 59, 1259–1265.

Trippett, T.M., Schwartz, C.L., Guillerman, R.P., Gamis, A.S., Gardner, S., Hogan, S., et al., 2015. Ifosfamide and vinorelbine is an effective reinduction regimen in children with refractory/relapsed Hodgkin lymphoma, AHOD00P1: a children's oncology group report. Pediatr. Blood Cancer 62 (1), 60–64.

Wolden, S.L., Chen, L., Kelly, K.M., et al., 2012. Long-term results of CCG 5942: a randomized comparison of chemotherapy with and without radiotherapy for children with Hodgkin's lymphoma—a report from the children's oncology group. J. Clin. Oncol. 30, 3174–3180.

Younes, A., Gopal, A.K., Smith, S.E., et al., 2012. Results of a pivotal phase II study of brentuximab vedotin for patients with relapsed or refractory Hodgkin's lymphoma. J. Clin. Oncol. 30, 2183–2189.

Web Resources:

Children's Oncology Group:

- http://www.childrensoncologygroup.org/index.php/hodgkindisease
- http://www.survivorshipguidelines.org
- National Cancer Institute PDQ: http://www.cancer.gov/cancertopics/pdq/treatment/childhodgkins/HealthProfessional

22

Non-Hodgkin Lymphoma

Mary S. Huang and Howard Weinstein

INTRODUCTION

Childhood non-Hodgkin lymphoma (NHL) is distinguished from adult NHL by differing frequencies of histopathologic types and by the greater frequency of extranodal presentations. With current combination chemotherapy regimens, survival is generally excellent (85 to over 90%) for all patients, including those with disseminated disease, bone marrow involvement, central nervous system (CNS) involvement, and high serum lactate dehydrogenase (LDH).

With improved survival in all subtypes and stages, recent efforts have focused on maintaining excellent event-free survival (EFS) with reduced late toxicity by incorporating more specific targeted therapies for patients with less favorable subgroups of NHL and identifying new prognostic factors.

INCIDENCE AND EPIDEMIOLOGY

Incidence

1. NHL represents approximately 6–8% of all malignancies in patients under 20 years of age.
2. Surveillance, Epidemiology and End Results data estimate an incidence of about 1 per 100,000, with an annual incidence of 750–800 cases per year in children up to 19 years of age in the United States.
3. There is geographic variation in the incidence of NHL. For example, in equatorial Africa, Burkitt lymphoma (BL) accounts for almost 50% of all childhood cancers. In this setting, endemic BL is invariably positive for Epstein–Barr virus (EBV), in contrast to about 10% of cases of sporadic BL.
4. There has been a gradual increase in incidence of NHL in the United States over the past 40 years, more pronounced in the 15–19 age group.

Epidemiology

- *Sex*: Male:female is 2–3:1.
- *Age*: Median age of presentation is 10 years. It is rare to have cases under 3 years of age.
- *Risk factors*: Inherited or acquired risk factors have been identified, including those listed below. NHL may develop as second malignancy after chemotherapy and/or radiation therapy or in the setting of congenital or acquired immunodeficiency.
  - *Genetics*: Immunological defects (Bruton type of sex-linked agammaglobulinemia, common variable agammaglobulinemia, severe combined immunodeficiency ataxia-telangiectasia, Bloom syndrome, Wiskott–Aldrich syndrome, autoimmune lymphoproliferative syndrome).
  - *Post-transplant immunosuppression*: Post bone marrow transplantation (especially with use of T-cell-depleted marrow), post-solid organ transplantation.
  - *Lymphomatoid papulosis* in children may evolve into or coexist with anaplastic large-cell lymphoma (ALCL).

Lanzkowsky's Manual of Pediatric Hematology and Oncology.
DOI: http://dx.doi.org/10.1016/B978-0-12-801368-7.00022-3

- *Drugs*: Infliximab and other immunosuppressive agents used in inflammatory bowel disease and autoimmune disease.
- *Viral*: EBV, human immune deficiency virus (HIV) and possible link to human T-lymphotropic virus.

TABLE 22.1   WHO Classification of Lymphoid Neoplasms (2008)

PRECURSOR LYMPHOID NEOPLASMS

B lymphoblastic leukemia/lymphoma NOS

B lymphoblastic leukemia/lymphoma with recurrent genetic abnormalities

B lymphoblastic leukemia/lymphoma with t(9;22); bcr-abl1

B lymphoblastic leukemia/lymphoma with t(v;11q23); MLL rearranged

B lymphoblastic leukemia/lymphoma with t(12:21); TEL-AML1 and ETV6-RUNX1

B lymphoblastic leukemia/lymphoma with hyperploidy

B lymphoblastic leukemia/lymphoma with hypodiploidy

B lymphoblastic leukemia/lymphoma with t(5;14); IL3-IGH

B lymphoblastic leukemia/lymphoma with t(1;19); E2A-PBX1 and TCF3-PBX1

T lymphoblastic leukemia/lymphoma

MATURE B-CELL NEOPLASMS

Chronic lymphocytic leukemia/small lymphocytic lymphoma

B-cell prolymphocytic leukemia

Splenic marginal zone lymphoma

Hairy cell leukemia

Lymphoplasmacytic lymphoma/Waldenstrom macroglobulinemia

Heavy chain disease

Plasma cell myeloma

Solitary plasmacytoma of bone

Extraosseous plasmacytoma

Extranodal marginal zone B-cell lymphoma of mucosa-associated lymphoid tissue type

Nodal marginal zone lymphoma

Follicular lymphoma

Primary cutaneous follicular lymphoma

Mantle cell lymphoma

Diffuse large B-cell lymphoma, NOS (T-cell/histiocyte-rich type; primary CNS type; primary leg skin type and EBV + elderly type)

Diffuse large B-cell lymphoma with chronic inflammation

Lymphomatoid granulomatosis

Primary mediastinal large B-cell lymphoma

Intravascular large B-cell lymphoma

ALK+ large B-cell lymphoma

Plasmablastic lymphoma

Large B-cell lymphoma associated with human herpes virus 8+ Castleman disease

Primary effusion lymphoma

(Continued)

TABLE 22.1 *(Continued)*

Burkitt lymphoma
B-cell lymphoma, unclassifiable, Burkitt-like
B-cell lymphoma, unclassifiable, Hodgkin lymphoma-like

MATURE T-CELL AND NK-CELL NEOPLASMS

T-cell prolymphocytic leukemia
T-cell large granular lymphocytic leukemia
Chronic lymphoproliferative disorder of NK cells
Aggressive NK-cell leukemia
Systemic EBV+ T-cell lymphoproliferative disorder of childhood
Hydroa vacciniforme-like lymphoma
Adult T-cell lymphoma/leukemia
Extranodal T-cell/NK-cell lymphoma, nasal type
Enteropathy-associated T-cell lymphoma
Hepatosplenic T-cell lymphoma
Subcutaneous panniculitis-like T-cell lymphoma
Mycosis fungoides
Sézary syndrome
Primary cutaneous CD30+ T-cell lymphoproliferative disorder
Primary cutaneous gamma-delta T-cell lymphoma
Peripheral T-cell lymphoma, NOS
Angioimmunoblastic T-cell lymphoma
Anaplastic large cell lymphoma, ALK+ type
Anaplastic large cell lymphoma, ALK− type

HODGKIN LYMPHOMA (HODGKIN DISEASE)

Nodular lymphocyte-predominant Hodgkin lymphoma
Classic Hodgkin lymphoma
Nodular sclerosis Hodgkin lymphoma
Lymphocyte-rich classic Hodgkin lymphoma
Mixed cellularity Hodgkin lymphoma
Lymphocyte depletion Hodgkin lymphoma

PTLD

Plasmacytic hyperplasia
Infectious mononucleosis like PTLD
Polymorphic PTLD
Monomorphic PTLD (B and T/NK cell types)
Classic HD type PTLD

HISTIOCYTIC AND DENDRITIC CELL NEOPLASMS

Histiocytic sarcoma
Langerhans cell histiocytosis

(Continued)

TABLE 22.1 *(Continued)*

Langerhans cell sarcoma

Interdigitating dendritic cell sarcoma

Follicular dendritic cell sarcoma

Fibroblastic reticular cell tumor

Indeterminate dendritic cell sarcoma

Disseminated juvenile xanthogranuloma

NOS, not otherwise specified; CNS, central nervous system; EBV, Epstein–Barr virus; PTLD, Post-transplant lymphoproliferative disorders; NK, natural killer; IGH, immunoglobulin heavy chain; MLL, mixed-lineage Leukemia.
Adapted from research originally published by Jaffe et al. (2008).

PATHOLOGIC CLASSIFICATION

Table 22.1 presents the World Health Organization (WHO) classification for lymphoid neoplasms from the International Lymphoma Study Group and incorporates histology as well as immunohistochemistry, gene expression profiling, cytogenetic, molecular and clinical features.

Pediatric NHL is mostly (more than 95%) high-grade and includes the following four major subtypes:

1. B- and T-lymphoblastic lymphoma (LL).
2. BL.
3. Diffuse large B-cell lymphoma (DLBCL).
4. ALCL.

Many of these high-grade lymphomas disseminate noncontiguously, evolve into a leukemic phase, and involve the CNS. The more common low-grade lymphomas seen in adults, such as follicular and marginal zone, are rare in children. T and B-LL comprise about 20% of NHL in childhood; T-cell is the more common type. Mature B-cell lymphoma includes both BL (19% of cases) as well as DLBCL and primary mediastinal large B-cell lymphoma (PMBL). Together the latter two comprise approximately 22% of NHL in childhood. In the current WHO classification, PMBL is considered distinct from other DLBCL based upon its unique clinical, histologic, and molecular features. Other mature B-cell lymphomas, including pediatric marginal zone lymphoma, pediatric-type follicular lymphoma (PFL), and mucosa-associated lymphoid tissue (MALT) lymphoma as well as rare cutaneous lymphomas are also recognized as distinct entities that rarely occur in childhood. Included in the most recent WHO classification, a subtype of follicular lymphoma, referred to as PFL has been described and mostly presents as Stage I or II disease. Within the category of mature T-cell lymphoma, ALCL accounts for approximately 10% of NHL in childhood.

CLINICAL FEATURES

The clinical manifestations of childhood NHL depend primarily on pathological subtype and sites of involvement. Tumors which grow rapidly can cause symptoms based on size and location. Approximately 70% of children present with advanced-stage disease, including extranodal disease with gastrointestinal, bone marrow, and CNS involvement.

Approximately 25% of children with NHL have an anterior mediastinal mass (usually T-LL or PMBL) and present with wheezing, orthopnea, and cough progressing to dyspnea. The majority of these patients are adolescents, and their presentation may manifest as superior vena cava (SVC) syndrome, an oncological emergency discussed in Chapter 32. Patients with large anterior mediastinal masses are at major risk of cardiac or respiratory arrest when laid flat during general anesthesia or deep sedation. A careful workup including a chest computed tomography (CT) scan with airway measurements is essential before attempting any procedures. The least invasive procedure (e.g., biopsy of a peripheral lymph node) should be carried out. If these procedures are not successful in providing a diagnosis, then a CT-guided needle biopsy of the mediastinal mass should be considered. In some clinical situations (e.g., orthopnea or significant airway narrowing), preoperative or preprocedure steroids should be considered for up to 48 h. The use of steroids has largely replaced localized irradiation in this setting.

Primary gastrointestinal involvement occurs in about 30% (usually Burkitt histology), commonly presenting as an abdominal mass with ascites, an "acute abdomen" from an intussusception, or rarely a malnutrition syndrome with colitis symptoms. The majority of children with BL presenting with an ileal—cecal intussusception have limited gastrointestinal involvement that is amenable to complete surgical resection (Murphy Stage 2 or group A).

In 20—30% of children, the head and neck, including Waldeyer's ring or cervical lymph nodes, is the site of origin. The remainder of patients have miscellaneous primary sites, including bone, breast, skin, epidural space, or non-cervical lymph nodes. Involvement of the bone marrow occurs at diagnosis in 10—30% of patients with BL and LL. Overt CNS involvement at diagnosis is not common but is mostly seen in children with advanced-stage BL and LL. Children who develop BL in endemic areas of the world often have a mass in the head or neck region (especially jaw) in contrast to the abdominal presentation typical of non-endemic BL. Both endemic and sporadic cases of BL have the same chromosomal translocations involving one of the loci encoding immunoglobulin heavy or light chains and c-myc oncogene. The exact role of EBV in the pathogenesis of BL and other malignancies is unknown.

DIAGNOSIS

Tissue is required for diagnosis. For some subtypes of NHL, architecture is critical to diagnosis, such that a fine-needle aspirate or core biopsy may not be sufficient to establish a diagnosis. In some cases, however, diagnostic samples may also be obtained from bone marrow, cerebrospinal fluid (CSF), or pleural/paracentesis fluid. In all cases, samples should be tested by flow cytometry for immunophenotype as well as cytogenetics and molecular assays. Recommended laboratory and radiologic testing includes: complete blood count with differential, electrolytes, uric acid, calcium, phosphorus, and creatinine, liver function tests, and LDH. Bone marrow aspiration and biopsy; lumbar puncture with CSF cytology, cell count, glucose, and protein are indicated in histologies in which bone marrow and CNS involvement are common. These studies can potentially be omitted in subsets of pediatric NHL, such as DLBCL, ALCL, and PMBL, in which bone marrow and/or CNS involvement are rare. Chest radiograph and neck, chest, abdominal and pelvic CT scans are recommended to define the extent of disease in most instances. Positron emission tomography (PET) scan, or combined PET/CT, may also be performed to help in initial staging as well as to measure tumor response during the course of therapy. Of note, in contrast to Hodgkin lymphoma (HL), the role of PET in initial staging and treatment response has not yet been well-established (see Chapter 21).

The possibility of inherited or acquired predisposition to NHL should be considered. In some patients, evaluation for specific infection, including HIV testing, or immune function may be appropriate.

STAGING

Staging of NHL requires an investigation to determine the clinical extent of the disease, the degree of organ impairment and biochemical disturbance present. Correct staging is critically important at the time of diagnosis in NHL due to the high number of patients who present with advanced disease. Prior staging systems for NHL entailed a modification of the Ann Arbor Staging System for HL while taking into consideration the common presentations of childhood NHL such as extranodal involvement and metastatic spread to the bone marrow and CNS. The Murphy Staging system has been widely accepted (see Table 22.2). A French,

TABLE 22.2   Murphy and St Jude Children's Research Hospital Staging System for Childhood Non-Hodgkin Lymphoma

Stage I	A single tumor (extranodal) or single anatomic area (nodal), with the exclusion of mediastinum or abdomen
Stage II	A single tumor (extranodal) with regional node involvement
	Two or more nodal areas on the same side of the diaphragm
	Two single (extranodal) tumors with or without regional node involvement on the same side of the diaphragm
	A primary gastrointestinal tract tumor, usually in the ileocecal area, with or without involvement of associated mesenteric nodes only[a]
Stage III	Two single tumors (extranodal) on opposite sides of the diaphragm
	Two or more nodal areas above and below the diaphragm
	All the primary intrathoracic tumors (mediastinal, pleural, thymic)

(Continued)

TABLE 22.2 (*Continued*)

All extensive primary intra-abdominal disease[a]

All paraspinal or epidural tumors, regardless of other tumor sites

Stage IV Any of the preceding stages with initial central nervous system or bone marrow involvement[b]

[a]*A distinction is made between apparently localized gastrointestinal tract lymphoma and more extensive intra-abdominal disease. Stage II disease typically is limited to a segment of the gut with or without the associated mesenteric nodes only, and the primary tumor can be completely removed grossly by segmental excision. Stage III disease typically exhibits spread to para-aortic and retroperitoneal areas by implants and plaques in mesentery or peritoneum or by direct infiltration of structures adjacent to the primary tumor. Ascites may be present, and complete resection of all gross tumor is not possible.*
[b]*If marrow involvement is present initially, the number of abnormal cells must be 25% or less in an otherwise normal marrow aspirate with normal peripheral blood picture.*

TABLE 22.3 Clinical Staging of B-Cell Lymphomas from LMB-96

Stage	Extent of tumor
A	Resected stage I and abdominal stage II
B	Multiple extra-abdominal sites; non-resected Stages I, II, III, and IV (CNS-, BM < 25%)
C	Intra-abdominal tumor Stage IV (CNS + or), BM >25% (Burkitt's leukemia)

BM, bone marrow; CNS, central nervous system.

American and British (FAB) childhood NHL risk stratification, used in the FAB Lymphoma Malignant B type (LMB)-96 (LMB) protocol, has also been developed for B-LL, BL, and DLBCL (Table 22.3), and incorporates standard staging along with the impact of dissemination to either the bone marrow or the CNS on prognosis.

PROGNOSIS

Children with Stage I to II disease have 2-year disease-free survival (DFS) rates of 85–98%. Children with Stage III disease have an 85–90% 2-year DFS. Those with Stage IV disease and marrow involvement have an 85–90% DFS, and those with CNS disease (BL) have an 80% EFS. Patients with PMBL have historically had an EFS of 60–70%. Newer treatment regimens for this subset of patients, however, may have an improve EFS. Deaths from second malignancies may occur after prolonged time intervals and modify survival curves.

MANAGEMENT

Emergency Treatment

The following clinical manifestations require immediate attention:

1. SVC syndrome secondary to a large mediastinal mass obstructing blood flow.
2. Respiratory airway compression.
3. Tumor lysis syndrome (TLS) secondary to severe metabolic abnormalities from massive lysis of tumor cells.

The management of these critical conditions is fully described in Chapter 32.

Chemotherapy

Because of the high likelihood of disseminated disease at presentation, all children with the high-grade sub-types of NHL, regardless of stage, receive combination chemotherapy. All four major histologic subtypes of childhood NHL respond to a wide range of drugs including steroids, anthracyclines, vinca alkaloids, topisome-rase 2 inhibitors, and alkylating agents, but different combinations and schedules are optimal for particular histologies. Tables 22.4 and 22.5 list some of the commonly used regimens and treatment outcomes. Historically, patients with localized Murphy Stages I and II lymphomas have a better prognosis than those with more

TABLE 22.4 Low-Stage/Low-Risk Patients

WHO	Regimen	EFS (%)
Lymphoblastic lymphoma, T and B	BFM-95; BFM-like	85–90
Burkitt	CHOP or COPAD (LMB-96)	90–98
Diffuse large B-cell	CHOP or COPAD (LMB-96)	90–98
Anaplastic large cell	CHOP	85–90
	BFM-NHL90	

BFM, Berlin–Frankfurt–Munster; NHL, non-Hodgkin lymphoma; EFS, event-free survival; LMB, Lymphoma Malignant B type.

TABLE 22.5 Advanced-Stage/High-Risk Patients

WHO	Regimen	EFS (%)
Lymphoblastic lymphoma, T and B	BFM-95	75–80
		50[a]
Burkitt	LMB-96	80–95
Diffuse large B-cell	LMB-96	80–95
Primary mediastinal large B-cell	LMB-96	60–70
	DA-EPOCH-R	>90[b]
Anaplastic large cell	APO	70
	ALCL 99	

[a]Patients deemed high risk by poor prednisone response.
[b]Phase II data in adults. Currently in pediatric trial.
BFM, Berlin–Frankfurt–Munster; EFS, event-free survival; LMB, Lymphoma Malignant B type; ALCL, anaplastic large-cell lymphoma.

extensive Stages III and IV. Stratified therapy, however, has resulted in excellent outcomes for both low- and high-stage disease.

Bone marrow transplantation is usually reserved for patients in second or subsequent remission. Patients undergoing transplantation early in their disease, after first relapse or second complete remission, have 2-year DFS approaching 50%, whereas those with refractory disease fare less well (5–20%).

In patients with an excellent prognosis for cure, potential late effects related to chemotherapy must be taken into consideration. Every effort has to be made to minimize impact on cardiac function, fertility, growth, development, and neurocognitive function (Chapter 34).

LYMPHOBLASTIC LYMPHOMA

Although there is evidence that this entity may be distinct from acute lymphoblastic leukemia (ALL) on a molecular level, current therapeutic strategies are based on ALL treatment, with approximately 85–90% DFS. The use of more intensive ALL chemotherapy regimens has also resulted in excellent outcomes for most children with advanced-stage disease (DFS of 80–85%). Similar to the case with ALL patients, there is evidence that poor initial response to treatment appears to confer a worse prognosis with EFS only in the 50% range. The least toxic and most effective method of CNS prophylaxis for these patients remains unclear. Both cranial irradiation and intrathecal (IT) methotrexate have been very effective but are associated with neurocognitive late effects. Data indicate that intensive IT methotrexate with or without systemic high-dose methotrexate is adequate therapy to prevent CNS relapse.

Currently in the United States, LL patients are eligible to be enrolled on ALL protocols through Children's Oncology Group (COG) institutions.

B LINEAGE NHL

For the purposes of treatment, high-grade mature B-cell lymphomas have been treated similarly. BL, in contrast to DLBCL, is much more likely to involve the bone marrow and CNS and much more likely to be complicated with a TLS.

Although histologically and clinically distinct, both BL and DLBCL respond well to treatments as outlined below. For low-stage BL as well as DLBCL, short-duration therapy with cyclophosphamide, doxorubicin, vincristine (Oncovin), and prednisone (CHOP); or COPAD result in excellent relapse-free survivals of 90−98% and 85−90%, respectively. For higher-stage and -risk groups, the LMB-96 protocol, a multiagent intensive chemotherapy regimen for B-cell lymphoma, has resulted in greater than 90% 4-year EFS. LMB-96 includes 3−5 months of intensive cycles of cyclophosphamide, vincristine, prednisone, high-dose methotrexate, doxorubicin, cytosine arabinoside (AraC), and etoposide.

In the setting of excellent cure rates, particularly in low-stage disease, potential late effects, comorbidities or specifics of a patient's clinical situation may direct a choice between acceptable regimens. A current effort to improve outcomes for high-risk BL and DLBCL builds on the experience of LMB-96 with the randomized addition of rituximab (anti-CD-20 targeted monoclonal antibody). The addition of rituximab to therapeutic regimens in adult mature B-cell lymphoma has been beneficial. How best to incorporate this agent in the treatment of pediatric mature B-cell lymphoma has not yet been well-established.

Primary Mediastinal Large B-Cell Lymphoma

As noted previously, when treated using regimens that have been successful for other mature B-cell lymphomas, patients with PMBL have stood out with worse prognosis, with significantly inferior EFS, <70% using LMB-96 therapy. There is increasing evidence that this disease represents a distinct entity requiring specific alternative therapy. Reports of improved outcomes with alternative regimens, most notably dose-adjusted EPOCH-R in adults with this disease, are promising (shown in Table 22.6). This approach is currently being evaluated in the pediatric age group as part of a collaborative study including both the European Intergroup for Children

TABLE 22.6 Dose Adjusted EPOCH-R for PMBL

Drug	Route	Dose	Day of cycle
Rituximab	IV	375 mg/m^2/dose	1
PredniSONE	PO	60 mg/m^2/dose BID (120 mg/m^2/day)	1−5
Etoposide	IV over 24 h	50 mg/m^2/day as initial dose	1−4
DOXOrubicin	IV over 24 h	10 mg/m^2/day as initial dose	1−4
VinCRIStine	IV over 24 h	0.4 mg/m^2/dose	1−4
Cyclophosphamide	IV bolus	750 mg/m^2/day as initial dose	5
G-CSF	SubQ	5 g/kg/dose	

DOSE ADJUSTMENTS

ANC nadir after previous course	Dose level for next course
0.5 × 10^9/l on all measurements	Increase 1 dose level above last course
0.5 × 10^9/l on 1 or 2 measurements	Same dose as last course
0.5 × 10^9/l on ≥3 measurements	Decrease 1 dose level below last course
Platelet nadir after previous course	**Dose level for next course**
25 × 10^9/l on ≥1 measurement	Decrease 1 dose level below last course

Repeat courses every 3 weeks (21 days) for a total of six courses.
If ANC ≥1000/μl and platelets ≥100,000/μl on day 21, begin next treatment course.
If ANC <1000/μl or platelets <100,000/μl on day 21, delay next treatment course up to 1 week. G-CSF is resumed until ANC >1000/μl and stopped 24 h before treatment. If counts still remain low after 1 week delay, reduce 1 dose level below last course.
CSF, cerebrospinal fluid; IV, intravenous; PO, by mouth.

Non-Hodgkin's Lymphoma and COG. This particular approach exposes patients to very high cumulative doses of anthracycline, albeit as continuous infusion. Particular attention to late cardiac effects will be of particular interest in pediatric patients treated with this regimen.

Anaplastic Large-Cell Lymphoma

ALCL consists of lymphoid cells that are usually large with abundant cytoplasm and pleomorphic, often horseshoe-shaped nuclei. The cells are CD30+, and the majority of pediatric cases are also positive for the anaplastic lymphoma kinase (ALK) protein and have the chromosomal translocation t(2;5)(p23;q35) involving the nucleophosmin and ALK genes.

For early-stage ALCL, CHOP is effective therapy. Excellent results have also been reported with NHL-Berlin–Frankfurt–Munster (BFM) 90 therapy. The APO regimen (Adriamycin, Prednisone, Vincristine) has also been effective for higher-stage ALCL in children (DFS 70–75%). Other regimens for ALCL have been modeled after B-cell protocols and report similar results.

Interestingly, vinblastine has been shown to have significant activity as a single agent in recurrent ALCL. Recent randomized clinical trials testing the addition of vinblastine to standard therapy have not shown significant impact on outcome. However, alternative strategies to integrate vinblastine into existing therapy may be more beneficial. In consideration of the unique molecular features of ALCL, agents that target ALK or CD30 are also of interest and are currently in clinical trials. In particular, trials involving brentuximab vedotin, an anti-CD30 antibody conjugated to a vinca alkaloid, and crizotinib, a small molecule inhibitor of ALK have shown significant promise.

Rare Pediatric NHL

Rare low-grade mature B-cell lymphomas, such as PFL, pediatric marginal zone lymphoma, and MALT lymphoma, occur in childhood. For the former two presenting with low-stage disease, many patients have done well with resection and observation only. PFL has been described with unique clinical and pathologic features compared with adult-type follicular lymphoma. For extranodal MALT lymphoma, treatment has consisted of resection followed by radiation therapy without adjuvant chemotherapy. Rare cutaneous lymphomas have also been described in the pediatric age group. Efforts to collect clinical information and pathologic specimens to further characterize these rare pediatric lymphomas and determine best approaches are in progress.

Radiation Therapy

The role of radiotherapy in the management of all childhood NHL has decreased as systemic chemotherapeutic regimens have become more effective. As the overall survival improved for children with all stages and subtypes of NHL, studies designed to assess the need for regional radiotherapy followed. This has evolved on a background of significant concerns for late effects of treatment, including neurocognitive effects of low-dose CNS radiation as well as the risk for second malignancy, cardiac and pulmonary toxicity in patients receiving radiation therapy to the mediastinum and neck combined with chemotherapy. Reports from the Pediatric Oncology Group demonstrated successful local and systemic control with chemotherapy alone for children with Murphy Stages I and II NHL regardless of histology. The addition of localized radiation therapy (RT) (e.g., mediastinal RT) to chemotherapy was also not shown to benefit patients with advanced-stage LL. In the past 10 years, several groups have reported that primary lymphoma of bone in children can be treated successfully with chemotherapy alone.

Prophylactic CNS treatment is needed in the majority of children with NHL in order to reduce the high risk of CNS relapse. The approach for all non-lymphoblastic histologies includes systemic and IT chemotherapy. It has become evident that there is no benefit to adding radiation therapy for CNS prophylaxis in BL. The current CNS preventive regimens for advanced-stage LL include either cranial irradiation and IT chemotherapy or IT chemotherapy and systemic chemotherapy only. The BFM group safely reduced the dosage of cranial radiation therapy to 12 Gy for LL and eliminated it entirely from BFM-95 without increasing the CNS relapse rate. Cranial irradiation (12–18 Gy) is still warranted for the rare patient with LL with initial CNS disease or a CNS relapse. Patients with cranial nerve palsies at diagnosis, independent of CSF findings, are treated in a similar fashion as children with initial CNS lymphoma.

The current indications for radiation therapy to involved sites of disease outside the CNS in NHL have become somewhat limited. Involved field radiation is considered for patients who do not achieve a complete remission after induction chemotherapy. This situation is relatively rare, but more common in PMBL. Given the increased rate of local failure compared to other pediatric NHL, the role of involved field radiation remains a topic of some debate.

Limited data are available as to the best therapy for patients with rare, low-grade lymphoma. Historically, extranodal MALT in pediatrics, such as of the ocular adenexa, is treated with low-dose superficial radiation therapy. No role for radiation therapy in low-stage PFL or pediatric marginal zone lymphoma has been established.

Involved field radiation may also be considered for palliation of pain or mass effect, and consolidation to regions of local disease before or after bone marrow transplantation in patients with recurrent disease. Care should be taken to use modest fractionation schedules and to avoid exceeding normal tissue tolerance dosages because patients may be eligible for subsequent bone marrow transplantation necessitating total body irradiation (TBI) as a component of the preparative regimen.

Children with refractory or relapsed NHL can experience prolonged DFS after treatment with high-dose chemotherapy or high-dose chemoradiotherapy followed by autologous or allogeneic bone marrow transplant. When TBI is a component of the preparatory regimen, fractionated courses to total dosages of 12–14 Gy are common. Disease recurrence rather than the morbidity of transplant is the predominant cause of failure in this setting. The strategy of irradiating the local sites of initial disease recurrence before or after transplant, whether or not TBI is used, has proved to be effective in adults and should be considered. Dosages to these sites (usually at least 20 Gy) are constrained by normal tissue tolerances.

Historically, emergency radiation therapy for SVC syndrome, acute airway compromise, or spinal cord compression has been used to provide rapid symptom relief. The response is particularly dramatic with LL and symptoms are usually improved within 48 h of treatment. Usually 1.5–2 Gy per fraction for a total dosage of 6–7.5 Gy is adequate to relieve symptoms. Hyperfractionated regimens (1.2–1.5 Gy per fraction twice a day for a total dosage of 6–10 Gy) can also be used. More recently, systemic steroids are more commonly used and equally effective, especially in patients with large mediastinal masses. Emergency radiotherapy or treatment with steroids may be appropriate in the absence of a histologic diagnosis. Despite the rapid response with either, it does not significantly compromise a tissue diagnosis if sampling occurs within 48 h of treatment.

Surgical Therapy

Although surgical biopsy is critical in the original diagnosis, as well as for evaluation of residual masses and response to therapy, the role of surgery in NHL as a treatment is limited. It should be performed on patients in whom there is good reason to believe that total resection can be achieved, such as in limited abdominal disease. Patients with widespread high-grade lymphoma are not eligible for surgical resection.

In patients with rare low-grade lymphoma surgical resection may eliminate the need for systemic therapy and surgical resection may be considered optimal therapy.

Management of Relapse

Most relapses occur within 12 months of diagnosis for BLs and within the first 2–4 years for most other histologic subtypes. Late relapse has been reported with ALCL and B-LL.

Relapse indicates a poor prognosis, regardless of the site of relapse, tumor histology, other original prognostic factors, prior therapy, or time from diagnosis to relapse. For this reason, the selection of most effective front-line treatment is critical. Relapsed patients are first treated to induce remission. After induction of complete or a very good partial remission is achieved, consolidation with stem cell transplantation is usually indicated. Patients who are chemotherapy-resistant are unlikely to be cured using autologous stem cell transplantation.

Patients with ALCL and DLBCL are more likely to have better outcomes after autologous hematopoietic stem cell transplant compared to those with lymphoblastic and BLs. Targeted therapies such as rituximab, brentuximab, and crizotinib, in combination with chemotherapy, are showing promising efficacy in the management of relapsed NHL.

In recent years, outcomes for patients with NHL have improved dramatically. Optimal combinations of agents, intensity, and duration of therapy have evolved for disease groups. In the future we anticipate further stratification of treatment regimens and the addition of disease-specific agents, such as rituximab, in mature B-cell lymphomas or agents targeting ALK and CD30 in ALCL. Nelarabine appears to be specifically effective in T-cell LL, and data indicate that vinblastine may be particularly effective, even as a single agent, in ALCL. How best to incorporate these agents into existing regimens will be a focus of future efforts. Further advances in treatment with consideration of late effects will likely be a focus of future clinical trials.

Further Reading and References

Alexander, S., Kraveka, J.M., Weitzman, S., Lowe, E., Smith, L., Lynch, J.C., et al., 2014. Advanced stage anaplastic large cell lymphoma in children and adolescents: results of ANHL0131, a randomized phase III trial of APO versus a modified regimen with vinblastine: a report from the chidlren's oncology group. Pediatr. Blood Cancer 61 (12), 2236−42.

Attarbaschi, A., Beishuizen, A., Mann, G., Rosolen, A., Mori, T., Uyttebroeck, A., et al., European Intergroup for Childhood Non-Hodgkin Lymphoma (EICNHL) and the international Berlin-Frankfurt-Münster (i-BFM) Study Group, 2013. Children and adolescents with follicular lymphoma have an excellent prognosis with either limited chemotherapy or with a "watch and wait" strategy after complete resection. Ann. Hematol. 92, 1537−1541.

Bluhm, E.C., Ronckers, C., Hayashi, R.J., Neglia, J.P., Mertens, A.C., Stovall, M., et al., 2008. Cause-specific mortality and second cancer incidence after non-Hodgkin lymphoma: a report from the Childhood Cancer Survivor Study. Blood 111, 4014−4021.

Bollard, C.M., Lim, M.S., Gross, T.G.; COG Non-Hodgkin Lymphoma Committee, 2013. Review Children's Oncology Group's 2013 blueprint for research: non-Hodgkin Lymphoma. Pediatr. Blood Cancer 60, 979−984.

Bradley, M.B., Cairo, M.S., 2008. Stem cell transplantation for pediatric lymphoma: past present and future. Bone Marrow Transplant. 41 (2), 149−158.

Cairo, M.S., Gerrard, M., Sposto, R., Auperin, A., Pinkerton, C.R., Michon, J., et al., FAB LMB96 International Study Committee, 2007. Results of a randomized international study of high-risk central nervous system B non-Hodgkin Lymphoma and B acute lymphoblastic leukemia in children and adolescents. Blood 109, 2736−2743.

Dunleavy, K., Pittaluga, S., Maeda, L.S., Advani, R., Chen, C.C., Hessler, J., et al., 2013. Dose-adjusted EPOCH-rituximab therapy in primary mediastinal B-cell lymphoma. N. Engl. J. Med. 368, 1408−1416.

Foyil, K.V., Bartlett, N.L., 2012. Brentuximab vedotin and crizotinib in anaplastic large-cell lymphoma. Cancer J. 18, 450−456.

Garner, R., Li, Y., Gray, B., Zori, R., Braylan, R., Wall, J., et al., 2009. Longer-term disease control of refractory anaplastic large cell lymphoma with vinblastine. J. Pediatr. Hematol. Oncol. 31, 145−147.

Gerrard, M., Waxman, I.M., Sposto, R., Auperin, A., Perkins, S.L., Goldman, S., et al., French-American-British/Lymphome Malins de Burkitt 96 (FAB/LMB 96) International Study Committee, 2013. Outcome and pathologic classification of children and adolescents with mediastinal large B-cell lymphoma treated with FAB/LMB96 mature B-NHL therapy. Blood 121, 278−285.

Jaffe, E.S., Harris, N.L., Stein, H., Isaacson, P.G., 2008. Classification of lymphoid neoplasms: the microscope as a tool for disease discovery. Blood 112, 4384−4399.

Kluge, R., Kurch, L., Montravers, F., Mauz-Körholz, C., 2013. FDG PET/CT in children and adolescents with lymphoma. Pediatr. Radiol. 43, 406−417.

Laver, J.H., Mahmoud, H., Pick, T.E., Hutchison, R.E., Weinstein, H.J., Schwenn, M., et al., 2002. Results of a randomized phase III trial in children and adolescents with advanced stage diffuse large cell non-Hodgkin's lymphoma: a Pediatric Oncology Group study. Leuk. Lymphoma 43, 105−109.

Le Deley, M.C., Rosolen, A., Williams, D.M., Horibe, K., Wrobel, G., Attarbaschi, A., et al., 2010. Vinblastine in children and adolescents with high-risk anaplastic large-cell lymphoma: results of the randomized ALCL99-Vinblastine trial. JCO 28 (25), 3987−3993.

Link, M.P., Donaldson, S.S., Berard, C.W., Shuster, J.J., Murphy, S.B., 1990. Results of treatment of childhood localized non-Hodgkin's lymphoma with combination chemotherapy with or without radiotherapy. N. Engl. J. Med. 322, 1169−1174.

Lones, M.A., Perkins, S.L., Sposto, R., Tedeschi, N., Kadin, M.E., Kjeldsberg, C.R., et al., 2002. Non-Hodgkin's lymphoma arising in bone in children and adolescents is associated with an excellent outcome: a Children's Cancer Group report. J. Clin. Oncol. 20, 2293−2301.

Louissant Jr, A., Ackerman, A.M., Dias-Santagata, D., Ferry, J.A., Hochberg, E.P., Huang, M.S., et al., 2012. Pediatric-type nodal follicular lymphoma: an indolent clonal proliferation in children and adults with high proliferation index and no BCL2 rearrangement. Blood 120 (12), 2395−2404.

Patte, C., Auperin, A., Gerrard, M., Michon, J., Pinkerton, R., Sposto, R., et al., FAB/LMB96 International Study Committee, 2007. Results of the randomized internation FAB/LMB96 trial for intermediate risk B-cell non-Hodgkin's lymphoma in children and adolescents: it is possible to reduce treatment for the early responding patients. Blood 109, 2773−2780.

Shamberger, R.C., Holzman, R.S., Griscom, N.T., Tarbell, N.J., Weinstein, H.J., Wohl, M.E., et al., 1995. Prospective evaluation by computed tomography and pulmonary function tests of children with mediastinal masses. Surgery 118, 468−471.

Swerdlow, S.H., Campo, E., Harris, N.L., et al., 2008. WHO classification of tumours for haematopoietic and lymphoid tissues, fourth ed.

Termuhlen, A.M., Smith, L.M., Perkins, S.L., Lones, M., Finlay, J.L., Weinstein, H., et al., 2012. Outcome of newly diagnosed children and adolescents with localized lymphoblastic lymphoma treated on Children's Oncology Group trial A5971: a report from the Children's Oncology Group. Pediatr. Blood Cancer 59 (7), 1229−1233.

Termuhlen, A.M., Smith, L.M., Perkins, S.L., Lones, M., Finlay, J.L., Weinstein, H., et al., 2013. Disseminated lymphoblastic lymphoma in children and adolescents: results of the COG A5971 trial: a report from the Children's Oncology Group. Br. J. Haematol. 162, 792−801.

Ward, E., DeSantis, C., Robbins, A., Kohler, B., Jemal, A., 2014. Childhood and adolescent cancer statistics, 2014. Cancer J. Clin. 64 (2), 83−103.

23

Central Nervous System Malignancies

Derek R. Hanson and Mark P. Atlas

EPIDEMIOLOGY

Tumors of the central nervous system (CNS) are the second most common group of cancers in childhood, accounting for 20% of all childhood malignancies, and affecting 35 per million children. A total of 60–70% of childhood CNS tumors arise from glial cells and tend not to metastasize outside the CNS. The relative frequency of specific CNS tumor types can be seen in Table 23.1.

PATHOLOGY

Supratentorial Lesions

1. Cerebral hemisphere: low- and high-grade glioma, ependymoma, meningioma, primitive neuroectodermal tumor (PNET).
2. Sella or chiasm: craniopharyngioma, pituitary adenoma, germ cell tumors, optic nerve glioma.
3. Pineal region: Pineoblastoma, pineocytoma, germ cell tumors, astrocytoma.

Infratentorial Lesions

1. Posterior fossa: medulloblastoma, glioma (low more frequent than high-grade), ependymoma, meningioma.
2. Brain stem tumors: low- and high-grade glioma, PNET.

Ventricular Lesions

1. Choroid plexus papilloma, choroid plexus carcinoma, neurocytoma.

   The most useful pathologic classification for brain tumors is based on embryonic derivation and histologic cell of origin. Tumors are further classified by grading the degree of malignancy within a particular tumor type. This grading is useful in astrocytoma and ependymoma. Criteria that are useful microscopically in grading the degree of malignancy include:
   - Cellular pleomorphism.
   - Mitotic index.
   - Anaplasia and necrosis.
   - MIB-1 index.

Molecular Pathology of CNS Neoplasms

The majority of CNS neoplasms are sporadic. Only a small percentage is associated with inherited genetic disorders.

Chromosomal abnormalities have been identified in many pediatric brain tumors. These chromosomal abnormalities can be helpful in determining the pathologic classification of certain tumors. For example, a mutation or deletion of the tumor suppressor gene *INI1* is often used to help distinguish atypical teratoid rhabdoid tumors from PNET or medulloblastoma. Discoveries regarding the molecular variation among tumors of the same

TABLE 23.1   Relative Frequency of Common Childhood Brain Tumors

Location and histology	Frequency (%)
Supratentorial	16
Low-grade glioma	3
Ependymoma	5
High-grade glioma	3
Primitive neuroectodermal	2
Tumor (primitive neuroectodermal tumor)	
Other	
Sella/chiasm	
Craniopharyngioma	6
Optic nerve glioma	5
Total	40
Infratentorial	
Cerebellum	
Medulloblastoma	20
Astrocytoma	18
Ependymoma	6
Brain stem	
Glioma	14
Other	2
Total	60

TABLE 23.2   Cytogenetic Loci Implicated in Malignant Brain Tumors

Tumor	Chromosome	Disorder
PNET/MB	5q21-22	Turcot syndrome[a]
	9q22.3	NBCCS
	17p13.3	
	17q	
Astrocytoma grades III–IV	7p12	
	17p13.1	Li–Fraumeni syndrome
	17q11.2	NF-1
	3p21, 7p22	Turcot syndrome[a]
Subependymal giant cell tumor	9q34	Tuberous sclerosis
	16p13	
Meningioma/ependymoma	22q12	NF-2
AT/RT	22q11.2	
Hemangioblastoma	3p25	von Hippel–Lindau

[a]A rare and distinctive form of multiple intestinal polyposis associated with brain tumors. Autosomal-recessive inheritance caused by mutation in one of the mismatch repair genes MLH1 or chromosome 3p, MPS2 on chromosome 7p or the adenomatous polyposis gene on chromosome 5q.
PNET/MB, primitive neuroectodermal tumor, medulloblastoma; AT/RT, atypical teratoid rhabdoid; NBCCS, nevoid basal cell carcinoma syndrome; NF-1, neurofibromatosis-1; NF-2, neurofibromatosis-2.
Adapted from Biegel (1999).

pathologic classification have caused a major shift in the treatment of these tumors, as targeted therapies and unique approaches to these molecular subtypes are being explored.

Table 23.2 describes common cytogenetic abnormalities identified in brain tumors.

CLINICAL MANIFESTATIONS

Intracranial Tumors

Symptoms and signs are related to the location, size, and growth rate of tumor:

- Slow-growing tumors produce massive shifts of normal structures and may become quite large by the time they first become symptomatic.
- Rapidly growing tumors produce symptoms early and present when they are relatively small.

The most common presenting signs and symptoms of an intracranial neoplasm are increased intracranial pressure (ICP) and focal neurologic deficits.

If symptoms and signs of increased ICP precede the onset of localized neurologic dysfunction, a tumor of the ventricles or deep midline structures is most likely. If localizing signs (seizures, ataxia, visual field defects, cranial neuropathies, or corticospinal tract dysfunction) are predominant in the absence of increased ICP it is more probable that the tumor originates in parenchymal structures (cerebral hemispheres, brain stem, or cerebellum).

General Signs and Symptoms of Intracranial Tumors

1. Headache: in young children headache can present as irritability; often worse in the morning, improving throughout the day.
2. Vomiting.
3. Disturbances of gait and balance.
4. Hemiparesis.
5. Cranial nerve abnormalities.
6. Impaired vision.
   a. Diplopia (sixth nerve palsy): in young children diplopia may present as frequent blinking or intermittent strabismus.
   b. Papilledema from increased ICP may present as intermittent blurred vision.
   c. Parinaud syndrome (failure of upward gaze and setting-sun sign, large pupils, and decreased constriction to light).
7. Mental disturbances: somnolence, irritability, personality or behavioral change, or change in school performance.
8. Seizures: usually focal.
9. Endocrine abnormalities: midline supratentorial tumors may cause endocrine abnormalities due to effects on the hypothalamus or pituitary and visual field disturbances as a result of optic pathway involvement.
10. Cranial enlargement in infants: characteristic of increased ICP.
11. Diencephalic syndrome: seen in patients aged 6 months to 3 years with brain tumors who present with sudden failure to thrive and emaciation. The syndrome is caused by a hypothalamic tumor in the anterior portion of the hypothalamus or the anterior floor of the third ventricle.

Spinal Tumors

Spinal tumors of children may be found anywhere along the vertebral column. They cause symptoms by compression of the contents of the spinal canal. Localized back pain in a child or adolescent should raise suspicion of a spinal tumor, especially if the back pain is worse in the recumbent position and relieved when sitting up. The major signs and symptoms of spinal tumors are listed in Table 23.3. Most spinal tumors have associated muscle weakness and the muscle group affected corresponds to the spinal level of the lesions.

Spinal tumors can be divided into three distinct groups:

- *Intramedullary*: these tumors tend to be glial in origin and are usually gliomas or ependymomas.
- *Extramedullary—intradural*: these tumors are likely to be neurofibromas often associated with neurofibromatosis. If they arise in adolescent females, meningiomas are more likely.
- *Extramedullary—extradural*: these tumors are most often of mesenchymal origin and may be due to direct extension of a neuroblastoma through the intervertebral foramina or due to a lymphoma. Tumors of the vertebra may also encroach upon the spinal cord, leading to epidural compression of the cord and paraplegia (e.g., PNET or Langerhans cell histiocytosis occurring in a thoracic or cervical vertebral body).

TABLE 23.3   Major Signs and Symptoms of Spinal Tumors

Back pain (in 50%, increased in supine position or with Valsalva maneuver)

Resistance to trunk flexion

Paraspinal muscle spasm

Spinal deformity (especially progressive scoliosis)

Gait disturbance

Weakness, flaccid or spastic

Reflex changes (especially decreased in arms and increased in legs)

Sensory impairment below level of tumor (30% of cases)

Decreased perspiration below level of tumor

Extensor plantar responses (Babinski signs)

Sphincter impairment (urinary or anal)

Midline closure defects of skin or vertebral arches

Nystagmus (with lesions of upper cervical cord)

TABLE 23.4   Genetic Syndromes Associated with Pediatric Brain Tumors

Syndrome	Chromosome	Protein	Inheritance	Brain tumor
Neurofibromatosis -1	17q11.2	Neurofibromin	Autosomal dominant	Optic glioma
Neurofibromatosis-2	22q12.2	Merlin	Autosomal dominant	Acoustic neuroma
Tuberous sclerosis	9q34	Hamartin	Autosomal dominant	Subependymal giant cell astrocytomas
	16p13.3	Tuberin	Autosomal dominant	
Li−Fraumeni	17p13.1	Tp53	Autosomal dominant	Choroid plexus carcinoma, also glioblastoma multiforme
VHL	3p25.3	VHL	Autosomal dominant	Hemangioblastoma
Turcot	5q21-22	APC	Autosomal dominant	Medulloblastoma
	3p21, 7p22		Autosomal recessive	Glioblastoma multiforme
Gorlin	9q22	PTCH1	Autosomal dominant	Medulloblastoma

GENETIC SYNDROMES ASSOCIATED WITH PEDIATRIC BRAIN TUMORS

There are a number of genetic syndromes associated with pediatric CNS tumors (Table 23.4).

Neurofibromatosis Type 1

Neurofibromatosis type 1 (NF-1) is an autosomal-dominant disorder with a worldwide incidence of 1 in 2500−3000 individuals. There is a high incidence of *de novo* mutations due to the large size of the NF-1 gene, which is 350,000 base pairs in length. The disorder is caused by a mutation on chromosome 17q11.2, which encodes for the neurofibromin protein, a negative regulator of Ras. Individuals with the condition are susceptible to the development of benign and malignant tumors. Other than neurofibromas, the most common tumors associated with this disorder are optic pathway gliomas. NF-1 is characterized by extreme clinical variability, even within a single family. Diagnosis is made when a patient exhibits two of the following seven criteria: six or more café-au-lait spots (>5 mm in prepubertal, >15 mm in postpubertal), two or more neurofibromas or one plexiform neurofibroma, axillary/inguinal freckling, optic pathway glioma, Lisch nodules, osseous lesions (kyphoscoliosis, sphenoid dysplasia, thinning of long bone cortex), and a first-degree relative with NF-1 by criteria.

Neurofibromatosis Type 2

Neurofibromatosis type 2 (NF-2) is an autosomal-dominant disorder caused by a mutation in the merlin gene located on chromosome 22q12.2. The syndrome is characterized by bilateral acoustic neuromas, meningiomas of the brain, spinal cord ependymoma, and schwannomas of the dorsal roots of the spinal cord.

Li—Fraumeni Syndrome

Li—Fraumeni syndrome is an autosomal-dominant disorder characterized by the early onset of tumors, multiple tumors within an individual, and multiple affected family members. The syndrome is caused by a mutation in the *p53* gene located on chromosome 17p13.1. A variety of tumor types are seen throughout families affected with this disorder, most commonly sarcomas, breast cancer, leukemia, brain tumors, and adrenocortical carcinoma. In pediatric patients choroid plexus carcinoma has been found to have a strong association with Li—Fraumeni. Glioblastoma multiforme is also a CNS tumor frequently seen in children with this disorder.

Von Hippel—Lindau

von Hippel—Lindau disease is a dominantly inherited familial cancer syndrome caused by a heterozygous mutation in the VHL gene found on chromosome 3p25.3. Affected patients are predisposed to a variety of malignant and benign neoplasms. The most frequent tumors seen are retinal, cerebellar, and spinal hemangioblastomas. Other tumors commonly seen in the condition are renal cell carcinoma, pheochromocytoma, and pancreatic tumors.

Tuberous Sclerosis

Tuberous sclerosis complex is an autosomal-dominant disorder caused by a mutation in either the *TSC1* gene on chromosome 9q34 or the *TSC2* gene on chromosome 16p13.3. Either of these mutations results in an overexpression of the mTOR complex 1. The syndrome is characterized by hamartomas in multiple organ systems including the brain, skin, heart, and kidneys. Patients are predisposed to developing subependymal giant cell astrocytomas. These tumors have been found to respond well to the mTOR inhibitor everolimus. Additional CNS manifestations include the formation of cortical tubers and learning difficulties. Skin findings include shagreen patches, ash leaf spots, and facial angiofibromas. Other tumors associated with the syndrome include angiomyolipomas of the kidney and rhabdomyomas of the heart.

Turcot Syndrome

Turcot syndrome is a disorder characterized by colorectal polyposis and primary CNS tumors. There are thought to be two distinct groups of patients with Turcot syndrome. One group of patients has an autosomal-recessive pattern with early onset of malignant gliomas and colorectal adenomas. These patients typically have mutations in DNA mismatch repair genes. The second group of patients is those with familial adenomatous polyposis who have autosomal-dominant inheritance of mutations in the *APC* gene. The brain tumor commonly associated with this group is medulloblastoma.

Gorlin Syndrome

Gorlin syndrome is an autosomal-dominant disorder most commonly involving a mutation in the *PTCH1* gene on chromosome 9q22. The disorder is characterized by the development of both basal cell carcinomas and medulloblastomas. Other clinical findings include odontogenic keratocysts, abnormal facies, intracranial calcifications, and rib, vertebrae, and other skeletal abnormalities.

DIAGNOSTIC EVALUATION

Computed Tomography

The computed tomography (CT) scan is an important procedure in the detection of CNS malignancies. Scans performed both with and without iodinated contrast agents detect 95% of brain tumors. However, tumors of the posterior fossa, which are common in children, are better evaluated with magnetic resonance imaging (MRI). CT scans should be performed using thin sections (usually 5 mm). Sedation is often necessary to avoid motion artifacts. CT is more useful than MRI in:

- Evaluating bony lesions.
- Detection of calcification in tumor.
- Investigating unstable patients because of the shorter imaging time.

Magnetic Resonance Imaging

MRI provides the following additional advantages:

- No ionizing radiation exposure (especially important in multiple follow-up examinations).
- Greater sensitivity in detection of brain tumors, especially in the temporal lobe and posterior fossa (these lesions are obscured by bony artifact on CT).
- Ability to directly image in multiple planes (multiplanar), which is of value to neurosurgical planning (CT is usually only in axial planes, though reconstructions may be performed).
- Ability to apply different pulse sequences which is useful in depicting anatomy (T1-weighted images) and pathology (T2-weighted images).
- Ability to map motor areas with functional MRI.

MRI specificity is enhanced with the contrast agent gadolinium diethylenetriaminepentaacetic acid dimeglumine (Gd-DTPA), which should be used in the evaluation of childhood CNS tumors and has the following advantages:

- Highlights areas of blood—brain barrier breakdown that occur in tumors.
- Useful in identifying areas of tumor within an area of surrounding edema.
- Improves the delineation of cystic from solid tumor elements.
- Helps to differentiate residual tumor from gliosis (scarring).

The major difficulty with MRI in infants and children is the long time required to complete imaging and for this reason adequate sedation is required.

Magnetic Resonance Angiography

Magnetic resonance angiography has been utilized in the preoperative evaluation of the normal anatomic vasculature (e.g., dural sinus occlusion) but has not been particularly useful in the assessment of tumor vascularity.

Magnetic Resonance Spectroscopy

Magnetic resonance spectroscopy may be helpful in both diagnosing pediatric brain tumors and during follow-up investigations. This technique has been shown to be able to distinguish between malignant tumors and areas of necrosis by comparing creatine/choline ratios with n-acetyl aspartate/choline ratios. This technique, in combination with tumor characteristics as identified by MRI, tumor site, and other patient characteristics, may be able to more accurately predict the tumor type preoperatively. In addition, it may be helpful in identifying postoperative residual tumor from postoperative changes.

Positron Emission Tomography

Positron emission tomography (PET) is a potentially useful technique for evaluating CNS tumors. The fluorine-18-labeled analog of 2-deoxyglucose is used to image the metabolic differences between normal and malignant cells. Astrocytomas and oligodendrogliomas are generally hypometabolic, whereas anaplastic astrocytomas and glioblastoma multiforme are hypermetabolic.

PET is useful to determine:

- Degree of malignancy of a tumor.
- Prognosis of brain tumor patients.
- Appropriate biopsy site in patients with multiple lesions, large homogeneous and heterogeneous lesions.
- Recurrent tumor from necrosis, scar, and edema in patients who have undergone radiation therapy and chemotherapy.
- Recurrent tumor from postsurgical change.

The utility of recently developed PET-MRI scanning requires evaluation, though preliminary data demonstrate utility in astrocytomas.

Evaluation of the Spinal Cord

MRI and Gd-DTPA has replaced myelography in the evaluation of meningeal spread of brain tumors in the spinal column and delineating spinal cord tumors.

Cerebrospinal Fluid Examination

The cerebrospinal fluid (CSF) should have the following studies performed:

- Cell count with cytocentrifuge slide examination for cytology of tumor cells.
- Glucose and protein content.
- CSF α-fetoprotein (AFP).
- CSF human chorionic gonadotropin (β-hCG).

Polyamine assays in the CSF are of value in the evaluation of tumors that are in close proximity to the circulating CSF (medulloblastoma, ependymoma, brain stem glioma). The assay is not useful in glioblastoma multiforme and not predictive in astrocytomas. AFP and β-hCG of the CSF may be elevated in CNS germ cell tumors.

Bone Marrow Aspiration and Bone Scan

These studies are indicated in medulloblastoma and high-grade ependymomas with evidence of cytopenias on the blood count because a small percentage of these patients have systemic metastases at the time of diagnosis.

TREATMENT

Surgery

The purpose of neurosurgical intervention is threefold.

1. To provide a tissue biopsy for purposes of histopathology and cytogenetics.
2. To attain maximum tumor removal with the fewest neurologic sequelae.
3. To relieve associated increased ICP due to CSF obstruction.

Use of preoperative dexamethasone can significantly decrease peritumoral edema, thus decreasing focal symptoms and often eliminating the need for emergency surgery. For patients with increased hydrocephalus that is moderate to severe, endoscopic or standard ventriculostomy can decrease ICP. Tumor resection is safer when performed 1–2 days following reduction in edema and ICP by these means.

Technical advances in neurosurgery that have improved the success of surgery include ultrasonic aspirators; image guidance (allows three-dimensional (3D) mapping of tumor); functional mapping and electrocorticography, which allow pre- and intraoperative differentiation of normal and tumor tissue; and neuroendoscopy. Intraoperative MRI scanning is useful in limited circumstances, when visualization by the neurosurgeon is compromised. Stereotactic biopsies allow biopsy of deep-seated midline tumors. The ideal goal is a gross-total resection of tumor (likely leaving microscopic residual) as removal of a margin of normal tissue would cause devastating neurologic sequelae.

Radiotherapy

Most patients with high-grade brain tumors require radiotherapy to achieve local control of microscopic or macroscopic residual. Radiation therapy for intracranial tumors consists of external beam irradiation using conventional fields or 3D-conformal radiotherapy. The latter decreases radiation to normal brain tissue by up to 30%.

Intensity-modulated radiation therapy (IMRT) uses more complex computerized planning and intensity modulation of the radiation to further decrease radiation to normal tissue under certain circumstances. Proton beam radiation virtually eliminates exit dose (the radiation dose that passes through the tumor and impacts normal tissues beyond), sparing some tissue from unwanted radiation.

It is particularly useful in tumor adjacent to sensitive areas such as the pituitary gland and spinal cord. Proton beam radiation is only available at relatively few centers nationwide.

The wider use of ionizing radiation in pediatric brain tumors has resulted in improved long-term survival. However, the significant long-term effects on cognition and growth, especially in patients requiring craniospinal irradiation (e.g., PNET) can be devastating.

The total dose of radiotherapy depends on:

- Tumor type (which also influences volume of treatment).
- Age of the child.
- Volume of the brain or spinal cord to be treated.

Current efforts seek to decrease radiation by conforming better to the tumor and by using chemotherapy in addition. Children 3 years of age are most drastically affected. In this group newer strategies that avoid or delay radiation therapy, by initial treatment with chemotherapy, have promising preliminary results. In medulloblastoma therapy, this approach allows up to 50% cure without ionizing radiation.

Brachytherapy, stereotactic radiosurgery, and fractionated stereotactic radiosurgery are alternatives to conventional radiation therapy presently under continuing study and may prove useful in relapsed patients.

Chemotherapy

Chemotherapy plays an important role in the management of recurrent disease and in many newly diagnosed patients.

Two anatomic features of the CNS make it unique with respect to the delivery of chemotherapeutic agents.

1. The tight junction of the endothelial cells of the cerebral capillaries—the blood—brain barrier.
2. The ventricular and subarachnoid CSF.

The blood—brain barrier inhibits the equilibration of large polar lipid-insoluble compounds between the blood and brain tissue, while small nonpolar lipid-soluble drugs rapidly equilibrate across the blood—brain barrier. The blood—brain barrier is probably not crucial in determining the efficacy of a particular chemotherapeutic agent, since in many brain tumors the normal blood—brain barrier is impaired. Factors such as tumor heterogeneity, cell kinetics and drug administration, distribution and excretion play a more significant role in determining the chemotherapeutic sensitivity of a particular tumor than the blood—brain barrier. Tumors with a low mitotic index and small growth fraction are less sensitive to chemotherapy; tumors with a high mitotic index and larger growth fraction are more sensitive to chemotherapy.

The CSF circulates over a large surface area of the brain and provides an alternate route of drug delivery; it can function as a reservoir for intrathecal administration or as a sink after systemic administration of chemotherapeutic agents. The rationale of instillation of chemotherapy into the CSF compartment is that significantly higher drug concentrations can be attained in the CSF and surrounding brain tissue. This mode of administration is most applicable in cases of meningeal spread or in those tumors in which the risk of spread through the CSF is high.

Adjuvant chemotherapy is used in select primary brain tumors in addition to recurrent disease. Tables 23.5, 23.6, 23.9, 23.10 and 23.12 list chemotherapy regimens commonly employed in brain tumors. In certain cases chemotherapy allows for decreased radiation doses with equal or improved cure rates. In others adjuvant chemotherapy improves outcome with standard radiation therapy. Trials of new agents, combinations of agents, and standard drugs as radio-sensitizing agents are ongoing.

TABLE 23.5 Chemotherapy Regimens for High-Grade Astrocytoma

Regimen	Dosage/Route	Frequency
CCNU	100 mg/m² PO, day 1	
Vincristine[a]	1.5 mg/m² IV, days 1, 8	Every 6 weeks for 8 cycles
Prednisone	40 mg/m² PO, days 1–14	
	OR	
CCNU	100 mg/m² PO, day 1	
Vincristine[a]	1.5 mg/m² IV, days 1, 8	Every 6 weeks for 8 cycles
Procarbazine	100 mg/m² PO, days 1–14	
	OR	
CCNU	75 mg/m² PO, day 1	
Vincristine[a]	1.5 mg/m² IV, days 1, 8, 14	Every 6 weeks for 8 cycles
Cisplatin	75 mg/m² IV, day 1 over 6 h	
	OR	
Temozolomide	90 mg/m²/day PO for 42 days	During radiation therapy
Temozolomide	200 mg/m²/day PO for 5 days	After radiation therapy every 4 weeks for 10 cycles
	OR	
Methotrexate	400 mg/kg IV day 1	
Vincristine[a]	0.05 mg/kg IV, days 1, 8, 15	
Cyclophosphamide	65 mg/kg IV days 5, 6	
Cisplatinum	3.5 mg/kg IV over 6 h day 4	
Etoposide	6.5 mg/kg IV over 1 h days 5, 6	

[a]*Maximum dose 2 mg.*
PO, per orum (by mouth); IV, intravenous.

ASTROCYTOMAS

Astrocytomas account for approximately 50% of the CNS tumors with peaks between ages 5–6 and 12–13 years. The WHO grades these tumors I–IV in order of increasing malignancy. Low-grade tumors (WHO grades I and II) are distinguished histologically from high-grade astrocytomas (WHO grades III and IV) by the absence of cellular pleomorphism, high cell density, mitotic activity, and necrosis. The following are the histologic subtypes:

- *Pilocytic astrocytoma* has a fibrillary background, rare mitoses, and classically, Rosenthal fibers. It usually behaves in a benign fashion. These tumors are well circumscribed and grow slowly (WHO grade I).
- *Diffuse or fibrillary astrocytoma* is more cellular and infiltrative and more likely to undergo anaplastic change (WHO grade II).
- *Anaplastic astrocytoma* is highly cellular with significant cellular atypia. It is locally invasive and aggressive (WHO grade III).
- *Glioblastoma multiforme* demonstrates increased nuclear anaplasia, pseudopalisading, and multinucleate giant cells (WHO grade IV).

Pleomorphic xanthoastrocytomas are usually classified as WHO grade II subtype, but often behave more aggressively.

Pilomyxoid astrocytomas have variable predictability and may present with diffuse disease. They generally respond as WHO grade II tumors.

The majority of cerebellar astrocytomas remain confined to the cerebellum. Very rarely do they have neuraxis dissemination.

Management of Astrocytomas

Low-Grade Astrocytomas (WHO Grades I and II)

Low-grade astrocytomas present with hydrocephalus, focal signs, or seizures.

Surgery

Surgical excision is the initial treatment. Gross-total resection is desirable. Pilocytic astrocytomas are slow-growing and well-circumscribed with a distinct margin. These features permit complete resection in 90% of patients with posterior fossa tumors and a majority with hemispheric tumors. By contrast, diffuse low-grade astrocytomas are infiltrative and are less often completely resected. Diencephalic tumors are amenable to gross-total resection in less than 40% of cases. If removal is complete, no further treatment is recommended. Pilomyxoid astrocytomas may be localized or have diffuse leptomeningeal spread. In the latter case, surgical resection of the primary lesion with adjuvant chemotherapy is warranted. Patients with significant residual tumor postoperatively may require further therapy if the risk of subsequent surgery to remove progressive tumor is too great.

Radiotherapy

The ability to re-resect hemispheric lesions usually allows postponement of adjuvant therapy. Radiotherapy is generally avoided in patients with low-grade gliomas due to its increased long-term toxicity compared to chemotherapy. Radiotherapy is reserved for cases where multiple chemotherapy treatments have failed in unresectable, symptomatic tumors, and is usually reserved for older patients. Dosing is 5000—5500 cGy, depending on age, to the original tumor bed with a 2-cm margin.

Chemotherapy

Carboplatin and vincristine (CV) is recommended in patients with newly diagnosed, progressive low-grade astrocytoma (Table 23.6).

TABLE 23.6 Chemotherapy for Low-Grade Astrocytomas and Optic Gliomas

A. PATIENTS YOUNGER THAN 5 YEARS OF AGE

Induction (one 12-week cycle)

Week	1	2	3	4	5	6	7	8	9	10	11—12
	V	V	V	V	V	V	V	V	V	V	
	C	C	C	C			C	C	C	C	Rest

MAINTENANCE (EIGHT 6-WEEK CYCLES)

Week	1	2	3	4	5—6
	V	V	V		
	C	C	C	C	Rest

(C) Carboplatin	175 mg/m^2 IV over 1 h
(V) Vincristine	1.5 mg/m^2 IV (maximum dose 2 mg)

B. PATIENTS OLDER THAN 5 YEARS OF AGE

TPCV (eight 6-week cycles)

Thioguanine	30 mg/m^2/dose orally every 6 h for 12 doses, days 1—3, hours 0—66
Procarbazine	50 mg/m^2/dose orally every 6 h for 4 doses, days 3—4, hours 60—78
CCNU (lomustine)	110 mg/m^2/dose once on day 3, hour 72
Vincristine	1.5 mg/m^2 (0.05 mg/kg in children <12 kg) (maximum dose 2 mg) intravenous push on days 14, 28

In a recent Children's Oncology Group (COG) study comparing CV with thioguanine, procarbazine, CCNU, and vincristine (TPCV), the 5-year event-free survival (EFS) was $39 \pm 4\%$ for CV compared to $52 + 5\%$ for TPCV. However, TPCV has slightly more toxicity but no carboplatin allergy. TPCV is a reasonable alternative regimen for older patients able to swallow oral medication (Table 23.6). A pilot study of vincristine, carboplatin, and temozolomide demonstrated a 2-year EFS of 80 ± 5 and 5-year EFS close to 60%.

Use of single-agent vinblastine demonstrated a 42% response rate in a phase II trial for recurrent disease and there are ongoing trials for its use in newly diagnosed patients.

Prognosis

The 5- and 10-year overall survival (OS) rates for completely resected low-grade supratentorial astrocytomas treated with surgery alone are 76–100% and 69–100%, respectively. In the posterior fossa these rates approach 100%. In patients with partially resected low-grade astrocytomas who are observed without treatment or who are treated with postoperative radiotherapy the 5- and 10-year survival rates are 67–87% and 67–94%, respectively.

Genomic studies can assist in the prognostication of outcomes. Genomic mutations resulting in activation of BRAF are very common in pediatric low-grade astrocytomas. BRAF is part of the Raf kinase family of growth signal transduction protein kinases, which regulate the MAP kinase/ERKs signaling pathway thus affecting cell division and differentiation. The most commonly seen mutation is the KIAA1549-BRAF fusion. The BRAF-KIAA1549 fusion is most commonly associated with cerebellar pilocytic astrocytomas (60–80%) and is also frequently seen in pilomyxoid astrocytoma.

BRAF fusion-positive patients with incomplete resections have been found to have better clinical outcomes than fusion-negative patients, with improved progression-free survival (PFS) and OS. BRAF^{V600E} point mutations are also commonly found in pediatric low-grade astrocytomas. While the mutation is less common in pilocytic astrocytomas (10–15%) than the truncated fusion alteration, it is frequently seen in gangliogliomas (30–60%) and pleomorphic xanthoastrocytomas ($\sim$60%). V600E-mutated tumors have a worse 5-year recurrence-free survival than those without the mutation (60% vs 80%).

Due to the inherent Ras activation in patients with NF-1, genomic alterations in BRAF are infrequently seen in the tumors of patients with NF-1.

Recurrence

Recurrent low-grade astrocytomas should be approached surgically when possible. If not completely resectable, chemotherapy should be given. Bevacizumab (10 mg/kg IV over 90 min) and irinotecan (125 mg/m^2 IV over 90 min) every 2 weeks are commonly used and successfully stabilize or shrink many tumors, but are associated with a high rate of progression upon cessation of bevacizumab. Some of this is related to rebound effect, but many patients continue to progress beyond the first 3 months off therapy. Single-agent vinblastine may be used. Trials of BRAF inhibitors have shown promising results in these tumors and may have a significant impact on outcomes, including debrafinib, a BRAF inhibitor targeting the BRAFV600E mutation. Clinical investigations using MEK inhibitor agents are also underway for patients with recurrent or refractory low-grade astrocytomas with activating BRAF mutation.

High-Grade Astrocytomas (WHO Grades III and IV)

Surgery

Complete surgical removal of these tumors is rarely accomplished because of their infiltrative nature. In about 20–25% of cases, the contralateral hemisphere may be involved, due to spread through the corpus callosum. The main purpose of surgery in these cases is to reduce the tumor burden. Patients who have total or near-total resection have longer survival than do patients who have partial resection or simple biopsy. Maximal surgical debulking is recommended unless the neurologic sequelae will be too devastating.

Radiotherapy

Postoperative irradiation increases survival in high-grade astrocytomas. The port should include the tumor bed and a 2-cm margin of normal surrounding tissue. The dose is 5000–6000 cGy in children over 5 years of age. The use of stereotactic radiotherapy is being investigated as a technique to increase local tumor doses.

Chemotherapy

In children, multiagent adjuvant chemotherapy added to postoperative radiation may result in a modest improvement in disease-free survival compared with postoperative radiation alone. Without a reasonable resection, chemotherapy is palliative. Procarbazine (or prednisone), CCNU, and vincristine (Table 23.5) has a 33% PFS at 5 years. This may be an overestimate because some low-grade astrocytomas may have been included in published reports. Temozolomide significantly improved survival in adults when given for 6 cycles after radiotherapy (Table 23.5) but did not change the outcome in pediatric patients (3-year EFS was $11 \pm 3\%$ and OS was $22 \pm 5\%$). In the temozolomide study, O^6-methylguanine-DNA-methyltransferase (MGMT) overexpression (which correlates with lack of MGMT methylation) worsened outcome (2-year EFS $17 \pm 5\%$ without MGMT overexpression; $5 \pm 4\%$ with MGMT overexpression). It is therefore reasonable to consider temozolomide for treatment in patients with hypermethylation of MGMT. The COG ACNS0432 studied the addition of lumustine to temozolomide. Preliminary data indicate a possible modest improvement in EFS. Bevacizumab plus irinotecan demonstrated superior survival in glioblastoma multiforme in adult studies. The recent COG ACNS0822 showed no advantage of vorinostat or bevacizumab as a radiosensitizer over the control of temozolomide. A recent study using erlotinib demonstrated no benefit in high-grade glioma.

MEDULLOBLASTOMA

Medulloblastoma is the most common CNS tumor in children, representing approximately 20% of all childhood brain tumors with 80% of the cases presenting before the age of 15 years. The tumor presents in the posterior fossa and widespread seeding of the subarachnoid space may occur. The frequency of CNS spread outside the primary tumor can be as high as 40% at the time of diagnosis. Extraneural spread to bone, bone marrow, lungs, liver, or lymph nodes rarely occurs.

Staging studies should include MRI of the spine (preferable preoperatively), lumbar CSF cytology, liver function tests and, if clinically symptomatic, bone scan and/or bone marrow examination. Histology and cytogenetics of the original tumor are essential to evaluate for large-cell anaplastic subtype and for INI-1 underexpression, which is characteristic of atypical teratoid/rhabdoid tumor. These more aggressive tumors fare poorly when treated as typical medulloblastoma. The Chang staging system (Table 23.7), which evaluates tumor size, local extension, and metastases is used to assess prognosis. Patients are divided into average- and high-risk categories based on extent of disease (i.e., CSF cytology or spinal cord involvement), volume of residual tumor, histology, and age at diagnosis (Table 23.8).

TABLE 23.7 Chang System for Posterior Fossa Medulloblastoma (Primitive Neuroectodermal Tumor)

Classification	Description
T1	Tumor less than 3 cm in diameter and limited to the classic midline position in the vermis, the roof of the fourth ventricle and less frequently to the cerebellar hemispheres
T2	Tumor 3 cm or greater in diameter, further invading one adjacent structure or partially filling the fourth ventricle
T3	Divided into T3a and T3b:
T3a	Tumor further invading two adjacent structures or completely filling the fourth ventricle with extension into the aqueduct of Sylvius, foramen of Magendie, or foramen of Luschka, producing marked internal hydrocephalus
T3b	Tumor arising from the floor of the fourth ventricle or brain stem and filling the fourth ventricle
T4	Tumor further spreading through the aqueduct of Sylvius to involve the third ventricle or midbrain or tumor extending to the upper cervical cord
M0	No evidence of gross subarachnoid or hematogenous metastasis
M1	Microscopic tumor cells formed in cerebrospinal fluid
M2	Gross nodular seeding demonstrated in cerebellar, cerebral subarachnoid space, or in the third or lateral ventricles
M3	Gross nodular seeding in spinal subarachnoid space
M4	Metastasis outside the cerebrospinal axis

From Laurent (1985) with permission.

Surgery

Surgical excision is employed as the initial therapy with an objective of gross-total resection or near-total resection with less than 1.5 cm² of residual tumor. Disease-free survival improves for patients who have had a radical resection of the tumor. Occasionally a child presents with life-threatening raised ICP due to obstruction of the fourth ventricle. In these cases, preoperative endoscopic third ventriculostomy or shunting should be performed to reduce the intraoperative mortality of definitive tumor resection in the presence of increased ICP.

Radiotherapy

Radiation therapy plays a critical role in treatment. Medulloblastomas are one of the most radiosensitive primary CNS tumors of childhood. The standard dose of radiotherapy is 5400–5580 cGy to the area of the primary tumor, with 2340 cGy given to the neuroaxis in average-risk medulloblastoma and 3600 cGy with high-risk disease. Use of IMRT to spare the cochlea is important in decreasing therapy-induced hearing loss. Proton beam radiation may reduce this further and decrease the exit dose to structures anterior to the spine.

Radiation therapy to children under 3 years of age with medulloblastoma is used only under rare circumstances because of significant neurocognitive, endocrinologic and growth side effects. Chemotherapy should be utilized so that radiation can either be omitted or postponed in this young age group.

Chemotherapy

The relatively rapid rate of growth and high mitotic index of medulloblastoma result in this tumor's responsiveness to a number of chemotherapeutic agents.

1. In patients who have high-risk disease (Table 23.8) adjuvant chemotherapy in combination with 36-Gy craniospinal radiation improves the disease-free survival for medulloblastoma. The use of high-dose chemotherapy with autologous stem cell rescue (Table 23.9) remains experimental.
2. Patients with average-risk disease (Table 23.8) have a greater than 80% 5-year disease-free survival with surgery and radiotherapy. The COG study utilizing craniospinal irradiation following surgery and adjuvant CCNU, vincristine, and cisplatinum (Table 23.5) chemotherapy demonstrated an 89% PFS rate in nondisseminated posterior fossa medulloblastoma with decreased craniospinal irradiation. Another study using cisplatin, etoposide, vincristine, cyclophosphamide, and carboplatin reduces the cisplatin and vincristine doses by 50% and the incidence of significant ototoxicity and neuropathy (Table 23.10).

TABLE 23.8   Medulloblastoma Risk Categories

	Average risk	High risk
Extent of disease	Negative cerebrospinal fluid (CSF) cytology	Positive CSF cytology
	Normal magnetic resonance imaging (MRI) of spine	Positive MRI of spine with gadolinium diethylenetriaminepentaacetic acid dimeglumine
Extent of residual tumor[a]	<1.5 cm² residual	>1.5 cm² residual
Histology	Undifferentiated	Large-cell anaplastic
Age at diagnosis	>3 years	<3 years

[a]On a two-dimensional measurement.

TABLE 23.9   Conditioning Regimen for Autologous Stem Cell Transplantation

Carboplatin	500 mg/m²/day IV[a]	Days −8 to −6
Thiotepa	300 mg/m²/day IV	Days −5 to −3
Etoposide	250 mg/m²/day IV	Days −5 to −3
Rest		Days −2 to −1
Stem cell infusion		Day 0
G-CSF	5 μg/kg/day	Day 1 until neutrophil engraftment

[a]Modified to area under the curve (AUC) of 7 mg/ml/min by pediatric Calvert formula.
IV, intravenous; G-CSF, granulocyte colony stimulating factor.

TABLE 23.10   Chemotherapy for Medulloblastoma

Cycles 1, 4, 7		
Cisplatin	20 mg/m^2 IV days 1–5	Infuse over 1 h
Etoposide	50 mg/m^2 IV days 1–5	Infuse over 1 h
Cycle 2, 5, 8		
Vincristine	1.5 mg/m^2 IV, days 1	IV over 1–15 min
Cyclophosphamide	900 mg/m^2 IV, days 1–2	Infuse over 1 h
Cycle 3, 6, 9		
Carboplatin	150 mg/m^2 PO, days 1, 8, 15, 22	Infuse over 1 h
Vincristine	1.5 mg/m^2 IV, days 1, 15	IV over 1–15 min

TABLE 23.11   Molecular Subgroups of Medulloblastoma

Subgroup	WNT	Sonic hedgehog	Group 3	Group 4
Age group	Older children, adolescents, and adults	Young children, adolescents, and adults	Children and infants	Infants, children, and adults
Gender	Equal	Equal	Male:female 2:1	Male predominant
Histology	Classic, rarely LCA	Desmoplastic/nodular, classic, LCA	Classic, LCA	Classic, LCA
Metastasis	Rarely M+	Uncommonly M+	Very frequently M+	Frequently M+
Prognosis	Very good	Infants good, others intermediate	Poor	Intermediate
Genetics	CTNNB1 mutation	PTCH1/PTCH2/SMO/SUFU mutation	i17q	i17q
		GLI2 amplification	MYC amplification	CDK6 amplification
		MYCN amplification		MYCN amplification
Gene expression	WNT signaling	SHH signaling	MYC+++	Minimal MYC/MYCN
	MYC+	MYCN+		

Prognosis

Survival for patients with average-risk medulloblastoma treated with adjuvant radiation and chemotherapy is at least 82% at 5 years. In high-risk patients using craniospinal radiation with chemotherapy, 45–50% of patients are disease-free at 5 years. When chemotherapy dose intensity is increased with more aggressive regimens and high-dose chemotherapy with autologous stem cell rescue (Table 23.9), 2-year disease-free survival data are encouraging at 73–78%.

RELAPSED MEDULLOBLASTOMA

Surveillance scanning is very important after therapy since it can detect most recurrences prior to the onset of symptoms. Studies have shown some improvement of survival in relapsed patients using high-dose carboplatin, thiotepa, and etoposide followed by autologous stem cell rescue (Table 23.9), with salvage rates up to 30%.

Molecular Subgroups

Gene expression profiling studies of medulloblastoma cohorts have demonstrated the existence of distinct molecular subgroups that differ in their demographics, genomics, and clinical outcomes. The discovery of novel driver mutations for each subgroup may offer a dramatic shift in the approach to this tumor as new diagnostic and therapeutic targets are identified. These findings are having a significant impact as new clinical trials and targeted therapies are being developed for the treatment of patients with these distinct molecular subgroups (Table 23.11).

WNT Tumors

Tumors of this subgroup demonstrate a WNT signaling gene expression signature and β-catenin nuclear staining. They are frequently of the classic medulloblastoma histologic subtype and rarely have a large-cell/anaplastic appearance. Metastases are infrequent at diagnosis. These tumors are characterized genetically by 6q loss and CTNNB1 mutations and have activated WNT signaling. MYC overexpression is occasionally seen. There is no gender predisposition for this subgroup and patients tend to be older children, adolescents, and adults. Overall, tumors of this subgroup are associated with very good outcomes. There are several clinical trials under development investigating whether this group of patients might be able to receive reduced therapy while maintaining their good outcomes.

Sonic Hedgehog Tumors

Tumors of this subgroup are characterized by classic or desmoplastic/nodular histology, chromosome 9q deletions, MYCN amplification, and mutations in sonic hedgehog pathway genes such as *SMO, PTCH1, PTCH2, SUFU*, and *GLI2*. Metastases are uncommon at diagnosis. There is no gender predisposition for this subgroup. Age distribution follows a bimodal pattern with cases seen primarily in children under 3 years of age and then again in older adolescents and adults. Outcomes are favorable for young children and intermediate for adolescents and young adults. Early-phase trials are examining the use of sonic hedgehog pathway inhibitors to treat these tumors.

Group 3 Tumors

Tumors in this group typically demonstrate classic or large-cell/anaplastic histology and are characterized by MYC amplification. Other mutations have been found in these tumors, including the presence of i17q. These tumors are frequently found to have metastasized at the time of diagnosis. Males tend to outnumber females by a ratio of 2:1. This subgroup primarily affects children and can also occur in infants. Prognosis is dependent on the patient's MYC status, as patients with MYC amplification or overexpression have a 5-year survival of less than 50%. Those without these characteristics have a prognosis similar to most patients with medulloblastoma.

Group 4 Tumors

These tumors are characterized by classic or large-cell/anaplastic histology and amplification of MYCN and CDK6. They may also have an i17q abnormality. These tumors are also commonly found to be metastatic at diagnosis, but not as frequently as group 3 tumors. This subgroup is found in infants, children, and adults, with a male predominance. Prognosis is intermediate.

BRAIN STEM TUMORS

Brain stem gliomas comprise 15–20% of all childhood CNS tumors. The median age at presentation is 6–7 years of age. Fifty percent of the patients present with cranial nerve and long-tract signs.

The majority of patients have diffuse, infiltrating pontine tumors, sometimes with an exophytic component. These are typically WHO grade II–IV gliomas with 1-year and 2-year survival rates of 10–20%. A subset of patients have focal lesions that are usually grade I tumors and have 2-year survival rates of about 60% or greater. Localized brain stem tumors may also be PNETs. While sensitivity to chemotherapy as well as radiation is common, prognosis remains poor, at around 10% 2-year survival.

Surgery

Surgical resection is not usually possible because of the proximity to vital structures, limited room for expansion and swelling, and possible damage to medullary structures. There is no apparent benefit from a surgical biopsy when the imaging and clinical picture are indicative of a diffuse infiltrating pontine glioma. For focal tumors (nontectal), complete resection may be safe and may not require any further therapy. Partial resection of exophytic tumors will establish the diagnosis and reduce the obstructing mass within the fourth ventricle. If hydrocephalus is present, a CSF diversion should be performed.

Radiotherapy

Limited field irradiation is standard palliative care in patients with infiltrative pontine gliomas. A tumor dose of approximately 5400 cGy is standard. The treatment field should include the extent of the defined tumor and a 2-cm margin around the tumor. In irradiating these tumors, precautions should be taken to minimize brain stem edema, especially in patients who have not been shunted. The use of high-dose steroids is valuable during treatment and may be required throughout the treatment period. Children with diffusely infiltrating pontine gliomas often initially respond to radiation therapy but progressive disease is usually seen within 8–12 months. Timing of radiation for partially resected focal lesions is unclear. It is not known whether immediate adjuvant therapy is superior to observation and therapy only at the time of progression.

Chemotherapy

The use of combination chemotherapy after radiotherapy has not improved the disease-free survival in brain stem tumors. Until some chemotherapeutic regimen confirms a survival advantage in newly diagnosed patients, chemotherapy currently plays a palliative role.

Prognosis

The overall prognosis for brain stem tumors is poor. Most centers report a 5–20% 2-year survival rate when high doses of irradiation are employed. Children with diffusely infiltrating pontine gliomas often respond initially to radiation therapy but progressive disease is usually seen within 8–12 months. Improved survival is seen in focal low-grade brain stem astrocytomas, especially those seen in patients with NF-1 and those in the tectal area.

EPENDYMOMAS

Fifty percent of all ependymomas occur during childhood and adolescence and they constitute approximately 9% of all primary childhood CNS tumors. The tumors occur either infratentorially or supratentorially. The fourth ventricle is the most common location. Ependymomas can also occur in the spinal cord and account for 25% of spinal cord tumors. Obstructive hydrocephalus is the most common presenting condition.

Surgery

Total removal of these tumors is difficult to accomplish, especially in tumors originating from the fourth ventricle. However, since gross-total tumor resection predicts a greatly improved outcome, this should be attempted. In the posterior fossa, the use of evoked potentials helps to make this process safer, but patients may still have multiple cranial nerve palsies and may require tracheotomy and gastrostomy for months until normal swallowing recovers. Patients with a subtotal resection may consider two courses of chemotherapy before potential second-look surgery.

Radiotherapy

The standard of care for grade II and grade III ependymoma is maximal surgical resection plus focal radiation therapy, as the recurrence rate with surgery alone is extremely high. Local disease without evidence of subarachnoid spread should receive local irradiation with a margin (that should include extension to lateral ventricles or superior cervical spine when appropriate). Use of 3D-conformal radiation or IMRT is appropriate and is currently under investigation. The dose is 5400–6000 cGy for intracranial lesions. Nonanaplastic supratentorial ependymoma can be observed if a true gross-total resection has been performed. Myxopapillary ependymoma can also be watched after surgery (even if metastatic at presentation, if lesions cause no symptoms) without further focal radiation therapy.

Chemotherapy

No advantage has been demonstrated from the use of adjuvant chemotherapy and although platinum agents are the most active they are not currently part of standard therapy. In infants, chemotherapy is used to postpone radiotherapy. There is some evidence that low-dose metronomic chemotherapy (i.e., oral etoposide at 50 mg/m^2/day for 21 or 28 days) can be effective in treating relapsed disease. The optimal length of therapy has not yet been determined.

Prognosis

The overall prognosis for ependymomas is dependent on the initial extent of resection and the presence of metastases. Grade III tumors tend to metastasize more often. They also tend to be more invasive and are less frequently completely resected. The survival outcome for nonmetastatic, completely resected ependymomas is the same for patients with grade II and grade III tumors. Patients with total resection who receive radiation have a 5-year survival of 67−86% compared to 22−47% 5-year EFS for patients who receive radiation after subtotal resection.

OPTIC GLIOMA

Optic gliomas constitute 5% of the primary CNS tumors in childhood and the majority are diagnosed by 5 years of age with almost all diagnosed by age 20. Involvement of the optic chiasm is usually seen in older children. Neurofibromatosis is present in up to 70% of patients with tumors of the optic nerve or chiasmatic tumors. Isolated optic nerve tumors are more commonly seen in patients with NF-1. Patients may present with decreased vision, proptosis, optic atrophy, and papilledema. In young children asymmetric nystagmus may be the presenting sign of a chiasmatic tumor. Tumors that extend beyond the optic pathway may be associated with hypothalamic dysfunction. Histologically, approximately 90% are low-grade astrocytomas.

Surgery

As surgical intervention may compromise vision, MRI should be used to make a clinical diagnosis and biopsy should be used for unusual circumstances. Surgery should be reserved for tumor extension into the optic canal or increasing visual compromise.

Chemotherapy

Chemotherapy has been used to treat progressive disease. Regimens used include:

- Carboplatin 175 mg/m^2 IV weekly for 4 weeks followed by a 2-week rest and then four more weekly doses of carboplatin. Vincristine 1.5 mg/m^2 (maximum dose 2 mg/weekly) for 10 weeks is given concurrently with the carboplatin course. If a response or stabilization is obtained, maintenance therapy is given consisting of courses of carboplatin 175 mg/m^2 weekly for 4 weeks with vincristine 1.5 mg/m^2 (maximum dose 2 mg) weekly for the first 3 weeks of each 6-week cycle. There is a 2-week rest between each maintenance cycle. In children with stable or improved disease the regimen is continued for 8 cycles (Table 23.12).
- Actinomycin D and vincristine. Actinomycin D 0.015 mg/kg/day for 5 days (0.5 mg maximum per dose) IV every 12 weeks for 6 cycles. Vincristine 1.5 mg/m^2 (2 mg maximum dose) IV weekly for 8 weeks with a 4-week rest between cycles.

Recommendations

Since optic glioma can run an indolent course, the decision for therapeutic intervention should be based on radiographic findings and assessment of visual acuity, visual fields, and visual-evoked response. The following treatment plan is recommended:

- *Evidence of optic nerve tumor and normal visual assessment*: no therapeutic intervention is recommended. MRI scan and visual assessments are performed every 6 months to monitor progression of disease.
- *Evidence of progression on visual assessment, with or without tumor extending posteriorly into the optic canal*: a trial of chemotherapy is recommended in an attempt to preserve useful vision.

TABLE 23.12   Chemotherapy for Optic Glioma

INDUCTION (ONE 12-WEEK CYCLE)

Week	1	2	3	4	5	6	7	8	9	10	11–12
	V	V	V	V	V	V	V	V	V	V	Rest
	C	C	C	C			C	C	C	C	

MAINTENANCE (EIGHT 6-WEEK CYCLES)

Week	1	2	3	4	5–6
	V	V	V		Rest
	C	C	C	C	

(C) Carboplatin	175 mg/m^2 IV
(V) Vincristine	1.5 mg/m^2 IV (maximum dose 2 mg)

Prognosis

PFS rates with CV are 68% at 3 years. With actinomycin D and vincristine, 4-year PFS is 62.5% but 7-year PFS is only about 33%. While both regimens have been utilized, the CV regimen is preferred.

CRANIOPHARYNGIOMAS

Craniopharyngiomas may involve the pituitary gland. They account for 6–9% of all CNS tumors in children. The peak incidence is 5–10 years of age. Patients present with symptoms of increased ICP, visual loss, and endocrine deficiencies. They typically require replacement therapy with cortisone, thyroxine, growth hormone, and/or sex hormones. These tumors are slow-growing benign lesions amidst vital anatomic structures.

Surgery

Craniopharyngiomas should be completely removed if possible without producing untoward neurologic sequelae. Complete excision is possible in 60–90% of cases. If radical excision can be accomplished without significant postoperative morbidity, a primary surgical approach is warranted. However, controversy exists over whether subtotal resection followed by radiation will produce less morbidity and be an equally effective strategy. Complete tumor removal is most easily accomplished in cystic tumors and is most difficult in solid or mixed tumors larger than 3 cm in size.

Radiotherapy

In tumors in which conservative surgery consisting only of drainage or subtotal removal is performed, the addition of radiotherapy reduces the local recurrence rate and improves long-term survival. For children older than 5 years, 5000–5500 cGy are given. In children less than 5 years of age the dose may be reduced to 4000–4500 cGy.

For patients with complete resections, radiation may be reserved for those who relapse.

Chemotherapy

At present there is no established role for systemic chemotherapy in craniopharyngioma. Interferon-α 2a systemically shrank the lesion in 25% of patients treated, primarily those with cystic lesions.

Prognosis

The long-term survival of patients treated with radical and total removal is 80–90% at 5 years and 81% at 10 years. The 5-year recurrence-free survival after subtotal removal is approximately 50%. Survival after partial removal and radiation therapy is 62–84%.

INTRACRANIAL GERM CELL TUMORS

Primary intracranial germ cell tumors comprise 1–3% of primary pediatric brain tumors. The peak age is between 10 and 21 years of age. Multiple tumor types can be seen: the majority are germinomas (~55%), teratomas, and mixed germ cell tumors (~33%), and the remaining 10% are malignant endodermal sinus tumors, embryonal cell carcinomas, choriocarcinomas, and teratocarcinomas. In all but the germinomas, serum and CSF AFP and β-hCG may be elevated. MRI of the spine with Gd-DTPA should be performed, as leptomeningeal spread is relatively common.

The outcome of patients with pure germinoma is considerably better than with mixed germ cell tumors.

Surgery

Surgical biopsy is indicated in all germ cell tumors to make an appropriate diagnosis unless elevated serum or CSF tumor markers establish the diagnosis. For patients with benign tumors such as teratomas or dermoids, surgery can be curative.

Radiotherapy

Conventional radiotherapy for a CNS germinoma includes doses to the primary tumor of 5000 cGy with 3000 cGy craniospinal therapy. However, there are several studies which demonstrate the ability to use chemotherapy to reduce the dose of radiation to between 3060 and 5040 cGy to gross tumor and the ventricles only (in nondisseminated disease) depending on response while maintaining outcomes of greater than 90% EFS. Nongerminoma germ cell tumors respond poorly to radiation therapy but the use of chemotherapy with radiation appears to improve survival substantially in preliminary studies. These data need to be replicated in larger studies.

Chemotherapy

In both germinomas, cycles of carboplatin 600 mg/m^2 IV for 1 day with etoposide 150 mg/m^2/day IV for days 1–3 of a 21-day cycle are often used. For nongerminoma germ cell tumors, cycles of carboplatin 600 mg/m^2 IV for 1 day with etoposide 150 mg/m^2/day IV for days 1–3 alternating with ifosfamide 1800 mg/m^2/day IV for days 1–5 with etoposide 100 mg/m^2/day IV for days 1–5 of a 21-day cycle are used. These approaches are generally combined with graded doses of radiation depending on initial risk and response.

Prognosis

The germinomas have the best OS rate (>95% disease-free survival) followed by teratomas and pineal parenchymal tumors. Nongerminoma germ cell tumors, such as embryonal carcinoma, endodermal, and choriocarcinoma have a worse survival. However, preliminary data from combined chemotherapy/radiation approaches demonstrate encouraging data, indicating that survival may now approach 60–80%.

MALIGNANT BRAIN TUMORS IN INFANTS AND CHILDREN LESS THAN 3 YEARS OF AGE

Infants and very young children with brain tumors have a worse prognosis than older children. They are also at higher risk for neurotoxicity including mental retardation, growth failure, and leukoencephalopathy. Due to these factors there is a reluctance to treat infants and young children with radiation therapy. Recent studies have been designed to use chemotherapy and to either withhold radiation therapy or postpone its use to a time when the patient is older. Postoperative chemotherapy cycles with high-dose methotrexate, cyclophosphamide, vincristine, cisplatin, and etoposide, followed by high-dose chemotherapy with autologous stem cell rescue are used in young children in an attempt to avoid radiation altogether or at least delay its use (Table 23.9). In children with nonmetastatic medulloblastoma treated on the Headstart I and Headstart II protocols, the 5-year EFS and OS rates for all patients were $52 \pm 11\%$ and $70 \pm 10\%$. Patients with gross-total resection and patients with residual tumor had outcomes of $64 \pm 13\%$ and $79 \pm 11\%$, and $29 \pm 17\%$ and $57 \pm 19\%$, respectively.

Atypical teratoid/rhabdoid tumors are very aggressive tumors of infancy that present similarly to medulloblastoma and are at times indistinguishable pathologically except for monosomy 22 (lack of expression of INI-1). While uniformly fatal if treated with conventional medulloblastoma therapy, these patients have a survival approaching 50% if treated with high-dose chemotherapy followed by autologous stem cell rescue and focal irradiation.

Trials of intensive chemotherapy followed by stem cell rescue have also been attempted in young children with malignant brain tumors with promising results with avoidance of radiation therapy. The current COG study uses three tandem auto stem cell rescues after 3 cycles of chemotherapy. These approaches require further research. Additionally, with 3D-conformal radiation and IMRT, the timing and use of radiation therapy in subsets of disease that do worse (i.e., subtotal resections) is being studied in children as young as 8 months with focal radiotherapy alone. Small studies using focal radiation in nondisseminated disease with gross-total resection demonstrate disease-free survival of 75% or greater. The use of second-look surgery is also being evaluated in current trials.

Further Reading and References

Allen, J., Finlay, J., 1998. Intensive chemotherapy and bone marrow rescue for young children with newly diagnosed malignant brain tumors. J. Clin. Oncol. 16, 210–221.

Balmaceda, C., Heller, G., Rosenblum, M., et al., 1996. Chemotherapy without irradiation—a novel approach for newly diagnosed CNS germ cell tumors: results of an international cooperative trial. J. Clin. Oncol. 14, 2908–2915.

Biegel, J.A., 1999. Cytogenetics and molecular genetics of childhood brain tumors. Neuro. Oncol. 1 (2), 139–151.

Blaney, S.M., Kun, L.E., Hunter, J., et al., 2006. In: Pizzo, P.A., Poplack, D.G. (Eds.), Tumors of the Central Nervous System in Principles and Practice of Pediatric Oncology, fourth ed. Lippincott-Williams & Wilkins, Philadelphia, PA.

Dasgupta, T., Haas-Kogan, D.A., 2013. The combination of novel targeted molecular agents and radiation in the treatment of pediatric gliomas. Front. Oncol. 3, 110.

Duffner, P.C., Horowitz, M.E., Krischer, J.P., et al., 2008. Myeloablative chemotherapy with autologous bone marrow rescue in children and adolescents with recurrent malignant astrocytoma: outcome compared with conventional chemotherapy: a report from the Children's Oncology Group. Pediatr. Blood Cancer 51 (6), 806–811.

Geyer, J.R., Sposto, R., et al., 2005. Multiagent chemotherapy and deferred radiotherapy in infants with malignant brain tumors: a report from the Children's Cancer Group. J. Clin. Oncol. 23 (30), 7621–7631.

Halperin, E.C., Constine, L.S., Tarbell, N.J., Kun, L.E., 2005. Pediatric Radiation Oncology, fourth ed. Lippincott-Williams & Wilkins, Philadelphia, PA.

Laurent, J., 1985. Classification system for primitive neuroectodermal tumors (medulloblastoma) of the posterior fossa. Cancer 56, 1807.

MacDonald, T.J., Rood, B.R., Santi, M.R., et al., 2003. Advances in the diagnosis, molecular genetics and treatment of pediatric embryonal CNS tumors. Oncologist 8, 174–186.

Mason, W.P., Growas, A., Halpern, S., et al., 1997. Carboplatin and vincristine chemotherapy for children with newly diagnosed progressive low-grade glioma. J. Neurosurg. 86, 747–754.

Taylor, M.D., Northcott, P.A., Korshunov, A., et al., 2012. Molecular subgroups of medulloblastoma: the current consensus. Acta Neuropathol. 123 (4), 465–472.

Thompson, D., Phipps, K., et al., 2005. Craniopharyngioma in childhood: our evidence-based approach to management. Childs Nerv. Syst. 21 (8–9), 660–668.

24

Neuroblastoma

Julie R. Park and Rochelle Bagatell

Neuroblastoma originates from primordial neural crest cells that normally give rise to adrenal medulla and the sympathetic ganglia.

EPIDEMIOLOGY

- Neuroblastoma is the most common extracranial solid tumor in children, accounting for 7% of all childhood malignancies. It accounts for 15% of childhood cancer mortality.
- Neuroblastoma is the most common malignancy in infants with an annual incidence of 10 per million live births.
- At diagnosis, 50% of patients are under 2 years of age, 75% under age 4, and 90% under age 10. Peak age of incidence is 2 years.
- *In situ* neuroblastoma is detected in one in 259 autopsies among infants less than 3 months of age. This 400-fold increase in the autopsy incidence of neuroblastoma compared with the clinical incidence of the tumor indicates that involution or maturation of the tumor occurs spontaneously in a moderate number of infants.

PREDISPOSITION

- Disorders involving the autonomic nervous system are occasionally seen in patients with neuroblastoma. Conditions associated with neuroblastoma are listed in Table 24.1.
- A family history of neuroblastoma is identified in 1–2% of cases and the tumors that arise in patients with familial disease are heterogeneous. The study of neuroblastoma predisposition has yielded remarkable insights into the biology of this disease.
- Mutations in PHOX2B (a key regulator of normal development of the autonomic nervous system) have been noted in a subset of patients with hereditary neuroblastoma.
- Heritable mutations in the anaplastic lymphoma kinase gene are an important cause of familial neuroblastoma.

PATHOLOGY AND BIOLOGY

The diagnosis of neuroblastoma is based on the presence of characteristic histopathologic features of tumor tissue or the presence of tumor cells in a bone marrow aspirate or biopsy accompanied by elevated levels of urinary catecholamines. Neuroblastoma is one of the small round blue cell tumors of childhood and tumors are classified as histologically favorable or unfavorable according to the International Neuroblastoma Pathology Classification (INPC) as shown in Table 24.2.

TABLE 24.1　Conditions Associated with Neuroblastoma

Neurofibromatosis type I
Hirschsprung disease with aganglionic colon
Congenital central hypoventilation syndrome
Pheochromocytoma
Heterochromia
Turner syndrome
Noonan syndrome
Cardiac malformations

TABLE 24.2　Prognostic Grouping of Neuroblastic Tumors According to Histologic Classification Systems: International Neuroblastoma Pathology Classification System (INPC) and Shimada Classification

International neuroblastoma pathology classification		Original Shimada classification	Prognostic group
Neuroblastoma	(Schwannian poor)[a]	Stroma-poor	
Favorable		Favorable	Favorable
<1.5 years	Poorly differentiated or differentiating and low or intermediate MKI		
1.5–5 years	Differentiating and low MKI		
Unfavorable		Unfavorable	Unfavorable
<1.5 years	(a) Undifferentiated tumor[b]		
	(b) High MKI		
1.5–5 years	(a) Undifferentiated or poorly differentiated tumor		
	(b) Intermediate or high MKI tumor		
≥5 years	All tumors		
Ganglioneuroblastoma, intermixed	(Schwannian stroma-rich)	Stroma-rich intermixed (favorable)	Favorable
Ganglioneuroma	(Schwannian stroma-dominant)		
Maturing		Well differentiated (favorable)	Favorable[c]
Mature		Ganglioneuroma	
Ganglioneuroblastoma, nodular	(Composite Schwannian stroma-rich/stroma dominant and stroma poor)	Stroma-rich nodular (unfavorable)	Unfavorable[c]

[a]Neuroblastoma subtypes detailed in Shimada et al. (1999).
[b]Rare subtype, especially diagnosed in this age group. Further investigation and analysis required.
[c]Prognostic grouping not related to age.
MKI, mitosis-karyorrhexis index.
Adapted from: Shimada et al. (1999).

CLINICAL FEATURES

Neuroblastoma usually presents with an adrenal mass or less commonly as a tumor arising anywhere along the sympathetic neural chain. The most common presenting feature is an asymptomatic abdominal mass, with distant metastases detected at the time of diagnosis in 75% of cases.

Anatomic Site

Clinical findings related to the anatomic site of the primary tumor are listed below:

- Head and neck
  - Unilateral palpable neck mass
  - Horner's syndrome (myosis, ptosis, enophthalmos, anhydrosis)
- Orbit and eyes
  - Orbital metastases with periorbital hemorrhage ("raccoon eyes")
  - Exophthalmos, palpable supraorbital masses, ecchymosis, edema of eyelids and conjunctiva, ptosis
  - Opsoclonus ("dancing eyes syndrome")
  - Heterochromia iridis, anisocoria
- Chest
  - Upper thoracic tumors: dyspnea, pulmonary infections, dysphagia, lymphatic compression, Horner syndrome
  - Lower thoracic tumors: usually no symptoms
- Abdomen
  - Anorexia, vomiting
  - Abdominal pain
  - Palpable mass
  - Massive involvement of the liver with metastatic disease; may result in respiratory insufficiency in the newborn
- Pelvis
  - Constipation
  - Urinary retention
  - Presacral mass palpable on rectal examination
- Paraspinal area (dumbbell or hourglass-shaped neuroblastoma)
  - Localized back pain and tenderness
  - Limp
  - Weakness of lower extremities
  - Hypotonia, muscle atrophy, areflexia, hyperreflexia
  - Paraplegia
  - Scoliosis
  - Bladder and anal sphincter dysfunction.
- Lymph nodes: enlarged
- Bone: pain, limping and irritability in the young child associated with bone and bone marrow metastasis
- Lung: pulmonary parenchyma or pleura may rarely be involved. Lung involvement should be proven by biopsy or MIBG (I^{131}-meta-iodobenzylguanidine (MIBG) scan) avidity due to its rarity in this disease
- Brain: metastatic involvement of the brain has been described, though it is more common in the setting of recurrent rather than newly diagnosed disease.

Other Symptoms

- Nonspecific constitutional symptoms include lethargy, anorexia, pallor, weight loss, abdominal pain, weakness, and irritability. These symptoms are more commonly seen in patients with metastatic disease at diagnosis.
- Intermittent attacks of sweating, flushing, pallor, headaches, palpitations, and hypertension related to catecholamine release can occur, but are relatively rare.
- Hypertension in children with neuroblastoma is usually due to renovascular compromise rather than catecholamine secretion.

Paraneoplastic Syndromes

1. Vasoactive intestinal peptide (VIP) secretion (Kerner–Morrison syndrome):
   a. Signs and symptoms include intractable watery diarrhea resulting in failure to thrive, associated with abdominal distention and hypokalemia

   b. Symptoms are due to the secretion of an enterohormone, VIP, by the neuroblastoma cells

   c. VIP-secreting tumors typically have favorable histology

2. Opsoclonus myoclonus ataxia syndrome (OMAS):

   a. Signs and symptoms include bursts of rapid involuntary random eye movements in all directions of gaze (opsoclonus); and motor incoordination due to frequent, irregular, jerking of muscles of the limbs and trunk (myoclonic jerking)

   b. Symptoms may or may not resolve after the tumor is removed

   c. Prognosis for survival is favorable because OMAS is typically seen in patients with low-stage, biologically favorable tumors

   d. There is a high incidence of chronic neurologic deficits, including cognitive and motor delays, language deficits, and behavioral abnormalities

   e. All children presenting with this syndrome should be evaluated for the presence of neuroblastoma.

DIAGNOSIS AND STAGING

When neuroblastoma is suspected, levels of the catecholamines homovanillic acid (HVA) and vanillylmandelic acid (VMA) should be analyzed to facilitate diagnosis. Tumor biopsies should be performed and should contain enough tissue to permit assessment of tumor histology and sufficient material for the molecular genetic analyses that are central to correct risk stratification and treatment planning. Open biopsies are preferred rather than needle biopsies. Criteria for neuroblastoma diagnosis are described in Table 24.3.

Once the diagnosis of neuroblastoma has been confirmed, a complete staging workup should be performed consisting of the following:

- Computed tomography (CT) or magnetic resonance imaging (MRI) for evaluation of tumors in the abdomen, pelvis, or mediastinum. MRI decreases imaging-associated radiation exposure, especially when serial imaging evaluations are required.
- MRI is essential when evaluating paraspinal lesions as it permits evaluation of intraforaminal extension with the potential for cord compression.
- I^{123}-MIBG scintigraphy is recommended for detection of bony metastases and occult soft tissue disease. MIBG is taken up by approximately 90% of neuroblastoma tumors.
- ^{18}F-fluorodeoxyglucose positron emission tomography (FDG-PET) may be useful to visualize primary and metastatic sites of disease, especially soft tissue disease, and should be utilized to diagnose and monitor response in patients whose tumors are not MIBG avid.
- Bone marrow disease should be assessed by performing bilateral bone marrow aspirates and biopsies. Standard histological analyses, including use of neuroblastoma immunohistochemical staining of biopsy specimens (e.g., CD56 or PGP9.5), should be performed.
- Table 24.4 shows the recommended tests for the assessment of extent of disease.

Staging Systems

Table 24.5 shows the International Neuroblastoma Staging System (INSS) and the more recently developed International Neuroblastoma Risk Group Staging System (INRGSS).

TABLE 24.3 Recommended Criteria for Neuroblastoma Diagnosis

Established if:

1. Unequivocal pathologic diagnosis[a] is made from tumor tissue by light microscopy (with or without immunohistology, electron microscopy)

OR

2. Bone marrow aspirate or trephine biopsy contains unequivocal tumor cells[a] (e.g., syncytia or immunocytologically positive clumps of cells) and increased urine or serum catecholamines or metabolites[b]

[a]*If histology is equivocal, karyotypic abnormalities in tumor cells characteristic of other tumors (e.g., t(11;22)) excludes a diagnosis of neuroblastoma, whereas genetic features characteristic of neuroblastoma (1p deletion, MYCN amplification) support this diagnosis.*
[b]*Includes dopamine, HVA, and/or VMA (levels must be >3.0 SD above the mean per milligram creatinine for age to be considered increased and at least two of these must be measured).*
From: Brodeur et al. (1993).

TABLE 24.4 Tests Recommended for Assessment of Extent of Disease

Tumor site	Recommended tests
Primary tumor	CT and/or MRI scan[a] with 3D measurements; MIBG scan[b]
METASTATIC SITES	
Bone marrow	Bilateral posterior iliac crest marrow aspirates and trephine (core) bone marrow biopsies required to exclude marrow involvement. A single positive site documents marrow involvement. Core biopsies must contain at least 1 cm of marrow (excluding cartilage) to be considered adequate
Bone	MIBG scan; if tumor does not take up MIBG use ^{18}F-FDG-PET
Lymph nodes	Clinical examination (palpable nodes), confirmed histologically. CT and/or MRI scan for non-palpable nodes (3D measurements)
Abdomen/liver	CT and/or MRI scan[a] with 3D measurements
Chest	CT/MRI necessary if abdominal mass/nodes extend into chest, or evidence of soft tissue MIBG uptake. Consider inclusion if patient has non-thoracic primary with disseminated disease to allow evaluation for distant nodal metastases

[a]Ultrasound considered suboptimal for accurate 3D measurements in complex locoregional tumors; may be considered for serial monitoring if expertise in US imaging exists and tumors are well circumscribed (INRGSS L1).
[b]The MIBG scan is applicable to all sites of disease. I-123 MIBG is the recommended tracer.
AP, anteroposterior.
From: Brodeur et al. (1993) and Sharp et al. (2009).

TABLE 24.5 Neuroblastoma Staging: INSS and INRGSS

	INSS		INRGSS
Stage 1.	Localized tumor with complete gross excision, with or without microscopic residual disease; representative ipsilateral lymph nodes negative for tumor microscopically (nodes attached to and removed with the primary tumor may be positive)	*Stage L1.*	Localized tumor not involving vital structures (as defined by the list of image-defined risk factors, Table 24.6) and confined to one body compartment
Stage 2A.	Localized tumor with incomplete gross excision; representative ipsilateral non-adherent lymph nodes negative for tumor microscopically	*Stage L2.*	Locoregional tumor with presence of one or more image-defined risk factors
Stage 2B.	Localized tumor with or without complete gross excision, with ipsilateral non-adherent lymph nodes positive for tumor. Enlarged contralateral lymph nodes must be negative microscopically		
Stage 3.	Unresectable unilateral tumor infiltrating across the midline,[a] with or without regional lymph node involvement; or localized unilateral tumor with contralateral regional lymph node involvement; or midline tumor with bilateral extension by infiltration (unresectable) or by lymph node involvement		
Stage 4.	Any primary tumor with dissemination to distant lymph nodes, bone, bone marrow, liver, skin, and/or other organs (except as defined for stage 4S)	*Stage M.*	Distant metastatic disease (except stage MS)
Stage 4S.	Localized primary tumor (as defined for stage 1, 2A, or 2B), with dissemination limited to skin, liver, and/or bone marrow[b] (limited to infants <1 year of age)	*Stage MS.*	Metastatic disease in children younger than 18 months with metastases confined to skin, liver, and/or bone marrow

[a]The midline is defined as the vertebral column. Tumors originating on one side and crossing the midline must infiltrate to or beyond the opposite side of the vertebral column. A grossly resectable tumor arising in the midline from pelvic ganglia or the organ of Zuckerkandl would be considered stage 1. A midline tumor that extended beyond one side of the vertebral column and was unresectable would be considered stage 2A. Ipsilateral lymph node involvement would make it stage 2B, whereas bilateral lymph node involvement would make it stage 3. A midline primary tumor with bilateral infiltration that was not resectable would be considered stage 3. A tumor of any size with malignant ascites or peritoneal implants would be stage 3 (but a thoracic tumor with malignant pleural effusion unilaterally would be stage 2B).
[b]Marrow involvement in stage 4S should be minimal, that is, <10% of total nucleated cells identified as malignant on bone marrow biopsy or on marrow aspirate. More extensive marrow involvement would be considered to be stage 4. The MIBG scan should be negative in the marrow.
Multifocal primary tumors (e.g., bilateral adrenal primary tumors) are staged according to the greatest extent of disease in both systems.
From: Brodeur et al. (1993), with permission and Monclair et al. (2009), with permission.

- The INSS uses extent of surgical resection, presence of lymph node involvement, and metastatic disease to determine disease stage
- The INRGSS uses presurgical assessments of extent of disease based upon results of imaging studies and bone marrow morphology. Radiologic features are used to distinguish locoregional tumors that do not involve local structures (INRGSS L1) from locally invasive tumors (INRGSS L2, exhibiting image-defined risk factors (IDRF)). Table 24.6 shows neuroblastoma IDRF. Stages M and MS refer to tumors that are metastatic, however the MS designation refers to patients under 18 months of age who have metastases restricted to skin, liver, and/or bone marrow.

TREATMENT

Treatment modalities used are surgery, chemotherapy, radiotherapy, and immunotherapy. The role of each is determined by clinical and molecular characteristics determined at diagnosis such as age, stage, and biological features.

TABLE 24.6   Neuroblastoma Image-Defined Risk Factors (IDRF) By Location

NECK:

Tumor encasing carotid and/or vertebral artery and/or internal jugular vein

Tumor extending to base of skull

CERVICO-THORACIC JUNCTION:

Tumor encasing brachial plexus roots

Tumor encasing subclavian vessels and/or vertebral and/or carotid artery

Tumor compressing the trachea

THORAX:

Tumor encasing the aorta and/or major branches

Tumor compressing the trachea and/or principal bronchi

Lower mediastinal tumor, infiltrating the costovertebral junction between T9 and T12

Significant pleural effusion with or without presence of malignant cells

THORACO-ABDOMINAL:

Tumor encasing the aorta and/or vena cava

ABDOMEN/PELVIS:

Tumor infiltrating the porta hepatis

Tumor infiltrating the branches of the superior mesenteric artery at the mesenteric root

Tumor encasing the origin of the celiac axis and/or of the superior mesenteric artery

Tumor invading one or both renal pedicles

Tumor encasing the aorta and/or vena cava

Tumor encasing the iliac vessels

Pelvic tumor crossing the sciatic notch

Ascites with or without presence of malignant cells

DUMBBELL TUMORS WITH SYMPTOMS OF SPINAL CORD COMPRESSION:

Any localization

INVOLVEMENT/INFILTRATION OF ADJACENT ORGANS/STRUCTURES:

Pericardium, diaphragm, kidney, liver, duodeno-pancreatic block, mesentery, and others

From: Monclair et al. (2009), with permission.

Surgery

The role of surgery is to establish the diagnosis, provide tissue for biologic studies, stage the disease, and excise the tumor (when feasible).

- Timing of surgical resection is dependent upon radiographic features of response to treatment and upon a patient's risk stratification.
- Residual tumor does not adversely affect prognosis in low- or intermediate-risk disease, while an attempt at complete surgical resection is recommended for patients with high-risk disease if resection can be performed without significant morbidity.
- Attempted surgical resection of localized tumors (INSS stage 1 or stage 2, INRGSS L1) is the mainstay of therapy. Recent studies, however, demonstrated that infants less than 6 months of age with small localized tumors (INSS stage 1, INRGSS L1) of the adrenal gland are likely to have spontaneous tumor regression. An alternative treatment option in this cohort of patients is to perform surgery only in those cases in which an increase by greater than 25% from baseline in tumor size is demonstrated by serial imaging studies.
- Surgical resection that may result in loss or damage to vital organs should be avoided, especially in the setting of low- and intermediate-risk neuroblastoma. In such cases, chemotherapy should be administered prior to attempted surgical resection.
- Delayed surgical resection of primary disease should be performed in patients with evidence of metastatic disease.

Radiation Therapy

Although neuroblastoma is an extremely radiosensitive tumor, the long-term risks of radiotherapy are generally not warranted in patients with low- and intermediate-risk disease. Radiotherapy is, however, an important component of therapy for patients with high-risk disease.

Radiotherapy may be warranted in the following clinical scenarios involving patients with low- or intermediate-risk neuroblastoma:

- Neonates with INSS stage 4S neuroblastoma who develop respiratory distress secondary to hepatomegaly or hepatic failure and for whom treatment with chemotherapy is not feasible or has been ineffective.
- Children with neuroblastoma-associated spinal cord compression and neurologic compromise for whom there is a contraindication to the use of emergent chemotherapy. Laminectomy has also been used to rapidly reduce cord compression. Both radiation and surgery are associated with a significant incidence of vertebral body damage and scoliosis.

PROGNOSIS, RISK STRATIFICATION, AND THERAPY

- Factors of prognostic significance are listed in Table 24.7. Table 24.8 shows the approach used by the Children's Oncology Group (COG) for classification for neuroblastoma risk group.
- Clinical variables utilized for risk stratification include stage and age; 18 months of age is used in risk designation.
- Biologic features of neuroblastoma tumors are of critical importance for risk assessment. Amplification of the *MYCN* oncogene (v-Myc avian myelocytomatosis viral oncogene neuroblastoma derived homolog) is detected in approximately 20% of primary tumors overall, and is strongly correlated with advanced-stage disease and treatment failure. Its association with poor outcome in patients with otherwise favorable disease patterns highlights its importance as a prognostic factor.
- The presence of segmental chromosome aberrations (SCA) within tumor cells, frequently involving loss of heterozygosity of chromosome 1q, 1p or gain of chromosome 17q is associated with an increased risk of disease recurrence and in some cohorts of patients, decreased survival.
- Diploid DNA content of tumor cells is associated with a less favorable outcome in subsets of children with neuroblastoma, while hyperdiploid DNA content appears to be associated with more favorable outcome.
- Response criteria are based upon disease status at primary and metastatic sites. Neuroblastoma-specific definitions of response to treatment are listed in Table 24.9. An international effort is underway to revise these criteria specifically in regard to response in metastatic sites including bone and bone marrow. In addition, the serial measurement of serum or urine catecholamine levels, while useful as a diagnostic tool, is no longer used in response assessment due to lack of international standardization and well-documented intrapatient variability.

TABLE 24.7 Factors of Prognostic Significance in Neuroblastoma

Prognostic factor	Comparison[a]	5-year overall survival
Pathology	Differentiated > undifferentiated	89% vs 72%
	Low MKI > high MKI	82% vs 44%
Age at diagnosis	<18 months >18 months or older	89% vs 49%
Ferritin level (ng/ml)	<92 ng/ml >92 ng/ml or greater	87% vs 52%
LDH	Low (<1500 u/ml) > high (>1500 u/ml)	85% vs 58%
Stage	1, 2, 3, 4s >4	91% vs 42%
MYCN gene amplification	Normal > amplified copy number	82% vs 34%
DNA index (DI)	Hyperdiploid (DI > 1) > diploid (DI = 1)	82% vs 60%
Segmental chromosomal Aberrations	Absent > present	
Chromosome 1p	Normal > ch1p aberration	83% vs 48%
Chromosome 11q	Normal > ch11q aberration	79% vs 57%
Chromosome 17q	Normal > Gain ch17q	74% vs 55%

[a]>, better prognosis.
MKI, mitosis-karyorrhexis index; ch, chromosome.
From: Cohn et al. (2009), with permission.

TABLE 24.8 Children's Oncology Group (COG) Risk Group Assignment for Neuroblastoma

Risk group	INSS	Age	*MYCN*	Ploidy	INPC pathology
Low risk	1	Any	Any	Any	Any
Low risk	2a/2b	Any	Not amp	Any	Any
High risk	2a/2b	Any	Amp	Any	Any
Intermediate risk	3	<547 days	Not amp	Any	Any
Intermediate risk	3	≥547 days	Not amp	Any	FH
High risk	3	Any	Amp	Any	Any
High risk	3	≥547 days	Not amp	Any	UH
High risk	4	<365 days	Amp	Any	Any
Intermediate risk	4	<365 days	Not amp	Any	Any
High risk	4	365 to <547 days	Amp	Any	Any
High risk	4	365 to <547 days	Any	DI = 1	Any
High risk	4	365 to <547 days	Any	Any	UH
Intermediate risk	4	365 to <547 days	Not amp	DI>1	FH
High risk	4	≥547 days	Any	Any	Any
Low risk	4s	<365 days	Not amp	DI>1	FH
Intermediate risk	4s	<365 days	Not amp	DI = 1	Any
Intermediate risk	4s	<365 days	Not amp	Any	UH
High risk	4s	<365 days	Amp	Any	Any

INSS, International Neuroblastoma Staging System; amp, amplified; DI, DNA index; FH, favorable histology; UH, unfavorable histology.
Adapted from: Park et al. (2008), with permission.

While current studies use the INRGSS for risk stratification, data generated from completed clinical trials made use of the INSS. For this reason, the following discussion refers to patient groups as designated using INSS nomenclature.

Low-Risk Group

The low-risk group consists of (see Table 24.8):

- All patients with INSS stage 1 and 2 neuroblastoma without *MYCN* amplification.
- Patients with INSS stage 4S disease with favorable International Neuroblastoma Pathology Committee (INPC) tumor histology, hyperdiploid tumor DNA content without *MYCN* amplification.

Treatment of Low-Risk Patients

- Two sequential multi-institutional clinical trials in COG (CCG3881 and COG P9641) have demonstrated that surgical resection is the mainstay of treatment of patients with INSS stage 1 or stage 2 neuroblastoma without *MYCN* amplification.
- An additional multi-institutional clinical trial in the COG (ANBL00P2) demonstrated that for patients less than 6 months of age with small (volume less than 16 ml) adrenal tumor it is safe to observe only, and surgery should be reserved only for those patients with disease progression.
- Chemotherapy is reserved for patients whose tumor cannot be approached surgically without damaging vital structures or who present with life-threatening symptoms or neurologic compromise.
- Incomplete surgical resection does not compromise relapse-free survival in localized disease and favorable biologic features.
- Chemotherapy is recommended if less than 50% of the tumor is resected at diagnosis. Even for those receiving chemotherapy, results from recent clinical trials indicate that excellent outcomes can be maintained while delivering less intense chemotherapy for those patients with localized tumors but less than 50% tumor resection at diagnosis.
- Reduction in chemotherapy is not recommended in patients with tumors exhibiting SCA (including loss of heterozygosity of chromosome 11q or 1p), as these patients have increased risk for recurrence.
- Disease recurrence is rare in low-risk patients regardless of extent of surgical resection. When recurrent disease is seen in this population, it typically occurs at the primary site or within regional lymph nodes. Treatment of recurrent disease is patient-specific and may include repeat surgical resection or chemotherapy. Decision-making is based upon the site of disease recurrence and associated symptoms.

TABLE 24.9 Modified INRC Definitions of Response to Treatment

Response	Primary tumor[a]	Metastatic sites[a]
CR	No tumor	No tumor; catecholamines normal; MIBG normal
VGPR[b]	Decreased by 90–99%	No tumor; catecholamines normal; residual ^{99}Tc bone changes allowed, MIBG normal
PR	Decreased by >50%	All measurable sites decreased by >50%. Bones and bone marrow: number of positive bone sites decreased by >50%; no more than one positive bone marrow site allowed[c]
MR	No new lesions; >50% reduction of any measurable lesion (primary or metastases) with <50% reduction in any other; <25% increase in any existing lesion	
SD	No new lesions; <25% increase in any existing lesion; exclude bone marrow evaluation	
PD	Any new lesion; increase of any measurable lesion by >25%; previous negative marrow positive for tumor	

[a]*Evaluation of primary and metastatic disease as outlined in Table 24.4.*
[b]*VGPR designation will be removed in the planned revision of the INRC.*
[c]*One positive marrow aspirate or biopsy allowed for PR if this represents a decrease from the number of positive sites at diagnosis. Of note, bone marrow designation will be revised in planned revision of the INRC.*
CR, complete response; VGPR, very good partial response; PR, partial response; MR, mixed response; SD, stable disease; PD, progressive disease.
From: Brodeur et al. (1993).

TABLE 24.10   Intermediate-Risk Chemotherapy Regimen (COG A3961 SCHEMA)

Course #	Cycle #	Day	Agents
I	1	0	Carboplatin and etoposide
		1	Etoposide
		1	Etoposide
	2	21	Carboplatin, cyclophosphamide, and doxorubicin
	3	42	Cyclophosphamide and etoposide
		43	Etoposide
		44	Etoposide
	4	63	Carboplatin, etoposide, and doxorubicin
		64	Etoposide
		65	Etoposide
II	5	84–86	Repeat chemotherapy drugs in cycle 3
	6	105	Repeat cycle 2
	7	126–128	Repeat cycle 1
	8	147	Cyclophosphamide and doxorubicin

For children <1 year of age or who are ≤12 kg in weight, the doses of chemotherapy will be adjusted and given as mg/kg. Each cycle is given at 3 weeks (21 days) interval

DOSE OF CHEMOTHERAPEUTIC AGENTS

Carboplatin 560 mg/m^2 or 18.6 mg/kg IV over 1 h for 1 day

Etoposide 120 mg/m^2 or 4 mg/kg IV over 2 h daily for 3 days

Cyclophosphamide 1000 mg/m^2 or 33.3 mg/kg over 1 h daily for 1 day

Doxorubicin 30 mg/m^2 or 1 mg/kg IV over 60 min daily for 1 day

Myeloid growth factors should be given for infants less than 60 days of age.

Patients with INSS Stage 1 and 2 Tumors

- Surgical removal of the primary tumor without damage to vital organs is followed by close observation for patients with localized, resectable tumors. Three-year event-free survival (EFS) and overall survival (OS) exceeds 90% following >50% resection in patients with tumors that have favorable biologic features.
- For patients with INSS stage 2 disease and no evidence of SCA in tumor tissue but for whom a >50% resection of the primary tumor cannot be achieved, two cycles of moderate-dose chemotherapy utilizing carboplatin, etoposide, cyclophosphamide, and doxorubicin are recommended. Chemotherapy agents and dosing (cycles 1–2) are according to the regimen detailed in Table 24.10.

Patients with INSS Stage 4S Disease

The majority of patients with INSS stage 4S disease fall into the low-risk category with EFS 86% and OS 92%.

- The majority of these tumors will undergo spontaneous regression. However, patients less than 2 months of age have a higher incidence of respiratory compromise and liver dysfunction due to diffuse infiltration of the liver with tumor.
- In the absence of life-threatening complications, no therapy is indicated.
- Surgical resection of the primary tumor is not usually necessary, although biopsy of primary or metastatic sites of disease should be undertaken to ascertain biologic characteristics in clinically stable patients.
- In very young infants with respiratory compromise or evolving hepatomegaly, surgical biopsies should be avoided and chemotherapy considered.

- Chemotherapy should be utilized in those patients with life-threatening complications including respiratory compromise and severe liver dysfunction. Studies have demonstrated that initiation of intermediate-risk chemotherapy (cycles 1–4, Table 24.10) can produce clinical stabilization. Chemotherapy should be discontinued once the patient's clinical condition has stabilized. Emergent low-dose radiotherapy can also be utilized (150 cGy two or three times to the anterior two-thirds of the liver through lateral oblique ports) in extreme clinical situations. This dose is generally sufficient to halt the progression of the tumor and induce regression.
- The rare INSS stage 4S patient with unfavorable biological features may be a candidate for more intensive treatment.

Intermediate-Risk Group

The intermediate-risk group (Table 24.8) consists of the following:

- Patients less than 18 months of age with INSS stage 3 tumors without *MYCN* amplification.
- Patients 18 months of age or older with INSS stage 3 tumors with favorable histology and without *MYCN* amplification.
- Patients less than 12 months of age with INSS stage 4 neuroblastoma whose tumors do not have *MYCN* amplification.
- Patients with INSS stage 4S MYCN non-amplified disease with either unfavorable INPC histology and/or diploidy.
- Patients between 12 months and 18 months of age with INSS stage 4 MYCN non-amplified neuroblastoma, whose tumors have favorable INPC histology and hyperdiploid DNA content.

Treatment of Intermediate-Risk Patients

Chemotherapy

- Chemotherapy regimens utilized include cyclophosphamide, carboplatin, etoposide, and doxorubicin. The clinical trial COG A3961 outlined in Table 24.10, utilized four cycles of chemotherapy (course I) for the following patients:
  - Intermediate-risk INSS stage-3 patients noted above.
  - Patients less than 12 months of age with INSS stage 4 neuroblastoma exhibiting favorable tumor characteristics, defined as favorable INPC histology and hyperdiploid DNA content.
    Eight cycles of the same chemotherapy agents (courses I and II) are administered to the following patients:
  - Patients less than 12 months with INSS stage 4 or with INSS stage 4S without *MCYN* amplification but with unfavorable tumor histology or diploid tumor cell DNA content at time of diagnosis.
  - Patients with residual metastatic disease or less than 90% reduction in primary tumor with chemotherapy and surgical approaches (very good partial response, VGPR).
- The COG clinical trial (ANBL0531) demonstrated that excellent outcome was obtained (OS exceeding 95%) for select intermediate-risk patients (INRG stage 2B disease with less than 50% resection at diagnosis and INSS stage 3 disease AND favorable tumor biologic features defined as favorable INPC histology, hyperdiploid DNA content and no SCA) with as few as two cycles of chemotherapy (e.g., cycles 1 and 2 of intermediate-risk chemotherapy per Table 24.10) and stopping chemotherapy once the patient has achieved at least 50% reduction in primary tumor volume (partial response, PR).
- Clinical trials performed in Germany have demonstrated spontaneous tumor regression in the majority of patients less than 12 months of age with localized intermediate-risk neuroblastoma (INSS Stage 3/INRGSS L2) and favorable tumor biologic features. Current clinical trials are underway in both North America and Europe to further explore whether chemotherapy can be eliminated in a select cohort of patients with INSS stage 3/INRGSS L2 whose tumors exhibit favorable INPC histology, hyperdiploid DNA content, and no SCA.
- A minimum of 4–8 cycles of chemotherapy should be administered to patients with intermediate-risk disease whose tumors exhibit SCA as these patients exhibit increased risk for recurrence compared to patients whose tumors lack these chromosomal aberrations. The decision regarding the precise number of cycles to be given depends upon additional biologic features noted above.
- Retrospective analyses of clinical trials performed by the Children's Cancer Group and Pediatric Oncology Group suggest that patients with INSS stage 4 neuroblastoma between the ages of 12 and 18 months with

favorable histology and hyperdiploid tumors without *MYCN* amplification have an excellent prognosis, similar to that of patients less than 12 months of age. Historically these patients have been treated using high-risk therapy algorithms, but should be treated as intermediate-risk patients.

- The COG ANBL0531 clinical trial assessed whether therapy can be significantly reduced while maintaining excellent survival in patients with INSS stage 4 neuroblastoma between the ages of 12 and 18 months whose tumors have favorable histology, lack *MYCN* amplification, and are hyperdiploid. In lieu of high-risk therapy, patients received eight cycles of intermediate-risk chemotherapy followed by six cycles of 13-cis retinoic acid therapy (80 mg/m^2/dose for patients >12 kg and 2.67 mg/kg/dose for patients ≤12 kg given twice per day for 14 consecutive days in each 28-day cycle for a total of six cycles) using a therapeutic endpoint of at least 90% reduction in tumor volume.

Surgery

- Surgical resection following preoperative chemotherapy for treatment of intermediate-risk neuroblastoma may be performed to achieve the desired primary tumor response, although the extent of resection required remains controversial.
- Clinical trials performed in Germany reveal excellent OS with limited surgical intervention for patients with stage 3 intermediate-risk disease.
- Mass screening studies suggest that tumors occurring in infancy can regress without chemotherapy or surgical resection.
- COG ANBL0531 demonstrated that a further reduction in therapy is appropriate for patients with INSS stage 3 disease and favorable tumor biology or patients less than 18 months of age with INSS stage 4 disease and favorable biology based on use of an endpoint of partial response (PR; at least 50% reduction in primary tumor) rather than the very good partial response (VGPR, at least 90% reduction in primary tumor) endpoint used in prior trials.

Radiation Therapy

Radiation therapy has a limited role in the treatment of intermediate-risk neuroblastoma.

- Radiotherapy is to be administered when there are life-threatening symptoms or a risk of organ impairment/neurologic compromise due to tumor bulk not responding to initial chemotherapy.
- Intraspinal neuroblastoma with neurologic deficits should be managed with chemotherapy alone when possible to avoid the late effects associated with laminectomy or radiation to the spine of a young child. Laminectomy or radiation should be used if there is progression of neurological signs despite chemotherapy.

High-Risk Group

The high-risk group (Table 24.8) consists of the following:

- INSS stage 4 greater than 18 months of age at diagnosis.
- INSS stage 4 with tumor *MYCN* amplification, regardless of age.
- INSS stage 4 between 12 and 18 months of age at diagnosis with unfavorable tumor histology and/or diploid tumor DNA content.
- INSS stage 2 or stage 3 with tumor *MYCN* amplification, regardless of age.
- INSS stage 4S with *MYCN* amplification.
- Classification of patients with INSS stage 3 and unfavorable histology, greater than 18 months of age at diagnosis is not standardized internationally. Survival for this patient cohort is inferior to that of patients with similar disease stage but age less than 18 months at diagnosis (OS <80% vs >95%). For this reason, these children are considered to have high-risk neuroblastoma according to the COG risk schema but are classified differently according to European classification systems.

The probability of long-term survival in this group of patients is approximately 45% following comprehensive treatment.

Treatment approaches for high-risk neuroblastoma include:

- Induction therapy that incorporates alternating cycles of dose-intensive multiagent chemotherapy and surgical resection of primary disease.
- Consolidation therapy that includes myeloablative doses of chemotherapy with autologous hematopoietic stem cell rescue followed by external beam radiotherapy to the primary tumor site.
- Postconsolidation therapy directed at the treatment of minimal residual disease.

TABLE 24.11 Induction Chemotherapy Regimens and Response Rates from Multicenter Clinical Trials

Study	Regimen[a]	Response rate (CR/VGPR/PR)[b]
CCG 3891	CDDP, DOX, ETOP, CPM	76%
	Repeat every 28 days for 5 cycles	
POG 9341	CDDP/ETOP: VCR/DOX/CPM; IFOS/ETOP; CARBO/ETOP; CDDP/ETOP	78%
	Five sequential cycles every 28 days	
COG A3973	VCR/DOX/CPM cycles 1, 2, 4, 6	78%
	CDDP/ETOP cycles 3, 5	
	Cycles given every 21 days	
ENSG-5	VCR/CARBO/ETOP cycle 1, 3, 5, 7	85%
	VCR/CDDP cycles 2, 4, 6, 8	
	Cycles given every 10 days	

[a]All drugs are given intravenously.
[b]Responses include surgery and chemotherapy.
CARBO, carboplatin, CDDP, cisplatin; CPM, cyclophosphamide; CR, complete response; DOX, doxorubicin; PR, partial response; VCR, vincristine; ETOP, etoposide; IFOS, Ifosfamide.
CCG 3891
Cisplatin 60 mg/m^2/dose on day 1, doxorubicin 30 mg/m^2/dose on day 3, etoposide 100 mg/m^2/dose on days 3 and 6 and cyclophosphamide 1000 mg/m^2/dose on days 4 and 5.
POG 9341
Cycles 1 and 5: cisplatin 40 mg/m^2/day on days 1–5, etoposide 100 mg/m^2/dose twice daily on days 1–3. *Cycle 2*: cyclophosphamide 1000 mg/m^2/dose on days 1 and 2, doxorubicin 60 mg/m^2/dose on day 1, vincristine 1.5 mg/m^2/dose on days 1, 8, and 15. *Cycle 3*: ifosfamide 2 g/m^2/dose on days 1–5, etoposide 75 mg/m^2/dose twice per day on days 1–3. *Cycle 4*: carboplatin 500 mg/m^2/dose on days 1 and 2, etoposide 75 mg/m^2/dose twice per day on days 1–3.
COG A3973
Vincristine 0.67 mg/m^2/24 h or 0.022 mg/kg/24 h (whichever is lower) by continuous infusion over 72 h on days 1, 2, and 3 (total dose 2 mg/m^2/over 72 h or 0.067 mg/kg/72 h with *maximum* total per cycle of 2 mg). For patients ≤12 kg, use 0.022 mg/kg/24 h by continuous infusion for 72 h (total dose 0.066 mg/kg/3 days). For infants <12 months, use 0.017 mg/kg/24 h by continuous infusion for 72 h, total dose 0.051 mg/kg/3 days. Cyclophosphamide 2.1 g/m^2/day (for patients ≤12 kg use 70 mg/kg/day) on days 1 and 2. Doxorubicin 25 mg/m^2/24 h (for patients ≤12 kg, use 0.83 mg/kg/24 h) by continuous infusion over 72 h on days 1, 2, and 3 (total dose 75 mg/m^2/72 h or 2.49 mg/kg/72 h if ≤12 kg). Etoposide 200 mg/m^2/day (for patients ≤12 kg use 6.67 mg/kg/day) on days 1, 2, and 3 (total dose 600 mg/m^2/3 days or 20 mg/kg/3 days if ≤12 kg). Cisplatin 50 mg/m^2/day (for patients ≤12 kg use 1.66 mg/kg/day) on days 1, 2, 3, and 4 (total dose 200 mg/m^2/4 days or 6.64 mg/kg/4 days).
ENSG-5
Cycles 1 and 5: vincristine 1.5 mg/m^2 on day 1, carboplatin 750 mg/m^2 on day 1, etoposide 175 mg/m^2 on days 1 and 2.
Cycles 2, 4, 6, and 8: vincristine 1.5 mg/m^2 on day 1, cisplatin 80 mg/m^2 over 24 h beginning on day 1.
Cycles 3 and 7: vincristine 1.5 mg/m^2 on day 1, etoposide 175 mg/m^2 on days 1 and 2, cyclophosphamide 1050 mg/m^2 on days 1 and 2.
Adapted with permission from: Matthay et al. (2009), Zage et al. (2008), and Pearson et al. (2008).

Induction Therapy

Neuroblastoma is usually responsive to chemotherapy even in the presence of *MYCN* amplification. The goal of induction therapy is to maximally reduce tumor burden by decreasing primary tumor bulk and eliminating metastatic sites of disease. The effectiveness of induction chemotherapy is assessed by the response rate.

- The quality of remission at the end of induction therapy is associated with long-term survival.
- Retrospective and prospective analyses have determined that persistent metastatic disease as detected by uptake of ^{123}I MIBG nuclear imaging is an adverse prognostic factor.
- There are many treatment protocols available for induction chemotherapy. The duration of induction therapy is approximately 6 months. Table 24.11 describes chemotherapy regimens currently being used in North America and Europe. The induction regimen currently being used by the COG is shown in Table 24.12.
- Despite marked increases in the dose intensity of active agents such as cisplatin, cyclosphosphamide, and etoposide, complete remission is achieved in less than 75% of patients at the end of induction. Approximately 10% of patients will have disease progression during induction.
- Delayed resolution of MIBG avid metastatic disease during induction therapy correlates with decreased EFS and OS. Although the ideal therapy for patients with multiple sites of persistent metastatic disease at completion of induction therapy has not been clearly delineated, clinicians may elect to administer alternative therapy to these patients prior to delivering consolidation therapy.

TABLE 24.12 Induction Therapy for High-Risk Group Neuroblastoma (Children's Oncology Group)

Cycle 1	Topotecan	1.2 mg/m^2/dose IV once daily for 5 doses
	Cyclophosphamide	>12 kg: 400 mg/m^2/dose IV once daily for 5 doses ≤12 kg: 13.3 mg/kg/dose IV once daily for 5 doses
Cycle 2	Same as cycle 1	
PERIPHERAL STEM CELL HARVEST		
Cycle 3	Cisplatin	>12 kg: 50 mg/m^2/dose IV once daily for 4 doses ≤12 kg: 1.66 mg/m^2/dose IV once daily for 4 doses
	Etoposide	>12 kg: 200 mg/m^2/dose IV once daily for 3 days ≤12 kg: 6.67 mg/kg/dose IV once daily for 3 days
Cycle 4	Cyclophosphamide	>12 kg: 2100 mg/m^2/dose IV once daily for 2 doses ≤12 kg: 70 mg/kg/dose IV once daily for 2 doses
	Doxorubin	>12 kg: 25 mg/m^2/dose IV once daily for 3 doses ≤12 kg: 0.83 mg/kg/dose IV once daily for 3 doses
	Vincristine	>12 kg and ≥12 months: 0.67 mg/m^2/dose IV once daily for 3 days ≤12 kg and ≥12 months: 0.022 mg/kg/dose IV once daily for 3 days <12 months: 0.017 mg/kg/dose IV once daily for 3 days
Cycle 5	Same as cycle 3	
SURGERY		
Cycle 6	Same as cycle 4, ongoing trails have eliminated cycle 6 and moved surgical resection to occur prior to cycle 5	

Consolidation Therapy

Patients without progressive disease during induction therapy proceed with consolidation therapy.

- The goal of consolidation therapy is to achieve a state of minimal residual disease. The efficacy of myeloablative chemotherapy with autologous hematopoietic stem cell rescue as consolidation therapy has been demonstrated in several multicenter randomized clinical trials.
- Peripheral blood stem cells (PBSC) are the recommended source for hematopoietic stem cell rescue following myeloablative chemotherapy. The timing of PBSC collection varies. In COG trials, PBSC are collected during early induction chemotherapy (usually after 2−3 cycles of chemotherapy) regardless of the presence of metastatic bone marrow disease while in European trials, PBSC collection is delayed until later in induction and with clearance of BM metastatic disease.
- A randomized clinical trial conducted by the COG (A3973) showed no benefit associated with *ex vivo* purging of PBSC.
- Data from a recently completed COG trial (ANBL0532) will determine whether sequentially administered (tandem) myeloablative chemotherapy regimens may provide a further decrement in risk of neuroblastoma relapse compared with single autologous transplant. Current standard of care is a single high-dose chemotherapy regimen.

Consolidation regimens differ and, in general, regimens containing total body irradiation as a component of high-dose therapy are not recommended due to associated late effects. Recent clinical trials in the COG have utilized a myeloablative chemotherapy regimen of carboplatin, etoposide, and melphelan (CEM). Dosing in patients with creatinine clearance or glomerular filtration rate ≥100 ml/min/1.73 m^2 includes:

- Carboplatin (425 mg/m^2/dose IV for patients >12 kg and 14.2 mg/kg/dose IV for patients ≤12 kg once daily for four doses).

- Melphalan (70 mg/m^2/dose IV for patients >12 kg and 2.3 mg/kg/dose IV for patients ≤12 kg once daily for three doses).
- Etoposide (338 mg/m^2/dose IV for patients >12 kg and 11.3 mg/kg/dose IV for patients ≤12 kg once daily for four doses).

Doses for patients with altered renal function (creatinine clearance or glomerular filtration rate ≥60 and <100 ml/min/1.73 m^2) should be carefully modified.

A European clinical trial demonstrated decreased risk of neuroblastoma relapse and improved survival for those children who received rapid COJEC induction (cisplatin, vincristine, carboplatin, etoposide, and cyclophosphamide) therapy followed by busulfan and melphalan myeloablative chemotherapy compared to those who received CEM myeloablative chemotherapy. Initial use of busulfan and melphalan myeloablative chemotherapy by North American centers raised concern regarding risk of sinusoidal obstructive syndrome. Ongoing trials are assessing the feasibility and tolerability of using myeloablative busulfan and melphalan within the context of standard induction and postconsolidation therapies used in COG trials.

Postconsolidation Continuation Therapy

The high rate of relapse observed in patients with high-risk neuroblastoma following treatment limited to induction and consolidation therapies suggested the existence of minimal residual disease at completion of consolidation therapy. The aim of postconsolidation therapy is to eradicate any residual tumor cells using non-cytotoxic agents that are active against persisting chemoresistant tumor cells. The following are used in postconsolidation treatment:

- Radiation therapy: the COG trials have prescribed radiation to the primary site and any residual sites of MIBG avidity following consolidation with high-dose chemotherapy and stem cell rescue. The dose is 2160 cGy.
- Isotretinoin (13-*cis*-retinoic acid, cisRA): The administration of cisRA (80 mg/m^2/dose for patients >12 kg and 2.67 mg/kg/dose for patients ≤12 kg given twice per day for 14 consecutive days in each 28-day cycle for a total of six cycles) following consolidation therapy has been shown to improve EFS in a randomized clinical trial (CCG3891) and has been incorporated into postconsolidation therapy for high-risk neuroblastoma.
- Anti-ganglioside (GD2) antibody therapy: The addition of anti-GD2 immunotherapy (chimeric 14.18 antibody) combined with cytokines granulocyte macrophage colony stimulating factor (GMCSF) and interleukin-2 (IL2) to standard cisRA has been shown to decrease the risk of recurrence and improve OS in children with high-risk neuroblastoma. Although clinical trials designed to optimize anti-GD2 antibody containing regimens for use in the postconsolidation setting are underway, the administration of anti-GD2 has become standard of care. *The acute toxicities of the ch14.18 antibody are significant, and include life-threatening anaphylaxis, capillary leak syndrome, and cardiac events, as well as severe pain amongst other acute toxicities. The antibody should only be administered in a closely monitored setting, by experienced and skilled personnel.*

The COG study, ANBL0032, administered the ch14.18 anti-GD2 antibody with GMCSF, IL-2 and *cis*-retinoic acid in six cycles.

- Cycles 1, 3, and 5:
  - GMCF 250 micrograms/day for 14 days from days 0 to 13
  - Ch14.18 antibody at 17.5 mg/m^2 per day for 4 days from days 3 to 6
  - *Cis*-retinoic acid 160 mg/m^2/day or 5.33 mg/kg/day if <12 kg from days 10 to 23 of the cycles.
- Cycles 2 and 4:
  - IL-2 3 MIU/m^2/day daily from days 0 to 3 and 4.5 MIU/m^2/day daily from days 7 to 10
  - Ch14.18 antibody at 17.5 mg/m^2 per day for 4 days from days 7 to 10
  - Cis-retinoic acid 160 mg/m^2/day or 5.33 mg/kg/day if <12 kg from days 14 to 27.
- Cycle 6:
  - *Cis*-retinoic acid 160 mg/m^2/day or 5.33 mg/kg/day if <12 kg from days 14 to 27.

POST-THERAPY MONITORING

Routine monitoring of patients during and following completion of therapy is required given the risk for recurrent disease. Table 24.13 lists the recommended disease status monitoring for high-risk neuroblastoma.

TABLE 24.13 Recommended Disease Status Monitoring for High-Risk Neuroblastoma

Diagnostic test	First year	Second year	After second year
FREQUENCY POST-THERAPY			
CT or MRI	3 months or as clinically indicated	3–6 months or as clinically indicated	6–12 months if clinically indicated
^{123}I-MIBG scintigraphy if MIBG avid at diagnosis	3 months or as clinically indicated	3–6 months or as clinically indicated	6–12 months if clinically indicated
^{18}F-FDG PET if not MIBG avid	3 months or as clinically indicated	3–6 months or as clinically indicated	6–12 months if clinically indicated
Bone marrow biopsy and aspirate (bilateral)	as clinically indicated	as clinically indicated	as clinically indicated

Patients must also be monitored for long-term complications of therapy including anthracycline-induced cardiomyopathy, platinum-induced ototoxicity and renal dysfunction, chemotherapy and radiation-induced risk for secondary malignancies, infertility, and growth and development deficiencies.

- The risk for disease recurrence is highest in the initial 2 years following completion of therapy but late relapses can also occur.
- Disease recurrence primarily occurs in metastatic sites. Bone and bone marrow are the most frequent sites of disease recurrence. Soft tissue disease (including nodal sites) can also occur.
- Primary site recurrence is less frequent, occurring in 10–15% of children following aggressive surgical resection and local radiotherapy.
- The incidence of recurrence in the central nervous system may be increasing as more effective systemic therapies are developed.

Post-Therapy Surveillance Recommendations

- I[123]-MIBG scan (if MIBG avid at diagnosis), bone marrow assessment and CT/MRI scans of primary site of disease are performed during the initial 5 years from completion of therapy.
- Surveillance (MIBG scans, CT/MRI scans, and bone marrow aspirations) should be carried out every 3–6 months for the initial 2 years as the risk for recurrence is greatest in the initial 2 years from diagnosis. For those patients who remain in complete remission after 2 years the frequency of surveillance scans can be extended to every 6–12 months until 5 years from diagnosis.
- Patients who survive long term from high-risk neuroblastoma should be monitored in a comprehensive survivorship program skilled at facilitating management of therapy-related morbidities.

Table 24.13 lists diagnostic evaluations that should be used to monitor disease status.

SPECIAL TREATMENT CONSIDERATIONS

Dumbbell Neuroblastoma and Spinal Cord Compression

Spinal cord decompression is a neurological emergency and various treatment modalities including chemotherapy, surgery, and radiation can be effective. Complete neurological recovery is observed in 30–40% of patients. Neurological improvement is observed in 65–70% of patients.

- *Decompression with chemotherapy combined with the use of corticosteroids (e.g., dexamethasone):* Dexamethasone should be administered immediately upon recognition of neuroblastoma-related spinal cord compression. Diagnosis and staging should be performed rapidly to avoid delay in initiation of chemotherapy. The intensity of chemotherapy is determined according to stage and age risk criteria as previously described but treatment should not be withheld if nuclear imaging cannot be performed promptly. The majority of patients (>85%) will have rapid tumor shrinkage and improvement in or prevention of neurologic compromise. The patient's neurologic status must be monitored every 4–6 h.

- Decompression by laminectomy is reserved only for patients who do not respond to initial chemotherapy and dexamethasone. These patients are at risk for the later development of kyphoscoliosis and spinal column instability following a laminectomy procedure.
- Decompression by radiotherapy is no longer indicated in the initial management of neuroblastoma but should be considered if the patient is not responding to initial chemotherapy.

Opsoclonus Myoclonus Ataxia Syndrome (OMAS)

Clinical features of the OMAS are detailed above. Clinical observations suggest an immune-mediated mechanism for OMAS.

- Patients with OMAS have serum and cerebrospinal fluid IgG and IgM auto-antibodies that bind to the cytoplasm of cerebellar Purkinje cells, central and peripheral axons, and neural proteins of the neurofilaments.
- Diffuse and extensive lymphocytic infiltration with lymphoid follicles is a characteristic histologic feature of neuroblastic tumors with associated opsoclonus-myoclonus and ataxia.

Treatment

Definitive treatment remains controversial; suppression of the immune system is the mainstay of therapy.

- Chemotherapy should be utilized as appropriate, based upon documented clinical risk group at the time of diagnosis.
- Immunomodulatory therapies such as corticosteroids and low-dose cyclophosphamide have been utilized for treatment.
- High-dose intravenous immunoglobulin (IVIG 150–500 mg/kg/day for 4–5 days) may benefit patients or provide a steroid-sparing effect.
- The use of an anti-CD20 antibody (rituximab) is being evaluated in patients who fail to respond to the above therapies. Definitive data regarding the effectiveness of this agent in OMAS are not yet available.

A stepwise approach to treatment of OMAS is being evaluated in an international multicenter trial.

NEUROBLASTOMA IN THE ADOLESCENT AND YOUNG ADULT

The distribution of primary neuroblastoma sites in this group of patients is similar to that seen in pediatric age group patients.

- Young and older adults with neuroblastoma tend to experience a more indolent disease course than younger patients but their disease is less likely to respond to initial cytotoxic therapy.
- The best treatment approach remains to be determined. Surgery with or without radiation therapy may be considered for localized disease. Intensive multiagent chemotherapy including use of autologous bone marrow transplantation may be considered, especially for those patients with advanced regional or metastatic disease.

Further Reading and References

Ambros, P.F., Ambros, I.M., Brodeur, G.M., Haber, M., Khan, J., Nakagawara, A., et al., 2009. International consensus for neuroblastoma molecular diagnostics: report from the International Neuroblastoma Risk Group (INRG) Biology Committee. Br. J. Cancer 100, 1471–1482.
Baker, D.L., Schmidt, M.L., Cohn, S.L., Maris, J.M., London, W.B., Buxton, A., et al., 2010. Children's Oncology Group. Outcome after reduced chemotherapy for intermediate-risk neuroblastoma. N. Engl. J. Med. 363 (14), 1313–1323.
Berthold, F., Boos, J., Burdach, S., et al., 2005. Myeloablative megatherapy with autologous stem-cell rescue versus oral maintenance chemotherapy as consolidation treatment in patients with high-risk neuroblastoma: a randomised controlled trial. Lancet Oncol. 6, 649–658.
Brisse, H.J., McCarville, M.B., Granata, C., Krug, K.B., Wootton-Gorges, S.L., Kanegawa, K., et al., 2011. Guidelines for imaging and staging of neuroblastic tumors: consensus report from the International Neuroblastoma Risk Group Project. Radiology 261 (1), 243–257.
Brodeur, G.M., Pritchard, J., Berthold, F., et al., 1993. Revisions of the international criteria for neuroblastoma diagnosis, staging, and response to treatment. J. Clin. Oncol. 11, 1466–1477.
Brodeur, G.M., 2003. Neuroblastoma: biological insights into a clinical enigma. Nat. Rev. Cancer 3, 203–216.
Cohn, S.L., Pearson, A.D., London, W.B., Monclair, T., Ambros, P.F., Brodeur, G.M., et al., 2009. The International Neuroblastoma Risk Group (INRG) classification system: an INRG Task Force report. J. Clin. Oncol. 27, 289–297.
George, R.E., Li, S., Medeiros-Nancarrow, C., et al., 2006. High-risk neuroblastoma treated with tandem autologous peripheral-blood stem cell-supported transplantation: long-term survival update. J. Clin. Oncol. 24 (18), 2891–2896.

Hero, B., Simon, T., Spitz, R., Ernestus, K., Gnekow, A.K., Scheel-Walter, H.G., et al., 2008. Localized infant neuroblastomas often show spontaneous regression: results of the prospective trials NB95-S and NB97. J. Clin. Oncol. 26 (9), 1504–1510.

Kreissman, S.G., Seeger, R.C., Matthay, K.K., London, W.B., Sposto, R., Grupp, S.A., et al., 2013. Purged versus non-purged peripheral blood stem-cell transplantation for high-risk neuroblastoma (COG A3973): a randomised phase 3 trial. Lancet Oncol. 14 (10), 999–1008.

Kushner, B.H., Kramer, K., LaQuaglia, M.P., et al., 2004. Reduction from seven to five cycles of intensive induction chemotherapy in children with high-risk neuroblastoma. J. Clin. Oncol. 22, 4888–4892.

Matthay, K.K., Blaes, F., Hero, B., et al., 2005. Opsoclonus myoclonus syndrome in neuroblastoma a report from a workshop on the dancing eyes syndrome at the advances in neuroblastoma meeting in Genoa, Italy, 2004. Cancer Lett. 228, 275–282.

Matthay, K.K., Reynolds, C.P., Seeger, R.C., Shimada, H., Adkins, E.S., Haas-Kogan, D., et al., 2009. Long-term results for children with high-risk neuroblastoma treated on a randomized trial of myeloablative therapy followed by 13-cis-retinoic acid: a Children's Oncology Group study. J. Clin. Oncol. 27, 1007–1013.

Modak, S., Cheung, N.K., 2007. Disialoganglioside directed immunotherapy of neuroblastoma. Cancer Invest. 25, 67–77.

Monclair, T., Brodeur, G.M., Ambros, P.F., Brisse, H.J., Cecchetto, G., Holmes, K., et al., 2009. The International Neuroblastoma Risk Group (INRG) staging system: an INRG Task Force report. J. Clin. Oncol. 27, 298–303, 2009.

Mossé, Y.P., Deyell, R.J., Berthold, F., Nagakawara, A., Ambros, P.F., Monclair, T., et al., 2014. Neuroblastoma in older children, adolescents and young adults: a report from the International Neuroblastoma Risk Group project. Pediatr. Blood Cancer 61 (4), 627–635.

Nickerson, H.J., Matthay, K.K., Seeger, R.C., et al., 2000. Favorable biology and outcome of stage IV-S neuroblastoma with supportive care or minimal therapy: a children's cancer group study. J. Clin. Oncol. 18, 477–486.

Nuchtern, J.G., London, W.B., Barnewolt, C.E., Naranjo, A., McGrady, P.W., Geiger, J.D., et al., 2012. A prospective study of expectant observation as primary therapy for neuroblastoma in young infants: a Children's Oncology Group study. Ann. Surg. 256 (4), 573–580.

Park, J.R., Eggert, A., Caron, H., 2008. Neuroblastoma: biology, prognosis and treatment. Pediatr. Clin. North. Am. 55 (1), 97–120.

Pearson, A.D., Pinkerton, C.R., Lewis, I.J., Imeson, J., Ellershaw, C., Machin, D., 2008. High-dose rapid and standard induction chemotherapy for patients aged over 1 year with stage 4 neuroblastoma: a randomised trial. Lancet Oncol. 9, 247–256.

Pritchard, J., Cotterill, S.J., Germond, S.M., Imeson, J., de Kraker, J., Jones, D.R., 2005. High dose melphalan in the treatment of advanced neuroblastoma: results of a randomised trial (ENSG-1) by the European Neuroblastoma Study Group. Pediatr. Blood Cancer 44 (4), 348–357.

Rubie, H., De Bernardi, B., Gerrard, M., Canete, A., Ladenstein, R., Couturier, J., et al., 2011. Excellent outcome with reduced treatment in infants with nonmetastatic and unresectable neuroblastoma without MYCN amplification: results of the prospective INES 99.1. J. Clin. Oncol. 29 (4), 449–455.

Schleiermacher, G., Mosseri, V., London, W.B., Maris, J.M., Brodeur, G.M., Attiyeh, E., et al., 2012. Segmental chromosomal alterations have prognostic impact in neuroblastoma: a report from the INRG project. Br. J. Cancer 107 (8), 1418–1422.

Sharp, S.E., Shulkin, B.L., Gelfand, M.J., et al., 2009. ^{123}I-MIBG scintigraphy and ^{18}F-FDG PET in neuroblastoma. J. Nucl. Med. 50, 1237–1243.

Shimada, H., Ambros, I.M., Dehner, L.P., et al., 1999. The international neuroblastoma pathology classification (the Shimada system). Cancer 86, 364–372.

Shimada, H., Umehara, S., Monobe, Y., et al., 2001. International neuroblastoma pathology classification for prognostic evaluation of patients with peripheral neuroblastic tumors: a report from the Children's Cancer Group. Cancer 92, 2451–2461.

Simon, T., Häberle, B., Hero, B., von Schweinitz, D., Berthold, F., 2013. Role of surgery in the treatment of patients with stage 4 neuroblastoma age 18 months or older at diagnosis. J. Clin. Oncol. 31 (6), 752–758.

Simon, T., Niemann, C.A., Hero, B., Henze, G., Suttorp, M., Schilling, F.H., et al., 2012. Short- and long-term outcome of patients with symptoms of spinal cord compression by neuroblastoma. Dev. Med. Child Neurol. 54 (4), 347–352.

Strother, D.R., London, W.B., Schmidt, M.L., Brodeur, G.M., Shimada, H., Thorner, P., et al., 2012. Outcome after surgery alone or with restricted use of chemotherapy for patients with low-risk neuroblastoma: results of Children's Oncology Group study P9641. J. Clin. Oncol. 30 (15), 1842–1848.

Yanik, G.A., Parisi, M.T., Shulkin, B.L., Naranjo, A., Kreissman, S.G., London, W.B., et al., 2013. Semiquantitative mIBG scoring as a prognostic indicator in patients with stage 4 neuroblastoma: a report from the Children's Oncology Group. J. Nucl. Med. 54 (4), 541–548.

Yu, A.L., Gilman, A.L., Ozkaynak, M.F., London, W.B., Kreissman, S.G., Chen, H.X., et al., 2010. Anti-GD2 antibody with GM-CSF, interleukin-2, and isotretinoin for neuroblastoma. N. Engl. J. Med. 363 (14), 1324–1334.

Zage, P.E., Kletzel, M., Murray, K., et al., 2008. Outcomes of the POG 9340/9341/9342 trials for children with high-risk neuroblastoma: a report from the Children's Oncology Group. Pediatr. Blood Cancer 51, 747–753.

25

Renal Tumors

Anne B. Warwick and Jeffrey S. Dome

WILMS' TUMOR

Incidence

- Renal tumors comprise approximately 6% of all childhood cancers and nearly 10% of all malignancies among children aged 1–4 years.
- Wilms' tumor (nephroblastoma) is the most common primary renal tumor of childhood and the sixth most common childhood malignancy in the United States.
- The male to female ratio 0.92:1.00 in unilateral Wilms' and 0.6:1.00 in bilateral Wilms' tumor.
- Wilms' tumor, clear cell sarcoma of the kidney, and malignant rhabdoid tumor of the kidney occur predominantly in younger children and are rarely seen after 10 years of age. The majority of renal tumors occurring in adolescents and young adults are renal cell carcinomas (RCCs).
- Seventy-eight percent of children with Wilms' tumor are diagnosed at 1–5 years of age, with a peak incidence occurring between 3 and 4 years of age. The median age of presentation is 44 months in unilateral disease and 32 months in bilateral disease.
- Wilms' tumor is usually sporadic, but 1% of cases are familial.

Associated Congenital Anomalies

Congenital anomalies occur in 12–15% of cases. The most frequently diagnosed congenital anomalies are aniridia, genitourinary anomalies, and hemihypertrophy. Table 25.1 lists the incidence of congenital anomalies in patients with Wilms' tumor.

The three best characterized syndromes associated with Wilms' tumor include:

- WAGR syndrome.
- Denys–Drash syndrome.
- Beckwith–Wiedemann syndrome.

Other syndromes with an estimated Wilms' tumor risk of at least 5% are listed in Table 25.2.

WAGR Syndrome

- The syndrome consists of Wilms' tumor, aniridia, genitourinary malformations, and mental retardation, but other associated anomalies are seen.
- A chromosomal deletion has been consistently found in the short arm of chromosome 11 (11p13 deletion). This area encompasses both the *WT1* and *PAX6* genes; *WT1* is implicated in Wilms' tumor and *PAX6* is implicated in aniridia.
- Aniridia is present in 1.2% of children with Wilms' tumor.
- A child with sporadic aniridia has a 5% chance of developing Wilms' tumor; most cases of sporadic aniridia have *PAX6* mutations/deletions but not *WT1* deletions. A child with *WT1* deletion has a 30–40% risk of Wilms' tumor.
- WAGR syndrome is associated with focal segmental glomerular sclerosis and renal failure in 30–40% of individuals 20 years following Wilms' tumor diagnosis.

Lanzkowsky's Manual of Pediatric Hematology and Oncology.
DOI: http://dx.doi.org/10.1016/B978-0-12-801368-7.00025-9

TABLE 25.1 Incidence of Congenital Anomalies in Patients with Wilms' Tumor

Anomaly	Incidence (%)
Genitourinary anomalies: horseshoe kidney, dysplasia of kidney, cystic disease of kidney, hypospadias, cryptorchidism, duplication of collecting system	4.4
Congenital aniridia	1.1
Congenital hemihypertrophy	2.9
Musculoskeletal anomalies: clubfoot, rib fusion, distal phocomelia, hip dislocation	2.9
Hamartomas: hemangiomas, "birthmarks," multiple nevi, café-au-lait spots	7.9

TABLE 25.2 Syndromes with Estimated Wilms' Tumor Risk ≥ 5%

Syndrome	Chromosome locus	Implicated gene(s)	Phenotype	Estimated Wilms' tumor risk%
WAGR	11p13	WT1	Aniridia, genitourinary anomalies, delayed-onset renal failure	30
Denys–Drash	11p13	WT1	Ambiguous genitalia, diffuse mesangial sclerosis	>90
Frasier	11p13	WT1	Ambiguous genitalia, streak gonads, focal segmental glomerulosclerosis	8
Beckwith–Wiedemann Isolated hemihypertrophy	11p15	IGF2, H19, KCNQ1 (KvLQT1), KCNQ1OT1 (LIT1), or CDKN1C (p57^{KIP2})	Organomegaly, large birth weight, macroglossia, omphalocele, hemihypertrophy, ear pits and creases, neonatal hypoglycemia	5
Perlman	2q37	DIS3L2	Prenatal overgrowth, facial dysmorphism, developmental delay, cryptorchidism, renal dysplasia	33
Mosaic variegated aneuploidy	15q15	BUB1B	Microcephaly, growth retardation, developmental delay, cataracts, heart defects	25
Fanconi anemia D1	13q12	BRCA2	Short stature, radial ray defects, bone marrow failure	20
Simpson–Golabi–Behmel	Xq26	GPC3	Overgrowth, course facial features	10

Denys–Drash Syndrome
- The syndrome consists of Wilms' tumor, early renal failure with mesangial sclerosis, and pseudohermaphroditism.
- Associated with point mutations of the WT1 gene.
- Associated with 90% risk of Wilms' tumor.

Beckwith–Wiedemann Syndrome
1. This syndrome consists of
   - Hyperplastic fetal visceromegaly involving the kidney, adrenal cortex, pancreas, gonads, and liver, hemihypertrophy, macroglossia, abdominal wall defects (omphalocele, umbilical hernia, diastasis recti), ear pits or creases, microcephaly, mental retardation, hypoglycemia, and postnatal somatic gigantism.
   - Predisposition to embryonal tumors (Wilms' tumor, hepatoblastoma, neuroblastoma, and rhabdomyosarcoma) and adrenocortical carcinoma. Risk of Wilms' tumor is approximately 5%.
2. The Beckwith–Wiedemann locus at chromosome 11p15.5 contains a cluster of imprinted genes (IGF2, H19, CDKN1C, KCNQ1OT1, and KCNQ1). Various genetic and epigenetic changes can lead to the phenotypic features of the syndrome. This locus is often referred to as WT2.

Screening of Children with WAGR, Denys–Drash, and Beckwith–Wiedemann Syndromes
Among individuals with Beckwith–Wiedemann syndrome and Wilms' tumor, 93% develop the tumor by 8 years of age. Therefore, children with Beckwith–Wiedemann syndrome should have an abdominal ultrasound every 3 months until 8 years of age to evaluate for abdominal malignancies. Serum α-fetoprotein level should be tested during infancy

TABLE 25.3   Initial Signs and Symptoms of Wilms' Tumor in Order of Frequency

Sign/symptom	Frequency (%)
Palpable mass in abdomen	60
Hypertension	25
Hematuria	15
Obstipation	4
Weight loss	4
Urinary tract infection	3
Diarrhea	3
Previous trauma	3
Other signs/symptoms: nausea, vomiting, abdominal pain, inguinal hernia, cardiac insufficiency[a], acute surgical abdomen, pleural effusion, polycythemia	8

[a]*Due to propagation of tumor clot from inferior vena cava into right atrium.*

and early childhood because of the association with hepatoblastoma. 11p15 mutation and methylation analysis should be considered because new data indicate that certain Beckwith–Wiedemann epigenetic changes are associated with a higher risk of tumor development than others. Among individuals with WAGR syndrome and Wilms' tumor, 90% develop the tumor by age 4 and 98% by age 7 years. Based on this information, children with WAGR and Denys–Drash syndromes should have screening abdominal ultrasounds every 3 months until 5 years of age, but screening may be extended if Wilms' tumor precursor lesions (nephrogenic rests) are identified. Children with sporadic aniridia should be referred for FISH (fluorescence *in situ* hybridization) of the *WT1* gene because if *WT1* is not deleted, the risk of Wilms' tumor is similar to the population risk.

Signs and Symptoms

- Initial signs and symptoms, in order of frequency, are listed in Table 25.3.
- Abdominal mass is the most common presenting symptom and sign. Occasionally, there is abdominal pain, especially when hemorrhage occurs in the tumor following trauma.
- Hematuria is not common but is more often seen microscopically than on gross examination.
- Hypertension is seen in approximately 25% of patients due to elaboration of renin by tumor cells or, less commonly, due to distortion or compression of renal vasculature.
- Polycythemia is occasionally present. Erythropoietin levels are usually increased but can also be normal. Polycythemia is usually associated with males, older age, and low clinical stage. All children with unexplained polycythemia should be investigated for Wilms' tumor.
- Bleeding diathesis can occur due to the presence of acquired von Willebrand disease (reduced von Willebrand factor antigen level, prolonged bleeding time, decreased factor VIII and factor VIII ristocetin cofactor activity). The frequency of this complication is unknown.

Diagnostic Studies

Table 25.4 lists the standard clinical, laboratory, and radiographic investigations required to evaluate a patient with a suspected renal tumor. Since the most common initial surgical approach is nephrectomy, the following information is required prior to surgery:

- Presence/absence of a functioning kidney on the contralateral side.
- Presence/absence of lung or liver metastases.
- Presence/absence of bone or brain metastases if indicated by history or physical examination.
- Presence/absence of tumor thrombus in the renal vein or inferior vena cava (IVC).

Staging System

Table 25.5 describes the current staging system for renal tumors utilized by the Renal Tumors Committee of the Children's Oncology Group (COG).

TABLE 25.4   Investigations to Evaluate Newly Diagnosed Wilms' Tumor

- *History*: family history of cancer, congenital defects, and benign tumors
- *Physical examination*: congenital anomalies (aniridia, hemihypertrophy, genitourinary anomalies), blood pressure, liver enlargement
- *Complete blood count*: presence/absence of polycythemia, anemia, thrombocytopenia
- *Urinalysis*: presence/absence of hematuria
- *Serum chemistries*: blood urea nitrogen, creatinine, uric acid, serum glutamic-oxaloacetic transaminases, serum glutamic pyruvic transaminases, lactic dehydrogenase, alkaline phosphatase
- *Assessment of coagulation factors*: prothrombin time, partial thromboplastin time, fibrinogen level, PFA-100 closure time, factor VIII level, von Willebrand factor antigen level, factor VIII ristocetin cofactor activity should be considered to screen for acquired von Willebrand disease
- *Assessment of cardiac status*: electrocardiogram and echocardiogram in all patients who will receive doxorubicin; echocardiogram may also be useful in detecting the presence of tumor in the inferior vena cava or right atrium
- *Abdominal ultrasound* including Doppler imaging of renal veins and inferior vena cava
- *Abdominal CT scan or MRI with special attention to*
  - Presence and function of the opposite kidney
  - Evidence of bilateral lesions
  - Evidence of involvement of renal vein or inferior vena cava with tumor
  - Lymph node involvement
  - Liver metastasis
- *Chest computed tomography scan*
- *Bone scintigraphy*: only in cases of clear cell sarcoma or rhabdoid tumor of kidney; may be considered for renal cell carcinoma if symptoms are present
- *Magnetic resonance imaging of brain*: only in cases of clear cell sarcoma or rhabdoid tumor of kidney; may be considered for renal cell carcinoma if symptoms are present
- *Peripheral blood for genetic analysis*: in cases of congenital anomalies, such as aniridia, Beckwith–Wiedemann syndrome, hemihypertrophy or family history of Wilms tumor

Pathology

Wilms' tumor is derived from primitive metanephric blastema and is characterized by histopathologic diversity. The classic Wilms' tumor is composed of persistent blastema, dysplastic tubules (epithelial), and supporting mesenchyme or stroma. The coexistence of epithelial, blastemal, and stromal cells has led to the term *triphasic* to characterize the classic Wilms' tumor. Each of the cell types may exhibit a spectrum of differentiation, generally replicating various stages of renal embryogenesis. The proportion of each cell type may also vary significantly from tumor to tumor. Some Wilms' tumors may be biphasic or even monomorphous in appearance. The absence of anaplastic features identifies Wilms' tumors as having favorable histology (FH). Clear cell sarcoma of the kidney, rhabdoid tumor of the kidney and RCCs have unique histologies and are not Wilms' tumor variants.

Anaplastic Wilms' Tumor

Histologic Findings

Anaplastic histology is identified by the presence of cells with nuclear enlargement, nuclear atypia, and irregular mitotic figures. Focal anaplasia is distinguished from diffuse anaplasia by the distribution of anaplastic elements. Anaplasia well-contained within a single region of the tumor is considered focal anaplasia and anaplasia found in multiple regions of the tumor or outside the renal parenchyma is considered diffuse anaplasia.

FH, as the name implies, is associated with the best prognosis, followed by focal anaplasia, followed by diffuse anaplasia, which has significantly worse prognosis.

A diagnosis of diffuse anaplasia is made if the following characteristics are present:

- Anaplasia in any extrarenal site, including vessels of the renal sinus, extracapsular infiltrates, or nodal or distant metastases.
- Anaplasia in a random biopsy specimen.
- Anaplasia unequivocally expressed in one region of the tumor, but with extreme nuclear pleomorphism approaching the criteria of anaplasia (extreme nuclear unrest) elsewhere in the lesion.
- Anaplasia in more than one tumor slide, unless (i) it is known that every slide showing anaplasia came from the same focused region of the tumor or (ii) anaplastic foci on the various slides are minute and surrounded on all sides by non-anaplastic tumor.

TABLE 25.5 Staging System for Renal Tumors (Children's Oncology Group Renal Tumors Committee)

STAGE I

- Completely resected tumor limited to kidney with intact capsule
- No biopsy or rupture of tumor prior to removal
- No involvement of vessels or renal sinuses
- No tumor at or beyond margins of resection
- Regional lymph nodes negative for tumor

STAGE II

- Completely resected tumor
- No tumor at or beyond margins of resection
- Regional lymph nodes negative for tumor
- One or more of the following findings:
  - Penetration of the renal capsule
  - Invasion of vasculature extending beyond renal parenchyma

STAGE III

- Residual tumor present after surgery, confined to abdomen, with one or more of the following present:
  - One or more regional lymph nodes positive for tumor
  - Tumor implanted on or penetrating through peritoneum
  - Presence of gross unresected tumor or tumor at margin of resection
  - Any tumor spillage occurring before or during surgery, including biopsy
  - Tumor removed in more than one piece

STAGE IV

- Presence of hematogenous metastasis (e.g., lung, liver, bone, brain)
- Presence of lymph node metastasis outside the abdomen and pelvis

STAGE V

- Wilms' tumor in both kidneys

TABLE 25.6 Prognostic Significance of Loss of Heterozygosity for 1p and 16q in Favorable-Histology Wilms' Tumor

Stage	LOH Status	4-year RFS (%)	RR	P	4-year OS (%)	RR	P
I or II	Neither	91.2			98.4		
I or II	Both	74.9	2.88	0.001	90.5	4.25	0.01
III or IV	Neither	83.0			91.9		
III or IV	Both	65.9	2.41	0.01	77.5	2.66	0.04

LOH, loss of heterozygosity; RFS, relapse-free survival; RR, relapse risk; OS, overall survival.
From: Grundy et al. (2005), with permission.

Loss of Heterozygosity as a Prognostic Factor for Wilms' Tumor

Loss of heterozygosity (LOH) for polymorphic DNA markers at both chromosomes 1p and 16q occurs in approximately 5% of FH Wilms' tumor cells, and has been shown to be associated with inferior relapse-free survival (RFS) and overall survival (OS) in patients with FH Wilms' tumor. Table 25.6 summarizes the prognostic significance of LOH for 1p and 16q in the context of therapy used on NWTS-5. Although LOH 1p/16q has been used for risk stratification and therapy determination in the most recent COG studies, it is not known at this time whether subsequent alteration of upfront therapy will improve the inferior RFS and OS conferred by LOH of 1p and 16q.

Gain of Chromosome 1q as a Prognostic Factor for Wilms' Tumor

Gain of chromosome 1q, one of the most common cytogenetic findings in Wilms' tumor, has recently been shown to be independently associated with a poorer RFS and OS in patients with FH Wilms' tumor. LOH 1p/16q and 1q gain are frequently, but not always, concurrent because common cytogenetic events can lead to

TABLE 25.7　Prognostic Significance of Gain of Chromosome 1q in Favorable-Histology Wilms' Tumor

1q Gain status	8-year EFS (%)	(99% CI)	P	8-year OS (%)	(99% CI)	P
Absent	93	(87–96%)		98	(94–99%)	
Present	76	(63–85%)	0.0024	89	(78–95%)	0.0075

EFS, event-free survival; CI, confidence interval; OS, overall survival.
Adapted from: Gratias et al. (2013).

TABLE 25.8　Treatment Recommendations for Previously Untreated Wilms' Tumor

Stage	Histology	Surgery	Radiation therapy	Chemotherapy		
				Agents	Dose and schedules	Duration (weeks)
I and II	FH	Yes	No	AMD + VCR	Table 25.9	18
III and IV	FH	Yes	Yes	AMD + VCR + DOX	Table 25.10	24
I-IV	FA	Yes	Yes	AMD + VCR + DOX	Table 25.10	24
I	DA	Yes	Yes	AMD + VCR + DOX	Table 25.10	24
II, III and IV	DA	Yes	Yes	VCR + DOX + CTX + ETOP	Table 25.11	24

FH, favorable histology; DA, diffuse anaplasia; FA, focal anaplasia; AMD, actinomycin-D; VCR, vincristine; DOX, doxorubicin; CTX, cyclophosphamide; ETOP, etoposide.
Infants less than 12 months of age should receive one-half the recommended dose of all drugs, as full doses lead to prohibitive hematologic toxicity in this age group.
Full doses of chemotherapeutic agents should be administered to children more than 12 months of age.

both genetic lesions. Table 25.7 summarizes the prognostic significance of 1q gain in FH Wilms' tumor. It is likely that future COG studies for FH Wilms' tumor will incorporate 1q gain into the risk-stratification schema.

Treatment

Standard treatment of Wilms' tumor in North America consists of surgery and chemotherapy as shown in Tables 25.8–25.11. Treatment of Wilms' tumor in other parts of the world generally involves prenephrectomy chemotherapy. The chemotherapy regimens use the same agents at slightly different doses and durations of therapy.

Surgery

- Complete exploration of the involved kidney should be undertaken.
- Manual exploration of the liver and contralateral kidney is not required if high-quality preoperative computed tomography (CT) imaging is available.
- In all cases, lymph nodes should be sampled from the perihilar and para-aortic/paracaval regions at a minimum, in addition to all visible abnormal-appearing nodes. All lymph nodes removed should be identified and the site marked.
- In unilateral renal tumor cases, radical excision, often nephrectomy, is advised so that all tumor tissue can be completely removed. The junction of suspicious abnormal areas with normal kidney should be removed to facilitate the accurate diagnosis of small lesions.
- In cases involving bilateral renal tumors, a renal-sparing approach is warranted. A renal-sparing approach may also be considered for patients with unilateral Wilms' tumor at high risk for a second primary tumor (i.e., Beckwith–Wiedemann, WAGR, and Denys–Drash).

Treatment of Tumors Considered Inoperable

With the North American approach to Wilms' tumor treatment, preoperative therapy is recommended only in carefully selected cases. Indications for preoperative therapy include bilateral Wilms' tumor, tumor thrombus in the IVC above the level of the hepatic veins, tumors that invade other organs if resection of the tumor would involve

TABLE 25.9 Chemotherapy for Previously Untreated Stages I–II Favorable-Histology Wilms' Tumor

Week 0		AMD
Week 1	VCR1	
Week 2	VCR1	
Week 3	VCR1	AMD
Week 4	VCR1	
Week 5	VCR1	
Week 6	VCR1	AMD
Week 7	VCR1	
Week 8	VCR1	
Week 9	VCR1	AMD
Week 10	VCR1	
Week 11		
Week 12	VCR2	AMD
Week 13		
Week 14		
Week 15	VCR2	AMD
Week 16		
Week 17		
Week 18	VCR2	AMD

DRUG DOSING INFORMATION (BASED ON NATIONAL WILMS' TUMOR STUDY-5)

VCR1	Vincristine	• 0.025 mg/kg/day for 1 day IV for infants <12 months • 0.05 mg/kg/day for 1 day IV for children ≥12 months and <30 kg • 1.5 mg/m^2/day for 1 day IV for children >30 kg (maximum dose 2 mg)
VCR2	Vincristine	• 0.034 mg/kg/day for 1 day IV for infants <12 months • 0.067 mg/kg/day for 1 day IV for children ≥12 months and <30 kg • 2 mg/m^2/day/day for 1 day IV for children >30 kg (maximum dose 2 mg)
AMD	Actinomycin-D	• 0.023 mg/kg/day for 1 day IV for infants <12 months • 0.045 mg/kg/day for 1 day IV for children ≥12 months and <30 kg • 1.35 mg/m^2/day for 1 day IV for children >30 kg (maximum dose 2.3 mg)

AMD, actinomycin-D; VCR, vincristine.
Granulocyte colony stimulating factor (filgrastim) is not recommended with this regimen.
Doses of actinomycin-D reduced 50% if given within 6 weeks of whole-lung or whole-abdominal radiation therapy.

resection of the other organ (excluding the adrenal gland), tumors that, in the surgeon's judgment, would result in excessive morbidity if removed before chemotherapy. The following management plan is recommended:

- Histologic diagnosis by biopsy should be made before starting treatment. An exception is made in patients with bilateral Wilms' tumor with typical presenting features.
- The chemotherapy regimen shown in Table 25.10 (vincristine, doxorubicin, actinomycin-D) should be administered.
- An abdominal CT or magnetic resonance imaging (MRI) scan should be performed after approximately 6 weeks of therapy.
- Shrinkage usually occurs in 6 weeks. Surgery is planned according to tumor shrinkage and when a radiographic assessment suggests that the tumor can be totally excised, and should be accomplished no later than week 12 of therapy.
- The primary tumor should be considered Stage III, regardless of the findings at surgery. Postoperative radiation therapy (RT) is required.
- Pathology at definitive surgery should dictate the chemotherapy approach utilized postoperatively.

TABLE 25.10 Chemotherapy for Previously Untreated Stages III—IV Favorable Histology, Stages I—IV Focal Anaplastic and Stage I Diffuse Anaplastic Wilms' Tumor

Week 0		AMD	
Week 1	VCR1		
Week 2	VCR1		
Week 3	VCR1		DOX1
Week 4	VCR1		
Week 5	VCR1		
Week 6	VCR1	AMD	
Week 7	VCR1		
Week 8	VCR1		
Week 9	VCR1		DOX1
Week 10	VCR1		
Week 11			
Week 12	VCR2	AMD	
Week 13			
Week 14			
Week 15	VCR2		DOX2
Week 16			
Week 17			
Week 18	VCR2	AMD	
Week 19			
Week 20			
Week 21	VCR2		DOX2
Week 22			
Week 23			
Week 24	VCR2	AMD	

DRUG DOSING INFORMATION (BASED ON NATIONAL WILMS' TUMOR STUDY-5)

VCR1	Vincristine	• 0.025 mg/kg/day for 1 day IV for infants <12 months
		• 0.05 mg/kg/day for 1 day IV for children ≥12 months and ≤30 kg
		• 1.5 mg/m^2/day for 1 day IV for children >30 kg (maximum dose 2 mg)
VCR2	Vincristine	• 0.034 mg/kg/day for 1 day IV for infants <12 months
		• 0.067 mg/kg/day for 1 day IV for children ≥12 months and ≤30 kg
		• 2 mg/m^2/day for 1 day IV for children >30 kg (maximum dose 2 mg)
AMD	Actinomycin-D	• 0.023 mg/kg/day for 1 day IV for infants <12 months
		• 0.045 mg/kg/day for 1 day IV for children ≥12 months and ≤30 kg
		• 1.35 mg/m^2/day for 1 day IV for children >30 kg (maximum dose 2.3 mg)
DOX1	Doxorubicin	• 0.75 mg/kg/day for 1 day IV for infants <12 months
		• 1.5 mg/kg/day for 1 day IV for children ≥12 months and ≤30 kg
		• 45 mg/m^2/day for 1 day IV for children >30 kg
DOX2	Doxorubicin	• 0.5 mg/kg/day for 1 day IV for infants <12 months
		• 1 mg/kg/day for 1 day IV for children ≥12 months and ≤30 kg
		• 30 mg/m^2/day for 1 day IV for children >30 kg

AMD, actinomycin-D; VCR, vincristine; DOX, doxorubicin.
Granulocyte colony stimulating factor (filgrastim) is not recommended with this regimen. Doses of doxorubicin and actinomycin-D reduced 50% if given within 6 weeks of whole-lung or whole-abdominal radiation therapy.

TABLE 25.11 Chemotherapy for Previously Untreated Stages II—IV Diffuse Anaplastic Wilms' Tumor

Week 1	DOX			
Week 2		VCR1		
Week 3		VCR1		
Week 4			CTX5	ETOP
Week 5		VCR1		
Week 6		VCR1		
Week 7	DOX	VCR1	CTX3	
Week 8		VCR1		
Week 9		VCR1		
Week 10			CTX5	ETOP
Week 11		VCR1		
Week 12		VCR1		
Week 13	DOX	VCR2	CTX3	
Week 14		VCR2		
Week 15				
Week 16			CTX5	ETOP
Week 17				
Week 18				
Week 19	DOX	VCR2	CTX3	
Week 20				
Week 21				
Week 22			CTX5	ETOP
Week 23				
Week 24				
Week 25	DOX	VCR2	CTX3	

DRUG DOSING INFORMATION (BASED ON NATIONAL WILMS' TUMOR STUDY-5)

DOX	Doxorubicin	• 0.75 mg/kg/day for 1 day IV for infants <12 months • 1.5 mg/kg/day for 1 day IV for children ≥12 months and ≤30 kg • 45 mg/m^2/day for 1 day IV for children >30 kg
VCR1	Vincristine	• 0.025 mg/kg/day for 1 day IV for infants <12 months • 0.05 mg/kg/day for 1 day IV for children ≥12 months and ≤30 kg • 1.5 mg/m^2/day for 1 day IV for children >30 kg (maximum dose 2 mg)
VCR2	Vincristine	• 0.034 mg/kg/day for 1 day IV for infants <12 months • 0.067 mg/kg/day for 1 day IV for children ≥12 months and ≤30 kg • 2 mg/m^2/day for 1 day IV for children >30 kg (maximum dose 2 mg)
CTX5	Cyclophosphamide	• 7.35 mg/kg/day IV for 5 days for infants <12 months • 14.7 mg/kg/day IV for 5 days for children ≥12 months and ≤30 kg • 440 mg/m^2/day IV for 5 days for children >30 kg
CTX3	Cyclophosphamide	• 7.35 mg/kg/day IV for 3 days for infants <12 months • 14.7 mg/kg/day IV for 3 days for children ≥12 months and ≤30 kg • 440 mg/m^2/day IV for 3 days for children >30 kg
ETOP	Etoposide	• 1.65 mg/kg/day for 5 days for infants <12 months • 3.3 mg/kg/day IV for 5 days for children ≥12 months and ≤30 kg • 100 mg/m^2/day IV for 5 days for children >30 kg

VCR, vincristine; ETOP, etoposide; DOX, doxorubicin.

Granuloctye colony stimulating factor (filgrastim) 5 μg/kg/day subcutaneous starting 24 h after cyclophosphamide until the ANC is >2000/μl after the expected nadir.

Doses of doxorubicin reduced 50% if given within 6 weeks of whole-lung or whole-abdominal radiation therapy.

Radiation Therapy

Table 25.8 indicates which patients should receive radiation.

- RT is usually begun shortly after primary tumor resection (within 10–14 days), with the aim of eradicating tumor cells that may have spilled during surgery. The size of the field depends on the findings at surgery, but in all cases, the liver, spleen, and opposite kidney should be carefully shielded.
- Abdominal Stage III FH and abdominal Stages I–III anaplastic Wilms' tumor, clear cell sarcoma and rhabdoid tumor require irradiation to the tumor bed at a dose of 1080 cGy as determined by the preoperative radiographic findings. The tumor bed is defined as the outline of the kidney and any associated tumor. The portals should be extended to include areas of more diffuse involvement (e.g., the para-aortic chains) when those nodes are found to be involved.
- When there is peritoneal seeding or gross tumor spillage, whole-abdominal irradiation should be given at the same doses as flank RT through anterior and posterior portals from the diaphragmatic domes to the inferior margin of the obturator foramen, sparing the femoral heads. The contralateral kidney should be shielded.
- Unresected abdominal lymph node metastasis requires RT at a dose of upto 1980 cGy.
- Actinomycin-D and doxorubicin doses should be decreased 50% if given within 6 weeks following whole-lung or whole-abdomen RT.
- Whole-abdominal radiotherapy is unnecessary for patients with tumor spills confined to the flank or for those who had prior biopsy of the neoplasm.
- Metastatic disease to lungs requires whole-lung RT 1200 cGy (1050 cGy if <12 months age).
- *Pneumocystis jiroveci* prophylaxis with trimethoprim/sulfamethoxazole or pentamidine should be instituted in patients receiving lung RT.

Bilateral Wilms' Tumor

Synchronous bilateral Wilms' tumor occurs in about 5% of all Wilms' tumor patients. A significant concern for patients with bilateral Wilms' tumor is a vast increase in risk for end-stage renal disease, 11.5% compared to 0.6% for unilateral Wilms' tumor patients. Historically, patients with bilateral Wilms' tumor have had poorer outcomes than patients of similar histology, age, and local stage with unilateral lesions.

Treatment Approach

The main objective in the treatment of bilateral Wilms' tumor is eradicating the tumor with maximum preservation of renal tissue. The goal is for patients to undergo unilateral or bilateral partial nephrectomies, if feasible.
Current therapy recommendations for bilateral Wilms' tumor include:

- Open or percutaneous biopsy of each lesion to determine that the tumor is indeed Wilms' tumor. Some physicians advocate not doing a biopsy because the overwhelming majority of bilateral renal tumors are Wilms' tumor. Moreover, biopsies are insensitive in the detection of anaplasia.
- Six weeks of preoperative chemotherapy are appropriate for the lesion with poorest prognosis stage and histology (Tables 25.9–25.11).
- Open wedge biopsy or nephrectomy should be performed at week 6 of therapy for patients whose disease has shrunk <50% during initial therapy, as the likelihood of histopathologic findings other than FH are increased in the setting of disease that is poorly responsive to 6 weeks of therapy.
- For patients who have experienced significant response but renal-sparing definitive surgery is still not possible, 6 additional weeks of chemotherapy may be given.
- Definitive surgery should not be delayed beyond week 12, as further delays may result in undertreatment of resistant disease and increased risk for relapse.
- Chemotherapy should be adjusted accordingly if anaplastic histology is identified at the time of definitive surgery. Due to increased risk of injury to the remaining renal parenchyma, RT is reserved for those patients with positive tumor margins present following definitive surgery.

TABLE 25.12 Four-Year Outcomes in National Wilms' Tumor Study-5

Stage/histology	Drug treatment	4-year RFS (%)	4-year OS (%)
I FH ($n = 75$) (age <24 m, tumor <550 g)	Nephrectomy alone	86.5	98.7
I FH ($n = 97$) (age <24 m, tumor <550 g)	VCR, AMD	97.9	99.0
I FH ($n = 243$) (age >24 m, tumor >550 g)	VCR, AMD	93.8	97.0
II FH ($n = 555$)	VCR, AMD	84.4	97.2
III FH ($n = 488$)	VCR, AMD, DOX	86.5	94.4
IV FH ($n = 198$)	VCR, AMD, DOX	75.1	85.2
V FH ($n = 68$)	VCR, AMD, DOX	66.4	87.7
I FA + DA ($n = 29$)	VCR, AMD	69.5	82.6
II DA ($n = 23$)	VCR, DOX, CTX, ETOP	82.6	81.5
III DA ($n = 43$)	VCR, DOX, CTX, ETOP	64.7	66.7
IV DA ($n = 15$)	VCR, DOX, CTX, ETOP	33.3	33.3
V DA ($n = 20$)	VCR, DOX, CTX, ETOP	25.1	41.6

FH, favorable histology; FA, focal anaplasia; DA, diffuse anaplasia, RFS, relapse-free survival; OS, overall survival; VCR, vincristine; AMD, actinomycin-D; DOX, doxorubicin; CTX, cyclophosphamide; ETOP, etoposide.
From: Gratias and Dome (2008), modified with permission.

Post-Therapy Follow-Up

The optimal imaging protocol for surveillance for recurrence after completion of therapy is an area of active study.

- In patients with FH Wilms' tumor chest radiographs and abdominal ultrasounds should be performed every 3 months during years 1 and 2, every 6 months during years 3 and 4, and then once at year 5 after completion of therapy.
- In patients who have an anaplastic histology Wilms' tumor (which has a higher risk of recurrence compared to FH), chest and abdominal CT scans, alternating with chest radiographs and ultrasounds is a reasonable approach. It is not clear, however, whether early detection of relapse affects long-term prognosis.

Prognosis

Table 25.12 lists the percentage of RFS and OS 4 years from diagnosis in relationship to the stage of disease, histology, and chemotherapy treatment.

Treatment in Relapse

Patients with relapsed Wilms' tumor may be divided into three risk categories: standard risk, higher risk, and very high risk.

Standard Risk Relapse

Patients with all the following factors are considered "standard risk" for relapse:

- FH Wilms' tumor at diagnosis.
- Stage I or II at time of diagnosis.
- Initial treatment with only actinomycin-D and vincristine.
- No radiotherapy given during initial therapy.

This group may be treated with the chemotherapy regimen shown in Table 25.10 and have 70% RFS and 80% OS post-relapse.

High-Risk Relapse

Patients with the following features are considered to have "high risk" for relapse:

- FH Wilms' tumor at diagnosis.
- Initial treatment with three chemotherapy drugs, typically actinomycin-D, doxorubicin, and vincristine, with or without RT.

This group may appropriately be treated with one of the following regimens:

- Cyclophosphamide/etoposide alternating with carboplatin/etoposide. With this regimen, there is a 42% event-free survival (EFS) and 48% OS.
- Ifosfamide/carboplatin/etoposide (ICE). This regimen is associated with a high objective response rate (82%) in recurrent Wilms' tumor.
- Topotecan and vincristine/irinotecan have been demonstrated to have activity in recurrent or high-risk Wilms' tumor. Consideration may be given to adding these agents to the more conventional therapy.
- High-dose chemotherapy with autologous stem cell rescue (HDC/ASCR) has been successfully used for high-risk relapsed Wilms' tumor, though it is not clear whether it provides a benefit compared to conventional-dose chemotherapy. A recent meta-analysis did not suggest a survival benefit for HDC/ASCR in the high-risk group.

Very High Risk of Relapse

Patients with the following features are considered to have "very high risk" of relapse:

- Initial treatment with four or more chemotherapy drugs, with or without RT.
- Multiply relapsed Wilms' tumor.

A recent meta-analysis suggested a benefit for HDC/ASCR for patients with very high risk of relapse. Stem cell transplant appears to be most effective in the setting of demonstrated chemosensitivity and minimal residual disease.

Intensive study of the biology of these lesions and development of novel therapeutic agents offers the best hope for improved prognosis for this group of patients. These patients should be considered for phase 1 and 2 studies.

Local Control for Patients with Relapsed Disease

Treatment for local control of patients with relapsed disease is as follows:

- *Pulmonary relapse*: radiation to both lungs, in addition to aforementioned chemotherapy.
- *Liver relapse*: surgery with partial hepatectomy if the tumor is localized to one lobe with or without radiotherapy, in addition to aforementioned chemotherapy.
- *Local relapse in tumor bed*: surgical excision and local radiotherapy, in addition to aforementioned chemotherapy.

NEPHROBLASTOMATOSIS

Nephrogenic rests are regions of persistent embryonal tissue in the kidney that represent potential precursors to Wilms' tumor. Nephroblastomatosis is the presence of multiple nephrogenic rests or diffusely distributed nephrogenic rests, giving a rind-like appearance to the kidneys. Management of nephroblastomatosis involves:

- Close surveillance with ultrasounds, CT, or MRI scans (preferred) to assess the pattern of nephroblastomatosis and growth.
- Conservative renal tissue-preserving surgery for growing lesions that could be representative of Wilms' tumor.
- Chemotherapy may be beneficial in shrinking nephroblastomatosis. Vincristine and actinomycin-D are used but the optimal duration of therapy is unknown.

CONGENITAL MESOBLASTIC NEPHROMA

- Generally a tumor of infancy, usually presenting in the first 3 months of life.
- Two main histologic types have been described: (i) classic congenital mesoblastic nephroma (CMN), which is histologically similar to infantile fibromatosis and (ii) cellular CMN, which is histologically similar to infantile fibrosarcoma and carries the characteristic t(12;15). Some tumors have mixed classic and cellular elements.
- Complete nephrectomy without adjuvant therapy is usually adequate treatment, though local recurrence and/or metastasis have been associated with cellular histology and Stage III disease. Patients who experience a local recurrence or metastases may require additional surgery or treatment with chemotherapy (vincristine, dactinomycin, and doxorubicin; or vincristine, dactinomycin, and cyclophosphamide; or vincristine, doxorubicin, and cyclophosphamide have been employed).

CLEAR CELL SARCOMA OF THE KIDNEY

- Represents 3–5% of pediatric renal tumors.
- Peak incidence between age 3 and 5 years.
- Male to female ratio is 2:1.
- Bone, lung, and liver are most common sites of metastasis at presentation, but the brain is the most common site of metastasis at recurrence.
- Treatment consists of nephrectomy, RT, and chemotherapy with cyclophosphamide, etoposide, vincristine, and doxorubicin for 24 weeks (see Table 25.11).
- EFS estimates are approximately 100% for Stage I, 85% for Stage II, 75% for Stage III, and 35% for Stage IV disease.

RHABDOID TUMOR OF THE KIDNEY

- Represents 2% of renal tumors, median age of presentation is 1 year old.
- Chromosomal deletion or mutations of the *SMARCB1/INI1* gene at chromosome 22q11-12 are present in the tumors of most patients.
- Some patients have constitutional *SMARCB1/INI1* mutations and multifocal disease in the brain (atypical teratoid/rhabdoid tumors) or rhabdoid tumors at other sites.
- Metastases common to lung, liver, and brain. Most patients present with Stage III or IV disease.
- Event-free and OS is only approximately 25%. The tumor responds poorly to current chemotherapy and radiotherapy, with high rates of recurrence. Multiple case series have shown some response to an aggressive approach using intensive chemotherapy (vincristine–doxorubicin–cyclophosphamide alternating with ICE plus RT and surgical resection of all possible sites of disease). The patients with the best chance of cure are those with localized disease (Stages I/II) and age over 3 years.
- Intensive study of the biology of this tumor and development of novel therapeutic agents offer the best hope for improved prognosis.

RENAL CELL CARCINOMA

- Second most common pediatric renal malignancy, representing 5–6% of primary renal tumors under 21 years of age.
- Most common renal neoplasm in adults, but pediatric subtypes of RCC are distinct from the clear cell subtype commonly seen in adults.
- "Translocation" histologic type (involving *TFE3* gene on X chromosome) is the most common type of pediatric RCC.
- Survival for Stage I disease >90%, 70–80% for Stages II and III, and <15% for Stage IV.
- Treatment: Surgery is mainstay of therapy. Anecdotal responses have occurred with cytotoxic chemotherapy and data are emerging to suggest vascular endothelial growth factor receptor inhibitors such as sunitinib, a multitargeted tyrosine kinase inhibitor approved for use in adult-type clear cell RCC, may have activity in the "translocation" type tumors. Future studies are planned in the COG to investigate these agents.

Further Reading and References

Abu-Gosh, A.M., Krailo, M.D., Goldman, S.C., et al., 2002. Ifosfamide, carboplatin and etoposide in children with poor risk relapsed Wilms' tumor: a Children's Cancer Group report. Ann. Oncol. 13, 460–469.

Dome, J.S., Cotton, C.A., Perlman, E.J., et al., 2006. Treatment of anaplastic histology Wilms' tumor: results from the Fifth National Wilms' Tumor Study. J. Clin. Oncol. 20, 2352–2358.

Dome, J.S., Fernandez, C.V., Mullen, E.A., et al., 2013. Children's Oncology Group's 2013 blueprint for research: renal tumors. Pediatr. Blood Cancer 60, 994–1000.

Dome, J.S., Perlman, E.J., Graf, N., 2014. Risk stratification for Wilms tumor: current approach and future directions. Am. Soc. Clin. Oncol. Educ. Book, 215–223.

Geller, J.I., Dome, J.S., 2004. Local lymph node involvement does not predict poor outcome in pediatric renal cell carcinoma. Cancer 101, 1575–1583.

Gooskens, S.L.M., Furtwängler, R., Vujanic, G.M., et al., 2012. Clear cell sarcoma of the kidney: a review. Eur. J. Cancer 28, 2219–2226.

Gratias, E.J., Dome, J.S., 2008. Current and emerging chemotherapy treatment strategies for Wilms tumor in North America. Pediatr Drugs 10 (2), 115–124.

Gratias, E.J., Jennings, L.J., Anderson, J.R., et al., 2013. Gain of 1q is associated with inferior event-free and overall survival in patients with favorable histology Wilms tumor: a report from the Children's Oncology Group. Cancer 119, 3887–3894.

Green, D.M., Cotton, C.A., Malogolowkin, M., et al., 2007. Treatment of Wilms' tumor relapsing after initial treatment with vincristine and actinomycin D: a report from the National Wilms' Tumor Study Group. Pediatr. Blood Cancer 48, 493–499.

Grundy, P.E., Breslow, N.E., Li, S., et al., 2005. Loss of heterozygosity for chromosomes 1p and 16q is an adverse prognostic factor in favorable-histology Wilms' tumor: a report from the National Wilms' Tumor Study Group. J. Clin. Oncol. 23, 7312–7321.

Ha, T.C., Spreafico, F., Graf, N., et al., 2013. An international strategy to determine the role of high dose therapy in recurrent Wilms' tumour. Eur. J. Cancer 49, 194–210.

Malogolowkin, M., Cotton, C.A., Green, D.M., et al., 2008. Treatment of Wilms' tumor relapsing after initial treatment with vincristine, actinomycin D and doxorubicin. A report from the National Wilms' Tumor Study Group. Pediatr. Blood Cancer 52 (2), 236–241.

Metzger, M.L., Stewart, C.F., Freeman, B.B., et al., 2007. Topotecan is active against Wilms' tumor: results of a multi-institutional phase II study. J. Clin. Oncol. 21, 3130–3136.

Scott, R.H., Stiller, C.A., Walker, L., Rahman, N., 2006. Syndromes and constitutional chromosomal abnormalities associated with Wilms tumour. J. Med. Genet. 43, 705–715.

Segers, H., van den Heuvel-Eibrink, M.M., Williams, R.D., et al., 2013. Gain of 1q is a marker of poor prognosis in Wilms' Tumors. Genes Chromosomes Cancer 52, 1065–1074.

Seibel, N.L., Li, S., Breslow, N.E., et al., 2004. Effect of duration of treatment on treatment outcome for patients with clear-cell sarcoma of the kidney: a report from the National Wilms' Tumor Study Group. J. Clin. Oncol. 22, 468–473.

Tomlinson, G.E., Breslow, N.E., Dome, J., et al., 2005. Age at diagnosis as a prognostic factor for rhabdoid tumor of the kidney: a report from the National Wilms' Tumor Study Group. J. Clin. Oncol. 23, 7641–7645.

26

Rhabdomyosarcoma and Other Soft-Tissue Sarcomas

Carolyn Fein Levy and Leonard H. Wexler

Soft-tissue sarcomas (STS) are a heterogeneous group of malignant tumors derived from primitive mesenchymal cells. These tumors arise from muscle, connective tissue, supportive tissue, and vascular tissue. As a group, they are locally highly invasive and have a high propensity for local recurrence. They usually metastasize via the bloodstream and, less commonly, via the lymphatics. The STS can be divided into two groups:

1. Rhabdomyosarcoma (RMS).
2. Nonrhabdomyosarcoma soft-tissue sarcomas (NRSTS).

INCIDENCE AND EPIDEMIOLOGY

STS account for 7% of childhood malignancies. RMS accounts for 40% of STS and RMS is the most common pediatric STS, with an incidence of 4–5 per million in children less than 15 years old.

- RMS is the third most common solid extracranial tumor, following neuroblastoma and Wilms' tumor.
- RMS accounts for 3% of all malignant neoplasms in children, with approximately 400 new cases diagnosed in the United States each year in children under 19 years of age.
- Approximately one-third of cases of RMS are in children less than 5 years of age and 60% of cases are diagnosed in children less than 10 of age.
- In RMS there is a slight male predominance with a male:female ratio of 1.4:1; however adolescent patients are disproportionately males.
- RMS incidence in African-American females is half that of Caucasian females. The incidence is lower in Asian populations residing in Asia or the West.

NRSTS comprise a diverse group of malignancies. Although each type is rare, together they constitute about 60% of all pediatric STS with an incidence of 6–8 cases per million in children less than 20 years of age. There are approximately 550–600 new NRSTS cases per year in the United States, which represents 4% of all pediatric malignancies. Among the NRSTS the most common subtypes include synovial sarcoma, malignant peripheral nerve sheath tumors (MPNST), dermatofibrosarcoma protuberans (DFSP), malignant fibrous histiocytoma, and fibrosarcoma. Tumors typically found in children under 5 years of age are infantile fibrosarcoma and infantile hemangioperiocytoma. Males are affected slightly more than females and African-Americans are affected slightly more often than Caucasians.

Most cases of STS occur sporadically but up to 30% of cases may have an underlying risk factor, which may include:

- Germline mutations of the *p53* suppressor gene, as in Li–Fraumeni familial cancer syndrome. There is an association between early-onset breast cancer, sarcomas, brain tumors, and adrenocortical tumors in family members.
- Ionizing radiation.
- Neurofibromatosis (*NF1*)—patients with *NF1* have up to a 15% lifetime risk of developing an MPNST associated with chromosome 17 deletions.

Lanzkowsky's Manual of Pediatric Hematology and Oncology.
DOI: http://dx.doi.org/10.1016/B978-0-12-801368-7.00026-0

- Syndromes such as Beckwith—Weidemann syndrome and Costello syndrome (a genetic disorder characterized by delayed development and mental retardation, unusually flexible joints, hypertrophic cardiomyopathy, short stature, and an increased risk of developing tumors, with the most frequent being RMS).
- *DICER1* mutations—familial pleuropulmonary blastoma predisposition syndrome increases the risk of developing tumors including embryonal RMS (ERMS).
- Maternal and paternal use of marijuana and cocaine and first-trimester prenatal X-ray exposure, possibly as an environmental interaction with a genetic trigger.

PATHOLOGIC AND GENETIC CLASSIFICATION

Immunohistochemistry, molecular diagnostics including reverse-transcriptase polymerase chain reaction and/or fluorescence *in situ* hybridization may be necessary to differentiate RMS and NRSTS from the other small round blue cell tumors of childhood (i.e., lymphoma, Ewing sarcoma, neuroblastoma). There are various chromosomal translocations that are characteristic of NRSTS which have led to the refinement of the histopathological classification of pediatric STS. Table 26.1 summarizes the clinical and biological features of NRSTS. Table 26.2 lists the histologic subtypes of RMS with reference to their morphology, site of origin, and age distribution.

Genetics of RMS

1. Alveolar rhabdomyosarcoma (ARMS) has a characteristic translocation of the *FOXO1* gene (previously known as Forkhead, or *FKHR*) at 13q14 with *PAX3* at 2q35 or less commonly *PAX7* at 1p36. The fusion protein functions as a transcription factor that activates transcription from PAX binding sites that are 10—100 times more active than wild-type *PAX7* and *PAX3*. This alteration in growth, differentiation, and apoptosis results in tumorigenic behavior. Approximately 75% of ARMS contain the *FOXO1-PAX 3* translocation; the remaining 25% contain the *FOXO1-PAX7* translocation. There are conflicting data with regard to whether translocation subtype is prognostically significant for any group of patients.
2. ERMS has a loss of heterozygosity (LOH) at the 11p15.5 locus. This LOH involves loss of the imprinted maternal genetic information and all that remains is expression of the paternal genetic material. This LOH includes loss of tumor suppressor genes that have been implicated in oncogenesis which results in the overproduction of insulin-related growth factor-II (IGF-II). IGF-II stimulates the growth of RMS and blockade of the IGF-II receptor inhibits RMS growth *in vivo*.
3. "Translocation-negative" ARMS (i.e., tumors that have an alveolar pattern on routine light microscopy but lack the defining *FOXO1-PAX* translocation) represent approximately 25% of cases of ARMS and have been demonstrated conclusively to cluster genomically and clinically with ERMS. These cases will be considered ERMS in future Children's Oncology Group (COG) clinical trials.
4. Spindle cell RMS, a less common subtype of ERMS, may be seen in children and adults and appears to have distinctive genetic features and clinical behavior in each group: children typically have a favorable prognosis and cases of recurrent *NCOA2* translocation have been described; conversely, adults typically have more aggressively behaving disease and there is growing evidence that many of these tumors contain mutations in the *MYOD1* gene.

CLINICAL FEATURES

Primary Sites

RMS may occur in any anatomic location of the body where there is skeletal muscle, as well as in sites where no skeletal muscle is found (e.g., urinary bladder, common bile duct). RMS in children under 10 years of age generally involves the head and neck or genitourinary areas. Adolescents more commonly develop extremity, truncal, or paratesticular lesions. Table 26.3 lists the relative frequency of the various primary sites and sites of regional spread and distant metastases.

NRSTS can arise in any tissue but are extremely uncommon in bone. Approximately half of NRSTS arise in extremities. The remaining NRSTS are divided between trunk, head and neck, and visceral/retroperitoneal sites.

TABLE 26.1 Clinical and Biological Features of the Nonrhabdomyosarcoma Soft-Tissue Sarcomas (NRSTs)

Tumor[a]	Cell origin/ cytogenetics/ product	Common sites	Common ages	Good prognostic factors[b]	Outcome	Therapy
Synovial sarcoma	Mesenchymal cells/ t(X;18) (p11q11)/ SSX1-SYT (seen in biphasic tumors) SSX2-SYT (seen in monophasic tumors)/ translocation present in >90%, mycn overexpression	Extremities (lower twice as common as upper extremity)	Adolescence/ young adulthood, accounts for 30% of pediatric NRSTS	Age ≤ 14 years, size <5 cm, calcification, chemosensitive	Stages I and II, 70%; stages III and IV, poor	WLE with/without RT Chemo: ifosfamide/ doxorubicin
Dermatofibrosarcoma Protuberans (DFDSP)	Dermis/t(17;22) (q21;q13) ring chromosome/ COL1A1-PDGFB	Trunk and proximal limbs, rare head and neck	20–50 years, rare in childhood	Complete excision, local recurrence 60% with incomplete resection		WLE (3 cm) margin pseudopod-like projections with Mohs micrographic surgery, RT has been used when WLE not possible, Imatinib for unresectable, locally advanced, recurrent or metastatic disease
Malignant fibrous histiocytoma (MFH aka undifferentiated pleomorphic sarcoma)	Unknown/19p+, complex abnormalities	Lower extremity, trunk, head and neck	In children, 10–20 years, 40–60 years Common radiation-induced sarcoma	Extremity site	5-year survival, 27–53%	WLE Chemo: ifosfamide/ doxorubicin
Angiomatoid fibrous histiocytoma	Fibroblast/t(2;22) (q34;q12) t(12;16) (q13;p11) t(12;22) (q13;q12)/EWSR1-CREB1 TLS-ATF1 EWSR1-ATF1	Extremity, trunk and head and neck (subcutis may infiltrate dermis or muscle)	Young children and young adults	Much less aggressive than MFH	Excellent with surgery alone	WLE
Malignant peripheral nerve sheath tumor (MPNST)	Schwann cell or fibroblast/17q, 22q loss or rearrangement, complex abnormalities in high-grade tumors	Extremity, retroperitoneum trunk	Younger in patients with Neurofibromatosis (NF1) Develop in 10% patients with NF1 and 20–60% cases of MPNST occur in association with NF1.	Size < 5 cm, No NF1	53% survival without NF, 16% with NF	WLE with/without RT Chemo: neoadjuvant role, ifosfamide/ doxorubicin
Fibrosarcoma	Fibroblast/t(X;18), t(2;5), t(7;22)	Truncal/ Proximal site	Adolescence		5-year survival 34–60%	WLE with/without RT Chemo: no established role
Infantile fibrosarcoma	Fibroblast/t(12;15) (p13;q25)/ETV6-NTRK3	Distal extremity	Most <2 years	<5 years	5-year survival 84%	WLE, RT/chemo if WLE not possible Neoadjuvant chemotherapy with VA ± C
Leiomyosarcoma	Deletion 1p, other complex abnormalities, Smooth muscle-uterine t(12;14) (q15;q24); HMGA2 rearrangement	Retroperitoneum GI tract, any soft tissue or vascular area	40–70 years, when in children, any age, associated with human immunodeficiency virus related to EBV infection, reported in patients who received RT for retinoblastoma and Carney triad[c]	<5 cm	33% disease-free survival at 1–5 years	WLE Chemo: ifosfamide/ doxorubicin or gemcitabine/ docetaxel

TABLE 26.1 *(Continued)*

Tumor[a]	Cell origin/ cytogenetics/ product	Common sites	Common ages	Good prognostic factors[b]	Outcome	Therapy
Alveolar soft part sarcoma	(?)Unknown/ t(X;17)(p11;q25)/ ASPSCR1-TFE3	Orbit, head and neck, lower extremity	15–35 years	Young age, orbital site, <5 cm	5-year survival 27–59% (indolent; death from disease after 10–20 years) 79% metastatic disease including brain	WLE Chemo or RT only after recurrence Chemo: no clear role. Possible role of vascular endothelial growth factor inhibitors being explored
Hemangiopericytoma infantile form (<1 year of age)	Pericytes/t(12;19), (q13;q13)/t(13;22) (q22;q11)	Extremity, retroperitoneum head and neck Extremity, trunk	20–70 years, when in children, 10–20 years Rare, but typically , 1 year	Low stage, <5 cm, infantile form	Stage I and II, 30–70% 5-year survival with adjuvant therapy Stages III and IV, poor Infantile— Excellent with surgery alone	WLE, with/ without RT Chemo: no established role but can be chemoresponsive, Infantile form responds more favorably to chemotherapy
Liposarcoma (myxoid)	Primitive mesenchyme/ t(12;16) (q13p11)/ FUS-DDIT3	Extremity, retroperitoneum	0–2 years and second decade; sixth decade most common	Child, myxoid type	Very good with WLE, rarely metastasizes	WLE, with/ without RT RT important in retroperitoneal lesions Chemo: no established role
Clear cell sarcoma	Mesoderm, melanin deposits t(12;22) (p13; q12)/EWSR1-ATF1	Tendons and aponeuroses of lower extremity	Young adults, females	<5 cm, no necrosis, nonmetastatic	Adverse prognosis; 5-year survival rates of 60–70%. However, only 30–40% are long-term survivors due to late recurrences	WLE with sentinel node biopsy No clear role for adjuvant chemotherapy. Potential role for immunotherapy (e.g., translocation-targeted vaccines, interferon, GM-CSF secreting vaccine)

[a]Listed in order of decreasing incidence.
[b]Low histologic grade and low stage are good prognostic factors.
[c]Carney triad: A condition consisting of gastric epithelioid leiomyosarcoma, pulmonary chondroma, functioning extra-adrenal paraganglioma.
WLE, wide local excision; RT, radiation therapy; VAC, vincristine, actinomycin D, cyclophosphamide.

TABLE 26.2 Histologic Subtypes of Rhabdomyosarcoma (RMS)

Pathologic subtype	Morphology	Usual site of origin	Usual age (years) distribution
Embryonal (ERMS) solid	Resembles skeletal muscle in 7- to 10-week fetus. Moderately cellular with loose myxoid stroma. Actin and desmin positive; myogenin scattered positivity <50%	Head and neck, orbit, genitourinary tract	3–12
Botryoid variant	Only one microscopic field of cambium layer necessary to diagnose as botryoid. Grossly presents with grape-like configuration	Bladder, vagina, nasopharynx, bile duct	0–8
Spindle cell variant[a]	Spindle-shaped cells with elongated nuclei and prominent nucleoli. Low cellularity. Collagen-rich and -poor variants	Paratesticular	2–12
Alveolar (ARMS)	Resembles skeletal muscle in 10- to 21-week fetus. Basic cell is round with scanty eosinophilic cytoplasm; alveolar pattern may be lost if densely packed; cross striations more common than embryonal variety. Up to one-third of ARMS-negative tumors are actually dense pattern ERMS. Diffuse myogenin positivity. ARMS requires confirmation of PAX3/7-FKHR (FOXO1) translocation	Extremities, trunk, perineum (adolescents)	6–21
RMS, NOS	Heterogeneous, unable to subtype due to paucity of tissue	Extremities, trunk	6–21

[a]Spindle cell RMS in adults is a distinct clinical entity with more aggressive behavior than in the pediatric population.
NOS, not otherwise specified.

TABLE 26.3 Frequency of Primary Sites, Sites of Regional Spread, and Distant Metastatic Sites in Rhabdomyosarcoma

Primary site	Relative frequency (%)	Regional spread and distant metastatic sites
Head and neck	**40**	
Orbit	8	Nodes rarely involved; rare lung metastasis
Parameningeal[a]	25	Regional spread to bone, meninges, brain; lung and bone metastases
Other[b]	7	Nodes rarely involved; lung metastases
Genitourinary tract	**29**	
Bladder, prostate	10	Nodes rarely involved; metastases to lung, bone, and bone marrow (primarily prostate primaries)
Vagina, uterus	5	Nodes rarely involved; metastases to retroperitoneal nodes (mainly from uterus)
Paratesticular	14	Retroperitoneal nodes in up to 50% of boys 10 or older; metastases to lung and bone
Extremities	**14**	Nodes involved in up to 50% of cases; metastases to lung, bone marrow, bone, central nervous system
Trunk	**12**	Nodes rarely involved; metastases to lung and bone
Other	**5**	Nodal involvement site-dependent (increased in perineal/perianal primaries); metastases to lung, bone, and liver

[a]*Parameningeal sites are adjacent to the meninges at the base of the skull; they consist of nasopharynx, middle ear, paranasal sinuses, and infratemporal and pterygopalatine fossae.*
[b]*Nonorbital, nonparameningeal sites consist of larynx, oropharynx, oral cavity, parotid, cheek, and scalp.*

Signs and Symptoms

Most STS present as painless masses. Symptoms depend on the location and invasion of the adjacent normal structures.

Specific clinical manifestations vary with the site of origin of the primary lesion and are outlined in Table 26.4. About 50% of RMS of head and neck primary tumors (nasopharynx, sinuses, middle ear) have infiltration of tumor through the skull base with intracranial extension of disease that may manifest as cranial nerve palsies with or without headache and/or other signs of raised intracranial pressure.

Rare primary sites for RMS include the gastrointestinal–hepatobiliary tract (3%), where it presents with obstructive jaundice and a large abdominal mass. These tumors arise in the common bile duct and may extend into both lobes of the liver. Other rare primary sites are the intrathoracic region (2%) and the perineal–perianal area (2%).

Approximately 20% of RMS have metastatic disease at diagnosis and the most common sites are bone marrow and lung, followed by lymph nodes and bone.

NRSTS rarely presents with systemic symptoms. Up to 15% of patients present with metastatic disease, most commonly to the lung. Nodal spread is rarely seen except with epithelioid sarcoma and clear cell sarcoma. Rarely bone, liver, subcutaneous and brain metastases are seen, and bone marrow involvement is exceedingly rare.

DIAGNOSTIC EVALUATION

The diagnostic evaluation should delineate the extent of the primary tumor and the location and extent of metastatic disease and should consist of the following:

- *Complete history and physical examination* including measurements of the primary tumor and assessment of regional lymph nodes.
- *Laboratory tests* including complete blood count, comprehensive metabolic panel including liver functions and urinalysis. LDH tends to be elevated in cases of advanced disease and its trend over time often correlates with response to treatment and/or progression of disease. In cases of advanced-stage ARMS, a coagulation profile should be performed as these patients may present with tumor-induced disseminated intravascular coagulation.
- *Primary tumor imaging*:
  - Magnetic resonance imaging (MRI) to assess primary tumor.
  - Ultrasound for paratesticular, bladder/prostate, or biliary tree.

TABLE 26.4 Clinical Manifestations of Rhabdomyosarcoma in Various Anatomic Locations

Location	Signs and symptoms
HEAD AND NECK[a]	
Neck	Soft-tissue mass
	Hoarseness
	Dysphagia
Nasopharynx	Sinusitis
	Local pain and swelling
	Epistaxis
Paranasal sinus	Sinus obstruction/sinusitis
	Unilateral nasal discharge
	Local pain and swelling
	Epistaxis
Middle ear/mastoid	Chronic otitis media—purulent blood stained discharge
	Polypoid mass in external canal
	Peripheral facial nerve palsy
Orbit	Proptosis
	Ocular palsies
	Conjunctival mass
GENITOURINARY	
Vagina and uterus	Vaginal bleeding
	Grapelike clustered mass protruding through vaginal or cervical opening (i.e., sarcoma botryoides)
Prostate	Hematuria
	Constipation
	Urinary obstruction
Bladder	Urinary obstruction
	Hematuria
	Tumor extrusion
	Recurrent urinary tract infections
Paratesticular	Painless scrotal or inguinal mass
Retroperitoneum	Abdominal pain
	Abdominal mass
	Intestinal obstruction
Biliary tract	Obstructive jaundice
Pelvic	Constipation
	Genitourinary obstruction
Extremity/trunk	Asymptomatic or painful mass

[a]*All can extend through multiple foramina and fissures into the epidural space and infiltrate the central nervous system with cranial nerve palsies, meningeal symptoms, and brain stem signs.*

- *Metastatic workup:*
  - Chest computed tomography (CT) to look for lung metastasis.
  - MRI/CT of draining lymph nodes. This is required in lower extremity and paratesticular primaries to optimally evaluate regional adenopathy.
  - ^{18}FDG-PET scan to assess primary and metastatic disease, as well as to monitor response to therapy.
  - Bone scan was traditionally performed to identify osseous metastases, but has been replaced by ^{18}FDG-PET scan, which has both greater sensitivity and specificity.
  - Bilateral bone marrow aspiration and biopsy are not necessary in patients with noninvasive, node-negative tumors and in patients with node-negative invasive ERMS with negative CT chest and is not indicated for NRSTS.
  - Brain MRI is recommended in patients with widespread metastatic disease in NRSTS.

- *Additional diagnostic evaluation depending on primary site:*
  - Dental evaluation with Panorex radiography for maxillary or mandibular disease.
  - Lumbar puncture for parameningeal head and neck disease.
  - Cystoscopy or vaginoscopy for bladder, prostate, or vaginal disease.

- *Surgical consultation* for:
  - Percutaneous, incisional, or excisional biopsy. An adequate biopsy is critical for accurate diagnosis. Incisional biopsy is the gold standard, however multiple core needle biopsies may be adequate. Core needle biopsies may not provide adequate tissue for molecular pathologic studies. Incisional and core biopsy tracks need to be resected at the time of definitive resection. Fine-needle aspiration is not acceptable. Sampling of suspicious lymph nodes. Sentinel lymph node evaluation is required for all RMS extremity primaries as well as in all epithelioid sarcoma and clear cell sarcomas.
  - Required ipsilateral retroperitoneal lymph node dissection for paratesticular RMS for boys 10 years and older.

STAGING

Staging for RMS

It is essential to fully stage RMS. The prognosis, selection of systemic therapy, and the design of optimal local therapy depend on the following:

- Primary site.
- Histologic type.
- Tumor size.
- Degree of regional spread.
- Nodal involvement.
- Distant metastatic disease.
- Extent of prechemotherapy tumor resection.

The RMS staging systems consist of:

Pretreatment clinical stage: the Intergroup Rhabdomyosarcoma Study (IRS) incorporated the most significant prognostic variables (primary site, invasiveness, size, regional lymph node involvement) into a tumor—node—metastasis. This soft-tissue sarcoma—Children's Oncology Group (STS-COG) staging system is provided in Table 26.5.

Postoperative clinical group: this is based on the extent of the surgical resection and takes into account the lymph node evaluation (Table 26.6). In some locations, when complete resection is possible with negative margins, overall survival is improved.

Risk group classification (Table 26.7): the risk group classification and treatment is based on the combination of pretreatment stage, postoperative clinical group, and histology.

The *low-risk group* includes all nonmetastatic favorable-site ERMS (stage 1, regardless of degree of initial surgical resection) and completely resected (groups I and II) nonmetastatic unfavorable-site ERMS. Girls with nonbladder genitourinary tract ERMS treated without local radiation have been included in the low-risk group, but are better categorized as intermediate-risk as discussed below.

TABLE 26.5   Soft-Tissue Sarcoma—Children's Oncology Group Pretreatment Tumor—Node—Metastasis Staging of Rhabdomyosarcoma

Stage	Sites	T Invasiveness	T Size	N	M
I	**Favorable sites** Orbit, head, and neck (excluding parameningeal), genitourinary (nonbladder/nonprostate)	T1 or T2	a or b	N0 or N1 or NX	M0
II	**Unfavorable sites** Bladder/prostate, extremity Cranial parameningeal Other (includes trunk, retroperitoneum, etc.)	T1 or T2	a	N0 or NX	M0
III	**Unfavorable sites** Bladder/prostate, extremity Cranial parameningeal Other (includes trunk retroperitoneum, etc.)	T1 or T2	a b	N1 N0 or N1 or NX	M0
IV	**Any sites**	T1 or T2	a or b	N0 or N1	M1

T = Tumor	N = Regional nodes	M = Metastasis[a]
T1 = Confined to anatomic site of origin	N0 = Not clinically involved	M0 = No distant metastasis
T2 = Extension and/or fixation to surrounding tissue	N1 = Clinically involved	M1 = Distant metastasis present
a ≤ 5 cm in diameter	NX = Clinical status unknown	M1 includes positive CSF cytology, pleural and/or peritoneal fluid and distant nodes
b > 5 cm in diameter		

[a]*Distant metastatic disease consists of lung, liver, bones, bone marrow, brain, and distant muscle and nodes. The presence of positive cytology in CSF, pleural, or abdominal fluids, as well as implants on pleural or peritoneal surfaces, also constitutes stage IV disease.*
Modified from Mandell (1993) with permission.

TABLE 26.6   Postoperative Clinical Grouping System (Intergroup Rhabdomyosarcoma Study)

Group	Definition (Incidence)
I.	No residual disease (16%)
	A. Localized, completely resected, confined to site of origin
	B. Localized, completely resected, infiltrated beyond site of origin
II.	Microscopic residual disease (20%)
	A. Margins positive, lymph nodes negative
	B. Margins negative, lymph nodes positive
	C. Margins positive, lymph nodes positive
III.	Gross residual disease (48%)
	A. Biopsy only
	B. Grossly visible disease after 50% resection of primary tumor
IV.	Distant metastasis present at diagnosis (16%)

The *intermediate-risk group* includes all nonmetastatic alveolar tumors, and embryonal tumors in unfavorable primary sites (stage 2 or 3) that have been incompletely resected (group III). Girls with nonbladder genitourinary tract ERMS treated without local radiation therapy have inferior outcomes, and should be considered to have *intermediate-risk disease*. Their prognosis improves when treated with either intensified systemic therapy (cyclophosphamide at a dose of 2.2 g/m²/cycle) or "standard-intensity" systemic therapy (cyclophosphamide at 1.2 g/m²/cycle) combined with appropriate local therapy consisting of some combination of conservative surgery with either vaginal brachytherapy or external beam radiation therapy. Children under the age of 10 years with ERMS and isolated lung metastases have been included in the high-risk group, but are better categorized as intermediate-risk as they have a more favorable prognosis than other patients with metastatic disease, and can be considered to have *intermediate-risk disease*.

TABLE 26.7 Rhabdomyosarcoma Risk Group Classification and Outcome for Rhabdomyosarcoma (RMS) from Soft-Tissue Sarcoma—Children's Oncology Group

Risk group	Histology	Pretreatment stage	Postoperative clinical group	EFS (%)
Low	Embryonal	1 (all favorable sites)	I or II	85–95
Subset 1		1 (orbit only)	III	
		2 (unfavorable sites ≤5 cm)	I or II	
Low	Embryonal	1 (nonorbit only)	III	70–85
Subset 2		3 (unfavorable sites <5 cm, or unfavorable site, any size, with regional nodal involvement)[a]	I or II	
Intermediate	Embryonal	2, 3	III	65–75
Intermediate	Alveolar	1, 2, 3[b]	I, II, III	50–60
High	Embryonal	4[c]	IV	20–40
High	Alveolar	4	IV	5–20

[a]See text on risk stratification for exceptions regarding girls with nonbladder genitourinary tract embryonal RMS treated without local radiation.
[b]See text on risk stratification for exceptions regarding children with alveolar RMS and regional lymph node involvement.
[c]See text on risk stratification for exceptions regarding children under 10 years of age with ERMS and isolated lung lesions.

The *high-risk group* includes all metastatic tumors, both alveolar and embryonal. Children under the age of 10 years with ERMS and isolated lung metastases are discussed in the intermediate-risk group. Patients with ARMS with regional nodal involvement have been included in the intermediate-risk group, but they are better categorized as high-risk as their survival is less than 50%.

Since prognosis is determined by the risk group, the risk group classification determines treatment.

Staging for NRSTS

Staging using the American Joint Committee on Cancer staging system incorporates tumor size (T), depth, nodal (N), and metastatic (M) involvement and histologic grade (G). Tumors with necrosis greater than 15% and mitotic rate greater than 5–10 per high power field are consider high grade. Risk stratification is based on tumor size, tumor grade, margins, resectability, and metastatic disease. In pediatric NRSTS, risk stratification with an emphasis on tumor size and resectability predicts outcome.

The *low-risk group* (~50% of patients) includes those whose primary tumor has been grossly resected, have nonmetastatic disease and whose pathology shows:

- Low-grade tumors regardless of margin status.
- High-grade tumors ≤5 cm in size.

The *intermediate-risk group* (~35% of patients) includes those with nonmetastatic disease and:

- High-grade tumors that are grossly resected and greater than 5 cm in size.
- High-grade tumors that are unresectable and where a delayed resection is planned.

The *high-risk group* (~15% of patients) includes those who have metastatic disease, irrespective of the pathological grade of the tumor.

PROGNOSIS

Prognosis for RMS

Overall, RMS is curable in the majority of children (70% survival 5 years after diagnosis). The type of treatment failure differs among different risk groups. The failures are typically:

- Local treatment failure for patients with nonmetastatic ERMS.
- Regional and distant treatment failure for patients with nonmetastatic ARMS.

- Distant treatment failure for patients with metastatic RMS.
  - Extent of disease is the most important prognostic factor. Children with localized, completely resected disease do better than those with widespread or disseminated disease.
  - Patients with gross residual disease after surgery (group III) have a statistically significantly lower 3-year FFS of 73% compared with 83% and 86% FFS for groups I and II, respectively.
  - Patients with smaller tumors (≤5 cm) have improved survival compared to children with tumors greater than 5 cm in size.
  - Those with metastatic disease at diagnosis have the worst prognosis. Among those with metastatic disease, two or less metastases are significantly better than three or more.
  - Among patients with intermediate-risk disease, patients with alveolar histology have a worse prognosis, and the group with regional nodal involvement has a prognosis that approaches that of patients with conventionally defined "distant metastases." The alveolar subtype is frequently associated with an extremity site, which is a poor prognostic factor.
  - Nonparameningeal head and neck sites, nonbladder/nonprostate male and female genitourinary tract sites and biliary tract are "favorable sites"; all other sites are "unfavorable."
  - Children between 1 and 9 years of age have a better prognosis than those older than 9 years of age (83% vs 68% 3-year FFS). This finding might be due to the higher incidence of advanced disease and alveolar type in older children.

The high-risk group of patients have metastatic disease and represent approximately 20% of patients with RMS. Within this group there is significant diversity of outcome. Patients with ERMS aged 1−9 years of age and lung-only metastases have 50% EFS and those with bone and bone marrow involvement at diagnosis having an overall survival less than 10%. The dominant risk of treatment failure is from the inability to control systemic disease, although local treatment failure also appears to be higher in patients with metastases at diagnosis.

Prognosis for NRSTS

NRSTS in children frequently have a better prognosis than in adults. In infants and young children, NRSTS often behave in a benign manner and have an excellent prognosis with surgery alone. The presence of specific cytogenetic changes in the NRSTS is the key to accurate diagnosis and in the future may be helpful to identify therapeutic targets.

NRSTS are broadly divided into low-grade and high-grade tumors based upon their propensity for local invasion and distant metastases. The single most important prognostic variable for nonmetastatic tumors remains initial surgical resection. Factors for poor outcome include:

- Metastatic disease.
- High histologic grade.
- Tumors greater than 5 cm in size.
- Unresectable disease.

The *low-risk group* has an 89% 5-year survival.
The *intermediate-risk group* has a 56% 5-year survival
The *high-risk group* has a 15% 5-year survival.

Factors affecting the risk of local recurrence overlap with those for metastases. Local recurrence and metastatic spread are substantially more likely to occur in the following conditions:

- Microscopically positive margins for high-grade tumors.
- An intra-abdominal primary tumor.
- Tumor size greater than 5 cm are more likely to metastasize.
- Tumors that have a high histologic grade.

High-grade tumors that recur locally following adequate local therapy will eventually metastasize in two-thirds of patients and curative treatment for such patients rarely occurs.

Molecular determinants of outcome for patients with high-grade NRSTS do not exist at present.

TREATMENT

Treatment for RMS

General Principles

Successful treatment of RMS requires both local and systemic control of disease.

Local Control

For all patients with ARMS and for the vast majority of patients with ERMS with microscopic or gross residual disease following initial surgery, local control is accomplished using radiotherapy.

Systemic Control

This is accomplished with chemotherapy. Patients with localized disease require systemic therapy to eradicate micrometastatic disease typically present at diagnosis.

Surgery

Surgery is most effective if the primary tumor can be completely excised with an adequate margin of uninvolved tissue. If the primary tumor cannot be resected because of proximity to blood vessels and nerves or because this would produce major functional or cosmetic sequelae, an incisional or debulking biopsy is performed. This is followed by induction chemotherapy and radiation to cytoreduce the tumor. Second-look surgery is limited to situations where complete surgical resection of the postinduction chemotherapy mass may result in a meaningful reduction in the dose of postoperative radiation. This is particularly important in infants, toddlers, and very young children where nonmutilating surgical removal of postchemotherapy gross residual disease (during second-look surgery) can result in a reduction in radiation dose from 50.4 to 36 Gy (if there are negative margins) or to 41.4 Gy (if there are positive margins) and, in selected instances, to no postoperative radiation being given. In patients with exceptionally large tumors in the thorax or retroperitoneum, surgical removal of postchemotherapy gross residual disease may permit the use of a lower (potentially more tolerable) dose of radiation. Surgical resection of postradiation residual masses is generally not indicated unless imaging studies (including PET scan) strongly suggest the presence of residual viable tumor.

Prior to starting chemotherapy there may be a role for primary re-excision under the following conditions:

- If the patient had a noncancer operation initially.
- If gross or microscopic residual disease is amenable to wide excision.
- If there is uncertainty about residual disease and/or margins.

For primary tumors arising in the orbit, head, and neck and certain extremity locations, aggressive surgical debulking, such as enucleation, head and neck dissection and amputation, are not indicated. Instead, chemotherapy and radiotherapy should be relied on to control the tumor at the primary site. For tumors arising in the bladder and prostate the goal is to preserve function. Radical surgical debulking is reserved for cases with residual disease after chemotherapy and radiation therapy.

The role of lymph node biopsy or dissection as part of the primary approach depends on the tumor site. Sites with a high incidence of regional node involvement include the extremity (30–50%), genitourinary (20%), perirectal (33%), and paratesticular (40%). Consideration should be given to lymph node biopsy, when management may be affected by the results of the biopsy or excision of clinically suspicious nodes at any site. Regional nodal exploration, with sentinel node mapping, is required in extremity lesions. Ipsilateral retroperitoneal lymph node dissection is mandatory for boys with paratesticular tumors who are 10 years of age and older (Table 26.8).

Radiotherapy

STS are only moderately radiation-sensitive. In general, NRSTS require higher radiation doses than do RMS to achieve local control. The treatment volume is determined by the extent of tumor at diagnosis prior to surgical resection, with an appropriate margin which may be influenced by the surrounding normal tissue structures. Radiation dose guidelines for RMS and NRSTS are outlined in Table 26.9. Use of intensity-modulated radiation

TABLE 26.8   Outcome of the Children's Oncology Group (COG) ARST-0332 NRSTS Trial

Arm	Risk group	Radiotherapy	N = number of patients (percent)	3-year EFS (%)	3-year OS (%)
A	Low	No	212 (38%)	91	99
B	Low	55.8 Gy	19 (4%)	79	100
C	Intermediate	55.8 Gy	120 (22%)	68	81
	High[a]				
D	Intermediate	Preoperative 45 Gy Postoperative 10.8 Gy to negative margins 19.8 Gy to positive margins	200 (36%)	52	66
	High[a]				

[a]*Both arms C and D included patients with metastatic disease. These patients had worse survival that those without metastatic disease.*

Arm A—low grade and grossly resected (either localized regardless of margins or metastatic)—treated with surgery only. The local relapse rates in these low-grade tumors were 2.5% with negative margins and 14.5% with positive margins. There were no patients on this trial with low-grade metastatic lesions.

Arm B—high grade, grossly resected, <5 cm in diameter, positive margins—surgery and radiation therapy.

Arm C—grossly resected, high grade, >5 cm or metastatic grossly resected surgery followed by chemoradiotherapy.

Arm D—unresectable or high-grade tumor >5 cm with planned delayed surgery—biopsy only, chemoradiotherapy with surgery at week 13 followed by chemotherapy and radiation depending on margins.

Chemotherapy: cumulative chemotherapy dose: ifosfamide = 9 g/m^2 × 6 cycles = 54 g/m^2; doxorubicin = 75 mg/m^2 × 5 cycles = 375 mg/m^2.

therapy should be considered in order to spare normal tissues. There is no added benefit to hyperfractionation and the current recommendation is fractionated daily doses of 180 cGy for most sites. Proton beam radiation therapy may offer advantages with regard to sparing critical normal surrounding structures, particularly for skull base, paraspinal/truncal, and pelvic tumors. Brachytherapy can be considered for vaginal tumors and for tumors where there would be value in a higher dose over a shorter period of time and a concentrated dose given to the tumor bed. For NRSTS, radiotherapy may be delivered preoperatively or postoperatively. Preoperative radiotherapy is recommended in NRSTS when tumor shrinkage will enable the surgeon to achieve negative margins or reduce morbidity. The dose of preoperative radiation is lower than postoperative radiation and if margins are positive then a boost can be given to the surgical bed. The downside of preoperative radiation is the increased risk of wound complications. All metastatic lesions should be radiated as well.

Chemotherapy

Patients can either receive "standard" chemotherapy or a treatment regimen determined by risk classification.

Standard Treatment of RMS

All patients with RMS regardless of their initial stage or group receive combination chemotherapy as "standard therapy" consisting of VAC:

- Vincristine (V) age-adjusted (0.025 mg/kg for infants under age 1 year; 0.05 mg/kg for toddlers 1—3 years of age; 1.5 mg/m^2, maximum dose 2 mg, for children 3 years of age and older; all vincristine doses are given as an IV push over 1 min (however some institutions have advocated administering vincristine as a short IV infusion over 10—15 min as a safety enhancement)).
- Actinomycin D (A) age-adjusted (0.025 mg/kg for infants under age 1 year; 0.045 mg/kg for children 1 year of age and older, maximum dose 2.5 mg; actinomycin D doses are given as an IV push over 1—10 min and because it is a radiosensitizer this agent is omitted during radiotherapy).
- Cyclophosphamide (C) age-adjusted (40 mg/kg for infants and toddlers <3 years of age; 1.2 g/m^2 for children 3 years of age and older; all cyclophosphamide doses should be administered IV over 30—60 min with appropriate pre- and postinfusion IV hydration and Mesna to protect the bladder from bleeding. Each dose of Mesna is 20% of the cyclophosphamide dose given IV every 3 h for 3—5 doses beginning just prior to the cyclophosphamide infusion or as a continuous 24-h IV infusion at the mg equivalent dose of the cyclophosphamide dose).

TABLE 26.9   Guidelines for Radiotherapy for Rhabdomyosarcoma (RMS) and Nonrhabdomyosarcoma Soft-Tissue Sarcomas (NRSTS)

RMS

Clinical group I Embryonal	0 Gy
Clinical group I Alveolar	36 Gy
Clinical group IIA (nodes negative)	36 Gy
Clinical group IIB/C (nodal sites)	41.4 Gy
Clinical group III (orbit only)	45 Gy
Clinical group III (other sites)	50.4 GY
Second-look surgery	
Negative margins	36 Gy
Positive margins	41.4 Gy
Clinical group IV (metastases)	50.4 Gy
Second-look surgery	
Negative margins	36 Gy
Positive margins	41.4 Gy

NRSTS

Low-grade tumors

Grossly resected (any margins-observation only)	0 Gy
Unresectable tumors	
Preoperative	45 Gy
Postoperative	
Negative margins	0 Gy
Positive margins	10.8 Gy
Macroscopic margins	19.8 Gy

HIGH-GRADE TUMORS

Grossly resected ≤5 cm	
Negative margins (observation only)	0 Gy
Positive margins	55.8 Gy
Grossly resected >5 cm	
Negative or positive margins	55.8 Gy
Unresectable tumors	
Preoperative RT	45 Gy
Postoperative	
Negative margins	0 Gy
Positive margins	10.8 Gy
Macroscopic margins	19.8 Gy
Inoperable tumors	63 Gy

Dactinomycin (actinomycin D) and cyclophosphamide are given every 3 weeks and vincristine is given weekly for 9–12 weeks; doses of all agents are age-adjusted because of the higher risk of hepatic sinusoidal obstructive syndrome (formerly veno-occlusive disease) in children under 3 years of age. Dactinomycin is omitted during radiation therapy; vincristine may be given weekly or every 3 weeks in combination with cyclophosphamide during radiation therapy. This regimen is typically given for 12–14 cycles lasting 8–10 months.

Patients with low-risk RMS may receive VA with or without C, depending on their specific risk factors.

Other effective agents in RMS include ifosfamide, etoposide, doxorubicin, irinotecan, topotecan, carboplatin, and vinorelbine.

Treatment Regimens by Risk-Adapted Classification

Low-Risk Group

Low-risk group patients are those with ERMS that has been completely resected (Group I or II) either in a favorable-site or gross-totally resected localized ERMS.

Radiation Therapy

1. Group I embryonal patients do not receive radiation therapy.
2. Patients with microscopic, locoregional, or gross residual tumor receive radiation therapy at week 13 of chemotherapy.

Chemotherapy

All low-risk patients receive the following chemotherapy:

- Vincristine IV 1 dose weekly for 9 weeks.
- Dactinomycin IV and cyclophosphamide ($1.2 \, g/m^2$) IV 1 dose each on the first week and then on weeks 4, 7, and 10.

Low-risk subset 1 patients receive the following additional treatment:

- Vincristine IV 1 dose weekly from week 13 to week 21.
- Dactinomycin IV 1 dose on weeks 13, 16, 19, and 22. (Radiation is started on week 13 when indicated; dactinomycin is omitted during radiation.)

Low-risk subset 2 patients receive the following additional treatment:

- Vincristine IV 1 dose weekly on weeks 13–21, 25–33, and 37–45 (total 36 doses).
- Dactinomycin IV 1 dose every 3 weeks starting on week 13 until week 46 (total 14–16 doses; 1–2 doses held during radiation).

Compared to the standard therapy discussed above, the risk-adapted therapy results in a lower cumulative dose of cyclophosphamide (4 doses, $4.8 \, g/m^2$) and a shorter duration of therapy for these low-risk patients with very favorable prognoses.

Intermediate-Risk Group

Intermediate-risk-group patients account for the majority of the patients with RMS (55%). This group includes unfavorable-site ERMS that have gross residual disease and nonmetastatic ARMS tumors. There has been no improvement in outcome for this group in over two decades. The COG intermediate-risk study (ARST-0531) compared 14 cycles of VAC (with a reduced cyclophosphamide dose of $1.2 \, g/m^2$) to 7 cycles of VAC alternating with 7 cycles of vincristine–irinotecan (VI). Irinotecan was given at a dose of $50 \, mg/m^2$ IV over 90 min daily for 5 days every 21 days. There was no difference in 2-year event-free survival between the two groups.

High-Risk Group

Given the relatively poor outcome for these patients using the standard "VAC" therapy discussed above, the COG high-risk study (ARST-0431) used a more intensive seven-drug regimen given on an interval-compressed (every 2 weeks) schedule. This regimen included:

Weeks 1–5: VI: vincristine (doses as above) weekly; irinotecan $50 \, mg/m^2$ (maximum 100 mg) IV over 60 min daily × 5 at weeks 1 and 4.

Weeks 6,7: VDC: vincristine weekly in combination with doxorubicin (37.5 mg/m^2/day × 2 days (50% dose reduction for infants <1 year)) and cyclophosphamide (dosed as above) at week 6.

Week 9: IE: ifosfamide (1800 mg/m^2 IV over 1 h daily × 5) and etoposide 100 mg/m^2 IV over 1 h daily × 5 (50% dose reduction of both agents for infants <1 year).

Weeks 11, 12: VDC.

Week 13: IE

Weeks 15, 16: VDC

Week 17: IE

Weeks 20−24: VI; local radiation therapy.

Week 26: IE

Weeks 28, 29: VDC

Week 30: IE

Weeks 32, 33; VDC

Weeks 35, 38: VAC (as above for ARST-0531 and ARST-0331)

Weeks 41−44: Vincristine weekly with AC at weeks 41 and 44

Weeks 47, 48, 50, 51: VI (weekly vincristine; irinotecan only at weeks 47 and 50).

This regimen produced a superior 18-month progression-free survival of 66% compared to the expected 45% of historical controls, although there was no clear improvement in overall survival. It remains unclear whether the apparent improvement in short-term progression-free survival justifies the use of this more toxic regimen relative to the results that are achieved with "standard" VAC therapy.

No improvement in outcome has been demonstrated for this group of patients with single or tandem high-dose chemotherapy plus autologous stem-cell transplant regimens.

Treatment of NRSTS

The mainstay of therapy for this histologically and biologically diverse group of tumors is surgical resection. Tumors are broadly divided into two main categories:

- Low-grade lesions which may recur locally but are unlikely to metastasize.
- High-grade lesions, which have a substantially greater risk of both local invasiveness and distant spread.

The role of chemotherapy remains unclear for many patients. Preoperative chemotherapy may improve resectability and some studies suggest an improvement in metastasis-free survival with chemotherapy. The majority of randomized trials do not show improved overall survival. Chemotherapy appears to be of more use in synovial cell sarcoma and in extremity tumors. Ifosfamide and doxorubicin or etoposide are the most active chemotherapy agents. The standard dose of ifosfamide is 1800 mg/m^2/day for 5 days or 3000 mg/m^2/day for 3 days every 3 weeks for 5−7 cycles with doxorubicin 75 mg/m^2 per cycle either as a continuous infusion or in a divided daily dose for 2 days every 3 weeks for 5 cycles. During radiation doxorubin should be omitted and can be replaced with etoposide 100 mg/m^2/day for 5 days. For unresectable tumors, the length of therapy depends upon response. In general, tumors that have not become resectable after 5 cycles of chemotherapy should receive radiation therapy and resectability should be reevaluated after 45 Gy.

Low-grade tumors are, for the most part, observed following total surgical resection. The approach to high-grade tumors is based on their size, respectability, and status of the margins. Small tumors (5 cm or less) that are completely resected with negative margins are generally observed without further treatment. Postoperative radiation is generally recommended for small tumors with positive margins and for all tumors more than 5 cm, even if margins are negative, because of the high risk of local recurrence. Neoadjuvant chemotherapy (i.e., chemotherapy before surgery or radiation therapy) is generally reserved for those patients with unresectable tumors and those with metastatic disease; adjuvant chemotherapy is typically recommended for patients with tumors greater than 5 cm in size even if completely resected. The aim is to reduce the size or extent of the cancer prior to surgery. This makes resection easier and more likely to be successful and reduces the consequences of more extensive surgery.

The COG trial ARST-0332 was developed as a risk-based treatment strategy for patients with NRSTS. The goal was to reduce therapy for patients with low-risk disease and to evaluate the use of chemoradiotherapy in intermediate- and high-risk disease. The patients were stratified into risk groups based on tumor grade and extent of disease.

ARST-0332 risk groups

- *Low-risk group*—(~50% of patients) was defined as tumors that were nonmetastatic grossly resected tumors. They were treated with surgery with or without adjuvant radiation therapy, depending on the histologic grade of the tumor and the surgical margin status. Low-grade tumors regardless of margins underwent observation only postresection. High-grade tumors that were less than 5 cm in size received radiation only if the margins were positive.
- *Intermediate-risk group*—(~35% of patients) was defined as those with nonmetastatic grossly resected tumor that was both high grade and greater than 5 cm in maximal diameter and those with nonmetastatic unresectable tumors. Intermediate-risk patients in whom the primary tumor has not been excised receive neoadjuvant combined chemoradiotherapy prior to definitive resection. If the primary tumor is able to be excised, they are treated with adjuvant chemotherapy with radiation. The chemotherapeutic regimen for all patients is doxorubicin with ifosfamide.
- *High-risk group*—(~15% of patients) was defined as patients with metastatic tumors including those with metastases restricted to regional lymph nodes. Patients with low-grade tumors with microscopic margins with all disease resected received observation only, while patients with low-grade tumors that were unresected and those with high-grade tumors received neoadjuvant or adjuvant chemoradiotherapy.

The regimen utilized on ARST-0332 is:

- Weeks 1, 4, 13, and 16: ifosfamide (3000 mg/m^2 IV over 3 h daily × 3) and doxorubicin (37.5 mg/m^2/day IV over 24 h × 2 days).
- Weeks 7 and 10: ifosfamide (3000 mg/m^2 IV over 3 h daily × 3).
- Week 19: doxorubicin (37.5 mg/m^2/day × 2 days).

(Note that if patients do not receive radiotherapy then weeks 16 and 19 doxorubicin should be given at weeks 7 and 10. Cytokine support is recommended with filgastrim or peg-filgastrim after each cycle.)

The outcomes of this trial are given in Table 26.8. Given the favorable results, the treatment strategy developed in this study is recommended for the care of patients with NRSTS.

FOLLOW-UP AFTER COMPLETION OF THERAPY

First Year After Completion of Therapy

1. Physical examination including complete blood count, kidney function, urinalysis, every 3 months.
2. Chest radiograph every 3 months.
3. Appropriate imaging studies (i.e., CT or MRI of the involved region) every 3 months.
4. Referral and follow-up with other subspecialists as indicated by primary site of disease (i.e., orbital lesion—ophthalmology, genitourinary—urology or gynecology, head and neck—otolaryngology or dental, bone—orthopedics).
5. Patients who received radiation should be followed-up annually with radiation oncology.

Second and Third Years After Completion of Therapy

1. Physical examination including complete blood count, kidney function, urinalysis, every 4 months.
2. Chest radiograph every 4 months.
3. Appropriate imaging studies (i.e., CT or MRI of the involved region) every 4 months.
4. Patients who receive cyclophosphamide: evaluation of gonadal function (FSH, LH, testosterone/estradiol, and anti-Mullerian hormone).
5. Echocardiogram to evaluate for anthracycline-related late effects.
6. Monitor for secondary malignancies, especially in cases with known familial risk factors.

Fourth and Fifth Years After Completion of Therapy

1. Physical examination including complete blood count, kidney function, urinalysis, every 6 months.
2. Chest radiograph every 6 months.

3. Appropriate imaging studies (i.e., CT or MRI of the involved region) every 4 months.
4. *Referral to a survivorship program is strongly recommended.* Because of the widespread need for radiotherapy to achieve local control and the young age of the patients, late-onset radiation-associated complications are a significant source of treatment-related morbidities after 5 years.

RECURRENT DISEASE

The overwhelming majority of recurrences of RMS occur within the first 2 years of completing therapy. Although some children attain durable remissions with secondary therapy, the long-term prognosis for children with progressive or recurrent disease is extremely poor. The 3-year survival following relapse for patients is less than 15%. When recurrent disease is suspected, a biopsy should be done and an extent-of-disease workup should be undertaken. In formulating a treatment plan, the following factors should be considered:

- Timing of the recurrence.
- Extent of disease at diagnosis and recurrence.
- Type of chemotherapy and radiation therapy previously given.

A durable remission is very difficult to attain in patients who have unresectable or widespread disease at diagnosis and in those who progressed on therapy or relapsed early.

Treatment

Complete surgical resection, when feasible, improves outcome, particularly if the recurrence is localized or within a prior radiation field. Surgery should be followed by adjuvant chemotherapy and radiation (if further radiation is feasible). Durable remissions in paratesticular and vaginal recurrences can be attained with this salvage approach. Disseminated recurrence is essentially incurable.

While surgical resection of metastatic lesions is not curative, it may be palliative. Even when followed with intensive multiagent therapy with vincristine/doxorubicin/cyclophosphamide and ifosfamide/etoposide (plus irinotecan for responding patients), the median postrelapse survival time is 1.1 years and fewer than 20% of patients are alive 3 years' postrelapse. Multiple trials have used intensive regimens followed by autologous stem-cell rescue, but no large studies have demonstrated any significant salvage rate.

A recently completed COG study (ARST-0921) compared postrelapse outcome in patients with first progression/recurrence receiving a standard chemotherapy backbone of cyclophosphamide (1200 mg/m^2 IV over 30–60 min (40 mg/kg for patients <10 kg) on day 1 of each 21-day cycle; all patients received mesna) and vinorelbine (25 mg/m^2 IV over 6–10 min (0.83 mg/kg for patients <10 kg) on days 1 and 8 of each 21-day cycle) and randomly assigned to receive either bevacizumab (Regimen A) at a dose of 15 mg/kg (0.5 m/kg for patients <10 kg) IV over 30–90 min on day 1 of each 21-day cycle or temsirolimus (Regimen B) at a dose of 15 mg/m^2 (maximum dose 30 mg) IV over 30–60 min on days 1, 8, and 15 of each 21-day cycle. Accrual to the study was ended early after an interim analysis following enrollment of 87 patients demonstrated significantly improved 6-month progression-free survival of 65% with Regimen B versus 50% in Regimen A. There was a threefold reduction in the risk of early disease progression from 26% to 9% with the use of Regimen B. No differences, however, were seen in 2-year overall survival between the two groups.

Active agents in relapsed RMS include combinations such as ifosfamide/carboplatin/etoposide, docetaxel/gemcitabine, and irinotecan in combination with vinorelbine or temozolomide. Chemotherapy regimens in use for recurrent STS are outlined in Table 26.10.

FUTURE PERSPECTIVES

Survival has dramatically increased over the past 30 years; however, improvements in disease control are still needed, especially for the majority of patients with gross residua after resection or metastatic disease at the time of diagnosis. Risk stratification of RMS into low-, intermediate-, and high-risk strata has already translated into the development of therapies that aim to maintain excellent outcome while simultaneously reducing the short- and long-term toxicities of therapy. The excellent outcome of patients with lower-risk RMS makes it unlikely that

TABLE 26.10 Alternative Chemotherapy for Recurrent Soft-Tissue Sarcomas

REGIMEN 1[a]		
Irinotecan IV	50 mg/m^2/day	Days 1–5
Temozolomide po	100 mg/m^2/day	Days 1–5
REGIMEN 2[a]		
Vinorelbine IV	20–30 mg/m^2/day	Days 1, 8
Irinotecan IV	20 mg/m^2/day	Days 1–5 and days 8–12
REGIMEN 3[a]		
Vinorelbine IV	25 mg/m^2/day	Days 1, 8, 15
Cyclophosphamide po	25 mg/m^2/day	Days 1–28

[a]*Personal communication with Leonard H. Wexler, MD. Regimens can be given for 6 cycles up to 26 cycles if tolerated.*

future clinical trials can be conducted to demonstrate further improvements in outcome, however, novel approaches are clearly needed for patients with intermediate- and high-risk disease. Based on the results of ARST-0531 and ARST-0921, the next-generation intermediate-risk study will use the less toxic VAC/VI regimen and randomize patients to receive the mTOR inhibitor, temsirolimus. Additional trials using angiogenesis inhibitors such as pazopanib, a small molecule multitargeted tyrosine kinase inhibitor which blocks signaling pathways and inhibits angiogenesis, tumor growth, and metastatic formation, in conjunction with the chemotherapy backbone of ARST-0431 are planned for patients with high-risk disease. For NRSTS, a number of clinical trials with pazopanib have demonstrated activity and have been shown to increase progression-free survival in recurrent and refractory sarcomas in adults. The current COG trial builds on the chemoradiotherapy backbone of the ARST-0332 and randomizes patients to receive treatment with ifosfamide/doxorubicin/radiation with or without pazopanib.

Further Reading and References

Arndt, C., Stoner, J.A., Hawkins, D.S., et al., 2009. Vincristine, actinomycin and cyclophosphamide compared with vincristine, actinomycin and cyclophosphamide alternating with vincristine, topotecan and cyclophosphamide for intermediate-risk rhabdomyosarcoma: Children's Oncology Group Study D9803. J. Clin. Oncol. 27, 5182–5188.

Casanova, M., Ferrari, A., Bisogno, G., et al., 2004. Vinorelbine and low-dose cyclophosphamide in the treatment of pediatric sarcomas. Pilot study for the upcoming European rhabdomyosarcoma protocol. Cancer 101, 1664–1671.

Dantonello, T.M., Int-Veen, C., Winkler, P., et al., 2008. Initial patient characteristics can predict pattern and risk of relapse in localized rhabdomyosarcoma. J. Clin. Oncol. 26, 406–413.

Davicioni, E., Anderson, M.J., Finckenstein, F.G., et al., 2009. Molecular classification of rhabdomyosarcoma—genotypic and phenotypic determinants of diagnosis. A report from the Children's Oncology Group. Am. J. Pathol. 174, 550–564.

Dharmarajan, K.V., Wexler, L.H., Gavane, S., et al., 2012. Positron emission tomography (PET) evaluation after initial chemotherapy and radiation therapy predicts local control in rhabdomyosarcoma. Intl. J. Radiation Oncol. Biol. Phys. 84, 996–1002.

Klem, M.L., Grewal, R.K., Wexler, L.H., et al., 2007. PET for staging in rhabdomyosarcoma: an evaluation of PET as an adjunct to current staging tools. J. Pediatr. Hematol. Oncol. 29, 9–14.

Mancini, B.R., Roberts, K.B., 2013. Pediatric non-rhabdomyosarcoma soft tissue sarcomas. J. Radiat. Oncol. 2, 135–148.

Mandell, L.R., 1993. Ongoing progress in the treatment of childhood rhabdomyosarcoma. Oncology 7, 71–84.

Mattke, A.C., Bailey, E.J., Schuck, A., et al., 2009. Does the time-point of relapse influence outcome in pediatric rhabdomyosarcomas?. Pediatr. Blood Cancer 52, 772–776.

Meza, J.L., Anderson, J., Pappo, A.S., et al., 2006. Analysis of prognostic factors in patients with nonmetastatic rhabdomyosarcoma treated on the intergroup rhabdomyosarcoma studies III and IV: the Children's Oncology Group. J. Clin. Oncol. 24, 3844–3851.

Million, L., Donaldson, S.S., 2012. Resectable pediatric nonrhabdomyosarcoma soft tissue sarcoma: which patients benefit from adjuvant radiation therapy and how much? ISRN Oncol. 2012, 1–6.

Oberlin, O., Rey, A., Lyden, E., et al., 2008. Prognostic factors in metastatic rhabdomyosarcoma: results of a pooled analysis from the United States and European cooperative groups. J. Clin. Oncol. 26 (14), 2384–2389.

Pappo, A.S., Devidas, M., Jenkins, J., et al., 2005. Phase II trial of neoadjuvant vincristine, ifosfamide, doxorubicin with GCSF support in children and adolescents with advanced stage NRSTSs: a Pediatric Oncology Group study. J Clin Oncol. 23, 18.

Puri, D.R., Wexler, L.H., Meyers, P.A., et al., 2006. The challenging role of radiation therapy for very young children with rhabdomyosarcoma. Int. J. Radiat. Oncol. Biol. Phys. 65, 1177–1184.

Rodeberg, D.A., Stoner, J.A., Hayes-Jordan, A., et al., 2009. Prognostic significance of tumor response at the end of therapy in Group III rhabdomyosarcoma. A report from the Children's Oncology Group. J. Clin. Oncol. 27, 3705–3710.

Rodeberg, D.A., Garcia-Henriquez, N., Lyden, E.R., et al., 2011. Prognostic significance and tumor biology of regional lymph node disease in patients with rhabdomyosarcoma: a report from the Children's Oncology Group. J. Clin. Oncol. 29, 1304–1311.

Skapek, S.X., Anderson, J., Barr, F.G., et al., 2013. *PAX-FOXO1* fusion status drives unfavorable outcome for children with rhabdomyosarcoma: A Children's Oncology Group Report. Pediatr. Blood Cancer 60, 1411–1417.

Spunt, S.L., Hill, D.A., Motosue, A.M., et al., 2002. Clinical features and outcome of initially unresected nonmetastatic pediatric nonrhabdo-myosarcoma soft tissue sarcoma. J. Clin. Oncol. 20, 3225–3235.

Spunt, S.L., Million, L., Anderson, J.M., et al., 2014. Risk-based treatment for nonrhabdomyosarcoma soft tissue sarcomas (NRSTS) in patients under 30 years of age: Children's Oncology Group study ARST0332. J. Clin. Oncol. (Meeting Abstract) 32 (15), 10008.

Walterhouse, D.O., Pappo, A.S., Meza, J.L., et al., 2014. Shorter-duration therapy using vincristine, dactinomycin, and lower-dose cyclophos-phamide with or without radiotherapy for patients with newly diagnosed low-risk rhabdomyosarcoma: a report from the Soft Tissue Sarcoma Committee of the Children's Oncology Group. J. Clin. Oncol. 55, 6787.

Weiss, A.R., Lyden, E.R., Anderson, J.R., et al., 2013. Histologic and clinical characteristics can guide staging evaluations for children and ado-lescents with rhabdomyosarcoma: a report from the Children's Oncology Group Soft Tissue Sarcoma Committee. J. Clin. Oncol. 31, 3226–3232.

Wexler, L.H., Ladanyi, M., 2010. Diagnosing alveolar rhabdomyosarcoma: morphology must coupled with fusion confirmation. J. Clin. Oncol. 28, 2126–2128.

27

Malignant Bone Tumors

Noah Federman, Elizabeth A. Van Dyne and Nicholas Bernthal

Malignant bone tumors constitute approximately 6% of all childhood malignancies. In the United States, the annual incidence in children under 20 years of age is 8.7 per million. Osteosarcoma (56%), the Ewing sarcoma family of tumors (34%), and chondrosarcoma (6%) comprise the most frequently encountered malignant bone tumors in children and adolescents. In the United States 650–700 children and adolescents under the age of 20 are diagnosed with malignant bone tumors each year.

Figure 27.1 shows the age-specific incidence rate of bone cancers for all races and both sexes.

The peak incidence of bone cancer steadily increases through childhood and peaks at age 15. This coincides with the adolescent growth spurt, after which rates decline. Table 27.1 summarizes the differences between osteosarcoma and the Ewing sarcoma family of tumors.

OSTEOSARCOMA

Epidemiology

The incidence of osteosarcoma is approximately 800 cases per year in the United States, with approximately 400 cases per year in children less than 20 years old. The incidence is slightly higher in males, and in African-Americans, Asians/Pacific Islanders, and Hispanics than in Whites. Although the etiology of osteosarcoma is unknown, certain factors correlate with occurrence:

- *Bone growth*: The development of osteosarcoma is correlated with increased growth velocity as suggested by:
  - The peak incidence occurs during the pubescent growth spurt. The age peak is 12 years in girls and 16 years in boys, correlating with the different average ages for pubertal development.
  - Patients with osteosarcoma are taller than average.
  - The most common sites are metaphyses of the most rapidly growing bones (distal femur, proximal humerus, proximal tibia).
- *Genetic factors*: Osteosarcoma is a genetically intricate tumor with complex karyotypes, numerous genetic abnormalities, and absence of characteristic chromosomal translocations. Osteosarcoma can be hereditary in some rare cases. Human predisposition syndromes, murine models, genetic analyses of osteosarcoma tumor specimens, and environmental factors may all help to understand the pathogenesis of this disease.
  - Human predisposition syndromes:
    - *Retinoblastoma*—In patients with early-onset bilateral retinoblastoma, who are likely to have germline alteration in the Rb gene, the development of a secondary osteosarcoma or soft-tissue osteosarcoma is independent of the therapeutic modalities or radiation field used in the treatment of the retinoblastoma. Approximately 40% of patients will develop this secondary malignancy by 40 years of age.
    - *P53*—Osteosarcoma is seen frequently in families with Li–Fraumeni syndrome, in which members of the affected families have breast cancer, brain tumors, soft tissue sarcomas, leukemia, adrenocortical carcinoma, and osteosarcoma.

- *Rothmund–Thomson syndrome*—An inherited autosomal recessive disorder consisting of skeletal abnormalities, short stature, underdeveloped or missing forearm bones or thumbs, cataracts, abnormalities of teeth and nails, sparse hair and red, and swollen patches on the skin.
- *Werner syndrome* (also known as progeria adultorum)—A genetic disorder of premature aging consisting of scleroderma-like, thin, wrinkled skin, loss of subcutaneous fat, graying and loss of hair, bilateral cataracts, hypogonadism, and premature menopause.
- *Diamond–Blackfan anemia*—A rare congenital bone marrow failure disorder characterized by pure red cell aplasia. About 50% of patients have mutations in the ribosomal protein subunit genes that are associated with increased incidences of osteosarcoma among other malignancies.

- Murine models that produce osteosarcoma include those with alterations in the p53, Rb, Wnt, Notch, IGF, mTor pathways, *MYC, FOS* as well as chronic parathyroid hormone exposure.
- Genetic analyses of osteosarcoma tumor specimens:
  - Osteosarcomas are genetically complex with numerous alterations in each tumor sample. These include a multitude of genes involved in cell cycle regulation, differentiation, oncogenic transformation, cell signaling, signal transduction, among many pathways. Consistent chromosomal gains and losses occur at genetic sites in which the gene(s) of interest have not been clearly identified.
  - Approximately 3% of patients with spontaneous osteosarcoma are found to have a germline alteration in *p53* (Li–Fraumeni syndrome).
  - There are non-germline mutations of both *p53* and *Rb* in osteosarcoma tumors. Twenty-five percent have point mutations in the *p53* gene, 10–20% have *p53* rearrangements, 75% have loss of one 17q allele, and 60% of OS tumors have a loss of heterozygosity at 13q of the site of the *Rb* gene.

- *Environmental factors*: The environmental factor most consistently associated with the development of osteosarcoma is radiation therapy. Bone turnover also may be related to osteosarcoma development. The evidence for these is:
- Osteosarcoma may arise in any previously irradiated bone.
- The interval between irradiation and the appearance of osteosarcoma has ranged from 4 to 40 years but is generally 10–20 years, suggesting it is not a major etiologic factor in the younger patients.
- Paget disease of bone and other disorders associated with rapid bone turnover are associated with a higher incidence of osteosarcoma.

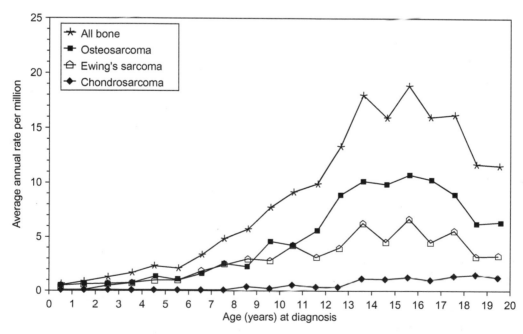

FIGURE 27.1 Bone cancer age-specific incidence rate by histology for all races and both sexes. SEER, 1976–1994 and 1986–1994, combined. *From: Gurney et al. (1999).*

TABLE 27.1　Summary of Differences Between Osteosarcoma and the Ewing Sarcoma Family of Tumors

	Osteosarcoma	Ewing sarcoma family of tumors
Epidemiology	Bimodal age distribution with peaks in early adolescence and adults over 60	In children and adolescents, rare in adults
	Males affected more often than females	Caucasians affected more than other races
	African-Americans and other races more affected than Caucasians	
Location	Metaphysis of long bones	Soft tissue mass
		Flat bones
		Diaphysis of bones
Metastatic disease	Lung, bone	Lungs, bone, bone marrow
Radiographic findings	Radial or "sunburst" pattern	"Permeative" or "moth-eaten"
	Codman triangle	Bone deposited in "onion peel" appearance
		Codman triangle

TABLE 27.2　Classification of Osteosarcoma

1. Classic
   a. Osteoblastic
   b. Chondroblastic
   c. Fibroblastic
2. Telangiectatic
3. Small cell
4. Periosteal
5. Parosteal

Pathology

The histologic diagnosis of osteosarcoma is dependent on the demonstration of malignant osteoid. Table 27.2 lists the classification of osteosarcoma. Although several of the rarer variants may be confused radiographically or histologically with other entities, the pathologic hallmark of osteosarcoma is the production of osteoid. Parosteal osteosarcoma is generally considered a low-grade variant of osteosarcoma and is treated with surgical resection alone.

Clinical Manifestations

Symptoms are usually present for several months before the diagnosis is made, with an average duration of approximately 3 months. Patients commonly present with the following symptoms:

- Pain at site of tumor (90%)
- Local swelling (50%)
- Decreased range of motion (45%)
- Pathologic fracture (8%)
- Joint effusion can occur rarely. Its presence may suggest an extension of the tumor into the joint space.

It is important to be aware of the sites and ages at which osteosarcoma presents as this helps distinguish it from other entities which occur more commonly in this population, such as growing pains and osteomyelitis. Osteosarcoma can arise in any bone but has a propensity for the metaphyseal portions of long bones, particularly in the lower extremity around the knee (>50% involving distal femur or proximal tibia). The most common presenting symptom is persistent pain and swelling at the primary site. Approximately 10−20% have evidence of metastatic disease at the time of diagnosis, with lungs and bone being the most common sites. Systemic manifestations of osteosarcoma are rare at initial presentation, and when they occur they are usually associated with significant metastatic disease.

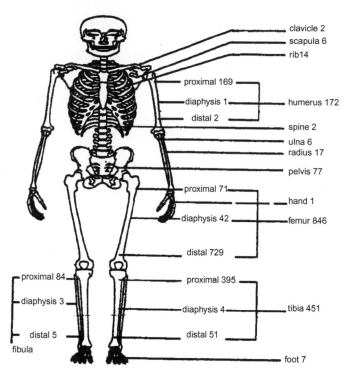

clavicle 2
scapula 6
rib14

proximal 169
diaphysis 1 — humerus 172
distal 2

spine 2
ulna 6
radius 17

pelvis 77

proximal 71
hand 1
diaphysis 42 — femur 846

distal 729

proximal 84 — proximal 395
diaphysis 3 — diaphysis 4 — tibia 451
distal 5 — distal 51
fibula

foot 7

FIGURE 27.2 Skeletal distribution of 1702 patients with osteosarcoma. *From: Bielack et al. (2002), with permission.*

Figure 27.2 shows the skeletal distribution of patients with osteosarcoma. The most common primary sites in individuals less than 25 years of age are:

- Lower long bones (74.5%)
- Upper long bones (11.2%)
- Pelvic region (3.6%)
- Face or skull (3.2%)
- Mandible (1.9%)
- Chest region (1.8%)
- Vertebral column (1.2%)
- Lower short bones (1.1%)
- Upper short bones (0.3%).

Diagnostic Evaluation

The diagnostic evaluation should delineate the extent of the primary tumor and the presence or absence of metastatic disease. All patients should have the following evaluation:

- *History.*
- *Physical examination.*
- *Blood count.*
- *Urinalysis.*
- *Blood urea nitrogen (BUN), creatinine, liver enzymes, alkaline phosphatase, total bilirubin, and lactate dehydrogenase*: 30–50% of patients have elevated alkaline phosphatase which has been associated with a worse prognosis.
- *Radiographs of the entire affected bone*—The Codman triangle, the formation of new periosteum and the elevation of the cortex, may be present.
- *Computed tomography (CT) and magnetic resonance imaging (MRI) of the entire affected bone.* Imaging of the primary tumor includes MRI, because the extent of soft tissue involvement, especially that of neurovascular structures, will dictate the feasibility of limb salvage surgery. MRI will also show the extent of marrow involvement, essential information in surgical planning.

- *Technetium-99 bone scan and/or whole body 18-fluoro-deoxyglucose positron emission tomography (FDG PET)/CT scan*—Approximately 10% of patients will have evidence of bone metastases on bone scan. Whole-body FDG PET/CT is used for diagnostic evaluation and may be more sensitive and specific for diagnosing bone metastases when compared to technetium-99 bone scan. Bone scan, however, is still considered standard for detection of osseous metastases.
- *High-resolution chest CT*—The use of chest CT is essential for staging patients, because it detects metastases not seen on chest radiographs. If a lesion detected on chest CT cannot be diagnosed as metastatic disease, histologic confirmation is indicated.
- *Assessment of renal function*—Institutional practice varies with some obtaining a creatinine clearance or radionuclide GFR study to assess renal function prior to chemotherapy whereas others rely solely on the serum creatinine with a calculated estimate of glomerular filtration rate.
- *Echocardiogram* to assess cardiac function prior to chemotherapy.
- *Baseline audiogram* prior to chemotherapy.
- *Fertility preservation*—sperm banking should be offered whenever feasible. If timing allows, oocyte cryopreservation is the standard of care for postpubertal young women who are not prepared to undertake embryo cryopreservation. Cryopreservation of ovaries has not become a clinical standard at present but interest in this approach as a means of preserving fertility continues to increase.
- *Biopsy of primary site and possibly secondary sites if questionable for other diagnosis*—Biopsy and pathologic review are essential to establish diagnosis of osteosarcoma. Biopsy should be performed by an orthopedic surgeon, radiologist, or interventional radiologist with expertise. Improperly performed biopsies can contaminate unaffected tissue and alter definitive surgical resection. Either an open biopsy or image-guided percutaneous core biopsy can be performed. The biopsy should be performed with future surgery in mind as a poorly placed biopsy may jeopardize future limb salvage surgery.

Radiographically, osteosarcoma typically shows destruction of normal trabecular bone in the metaphysis of the affected long bone along with the following findings (Figure 27.3):

- Soft-tissue extension (75%)
- Radiating calcification, or sunburst (60%)
- Osteosclerotic lesion (45%)
- Lytic lesion (30%)
- Mixed lesion (25%).

Treatment of Localized Osteosarcoma

Treatment relies on early and proper diagnosis, neoadjuvant and adjuvant multiagent chemotherapy, and aggressive surgery, with limb-preserving therapy when possible. The tumor is generally viewed as radioresistant and radiation therapy only plays a role in initial management in a case-by-case basis when positive margins are unable to resect. The survival rate of non-metastatic osteosarcoma is approximately 70%.

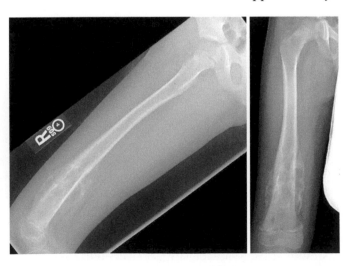

FIGURE 27.3   Anteroposterior/lateral radiographs of an aggressive osteosarcoma lesion in the right distal femur in a skeletally immature child. Note the periosteal reaction and associated soft tissue mass with calcification and mineral deposition classic for osteosarcoma.

Surgery

The surgical approach to osteosarcoma includes surgical biopsy, followed by either limb salvage surgery or amputation. Complete resection is a prerequisite for cure.

Surgery should remove all gross and microscopic tumor with a margin of normal tissue to prevent local recurrence as positive margins and poor necrosis lead to a higher risk of recurrence. Tumor involvement of neurovascular structures, seeding of surrounding compartments from a pathological fracture, and inadequacy of surrounding soft tissue for closure are the common contraindications for limb salvage. In recent decades, approximately 85% of patients with osteosarcoma have been able to be surgically managed with limb salvage surgery. If adequate surgical margins are achieved, the rate of local recurrence and survival after amputation and limb salvage surgery are equivalent.

Definitive surgery is generally performed following preoperative (neoadjuvant) chemotherapy. Delaying definitive surgery until after neoadjuvant chemotherapy is the standard treatment approach, though there has been no defined survival benefit.

Neoadjuvant chemotherapy offers several theoretical advantages:

- Decrease in tumor-related edema and shrinkage of the tumor at the primary site so as to facilitate less radical surgery.
- Initial chemotherapy directed toward the assumed micrometastases which are present in 80–90% of patients.
- Assessment of sensitivity of primary tumor to chemotherapy which has prognostic significance. Greater than 90% necrosis is considered a good response and portends an improvement in overall survival.

The type of surgery is determined by tumor location, size, extramedullary extent, the presence of metastatic disease, age, skeletal development, and the patient's lifestyle choices. If complete excision cannot be accomplished with limb-sparing surgery, amputation is indicated. Guidelines for the use of limb-sparing surgery include:

- No major neurovascular involvement by tumor.
- Ability to have a wide resection of affected bone with a normal muscle cuff in all directions.
- Ability to perform en bloc removal of all previous biopsy sites and all potentially contaminated tissue.
- Adequate motor reconstruction.

Chemotherapy

The use of chemotherapy results in significant improvement in disease-free survival. More than 80–90% of patients treated with surgery alone will develop metastatic disease. The standard components of multiagent chemotherapy for osteosarcoma are high-dose methotrexate (HDMTX), doxorubicin (Adriamycin, ADM), and cisplatin (CDDP), a regimen often referred to as MAP. Figure 27.4A shows the schedule and doses of MAP regimens.

Ifosfamide (IFO) and etoposide (VP16) are other agents with demonstrated activity in osteosarcoma, however, the use of these agents upfront in the management of localized osteosarcoma has not shown a survival advantage at this time.

Specific HDMTX administration guidelines must be followed to prevent renal injury and life-threatening toxicity. Normal renal function should be documented prior to administration. Alkalinizing intravenous fluids should be administered with HDMTX and continued until a sufficient decrease in MTX blood levels has occurred. MTX levels should be monitored starting 4 h after the beginning of MTX administration and then every 24 h from the starting time of the MTX infusion. Leucovorin should be initiated 24 h after HDMTX administration in a dose of 15 mg/m^2 oral or IV route every 6 h until the MTX level is less than 1.0×10^7 mol/l. Patients whose MTX excretion is delayed require an increased dose or duration of leucovorin, according to Table 27.3. Glucarpidase (Voraxaze) is now FDA-approved for patients with toxic plasma methotrexate levels in patients with delayed clearance due to renal dysfunction. It should be administered at 50 units/kg IV once.

A study conducted by the Pediatric Oncology Group and Children's Cancer Group demonstrated that the addition of muramyl-tripeptide phosphatidylethanolamine (MTP-PE) to standard chemotherapy significantly improved 6-year overall survival, from 70% to 78%, but did not significantly improve event-free survival. MTP-PE, a liposomal encapsulated bacille Calmette–Guerin (BCG) cell wall component, is approved for use in Europe but not in the United States, and prior to standard use will need further confirmatory trials.

The degree of tumor necrosis following neoadjuvant chemotherapy is an independent predictor of event-free and overall survival, presumably reflecting tumor resistance to chemotherapy. The histologic response of osteosarcoma to neoadjuvant (preoperative) chemotherapy is graded I to IV (grade I ≤50%, grade II >50%, grade III >90%, grade IV 100% tumor necrosis). Patients whose tumors exhibit a good response (90% necrosis or grades III and IV) have superior event-free and overall survival compared with poor responders.

(A) Schema of typical administration scheduled for MAP therapy for osteosarcoma

Week	1	2	3	4	5	6	7	8	9	10	11	12	13	14	15	16	17	18	19	20	21	22	23	24	25	26	27	28	29
Chemo	A			M	M	A			M	M	surgery	A			M	M	A			M	M	A		M	M	M	A		M	M
	P					P						P					P												

A = Doxorubicin 75 mg/m² continuous infusion over 48 h
P = Cisplatin 60 mg/m²/day × 2 days
M = Methotrexate 12 gm/m² × 1 dose, maximum dose 20 gm*

(B) Schema of MAP chemotherapy plus ifosfamide and etoposide

Week	1	4	5	6	9	10	11	12	15	16	19	20	23	24	27	28	31	32	35	36	39	40
Chemo	A	M	M	A	M	M	surgery	A	M	I	M	A	M	I	M	A	M	I	M	A	M	M
	P			P				P		E		i		E		P		E		i		

A = Doxorubicin 75 mg/m² continuous infusion over 48 h
P = Cisplatin 60 mg/m²/day × 2 days
M = Methotrexate 12 gm/m² × 1 dose, maximum dose 20 g
I = Ifosfamide 2.8 gm/m²/day × 5 days
i = Ifosfamide 3 gm/m²/day × 3 days
E = Etoposide 100 mg/m²/day × 5 days

* Methotrexate is administered in 1 l of D5W ¼ NS with 50 m Eq of sodium bicarbonate per liter over 4 h
Hydration is administered to achieve a urine output of 1400–1600 ml/m² for each 24-h period and a urine pH greater than 7.0

FIGURE 27.4 (A, B) Protocol schema for the ongoing EURAMOS clinical trial, including standard arm with methotrexate, doxorubicin and cisplatin (MAP) and experimental arm for patients with poor histologic necrosis with MAP plus ifosfamide and etoposide (MAPIE). *From: Protocol document and with permission of Mark Bernstein.*

TABLE 27.3 Methotrexate Toxicity and Recommendations for Management, MTX Level, Hours After Start Time of Infusion[a]

	Expected excretion	Delayed excretion	Grade I toxicity (mild)	Grade II toxicity (moderate)	Grade III toxicity (severe)	Grade IV toxicity (life-threatening)
24	$\leq 10\,\mu M$ (1×10^{-5} M)		>10 to $<50\,\mu M$ ($1-5 \times 10^{-5}$ M)	>10 to $<50\,\mu M$ ($1-5 \times 10^{-5}$ M)	$\geq 50\,\mu M$ to $<500\,\mu M$ ($5-50 \times 10^{-5}$ M)	$\geq 500\,\mu M$ (5×10^{-4} M)
48	$\leq 1\,\mu M$ (1×10^{-6} M)		$\geq 1\,\mu M$ to $<5\,\mu M$ ($1-5 \times 10^{-6}$ M)	$\geq 1\,\mu M$ to $<5\,\mu M$ ($1-5 \times 10^{-6}$ M)	$\geq 5\,\mu M$ to $<100\,\mu M$ (5×10^{-6} M) $-(1 \times 10^{-4}$ M)	$\geq 100\,\mu M$ (1×10^{-4} M)
72	$\leq 0.1\,\mu M$ (1×10^{-7} M)	≥ 0.1 to $<0.5\,\mu M$ ($1-5 \times 10^{-7}$ M)	≥ 0.5 to $<5\,\mu M$ ($0.5-5 \times 10^{-6}$ M)	≥ 0.5 to $<5\,\mu M$ ($0.5-5 \times 10^{-6}$ M)	≥ 5 to $<50\,\mu M$ ($0.5-5 \times 10^{-5}$ M)	$\geq 50\,\mu M$ (5×10^{-5} M)
Increase in baseline creatinine[a]			25–50%	50–100%	>100%	
Other[a]			Grade I–II stomatitis	On previous MTX course: Grade III–IV stomatitis, myelosuppression		
Leucovorin	15 mg/m² q6h PO/IV Until MTX level $<0.1\,\mu M$ (1×10^{-7} M)	15 mg/m² q6h PO/IV Recheck at 96 h if $<0.08\,\mu M$ (8×10^{-8} M) discontinue leucovorin if $\geq 0.08\,\mu M$ discontinue leucovorin when level $<0.05\,\mu M$ ($5 \times 10{-8}$ M)	15 mg/m² q6h PO/IV[b]	15 mg/m² q3h IV Until MTX level[b]	150 mg/m² q3h IV Until MTX level[b]	1500 mg/m² q6h IV (maximum dose 1500 mg)[b]
Other interventions	Hydration 125 ml/m²/h	Hydration 125 ml/m²/h	Increase hydration 200 ml/m²/h	Increase hydration 200 ml/m²/h	Consider glucarpidase increase hydration 200 ml/m²/h	Consider glucarpidase Increase hydration 200 ml/m²/h

[a]Classify grade of toxicity based on highest grade: elevated MTX, creatinine, OR other factor listed.
[b]Until MTX level $<0.1\,\mu M$ (1×10^{-7} M) or until criteria for delayed excretion met.
Adapted from: COG appendix for AOST0331 A Randomized Trial of the European and American Osteosarcoma Study Group to Optimize Treatment Strategies for Resectable Osteosarcoma Based on Histological Response to Pre-Operative Chemotherapy.

In patients whose response is suboptimal, intensification or alteration of chemotherapy has been the subject of the Intergroup European and American Study Group COG (Children's Oncology Group) AOST0331/EURAMOS trial, the largest randomized international prospective trial performed to date in osteosarcoma, registering over 2200 patients. Osteosarcoma patients with a good response (>90% necrosis) were randomized to continue standard MAP versus receiving MAP with addition of pegylated interferon alfa-2B. Patients that had a poor response were randomized to either receive standard MAP or MAP with the addition of ifosfamide and etoposide (Figure 27.4). The results from the trial are disappointing, showing neither an event-free survival benefit to the addition of interferon to the good responders nor a benefit to the addition of ifosfamide and etoposide to the poor responder cohort, while clearly causing increased toxicity in both groups. Long-term survival data analysis is ongoing.

Treatment of Metastatic Osteosarcoma at Presentation

Approximately 20% of patients with osteosarcoma present with metastatic disease, and survival is 10–50%. Because of the poor prognosis, metastatic osteosarcoma is often treated with MAP chemotherapy plus ifosfamide and etoposide (Figure 27.4B). In metastatic disease, the addition of ifosfamide and etoposide increases response rates. However, a survival advantage has not been shown to date. Surgical management of the primary tumor is performed as it would be for non-metastatic osteosarcoma. When the tumor arises in a site that is unresectable, radiation may be used for local control even though the tumor is usually not radiosensitive. Radiation is more commonly used as a local control modality in osteosarcoma of the head and neck.

Complete surgical removal of all metastatic sites, if feasible, is an important component of the treatment of metastatic osteosarcoma. About one-third of patients with pulmonary metastases from osteosarcoma have pulmonary nodules that are not visible on CT but can be identified by intraoperative palpation. Because it allows the surgeon to identify these nodules by palpation, open thoracotomy is the recommended approach for resection of pulmonary metastases. Patients with unilateral lung metastases should have an open thoracotomy on the contralateral side to permit identification and resection of palpable, non-visible pulmonary metastases.

Treatment of Relapsed Osteosarcoma

More than 85% of osteosarcoma recurrences are in the lung and in a significant proportion of patients the lungs are the only site of recurrence. Patients with recurrent osteosarcoma have a poor prognosis. In retrospective studies, patients who have complete resection of all sites of recurrent disease have a better prognosis than patients who do not have complete resection. Therefore, surgical resection of all sites of recurrent disease is recommended, if feasible. Whether chemotherapy improves outcome in recurrent osteosarcoma is not clear. Common salvage regimens include high-dose ifosfamide with or without etoposide and gemcitabine and docetaxel (Taxotere). Radiation can be used for palliation, typically with painful bony metastasis. Stereotactic beam body radiotherapy (SBRT) has been used successfully for local control of multiple lung metastases when thoracotomy is not feasible without major morbidity.

Newer Agents Under Investigation

Intensification of postoperative treatment to improve the outcome of patients who have a poor response (<90% necrosis) to neoadjuvant chemotherapy is under investigation in the COG/EURAMOS phase III clinical trial which is closed to accrual. Intensification of neoadjuvant chemotherapy has also attempted to improve outcome. Although the percentage of good responders increased, the event-free and overall survival rates have remained unchanged. Targeted or biologically based therapies such as bisphosphonates, denosumab, interferon, trastuzumab, insulin-like growth factor receptor-1 antibody, and granulocyte macrophage stimulating factor have been and are currently being investigated. Human epidermal growth factor receptor 2 (HER2) overexpression is present in many cancers including breast cancer and osteosarcoma. The monoclonal antibody trastuzumab targets HER2, and has shown survival benefits when administered with adjuvant chemotherapy in breast cancer; although in trials in metastatic osteosarcoma, the benefit is controversial. Immunostimulant liposomal muramyl-tripeptide phosphatidyl-ethanolamine (MTP-PE) had shown increased overall survival when added to the standard regimen but no improvement in event-free survival, and with several other confounding variables the approval of MTP-PE in the United States was denied by the FDA. Vaccines to generate T-cell responses are being developed. Gemcitabine and docetaxel have demonstrated antitumor activity in refractory tumors and have been used as a standard salvage regimen though response rates in small series have not been impressive. Tyrosine kinase inhibitors are also being investigated in recurrent disease.

Posttreatment Surveillance

1. Non-contrast CT of the chest every 3 months for 2 years after completion of therapy, every 6 months for 1 year, then every year for 2 years.
2. Chest radiograph yearly for 5 years starting after the last CT of the chest.
3. Evaluation of the primary site every 3 months for 2 years after completion of therapy, every 6 months for 3 years, then every year for 5 years.

Prognosis

The actuarial disease-free survival for patients with non-metastatic osteosarcoma is approximately 70%. Patients who present with metastatic disease have a significantly worse prognosis with 2-year survival of 10–30%. Key prognostic factors are surgical remission and response to chemotherapy (Figure 27.5). For newly diagnosed osteosarcoma, the following factors have been associated with a poor prognosis:

- Detectable metastatic disease at diagnosis
- Incomplete resection
- Poor response to chemotherapy (<90% necrosis)
- Axial skeletal primaries
- Larger tumor size
- Proximal location within the limb.

(A)

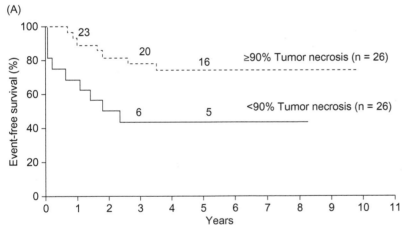

(B)

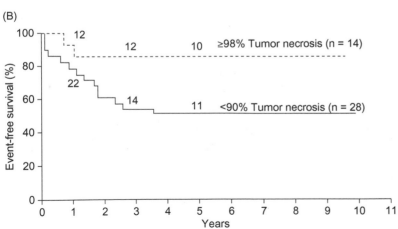

FIGURE 27.5 Event-free survival comparing pathologic response at (A) less than 10% residual viable tumor versus greater than 10% residual viable tumor and (B) less than 2% residual viable tumor versus greater than 2% residual viable tumor. Numbers on the curves represent patients still at risk. *From: Goorin et al. (2003), with permission.*

For recurrent osteosarcoma, the following factors have been associated with a poor prognosis:

- Incomplete surgical resection
- Early (less than 2 years posttreatment) relapse
- More than one or two pulmonary nodules
- Bilateral pulmonary involvement
- Pleural disruption by metastatic disease.

EWING SARCOMA FAMILY OF TUMORS

Ewing sarcoma family of tumors (EFT) includes Ewing sarcoma (ES, osseous), extraosseous Ewing sarcoma (EES) and atypical Ewing sarcoma, primitive neuroectodermal tumor (PNET), and Askin tumors (chest wall and pulmonary region). EFT are malignant tumors of bone or soft tissue harboring the Ewing sarcoma translocation.

Epidemiology

Ewing sarcoma is the second most common malignant primary bone tumor of childhood. The etiology of EFT is unknown. Certain epidemiologic association studies have indicated higher rates of EFT in children with a history of inguinal hernia and in children of farm workers; however, they are not considered risk factors. Eighty percent of patients with EFT are younger than 20 years at diagnosis. EFT has a striking difference in racial incidence with a low incidence in Black and Asian children. It is not usually associated with familial cancer syndromes. Unlike osteosarcoma, the risk of EFT has not been shown to increase following radiation exposure.

Pathology

EFT is one of the small round blue cell tumors of childhood, with unknown histogenesis.
The pathologic features of EFT are:

- Highly cellular aggregates of small round cells compartmentalized by strands of fibrous tissue.
- Tumor cells have round or oval nuclei, occasionally with indistinct nucleoli.
- Nuclei containing finely dispersed chromatin (ground-glass appearance).
- Less than two mitotic figures per high-power field.

In order to differentiate EFT from other small round blue cell tumors, immunohistochemical studies are necessary. Table 27.4 shows the immunohistochemistry of EFT and other small round blue cell tumors of childhood. Glycoprotein p30/32mic2 (CD99) has proven to be one of the most useful immunohistochemical stains in supporting the diagnosis of EFT. CD99 reacts with the cytoplasmic membrane of Ewing sarcoma cells, giving a honeycomb staining pattern. It is also expressed by other tumors and normal tissues including non-blastematous portion of Wilms' tumor, clear cell sarcoma of the kidney, T-cell lymphoma and leukemia, pancreatic islet cell tumors, immature teratomas, testicular embryonal carcinoma, ependymomas, and rhabdomyosarcoma. The pattern of CD99 staining in other tissues is often cytoplasmic, rather than the membranous pattern seen in Ewing sarcoma.

Molecular Genetics

EFT is characterized by the presence of one of several chromosomal translocations (Table 27.5). The t(11;22) translocation is the most common and has been found to involve the juxtaposition of two normally distinct genes, *FLI1* on chromosome 11 and *EWS* on chromosome 22 (Figure 27.6). The fused gene results in an EWS/FLI1 chimeric protein. Other variant translocations have been described that fuse EWS to other ETS family transcription factors, such as t(21;22) and EWS/ERG fusion. While the exact role of the fusion proteins remains unknown, they are felt to act primarily as aberrant transcription factors that play an important role in growth deregulation.

TABLE 27.4   Immunohistochemistry of EFT and Other Small Round Blue Cell Tumors of Childhood

Marker	EFT	Neuroblastoma	Rhabdomyosarcoma	Lymphoma[a]
NSE	+/−	+	+/−	−
S-100	+/−	+	+/−	−
NFTP	+/−	+	+/−	−
Desmin	+/−	−	+	−
Actin	+/−	−	+	−
Vimentin	+/−	−	+	+
Cytokeratin	+/−	−	+/−	−
LCA	−	−	−	+
HNK-1	+/−	+	+	−
b2-Microglobulin	+	−[b]	+/−	+
CD99	+	−	+[c]	+/−[d]

[a]Non-histiocytic.
[b]Only in ganglion cells, Schwann cells, and stage IVS neuroblastoma.
[c]Only well-differentiated rhabdomyoblasts.
[d]Positive in T-cell lymphoblastic lymphoma.
−, negative; +/−, positive in greater than 10% of cases or variably positive; +, positive.
Adapted from: Ginsberg JP, Woo SY, Johnson ME, et al. Ewing sarcoma family of tumors: Ewing sarcoma of bone and soft tissue and the peripheral primitive neuroectodermal tumors. In: Pizzo PA, Poplack DG, editors. Principles and Practice of Pediatric Oncology. 4th ed. Philadelphia: Lippincott–Raven, 2003, with permission.

TABLE 27.5   Frequency of Chromosomal Translocations in EFT

Chromosomal abnormality	Fusion	Frequency (%)
t(11;22)(q24;q12)	EWS-FLI1	90−95
t(21;22)(q22;q12)	EWS-ERG	5−10
t(7;22)(p22;q12)	EWS-ETV1	<1
t(17;22)(q12;q12)	EWS-ETV4	<1
t(2;22)(q33;q12)	EWS-FEV	<1
t(16;21)(p11;q22)	FUS-ERG	<1
t(2;16)(q35;p11)	FUS-FEV	<1
t(1;22)(p36;q12)	EWS-ZSG	<1
t(20;22)(q13;q12)	EWS-NFATc2	<1

Most of these translocations can be detected by reverse transcription polymerase chain reaction (RT-PCR) and fluorescent in situ hybridization (FISH) for EWS rearrangement.

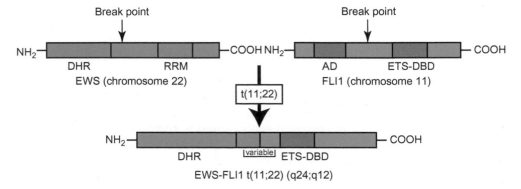

FIGURE 27.6   Schematic of EWS-FLI1 t(11;22)(q24;q12) translocation. Variation in the sites of chromosomal break points leads different fusion types. *Anderson et al. (2012), with permission.*

TABLE 27.6 Distribution of Primary Site for 303 Patients with Ewing Sarcoma of Bone

Primary site	Percentage
Pelvic	20.0
Ilium	12.5
Sacrum	3.3
Ischium	3.3
Pubis	1.7
Lower extremity	45.6
Femur	20.8
Fibula	12.2
Tibia	10.6
Feet	2.0
Upper extremity	12.9
Humerus	10.6
Forearm	2.0
Hand	0.3
Axial skeleton/ribs	12.9
Face	2.3

From: Grier (1997), with permission.

Clinical Features

Patients commonly present with the following symptoms:

- Pain at site of tumor (96%)
- Local swelling and/or palpable mass (61%)
- Fever (21%)
- Pathologic fracture (16%).

Fever is a frequent symptom of Ewing sarcoma, increasing the risk that the tumor may be initially mistaken for osteomyelitis. Patients may have symptomatology for 3–6 months before a diagnosis is established. A small proportion of patients may have symptoms for longer than 6 months before the tumor is identified. The length of symptoms does not necessarily correlate with the presence of metastases, since these appear to be related to the biological behavior of the tumor rather than to the length of symptoms. The pain may be intermittent in nature.

Ewing sarcoma most commonly involves the lower extremity, with the pelvis being the next most common site. Table 27.6 lists the distribution of the primary sites of origin in Ewing sarcoma of bone. In long bones, Ewing sarcoma is generally of diaphyseal origin. EFT of the thoracopulmonary region involving the chest wall has been referred to as an Askin tumor.

Diagnostic Evaluation

The diagnostic evaluation typically begins with a history, physical examination, and plain radiographs of the affected bone. Radiographically, Ewing sarcoma produces changes in the diaphysis of long bones with extension toward the metaphysis (Figure 27.7). Table 27.7 lists the radiographic findings in Ewing sarcoma. Patients with soft-tissue primary tumors may have normal plain radiographs.

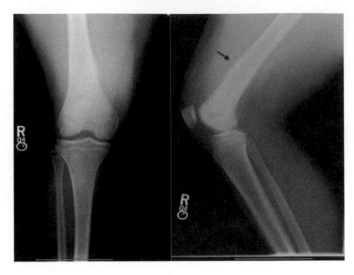

FIGURE 27.7   AP/lateral radiographs of Ewing sarcoma of the femur in an adolescent. Note the periosteal reaction at the femoral diaphysis (arrow) causing the "onion-skin" appearance classic for Ewing sarcoma.

TABLE 27.7   Radiographic Findings in Ewing Sarcoma

Finding	Percentage
Bone destruction	75
Soft-tissue extension	64
Reactive bone formation	25
Lamellated periosteal reaction (onion skin)	23
Radiating spicules (sunburst)	20
Periosteal thickening	19
Sclerosis	16
Fracture	13

From Green (1985), with permission.

The remainder of the staging studies for EFT should consist of the following:

- Blood count
- Urinalysis
- BUN, creatinine, liver enzymes, alkaline phosphatase, serum LDH
- MRI of primary site with gadolinium contrast
- Chest CT scan
- Bilateral bone marrow aspirate and biopsy (for pelvic/ilium primary unilateral biopsy in the unaffected iliac crest)
- Bone scan
- FDG PET/CT recommended.

The role of PET scans in the management of EFT remains unclear. Studies have suggested that these scans may be useful for evaluating the presence of metastatic disease and for monitoring response to therapy and may be more sensitive and specific for detecting osseous metastases and bone marrow metastases. However, their role in the staging of EFT remains controversial. Bone marrow biopsy and aspiration is still considered standard in staging of EFT, although this may change with the increased study of PET/CT and MRI in staging and detection of marrow involvement.

Approximately 10–30% of patients have detectable metastases at the time of diagnosis. The most common metastatic sites are:

- Lung (38%)
- Bone (31%)
- Bone marrow (11%).

The site of the primary tumor is related to the incidence of metastases at diagnosis:

- Central primary—40% incidence of metastases
- Proximal primary—30% incidence of metastases
- Distal primary—15% incidence of metastases.

Treatment

The treatment of EFT involves a multimodality therapy, using surgery and/or radiation therapy for local control of the primary lesion and chemotherapy for decreasing the size of the primary lesion and eradication of subclinical micrometastases. Most current treatment protocols utilize neoadjuvant chemotherapy, followed by local control, usually with surgery, and then adjuvant chemotherapy.

Surgery

Biopsy should ideally be performed via open technique or percutaneous core needle at the center at which ultimate surgical treatment will be rendered. The biopsy site should be chosen carefully and placed in line with the potential resection site or radiation portals. The biopsy should be preferably taken from the extraosseous component to prevent pathologic fracture. The biopsy site must be included *en bloc* with the resection or in the planned radiation field.

The preservation of function and the reduction of long-term sequelae should be taken into consideration in determining the mode of local control selected for an individual patient. For those tumors felt to be resectable with negative margins and acceptable long-term physical function, definitive surgical local control is generally preferred to radiation. This preference is largely due to the increased risk of second malignancy following radiation as well as to the improved surgical techniques. However, local control rates and overall survival are similar for definitive surgery alone, surgery with neoadjuvant or adjuvant radiation, or definitive radiation. Definitive surgical resection typically takes place after several cycles of chemotherapy. Tumors that initially appear unresectable may become resectable with chemotherapy. Given that EFT is radioresponsive, amputation for local control is used less commonly than in osteosarcoma. Decisions regarding local control planning should take place with an experienced and integrated multidisciplinary musculoskeletal team that involves orthopedic oncology surgeons, radiation oncologists, pediatric oncologists, radiologists, and other members of the patient's healthcare team.

Radiation

Radiation plays an important role in the treatment of EFT when function-preserving surgery cannot be obtained, when margins are positive, or used in a neoadjuvant setting to improve the resectability of a tumor. The decision to use radiation should be made with the understanding of the risk of secondary radiation-induced malignancies and the possibility of skeletal development retardation and deformity. With modern radiation techniques the successful local control and overall survival appear to be similar between surgical and radiation modalities. However, no randomized clinical trial has been performed comparing the two, and the use of radiation definitively or as an adjunct has been largely institutionally driven. When a tumor is unresectable (such as large sacral and pelvic primaries) or there are positive margins postoperatively, radiation therapy is indicated. Radiation is also used for metastatic disease and palliation of recurrent disease. Whole-lung radiation therapy with a dose of 12–21 Gy for patients with pulmonary metastatic EFT has been recommended as standard. However, the improvement in event-free and overall survival is not entirely clear, nor has a randomized trial been performed.

Current recommendations for definitive radiation of the primary site utilize doses ranging from 45–60 Gy. In the postoperative setting, gross disease standard dosing is 55.8 Gy and microscopic disease is 45 Gy.

Chemotherapy

The major impact on improved long-term survival in EFT has been the initiation of systemic chemotherapy. Chemotherapy can rapidly reduce tumor burden, even to the extent that there may be signs of tumor lysis (such as elevated LDH and uric acid). This reduction in tumor volume can facilitate surgical local control, and allow for limb-sparing procedures when there is extremity involvement.

Current standard chemotherapy includes vincristine, doxorubicin, and cyclophosphamide (VDC) alternating with ifosfamide and etoposide (IE) for 12 weeks (six cycles) of induction given every 2 weeks:

- Vincristine 1.5 mg/m^2 (maximal dose 2 mg) and cyclophosphamide 1200 mg/m^2 for 1 day, doxorubicin 37.5 mg/m^2/day for 2 days; Ifosfamide 1,800 mg/m^2/day for 5 days and etoposide 100 mg/m^2/day for 5 days.
- Children's Oncology Group trial AEWS1031 is evaluating the use of alternating VDC (two cycles), IE (two cycles) and vincristine, topotecan (0.75 mg/m^2/dose IV once daily for 5 days), and cyclophosphamide (VTC, two cycles) in induction. Topotecan was added to the study regimen as it demonstrated activity in recurrent and metastatic disease and has been shown to be synergistic with vincristine. After local control (by surgery or radiation), consolidation consists of 22 weeks (12 cycles), with standard VDC/IE/VC (VC = vincristine and cyclophosphamide). On the study regimen, consolidation consists of 22 weeks of VTC/IE/VDC. Patients with localized extraosseous tumors fare well with the same treatment as with those with osseous disease. Intervals of chemotherapy have been decreased to 2 weeks as interval compression was demonstrated to be superior to previously standard 3-week timing. The use of filgrastim is critical in the ability to compress chemotherapy timing. Long-term cardiotoxicity has been recognized as a complication of anthracycline-based chemotherapy, so dexrazoxane has been added to courses containing above 225 mg/m^2 cumulative dose of doxorubicin for cardioprotection.

Treatment of Metastatic EFT

Patients with metastatic EFT have a poor outcome, with a 5-year event-free survival of 10–30%. These patients appear to have more chemotherapy-resistant disease, and cure rates are variable based on disease location. Typically cure rates in lung and pleural metastasis are higher than in metastatic disease to the bone and bone marrow. Superiority of chemotherapy regimens is difficult to determine based on studies available. The addition of ifosfamide and etoposide to vincristine, doxorubicin, and cyclophosphamide does not result in an improved outcome for these patients, although it is still used as a standard regimen in patients with metastatic disease. The impact of interval compression of chemotherapy cycles has not yet been evaluated in this subgroup of patients. VDC/IE is used, in some cases with the addition of actinomycin D or higher doses of the alkylating agents cyclophosphamide and ifosfamide. Because of poor outcomes with current regimens, some treatments incorporate high-dose myeloablative chemotherapy (such as busulfan, melphalan, thiotepa, etoposide, carboplatin) and total-body irradiation followed by autologous hematopoietic cell support. The use of high-dose therapy with autologous stem cell rescue for metastatic EFT is being analyzed in the Euro-Ewing 99 phase III trial to compare the event-free and overall survival of patients with EFT treated with standard induction chemotherapy comprising vincristine, ifosfamide, dactinomycin, and etoposide (VIDE) followed by consolidation chemotherapy comprising vincristine, dactinomycin, and ifosfamide versus high-dose busulfan and melphalan (Bu-Mel) followed by autologous peripheral blood stem cell (PBSC) transplantation with or without radiotherapy and/or surgery. This study is currently closed to accrual within the United States. In metastatic disease, radiation and/or surgery is used for control in multiple sites. Radiation is used more often when there is metastatic disease compared to local disease. Surgical resection can be performed when sites are amenable to resection.

Treatment of Relapsed EFT

The prognosis for patients with recurrent disease is poor. Early relapse, typically less than 2 years, elevated LDH, and having combined local and distant sites of disease are poor prognostic factors. Full evaluation for metastatic disease must take place as a majority of patients with local disease have gross or microscopic metastatic disease. Therapy for these patients must be individualized. Management of local recurrence usually involves surgery or radiation. Surgery can be used in isolated lung metastasis. Radiation is beneficial for palliation of painful lesions.

Common salvage chemotherapy regimens include irinotecan (e.g., 50 mg/m^2 IV for 5 days) and temozolomide (e.g., 100 mg/m^2 by mouth for 5 days) or topotecan/cyclophosphamide (cyclophosphamide 1200 mg/m^2 for 1 day, topotecan 0.75 mg/m^2/dose IV once daily for 5 days, with or without vincristine 1.5 mg/m^2 (maximal dose 2 mg)); cisplatin has also been used in treatment protocols. Ifosfamide and etoposide (ifosfamide 1800 mg/m^2/day for 5 days and etoposide 100 mg/m^2/day for 5 days) may also be given to patients who did not receive in their initial regimen, or at escalating doses. Those that have gone longer durations with disease control prior to recurrence tend to respond better to chemotherapy than those with primary refractory disease. In recent years, insulin-like growth factor receptor monoclonal antibodies have been investigated in clinical trials with Ewing sarcoma. While early results showed some impressive anecdotal complete responses in recurrent and refractory disease, larger clinical trials had disappointing overall response rates though some Ewing patients did have meaningful benefit. Other small-molecule inhibitors are being evaluated in this population with some early indications of activity. PARP inhibitors, pharmacological inhibitors of the enzyme poly ADP ribose polymerase are also under current clinical trial investigation having shown exciting preclinical activity in EFTs. Retrospective studies have evaluated the role of myeloablative therapy and autologous stem cell rescue for relapsed EFT with equivocal results.

Prognosis

The overall 5-year disease-free survival for localized EFT treated with surgery, radiation, and multiagent chemotherapy is approximately 60–70%. Patients who present with metastases at diagnosis have a 5-year survival rate of 10–30%, though there are some indications that patients with lung metastases only may have an improved outcome compared to patients with more widespread metastatic disease. In recurrent disease, survival is dismal, but can be up to 15–20% in those who relapse later.

While the presence of metastatic disease appears to be the most important prognostic factor, other commonly reported adverse prognostic factors include:

- Older age
- Axial tumor location
- Larger tumor size
- Elevated serum LDH level.

Other prognostic factors that have been reported include fever, anemia, proliferative index, and chemotherapy-induced necrosis. Whereas EFT translocation type had been implicated as a prognostic factor, recent studies re-examining this issue have shown that current treatment protocols have erased the prognostic significance of translocation type on overall outcome.

Although the time of the onset of symptoms to diagnosis is long in Ewings tumors (especially for adolescents and for tumor sites such as the pelvis, or hand and foot), it does not impact survival.

OTHER BONE TUMORS

Chondrosarcoma

Chondrosarcomas are a group of malignant bone tumors that produce a cartilaginous matrix, typically in bones that undergo endochondral ossification. Less than 10% of all chondrosarcomas are in children. Greater than 90% are slow-growing with low metastatic potential. High-grade chondrosarcomas (5–10%) have a high metastatic potential and require more aggressive treatment. A diagnostic biopsy is required to evaluate histopathologic grade and make treatment decisions. Chondrosarcomas can be primary or secondary (arise from precursor cartilaginous lesions).

Precursor lesions, osteochondromas and enchondromas, have a possible but low propensity for malignant transformation to chondrosarcoma.

1. Osteochondroma is a bony projection with cartilaginous cap, typically present around the knee. Hereditary multiple exostoses is an autosomal dominant genetic condition, most commonly a germline mutation of the tumor suppressor genes *EXT1* or *EXT2*, characterized by two or more osteochondromas.
2. Enchondroma is a benign tumor of the medulla associated with enchondromatosis (such as Ollier disease) when multiple lesions are present.

Pathology

Chondrosarcomas are histologically graded from 1 to 3 based on staining pattern, mitotic activity, and degree of cellularity. Surgical resection is required and complete excision is ideal. Grade 1 has marginally enlarged chondrocytes with a predominance of chondroid matrix, high survival rates of 83–95%, and almost never metastasizes. Grade 3 chondrosarcomas are highly cellular, with decreased chondroid matrix, necrosis, and mitoses easily detectable.

Chondrosarcomas can arise from the medullary cavity (central), cartilage cap (peripheral), and the surface of the bone (periosteal). Mesenchymal chondrosarcomas are highly malignant tumors, and the most common in the pediatric and adolescent population. Unlike other types of chondrosarcomas, these tumors are more common in axial bones (craniofacial, pelvis, vertebra) and approximately one-third are extraosseous. The meninges are the most common extraosseous location. Approximately 20% are metastatic at diagnosis, and overall survival is 10–20%. Clear cell, myxoid, and dedifferentiated chondrosarcoma (well-differentiated cartilage tumor and high-grade non-cartilaginous sarcoma) are other subtypes.

The following genetic and signaling abnormalities have been described in chondrosarcomas:

1. Parathyroid hormone-related protein (PTHRP) signaling in enchondromas
2. Hedgehog signaling in central chondrosarcomas
3. Activation and overexpression of platelet-derived growth factor receptor-alpha (PDGFRA) and beta (PDGFRB) in conventional primary chondrosarcomas
4. Point mutations in isocitrate dehydrogenase-1 and isocitrate dehydrogenase-2 genes (*IDH1* and *IDH2*) in 40–56% of enchondromas and chondrosarcomas
5. Intrachromosomal rearrangement of chromosome arm 8q creating HEY1-NCOA2 fusion in a majority of mesenchymal chondrosarcomas.

Clinical Features

Typically chondrosarcomas present with insidious pain and local swelling. A total of 95% of patients with chondrosarcomas report slowly progressive pain. Vague symptoms can last 1–2 years prior to diagnosis. Approximately 20% present with a pathologic fracture. Most tumors arise from the shoulder, pelvis, or proximal femur and approximately 50% involve the metaphysis. In children there is a propensity to arise from other locations, including axial sites such as the ribs.

Diagnostic Evaluation

Preoperative evaluation includes imaging and diagnostic biopsy:

1. Radiographs (mixed sclerosis, lucent central lesion, flocculent calcifications in "ring and arc" pattern).
2. MRI and/or CT imaging. MRI is the preferred imaging modality for chondrosarcomas due to the ability to see soft tissue extension and marrow involvement.
3. Bone scan and/or whole-body FDG PET/CT.
4. Biopsy, directed toward the area of the lesion that appears more aggressive on imaging.
5. Chest CT scan.

Treatment

Surgical treatment is the standard of care in all grades of chondrosarcoma. In intermediate to high-grade tumors, a wide *en bloc* local excision is preferred. In low-grade chondrosarcoma, intralesional and wide resection are both considered acceptable treatment modalities. Intralesional curettage is usually supplemented with local adjuvant therapy—cryotherapy, high-speed burr margin expansion, phenol, hydrogen peroxide, or thermoablation.

Chondrosarcomas are generally radioresistant; however, radiotherapy is used when resection is not feasible, resection is incomplete, or for palliation. Chondrosarcomas are also generally chemoresistant, although there is suggested benefit in dedifferentiated and mesenchymal chondrosarcoma. Chemotherapy is typically with doxorubicin and cisplatin-based regimens. Osteosarcoma protocols have also been used. Neoadjuvant chemotherapy has been shown to decrease tumor volume. Novel therapies have been trialed, such as tyrosine kinase inhibitors, dasatinib and imatinib. Also under investigation are aromatase inhibitors, histone deacetylase inhibitors, angiogenesis inhibitors, vascular endothelial growth factor antisense molecules, recombinant human Apo2L/TRAIL, and IgG1 monoclonal antibodies to trigger apoptosis.

Posttreatment Surveillance

Posttreatment surveillance is generally conducted in the manner of other malignant bone tumors with chest CT every 3–6 months, plain films, and MRI of the primary site and spaced out to yearly intervals up to 10 years as late local and distant recurrences can occur.

Giant Cell Tumor of Bone

Giant cell tumor of bone (GCTB) is a rare neoplasm typically in adolescents and young adults, and most common after skeletal maturity and closure of the growth plates. Less than 3% of cases of GCTB are in patients under the age of 14. There is a higher predominance in females (slight predilection), those with Paget disease of bone (where disease location is more common skull, facial bones, pelvis, spine), and Asian populations. GCT is composed of mononuclear ovoid and spindle-shaped cells and characteristic multinucleated giant cells that promote osteolysis. In general most cases are benign, but can vary from static tumors to aggressive lesions with bone destruction and soft tissue extension. Distant metastasis occur in about 1–3% of cases, most commonly in the lung, and local recurrence is possible. There is also the rare possibility of development of a high-grade sarcoma adjacent to the site of GCTB, known as a primary malignant giant cell tumor (PMGCT). If the malignancy occurs after treatment, typically after radiotherapy, it is a secondary malignant giant cell tumor (SMGCT).

Clinical Features

GCTB typically presents in the epiphyses of long bones (distal femur, proximal tibia, and distal radius) with pain, swelling, and limitation of movement. Ten to thirty-five percent present with a pathologic fracture. Over 99% of cases are solitary and those with multiple lesions are typically younger in age. Lung metastasis can occur though uncommonly.

Diagnostic Evaluation

On radiographic imaging, the lesion is lytic, eccentric in location, and well-defined with a non-sclerotic margin (Figure 27.8). Most commonly in the epiphysis, it can extend to subchondral bone and surrounding tissues. New bone formation and matrix calcification are absent. GCTB contain fluid–fluid levels in up to 14% of cases due to secondary aneurysmal bone cyst formation. MRI/CT of the lesion, CT of chest to evaluate for metastatic disease, PET/CT and radionuclide bone scan for multicentric disease (especially in younger patients) are performed for staging. Typically MRI is more helpful for soft tissue extensions, and CT for bony components. On MRI there is decreased T1 intensity of the lesion and increased signal intensity on images obtained with fluid-sensitive sequences.

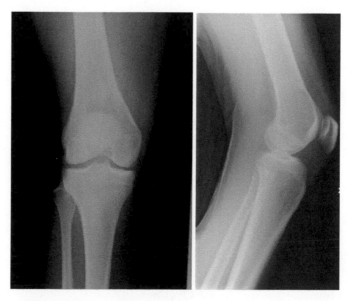

FIGURE 27.8 AP/lateral radiographs of giant cell tumor of bone of right distal femur. This well-defined and eccentric epiphyseal-based lesion crosses the physeal scar into the metaphysis.

Treatment

Surgery continues to be the treatment of choice for resectable disease, and type of surgery is dependent on location, extent, grade of tumor (biologic aggressiveness), whether a pathologic fracture has occurred, and if the lesion is *de novo* or recurrent. Curettage, marginal excision, wide local excision, and *en bloc* resection are all options. Complete resection of tumor is the best predictor of outcome. Therefore, wider resections have lower recurrence rates, but this must be balanced with the possible morbidity of the surgical intervention. Curettage is typically the treatment of choice and has good functional outcomes. Multiple different local adjuvants have been used with curettage, including phenol, polymethylmethacrylate, liquid nitrogen, and argon beam laser. Bony defects are often reconstructed with either bone graft or bone cement (polymethylmethacrylate). Bone cement carries the added benefit of emanating heat in the process of setting, providing an adjuvant thermoablation therapy to residual microscopic disease.

GCTB is radiosensitive, though reserved for use in recurrent disease, unresectable disease, and disease with positive margins after surgery. Sacral GCTB is a challenging disease location because of the high morbidity associated with resection. Serial arterial embolization, cryotherapy, and more recently denosumab have been used with success.

Local recurrence can be managed with surgery and/or radiotherapy. With local recurrence there is a higher incidence of pulmonary involvement so chest imaging should be performed. The clinical course of lung metastasis ranges from mortality to self-resolution. For metastatic disease surgical resection, observation, radiation, chemotherapy, and denosumab have all been used either alone or in combination. Chemotherapy regimens have been based on an anthracycline and alkylating agent backbone and the decision to use chemotherapy in malignant GCTB should be made on a case-by-case basis.

RANK-L Inhibition

Recent developments in the understanding of the underlying molecular pathogenesis of GCTB have pointed to interactions between the stromal component producing receptor activator of nuclear factor-kappa B (RANK) and RANK-ligand (RANKL) causing the formation of osteoclast-like giant cells that drive bone destruction. The development of a monoclonal humanized antibody to RANKL, denosumab, has been shown to reduce skeletal-related events from osteoporosis and from bony metastases from solid tumors. In recent clinical trials in skeletally mature adolescents over age 12 and adults with recurrent or unresectable GCTB, denosumab has shown remarkable outcomes regarding its efficacy, leading to accelerated approval in the United States. In the skeletally immature pediatric population, only anecdotal use has been published. Clearly more data will be required to better assess its short- and long-term toxicities in this vulnerable age group, but early results are promising and may change the landscape of surgical recommendations for GCTB.

Further Reading and References

Anderson, J.L., Denny, C.T., Tap, W.D., Federman, N., 2012. Pediatric sarcomas: translating molecular pathogenesis of disease to novel therapeutic possibilities. Pediatr. Res. 72 (2), 112–121.

Balamuth, N.J., Womer, R.B., 2010. Ewing's sarcoma. Lancet Oncol. 11 (2), 184–192.

Bielack, S.S., Kampf-Bielack, B., Delling, G., et al., 2002. Prognostic factors in high grade osteosarcoma of the extremities or trunk: an analysis of 1,702 patients treated on neoadjuvant cooperative osteosarcoma study group protocols. J. Clin. Oncol. 20 (3), 776–790.

Bielack, S.S., Kempf-Bielack, B., Branscheid, D., et al., 2009. Second and subsequent recurrences of osteosarcoma: presentation, treatment and outcomes of 249 consecutive cooperative osteosarcoma study group patients. J. Clin. Oncol. 27 (4), 557–565.

Bielack, S.S., Smeland, S., Whelan, J.S., et al., 2015. Methotrexate, doxorubicin, and cisplatin (MAP) plus maintenance pegylated interferon alfa-2b versus MAP alone in patients with resectable high-grade osteosarcoma and good histologic response to preoperative MAP: first results of the EURAMOS-1 Good Response Randomized Controlled Trial. J. Clin. Oncol. Available from: http://dx.doi.org/10.1200/JCO.2014.60.0734.

Chakarun, C.J., Forrester, D.M., Gottsegen, C.J., Patel, D.B., White, E.A., Matcuk Jr., G.R., 2013. Giant cell tumor of bone: review, mimics, and new developments in treatment. Radiographics 33 (1), 197–211.

Donaldson, S.S., 2004. Ewing sarcoma: radiation dose and target volume. Pediatr. Blood Cancer 42, 471–476.

Federman, N., Bernthal, N., Eilber, F.C., Tap, W.D., 2009. The multidisciplinary management of osteosarcoma. Curr. Treat Options Oncol. 10 (1–2), 82–93.

Federman, N., Brien, E.W., Narasimhan, V., Dry, S.M., Sodhi, M., Chawla, S.P., 2014. Giant cell tumor of bone in childhood: clinical aspects and novel therapeutic targets. Paediatr. Drugs 16 (1), 21–28.

Gaetano, B., Ferrari, S., Bertoni, F., et al., 2000. Prognostic factors in nonmetastatic Ewing Sarcoma of bone treated with adjuvant chemotherapy: analysis of 359 patients at the Istituto Ortopedico Rizzoli. J. Clin. Oncol. 18 (1), 4–11.

Giuffrida, A.Y., Burgueno, J.E., Koniaris, L.G., et al., 2009. Chondrosarcoma in the United States (1973 to 2003): an analysis of 2890 cases from the SEER database. J. Bone Joint Surg. Am. 91, 1063.

Goorin, A.M., Schwartzentruber, D.J., Devidas, M., et al., 2003. Presurgical chemotherapy compared with immediate surgery and adjuvant chemotherapy for non-metastatic osteosarcoma: Pediatric Oncology Group Study POG-8651. J. Clin. Oncol. 21 (8), 1574–1580.

Gorlick, R., Bielack, S., Teot, L., Meyer, J., Randall, R.L., Marina, A., 2010. Osteosarcoma: biology, diagnosis, treatment, and remaining challenges. In: Pizzo, P.A., Poplack, D.G. (Eds.), Principles and Practice of Pediatric Oncology, sixth ed. Lippincott Williams & Wilkins, Philadelphia.

Gorlick, R., Janeway, K., Lessnick, S., Randall, R.L., Marina, N., C.O.G. Bone Tumor Committee, 2013. Children's Oncology Group's 2013 blueprint for research: bone tumors. Pediatr. Blood Cancer 60 (6), 1009–1015.

Green, D.M., 1985. Diagnosis and Management of Solid Tumors in Infants and Children. Nijhoff, Boston.

Grier, H.E., 1997. The Ewing Family of tumors: Ewing sarcoma and primitive neuroectodermal tumors. Pediatr. Clin. N. Am. 64, 991–1004.

Gurney, J.G., Swensen, A.R., Bukterys, M., 1999. Malignant Bone Tumors. Cancer Incidence and Survival Among Children and Adolescents: United States SEER Program 1975–1995. National Cancer Institute, Bethesda, MD, SEER Program. NIH Pub No 99-4649, pp. 99–110.

Hawkins, D.S., Bölling, T., Dubois, S., Hogendoorn, P.C.W., Jürgens, H., Paulussen, M., et al., 2010. Ewing sarcoma. In: Pizzo, P.A., Poplack, D.G. (Eds.), Principles and Practice of Pediatric Oncology, sixth ed. Lippincott Williams & Wilkins, Philadelphia.

Leddy, L.R., Holmes, R.E., 2014. Chondrosarcoma of Bone. Peabody TD and Attar S (eds.), Orthopaedic Oncology. Cancer Treat. Res. 162, 117–130.

Mendenhall, W.M., Zlotecki, R.A., Scarborough, M.T., Gibbs, C.P., Mendenhall, N.P., 2006. Giant cell tumor of bone. Am. J. Clin. Oncol. 29 (1), 96–99.

Mirabello, L., Troisi, R.J., Savage, S.A., 2009. Osteosarcoma incidence and survival rates from 1973 to 2004: data from the Surveillance, Epidemiology, and End Results Program. Cancer 115, 1531–1543.

van Maldegem, A.M., Bovée, J.V., Gelderblom, H., 2014. Comprehensive analysis of published studies involving systemic treatment for chondrosarcoma of bone between 2000 and 2013. Clin. Sarcoma Res. 4, 11. Available from: http://dx.doi.org/10.1186/2045-3329-4-11.

Womer, R.B., West, D.C., Krailo, M.D., Dickman, P.S., Pawel, B.R., Grier, H.E., et al., 2012. Randomized controlled trial of interval-compressed chemotherapy for the treatment of localized Ewing sarcoma: a report from the Children's Oncology Group. J. Clin. Oncol. 30 (33), 4148–4154.

28

Retinoblastoma

Ann Leahey

Retinoblastoma (RB) is a malignant tumor of the embryonic retina. It affects young children, usually under the age of 5 years. Tumors may be unilateral or bilateral, unifocal or multifocal. There are hereditary and nonhereditary forms of the disease and the disease can be sporadic or familial. Intraocular growth occurs first, prior to invasion of structures within the globe or spread to metastatic sites. In developed nations, presentation with metastatic disease is unusual. However, metastatic disease is not uncommon in developing countries, where it is a significant cause of morbidity and mortality. RB is the paradigm for a genetically inherited cancer and provides the basis for Knudson's two-hit hypothesis of carcinogenesis.

INCIDENCE

1. RB is the most common intraocular malignancy of childhood, occurring at a rate of approximately 1 in 20,000 live births.
2. The annual incidence is 11 per million under 5 years of age.
3. Approximately 200–300 new cases of RB occur each year in the United States.
4. RB accounts for 11% of cancers developing in the first year of life, but for only 3% of all cancers diagnosed in children younger than 15 years of age.
5. The median age at diagnosis is 2 years.
6. Approximately 40% of cases are hereditary. The majority of these patients present with bilateral disease with an average of three tumors per eye and present in the first year of life. Of these, only 15% have an established family history of RB.
7. The remaining 60% of cases are nonhereditary, most often presenting with unilateral and unifocal disease in the second and third years of life.

CLASSIFICATION

Laterality

Unilateral Tumors

Patients can have one or more tumors in one eye. Median age of presentation is 23 months, with few presenting under a year of age. Tumor development within the affected eye can be synchronous or metachronous. For those with unifocal unilateral disease, it is likely that these patients have the nonhereditary form of the disease. Patients with multifocal unilateral disease or those that present at a younger age are more likely to have the hereditary form of the disease.

Bilateral Tumors

Patients have one or more tumors in both eyes. Tumor development can be synchronous or metachronous. It is presumed that these patients have the hereditary form of the disease, even in the absence of a positive family history. These patients almost always present prior to age 2 years and most present in the first year of life.

The severity of tumors may be variable in the two eyes, and this becomes important in weighing up potential treatment options.

Focality

If a single tumor focus exists, these tumors are more likely to be nonhereditary. In contrast, multiple tumor foci can be noted in one or both eyes. These patients are more likely to harbor a germline mutation in the *RB1* gene.

Genetics

The Two-Hit Hypothesis

Knudson's two-hit hypothesis proposes that as few as two stochastic events are required for tumor initiation. The first can be either germline or somatic and the second occurs somatically in the individual retinoblast cells.

The RB1 Gene

RB occurs as a result of mutations of the *RB1* gene located on chromosome 13q14. This was the first tumor suppressor gene cloned. It spans 183 kilobases of genomic DNA, consisting of 27 exons, and coding for a 110-kd protein p110, with 928 amino acids. Regulation of transcription, and thus, cell proliferation are linked to the phosphorylation of the RB protein. Involved in this process are E2F1, a transcription factor that regulates the cell cycle during G1, histone deacetylase 1, and downstream cell cycle-specific kinases. Loss of function is the initiating event for RB. Mutations may be germline or somatic and there are a broad array of types and locations of mutations, ranging from single base changes to large deletions. No hot spots of common mutation have been identified.

Hereditary RB

A "two-hit" model has been proposed to explain the different clinical features of hereditary and nonhereditary cases of RB. These patients inherit a germline mutation in the *RB1* gene that is present in every cell of the body. Eighty-five percent of these are new spontaneous mutations (sporadic), while the remaining 15% have a positive family history (familial). For the sporadic cases, the paternal allele is affected in approximately 94% of cases. Postconception, a second somatic mutation occurs in the remaining *RB1* gene in one or more retinoblasts and tumor results. These cases are at risk of multifocal and bilateral tumors.

Nonhereditary RB

These patients inherit two normal copies of the *RB1* gene. Postconception two somatic mutations occur, one in each copy of the *RB1* gene in a retinoblast and tumor results. These cases are usually unifocal or unilateral tumors.

Genetic Counseling

The *RB1* gene is inherited in an autosomal dominant fashion. Penetrance is high, at approximately 95%. Genetic counseling should be an integral part of the therapy for a patient with RB, whether unilateral or bilateral. Genetic counseling, however, is not always straightforward. A significant proportion (10–18%) of children with RB have somatic genetic mosaicism, making the genetic story more complex and contributing to the difficulty of genetic counseling.

Testing, including sequencing, duplication and deletion testing, MYCN copy number and promoter methylation analysis, is done on the tumor and then compared to somatic tissue (usually blood). If the globe is salvaged, then testing proceeds with a blood sample alone. Of note, in situations where the tumor is not available a negative peripheral blood test does not conclusively rule out that the patient has heritable RB, although it makes it significantly less likely. Environmental exposures associated with development of RB have been inconsistent in epidemiologic studies.

Prenatal Diagnosis and Further Genetic Counseling

With sequencing of the *RB1* gene available, prenatal diagnosis can be undertaken, particularly if there is a family history and the mutation has been identified. However, a negative result cannot categorically rule out disease, as 100% of mutations are not picked up by current screening methods. The positive predictive value of screening, however, is rapidly improving.

13Q DELETION SYNDROME

This syndrome is associated with an increased risk of RB as 13q14 is the genomic location of the RB gene. Physical findings include microcephaly, a high forehead, short nose, hypertelorism, and micrognathia.

There are reports which correlate the size of the deletion with the degree of the attendant developmental delay, although this correlation can be inexact. Patients identified with this syndrome in the neonatal period should be sent to an oncologic ophthalmologist for serial ocular screenings.

RISK FOR SECOND MALIGNANT NEOPLASMS

1. As the *RB1* gene is a tumor suppressor gene, individuals with heritable RB are at elevated risk for the development of second and subsequent malignancies.
2. The most common second malignancies reported have been osteosarcoma, followed by soft tissue sarcomas and melanoma. Leukemia, lymphoma, and breast cancer are also reported in excess of that expected.
3. The risk is highest for those children with the heritable form of the disease who are treated with full-dose, external beam radiotherapy delivered without use of conformal fields under the age of 12 months.
4. The 40-year risk is approximately 35% for those treated with radiotherapy and 10% for those treated without radiotherapy.
5. The risk for those patients with unilateral disease is approximately 5%, likely representative of the small fraction of genetic cases with only a single eye affected.
6. For those who survive an second malignant neoplasm (SMN), the risk of developing yet another primary malignancy is about 2% per year.
7. Approximately 60% of SMNs occur within the radiotherapy field, with the remainder occurring outside the field.
8. Patients with RB should be counseled carefully regarding their increased risk of SMN and should be followed by routine clinical evaluation. Any signs or symptoms potentially referable to a SMN should be promptly evaluated. It is important to counsel against smoke exposure and in the proper use of sunscreen.

PATHOLOGY

Retinoblastoma

The tumor is composed mainly of undifferentiated anaplastic cells that arise from the nuclear layers of the retina. Histology shows similarity to other embryonal tumors of childhood, such as neuroblastoma and medulloblastoma, including features such as aggregation around blood vessels, necrosis, calcification, and Flexner–Wintersteiner rosettes. RBs are characterized by marked cell proliferation as evidenced by high mitosis counts. Tumors can undergo spontaneous regression.

Retinocytoma

This is a variant of RB referred to as either retinoma or retinocytoma. These tumors are composed of benign-appearing cells with a high degree of photoreceptor differentiation. Eyes with such tumors have normal vision. The main issue is that these tumors do not regress in the same way as RB, when treated with chemotherapy or local ophthalmic therapy. Thus, differentiating these benign variants from the malignant form of the tumor is essential, so unnecessary treatment is not employed.

CLINICAL FEATURES

Presenting Signs and Symptoms

1. Leukocoria.
2. Strabismus.
3. Decreased visual acuity.

4. Inflammatory changes.
5. Hyphema.
6. Vitreous hemorrhage, resulting in a black pupil.

Differential Diagnosis

Pseudoretinoblastoma

The diagnosis of RB is made visually. There are multiple conditions that can mimic RB and in a review of almost 3000 patients referred to a major center with the diagnosis of RB, 22% had a simulating lesion referred to as pseudo RB. In children under the age of 2 years, these most commonly included persistent fetal vasculature (also termed persistent hyperplastic primary vitreous), Coats disease, and vitreous hemorrhage. Older children may have toxocariasis or familial exudative vitreoretinopathy.

Patterns of Spread

Intraocular

1. With endophytic growth, there is a white hazy mass.
2. With exophytic growth, there is retinal detachment.
3. Most tumors have combined growth.
4. Retinal cells frequently break off from the main mass and seed the vitreous, or new locations on the retina.
5. Glaucoma may result from occlusion of the trabecular network or from iris neovascularization.

Extraocular

1. RB spreads first to soft tissues surrounding the eye or invades the optic nerve. From there it can spread directly along the axons to the brain, or may cross into the subarachnoid space and spread via the cerebrospinal fluid to the brain.
2. Hematogenous spread leads to metastatic disease, most commonly to brain, lungs, bone marrow, or bone.
3. RB can spread lymphatically to the preauricular and submandibular nodes.

In patients with the genetic form of RB, central nervous system (CNS) disease is less likely the result of metastatic or regional spread than pineoblastoma, which is also termed trilateral RB.

Trilateral RB

Trilateral RB is a syndrome that occurs in children under the age of 5 years. It consists usually of bilateral hereditary RB associated with an intracranial neuroblastic tumor of the pineal gland, and occurs in less than 10% of children with familial, multifocal, or bilateral RB. With the onset of systemic neoadjuvant chemotherapy, the incidence of this syndrome is decreasing.

DIAGNOSTIC PROCEDURES

Screening

All children should have screening performed as part of well-child check-ups, primarily by eliciting red reflexes in the eye. However, most cases of RB are diagnosed after a parent or other relative notices an abnormality of the eye and this prompts further evaluation.

Siblings of children with RB should be screened by ophthalmology at regular intervals at least through age 2 to 3 years unless their genetic testing has revealed they do not carry the *RB1* gene mutation.

Diagnosis of Intraocular RB

Diagnosis is made by ophthalmologists, retinal specialists, or ocular oncologists using a combination of an ophthalmologic examination generally performed under sedation or anesthesia, together with retinal camera (RetCam) imaging, ultrasound, computed tomography, or magnetic resonance imaging (MRI). Calcification of the

TABLE 28.1   Investigations for Diagnosis of Retinoblastoma

Examinations	Imaging studies	Laboratory evaluations	Diagnostic studies
• Examination under anesthesia by pediatric ophthalmologist • Examination and consultation with pediatric oncologist • Audiology evaluation if newborn hearing screen was abnormal or family history of hearing loss	• MRI scan of brain and orbits (should include the pineal gland to exclude trilateral RB)	• Complete blood count with white blood cell differential • Chemistry panel, if systemic chemotherapy is considered	• Lumbar puncture only when there is radiographic or clinical suspicion of CNS disease • Bone scan only when bone pain or other extraocular disease • Bone marrow biopsy only when there is an abnormal blood count (without alternative explanation) or other extraocular disease • If enucleation is performed, subsequent evaluation by a pathologist experienced in the evaluation of RB

CNS, central nervous system; RB, retinoblastoma; MRI, magnetic resonance imaging.

lesion(s) may be seen. Due to concern about rupturing the tumor and causing both intraocular and extraocular spread, surgical biopsies are not performed for confirmation.

Defining Extent of Disease

The ophthalmologic examination will determine the extent of intraocular tumor and presence or absence of orbital extension. It is crucial that intraocular examination with a binocular indirect ophthalmoscope be performed on both eyes with the pupils maximally dilated. The ophthalmologist should make diagrams of the retina to show the number, size, and location of all tumors. These diagrams are now being supplemented by the use of RetCam images, which are helpful in determining not only the extent of disease, but response to treatment. When the binocular indirect ophthalmoscope is used, the location of the tumor(s) should be related to specific landmarks such as the optic nerve head, fovea, and ora serrata. Size of the lesion is estimated by comparison with the optic nerve head diameter.

Extraocular Extent of the Disease

All children with RB should be referred to a pediatric oncologist for evaluation of extraocular disease.
Table 28.1 summarizes the investigations to be performed for diagnosis of RB.

Classification

The Reese–Ellsworth classification is no longer employed because it is not useful in stratifying patients with respect to outcome following modern treatment modalities. The International Classification System for Intraocular RB, which is based on the extent and location of intraocular RB (Table 28.2), has replaced it as the standard classification system.

TREATMENT

In order to maximize the preservation of useful vision, treatment should be undertaken in specialized centers, where there is close collaboration between pediatric oncology and pediatric ophthalmology. The initial therapy for RB is dependent on both the intraocular and extraocular extent of the disease. Therapeutic modalities include the following and a combined modality approach is not uncommon:

1. Systemic chemotherapy.
2. Local ophthalmic therapy, including local administration of chemotherapy including subtenon or intravitreal chemotherapy.
3. Intra-arterial chemotherapy (IAC), using an interventional neuroradiology approach.
4. Radiotherapy including external beam (photons or protons) and plaque brachytherapy.
5. Enucleation: The need for bilateral enucleation, with optimum treatment, is <1%.

TABLE 28.2   International Classification System for Intraocular Retinoblastoma

---

GROUP A: SMALL INTRARETINAL TUMORS AWAY FROM FOVEOLA AND DISC

- All tumors are 3 mm or smaller in greatest dimension, confined to the retina *and*
- All tumors are located further than 3 mm from the foveola and 1.5 mm from the optic disk

GROUP B: ALL REMAINING DISCRETE TUMORS CONFINED TO THE RETINA

- All other tumors confined to the retina not in Group A
- Tumor-associated subretinal fluid less than 3 mm from the tumor with no subretinal seeding

GROUP C: DISCRETE LOCAL DISEASE WITH MINIMAL SUBRETINAL OR VITREOUS SEEDING

- Tumors are discrete
- Subretinal fluid, present or past, without seeding involving one-fourth of the retina
- Local fine vitreous seeding may be present close to discrete tumor
- Local subretinal seeding less than 3 mm (2 disk diameters) from the tumor

GROUP D: DIFFUSE DISEASE WITH SIGNIFICANT VITREOUS OR SUBRETINAL SEEDING

- Tumors may be massive or diffuse
- Subretinal fluid present or past without seeding, involving up to total retinal detachment
- Diffuse or massive vitreous disease may include "greasy" seeds or avascular tumor masses
- Diffuse subretinal seeding may include plaques or tumor nodules

GROUP E: PRESENCE OF ANY ONE OR MORE OF THESE POOR PROGNOSTIC FEATURES

- Tumor touching the lens
- Tumor anterior to the anterior vitreous face involving ciliary body or anterior segment
- Diffuse infiltrating retinoblastoma
- Neovascular glaucoma
- Opaque media from hemorrhage
- Tumor necrosis with aseptic orbital cellulites
- Phthisis bulbi

Treatment of Intraocular RB

Systemic Chemotherapy

For intraocular RB, chemotherapy is used in a neoadjuvant setting, often in combination with local ophthalmic therapies, where its purpose is to decrease tumor size and volume to allow the successful utilization of local ophthalmic therapies. In this setting, it is referred to as chemoreduction. Systemic chemotherapy can also be used in combination with enucleation, where one eye is enucleated and chemotherapy is delivered to treat tumors in the remaining eye.

The goals of treatment are:

1. To save the child's life.
2. To preserve as much vision as possible.
3. To avoid external beam radiotherapy in patients with heritable RB.

Chemotherapy regimens generally include carboplatin, etoposide, and vincristine (CEV). Treatment is often given monthly for a total of six cycles. Some investigators have employed cyclosporine in addition to CEV in attempts to modulate drug resistance. Another frontline regimen used is vincristine and topotecan (Table 28.3). The success of either regimen is proportional to the grouping of each eye involved as well as the quality of the focal ophthalmic therapy administered in conjunction with any chemotherapy. Group A tumors can be successfully managed with focal ophthalmic therapies alone. Group B and C eyes are salvaged in the majority of cases. Group D eyes can be salvaged in approximately half the cases with systemic chemotherapy. Rates of 80–90% salvage with IAC in Group D eyes have been reported in single-institution studies with limited follow-up. Significantly fewer Group E eyes can be salvaged and the concern is that they have the greatest risk of features portending metastatic disease. Enucleation of a unilateral Group E eye can be lifesaving. One third of the patients who undergo unilateral enucleation may require adjuvant chemotherapy to prevent metastatic disease based on high-risk pathologic features. In the Children's Oncology Group (COG) these include massive (>3 mm) choroidal involvement, post-laminar optic nerve involvement (ONI) or any uveal involvement in the presence of ONI.

TABLE 28.3　Neoadjuvant Chemotherapy Protocols for Intraocular Retinoblastoma[a]

- A: CEV: Six cycles of the following are given every 28 days
  - Vincristine 0.05 mg/kg (1.5 mg/m^2 if >age 3 years) Day 1
  - Carboplatin 18.6 mg/kg (560 mg/m^2 if >age 3 years) Day 1
  - Etoposide 5 mg/kg (150 mg/m^2 if >age 3 years) Days 1 and 2
- B: Vincristine and Topotecan
  - Vincristine 0.05 mg/kg (1.5 mg/m^2 if age greater than or equal to 3 years) on Day 1
  - Topotecan 0.066 mg/kg (2 mg/m^2 if greater than or equal to 3 years of age) on Days 1–5

[a]There are protocols utilizing higher doses of carboplatin and etoposide for patients with Group D disease. There are other protocols that include cyclosporine A aimed at deceasing drug resistance.
CEV, carboplatin, etoposide, and vincristine.

Concerns about systemic chemotherapy have included hearing loss, secondary leukemia, and infertility. A recent prospective trial showed no hearing loss found on serial audiograms with CEV. Patients with abnormal hearing prior to treatment can worsen. Thus, it is important to establish normal hearing prior to commencement of therapy. Additionally, the dosing strategy of the chemotherapy is crucial. Infants, especially those under 10 kg, should be based on weight, not body surface area.

Secondary acute myeloid leukemia (AML) has not been reported in the context of modern treatment regimens, and a Surveillance Epidemiology and End Result report more recently calculated an observed to expected rate of secondary AML of zero. IAC utilizing melphalan theoretically carries a risk of AML as melphalan is a leukemogen.

CEV systemic chemotherapy does not cause infertility.

External Beam Radiotherapy

Standard of care for intraocular RB for many years has included external beam radiotherapy with doses of 40–45 Gy. This resulted in a number of late effects, and increased the risk of second malignancies both in and out of the field of radiotherapy. Newer methods of delivering radiotherapy with more conformal fields are being used at present, so that the normal structures receive less scatter and thus the risk for adverse long-term outcomes would be hypothesized to be less. Conformal radiotherapy, stereotactic radiotherapy, proton-beam radiotherapy, as well as intensity-modulated radiation therapy all use technology that minimizes doses to non-target structures. In addition, there are current protocols testing doses of 23–36 Gy with encouraging results.

Local Ophthalmic Therapies

Local therapy is used to eradicate local disease after reduction of the tumor volume by chemotherapy and may include cryotherapy, green laser, infrared laser, and/or radioactive plaque. The goal of local therapy is to achieve Type I regression pattern with calcification or Type IV with flat chorioretinal scars, or avascular, linear, white gliosis.

Cryotherapy

Cryotherapy can be utilized for ablating tumor remnants/recurrences after chemotherapy up to 3 mm in height and 6 mm in diameter that are located at or anterior to the equator. It is recommended that no more than four different sites be frozen in one eye at a single session. There is less likelihood of creating vitreous seeds with cryotherapy if a tumor has been previously treated with chemotherapy. Extensive cryotherapy has been associated with significant persistent retinal detachment, particularly if the retina was originally detached prior to chemotherapy.

Laser Photoablation

Laser (using argon laser, diode laser, or xenon arc photocoagulation) can be used to directly coagulate tumors up to 8 mm in thickness, especially posterior to the equator following chemotherapy. Laser-induced hemorrhage has been associated with vitreous seeding. Inadequate dilation of the pupil can cause laser burns to the iris.

Transpupillary Thermotherapy

Thermotherapy uses infrared radiation (or alternatively ultrasound or microwaves) to deliver heat to the eye with the goal of achieving a temperature of 42–60°C. In general, small tumors require approximately 300 mW of power for 10 min or less repeated for three times at 1-month intervals. Tumors up to 15 mm in base can be adequately treated if the patient is receiving chemoreduction.

Intravitreal Chemotherapy

Presently this treatment modality is reserved for recurrent vitreous seeds. Injections are given on a weekly to monthly schedule for three to six consecutive injections. Doses of 50 µg of melphalan can lead to phthisis bulbi and thus many investigators prefer 10–30 micrograms per injection.

There is concern that tumor seeding of the orbit could occur during this procedure but a recent review of all 304 published cases demonstrated only one case who developed metastatic disease.

Plaque Radiotherapy

Episcleral plaque radiotherapy (I-125 brachytherapy) may be used to treat local recurrences up to 8 mm in thickness and 15 mm in base. Plaque radiotherapy involves securing a radioactive plaque to the eye wall at the apex of the tumor. It may be used as a primary treatment or as an adjunct to surgery. Iodine-125 or Ruthenium-106 can be utilized for plaques. The tumor dose is prescribed at the apex of the tumor and is 30–35 Gy. Proliferative retinopathy secondary to plaque can occur.

Subtenon (Subconjunctival) Chemotherapy

For patients with more advanced intraocular disease (International classification groups D and E), pilot studies have been conducted using subtenon carboplatin, in addition to systemic chemotherapy and other local ophthalmic therapies. The ophthalmologist makes a 3-mm incision in the conjunctiva and anterior tenons. A 5-cc syringe containing the carboplatin is fitted with an irrigating cannula. The cannula is placed through the conjunctival incision and is gently and bluntly passed through the posterior tenon's capsule while maintaining constant contact with the globe. Once the irrigating cannula has been passed posteriorly to its full extent, the drug is delivered into the retrobulbar space.

Intra-Arterial Chemotherapy

With the goal of lessening systemic exposure to chemotherapy, IAC involves delivery of a single or double drug regimen via a microcatheter that is guided under fluoroscopy to the ostium of the ophthalmic artery. The regimens reported to date have included melphalan, topotecan, and/or carboplatin monthly for a total of three sessions. The medication is delivered in a pulsatile fashion to minimize reflux into the internal carotid artery and so as not to occlude the ophthalmic artery itself. Response is assessed via a separate evaluation under anesthesia (EUA).

Some teams dose both eyes in a single procedure by withdrawing the catheter after the first eye is treated and repositioning it on the other side. This can double the systemic exposure. Children under approximately 4 months of age are often not treated due to lack of availability of appropriately sized microcatheters.

Although clinically apparent strokes have rarely been reported, carotid artery spasm has been seen. Additionally, approximately one quarter of patients will experience an autonomic cardiorespiratory reflex evidenced as hypoxia, bradycardia, and hypotension. This may be life-threatening, requiring epinephrine administration.

Choroidal atrophy has been detected post-procedure and histopathologic examination of a non-human primate model revealed toxic effects in the ocular and orbital vasculature. The COG has recently launched a limited institution IAC feasibility trial of monthly melphalan.

In comparison to systemic chemotherapy which has been employed for over 20 years there is less than a decade of American experience with IAC. Nevertheless, it can provide complete tumor control, even in advanced eyes. Additionally, it can provide ocular salvage in approximately half of patients whose tumors progressed following systemic chemotherapy.

Enucleation

Enucleation of the eye is recommended when there is no chance for useful vision even if the entire tumor is destroyed. It is also indicated where risk of development of extraocular or metastatic disease is high.

Careful examination of the enucleated specimen by an experienced pathologist is necessary to determine whether high-risk features for metastatic disease are present. These include

1. Anterior chamber seeding (although this is unlikely to be detected in the absence of other high-risk features).
2. Massive choroidal involvement (>3 mm).

3. Tumor beyond the lamina cribrosa.
4. Any choroidal involvement with concomitant ONI.
5. Disease at the cut edge of the optic nerve.
6. Scleral and extrascleral extension.

External beam radiotherapy or systemic adjuvant therapy is generally required in patients exhibiting features (5) and (6) listed above. An orbital implant (typically hydroxyapatite) is placed at the time of enucleation.

Treatment of Extraocular RB

There is no single clearly proven effective therapy for the treatment of extraocular RB. Those with CNS metastases appear to do worse than those with other forms of extraocular disease. The COG presently is conducting a trial for patients with metastatic RB.

Generally, induction chemotherapy with CEV is given for four cycles and this is followed by high-dose chemotherapy followed by stem cell rescue. Following recovery from stem cell transplantation, radiotherapy is generally given to sites of initial bulky disease although this has been associated with the development of secondary solid tumors following stem cell transplantation.

Treatment of Recurrent RB

The prognosis for a patient with recurrent or progressive RB depends on the site and extent of the recurrence or progression. Recurrence is not uncommon for those individuals treated with systemic chemotherapy without radiation therapy or enucleation. For these individuals, recurrence typically occurs in the first 6 months post chemotherapy. If the recurrence or progression of RB is confined to the eye and is small, the prognosis for sight and survival may be excellent with local ophthalmic therapy only. If the recurrence or progression is confined to the eye but is extensive, the prognosis for sight is poor; however, the survival remains excellent. If the recurrence or progression is extraocular, the prognosis is more guarded and the treatment depends on many factors and individual patient considerations.

POSTTREATMENT MANAGEMENT

Disease-Related Follow-Up

Patient should be monitored for follow-up of the primary tumor. The majority of recurrences appear within 3 years of diagnosis and recurrences are extremely rare after 5 years of age. The follow-up schedule is largely dependent on the therapy used and the extent of original disease.

For patients treated with chemotherapy for intraocular or extraocular disease, an EUA is recommended to be performed every month until there is no active tumor seen on a minimum of three EUAs, then approximately every 2–3 months until age 3 years and then every 6 months to age 10 years. Patients should be examined without anesthesia when old enough to cooperate.

For patients who have undergone an enucleation and/or received external beam radiotherapy, an evaluation should be performed every 2–3 months for the first year post therapy, every 3–4 months the following year, every 6 months until age 5 years and then annually. For patients with extraocular disease, assessment for recurrent metastatic disease should also be performed, which should include physical examination, MRI of the brain and orbits as well as a complete blood count (CBC) and chemistry panel. Spinal tap evaluation and bone marrow aspirates are done for patients with signs or symptoms of abnormalities found on MRI or CBC.

To screen for trilateral RB, patients diagnosed with bilateral RB, especially those diagnosed at less than 1 year of age, or those with positive family history are recommended at many centers to have an MRI every 6 months from the end of therapy until 5 years of age.

Toxicity-Related Follow-Up

This is largely dependent on the therapy received. For those treated with chemotherapy or radiation therapy, a history and physical examination should be performed at least every 3 months until 2 years off therapy, then every 6 months until 5 years off therapy and yearly thereafter. A visual acuity assessment should be conducted at least yearly. Attention should be paid to late effects of chemotherapy and radiotherapy, growth and development and surveillance for SMNs. Parents (and patients as they get older) should be counseled regarding their risk of SMN and should have a comprehensive risk-directed evaluation annually lifelong.

If systemic chemotherapy has been used, the COG Long-Term Follow-Up Guidelines (www.survivorshipguidelines.org) have the most comprehensive guidelines for follow-up based on therapeutic exposures.

There is no clear screening known to be effective for SMNs. Patients who have received chemotherapy associated with the occurrence of secondary leukemia (etoposide, alkylating agents, doxorubicin, nitrosoureas) should have annual blood counts with differential white cell count performed for 10–15 years from exposure. Patients should be made aware of signs and symptoms of the most common second malignancies following RB, osteosarcoma, and soft tissue sarcomas, as well as skin cancers and breast cancer. No routine imaging studies are recommended for surveillance for skin and solid organ tumors unless there is a clinical indication. However, careful physical examinations by the patient and physician are clearly indicated for early detection and successful treatment.

FUTURE PERSPECTIVES

Until recently, RB was largely a disease managed by ocular oncologists, pediatric ophthalmologists, and radiation oncologists. With the advent of neoadjuvant chemotherapy for intraocular disease and dose-intensive myeloablative protocols for extraocular disease, the role of the pediatric oncologist in the management of this disease is growing. Pediatric oncologists are also well versed in the late effects of therapy and surveillance for SMNs. Standardized databases regarding diagnosis, group and stage determination, treatment, and late effects should be established. The goal of neoadjuvant systemic chemotherapy is to avoid high-dose external beam radiotherapy and enucleation, and the long-term effects associated with both. It is only with these cooperative group studies that long-term data can be systematically collected to demonstrate that changes in approach toward eye salvage provides an acceptable toxicity profile. Biologic samples could be obtained to better understand RB genetics, and to test new treatment strategies.

Further Reading and References

Aerts, I., Sastre-Garau, Y., Savignoni, A., et al., 2013. Results of a multicenter prospective study on the postoperative treatment of unilateral retinoblastoma after primary enucleation. J. Clin. Oncol. 31 (11), 1458–1463.

Dunkel, I.J., Khakoo, Y., Kernan, N.A., et al., 2010. Intensive multimodality therapy for patients with stage 4A metastatic retinoblastoma. Pediatr. Blood Cancer 55 (1), 55–59.

Gallie, B.L., Budning, A., DeBoer, G., et al., 1996. Chemotherapy with focal therapy can cure intraocular retinoblastoma without radiotherapy. Arch. Ophthalmol. 114, 1321–1328.

Ghassemi, F., Shields, C.L., 2012. Intravitreal melphalan for refractory or recurrent vitreous seeding from retinoblastoma. Arch. Ophthalmol. 130, 1268–1271.

Gobin, Y.P., Dunkel, I.J., Marr, B.P., et al., 2011. Intra-arterial chemotherapy for the management of retinoblastoma: four-year experience. Arch. Ophthalmol. 129, 732–737.

Gombos, D.S., Hungerford, J., Abramson, D.H., et al., 2007. Secondary acute myelogenous leukemia in patients with retinoblastoma. Is chemotherapy a factor? Ophthalmology 114, 1378–1383.

Hurwitz, R.L., Shields, C.L., Shields, J.A., et al., 2006. Retinoblastoma. In: Pizzo, P.A., Poplack, D.G. (Eds.), Principles and Practice of Pediatric Oncology, fifth ed. Lippincott, Williams and Wilkins, Philadelphia, PA, pp. 865–886.

Kaliki, S., Shields, C.L., Shah, S.U., et al., 2011. Postenucleation adjuvant chemotherapy with vincristine, etoposide, and carboplatin for the treatment of high-risk retinoblastoma. Arch. Ophthalmol. 129, 1422–1427.

Kaliki, S., Shields, C.L., Rojanaporn, D., 2013. High-risk retinoblastoma based on International Classification of Retinoblastoma: analysis of 519 enucleated eyes. Opthalmology 120, 97–1003.

Kivela, T., 2009. The epidemiological challenge of the most frequent eye cancer: retinoblastoma, an issue of birth and death. Br. J. Ophthalmol. 93, 1129–1131.

Kivela, T., 2012. Alive with good vision: the ultimate goal in managing retinoblastoma. Clin. Exp. Ophthalmol. 40, 655–656.

Kivelä, T., Eskelin, S., Paloheimo, M., 2011. Intravitreal methotrexate for retinoblastoma. Ophthalmology 118, 1689.

Kleinerman, R.A., Tucker, M.A., Tarone, R.E., et al., 2005. Risk of new cancers after radiotherapy in long-term survivors of retinoblastoma: an extended follow up. J. Clin. Oncol. 23 (101), 2272–2279.

Knudson, A.G., 1971. Mutation and cancer: statistical study of retinoblastoma. Proc. Natl. Acad. Sci. USA 68 (4), 820–823.

Lambert, M.P., Shields, C.L., Meadows, A.T., 2008. A retrospective review of hearing in children with retinoblastoma treated with carboplatin-based chemotherapy. Pediatr. Blood Cancer 50, 223–226.

Leahey, A., 2012. A cautionary tale: dosing chemotherapy in infants with retinoblastoma. J. Clin. Oncol. 30, 1023–1024.

Marr, B.P., Brodie, S.E., Dunkel, I.J., et al., 2012. Three-drug intra-arterial chemotherapy using simultaneous carboplatin, topotecan and melphalan for intraocular retinoblastoma: preliminary results. Br. J. Ophthalmol. 96, 1300—1303.

Mulvihill, A., Budning, A., Jay, V., Vandenhoven, C., et al., 2003. Ocular motility changes after subtenon carboplatin chemotherapy for retinoblastoma. Arch. Ophthalmol. 121, 1120—1124.

Murphree, A.L., 2005. Intraocular retinoblastoma: the case for a new group classification. Ophthalmol. Clin. North Am 200, 41—53 (viii).

Phillips, T.J., McGuirk, S.P., Chahal, H.K., et al., 2013. Autonomic cardio-respiratory reflex reactions and super selective ophthalmic arterial chemotherapy for retinoblastoma. Paediatr. Anaesth. 23, 940—945.

Qaddoumi, I., Bass, J.K., Wu, J., et al., 2012. Carboplatin-associated ototoxicity in children with retinoblastoma. J. Clin. Oncol. 30 (10), 1034—1041.

Ramasubramanian, A., Shields, C.L. (Eds.), 2012. Retinoblastoma. Jaypee Brothers Medical Publishers, New Delhi, India.

Ramasubramanian, A., Kytasty, C., Meadows, A.T., et al., 2013. Incidence of pineal gland cyst and pineoblastoma in children with retinoblastoma during the chemoreduction era. Am. J. Ophthalmol. 156, 825—829.

Rihani, R., Bazzeh, F., Faqih, N., et al., 2010. Secondary hematopoetic malignancies in survivors of childhood cancer: an analysis of 111 cases from the Surveillance, Epidemiology and End Result-9 Registry. Cancer 116 (18), 4385—4394.

Rushlow, D.E., Mol, B.M., Kennett, J.Y., et al., 2013. Characterisation of retinoblastomas without RB1 mutations: genomic, gene expression and clinical studies. Lancet Oncol. 14 (4), 327—334.

Shields, C.L., Shields, J.A., 2006. Basic understanding of current classification and management of retinoblastoma. Curr. Opin. Ophthalmol. 17, 228—234.

Shields, C.L., Meadows, A.T., Shields, J.A., et al., 2001. Chemoreduction for retinoblastoma may prevent intracranial neuroblastic malignancy (trilateral retinoblastoma). Arch. Ophthalmol. 119, 1269—1272.

Shields, C.L., Mashayekhi, A., Sun, H., et al., 2006a. Iodine 125 plaque radiotherapy as salvage treatment for retinoblastoma recurrence after chemoreduction in 84 tumors. Ophthalmology 113 (11), 2087—2097.

Shields, C.L., Mashayekhi, A., Au, A.K., et al., 2006b. The International Classification of Retinoblastoma predicts chemoreduction success. Ophthalmology 113, 2276—2280.

Shields, C.L., Bianciotto, C.G., Ramasubramanian, A., et al., 2011. Intra-arterial chemotherapy for retinoblastoma. Report #1: control of tumor, subretinal seeds, and vitreous seeds. Arch. Ophthalmol. 129, 1399—1406.

Shields, C.L., Schoenfeld, E., Kocher, K., et al., 2013. Lesions simulating retinoblastoma (pseudoretinoblastoma) in 604 cases. Ophthalmology 120, 311—316.

Shields, C.L., Lally, S.E., Leahey, A.M., et al., 2014. Targeted retinoblastoma management: when to use intravenous, intra-arterial, periocular, and intravitreal chemotherapy. Curr. Opin. Ophthlamol. 25 (5), 374—385.

Smith, S.J., Smith, B.D., 2013. Evaluating the risk of extraocular tumour spread following intravitreal injection therapy for retinoblastoma: a systematic review. Br. J. Ophthalmol. 97, 1231—1236.

Superstein, R., Lederer, D., Dubois, J., et al., 2013. Retinal vascular precipitates during administration of melphalan into the ophthalmic artery. JAMA Ophthalmol. 131, 963—965.

Tse, B.C., Steinle, J.J., Johnson, D., et al., 2013. Superselective intraophthalmic artery chemotherapy in a nonhuman primate model: histopathologic findings. JAMA Ophthalmol. 131, 903—911.

Turaka, K., Shields, C.L., Leahey, A., et al., 2012. Second malignant neoplasms following chemoreduction with carboplatin, etoposide, and vincristine in 245 patients with intraocular retinoblastoma. Pediatr. Blood Cancer 59, 121—125.

Yousef, Y.A., Halliday, W., Chan, H.S., et al., 2013. No ocular motility complications after subtenon topotecan with fibrin sealant for retinoblastoma. Can. J. Ophthalmol. 48, 524—528.

29

Germ Cell Tumors

Thomas A. Olson

Germ cell tumors (GCTs) are neoplasms that develop from primordial germ cells of the human embryo, which are normally destined to produce sperm or ova. Primordial germ cells appear to originate in the yolk sac endoderm and migrate around the hindgut to the genital ridge on the posterior abdominal wall where they become part of the developing gonad. Figure 29.1 depicts the histogenesis of tumors of germ cell origin. Viable germ cells arrested along this path of migration may form neoplasia in midline sites, such as the pineal region (6%), mediastinum (7%), retroperitoneum (4%), sacrococcygeal region (42%) or in the ovary (24%), testis (9%), and other sites (8%).

INCIDENCE

Tumors of germ cell origin account for approximately 2–3% of childhood malignancies. The incidence of GCTs is 2.5 per million in white children and 3.0 per million in African-American children under 15 years of age with a male:female ratio of 1.0:1.1. GCTs are more common in the ovaries and testes than in extragonadal sites.

PATHOLOGY

The germ cells are the precursors of sperm and ova and retain the potential to produce all the somatic (embryonic) and supporting (extraembryonic) structures of a developing embryo. *Yolk sac tumors (YSTs) (endodermal sinus tumors)* are derived from a totipotential germ cell that differentiates to extraembryonic structures. *Teratomas* are embryonal neoplasms that contain tissues from all three germ layers (ectoderm, endoderm, and mesoderm). Teratomas are mature or immature and may occur with or without malignant germ cell elements (endodermal sinus tumors, choriocarcinomas, embryonal carcinomas, or germinomas) or rarely malignant somatic elements (such as primitive neuroectodermal tumors in children).

The malignant histologic variants of GCTs are:

1. Germinoma.
   a. Dysgerminoma (ovary).
   b. Seminoma (testis).
2. Immature teratoma.
3. Embryonal carcinoma.
4. YST (endodermal sinus tumor).
5. Choriocarcinoma.

Table 29.1 describes the pathology of GCTs and associated biochemical markers.

Cytogenetic analysis has identified isochromosome 12p (i12p) as a specific abnormality in more than 80% of GCTs in postpubertal patients. Isochromosome 12p has also been identified in one-third of GCT in children under 5 years of age, suggesting that similar mechanisms for GCT generation may exist.

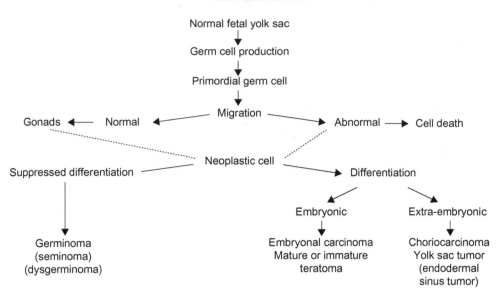

FIGURE 29.1 Histogenesis of tumors of germ cell origin.

TABLE 29.1 Pathology of Germ Cell Tumors and Associated Markers

Histologic variant	Morphology	Common sites of origin	Markers[a]		
			AFP	β-hCG	PLAP
Germinoma	• Cells are round with discrete membranes, abundant clear cytoplasm. • Round nucleus with one to several prominent nucleoli. Cells arranged in nest or lobules separated by fibrous stroma. Areas of necrosis with granulomatous reaction. Multinucleated giant cells may be present	• Ovary: dysgerminoma • Testis: seminoma • Anterior mediastinum	−	−	+

Mature teratoma	• Mature tissue derived from:
	  • Ectoderm: squamous epithelium, neuronal tissue
	  • Mesoderm: muscle, teeth, cartilage, bone
	  • Endoderm: mucous glands, gastrointestinal and respiratory tract Lining
	• No mitoses seen
		• Sacrococcygeal/presacral
		• Gonads
		• Mediastinum
			−	−	−

Immature teratoma	• Immature tissue derived from the three germinal layers. Lesions are graded histologically:
	  • Grade 1: Some immaturity with neuroepithelium absent or limited to a rare low magnification (×40) field and not more than one such field in any slide
	  • Grade 2: Immaturity and neuroepithelium present but does not exceed three low-power microscopic fields in any slide
	  • Grade 3: Immaturity and neuroepithelium prominent; occupying four or more low magnification microscopic fields
		• Sacrococcygeal/presacral
		• Gonads
		• Mediastinum
			−	−	−

Embryonal carcinoma	Tumor cells polygonal with abundant pink vacuolated cytoplasm with ill-defined cellular membranes. Nucleus irregular and pleomorphic with prominent nucleoli. Cells arranged in solid sheets with scanty stroma or in tubules, acini, or papillary structures. Typical and atypical mitoses present	Testis (young adult)	−	±	±

Yolk sac tumor (endodermal sinus)	• Characteristics as seen microscopically:
	  • Aggregates of small undifferentiated embryonal cells
	  • Areas of stellate mesodermal cells
	  • Areas of perivascular formation consisting of a mesodermal core with a capillary in the center lined by columnar cells (Schuller–Duvall body)
		• Testis (infant)
		• Ovary
		• Presacral
			+	−	−

AFP, α-Fetoprotein; β-hCG, beta subunit of human chorionic gonadotropin; PLAP, placental alkaline phosphatase.

CLINICAL FEATURES

The signs and symptoms of GCTs are dependent on the site of origin. Table 29.2 lists the clinical features of GCTs at different sites. In addition, certain histologic variants may have associated clinical findings, as listed in Table 29.3.

Although GCTs are a diverse group histologically, all originate from primordial germ cells and have a common pattern of spread, irrespective of the primary site as follows:

1. Lungs.
2. Liver.
3. Regional nodes.
4. Central nervous system (CNS).
5. Bone and bone marrow (less commonly).

TABLE 29.2 Clinical Features of Germ Cell Tumors

Tumor type	Median age year	Relative frequency (%)	Features
PEDIATRIC OVARIAN TUMORS			
Dysgerminoma	16	24	Rapidly developing; 14–25% with other germ cell elements; very radiosensitive
Yolk sac tumor (endodermal sinus tumor)	18	16	↑ AFP; 75% Stage I; all patients require chemotherapy because of high risk of relapse even in low-stage disease
Teratoma mature (solid, cystic)	10–15	31	Neuroglial implants may occur with cystic or solid teratomas, but do not affect prognosis; surgery is mainstay of treatment
Immature inversely	11–14	10	Grading system based on amount of neuroepithelium present; prognosis related to stage and grade; 30% with ↑ AFP
Embryonal carcinoma	14	6	47% prepubertal; ↑β-hCG and precocious puberty common; chemotherapy indicated
Malignant mixed germ cell tumor	16	11	40% premenarchal; 30% sexually precocious; AFP/β-hCG may be increased
Gonadoblastoma both	8–10	1	Associated with dysgenetic gonads and sexual maldevelopment; removal of gonads is treatment of choice
Other (polyembryoma, choriocarcinoma)	NA	<1	Rare in children
PEDIATRIC TESTICULAR TUMORS			
Yolk sac tumor (endodermal sinus tumor)	2	26	Most common of malignant germ cell tumors of the testes; ↑AFP; compared to adult cases, pediatric tumors are pure histologically; 85% Stage I; chemotherapy reserved for higher-stage or recurrent disease
Teratoma	3	24	Poorly differentiated histologic features do no impart a malignant course in children. Surgery alone is usually sufficient treatment
Embryonal carcinoma on stage	Late teens	20	Uncommon in young children; ↑AFP ± β-hCG; managed as for adults, with retroperitoneal lymphadenectomy ± chemotherapy ± irradiation based
Teratocarcinoma	Late teens	13	80% stage I with 75% survival after surgery alone; more advanced disease requires multimodality therapy
Gonadoblastoma	5–10	<1	Associated with sexual maldevelopment syndromes; bilateral involvement in 30%; bilateral removal of gonads is treatment of choice
Other (polyembryoma choriocarcinoma)	NA	16	Rare in children

AFP, α-fetoprotein; β-hCG, beta subunit of human chorionic gonadotropin.
Source: Pizzo and Poplack (2002), with permission.

TABLE 29.3   Clinical Association of Different Histologic Variants of Germ Cell Tumors

Histology	Clinical association
Teratoma	Musculoskeletal anomalies
	Rectal stenosis
	Congenital heart disease
	Microcephaly
Ovarian dysgerminoma (postpubertal)	Amenorrhea
	Menorrhagia
	46XY (male pseudohermaphrodite)
Ovarian embryonal carcinoma	Precocious puberty
	Amenorrhea
	Hirsutism

DIAGNOSTIC EVALUATION

The following evaluations should be carried out:

1. History.
2. Physical examination.
3. Complete blood count.
4. Liver function tests, electrolytes, blood urea nitrogen, creatinine.
5. α-Fetoprotein (AFP).
6. Human chorionic gonadotropin (β-hCG).
7. Lactic dehydrogenase (LDH) isoenzyme 1. LDH may also correlate with disease activity such as:
   a. Tumor bulk.
   b. Residual tumor after surgery.
   c. Response to chemotherapy and radiotherapy.
   d. Tumor recurrence.
8. Radiographic evaluation of primary site and regional disease.
   a. Mediastinum: chest and upper abdominal computed tomography (CT).
   b. Ovary: ultrasound, pelvic and abdominal CT or magnetic resonance imaging (MRI).
   c. Sacrococcygeum: pelvic and abdominal CT or MRI.
   d. Testis: ultrasound, pelvic and abdominal CT or MRI.
9. Radiographic evaluation for distant metastases.
   a. Chest radiographs (posteroanterior and lateral).
   b. Chest CT.
   c. Bone scan (may be reserved for patients with pure choriocarcinoma or known Stage IV disease).

TUMOR MARKERS

Certain histologic variants of GCTs secrete the tumor markers AFP and beta subunit of human chorionic gonadotropin (β-hCG). The production of these markers can be assessed by immunohistologic staining of tissue sections or measurement of blood levels, and has important diagnostic value and can be used to assess disease activity.

AFP is a major serum protein of the human fetus. It is produced in the embryonic liver, in the yolk sac, and, in smaller amounts, in the gastrointestinal tract. In general, the highest AFP levels are seen in YST, with embryonal carcinoma exhibiting intermediate levels.

The β-hCG can serve as a tumor marker when positive. It may be positive when syncytiotrophoblasts are present in the tumor, and it is found to be elevated in embryonal carcinomas and choriocarcinomas. In choriocarcinoma, there is generally a marked elevation of β-hCG.

Pure teratomas are not associated with excessive AFP or β-hCG production. Elevation of either marker in association with teratoma may indicate the presence of more malignant germ cell elements and requires review of the histologic material or study of more histologic sections. AFP is difficult to evaluate as an indicator of residual or recurrent GCTs in infants less than 1 year of age. The half-life of AFP may also vary with age. The wide range of normal AFP levels in infants is shown in Appendix 2, but these data are based on small numbers of patients. It is essential to monitor AFP levels over the first year of life in these patients to ensure that AFP levels continue to fall. The half-life of AFP (beyond 8 months of age) and β-hCG is 5 days and 16 h, respectively.

Increasing levels of serum AFP are not necessarily indicative of tumor progression because an abrupt escalation in serum AFP can occur after chemotherapy-induced tumor lysis. Elevations of serum AFP could also be caused by alterations in hepatic function such as viral hepatitis, cirrhosis, hepatoblastoma, pancreatic and gastrointestinal malignancies, and lung cancers. β-hCG can also experience sudden increases secondary to cell lysis during chemotherapy or can be increased with malignancies of the liver, pancreas, gastrointestinal tract, breast, lung, and bladder. The utility of CA-125 in monitoring ovarian GCTs in children has not been thoroughly evaluated.

STAGING

The staging systems for pediatric GCTs differ depending on the site of origin and are listed in Table 29.4.

TABLE 29.4 Staging Systems for Pediatric Germ Cell Tumors

Stage	Extent of disease
EXTRAGONADAL	
I	Complete resection at any site, including coccygectomy for sacrococcygeal site; negative tumor margins; tumor markers positive but fall to normal or markers negative at diagnosis; lymphadenectomy negative for tumor
II	Microscopic residual disease; lymph nodes negative; tumor markers positive or negative
III	Gross residual or biopsy only; retroperitoneal nodes negative or positive; tumor markers positive or negative
IV	Distant metastases, including liver
OVARIAN	
I	Limited to ovary or ovaries; peritoneal washings negative for malignant cells; no disease beyond ovaries; presence of gliomatosis peritonei[a] does not result in changing Stage I disease to a higher stage; tumor markers normal after appropriate half-life decline (AFP, 5 days; β-hCG, 1616 h)
II	Microscopic residual or positive lymph nodes (≤2 cm); peritoneal washings negative for malignant cells; presence of gliomatosis peritonei does not result in changing Stage II disease to a higher stage; tumor markers positive or negative
III	Lymph node involvement (metastatic nodule) >2 cm; gross residual or biopsy only; contiguous visceral involvement (omentum, intestine, bladder); peritoneal washings positive for malignant cells; tumor markers positive or negative
IV	Distant metastases, including liver
TESTICULAR	
I	Limited to testes; complete resection by high inguinal orchiectomy or trans-scrotal orchiectomy with no spill at surgery; no clinical, radiographic, or histologic evidence of disease beyond the testes; tumor markers normal after appropriate half-life decline; patients with normal or unknown tumor markers at diagnosis must have a negative ipsilateral retroperitoneal node dissection to confirm Stage I disease
II	Trans-scrotal orchiectomy with gross spill of tumor; microscopic disease in scrotum or high in spermatic cord (≤5 cm from proximal end); retroperitoneal lymph node involvement (≤2 cm); increased tumor markers after appropriate half-life decline
III	Retroperitoneal lymph node involvement (>2 cm) but no visceral or extra-abdominal involvement
IV	Distant metastases, including liver

[a] *Peritoneal implants that contain only mature glial elements with no malignant elements.*
AFP, α-fetoprotein; β-hCG, beta subunit of human chorionic gonadotropin.

TREATMENT

The treatment strategies for GCTs depend on:

- Histologic subtype.
- Site of origin.
- Stage of disease.

Surgical resection is essential for treatment of most germ cells. It is the treatment of choice for benign teratomas, immature teratomas, and low-stage malignant GCTs. Chemotherapy has significantly improved outcome for children with malignant GCTs.

Teratoma

Pathologically, teratomas can be divided into three main groups for the purpose of therapeutic strategies.

Mature Teratoma

These tumors contain mature tissue, and the treatment is surgical excision, irrespective of the site. It is critically important that multiple histologic sections be examined to exclude the presence of focal immature tissue and malignant germ cell elements.

Immature Teratoma

These tumors contain elements of immature tissue and are graded histologically (see Table 29.1). The treatment of choice is surgical resection. Most immature teratomas in prepubertal children do not respond to chemotherapy. However, if AFP is increased, a malignant germ cell component may be present and chemotherapy may be indicated.

Teratoma with Malignant Germ Cell Elements

These tumors contain foci of malignant tissue, usually embryonal carcinoma, YST, or choriocarcinoma. These lesions are treated in an aggressive multimodal fashion. Chemotherapy improves local and metastatic control. Specific chemotherapy guidelines are discussed later in this chapter under GCTs with malignant elements. The benign teratomatous portion will not respond to chemotherapy or radiation therapy and must be treated surgically.

Sacrococcygeal Teratoma

Sacrococcygeal teratomas (SCTs) are the most common GCTs of childhood and account for 44% of all GCTs and 78% of extragonadal tumors. These tumors are relatively rare with an incidence of 1 in 40,000 live births. Females are more frequently affected. As most of these neoplasms are exophytic and are visible externally, approximately 80% are diagnosed within the first month of life. Sacrococcygeal tumors can be diagnosed antenatally on fetal ultrasound. Large tumors can be associated with congestive cardiac failure and hydrops fetalis due to arteriovenous shunting within the tumor. Tables 29.5 and 29.6 describe, respectively, the anatomic location and histologic grading of SCTs.

Though the majority of SCTs are benign, it is difficult to separate treatment discussion into benign and malignant tumors. Seventeen percent are malignant at diagnosis and 5% have distant metastases. The histology of the

TABLE 29.5   Anatomic Location of Sacrococcygeal Teratoma

Type	Location	Malignant histology (%)
I	Tumor predominantly external (sacrococcygeal) with only a minimal presacral component (the most common type)	8
II	Tumor presenting externally but with a significant intrapelvic extension (second in frequency to type I)	21
III	Tumor minimal external component but with the predominant mass extending into the pelvis and abdomen	34
IV	Tumor internalized presacral tumor without external presentation	38

malignant component of SCT is almost always YST. The incidence of a malignant component of the tumor is related to the type of the tumor, 38% in type IV versus 8% in type I. The average age at diagnosis in patients with metastatic disease is 22 months of age. Congenital anomalies, musculoskeletal and CNS defects are seen in up to 18—20% of patients.

Treatment

Surgery

Complete surgical excision should be performed, which necessitates removal of the entire coccyx. In almost all cases, the tumor is attached to the coccyx and failure to remove the coccyx results in local tumor recurrence in 30—40% of cases.

Chemotherapy

In patients with complete removal with grade 0 or 1 histology, no further therapy is required. However, continuous monitoring of these patients is required because malignant GCT can recur either from missed malignant tissue in the original tumor or malignant conversion of residual benign teratoma. In patients with histologic grade 3, malignant foci, or elevated tumor markers for age, postsurgical chemotherapy may be required (Tables 29.7 and 29.8). However, the benefit of chemotherapy in immature teratomas has not been validated. The chemotherapy regimens described in these tables are applicable to all malignant GCTs. In patients with Stage I SCT that contain <5% YST elements careful observation for 2 years with monthly AFP levels is a reasonable approach. However, chemotherapy as an option should be discussed with the family.

In patients with elevated AFP and a tumor that cannot be primarily resected, neoadjuvant chemotherapy followed by surgery is indicated.

TABLE 29.6 Histologic Grading of Sacrococcygeal Teratoma

Grade	Description
0	All tissue mature; no embryonal tissue
1	Rare foci of embryonal tissue not exceeding one low-power field per slide
2	Moderate quantity of embryonal tissue and some atypia but not exceeding three lower-power fields per slide
3	Large quantity of immature tissue exceeding three lower-power fields per slide, with abundant mitoses and cellular atypia

TABLE 29.7 Various Chemotherapy Regimens for Germ Cell Tumors

Regimen 1 (PEB)	<12 Months of Age (Every 3 Weeks)	≥12 Months of Age (Every 3 Weeks)
Cisplatin (P)	0.7 mg/kg/IV days 1—5	20 mg/m^2/IV on days 1—5
Etoposide (E)	3 mg/kg/IV days 1—5	100 mg/m^2/IV over days 1—5
Bleomycin (B)	0.5 units/kg/IV day 1	15 units/m^2/IV on day 1[a]

OR

REGIMEN 2 (JEB)	
Carboplatin	600 mg/m^2/IV day 1
Etoposide	120 mg/m^2/IV days 1—3
Bleomycin	15 units/m^2 day 3

[a]Adult treatment regimens include weekly bleomycin (30 units) during chemotherapy.
PEB, cisplatin, etoposide and bleomycin; JEB, carboplatin, etoposide and bleomycin.
Disease is evaluated in week 12 after four cycles; if complete remission, then discontinue chemotherapy; if partial remission, then surgery (week 12); if residual viable disease, consider relapse therapy.

TABLE 29.8   Chemotherapy Regimens for Malignant Germ Cell Tumors

Site	Histology	Therapy
Testicular	Stage I	Radical inguinal orchiectomy and observation[a]
Testicular	Stage II	PEB or JEB every 3 weeks for four cycles
Testicular	Stages III and IV	PEB or JEB every 3 weeks, for 4 cycles followed by surgery.
		If no disease (CR), discontinue therapy; if PR, if viable tumor consider relapse therapy
Ovarian	Stage I	Unilateral salpingo-oophorectomy and observation possible[b]
Ovarian	Stages I and II	PEB or JEB every 3 weeks for four cycles
Ovarian	Stages III and IV	PEB every 3 weeks, for four cycles followed by surgery
		If CR, discontinue therapy; if viable tumor consider relapse therapy
Malignant extragonadal germ cell tumor	Stages I–IV	PEB every 3 weeks, for four cycles followed by surgery[c]
		If CR, discontinue therapy; if viable tumor consider relapse therapy
Immature teratoma[a]		Surgery alone[d]

[a]If AFP is increased, chemotherapy as per Table 29.7.
[b]There is option to give chemotherapy.
[c]Observation for Stage I extragonadal tumors has not been studied.
[d]Chemotherapy has been used in adolescent girls and women with Stages II–IV immature teratoma. Data are limited.
CR, complete remission; PR, partial remission.

Prognosis

The prognosis of SCT depends on the following factors:

- Age.
- Surgical resectability.
- Histological grading.

Patients under 2 months of age have a favorable outlook, because only 7% of tumors in females and 10% in males are malignant at that age. Patients over 2 months of age have a less favorable outlook; the incidence of malignancy is 48% in females and 67% in males.

Patients whose tumors are initially resectable have a more favorable outcome. Patients whose tumor cannot be resected do poorly, even though these lesions are initially responsive to chemotherapy.

The histologic grading is extremely important. Patients with grade 0, 1, and 2 lesions have a 90–95% cure rate with complete surgical excision. Patients with grade 3 or malignant elements have an approximately 45–50% 2-year disease-free survival when treated with surgery and chemotherapy.

Ovarian Teratoma

These tumors comprise 40–50% of the ovarian tumors seen in childhood and adolescence. Many are classified as immature teratomas. The clinical behavior of immature ovarian teratoma correlates well with the histologic grading but poorly with the stage.

Treatment

Surgery

Surgery is essential, especially for immature teratomas that do not have malignant elements. Pure teratomas and immature teratomas do not usually respond to chemotherapy. The surgical approach is similar to that for other ovarian tumors. The incidence of bilateral involvement is unusual, but the contralateral ovary should be inspected and biopsied if abnormalities are found on inspection and palpation.

Chemotherapy

There are rare reports in adolescents that high grade (2 and 3) immature teratomas may respond to chemotherapy as described below for malignant GCT.

Prognosis

Outcomes are generally very good, although evidence of overt malignant elements (YST, embryonal carcinoma, choriocarcinoma) requires intensive chemotherapy to be administered. Subsequent surgery may be needed to remove residual nonmalignant elements.

Mediastinal Teratoma

Mediastinal teratomas are located in the anterior mediastinum. The average age of the pediatric patient is 3 years. They are more common in males. Teratoma subtypes (mature or immature) comprise the bulk of the tumor but YST and choriocarcinoma may occasionally be seen. Although most of these tumors are benign lesions, mediastinal teratomas occasionally have sarcomatous foci resembling rhabdomyosarcoma or undifferentiated sarcoma. These foci are extremely aggressive, therefore appropriate metastatic survey including serum markers (AFP and β-hCG and LDH) is recommended.

Treatment

Surgery

Patients with benign lesions are treated successfully with surgical excision. Patients who have malignant lesions (grade 2 or 3) have tumors that are not generally amenable to complete surgical excision because of infiltration of surrounding vital structures. Surgical debulking must be attempted after reduction with chemotherapy.

Chemotherapy

Aggressive chemotherapy is used for those with foci of malignant elements. The combinations appropriate for patients with malignant germ cell elements are shown in Table 29.7. If other elements are present, such as rhabdomyosarcoma, therapy must be adjusted to treat these more malignant elements.

Prognosis

The two critical prognostic factors in mediastinal GCTs are age and histology. Mediastinal teratomas in patients under 15 years of age tend not to have malignant elements or metastasize. In older patients, these tumors often contain YST, embryonal carcinoma, or choriocarcinoma and have a worse prognosis, with 50% of patients developing metastases within 1 year.

GCTs with Malignant Elements

Chemotherapy has an important role in the treatment of pediatric GCTs. Most of the treatment recommendations come from pediatric trials (age <15 years) and adult testicular GCT trials. Tables 29.7 and 29.8 reflect treatment recommendations for children and adolescents under 15 years of age. The treatment of those over 15 years is more complex and may require adherence to adult GCT guidelines. Chemotherapy also has a role in a neoadjuvant setting where primary surgical excision is not possible.

Germinoma

Germinoma is the most common pure malignant GCT. When it arises in extragonadal sites such as the pineal region, anterior mediastinum, and retroperitoneum, it is termed germinoma. It has been designated seminoma when it occurs in the testes and dysgerminoma when it involves the ovary. Germinomas comprise 10% of ovarian tumors in children and 15% of all GCTs. Seminomas are the most common malignancy found in undescended testes. Ovarian dysgerminomas are sometimes associated with precocious sexual development, but the majority of patients are developmentally normal.

Ovarian Dysgerminoma

Rapid development of signs and symptoms of an abdominal mass is the usual presentation. Abdominal pain is not common unless torsion is present. Seventy-five percent of patients have Stage I disease at presentation. Patterns of spread include contiguous extension, metastasis to regional lymph nodes and rarely to liver and lungs.

Treatment

Surgery

Dysgerminomas have a high incidence of bilateral ovarian involvement (5–10%). Bilateral involvement is particularly common in women with a Y chromosome and gonadal streaks.

Conservative surgery consisting of unilateral salpingo-oophorectomy and wedge biopsy of the contralateral ovary, as well as sampling of regional lymph nodes, is recommended for patients with the following criteria:

1. Stage I unilateral encapsulated tumor less than 10 cm in diameter.
2. Normal contralateral ovary.
3. No evidence of retroperitoneal lymph node metastases.
4. No ascites; negative cytology of peritoneal washing.
5. Normal female 46XX karyotype.

Patients with Stage I disease require no further postsurgical therapy.

In the past, the standard surgical management of patients with Stages II and III disease had been total abdominal hysterectomy and bilateral salpingo-oophorectomy followed by postoperative radiotherapy. However, because the tumor most commonly affects children and young women, a program employing limited surgery and chemotherapy is more appropriate.

Chemotherapy

Dysgerminoma and seminoma (testicular tumor analogous to dysgerminoma) are responsive to combination chemotherapy. Tables 29.7 and 29.8 list the therapy regimens used for the treatment of GCTs. When dysgerminomas occur in combination with other malignant germ cell elements such as malignant teratomas, YST, or embryonal carcinoma they are called mixed GCTs. Management should be based on the most malignant component present.

The primary chemotherapy approach in patients with Stages II and III disease, combined with limited surgery, has been effective in children. This approach has the advantage of preserving as much reproductive and endocrine function as possible without compromising long-term survival. It is recommended for pediatric and adolescent patients with 46XX karyotype.

Radiotherapy

Dysgerminoma is the most radiosensitive of the ovarian GCTs. However, radiation therapy is reserved for patients with persistent disease after chemotherapy.

Prognosis

The prognosis for ovarian dysgerminoma correlates with extent of disease at diagnosis as shown in following table:

Extent of disease	5-year survival (%)
Stage I	95
Stage II	75
Stage III	60
Stage IV	33

Extragonadal Germinoma

Incomplete migration of germ cells seems to be responsible for the origin of extragonadal GCTs. Most extragonadal germinomas occur in the mediastinum. In contrast to other GCTs in the mediastinum, hematogenous metastases are rare and the primary therapeutic strategy is aimed at local control.

Treatment

Surgery

Although excellent cure rates of 70–80% have been reported with mediastinal germinomas treated by surgical excision, most tumors are not resectable because of their size and/or proximity to great vessels and vital structures. In most cases, biopsy or minimal debulking is the only feasible surgical approach.

Radiotherapy

These tumors are radiosensitive and radiation to the mediastinum in the range of 4500–5000 cGy cures approximately 50–60% of patients. Again, radiation may be an option for patients who progress on chemotherapy.

Chemotherapy

Combination chemotherapy has been employed in advanced extragonadal germinomas. Cisplatin is particularly effective with excellent results. The chemotherapy regimens employed are shown in Table 29.7. In patients with large bulky mediastinal masses, an approach using combination chemotherapy with or without radiotherapy (depending on the response to two to three courses of chemotherapy) may improve the results compared to patients treated with radiotherapy alone.

Endodermal Sinus Tumor (YST)

YST is the most common histology in children. It is often the only malignant element and is frequently found in immature teratomas. When present in an extragonadal site, YST behaves in a highly malignant fashion. It is important to recognize that foci of YST present in immature teratoma require aggressive chemotherapy similar to that for pure YST. The sacrococcygeal area is the major site of involvement in the newborn and infant; the ovary is the most common location for YSTs in older children and adolescents. Testicular YSTs have two peaks of incidence in infancy and adolescence. They are the most frequent malignant testicular tumors in young boys.

Testicular YST

This tumor is localized (Stage I) in 85% of cases. The overall survival rate is higher than 85% which seems to correlate with age. The management of this tumor is considered separately in two age groups: prepubertal and postpubertal. Tumors in these two age groups behave differently clinically and have distinct molecular profiles.

Prepubertal Males

Surgery

The definitive surgical treatment is inguinal orchiectomy with high ligation of the spermatic cord. Retroperitoneal lymph node dissection is not indicated in young patients with disease limited to the testes because less than 5–6% will have positive retroperitoneal lymph nodes. Patients should be staged with CT of chest and abdomen. Any abdominal nodal or metastatic disease can be treated successfully with chemotherapy alone. Infants with a negative metastatic workup and normal AFP and β-hCG require no further therapy following orchiectomy. Close observation with serum AFP for 2 years is indicated. Radiologic evaluation of chest and abdomen should be used sparingly in this population, as AFP is an excellent marker.

Prognosis

In prepubertal boys with normal AFP and β-hCG following orchiectomy, the cure rate is 70–85% with surgery alone. In the small number of patients with recurrence, combination chemotherapy can produce subsequent cure. In higher-stage disease treated with chemotherapy, overall survival at 5 years approaches 100%.

Postpubertal Males

Surgery

Radical orchiectomy should be combined with appropriate staging. Historically, in patients with no evidence of retroperitoneal adenopathy, retroperitoneal lymphadenectomy (RPLND) has been recommended. There is controversy concerning the need for RPLND which has been well documented in adult literature. Some institutions observe CT-negative patients who did not undergo RPLND, while others administer two cycles of chemotherapy. In patients with Stage II or III disease, initial chemotherapy followed by surgical debulking and possibly retroperitoneal lymphadenectomy should be done. Adolescent testicular GCTs may have residual benign lesions after chemotherapy. These lesions should be removed as they pose a risk for recurrence as malignant GCT.

Chemotherapy

In patients with Stage I disease, no further therapy is given. In patients with Stage II, III, or IV disease, combination chemotherapy is administered. Tables 29.7 and 29.8 describe the dose for prepubertal patients. Adolescents may benefit from additional bleomycin (weekly).

Ovarian Yolk Sac Tumor

Ovarian YSTs occur in both prepubertal and postpubertal females and are characterized by rapid growth with intrapelvic and intra-abdominal spread with widespread involvement of the peritoneal surfaces. YSTs spread rapidly to lymphatic and peritoneal structures with a short duration of symptoms, high frequency of abdominal pain, and high frequency of high-stage disease at diagnosis. Distant metastases are seen in liver, lungs, lymph nodes, and rarely in bones. Most patients have elevated AFP levels.

Treatment

Surgery

Unilateral salpingo-oophorectomy is performed for patients with Stage I disease. In patients with more extensive disease, debulking should be performed, including omentectomy and retroperitoneal lymph node sampling. For unilateral disease, unilateral oophorectomy with biopsy of uninvolved ovary should be performed. Bilateral disease evaluation includes bilateral oophorectomy. Peritoneal washings should be obtained in all cases.

Chemotherapy

Chemotherapy is currently indicated in most cases postoperatively (Tables 29.7 and 29.8). Stage I ovarian YST may be observed. In a recent Children's Oncology Group trial, approximately 50% were treated successfully with surgery and observation. The salvage rate was near 98%. The option of minimal chemotherapy or observation can be discussed with the family. In patients who have unresectable gross disease at initial surgery, four cycles of chemotherapy should be followed by second-look surgery to confirm eradication of disease and to remove residual disease if present.

Radiotherapy

Radiotherapy does not play a major role in management.

Prognosis

The prognosis depends on the stage and the chemotherapeutic regimen employed, but with platinum-based combination therapy disease-free survival may be as high as 80%.

Embryonal Carcinoma

Embryonal carcinoma occurs either in pure form or in combination with a mixed GCT (malignant teratoma). In either case, the management is the same.

Testicular Embryonal Carcinoma

Embryonal carcinoma occurs more commonly in late adolescence or in early adulthood. The presenting symptoms include enlarging scrotal mass, metastatic abdominal or mediastinal disease, or localized peripheral lymphadenopathy. Serum AFP or β-hCG may be elevated. The therapy required for cure depends on stage of disease and quantitation of AFP.

Treatment

Surgery

The standard surgical approach for patients with clinical Stage I and II disease is radical orchiectomy with high ligation of the spermatic cord and retroperitoneal lymphadenectomy as described previously for postpubertal males.

Prognosis

The cure rate for patients with testicular embryonal carcinoma is excellent, provided that the appropriate chemotherapeutic regimens are employed and that patients who are treated with surgery alone have careful follow-up evaluation, including radiographic evaluation and serial AFP and β-hCG determinations to detect recurrence.

Ovarian Embryonal Carcinoma

Ovarian embryonal carcinoma is distinct from YST of the ovary. This rare tumor resembles embryonal carcinoma of the testis histologically and resembles it biochemically (with secretion of AFP and β-hCG). It occurs predominantly in adolescent females with a mean age of 15 years. The treatment approach is the same as that of an ovarian YST.

Relapsed and Resistant GCTs

The prognosis for children with malignant GCTs is excellent. However, there are patients who do not respond or relapse after therapy. They are usually found to have persistent marker elevation or significant marker elevation after treatment. Recurrent GCTs often respond to various chemotherapy regimens, such as carboplatin, ifosfamide, and etoposide or paclitaxel, ifosfamide, and carboplatin. However, surgical resection is essential. Though there are minimal data from pediatric trials, some patients whose recurrent disease is responsive to chemotherapy may be candidates for ablative chemotherapy and autologous hematopoietic stem cell transplantation.

Further Reading and References

Balmaceda, C., Heller, G., Rosenblum, M., et al., 1996. Chemotherapy without irradiation a novel approach for newly diagnosed CNS germ cell tumors: results of an international cooperative trial. J. Clin. Oncol. 12, 2908–2915.

Billmire, D.F., Cullen, J.W., Rescorla, F.J., Davis, M., Schlatter, M.G., Olson, T.A., et al., 2014. Surveillance after initial surgery for pediatric and adolescent girls with Stage I ovarian germ cell tumors: report from the Children's Oncology Group. J. Clin. Oncol. 32 (5), 465–470.

Cushing, B., Perlman, E., Marina, N., Castleberry, R.P., 2002. Germ cell tumors. In: Pizzo, P.A., Poplack, D.G. (Eds.), Principles and Practice of Pediatric Oncology, fourth ed. Lippincott–Raven, Philadelphia.

Cushing, B., Giller, R., Cullen, J., et al., 2004. Randomized of combination chemotherapy with etoposide, bleomycin and either high-dose or standard-dose Cisplatin in children and adolescents with high-risk malignant germ cell tumors: a pediatric intergroup study—Pediatric Oncology Group 9049 and Children's Cancer Group 8882. J. Clin. Oncol. 22 (13), 2691–2700.

De Backer, A., Madern, G.C., Oosterhuis, J.W., Hakvoort-Cammel, F.G., Hazbroek, F.W., 2006. Ovarian germ cell tumors in children: a clinical study of 66 patients. Pediatr. Blood Cancer 46 (4), 459–464.

Einhorn, L.H., Williams, S.D., Chamness, A., Brames, M.F., Perkins, S.M., Abonour, R., 2007. High-dose chemotherapy and stem cell rescue for metastatic germ-cell tumors. N. Engl. J. Med. 357 (45), 340–348.

Hawkins, E.P., Perlman, E., 1996. Germ cell tumors in childhood: morphology and biology. In: Parham, D.M. (Ed.), Pediatric Neoplasia: Morphology and Biology. Raven Press, New York. pp. 297.

International Germ Cell Cancer Collaborative Group, 1997. International Germ Cell Consensus Classification: a prognostic factor-based staging system for metastatic germ cell tumors. J. Clin. Oncol. 15 (2), 594–603.

Mann, J.R., Raafat, F., Robinson, K., et al., 2000. The United Kingdom's Cancer Study Group's second germ cell study: carboplatin, etoposide, and bleomycin are effective treatment for children with malignant extracranial germ cell tumors, with acceptable toxicity. J. Clin. Oncol. 18 (22), 3809–3818.

Mann, J.R., Gray, E.S., Thornton, C., et al., 2008. Mature and immature extracranial teratomas in children: The UK Children's Cancer Study Group experience. J. Clin. Oncol. 26 (21), 3590–3597.

Marina, N., London, W.B., Frazier, A.L., et al., 2006. Prognostic factors in children with extragonadal germ cell tumors: a pediatric intergroup study. J. Clin. Oncol. 24 (16), 2544–2548.

Marina, N.M., Cushing, B., Giller, R., Cohen, L., et al., 1999. Complete surgical excision is effective treatment for children with immature teratomas with or without malignant elements: a Pediatric Oncology Group/Children's Cancer Group intergroup study. J. Clin. Oncol. 17, 2137–2143.

Motzer, R.J., Nichols, C.J., Margolin, K.A., et al., 2007. Phase III randomized trial of conventional-dose chemotherapy with or without high-dose chemotherapy with autologous hematopoietic stem-cell rescue as first-line treatment for patients with poor-prognosis metastatic germ cell tumors. J. Clin. Oncol. 25 (3), 247–256.

Nichols, C.G., Fox, E.P., 1990. Extragonadal and pediatric germ cell tumors in children. Hematol. Oncol. Clin. 5, 1189–1209.

Pizzo, P.A., Poplack, D.G., 2002. Principals and Practice of Pediatric Oncology, fourth ed. Lippincott–Raven, Philadelphia.

Rogers, P.C., Olson, T.A., Cullen, J.W., et al., 2004. Treatment of children and adolescents with stage II testicular and stages I and II ovarian malignant germ cell tumors: A pediatric intergroup study—Pediatric Oncology Group 9048 and Children's Cancer Group 8891. J. Clin. Oncol. 22 (17), 3563–3569.

Schlatter, M., Rescorla, F., Giller, R., et al., 2003. Excellent outcome in patients with stage I germ cell tumor of the testes: a study of the Children's Cancer Group/Pediatric Oncology Group. J. Pediatr. Surg. 38 (3), 319–324.

30

Hepatic Tumors

Louis B. Rapkin and Thomas A. Olson

Primary hepatic neoplasms are rare and account for only 1–2% of all childhood cancers. Hepatoblastoma accounts for approximately two-thirds of liver tumors in children. Hepatoblastoma and hepatocellular carcinoma (HCC) are the two most common malignancies that arise de novo in the liver. Table 30.1 lists the most prevalent malignant and benign liver tumors in children.

INCIDENCE

Table 30.2 lists the demographic features associated with hepatoblastoma and HCC. Hepatoblastoma occurs primarily in young children, with 80% of cases reported before 3 years of age. The incidence of hepatoblastoma has increased over the last 25 years. The cause for this increase is unknown, but increased survival of very-low-birth-weight infants may be a contributing factor. The association between the development of hepatoblastoma and low birth weight has been well documented.

EPIDEMIOLOGY

The etiology of hepatoblastoma and HCC is unknown. Certain disorders increase the risk of liver cancer (see Table 30.3). Congenital anomalies which have been reported with hepatoblastoma include:

- Hemihyperplasia syndromes (formerly hemihypertrophy) including Beckwith–Wiedemann syndrome.
- Meckel's diverticulum.
- Congenital absence of adrenal gland.
- Congenital absence of kidney.
- Umbilical hernia.

Patients with Beckwith–Wiedemann syndrome and isolated hemihyperplasia should be screened every 3 months for hepatoblastoma with measurements of α-fetoprotein (AFP) tumor markers and abdominal ultrasounds until age 7 years. Though there is an association between hepatoblastoma and familial adenomatous polyposis (FAP), hepatoblastoma occurs in less than 1% of members of families with FAP. Screening in these families is controversial, but in children with *APC* gene (adenomatous polyposis coli) mutations it may be warranted.

Hepatic tumors have a wide geographic variation in incidence:

1. They are the third most common abdominal cancer in Japan.
2. They are seen more frequently in Asian and African children.

The geographic variation is thought to reflect the etiologic role of environmental conditions.

Lanzkowsky's Manual of Pediatric Hematology and Oncology.
DOI: http://dx.doi.org/10.1016/B978-0-12-801368-7.00030-2

TABLE 30.1   Malignant and Benign Liver Tumors in Children

Malignant	Benign
Hepatoblastoma	Hemangioma
Hepatocellular carcinoma	Hemangioendothelioma
Rhabdomyosarcoma	Angiomyolipoma
Undifferentiated embryonal sarcoma	Hamartoma
Angiosarcoma	Biliary cyst
Mesenchymal (mixed)	Adenoma
Sarcoma	Teratoma
Rhabdoid tumor	Myofibroblastic tumor
Yolk sac tumor	
Leiomyosarcoma	

TABLE 30.2   Demographic Features Associated with Hepatoblastoma and Hepatocellular Carcinoma in Children

Host factor	Hepatoblastoma	Hepatocellular carcinoma
Incidence		
White males	1.4 per million	0.5 per million
African American males	0.9 per million	0.0 per million
White females	0.5 per million	0.9 per million
African American females	0.0 per million	0.0 per million
Median age	1 year	12 years
Male:female ratio	1.7:1.0	1.0:1.1

TABLE 30.3   Disorders Associated with Increased Risk of Hepatoblastoma and Hepatocellular Carcinoma

Hepatoblastoma	Hepatocellular carcinoma
Low-birth-weight infant	Familial cholestatic cirrhosis of childhood
Von Gierke disease	Ataxia telangiectasia
Congenital cystathioninuria and hemihyperplasia	Biliary cirrhosis due to bile duct atresia
Maternal use of hormonal therapy	Cirrhosis following giant cell hepatitis
Exposure to metals such as in welding and soldering fumes	Chronic carrier of hepatitis B virus
Beckwith–Wiedemann syndrome	Hereditary tyrosinemia
Li–Fraumeni syndrome	Androgen therapy
Trisomy 18	Methotrexate therapy
Fetal alcohol syndrome	α_1-Antitrypsin deficiency
Gardner syndrome[a]	Fanconi anemia
Type I glycogen storage disease	Type 1 glycogen storage disease
Prader–Willi syndrome	Neurofibromatosis
	Soto syndrome[b]
	Familial adenomatous polyposis (FAP)
	Hepatoblastoma
	Alagille syndrome
	Wilms' tumor with liver metastases treated with hepatic radiation

[a]Gardner syndrome, a variant of familial adenomatous polyposis, is an autosomal syndrome associated with deletions on the long arm of chromosome 5. It is characterized by colonic polyps that undergo malignant change and benign and malignant extracolonic lesions. Tumors frequently associated with Gardner syndrome include carcinoma of the ampulla of Vater, papillary carcinoma of the thyroid, and, in children, hepatoblastoma. The childhood malignancies often precede the appearance of other manifestations by several years.
[b]A rare genetic disorder characterized by excessive physical growth during the first 2–3 years of life. They tend to be larger at birth. May have mild mental retardation, delayed development, and hypotonia.

PATHOLOGY

Hepatic tumors are divided into two major histologic types: hepatoblastoma and HCC. Other less frequent liver tumors include transitional liver cell tumor, undifferentiated embryonal sarcoma of liver, and infantile choriocarcinoma. The pathologic classifications for hepatoblastoma and HCC are as follows:

1. Hepatoblastoma.
   a. Epithelial type.
      i. Embryonal pattern.
      ii. Pure fetal pattern.
      iii. Macrotrabecular type.
      iv. Small cell undifferentiated type or anaplastic
   b. Mixed epithelial and mesenchymal type.
2. HCC.
3. Fibrolamellar HCC (a histologic variant of HCC and has a similar prognosis when adjusted for stage).

CLINICAL FEATURES

The most common sign of primary liver malignancy is an upper abdominal mass or generalized abdominal enlargement. Table 30.4 outlines the clinical features of the various pathologic types of hepatic tumors. Table 30.5 shows the frequency of signs and symptoms at diagnosis in children with hepatoblastoma and HCC. The clinical manifestations are similar in both.

TABLE 30.4 Clinical Features of the Various Pathological Types of Hepatic Tumors

Feature	Hepatoblastoma	Hepatocellular carcinoma	Fibrolamellar variant
Usual age of presentation	0–3 year	5–18 year	10–20 year
Associated congenital anomalies	Dysmorphic features	Metabolic	None
	Hemihyperplasia		
	Beckwith–Wiedemann syndrome		
Advanced disease at presentation	40%	70%	10%
Usual site of origin	Right lobe	Right lobe, multifocal	Right lobe
Abnormal liver function tests	15–30%	30–50%	Rare
Jaundice	5%	25%	Absent
Elevated AFP	60–70%	50%	10%
Positive hepatitis B serology	Absent	Present in some	Absent
Abnormal B12-binding protein	Absent	Absent	Present
Chromosomal abnormalities	11p15.5, 18, 17p13 occasional loss of heterozygosity	17q11.2, 20p12, 11q22-23 rare presence of TP53 mutation	Not reported
Distinctive radiographic appearance	None	None	None
Pathology hepatocytes	Fetal and/or embryonal cells + mesenchymal component	Large pleomorphic tumor cells and tumor giant cells	Eosinophilic with dense fibrous stroma

AFP, α-fetoprotein.
Source: *From Pizzo and Poplack (2002), with permission.*

TABLE 30.5　Frequency of Signs and Symptoms in Children with Hepatoblastoma and Hepatocellular carcinoma

Sign/symptom	Hepatoblastoma (%)	Hepatocellular carcinoma (%)
Abdominal mass	80	60
Abdominal distention	27	34
Anorexia	20	20
Weight loss	19	19
Abdominal pain	15	21
Vomiting	10	10
Pallor	7	Rare
Jaundice	5	10
Fever	4	8
Diarrhea	2	Rare
Constipation	1	Rare
Pseudoprecocious puberty	Occasional	Not reported

DIAGNOSTIC EVALUATION

Patients should have the following evaluations:

- History
- Physical examination
- Complete blood count (anemia with moderate leukocytosis is commonly seen; thrombocytosis with platelet counts greater than 500,000/mm^3 is the most frequent abnormality)
- Urinalysis
- Liver profile and electrolytes
- Fibrinogen, partial thromboplastin time, and prothrombin time
- Hepatitis B surface antigen (HB$_s$Ag), core antigen (HB$_c$Ag), and core antibody
- AFP
- Beta subunit of human chorionic gonadotropin (β-hCG)
- Carcinoembryonic antigen
- Radiographic evaluation of intrahepatic disease
  - Sonogram
  - Abdominal computed tomography (CT)
  - Magnetic resonance imaging (MRI)
  - MRI angiogram
  - MRI cholangiogram
- Radiographic evaluation of extrahepatic disease
  - Chest radiographs (posteroanterior and lateral)
  - Chest CT
  - Bone scan
- Bone marrow aspirate/biopsy.

The diagnostic evaluation should determine:

- Extent of intrahepatic disease.
- Potential for hepatic respectability.
- Presence or absence of extrahepatic disease.

MRI angiography and MRI cholangiogram are useful complements to CT for evaluation of intrahepatic disease because they aid in determining surgical resectability. In hepatoblastoma, the right lobe of the liver is more commonly involved than the left, but the tumor involves both lobes in 30% of patients.

Approximately 10−20% of patients with hepatic tumors have demonstrable pulmonary metastases on chest CT.

TABLE 30.6 Staging of Hepatic Tumors (Based on Postsurgical Findings) in Children and Percentage of Cases

Stage	Description	Percentage of cases by stage
I	Complete resection of tumor by wedge resection lobectomy or by extended lobectomy as initial treatment	25
II	Tumors rendered completely resectable by initial radiotherapy or chemotherapy	4
	Residual disease confined to one lobe	
III	A. Gross residual tumor involving both lobes of liver	48
	B. Regional lymph node involvement	
IV	Metastatic disease irrespective of extent of liver involvement	23

TABLE 30.7 PRETEXT (based on presurgical findings) Staging for Hepatoblastoma

PRETEXT stage	Description
Stage 1	Tumor involves one quadrant. Three adjoining quadrants are free of disease
Stage 2	Tumor involves two adjoining quadrants with remaining two free of disease
Stage 3	Tumor involves three adjoining quadrants or two nonadjoining quadrants. One quadrant or two nonadjoining quadrants are free of disease
Stage 4	Tumor involves all four quadrants. No quadrant is free of disease.

STAGING

Staging of hepatic tumors is complex. Both presurgical (PRETEXT) and postsurgical staging systems are described below:

- System based on postsurgical findings: Based on the degree of resectability of the primary lesion and the presence or absence of metastatic disease. This staging system for hepatic tumors, used in the United States, is shown in Table 30.6.
- System based on presurgical findings: Extent of disease (PRETEXT) at diagnosis, used in European protocols, is shown in Table 30.7. This system was designed for international treatment programs in which only patients with PRETEXT stage 1 would undergo an initial attempt at resection. All other patients would receive chemotherapy prior to surgery. There is considerable interobserver variability and some patients with potential postsurgical (United States) stage 1 would be upstaged in PRETEXT system and have surgery after chemotherapy. Whichever approach is used, the outcomes are similar.

TREATMENT

Hepatoblastoma

Surgery

Only patients in whom complete resection can be achieved have a reasonable chance of cure. There are three appropriate surgical options: initial surgical resection; delayed surgical resection after chemotherapy; and orthotopic liver transplantation. Chemotherapy plays an important role, not only in eradicating subclinical metastases in completely resected disease, but also may allow unresectable disease to become resectable. Timing of surgery is essential as is early referral to a pediatric liver surgeon (with liver transplant experience). In the United States, an initial surgical resection in appropriate patients is preferred (see below).

Patients who are suitable candidates for complete resection include those with:

1. Tumors confined to the right lobe.
2. Tumors originating in the right lobe that do not extend beyond the medial segment of the left lobe.
3. Tumors confined to the left lobe.

Patients who have tumor involvement of both lobes are not candidates for curative surgical resection. In these cases, biopsy should be performed. The European approach is different. Patients are staged by PRETEXT and neoadjuvant chemotherapy is administered prior to surgery to all patients except PRETEXT stage 1. Surgery is recommended for limited metastatic disease (especially in lung). This surgery is often done at time of liver tumor resection.

Radical hepatic resection results in many potential postoperative complications, including the following:

- Hypovolemia
- Hypoglycemia
- Hypoalbuminemia
- Hypofibrinogenemia and deficiency of coagulation proteins
- Hyperbilirubinemia persists for 2–4 weeks after resection, and hepatic regeneration is complete by 1–3 months postsurgery.

Chemotherapy

Stages I and II: Completely resected with pure fetal histology require no chemotherapy. Other histologic patterns require combination chemotherapy.
Stage III: Chemotherapy followed by resection (including orthotopic liver transplantation if not resectable).
Stage IV: Same as for Stage III and resection of pulmonary metastases.

Combination chemotherapy plays several roles in the management of hepatoblastoma:

1. Adjuvant therapy for patients who have undergone complete resection, because its use improves disease-free survival.
2. Preoperative therapy for patients who have initially unresectable disease to shrink the primary tumor.
3. Palliative therapy for patients with metastatic disease at diagnosis.

Table 30.8 lists three commonly used chemotherapy regimens for liver tumors. Regimen 2 has recently been shown to be as effective as regimen 1 with an equivalent survival. Regimen 2 eliminates the use of doxorubicin

TABLE 30.8　Various Combination Chemotherapy Regimens for Liver Tumors

Regimen 1 (every 3–4 weeks)
　　Doxorubicin 25 mg/m^2/day × 3 days continuous infusion by central line:
　　Cisplatin 20 mg/m^2/day × 5 days continuous infusion, days 0–4 (six cycles of chemotherapy)
　　Patients less than 10 kg receive chemotherapy according to weight:
　　Doxorubicin 0.83 mg/kg × 3 days; cisplatin 0.66 mg/kg/day for 5 days (six cycles of chemotherapy)
　　　　OR
Regimen 2 (every 3–4 weeks)
　　Cisplatin 100 mg/m^2 IV infused over 6 h, day 1
　　Vincristine 1.5 mg/m^2 IV (max 2 mg), day 3, 10, 17
　　Fluorouracil 600 mg/m^2 IV, day 3
Under 1 year of age:
　　Cisplatin 3.3 mg/kg and vincristine 0.05 mg/kg
　　After four cycles, attempt surgical resection, followed by four more cycles
　　　　OR
Regimen 3 (every 3–4 weeks)
　　Cisplatin 90 mg/m^2 IV infused over 6 h (hours 0–6), day 0
　　Doxorubicin 20 mg/m^2/day continuous infusion × 4 days, days 0–3
Under 1 year of age:
　　Cisplatin 3 mg/kg
　　Doxorubicin 0.66 mg/kg
After four cycles, attempt surgical resection followed by four additional courses of chemotherapy

and has minimal toxicity. A recent pilot study by International Society of Paediatric Oncology intensified cisplatin (every 2 weeks) in patients with high-risk disease. Overall survival was 83%. However, controversy still remains as to the most effective regimen, and international co-operative trials are needed. Chemotherapy has been administered postsurgically after resection (including post-transplantation) as part of ongoing clinical trials.

Radiation

Radiotherapy is not curative for intrahepatic disease because the effective tumor dose exceeds hepatic tolerance. However, radiotherapy is useful in shrinking unresectable disease or microscopic residual disease. Radiation dosages used to treat hepatic tumors range from 1200 to 2000 cGy. Occasionally, higher doses to localized areas of tumor have been associated with tumor regression. Radiotherapy immediately after hepatic resection will limit hepatic regeneration.

Hepatocellular Carcinoma

Surgery

Complete surgical resection is essential for survival in HCC. Chemotherapy (cisplatin and doxorubicin) may be given in some cases to make surgical removal possible, although clinical trials have not shown that chemotherapy has improved outcomes in HCC.

Chemotherapy

Stages I and II: Adjuvant chemotherapy as described in Table 30.8.
Stage III: Chemotherapy followed by resection (including orthotopic liver transplantation if not resectable).
Stage IV: No chemotherapy combination has been effective, though cisplatin and doxorubicin have been used.

Liver Transplantation

Liver transplantation has been used successfully in selected cases of otherwise unresectable HCC.

FOLLOW-UP

Approximately 90% of children with hepatoblastomas and 50% of children with HCC have elevated AFP. AFP is a valuable marker for monitoring residual or metastatic disease following resection of the primary tumor or for monitoring the response of an unresectable primary tumor to therapy. AFP is normally elevated in the newborn period and then declines (see Appendix 2). Concentrations of AFP greater than 500,000 ng/ml are not unusual in hepatoblastoma. Failure of AFP to return to normal after surgery is an indication of incomplete tumor removal or metastases. Serial determinations of AFP are the most precise measurement of the effectiveness of hepatoblastoma therapy.

PROGNOSIS

The factors that determine prognosis include staging, histology, and tumor- and treatment-related factors. The overall survival rate is 70% for hepatoblastoma and 25% for HCC. Using the staging system shown in Table 30.6, patients with HCC Stage I have a good outcome and Stage IV are usually fatal. The 5-year survival rates for hepatoblastoma are as follows:

Stage	Percentage
I/II	90
III	60
IV	20

Patients with gross resection with pure fetal histology have an excellent prognosis without chemotherapy. The embryonal hepatoblastoma pattern requires chemotherapy. The small cell variant of hepatoblastoma has a poor prognosis. Hepatocellular carcinoma and the fibrolamellar variant have an unfavorable prognosis. Studies of DNA content have shown that diploid tumors with low proliferation index have a better prognosis than aneuploid tumors and high proliferative index. Patients who show a decline in AFP by 2 logs in the initial four courses of chemotherapy have a 75% chance of survival. An elevated β-hCG level is found in the occasional patient with HCC who has associated precocious puberty. However, β-hCG levels do not necessarily reflect the clinical course of the tumor.

Further Reading and References

Hepatic Tumors

Bilk, P., Superina, R., 1997. Transplantation for unresectable liver tumors in children. Transplan. Proc. 29, 2834–2836.

Buckley, J.D., Sather, H., Ruccione, K., Rogues, P.C.J., Haas, J.G., Henderson, B.E., Hammond, G.D., 1989. A case control study of risk factors for hepatoblastoma: a report from the Children's Cancer Study Group. Cancer 64, 1169–1176.

Douglass, E.C., Reynolds, M., Finegold, M., Cantor, A.B., Glicksman, A., 1993. Cisplatin, vincristine and fluorouracil therapy for hepatoblastoma: a Pediatric Oncology Group study. J. Clin. Oncol. 11, 96–99.

Katzentein, H.M., Krailo, M.D., Malogolowkin, M.H., et al., 2002. Hepatocellular carcinoma in children and adolescents: results from the Pediatric Oncology Group and the Children's Cancer Group intergroup study. J. Clin. Oncol. 20 (12), 2789–2797.

Litten, J.B., Tomlinson, G.E., 2008. Liver tumors in children. Oncologist 13 (7), 812–820.

Ortega, J.A., Krailo, M.D., Haas, J.E., King, D.R., Ablin, A.R., Quinn, J.J., et al., 1991. Effective treatment of unresectable or metastatic hepatoblastoma with cisplatin and continuous infusion doxorubicin chemotherapy: a report from the Children's Cancer Study Group. J. Clin. Oncol. 9, 2167–2176.

Ortega, J.A., Douglass, E., Feusner, J., et al., 2000. Randomized comparison of cisplatin/vincristine/5-flourouracil and cisplatin/continuous infusion doxorubicin for treatment of pediatric hepatoblastoma: a report from the Children's Cancer Group and the Pediatric Oncology Group. J. Clin. Oncol. 18 (14), 2665–2675.

Otte, J.B., Pritchard, J., Aronson, D.C., et al., 2004. Liver transplantation for hepatoblastoma: results from the International Society of Paediatric Oncology (SIOP) study—SIOPEL 1 and review of world literature. Pediatr. Blood Cancer 42 (1), 74–83.

Perilongo, G., Shafford, E., Mailbach, R., et al., 2004. Risk-adapted treatment for childhood hepatoblastoma. Final report of the second study of the International Society of Paediatric Oncology—SIOPEL 2. Eur. J. Cancer 40 (3), 411–421.

Pizzo, P.A., Poplack, D.G. (Eds.), 2002. Principles and Practice of Pediatric Oncology. fourth ed. JB Lippincott, Philadelphia.

Raney, B., 1997. Hepatoblastoma in children: a review. J. Pediatr. Hematol. Oncol. 19, 418–422.

Tomilinson, G.E., Finegold, M.J., 2002. Tumors of the liver. In: Pizzo, P.A., Poplack, D.G. (Eds.), Principles and Practice of Pediatric Oncology, fourth ed. Lippincott–Raven, Philadelphia.

Zsiros, J., Brugieres, L., Brock, P., et al., 2013. Dose-dense cisplatin-based chemotherapy and surgery for children with high-risk hepatoblastoma (SIOPEL-4): a prospective, single-arm, feasibility study. Lancet Oncol. 14 (9), 834–842.

31

Hematopoietic Stem Cell Transplantation

Indira Sahdev and Hisham Abdel-Azim

Hematopoietic stem cell transplantation (HSCT) has become an accepted therapeutic modality for the treatment of malignant and nonmalignant disorders. Tables 31.1 and 31.2 list the indications for allogeneic and autologous HSCT. Most allogeneic transplants performed in patients <20 years old are for acute leukemias (43%) or nonmalignant indications (35%). Preparation, or "conditioning," for HSCT involves the delivery of high-dose chemotherapy (HDC) with or without radiation to ablate or reduce hematopoiesis and to provide sufficient immunosuppression to allow donor cell engraftment.

The rationale for HDC involves the theory of the steep dose–response curve for many chemotherapeutic agents. Most drugs exhibit a log-linear relationship between tumor cell kill and dose over a certain range, followed by flattening of the curve in the upper dose ranges. For this reason, small changes in dose can produce a significant change in response to chemotherapeutic agents. For many chemotherapeutic agents, the major dose-limiting toxicity is myelosuppression. Sufficiently high doses of chemotherapy cannot be delivered out of concern for permanent damage to the hematopoietic system. While hematopoietic growth factor support offers the potential to maximize the dose–response of standard-dose chemotherapy, allogeneic HSCT or autologous hematopoietic stem cell (HSC) rescue offer the opportunity to exceed marrow tolerance. This permits the delivery of a higher dose of chemotherapeutic agents, thus achieving a higher peak on the tumor-kill versus drug dose curve. A 3- to 10-fold increase in drug dose may result in a multiple log increase in tumor cell killing. Generally, multiple drugs are used for the conditioning regimen in order to overcome tumor heterogeneity and drug resistance.

The most common type of HSCT to treat hematological malignancies and nonmalignant disorders is allogeneic, using a human leukocyte antigen (HLA)-matched histocompatible donor. Solid tumors have been treated with HDC followed by autologous HSC rescue with the concomitant use of hematopoietic growth factors such as granulocyte colony-stimulating factor (G-CSF) to reduce the duration of neutropenia caused by escalating doses of chemotherapy.

Table 31.3 lists the different sources of HSCs used for transplantation. Tables 31.4 and 31.5 list the advantages and disadvantages of allogeneic and autologous HSCT, respectively.

ALLOGENEIC STEM CELL TRANSPLANTATION

The single most important predictive factor for the outcome of an allogeneic HSCT is the degree of immunologic match between the donor and recipient. This is dependent on the degree of match between the donor and recipient major histocompatibility complex (MHC).

Histocompatibility Testing

The MHC genes are mapped within a region called HLA (human leukocyte antigen system A) located on the short arm of chromosome 6. HLA antigens are responsible for rejection of foreign objects from the body. There are six major loci—A, B, C, DR, DQ, and DP, which are divided into two groups. The class I antigens are HLA-A, HLA-B, and HLA-C and the class II antigens are HLA-DR, HLA-DQ, and HLA-DP. They are segregated by haplotype from father and mother. The class I molecules are composed of an α chain and $\beta2$ microglobulin, are highly

Lanzkowsky's Manual of Pediatric Hematology and Oncology.
DOI: http://dx.doi.org/10.1016/B978-0-12-801368-7.00031-4

TABLE 31.1   Indications for Allogeneic Hematopoietic Stem Cell Transplantation in Children

---

MALIGNANT DISORDERS

This group includes patients under 21 years of age with any of the following:

Acute Myelogenous Leukemia (AML)
1. High-risk first complete remission (CR1)[a], defined as:
   a. Having preceding myelodysplasia (MDS) or
   b. High risk cytogenetics: del (5q) -5, -7, abn (3q), t(6;9) complex karyotype (≥ 5 abnormalities) or
   c. Requiring more than 1 cycle of chemotherapy to obtain CR (this includes any clinical or radiographic evidence of progressive extramedullary AML).
   d. FAB classification: M6
   e. The presence of high (>0.4) allelic ratio FLT3-ITD
   f. Patients with therapy-related AML
2. Second or greater CR

Acute Lymphocytic Leukemia (ALL)
1. High-risk first complete remission (CR1)[b], defined as:
   a. Infants with MLL (myeloid/lymphoid or mixed lineage leukemia) rearrangements <6 month of age with high risk characteristics; or
   b. Severe hypodiploidy (<44 chromosomes and/or a DNA index of <0.81); or,
   c. M3 bone marrow at end of induction, or
   d. M2 bone marrow at end of induction with M2-3 at Day 42.
2. High-risk second remission, defined as:
   a. Ph + ALL or
   b. Bone marrow relapse <36 months from induction or
   c. T-lineage relapse at any time or
   d. Early isolated CNS relapse (<18 months from diagnosis); or
   e. Slow re-induction (M2-3 bone marrow at Day 28) after relapse at any time.
3. Any third or subsequent CR

Chronic Myelogenous Leukemia Ph + (CML)
1. Chronic unstable phase[c]
2. Early accelerated phase or
3. Chronic phase after treatment for blast crisis

Juvenile Chronic Myeloid Leukemia (JCML)

Juvenile Myelo-Monocytic Leukemia (JMML)

Lymphomas (Second/Subsequent Complete Remission, or Partial Remission)
1. Hodgkin
2. Non-Hodgkin

Myelodysplasia

Myelofibrosis

Familial Hemophagocytic Lymphohistiocytosis (HLH)

HLH Relapse After Initial Therapy

NON-MALIGNANT DISORDERS

Congenital:
1. Immunodeficiency syndromes:
   a. Severe combined immunodeficiency syndromes
   b. Wiskott–Aldrich syndrome (WAS)[d]
2. Hematologic disorders:
   a. Hemoglobinopathies:
      Sickle cell anemia (selected cases)[e]
      Thalassemia
   b. Fanconi anemia[f] (with evidence of marrow failure or myelodysplasia)
   c. Shwachman–Diamond syndrome[f] (with evidence of marrow failure or myelodysplasia)
   d. Kostmann agranulocytosisc[g]
   e. Diamond–Blackfan anemia[h]
   f. Dyskeratosis congenita[f]
   g. Congenital amegakaryocytic thrombocytopenia (CAMT)
   h. Chronic granulomatous disease
   i. Chediak–Higashi syndrome
   j. Leukocyte adhesion deficiency type I (LAD I): severe phenotype defined by absence ($\leq 1\%$) of CD11b/CD18 expression on neutrophils.
   k. Neutrophil actin defects

(Continued)

TABLE 31.1 (*Continued*)

3. Lysosomal storage diseases[i]
   a. X-ALD with all of the following:
      - Evidence of brain demyelination (i.e. measurable MRI severity score ≥ 1) and/or gadolinium enhancement by MRI brain.
      - Brain demyelination MRI severity score <9
   b. MLD with all of the following:
      - Confirmed with both leukocyte arylsulfatase A deficiency and increased urinary sulfatides to exclude pseudo-deficiency.
      - Presymptomatic late infantile, juvenile or adult form of the disease.
   c. Infantile GLD (Krabbe disease) must be asymptomatic.
   d. Gaucher disease (Type III) : with new or deteriorating neurological and/or pulmonary symptoms while on enzyme replacement therapy.
4. Osteopetrosis: Malignant Infantile Osteopetrosis or Osteopetrosis with Carbonic Anhydrase (CA) II deficiency.
5. Congenital hemolytic anemia patients, who are transfusion dependent.

Acquired:
1. Severe aplastic anemia
2. Paroxysmal nocturnal hemoglubinuria

[a]The high-risk group as defined represents approximately 30% of patients and has a predicted overall survival (OS) <35%. This group additionally, may include non-low risk cytogenetics with positive (≥ 0.1%) positive minimal residual disease (MRD) at the end of Induction I[low risk cytogenetics is defined as inc(16)/t(16;16) or (8;21) cytogenetics or NPM or CEBPα mutations].This group of patients should optimally receive HSCT, if possible, after Intensification I.

[b]Investigators differ regarding the optimum treatment and may consider HSCT for ALL in CR1 for: Ph + ALL with available HLA matched sibling; MLL rearrangement with slow early response [defined as having M2 (5–25% blasts) or M3 (>25% blasts on bone marrow examination on Day 14 of induction therapy)]; intrachromosomal amplification of chromosome 21(iAMP21); Early T-cell precursor (ETP)-ALL, ALL with positive > or = 0.01% MRD especially at the end of consolidation.

[c]Unstable phase as defined with longitudinal Q RT-PCR for BCR-ABL in response to tyrosine kinase inhibitors (TKIs) treatment; unlike adults, in children, investigators differ regarding the optimum treatment and may consider HSCT, if optimal, donor is available, regardless of the response to TKIs treatment.

[d]Indications for HSCT using unrelated donor (or cord blood) in WAS include; age ≤5 years old, refractory or symptomatic transfusion dependent thrombocytopenia, severe auto-immune manifestations not controlled on medical therapy, life-threatening or recurrent serious infections in the past requiring hospital admissions, while on standard prophylaxis.

[e]Indications for HSCT in sickle cell disease (SCD) may include history of one or more of the following: Recurrent vaso-occlusive crisis requiring hospitalizations or emergency room visits and narcotic use to control pain; evidence of SCD ischemia or pathology by cerebral magnetic resonance imaging (MRI) or cerebral magnetic resonance angiography (MRA) scan; history of elevated trans-cranial flow Doppler studies; recurrent acute chest syndrome, sickle nephropathy; Grade ≥1 avascular necrosis of 1≥ joint(s); red-cell alloimmunization during transfusion therapy interfering with the ability to use transfusion therapy.

[f]Reduced intensity conditioning used due to poor tolerance to chemotherapy.

[g]G-CSF may be successful in treatment and avoiding HSCT. Requiring GCSF therapy ≥10 ug/kg/day or marrow failure resulting in pancytopenia requiring transfusions are clear indications to consider HSCT from any suitable unrelated donor or cord blood.

[h]Transfusion dependance or intolerance to steroids are clear indications to consider HSCT from any suitable unrelated donor or cord blood.

[i]HSCT provides a population of cells with the capacity to produce the missing enzyme. Early transplantation is the goal so that enzyme replacement may occur before extensive central nervous system injury becomes evident.

TABLE 31.2 Indications for Autologous Hematopoietic Stem Cell Transplantation

HEMATOLOGIC MALIGNANCIES

1. Non-Hodgkins lymphoma[a]
2. Hodgkins disease[a]

SOLID TUMORS

1. Neuroblastoma
2. Ewing's sarcoma, primitive neuroectodermal tumor[b]
3. Rhabdomyosarcoma[a]
4. Germ cell tumor[a]
5. Brain tumors[a]
6. Germ cell tumors[a]
7. Wilms tumor[a]

[a]Relapsed or refractory disease but achieved at least partial response with conventional chemotherapy (i.e., chemosensitive disease).
[b]Metastatic disease at presentation but achieved partial remission with conventional chemotherapy or relapsed but has chemosensitive disease.

TABLE 31.3   Sources of Hematopoietic Stem Cells for Transplantation

1. Allogeneic bone marrow or PBSC using HLA-matched sibling
2. Allogeneic bone marrow or PBSC using matched (or partially matched (7 out of 8 allele matched)) HLA family members if available
3. Syngeneic bone marrow or PBSC (an identical twin)
4. Allogeneic bone marrow or PBSC using HLA-matched unrelated donors (mismatched or haplo donors may be used in certain conditions but require CD34 selection or *in vivo* T-cell depletion pre- or post-HSCT)
5. Autologous bone marrow: Patient's own marrow is cryopreserved and reinfused after patient has received aggressive chemotherapy and/or radiation therapy to treat the underlying malignancy *in vivo* purging or *ex vivo* purging[a]
6. Autologous PBSCs: The number of circulating stem cells can be increased by the use of recombinant growth factors, which can increase stem cell numbers in peripheral blood prior to apheresis. This induces a more rapid hematopoietic recovery after transplantation[b]
7. UCB stem cells (always allogeneic). using HLA-matched sibling or four or five out of six HLA-matched unrelated cord blood (single or double units). *Ex vivo* expansion of cells to increase the number of CD34 + cells available for transplant has been employed experimentally
8. Fetal liver stem cells administered to patients in utero (26- to 30-week gestation) with severe immunodeficiency and inborn errors of metabolism (experimental)

[a]*Ex vivo purging for removing malignant cells from peripheral blood stem cells or marrow rely upon physical (density or velocity sedimentation, filtration), pharmacologic (e.g., chemotherapy) or immunologic (monoclonal antibodies) principles.*
[b]*This is a useful method in patients who have received previous pelvic radiation therapy or because of tumor in the marrow. G-CSF 10 μg/kg/day or GM-CSF 500 mg/m² for 5–7 days can be used to mobilize progenitor cells. Blood progenitor cells can be collected for 2–5 days beginning on day 4 or 5 after initiation of growth factor.*
HLA, human leukocyte antigen; PBSC, peripheral blood stem cell; HSCT, hematopoietic stem cell transplantation; UCB, umbilical cord blood.

TABLE 31.4   Advantages and Disadvantages of Allogeneic Hematopoietic Stem Cell Transplantation

Advantages	Disadvantages
1. Lower relapse rate 2. Graft versus leukemia effect[a]	1. Compatible donor availability limited (approximately 25–30% of siblings are compatible) 2. Graft-versus-host disease (GVHD), except in syngeneic (identical twins) 3. Increased risk of viral infections such as cytomegalovirus or adenovirus and interstitial pneumonia (this risk is 2.5 times greater in unrelated matched donors compared to matched siblings) 4. Risk of veno-occlusive disease greater in unrelated donors 5. Immune suppression due to GVHD prophylaxis or treatment drugs

[a]*The relapse rate is 2.5 times lower in allogeneic recipients who have grade II–IV acute GVHD compared to recipients without GVHD. Leukemic cells have been reported to disappear during episodes of acute GVHD.*

TABLE 31.5   Advantages and Disadvantages of Autologous Hematopoietic Stem Cell Transplantation

Advantages	Disadvantages

1. No need for an allogeneic donor
2. No GVHD
3. Lower risk of opportunistic infection (e.g., CMV, Pneumocystis jiroveci)
	1. creased risk of relapse (approximately 50%)
	2. No graft versus leukemia effect
	3. Graft failure due to:
	  a. Effect of in vitro purging
	  b. Effect of previous chemotherapy and radiotherapy

polymorphic, and are expressed on most nucleated cells and on platelets. The HLA class II loci are involved in exogenous antigen processing. The HLA class II antigens are heterodimeric cell surface molecules formed by the α and β chains, each of which is polymorphic. HLA terminology is designated by the World Health Organization (WHO) nomenclature committee for factors of the HLA system and is updated at regular intervals.

To assess the degree of HLA compatibility, donor and recipient need to be HLA "typed." Typing methods are serological (antigen) or molecular (allele) (with low, intermediate, or high resolution). The broadest designation of HLA type is based on serologic typing, with the highest specificity based on the actual DNA sequence.

The following different techniques have been used for the identification of class I and II types:

- Sequence-specific oligonucleotide probe.
- Sequence-specific primer.
- DNA sequencing (usually reserved to resolve any ambiguity that is not resolved by the above methods).

HLA matching for the purpose of HSCT is generally confined to major class I and II loci, although it is increasingly appreciated that minor histocompatibility antigens also play important roles in the outcome of allogeneic HSCT.

Any disparity at major HLA loci between the donor and the recipient requires vigorous immunologic intervention to avoid rejection, or, should engraftment occur, subsequent graft-versus-host disease (GVHD).

Donor Selection

Any HLA-identical sibling, who does not have an anticipated risk for the underlying condition for which their sibling is undergoing transplant, should be considered a potential donor. ABO mismatch is not a contraindication, but if multiple donors are available, a donor with the same ABO type is preferred (major ABO mismatch (A or B or O) requires the removal of the red cells and/or plasma from the bone marrow graft prior to infusion). The HLA match can be phenotypic or genotypic. Despite the preference for a sibling donor, 70–75% of patients have no acceptable donors within their families. In these cases, volunteer unrelated donors who are HLA, A, B, C, and DR compatible or minimally mismatched with the recipients are sought through a search of the US National Marrow Donor Program and Bone Marrow Donors Worldwide.

An acceptable adult unrelated donor donating bone marrow or peripheral blood stem cells (PBSCs) should be matched via high-resolution DNA typing at least seven of eight alleles (HLA-A, B, C, and DRB1).

An acceptable umbilical cord blood (UCB) donor should be matched via low-resolution DNA typing (or serology with high-resolution DRB1 typing) at least four of six antigens (HLA-A, B, and DRB1). For double UCB donations (used as needed to overcome cell dose limitation), each should be matched to at least three of six HLA-A, B, and DRB1 antigens with the other unit, and each unit must match independently with at least four of six HLA antigens with the recipient.

Suggested Strategy for Donor Selection and Prioritization

1. Priority of donor by HLA disparity for a related donor is
   a. HLA-identical sibling.
   b. HLA-matched relative (fully matched 10/10).
   c. Single HLA class I antigen or single class II allele mismatched relative. The class II antigen matching must be confirmed by high-resolution molecular typing.
2. The priority of donor by HLA disparity for an unrelated, volunteer donor is
   a. HLA alleles matched at all loci, 10/10 match (A, B, C, DRB1, DQB1).
   b. Single HLA-A or HLA-B allele mismatch, 9/10 match.
   c. HLA alleles match at A, B, and DRB1 with a micromismatch or major mismatch at C loci (8/10 or 9/10 match).
   d. HLA A, B allele match but one allele mismatch at DRB1, 9/10 match.
   e. One antigen mismatch at HLA A or B in the cross-reactive group with allele match at DRB1 (high-resolution typing for all class I and class II antigens are required).
3. The priority of donor for UCB units is
   a. A 6/6 HLA-matched unit.
   b. A unit with one minor or major mismatch at class I (5/6).
   c. Two minor or major mismatches at class I (4/6 match) or one minor or major mismatch at class I and a micromismatch at class II (4/6 match). Recent data support the use of allele-level HLA matching in the selection of single UCB units.

Secondary Donor Prioritization (After HLA Matching Prioritization)

1. *HSC source.* For recipients of HLA-matched related donor allografts, the preferred source of HSC, PBSC, bone marrow, or UCB, will be determined by the transplant protocol on which the recipient is enrolled, the disease status of the recipient, and/or the medical condition of the donor. Bone marrow is the preferred source for most allogeneic pediatric transplants. This is discussed in more detail below.
2. *Cell dose.* When using bone marrow or peripheral blood as the HSC source, the donor's size must be adequate to allow safe donation of the requested cell dose. This is more likely to be a consideration when the recipient is large-sized or in circumstances where bone marrow harvest is the preferred source of graft. While the cell dose is generally controllable when bone marrow or peripheral blood is the HSC source, UCB units have fixed cell doses.
3. *Donor gender* (male preferred).

TABLE 31.6　Blood Bank Support for Hematopoietic Stem Cell Transplantation

Problem	Solution
Transfusion associated GVHD	Standard: gamma radiation (2,500–3,000 cGy)
CMV transmission (CMV negative donor and recipient)	*Standard:* use of only CMV-negative blood products or through an in-line filter to remove lymphocytes and leukocytes that may harbor latent viruses
	Alternative: leukocyte-depleted blood products (most commonly used)

Alloimmunization

1. Graft failure or rejection (especially aplastic anemia)

    Avoidance or minimizing of pretransplant transfusions; avoidance of transfusions from family members, especially potential donors

2. Refractory thrombocytopenia[a]

    Most effective: leukocyte-depleted blood products
    a. Single-donor platelets
    b. HLA-matched donors
    c. Cross-matched platelets

ABO Incompatibility

1. ABO Incompatibility

		Donor			
		O	A	B	AB
Recipient	O	Compatible	Major	Major	Major
	A	Minor	Compatible	Bidirectional	Major
	B	Minor	Bidirectional	Compatible	Major
	AB	Minor	Minor	Minor	Compatible

2. Suggested processing for bone marrow grafts*

		Recipient Blood Group			
		A	B	AB	O
Donor Blood Group	A	None	RBCS & Plasma depletion	Plasma depletion	RBCS depletion
	B	RBCS & Plasma depletion	None	Plasma depletion	RBCS depletion
	AB	RBCS depletion	RBCS depletion	None	RBCS depletion
	O	Plasma depletion	Plasma depletion	Plasma depletion	None

* Isohemagglutinin titers in both donor and recipient may affect the specific requirements for processing with regard to need for RBCS and/or plasma depletion.

Donor red cell transfusion requirements	Autologous red blood cell salvage
	Predeposit autologous red blood cell units

[a]*Non-alloimmunization causes of refractory thrombocytopenia include drugs, hepatic veno-occlusive disease (VOD) or Sinusoidal obstruction syndrome (SOS), hypersplenism, or sepsis. Another more unusual cause of refractory thrombocytopenia is a syndrome resembling thrombotic thrombocytopenic purpura, often associated with the use of Calcineurin inhibitors, total-body irradiation, or the development of acute GVHD.*

4. *Donor age group.* The younger donor is preferred unless the donor's size would prohibit collection of sufficient cells for transplantation.
5. *ABO compatibility* (ABO match followed by minor mismatch followed by major mismatch and least preferred bidirectional mismatch).
6. *Donor parity* (lower parity preferred).
7. *Cytomegalovirus (CMV) serology.* A CMV-negative donor is preferred for a CMV-negative recipient.

Table 31.6 lists the blood bank support required for HSCT.

HSC SOURCES, COLLECTION, AND MANIPULATION

Bone Marrow

Bone marrow harvesting is carried out using general or epidural anesthesia under sterile conditions.

The recommended cell dose and volume of marrow to be collected is determined by the recipient's body weight and diagnosis, the type of graft manipulation (if any) that will occur, and the size of the donor.

The iliac crests (most common posterior and rarely anterior if an adequate number of cells are not obtained from posterior iliac crests) of the donor are prepared and draped. Approximately 2 ml of bone marrow is aspirated from each site, avoiding dilution with blood by taking multiple aspirates from the iliac crests. The marrow is collected in a heparinized collection bag. The minimum concentration of heparin (preservative-free) is 3−5 units/ml of bone marrow. The quantity of nucleated bone marrow cells required to ensure engraftment is $2-5 \times 10^8$ cells/kg of recipient body weight. The usual volume of marrow required to achieve this cell yield is 10−20 ml/kg of recipient body weight. Marrow from children, especially infants, has a higher proportion of marrow-repopulating cells than marrow from older donors. The marrow is filtered through a filtering apparatus to remove bone and tissue fragments and placed in a blood transfer pack. In allogeneic transplants, the pack containing the bone marrow is then given to the recipient as an intravenous (IV) infusion over a period of a few hours.

For autologous bone marrow transplantation, some form of processing to remove the erythrocytes is necessary prior to the collected marrow being cryopreserved in liquid nitrogen. At this temperature, the marrow can be preserved for periods of up to 10 years or possibly longer. The cryopreserved bone marrow is thawed at the bedside in a 37°C water bath and given intravenously to the recipient over a period of a few minutes to an hour, depending on the volume infused. Generation of a "buffy coat" or sedimentation at unit gravity in hetastarch both produce an erythrocyte-poor product that contains most of the nucleated cells, including mature granulocytes. Most investigators freeze cells in tissue culture medium (e.g., PlasmaLyte) containing 5−10% dimethyl sulfoxide and a colloid (autologous plasma or human serum albumin are common); use of bags for marrow cryopreservation simplifies thawing and allows direct infusion afterwards. Because of the possibility of bag breakage, it is recommended that at least two bags (preferably more) be used for each patient.

Peripheral Blood Stem Cells

PBSCs are the most commonly used source for an autologous graft, and bone marrow is reserved for cases where PBSC collection or mobilization are not feasible.

PBSCs are collected from donors after stimulation with G-CSF (10 micrograms/kg/day) and have been successfully used to reconstitute hematopoiesis in recipients of autologous, syngeneic, and allogeneic grafts. Patients who receive more than 5×10^6 CD34 + cells/kg engraft satisfactorily. In autologous donors, a combination of myelosuppressive chemotherapy and G-CSF administration is more effective at mobilizing CD34 + cells. Plerixafor alone or in combination with G-CSF may be used in certain conditions for mobilization.

Umbilical Cord Blood

From 2007 to 2011 UCB was the most common (45%) graft source, followed by bone marrow grafts (35%) for patients <21 years of age. UCB has been collected, cryopreserved, and is used as the source of pluripotential HSCs when a related or unrelated stem cell donor is not available. UCB cells have increased proliferative capacity and decreased alloreactivity, with a lower incidence of GVHD. This property, coupled with an absence of donor risk, is an advantage in the use of this source of HSC, but a disadvantage is the inability to evaluate the genetic history of these donors. For a single UCB transplant, the minimum cell dose must be equal to or greater than 2.5×10^7 total nucleated cells (TNC) per kilogram of recipient weight. If the transplant is performed with two cord blood units, the preferred dose of the combined units is greater than or equal to 2.5×10^7 TNC per kilogram of recipient weight. A higher cell dose is preferred when there a greater degree of HLA mismatch between the cord unit(s) and the intended recipient. For cord blood the interaction between the cell dose available and HLA matching affects the selection strategy as follows:

1. Single unit:
   a. The UCB unit is a 6/6 match with the recipient with a cell dose greater than or equal to 2.5×10^7 TNC/kg.
   b. The UCB unit is a 5/6 matched with a cell dose greater than or equal to 4.0×10^7 TNC/kg.
   c. The UCB unit is a 4/6 matched with a cell dose greater than or equal to 5.0×10^7 TNC/kg.
2. Double units should be adequately matched with each other and the recipient as discussed above.

Graft Manipulation Post-collection

Several manipulations can be performed on HSC post-collection to reduce the risk of red blood cell (RBC) hemolysis due to ABO incompatibility, GVHD, and the reinfusion of malignant cells.

ABO Incompatibility

ABO incompatibility between donor and recipient is encountered in 25—30% of allogeneic transplants. A major incompatibility occurs when the recipient plasma contains isohemagglutinins directed against the donor RBC antigens (e.g., group O recipient, group A donor) and minor incompatibility occurs when the donor plasma contains isohemagglutinins directed against recipient RBC antigens (e.g., group A recipient, group O donor). In both instances, appropriate anti-A or anti-B isohemagglutinin titers must be determined before infusion of the cells. After the transplant procedure, all patients should have immunohematologic testing for the appearance of donor-derived RBCs and changes in recipient isohemagglutinin titers.

Graft-Versus-Host Disease

The removal of T lymphocytes from allogeneic HSC post-collection decreases the risk of GVHD. This can be achieved through various techniques, including monoclonal antibodies accompanied by complement-mediated lysis, immunotoxins, or immunomagnetic beads (complete or partial *in vivo* T-cell depletion may include administration of ATG or alemtuzumab (Campath)). Physical methods of T-cell depletion include counter-elutriation and soybean lectin agglutination plus E-rosette depletion.

Purging of Malignant Cells

Purging can be performed to remove malignant cells. This can be achieved through the use of monoclonal antibodies combined with complement, monoclonal antibodies linked to toxins and various chemotherapy drugs.

Medical Evaluation of HSC Donors

Donor evaluation procedures protect the safety of the HSC donor and recipient. A standardized, comprehensive evaluation of donors identifies potential medical risks to the donor and the recipient.

The medical history should focus on indicators of active disease processes and the risk of an indolent infectious disease. The history should include details of vaccination, travel, blood transfusions, questions to identify persons at high risk for the transmission of communicable or inherited, hematological or immunological diseases, and questions to identify any past history of malignant diseases.

Donor assessment should also include pregnancy testing, prior deferrals from blood donation, contraindications to blood donation, and findings that would increase the anesthesia risk (an electrocardiogram and chest x-ray should be performed on adult donors). PBSC donors should be evaluated for potential contraindications to central venous access catheter placement and G-CSF for HSC mobilization.

Laboratory evaluation for potential donors should include

- Complete blood count.
- Comprehensive metabolic panel (including electrolytes, glucose, blood urea nitrogen and creatinine, serum protein, serum albumin, AST, and ALT).
- ABO group and Rh type, red cell antibody screen.
- Confirmatory HLA typing.
- Human immunodeficiency virus (HIV), type 1 (HIV-1).
- HIV-2.
- Hepatitis B virus.
- Hepatitis C virus.
- *Treponema pallidum* (syphilis).
- Human T-cell lymphotrophic virus I (HTLV-1).
- HTLV-2.
- CMV.
- Ebstein—Barr virus (EBV).
- Herpes viruses.

TABLE 31.7 Pretransplantation Evaluation of the Recipient[a]

History, physical examination, weight, height, body surface area (BSA) and head circumference (if appropriate for age)
Complete blood count
Comprehensive metabolic panel
24-hour urine for creatinine clearance or glomerular filtration rate (GFR)
ABO blood group, Rh typing and isohemagglutinin titer (if indicated depending on donor blood group)
HIV-1 and -2, HTLV I and II[b]
CMV, hepatitis B and C, EBV and HSV 1/2[b]
Syphilis[b]
Echocardiogram (or MUGA) with ejection fraction (or shortening fraction) as appropriate and EKG
Pulmonary function test (PFT) (if age appropriate and feasible) or O_2% saturation.
Imaging: Chest x-ray at minimum (additional imaging requirements vary depending on the specific transplant program)[c]
Dental and eye evaluation
Disease-specific manifestation assessments at baseline as clinically indicated (e.g., bone marrow aspiration/biopsy if applicable)
Karnofsky or Lansky score (age appropriate)
Psychosocial evaluation

[a] *Additional testing may be needed as indicated to evaluate any existing comorbidities.*
[b] *Antigen detection or molecular methods may need to be used for testing especially if primary disease involves immune deficiency. All HSCT recipients require insertion of vascular access device (VAD) (i.e., central line), the specific VAD type used is different depending on patient, disease, transplant center, preference, and availability.*
[c] *CT scan chest, abdomen, and pelvis and CT of the neck; MRI of the brain for all to establish a baseline for future evaluation of possible CNS toxicities may be done based on the specific transplant program standards.*

TABLE 31.8 Commonly Used Stem Cell Transplantation Preparative Regimens

Drug	Total dose	Cyclophosphamide total dose	Total body irradiation (cGy) (fractionated)
Cyclophosphamide	120 mg/kg		1200–1350
Etoposide	60 mg/kg		1200–1350
Etoposide	30 mg/kg	120 mg/kg	1200
Busulfan[a]	16 mg/kg	120–200 mg/kg	
Melphalan	140 mg/m²		1200–1350

[a] *Busulfan is commonly administered as IV medication and is usually dosed initially as follows: weight <10 kg: 0.8 mg/kg/dose; weight ≥10 kg and age <4 years: 1 mg/kg/dose; age ≥4 years: 0.8 mg/kg/dose. Subsequent doses are usually adjusted to achieve target of AUC 900–1500 μmol/l/min (or net steady-state concentration (SSc) of 800–1200 ng/ml) (target can be slightly variable depending on local center and treatment protocol).*

Additional testing may be recommended based on local regulations or as clinically indicated (e.g., West Nile virus, *Trypanosoma cruzi* (Chagas' disease), screening for hemoglobin S).

Table 31.7 lists the pretransplantation evaluation of the HSCT recipient.

Pretransplantation Preparative Regimens (Conditioning)

Pretransplantation conditioning is used both to eradicate disease and as a means of immunosuppressing the recipient sufficiently to allow the acceptance of a new immunologically disparate hematopoietic system. Table 31.8 shows examples of various preparative regimens. The current mainstay preparatory regimen for traditional myeloablative HSCT consists of ablative doses of either total body irradiation (TBI) or busulfan (Bu) together with either one or two chemotherapy agents. On completion of the preparative regimen, the donor marrow, PBSC, or UCB is infused.

Commonly used Preconditioning Regimens

Leukemia

Two preconditioning regimens (shown in Table 31.8), which have antitumor as well as immunosuppressive properties, are commonly used prior to stem cell transplantation. TBI (fractionated doses) and cyclophosphamide (or etoposide) are utilized in acute lymphocytic leukemia (ALL). A combination of TBI and cyclophosphamide or

TABLE 31.9 CBV Regimen[a] for Stem Cell Transplantation in Non-Hodgkin Lymphoma

Days	Chemotherapy
$-8, -7, -6$	BCNU: 100 mg/m^2/day IV over 3 hours (total dose 300 mg/m^2)
	Etoposide 800 mg/m^2/day IV as a 72-hour continuous infusion (total dose 2400 mg/m^2)
$-5, -4, -3, -2$	Cyclophosphamide: 1500 mg/m^2/day IV over 1 h. Use MESNA (total dose 2200 mg/m^2 IV every 24 h)
-1	No treatment
0	Stem cell infusion

[a]*Various dosing schedules for CBV regimen exist.*
Methylprednisolone 1 mg/kg/day in divided doses given every 6 hours for pulmonary protection from BCNU toxicity on days -9 to -2. Wean by 20% every 2 days, taper.

TABLE 31.10 BEAM Regimen for Stem Cell Transplantation in Non-Hodgkin Lymphoma

Day	Chemotherapy
-6	BCNU: 300 mg/m^2 IV over 3 hours
$-5, -4, -3, -2$	Etoposide 200 mg/m^2/day IV over 1 h (total dose 800 mg/m^2)
$-5, -4, -3, -2$	Cytosine arabinoside: 400 mg/m^2/day IV over 1 h (total dose 1600 mg/m^2)
-1	Melphalan 140 mg/m^2 IV over 30 minutes
0	Stem cell infusion

Methylprednisolone 1 mg/kg/day in divided doses given every 6 hours for pulmonary protection from BCNU toxicity on days -7 to -2. Wean by 20% every 2 days for an 8-day taper.

Bu and cyclophosphamide in acute myeloid leukemia (AML), chronic myeloid leukemia (CML), myelodysplasia, juvenile chronic myeloid leukemia, and juvenile myelomonocytic leukemia. (See also Chapters 18 and 19.)

Non-Hodgkin Lymphoma (NHL)

Myeloablative chemotherapy regimens such as cyclophosphamide, Bis-Chloroethyl-Nitroso-Urea (BCNU) or Carmustine, etoposide (VP 16) (CBV) (Table 31.9) or BCNU, etoposide, cytosine arabinoside, and melphalan (BEAM) (Table 31.10) are preconditioning regimens utilized prior to stem cell transplantation in relapsed non-Hodgkin lymphoma (NHL). (See also Chapter 22.)

Solid Tumors

HDC with autologous stem cell rescue has become the standard of care for neuroblastoma and medulloblastoma, and has been utilized in the setting of other relapsed solid tumors.

Severe Aplastic Anemia (SAA)

Patients who have received less than five blood transfusions usually receive cyclophosphamide 50 mg/kg/day for 4 days, followed by 1 day of rest and stem cell transplantation on the following day.

In multiply transfused patients, rejection of the bone marrow is a major problem (can be as high as 20–25%). To improve long-term survival in these patients, pretransplantation conditioning requires more intensive immunosuppression to be employed. (See also Chapter 8.)

As graft rejection is the major cause of morbidity and mortality in HSCT for severe aplastic anemia (SAA) the pretransplantation preparative regimen of cyclophosphamide (Cytoxan) and antithymocyte globulin, ATG (ATGAM/thymoglobulin)[1] with the inclusion of cyclosporin A, (Sandimmune) in the GVHD prophylaxis

[1]Throughout this chapter, ATG refers to horse ATG, unless noted otherwise.

TABLE 31.11 Hematopoietic Stem Cell Transplantation Preparative Regimen for Severe Aplastic Anemia

The following regimen is commonly used for non-sensitized (not heavily transfused) matched sibling donors[a]:	
Day 5	*Morning:* Cyclophosphamide, 50 mg/kg IV over 1 hour
	Afternoon: ATG (antithymocyte globulin), 30 mg/kg IV. First dose of ATG is given over 8 hours; subsequent doses are given over 4 hours
Day 4	(Same as for Day 5)
Day 3	(Same as for Day 5)
Day 2	*Morning:* Cyclophosphamide, 50 mg/kg IV over 1 hour
Day 1	Rest; cyclosporine A, 10 mg/kg/d PO daily adjusted for serum levels
Day 0	Marrow infusion

[a]*The following regimen is commonly used for sensitized (heavily transfused) matched sibling, or unrelated donors: Rabbit (r)ATG 3 mg/kg Day −4 to −2, Fludarabine 30 mg/m2 Day −5 to −2 and TBI 200 cGy Day −1.*
Adapted from: Vlachos and Lipton (2002), with permission.

regimen is designed to be highly immunosuppressive. Table 31.11 describes a commonly used immunosuppressive HSCT preparative regimen for SAA. Even when an identical twin is used as a donor, a similar preparatory regimen is recommended. Long-term survival in the range of >90% can be expected with HSCT using histocompatible related donors. The overall risk of unrelated or mismatched related donor transplantation precludes its use as front-line therapy for SAA at this time. As improved HLA typing, preparative regimens and GVHD prophylaxis are utilized, HSCT is becoming available to a wider group of patients with SAA.

Fanconi Anemia

Due to impaired tissue repair and increased risks of adverse effects, reduced-intensity conditioning regimens are employed. (See also Chapter 8.)
Matched sibling graft:

Cyclophosphamide 10 mg/kg/day × 4 days
Fludarabine 35 mg/m^2/day × 4 days
ATG (rabbit) 2.5 mg/kg/day × 4 days
Methylprednisolone 2 mg/kg/day though +20 then taper

Unrelated bone marrow graft[2]:

TBI 450 cGy (single fraction)
Cyclophosphamide 10 mg/kg/day × 4 days
Fludarabine 35 mg/m^2/day × 4 days
ATG (rabbit) 2.5 mg/kg/day × 4 days
Methylprednisolone 2 mg/kg/day though +20 then taper

Cord graft:

Bu 1 mg/kg × 2 doses/day × 2 days
Cyclophosphamide 10 mg/kg/day × 4 days
Fludarabine 35 mg/m^2/day × 4 days
ATG (rabbit) 3 mg/kg/day × 4 days
Methylprednisolone 2 mg/kg/day though +20 then taper

Miscellaneous Conditions

In hemoglobinopathies, selected immunodeficiency, and other inherited disorders, the conditioning regimen commonly used is Bu and cyclophosphamide with or without ATG.

[2]CD34 selected graft may be used to minimize risks for GVHD.

Nonmyeloablative (Reduced-Intensity) Regimens

These regimens are based on the ability of donor T cells to ablate the host hematopoiesis or malignancy or both. The object is to use the donor lymphoid cells to eradicate the disease in the host. This allows the use of decreased doses of myeloablative chemotherapy with less toxicity. However, aggressive immunosuppressive therapy is still required to eliminate the potential host rejection of the donor cells. This approach is referred to as nonmyeloablative or reduced-intensity transplantation. This approach is traditionally associated with decreased morbidity and mortality of the transplant procedure itself, although it is also associated with increased chances of disease recurrence and graft rejection. It is most commonly utilized if the patient cannot tolerate a myeloablative HSCT, for example decreased organ functions or comorbidities. PBSC are often utilized as a graft source for this approach and are associated with increased risks for GVHD.

The most common conditioning regimens used for this approach include:[3]

- Fludarabine: $30 \, \text{mg/m}^2$/day on days -5 to -2 (total dose of $120 \, \text{mg/m}^2$) and Melphalan: $140 \, \text{mg/m}^2$ on day -2.

  OR

- Fludarabine: $30 \, \text{mg/m}^2$/day on days -6 to -2 (total dose of $150 \, \text{mg/m}^2$): Bu: $4 \, \text{mg/kg}$/day or $3.2 \, \text{mg/kg}$/day (total dose of $8 \, \text{mg/kg}$ or $6.4 \, \text{mg/kg}$, respectively) on days -5 to -4.

  OR

- Fludarabine ($30 \, \text{mg/m}^2$ IV per day for 3 days)
- 2 Gy TBI.

At present this approach is being used in both selected cases of malignant and nonmalignant diseases such as:

- Malignant diseases
  - CML.
  - AML.
  - NHL.
  - Hodgkin lymphoma.
- Nonmalignant diseases:
  - Immune deficiency disorders (e.g., severe combined immunodeficiency (SCID)).
  - Bone marrow failure syndromes (e.g., Fanconi anemia, dyskeratosis congenita).
  - Hemophagocytic lymphohistiocytosis.
  - Hemoglobinopathies.

Second Transplantation Regimens

The following regimens are being used for second transplantation following graft rejection with autologous recovery or relapse of disease:

- Fludarabine: $30 \, \text{mg/m}^2$ IV daily for 4 days.
- Melphalan: $70 \, \text{mg/m}^2$ IV daily for 2 days.
- Antithymocyte globulin (horse):[4] $30 \, \text{mg/kg}$ IV for 3 days.

  OR

- Fludarabine: $30 \, \text{mg/m}^2$ IV daily for 4 days.
- Bu: $4 \, \text{mg/kg}$/day IV daily for 2 days.
- Antithymocyte globulin(horse) $15 \, \text{mg/kg}$ IV daily for 3 days.

  OR

- Cyclophosphamide: $50 \, \text{mg/kg}$ IV for 1 day.
- Fludarabine: $35 \, \text{mg/m}^2$ IV daily for 4 days.
- Antithymocyte globulin (horse) $30 \, \text{mg/kg}$ IV daily for 3 days.

---

[3]ATG is occasionally added to this regimen in unrelated donor or nonmalignant conditions.

[4]Horse ATG can be substituted with rabbit ATG using appropriate dosing.

For graft failure without autologous recovery, the following regimen is commonly used:[5]

- Fludarabine 30 mg/m^2 IV daily for 4 days.
- ATG (rabbit) 2.5 mg/kg IV daily for 3 days.
- TBI 200 cGy for 1 day.

ENGRAFTMENT

Engraftment is defined by neutrophil engraftment, which is the first day of three consecutive days when the absolute neutrophil count is >500/mm^3.

Platelet engraftment is defined as the first day of a minimum of three consecutive measurements on different days when: platelet count >50,000/mm^3, and the patient is platelet transfusion-independent for a minimum of 7 days.

Evidence of Engraftment

Evidence of engraftment is usually seen within 3 weeks following stem cell transplantation and is identified as follows:

1. Direct evidence
   a. Fluorescence *in situ* hybridization in sex-mismatched transplantation.
   b. Variable number of tandem repeats in same sex-matched transplantation.
   c. HLA typing: mismatched transplantation.
2. Indirect evidence
   a. Aplastic anemia and leukemia:
      i. Adequate bone marrow cellularity.
      ii. Increase in white blood cells with normal morphology and increase in platelet counts.
3. SCID: improved immunologic parameters.

Causes of Failure to Engraft

- Increased HLA genetic disparity.
- Inadequate T-cell content of the graft.
- Inadequate stem cell dose.
- History of multiple transfusions in aplastic anemia and hemoglobinopathies leading to alloimmunization with HLA antibodies.
- Disordered microenvironment in storage disorders and osteopetrosis.

COMPLICATIONS OF HSCT

The risks of HSCT are related to the underlying disease, the pretransplantation conditioning with cytotoxic drugs or irradiation or both, posttransplantation immunosuppression (infections), and GVHD. The early mortality can be as high as 20%, depending on the conditioning regimens, the disease, and the patient's status.

The potential complications post transplantation of conditioning regimens include the following:

- Toxic vasculitis.
- Severe mucositis.
- Veno-occlusive disease (VOD) of the liver.
- Acute alveolitis.
- Obliterative bronchiolitis.
- Hemolytic–uremic syndrome.
- Multiple organ failure.

[5]CD34 + selected graft can be used to minimize risk of GVHD.

Despite quantitative myeloid (neutrophil) recovery after bone marrow transplantation, the functional recovery of humoral and cellular immunity may take a year or longer. The type of graft used (autologous or allogeneic), the type and method of administration of immunosuppressive therapy after transplant and whether GVHD has occurred (especially chronic GVHD) all influence the rate of lymphoimmunologic reconstitution.

Immunodeficiency

After stem cell transplantation, a transient state of combined immunodeficiency develops in all patients. The natural history of immune reconstitution is similar for autologous and allogeneic transplants but often is altered in allogeneic transplants by GVHD as well as by immunosuppressive therapy. Although natural killer (NK) cells reconstitute to normal levels usually within the first month after transplantation, other immunologic defects may persist.

The importance of the recognition of the delayed overall immunologic reconstitution relates to the clinically observed incidence of recurrent bacterial (*Streptococcus pneumoniae*) and opportunistic infections (*Pneumocystis jiroveci* (formerly, *Pneumocystis carinii*), fungal, herpes zoster, CMV) that can occur many months after transplantation. In addition to the use of prophylaxis against *Pneumocystis jiroveci* infection, many centers add penicillin (or a first-generation cephalosporin or fluoroquinolone) as posttransplant prophylaxis against *Streptococcus pneumoniae* infection, particularly if the patient has chronic GVHD. The suppressed immunity also has major practical implications for consideration of the timing of vaccinations after bone marrow transplantation. The following approach for the timing of vaccinations should be considered:

- At 6−12 months (depending on immune recovery), individuals who do not have chronic GVHD should receive killed influenza (yearly) and pneumococcal polysaccharide vaccines and should also have vaccinations with inactivated poliovirus and diphtheria−acellular pertussis−tetanus. It is probably also safe to administer hepatitis B and *Haemophilus influenzae* conjugate vaccines.
- Immunosuppressed patients with chronic GVHD are less likely to develop an adequate antibody response after vaccinations. If vaccines are administered, antibody titers should be checked after vaccination to determine the efficacy of the response.
- For patients who are 2 years post-stem cell transplantation, have no evidence of GVHD, and are not receiving immunosuppressive therapy, measles, mumps, and rubella live vaccines should be given. Table 31.12 lists a reimmunization schedule after HSCT.

Infections

Patients undergoing stem cell transplantation have increased risks of infection related to their disease and its status. Patients with more than two relapses who undergo stem cell transplantation develop more infectious complications than those who receive transplants during the first remission. The adequacy of treatment of any infection in the patient prior to marrow ablation influences the risk of subsequent complications. Prolonged duration of neutropenia, use of central venous catheters, and slow speed of marrow engraftment are important risk factors.

Table 31.13 lists prophylaxis and supportive care for stem cell transplantation. Table 31.14 lists the bacterial and fungal infections most frequently seen at different times post transplantation.

During the first 30 days, the most frequently documented infection is coagulase-negative *Staphylococcus epidermidis* infection associated with the use of indwelling central venous catheters.

After 120 days post bone marrow transplantation, the most common infections are sino pulmonary infections with encapsulated organisms and cutaneous infections with herpes zoster.

Other localized or disseminated viral infections that occur include adenovirus, herpes simplex virus, varicella virus, respiratory syncytial virus, para-influenza, and influenza A or B.

Table 31.15 outlines the management of infection during and after stem cell transplantation.

Cytomegalovirus (CMV)

During the second and third months, infections with CMV and interstitial pneumonia occur. CMV infection occurs within 20−100 days after transplantation, with the highest infection rate in patients who are CMV-seropositive before transplantation or seronegative recipients receiving a seropositive graft. CMV infection can also arise from the exogenous introduction of virus in blood products.

The clinical manifestations of CMV infection are variable, ranging from asymptomatic viral excretion to fever, arthralgia, arthritis, hepatitis, secondary bone marrow hypoplasia with thrombocytopenia and leukopenia, retinitis, esophagitis, gastroenteritis, and pneumonia with up to 80% mortality.

Early treatment with the antiviral drug ganciclovir plus high-titer CMV IV gammaglobulin has reduced morbidity to less than 20%. Prophylactic or preemptive therapy in seropositive patients or those receiving a seropositive graft is an effective way to prevent the development of CMV infection.

The following guidelines for the prophylaxis and treatment of CMV infections should be carried out (as recommended by the US Center for Disease Control and Prevention along with the American Society of Blood and Marrow Transplantation):

- Allogeneic transplantation patients who are seronegative at the time of transplant and have a seronegative donor may not require additional CMV prophylaxis other than the strict use of CMV-negative blood products.

TABLE 31.12 Reimmunization Schedule after Hematopoietic Cell Transplantation (HSCT)

	Month 0[a] (6–12 months post-HCT)	Month 2[a]	Month 6	Month 12[a]
Diphtheria/tetanus toxoid (Td) or DTaP/Tdap[b]	X	X		X
Inactivated poliovirus (IPV)[c]	X	X	X	
Haemophilus influenza (HiB)	X	X	X	
Hepatitis A[d] (see footnotes for pediatric schedule)	X	X (skip this dose if using single HepA)	X	
Hepatitis B[d]	X	X	X	
Human papillomavirus (HPV)[e]	X	X	X	
Pneumococcal[f]	Total 3 doses PVC (1-month interval) then 4th dose with PPSV 7 months later			
Meningococcal vaccine (MCV4 or MPSV4)[g]	X			
Influenza vaccine (inactivated)[h]	Seasonal; Lifelong administration; start before HCT then resuming ≥6 months after HSCT			
Influenza vaccine (live, nasal)	Contraindicated			
MMR (live, Measles, Mumps, Rubella)[i]	Contraindicated in HSCT recipients with GVHD or on immunosuppressive medications			X[i]
Varicella (live)[j]	Limited data for use after HSCT			
Zoster (live)	Contraindicated			

[a]Vaccinations may be administered all at one session or separated by 2–4 weeks for each time point post-HSCT. May be initiated 6–12 months post-HSCT, keeping intervals between vaccinations similar to schedule noted in chart above.

[b]DTaP (Diptheria toxoid, Tetanus toxoid and Pertussis) vaccine should be administered to children <7 years old. Tdap (Diptheria toxoid, tetanus toxoid and pertussis) for patients >7 years old. Tdap should replace a single dose of Td as booster for adults >19 years who have not previously received a dose of Tdap >10 years. Then boost with Td every 10 years.

[c]Oral polio vaccine is contraindicated in HSCT receipients.

[d]Hepatitis A: Give with Hepatitis B in a combination vaccine HepA/B at Hepatitis B schedule in >18 years as noted in chart above. For people 1 to 18 years old, single-antigen hepatitis A vaccine formulations should be administered in a 2-dose schedule at 12 and 24 months post-HSCT. Hepatitis A is recommended in endemic areas.

[e]HPV is recommended in all females at ages 11–12 years (range: 9–26 years) who have not completed the series. Series consists of 3 doses, second dose 2 months after the first, and 3rd dose administered 6 months after the first.

[f]Pneumococcus: All ages: PVC13 (pneumoccocal conjugate vaccine 13 valent) total 3 doses 1 month apart. A fourth dose with PPSV (pneumococcal polysaccharide) 7 months after the last PVC13 may be given to broaden immune response. In patient with chronic GVHD, may have poor response to PPSV, fourth dose with PVC13 may be considered.

[g]Meningococcal: Indicated in anatomic or functional asplenia, terminal complement component deficiencies, travel to endemic or epidemic areas, all college students living in dorms who have not been previously vaccinated. MCV4 preferred in patients >2 years old. Polysaccharide vaccine (MPSV4) is an acceptable alternative in that age group. MPSV4 is recommended for children 3 months to 2 years under certain circumstances.

[h]Influenza: It is strongly recommended that all household members of a transplant patient receive the influenza vaccine on a yearly basis. Transplant patients themselves should also receive this vaccination on a yearly basis beginning before HSCT and then resuming at least 6 months post-HSCT. For children <9 years, first year post-HSCT, 2 doses are recommended administered 1 month apart, then 1 dose there after annually.

[i]MMR: Contraindicated for patients <24 months post-HCT, with chronic GVHD or on immunosuppression. Not generally recommended for all transplant recipients, although should generally be administered to children. In children, 2 doses are favored, at least 28-day interval. Should be considered on an individual basis for patients from high-prevalence geographic areas or where risk is increased for these diseases (i.e., travel to foreign countries), and if >24 months post-HSCT, no GVHD, no immunosuppression. Rubella vaccine should be given to females with potential to become pregnant.

[j]Varicella vaccine: NOT for adults. Limited data on safety and efficacy. May be considered as optional in pediatric patients on case-by case basis and only if >24 months post-HSCT, no active GVHD, and no steroids/immunosuppression.

TABLE 31.13   Prophylaxis and Supportive Care for Stem Cell Transplantation

1. Mouth care: Peridex or Biotene and mycostatin every 3–4 h daily
2. Recombinant hematopoietic growth factors: G-CSF[a]
3. Prophylaxis against *Pneumocystis jiroveci*: trimethoprim/sulfamethoxazole 5 mg/kg in two divided doses (three times per week) starting after engraftment; continue if chronic GVHD is present[b,c]
4. Prophylaxis against *Candida* and other fungal infections is used. The choice of appropriate agent depends on several factors[d,e]
5. Prophylaxis for herpes simplex (HSV) and CMV: IV acyclovir 250 mg/m^2 for HSV; or 500 mg/m^2 for CMV every 8 hours for first 3–6 months
6. Additional prophylaxis for CMV:
   - CMV-negative blood products should be used if patient and donor are CMV-negative; if CMV-negative blood is not available, leukocyte filters to remove leukocytes should be employed
   - Foscarnet may be used to suppress CMV during neutropenia (in certain cases with history of CMV reactivation during conditioning or before engraftment)
7. Intravenous immunoglobulin (mostly in recipient of allogeneic stem cell transplant)[f]: 400 mg/kg every 4 weeks for first 3–6 months after transplantation and then check trough immunoglobulin G levels periodically if less than within normal limits then infuse accordingly. May consider to restart or continue for patients with GVHD
8. Nutrition: total parenteral nutrition given if enteral nutrition is not possible or inadequate

[a]*Does not increase relapse and GVHD rates. Stimulates hematopoietic recovery.*
[b]*If patients are allergic to trimethoprim/sulfamethoxazole or if severe cytopenia is present, aerosolized pentamidine or intravenously 4 mg/kg once a month can be used; other alternatives may include dapsone or atovaquone.*
[c]*Drug-related neutropenia is a major side effect.*
[d]*Use of azoles is usually avoided during conditioning chemotherapy and time interval of increased risk for SOS(VOD) after HSCT and commonly employed for prophylaxis after then. Lipid complex (or liposomal) amphotericin: 1–2 mg/kg IV/day. Micafungin IV (once daily), dose is based on age: for Candida prophylaxis: 6 mg/kg/dose (<2 Years); for age: 2 mg/kg (2–8 years) (maximum 50 mg/kg/day); 1 mg/kg (≥18 years). For Aspergillous (>2 years old): 4 mg/kg (maximum 200 mg/day). Fluconazole PO or IV (Candida only), dose is based on age and weight: <12 years (<20 kg, 100 mg PO or 6 mg/kg IV daily; >20 kg, 200 mg PO or 6 mg/kg IV daily; >12 years old 400 mg PO or IV daily. Voriconazole* (Candida and aspergillus), dose is based on age and weight: <12 years: PO (<20 kg, 100 mg q12 h; >20 kg, 200 mg q12 h); IV (<20 kg 6 mg/kg q12 h (max 100 mg/dose); >20 kg, 6 mg/kg q12 h (max 200 mg/dose); >12 years and >20 kg 400 mg q12 h). Posaconazole* PO (Candida, aspergillus and Zygomycetes) prophyalxis: for <13 years (<20 kg: 10 mg/kg BID; 20–34 kg: 8 mg/kg BID; >34 kg: 400 mg BID; (maximum 400 mg/dose for all)); for 13 years and older 200 mg TID.*
[e]*Trough concentration (μg/ml) is used for therapeutic drug monitoring as follows; Voriconazole: >0.5 prophylaxis; >2 treatment (keep level <5) Posaconazole (best absorbed with high fatty meals): >0.7 prophylaxis, >0.7 or 1.5 treatment.*
[f]*For its immunomodulatory (decreased incidence of GVHD) effect and antimicrobial activity, especially CMV. It also decreases Gram-negative sepsis.*

TABLE 31.14   Infections Most Frequently Seen at Different Times Posttransplantation

1. Infections in first 30 days posttransplantation
   a. Bacteremia
      i. Gram-positive organisms: *Staphylococcus epidermidis*
      ii. Gram-negative aerobes and anaerobes
   b. Invasive fungal infections: *Aspergillus, Candida*
   c. Reactivation of herpes simplex I
2. Infections 30–120 days posttransplantation
   a. Protozoal infections
      i. *Pneumocystis jiroveci*
      ii. *Toxoplasma*
   b. Viral infections
      i. Cytomegalovirus (CMV)
      ii. Adenovirus
      iii. Epstein–Barr virus (EBV)
      iv. Human herpes virus 6 (HHV-6)
   c. Fungal infections
      i. *Candida* (*C. albicans* and *C. tropicalis*)
      ii. *Aspergillus*
      iii. *Trichosporon*
      iv. *Fusarium*
      v. *Candida kruseii*
3. Infections after 120 days posttransplantation
   a. Sinopulmonary infections with encapsulated organisms
   b. Viral infections
      i. Cutaneous herpes zoster

TABLE 31.15 Management of Infections During and After HSCT

1. Gram-negative infections:
   a. Empiric combination broad-spectum therapy (e.g., fourth generation cephalosporins or carbapenem, and an aminoglycoside)
2. Gram-positive infections:
   a. Vancomycin
   b. Linezolid (if vancomycin-resistant enterococci)
3. Fungal infections[a]:
   a. Liposomal amphotericin
   b. Voriconazole
   c. Micafungin (or caspofungin)
   d. Posaconazole
   e. Removal of indwelling catheter
4. Interstitial pneumonia:
   a. Bronchoalveolar lavage (BAL)
   b. Open lung biopsy
   c. Treatment depends on etiology
      i. Pneumocystis: bactrim, pentamidine
      ii. CMV: ganciclovir (foscarnet before engraftment) plus CMV high-titer gammaglobulin
      iii. Idiopathic: supportive care, steroids
      iv. Fungal: antifungals
5. Herpes zoster:
   a. Acyclovir

[a]See Table 31.13 for detailed dosing and therapeutic monitoring. It is highly recommended that all patients undergoing HSCT with proven, probable, or possible aspergillosis infection (including past history with complete radiological resolution at time of transplant) receive double antifungal coverage until engraftment or radiological and laboratory resolution as well as patients with evidence of new aspergillosis infection after HSCT. It is recommended to screen for aspergillosis infection with galactomannan (GM) during HSCT and after then for selected high-risk patients.

- Allogeneic transplantation patients who are CMV-positive at transplantation or those who received transplantation from a CMV-positive donor should receive CMV safe, leukocyte-filtered blood products.
- Allogeneic transplantation patients should have surveillance for CMV antigenemia assay or a polymerase chain reaction (PCR) assay for blood and urine obtained weekly (or twice weekly) for the first 120 days posttransplantation. Patients CMV-positive from blood, bronchial washings, or urine should be treated with ganciclovir (foscarnet is used before engraftment or if indicated) and gammaglobulin at least until day 100 after transplantation or for 2–3 weeks post the latest date of positive PCR.
- Autologous transplantation patients who are seronegative at the time of transplantation should preferably receive CMV-negative blood products but may, as an alternative, receive leukocyte-filtered blood products.

Interstitial Pneumonitis

Interstitial pneumonitis is an important complication and may be due to any of the following causes:

- CMV.
- *Pneumocystis jiroveci.*
- Undiagnosed infections (e.g., fungal).
- Drugs.
- Radiation toxicity.
- Immune reactions involving the lung.
- Idiopathic (16%).

The mortality rate for interstitial pneumonitis is 60%. The use of prophylactic bactrim has reduced *Pneumocystis jiroveci* infection to less than 2%.

Pancytopenia

In addition to the corrections of anemia and thrombocytopenia by the use of irradiated leukocyte-depleted packed red cells and platelets, recombinant hematopoietic growth factor plays the following roles:

- It is safe and effective to administer growth factors to allogeneic transplantation patients. All allogeneic and autologous transplantation patients should receive G-CSF therapy following stem cell transplantation.

- With the use of ganciclovir to treat CMV infections and other myelosuppressive supportive therapies, growth factors may play an added role in the prevention or treatment of myelosuppression caused by these agents after stem cell transplantation.

Graft-Versus-Host Disease

GVHD is caused by donor alloreactivity in an immunocompromised host and results from T lymphocytes contained in the stem cell graft that proliferate and differentiate in the host. These T cells recognize host alloantigens as foreign and through both direct effector mechanisms and by inflammatory mediators released by T cells, monocytes, and production of cytokines in the host may cause tissue damage.

GVHD is divided into two stages:

- Acute GVHD manifests in the first 100 (most common within 30–40) days after stem cell transplantation.
- Chronic GVHD manifests 100 days after stem cell transplantation.

The incidence of acute GVHD varies among different transplantation centers and depends on the primary diagnosis. In patients with primary immunodeficiency and aplastic anemia, the incidence of acute GVHD may vary from 10–35%, whereas in patients with acute leukemia, it may be approximately 50%. The probability of acute GVHD grade II–IV is approximately 70% in unrelated matched donors compared to 30% in patients with HLA-A, -B, and -DR matched sibling transplantation. Approximately 10–30% of all patients with sustained allogeneic engraftment die of acute GVHD or its complications.

Chronic GVHD (cGVHD) is the primary cause of late non-relapse morbidity and mortality after allogeneic HSCT and is the major determinant of long-term quality of life in HSCT patients. The greatest risk factor for the development of cGVHD is preceding acute GVHD; therefore effective prophylaxis for acute GHVD is the best strategy for prevention of cGVHD. Chronic GVHD occurs in 30–80% of allogeneic HSCT patients with a higher incidence in recipients of HLA-mismatched or unrelated donor allografts. The risk factors predicting poor outcome for cGVHD are: thrombocytopenia (platelets <100,000), progressive-onset cGVHD (acute GVHD evolving into cGVHD without a disease-free interval), extensive skin involvement (>50% body surface area) and Lansky/Karnofsky performance status <50%.

The pathophysiology of this disease process includes the formation of autoantibodies and the inability to produce protective antibodies against environmental pathogens. Patients with extensive, multiorgan cGVHD have a poor prognosis with a high mortality rate. The primary cause of death in these patients is infection, hence aggressive prophylaxis and treatment of infection is an important component of therapy for cGVHD.

GVHD Prophylaxis

Methotrexate, calcineurin inhibitors (cyclosporine/tacrolimus), mycophenolate mofetil (MMF), sirolimus, and methylprednisolone are the most common drugs used singularly or in combination for GVHD prophylaxis in HSCT. The most commonly used combinations are a calcineurin inhibitor plus methotrexate or methylprednisolone. The drugs are used in the following dosages.

Methotrexate (Short Course)

Methotrexate 15 mg/m^2 IV on the day following transplantation and 10 mg/m^2 IV on days 3, 6, and 11 post transplantation (alternatively mini-methotrexate 5 mg/m^2 is used for all doses).

Cyclosporine

Cyclosporine is given IV, starting on the day before transplantation, based on body weight at age ≤6 years of age 6 mg/kg IV/day in divided doses (e.g., 2 mg/kg q8h); age >6 years of age: 3 mg/kg IV/day in divided doses (1.5 mg/kg q12h). Cyclosporine can be switched to an oral equivalent dose when the patient can take oral medication. Serum levels are usually kept between 200 and 250 mg/ml. Cyclosporine is slowly tapered after HSCT. The taper schedule is usually tailored to the treatment plan, indication for HSCT and type of donor (e.g. leukemia taper starts usually around day 40–60 for related donor and day 100–180 for unrelated donor) by 10% weekly.

Tacrolimus

The usual dose of the drug is 0.05 mg/kg to 0.1 mg/kg/day as IV continuous infusion to achieve the trough level of 5–12 mg/ml. Tacrolimus can be switched to oral equivalent dose when the patient can take oral medication.

Sirolimus

Sirolimus is an mTOR kinase inhibitor which can be used alone or commonly in combination with a calcineurin inhibitor. It is available only as oral medication. The starting dose is usually 2.5 mg/m^2/day (4 mg maximum). A loading dose is sometimes used, especially in older children and young adults. The dose is adjusted to achieve a target serum trough of 3–12 ng/ml. The sum of sirolimus and tacrolimus serum concentrations should be between 10–16 ng/ml for recipients of fully matched related/unrelated donors and 14–16 ng/ml for recipients of mismatched related/unrelated/CB donors.

Mycophenolate Mofetil

MMF is dosed at 15 mg/kg, based on adjusted body weight, every 8 h (45 mg/kg/day; max. 3 g/day) PO, or IV if indicated, after HSCT infusion (i.e., first dose at least 4–6 h following stem cell infusion). MMF can be tapered or stopped at once, depending on the treatment plan.

Methylprednisolone

Methylprednisolone is often used in combination with other immunosuppressive medications for high-risk patients with liver disease and with cord blood transplantation. Administration, dose schedule, and the taper are variable. The dose can range from 1–2 mg/kg/day posttransplantation and thereafter it is tapered slowly if there is no evidence of GVHD.

The other major technique to prevent GVHD is purging of the donor marrow of T lymphocytes as discussed above.

Acute GVHD

Despite GVHD prophylaxis, 20–80% of recipients of allogeneic HSCT will develop acute GVHD, with the incidence and severity dependent on the presence of various risk factors. Acute GVHD occurs more frequently and is more severe after HSCT from HLA-nonidentical or unrelated donors.

The clinical manifestations of acute GVHD range from a mild maculopapular eruption to generalized erythroderma, hepatic dysfunction, gastroenteritis, stomatitis, and lymphocytic bronchitis. Thrombocytopenia and anemia have been reported in GVHD. Ocular symptoms, including photophobia, hemorrhagic conjunctivitis, and pseudomembrane formation, have also been observed. It is occasionally difficult to separate the clinical manifestations of GVHD from other disorders in the posttransplant patient and, in these cases, a biopsy of the skin or liver may be required. Table 31.16 lists the clinical manifestations, stages, and grades of acute GVHD and Table 31.17 describes the histologic grades of acute GVHD. Table 31.18 presents a comparison between acute and chronic GVHD.

Treatment

Prophylaxis

Complete matching of the donor and recipient at the level of HLA class I and II by high-resolution DNA typing is the most important factor in preventing GVHD. Consideration of the sex and parity of the donor are advisable if there is a choice of donor. If the patient is CMV seronegative, use of a CMV-negative donor appears to reduce the risk of CMV infection as well as risk of GVHD. Table 31.19 lists additional measures that can be employed in the prophylaxis of GVHD, as mentioned earlier in this chapter. Prophylactic drugs have been used in an attempt to inhibit the T-cell response by relying on *in vivo* immunosuppression. Currently all agents are used in a combination that targets different molecular intermediates of T-cell signals.

TABLE 31.16 Clinical Manifestations, Stages and Grades of Acute Graft-versus-Host Disease (GVHD)

ACUTE GVHD STAGING

Clinical stage	Skin[a]	Liver (serum total bilirubin, mg/dl)[b]	Gut (diarrhea)[c]	
			Children (ml/kg/day)	Adults (ml/day)
I (mild)	Maculopapular rash, <25% body surface. May be pruritic or painful	1.5–3.0	10–15	500–1000 nausea and vomiting
II (moderate)	Maculopapular rash, 25–50% body surface	3.0–6.0	16–20	1000–1500 nausea and vomiting
III (severe)	Maculopapular rash, >50% body, or generalized erythroderma	6.0–15.0	21–25	>1500 nausea and vomiting
IV (life-threatening)	Desquamation and bullae	>15	>25 pain or ileus	>2500 pain or ileus

ACUTE GVHD GRADING

Clinical grade	Skin	Liver	Gut
I	Stage 1–2	None	None
II	Stage 3 and/or	Stage 1 and/or	Stage 1
III	None or Stage 3	Stage 2–3 or	Stage 2–4
IV	Stage 4 or	Stage 4	NA

[a]*Differential diagnosis includes chemoradiotherapy-induced rash, drug allergy, and viral exanthema.*
[b]*The degree of hyperbilirubinemia does not correlate well with clinical outcome. Differential diagnosis includes hepatotoxic drug reactions and viral infections. Liver biopsy is helpful in establishing the diagnosis.*
[c]*Diarrhea may be severe and associated with crampy abdominal pain. The diarrhea is green, mucoid, watery, and mixed with exfoliated cells that may form fecal casts. It may progress to bleeding and ileus. Intestinal radiographs show mucosal and submucosal edema with rapid barium transit time and loss of haustral folds. A variant of enteric GVHD in 13% of patients has presenting features of anorexia and dyspepsia. Patients with upper gastrointestinal disease may not manifest lower tract involvement. Gastrointestinal endoscopy and biopsy are mandatory for the diagnosis of upper gastrointestinal tract disease. Lower tract disease may be diagnosed with rectal biopsy, which shows necrosis of individual cells in crypts that may progress to dropout of entire crypts and loss of epithelium.*

TABLE 31.17 Histologic Grades of Acute Graft-versus-Host Disease (GVHD)

Grade	Skin	Liver	Gut
I	Epidermal base cell vacuolar degeneration	Fewer than 25% small interlobular bile ducts abnormal (degeneration or necrosis)	Single-cell necrosis of epithelial cells
II	Grade I changes plus "eosinophilic bodies"	25–50% bile ducts abnormal	Necrosis
III	Grade II changes plus separation of the dermal–epidermal junction	50–75% bile ducts abnormal	Focal microscopic mucosal denudation
IV	Frank epidermal denudation	More than 75% bile ducts abnormal	Diffuse mucosal denudation

Therapy

Grades II, III, and IV GVHD require therapeutic intervention with high-dose methylprednisolone 2–5 mg/kg/day for 7 days, after which the dose is tapered. Basiliximab (Anti-IL-2 receptor) is often used as a steroid-sparing agent. Topical steroids are frequently used for gastrointestinal (GI) GVHD (e.g., beclomethasone or budesonide).

The progression of GVHD despite steroid therapy in these doses requires an increase in the dose of methyl-prednisolone to 20 mg/kg/day for 3 days, 10 mg/kg/day for 3 days, 5 mg/kg/day for 3 days, 3 mg/kg/day with tapering doses of steroids, depending on the response or the use of ATG. ATG is usually given in doses of 10–15 mg/kg every other day for 7–14 days. Newer immunosuppressive agents like tacrolimus, sirolimus, MMF, rituximab, cyclophosphamide (post-HSCT), and monoclonal antibodies targeting CD3, CD52, IL-2 receptor, and TNFα have been used with some success. Extracorporeal photopheresis (ECP) has also been utilized for the treatment of aGVHD.

TABLE 31.18 Comparison of Acute and Chronic Graft-versus-Host Disease (GVHD)

Characteristic	Acute	Chronic
Incidence	40–60%	20–40%
Onset, days	7–60 (up to 100)	>100
Clinical Manifestations		
Skin	Erythematous rash	Sclerodermatous changes
Gut	Secretory diarrhea	Dry mouth, esophagitis, malabsorption
Liver	Hepatitis	Cholestasis
Lung	Diffuse alveolar hemorrhage	Pulmonary dysfunction
Other	Fever	Contractures (due to sclerodermatous skin damage)
		Alopecia
		Thrombocytopenia
Target cells	Epidermal	Mesenchymal
Basis of clinical grading	Severity of disease	Extent of disease

TABLE 31.19 Prophylaxis of Graft-versus-Host Disease (GVHD)

1. Complete matching of the donor and recipient at the level of HLA class I and II by high-resolution DNA typing
2. Prevention of infection
3. *In vivo* treatment of recipient[a]
   a. Cyclosporine or tacrolimus
   b. Methotrexate
   c. Corticosteroids
   d. MMF
   e. Sirolimus
   f. Basiliximab (rarely used for primary prophylaxis)
4. Depletion of donor T cells from graft such as:
   a. CD34 selection of the graft
   b. *In vivo* T-cell depletion: ATG, alemtuzumab

[a]*Current prevention in many centers consists of methotrexate and cyclosporine (or tacrolimus).*

Chronic GVHD

Chronic GVHD occurs 100 days after transplantation. It may:

- Follow a progressive extension of acute GVHD, or
- Follow a quiescent period after acute GVHD has resolved, or
- Occur in a patient who did not have acute GVHD.

In matched sibling grafts, the incidence of chronic GVHD is 13% for children under 10 years of age and 28% for children 10–19 years of age. In unrelated donors or mismatched-family member donors, the incidence of chronic GVHD ranges from 42–56%.

The predictors for chronic GVHD are:

- Donor–recipient HLA disparity.
- Increasing patient/donor age.
- Latent viral infection.
- Female donor with a history of parity (sensitized).
- Source of allogeneic cells, number of T cells collected during pheresis is greater than that collected during bone marrow harvest.

TABLE 31.20   Classification of Chronic Graft-Versus-Host Disease (cGVHD)

Limited:	• Either or both • Localized skin involvement • Hepatic dysfunction

Extensive	• Either
	• Generalized skin involvement
	• Hepatic dysfunction
	• PLUS any of the following:
	  ○ Ocular involvement (Schirmer's test, 5 mm wetting)
	  ○ Oral mucosal involvement (lip biopsy positive) or involvement of minor salivary glands
	  ○ Liver histology showing chronic progressive hepatitis, bridging necrosis or cirrhosis
	  ○ Any other target organ

The overall severity of cGVHD can be assigned as follows[a]:

• Mild	Signs and symptoms of cGVHD do not interfere substantially with function and do not progress once appropriately treated with local therapy or standard systemic therapy (steroids and/or cyclosporine or tacrolimus)
• Moderate	Signs and symptoms of cGVHD interfere somewhat with function despite appropriate therapy or are progressive through first line systemic therapy defined as steroids and/or cyclosporine or tacrolimus
• Severe	Signs and symptoms of cGVHD limit function substantially despite appropriate therapy or are progressive through second line therapy

[a]CIBMTR utilizes a global severity scores (where eight organs are scored on a 0−3 scale to reflect degree of cGVHD involvement, as well as liver and pulmonary function results) to characterize cGVHD as mild, moderate or severe.

One-third of patients who develop chronic GVHD are at risk for bacterial and opportunistic infections. The duration of chronic GVHD varies and the morbidity and mortality range from 10 to 15%. Chronic GVHD can be divided into two groups (limited or extensive) (Table 31.20).

Limited (Usually Involving Only One Organ)

• Localized skin involvement: erythema, hyperkeratosis, patchy scleroderma-like lesions, reticular hyperpigmentation, desquamation, alopecia, nail loss; rash initially involving palms and soles, back of neck and ears and later the trunk and extremities.
• Hepatic dysfunction: predominantly cholestatic abnormalities.

Extensive

• Generalized skin involvement: resembling lichen planus or scleroderma; atrophy of skin with ulceration; fibrosis of skin with or without limitation of joint movement.
• Hepatic dysfunction: liver histology showing chronic aggressive hepatitis, bridging necrosis, or cirrhosis.
• Ocular symptoms: dry eyes, Schirmer's test with less than 5 mm wetting; keratoconjunctivitis sicca including burning, irritation, photophobia, and pain.
• Buccal cavity: dry mouth, sensitivity to acidic or spicy foods and pain; oral atrophy with depapillation of tongue; erythema and lichenoid lesions of buccal and labial mucosa; involvement of minor salivary glands or oral mucosa demonstrated on labial biopsy specimen.
• GI tract: malabsorption due to submucosal fibrosis of GI tract; dysphagia, pain, and weight loss.
• Genitourinary tract: vaginitis and vaginal strictures.
• Polyarthritis, myositis, and fasciitis leading to contractures.
• Pulmonary complications: interstitial pneumonia, and bronchiolitis obliterans (BO).

Prognosis

Morbidity and mortality are highest in patients with progressive-onset chronic GVHD that directly follows acute GVHD, are intermediate in patients after resolution of acute GVHD and are lowest in those with de novo onset. History of previous acute GVHD independently predicts a threefold increase in the relative risk of subsequent chronic GVHD.

Overlap Syndrome

Manifestations of both acute and chronic GVHD are present. Features of acute GVHD that usually exist in overlap syndrome include changes in the skin (erythema, rash, pruritus), mouth (gingivitis, mucositis, oral erythema, and pain), GI symptoms (anorexia, nausea, vomiting, diarrhea, weight loss), liver dysfunction (elevations in bilirubin, alkaline phosphatase, ALT, or AST), and BO.

Treatment

Methylprednisolone 1−2 mg/kg (or prednisone) alone or in combination with PO cyclosporine 10−12 mg/kg/day or tacrolimus 0.15−0.3 mg/kg/day in two divided doses. Cyclosporine and tacrolimus are adjusted as needed to achieve therapeutic drug levels or in response to drug toxicity. After 2 weeks of therapy if chronic GVHD is responding, the steroids are tapered, as tolerated. After the steroid taper is completed, if the chronic GVHD is stable to improving, the cyclosporine or tacrolimus are tapered. The goal is to taper the immunosuppressive medication to a regimen in which steroids are given on alternate days to the cyclosporine or tacrolimus. Usually immune suppression is continued for 6 months. If patients are responding to primary therapy at 3 months, the same regimen may continue until maximal response is achieved and then the immunosuppression can be weaned with close monitoring. Patients who do not respond to primary therapy or show progression should be treated with an immunosuppressive regimen, as described below.

Salvage therapy for cGVHD may include:

1. MMF 15 mg/kg IV or PO (2−3 times/day):
   a. The combination of MMF plus tacrolimus was reported to result in objective responses in 46% of steroid-refractory cGVHD. Toxicities of MMF include: reversible cytopenias, especially leukopenia, when the drug is used early posttransplant and GI side effects.
2. Sirolimus:
   a. The combination of sirolimus and tacrolimus was reported to result in objective responses in 62% of steroid-refractory cGVHD. For patients <40 kg, a loading dose of 3 mg/m^2 on day 1 should be followed by a once-daily dose of 1 mg/m^2. For patients >40 kg, a loading dose of 6 mg sirolimus on day 1 of treatment should be followed by a once-daily dose of 2−4 mg.
   b. The dose should be adjusted to achieve a whole blood trough concentration of 3−12 ng/ml. Trough levels should be measured 20−24 h after the last dose.
   c. When used in combination with tacrolimus the total sum of tacrolimus and sirolimus levels should not exceed 12 ng/ml to minimize chances for toxicity.
3. Hydroxychloroquine (HCQ):
   a. Hydroxychloroquine (HCQ) is an antimalarial drug used for the treatment of autoimmune disease. HCQ interferes with antigen processing and presentation, cytokine production and cytotoxicity and is synergistic with cyclosporine and tacrolimus *in vitro*. Common toxicities are nausea, diarrhea, and cramps. HCQ has also been reported to rarely cause neuropathy or myopathy. G6PD testing is recommended before starting therapy with HCQ. HCQ therapy is used in steroid-refractory cGVHD with reported response in cGVHD of skin, liver, and oral mucosa as the most responsive sites. The dose used for cGVHD is 800 mg/day or 12 mg/kg/day if weight is <50 kg. Recommended dose adjustments are: decrease dose by 25% for bilirubin levels >6 times the upper limit of normal and decrease by 50% for bilirubin levels >12 times the upper limit of normal. Decrease dose by 25% for creatinine values greater than or equal to three times the upper limit of normal. Increase dose by 50% for documented malabsorption or severe diarrhea. Patients on HCQ should be monitored for retinal, hepatic, and renal toxicity.
4. Thalidomide: Thalidomide is most efficacious in treating mucosal and chronic GVHD of skin. Patient tolerance of thalidomide varies widely; thus the dose prescribed can vary from 50 mg to 600 mg per day. Toxicities include sedation, constipation, neuropathy, marrow suppression, and occasionally severe skin reactions. Thalidomide should not be prescribed for patients who have preexisting neuropathy and is contraindicated in any female patient of childbearing age who is not using adequate contraception. It is recommended to start at a low dose of 100 mg/day and escalate the dose as tolerated. Thalidomide can be given as a single bedtime dose or divided into three to four doses per day. Most patients cannot tolerate a dose exceeding 400−600 mg/day.

5. Azathioprine: 1.5 mg/kg once a day may also be utilized. The azathioprine dose is adjusted to keep the white blood cell count between 2500 and 3500/mm^3. This can have a steroid-sparing effect if unable to taper steroids.
6. PUVA: Photochemotherapy with psoralen plus ultraviolet A irradiation (PUVA) prevents or inhibits allorecognition between donor and recipient cells and tissues through a variety of immunomodulatory mechanisms including alterations in cell surface antigens and interference in the processes of transmembrane signaling and intracellular activation. Psoralen covalently bonds to pyrimidine bases in DNA and thereby suppresses DNA synthesis and cell division. PUVA has been used to treat skin and oral cGVHD. Side effects of psoralen include nausea and vomiting, elevation of liver function tests and pruritis as well as extreme photosensitivity. Side effects of PUVA include skin erythema, tenderness, blistering and dryness, hyperpigmentation and actinic keratoses, and secondary malignancy.
7. ECP has also been found to be effective in controlling chronic GVHD and in reducing the need for prednisone in patients with refractory GVHD. The ECP procedure involves collecting the patient's donor-derived white cells and treating the cells with the drug uvadex and exposing them to ultraviolet light. The treated cells are then given back to the patient.

The following supportive care should be given in addition to the primary therapy:

- IV gammaglobulin 500 mg/kg every 2–4 weeks (as per IgG levels).
- Bactrim prophylaxis for *Pneumocystis jiroveci* infection.
- Antibiotics for chronic sinopulmonary infection.
- Acyclovir or valganciclovir prophylaxis.
- Antifungal prophylaxis.

Therapy should be continued for a minimum of 6–9 months or until all clinical and pathologic evidence of chronic GVHD has cleared then taper medical treatment gradually.

Sinusoidal Obstruction Syndrome (VOD)

Veno-occlusive disease (sinusoidal obstruction syndrome (SOS)) occurs as a complication of chemotherapy and/or radiation in allogeneic stem cell transplant. It is a common life-threatening complication of preparative-regimen-related toxicity of stem cell transplantation. It occurs after approximately 20% of allogeneic stem cell transplantations and after about 10% of autologous stem cell transplantations. SOS is characterized by fibrous obliteration of small hepatic vessels. It usually occurs within the first 30 days.

SOS incidence is reported between 5 and 50% with mortality as high as 15% in HSCT recipients. Clinically, SOS is defined as jaundice, tender hepatomegaly, and weight gain (>10% baseline weight) following the high-dose conditioning regimen, in the absence of other explanations for these symptoms. It is characterized histologically by necrosis of zone III hepatocytes. SOS is usually a clinical diagnosis and is more common after conditioning regimens that include high-dose cyclophosphamide, TBI, and Bu and in patients with a history of treatment with gemtuzumab ozogamicin. Use of sirolimus-based GVHD prophylaxis regimens has also been associated with a higher rate of SOS in patients receiving myeloablative conditioning regimens. Since treatment of established SOS is challenging, it is preferable to prevent this complication. Various measures have been utilized for SOS prophylaxis including continuous IV unfractionated heparin, low-molecular weight heparins, and ursodiol. Treatment of established SOS has been unsatisfactory and typically conservative supportive measures and strict fluid are utilized. Early initiation of continuous veno-venous hemofiltration (CVVH) for treatment of SOS for both fluid management and cytokines removal has been used with successful outcomes for treatment of SOS. CVVH is used as salvage therapy for SOS by many transplant centers.

Defibrotide (not FDA-approved for use in the United States) has been associated with some success in treatment and prophylaxis of SOS.

Clinical Manifestations

1. Jaundice.
2. Weight gain.
3. Ascites.
4. Hepatomegaly.

5. Right upper quadrant pain.

Because not all patients exhibit the full spectrum of the syndrome, a common clinical definition requires the presence of any two of the above-listed features with the onset occurring between 7 and 21 days from stem cell infusion. Because these clinical manifestations are not specific to SOS, all other causes of hepatic dysfunction must be excluded. Another condition that may mimic hepatic SOS is nodular regenerative hyperplasia of the liver, a diffuse nonfibrotic nodulation of the liver with areas of regenerative activity alternating with areas of atrophy. This syndrome affects the same transplantation population at risk for hepatic SOS and both conditions are frequently associated with the development of ascites. Although no specific treatments are available for either entity, the mortality rate is considerably higher for patients who develop hepatic SOS than for those who develop nodular regenerative hyperplasia.

Predisposing Factors

The cause of SOS remains unknown; however, patients with the following conditions have an increased risk:

- Preexisting hepatitis (one of the most strongly predictive risk factors).
- Elevated liver function tests.
- Conditioning agents (such as Bu based, TBI based, and cyclophosphamide).
- Mismatched or unrelated allogeneic transplants

Prophylaxis

The following measures have been employed as prophylaxis in SOS:

- Lower or fractionated doses of TBI.
- Targeted Bu levels.
- Strict fluid management.
- Use of ursodeoxycholic acid.
- Continuous infusion of low-dose heparin has been used to reduce toxicity.[6]
- Defibrotide has shown promising results (not FDA-approved for use in the United States).

Treatment

1. Supportive care.
2. High-dose spironolactone therapy.
3. CVVH.
4. Defibrotide (not FDA-approved for use in the United States).
5. Antithrombin III (AT III) infusion if AT III levels are low.
6. Transjugular intrahepatic portosystemic shunt procedure.

LATE SEQUELAE OF STEM CELL TRANSPLANTATION

Table 31.21 lists the late sequelae of allogeneic stem cell transplantation in children.

Recent Advances in HSCT

Minimal Residual Disease Assessment Pre-HSCT

In patients diagnosed with ALL in morphological remission, who proceed to HSCT, there is strong evidence that pre-HSCT minimal residual disease (MRD) (defined as any detection of disease $\geq 0.01\%$) is associated with a higher risk of relapse after HSCT. In one study, the estimated 2-year relapse rate was 27.4% (17.4–43.1) in

[6]Heparin has an effect on the overall incidence of VOD and may lessen the severity of the clinical illness. A dose of 100 units/kg/day of heparin has been shown to be safe. A combination of heparin, glutamine, and ursodeoxycholic has been shown to decrease the incidence of VOD.

TABLE 31.21	Late Sequelae of Allogeneic Stem Cell Transplantation in Children

1. Chronic GVHD
2. Multiple endocrine disorders (hypothyroidism, delayed pubescence, gonadal failure, growth hormone deficiency)[a]
3. Second malignancies[b]
4. Sterility due to gonadal failure
5. Cataracts (secondary to radiation therapy); 20–50% incidence at 5–6 years
6. Renal insufficiency (nephrotoxic antibiotics and cyclosporine)
7. Obstructive and restrictive pulmonary disease (occurs in 10–15% of patients with chronic GVHD)
8. Cardiomyopathy
9. Aseptic necrosis of bone
10. Leukoencephalopathy (especially with IT MTX)
11. Immunologic dysfunction
12. Disturbance in dental development (from TBI) particularly if HSCT takes place in patients less than 6 years of age

[a]Associated with the use of TBI.
[b]The frequency and severity of these sequelae vary considerably with the conditioning required preceding SCT.

MRD-negative patients and 70.8% (50.4–99.4) in MRD-positive patients. The level of MRD detection also correlates with relapse risk, with higher levels of MRD being associated with higher relapse rates especially when the MRD level is >0.1%.

In AML, a negative impact of pre-HCT MRD positivity was also observed.

The impact of pretransplant MRD reduction on post-HCT outcomes has not been prospectively determined. Achieving an MRD-negative complete remission prior to HCT is not always possible. Furthermore, attempts to reduce the level of or eradicate MRD are not without risk. For example, treatment-associated toxicities, progressive disease, or delays in the time to HSCT with further attempts to treat MRD could compromise transplant efficacy, safety, or feasibility.

It is confirmed in several studies that not all patients who undergo HSCT with MRD-positive disease will relapse. A recent study of children undergoing HSCT for ALL or AML demonstrated that the survival rates increased as pre-HSCT MRD levels decreased, being 29, 52, and 68% for patients with high, low, and no MRD positivity, respectively. These results suggest that more than half of patients with low levels of MRD prior to HSCT will have long-term survival after HSCT and that MRD positivity alone should not eliminate HSCT as a treatment option. Therefore, the risk between delaying HSCT to administer additional therapy pre-HSCT to achieve negative MRD, and proceeding to HSCT with MRD-positive disease needs to be carefully evaluated and individually decided.

A possible approach for MRD-positive disease is early weaning off immune suppression and/or use of novel therapies (e.g. immune- or cell-based) in the peri-HSCT setting.

Adoptive Cell Therapy

Adoptive cell transfer therapy uses patient-specific autologous or HLA-matched allogeneic lymphocytes to elicit an antitumor immune response. Several approaches are currently being tested in clinical trials.

Dendritic cells (DC) are the most efficient antigen-presenting cells *in vivo*. Therefore, DC-based vaccines have been used to mediate interaction between innate and adaptive immune responses. Mature DC can stimulate activation of autologous tumor-specific CD8 + T cells to reduce the tumor mass. DC can be isolated from peripheral blood mononuclear cells, expanded *in vitro* and challenged with a wide variety of cancer-specific antigens.

NK cells can lyse target cells without MHC restriction. NK cell cytotoxicity is mainly dependent on the balance between activating and inhibitory signals. NK cells are activated to kill target cells which have downregulation of MHC-I expression. Therefore, tumor cells that express low MHC-I molecules to evade immunosurveillance are the ideal target cells for NK cells to exert antitumor effects. NK-cell-mediated cell lysis can be enhanced by using antibodies blocking NK-inhibitory receptors or antibodies targeting activating receptors. For example, antibody blocking KIR significantly promoted NK cell Ab-dependent cellular cytotoxicity responses in a human cancer. NK cells can be derived from several sources including autologous NK cells, allogeneic NK cells, NK cell lines, genetic modified NK cells, HSCs, and induced pluripotent stem cells. By cytokine stimulation, autologous NK cells can be transformed into lymphokine-activated killer cells and exhibit greater cytotoxicity against tumor cells.

T-cell-based transfer, either through cytokine-activated T cells or through genetically modified T cells, is currently being tested in clinical trials.

T cells can be equipped with genes that encode receptors capable of recognizing cancer-specific antigen. Most recently several studies are investigating the use of autologous chimeric antigen receptor T cells directed at CD19 for treatment of B-precursor ALL, and have shown promising results.

Outcomes

Table 31.22 lists the probability of survival following allogeneic stem cell transplantation. Disease-free survival varies among different published series.

TABLE 31.22 Probability of Survival Following Allogeneic Stem Cell Transplantation[a]

Disease	Probability of survival (%)	
	Identical sibling	Unrelated donor
AML		
Disease status		
Early	66 ± 1	48 ± 1[b]
Intermediate	57 ± 4	46 ± 1[b]
Advanced	35 ± 3	26 ± 1[b]
ALL		
Disease status		
Early	67 ± 2	64 ± 2
Intermediate	55 ± 2	47 ± 1
Advanced	26 ± 3	31 ± 3
Aplastic anemia	88 ± 1	70 ± 2

[a]*Between 2001 and 2011, the 3-year probabilities of survival following transplant from HLA-identical sibling for patients <20 years old.*
[b]*Includes all ages.*
AML and ALL where classified as: early phase (first complete remission (CR1)), intermediate phase (second or subsequent CR), or advanced phase (primary induction failure, active disease).
AML, acute myeloid leukemia; ALL, acute lymphocytic leukemia.

Further Reading and References

Barrett, D., Fish, J.D., Grupp, S.A., 2010. Autologous and allogeneic cellular therapies for high-risk pediatric solid tumors. Pediatr. Clin. North Am. 57 (1), 47–66.

Coppell, J.A., Richardson, P.G., Soiffer, R., Martin, P.L., Kernan, N.A., Chen, A., et al., 2010. Veno-occlusive disease following stem cell transplantation: incidence, clinical course and outcome. Biol. Blood Marrow Transplant. 16 (2), 157–168.

Grupp, S.A., Kalos, M., Barrett, D., Aplenc, R., Porter, D.L., Rheingold, S.R., et al., 2013. Chimeric antigen receptor-modified T cells for acute lymphoid leukemia. N. Engl. J. Med. 368 (16), 1509–1518.

Inamoto, Y., Kim, D.D., Storer, B.E., Moon, J.H., Lipton, J.H., Kuruvilla, J., et al., 2014. Application of CIBMTR risk score to NIH chronic GVHD at individual centers. Blood 123 (3), 453–455.

Lakshminarayanan, S., Sahdev, I., Goyal, M., Vlachos, A., Atlas, M., Lipton, J.M., 2010. Low incidence of hepatic veno-occlusive disease in pediatric patients undergoing hematopoietic stem cell transplantation attributed to a combination of intravenous heparin, oral glutamine, and ursodiol at a single transplant institution. Pediatr. Transplant. 14 (5), 618–621.

Leung, W., Pui, C.H., Coustan-Smith, E., Yang, J., Pei, D., Gan, K., et al., 2012. Detectable minimal residual disease before hematopoietic cell transplantation is prognostic but does not preclude cure for children with very-high-risk leukemia. Blood 120 (2), 468–472.

Li, M., Sun, K., Welniak, L.A., Murphy, W.J., 2008. Immunomodulation and pharmacological strategies in the treatment of graft-versus-host disease. Expert Opin. Pharmacother. 9 (13), 2305–2316.

Locatelli, F., Pagliara, D., 2012. Allogeneic hematopoietic stem cell transplantation in children with sickle cell disease. Pediatr. Blood Cancer 59 (2), 372–376.

Luke, J.L., McDonald, L., Jude, V., Chan, K.W., Cuellar, N.G., et al., 2013. Clinical practice implications of immunizations after pediatric bone marrow transplant: a literature review. J. Pediatr. Oncol. Nurs. 30 (1), 7–17.

Martin, P.J., Inamoto, Y., Flowers, M.E., Carpenter, P.A., 2012. Secondary treatment of acute graft-versus-host disease: a critical review. Biol. Blood Marrow Transplant. 18 (7), 982–988.

Pizer, B., Donachie, P.H., Robinson, K., Taylor, R.E., Michalski, A., Punt, J., Ellison, D.W., 2011. Treatment of recurrent central nervous system primitive neuroectodermal tumours in children and adolescents: results of a Children's Cancer and Leukaemia Group study. Eur. J. Cancer 47 (9), 1389–1397.

Prasad, V.K., Kurtzberg, J., 2010. Transplant outcomes in mucopolysaccharidoses. Semin Hematol. 47 (1), 59–69.

Pui, C.H., Carroll, W.L., Meshinchi, S., Arceci, R.J., 2011. Biology, risk stratification, and therapy of pediatric acute leukemias: an update. J. Clin. Oncol. 29 (5), 551–565.

Pulsipher, M.A., Peters, C., Pui, C.H., 2011. High-risk pediatric acute lymphoblastic leukemia: to transplant or not to transplant? Biol. Blood Marrow Transplant. 17 (1 Suppl), S137–S148.

Tomblyn, M., Chiller, T., Einsele, H., Gress, R., Sepkowitz, K., Storek, J., et al., 2009. Guidelines for preventing infectious complications among hematopoietic cell transplant recipients: a global perspective. Biol. Blood Bone Marrow Transplant. 15 (10), 1143–1238.

Vlachos, A., Lipton, J.M., 2002. Conn's Current Therapy. W.B. Saunders Company, Philadelphia, PA.

Management of Oncologic Emergencies

Jason L. Freedman and Susan R. Rheingold

Survival in children with cancer has increased dramatically during the past five decades. This progress is due to advances in specific oncologic therapies as well as supportive care, and an improved ability to manage life-threatening complications. Oncologic emergencies can occur as an initial manifestation of cancer, as a side effect of therapy, or at the time of progression or recurrence of the disease.

Excellent cancer management requires mastery of the following:

- Metabolic emergencies including hyperleukocytosis, tumor lysis syndrome (TLS), and associated electrolyte derangements.
- Cardiothoracic emergencies including superior vena cava syndrome and mediastinal masses.
- Acute abdominal processes.
- Renal dysfunction and hypertension.
- Neurologic emergencies.
- Endocrine emergencies.
- Treatment-related emergencies.

METABOLIC EMERGENCIES

Hyperleukocytosis

Hyperleukocytosis is defined as a total white cell count greater than $100,000/mm^3$. Hyperleukocytosis is seen at presentation in 9–13% of children with acute lymphocytic leukemia (ALL) and 5–22% of children with acute myeloid leukemia (AML). It occurs in an even higher percentage of patients with chronic myeloid leukemia (CML).

The presence of a high number of blasts in the microcirculation leads to sludging, which interferes with oxygenation of local tissue, ultimately leading to tissue ischemia. This in turn leads to an adhesive reaction between abnormal vascular endothelium and the circulating blasts worsening leukostasis, thrombosis, and leading to secondary hemorrhage. The higher metabolic rate of the blasts and the local production of cytokines also contribute to tissue hypoxia. Thrombi in the circulation lead to vascular damage and parenchymal ischemia manifested as pulmonary or cerebrovascular hemorrhage and edema.

Myeloblasts are larger, less deformable, and more adherent to vasculature than lymphoblasts. Due to these intrinsic properties, leukostasis and thrombosis are far more prevalent in AML than in ALL. At presentation, patients with AML are more likely to have intracranial hemorrhage or thrombosis or pulmonary hemorrhage and leukostasis, whereas ALL is more likely to lead to metabolic disturbance from tumor lysis syndrome.

As leukostasis is associated with early morbidity and mortality, any patient presenting with a white blood cell count greater than $50,000/mm^3$ should be evaluated closely for clinical signs and symptoms of leukostasis.

Lanzkowsky's Manual of Pediatric Hematology and Oncology.
DOI: http://dx.doi.org/10.1016/B978-0-12-801368-7.00032-6

Clinical Features

- Central nervous system (CNS): blurred vision, confusion, somnolence, delirium, stupor, coma and papilledema.
  - Computed tomography (CT) may reveal hemorrhage or leukemic infiltrate.
- Pulmonary: tachypnea, dyspnea, hypoxia.
  - Chest radiograph may reveal varying degree of diffuse interstitial or alveolar infiltrates.
- Genitourinary: oliguria, anuria, rarely priapism.
- Vascular symptoms which include disseminated intravascular coagulation (DIC), retinal hemorrhage, myocardial infarction, and renal vein thrombosis.

Risk Factors

- WBC counts >200,000/mm^3 in AML, and >300,000/mm^3 in ALL and CML.
- Age less than 1 year.
- M4, M5 AML (higher lysozyme activity).
- Cytogenetic abnormalities including MLL 11q23, t(4:11), inv16, Philadelphia positive, and Philadelphia-like ALL and FLT3-ITD.

Tumor Lysis Syndrome

TLS arises due to the rapid release of intracellular metabolites (such as phosphorous, potassium, and uric acid) from dying tumor cells in quantities that exceed the excretory capacity of the kidneys.

In patients with a high tumor burden or rapid cell proliferation, such as Burkitt or Burkitt-like lymphoma, B-cell ALL, and T-cell leukemia or lymphoma, significant cell death and release of intracellular ions, even prior to chemotherapy initiation, may result in the following metabolic complications:

- Hyperuricemia.
- Hyperkalemia.
- Hyperphosphatemia.
- Hypocalcemia.
- Renal insufficiency/failure.

If not successfully treated, TLS can result in cardiac arrhythmias, renal failure, seizures, coma, DIC, and death.

Diagnostic Criteria for Laboratory TLS and Clinical TLS

Laboratory TLS (LTLS): The presence of two or more abnormal serum values at presentation (i.e., uric acid ≥8 mg/dl, potassium ≥6 mg/dl, phosphate ≥2.1 mmol/l, calcium ≤1.75 mmol/l, creatinine >normal).

Clinical TLS (CTLS): Presence of LTLS and one or more of the following clinical complications: renal insufficiency, cardiac arrhythmias, seizures or sudden death.

Recognition of risk factors, close monitoring, and appropriate preventative intervention are vital in managing TLS. Table 32.1 lists patient stratification by risk for TLS for various types of cancer.

TABLE 32.1 Patient Stratification by Risk for TLS for Various Types of Cancer

Type of cancer	Risk		
	High	Intermediate	Low
NHL	Burkitt, lymphoblastic	DLBCL	Indolent NHL
ALL	WBC ≥100,000/mm^3	WBC 50,000–100,000/mm^3	WBC ≤50,000/mm^3
AML	WBC ≥50,000/mm^3	WBC 10,000–50,000/mm^3	WBC ≤10,000/mm^3
Other hematologic malignancies (including CML) and solid tumors		Rapid proliferation with expected rapid response to therapy	Remainder of patients

NHL, non-Hodgkin's lymphoma; DLBCL, diffuse large B-cell lymphoma; ALL, acute lymphoblastic leukemia; AML, acute myeloid leukemia; CML, chronic myeloid leukemia.
Source: Reproduced with permission from Coiffier et al., 2008.

Prevention and Management of TLS

Prevention

1. Fluids and hydration
   a. Promote excretion of uric acid and phosphorus; increase glomerular filtration rate (GFR) and renal blood flow.
   b. Hydration at a rate of $>2 \, l/m^2/day$ should start 24 h before chemotherapy.
   c. Urine output goal of 3 ml/kg/h should be maintained. Diuretics may be needed to maintain this urine output.
2. Allopurinol
   a. A xanthine analog which blocks conversion of xanthine and hypoxanthine to uric acid; works as a competitive inhibitor of xanthine oxidase.
   b. Does not remove preformed uric acid.
   c. Slow reduction in uric acid levels.
   d. Decreased risk of uric acid crystallization in kidney tubules.
   e. Adverse effects include hypersensitivity reaction and a build up of xanthine and hypoxanthine, which may lead to renal insufficiency as well.
3. Rasburicase
   a. Indicated for patients at high risk of TLS.
   b. Assess patient's G6PD status or risk of having G6PD, as may cause methemoglobinemia or severe hemolytic anemia.
   c. May require reinitiation of allopurinol several days after rasburicase.

Active management

Table 32.2 outlines the management of hyperleukocytosis and TLS.

TABLE 32.2 Management of Hyperleukocytosis and TLS

Management/objective	Guidelines
Aggressive hydration	Usually 2–4 times maintenance (alkalinization is avoided so as not to precipitate xanthine calculi and cause acute kidney injury)
Diuresis	Furosemide 0.5–1 mg/kg Mannitol 0.5 g/kg can be used if patient has oliguria unresponsive to increased hydration and furosemide
Uric acid reduction	(1) Allopurinol 300 mg/m²/day or 10 mg/kg/day PO (maximum dose 800 mg/day) or 200 mg/m²/day IV (maximum dose 600 mg/day) may be used if available; (2) Rasburicase (recombinant urate oxidase) 0.15–0.2 mg/kg/day IV, the dose can be repeated
Leukocyte reduction	Leukapheresis or exchange transfusion (for infants) can be used at any WBC count if the patient is symptomatic. In asymptomatic patients, leukapheresis should be considered if the initial white cell count is greater than 200,000/mm³ in AML or 300,000/mm³ in ALL or CML
Transfusion	Platelet transfusion to keep platelet count over 20,000/mm³ to decrease the risk of intracranial hemorrhage. Avoid packed RBC transfusions if cardiovascularly stable, because it increases blood viscosity. Fresh frozen plasma transfusion and administration of vitamin K can be considered if coagulopathy is present prior to any invasive procedures
Chemotherapy	Chemotherapy should be started when patient is stabilized and has adequate urine output
Dialysis	Dialysis is indicated for progressive renal failure with potassium >6 mEq/l, phosphate >10 mg/dl, oliguria, anuria, or volume overload unresponsive to the above measures
Monitor	Electrolytes; calcium, phosphorus, potassium, uric acid, BUN, and creatinine every 4–12 h depending upon risk of TLS. Complete blood counts 1–2 times per day. Respiratory, CNS, and cardiac monitoring if hyperkalemia or hypocalcemia are present
Imaging	Brain CT with contrast if no renal insufficiency in the presence of neurologic symptoms or signs. MRI, MRA, or MRV should be carried out if thrombosis is suspected

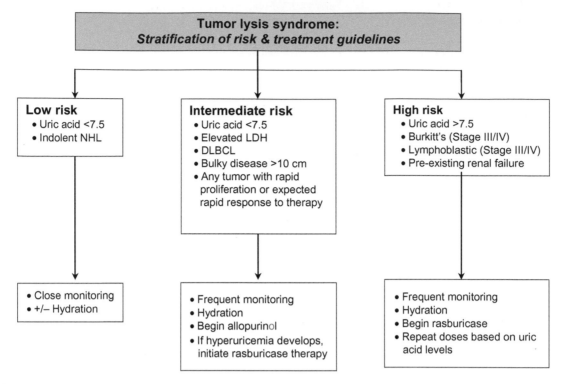

FIGURE 32.1   Treatment algorithm for the prevention and management of hyperuricemia. The treatment approach for low-risk patients is close observation with or without hydration. For those with intermediate risk, rasburicase is recommended if hyperuricemia develops despite prophylactic treatment with allopurinol. Vigorous hydration is recommended for all patients in the intermediate-to-high-risk groups, or those with diagnosed TLS. The use of rasburicase is recommended for the treatment of patients with hyperuricemia associated with diagnosed TLS, or in the initial management of patients considered to be at high risk of developing TLS. TLS, tumor lysis syndrome; NHL, non-Hodgkin lymphoma; LDH, lactate dehydrogenase; DLBCL, diffuse large B-cell lymphoma. *Adapted from Hochberg and Cairo (2008). Reprinted with permission.*

Management of the Various Metabolic Derangements in TLS

Hyperuricemia

1. Pharmacologic treatment should be based on risk stratification as outlined in Figure 32.1, which shows a treatment algorithm for the prevention and management of hyperuricemia.
   a. For low-risk patients: Observation and frequent laboratory monitoring are required. Allopurinol can be initiated as first-line therapy at diagnosis.
   b. For intermediate-risk patients: Allopurinol should be initiated as first-line therapy at diagnosis. Rasburicase should be considered if laboratory or clinical evidence of TLS is present despite allopurinol.
   c. For high-risk patients: Consider the use of rasburicase at diagnosis for tumors with high proliferative index, large tumor burden, or tumors that are highly chemo-sensitive. Rasburicase should also be considered as first-line therapy for patients with clinical evidence of TLS despite risk stratification. Rasburicase must be used with caution in patients with unknown G6PD status and/or at high risk for having G6PD deficiency.
2. Hydration $>2\ l/m^2/day$ should start before chemotherapy and urine output goal of $>100\ ml/m^2/h$ should be maintained during chemotherapy. Strict measurement of intake and output should be carried out every 2–4 h. Diuretics may be needed to maintain this urine output (i.e., furosemide 0.5–1 mg/kg/dose or mannitol 0.5 g/kg of the 25% solution over 5–10 min repeated every 6 h as needed).

Hyperkalemia

1. Potassium should be avoided unless dangerously low until tumor lysis is controlled.
2. Mild (<6 mEq/l) and asymptomatic:
   a. Hydration and diuretics as described in hyperuricemia.
   b. Sodium polystyrene sulfonate (kayexelate) at dose of 1 g/kg every 6 h with sorbitol 50–150 ml will remove 1 mEq of potassium per liter per gram of resin over 24 h.

3. Moderate to severe hyperkalemia:
   a. Obtain an EKG.
   b. Rapid insulin (0.1 U/kg/h) plus glucose (dextrose 0.5 g/kg/h); in emergency cases, 50% dextrose can be used at 1 ml/kg through a central line. Monitor serum glucose closely.
   c. For life-threatening arrhythmias IV calcium gluconate (100–200 mg/kg) or calcium chloride (10 mg/kg) slow IV infusion rate. The onset of action is within minutes and duration of activity lasts about 30 min (cannot be administered in the same line as $NaHCO_3$).
   d. $NaHCO_3$ to stabilize myocardial cell membrane and to reverse acidosis; 1–2 mEq/kg IV. For every increase of 0.1 pH unit, potassium is decreased about 1 mEq/l. Onset of action is within 30 min and duration of action lasts several hours.
   e. Dialysis should be considered while these rescuing approaches are underway.

Hyperphosphatemia

1. Lymphoblasts contain four times the amount of phosphate present in normal lymphocytes; consider a low-phosphate diet.
2. Aluminum hydroxide 150 mg/kg/day divided into doses every 4–6 h should be administered. This will prevent absorption of oral phosphate intake, but has little direct effect on lowering serum phosphate. Sevelamer hydrochloride (RenaGel), a noncalcium phosphate binder (400 mg twice daily for older children) can also be used.
3. Hydration necessary to maintain a urine output of ≥3 ml/kg/h.
4. Dialysis may be necessary to lower serum phosphate and prevent metastatic calcium deposition.

Hypocalcemia

1. As the product of serum calcium and phosphate increases over 60 due to hyperphosphatemia, a compensatory hypocalcemia may occur to maintain the calcium phosphate product at 60. At a calcium × phosphate product of >60 renal calcification can occur and exacerbate renal damage.
2. For symptomatic hypocalcemia (e.g., tetany), 10 mg/kg of elemental calcium (i.e., 0.5–1.0 ml/kg of 10% calcium gluconate) should be given. Calcium administration should be discontinued when symptoms resolve. Dialysis should be carried out if hyperphosphatemia persists. (Caution: Do not administer calcium in the same line as $NaHCO_3$.)

Renal Dysfunction from Tumor Lysis

Mechanisms of Renal Dysfunction
1. Precipitation of urates in the acid environment of the renal tubules.
2. Precipitation of hypoxanthine when the urine pH exceeds 7.5.
3. Increase in the hypoxanthine levels after starting treatment with allopurinol.
4. Precipitation of calcium phosphate in renal microvasculature and renal tubules when the product of serum calcium and phosphate values exceeds 60.

Indications for Dialysis Include the Following
1. Presence of hyperphosphatemia (>6 mg/dl) and hypercalcemia which promotes deposition in renal interstitium and tubular system, exacerbating kidney damage.
2. An estimated GFR less than 50%.
3. Persistent hyperkalemia with QRS interval widening and/or level exceeding 6 mEq/l.
4. Severe metabolic acidosis.
5. Volume overload unresponsive to diuretic therapy.
6. Anuria and overt uremic symptoms (i.e., encephalopathy).
7. Severe symptomatic hypocalcemia.
8. Hypertension (BP >150/90) and inadequate urine output at 10 h from start of treatment.
9. Congestive heart failure.

Hemodialysis or hemofiltration should be used when renal failure occurs. Continuous renal replacement therapy (CRRT) may be used for hemodynamically unstable patients because it is less likely to exacerbate hypotension. CRRT is not as effective for the treatment of hyperphosphatemia. Peritoneal dialysis should not be used.

Dialyzable chemotherapy, such as cyclophosphamide, is given immediately after dialysis and not before. Renal dialysis usually needs to be repeated every 12 h while there is continuous rapid tumor lysis.

CARDIOTHORACIC EMERGENCIES

Superior Vena Cava Syndrome and Superior Mediastinal Syndrome

Superior vena cava syndrome (SVCS) consists of the signs and symptoms of superior vena cava (SVC) obstruction due to compression or thrombosis. This condition is frequently due to a large anterior mediastinal mass compressing the SVC. Rapid growth of the mediastinal mass does not permit the development of effective collateral circulation to compensate and results in the signs and symptoms of compression of the SVC.

Superior mediastinal syndrome (SMS) consists of SVCS with tracheal compression.

The trachea and main stem bronchus are more compressible in children, making them more susceptible to SMS. In pediatrics, the terms SVCS and SMS are often used synonymously.

Etiology

1. Intrinsic causes: vascular thrombosis following the introduction of a catheter, intravascular tumor thrombosis (Wilms, lymphoma).
2. Extrinsic causes: malignant anterior mediastinal tumors
   a. Hodgkin lymphoma.
   b. Non-Hodgkin lymphoma.
   c. Teratoma or other germ cell tumor.
   d. Thyroid cancer.
   e. Thymoma.

Clinical Features

1. Superior vena cava syndrome:
   a. Swelling, plethora, and cyanosis of the face, neck, and upper extremities.
   b. Suffusion of the conjunctiva.
   c. Engorgement of collateral veins.
   d. Altered mental status.
2. Superior mediastinal syndrome:
   a. Respiratory symptoms: cough, hoarseness, dyspnea, orthopnea, wheezing, and stridor. Supine position worsens symptoms.
   b. Dysphagia.
   c. Chest pain.
   d. Altered mental status and syncope.

Management

1. Extreme care is required in handling the patient. The following may precipitate respiratory arrest:
   a. Supine position (as for CT or operative procedures)—Do not place patient in recumbent position.
   b. Medications that cause intercostal muscle relaxation.
   c. Stress.
   d. Sedation (conscious sedation, anxiolytics, or general anesthesia). The patient may have to be intubated. Extubation may not be possible until the anterior mediastinal mass has significantly decreased in size. Extracorporeal membrane oxygenation may be required if intubation is not possible.
2. Diagnosis should be made quickly in the least invasive manner.
   a. Radiograph of the chest and CT (if tolerated).
   b. Screening blood work such as CBC, LDH, uric acid, α-fetoprotein, and β-hCG (to screen for germ cell tumors and lymphoma).
   c. Echocardiogram, to assess anesthesia risk, cardiac function and for possible intravascular thrombus if no evidence of mass on chest radiograph.
   d. Determine anesthesia risk. If high risk, perform the least invasive technique with local anesthesia (bone marrow, pleurocentesis, pericardiocentesis, lymph node biopsy, or fine-needle aspirate). If low risk use sedation or anesthesia and monitor closely.

3. Therapy:
   a. Establishing a tissue diagnosis may not be possible and patients may need empiric treatment as a life-saving measure. First-line treatment in emergent situations is high-dose steroids, although they may confound the diagnosis. Prednisolone 60 mg/m^2/day (2 mg/kg/day) or methylprednisolone 48 mg/m^2/day (1.6 mg/kg/day) divided into two daily doses should be employed. This will treat hematologic malignancies and decrease airway edema. The patient should undergo biopsy as soon as the mass shrinks and the patient is stable.
   b. If poor response to steroids, chemotherapy such as vincristine, cyclophosphamide with or without an anthracycline can be added. Tumor-specific chemotherapy should be instituted after a biopsy has been obtained.
   c. If a solid tumor not responsive to steroids or chemotherapy, emergent radiation can be performed.
   d. For symptomatic venous thrombosis with no evidence of hemorrhage, anticoagulation can be initiated using systemic or low-molecular-weight heparin (LMWH):
      i. Unfractionated heparin can be started with a 75 U/kg bolus followed by 18 U/kg/h (for children) to 28 U/kg/h (infants) continuous infusion. Titrate to a goal-activated partial thromboplastin time of 60–85 s or anti-Xa level of 0.3–0.7 U/ml.
      ii. LMWH 1 mg/kg every 12 h. Titrate to a goal anti-Xa level of 0.5–1 U/ml.

ABDOMINAL EMERGENCIES

1. Esophagitis: the most common gastrointestinal (GI) problem in oncology patients.
2. Gastric hemorrhage: especially in patients on corticosteroid therapy.
3. Typhlitis: seen primarily in patients with prolonged neutropenia or at new diagnosis of leukemia.
4. Perirectal abscess: in prolonged neutropenia.
5. Hemorrhagic pancreatitis: especially in patients on asparaginase therapy.
6. Massive hepatic enlargement from tumor: especially in infants with stage IVS neuroblastoma.

Table 32.3 provides an evaluation and management of common causes of abdominal pain.

TABLE 32.3 Evaluation and Management of Common Causes of Abdominal Pain

Diagnosis	Signs and symptoms	Clinical set-up	Evaluation	Management
Bowel obstruction	Pain Decreased bowel sounds	Tumor Adhesions	Abdominal radiograph	Bowel rest/NG Surgical consultation
Constipation/ileus	Hard stools or no stool Pain	Narcotics Postoperative Vincristine Dehydration	Abdominal radiograph	Stool motility agents Stool softeners, bulking agents IV hydration
Gastritis/esophagitis	Gastric/throat pain Metallic taste	Chemotherapy, steroids *Candida*	Oral examination	Oral or IV antacid Antifungal
Hepatic enlargement/ VOD	RUQ mass	Neuroblastoma Stage 4S	CT, urine VMA/HVA	Chemotherapy
Pancreatitis	RUQ pain Emesis	Asparaginase Steroids	Amylase, lipase US/CT for pseudocyst	Bowel rest/NG
Perirectal abscess	Erythema Induration Pain with defecation	Severe myelosuppression	Perirectal examination	Broad-spectrum antibiotics (Gram-negative and anaerobes) Sitz baths (4 times a day)
Typhlitis/colitis	Acute abdomen/RLQ pain Diarrhea Hypotension Bloody diarrhea	Severe myelosuppression Acute leukemia	Abdominal radiograph or CT (free air or thickened bowel wall) Stool cultures	Nothing by mouth Bowel rest/NG Surgical consultation Broad-spectrum antibiotics (Gram-negative, anaerobes, fungus)
Veno-occlusive disease	RUQ mass Edema Weight gain Jaundice	Post-transplant 6-Thioguanine, Actinomycin Gemtuzumab	US for reversal of flow Bilirubin Weight checks	Stop inciting agents Defibrotide

RUQ, right upper quadrant of abdomen; RLQ, right lower quadrant of abdomen; US, ultrasound; NG, nasogastric tube suctioning.

Evaluation and Diagnosis of Abdominal Emergencies

1. History regarding onset, timing, location, and radiation of pain.
2. Observation and gentle examination including mouth and perirectal area in particular. If rectal examination is deemed necessary it should be performed very gently.
3. The classic signs of an acute abdomen may be muted in a neutropenic patient or a patient on steroids.
4. Serial blood counts to evaluate for hemorrhage, neutropenia, infection.
5. Blood, stool, and urine cultures as indicated.
6. Laboratory tests, liver enzymes, bilirubin, amylase, lipase, electrolytes.
7. Vital signs monitoring.
8. Abdominal radiography: ultrasonography, CT, and magnetic resonance imaging (MRI) as indicated.

Typhlitis

Typhlitis, a necrotizing colitis often localized in the cecum, occurs in the setting of severe neutropenia, particularly in patients with leukemia and in stem cell transplant recipients. It should be strongly suspected in patients with right lower quadrant pain or the development of a partially obstructive right lower quadrant mass. Typhlitis is the result of bacterial or fungal invasion of the mucosa and can quickly progress from inflammation to full-thickness infarction to perforation, peritonitis, and septic shock.

Etiology

1. The responsible pathogens include *Pseudomonas* species, *Escherichia coli*, other Gram-negative bacteria, *Staphylococcus aureus*, α-hemolytic *Streptococcus*, *Clostridium*, *Aspergillus*, and *Candida*.
2. Typhlitis in patients receiving chemotherapy is linked to mucosal injury caused by cytotoxic chemotherapeutic agents.

Diagnosis

Typhlitis is usually diagnosed clinically when a neutropenic patient presents with:

1. Right lower quadrant pain.
2. Physical examination may reveal an absence of bowel sounds, bowel distention, tenderness on palpation maximal in the right lower quadrant, or a palpable mass in the right lower quadrant. Serial abdominal examinations are required.
3. Imaging studies may aid in the diagnosis of typhlitis:
   a. Radiograph of the abdomen may reveal pneumatosis intestinalis, free air in the peritoneum or bowel wall thickening.
   b. Ultrasonography may reveal thickening of the bowel wall in the region of the cecum and is becoming a more commonly used nonradiation modality to image for typhlitis.
   c. CT scan is the definitive imaging study and may demonstrate diffuse thickening of the cecal wall.

Treatment

1. Medical management is the initial treatment, consisting of:
   a. Discontinuation of oral intake.
   b. Nasogastric tube to suction.
   c. Broad-spectrum antibiotics (anaerobic and Gram-negative coverage) and anti-fungals.
   d. Intravenous fluid and electrolytes.
   e. Packed red cell and platelet transfusions, as indicated.
   f. Vasopressors, as needed (hypotension is associated with a poor outcome).
2. Indications for surgical intervention:
   a. Persistent GI bleeding despite resolution of neutropenia and thrombocytopenia.
   b. Evidence of free air in the abdomen on abdominal radiograph (indicating perforation).
   c. Clinical deterioration requiring fluid and pressor support, indicating uncontrolled sepsis from bowel infarction.
   d. Surgery consists of removing necrotic portions of the bowel and diversion via colostomy. Healing can occur with fibrosis and stricture formation.
   e. Mortality is related to bowel perforation, bowel necrosis, and sepsis.

Perirectal Abscess

Inflammation and infection of the rectum and perirectal tissue occur commonly in patients receiving chemotherapy or radiation therapy, especially in patients with prolonged neutropenia. Most abscesses are caused not by a single organism but rather by a combination of aerobic organisms, such as staphylococci, streptococci, *E. coli*, *Pseudomonas*, and anaerobic Gram-positive and -negative organisms.

Presentation includes anorectal pain, tenderness, and painful bowel movements. An abscess or draining fistula may be present; however, in the neutropenic patient, pus will be absent and the patient will present with a brawny edema and dense cellulitis.

Management

1. Initial therapy with intravenous antibiotics to cover Gram-negative organisms and anaerobes.
2. Granulocyte colony-stimulating factor to shorten period of neutropenia.
3. Sitz baths four times a day, and meticulous attention to perirectal hygiene.
4. Surgical incision and drainage of obviously fluctuant areas or draining fistulas that do not resolve with medical management.

RENAL EMERGENCIES

Renal complications can be caused by the tumor itself (urinary tract obstruction and renal vein thrombosis) or as a result of cancer therapy (hemorrhagic cystitis, acute renal failure, or hypertension).

Oliguria/Anuria

Differential Diagnoses

- Prerenal: septic shock, dehydration, emesis, diarrhea, decreased oral intake, metabolic abnormalities from tumor lysis.
- Postrenal: bulky abdominopelvic tumors with obstruction, hemorrhagic cystitis.
- Renal insufficiency: chemotherapy agents, contrast dyes, anti-infectives.

Evaluation

- BUN, creatinine, electrolytes.
- Close monitoring of all intake and output.
- CT or ultrasound (US) of abdomen/pelvis.

Therapy

- Vigorous hydration for prerenal etiologies including BK viremia.
- Decompression of obstructed kidney with stenting or catheter placement in postrenal etiologies.
- Treatment of underlying tumor with chemotherapy, surgery, or radiation to decrease outflow obstruction.
- Avoidance of nephrotoxic agents, IV contrast, and appropriate renal dosing of medications.
- With significant electrolyte abnormalities, fluid overload or true anuria, consideration of dialysis may be indicated.

Hypertension

Definition

- Systolic or diastolic blood pressure outside the 95th percentile for age, gender, and height.

Etiology

- Secondary to pain or anxiety, and often transient.
- Secondary to tumor compression of renal parenchyma leading to increased renin production (secondary hyperaldosteronism).

- Secondary increased renin production from tumors (e.g., pheochromocytoma, Wilms' tumor, neuroblastoma).
- Secondary to medications: steroids, anti-infectives, calcineurin inhibitors.
- Renal vein thrombosis.
- Increased intracranial pressure (ICP).

Symptoms

- Symptoms may include headache, irritability, lethargy, confusion, and, if untreated, seizures, posterior reversible encephalopathy syndrome (PRES).

Treatment

- Acute hypertension:
  - Hydralazine (0.2–0.6 mg/kg/dose IV).
  - Nicardipine (0.5–1 μg/kg/min infusion to be titrated to desired blood pressure).
  - Labetalol (0.2–1 mg/kg/dose IV, to be avoided in patients with bronchospasm or diabetes).
  - Sublingual nifedipine (5–10 mg/dose for children weighing >10 kg).
  - Oral clonidine (0.05–0.1 mg/dose).
- Chronic hypertension:
  - Amlodipine (0.1 mg/kg per dose, or 2.5–5 mg/day).
  - Consider ACE inhibitors or beta-blockers.
- Hypertension from fluid overload:
  - Furosemide (0.5–1 mg/kg).

NEUROLOGIC EMERGENCIES

Neurologic emergencies, such as seizure, alterations in mental status (AMS), cerebrovascular accidents (CVAs), spinal cord compression, and increased ICP occur in over 10% of pediatric oncology patients. Initial intervention for acute neurologic deterioration requires immediate stabilization of the patient.

Evaluation and Diagnosis of Neurologic Emergencies

1. Detailed history including current medications, recent chemotherapy and radiation therapy, prior events.
2. Thorough physical and neurologic examination, pulse oximetry, vital signs.
3. Laboratory tests including blood count, electrolytes, ammonia, liver, toxicology screen, and renal function tests.
4. Head CT acutely, MRI/magnetic resonance angiography (MRA).
5. Spinal tap.
6. Electroencephalogram.

Differential Diagnosis

- *Seizures:* CNS tumors, intrathecal chemotherapy, metabolic derangements.
- *Raised ICP:* CNS tumors, shunt obstruction, pseudotumor cerebri, infection, CVA.
- *CVA:* Asparaginase, hyperleukocytosis, coagulopathy/hemorrhage, radiation-induced vasculopathy.
- *AMS:* CNS tumor, opiates, benzodiazepines, steroids, intrathecals, high-dose cytarabine or methotrexate, ifosfamide, nelarabine, postictal, CVA, CNS infection, metabolic derangements, postradiation somnolence.

Management

1. Stabilize the patient: oxygen and hydration as needed.
2. Stop any intravenous infusions, especially chemotherapy, narcotics.
3. Transfuse platelets if thrombocytopenic.
4. Start broad-spectrum antibiotics after blood culture is obtained, a spinal tap can be performed once the CT reveals normal ventricular size.

5. If actively having seizures administer ativan or dilantin load.
6. Emergent head CT to look for CNS bleed, hydrocephalus, herniation, or mass, MRI/MRA/magnetic resonance venography (MRV), when available, to look for white matter changes and vascular issues. If nonhemorrhagic stroke is found on MRI, thrombolysis and anticoagulation should be considered.

Spinal Cord Compression

Incidence and Etiology

- Three to five percent of children with cancer develop acute spinal cord or cauda equina compression. Sarcomas account for about half of the cases of spinal cord involvement in childhood. The remainder are caused by neuroblastoma, germ cell tumor, lymphoma, leukemia, and drop metastasis of CNS tumors.
- The spinal cord can be compressed by tumor in the epidural or subarachnoid space or by metastases within the cord parenchyma.

Pathophysiology

- Direct extension of the tumor.
- Metastatic spread to the vertebrae with secondary cord compression.
- Spread to the epidural space via infiltration of the vertebral foramina.
- Subarachnoid spread down the spinal cord from primary CNS tumor (such as medulloblastoma).

Clinical Presentation

- Back pain with localized tenderness occurs in 80% of patients.
- Incontinence, urinary retention, and other abnormalities of bowel or bladder function are not frequent if spinal compression is diagnosed early.
- Loss of strength and sensory deficits with a sensory level may also occur.
- Any child with cancer and back pain should be presumed to have spinal cord involvement until further workup indicates otherwise.

Evaluation

- A thorough history and neurologic examination should be included in the evaluation.
- Spinal radiographs are useful if the compression is due to vertebral metastases but they will miss epidural disease in 50% of cases.
- MRI with and without gadolinium is necessary to detect the presence and extent of epidural involvement.
- Cerebrospinal fluid analysis is important in the evaluation of subarachnoid disease, but it is not helpful in localizing epidural disease.

Treatment

- Because the potential for permanent neurologic damage is high, it is crucial to initiate treatment immediately.
- Dexamethasone is initiated to decrease local edema, prior to diagnostic studies.
  - In the presence of neurologic abnormalities, immediately start dexamethasone 1−2 mg/kg/day loading dose. Follow with 1.5 mg/kg/day divided every 6 h and obtain emergent MRI.
  - With back pain and the absence of neurologic symptoms, start dexamethasone 0.25−1 mg/kg/dose every 6 h and perform MRI within 24 h.
- If an epidural mass is identified, treatment is aimed at rapid decompression. Chemotherapy, radiation therapy, or surgical decompression may be used.
  - Specific chemotherapy can be instituted in addition to the use of dexamethasone in lymphoma, leukemia, and neuroblastoma.
  - If tumor is known to be radiosensitive, give local radiation including the full volume of the tumor plus one vertebra above and below the lesion. Consult a radiation oncologist for daily dosing fractions and total dose.
  - Surgical emergent laminotomy or laminectomy may be indicated for paralysis requiring rapid decompression, especially in lesions expected to be less radioresponsive (e.g., sarcoma) or no symptomatic improvement with emergent steroids, chemotherapy, and/or radiation.

ENDOCRINE EMERGENCIES

Syndrome of Inappropriate Antidiuretic Hormone Secretion (SIADH)

Etiology

1. Involves continuous pituitary release of antidiuretic hormone (ADH), irrespective of plasma osmolality.
2. Leads to significant hyponatremia, serum hypo-osmolality, and water intoxication.
3. Results from physiologic stress, pain, surgery, mechanical ventilation, infections, CNS and pulmonary lesions, lymphomas, and leukemias.
4. Occurs as a side effect of vincristine, vinblastine, cyclophosphamide, ifosfamide, cisplatin, and melphalan.
5. Occurs in overhydration with hypotonic fluids, diabetes insipidus with free water replacement, and cerebral salt wasting.

Clinical Features

- Oliguria.
- Weight gain.
- Often asymptomatic.
- Early symptoms can include fatigue, headache, and nausea.
- Late manifestations include lethargy, confusion, hallucinations, seizures, and coma.

Laboratory Features of SIADH

- Low serum osmolality ($<$280 mOsm/l).
- High urine osmolality ($>$500 mOsm/l).
- Urine to serum osmolality ratio $>$1.
- Hyponatremia (sodium, $<$130 mEq/l).
- Increased urine specific gravity.

Treatment

1. Fluid restriction.
2. Furosemide 1 mg/kg should be administered to increase diuresis of free water and reduce the impact of the excess ADH.
3. Hydration with normal saline limited to insensible losses (500 ml/m^2/24 h) plus ongoing losses.
4. In cases of severe neurologic involvement (seizures or coma), hydrate carefully with hypertonic saline 3%. The rate of sodium correction should be limited to 2 mEq/l/h over the first 2 h. Subsequent to that the rate of correction should target a change of serum sodium by 10 mEq over the first 24 h and 18 mEq by 48 h of treatment. Too rapid correction may lead to further permanent neurologic sequelae.

Hypercalcemia of Malignancy

Etiology

1. Osteolytic bone lesions (particularly in T-cell leukemia and lymphoma).
2. Bone demineralization secondary to parathyroid-like hormone (PTHrP) produced by tumors (paraneoplastic syndrome).
3. Immobilization.
4. Defect in renal excretion.

Clinical Features

Patients typically become symptomatic when serum calcium exceeds 12 mg/dl.

- Anorexia, nausea, vomiting, and constipation.
- Weakness.
- Coma.
- Pruritis.
- Bone pain.
- Polyuria, polydipsia, nephrogenic diabetes insipidus.

- Bradycardia, arrhythmias.
- Dehydration, impaired renal function.
- Disseminated intravascular coagulation.

Treatment

1. Dehydration and electrolyte disturbances should be corrected. Stop calcium-containing medications.
2. Renal calcium excretion should be increased by inducing diuresis with normal saline at two- to threefold maintenance and furosemide 1−2 mg/kg/dose every 6 h.
3. Calcium mobilization from bone should be decreased by
   a. Bisphosphonates, such as pamidronate 0.5−1 mg/kg IV over 4−6 h with very close monitoring of serum calcium, phosphate, and magnesium for 2 weeks.
   b. Prednisone 1.5−2.0 mg/kg daily (in lymphoproliferative disorders).

Adrenal Insufficiency

Etiology

1. Secondary to significant prior corticosteroid exposure.
2. During periods of critical illness, trauma, surgery, or infection.
3. Tissue resistance to steroids.
4. Adrenal gland failure.
5. Radiation or surgically induced impairment of cortisol synthesis due to injury to the adrenal glands.

Clinical Features

1. Fatigue, dizziness, weakness, myalgia, nausea/vomiting.
2. Severe hypotension, shock.
3. Hyponatremia, hyperkalemia, metabolic acidosis with normal anion gap.

Treatment

1. Glucocorticoid replacement therapy (hydrocortisone):
   a. Hydrocortisone at 100 mg/m^2 and then 25 mg/m^2 per dose given every 6 h for 7 days without taper.
   b. Fludrocortisone.
   c. Strict monitoring for hyperglycemia is critical during stress dosing of steroids.
2. Treatment includes interventions for sepsis and hypotension if warranted.
3. Consider checking cortisol levels as patient may need physiologic replacement once stress dosing is complete.

TREATMENT-ASSOCIATED EMERGENCIES

Anaphylaxis and Hypersensitivity to Chemotherapeutic Agents

Etiology

1. Asparaginase.
2. Platinum agents.
3. Etoposide.
4. Monoclonal antibodies.

Presentation

1. Bronchospasm.
2. Dyspnea.
3. Wheezing.
4. Angioedema.
5. Flushing.
6. Nausea.
7. Rhinitis.

8. Pruritis.
9. Urticaria.
10. Hypotension, shock, fluid overload.

Treatment

1. Discontinue inciting agent.
2. Maintain vascular access.
3. Assess respiratory status.
4. If bronchospasm or hypotension:
   a. IM epinephrine (0.01 mg/kg of 1:1000 dilution (if <10 kg), or autoinject 0.15 mg if 10–24 kg or 0.3 mg if ≥ 25 kg).
   b. IV fluid boluses with normal saline are recommended.
   c. Persistent wheezing, consider albuterol and prepare for intubation.
5. H1 blockers (diphenhydramine or hydroxyzine).
6. H2 antagonists (ranitidine).
7. Parenteral steroids, typically methylprednisolone or hydrocortisone, are recommended to dampen the late allergic symptoms.

APL Differentiation Syndrome

Attributed to maturation of promyelocytes and associated hyperinflammatory response with vasoactive cytokines.

Presentation

1. Respiratory distress, edema, weight gain, hypotension, fever, renal failure.
2. Onset of symptoms has a bimodal pattern, often within 1 week of starting ATRA therapy, while others manifest symptoms 3–4 weeks after starting treatment.

Treatment

1. Initiate dexamethasone 0.5–1 mg/kg (max 10 mg per dose) IV every 12 h.
2. Hold ATRA if symptoms are life-threatening.

Veno-Occlusive Disease

- Also referred to as hepatic sinusoidal obstruction syndrome.
- Characterized by rapid and often massive hepatic enlargement with resultant right-upper-quadrant pain, liver tenderness, jaundice, weight gain, and ascites.
- Ultrasonography shows ascites and reversal of flow in hepatic vessels.

Etiology

Preexisting hepatic disease, radiation, allogeneic transplant, vincristine, actinomycin-D, and 6-thiguanine, busulfan with or without cyclophosphamide, and younger age have all been identified as risk factors for veno-occlusive disease (VOD).

Treatment

1. Primarily supportive in the nontransplant setting including discontinuation of the offending agent(s).
2. Studies evaluating defibrotide as a treatment option are ongoing in the transplant setting. Defibrotide is also available on a compassionate basis.

Further Reading and References

Adelberg, D.E., Bishop, M.R., 2009. Emergencies related to cancer chemotherapy and hematopoietic stem cell transplantation. Emerg. Med. Clin. North Am. 27 (2), 311–331.

Albanese, A., Hindmarsh, P., Stanhope, R., 2001. Management of hyponatraemia in patients with acute cerebral insults. Archiv. Dis. Child. 85 (3), 246–251.

Anghelescu, D.L., et al., 2007. Clinical and diagnostic imaging findings predict anesthetic complications in children presenting with malignant mediastinal masses. Paediatr. Anaesth. 17 (11), 1090–1098.

Antunes, N.L., DeAngelis, L.M., 1999. Neurologic consultations in children with systemic cancer. Pediatr. Neurol. 20 (2), 121–124.

Bowers, D.C., Liu, Y., Leisenring, W., McNeil, E., Stovall, M., Gurney, J.G., et al., 2006. Late-occurring stroke among long-term survivors of childhood leukemia and brain tumors: a report from the Childhood Cancer Survivor Study. J Clin Oncol 24 (33), 5277–5282.

Coiffier, B., Altman, A., Pui, C.H., Younes, A., Cario, M.S., 2008. Guidelines for the management of pediatric and adult tumor lysis syndrome: an evidence-based review. J. Clin. Oncol. 26 (16), 2767–2778.

Cole, J.S., Patchell, R.A., 2008. Metastatic epidural spinal cord compression. Lancet Neurol. 7 (5), 459–466.

Demshar, R., Vanek, R., Mazanec, P., 2011. Oncologic emergencies: new decade, new perspectives. AACN Adv. Crit. Care 22 (4), 337–348.

DiMario Jr., F.J., Packer, R.J., 1990. Acute mental status changes in children with systemic cancer. Pediatrics 85 (3), 353–360.

Ganzel, C., et al., 2012. Hyperleukocytosis, leukostasis and leukapheresis: practice management. Blood Rev. 26 (3), 117–122.

Ho, V.T., Linden, E., Revta, C., Richardson, P.G., 2007. Hepatic veno-occlusive disease after hematopoietic stem cell transplantation: review and update on the use of defibrotide. Semin. Thromb. Hemost. 33 (4), 373–388.

Hochberg, J., Cairo, M.S., 2008. Rasburicase: future directions in tumor lysis management. Expert Opinion on Biological Therapy 8 (10), 1595–1604.

Howard, S.C., Jones, D.P., Pui, C.H., 2011. The tumor lysis syndrome. N. Engl. J. Med. 364 (19), 1844–1854.

Ilgen, J.S., Marr, A.L., 2009. Cancer emergencies: the acute abdomen. Emerg. Med. Clin. North Am. 27 (3), 381–399.

Kaste, S.C., Rodriguez-Galindo, C., Furman, W.L., 1999. Imaging pediatric oncologic emergencies of the abdomen. AJR Am. J. Roentgenol. 173 (3), 729–736.

Kearney, S.L., et al., 2009. Clinical course and outcome in children with acute lymphoblastic leukemia and asparaginase-associated pancreatitis. Pediatr. Blood Cancer 53 (2), 162–167.

Kerdudo, C., Aerts, I., Fattet, S., Chevret, L., Pacquement, H., Doz, F., et al., 2005. Hypercalcemia and childhood cancer: a 7-year experience. J. Pediatr. Hematol. Oncol. 27 (1), 23–27.

Leslie, J.A., Cain, M.P., 2006. Pediatric urologic emergencies and urgencies. Pediatr. Clin. North Am. 53 (3), 513–527, viii.

Lewis, M.A., Hendrickson, A.W., Moynihan, T.J., 2011. Oncologic emergencies: pathophysiology, presentation, diagnosis, and treatment. CA Cancer J. Clin. 61, 287–314.

Lowe, E.J., Pui, C.H., Hancock, M.L., Geiger, T.L., Khan, R.B., Sandlund, J.T., 2005. Early complications in children with acute lymphoblastic leukemia presenting with hyperleukocytosis. Pediatr. Blood Cancer 45 (1), 10–15.

McCarville, M.B., et al., 2005. Typhlitis in childhood cancer. Cancer 104 (2), 380–387.

McCurdy, M.T., Shanholtz, C.B., 2012. Oncologic emergencies. Crit. Care Med. 40 (7), 2212–2222.

Perger, L., Lee, E.Y., Shamberger, R.C., 2008. Management of children and adolescents with a critical airway due to compression by an anterior mediastinal mass. J. Pediatr. Surg. 43 (11), 1990–1997.

Pitfield, A.F., Carroll, A.B., Kissoon, N., 2012. Emergency management of increased intracranial pressure. Pediatr. Emerg. Care 28 (2), 200–204, quiz 205–7.

Taub, Y.R., Wolford, R.W., 2009. Adrenal insufficiency and other adrenal oncologic emergencies. Emerg. Med. Clin. North Am. 27 (2), 271–282.

Wilson, F.P., Berns, J.S., 2014. Tumor lysis syndrome: new challenges and recent advances. Adv. Chronic Kidney Dis. 21 (1), 18–26.

Wilson, L.D., Detterbeck, F.C., Yahalom, J., 2007. Clinical practice. superior vena cava syndrome with malignant causes. N. Engl. J. Med. 356 (18), 1862–1869.

33

Supportive Care of Patients with Cancer

Anurag K. Agrawal and James Feusner

Increasing intensity of chemotherapy has improved pediatric oncology outcomes concomitant with recognition and augmentation of supportive care practices. Prophylaxis and treatment of infection remains the cornerstone of supportive care along with transfusion practice (see Chapter 36). Additional considerations include the recognition and management of nausea and vomiting, mucositis and pain, nutritional status of the oncology patient, utilization of hematopoietic growth factors, acute radiation side effects, management of central venous catheters (CVCs), posttreatment immunizations, and palliative care.

MANAGEMENT OF INFECTIOUS COMPLICATIONS

Oncologic treatment affects both innate immunity, such as skin and mucosal barriers, as well as adaptive immunity, such as pathogen-specific B- and T-cell response. Factors leading to infection susceptibility include:

- Underlying disease: hematologic malignancy, advanced-stage lymphoma, progressive disease and patients undergoing hematopoietic stem cell transplant (HSCT) being at highest risk.
- Type of therapy: dose-intensive therapies such as high-dose cytarabine, acute myelogenous leukemia (AML) induction, and HSCT.
- Degree and duration of neutropenia with profound neutropenia defined as absolute neutrophil count (ANC) $\leq 0.1 \times 10^9$/l and prolonged neutropenia defined as lasting >7 days.
- Disruption of normal skin and mucosal barriers.
- Malnutrition.
- Defects in humoral immunity leading to risk of encapsulated bacteremia.
- Defects in cellular immunity (either at baseline or secondary to therapy) leading to susceptibility to viral, fungal, and some bacterial infections (especially those replicating intracellularly).
- Colonizing microbial flora.
- Foreign bodies; for example, CVCs and ventriculoperitoneal (VP) shunts.

Critical and emergent assessment will direct the initial risk stratification and subsequent diagnostic evaluations. Important initial questions and examinations include:

- Height of fever: fever >39.0°C has been noted as an independent risk factor for serious bacterial infection.
- Presence of rigors or chills with central line flushing.
- Recent chemotherapy or radiation therapy (RT).
- Current medications (i.e., antimicrobial prophylaxis).
- Possible infectious exposures at home or school or with recent travel.
- Prior history of documented infections.
- Thorough physical examination with particular consideration for oral and perirectal mucosa, CVC access site, skin, and sites of any invasive procedure.

Common organisms that must be considered are listed in Table 33.1.

TABLE 33.1 Common Organisms Causing Bacteremia and Sepsis

Gram-positive bacteria	Staphylococci	Coagulase-negative (i.e., *S. epidermidis*)
		S. aureus (including MRSA)
	Streptococci	Alpha-hemolytic (i.e., *S. viridians, S. mitis*)
Gram-negative bacteria	Enterobacteriaceae	*E. coli, Enterobacter, Klebsiella, Serratia*
	Pseudomonas aeruginosa	
	Stenotrophomonas maltophilia	
	Acinetobacter sp.	
Anaerobic bacteria	*Clostridium difficile*	
	Bacteroides sp.	
	Propionibacterium acnes	
Fungi	*Candida* sp.	
	Aspergillus sp.	
	Zygomycetes	
	Cryptococci	
	Pneumocystis jiroveci	
Viruses	Herpes simplex virus	
	Varicella zoster virus	
	Cytomegalovirus	
	Epstein–Barr virus	
	Respiratory syncytial virus	
	Adenovirus	
	Influenza	
	Parainfluenza	
	Human herpesvirus 6	
Other	*Toxoplasma gondii*	
	Strongyloides stercoralis	
	Cryptosporidium	
	Bacillus sp.	
	Atypical mycobacterium	

Febrile Neutropenia

Febrile Neutropenia (FN) is defined by the following criteria:

- A single oral temperature ≥38.3°C (101.0°F) or an oral temperature ≥38.0°C (100.4°F) sustained for >1 h or that occurs twice within a 24-h period.
- An ANC $<0.5 \times 10^9/l$ or ANC $<1.0 \times 10^9/l$ expected to decrease to $<0.5 \times 10^9/l$ over the subsequent 48 h.

Families should be advised against taking rectal temperatures. Recent consumption of hot or cold beverages should not alter management if the patient has had a documented oral temperature taken. Alternate routes for fever measurement, including axillary, otic, and temporal, should be discouraged but all should be managed in the same manner if there is a documented fever.

Initial FN evaluation should include the following:

- Complete blood count with differential.
- Complete metabolic panel.
- Blood cultures from each lumen of the CVC or peripheral cultures if without a CVC (≥1 ml of blood).
- Clean-catch bacterial urine cultures (urine catheterization should not be done, especially in the neutropenic patient).
- Gram stain and culture from suspicious skin, oropharyngeal, or CVC sites.

Additional measures that may be considered but are not routinely recommended include:

- Peripheral blood cultures in addition to central cultures can be considered as a means to determine bacteremia versus CVC infection based on the differential time to positivity, although the impact of this measure on treatment decision-making in FN is unclear.
- Coagulation studies in the patient with bleeding.
- Chest radiography is not routinely recommended and should only be done in the patient with respiratory compromise, symptoms of pulmonary infection, or auscultatory signs.
- Patients with sinus tenderness should have computed tomography (CT) of the sinuses.
- Patients with esophagitis should be considered for endogastroduodenoscopy with biopsy and culture to rule out viral and fungal causes.
- Patients with diarrhea should have a stool sample sent for culture, rotavirus, and *Clostridium difficile* testing.
- Lumbar puncture is rarely indicated but if the patient has central nervous system (CNS) signs a head CT should be performed first to rule out mass lesions or hemorrhagic stroke which may lead to increased intracranial pressure.
- Shunt fluid examination from implanted devices such as VP shunts or Ommaya reservoirs is rarely indicated.

Management of Febrile Neutropenia

The initial management of pediatric FN includes rapid assessment of the patient, recognition of those exhibiting signs and symptoms of sepsis, rapid initiation of broad-spectrum antibiotics and other necessary supportive care measures, and subsequent admission to hospital. Although multiple risk stratification models have been published, none has been validated across varied pediatric oncology cohorts. Therefore, the current recommendation in pediatric patients is for admission with all cases of FN. Although early discharge may be considered in certain cases, this is not routinely recommended and should be done under carefully determined and monitored institutional guidelines. See Figure 33.1 for an algorithmic approach to initial management of FN.

Antibiotic selection should be based on microbial prevalence and sensitivity patterns at individual institutions. Due to the acute risk of Gram-negative sepsis, empiric coverage must include these organisms, including *Pseudomonas*. Multiple empiric regimens are acceptable, although in general monotherapy has supplanted dual therapy as the regimens of choice:

- Monotherapy
  - Fourth-generation antipseudomonal β-lactam cephalosporin; cefepime 150 mg/kg/day IV divided q8h (max 2 g/day).
  - Carbapenem
    - Imipenem/cilastatin 60−100 mg/kg/day (imipenem component) IV divided q6h (max 4 g/day).
    - Meropenem 60 mg/kg/day IV divided q8h (max 3 g/day) (can be increased to 120 mg/kg/day IV divided q8h with max 6 g/day in severe infection).
  - Piperacillin/tazobactam (Zosyn) 240−300 mg/kg/day (piperacillin component) IV divided q8h (max 16 g/day).
- Dual therapy (antipseudomonal β-lactam plus an aminoglycoside)
  - Ceftazidime 150 mg/kg/day IV divided q8h (max 6 g/day) plus tobramycin 7.5 mg/kg/day IV divided q8h or 7−9 mg/kg/dose IV daily.

Multiple meta-analyses have shown that monotherapy with broad-spectrum, antipseudomonal β-lactams is non-inferior to dual therapy. The empiric utilization of anti-Gram-positive antibiotics without a documented infection does not improve outcomes. Patients on aminoglycosides or vancomycin should have trough levels monitored weekly due to risks of nephrotoxicity and ototoxicity with frequent monitoring of renal function.

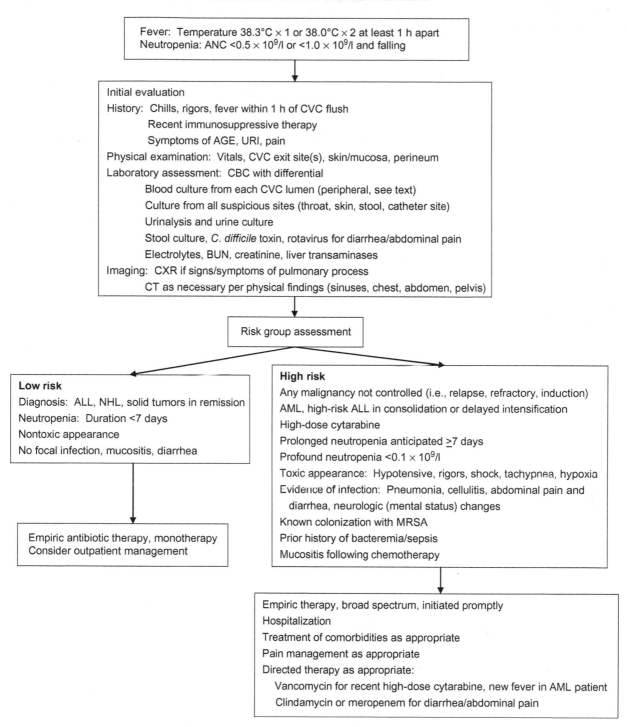

FIGURE 33.1 Evaluation and initial management of febrile neutropenia in the child with cancer. CVC, central venous catheter; AGE, acute gastroenteritis; URI, upper respiratory infection; CBC, complete blood count; BUN, blood urea nitrogen; CXR, chest radiography; CT, computed tomography; ALL, acute lymphoblastic leukemia; NHL, non-Hodgkin lymphoma; AML, acute myelogenous leukemia; MRSA, methicillin-resistant *Staphylococcus aureus*. From: Hastings et al. (2012). *Used with permission.*

Vancomycin trough levels are also utilized to determine antibiotic efficacy with a documented Gram-positive infection. Although not well studied, daily dosing of aminoglycosides may improve efficacy and decrease nephrotoxicity as compared to divided dosing throughout the day. Trough levels can similarly be monitored with daily dosing, although this methodology requires further validation. Patients with underlying renal dysfunction should receive renal dosing with more frequent trough monitoring. Dual therapy can be considered in the following clinical scenarios:

- Patient instability (e.g., hypotension, altered mental status, oliguria, moderate to severe respiratory distress).
- Concern for resistant pathogens (e.g., extended-spectrum β-lactamase (ESBL)-producing *Serratia, Pseudomonas, Acinetobacter, Citrobacter, Enterobacter, Klebsiella* spp.).
- Need for synergism for specific pathogens (e.g., *Enterococcus, Mycobacterium* spp., MRSA).
- Need for synergism with specific infections (e.g., endocarditis, cryptococcal meningitis).

Vancomycin should be considered in the following situations at a dose of 60 mg/kg/day IV divided q8h (max 4 g/day):

- Patients with AML receiving high-dose cytarabine due to risk for *S. viridians* infection with associated septic shock and acute respiratory distress syndrome.
- Presentation with hypotension or other evidence of shock.
- Mucositis.
- Prior history of alpha-hemolytic *Streptococcus* infection.
- Skin breakdown or catheter site infection.
- Colonization with resistant organisms treated only with vancomycin.
- Vegetations on echocardiogram.
- Severe pneumonia.

Anaerobic drugs including clindamycin at a dose of 40 mg/kg/day IV (max 2.7 g/day) divided q6−8h, metronidazole 30 mg/kg/day IV divided q8h (max 1.5 g/day), or oral vancomycin 40 mg/kg/day orally divided q6−8h should be considered in the following situations:

- Typhlitis (neutropenic colitis).
- Significant mucosal breakdown.
- Perianal skin breakdown.
- Peritoneal signs or other abdominal pathology.
- *C. difficile* infection.

Patients should be monitored closely for signs of sepsis and treated accordingly with fluid resuscitation, vasopressor support, and management in an intensive care setting, as required. Post-HSCT patients are at risk for particular infections based on the time point after transplant as summarized in Table 33.2.

TABLE 33.2 Common Infections Seen at Different Time Points Following Hematopoietic Stem Cell Transplantation

First 30 days	Bacterial	Gram-negative aerobes and anaerobes
		Staphylococcus epidermidis
	Fungal	*Aspergillus* sp.
		Candida sp.
	Viral	Herpes simplex type I reactivation
30−120 days	Fungal	*Candida albicans* and *C. tropicalis*
		Aspergillus sp.
		Other *Candida* sp., *Trichosporon* sp., *Fusarium* sp.
		Pneumocystis jiroveci
	Viral	Cytomegalovirus
		Adenovirus
		Epstein−Barr virus
		Human herpesvirus 6
	Protozoal	*Toxoplasma* sp.

From: Hastings et al. (2012). Used with permission.

Alterations in Initial FN Management

Modifications to the initial empiric regimen should be made based on the patient's clinical course, any positive cultures, and total time of FN. Patients who are initially treated with dual antibiotic therapy or have had vancomycin added can have these secondary agents discontinued 24–72 h after initial presentation if they have defervesced, have no new clinical signs, and all cultures are negative. Patients with continued fevers should have daily blood cultures from each CVC lumen and a daily CBC to monitor for count recovery with continuation of empiric antibiotics. Resolution of neutropenia occurs once the ANC is $\geq 0.5 \times 10^9/l$. An ANC $\geq 0.2 \times 10^9/l$ and rising on two consecutive days is a sign of impeding count recovery as well as bacterial protection. Once the patient has defervesced with negative cultures and neutrophil recovery, all antibiotics can be discontinued and the patient discharged. Patients with positive cultures require continuation of antibiotics to complete an appropriate course, generally 7–10 days for Gram-positive organisms and 10–14 days for Gram-negative organisms, assuming the infection can be cleared after initiation of antibiotics appropriate for the organism sensitivity (see the section below on Management of CVCs for further information). Viral studies can also be considered, especially with seasonal viruses such as RSV, influenza, and enterovirus and for HSV in those with mucocutaneous lesions.

Fungal Infection

Patients with prolonged (i.e., >7–10 days) and profound (i.e., ANC $<0.1 \times 10^9/l$) neutropenia are at highest risk for developing an invasive fungal infection (IFI). Patients, especially those with hematologic malignancy and those not anticipated to have prompt neutrophil recovery, should have the addition of an empiric antifungal approximately 3–5 days after presentation for FN, generally with an echinocandin (i.e., micafungin, caspofungin, anidulafungin) or azole (i.e., voriconazole, posaconazole). Possible drug interactions with azole agents must be considered. Once empiric antifungal agents have been initiated, chest CT and an abdominal ultrasound should be considered to rule out occult fungal infection once the patient has experienced neutrophil recovery if with persistent or recrudescent fever. CT of the sinuses can be considered in the patient with sinus symptoms. CT of the abdomen and pelvis is generally not indicated unless the patient has localizing signs and a negative abdominal ultrasound. See Figure 33.2 for an algorithmic approach to the continued management of FN.

Biomarkers of IFI have been utilized in adult patients and may also be beneficial in pediatric patients. Galactomannan is a cell wall component of growing hyphae and can be detected from serum, urine, or bronchoalveolar lavage (BAL) fluid. Galactomannan can be a useful marker for aspergillosis although false-positive results can occur; serial repetition in conjunction with corroborative clinical and radiographic findings makes the test most useful. Testing of serum β-D-glucan, a cell wall component of most fungi, as well as polymerase chain reaction (PCR), may improve detection of IFI although data are lacking in pediatric patients.

Risk factors for fungal disease include the following:

- Recrudescence of fever after recovery of neutrophils.
- Persistent fevers.
- Active graft-versus-host disease (GVHD).
- Prolonged recent corticosteroid usage.
- Development of lower respiratory symptoms (cough, chest pain, hemoptysis, dyspnea).
- Development of new, focal papulonodular skin rash or eschar.
- Upper respiratory symptoms (nasal discharge), nasal eschar, periorbital swelling, perforation of hard palate.
- Sinus tenderness with concomitant findings on CT sinuses (i.e., erosion of sinus walls, skull base destruction).
- Findings on CT chest imaging (i.e., nodules, infiltrates, halo or crescent sign, cavitation).
- Shoulder pain.
- Focal neurologic findings with concomitant finding of mass lesion, mastoiditis, or empyema on CT head.
- Galactomannan positivity (serum, BAL).
- Positive fungal culture (blood, urine).

Invasive aspergillosis (especially *A. fumigatus*) is being seen with increasing frequency and is often found as isolated pulmonary disease. CT chest findings most often show nodules and cavitation with the halo and crescent signs being much less common as compared with adults. Patients with concern for invasive candidiasis (often through blood or urine culture positivity) should have CT imaging of the head, chest, abdomen, and pelvis and ophthalmic evaluation to rule out potential sites of disseminated disease. Patients with sinus disease are at increased risk of zygomycosis, especially mucormycosis (i.e., *Mucor*, *Rhizomucor*, *Rhizopus*).

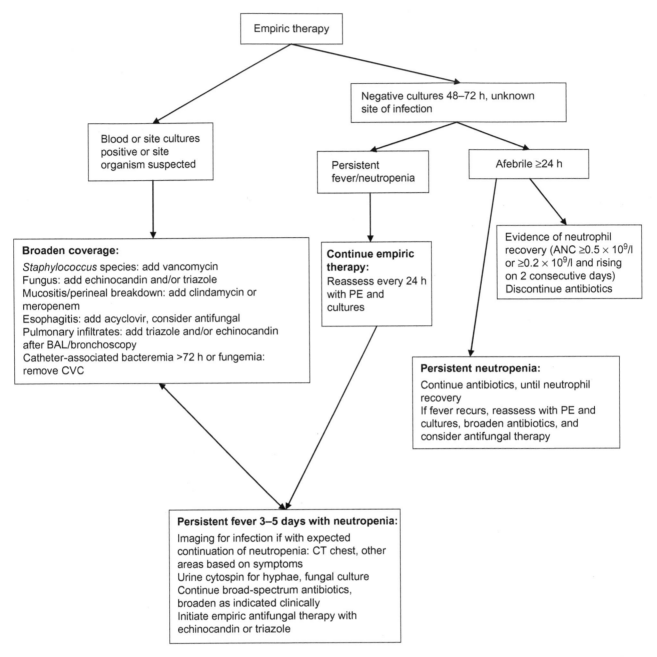

FIGURE 33.2 Ongoing management of fever and neutropenia. ANC, absolute neutrophil count; PE, physical examination; BAL, bronchoal-veolar lavage; CVC, central venous catheter; CT, computed tomography. *From: Hastings et al. (2012). Used with permission.*

Treatment of IFI includes the following:

- Invasive aspergillosis
  - IV voriconazole
    - 2 to <12 years: load 7 mg/kg/dose q12h × 2 doses (max 400 mg/dose) followed by 7 mg/kg/dose q12h (max 200 mg/dose).
    - ≥12 years: load 6 mg/kg/dose q12h × 2 doses followed by 4 mg/kg/dose q12h.
  - PO voriconazole
    - 2 to <12 years: load 8 mg/kg/dose (max 400 mg/dose) BID ×2 doses followed by 7 mg/kg/dose (max 200 mg/dose) BID.
    - ≥12 years:
      <40 kg: 100 mg BID (max dose 150 mg).
      ≥40 kg: 200 mg BID (max dose 300 mg).

- Posaconazole
  - For children >13 years of age, 400 mg PO BID with meals.
- Micafungin
  - 4 mg/kg (max 150 mg) IV q24h.
  - Can be given in conjunction with voriconazole although data on treatment synergy are lacking.
- Caspofungin
  - 70 mg/m^2 IV loading dose (max 70 mg/dose) followed by 50 mg/m^2 IV q24h (max 70 mg/dose).
  - Can be given in conjunction with voriconazole although data on treatment synergy are lacking.
- Mucormycosis
  - Liposomal amphotericin B (Ambisome)
    - 5 mg/kg/dose IV q24h.
  - Combination therapy has not shown benefit.
  - Surgical resection with soft tissue and rhino-orbito-cerebral diseases.
- Invasive candidiasis
  - Treatment based on sensitivities, often sensitive to fluconazole.
  - Fluconazole 12 mg/kg IV loading dose followed by 6–12 mg/kg IV q24h (max 400 mg/dose).
  - Echinocandins have been found to be at least non-inferior to fluconazole.

Fever in the Non-Neutropenic Oncology Patient

The non-neutropenic oncology patient remains susceptible to infection secondary to the presence of a CVC as well as immune dysfunction secondary to chemotherapy effect or effect of the underlying malignancy. Evaluation of the non-neutropenic patient should mirror that of the neutropenic patient with a careful history and physical as well as blood culture and CBC. Patients without concerning findings can be managed in the outpatient setting as long as close follow-up can be ensured. Patients should receive daily ceftriaxone for at least the first 48 h while awaiting results of the initial blood cultures. Patients in whom close follow-up or rapid return to hospital is unreliable, those with concerning findings on examination and those with dropping blood counts with the potential for severe neutropenia in the subsequent 48 h should be admitted.

INFECTION PROPHYLAXIS

Strategies to prevent infection are not as well established as treatment of infection, especially in pediatric patients. Ongoing studies are attempting to determine whether antibiotic and antifungal prophylaxis are beneficial and whether certain agents are superior. Risk stratification to determine which pediatric populations should receive prophylaxis have shown that those with the longest periods of chemotherapy-related neutropenia and therefore those receiving the most intensive myelosuppressive regimens, namely AML and relapsed acute lymphoblastic leukemia, are at the highest risk.

Antibacterial Prophylaxis

Adult data have shown benefit of antibacterial prophylaxis, especially with fluoroquinolones in high-risk patients, and should be considered in patients being treated for AML. Pediatric evidence is extremely limited though small studies have shown potential benefit. Whether antibacterial prophylaxis increases the rate of antibiotic-resistant organisms remains unclear. Risk of *C. difficile*-associated diarrhea has been shown to increase in adult patients receiving antibacterial prophylaxis.

CVCs are a potential source of infection and guidelines emphasize the importance of standardized central line care with institutional systems to ensure compliance. Additional strategies that can be considered include:

- Antibiotic lock therapy.
- Ethanol lock therapy.
- Chlorhexidine gluconate cleansing.

Antifungal Prophylaxis

Multiple defects in host defense secondary to intensive myelosuppressive chemotherapy regimens increase the risk of fungal infection, especially in the following patient groups:

- Patients undergoing HSCT, especially those with an alternative allogeneic donor.
- Treatment for AML.
- Treatment of relapsed acute lymphoblastic leukemia.
- Patients with severe aplastic anemia.

Multiple agents have been studied including fluconazole, extended-spectrum azoles (i.e., itraconazole, voriconazole, posaconazole), and echinocandins (i.e., micafungin, caspofungin) although no one agent has been shown to be consistently superior. Adult guidelines recommend antifungal prophylaxis in patients undergoing HSCT and in those with hematologic malignancy receiving intensive therapy. Any of the four azoles and both echinocandins are considered acceptable choices. Pediatric consensus guidelines recommend fluconazole 6–12 mg/kg/day (maximum 400 mg/day) for children with AML or myelodysplastic syndrome with posaconazole 200 mg three times daily as an alternative for those ≥13 years of age in settings with high local mold incidence. No such similar guidance is given regarding potential mold infection in those <13 years of age although current pediatric studies are attempting to answer this question in high-risk populations.

Pneumocystis jiroveci Pneumonia Prophylaxis

Pneumocystis jiroveci (formerly *Pneumocystic carinii*), is a yeast-like fungal species which causes pneumonia in patients with underlying T-cell immunosuppression. Prophylaxis with trimethoprim-sulfamethoxazole (TMP-SMX) remains the standard of care and is generally continued for 3 months after the completion of chemotherapy. TMP-SMX is given at a dose of 5 mg/kg/day divided BID (TMP component; maximum 320 mg TMP/day) on either 2 or 3 days per week. TMP-SMX may lead to myelosuppression or be poorly tolerated, although this is more likely in adults than children. The optimal second-line prophylactic agent is not well-defined and all appear inferior to TMP-SMX but include oral dapsone, IV or inhaled pentamidine, and oral atovaquone.

Antiviral Prophylaxis

Effective strategies to prevent viral infection in oncology patients are lacking due to a lack of risk stratification, wide variety of viruses with variable modes of transmission, and lack of effective antiviral prophylactic agents. Multiple other strategies can be utilized to prevent viral infection including:

- Preexposure prophylaxis (i.e., vaccination).
- Postexposure prophylaxis (i.e., immunoglobulin [Ig]).
- Chemoprophylaxis.
- Suppressive therapy.
- Hospital infection control practices.
- Anticipatory guidance for patient and family.

Preexposure Prophylaxis

Evidence-based guidelines are lacking on the utility of vaccination before and during chemotherapy. Increased rates of immunization against varicella zoster virus (VZV) in the United States have led to protection of the immunocompromised through herd immunity, mitigating the benefit of varicella vaccination before or during chemotherapy, especially given the risks of delaying therapy with the use of a live attenuated vaccine. In areas of high prevalence, VZV and hepatitis B virus (HBV) vaccination can be considered prior to and during chemotherapy. The appropriate timing and schedule for HBV vaccination are yet to be determined. VZV vaccination must follow rules including:

- ALL in continuous clinical remission for 1 year.
- Lymphocyte count ≥700 cells/μl.
- IgG level ≥100 mg/dl.
- Response to at least one mitogen (i.e., phytohemagglutinin or pokeweed mitogen) as a measure of T-cell function.

Risk from influenza is well-documented in immunocompromised children and the general consensus is that the benefit of inactivated influenza vaccination during therapy outweighs cost and other potential risks even if the seroresponse is blunted. Although small studies have shown safety of the intranasal live attenuated influenza vaccine (LAIV), with limited safety data and no evidence of increased immunogenicity, LAIV remains relatively contraindicated in pediatric oncology patients.

Postexposure Prophylaxis

Exposure to either VZV or measles should lead to postexposure prophylaxis to mitigate the risk of disease. In both cases live virus vaccination is contraindicated. VZV exposure is defined as contact from 2 days prior to rash development up to the time when all lesions crust over. Multiple studies have shown the potential benefit of varicella zoster immune globulin (VariZIG; VZIG) in immunocompromised children. Although VZIG may not prevent disease occurrence, it has been shown to decrease disease severity in the majority of cases, especially if given within 72 h of exposure. As an investigational agent, VariZIG may not be obtainable; in such cases intravenous immunoglobulin (IVIG) should be given instead. Oral acyclovir has also shown benefit with guidelines as below:

- *If within 4—10 days of exposure:*
  - VZIG 125 units/10 kg for the first 10—40 kg; >40 kg, 625 units IM (max 2.5 ml per injection site) OR
  - IVIG 400 mg/kg IV.
- *If within 7—10 days of exposure and neither VZIG nor IVIG administered:*
  - Acyclovir 80 mg/kg/day PO div QID (max dose 800 mg QID), for 7—14 days.

A formal comparison between VZIG and acyclovir efficacy is lacking.

Measles exposure is defined as contact 5 days prior through 4 days after onset of rash in the infectious contact. Ideally there should be virologic confirmation of exposure. Passive immunization with immunoglobulin (Ig) should be utilized as below:

- *If within 6—14 days of exposure:*
  - Immunoglobulin 0.5 ml/kg IM (max dose 15 ml; max 3 ml per injection site in children) OR
  - IVIG 400 mg/kg IV.

In settings where Ig is unavailable, ribavirin for treatment or postexposure prophylaxis of measles can be considered. Of note, a 6-month washout period after Ig is required prior to administration of measles vaccine.

Randomized studies have shown the benefit of postexposure prophylaxis in the immunocompetent patient after direct exposure with an influenza-infected person. Although studies in the immunocompromised are lacking, the Advisory Committee on Immunization Practices recommends antiviral therapy within 48 h of exposure for 10 days. Neuraminidase inhibitors are first-line agents dependent on seasonal and regional resistance patterns and are dosed as follows:

- Oseltamivir
  - 3—11 months: 3 mg/kg/dose once daily.
  - 1—12 years:
    - ≤15 kg: 30 mg once daily.
    - >15 to ≤23 kg: 45 mg once daily.
    - >23 to ≤40 kg: 60 mg once daily.
    - >40 kg: 75 mg once daily.
  - >12 years: 75 mg once daily.
- Zanamivir
  - ≥5 years: two inhalations (10 mg) once daily.

Suppressive Therapy for Viral Infections

Viral suppressive therapy is generally considered for reactivation of herpes viruses including cytomegalovirus (CMV), herpes simplex virus (HSV), and VZV after allogeneic HSCT; therefore it is important to know the patient's exposure status prior to HSCT preparative therapy. Although CMV reactivation has been reported in children after chemotherapy, data are lacking to support suppressive therapy. Ganciclovir is effective in preventing CMV reactivation post-transplant but is myelosuppressive. Additionally, prevention of reactivation with ganciclovir has not been shown to be more effective than preemptive therapy (i.e., initiation with CMV PCR positivity) to prevent symptoms of infection. In the patient with CMV reactivation post-transplant, ganciclovir should be given at a dose of 5 mg/kg IV BID for 1 week followed by 5 mg/kg/day 5 days per week.

HSV reactivation is common in adult patients although the mortality risk is low. Pediatric data are lacking and it is not recommended to routinely administer acyclovir prophylaxis in children receiving chemotherapy. In patients with breakthrough infection, acyclovir or valacyclovir therapy can be utilized. In those with recurrent infection, antiviral prophylaxis can be considered although evidence-based guidelines are lacking. Data are lacking on the use of acyclovir to prevent VZV reactivation with chemotherapy in pediatric patients. Acyclovir prophylaxis should be given to patients undergoing HSCT with a history of either HSV or VZV exposure to prevent reactivation. Acyclovir dosing is as follows:

- Prophylaxis for history of HSV or VZV exposure (post-transplant)
  - Patients >35 kg: generic acyclovir 800 mg PO BID or valacyclovir 500 mg PO BID.
  - Patients <35 kg: acyclovir oral suspension 600 mg/m^2 PO BID or valacyclovir 250 mg PO BID.
  - Patients without oral intake: acyclovir 250 mg/m^2 IV q12h.
- Treatment of symptomatic HSV infection (all dosing for 7-day duration)
  - Patients >35 kg: valacyclovir 500 mg PO TID.
  - Patients <35 kg: valacyclovir 500 mg PO BID or acyclovir suspension 600 mg/m^2 QID.
  - Patients without oral intake: acyclovir 250 mg/m^2 IV q8h.

Hospital Infection Control Practices

Multiple hospital-based infection control practices are vital to protect immunocompromised patients from nosocomial infection. Important interventions include the following:

- Hand hygiene.
- Mandatory vaccination of healthcare workers.
- Isolation of immunocompromised patients.
- Isolation of patients with communicable diseases.
- Visitor screening.
- Healthcare work restriction.

Healthcare workers are a significant potential reservoir of infection for patients and healthcare workers have been shown to not restrict themselves from work, especially those with a viral upper respiratory infection. Therefore, institutional standards must be in place to enforce restriction of healthcare workers from attending to high-risk patients if they develop upper respiratory symptoms.

Anticipatory Guidance

Patients undergoing chemotherapy and HSCT and their families must be advised as to the potential sources of infection outside of the hospital setting. As a means to prevent infection, patients should be advised to avoid crowded places such as movie theaters, shopping malls, and grocery stores where they may be exposed to viral pathogens. Similarly, families should employ screening at home to ensure that visitors are free of respiratory symptoms. It is often best to avoid exposure to young children who may be reservoirs of viral disease.

Household contacts should receive yearly inactivated influenza vaccine and young, susceptible contacts should be immunized against varicella. Those who develop a post-vaccination rash are recommended to be separated from susceptible individuals until the lesions have crusted over due to the theoretical risk of transmission even though no case of transmission of vaccine strain varicella to the immunocompromised has been reported. Additional live virus vaccines such as measles-mumps-rubella (MMR) and rotavirus have been deemed safe. Oral poliovirus is contraindicated and LAIV is relatively contraindicated.

RECOGNITION AND MANAGEMENT OF NAUSEA AND VOMITING

Chemotherapy-induced nausea and vomiting (CINV) can lead to:

- Decreased quality of life.
- Metabolic imbalances.
- Anorexia resulting in malnutrition.
- Prolonged hospitalizations.
- Potential delay or discontinuation of subsequent chemotherapy cycles.

Factors which may influence the incidence of CINV include:

- Type, dose, and schedule of chemotherapy.
- Target of RT.
- Individual patient variability based on age, gender, prior chemotherapy.

CINV is classified as the following:

- *Anticipatory nausea* occurs in patients with a history of significant CINV and may be triggered by multiple stimuli including odors and visual and auditory stimuli.
- *Acute* CINV occurs during drug administration and resolves within 24 h after completion of therapy with peak symptoms occurring 4–6 h after chemotherapy commencement.
- *Delayed* CINV begins >24 h after the completion of therapy.
- *Breakthrough* CINV occurs despite utilization of multiple antiemetics.
- *Refractory* CINV refers to CINV not responding to multiple antiemetics.

Breakthrough and refractory CINV require use of adjuvant agents with subsequent treatment cycles; refractory CINV may also benefit from the utilization of complementary therapies.

Nausea is mediated through the autonomic nervous system and vomiting is mediated by stimulation of the vomiting center which receives input from neuronal pathways including:

- Chemoreceptor trigger zone (CTZ).
- Peripheral stimuli from the gastrointestinal (GI) tract via vagal and splanchnic nerves.
- Cortical pathways (midbrain receptors, limbic system).
- Vestibular labyrinthine apparatus of the inner ear.

The CTZ is located in the area postrema in the floor of the fourth ventricle. Several receptors have been identified in the CTZ:

- Muscarinic.
- Dopamine (D_2).
- Serotonin (5-HT_3).
- Neurokinin-1 (NK-1).
- Histamine (H_1).

The emetic center is located in the nucleus tractus solitarii in the brainstem and coordinates afferent signaling from the GI tract and efferent signaling to the salivation and respiratory centers, abdominal muscles, and autonomic nerves. Cortical involvement is also a likely efferent signal which results in anticipatory CINV. Emesis results from the release of neurotransmitters including serotonin from intestinal enterochromaffin cells.

Antiemetic Agents

Effective antiemetic agents provide control of vomiting by blocking neurochemical receptors and thus inhibiting stimulation of the CTZ. Those agents with proven antiemetic activity in children include the following:

- *Dopamine receptor antagonists*—While these have been replaced by 5-HT_3 receptor antagonists as the primary antiemetic of choice secondary to their side effect profile, metoclopramide is commonly utilized in pediatric patients and must be used in conjunction with an antihistamine secondary to the risk of dystonic reaction and extrapyramidal symptoms.

- *Corticosteroids* are effective through unclear mechanisms but are most beneficial when initiated prior to the onset of chemotherapy in regimens which do not utilize steroids as part of treatment. Steroids should also be avoided in patients with CNS malignancy due to concern for dexamethasone decreasing influx of chemotherapeutic agents into the brain by altered permeability of the blood—brain barrier.
- *5-HT₃ receptor antagonists*—Ondansetron is the most widely utilized 5-HT$_3$ receptor antagonist although granisetron, dolasetron, and palonosetron are available in the United States, with tropisetron available internationally but not yet approved by the US Food and Drug Administration (FDA). Ondansetron is dosed as 0.15 mg/kg q8h to a maximum of 8 mg, although it is equally effective when given as a single daily dose of 0.45 mg/kg or 16 mg/m^2 to a maximum of 24 mg. Granisetron has clinically been shown to be equally efficacious and safe and is dosed as a single daily 40 μg/kg IV dose. Palonosetron has a higher binding affinity to the 5-HT$_3$ receptor and 5—10 times longer half-life than first-generation agents. Further research is required in pediatric oncology patients to determine the optimal dose, cost-effectiveness, and relative efficacy of palonosetron as compared to first-generation agents.
- *Neurokinin-1 receptor (substance P) antagonists* (aprepitant, fosaprepitant) are used in combination with a 5-HT$_3$ receptor antagonist and dexamethasone to prevent acute and delayed CINV in adult oncology patients receiving highly emetogenic chemotherapy. Pediatric studies are lacking; some centers are utilizing adult dosing (125 mg on day 1, 80 mg on days 2 and 3) in children ≥12 years of age. Variable dosing regimens have been utilized in children <12 years of age although optimal dosing is yet to be determined. Aprepitant is a moderate inhibitor of CYP3A4 and thus drug interactions are an important consideration, especially in patients receiving etoposide, ifosfamide, imatinib, irinotecan, paclitaxel, and vinca alkaloids (in addition to steroids). The antiemetic dose of dexamethasone should be halved when given with aprepitant. Additional NK-1 receptor antagonists casopitant and rolapitant have not been studied in pediatric patients.
- *Cannabinoids*—Those approved for CINV include dronabinol and nabilone and are thought to work by targeting cannabinoid-1 (CB-1) and CB-2 receptors in the brain. These agents have demonstrated modest efficacy for CINV in children, although side effects including euphoria, dizziness, and hallucinations must be considered.

Antihistamines, such as diphenhydramine, affect histaminergic receptors in the CTZ and are empirically effective but have not been systematically studied. Similarly benzodiazepines and anticholinergics are widely utilized and empirically beneficial but not well studied. Benzodiazepines are useful adjuncts especially in the patient with anticipatory CINV. See Table 33.3 for a description of agents, proposed mechanism of action and suggested dosage.

TABLE 33.3 Common Pediatric Antiemetic Agents

Agent	Mechanism of action	Dose	Comments
Ondansetron	5-HT$_3$ receptor antagonist	0.15 mg/kg q8h IV/PO (max 32 mg/day)	Well tolerated; common side effects include headache, fatigue, constipation, diarrhea. Also available as orally disintegrating tablet
Lorazepam	Interaction with GABA receptor; poorly understood antiemetic effects	0.25—0.5 mg q4—6h IV/ PO, max dose 2 mg	Used as adjunctive; can be utilized for anticipatory nausea. At higher doses has more sedation/anxiolytic effect than antiemetic effect
Diphenhydramine	H$_1$ histamine receptor antagonist	0.5—1 mg/kg q6h IV/ PO, max dose 50 mg	Used as adjunctive; can be utilized for anticipatory nausea. Higher dose used for prevention of dystonic reaction with metoclopramide
Metoclopramide	Dopamine antagonist	1 mg/kg q4—6h IV/PO, max dose 50 mg	Used as adjunctive; higher dose required for antiemetic effect as compared to prokinetic. Must be given with diphenhydramine at higher dose
Decadron	Poorly understood	5 mg/m^2 q6h IV/PO	Used as adjunctive; cannot be used in malignancies where steroids are part of the treatment regimen or in brain tumor regimens
Scopolamine	Anticholinergic	Transdermal patch for adolescents/adults	Must be changed q72h; patient must be advised to not touch patch and then rub eyes as this will lead to mydriasis
Dronabinol	Cannabinoid; agonist antiemetic effect	5 mg/m^2 q2—4h, max dose 15 mg/m^2 in adults	No established pediatric dosing. Teens and young adults should be advised to not smoke cannabinoids which can contain impurities or increase the risk of fungal infection
Aprepitant	Neurokinin-1 receptor antagonist	80—125 mg daily PO in adults	Pediatric dosing not established; insufficient studies in pediatric patients to date

From: Hastings et al. (2012). Used with permission.

TABLE 33.4 Emetogenic Potential of Common Pediatric Chemotherapeutic Agents

High (>90%)	Moderate–high (60–90%)	Moderate (30–60%)	Moderate–low (10–30%)	Minimal (<10%)
Cytarabine (>1000 mg/m^2)	Cytarabine ($250-1000$ mg/m^2)	Carboplatin	Cytarabine (<250 mg/m^2)	Bleomycin
Cisplatin (>50 mg/m^2)	Cisplatin (<50 mg/m^2)	Cyclophosphamide (<750 mg/m^2)	Etoposide	Decadron
Cytoxan (>1500 mg/m^2)	Cytoxan ($750-1500$ mg/m^2)	Daunomycin	Mercaptopurine	Prednisone
	Dactinomycin	Doxorubicin (<60 mg/m^2)	Methotrexate ($50-250$ mg/m^2)	Fludarabine
	Doxorubicin (>60 mg/m^2)	Idarubicin	Topotecan	Methotrexate (<50 mg/m^2)
	Methotrexate (>250 mg/m^2)	Ifosfamide	Vinblastine	Thioguanine
		Irinotecan	Radiation therapy	Vincristine

From: Hastings et al. (2012). Used with permission.

Alternative therapies including ginger, acupressure/acupuncture, hypnosis, and other behavior modification techniques may be beneficial but have not been studied systematically, especially in pediatric patients.

Although patient factors are important when considering the risk of CINV, chemotherapy agents have direct emetogenic risk as described in Table 33.4. When patients are undergoing therapy with multiple agents, the emetogenicity of the regimen is determined based on the agent with the highest risk. Management of CINV should follow the algorithm below:

- Moderate–high to high emetogenic risk
  - 5-HT$_3$ receptor antagonist.
  - Dexamethasone (if allowed).
  - If steroids not allowed, additional breakthrough agent(s).
  - Aprepitant (if ≥12 years of age).
- Moderate emetogenic risk
  - 5-HT$_3$ receptor antagonist.
  - Dexamethasone (if allowed).
  - If steroids not allowed, additional breakthrough agent.
- Moderate–low emetogenic risk
  - 5-HT$_3$ receptor antagonist.
- Minimal emetogenic risk
  - No routine prophylaxis.

MUCOSITIS

In addition to CINV, oral mucositis is one of the most common and distressing side effects of cancer therapy. Mucositis occurs secondary to damage of the GI mucosal lining from chemotherapy and RT with a continuum from limited mildly sore erythematous mucosae to diffuse areas of painful ulceration with pseudomembrane formation. Potential effects of mucositis include:

- Fever.
- Pain.
- Dysphagia.
- Delay in delivery of chemotherapy and RT.
- Anorexia resulting in malnutrition and need for nasogastric tube feeding or total parenteral nutrition (PN).
- Increased hospitalization.
- Overall decreased quality of life.

Risk factors for the development of mucositis include:

- HSCT (allogeneic > autologous transplantation).
- RT, especially high dose (i.e., >50 Gy) given to the head and neck.
- Combination chemotherapy and RT.
- Existing oral or dental disease.
- Altered nutritional status.
- Previous history of mucositis.

Mucositis is a complex physiologic process with five stages:

- Initiation.
- Primary damage response.
- Signal amplification.
- Ulceration.
- Healing.

Initiation occurs with the onset of chemotherapy or RT and subsequent direct cell damage as well as development of reactive oxygen species which cause further and more significant cell damage. The primary damage response leads to expression of NF-κB, which subsequently stimulates pro-inflammatory cytokines including TNF-α, IL-6, and IL-1β and results in apoptosis of the epithelial basal cells. Cytokine mediators then cause further damage for days after the initial chemotherapy or RT through signal amplification. Finally, clinical signs of mucositis are seen as ulceration with loss of mucosal integrity and pain. Pseudomembrane formation is a potential source of bacterial colonization and sepsis. Healing occurs over 2–3 weeks with migration, differentiation, and proliferation of new tissue and is somewhat dependent on neutrophil count recovery.

Appropriate assessment of oral mucositis is vital in order to facilitate management. Although multiple mucositis scales exist for adult cancer patients, there are limited data in pediatric patients. Additionally there are difficulties in assessing the young, uncooperative child. Grades of oral mucositis based on the National Cancer Institute Common Terminology Criteria for Adverse Events (CTCAE v4.0) include:

- Grade 1: Asymptomatic or mild symptoms not requiring intervention.
- Grade 2: Moderate pain not interfering with oral intake or requiring dietary modifications.
- Grade 3: Severe pain interfering with oral intake.
- Grade 4: Life-threatening consequences requiring urgent intervention.
- Grade 5: Death.

Prevention and Treatment of Oral Mucositis

Evidence-based interventions which are effective in the prevention and treatment of oral mucositis, especially in pediatric patients, are extremely limited. Palifermin, a recombinant keratinocyte growth factor-1 (KGF-1), has been shown to increase cellular proliferation and mediate epithelial cell repair with a decrease in duration and severity of oral mucositis in adult patients undergoing HSCT. Extremely limited pediatric data have shown potential benefit in HSCT at a dose of 90 μg/kg/day. Low-level laser therapy (local application of a monochromatic, narrow-band, coherent light source) has shown potential benefit in the prevention and treatment of oral mucositis though pediatric data are limited and somewhat conflicting. Other preventative and treatment modalities that have shown inconsistent and limited pediatric evidence include glutamine and cryotherapy. Additional treatment modalities which are vital include pain control as well as continued oral care with saline rinses at a minimum. Oral hygiene including gentle brushing and flossing should ideally be continued if tolerable.

PAIN MANAGEMENT

Pain in pediatric oncology patients can be secondary to the underlying tumor, due to treatment-related effects of chemotherapy and RT or procedure-related. Pain is a common, underreported, and underdiagnosed problem in hospitalized children. A number of studies have demonstrated that effective pain management not only increases a patient's comfort level but can also affect long-term changes in a patient's pain threshold, and, in critically ill patients, it has been demonstrated to improve morbidity and mortality. Despite this, other studies have demonstrated that pediatric pain management is often suboptimal. Multiple factors have been noted in the inadequate control of pediatric cancer pain:

- Persistent misconceptions about pediatric pain
  - Perception that pain is less severe or less frequent than is the case.
  - Infant hypoalgesia.
  - Concern of causing dependence or addiction.
- Challenges in pain assessment due to variable cognitive and developmental stage.
- Lack of pediatric pharmacokinetic data.
- Lack of access to pediatric pain and palliation specialists.
- Limited training in palliation amongst pediatric oncologists.

Developmental Issues in Pediatric Pain Management

Undertreatment of pain in young children is known to have short-term physiologic effects and may influence later pain behaviors including those children with newly diagnosed cancer. A child's development stage must be considered in the assessment of pain; children under 3 years of age and those with developmental delay must have pain intensity measured by behavioral observation scales while older children (e.g., >8 years) can utilize adult scales. Physiologic aspects can affect the pharmacokinetics and pharmacodynamics of analgesic drug delivery including increased potential for sedatory hypoventilation in infants and increased drug clearance in young children as compared to adults due to a proportionally larger liver mass.

Assessment of Pain

Multiple pain assessment tools are available in the medical literature although none has been specifically validated for pediatric oncology patients. In the 2007 National Comprehensive Cancer Network Pediatric Cancer Pain Guidelines, the Wong-Baker FACES® Pain Rating Scale (Figure 33.3) for self-report and the FLACC (face, legs, activity, crying, consolability) scale for preverbal children (Table 33.5) were utilized. Due to the number of potentially useful measures, it is important for institutions to determine an appropriate interdisciplinary pain management plan including which pain tools are preferred for systematic pain assessment in pediatric oncology patients. Behavioral and psychological assessment is a necessity in young, nonverbal children with the understanding that self-report is the best indicator of pain in all verbal patient populations (i.e., usually children ≥3 years).

Assessment of pain should be made for all patients, whether in the hospital or clinic. Patients undergoing specific interventions may require more frequent monitoring of pain; similarly, patients in pain requiring analgesic support must be closely monitored for the appropriate response as well as the development of side effects. An assessment of pain must also include documentation of an appropriate plan and follow-up to address pain issues. Pain assessment in infants with cancer is especially difficult and further research is required.

FIGURE 33.3   The Wong-Baker FACES™ Pain Rating Scale. *From: Hockenberry et al. (2005). Used with permission. Copyright Mosby.*

TABLE 33.5 The Face, Legs, Activity, Cry, Consolability (FLACC) Scale

Parameter	0 Points	1 Point	2 Points
Face	No particular expression or smile	Occasional grimace, frown, withdrawn or disinterested	Frequent to constant frown, clenched jaw, quivering chin
Legs	Normal position or relaxed	Uneasy, restless, or tense	Kicking or legs drawn up
Activity	Lying quietly, normal position, moves easily	Squirming, shifting back and forth, or tense	Arched, rigid, or jerking
Cry	No cry	Moans, whimpers, or occasional complaint	Crying steadily, screams or sobs, frequent complaints
Consolability	Content, relaxed	Reassured by occasional touching, hugging, or being talked to; distractible	Difficult to console or comfort

From: Merkel et al. (2004). Used with permission.

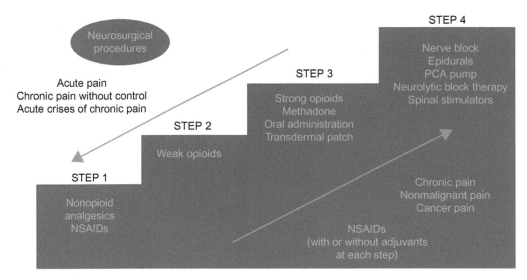

FIGURE 33.4 Adaptation of the World Health Organization analgesic ladder. NSAID, nonsteroidal anti-inflammatory drug; PCA, patient-controlled analgesia. *From: Hastings et al. (2012). Used with permission.*

Treatment of Pain

In 1986 the World Health Organization presented the analgesic ladder as a framework that physicians could use when developing treatment plans for pain. A recent adaptation is presented in Figure 33.4.

The cornerstone of the ladder rests on five simple recommendations:

1. The oral form of medication should be chosen whenever possible.
2. Analgesics should be given at regular intervals. It is necessary to respect the duration of the medication's efficacy and to prescribe doses at specific intervals in accordance with the patient's level of pain. The dose should be adjusted until the patient is comfortable.
3. Analgesics should be prescribed according to pain intensity as evaluated by a pain scale. The prescription must be given according to the level of the patient's pain and not according to the medical staff's perception of the pain.
4. Dosing of pain medication should be adapted to the individual. The correct dosage is one that will allow adequate relief of pain. The dosing should be adapted to achieve the best balance between the analgesic effect and the side effects.
5. Analgesics should be prescribed with a constant concern for detail. The regularity of analgesic administration is crucial for the adequate treatment of pain.

Based on level of symptoms, patients should be initiated with the appropriate type of pain control (i.e., step 1, 2, or 3) and then titrated up or down based on the improvement or worsening of their symptoms.

Step 1 Therapy: Non-Opioid Analgesics

Drugs that are commonly used for step 1 therapy of pain in children are presented in Table 33.6. These should be first-line therapy for pain unless the clinical situation dictates that therapy should begin at a higher step. Although nonsteroidal anti-inflammatory drugs are generally not utilized in oncology patients because of the theoretical risk of reversible platelet dysfunction in the setting of potential thrombocytopenia, there are no large, controlled trials demonstrating the validity of this concern.

Step 2 Therapy: Weak Opioids

Patients whose pain is inadequately controlled with step 1 therapy or who are being weaned from step 3 therapy require treatment with step 2 agents (Table 33.7). Note that many patients may not metabolize codeine to morphine and thus it will always provide inadequate pain relief in these cases. The use of codeine can cause the same side effects seen with stronger opioids: respiratory depression, hypoxia, nausea, vomiting, pruritus, constipation, physical tolerance, and dependence; therefore, the patient must be monitored closely.

TABLE 33.6 Step 1 Pain Therapy Medications

Drug	Dose	Comment
Ibuprofen	1–3 months: 5 mg/kg 3–4 times daily 3 months–1 year: 50 mg 3 times daily 1–4 years: 100 mg 3 times daily 4–7 years: 150 mg 3 times daily 7–10 years: 200 mg 3 times daily 10–12 years: 300 mg 3 times daily 12–18 years: 300–400 mg 3 times daily **Maximum dose 30 mg/kg/day; 2.4 g/day**	• Block conversion of arachidonic acid into prostaglandins and thromboxanes • Nonselective; variably impair platelet function • Ketorolac can affect renal function or bone growth with prolonged use
Naproxen	5 mg/kg/dose 2 times daily **Maximum dose 15 mg/kg/day**	
Ketorolac	0.5 mg/kg/dose IV/IM every 6 h **Maximum dose 30 mg** **Maximum 20 doses**	
Acetaminophen	15 mg/kg/dose every 4–6 h **Maximum 5 doses/day; maximum 4 g/day**	• Has no antiplatelet effect • Can be hepatotoxic • Avoid rectal dosing in neutropenic patients • Parenteral form recently licensed in US

IV, intravenous; IM, intramuscular.
From: Hastings et al. (2012). Used with permission.

TABLE 33.7 Step 2 Pain Therapy Medications

Drug	Dose	Comments
Codeine	0.5–1 mg/kg q4–6h Max: 60 mg/dose	• Comes as 30 and 60 mg tablets • Up to 35% of children are inefficient metabolizers of codeine to morphine; they will achieve minimal benefit from this product
Acetaminophen with codeine	0.5–1.0 mg/kg/dose of codeine q4–6h Max: 2 tablets/dose; 15 ml/dose	• Tablet: 300/15 mg, 300/30 mg • Liquid: 120/12 mg per 5 ml • Up to 35% of children are inefficient metabolizers of codeine to morphine; they will achieve minimal benefit from this product
Acetaminophen with hydrocodone	>2 years: 0.135 mg/kg/dose hydrocodone <40 kg: do not exceed 5 mg hydrocodone per dose >40 kg: do not exceed 7.5 mg hydrocodone per dose	• Tablet: 5/500 mg • Liquid: 7.5/500 mg per 15 ml • Elixir contains 7% alcohol

From: Hastings et al. (2012). Used with permission.

Step 3 Therapy: Strong Opioids

Patients not responding to step 2 therapy or whose condition indicates the need for stronger pain therapy at the outset are candidates for step 3 therapy (Table 33.8). Strong opioids act by binding to μ-receptors in the spinal cord and CNS. These receptors are also found throughout the body, and binding to peripheral sites accounts for many of the side effects seen with opiate therapy.

Oral and intravenous dosing of step 3 medications is presented in Table 33.9. It is important to remember that patients with severe chronic pain will often be on doses much larger than this due to tolerance. Patients should be monitored closely for medication effectiveness and titrated accordingly.

Side effects including constipation, nausea, vomiting, pruritus, and respiratory depression must be considered and treated accordingly. Opioid rotation can be considered in patients who develop tolerance and require large medication dosages for pain control or have troubling side effects. Additionally transdermal fentanyl can be considered in patients whose pain is poorly controlled with intravenous opioids especially if trying to return home and at the end of life. The patch strength should be individualized based on the amount of narcotic being utilized at the time of patch initiation. Each patch has a duration of action of 72 h.

Patient-controlled analgesia (PCA) provides a constant amount of opioid (basal rate) in addition to immediate pushes that the patient can administer if with breakthrough pain (usually for children ≥7 years of age). Generally the patient's previous 24-h opioid requirement is a good starting place when determining the necessary basal rate and push doses (e.g., if the patient required 48 mg of morphine in the previous 24 h, a 1.5−2 mg basal rate would be recommended with a similar push dose). A 10-min lockout period is traditionally utilized with a PCA as this is the time of onset for intravenous opioids. Patients should be continually monitored and the reversal agent naloxone available in case of respiratory depression.

Children who are too young to utilize a PCA or are developmentally delayed will benefit from authorized agent-controlled analgesia (AACA). AACA can include either nurse-controlled or caregiver-controlled analgesia as long as the caregiver is consistently available, competent, and properly educated and the authorized agent is designated in the medical order. Only one authorized agent should be utilized at a given time to avoid overdosing. Studies have shown a low complication rate with AACA but emphasize the need for institutional standards and guidelines, caregiver educational materials, and frequent nursing bedside assessment.

Step 4 Therapy: Adjuvant Modalities

Patients who fail therapy with high-dose opioids and, potentially, adjuvant anesthetic agents, develop intolerable side effects, or have severe but localized pain should be considered for regional anesthesia interventional procedures under consultation with palliative and pain management services. Most pediatric oncology patients will not need interventional management for pain. Risk factors for needing interventions include:

- Patients with solid tumors.
- Patients with metastases to the spinal nerve roots, nerve plexus, or large peripheral nerves.

TABLE 33.8   Step 3 Pain Therapy Medications

Drug	Oral dose	Comments
Oxycodone	• Instant release: 0.05−0.15 mg/kg/dose up to 5 mg/dose q4−6h • Sustained release: for patient taking >20 mg/day of oxycodone can administer 10 mg q12h	
Morphine	• 0.3−0.6 mg/kg/dose q12h for sustained release • 0.2−0.5 mg/kg/dose q4−6h prn for immediate release tablets or solution	• Injection (mg/ml): 1, 2, 4, 5, 10 • Injection, preservative free (mg/ml): 1, 5 • Oral solution (mg/ml): 2, 4, 20 • Tablet (IR) (mg): 10, 15, 30 • Tablet (ER) (mg): 15, 30, 60, 100, 200
Hydromorphone	0.03−0.08 mg/kg/dose PO q4−6h; max: 5 mg/dose	
Methadone	0.03−0.08 mg/kg/dose PO q4−6h; max: 5 mg/dose	

IR, immediate release; ER, extended release.
From: Hastings et al. (2012). Used with permission.

TABLE 33.9 Strong Opioid Characteristics

Medication	Equianalgesic doses (mg)		Pharmacokinetic profile (oral formulations unless specified)	
	IV	PO	Onset	Duration
Morphine sulfate	10	30	IV: 2–4 min	IV: 2–4 h
IR			20–30 min	3–6 h
CR (MS Contin/Oramorph)			2–4 h	8–12 h
ER (Kadian)			1–2 h	12–24 h
Hydromorphone	1.5	4.5	20–30 min	2–4 h
Oxycodone		20		
IR			20–30 min	4–6 h
CR (for regular, not prn use)			2–4 h	8–12 h
Fentanyl	180 mg oral morphine/24 h = 100 μg transdermal fentanyl/hour 1 mg IV morphine = 10 μg IV fentanyl		TD: 12–16 h IV: 1–5 min	TD: 48–72 h IV: 0.5–2 h
Methadone	Conversion ratios: • Oral: IV = 2:1 • Oral morphine: methadone based on 24-h morphine total		• Interindividual variability exists; methadone should be used by experienced clinicians only • Doses may need to be decreased after several days of administration; monitor vital signs daily and consult a specialist • May cause QT interval prolongation at higher doses	

24 h oral morphine total (mg)	Oral morphine: methadone ratio
<30	2:1
31–99	4:1
100–299	8:1
300–499	12:1
500–999	15:1
>1000	20:1

IV, intravenous; IR, immediate release; CR, controlled release; ER, extended release; TD, transdermal.
From: Hastings et al. (2012). Used with permission.

Practitioners must be aware of additional options including anesthetic agents, interventional procedures, complementary and alternative therapies, and end of life measures including palliative radiation and sedation for which there is an evidence basis in pediatric oncology patients. End-stage cancer pain may require the consideration of such treatment options. Evidence for neuroablation and neurostimulation in refractory pediatric cancer pain is lacking.

Intravenous anesthetic agents including ketamine, propofol, and lidocaine have all been reported effective in pediatric oncology patients. As an *N*-methyl-D-aspartate receptor antagonist, ketamine may have a beneficial effect on opioid tolerance and hyperalgesia. Low-dose intravenous propofol and lidocaine infusions have also been noted to be beneficial for end-stage pediatric cancer pain not well-controlled with opioids and benzodiazepines.

Procedural Pain Management in Children with Cancer

Institutional guidelines for pain management in children with cancer undergoing procedures are vital and components should include:

- Parental education and preparation.
- Supportive and normative parental–child communication to positively influence distress.
- Child life specialist support to help provide developmentally appropriate preparation for the child.
- Cognitive-behavioral interventions such as distraction.
- Sedation for bone marrow procedures and lumbar puncture.
- Topical anesthetics for intravenous and implanted CVC access.

Neuropathic Cancer Pain

Neuropathic cancer pain in pediatric oncology patients is poorly quantified as compared to nociceptive pain and neuropathic pain in adults. Neuropathic pain can be secondary to tumor invasion of the spinal cord or nerve roots, limb-sparing surgery, amputation, chemotherapy, RT, or HSCT. It is difficult to treat and patients often receive a combination of opioids, anticonvulsants (e.g., gabapentin), tricyclic antidepressants (e.g., amitriptyline) as well as psychological interventions and physical therapy. Systematic evidence on the utility of these agents in pediatric patients is lacking and utilization of a pain clinic, if available, may be beneficial.

Complementary Therapies

Cognitive-behavioral interventions and complementary therapies may improve pain and quality of life for pediatric oncology patients both during disease treatment and at the end of life. Methods with limited evidence include massage, hypnosis, acupuncture, music therapy, virtual reality, yoga, as well as biofeedback and relaxation techniques. Many of the reports are in relation to procedure-related pain but can likely be applied to treatment-related pain, pain from the underlying disease, and pain at the end of life. Generally, except for acupuncture, complementary therapies come with little risk in children with cancer and may provide some pain relief; therefore, resources for such modalities should be routinely provided to patients and their families both during treatment of disease and at the end of life as available.

NUTRITIONAL STATUS OF THE ONCOLOGY PATIENT

Malnutrition (both overnutrition and undernutrition) are affected by the underlying diagnosis, stage of disease, therapy intensity, and baseline socioeconomic factors. Poor nutritional status has been shown to correlate with increased treatment-related side effects and reduced survival. Studies have shown decreased survival with overnutrition and undernutrition in a variety of malignancies emphasizing the importance of timely and effective nutritional interventions.

Nutrition Assessment

Nutritional assessment should commence at diagnosis and then be carried out longitudinally during treatment as well as during survivorship. Body mass index is the simplest, normalized value to follow in children ≥ 2 years of age. Basic laboratory assessment of nutritional status should include liver and renal function, glucose measurement, and lipids. The practitioner should be cognizant of direct chemotherapeutic effects which can affect nutritional status such as hyperglycemia secondary to steroids or decreased liver protein synthesis after asparaginase administration. Chemotherapy, RT, and periods of infection all lead to a catabolic state with nutrient depletion, exacerbated by decreased oral intake and subsequent micronutrient deficiencies.

Nutrition Intervention

Maintenance of growth and development can and should occur during anticancer therapy. Interventions should be proactive and occur prior to the development of malnutrition; if malnutrition occurs, interventions should be implemented to both treat the malnutrition and prevent future recurrence. Weight loss >5% from baseline weight should lead to proactive interventions; weight loss >10% should be avoided and requires more intensive nutritional intervention and close follow-up. Facets of nutritional intervention include the following:

- *Dietary counseling* should begin at diagnosis to aid in the recognition and prevention of malnutrition.
  Although special diets such as the neutropenic or low microbial diet are often recommended in severely immunosuppressed patients such as those post-HSCT, adherence to this diet is difficult, further limits dietary intake and has not been shown to decrease risk of infection as compared to food safety guidelines alone.
  In patients with inadequate intake, decreasing weight or a lack of appropriate growth and development, the initial strategies entail increasing enteral caloric intake. Nutritionally fortified drinks such as Boost®, Boost®

Breeze, Pediasure®, and Ensure® may be an option in patients with difficulty in taking the appropriate calories. Utilization of additional fats such as oil and butter can be a simple way for the family to augment caloric intake. Medium-chain triglyceride oil may also provide additional calories.

- *Appetite stimulants*—Although megestrol acetate (Megace) has been shown to improve weight gain, adrenal suppression is a common and severe potential side effect. Cyprohepatidine has been reported to be effective in small studies with drowsiness as the main reported side effect. Other agents including cannabinoids, such as dronabinol, and mirtazapine, a noradrenergic and serotonergic antidepressant, have been empirically found effective but without reported systematic review.
- *Enteral tube feeding*—In patients unable to maintain adequate oral caloric intake after nutritional counseling and addition of high-calorie foods and nutritional supplements, enteral tube feedings should be initiated.
- *PN*—ideally PN should be avoided in all children with an intact GI system and ability to tolerate enteral feeds.

Enteral feedings have multiple potential benefits over PN:

- Maintenance of GI function.
- Cost-efficiency.
- Avoidance of potential PN complications
  - Risk of bacterial infection.
  - Cholestasis.
  - Thrombosis.
  - Hepatotoxicity.
  - Metabolic derangements.
- Method for medication administration.

Although more ideal than PN, enteral tube feedings are often perceived by the patient and family to be less ideal; the practitioner should be aware of the following potential pitfalls that should be overcome in presenting enteral feeds as a positive intervention to ensure appropriate nutrition:

- Representation of enteral feeds as a punishment for not eating.
- Familial perceptions
  - Inconvenience.
  - Discomfort in nasogastric tube placement.
  - Poor body image.
- Perceived inability to utilize enteral feeding based on underlying medical condition
  - Mucositis (mild to moderate).
  - Severe neutropenia.
  - Thrombocytopenia.

With assistance from the nutritionist or dietician, an appropriate schedule for enteral feeding as well as type and volume of formula should be determined based on the patient's caloric needs as well as history of recent oral intake. Factors that should be considered when making the decision on implementation of continuous versus bolus feeds include the following:

- Baseline considerations
  - Patient's oral intake.
  - Sleep patterns.
  - Lifestyle.
  - Food allergies/intolerances.
  - Underlying GI conditions.
- Consideration for continuous feeds
  - Poor oral intake.
  - High risk for nausea and vomiting.
  - Risk for diarrhea or constipation.
  - Ability to give nocturnal feeds with normal daytime oral intake.
- Consideration for bolus feedings
  - Ability to tolerate oral intake.
  - Lifestyle—desire to be off pump as much as possible.

As a general guideline, continuous feeds can be started at 1 ml/kg/h and increased by 1–2 ml/kg/h as tolerated to reach the goal rate. In patients with persistent nausea and vomiting with continuous feeds, post-pyloric (i.e., nasoduodenal, nasojejunal) tube placement may facilitate tolerance. Formula choice may also be altered in patients with poor tolerance. Generally, an unflavored milk-based formula is initiated first; patients with lactose intolerance should be initiated on a soy-based or lactose-free formula. Additional interventions which may improve tolerance to enteral feeds include elevation of the head of the bed and use of prokinetic medications such as metoclopramide and erythromycin. Although rarely indicated in pediatric oncology patients, percutaneous endoscopic gastrostomy placement should be considered in the patient with need for long-term nutritional support and inability to take oral feeds due to oral aversion or developmental delay.

Some potential indications for PN include the following:

- Post-HSCT.
- Neutropenic enterocolitis.
- Bowel obstruction.
- Severe mucositis.
- Postsurgical ileus.
- Initial management of pancreatitis.

Utilization of PN should occur with guidance from a nutritionist or dietician. It should be noted that although PN is utilized in most centers after HSCT, enteral feeds have been shown to be safe and feasible. Potential drug interactions between PN and other medications should be reviewed as well as medication compatibility in the infusing central line lumen. The potential for side effects with PN is directly related to the time on PN and therefore efforts should be made to wean to enteral feeds slowly as tolerated.

Nutrition and Survivorship

Long-term studies are showing the risk of nutrition-related conditions in survivors of pediatric cancers including:

- Obesity.
- Metabolic syndrome.
- Heart disease.
- Osteopenia/osteoporosis.

Risk of these conditions is due to a complex interplay of medical and psychosocial issues that is yet to be clearly delineated. Periods of malnutrition or anorexia during cancer therapy may subsequently lead to later poor dietary choices and overeating. Utilization of steroids as a part of therapy may similarly alter normal homeostatic mechanisms related to satiety and lead to metabolic syndrome. Heart disease is a complex interplay of inactivity, poor diet, as well as utilization of anthracyclines as a part of therapy. The practitioner must be aware of these longer-term survivor issues and counsel the patient and family on healthy nutrition and regular exercise as well as monitoring closely for incremental gains in weight and BMI.

UTILIZATION OF HEMATOPOIETIC GROWTH FACTORS

Several recombinant human hematopoietic growth factors are approved and in clinical use but not all are approved for use in children with cancer. Specifically, neutrophil stimulating factors, granulocyte colony-stimulating factor (G-CSF) and granulocyte-macrophage colony-stimulating factor (GM-CSF), are approved, while additional agents discussed here including erythropoietins (EPOs) and platelet growth factors interleukin-11 (IL-11) and thrombopoietin (TPO)-receptor agonists are not.

Granulocyte Colony-Stimulating Factors

GM-CSF (sargramostim) was the first FDA–approved cytokine for stimulation of myelopoiesis in the post-transplant setting and has activity on multiple cell lineages including monocytes and neutrophils. GM-CSF is not FDA-approved for treating chemotherapy-induced myelosuppression or FN as it has been shown to increase the rate of fever and to be ineffective compared to placebo in reducing the rate of FN. GM-CSF was subsequently followed by G-CSF for both chemotherapy-induced neutropenia and in the post-transplant setting. A longer-lasting pegylated form of G-CSF, pegfilgrastim, is FDA-approved in adults for stimulation of granulopoiesis after myelo-suppressive chemotherapy in non-myeloid malignancies but is yet to be approved in pediatric patients.

Utilization of myeloid CSFs has been reported for the following clinical situations:

- Treatment of myelosuppression after chemotherapy (primary prophylaxis).
- Prevention of FN and delay or dose reduction in subsequent chemotherapy delivery (secondary prophylaxis).
- Treatment of neutropenia to prevent infection.
- Treatment of infection with neutropenia.

As pediatric data are limited, adult guidelines must be evaluated for applicability to pediatric patients. The following recommendations for CSFs in pediatric patients seem reasonable:

- Children expected to have a ≥20% risk of chemotherapy-induced neutropenia.
- Children who have previously suffered chemotherapy-induced FN and delay or dose reduction in chemotherapy delivery which may affect treatment outcome.
- As supportive treatment in high-risk FN, variably defined as >7−10 days of neutropenia with uncontrolled primary disease, hypotension, profound neutropenia (i.e., ANC $<0.1 \times 10^9$/l), sepsis, pneumonia, or fungal infection.

Although treatment of myelosuppression has not shown benefit in infection-related mortality, meta-analyses in pediatric patients in general have shown significant reduction in FN, length of hospital stay, and documented infection. Whether these findings hold specifically for pediatric patients with AML remains unclear. Data regarding the benefit of CSFs in pediatric patients to prevent delay or dose reduction of chemotherapy and subsequently impacting treatment outcome are unclear; the published improved survival in pediatric patients with localized Ewing sarcoma treated on a compressed arm with G-CSF support is one example of benefit. Finally, utilization of CSFs as treatment of FN in pediatric patients has shown shortened median hospital stays, days of antibiotic use, cost of treatment and a reduction in duration of the FN episode though without impact on infection-related mortality. Based on these findings pediatric guidelines are similar to adult guidelines and recommend CSFs in patients with pneumonia, hypotension, multiorgan dysfunction, and fungal infection as well as, potentially, prolonged neutropenia (i.e., >28 days), bacterial sepsis, and age <12 months.

The optimal dose of G-CSF in pediatric patients is 5 μg/kg of filgrastim and potentially 100 μg/kg of pegfil-grastim. Higher dose G-CSF (i.e., 10 μg/kg) has not shown benefit in time to ANC $\geq1.0 \times 10^9$/l, incidence of infection, febrile days, incidence of hospitalization, or overall survival as compared with 5 μg/kg. Although package inserts for CSFs suggest equipotency between subcutaneous and intravenous dosing, adult studies have shown intravenous administration to be less efficacious and thus adult guidelines recommend subcutaneous dosing. Additionally, intravenous dosing is recommended to be given over 15−30 min rather than as a push. Pediatric data are lacking and institutional practice is often to give intravenous CSFs while hospitalized for patient comfort. Timing of CSF administration is also unclear with pediatric guidelines recommending CSF initiation 1−5 days after chemotherapy completion. Limited studies have shown no difference in mean duration of neutropenia, number of hospital days on parenteral antibiotics, or number of FN episodes based on CSF start time. Similarly, the need for daily dosing versus other schedules (e.g., every other day) as well as the optimal stopping time or ANC threshold is not clear. CSF discontinuation at least 24 h prior to the next cycle of chemotherapy is important as CSF-stimulated precursors appear more sensitive to chemotherapy with risk of enhanced myelosuppression with concomitant CSF usage.

Erythropoietin

EPO is produced in the cortical region of the kidney and stimulates proliferation and terminal differentiation of erythroid precursors. While recombinant human erythropoietin (rhEPO) and secondarily darbepoetin alfa (EPO-stimulating agents; ESAs) increase hemoglobin, reduce red blood cell transfusion requirements, and improve quality of life with chemotherapy- or RT-induced anemia, meta-analyses have repeatedly shown that survival may be worsened with utilization of ESAs in adults. It is unclear whether this effect is secondary to an increase in venous thromboembolism with high hemoglobin goals in ESA-treated patients or due to the promotion of antiapoptotic genes and angiogenic growth factors (and subsequent tumor proliferation) by rhEPO. Adult guidelines suggest cautious consideration for ESAs with hemoglobin <10 g/dl in non-myeloid malignancies. Data are lacking to support ESA use in pediatric oncology patients due to unclear quality of life benefit and cost-effectiveness as well as secondary to potential risks in regards to thromboembolism, tumor progression, and overall survival.

Platelet Growth Factors

Platelet transfusion remains the only method for treatment of clinically significant thrombocytopenia in pediatric oncology patients. Multiple growth factors have *in vitro* stimulatory effects on platelet production but only IL-11, stem cell factor, and TPO have shown *in vivo* benefit. Only IL-11 is approved for chemotherapy-induced thrombocytopenia, and only in adult patients. TPO-receptor antagonists have been approved for the treatment of adult immune thrombocytopenic purpura but as yet have not been found effective for the treatment of chemotherapy-induced thrombocytopenia. Multiple additional agents have been studied and been found to be noneffective, to have unacceptable toxicity, or lead to antibody development.

Interleukin-11

IL-11 stimulates megakaryocyte maturation in addition to effects on bone, chondrocytes, neurons, adipocytes, as well as GI and bronchial epithelium. Studies in adult solid tumor patients have shown benefit in platelet count and subsequent need for platelet transfusion. Pediatric data are extremely limited, with potential benefit in time to platelet recovery and need for platelet transfusion but with significant noted side effects due to the multiple sites of action of this cytokine.

TPO-Receptor Agonists

TPO is the primary regulator of megakaryopoiesis and has shown both *in vitro* and *in vivo* effects on platelet counts. Second-generation TPO-receptor antagonists, notably TPO peptide mimetic romiplostim and non-peptide mimetic eltrombopag, have shown dose-dependent increases in platelet count without the development of neutralizing antibodies and both drugs have been FDA-approved for the treatment of immune thrombocytopenic purpura in adult patients. No study has been published using second-generation TPO-receptor antagonists in chemotherapy-induced thrombocytopenia.

MANAGEMENT OF ACUTE RADIATION SIDE EFFECTS

Potential complications from RT requiring supportive care measures include acute, subacute (also called early delayed), and chronic (late delayed) etiologies and are directly related to the site of radiation, dose, and volume of tissue exposed. Here we specifically concentrate on acute and, to a lesser extent, subacute issues.

Hematologic Toxicity

The hematopoietic system is extremely radiosensitive and therefore patients receiving RT commonly experience myelosuppression which can lead to treatment interruptions and increase the risk of infection and bleeding. Similar to chemotherapy, patients experience a rapid decline in lymphocytes followed by neutrophils, platelets, and finally erythrocytes.

- Myelosuppresion increases for each additional 20% of marrow irradiated, synergistic with concomitant chemotherapy administration. Children, unlike adults, have functional bone marrow in the appendicular skeleton, although the axial skeleton is most prominent with the pelvis and vertebrae serving the largest roles.
- Treatment interruption should occur when the ANC is between 0.3 and $0.5 \times 10^9/l$ and platelet counts between 25 and $50 \times 10^9/l$. CSFs should be used to both prevent the development of neutropenia in high-risk cases and treat neutropenia once it has occurred. Platelet transfusion remains the mainstay of treatment for radiation-induced thrombocytopenia.

- Practitioners should increase the platelet transfusion thresholds to $>30-50 \times 10^9/l$ for radiation fields with a high degree of marrow involvement such as a pelvic or craniospinal field, or those potentiating the risk of mucositis.
- It remains unclear whether maintenance of higher hemoglobin thresholds during RT to solid tumors serves as a radiosensitizer and improves patient outcomes, and therefore evidence-based guidelines are lacking in both adult and pediatric patients.

CNS Complications

Risk factors for radiation-induced brain and spinal injury include:

- Higher total radiation dose.
- Increased dose fractions (e.g., $>180-200$ cGy/dose).
- Extended radiation field volume.
- Concomitant usage of CNS toxic drugs such as intrathecal methotrexate.

Acute neurologic complications include parasthesias, seizures, encephalopathy, myelopathy, paralysis, and coma and are most likely secondary to underlying brain and spinal pathology and the resulting alteration in the blood–brain barrier and tumor edema (and potential mass effect) which occurs with RT. Unlike adults, fatigue due to radiation has not been reported in pediatric patients and is likely quite rare in this population.

Somnolence syndrome, a subacute toxicity, has been reported in pediatric patients undergoing CNS irradiation. Symptoms of this syndrome include lethargy, excessive sleeping (up to 20 h per day), anorexia, headache, irritability, fever, nausea and vomiting, and transient cognitive dysfunction and are thought due to demyelination injury of oligodendrocytes. Concomitant administration of intrathecal chemotherapy has not been shown to increase the risk of somnolence syndrome and long-term cognitive function does not seem affected. Steroid treatment at the onset of symptoms has shown, in case reports, benefit in reducing the duration of illness.

Skin Complications

Acute skin changes with RT doses >20 Gy are common, presenting days to weeks after the commencement of therapy. Areas most affected include those containing skin folds, such as the:

- Axillae.
- Groin.
- Inframammary region.

The earliest skin changes are pruritus, mild erythema, anhydrosis, and dry desquamation progressing to tender erythema, edema, and moist desquamation and, in severe cases, ulceration and necrosis. Data on the incidence of dermatitis in children are lacking. Radiation recall, which can be precipitated by multiple agents and can occur days to years after RT, most often presents as low-grade dermatitis in a previously irradiated region although more severe reactions can also occur. General management recommendations include:

- The use of loose-fitting clothing.
- Prevention of scratching or other abrasive activities.
- Protection from the sun with hats and sunscreen.
- Avoidance of temperature extremes.
- Avoidance of cornstarch or baby powder especially to skin folds.
- Use of an electric razor rather than a straight blade.
- Use of nonaluminum-based deodorant on intact skin.
- Avoidance of cosmetic products in the treatment field.
- Avoidance of swimming in lakes or chlorinated pools.
- The use of gentle, non-perfumed soaps and lotions.

It is vital that the irradiated areas be dry at the time of treatment to prevent increasing the RT dose to the skin surface. Multiple topical agents for the prevention and treatment of radiation dermatitis have been studied with no clear benefit. Similarly, the use of oral agents and dressings for management of moist desquamation are of unclear benefit.

Oral Mucositis

Mucositis is a common and debilitating complication of RT to the head and neck occurring several days to weeks after RT initiation and leading to potential treatment interruption, dose-limiting toxicity, and decreased patient quality of life. Treatment of radiation-induced oral mucositis is similar to chemotherapy-induced mucositis as discussed earlier with potential benefit with palifermin and low-level laser therapy. As previously discussed, maintenance of good oral hygiene and bland rinses has been shown to be beneficial in the prevention of mucositis; nutritional status should be monitored closely.

Dysgeusia

Patients undergoing radiation for head and neck tumors are at risk for altered taste due to direct radiation effect on the fungiform papillae and taste buds. Taste loss can precede mucositis with histologic signs of degeneration and atrophy occurring after 10 Gy. Taste loss increases exponentially with cumulative doses of 30 Gy with bitter and acid flavors being most affected. Pediatric patients may develop anorexia due to dysgeusia and therefore nutritional status must be monitored closely. Dysgeusia usually returns to normal weeks to months after the completion of RT though it may persist in a small subsegment of patients.

Xerostomia

A reduction in salivary function is most commonly seen with head and neck RT and has also been noted post-HSCT. Xerostomia risk is dose-related, with minimal xerostomia risk at mean doses of 10–15 Gy to the parotid gland and a decrease in glandular function by more than 75% with mean doses >40 Gy. Recommendations in adult patients include the use of intensity-modulated RT (IMRT) with salivary gland sparing when oncologically feasible, muscarinic agonist stimulation (pilocarpine over newer and less well-studied agents cevimeline and bethanechol) after RT completion (but not during RT), oral mucosal lubricants/salivary substitutes, salivary gland transfer in strictly selected cases, and acupuncture to stimulate salivary gland secretion.

GI Complications

Acute GI RT-induced side effects include nausea, vomiting, and anorexia immediately after treatment as well as dysphagia, esophagitis, dyspepsia, ulceration, bleeding, enteritis (GI mucositis manifesting as cramping, diarrhea, and malabsorption), and proctitis within the first few weeks of therapy. Radiation-induced nausea and vomiting (RINV) is dependent on the radiation field, RT dose, and use of concurrent chemotherapy. Generally, emetic prophylaxis should be given according to the chemotherapy-related antiemetic schedule unless the risk of emesis is higher with RT. Adult guidelines suggest the following stratification for risk of RINV:

- High risk in those receiving TBI.
- Moderate risk with RT to the upper abdomen.
- Low risk for cranial, craniospinal, head and neck, lower thorax, and pelvic RT.
- Minimal risk with extremity and breast RT.

Pneumonitis

Lung tissue is extremely sensitive to radiation with increased risk of long-term damage for fractionated lung irradiation to total doses >20 Gy. Chemotherapeutic agents including bleomycin, methotrexate, alkylating agents, dactinomycin, anthracyclines, and vinca alkaloids can synergistically add to lung injury. Pneumonitis is usually seen 1–3 months after completion of RT although it can be seen more acutely. A decreased incidence has been noted with IMRT compared to external beam RT. Studies have described the long-term risks in regards to total lung capacity and diffusion capacity. Common symptoms of acute radiation pneumonitis include cough, dyspnea, low-grade fever, and pleuritic chest pain, with minimal physical signs. Acute radiation pneumonitis is a risk factor for the development of chronic changes and patients should be followed clinically and with serial pulmonary function testing. Treatment is empiric: steroids at a dose of 1–2 mg/kg/day for several weeks followed by a slow taper have been shown to be beneficial.

Additional organ toxicity including pericarditis, hepatitis, cystitis, and nephropathy are possible but have been rarely reported in pediatric patients.

MANAGEMENT OF CVCs

CVCs are an essential component of care in children and adolescents with cancer and allow for safe and compassionate administration of chemotherapy and supportive medications, infusions, and transfusions in an efficient and cost-effective manner. Infection and occlusion remain the major risk factors with CVCs.

Types of CVCs

Types of CVCs include the following:

- Peripherally inserted central catheter (PICC).
- External tunneled CVC.
- Implanted port.

The decision on what type of catheter to utilize depends on the following:

- Age of the child.
- Body habitus.
- Length and intensity of therapy.
- Frequency of blood sampling.
- Anticipated supportive care needs (i.e., transfusions, TPN).
- Level of patient activity/lifestyle.
- Body image.
- Family/patient ability to care for the CVC.

Peripherally Inserted Central Catheters

PICCs are non-tunneled CVCs for short-term use (i.e., weeks to months) typically inserted into the basilic vein and threaded into a larger vessel, typically the distal superior vena cava (SVC). In pediatric oncology patients, a PICC should be considered in the patient too sick or unstable to receive general anesthesia (e.g., pulmonary leukostasis with hyperleukocytosis, mediastinal mass, airway compromise). Advantages of a PICC include:

- Ability to be inserted at the bedside or in interventional radiology.
- Decreased cost.
- Decreased potential complications.
- Blood sampling (compared with implanted device).
- Ability to be removed at the bedside.

Disadvantages include:

- Need for frequent (i.e., BID) flushing (compared with implanted device).
- Frequent sterile dressing changes (compared with implanted device).
- Risk of phlebitis.
- Increased risk of dislodgement due to lack of a cuff.

External Tunneled CVC

Tunneled CVCs are typically placed in the subclavian or internal jugular vein with the tip at the SVC/RA (right atrial) junction. The external portion of the catheter is tunneled under the skin and secured for long-term use with a Dacron cuff. Thicker, pheresis-grade catheters are a type of tunneled CVC that can be used for rapid injection of CT contrast and for stem cell harvesting. Advantages of external CVCs include:

- Convenient and safe delivery of intravenous therapies.
- If desired, multiple lumens for delivery of complex chemotherapeutic regimens and supportive care infusions.

- Painless blood sampling.
- Ability to repair external breaks (except pheresis-grade catheters).

  Disadvantages include:

- Surgical placement with anesthesia.
- Increased risk of infection and thrombosis compared to implanted catheters.
- Sterile dressing changes.
- Daily heparin flushes.
- Activity limitations (e.g., swimming, bathing).
- Impact on body image.
- Potential for self-removal.

Implanted Ports

A port is a tunneled CVC with a reservoir placed beneath the skin in a surgically created pocket. Non-coring Huber needles are utilized to access the port. Advantages to port placement include:

- Decreased risk of infection and thrombosis.
- Ease of blood sampling when accessed.
- Ability to swim and bathe.
- No daily cleaning.
- Less impact on body image.
- No potential for self-removal.

  Disadvantages include:

- Needle access through the skin.
- Inability to do painless blood sampling.
- Potential for dislodgement of the Huber needle and subsequent extravasation/infiltration of infusions.
- Need for reaccessing with a fresh Huber needle every 7 days.
- Need for monthly heparin flushes.
- Difficulty using multilumen devices in small children.

CVC Complications

Catheter insertion comes with rare potential risks including:

- Pneumothorax.
- Hemothorax.
- Chylothorax.
- Malpositioning.
- Arterial puncture.

Complications are more likely in patients who have had previous CVCs, prior surgery or radiation to the insertion area, obesity, or in those with unusual anatomy. Lung complications, malpositioning, and failure to place are all more likely with placement into the subclavian vein.

Infection

Infection from an indwelling CVC remains the most frequent significant complication of CVCs and either occurs secondary to bacterial colonization at the exit site or from catheter hub contamination. Antiseptic-impregnated catheters have not been shown to be beneficial in preventing infection. Institutional standards are required to provide education for personnel who insert and maintain catheters using maximal sterile barrier precautions. Catheter-related infections may be cleared and the CVC salvaged with appropriate antimicrobial therapy except in the following instances which may require CVC removal:

- Persistent positive blood cultures (i.e., ≥72 h).
- Tunnel infection.

- Port pocket site infection.
- Septic thrombosis.
- Endocarditis.
- Sepsis with hemodynamic instability poorly responsive to appropriate therapies.
- Infection with *S. aureus*, *P. aeruginosa*, fungi, or mycobacterium.

Exit site infections present with erythema, tenderness, induration, or drainage around the catheter insertion site. Culture of the site should be obtained and antimicrobial therapy tailored accordingly. Children with neutropenia or Gram-negative organisms should receive parenteral therapy while Gram-positive organisms can often be managed with oral agents. Antibiotic ointment should not be used alone as it can increase the risk of *Candida* spp. infection and promotes bacterial resistance. Tunnel infections occur with similar signs as an exit site infection but extend for >2 cm from the catheter exit site along the subcutaneous catheter tract. Pocket infections occur in the surgically created pocket for port placement and present with similar signs. Potential methods to decrease risk of CVC infection were discussed earlier in the chapter.

Occlusions

After infection, occlusion is the most common CVC complication. Potential causes of occlusion include:

- Catheter malposition or migration.
- Mechanical occlusion.
- Drug precipitates/lipid residue.
- Thrombotic occlusion.

Each of these problems can lead to partial or complete CVC occlusion. Partial occlusion often allows for fluid infusion but either sluggish or complete inability to withdraw blood due to a ball-valve effect. Complete occlusion presents with inability to draw blood or flush the catheter. Catheter malposition or migration should be assessed by history, increased external catheter length, visibility of the Dacron cuff, and chest radiography. Mechanical obstruction typically occurs due to a closed clamp, kink in the external portion of the line, or from compression by an external suture. For an implanted port the Huber needle should be withdrawn and the patient re-accessed. Incompatible medications or lipid residue can also lead to partial or complete CVC occlusion. Residues that are soluble at low pH may be cleared with 0.1% hydrochloric acid though this may lead to catheter wall damage; high pH soluble residues can be cleared with sodium bicarbonate and lipid residues with 70% ethanol solution.

Thrombotic Occlusion

Fibrin will begin forming at the external catheter wall within 24 h of insertion and can lead to thrombotic occlusions including:

- Fibrin sheath.
- Intraluminal thrombus.
- Mural thrombus.

Risk factors for thrombus formation include:

- Prior catheterization into the same vessel.
- Difficult insertion.
- Poor catheter tip positioning.
- High catheter-to-vessel ratio.
- Suboptimal catheter care.
- Underlying malignancy.
- Type of chemotherapy (i.e., asparaginase, steroids).

A fibrin sheath occurs at the external catheter wall, covering the catheter tip and resulting in inability to withdraw blood. Intraluminal occlusion occurs due to a buildup of fibrin and blood products but can be simply avoided by following flush guidelines. A mural thrombus is most significant as fibrin forms on the vessel wall and subsequently leads to venous thrombosis. If a mural thrombus is suspected due to symptoms such as distal pain, erythema, or swelling, ultrasound with Doppler flow or CT imaging is required. A dye study may not diagnose a mural thrombus. In the asymptomatic patient with only the inability to withdraw blood, the development of a fibrin sheath is the most likely cause and can generally be safely treated with a tissue plasminogen activator (tPA) dwell:

- Instill one dose into each catheter lumen.
- Allow to dwell for 30 min.
- If ineffective (i.e., no brisk blood return) a second dwell is indicated.
- Dosing of tPA is as follows:
  - External tunneled CVCs
    - Children <3 months: dilute 0.5 ml with 0.5 ml 0.9% NaCl; instill 1 ml (0.5 mg) in each lumen.
    - Children between 10−30 kg: instill 1 ml (1 mg) in each lumen
    - Children ≥30 kg: instill 2 ml (2 mg) in each lumen.
  - Implanted ports
    - Children <10 kg: dilute 0.5 ml with 2.5 ml 0.9% NaCl; instill 3 ml (0.5 mg).
    - Children ≥10 kg: dilute 2 ml with 1 ml 0.9% NaCl; instill 3 ml (2 mg).

If the second dwell is not effective the patient should undergo radiographic imaging to rule out a mural thrombus and would likely need a low-dose tPA infusion for thrombolysis:

- Initial infusion of 0.01−0.03 mg/kg/h for 6 h.
- If no improvement, sequentially increase to 0.03−0.06 mg/kg/h (max dose 2 mg/h).
- Do not exceed 48−72 h of infusion.
- Close monitoring of fibrinogen, platelets, and plasminogen.
- Repletion of plasminogen with 10 ml/kg fresh frozen plasma (FFP) if plasma concentration <50%.
- Repletion of fibrinogen with FFP or cryoprecipitate (1 unit/5 kg) for levels <100 mg/dl.
- Thrombolysis is contraindicated if there are risk factors for bleeding such as recent surgery.

Catheter-Related Thrombosis

Catheter-related thrombosis typically occurs when a mural thrombus has enlarged to occlude the vein. Thrombophilia work-up is rarely indicated as the presence of a CVC is the single greatest risk factor for venous thromboembolism. Chemotherapeutic medications, including steroids and asparaginase, may contribute to thrombotic risk. Initial management is as above with low-dose tPA infusion to attempt clot dissolution. Often clot dissolution is not possible and in such cases CVC removal is warranted. Though evidence is lacking, anticoagulation, usually with low-molecular-weight heparin (LMWH; 1 mg/kg q12h) in pediatric patients, is recommended to be given until clot dissolution or for 3 months after catheter-related thrombosis. Beyond this time period, prophylactic LMWH (0.5 mg/kg q12h) should be continued until CVC removal (Figure 33.5). LMWH should be held during periods of thrombocytopenia (i.e., platelets $<50 \times 10^9$/l), for 24 h prior and 12 h after a lumbar puncture, and for 24 h prior and 24 h after a minor surgical procedure.

Catheter Maintenance

Important components in CVC maintenance include:

- Skin antisepsis.
- CVC dressing changes.
- Hub care.
- CVC flushing and locking.
- Strict adherence to sterile technique with any line manipulation.

Skin antisepsis with chlorhexidine has been found to be the superior cleansing agent both prior to CVC placement as well as with routine dressing changes. Dressing changes are required every 5−7 days in order to assess the site and keep it free from bacterial colonization. The type of dressing is based on patient, family, and institutional preference as neither semipermeable transparent nor sterile gauze has been found to be superior. Hub decontamination should also occur frequently with either 70% isopropyl alcohol or 2% chlorhexidine in 70%

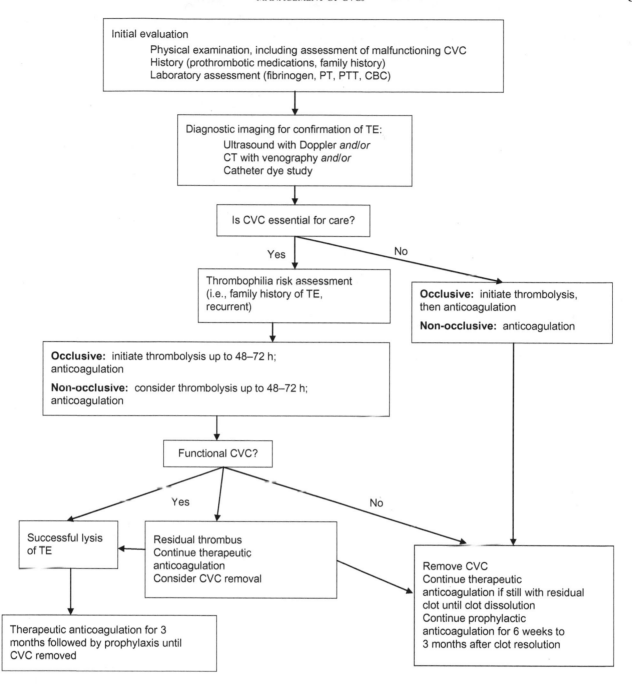

FIGURE 33.5 Evaluation of suspected thromboembolism in patients with tunneled central venous catheters. CVC, central venous catheter; PT, prothrombin time; PTT, partial thromboplastin time; CBC, complete blood count; TE, thromboembolism; CT, computed tomography. *From: Hastings et al. (2012). Used with permission.*

isopropyl alcohol. CVC flushing with normal saline is required to ensure line patency and to clear the line after medication administration or blood sampling. Finally, CVC locking with heparin to prevent reflux of blood into the catheter should occur as follows:

- PICC
  - 2 Fr: 1 ml heparinized saline (10 U/ml) every 6 h.
  - 2.6 Fr and larger: 2–3 ml heparinized saline (10 U/ml) every 12 h.
- Tunneled external CVC
  - 2 ml heparinized saline (10 U/ml) every 24 h.

- Implanted port
  - Daily to monthly flush: 5 ml heparinized saline (100 U/ml).
  - More frequent flushing: 5 ml heparinized saline (10 U/ml).

POSTTREATMENT IMMUNIZATIONS

The risk from vaccine-preventable disease is unclear after anticancer treatment although decreased immunization rates in the general community will continue to decrease herd protection. Immune reconstitution is variable after the completion of chemotherapy and HSCT and therefore inconsistent guidelines exist for (re)vaccination. Generally, patients who have received treatment for solid tumors will immune reconstitute first followed by those with hematologic malignancy and finally patients status-post HSCT. Guidelines suggest waiting 3–6 months after the completion of chemotherapy prior to inactivated/killed immunizations and 6–12 months for live virus vaccines. It is unclear as to how patients that interrupted their primary immunization series should be revaccinated but it is reasonable to continue with the primary vaccination series. Consideration can be given to reimmunizing with booster vaccines in patients that previously completed their primary vaccination series, especially for diphtheria-tetanus-acellular pertussis (DTaP), inactivated poliovirus (IPV), hepatitis B virus (HBV), MMR, and VZV.

Immune reconstitution after HSCT occurs after the patient is off all immunosuppressant therapy and therefore will occur sooner for patients' status-post autologous transplantation. Studies of immune reconstitution should occur once the absolute lymphocyte count is $>1 \times 10^9/1$ and are listed in Table 33.10. Patients with signs of immune reconstitution should first be revaccinated with diphtheria and tetanus and then subsequently have measurement of antibody response. Those with a positive response can subsequently commence revaccination with additional inactivated/killed immunizations including pneumococcus, *H. influenzae* type B, HBV, IPV, and meningococcus. Live virus vaccines (MMR and VZV) should be postponed until at least 2 years status-post HSCT. All patients should continue with yearly inactivated influenza vaccination.

PALLIATIVE CARE

The World Health Organization defines pediatric palliative care as care that "aims to improve the quality of life of patients facing life-threatening illnesses, and their families, through the prevention and relief of suffering by early identification and treatment of pain and other problems, whether physical, psychosocial, or spiritual." Ideally palliative care should be introduced when a potential life-limiting illness is diagnosed and continue whether or not a child receives disease-directed treatment. To be effective, palliative care requires a broad multidisciplinary approach that includes the family and makes use of available community resources. During any hospitalization, an important responsibility of the medical team is to ensure that each child has access to all needed services and is being supported to the greatest extent possible.

TABLE 33.10　Immune Reconstitution Studies Following Bone Marrow Transplantation

Test		Interpretation
Blastogenesis	Viral panel (CMV, HSV, VZV)	Measures response to CMV, HSV, and VZV and defines length of acyclovir prophylaxis
	Mitogens (concanavalin A, pokeweed, phytohemagglutinin)	Concanavalin A measures T- and B-cell function Pokeweed measures B-cell function Phytohemagglutinin measures T-cell function, must be reactive before PCP prophylaxis is discontinued
	Antigens (tetanus toxoid, *Candida albicans*)	Lack of response to tetanus toxoid indicates need for revaccination Response to *Candida* defines length of fluconazole prophylaxis
PRP	(If done by institution)	Measurement of T- and B-cell function; indicates need for PCP prophylaxis as well as ability to respond to conjugated vaccines

CMV, cytomegalovirus; HSV, herpes simplex virus; VZV, varicella zoster virus; PCP, *Pneumocystis jiroveci* pneumonia; PRP, polyribose phosphate.
From: Hastings et al. (2012). Used with permission.

Early introduction of palliative care principles allows for comfort-directed goals of decreasing suffering while still maintaining care-directed goals regarding treatment and cure. The palliative care team will aid in the development of a close and caring relationship with the patient and family, clarify patient and family values and priorities at different stages of treatment, and continually evaluate and reestablish the goals of care based on the current medical situation.

End of Life Care

The death of a child affects the physical and psychological wellbeing of family members for the rest of their lives. Events occurring around the time of death, both positive and negative, play a critical role in defining how family members grieve and ultimately come to terms with the event. Families who have lost a child have identified several needs regarding end of life care:

- The need to have complete information honestly communicated.
- Easy access to essential staff members who will be supportive.
- Assistance in coordinating necessary services.
- To have their relationship with their child maintained as much as possible.
- To be allowed to feel that their child's life and death has meaning.

Research studies involving pediatric palliative care are limited but have identified several important shortcomings in the level of palliative care provided to children:

- Lack of successful management of pain and other distressing symptoms.
- Significant levels of parental dissatisfaction with hospital staff arising from confusing, inadequate, or uncaring communication regarding treatment and prognosis.
- Discrepancies between parents and care providers in understanding the terminal condition.
- Problems associated with a language barrier.

Communicating effectively at the end of life requires that the practitioner listen to the unique needs of each patient and family, understand these needs and their goals, and provide honest and realistic information gently, accurately, and repeatedly. Families want to share the burden of decision-making while maintaining control and ensuring that their particular cultural, religious, and familial values are met.

Common Symptoms at the End of Life

Parents and caregivers report that their child's pain is the symptom they fear most. The desire to maintain alertness and interaction early in treatment may give way to a desire to keep the patient comfortable, even if it means the patient will be more sedated and less aware. Additional symptoms to consider include:

- Nausea and vomiting.
- Constipation.
- Weight loss and anorexia.
- Fatigue.
- Dyspnea.

Treatment of End of Life Symptoms

Practitioners must be aware of potential end of life symptoms and be willing to treat accordingly. Pain must be alleviated as outlined. The response to pain control therapy should be monitored closely and the goals of therapy frequently discussed with the patient and family. Symptoms of nausea and vomiting can often be alleviated with smaller, more frequent meals, elimination of aggravating odors or foods, and use of antiemetic medications. Constipation is a common side effect and is best treated by osmotic agents and bowel stimulants. Weight loss, anorexia, fatigue, and dyspnea are end-stage conditions. Family members should be made aware that patients lose the desire to eat at the end of life but do not suffer from hunger and secondary weight loss. Fatigue is often difficult to treat but is usually less significant in children as compared to adults. Fatigue can be alleviated by continuing activities of daily living as much as possible. Dyspnea can often be relieved by the use of strong opioids such as morphine as well as benzodiazepines.

Palliative RT

Although palliative RT is a routine part of adult oncology care, the usage of this treatment modality for end-stage pediatric cancer pain is likely underutilized and should be considered in patients with pain from soft tissues masses and metastases to bone, brain, and liver. The evidence in the medical literature for palliative radiotherapy in pediatric patients is limited. Multiple obstacles which must be overcome include lack of physician education and training, poor communication with patients, families, and interdisciplinary teams, poor healthcare coordination, lack of resources, and lack of treatment guidelines.

Palliative Sedation Therapy

Pain control is often suboptimal in the last days of life. Palliative sedation therapy (PST) should be utilized as a last resort in terminally ill patients with severe symptoms that persist despite the intensive efforts of an interdisciplinary team. Due to a lack of evidence-based research, an international panel of palliative care experts made the following recommendations for adults which can be extrapolated to pediatric oncology patients:

- The decision to utilize PST should arrive through interdisciplinary discussion with the healthcare team, family, and patient if developmentally appropriate.
- Sedative drugs should be titrated to the cessation of refractory symptoms.
- In the case of continuous deep PST, disease should be irreversible with death expected in hours to days.
- Midazolam should be considered first-line sedative therapy.
- Phenobarbital and propofol are reasonable second-line therapies.

Pediatric data are lacking.

Multidisciplinary Care

Patients and families will need assistance with multiple additional facets of care beyond physical end of life symptoms. This involves psychosocial, emotional, and spiritual support which can be provided by multiple additional caregivers such as nurses, social workers, psychologists, spiritual leaders, and family members. Incorporation of a multidisciplinary approach is vital to address the whole patient and family. Family members may also need assistance with organization of burial services and with bereavement. See Chapter 35 for a further discussion of the psychosocial aspects of cancer for patients and their families.

Further Reading and References

Anghelescu, D.L., Berde C., Cohen K.J., et al., 2007. NCCN Clinical Practice Guidelines in Oncology: Pediatric Cancer Pain v.1.2007. Available from: www.nccn.org.

Chopra, R.R., Bogart, J.A., 2009. Radiation therapy-related toxicity (including pneumonitis and fibrosis). Emerg. Med. Clin. North Am. 27, 293–310.

Dupuis, L.L., Boodhan, S., Sung, L., et al., 2011. Guideline for the classification of the acute emetogenic potential of antineoplastic medication in pediatric cancer patients. Pediatr. Blood Cancer 57, 191–198.

Dupuis, L.L., Boodhan, S., Holdsworth, M., et al., 2013. Guideline for the prevention of acute nausea and vomiting due to antineoplastic medication in pediatric cancer patients. Pediatr. Blood Cancer 60, 1073–1082.

Esposito, S., Cecinati, V., Brescia, L., et al., 2010. Vaccinations in children with cancer. Vaccine 28, 3278–3284.

Feudtner, C., 2007. Collaborative communication in pediatric palliative care: a foundation for problem-solving and decision-making. Pediatr. Clin. N. Am. 54, 583–607.

Friedrichsdorf, S.J., Kang, T.I., 2007. The management of pain in children with life-limited illnesses. Pediatr. Clin. N. Am. 54, 645–672.

Glenny, A.M., Gibson, F., Auld, E., 2010. The development of evidence-based guidelines on mouth care for children, teenagers and young adults treated for cancer. Eur. J. Cancer 46, 1399–1412.

Hastings, CH, Torkildson, JC, Agrawal, AK., 2012. Pediatric Hematology/Oncology Handbook. Children's Hospital and Research Center Oakland, second ed. Wiley Blackwell, Hoboken, NJ.

Hockenberry, M.J., Wilson, D., Winkelstein, M.L., 2005. Wong's Essentials of Pediatric Nursing, seventh ed. Mosby, St. Louis, p. 1259.

Kiman, R., Wuiloud, A.C., Requena, M.L., 2011. End of life care sedation for children. Curr. Opin. Support. Palliat. Care 5, 285–290.

Ladas, E.J., Sacks, N., Brophy, P., Rodgers, P.C., 2006. Standards of nutritional care in pediatric oncology: results from a nationwide survey on the standards of practice in pediatric oncology. A Children's Oncology Group Study. Pediatr. Blood Cancer 46, 339–344.

Lalla, R.V., Sonis, S.T., Peterson, D.E., 2008. Management of oral mucositis in patients who have cancer. Dent. Clin. North Am. 52, 61–77, viii.

Lehrnbecher, T., Phillips, R., Alexander, S., et al., 2012. Guideline for the management of fever and neutropenia in children with cancer and/or undergoing hematopoietic stem-cell transplantation. J. Clin. Oncol. 30, 4427–4438.

Lehrnbecher, T., Welte, K., 2002. Haematopoietic growth factors in children with neutropenia. Br. J. Haematol. 116, 28–56.

Marec-Berard, P., Chastagner, P., Kassab-Chahmi, D., et al., 2009. 2007 Standards, Options, and Recommendations: use of erythropoiesis-stimulating agents (ESA: epoetin alfa, epoetin beta, and darbepoetin) for the management of anemia in children with cancer. Pediatr. Blood Cancer 53, 7–12.

Merkel, S.I., Voepel-Lewis, T., Shayevitz, J.R., Malviya, S., 2004. The FLACC: a behavior scale for scoring postoperative pain in young children. Pediatr. Nurs. 23, 293–297.

Mermel, L.A., Allon, M., Bouza, E., et al., 2009. Clinical practice guidelines for the diagnosis and management of intravascular catheter-related infection: 2009 Update by the Infectious Diseases Society of America. Clin. Infect. Dis. 49, 1–45.

Monagle, P., Chan, A.K., Goldenberg, N.A., et al., 2012. Antithrombotic therapy in neonates and children: antithrombotic therapy and prevention of thrombosis, 9th ed: American College of Chest Physicians evidence-based clinical practice guidelines. Chest 141, e737S–e801S.

Royal College of Paediatrics and Child Health (RCPCH), 2002. Immunisation of the Immunocompromised Child—Best Practice Statement. RCPCH, United Kingdom.

Science, M., Robinson, P.D., MacDonald, T., et al., 2014. Guideline for primary antifungal prophylaxis for pediatric patients with cancer or hematopoietic stem cell transplant recipients. Pediatr. Blood Cancer 61, 393–400.

Sung, L., Heurter, H., Zokic, K.M., et al., 2001. Practical vaccination guidelines for children with cancer. Paediatr. Child Health 6, 379–383.

Wittman, B., Horan, J., Lyman, G., 2006. Prophylactic colony-stimulating factors in children receiving myelosuppressive chemotherapy: a meta-analysis of randomized controlled trials. Cancer Treatment Rev. 32, 289–303.

34

Evaluation, Investigations, and Management of Late Effects of Childhood Cancer

Julie I. Krystal and Jonathan D. Fish

Since the 1970s, outcomes for childhood cancer have shown remarkable and steady improvements. Five-year overall survival from childhood cancer now exceeds 80% and as many as 1 in 500 young American adults are survivors of childhood cancer. Despite these successes, nearly a quarter of survivors will have multiple severe, disabling or life-threatening conditions by age 50, over 90% have measurable end-organ damage and almost 20% will die within 30 years of diagnosis. The complications faced by survivors, and the care they require, are multisystem and complex. The most common complications include endocrine, cardiac, musculoskeletal and pulmonary complications, as well as treatment-related secondary neoplasms. Survivors also often face societal, emotional and psychological barriers, such as learning challenges, school difficulties, and problems obtaining insurance, and almost one in five suffer from stress-related mental disorders such as post-traumatic stress symptoms. Extensive, intricate, long-term medical and psychological follow-up is required to maintain survivors' health and quality of life.

Chemotherapy, radiation therapy, and surgery may all cause "late effects" involving any organ system. A definition of "late effects" includes any physical or psychological outcome that develops or persists beyond 5 years from the diagnosis of cancer, although there are many possible definitions. The field of "survivorship" has developed as a medical subspecialty that provides tailored and focused long-term follow-up for cancer survivors while conducting research to better understand, address and prevent the long-term complications of cancer therapy. Some late effects of therapy identified during childhood and adolescence resolve without consequence, while other late effects become chronic and may progress to become adult medical problems. The ability to predict and ameliorate late effects based on exposures has resulted in the development of screening guidelines. The following organizations have produced guidelines:

- Scottish Intercollegiate Guideline Network (http://www.sign.ac.uk/pdf/sign76.pdf)
- Late Effects Group of the United Kingdom Children's Cancer Study Group (http://www.ukccsg.org/public/followup/PracticeStatement/index.html)
- Children's Oncology Group (http://www.survivorshipguidelines.org)

These guidelines are a hybrid of evidence-based risk assessment with expert opinion screening recommendations that can be individually tailored based on exposures. It is critical to be familiar with the patient's exposures and to be able to elicit relevant information from the survivor. Important information to be obtained in follow-up of childhood cancer survivors is listed in Table 34.1. Table 34.2 lists selected late effects associated with common chemotherapeutic exposures and suggested screening. Table 34.3 lists selected late effects associated with organ exposure to radiation and suggested screening.

MUSCULOSKELETAL SYSTEM

Surgery

Surgery remains the primary therapy for many musculoskeletal tumors and the most visible late effect is amputation. As internal prostheses have become more refined, the ability to perform limb-salvage procedures

TABLE 34.1 Information to Be Elicited in Follow-up of Survivors of Childhood Cancer

1. History of previous cancer treatment:
   a. Cumulative doses of chemotherapeutic agents
   b. Doses and sites of radiotherapy
   c. Surgeries
2. History of intercurrent illnesses
3. Review of systems
4. Development of any benign tumors or other cancers
5. Medications (including prophylactic antibiotics and hormone replacement)
6. Educational status
   a. Highest grade completed and grade point average
   b. Results of neurocognitive evaluations
   c. Is there an individualized education plan in place at the school?
7. Employment status
8. Insurance coverage
   a. Does the patient have an individual policy, or coverage through his/her parents?
   b. Has the patient had difficulties obtaining insurance?
9. Marital history
10. Menses, libido, sexual activity
11. Pregnancy outcome (patient or spouse)
12. High-risk behaviors (smoking, alcohol or drug use, multiple sexual partners)

Adapted from: Pizzo and Poplock (2002), with permission.

TABLE 34.2 Late Effects and Screening for Chemotherapeutics

Exposure	Late effect	Suggested screening
Any chemotherapy exposure	1. Mental health disorders 2. Chronic pain 3. Fatigue 4. Difficulty accessing health care and insurance	1. Psychosocial assessment 2. Social work follow-up
Anthracyclines	1. Cardiomyopathy 2. Secondary leukemia	1. ECHO (see Table 34.4) 2. EKG 3. CBC with differential
Alkylators	1. Gonadal dysfunction (female) 2. Gonadal dysfunction (Male) 3. Pulmonary fibrosis 4. Renal/bladder toxicity 5. Acute leukemia, myelodysplasia	1. FSH 2. LH 3. Estradiol, AMH (female) 4. Testosterone, sperm count (male) 5. PFTs 6. CMP 7. Urinalysis 8. Blood pressure 9. CBC
Bleomycin	Pulmonary toxicity	1. PFTs
Corticosteroids	1. Reduced bone mineral density 2. Osteonecrosis 3. Cataracts	1. DXA scan 2. Musculoskeletal examination 3. Eye examination
Epipodophyllotoxins	Secondary leukemia	1. CBC with differential
Heavy metals (cisplatin/carboplatin)	Ototoxicity	1. Audiologic examination 2. Otoscopic examination
Methotrexate/cytarabine (high dose and intrathecal)	Neurocognitive deficits	1. Neurocognitive evaluation
Plant alkaloids	Peripheral neuropathy	1. Neurological examination

CMP, complete metabolic panel; ECHO, echocardiogram; EKG, electrocardiogram; FSH, follicle-stimulating hormone; LH, luteinizing hormone; PFT, pulmonary function test; AMH, anti-Mullerian hormone.

has dramatically improved. Limb-sparing procedures are now preformed in over 90% of osteosarcoma cases, without a deleterious effect on outcome. Although there are many advantages to a limb-salvage approach to extremity tumors, there are significant disadvantages as well. In particular, the prostheses generally cannot tolerate extreme stress, so sport participation is limited. Further, prosthesis infection, deterioration, and failure occur in as many as 20% of patients, which may require multiple additional surgeries.

TABLE 34.3   Late Effects and Screening for Radiation Exposures

Radiation exposure	Late effect	Suggested screening
Any site	1. Secondary malignancy 2. Dysplastic nevi/skin cancer	1. Target physical examination/screening 2. Dermatologic examination
Abdomen	1. Functional asplenia 2. Intestinal strictures/fibrosis/obstruction 3. Cholelithiasis 4. Renal/bladder toxicity	1. Blood culture when febrile 2. LFTs 3. Urinalysis 4. Electrolytes
Brain	1. Neurocognitive deficits 2. Leukoencephalopathy 3. Stroke 4. Craniofacial abnormalities 5. Overweight/obesity 6. Cataracts 7. Ototoxicity 8. Dental abnormalities	1. Neurocognitive evaluation 2. Neurologic examination 3. BMI, blood pressure 4. Fundoscopic examination 5. Audiologic examination 6. Dental examination
Breast tissue	1. Breast cancer 2. Breast tissue hypoplasia	1. Breast examination 2. Mammogram/breast MRI
Heart	1. Cardiomyopathy 2. Valvular disease 3. Coronary artery disease	1. ECHO (see Table 34.4) 2. EKG 3. Lipid profile 4. Blood pressure
Lungs	1. Pulmonary Fibrosis 2. Interstitial pneumonitis 3. Restrictive lung disease 4. Obstructive lung disease	1. PFTs 2. Lung examination
Ovaries	1. Delayed puberty 2. Premature menopause 3. Infertility	1. Tanner staging 2. FSH, LH 3. Estradiol
Pituitary (> 1800 cGy)	1. Growth hormone deficiency 2. Precocious puberty 3. Hypo/hyperthyroidism 4. Hyperprolactinemia 5. Gonadotropin deficiency	1. Height, weight, BMI 2. Tanner staging 3. Testicular volume (male) 4. FSH, LH (female) 5. Bone age 6. TFTs 7. Prolactin level
Testicles	1. Delayed puberty 2. Oligospermia/azoospermia 3. Testosterone deficiency	1. Tanner staging 2. FSH 3. Semen analysis 4. Testosterone level

MRI, magnetic resonance imaging; ECHO, echocardiogram; EKG, electrocardiogram; FSH, follicle-stimulating hormone; LH, luteinizing hormone; PFT, pulmonary function test; LFT, liver function test; TFT, thyroid function test.

Corticosteroids such as dexamethasone and prednisone used in cancer therapy (especially for acute lymphoblastic leukemia (ALL)) have resulted in avascular necrosis (AVN), with adolescents at highest risk. Nearly 3% of children treated for ALL will develop vascular necrosis, with much higher rates seen in those undergoing transplant. Protocols have attempted to decrease this incidence by decreasing the number of continuous weeks of exposure to dexamethasone and by utilizing more prednisone than dexamethasone. The incidence of reduced bone mineral density in survivors of childhood cancer is markedly elevated. Routine recommendations include adequate daily intake of calcium (1000–1500 mg) and vitamin D (400 IU daily) and weight-bearing exercise. The use of bisphosphonates and other absorption-reducing therapies in childhood cancer survivors remains investigational.

Radiation

For many musculoskeletal tumors, especially those not surgically resectable, radiation remains a keystone of therapy. Despite its effectiveness, radiation therapy carries with it the risk of significant consequences, which vary by site. Spine and extremity radiation can cause scoliosis, atrophy, or hypoplasia of muscles, AVN, reduced bone mineral density, discrepancy in extremity length, and alteration in sitting-to-standing height ratio.

Brain and head radiation may affect tooth enamel and formation, as well as cause learning challenges, memory loss, and personality changes. Secondary neoplasms are a risk of radiation to any site.

Women treated with chest radiation are at greatly increased risk of breast cancer, at rates similar to those of women carrying *BRCA* mutations. By the age of 50, survivors treated with chest radiation have a 30% incidence of breast cancer, with the highest risk being in those exposed to >20 Gy. Breast cancer screening in survivors is now recommended for women who received 10−19 Gy and required for those who received >20 Gy.

These late consequences are related to dose, site, volume, and age of the child at the time of radiation. The higher the dose and the younger the child, the more pronounced the late effects. Some protocols have been designed to reduce potential radiation-associated late effects through the use of:

- Low-volume, low-dose radiation combined with effective systemic therapy.
- Improvements in the delivery of radiation, including proton radiation.
- Dental prophylaxis prior to radiotherapy in maxillofacial sites.

Screening and Management

Regular physical examinations with appropriate imaging (e.g., magnetic resonance imaging (MRI) for suspected AVN) can identify musculoskeletal problems early and potentially reduce their impact on a survivor's quality of life. Survivors who have been exposed to high doses of steroids, especially in conjunction with radiation or high-dose methotrexate (MTX), may benefit from an evaluation of bone mineral density at entry into long-term follow-up. Survivors should be counseled to engage in routine weight-bearing exercise and to consume the recommended daily allowance of vitamin D (400 IU daily) and calcium (1000−1500 mg daily).

CARDIOVASCULAR SYSTEM

Chemotherapy, radiotherapy, and especially their combined use, have the potential to cause both early and late cardiac complications. After cancer recurrence and secondary neoplasms, the leading cause of morbidity and mortality in long-term survivors of childhood cancer is cardiac-related disease. The primary cardiac late effect is the development of cardiomyopathy and congestive heart failure. Other late cardiac complications include valvular heart disease and the onset of adult coronary artery disease at a younger age. Despite a focus on identification of risk factors and monitoring, cardiac toxicity remains an idiosyncratic event. Recent publications have begun to examine the role of genetic polymorphisms (e.g., those involved in anthracycline metabolism) in the development of late cardiovascular disease.

Chemotherapy

Anthracyclines

Anthracycline-induced myocyte death results in hypertrophy of existing myocytes, reduced thickness of the wall of the heart, ischemia, and interstitial fibrosis. The heart is unable to compensate adequately to meet the demands of growth, pregnancy, or other cardiac stress, which results in late-onset anthracycline-induced cardiac failure. The incidence of cardiomyopathy is related to the cumulative dose of anthracyclines, occurring in ~10% of survivors after a cumulative dose of less than $400 \, mg/m^2$, ~20% after $400-599 \, mg/m^2$, ~50% after $600-799 \, mg/m^2$, and almost 100% after $800 \, mg/m^2$. While toxicity with higher doses of anthracyclines is well documented and studied, recent data suggest that even patients exposed to $<100 \, mg/m^2$ may have long-term cardiac dysfunction.

Anthracycline-induced cardiomyopathy is a progressive disorder. It ultimately manifests with signs of congestive heart failure, including exercise intolerance, dyspnea, peripheral edema, pulmonary rales, S3 and S4 heart sounds and hepatomegaly. Rapid progression of symptoms may occur with pregnancy, anesthesia, isometric exercise, the use of illicit drugs (e.g., cocaine), prescription drugs, or alcohol. Early cardiomyopathy may be influenced by pharmacotherapy, such as with angiotensin-converting enzyme (ACE) inhibitors. Once it progresses to florid cardiac dysfunction, the only definitive treatment remains cardiac transplant, which exposes survivors to a new set of health risks and long-term consequences.

Anthracycline exposure is also associated with prolonged QTc intervals, sinus node dysfunction, and premature ventricular contractions. While arrhythmias and conduction abnormalities may be self-limited, some survivors may require pacemakers for persistent heart blocks.

Cyclophosphamide

Cyclophosphamide-induced cardiac effects occur primarily with high-dose preparatory regimens for stem cell transplantation. Cyclophosphamide causes intramyocardial edema and hemorrhage, often in association with serosanguineous pericardial effusion and fibrous pericarditis. This cardiotoxicity is usually reversible.

Radiation

Radiation is toxic to cardiomyocytes through multiple mechanisms, including ischemia, chronic inflammation, and fibrosis. In addition to acting in concert with anthracyclines to increase the risk of cardiomyopathy, radiation to the heart (e.g., mantle radiation for Hodgkin lymphoma or total body irradiation as conditioning for bone marrow transplantation) can also induce valvular damage, pericarditis, or coronary vessel damage, increasing the risk of ischemic heart disease. Cardiac doses over 20 Gy confer the highest risk. Radiation to the neck can damage the carotid vessels, increasing the risk of stroke.

Screening and Management

In addition to a good interval history and physical examination, important screening tools for evaluating cardiac function following exposure to anthracyclines or radiation include:

- Electrocardiogram (EKG).
- Echocardiogram (ECHO).
- Radionuclide angiocardiography.

EKG findings include prolonged QT_c (0.45 or longer), second-degree atrio-ventricular (A−V) block, complete heart block, ventricular ectopy, ST elevation or depression, and T-wave changes. An association between prolongation of QT_c interval and anthracycline dose over 300 mg/m^2 has been described in childhood cancer survivors.

ECHO findings include shortening fraction (SF) and velocity of circumferential fiber shortening for the measurement of left ventricular contractility.

Radionuclide cardiac cineangiocardiography (multigated acquisition (MUGA)) determines the ejection fraction and is useful for those patients in whom a good ECHO cannot be obtained. A left ventricular ejection fraction (LVEF) of 55% or more indicates normal systolic function.

The following criteria define progressively deteriorating cardiac function:

- A decrease in the SF by an absolute value of 10% from the previous test.
- SF less than 29%.
- A decrease in the MUGA LVEF by an absolute value of 10% from the previous test.
- MUGA LVEF less than 55%.
- A decrease in the MUGA LVEF with stress.

Management of anthracycline and radiation-induced cardiomyopathy should include:

- Digoxin to improve ventricular contractility.
- Diuretics to decrease sodium and water retention.
- ACE-inhibiting agents (e.g., enalapril) to decrease sodium and water retention and decrease afterload.

Prognosis of progressive late-onset heart failure is poor. Cardiac transplantation should be considered. The actuarial survival rate at 5 years after cardiac transplant is 77%.

The Children's Oncology Group provides guidelines for long-term follow-up of cardiac function following anthracycline and radiation exposure, the frequency of which is determined by the cumulative anthracycline doses and exposure to radiation (Table 34.4).

After completion of therapy, all patients with any amount of anthracycline exposure or thoracic radiation therapy should have an EKG and ECHO or MUGA. Those with normal studies at that point should be followed as per the recommended guidelines, while patients with an abnormal study either at the end of therapy or at the

TABLE 34.4   Recommended Frequency of Cardiac Screening

Age at treatment	Radiation with potential impact to the heart	Anthracycline dose	Recommended frequency
1 year old	Yes	Any	Every year
	No	<200 mg/m^2	Every 2 years
		≥200 mg/m^2	Every year
1–4 years old	Yes	Any	Every year
		<100 mg/m^2	Every 5 years
	No	≥100 to <300 mg/m^2	Every 2 years
		≥300 mg/m^2	Every year
≥5 years old	Yes	<300 mg/m^2	Every 2 years
		≥300 mg/m^2	Every year
		<200 mg/m^2	Every 5 years
	No	≥200 to <300 mg/m^2	Every 2 years
		≥300 mg/m^2	Every year
Any age with decrease in serial function			Every year

Adapted from: COG Survivorship Guidelines. http://www.survivorshipguidelines.org, with permission.

time of initial long-term follow-up (5 years after diagnosis) should have more frequent evaluations. It is extremely important to counsel patients to avoid tobacco smoke, and to encourage the maintenance of a healthy weight and physical fitness.

RESPIRATORY SYSTEM

Both chemotherapy and radiation therapy can cause acute and chronic lung injury. Younger children are at more risk than adolescents or adults for the development of chronic respiratory damage.

Chemotherapy

The primary offending agents include bleomycin and nitrosourea, with clinical manifestations usually occurring months after a critical cumulative dose is reached or exceeded. Busulfan is also associated with pulmonary fibrosis, with the greatest risk in patients who have received more than 500 mg. Pulmonary toxicity is increased when any of these agents are given in combination with radiation therapy to the lungs or chest. Children who have undergone transplant are at risk for chronic pulmonary issues such as graft versus host disease (GVHD) of the lungs and bronchiolitis obliterans organizing pneumonia, known as BOOP.

Bleomycin

The critical cumulative dose of bleomycin is 400 units, after which 10% of patients experience fibrosis. However, lung injury has been observed in children receiving 60–100 units/m^2. Bleomycin pulmonary toxicity manifests as dyspnea, dry cough, and rales. Radiographic findings include interstitial pneumonitis with reticular or nodular pattern and pulmonary function tests (PFTs) show a restrictive ventilatory defect with hypoxia, hypercapnia, and chronic hyperventilation. Assessment includes respiratory examination and chest radiography, however diffusion capacity of carbon monoxide is considered to be the most sensitive test. Radiation therapy, renal insufficiency, cisplatin, cyclophosphamide, exposure to high levels of oxygen and pulmonary infections can exacerbate the effects of bleomycin.

Nitrosourea

The greatest risk for those exposed to carmustine and lomustine occurs at doses greater than $600\,mg/m^2$. The clinical manifestations of nitrosourea toxicity are the same as bleomycin, although pulmonary fibrosis is more commonly associated with Carmustine.

While bleomycin and nitrosourea are most commonly associated with long-term pulmonary toxicity, other agents such as MTX, 6-mercaptopurine, and procarbazine have been associated with an acute hypersensitivity reaction, which can result in long-term pulmonary function changes.

Cytosine arabinoside, MTX, ifosfamide, and cyclophosphamide have been associated with non-cardiogenic pulmonary edema. This complication occurs within days of the beginning of treatment and can also result in long-term pulmonary function changes.

Other host factors that can contribute to chronic pulmonary toxicity include asthma, infection, smoking, and having had a history of assisted ventilation.

Radiation

Radiation therapy in children younger than 3 years of age results in increased pulmonary toxicity. Radiation therapy to the lungs can cause impairment of the proliferation and maturation of alveoli, resulting in chronic respiratory insufficiency. This is thought to be consistent with a proportionate interference with the growth of both the lungs and chest wall. Radiation also causes damage to the type II pneumocyte, which is responsible for the production of surfactant and the maintenance of patency and surface tension of the alveoli. As a result of changes in surfactant production, there is a decrease in alveolar surface tension and compliance. These children exhibit decreased mean total lung volumes and DLCO (diffusion capacity of carbon monoxide) that is approximately 60% of predicted values.

The effects of direct radiation to the lungs also include damage to the endothelial cells of the capillaries resulting in alterations of perfusion and permeability of the vessel wall, likely mediated by cytokine production that stimulates septal fibroblasts increasing collagen production and pulmonary fibrosis. Pulmonary toxicity for radiation can manifest as pulmonary fibrosis, interstitial pneumonitis, restrictive lung disease, or obstructive lung disease.

The late radiation injury to the lung is characterized by the presence of progressive fibrosis of alveolar septa and obliteration of collapsed alveoli with connective tissue. In asymptomatic adolescents treated for Hodgkin lymphoma, chest radiographic findings or PFTs consistent with fibrosis have been found in over 30% of patients and these changes have been detected months to years after radiation therapy. The incidence of radiation-induced fibrosis has decreased over time due to refinements in radiation therapy.

Children who receive craniospinal radiation (either for leukemia or malignant brain tumors) also have a significant risk of developing late restrictive lung disease.

Radiation to the chest, thorax, axilla, mantle, or mediastinum puts patients at risk for pulmonary toxicity, with the greatest risk seen in those who received $\geq 15\,Gy$ to these fields, or TBI with $\geq 6\,Gy$ in a single fraction or $\geq 12\,Gy$ in fractionated doses.

Screening and Management

It is critical to quickly identify patients developing pulmonary toxicity, obtain appropriate diagnostic tests, and implement interventions to prevent the toxicity from worsening. Concerning symptoms include cough, shortness of breath, and dyspnea on exertion. In addition to a good interval history and physical examination, important diagnostic tests for possible pulmonary toxicity include:

- PFTs: These should be obtained immediately following completion of therapy, at entry to long-term follow-up and as needed based on symptoms.
- Chest radiograph.
- Computed tomography (CT) scan of the chest based on symptoms and findings on PFTs and chest radiograph.

Any patient with symptoms or diagnostic testing consistent with pulmonary toxicity should be referred to a pulmonologist. In the interim, bronchodilators, expectorants, antibiotics, and oxygen can be used for symptomatic relief. It is extremely important to counsel patients to avoid tobacco smoke and on the importance of receiving the annual influenza vaccination.

CENTRAL NERVOUS SYSTEM

Understanding the risks of central nervous system (CNS) toxicity is critically important in long-term follow-up of childhood cancer survivors. Many survivors experience learning difficulties in school, which can be ameliorated with an appropriate individualized education plan (IEP) implemented by the school. Survivors qualify for school accommodations through section 504 of the federal Rehabilitation Act of 1973. A neuropsychologist familiar with the CNS effects of treatment for childhood cancer can perform a neurocognitive assessment and identify specific strengths and weaknesses in the childhood cancer survivor. These can be used to formulate interventions that can compensate for any weaknesses. Such interventions may include extra time on tests to compensate for reduced processing speed, sitting at the front of the class, and having tests taken in isolation to compensate for attention deficits, or having tutoring to bolster a specific weakness. Without the knowledge of the risks of therapy, subtle learning problems may be easily overlooked, potentially leading to poor school performance, reduced self-confidence, lower education achievement, lower earning potential, and a lower quality of life. Children treated before 5–6 years of age, especially those treated before 3 years of age, are at a higher risk for developing cognitive impairments than those treated after the age of 8–10 years.

Surgery

Survivors of childhood brain tumors usually have had surgery to remove the tumor. Depending on the location of the tumor and the degree of resection, almost any aspect of neurological function can be impaired. It is critically important to understand the location and extent of the surgery, as well as any perisurgical deficits that have persisted beyond therapy.

Chemotherapy

The pathogenesis of chemotherapy-induced CNS toxicity is not well understood. In addition to damage to the glial and capillary tissue, the neurotransmitter function of the brain may also be impaired. The primary offending chemotherapeutic agents leading to CNS toxicity include:

- Intrathecal (IT) chemotherapy (primarily MTX, cytosine arabinoside, and hydrocortisone).
- High-dose IV MTX, defined as single doses ≥ 1000 mg/m^2 (crosses the blood–brain barrier),
- High-dose IV cytosine arabinoside, defined as single doses ≥ 1000 mg/m^2 (crosses the blood–brain barrier).

These agents have more toxicity when used in combination with each other, and with steroids and radiation. Younger age at the time of therapy and the presence of CNS leukemia/lymphoma put patients at greater risk. Current trends in the use of higher doses of MTX, as well as the increased frequency with which MTX is used in IT therapy, have resulted in a rise in cognitive and neurologic sequelae.

Radiation

Radiation to a child's brain can result in altered capillary wall permeability, resulting in alterations in cerebral blood flow, primary damage to glial cells, demyelination of glial tissue, focal white matter destruction, and impaired neuronal differentiation, including dendritic formation and synaptogenesis. Children who receive cranial, craniospinal, or TBI at ages less than 3 years are at the greatest risk. In recognition of the serious consequences of brain radiation to children younger than 3 years old, protocols for the treatment of these young children have been developed specifically to avoid radiation.

Radiation to the brain over 20 Gy has been associated with pathological changes and learning deficits. Neurocognitive deficits may include issues with executive functioning, sustained attention, memory, processing speed, visual–motor integration, fine motor dexterity, and language. Although chemotherapy or radiation therapy alone can be sufficient to induce CNS changes, the combination can have an even more profound effect. Twenty-Gray cranial RT plus IT MTX and systemic MTX (40 mg/m^2 weekly) has been shown to be the most neurotoxic.

The pathologic findings as a result of radiation, IT chemotherapy, and systemic chemotherapy are:

- Leukoencephalopathy.
- Mineralizing microangiopathy.
- Subacute necrotizing leukomyelopathy.

Severe leukoencephalopathy may manifest with seizures, ataxia, lethargy, slurred speech, spasticity, dysphagia, lowered IQ scores, memory impairment, and confusion. Radioimaging findings include dilatation of the ventricles and subarachnoid space indicative of cerebral atrophy, white matter hypodensity by CT scan, and hyperdensity on MRI. With the use of current treatments, this severe form of leukoencephalopathy is seen infrequently. The subclinical form of leukoencephalopathy is more common, characterized by radiological abnormalities on CT scan and MRI without clinical symptoms and occurs in approximately 55–60% of patients receiving CNS prophylaxis. The more severe form is seen more often in patients treated with cranial radiation, IT MTX, and systemic MTX.

Mineralizing microangiopathy may manifest with seizures, electroencephalogram (EEG) abnormalities, incoordination, gait abnormalities, memory deficits, learning disabilities, decrease in IQ scores, and behavioral problems. The onset of symptoms usually occurs 10 months to several years after CNS exposure and the risk of these complications increases proportionately with higher doses of IT MTX and radiation. Imaging findings include changes in the gray matter of the brain, mainly in the region of basal ganglia and less frequently in the cerebellar gray matter. Histologically, there is deposition of calcium in the small blood vessels, causing lumen occlusion by mineralized debris. There may be dystrophic calcification of the surrounding neural tissue.

Subacute necrotizing leukomyelopathy is an unusual complication occurring after the use of cranial or craniospinal irradiation combined with IT MTX. Histologically, it shows focal myelin necrosis on the posterior and/or lateral columns of the spinal cord.

Table 34.8 lists the clinical manifestations of neurotoxicity associated with radiation therapy.

Screening and Management

In addition to a good interval history and physical examination, important diagnostic tests for possible CNS toxicity include:

- Neurocognitive assessment.
- Neuroradiologic studies.
- EEG.

Neurocognitive evaluation is essential to detect learning disability. It should be performed in all school-aged children who have received cranial irradiation, IT chemotherapy, or high-dose systemic MTX therapy. It should be performed early in therapy to establish a baseline, at the end of therapy and at entry into long-term follow-up. If any abnormalities are noted, more frequent assessment may be needed to maintain up-to-date recommendations for the school's IEP.

Neuroradiologic studies should include either CT or MRI as indicated based on physical examination, history, and neurocognitive assessment. MRI is the preferred imaging modality because of its greater sensitivity in delineating white matter damage. Mild change presents as occasional punctate areas of signal abnormality, moderate as large or multiple areas of damage, and severe as confluent areas of white matter damage. Patients with severe white matter damage manifest impairment in mentation, motor deficits, and seizures.

EEG abnormalities are relatively nonspecific, but can be useful to identify seizure activity should that be suspected based on history or physical examination.

Management of any CNS deficits should include working directly with the school to produce a functional IEP designed to enhance the strengths and ameliorate the weaknesses identified in each child.

ENDOCRINE SYSTEM

The primary endocrine systems that can sustain injury during therapy for childhood cancer include the hypothalamic–pituitary axis, the thyroid gland, the pancreas, and the gonads. More than 50% of survivors experience an endocrine disorder as a result of their treatment.

Surgery

The primary endocrine morbidities of surgery include pituitary dysfunction as a consequence of suprasellar brain tumors (i.e., germinomas and craniopharyngiomas) and gonadal dysfunction if the testes or ovaries are removed as part of therapy (i.e., for germ cell tumors).

Chemotherapy

The effects of systemic chemotherapy primarily affect the gonads. There are two aspects to gonadal function that require consideration:

- Fertility.
- Hormone production.

While these can be separated in males, they cannot be easily separated in females. The primary offending agents regarding gonadal dysfunction are the alkylating agents, such as procarbazine, cyclophosphamide, ifosfamide, and busulfan.

Female Gonadal Function

Female fertility and estrogen production are inexorably linked to the number of remaining oocytes. Girls are born with all of the oocytes they will have in their lifetime, approximately one million oocytes, which decline in an accelerating manner over time. When only 1000 oocytes remain, menopause ensues. Chemotherapy exerts a downward shift in the curve, leading to acute ovarian failure if <1000 oocytes remain after treatment or premature menopause if >1000 oocytes remain. Although anti-Mullerian hormone is being studied as a potentially reliable indicator of ovarian reserve in female survivors of childhood cancer, any girl treated with alkylating agents should be counseled that they may have a shortened fertility window. As estrogen production from follicles decreases, negative feedback to the pituitary is decreased with a resultant rise in luteinizing hormone (LH) and follicle-stimulating hormone (FSH). The decrease in viable oocytes and resulting decrease in estrogen also places women at risk for osteoporosis and sexual issues.

Male Gonadal Function

Unlike females, fertility and hormone (testosterone) production can be separated by sensitivity to treatment. While sperm are continuously produced from spermatic stem cells, testosterone is produced in the Leydig cell. The Leydig cell tends to be quite resistant to chemotherapy, so testosterone production can be conserved after therapy. In order to cause permanent male infertility, chemotherapy would have to completely eliminate all spermatic stem cells, as remaining stem cells can, over time, repopulate the testes and reestablish fertility. Sperm production can be assessed through semen analysis and testosterone production can be directly measured in the serum. As Leydig cells fail, the anterior pituitary is forced to increase the production of FSH and LH to stimulate the gonads to increase testosterone production. Any male who has been found to have elevated FSH and LH should be considered to have early Leydig cell failure. Generally, doses over $10 \, g/m^2$ of cyclophosphamide equivalents should raise concern of possible infertility and possible Leydig cell dysfunction.

Radiation

Any endocrine organ can be adversely affected by radiation. Radiation to the gonads can lead to sterility and hormone production failure as discussed above. Table 34.5 shows the effects of fractionated testicular irradiation on spermatogenesis and Leydig cell function.

Radiation to the neck can lead to thyroid dysfunction and radiation to the brain can damage the hypothalamic–pituitary axis.

The hypothalamic–pituitary axis plays a central role in translating neurologic and chemical signals from the brain into endocrine responses through its neuronal circuitry with the brain involving afferent and efferent pathways. It also produces peptide hormones and biogenic amines that serve as regulators of anterior pituitary hormones. These hypothalamic factors are supplied to the anterior pituitary gland by way of the portal venous system.

TABLE 34.5   Effect of Fractionated Testicular Irradiation on Spermatogenesis and Leydig Cell Function

Testicular dose (100 Rad) (Gy)	Effect on spermatogenesis	Effect on Leydig cell function
<0.1	No effect	No effect
0.1–0.3	Temporary oligospermia Complete recovery by 12 months	No effect
0.3–0.5	Temporary azoospermia at 4–12 months following irradiation 100% recovery by 48 months	
0.5–1.0	100% temporary azoospermia for 3–17 months from irradiation Recovery beginning at 8–26 months	Transient rise in FSH with eventual normalization
1–2	100% azoospermia from 2 months to at least 9 months Recovery beginning at 11–20 months with return of sperm counts at 30 months	Transient rise in FSH and LH No change in testosterone
2–3	100% azoospermia beginning at 1–2 months Some will suffer permanent azoospermia; others show recovery starting at 12–14 months Reduced testicular volume	Prolonged rise in FSH with some recovery Slight increase in LH No change in testosterone
3–4	100% azoospermia No recovery observed up to 40 months All have reduced testicular volume	Permanent elevation in FSH Transient rise in LH Reduced testosterone response to hCG stimulation
12	Permanent azoospermia Reduced testicular volume	Elevated FSH and LH Low testosterone Decreased or absent testosterone response to hCG stimulation Testosterone replacement may be needed to ensure pubertal changes
>24	Permanent azoospermia Reduced testicular volume	Effects more severe and profound than at 12 Gy Prepubertal testes appear more sensitive to the effects of radiation Replacement hormone treatment probably needed in all prepubertal cases

FSH, follicle-stimulating hormone; LH, luteinizing hormone; hCG, human chorionic gonadotropin.
From Leventhal et al. (1994), with permission.

Table 34.6 shows the anterior pituitary hormones, their physiologic functions, and their hypothalamic regulatory hormones.

Screening and Management

Gonadal Dysfunction

In addition to an interval history (especially menstrual and pregnancy history in females) and physical examination, important diagnostic tests for possible gonadal toxicity include:

- Serum FSH, LH, and estrogen or testosterone.
- Semen analysis.

For gonadal dysfunction, all patients who received alkylating agent and/or radiation to the gonad should be counseled about possible infertility and hormonal failure. Males should be offered the opportunity to have a

TABLE 34.6 Anterior Pituitary Hormones, Their Physiologic Functions and Their Hypothalamic Regulatory Hormones

Pituitary hormone	Functions of pituitary hormone	Hypothalamic regulatory factor
GH	Bone and soft-tissue growth through insulin-like GHs (ISF-1)	GHRH (+)
		Somatostatin (−)
PRL	Induction of lactation and interruption of ovulation and menstruation during postpartum period	Dopamine (−)
Gonadotropins		
LH	In males, LH stimulates Leydig cells to produce testosterone; in females, LH is responsible for normal steroidogenesis and ovulation	GnRH (+)
FSH	In males, FSH stimulates spermatogenesis; in females, FSH is responsible for normal steroidogenesis and ovulation	GnRH (+)
TSH	TSH regulates thyroid hormone production for thyroid gland	Thyrotropin-releasing hormone (+)
ACTH	ACTH regulates adrenal steroidogenesis	Corticotropin-releasing hormone (+)

(+), stimulating effect; (−), inhibitory effect.
FSH, follicle-stimulating hormone; LH, luteinizing hormone; TSH, thyroid-stimulating hormone; ACTH, adrenocorticotropin; PRL, prolactin; GH, growth hormone; GHRH, Growth-hormone-releasing hormones; GnRH, Gonadotropin-releasing hormone.

semen analysis performed and females should be counseled regarding the risk of premature menopause. All sexually mature males prior to therapy and males post-therapy who are still producing sperm but have elevated FSH and LH, should be offered sperm banking or testicular biopsy with cryopreservation. Studies using cryopreservation of testicular tissue in prepubertal boys are ongoing, and may become a viable option for fertility preservation in the future. For women, embryo and oocyte cryopreservation are available. While embryo cryopreservation requires an available partner, oocyte preservation allows women without a partner to increase their fertility options. Both options, however, require a delay in treatment in order to harvest oocytes and remain very costly. In addition to gonadal damage affecting fertility, the uterus and vagina can also be damaged, impacting pregnancy and delivery. Factors contributing to high-risk pregnancy and delivery after radiation to the abdomen during childhood include:

- Damage to elastic properties of the uterine musculature.
- Damage to the vasculature of the uterus.

Although the rate of birth defects among childhood cancer survivors is the same as in the general population, the rate of perinatal mortality is higher than in the general population and there is a four-fold increase in risk for low-birth-weight infants in women who have received abdominal radiation for the treatment of Wilms tumor during their childhood. These women are also at risk for premature labor and fetal malposition.

Thyroid Dysfunction

Hypothyroidism is the most common sequela of radiotherapy to the neck, occurring most commonly in those who received greater than 20 Gy of neck radiation. Elevated TSH levels with normal T3 and T4, indicative of subclinical hypothyroidism, can be detected in up to two-thirds of patients treated with mantle field radiation greater than 2600 cGy in Hodgkin lymphoma. Radiation dose-reduction lowered the incidence of hypothyroidism to 10–28%.

In addition to an interval history and physical examination, important diagnostic tests for possible thyroid toxicity include:

- Annual thyroid function tests for those who received radiation to the neck, either direct or as scatter from mantle or spinal radiation.
- Thyroid ultrasound to assess for possible secondary thyroid malignancies for those with a palpable nodule or abnormal thyroid function tests.

Patients with abnormal thyroid function tests should be referred to an endocrinologist for management and those with a nodule on physical examination or ultrasound should be referred to a neck surgeon for biopsy. Secondary papillary thyroid carcinoma can be treated with total thyroidectomy and radioactive iodine.

Anterior Pituitary Dysfunction

In addition to a good interval history and physical examination, important diagnostic tests for possible anterior pituitary toxicity include:

- Close monitoring of growth velocity with the use of a growth chart.
- Pubertal history and examination to ensure appropriate pubertal progression.
- Serum FSH, LH, and estrogen or testosterone.
- First morning cortisol level if indicated by history.

Growth Hormone Deficiency

If growth failure is due to documented growth hormone (GH) deficiency, then GH is appropriate treatment. Growth failure can occur due to the systemic effects of cancer or cancer treatment, and not necessarily due to GH deficiency, so due diligence is required to make the diagnosis of true GH deficiency.

Some considerations in deciding to use GH include:

- A worsened degree of scoliosis, although not an increased incidence.
- It may cause benign intracranial hypertension (pseudotumor cerebri), which resolves with medical management (i.e., acetazolamide and dexamethasone) and discontinuation of GH therapy.
- Children with prior neoplastic disease may have an increased risk of developing a second cancer following GH therapy, although the risk is clinically negligible.

Luteinizing and Follicle-Stimulating Hormone Deficiency

Treatment should include:

- Estrogen and progestin therapy in females.
- Androgen therapy in males.

Precocious Puberty

Treatment should include:

- Gonadotropin-releasing hormone (GnRH) analogs to suppress puberty.
- GnRH plus GH for patients with coexistent GH deficiency.

Thyroid-Stimulating Hormone Deficiency

Treatment should include daily thyroxine therapy.

Adrenocorticotropin Deficiency

Treatment should include

- Low-dose hydrocortisone therapy daily.
- Stress doses of hydrocortisone during febrile illness or under anesthesia.

Hyperprolactinemia

Treatment should include bromocriptine or related dopamine agonists are used to reduce prolactin levels in young women with amenorrhea and infertility as a result of hyperprolactinemia.

Table 34.7 describes the neuroendocrine complications, their clinical manifestations, standard fractionated radiation dose limits, diagnostic studies, and treatment.

Table 34.8 presents an evaluation of the hypothalamic–pituitary axis using an analysis of pertinent hormonal assays.

TABLE 34.7 Summary of Neuroendocrine Complications, Their Clinical Manifestations, Standard Fractionated Radiation Dose Limits to Brain Responsible for Damage, Diagnostic Studies, and Treatment

Disorder	Clinical presentation	Radiation dose (Gy)	Diagnostic studies	Treatment
GH deficiency	Short stature	≥18–20	Plotting growth on growth velocity chart GH stimulation test Bone age Frequent sampling for 12–24 h GH	Recombinant GH[a]
Gonadotropin deficiency	In young child, failure to enter puberty and primary amenorrhea; in adults, infertility, sexual dysfunction and decreased libido	>30	Basal serum concentration of LH, FSH, estradiol, or testosterone GnRH stimulation test	Estrogen/progestin (women) Depot testosterone (men)
Precocious puberty	Female 8 years of age or younger with appearance of breast and genital development and male 9 years of age or younger with testicular development Accelerated skeletal growth and premature epiphyseal fusion	>10(?)	GnRH stimulation test Estradiol or testosterone concentration Bone age Pelvic ultrasound (women) GH stimulation tests	GnRH agonists plus recombinant GH (if GH deficient)
TSH deficiency	Often subclinical	>30	Basal serum T3 uptake, T4 and TSH TRH stimulation test	L-thyroxine
ACTH deficiency	Decreased stamina, lethargy, fasting hypoglycemia, dilutional hyponatremia	>30	Basal serum cortisol concentration Adrenal stimulation test (e.g., insulin, ACTH)	Hydrocortisone

[a]See text discussion on use of growth hormone.
GH, growth hormone; TSH, thyroid-stimulating hormone; ACTH, adrenocorticotropin; LH, luteinizing hormone; FSH, follicle-stimulating hormone; GnRH, gonadotropin-releasing hormone; TRH, thyrotropin-releasing hormone.
Modified from Sklar (1994), with permission.

TABLE 34.8 Evaluation of Hypothalamic–Pituitary Axis

	Testosterone	FSH	LH	Response to GnRH	hCG
Primary Leydig cell disease	nl/lo	hi	hi	—	lo
Primary disease of germinal epithelium	nl	hi	nl	—	—
Hypothalamic disease	nl/lo	nl/lo	nl/lo	nl	—
Pituitary disease	nl/lo	nl/lo	nl/lo	lo	—

nl, normal; lo, low; hi, high; GnRH, gonadotropin-releasing hormone; hCG, human chorionic gonadotropin.
From Leventhal et al. (1994), with permission.

GENITOURINARY SYSTEM

Surgery

Surgery used to treat childhood tumors affecting the kidney (e.g., Wilms tumor and neuroblastoma) and bladder (e.g., rhabdomyosarcoma) can result in permanent damage to the genitourinary system. Other surgeries can also contribute, for example, damage to the thoracolumbar sympathetic plexus during retroperitoneal lymph node dissection can result in ejaculatory failure. Patients who have had a nephrectomy need to be counseled about the risks of trauma to their remaining single kidney (e.g., bicycle handlebar injuries or seatbelt injuries).

Chemotherapy

Ifosfamide, carboplatin, and cisplatin can cause renal complications, including acute tubular dysfunction, although carboplatin is less nephrotoxic than cisplatin. The acute tubular dysfunction can be manifested by increased excretion of potassium, phosphorus, and magnesium. Ifosfamide can also cause hypophosphatemic rickets. Risk factors for ifosfamide nephrotoxicity include:

- Younger age group, especially patients treated at less than 4 years of age.
- Hydronephrosis.
- Prior administration of cisplatin.
- Cumulative dose of ifosfamide greater than 60 g/m².
- Renal radiation ≥ 15 Gy.

Indicators of significant Fanconi syndrome include a serum glucose less than 150 mg/dl and ratio of urine protein to urine creatinine less than 0.2, as well as serum phosphate, <3.5 mg/dl; K +, <3 mEq/l; bicarbonate, <7 mEq/l; and 1 + glycosuria.

Hemorrhagic cystitis and fibrosis of the bladder can occur often after treatment with cyclophosphamide and ifosfamide, especially when the bladder is included in the radiation field.

Radiation

Hypoplastic kidney and renal arteriosclerosis can occur after the use of a combination of radiation therapy in a dose of 10−15 Gy with chemotherapy, or radiation therapy alone in a dose of 20−30 Gy. Nephrotic syndrome can occur after radiation therapy in a dose of 20−30 Gy of RT. Hypertension can be a consequence of radiation, and can exacerbate other renal dysfunction.

Abnormal bladder function is seen in children with pelvic rhabdomyosarcoma in whom radiation has been used. Thirty-two to one hundred percent of these children experience dribbling and nocturnal enuresis.

Screening and Management

In addition to a good interval history (especially sexual function in boys) and physical examination, important diagnostic tests for possible genitourinary toxicity include:

- Serum BUN and creatinine annually.
- Urinalysis annually.
- 24-h creatinine clearance or nuclear glomerular filtration rate in those with abnormal renal function.
- Blood pressure annually.
- Electrolytes annually.

OCULAR SYSTEM

Chemotherapy

Busulfan increases the risk of early cataract development, especially in conjunction with corticosteroids. Radiation alone, or in combination with chemotherapy, exacerbates these affects.

Radiation

Several areas of the eye can be damaged by radiation. In particular:

- *Lacrimal glands:* radiation >50 Gy can result in decreased tearing or fibrosis.
- *Cornea:* radiation >40 Gy can lead to ulceration, neovascularization, keratinization, or edema.
- *Lens:* radiation >10 Gy can cause cataracts.
- *Iris:* radiation >50 Gy can lead to neovascularization, glaucoma, and atrophy.
- *Retina:* radiation >50 Gy can cause infarction, exudates, hemorrhage, telangiectasia, neovascularization, and macular edema.
- *Optic nerve:* radiation >50 Gy can cause optic neuropathy.

Screening and Management

In addition to a good interval history and physical examination with importance given to visual changes or corrective lens changes, important diagnostic tests for possible ocular toxicity include:

- An annual eye examination by an ophthalmologist or optometrist, including a slit-lamp examination and dilation, for patients with exposure to corticosteroids or radiation to the eye.

AUDITORY SYSTEM

Chemotherapy

Cisplatin or carboplatin exposure can cause a mild to severe sensorineural hearing loss, often necessitating hearing aids. Both agents can also cause tinnitus and vertigo. Cisplatin is significantly more ototoxic than carboplatin. Children less than 1 year of age at the time of diagnosis are at risk of hearing loss regardless of the doses of cisplatin. Recent data have supported the use of vitamin E to reduce ototoxicity, during or after chemotherapy. The impact of vitamin E on risk of cancer relapse (given its role as an antioxidant) has not been definitively explored. Aminoglycosides used to treat infections or provide infectious prophylaxis also contribute to hearing loss.

Radiation

Radiation exposure can lead to both sensorineural and conductive hearing loses at doses of greater than 30 Gy to the cranium, infratentorium, nasopharyngeal tissue, or Waldeyer's ring. Tympanosclerosis, otosclerosis, and eustachian tube dysfunction can also occur.

Screening and Management

In addition to an interval history and physical examination, important diagnostic tests for possible audiological toxicity include:

- A hearing test for patients with exposure to cisplatin or carboplatin or radiation to the auditory system. For patients too young to perform a behavioral hearing test, a brainstem auditory-evoked potential can be performed under sedation.

GASTROINTESTINAL SYSTEM

Chemotherapy

Hepatic fibrosis/cirrhosis can occur following the use of MTX, actinomycin D, 6-mercaptopurine (6-MP), and 6-thioguanine (6-TG). 6-TG can also cause a mild to moderate sinusoidal obstructive syndrome of the liver. Viral hepatitis can exacerbate hepatic dysfunction.

Radiation

Radiation exposure to many fields, including hepatic, renal, splenic, para-aortic, extended mantle, bladder, cervical, vaginal, prostrate, spine, and flank, can lead to gastrointestinal effects. Fibrosis can occur anywhere along the gastrointestinal tract, including the liver, at doses greater than 40 Gy. Choleliathisis and bowel obstruction can occur following radiation therapy. Radiation of greater than 45 Gy to fields involving bowel can result in chronic enterocolitis, fistulas, and strictures. Radiotherapy-sensitizing agents, such as doxorubicin or actinomycin, can enhance radiation effects if given simultaneously. Patients who received greater than 30 Gy are at increased risk for colorectal cancer; however, those treated at ages greater than 50 are at the highest risk.

Transfusions

Many children being treated for cancer receive repeated transfusions of packed red blood cells for management of chemotherapy-induced anemia. This can lead to iron overload, especially liver iron overload, which in turn may exacerbate endocrinopathies, fertility, cardiac dysfunction, and hepatopathies associated with the prior cancer treatment.

Screening and Management

In addition to an interval history and physical examination, important diagnostic tests for possible gastrointestinal toxicity include:

- Liver function tests should be performed during follow-up for late effects.
- Serum ferritin level, as this correlates with liver iron overload in survivors of childhood cancer.
- Exposure to greater than 30 Gy of radiation should be screened for secondary colon cancer via colonoscopy starting at age 35, or 10 years after radiation, whichever occurs later.

A thorough family history should be obtained, as those with family histories of familial polyposis, inflammatory bowel disease, or other colorectal cancers or disorders, may require earlier or more frequent screening. Patients who may have received blood transfusions prior to 1972 may be at risk for transfusion-acquired hepatitis B and C, and screening should be done on entry to survivorship.

IMMUNOLOGIC SYSTEM

For patients who received conventional chemotherapy (i.e., not hematopoietic stem cell transplant) defects in both normal and cellular immunity generally resolve within 6 months to 1 year after cessation of chemotherapy. Although some immune deficiency may uncommonly persist for more than 2 years after completion of therapy, more than half of the children have no protective antibodies to one or more previously administered vaccines or related infections. Although most of them are able to produce antibodies after reimmunization, some patients repeatedly are unable to make protective antibodies after reimmunization or despite natural disease.

Current guidelines recommend that children resume their vaccine schedule 3 months after chemotherapy is completed, and should receive any vaccines they missed during their therapy, but should not repeat vaccinations already given. Patients who underwent hematopoietic stem cell transplantation (HSCT) should be given three doses of PCV13, three doses of Hib, two doses of MCV4 (for persons aged 11–18 years), and three doses of tetanus-diphtheria-containing vaccine, three doses of HepB, and three doses of IPV, starting 3–6 months posttransplant, and one dose of PPSV23 should be given 1 year after transplant in patient who do not have chronic GVHD. Two doses of both MMR and VAR should be given to seronegative patients who underwent HSCT, provided they are off immunosuppression drugs and do not have GVHD, 24 months following transplant. Annual influenza vaccination is recommended for all survivors.

In addition to the impact of chemotherapy and radiation on humoral and adaptive immunity, radiation to the mucosa (i.e., sinuses, nose, lungs) can cause defects in barrier immunity such as immotile cilia. This can lead to recurrent sinopulmonary infections in survivors of childhood cancer.

OBESITY

Adult survivors of childhood ALL, especially those who received cranial radiation, have an increased risk for obesity. Adult leukemia survivors are more physically inactive and have reduced exercise capacity which further increases their risk for obesity. The pathophysiology of the metabolic impacts of treatment for childhood cancer remains under investigation.

Interventions to prevent obesity and promote physical activity may decrease cardiovascular morbidity and improve the quality of life in this population. Follow-up with assessment of weight and physical activity is important in the evaluation of childhood cancer survivors.

SECOND MALIGNANT NEOPLASMS

Second malignant neoplasms are second cancers occurring in survivors of childhood cancer, caused by treatments administered for the first cancer or a heritable disposition, but otherwise unrelated to the first cancer. The risk of second malignant neoplasms varies with the type of therapy received, with some groups at significantly elevated risk. Within the first 20 years after diagnosis, the cumulative incidence of second neoplasm is between 3% and 12%. Childhood cancer survivors have 10−20 times the lifetime risk of second malignant neoplasms as compared to age-matched controls, and these second neoplasms are the most common cause of death in long-term survivors after recurrence of the primary cancer.

Patients with retinoblastoma, Hodgkin disease and Ewing sarcoma are at the greatest risk of developing second malignant neoplasms, due to their genetic makeup and therapy. Long-term survivors who have an underlying inherited susceptibility to cancer have a second malignant neoplasm rate that approaches 50%. Such inherited predispositions include:

- Retinoblastoma.
- Neurofibromatosis.
- Li−Fraumeni syndrome.
- Familial polyposis.

Both chemotherapy and radiation can predispose patients to second malignancies.

Chemotherapy

Most chemotherapy-induced second malignancies are acute leukemias and myelodysplastic syndromes. Three classes of chemotherapeutic agents increase the risk of secondary leukemias:

- Epipodophyllotoxins (e.g., etoposide).
- Anthracyclines (e.g., doxorubicin).
- Alkylating agents (e.g., cyclophosphamide).

Epipodophyllotoxins

This class of chemotherapeutics is topoisomerase II poisons. In the presence of a topoisomerase II poison, the topoisomerase enzyme complex can cleave the double-stranded DNA, but cannot re-anneal the cleaved ends. This initiates a DNA repair process, called nonhomologous end joining, that carries a high incidence of repair errors and therefore increases the risk of oncogenesis. As the topoisomerase II complex tends to function at the 11q23 location (the site of the MLL gene), MLL rearranged leukemias are prototypical for this agent. These leukemias are most commonly myeloid, but can also be lymphoid. The incidence is approximately 1−10% and may be dose- and schedule-dependent, although the specifics of these relationships remain under investigation. The peak occurrence is approximately 2−3 years following epipodophyllotoxin therapy, and these cancers are rarely seen more than 5 years following therapy.

Anthracyclines

This class of chemotherapeutics includes weak topoisomerase II poisons. The consequent secondary leukemias are similar to those seen with the epipodophyllotoxins, but occur less frequently.

Alkylating Agents

Alkylating agents increase the risk of a secondary acute myeloid leukemia, usually preceded by a myelodysplastic phase. The risk has been correlated with the total dose of alkylating agents. Typical cytogenetic abnormalities include aberrations in the long arm of chromosomes 5, 7, or both and the latency period is typically 3.5−5.5 years. The cumulative risk is 1% at 20 years.

The prognosis for patients with secondary acute leukemia is poor. Allogeneic hematopoietic stem cell transplantation is the treatment of choice. Radiotherapy does not contribute significantly to the development of secondary leukemias.

Radiation

Radiation therapy increases the childhood cancer survivors' risk of secondary solid tumors. The most common second tumor types include:

- Bone sarcoma.
- Breast cancer.
- Thyroid cancer.
- CNS cancer (e.g., meningioma).
- Skin cancers.

The risk of developing a radiation-induced second malignancy depends on the dose, site of radiation, first cancer diagnosis, age at the time of radiation, and underlying genetic risk factors. The median time to the development of a second malignant neoplasm is approximately 7 years, with a range from 1 to 15 years from therapy. Genetic abnormalities, such as neurofibromatosis or family histories suggestive of Li–Fraumeni syndrome, appear to predispose to the development of second solid neoplasms.

Screening and Management

In addition to an interval history and physical examination, important diagnostic tests for possible toxicity include:

- Annual complete blood count for those exposed to topoisomerase II poisons, anthracyclines, or alkylating agents.
- Annual dermatology checks for those exposed to radiation.
- Annual thyroid function tests (with ultrasounds if abnormal) for those exposed to thyroid radiation.
- Carotid Doppler evaluation starting at age 30.
- Research on optimal screening modalities for breast cancer are ongoing, however current recommendations for survivors treated with chest radiation include annual mammography in conjunction with breast MRI for female survivors who received chest radiation starting at age 25 or 8 years off-therapy, whichever occurs later.

Patients exposed to radiation should be counseled to apply SFP 45 sunscreen liberally and to avoid tobacco smoke.

PSYCHOSOCIAL ISSUES

Survivors of childhood cancer are at increased risk for a variety of psychosocial issues. Social withdrawal, marital issues, under/unemployment, depression, anxiety, and post-traumatic stress symptoms, all occur more frequently in survivors then in peer groups. Female survivors, and those with a family history of similar disorders, appear to be at the greatest risk. Chronic pain and chronic fatigue may have a substantial impact on survivor's quality of life, and those who have undergone amputation or major surgical procedures are most at risk.

Survivors have substantial health care needs; however access to care and compliance with follow-up remain significant issues in this population. Fewer than half of adult survivors have had a cancer-related health visit in the past 2 years, with even lower rates among those, in the United States, who are uninsured. Among patients who have had cancer-directed follow-up, many do not obtain the recommended screening tests. Multiple barriers to care exist, including lack of knowledge by general physicians, and patients own lack of knowledge about the care they received and the long-term consequences. In the United States, childhood cancer survivors are ten times more likely to have difficulty obtaining health insurance than their siblings. The recent implementation of the Affordable Care Act may result in increased insurance options for survivors, specifically since they can no longer be denied coverage because of preexisting medical conditions.

Survivors should be educated at each visit about the treatment they received and its potential long-term effects. The rationale and importance of each recommended screening test should be emphasized. Counseling on lifestyle choices, such as avoiding tobacco and maintaining a healthy weight, is part of every survivorship visit. If possible, a social worker should be available to survivors to ensure that issues such as insurance, transportation, and housing are addressed. A psychologist should be available to screen survivors for symptoms of depression, anxiety, and post-traumatic stress disorder, and recommended further counseling or treatment when needed.

Further Reading and References

Armenian, S.H., Kremer, L.C., Sklar, C., 2015. Approaches to reduce the long-term burden of treatment-related complications in survivors of childhood cancer. Am. Soc. Clin. Oncol. Educ. Book 35, 196–204.

Armstrong, G.T., Liu, Q., Yasui, Y., et al., 2009. Late mortality among 5-year survivors of childhood cancer: a summary from the Childhood Cancer Survivor Study. J. Clin. Oncol. 27, 2328–2338.

D'Angio, G.J., 1975. Pediatric cancer in perspective: cure is not enough. Cancer 35, 866–870.

Diller, L., Chow, E.J., Gurney, J.G., et al., 2009. Chronic disease in the Childhood Cancer Survivor Study cohort: a review of published findings. J. Clin. Oncol. 27, 2339–2355.

Fish, J.D., 2012. Fertility preservation for adolescent women with cancer. Adolesc. Med. State Art Rev. 23, 111–122, xi.

Fish, J.D., Ginsberg, J.P., 2009. Health insurance for survivors of childhood cancer: a pre-existing problem. Pediatr. Blood Cancer 53, 928–930.

Hudson, M.M., Ness, K.K., Gurney, J.G., et al., 2013. Clinical ascertainment of health outcomes among adults treated for childhood cancer. JAMA 309, 2371–2381.

Landier, W., Armenian, S., Bhatia, S., 2015. Late effects of childhood cancer and its treatment. Pediatr. Clin. North Am. 62, 275–300.

Leventhal, B.G., Halperin, E.C., Torano, A.E., 1994. The testes. In: Schwartz, C.L., Hobbie, W.L., Constine, L.S., Ruccione, K.S. (Eds.), Survivors of Childhood Cancer: Assessment and Management. Mosby, St. Louis, pp. 225–244.

Lipshultz, S.E., Cochran, T.R., Franco, V.I., Miller, T.L., 2013. Treatment-related cardiotoxicity in survivors of childhood cancer. Nat. Rev. Clin. Oncol. 10, 697–710.

Metzger, M.L., Meacham, L.R., Patterson, B., et al., 2013. Female reproductive health after childhood, adolescent, and young adult cancers: guidelines for the assessment and management of female reproductive complications. J. Clin. Oncol. 31, 1239–1247.

Moskowitz, C.S., Chou, J.F., Wolden, S.L., et al., 2014. Breast cancer after chest radiation therapy for childhood cancer. J. Clin. Oncol. 32, 2217–2223.

Nathan, P.C., Patel, S.K., Dilley, K., et al., 2007. Guidelines for identification of, advocacy for, and intervention in neurocognitive problems in survivors of childhood cancer: a report from the Children's Oncology Group. Arch. Pediatr. Adolesc. Med. 161, 798–806.

Oeffinger, K.C., Mertens, A.C., Sklar, C.A., et al., 2006. Chronic health conditions in adult survivors of childhood cancer. N. Engl. J. Med. 355, 1572–1582.

Park, E.R., Li, F.P., Liu, Y., et al., 2005. Health insurance coverage in survivors of childhood cancer: the Childhood Cancer Survivor Study. J. Clin. Oncol. 23, 9187–9197.

Pizzo, P.A., Poplock, D.G. (Eds.), 2002. Principles and Practice of Pediatric Oncology. fourth ed. Lippincott Williams and Wilkins, Philadelphia.

Rutter, M.M., Rose, S.R., 2007. Long-term endocrine sequelae of childhood cancer. Curr. Opin. Pediatr. 19, 480–487.

Sklar, C.A., 1994. Neuroendocrine complications of cancer therapy. In: Schwartz, C.L., Hobbie, W.L., Constine, L.S., Ruccione, K.S. (Eds.), Survivors of Childhood Cancer, Assessment and Management. Mosby, St. Louis, pp. 97–110.

35

Psychosocial Aspects of Cancer for Children and Their Families

Meg Tippy

Throughout the entire course of care, it is of paramount importance to address the psychosocial and educational needs of patients and their families.

The diagnosis and treatment of childhood cancer is extremely stressful and frightening for families. As with any life-threatening illness, childhood cancers interfere with the normal developmental milestones of childhood and adolescence, and the ability of families to function effectively. Understanding common reactions of children and families at various stages of the illness helps the practitioner communicate more effectively about the diagnosis, treatment, and prognosis, and to support the family throughout the process. While no single professional can meet a family's needs completely, collaborative and multidisciplinary health care team efforts can help family members enhance adaptive coping skills and mobilize their support systems to effectively engage in medical care.

TIME OF DIAGNOSIS

A diagnosis of childhood cancer is an acute traumatic event for a family. Many families recall the conversation confirming diagnosis as a turning point in their lives. Preconceptions or past experiences with cancer may lead to immediate thoughts that the child will die. Even if fears about death are not at the forefront, thoughts about a long, painful and invasive treatment course may be overwhelming. The way in which the diagnosis of cancer is presented significantly influences the family's initial reactions and sets the stage for collaboration with the medical team. The meeting with the family should be held as rapidly as possible once a diagnosis is established. They should be told in advance about the purpose of this meeting so that they can involve those individuals they regard as important to the conversation. The meeting should be held in a private room with seats for all attendees and interruptions should be minimized. Adolescent patients may be invited and included in this meeting. If not, an additional meeting should be set up between the health care team, parents, and the patient to discuss the diagnosis and treatment in a developmentally appropriate way, with time provided to answer the patient's questions. The oncologist must present the diagnosis of cancer in an empathic and unequivocal way, tailored to the cultural and educational characteristics of the family, patient, and situation. A member of the treating facility's mental health team, such as a pediatric psychologist, child life specialist or social worker, should attend. If the family has developed a particularly strong and trusting relationship with another staff member, such as a nurse, they should be invited for additional support. Information should be accurate and clearly communicated without the use of medical jargon or euphemisms. Having basic information prepared in written form to review together can be helpful in allowing them to listen and take in the message rather than only focusing on trying to memorize specific details. This summary should be specific, brief and written in easily understood terms, and include wide margins for notetaking.

Parents are often uncertain about how to talk with their younger children about the diagnosis, treatment, and prognosis. Parents must talk to their children openly and honestly about the cancer and treatment that they will receive in order to increase trust within their relationship and between the child and medical team. Open communication helps their children to feel more secure and less anxious. It is typically not necessary to include information on the detailed biological processes, statistics, or in-depth medical terminology. Children should be told the truth in developmentally appropriate terms, focusing on what their experience is likely to be. The word "cancer" and the specific name of the diagnosis should be used in order to demystify the terminology and save the patient from hearing them from others and feeling as if this information had been hidden from them. The conversation should not be rushed, and any questions should be answered honestly and gently. If the parent does not know the answer, they should say so and tell the child that they will consult with staff to get an answer. Parents may need to be coached as to when and what to say to be most effective. Oftentimes, families ask that a staff member such as a child life specialist or pediatric psychologist coach them or join them in telling the child. Whether the staff participates directly or helps the parents in preparing for the conversation, the health care team needs to support and reinforce their competence as parents, while helping them to develop their ability to support their child through the medical crisis with honesty and trust.

The family's reactions will determine the pace and flow of the conversation. Table 35.1 provides common reactions and suggestions for working with children at different developmental levels at the time of diagnosis and throughout cancer care. Table 35.2 provides common parental reactions and suggestions for psychosocial intervention throughout the care of the patient.

Despite knowing that cancer is a possibility prior to the family meeting, most family members report feeling shocked at the news confirming diagnosis. Disbelief, denial, confusion, sadness, grief, guilt, anger, helplessness, anxiety, and fear are all common reactions. Parents perceive their role as providing for their children and protecting them from fear, hurt, pain, and death. The diagnosis of cancer threatens that identity and sense of adequacy as a parent. Many family members report that they were unable to hear or comprehend any information from this first meeting beyond the naming of the diagnosis, as they were lost in their own thoughts and emotions. Some report feeling as if they were watching the meeting happen from a distance. Severe emotional reactions are normal and should be expected. As the cancer diagnosis is accepted, these emotions go through a process of evolution. Anger can become a significant emotion and it is common for someone feeling so overwhelmed to want to regain a sense of control. Parental anger may even be directed at the treatment team. Guilt is also common, and family members may pass through a period of self-blame during which they focus on transgressions they feel they may have committed and for which they feel they are now being punished. Careful listening and supportive attention by the team are important to reassure family members that they did not cause the cancer. This includes siblings, grandparents, and other relatives.

Treatment decisions need to be made quickly after the diagnosis is known. This requires the family to absorb and understand significant amounts of medical information, including weighing the risks and benefits of treatment research protocols. Families may want time to seek other medical opinions, consult with additional family members and friends, and conduct their own searches for information. These efforts may be appropriate to ensure that they are doing what is best for their child and they should be encouraged to seek a second opinion from a qualified and reputable oncologist if they so desire. It is likely that the second opinion will confirm the proposed treatment plan, and the family will gain confidence in their decision to proceed with the team's recommendations. They should be discouraged from selectively talking to those who promote unrealistic hope or provide inaccurate information. The best way to support the family during this time is to repeat and reinforce what is known in regard to the diagnosis and treatment options, build a partnership through rapport and the encouragement of family participation in the conversation, and recognize and attend to the family's level of distress, while gently encouraging their competence and ability to make a good decision. It is important that multiple conversations occur with the family in which different members of the team provide the same information to ensure the family understands the information and has opportunities to express their questions and concerns.

Psychological problems and/or developmental differences in the patient or any family member add significantly to the burden of dealing with cancer and can exhaust a family's time and emotional and financial resources. Even minor symptoms can be reactive, at least in part, to the stresses and disruptions of the cancer diagnosis and treatment course. An assessment of the family's strengths and vulnerabilities, psychosocial resources, and preexisting problems, can help the team anticipate the psychological adjustment and provide a means for quickly and efficiently providing psychosocial care based on their need. Several factors such as prior diagnosis of depression or anxiety, general "poor coping," or nonadherence to treatment indicate that it is appropriate to make a referral for psychological treatment. Other families seek psychological

TABLE 35.1   Reactions to Cancer by Developmental Level and Suggestions for Psychosocial Support

Developmental stage	Common reactions	Suggestions for psychosocial support
Preschoolers	May exhibit fears regarding unfamiliarity with hospital. Regression (i.e., loss of developmental milestones or return to previously discarded behavior) is common, as is a slow-down in meeting new developmental milestones. Child may experience a heightened dependence on parent (i.e., wanting to sleep with parents). Fears of abandonment may be seen. Many young children perceive treatment (i.e., painful procedures) as punishment or retribution for wrong doing. Sleep disturbance (i.e., refusal to go to bed, fitful sleep, nightmares) may also occur	Address concerns about all procedures, even temperature taking, by giving a brief, honest description explaining the treatment. Provide positive incentives for cooperation with procedures. Maintain gentle and positively focused discipline. Teach breathing exercises and developmentally appropriate distraction techniques during procedures, such as storytelling, fantasy play, and puzzles. Engage child life specialist in preparing the child for, and supporting the child and parents through, procedures. The child should be allowed to engage in medical play and to visit operating and recovery rooms if undergoing surgery to become familiar with surroundings. Developmentally appropriate toys and activities should be provided throughout inpatient stays and outpatient visits as young children benefit greatly from free play
School age	Reaction to diagnosis may be delayed or immediate and anxiety-ridden. Many children exhibit adaptive and efficient coping styles. Younger children may display any or all reactions listed for preschoolers. Psychosomatic complaints, nightmares, stoic-like acceptance, and labile emotions may all be seen. Child may pose difficult questions such as reasons and cause for treatment and diagnosis	School-age children may be interested to learn about various aspects of their illness and treatment at a developmentally appropriate level. They can be taught how to manage stress associated with procedures by developing self-regulating skills through breathing and imagery. The physician may forge a strong positive working relationship with school-age children by taking time to sit and answer questions. The child should be educated that their emotional responses are common and acceptable. Refusal to cooperate with treatments or restrictions should be met with consistent and lovingly applied discipline
	Sadness, fear, and worries of being different from peers due to physical changes and school absences may be expressed. Lowered self-esteem and confidence levels may initially be noted. The child may ask, "Why me?" and complain, "It's not fair!"	Maintaining friendships, continuing academic progress, school reentry, and other issues should be actively addressed by psychosocial staff
Adolescents	Many adolescents exhibit adaptive and efficient coping styles. Initial questions may be general (i.e., geared towards causation and prevention of disease) or more personal (i.e., individual treatment plan; disruption to social life). They experience a loss of independence, which can threaten their developing autonomy. Nonadherence may occur as an attempt to regain a sense of control. Many feel self-conscious about appearance and emerging sexuality (i.e., hair loss, treatment-related delays in puberty, scars). Fear of being different from peers may interfere with school/social scene reentry. May avoid intimate relationships with peers. May reconsider long-term career and family plans and expectations. Concerns about fertility are likely. Feelings of isolation and vulnerability may lead to social withdrawal, hesitancy, and sadness	Allow and encourage adolescents to participate in medical decisions (i.e., signing consents/assents when appropriate, control with scheduling, permit them to see tests, involve them in discussions of alternative treatments). Offer group support and individual counseling by psychosocial staff whenever possible. Encourage patient to maintain active participation in daily activities. Ensure academic program is appropriate to help adolescent reach their long-term life goals. Teach special skills for stress reduction, such as breathing, imagery, and relaxation training. When possible, arrange schedule to allow participation in major life/social events (i.e., prom, graduation service). Family therapy may be appropriate to address issues of autonomy

services as a "preventative" measure for their children, in order to have additional support and to help guide them as parents in this arena. Potential therapeutic interventions include individual expressive and/or play-based psychotherapy or cognitive-behavioral therapy (CBT) for the child, support groups and family therapy for members of the family, or a combination of approaches. Other therapeutic approaches appropriate for the early stages of treatment and beyond include art and music therapy. While not all families want or need such services, they often add a crucial dimension to comprehensive care.

TABLE 35.2 Parental Reactions to Cancer and Suggestions for Psychosocial Support

Treatment stage	Common reactions	Suggestions for psychosocial support
Diagnosis and initial treatment	Most experience shock, disbelief, fear, guilt, and anxiety. It is common to feel overwhelmed with new medical terminology and information and the need to make rapid treatment decisions. May experience frequent crying, difficulty sleeping, intrusive thoughts, and/or continued sense of stress. Often seek reassurance that their child "will survive"	Be factual and empathic. Allow time for questions and answers. Educate family on common reactions and feelings. Support parents in their request for time to talk with other family members and to seek a second opinion. Be sensitive to cultural beliefs and norms. Provide verbal and written information. Repeat information with compassion as parents/caregivers will need to review information in order to assimilate it
		Provide guidance on ways to communicate diagnosis to the child with cancer, the sibling(s), family, and friends. Prepare parents for possible delay in emotional reaction (themselves or their child(ren)). Help promote open communication in the family. Encourage parents to have their child, particularly adolescents, participate in medical decisions (i.e., signing consents; when appropriate, control with scheduling, permit them to see tests). Address logistical issues such as taking leave from job, childcare for siblings, financial issues, etc. With parent's permission, develop relationship with patient's school and begin to develop new academic plan with regard to home/hospital instruction, attendance at school, grade promotion vs. retention, assessment for special services, etc.
Remission	Fear of relapse is common, often parental anxiety increases in phases of less intensive treatment/end of treatment. Financial strains may become a key issue. Exhaustion as parents try to return to work and family activities while caring for the child with cancer. May become concerned about strain on relationships with other children, spouse/partner, etc. as the early crisis phase of treatment required their full attention. May become more lax in following through with medical appointments and medication schedule	Monitor family communication and coping. Encourage parents to maintain active participation in healthy daily activities. Educate regarding typical emotional shifts. Teach stress reduction such as breathing, imagery, and relaxation training. Refer for psychosocial support services and organizational support with regard to financial needs. Family therapy may be appropriate to address communication, behavior, or relationship issues. Suggest parents monitor adherence
Recurrence	Fear, anxiety, sadness, tearfulness, anger, difficulty sleeping. An urgent need to explore all treatment options. Concern how to best communicate information to the child with cancer and/or siblings	Monitor symptoms and offer counseling, particularly when symptoms interfere with ability to cope. Discuss treatment options openly and support family in obtaining second opinion as desired. Assess family communication. Encourage parents to include their child in medical decisions and, when appropriate, including end-of-life decisions as needed. Balance additional treatment with quality of life. When indicated, shift focus from "cure" to symptom control/palliative care. Refer for hospice services when end-of-life discussions begin. Encourage siblings to attend family conferences. Assist family in examining what is meaningful about the patient's goals in general terms (i.e., "go away to college" might be reframed as "I want to travel and learn") and identifying ways to continue to provide opportunities for growth and development rather than focus on activities/events/specific goals that may not be realistic and therefore experienced as a "failure." Offer psychosocial support to all family members

(Continued)

TABLE 35.2   (*Continued*)

Treatment stage	Common reactions	Suggestions for psychosocial support
Bereavement	Most families wish to maintain a relationship with the primary oncology team	Send a personal note to the family. Staff members that attend service/funeral can represent entire team. Provide selected reading materials about bereavement to parents, siblings, grandparents, and extended family members, classroom teachers, and others who knew the child well. Be aware of differences between "normal" grief reactions and those that are nonresolving and require counseling. Offer information on how to connect with other families who have lost a child, if the family desires. Support family's efforts to give back to the community
	Families appreciate the opportunity to keep their child's memory alive by sharing experiences with those who knew their child during his/her illness. Many families choose to facilitate donations to treatment/research/support organizations or start a charity in their child's memory	
Survivorship	Sense of relief along with ambivalence and fear of recurrence. Anxiety about decreased contact with medical team	Normalize emotional responses to end of treatment. Address realistic expectations regarding a quick "return to normal." Monitor family adjustment. Assess for psychological distress and offer psychological services when indicated. Have follow-up care arranged and provide education on potential late effects of cancer treatment. Monitor academic, social, and emotional functioning at follow-up appointments, and intervene where needed

TREATMENT INITIATION

The initiation of medical treatment often provides a sense of relief to families because the fight against cancer has begun. While it is reassuring that something is actively being done to help the child, tremendous burdens are placed on the family during this time. Daily lives are disrupted due to the demands of treatment, roles and responsibilities within the family need to be renegotiated to ensure that the basic needs of the family continue to be met (e.g., working to retain medical insurance and pay bills, caring for siblings), and highly technical medical information needs to be understood. Parents are in the position of consenting to treatment which is likely to be invasive, painful, and may create short-term and/or long-term side effects. While the ultimate goal is to restore the child's health and allow them to enjoy a long and productive life, many parents grapple with a deep sense of guilt for "subjecting" their child to treatment. This can be especially painful when a young child clings to a parent and pleads, "Don't let them hurt me!" Parents frequently report that the primary difficulty during treatment initiation is dealing with their own intense emotions. Within the first weeks following initiation of treatment, nearly all parents report clinically significant psychological distress in the form of anxiety, depression, or traumatic stress reactions. This distress can make it difficult for parents to establish consistently effective working relationships and open lines of communication with the medical team. It is vital to understand that parents have their child's best interests at heart but are struggling emotionally during this critical time. For the first time in the child's life they are not the sole caretakers or the ones who know the most about the child and his/her needs. Furthermore, parental marital relationships are often challenged as the stress can exacerbate previous marital problems, disrupt comfortable patterns of relating, or bring to light differences in beliefs and coping styles that make working together as a parenting team difficult. Parents who are separated or divorced, or whose relationship is already strained, face larger complications in being able to enter into treatment as a team. Parents who have open communication with one another and are mutually supportive, and trusting in their relationship, can surmount these challenges.

In addition to open communication with parents, providing a roadmap of treatment can help the family focus on what needs to be done and provide some sense of relief, optimism, and improved mood. This information needs to be presented carefully and repeatedly to the family as misunderstandings are common during this time. It is important for the treatment team to identify all involved caregivers within the family and to ensure that their questions are adequately answered. It is best to assess what the family understands or believes about the treatment plan and to correct erroneous perceptions. As the family comes to trust and rely on the medical team, it is also important for the physician and other members of the team to be clear about their roles and involvement in the child's care throughout the planned treatment course.

For the diagnosed child, the most significant aspects of initiating treatment are the physical effects. As treatment begins, anticipation and endurance of repeated painful procedures and side effects are some of the most distressing and traumatizing aspects of the cancer experience. Negotiating the impact of these physical changes on the patient's participation in normative social activities is also highly distressing. To help the child through this difficult aspect of treatment, a member of the team should explain procedures in developmentally appropriate terms, minimize the degree to which procedures are painful, and effectively manage chemotherapy-induced nausea and vomiting. Interventions that combine aspects of CBT and basic behavioral strategies (e.g., preparation, desensitization, modeling, and positive reinforcement) can be delivered by psychosocial specialists working as part of the multidisciplinary team. Parents can also learn to effectively use these techniques to help their children cope with procedures, boosting their self-efficacy as parents and allowing them to feel less helpless during treatment.

ILLNESS STABILIZATION

Parental distress present throughout the early stages of the illness and treatment starts to dissipate with illness stabilization. Some anxiously await the day their child is well enough to return home, especially if the initial hospitalization is long. Others find the return to home particularly stressful because they no longer have the expertise of the hospital staff monitoring and addressing the medical needs of the child. As treatment progresses and remission is established, most families develop ways of coping effectively with the cancer and its treatment, which some have described as "coming to terms" with the cancer routine. Distress decreases over time with few parents reporting clinically significant levels of anxiety and depression by 1–2 years post-diagnosis.

Still, there are moments of increased anxiety that surround specific stressful cancer-related events (e.g., emergency room visits, severe side effects, follow-up scans, and tests), and symptoms of cancer-related traumatic stress (e.g., intrusive thoughts about cancer, hypervigilance, and physiological arousal). This distress is normal. Parents continue to struggle to make sense of the diagnosis, and experience grief and mourning related to the loss of their child's health and prior lifestyle. Episodes of helplessness, fear, and crises of confidence are common and can influence emotional reactions and behavioral patterns. Maladaptive coping may manifest as excessive concern about relapse and death, refusal to allow the child to return to everyday activities, difficulties regularly attending scheduled clinic visits, or persistent and/or escalating anxiety or depression. Interestingly, the extent of distress in response to diagnosis and treatment is not consistently related to objective measures of the severity of illness or intensity of treatment.

Regarding the child with cancer, one of the most common concerns as treatment progresses is depression. Children with cancer, particularly adolescents, may display increased sleep, loss of interest in activities and social withdrawal among other symptoms that are consistent with depressed mood. Most research, however, reveals that children with cancer do not evidence greater rates of ongoing clinical depression, especially if their psychological needs are tended to during the cancer treatment.

One of the major tasks for the child and family during the illness stabilization phase of cancer treatment is to reestablish the patterns and routines of daily life that were disrupted due to diagnosis and initiation of treatment. A new day-to-day routine needs to be established that encompasses the needs of the ill child as well as those of the well siblings and the adults in the family. This "new normal" includes medical appointments and tending to the child treatment needs, but is balanced with activities of daily life with which they are familiar from prior to the onset of the illness. Parents should be encouraged to take some time to recharge themselves and their relationships and be cautioned against becoming overprotective or overindulgent with their children. Families of younger children should reestablish expectations and rules regarding bedtime, feeding, naps, and playtime. Families of older children need to reestablish their child's roles and responsibilities within the family, peer relationships, and school activities to the greatest extent possible. The maintenance of friendships and some level of independence from the family are important for the child, and especially the adolescent, with cancer. Contact with friends can be made through short visits, phone calls, social media, and video chats. In general, most children with cancer have been found to be quite resilient and to maintain good peer relationships, although they may be perceived as more socially withdrawn. Patients at highest risk for peer difficulties are those who had social difficulties prior to diagnosis, those whose treatment affected the central nervous system (CNS), and those who have more obvious changes in physical appearance. Interventions to improve socialization can be useful for these children.

School Reentry

The return to school may be particularly difficult for some families but is an important step toward psychosocial recovery. The parent may fear for the physical well-being of their child and the child or adolescent may feel embarrassment related to physical changes. Nonetheless, school is exceptionally important to child development and for normalizing the life of the child with cancer. While still in the hospital, the child should receive school services, and when home but not well enough to attend school, homebound services should be arranged. During these times, webcams to the classroom can be particularly helpful to maintain connection. In order for the patient's transition back to the classroom to be optimized, the classroom teacher should be asked to provide materials to supplement the hospital/home instruction materials. In this way, the child's work will parallel their classmates' in terms of content. The child's desk should remain in the classroom and their work from hospital/home instruction should be sent in and hung in the classroom alongside their peers' work. Prior to returning to school, the child, classmates, and school personnel need to be prepared. School reintegration programs, where the family, hospital, and school work together to ensure open communication and appropriate accommodations and services for the child, can help the family surmount this hurdle.

There are two federal laws which protect the educational rights of a child with a disabling condition. In the United States in 1973, Congress passed Section 504 of the Rehabilitation Act, a civil rights law which specifies that no one with a disability can be excluded from participating in elementary, secondary and postsecondary schooling and that modifications and accommodations have to be made so that they have the opportunity to perform at the same level as their peers. In 1997, Public Law 94-142, or the Individuals with Disabilities Education Act (IDEA), was enacted to provide federal funds to state and local agencies to guarantee special education and related services to children with one of 13 identified disabilities that impede learning to the point that specialized instruction is needed. A child or adolescent may qualify for either a Section 504 Accommodation Plan or an Individualized Education Plan via IDEA, and their needs may change over the years. While many schools offer to provide accommodation and/or special services without a formal plan, it is preferable to document the child's needs, services received, and progress made so that future education plans can follow suit.

A meeting should be scheduled for the hospital staff and parents to jointly educate the school staff about the child, ways in which their treatment may affect them in school, and potential late effects. This meeting should address potential medical, emotional, and academic challenges. A developmentally appropriate conversation with the class (or key classmates and friends in the case of a high school or college student) should also be held to demystify the illness and treatment process, abolish rumors and misunderstandings that often lead to anxiety and stress, and invite the class to participate in creating an open and accepting environment where no one is rejected because of their personal differences.

Medical Adherence

Medication adherence is a vital and often unrecognized clinical issue in pediatric oncology. While families may initially focus on the child's compliance with major procedures, careful and sustained adherence to the treatment regimen as prescribed, including exact dose and timing of medication administration across the entire length of treatment, is equally important to ensure success. Tools for evaluating and monitoring adherence should include asking at every visit how/when/where medications are taken, periodically checking pill bottles, and reviewing pharmacy reports. In addition, standardized scales can be used, and in some cases biochemical tests are appropriate. Evaluations of adherence should include a conversation about the barriers to absolute adherence, as well as what the family feels would be helpful to them to increase the fidelity of care. Poor adherence can be related to social/economic factors, health care system factors, patient-related factors, condition-related factors, treatment-related factors, or any combination of these. The cause of poor adherence must be thoughtfully considered (i.e., cost of medication, misunderstanding on the part of the caregiver regarding instructions, complexity of medication delivery, family systems and relationships), as addressing the issue often leads to improved adherence. For example, if a child is adherent to all medications except one, which they describe as having a horrible taste, the delivery method of that medication might be reconsidered (i.e., capsule to be swallowed rather than liquid). The clinical consequences of nonadherence can be significant, including lower survival rates and increased mortality due to relapse. Consequently, the need to anticipate, assess, monitor, and treat adherence to the treatment regimen is a critical component of the comprehensive care of the child with cancer.

DURING RELAPSES AND RECURRENCES

A relapse or recurrence of a child's cancer is a crisis point for the patient and family because the threat to the child's life is renewed. In response, the family commonly experiences anxiety, fear, anger, and sadness. For the child, the worst memories of earlier treatment may reemerge and they must reappraise each of these in order to prepare for another round of therapy. Families gain a sense of hope when new treatment options are offered as this implies further action will be taken against the disease, but they often realize that the chances for long-term cure are slimmer after relapse. Encouraging the child and family to once again adopt a positive attitude toward treatment can be challenging for the oncology staff, especially as parents require a heightened level of emotional availability and reassurance from the team members.

Most families manage to cope adequately following relapse or recurrence, with proper support. Development of a new sense of normality and stability in their lives will depend on the following:

- Treatment side effects
- The length of time the child needs to remain hospitalized or away from home and
- Concurrent stresses, such as financial pressures, career obligations, and family problems.

Assessing for maladaptive coping is essential. An overly pessimistic attitude about the future may immobilize parents in their day-to-day functioning, with each of the following requiring immediate psychosocial intervention:

- Emotional or physical withdrawal from the child.
- An inability to normalize the child's life.
- Refusal to follow through with medical care.

Similarly, an overly optimistic attitude or a desire to pursue every last medical option without concern for the child's quality of life can be problematic and should be discouraged. Individual or family sessions can help the family strengthen coping skills, openly discuss their hopes, fears, and desires for treatment options, and provide much needed support and guidance.

Whether or not a treatment response is obtained, families often feel an urgency to search for other curative treatment options, including second opinions. When appropriate, phase I trials, alternative therapies, and/or the choice for no further cure-focused treatment interventions should be presented as options, while reassuring the family that contact with the medical team will be maintained. If medically indicated, this might be a time to shift communication from "cure at all costs" to palliation, slowing the progress of the disease and skillfully managing symptoms while carefully balancing additional treatment with quality of life for the child.

TREATMENT OUTCOMES: THE UNSUCCESSFUL COURSE

Termination of Treatment

Maintaining adequate communication with both children and their parents is of particular importance when decisions to withhold or withdraw cure-focused and/or other life-prolonging therapies must be made. A series of meetings with the family should be held to discuss the transition in care associated with palliation-focused services. Ideally, the staff members who have been most intimately involved with the child and family during the course of the disease should be present. The child may or may not attend these meetings, depending on his or her age, developmental stage, and other circumstances. If a decision is made for the child not to be present, relevant information should be communicated to the child in a developmentally appropriate manner.

Once parents understand that treatment is no longer effective, they will begin the process of accepting that their child will die, while continuing to hope for cure or recovery. Parents are increasingly vulnerable during this period to the promises of nontraditional or even fraudulent healers as well as to misguided advice from internet sites and "chat rooms." Therefore, in order to preserve a relationship that is built on trust, the physician should openly and respectfully explore all options with the family, including resuscitation status, in-hospital palliative care programs and home care with or without the support of a hospice team. Advanced discussion about resuscitation status, when the physician can reassure the family that a life-threatening event is not imminent, is beneficial to the child and to the parent as it is very difficult to make a nonemotionally based reactive decision in the midst of a medical crisis. Parents usually respond best to an approach that is framed "in a worst case scenario" discussion. Once on a ventilator, the likelihood of being extubated is rare for the child with far-advanced cancer and therefore it is recommended that the phrase: "do not attempt resuscitation" and/or "allow a natural death" be used to help the family frame their decision.

Palliative/Hospice Care

The terms palliative care and hospice care are often used interchangeably, but this is an error. While they are distinct in their goals, the services are frequently offered by the same providers, and for many families, end-of-life is the first time that they have heard either phrase. Palliative care, in its purest sense, is defined as the active total care of the child's body, mind, and spirit, and also involves giving support to the family. Palliative care is, therefore, an essential part of treatment from the time of diagnosis as it involves evaluating and alleviating a child's physical, psychological, and social distress in an ongoing manner. Typically, at diagnosis, the primary intention of treatment is cure, and palliative care runs parallel in support of the child and family as they pursue this objective. When a change in treatment goal occurs so that the primary focus is quality of life for the extent of life that remains, palliative care becomes the primary focal point. Rather than a sudden transfer in focus and treatment team from "curative" to "palliative," it is preferable for the palliative care team to be known to the family and increase their services to the point where they become primary. A referral for hospice care services should ideally occur early in a child's treatment, as soon as it becomes apparent that cure is unlikely, so that the family and professional caregivers can derive maximum benefit from all that hospice services have to offer. Unfortunately, often a family's first contact with the hospice/palliative-focused care team comes when all treatment options have been exhausted, and this can be perceived as the treatment team "giving up" on the child. The transition to hospice/palliative-focused care can be much less traumatic when the family is reassured that they will not be cut off from the treatment team with whom they have developed a close relationship over the years, regardless of whether the child is at home or in the hospital. Hope should be encouraged by redirecting energies into providing as good a quality of life as possible for as long as possible, followed with as good a quality of death (typically described as absence of pain combined with the presence of loved ones) as possible. Access to care aimed at promoting optimal physical, psychological, and spiritual well-being is of utmost importance. It is especially critical that providers are responsive and attuned to questions the child may have and to the wishes of the parents and child.

Promoting open communication between child and parents about emotional concerns, needs, and desires is especially vital as the end of life nears. The child should be given honest and accurate information about end-of-life care and death, appropriate to his or her developmental age. A child's fears of separation, being alone, suffering, and being forgotten are not unusual and should be openly discussed. Many children need reassurance that they will not be alone, while others wish to discuss closure activities, such as funeral arrangements, giving away of belongings, and how they wish to be remembered. They need the opportunity to be involved in decision-making about how they will live the life they still have. Focusing on the quality of their child's life and death, adequate pain management, strengthening relationships with loved ones, and facilitating memory building can be the greatest gift to the child and family.

As death approaches, parents often question whether they have done all they could for their child. They may describe what feels like overwhelming anxiety as they try to hold on to any semblance of control or normalcy in their relationships and day-to-day lives. Such responses are appropriate given the sequence of experiences leading to the terminal phase of illness and the physical and emotional changes that their child is undergoing. Living with a dying child is all-consuming and it has a significant impact on each family member. Fluctuations in a parent's capacity to accept the idea of their child's death are common and enable them to emotionally be with their child in the present moment. Children, too, may talk about their funeral at one moment and soon after inquire about a future family vacation. Siblings cope similarly, gradually accepting the declining health of their brother or sister while appearing to deny the impending death. One needs to respect each family's readiness to discuss end-of-life events, as they delicately attempt to balance life issues with those related to palliative care, death, and loss.

The management of the terminal phase of illness has a dramatic effect on the psychosocial recovery of the family. The medical team's participation and investment in caring for the dying child is extremely important and greatly appreciated by all families, even those who appear to demonstrate strong skills in coping and adaptation. Entering a family's world and bearing witness to a parent's, sibling's or grandparent's anger, grief, and pain is emotionally demanding. When done well, being present at the time of death is also exceptionally intimate and rewarding.

Awareness of the needs of the medical staff caring for the dying child is critically important. Opportunities for staff to reflect with other staff or professionals, especially when facing multiple losses, doubts, guilt, or emotional exhaustion, are key. The impact of burnout can be profound on the individual, the team, and on the patients and families for whom health care is provided.

Bereavement

The pediatric cancer experience is an intimate, long-term relationship that does not need to be severed with the child's death. Following the immediate bereavement period, families often benefit from being able to discuss their cancer journey with the primary oncology team, including a review of autopsy findings, if such was performed. This may need to occur at a location outside of the oncology ward, if parents are not ready to return. Families benefit from receiving cards and notes that remind them that their child continues to remain alive in the thoughts of those who helped care for them. Having created a bond with staff over the shared love for your child is a powerful experience.

For years after the death of a child, parents try to make meaning out of the loss. Post-death adjustment is different for each family. Family members vividly recollect specific events in the course of their child's illness, some of which may be persistent and disturbing images of their child's last moments. They often have a need to review these, over and over again, until they find solace in their remembrance, and are able to integrate memories of the child and family from happier times. Parents, siblings, and others close to the child can benefit greatly from having someone available with whom to share their memories, grief, and concerns. If a hospice team is not involved for bereavement counseling, a psychologist, social worker, grief counselor, or bereavement group can be helpful. Some families find it too painful to ever return to the hospital, while others crave the caretaking and relationship with staff. It is not uncommon for parents to choose to devote their time and energy into "the cause" by developing foundations, promoting research, organizing blood drives, etc. This can establish a legacy for their child, provide hope for the future, and/or fulfill their parental caretaker role which was severed.

TREATMENT OUTCOMES: THE SUCCESSFUL COURSE

End of Treatment

The successful completion of therapy and discharge from treatment is a significant milestone, but one often accompanied by feelings of ambivalence within the family. The family often experiences a sense of accomplishment, relief, joy, and renewed hope for the future. Yet, at the same time, they might also have underlying anxiety about the possibility of relapse or recurrence and a sense of sadness as they lose the routine of coming to the clinic and receiving support from the health care team. An end to the active fight against the disease can also feel unsettling. These feelings are to be expected and should be normalized for the family. Extra support and education are important for the family at this juncture. Many families are surprised at the reactions of friends and family who expect them to "get over" their experience now that, "things are back to normal." This occurs because while their lives may return to a typical pattern of daily life, in the time that has passed each family member has been changed in some way by the experience, and so it is not their "old life" they return to. The signs and symptoms of possible relapse or recurrence should be reviewed along with remaining treatment options. The importance of monitoring for possible late effects of treatment should be explained and the family should be given a summary of the treatment received and written guidelines for self-care. Finally, the challenges that they have met and surmounted should be reviewed and their accomplishments lauded as they prepare for survivorship. Continued support and focus on the child's development, education, social, and emotional well-being should be stressed.

Survivorship

With approximately 80% of children diagnosed with cancer being cured, worries about relapse and recurrence tend to decrease as time passes from the end of treatment, and concerns about possible long-term effects of treatment begin to rise. At least two-thirds of childhood cancer survivors experience at least one late effect of treatment, and over one-fourth of survivors experience severe or life-threatening late effects (see Chapter 34). Psychosocial late effects in the form of neurocognitive deficits and post-traumatic stress reactions are prevalent.

Neurocognitive deficits are common late effects of childhood cancer treatment. While the timing of the emergence of these deficits is not always predictable, there is evidence that CNS-related cancers and treatments can result in deficits of memory, processing speed, visual-spatial and visual-motor skills, and executive functioning. Subsets of survivors learn new material at slower rates than peers and have poorer academic achievement. Children treated before the age of seven and those treated for brain tumors are at the highest risk for neurocognitive late effects. To address these difficulties, the family should be encouraged to develop a collaborative working relationship with the schools and advocate for individualized educational plans to meet their child's specific needs. A thorough evaluation of the child's cognitive strengths and weaknesses should be conducted via neuropsychological, psychological, or psychoeducational testing, and repeated as the child ages. Specific cognitive remediation programs for childhood cancer survivors have efficacy in improving academic achievement, but are not yet widely available. Pharmacological interventions, such as stimulants, have been used with some success in addressing attention difficulties.

Most survivors of childhood cancer and their family members emerge as psychologically competent and well-functioning. Still, the experience of childhood cancer may have long-lasting psychosocial effects. For example, the health-related quality of life of survivors, a multidimensional construct that encompasses physical functioning, psychological adjustment, social functioning, and an overall sense of well-being, seems to be poorer for childhood cancer survivors compared to peers. Furthermore, subsets of survivors do experience more serious problems. There is emerging evidence that survivors of CNS-related cancers and treatments experience some difficulties in social competence and peer interactions, exhibiting greater social isolation, reporting fewer best friends, and participating in fewer peer activities than other children. Also, the transition from adolescence to young adulthood is a time of psychosocial vulnerability for childhood cancer survivors. In addition to reports of increased distress during this time, milestones of adulthood (e.g., dating, employment, moving away from home, starting a family) seem to be achieved at a slower rate than for healthy peers.

A post-traumatic stress model has emerged as helpful in understanding the long-term cancer-related distress experienced by some childhood cancer survivors and their family members. Among child and adolescent survivors, rates of post-traumatic stress disorder (PTSD) are low, in the range of 5–20%, although a much higher percentage of survivors report experiencing post-traumatic stress symptoms, including intrusive thoughts about cancer, physiological arousal when reminded of cancer or their experience, and a desire to avoid anything linked to their cancer experience. PTSD rates are nearly double for young adults compared to adolescent survivors of childhood cancer. Remarkably, parents and siblings seem to be more likely than survivors to report cancer-related PTSD and post-traumatic stress symptoms. Across studies, approximately 40% of mothers, 35% of fathers and one-third of siblings report moderate to severe cancer-related post-traumatic stress reactions.

Through the survivorship phase of cancer, follow-up medical appointments should routinely screen for psychosocial late effects of treatment for the survivor and the family members. Education about the cancer treatment and the possible medical late effects will help to close gaps in knowledge about the experience and potential vulnerabilities. The provider should keep in mind that it was the parents who consented to treatment, and so the child may not have been given complete and detailed information at the time of treatment. This is particularly important to remember with regard to issues such as fertility. The fact that the parents were informed of potential risks or lack thereof, does not mean that the minor patient was given this information; nothing should be assumed to be understood unless specifically reviewed. The adolescent or young adult patient who had their illness explained in a developmentally appropriate manner during their childhood treatment may experience the adult explanation and accompanying information about risks and side effects in quite the same way as someone being told of their diagnosis for the first time. New questions and concerns may rise as the patient enters into new developmental phases across their lifespan. During these conversations, it is important to address any anxiety that arises, provide anticipatory guidance about possible traumatic stress symptoms, and ensure that psychological reactions do not interfere with follow-up care. When psychological symptoms or signs of poor adjustment are present, the family should be referred to an experienced psychologist, clinical social worker, or psychiatrist for consultation.

CONCLUSION

Comprehensive psychosocial support is an essential component of cancer care that begins with early assessment of family strengths and vulnerabilities, and emphasizes the importance of the child becoming medically stable and the child and family remaining socially and emotionally intact. The health care team needs to join together with families to form a therapeutic alliance and promote normalcy as soon as the child is medically capable. The treatment team needs to respect the cultural differences of the family and maintain open communication with them across the trajectory of illness whether the outcome is cure or end-of-life care. No child's illness-related distress should go unrecognized and untreated, and the strain on the professional staff needs to be equally recognized and addressed. In doing so, providers reduce the trauma associated with pediatric cancer and simultaneously promote hope, courage, and emotional growth.

Further Reading and References

Armstrong, G.T., Liu, Q., Yasui, Y., et al., 2009. Long-term outcomes among adult survivors of childhood central nervous system malignancies in the Childhood Cancer Survivor Study. J. Natl. Cancer Inst. 101, 946–958.

Brinkman, T.M., Zhu, L., Zeltzer, L.K., et al., 2013. Longitudinal patterns of psychological distress in adult survivors of childhood cancer. Br. J. Cancer 109 (5), 1373–1381.

Brown, R. (Ed.), 2006. Comprehensive Handbook of Childhood Cancer and Sickle Cell Disease: A Biopsychosocial Approach. Oxford University Press, New York.

Jankovic, M., Spinetta, J., Masera, G., et al., 2008. Communicating with the dying child: an invitation to listening—a report of the SIOP working committee on psychosocial issues in pediatric oncology. Pediatr. Blood Cancer 50 (5), 1087–1088.

Kazak, A.E., Abrams, A.N., Banks, J., et al., 2015. Psychosocial assessment as a standard of care in pediatric oncology. Pediatr. Blood Cancer 62 (S5), S426–S459.

Lown, E.A., Phillips, F., Schwartz, L.A., et al., 2015. Psychosocial follow-up in survivorship as a standard of care in pediatric oncology. Pediatr. Blood Cancer 62 (Suppl. 5), S531–s601.

Mittby, P.A., Robison, L.L., Whitton, J.A., et al., 2003. Utilization of special education services and educational attainment among long-term survivors of childhood cancer: a report from the Childhood Cancer Survivor Study. Cancer 97 (4), 1115–1126.

Oeffinger, K.C., Mertens, A.C., Sklar, C.A., et al., 2006. Chronic health conditions in adult survivors of childhood cancer. N. Engl. J. Med. 355, 1572–1582.

Pai, A.L., McGrady, M.E., 2015. Assessing medication adherence as a standard of care in pediatric oncology. Pediatr. Blood Cancer 62 (Suppl. 5), S696–s706.

Spinetta, J.J., Jankovic, M., Masera, G., et al., 2009. Optimal care for the child with cancer: a summary statement from the SIOP Working Committee on Psychosocial Issues in Pediatric Oncology. Pediatr. Blood Cancer 52 (7), 904–907.

Steele, A.C., Mullins, L.L., Mullins, A.J., Muriel, A.C., 2015. Psychosocial interventions and therapeutic support as a standard of care in pediatric oncology. Pediatr. Blood Cancer 62 (S5), S585–S618.

Thompson, A.L., Christiansen, H.L., Elam, M., et al., 2015. Academic continuity and school reentry support as a standard of care in pediatric oncology. Pediatr. Blood Cancer 62 (S5), S805–S817.

Tonorezos, E.S., Oeffinger, K.C., 2008. Survivorship after childhood, adolescent and young adult cancer. Cancer J. 14 (6), 388–395.

Wiener, L.S., Pao, M., Kazak, A.E., Kupst, M.J., Patenaude, A. (Eds.), 2009. Quick Reference for Pediatric Oncology Clinicians: The Psychiatric and Psychological Dimensions of Pediatric Cancer Symptom Management. IPOS Press, Charlottesville.

Zeltzer, L.K., Lu, Q., Leisenring, W., et al., 2008. Psychosocial outcomes and health-related quality of life in adult childhood cancer survivors: a report from the Childhood Cancer Survivor Study. Cancer Epidemiol. Biomarkers Prev. 17, 435–446.

Blood Banking Principles and Practices

Edward C.C. Wong and Naomi L.C. Luban

INTRODUCTION

The discipline of Transfusion Medicine (also known as "Blood Banking") includes the:

1. Collection, testing, processing, and preparation of blood and blood components.
2. Selection of the most appropriate products and transfusion practice based on laboratory findings and patient need.
3. Monitoring of the effectiveness of transfusion as modified by disease, physiological status, or the procedure(s) performed.

Certain patient groups (e.g., neonates, oncology patients, hematopoietic progenitor cell transplant recipients, those with sickle cell disease, and others); often require complex pretransfusion processing and specialized product selection and modification.

DONOR SELECTION AND COLLECTION

The US Food and Drug Administration (FDA) regulates donor selection, collection, testing, component preparation, storage, and distribution of blood and blood components through the Code of Federal Regulations. The American Association of Blood Banks regularly publishes standards for blood banks and transfusion services, technical bulletins, and a technical manual, which direct donor and patient transfusion practices and procedures. Outside of the United States, similar regulatory agencies oversee production, transfusion, and posttransfusion complications through biovigilance networks.

Infectious disease testing should include serological testing for human immunodeficiency virus (HIV), hepatitis B and C, human T lymphotrophic virus I/II, and syphilis and nucleic amplification testing for HIV, hepatitis C, and West Nile virus. Although not currently required by the FDA, testing for Chagas disease (*Trypanosoma cruzi*) is also performed. Cytomegalovirus (CMV) testing is not routinely performed. Because of the possibility of bacterial contamination of platelet products, different methods of bacterial testing for platelets are also performed. Donor deferrals or other testing may be indicated, depending on national or international infectious disease prevalence. An example includes variable and changing donor travel deferrals for malaria and Chikungunya virus. Testing for Dengue virus and babesia may be performed if these disorders are endemic to a specific geographical location.

COMPONENT PREPARATION

Whole-Blood-Derived Products

In the United States, the majority of red blood cells (RBCs) and plasma products (including fresh frozen plasma (FFP), thawed plasma, and other plasma products) are derived from the donation of whole blood. In Canada and Europe, whole-blood-derived platelets are prepared using an alternative "buffy coat" method.

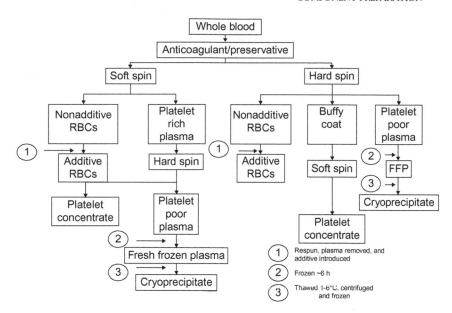

FIGURE 36.1 Methods of whole-blood separation. *Modified from Figure 2 in Wong (2013).*

Figure 36.1 outlines different whole-blood-derived blood components. Depending on the type of processing, one whole-blood unit can generate one unit each of RBCs, platelet concentrate, and plasma. In addition, subsequent thawing and precipitation of FFP at 1−6°C generates a unit of cryoprecipitate.

Apheresis-Derived Products

The same products obtained from whole-blood processing may also be obtained using apheresis techniques. Apheresis techniques provide the ability to obtain up to 12 equivalent donor whole-blood platelet concentrates, a double unit of RBCs, or up to 500 or 600 ml of plasma depending on donor weight. Granulocytes are also obtained using apheresis methods. The use of sterile connecting devices and specialized bag configurations permit these apheresis products to be divided into standard or even smaller units, especially valuable for infants and children.

RBC and Whole-Blood Products

Whole blood is collected into bags containing one of several anticoagulant preservative solutions; these include acid citrate dextrose solution, citrate-phosphate-dextrose solution, citrate-phosphate-double-dextrose solution, and citrate-phosphate-dextrose-adenine solution. Additive solutions are often added to the anticoagulated RBC product to increase shelf life. These solutions are each slightly different in their additive constituents and include AS-1, AS-3, AS-5, and AS-7 (Table 36.1). In Europe, Canada, and other countries, an alternative additive solution, SAG-M, may be used. Toxic levels of these anticoagulant preservative and additive solutions are only reached in patients undergoing massive transfusions or extracorporeal membrane oxygenation (ECMO) (Table 36.2). Different methods to reduce the anticoagulant additive solutions are utilized by transfusion services.

In some situations whole-blood units (reconstituted from RBCs and compatible plasma) may be necessary. Reconstituted whole blood may be needed as a prime for apheresis procedures in very small patients or for whole-blood exchange for hemolytic disease of the fetus and newborn (HDFN) in order to avoid iatrogenic dilution-induced coagulopathy. There are no platelets in reconstituted whole-blood units. To monitor bleeding risk associated with thrombocytopenia, a platelet count should be obtained.

Plasma Products

The following types of plasma are available from transfusion services.

1. FFP, prepared after separation of whole blood or apheresis and frozen at ≤−18°C within 6 h of preparation.
2. PF24 (plasma frozen within 24 h after phlebotomy), prepared from a whole blood or apheresis collection and stored at 1−6°C within 8 h of collection and frozen at ≤−18°C.
3. PF24RT24 (plasma frozen within 24 h after phlebotomy held at room temperature up to 24 h after phlebotomy), prepared from apheresis collection.

TABLE 36.1 Contents of Anticoagulant/Preservative Solutions (Based on Manufacturer's Data)

Constituent	CPDA-1[a]	CPD	AS-1[b]	AS-3[b]	AS-5[b]	AS-7[b]	SAG-M[b]
Volume (ml)	100	63	100	100	100	100	100
Sodium chloride (mg)	0	0	900	410	877	0	877
Dextrose (mg)	2000	1610	2200	1100	900	1585	820
Adenine	17.3	0	27	30	30	27	17
Mannitol	0	0	750	0	525	1000	0
Trisodium citrate (mg)	1660	1660	0	588	0	0	0
Citric acid	206	206	0	42	0	0	0
SODIUM PHOSPHATE							
Monobasic (mg)	140	140	0	276	0	0	0
Dibasic (mg)	0	0	0	0	0	170	0
Shelf life of WB units	35						
Shelf life of RBC units[c]	35	28	35–42	35–42	35–42	35–42	35–42

[a]*Approximately 450 ml of donor blood is drawn into 63 ml CPDA-1. A unit of RBCs (Hct ~70%) is prepared after centrifugation and removal of most plasma.*
[b]*When AS-1, AS-5, AS-7 or SAG-M is used, ~450 ml of donor blood is drawn into CPD. When AS is used, ~450 ml of donor blood is drawn into citrate-phosphate-double-dextrose solution which differs from CPD in containing, double the amount of dextrose. After centrifugation and removal of plasma, the RBCs are resuspended in 100 ml of the additive solution with final Hct 55–60%.*
[c]*Approved shelf life varies depending on country's regulatory agency.*
AS, additive solution; CPDA-1, citrate-phosphate-dextrose-adenine anticoagulant; CPD, citrate-phosphate-dextrose; SAG-M, sodium chloride, adenine, glucose, mannitol; RBC, red blood cell.
Note: CPD solution used to make SAG-M units contains 525 mg sodium chloride/63 ml.

TABLE 36.2 Quantity of Additives Infused During Transfusion of 15 ml/kg of Representative CPDA-1, AS-1, AS-3, or SAG-M Units in Comparison to Two Units (AS-1) for Pump Prime During ECMO[a]

Additive	CPDA-1 mg/kg	AS-1 mg/kg	AS-3 mg/kg	SAG-M mg/kg	ECMO 2 units	Toxic dose[b] mg/kg/h
NaCl	0	28	5	35.6	641	137
Dextrose	13	86	15	20.4	1982	240
Adenine	0.2	0.4	0.4	0.2	18	15
Citrate	12	6.5	8.4	6.5	148	180
Phosphate	9	1.3	3.7	1.3	31	>60
Mannitol	0	22	0	0	500	360

[a]*Assuming 60% Hct for CPDA-1, AS-1, AS-3, and SAG-M units.*
[b]*Toxic dose actually difficult to predict since adenine, dextrose, and phosphate enter RBCs for metabolism.*
ECMO, extracorporeal membrane oxygenation; CPDA-1, citrate-phosphate-dextrose-adenine anticoagulant; AS, additive solution; SAG-M, sodium chloride, adenine, glucose, mannitol.
Based on manufacturer's data and modified from Luban et al. (2012) with permission.

Each of these plasma products is considered to be equivalent in terms of clinical efficacy, with similar concentrations of coagulation and anticoagulation proteins. Delayed freezing reduces the concentration of factors V and VIII, protein C (PF24 only), and protein S (PF24RT24 only) making their use in clinical circumstances where these proteins would be beneficial contraindicated.

All plasma products thawed in an "open system" must be discarded after 24 h to minimize bacterial contamination. If the plasma product has been thawed in a closed system, storage for up to 5 days at 1–6°C is allowed, but the product must be relabeled as thawed plasma. There are no well-designed clinical trials in neonates or critically ill children demonstrating risk–benefit of one plasma product over another.

Platelet Products

Platelets are derived from whole blood or by apheresis. In the United States, whole-blood-derived platelets are prepared from a second "hard spin" of platelet-rich plasma. Platelets have a shelf life of 5 days (in some countries up to 7 days) and must be stored at 20–24°C. In Europe, Canada, and some other countries platelets may be prepared from whole blood using a "buffy coat" method (Figure 36.1; Table 36.3). Clinical studies comparing different platelet preparations have evaluated posttransfusion platelet increments, adjusting for patient blood volume and platelet product content. Few pediatric studies have been performed to assess efficiency, especially in the bleeding patient.

In a large multicenter trial to evaluate platelet dosing, it has been shown that there are differences in bleeding risk between adults and children receiving platelets for hypoproliferative thrombocytopenia. Children, particularly those undergoing hematopoietic stem cell transplantation (HSCT), were at a higher risk of bleeding than adults over a wide range of platelet count. Grade 3 or greater bleeding occurred most commonly in children 6–12 and 13–18 years of age regardless of HSCT.

The FDA recently approved the use of a platelet additive solution (PAS) for apheresis platelets. PAS solutions contain acetate or glucose as a substrate for platelet metabolism, phosphate to buffer lactate production, citrate to prevent coagulation and lactate production and potassium and magnesium to improve platelet function during storage. Theoretical advantages include improved *in vivo* and *in vitro* platelet quality, increased duration of storage, reduction of allergic reactions, and ability to use out-of-group platelets.

Cryoprecipitate

Cryoprecipitate is prepared by thawing whole-blood-derived FFP between 1 and 6°C and recovering the precipitate through centrifugation. The cold-insoluble precipitate is placed in the freezer within 1 h after removal from the refrigerated centrifuge then thawed for use. Cryoprecipitate contains high levels of fibrinogen (>150 mg), factor VIII (>80 IU), factor XIII, vWF, and fibronectin. Current indications include congenital and acquired hypofibrinogenemia, and factor XIII deficiency when viral-inactivated, monoclonal or otherwise viral-inactivated manufactured factor concentrates are not available or are contraindicated.

TABLE 36.3 Characteristics of Different Platelet Products

	Apheresis platelets	Whole-blood-derived (platelet-rich plasma)	Whole-blood-derived (buffy coat method)[a]
Donor exposure	Lower	Higher	Higher
Percentage of repeat donors	Higher	Lower	Lower
Compatibility with cost-effective, sensitive bacterial detection methods	Yes	No (unless prestorage pooling)	No (unless prestorage pooling)
Compatibility with pathogen reduction technologies	Yes	No (unless prestorage pooling)	No (unless prestorage pooling)
Hospital preparation (pooling, point-of-care bacterial testing)	Less	More	More
Transfusion service paperwork	Less	More	More
Wastage	Less	More	More
Febrile, nonhemolytic transfusion reactions	Fewer	More	More
Platelet activation	Lower	Highest	Lowest
Single donor plasma exposure	Higher than whole blood derived	Reduced	Reduced
Cost of production	Higher	Medium	Low
Availability	Low	High	High

[a]Not a licensed product in the United States.
Modified from Table 3 in Wong (2013) with permission.

Granulocytes

Granulocytes are collected using apheresis technology and are not an FDA-licensed product. Donors may be stimulated with oral dexamethasone or subcutaneous granulocyte colony-stimulating factor (G-CSF) and undergo apheresis to collect at least 1×10^{10} white blood cells (WBCs) per unit. Indications include a febrile, neutropenic patient with severe bacterial or fungal infection unresponsive to antifungals or to antibiotics and whose anticipated course will result in neutrophil recovery. Granulocytes may also be indicated in patients with neutrophil function defects (e.g., chronic granulomatous disease, leukocyte adhesion deficiency), who are unresponsive to antibiotics or antifungal agents for sepsis or acute infectious complications. Rigorous clinical trials for dosing are not available. Specialized consent and approval by the institutional review board may be required for use of this product as a clinical trial due to unresolved issues of risk–benefit for this product. The Resolving infection in Neutropenia with Granulocytes Study is a phase IIII randomized study of adjunctive granulocyte transfusion in septic patients undergoing standard antimicrobial therapy, which was published in 2015, was under powered and did not answer whether transfusion of dexamethasone or G-CSF-stimulated granulocytes has added benefit in septic children and adults. The decision to use granulocyte transfusion should be taken on a case by case basis, weighing the potential risks and benefits.

Complications of granulocyte transfusion include acute and chronic CMV and other viral infections for which testing is not performed or is unavailable, as well as human leukocyte antigen (HLA) alloimmunization since the product cannot be leukoreduced. Exposure to a large burden of foreign HLA antigens through WBC transfusion may contribute to the development of subsequent platelet refractoriness and a reduced likelihood of hematopoietic stem cell engraftment. Granulocytes contain a significant number of RBCs (>30 ml with a hematocrit generally between 5% and 15%), and therefore must be crossmatch compatible against a patient's specimen and can, if administered frequently, result in iatrogenic polycythemia.

Granulocyte concentrates have historically been irradiated to obviate the risk of graft-versus-host disease (GVHD). A study of adult acute myeloid leukemia (AML) patients has proposed that nonirradiated granulocytes are superior to irradiated granulocytes for the control of bacterial infections. The shelf life of the product is limited to only 24 h. This increases the risk of transfusion-transmitted infections as time-consuming donor infectious disease screening is often not available when the product is transfused.

BLOOD COMPONENT MODIFICATIONS

Leukoreduction

In Europe, Canada, and several other countries, universal leukocyte reduction (LR) has been adopted as a standard for all cellular blood components except granulocytes. RBCs typically contain up to 10×10^9 WBC and can be leukoreduced either by filtering whole blood at the beginning of processing (prestorage LR) or by attaching a filter with a sterile connecting device at the bedside to a unit of RBCs pretransfusion (Table 36.4). There are different approaches to the timing of filtration; at the bedside ensures that the filtration process has been observed while filtration in the blood bank allows for the monitoring and to control the WBC content postfiltration. Current LR filters reduce the number of WBCs a thousandfold so that the residual WBC count does not exceed 5×10^6 per packed RBC unit (United States) or 1×10^6 (Europe) while retaining 85% of the original red cell content. Transfusion services have developed quality control methods to ensure meeting these standards. LR should never be used as a strategy for the prevention of transfusion-associated graft-versus-host disease (TA-GVHD).

Irradiation

Irradiation prevents proliferation of donor lymphocytes which may recognize recipient tissues as foreign and initiate transfusion-acquired graft-versus-host disease (Table 36.5). Irradiation practice varies among institutions. Some institutions selectively irradiate blood products on the basis of diagnosis, age (or weight) of the patient, or the relationship of the donor to the recipient, while other institutions irradiate all products. Much of the practice variability occurs because of logistical issues such as location of the irradiator, real or estimated wastage of the RBC products and, in particular, technologist availability and patient population. Irradiation of blood at the point of distribution in the blood bank is a reliable method for ensuring that inadvertently nonirradiated units are not administered. FFP and cryoprecipitate are derived from thawed frozen plasma which damages WBCs during the freezing process and do

TABLE 36.4 Indication and Storage for Selected Blood Products

Component/product (volume)	Composition	Storage length	Storage conditions	Usual indications
Red blood cells (RBCs) (in citrate-phosphate-dextrose solution, citrate-phosphate-double-dextrose solution, citrate-phosphate-dextrose-adenine anticoagulant (CPDA-1), ~250 ml)	RBCs (~75% Hct); 50–70 ml plasma, 10^9–10^{10} white blood cells (WBCs) and platelets (nonfunctioning)	28 days	1–6°C	Increase RBC mass in symptomatic anemia
RBCs (additive, ~300 ml)	RBCs (~60% Hct); 40 ml plasma, 10^9–10^{10} WBCs and platelets (nonfunctioning); 100 ml additive solution	35–42 days	1–6°C	Increase RBC mass in symptomatic anemia
RBCs, leukocytes reduced (~300 ml with additive; ~225 ml without additive	RBCs with Hcts as above; $<5\times10^6$ WBCs (US) $<1\times10^6$ WBCs (Europe), few dysfunctional platelets, 40 ml plasma; should contain ≥85% of original red cell content	35–42 days	1–6°C	Increase RBC mass in symptomatic anemia, decrease likelihood of febrile reaction, alloimmunization to leukocyte antigens, reduce cytomegalovirus (CMV) transmission
Washed RBCs (~180 ml)	RBCs with ~Hct 75%; $<5\times10^6$ WBCs, no plasma	24 h	1–6°C	Increased RBC mass in symptomatic anemia; reduced potassium, additives and pathologic antibodies
Whole blood (~500 ml)	RBCs (~Hct 40%); 200 ml plasma, (deficient in labile clotting factors V and VIII, platelets (dysfunctional at 72 h); 10^9–10^{10} WBCs	28 days	1–6°C	Pump/other mechanical prime; hemolytic disease of the fetus and newborn, assists in hemostasis, if given 48 h or less after collection
Whole-blood-derived platelets (~50 ml)	Platelets $>5.5\times10^{10}$/unit, ~50 ml plasma; RBCs; WBCs	5 days	Agitated, 20–24°C	Bleeding caused by thrombocytopenia or thrombocytopathy
Apheresis platelets (~300 ml)	Platelets ($>3\times10^{11}$/unit); ~300 ml plasma; RBCs; WBCs	5 days	Agitated, 20–24°C	As above for whole-blood-derived platelets
Platelets, leukocytes reduced (~300 ml)	Platelets as above; $<5\times10^6$ WBCs	5 days	Agitated, 20–24°C	As above for whole-blood-derived platelets; decrease in febrile reaction, alloimmunization to leukocyte antigens and reduce CMV transmission
Fresh frozen plasma (FFP) (~250 ml)	Plasma, all coagulation factors and complement (no platelets)	1 year	≤−18°C	Treatment of coagulation factor deficiency when a suitable recombinant or derivative product is neither available or suitable; treatment for thrombotic thrombocytopenic purpura (TTP)
PF24 (plasma frozen within 24 h after phlebotomy, ~250 ml)	Plasma, all stable coagulation factors and complement (no platelets), decreased level of labile coagulation factors	1 year	≤−18°C	Clinically significant coagulation factor deficiencies; treatment for TTP. Note: not to be used for deficiencies of labile clotting factors such as factor V and VIII; TTP
PF24RT24 (plasma frozen within 24 h after phlebotomy, held at room temperature up to 24 h after phlebotomy, ~250 ml)	Plasma, all stable coagulation factors and complement (no platelets), decreased level of labile coagulation factors	1 year	≤−18°C	Clinically significant coagulation factor deficiencies; note: not to be used for deficiencies of labile clotting factors such as factors V and VIII and protein S; TTP

(Continued)

TABLE 36.4　(*Continued*)

Component/product (volume)	Composition	Storage length	Storage conditions	Usual indications
Thawed plasma (derived from FFP, PF24, or PF24RT24, ~250 ml)	Plasma, all stable coagulation factors and complement (no platelets), variable level of other coagulation factors	See next entry	Thawed at 30–37°C and maintained at 1–6°C for 4 days postinitial 24-h period	Clinically significant coagulation factor deficiencies; note: not to be used for deficiencies of labile clotting factors such as factors V and VIII (and protein S, if prepared from PF24RT24); TTP
Cryoprecipitate (~15 ml)	Fibrinogen ≥ 150 mg/unit; factor VIII ≥ 80 IU/unit; factor XIII, fibronectin, von Willebrand factor	1 year	$\leq -18°C$	Deficiency or dysfunction of fibrinogen or factor XIII, von Willebrand disease, or hemophilia A when a recombinant product or derivative is not available
Apheresis granulocytes (~200–300 ml)	Granulocytes ($>1 \times 10^{10}$ PMN/unit); lymphocytes and monocytes; platelets ($>2.0 \times 10^{11}$/unit), some RBCs, 3–8× greater content when G-CSF and dexamethasone-stimulated donors) are used.	24 h	Agitated 20–24°C	Provide functional granulocytes for febrile, septic patients, with anticipated WBC recovery and unresponsive to antibiotics/antifungals

Modified from Table 4 in Wong (2013) with permission.

TABLE 36.5　Clinical Indications for Irradiated Products

FETUS/INFANT

Intrauterine transfusion

Premature infants

Congenital immunodeficiency (identified or suspected)

Those undergoing exchange transfusion for erythroblastosis

CHILD/ADULT

Congenital immunodeficiency (identified or suspected)

Hematological malignancy or solid tumor (neuroblastoma, sarcoma, Hodgkin's disease receiving ablative chemo/radiotherapy)

Recipient of peripheral blood stem cells, marrow, cord blood, or cytotoxic T lymphocytes (CTL)

Recipient of familial blood donation

Recipient of HLA matched products

Lupus or any other condition requiring fludarabine, cyclophosphamide or combination myeloablative therapy

POTENTIAL INDICATIONS

Term infant

Recipients of solid organ transplants

Recipient and donor pair from a genetically homogeneous population

Other patients with hematological malignancy or solid tumor receiving immunosuppressive agents

Modified from Box 1 in Wong (2013) with permission.

not require irradiation unless fresh unfrozen plasma is being used. Irradiation reduces RBC storage shelf life to not >28 days postirradiation depending on the time of irradiation relative to collection. Irradiation followed by refrigerator storage causes a higher concentration of potassium in the product. To avoid wastage of RBC units, irradiation prior to issue is optimal. Instrumentation used for blood irradiation includes:

1. Free-standing cesium or cobalt gamma source irradiators (^{137}Cs or ^{60}Co).
2. Linear accelerators and other devices which use X-ray sources.

In the United States, the majority of facilities use free-standing lead shielded ^{137}Cs gamma irradiators which allow irradiation within minutes and have substantive controls to ensure that the bag or bags of blood products receive adequate dosing.

Washed Cellular Components

Washing of RBCs and platelets is used to reduce plasma proteins and cytokines and, in the case of RBCs, to reduce potassium, cytokines, and other by-products of RBC storage. Indications for washed products include:

1. Severe or recurrent allergic reactions to plasma proteins in cellular blood products.
2. Anaphylactic reactions due to IgA antibodies in IgA-deficient recipients.
3. Infants with neonatal alloimmune thrombocytopenia (NAIT) who receive platelets from their mother where washing removes the circulating platelet-specific antibodies responsible for NAIT.

In recurrent, severe allergic reactions, RBC additive units (which contain almost no plasma) and premedication with one or more antihistamines may be considered in lieu of washing. Potassium should be removed from large volume (>20–25 ml/kg) RBC transfusions in at-risk infants or children. In patients with T-cell activation due to the exposure of the cryptic T-antigen by neuraminidase from bacterial infection, IgM T antibody, which is present in all adult plasma, can be removed by washing. Washing has also been used to remove the small amount of anti-A and anti-B in donor blood to be transfused to non-O patients with paroxysmal nocturnal hemoglobinuria and rarely for incompatible blood group organ and HSCT.

Washed cellular products expire in 24 h for RBCs and in 4 h for platelets. Coordination is required between the ordering physician and the blood bank when a washed product is needed. Washing causes a significant loss of platelets during processing and the platelets may become activated with decreased platelet aggregation and potentially worse *in vivo* survival and function.

Volume Reduction

The following are indications for volume reduction procedures designed to reduce the plasma volume in a product:

1. Patients who have substantive renal or cardiorespiratory dysfunction (e.g., RBC and platelet transfusions in patients with renal failure).
2. Patients who would benefit from removal of potentially pathologic antibodies in plasma (e.g., ABO-incompatible platelets).

In the case of ABO-incompatible platelets, some institutions measure isoagglutinin titers and establish specific allowable cutoffs instead of washing or volume-reducing platelet products. Volume reduction is associated with significant loss of product and increased activation postprocessing and should be used only in situations with well-defined indications. Small-volume storage of platelets in syringes for administration in neonates by syringe pumps has been recently studied.

SPECIAL PEDIATRIC POPULATION NEEDS

Neonates

Neonates requiring transfusion utilize multiple small volumes of blood products issued in syringes or transfer bags. Small-volume transfusions (≤15 ml/kg) of additive or nonadditive RBC units are the primary product used. Such RBC products may be used up to the date of expiration without the risk of hyperkalemia. However, infants requiring exchange transfusion, extracorporeal support for cardiovascular surgery, and ECMO are transfused at rates >25 ml/kg and often require multiple transfusions. In emergent circumstances of unexpected massive bleeding, when fresh (<7–10 days old), washed, or volume-reduced RBC units are not available, the infusion rate should not exceed 0.5 ml/kg/min.

Given the short shelf life of an RBC unit once accessed and the small volumes of blood used to transfuse neonates, blood banks will access the same unit to provide RBC transfusions to multiple neonates. Additionally, the blood bank will access the same unit for multiple transfusions in the same neonate if there are short intervals between the transfusions. These strategies reduce both RBC wastage as well as exposure to multiple blood donors.

Neonates with hemolytic disease who require exchange transfusion postdelivery or following intrauterine fetal transfusion may require special blood products. Blood grouping of these infants can be confounded by intrauterine transfusion, Rhogam administration to the mother and IVIG administration to the infant at delivery (see Chapters 5 and 9). Selection and preparation of reconstituted RBCs for exchange is provided in Table 36.6.

Liberal transfusion criteria in the neonatal period may provide some protection from CNS hemorrhage, periventricular leukomalacia, and other known postnatal complications of prematurity. However, In terms of platelet

TABLE 36.6 Blood Product Characteristics Needed for Exchange Transfusion

Blood product	Citrate-phosphate-dextrose solution or citrate-phosphate-dextrose-adenine anticoagulant (CPDA-1) red blood cells (RBCs)
	Additive solutions can be used if they are hand packed or washed to remove additive solution
Hematocrit	45–60% depending on desired end result
Age	Should be as fresh as possible, <7 days
Irradiation	Required to prevent transfusion-associated graft-versus-host disease (TA-GVHD) and performed as close to the transfusion as possible (<24 h)
Sickle test	Negative
With severe hemolytic disease of the fetus and newborn	O Rh (D) negative RBCs crossmatched against mother's plasma and negative for any other antibody besides anti-D prior to delivery
	Donor blood prepared after infant's birth should be antigen-negative and crossmatched-negative against infant's specimen
For ABO HDN	RBCs must be type O and either Rh (D) negative or Rh (D) compatible with mother and infant. The blood should be washed free of plasma or have low titer anti-A or anti-B antibodies. Reconstitution with AB plasma may result in two donor exposures
For Rh HDN	RBCs should be Rh (D) negative and either Group O or same group as infant. Reconstitution with AB plasma may result in two donor exposures
For polycythemia	Dilution with saline or albumin is recommended

VOLUME OF BLOOD REQUIRED FOR EXCHANGE TRANSFUSION

Single volume exchange	Exchanges 60% of infant's blood volume
Double volume exchange	Exchanges 85% of infant's blood volume
	Example: 2 × infant's blood volume = 2 × 85 = 170 ml
Equations	**Partial exchange for severe anemia:** Volume (ml) = Infant's blood volume × (Hb desired − Hb initial)/(pRBC Hb − Hb initial) **Single volume or partial exchange for correction of polycythemia:** Volume (ml) = Infant's blood volume × Hct change/Initial Hct **Final or Target Hb** $$\text{Final Hb} = \frac{\% \text{ Exchanged} \times \text{Hb (donor)} + \% \text{ Unexchanged} \times \text{Hb (recipient)}}{100}$$ Note: % Exchanged = % effectively exchanged or target % exchanged % Unexchanged = 100 − % Exchanged Hb(donor) = Hemoglobin of donor blood Hb(recipient) = Hemoglobin of recipient blood **Volumes of RBCs and plasma needed for reconstitution:** **Example: Double volume exchange in an infant (85 ml/kg), desired Hct 0.45% in product.** Total volume = infant weight/kg × 2 = 85 ml/kg × 2 = 170 ml Absolute volume of RBCs needed (ml) = Total volume × desired Hct (0.45) = 170 ml × 0.45 = 76.5 ml Assuming Hct (unit) = 0.75 Actual volume of RBCs needed = Absolute volume/Hct (unit) = 76.5 ml/0.75 = 100.8 ml Volume of FFP needed = Total volume − Actual volume of RBCs needed = 170 ml − 100.8 ml = 69.2 ml **This will result in a double volume (170 ml) product at Hct of 0.45 needed for exchange**

transfusions, a multiinstitutional observational study found that infants with platelet counts $<20,000/mm^3$ were not at increased risk for major hemorrhage.

Patients on Extracorporeal Life Support and Other Critically Ill Patients

There has been a controversy over whether there is an increase in morbidity and mortality associated with the use of old versus fresh RBCs. Theoretically, older RBC products suffer from the RBC storage lesion, which entails biochemical changes that occur to stored RBCs that have potential clinical consequences (Table 36.7). Studies of oncology patients, patients undergoing cardiac surgery, and trauma patients have shown an increase in morbidity and mortality when "older" RBCs are transfused. Recently, the Age of Red Blood Cells in Premature Infants study, a double-blind, randomized clinical trial of transfusion of RBCs <7 days old versus standard-age RBCs in very low birth weight demonstrated no difference in clinical outcomes between the two groups.

TABLE 36.7 The RBC Storage Lesion

Biochemical and biochemical changes	*In vitro* findings	*In vivo* correlations with stored RBCs/potential clinical significance
2,3-diphosphoglycerate (2,3-DPG) depletion	Left shift of oxygen–hemoglobin (Hb) curve	Restoration of RBC 2,3-DPG within 24–72 h of transfusions; delay in peripheral oxygen offloading
adenosine triphosphate (ATP) depletion	Reversible red blood cell (RBC) spheroechinocyte formation	Poor correlation of ATP levels with 24 h posttransfusion RBC survival. Splenic trapping
Increased intracellular calcium	RBC dehydration with echinocytosis and microvesiculation	Increased viscosity, reduced flow in capillary systems, microparticles
Metabolic modulation	Increased glucose consumption via pentose phosphate pathway and increased O_2 saturation of Hb; progressive acidification and lactate accumulation	Metabolic abnormalities in recipient
Decreased Na–K pump activity	Leakage of K from RBCs not returned to RBCs	Increased K levels in older RBCs, clinically significant hyperkalemia; worsened by RBC irradiation and refrigerator storage
Depletion of SNO-Hb	Loss of SNO-Hb after 3 days of storage	No off-loading of NO under hypoxic conditions; less vasodilation and blood flow in capillary systems
Low-level hemolysis	Free Hb scavenges nitric oxide (NO); break down to free heme and iron	Decreased NO causes vasoconstriction; free heme causes pulmonary hypertension and acute tubular injury in the kidney
Membrane phospholipid loss	Deduction of RBC microvesicles correlates with spheroechinocyte formation and increased osmotic fragility	Less vesiculation associated with prolongative of storage time and increased survival of AS-1 preserved RBCs
Abnormal membrane phospholipid distribution	Phosphatidyl-serine (PS) accumulates on outer membrane in acid pH or in presence of aminophopholipid translocase inhibitor. This loss is seen less with additive solutions and hypotonic solutions. However, no change during routine blood bank storage	Senescent RBCs and sickle RBCs accumulate PS on outer membrane. Animal models show PS accumulation on RBCs associated with increased clearance Exposed PS potentially prothrombotic and proinflammatory
Lipid peroxidation and protein oxidation	Depletion of glutathione Spectrin oxidation Abnormal spectrin–actin–protein 4.1 complex binding associated with phospholipid loss. Rate of RBC lipid peroxidation slowed in donors taking antioxidants or addition of antioxidants to stored blood Addition of antioxidants to stored blood decreases lipid peroxidation, decreased osmotic fragility and increased RBC deformability	Decreased lipid peroxidation associated with prolongation of viable storage time and increased survival of AS-1 preserved RBCs. Oxidative damage to lipids can result in lysophopholipid-induced transfusion-related acute lung injury and posttransfusion thrombotic events

Modified from Table 5 in Wong (2013) with permission.

Hematopoietic Progenitor Cell Transplant Recipients

ABO minor and major mismatch hematopoietic progenitor cell transplant patients are subject to potential changes in ABO/Rh type which can result in significant hemolysis and prolonged red cell aplasia. When to change a recipient's blood type depends on the patient's forward and reverse type and the direct antiglobulin test (DAT). Other, less well-defined parameters include the degree of engraftment and risk of relapse. Blood type changes should be investigated by the blood bank and agreed upon jointly by the transfusion medicine and stem cell transplant physicians. Guiding principles include:

1. The avoidance of both donor graft and recipient hemolysis.
2. Protection against delayed RBC engraftment by the recipient (Table 36.8).

Because platelets are frequently in short supply and have short outdates, Rh-positive (D+) platelets may be given to Rh-negative (D−) HSCT or oncology patients. Studies have shown that the use of Rh-positive (D+) apheresis platelets which contain small numbers of RBCs is not associated with risk for anti-D alloimmunization.

Sickle Cell Disease Patients

Patients with sickle cell disease often require transfusions for acute and chronic complications and have a high prevalence of RBC alloantibody development. The causes of alloimmunization are related to RBC antigen discrepancies between the donor and the recipient and predisposing genetic risk factors, including inheritance of Rh variant alleles. It is helpful for the transfusion service to have an extended RBC phenotype (ABO, Rh, Kell, Duffy, Kidd, Lewis, Lutheran, P, and MNS) on hand to assist in appropriate transfusion management in cases of alloimmunization. Many centers only transfuse leukocyte-reduced RBCs phenotypically matched for C, E, and K and antigen negative for any other demonstrated antibodies or those historically identified. Serological and FDA licensed molecular RBC phenotyping is now available to guide RBC selection.

TABLE 36.8 Transfusion Support for Patients Undergoing ABO Mismatched Allogeneic Hematopoietic Stem Cell Transplantation

Recipient	Donor	Mismatch type	Pretransplant (phase I)[a] All components	Peritransplant (phase 2)[b] red blood cells	Peritransplant (phase 2)[b] First choice platelets	Peritransplant (phase 2)[b] Next choice platelets	Peritransplant (phase 2)[b] FFP	Posttransplant (phase 3)[c] All components
A	O	Minor	Recipient	O	A	AB,B,O	A,AB	Donor
B	O	Minor	Recipient	O	B	AB,A,O	B,AB	Donor
AB	O	Minor	Recipient	O	AB	A,B,O	AB	Donor
AB	A	Minor	Recipient	A	AB	A,B,O	AB	Donor
AB	B	Minor	Recipient	B	AB	B,A,O	AB	Donor
O	A	Major	Recipient	O	A	AB,B,O	A,AB	Donor
O	B	Major	Recipient	O	B	AB,A,O	B,AB	Donor
O	AB	Major	Recipient	O	AB	A,B,O	AB	Donor
A	AB	Major	Recipient	A	AB	A,B,O	AB	Donor
B	AB	Major	Recipient	B	AB	B,A,O	AB	Donor
A	B	Minor and major	Recipient	O	AB	A,B,O	AB	Donor
B	A	Minor and major	Recipient	O	AB	B,A,O	AB	Donor

[a]Phase 1: Patient is prepared for transplantation until myeloablation (Day −1).
[b]Phase 2: From the Start of myeloablative therapy (Day 0) until.
[c]Phase 3: After the forward and reverse type are consistent with the donor's ABO group.
For RBCs: Direct antiglobulin test is negative and antidonor isohemagglutinins are no longer detectable.
For FFP: Recipient's erythrocytes are no longer detectable (i.e., forward typing is consistent with donor's ABO group.
Modified from Table 7 in Wong (2013) with permission.

Many patients on chronic transfusion protocols receive RBC products from dedicated donor programs which target African-American donors. When a rare alloantibody or combinations of alloantibodies are identified, antigen-negative units may be obtained through the Rare Donor Registry. However this approach is expensive and often causes substantive delays.

Please see Chapter 11 for further details of the use of transfusion therapy in patients with sickle cell disease.

Special Blood Banking Products

1. IgA-deficient products

Products obtained from IgA-deficient donors may be indicated when there is evidence for posttransfusion anaphylaxis due to the development of anti-IgA antibodies reacting with IgA present in the plasma of blood components. Any blood product may induce such a reaction. Diagnosis requires documentation of anti-IgA antibodies, only available from reference laboratories. An alternative to use of IgA-deficient blood products is to wash platelets and RBCs. However, if plasma or cryoprecipitate is indicated, IgA-deficient products must be obtained or alternative products should be used. Other more rare causes of anaphylaxis include:

- Anti-Rg^a.
- Anti-Ch^a.
- Antihaptoglobin (especially in the Japanese population).

The Chido (Ch^a) and Rogers (Rg^a) antigens are distinct antigenic determinants on the C4 molecule, antibodies to which may result in anaphylaxis. Special testing is only available in reference laboratories. Consultation with the transfusion medicine specialist is necessary to appropriately provide products for patients with a known or predicted anaphylaxis risk.

2. HLA-matched/crossmatch-compatible platelets

Pediatric patients recovering from chemotherapy or undergoing allogeneic HSCT who require frequent platelet transfusions may become alloimmunized to HLA class I or platelet-specific antigens. Platelet refractoriness must be distinguished from consumptive thrombocytopenia due to fever, infection, acute blood loss, splenomegaly, or disseminated intravascular coagulation. A safe way to differentiate these two conditions is to calculate the correct count increment (CCI). A CCI $<7500/mm^3$ on two consecutive occasions despite using platelets that are ABO identical and 3 days or less in age suggests alloimmunization. The CCI is defined by the equation below.

$$CCI = \frac{\text{platelet count increase per } mm^3 \times \text{body surface area } (m^2) \times 10^{11}}{\text{total platelets transfused}}$$

For example, a 1-h posttransfusion platelet count increase of $40,000/mm^3$ in a patient with a BSA of $0.9\ m^2$ who received three "equivalent" whole-blood-derived platelet units ($3 \times 5.5 \times 10^{10}$) platelets would have a CCI of 21,818. Consecutive ABO identical platelet transfusions with CCIs <7500 are suggestive of platelet refractoriness and suggest that HLA-matched or crossmatched platelets might be a valuable alternative. Acquisition of these products requires close coordination with the transfusion service and blood center provider.

HLA matchmaker software program incorporates both the HLA type of the recipient and the anti-HLA antibodies identified can be useful to improved conventional cross-reactive epitope group matching. This usually requires coordination with a blood center provider.

3. Antigen-negative platelet products for neonatal alloimmune thrombocytopenia (NAIT)

NAIT develops when an infant inherits paternally derived platelet antigens that are recognized as foreign by the mother's immune system with the resultant production of IgG antibodies that cross the placenta and cause fetal or neonatal thrombocytopenia (see Chapter 14). If platelets are urgently needed for acute bleeding or in an infant at high risk of critical hemorrhage, nonantigen-matched leukoreduced platelets should be used.

On occasion when nonantigen-matched leukoreduced platelets are ineffective, antigen-negative platelets may be obtained from the mother, a known platelet antigen-negative family member, or pretested platelet donor. Platelet products from the mother must be washed, irradiated, and leukoreduced. Those from a blood relative must be irradiated and leukoreduced. Infectious disease testing may have to be waived if testing cannot be completed in time for transfusion need. Caution must be exercised when related donors are suggested in a patient being considered for HSCT as this may risk sensitization with a resultant increased risk of graft rejection.

4. Pathogen-inactivated blood products

Pathogen-inactivated products are new to the United States but have been used in Europe to reduce the risk of emerging infectious agents. Four processes (methylene blue plus visible light, amotosalen plus ultraviolet (UV) light, riboflavin plus UV light, and solvent detergent treatment) have received the CE mark in Europe for plasma, and two of these (amotosalen and riboflavin processes) have also received the CE mark for platelets (CE marking is the manufacturer's declaration that the product meets the requirements of the applicable European Council directives). Currently only the solvent detergent treated plasma product (Octaplas®, Octapharma AB Sweden) and pathogen inactivation methodology using amotosalen and UV light for platelets and plasma (Cerus Corporation, Concord, CA) are FDA cleared and licensed in the United States. Altered potency and toxicity have been reported for some of these products.

5. Frozen deglycerolized RBCs

RBCs can be frozen by adding the cryoprotective agent glycerol to the RBCs before freezing. RBCs are then thawed and washed with successively lower concentrates of sodium chloride to remove the glycerol. Recovery of 80% of the red cells present from the original unit of blood is expected. The deglycerolization process is not universally available and can take several hours to perform. The RBC product must be transfused within 24 h postthawing if prepared in an open system and within 2 weeks if a closed system is used. A major indication for this product is the need for rare, multiple RBC antigen-negative products.

6. CMV-negative products

CMV is a DNA virus of the Human Herpes Group known to asymptomatically infect children and young adults. CMV can be reactivated in some immunocompromised individuals and is transfusion transmitted. Prestorage LR in a blood donor center or blood bank using quality control methods reduces the risk of posttransfusion CMV. LR does not remove virions from those few donors who are asymptomatic, seronegative, and in the early phase of primary infection, accounting for the few cases of CMV infection reported. There are no clinical trials that have compared LR versus seronegative LR versus seropositive untested LR products in "at-risk" populations nor are there likely to be any given the rarity of transfusion-associated CMV infection. It is likely that universal LR of cellular components obviates the redundancy to request both LR and CMV-negative products for neonates, oncology, hematopoietic progenitor cell transplant, and solid organ transplant recipients. In summary, the use of "CMV-negative" blood products is unwarranted.

ADMINISTRATION

RBCs and Whole Blood

Transfusion reactions and their management are summarized in Table 36.9. Hemolytic reactions, both acute and delayed, are much more common than posttransfusion infections. Hemolysis can occur due to errors in the collection of a sample for crossmatch, bag labeling, crossmatch itself, nursing errors, infusion device failures, mechanical pump failure, and coadministration of inappropriate fluids.

Informed consent for blood and blood products must include risks, benefits, and alternatives to transfusion. Blood or blood products require a detailed written order specifying the product, volume, and any special requirements for administration including pretransfusion medication (if indicated), rate and duration of administration, special processing needs (irradiation, washed, etc.), and the use of a blood warmer or electromechanical device. Specific written procedures which guide transfusion practice should exist.

The following formula is used to determine RBC transfusion volume:

$$\text{Volume (ml)} = \frac{(\text{Hb}_{\text{target}} - \text{Hb}_{\text{observed}}) \times \text{weight} \times \text{blood volume}}{\text{Hb}_{\text{RBC}} \text{ unit}}$$

$\text{Hb}_{\text{target}}$ is the hemoglobin (Hb) concentration targeted posttransfusion (e.g., 10 g/dl), $\text{Hb}_{\text{observed}}$ is the most recently measured Hb concentration in the patient (g/dl), weight is expressed in kilograms, blood volume is expressed in ml/kg (80 ml/kg if child is <2 years old and 75 ml/kg if child is 2–14 years old) and Hb_{RBC} unit is the average Hb concentration in the RBC (g/dl) prepared by the transfusion service. In situations where additive units are used, the mean Hb concentration of the additive RBC is critical to the calculation.

TABLE 36.9 Management of Transfusion Reactions

Type	Etiology	Presentation	Evaluation	Treatment and prevention
Hemolytic (immune)	RBC incompatibility: BO incompatibility usually caused by clerical error; undetected alloantibody	• Fever/chills • Hemoglobinemia/hemoglobinuria • Anxiety • Nausea/vomiting • Pain • Hypotension • Renal failure with oliguria • DIC	• DAT • Visual inspection for plasma Hb • Clerical checks • Tests as indicated to determine RBC incompatibility • Tests to monitor hemolysis (Hb/Hct, haptoglobin, bilirubin, plasma-free Hb, LDH, urinalysis as indicated) • Tests to monitor renal function if urinanalysis indicates hemoglobinuria (BUN, creatinine) • Tests to monitor coagulation status if DIC is clinically suspected (PT/PTT, platelet count, fibrinogen)	Treatment: Supportive treatment for hypotension and renal perfusion (maintain output $\gg$ 1 ml/kg/h) may include • IV crystalloid (i.e., 10–20 ml/kg normal saline) • Furosemide[a] Neonate: 0.5–1 µg/kg/dose IV every 8–24 h (maximum dose IV: 2 µg/kg). Premature infants will require less frequent dosing Children: 0.5–2 µg/kg/dose IV every 6–12 h (maximum dose 6 µg/kg/dose) or continuous infusion 0.05 µg/kg/h titrating dose to clinical effect • Low-dose dopamine (2–5 µg/kg/min) Supportive treatment of DIC with active bleeding • Hemostatic components: platelets, plasma, cryoprecipitate Prevention: • Ensuring proper sample and recipient identification • Reviewing historical records • Providing antigen-negative units as appropriate
Hemolytic (nonimmune)	Physical or chemical destruction of blood (thermal, drugs, solutions added)	• Hemoglobinuria	• Plasma-free Hb • DAT (should be negative) • Visual inspection of unit for hemolysis	Treatment: • Hydrate Prevention: • Proper administration of blood components • Identify and eliminate cause
Fever/chill (nonhemolytic)	Antibody to donor WBC/plasma proteins Passive cytokine infusion	• Chills/rigors • Temperature rise (>1°C) not explained by condition • Headache • Nausea/vomiting Symptoms often occur near the end or after the completion of the transfusion	• Rule out hemolysis (DAT, inspection for plasma Hb per blood bank protocol) • Rule out bacterial contamination (culture patient and unit) • Diagnosis of exclusion	Treatment: • Antipyretic, for example, acetaminophen 10–15 mg/kg/dose PO (avoid aspirin) Prevention: • Leukocyte-reduced component if recurrent • Premedicate with antipyretic, i.e., acetaminophen 10–15 mg/kg/dose PO, 30–60 min before transfusion in patients with recurrent reactions
Allergic (mild)	Antibody to plasma protein	• Urticaria • Pruritis • Flushing	• Response to antihistamines	Treatment: • Antihistamine: children—diphenhydramine 1–1.5 mg/kg/dose PO/IV (administer 1 h before transfusion for prevention of recurrent hives) • May restart unit if symptoms are resolved

(Continued)

TABLE 36.9   (*Continued*)

Type	Etiology	Presentation	Evaluation	Treatment and prevention

Allergic (moderate to severe)		• Respiratory distress
		• Wheezing
		• Laryngeal edema
		• Nausea/vomiting
		• Hypotension
		
		Symptoms usually begin shortly after the start of transfusion
			• Rule out other etiologies for respiratory distress, for example, transfusion-associated circulatory overload (TACO), TRALI, anaphylactic reaction
			• IgA levels
			• Anti-IgA
				Treatment: (for anaphylaxis)
				
				• Epinephrine 0.01 ml/kg (1:1000) SC or IM (maximum 0.5 ml); can repeat in 15 min
				• (a) For management of anaphylaxis in children not responding to initial epinephrine IM or subQ injections and volume resuscitation, and where there is inadequate time for emergency transport or prolonged transport is required, recommend intravenous epinephrine 0.01 mg/kg (0.1 ml/kg of a 1:10,000 solution up to 10 μg/min), with a maximum dose of 0.3 mg
				• (b) For cardiopulmonary arrest during anaphylaxis epinephrine 0.01 mg/kg (0.1 ml/kg of a 1:10,000 solution), up to 10 μg/min rate of infusion, repeated every 3−5 min for continued cardiopulmonary arrest. Higher subsequent doses (0.1−0.2 mg/kg; 0.1 ml/kg of a 1:1000 solution) may be considered for unresponsive asystole or pulseless electrical activity
				• Diphenhydramine 1 mg/kg IV/IM/PO (maximum 50 mg); can repeat in 15 min
				• Albuterol nebulizer 0.05−0.15 mg/kg in 3 ml normal saline (estimate 2.5 mg <30 kg, 5.0 mg >30 kg)
				• Methylprednisolone 1−2 mg/kg/dose IV
				
				Prevention: (for moderate and recurrent reactions)
				
				• Diphenhydramine 1 mg/kg/dose IV/PO 1 h before transfusion
				• Corticosteroids 2−6 h before transfusion either one of the following:
				  Methylprednisolone 1 mg/kg/dose IV
				  Hydrocortisone 1 mg/kg/dose IV
				  Prednisone 1 mg/kg/dose PO
				• Washed RBCs/platelets
				• IgA-deficient components if appropriate

(Continued)

TABLE 36.9 (*Continued*)

Type	Etiology	Presentation	Evaluation	Treatment and prevention
Circulatory overload (TACO)	Too rapid or excessive blood transfusion or both	• Dyspnea • Cough • Rales • Rapid increase in systolic pressure • Headache • Cardiac arrhythmia	• Chest X-ray—bilateral infiltrates, may progress to complete "white out" • Rule out other etiologies for respiratory distress, for example, anaphylactic reaction, TRALI	Treatment: • Slow rate/stop transfusion • Upright position • Oxygen • Diuretics, i.e., furosemide Neonates: 0.5–1.0 mg/kg/dose every 8–24 h (maximum IV dose 2 mg/kg/dose)Children: 0.5–2.0 mg/kg/dose every 6–12 h (maximum dose 6 mg/kg/dose) Prevention: • Transfuse blood slowly • Aliquot small volumes
TRALI	Passive infusion of donor HLA/leukocyte antibody through plasma-containing components; neutrophil-priming lipid mediator; recipient antibody to donor white cells—these mechanisms lead to microvascular injury in the lung	• Fever • Dyspnea • Hypoxemia, may be severe • Hypotension • Pulmonary edema • Normal pulmonary capillary wedge pressure	• Rule out hemolytic transfusion reaction (visual inspection plasma Hb, DAT, and other clinical and laboratory findings) • Culture recipient • Culture blood component • Rule out TACO by obtaining chest X-ray and/or brain natriuretic peptide level • Test donor/recipient for white-cell-related antibody • Diagnosis of exclusion	Treatment: • Respiratory support may include oxygen, intubation/mechanical ventilation depending on severity of hypoxia • Blood pressure support (see above) Prevention: • If leukocyte antibody is present in recipient, use leukocyte-reduced blood components • If donor antibody is implicated, no special measures are needed • May want to prevent the use of plasma-containing components from implicated donors in the future—inform blood donor center
Transfusion-associated dyspnea	May occur with any product, especially platelets	• Dyspnea is the major symptom, otherwise no distinguishing features	• Primarily a clinical diagnosis after exclusion of TRALI, TACO, and allergic reactions	Treatment: Respiratory support may include oxygen, intubation/mechanical ventilation depending on severity of hypoxia
Bacterial contamination	Contaminated blood component (frequency is higher with platelet components than red cells)	• Rigors • Chills • Fever • Shock	• Culture recipient • Culture blood component • Rule out immune hemolytic transfusion reaction	Treatment: • Support blood pressure (see above) • Administer appropriate broad-spectrum antibiotics Prevention: • Attention to arm preparation for donor phlebotomy • Prevention of contamination during blood collection and storage • Screening of platelet components for bacterial contamination

(*Continued*)

TABLE 36.9 (*Continued*)

Type	Etiology	Presentation	Evaluation	Treatment and prevention
Hypotension	May occur with any product (associated with use of ACEi and negatively charged filters	Hypotension is the major symptom	Exclusion of other etiologies for hypotension	Treatment Support blood pressure (see above)
Hypothermia	Rapid infusion of cold blood	• Chills • Low temperature • Irregular heart rate • Possible cardiac arrest • Neonatal apnea		Treatment: • Slow infusion rate • Use blood warmer Prevention: • Transfusion with an approved blood warming device • Keep patient warm
Hyperkalemia (elevated serum potassium)	Hemolysis of red cells (immunologic/ nonimmunologic); massive/ rapid infusion of blood with high potassium level	• Nausea/diarrhea • Muscle weakness • Cardiac arrhythmias • Cardiac arrest	• Serum potassium (K^+) level • Monitor EKG for peaked T waves, loss of P wave, widening QRS complex, ST depression, bradycardia/ asystole	Treatment: Depends on severity • Cardiac monitor • Mild to moderate levels ($K^+ = 6.0–7.0$ mEq/l), try to enhance excretion; Kayexalate resin 1 g/kg/dose every 2–6 h • Severe ($K^+ > 7.0$ mEq/l), try to move K^+ into cell acutely: Regular insulin 0.1 unit/kg with glucose 0.5 g/kg over 30 min, sodium bicarbonate 1–2 mEq/ kg IV given over 5–10 min • If EKG changes are present then urgent reversal of membrane effects is required: Calcium gluconate (10%) 100 mg/kg/dose (1 ml/kg/ dose) over 3–5 min • If unsuccessful—dialysis recommended Prevention: Preventive measures for patients at risk for hyperkalemia: • Fresher red cells • Red cells with supernatant removed • Washed red cells
Hypocalcemia (low ionized calcium)	Massive transfusion of citrated blood particularly in clinical setting where citrate metabolism is dysfunctional	Neonates: • Jitteriness • Poor feeding • Apnea • Seizures • Arrhythmia • Increased irritability Older children: • Paresthesia • Tetany • Arrhythmia • Increased irritability • Vomiting	• Ionized calcium level • Prolonged Q-T interval on EKG	Treatment: • Slow rate of transfusion • Prophylactic calcium during exchange transfusion is not routinely performed • Seizure activity associated with hypocalcemia should be treated: calcium should be infused IV slowly with constant monitoring of heart rate • Calcium gluconate (10%) 100 mg/kg/dose slowly • If magnesium is low, correct with magnesium sulfate 25–50 mg/kg IV/IM every 4–6 h × 3–4 doses, prn

(*Continued*)

TABLE 36.9 (*Continued*)

Type	Etiology	Presentation	Evaluation	Treatment and prevention
Hypoglycemia	Discontinuing dextrose infusion during transfusion; rebound phenomena after exchange transfusion	• Jitteriness • Tremors • Seizure • Apnea/cyanosis	• Glucose level	Prevention: Monitor ionized calcium during procedure and administer calcium as necessary Treatment: • If asymptomatic, maintenance glucose either PO (D5W) or IV 4–8 mg/kg/min as required • If seizure activity associated with hypoglycemia: 5–10 mg/kg IV bolus (10% or 15% dextrose) followed by 8–10 mg/kg/min IV drip as required Prevention: Frequent glucose monitoring (30–60 min) during the transfusion; point-of-care glucose monitoring is optimal

[a]*Doses and frequency may be altered depending on gestational ages as well as clinical condition of neonate/infant/child. BUN, blood urea nitrogen; DAT, direct antiglobulin test; DIC, disseminated intravascular coagulation; Hb, hemoglobin; IV, intravenous; IM, intramuscular; PO, by mouth; PT, prothrombin time; PTT, partial thromboplastin time; RBCs, red blood cells; SC, subcutaneous; TRALI, transfusion-related acute lung injury.*
Immediate steps for all reactions: (1) Stop transfusion and clamp at hub; (2) keep IV open with 0.9% NaCl; (3) verify at bedside that the correct unit was administered to the correct patient; (4) notify patient's physician and blood bank that transfusion has been terminated; (5) send any necessary urine/blood samples to appropriate laboratories; (6) send blood unit and administration set to the blood bank.
Modified from Table 27 in Wong (2015), with permission.

For example, if the $Hb_{observed}$ in a 3-week-old baby weighing 3.5 kg is 6.0 g/dl, blood volume is 80 ml/kg and the Hb_{RBC} unit is 19.5 g/dl (usual Hb for an AS-3 unit), and the desired increase in the Hb concentration (Hb_{target}) is 10 g/dl.

The volume of an AS-3 RBC unit is calculated below:

$$\text{Volume} = \frac{(10.0 - 6.0 \, g/dl) \times 3.5 \, kg \times 80 \, ml/kg}{19.5 \, g/dl} = 57\text{ml or } 16.4 \, ml/kg$$

An alternative method in patients <50 kg, is to utilize the following rule: 10–15 ml/kg RBCs will increase Hb levels between 2 and 3 g/dl. In older children or adolescents (≥50 kg), transfusion of 1 RBC unit (~300 ml) may increase Hb by ~1 g/dl or increase hematocrit by 3%.

All RBC products must be transfused through a standard blood filter (150–200 μm), and completed within 4 h after release from the blood bank. Only normal saline or other FDA-approved products may be given through the same line where RBCs are transfused.

Plasma Products

While compatibility testing using crossmatching is not needed for plasma products, plasma-containing products must be ABO-compatible with the recipient's RBCs. In patients <50 kg, 10–20 ml/kg of frozen, thawed plasma should increase most coagulation factors by 15–20%. In older children and adolescents (≥50 kg), transfusion of two units of frozen, thawed plasma (~500 ml) should increase coagulation factors by a similar amount. Frozen, thawed plasma must be transfused using a microaggregate blood filter and completed within 4 h. AB plasma, thawed in advanced, is often available in institutions where there are active trauma units during massive transfusion episodes. Some institutions have initiated massive trauma protocols with fixed ratios of red cell:plasma:platelet concentrates.

Platelet Products

Platelets can be administered rapidly in acute bleeding situations. A specialized 150–200-μm platelet filter should be used which minimizes platelet loss. Approximately 10 ml/kg volume can be administered in 30 min and is usually well tolerated. Slower administration or the use of a diuretic can be used for volume-overloaded and cardiac-compromised patients. Five to ten ml/kg dose should increase platelet counts by $50,000–100,000/mm^3$. An alternative dosing scheme is to dose at 1 equivalent unit of platelets (5×10^{10} platelets) per every 5–10 kg body weight to achieve the same or similar increments. This calculation takes into account the actual number of platelets that are in the product.

Cryoprecipitate

Cryoprecipitate is not a common product for increasing fibrinogen or factor XIII since commercial and solvent detergent treated products are now available. Cryoprecipitate is dispensed in units. The equation below may be helpful.

$$\frac{\text{Desired fibrinogen (mg/dl)} - \text{Initial fibrinogen (mg/dl)} \times \text{plasma volume} \div 100 \text{ mg/dl}}{250 \text{ mg fibrinogen/unit}}$$

Alternatively, one can estimate a rise of 60–100 mg/dl by transfusion of 1–2 cryoprecipitate units/10 kg.

Granulocytes

Granulocytes are transfused over 1–2 h using a standard 150-μm blood filter with intermittent agitation of the product to avoid settling of the granulocytes. Granulocyte transfusion is frequently accompanied by fever, chills, and allergic reactions and should be discontinued in the case of severe pulmonary reactions. Premedication with an antipyretic and/or an antihistamine should be used unless contraindicated. Typical dosing is $\geq 1 \times 10^9$ neutrophils/kg in volumes of 10–15 ml/kg. Overall efficacy of granulocyte transfusion can be ascertained either by clinical improvement over time or assessment of serial absolute neutrophils counts.

Neonatal Immunohematological Issues

Neonates do not develop significant isohemagglutinins until ~4 months of age. Therefore, neonatal samples are not crossmatched against donor RBCs in many centers unless there have been multiple transfusions of plasma and platelets or massive RBC transfusion has occurred. Initial blood group should include two samples for ABO/D typing. ABO/D type (red cell, forward group) is performed to document there has been no misidentification of the sample; however, plasma typing (reverse group) is not required. In addition, antibody screen and testing for maternal IgG ABO isohemagglutinins may be performed. Sequential testing for ABO/D type, antibody screen, or crossmatch is not generally required. If circulating maternal antibody is present or abnormal antibody screen is identified, antigen-negative and/or crossmatch-compatible RBC units are required.

Immunohematology Techniques

Tables 36.10 and 36.11 identify specialized immunological techniques used in the transfusion service with a description of indications, methods, and turnaround time.

TABLE 36.10 Commonly Used Blood Bank Techniques

Test	Principle of the test	Purpose
Group and type TAT: 5–20 min	Mixing of known anti-A, anti-B, anti-AB reagent with patient red cells (forward type) and mixing of patient plasma or serum with red cells of known ABO and Rh (D) type (reverse type)	To determine ABO and Rh (D) type
Antibody screen (IAT) TAT: 45–60 min	Mixing of patient plasma or serum with donor reagent red cells with known antigenic phenotype (usually 2 or 3 reagent cells used). Used with or without enhancement media (e.g., low ionic strength solution, polyethylene glycol). Performed at various incubation phases (e.g., immediate spin, 37°C, antihuman globulin (AHG) and with various incubation times	To detect allo and/or auto IgG antibodies in the patient's plasma and/or serum
Antibody panel TAT: 45–60 min	Same as *Antibody screen*, but with a larger group of commercial reagent red cells (11 or more reagent cells)	To better delineate and classify allo and/or auto IgG antibodies in the patient's plasma and/or serum
Crossmatch TAT: 5–60 min	Mixing of patient plasma or serum with selected donor red cells. Used with or without enhancement media. Performed at various incubation phases	To detect incompatibility between patient and donor red cells (e.g., ABO and/or other antigens)
Direct antiglobulin test TAT: 20–30 min	Mixing of patient red cells from EDTA sample with various AHG antisera (e.g., anti-C3b, anti-C3d, anti-C3b,d, anti-IgG)	To detect *in vivo* sensitization of patient red cells with IgG and/or complement
Elution TAT: 60–90 min	To release bound antibodies from patient red cells to identify the antibody	To delineate the specificity of one or more antibodies bound to the patient's red cells and/or to phenotype patient's red cells after removal of bound antibodies
Cold antibody screen TAT: 60–90 min	Mixing of patient plasma or serum with donor reagent red cells with known antigenic phenotype (usually screening cells, A- and B-cells, and cord blood cells). Performed at various incubation phases (i.e., immediate spin, room temperature, 15°C, 4°C) and various incubation times	To detect allo and/or auto IgM antibodies in the patient's plasma and/or serum
Cold antibody panel TAT: 60–90 min	Same as *Cold antibody screen*, but with a larger group of reagent red cells (11 or more reagent cells)	To detect allo and/or auto IgM antibodies in the patient's plasma and/or serum
Serologic red blood cell antigen typing TAT: 30–90 min (may take longer if full phenotyping required)	Mixing of patient or donor red cells with known antisera (e.g., anti-A, anti-C, anti-D, etc.)	To detect the presence or absence of specific antigens on patient or donor red cells

Modified from Table 9 in Wong (2013) with permission.

TABLE 36.11 Uncommonly Used (Specialized) Blood Bank Techniques

Test	Principle of the test	Purpose
Donath–Landsteiner test TAT: 2–3 h	Mixing of patient serum with known P-positive reagent cell at cold temperature (ice water bath) and then incubating at warm temperature (37°C) resulting in visible hemolysis	To detect Donath–Landsteiner antibody (anti-P) and its unique hemolytic property
Minor Crossmatch TAT: 5–60 min	Mixing of donor plasma with patient red cells	To detect incompatibility between patient and donor plasma
Lectin panel TAT: 30–60 min	Mixing of patient red cells with known lectins (i.e., *Arachis hypogaea*, glycine maximum).	To detect the exposure of crypt T-antigens on the patient's red cells (polyagglutination)
Molecular red blood cell phenotyping TAT: 8 h to 1 week (depending on methodology and availability)	Extract DNA from patient or donor white blood cells, amplify specific targets, and detecting the amplified targets using US Food and Drug Administration-approved or research assay systems	To detect the presence or absence of specific polymorphisms to identify the patient or donor red cell antigenic phenotype
Adsorption for cold or warm antibodies TAT: ~8 h–days	Mixing of patient plasma or serum with autologous treated red cells or reagent cells with known phenotype, followed by incubation to adsorb out the autoantibodies. May need to perform 2 or 3 times	To remove unwanted autoantibodies in order to facilitate the detection of alloantibodies

Modified from Table 9 in Wong (2013) with permission.

SUMMARY

Blood component therapy plays a critical role in the treatment of pediatric hematology and oncology patients and those with critical illness. Product modifications should be based on a firm understanding of the risks and benefits of transfusion. Newer blood and blood products not yet licensed in the United States are likely to have an improved safety profile. Molecular immunohematologic testing for sickle cell disease patients and those with hemolytic anemia as well as improved devices for the infusion of blood and blood products also likely to provide improved patient transfusion safety.

Further Reading and References

Diab, Y., Wong, E., Criss, V.R., et al., 2011. Storage of aliquots of apheresis platelets for neonatal use in syringes with and without agitation. Transfusion 51, 2642–2646.

Diab, Y.A., Wong, E.C., Luban, N.L., 2013. Massive transfusion in children and neonates. Br. J. Haematol. 161, 15–26.

Dunbar, N.M., Omstein, D.L., Dumont, L.J., 2012. ABO incompatible platelets: risks versus benefit. Curr. Opin. Hematol. 19, 475–479.

Fasano, R.M., Paul, W.M., Pisciotto, P.T., 2012. Complications of neonatal transfusions. In: Popovsky, M.A. (Ed.), Transfusion Reactions, fourth ed. AABB Press, Bethesda, MD.

Fergusson, D.A., Hébert, P., Hogan, D.L., et al., 2012. Effect of fresh red blood cell transfusions on clinical outcomes in premature, very low-birth-weight infants: the ARIPI randomized trial. JAMA 308, 1443–1451.

Fung, M.B., Grossman, B., Hillyer, C.B., Westhoff, C.M. (Eds.), 2014. AABB Technical Manual. eighteenth ed. AABB Press, Bethesda, MD.

Josephson, C.D., 2011. Neonatal and pediatric transfusion practice. In: Roback, J.D., Grossman, B.J., Harris, T., Hillyer, C.D. (Eds.), Technical manual, seventeenth ed. AABB Press, Bethesda, MD, pp. 645–670.

Karam, O., Tucci, M., Combescure, C., Lacroix, J., Rimensberger, P.C., 2013. Plasma transfusion strategies for critically ill patients. Cochrane Database Syst. Rev. 12, CD010654.

Luban, N.L., McBride, E., Ford, J.C., Gupta, S., 2012. Transfusion medicine problems and solutions for the pediatric hematologist/oncologist. Pediatr. Blood Cancer 58, 1106–1111.

Luban, N.L.C., Strauss, R.G., Hume, H.A., 1991. Commentary on the safety of red cells preserved in extended storage media for neonatal transfusions. Transfusion 31, 229–235.

Neal, M.D., Raval, J.S., Triulzi, D.J., et al., 2013. Innate immune activation after transfusion of stored red cells. Transfus. Med. Rev. 27, 113–118.

Pavenski, K., Saidenberg, E., Lavoie, M., et al., 2012. Red blood cell storage lesions and related transfusion issues: A Canadian Blood Services research and development symposium. Transfus. Med. Rev. 26, 68–84.

Adverse effects of blood transfusion. In: Roseff, SD Pediatric Transfusion: A Physician's Handbook. third ed. AABB Press, Bethesda, MD.

Strauss, R.G., 2010. RBC storage and avoiding hyperkalemia from transfusions to neonates and infants. Transfusion 50, 1862–1865.

Tatari-Calderone, Z., Tamouza, R., Le Bouder, G.P., et al., 2013. The association of CD81 polymorphisms with alloimmunization in sickle cell disease. Clin. Dev. Immunol. 2013, 937846.

Veeraputhiran, M., Ware, J., Dent, J., et al., 2011. A comparison of washed and volume-reduced platelets with respect to platelet activation, aggregation, and plasma protein removal. Transfusion 51, 1030–1036.

Wong, E.C.C. (Ed.), 2015. Pediatric Transfusion Medicine Handbook. fourth ed. AABB Press, Bethesda, MD (Co-edited with Drs. Roseff, Sloan, Punzalan, Sesok-Pizzini, Josephson, Strauss, and King).

Wong, E.C.C., 2013. Blood banking/immunohematology: special considerations in pediatric patients. Pediatr. Clin. North Am. 60, 1541–1568.

Wong, E.C.C., Paul, W., 2010. Intrauterine, neonatal, and pediatric transfusion. In: Mintz, P.D. (Ed.), Transfusion Therapy: Clinical Principles, third ed. AABB Press, Bethesda, MD.

Yawn, B.P., Buchanan, G.R., Afenyi-Annan, A.N., et al., 2014. Management of sickle cell disease: summary of the 2014 evidence-based report by expert panel members. JAMA 312, 1033–1048.

1

Hematological Reference Values

The normal range for most hematologic parameters in infancy and childhood is different from that in adults. Dramatic changes occur during the first few weeks of life. The recognition of variables in the pediatric age group prevents unnecessary medical and laboratory investigations.

FETAL AND CORD BLOOD HEMATOLOGIC VALUE

TABLE A1.1 Hematologic Values[a] in Normal Fetuses at Different Gestational Ages

Week of gestation	Hemoglobin (g/dl)	RBCs ($\times 10^6$/ml)	Hematocrit (%)	Mean corpuscular volume (fl)	Total WBCs ($\times 10^6$/μl)	Corrected WBCs ($\times 10^6$/μl)	Platelets ($\times 10^6$/μl)
18–21 (N = 760)	11.69 ± 1.27	2.85 ± 0.36	37.3 ± 4.32	131.1 ± 11.0	4.68 ± 2.96	2.57 ± 0.42	234 ± 57
22–25 (N = 1200)	12.2 ± 1.6	3.09 ± 0.34	38.59 ± 3.94	125.1 ± 7.8	4.72 ± 2.82	3.73 ± 2.17	247 ± 59
26–29 (N = 460)	12.91 ± 1.38	3.46 ± 0.41	40.88 ± 4.4	118.5 ± 8.0	5.16 ± 2.53	4.08 ± 0.84	242 ± 69
>30 (N = 440)	13.64 ± 2.21	3.82 ± 0.64	43.55 ± 7.2	114.4 ± 9.3	7.71 ± 4.99	6.4 ± 2.99	232 ± 87

[a]Hematologic data obtained with a Coulter S plus II instrument. Total WBC count included nucleated red blood cells. Corrected WBC count included only WBCs, after subtracting the nucleated red cell component, based on a 100-cell manual differential.
RBCs, red blood cells; WBCs, white blood cells.
Modified from Forestier, F., Daffos, F., Catherine, N., et al., 1991. Developmental hematopoiesis in normal human fetal blood. Blood 77, 2360.

TABLE A1.2 WBC Manual Differential Counts in Normal Fetuses at Different Gestational Ages

Week of gestation	Lymphocytes (%)	Neutrophils (%)	Eosinophils (%)	Basophils (%)	Monocytes (%)	Nucleated RBCs (% of WBCs)
18–21 (N = 186)	88 ± 7	6 ± 4	2 ± 3	0.5 ± 1	3.5 ± 2	45 ± 86
22–25 (N = 230)	87 ± 6	6.5 ± 3.5	3 ± 3	0.5 ± 1	3.5 ± 2.5	21 ± 23
26–29 (N = 144)	85 ± 6	8.5 ± 4	4 ± 3	0.5 ± 1	3.5 ± 2.5	21 ± 67
>30 (N = 172)	68.5 ± 15	23 ± 15	5 ± 3	0.5 ± 1	3.5 ± 2	17 ± 40

RBCs, red blood cells; WBCs, white blood cells.
From Forestier, F., Daffos, F., Catherine, N., et al. 1991. Developmental hematopoiesis in normal human fetal blood. Blood 77, 2360.

TABLE A1.3 Hematologic Values for Cord Blood (Vaginal Delivery and Cesarean Section[a])

Characteristic	Study sample (N = 167) Median	Range	Vaginal delivery (N = 63) Median	Range	Cesarean section (N = 104) Median	Range	P-value[2]
WBC ($\times 10^9$/l)	15.1	5.54−39.7	18.4	12.0−34.1	13.6	8.54−39.7	<0.0001
RBC ($\times 10^{12}$/l)	4.7	3.46−6.62	4.78	3.89−6.30	4.62	3.46−6.62	NS
Hb (g/l)	174	130−234	176	140−230	171	130−234	NS
Hct (%)	53.6	40.1−73.1	54.7	41.9−73.1	52.6	40.1−72.2	NS
MCV (fl)	112	97.7−127	114	105−127	112	97.7−125	NS
MCH (pg)	36.5	31.4−41	36.5	31.4−41	36.6	32−39.9	NS
MCHC (g/l)	324	303−359	323	308−359	324	303−344	NS
RDW (%)	17.4	14.2−23.6	17.4	14.9−23.6	17.4	14.2−23.3	NS
PLT ($\times 10^9$/l)	270	161−607	297	169−607	254	161−424	0.0053
MPV (fl)	8.7	7.5−11.5	8.7	7.7−11.4	8.8	7.5−11.5	NS
Plateletcrit (%)	0.24	0.15−0.48	0.26	0.15−0.48	0.23	0.15−0.36	0.0056
CD34 + cells ($\times 10^6$/l)	43.9	7.14−253	47.7	15.9−253	39.9	7.14−120	0.007

[a]P-values of the differences between vaginal delivery and cesarean section. The concentrations were standardized to exclude the varying effect of the anticoagulant.
Hb, hemoglobin; Hct, hematocrit; MCH, mean corpuscular hemoglobin; MCHC, mean corpuscular hemoglobin concentration; MCV, mean corpuscular volume; MPV, mean platelet volume; NS, not significant; plateletcrit, MPV × PLT; PLT, platelet; RBC, red blood cell; RDW, red blood cell distribution width; WBC, white blood cell.
From Eskola, M., Juutistenaho, S., Aranko, K., et al. 2011. J. Perinatol. 258−262. Data obtained with Sysmex K-1000 analyzer (Sysmex, Kobe Japan).

TABLE A1.4 Red Cell and Reticulocyte Indices, Serum Iron Status Markers in Cord Blood

Parameter	Mean	SD	Reference range[a]
BLOOD COUNT AND CELLULAR INDICES			
Hb (g/l)	159	15	146−189
HCT (l/l)	0.49	0.05	0.44−0.58
MCV (fl)	109	4	102−118
MCVr (fl)	124	6	115−136
MCH (pg)	35	1	33−38
MCHC (g/l)	325	10	306−342
%Retic (%)	4.0	0.8	2.6−5.4
IRF-H (%)	24.1	7.8	10.2−40.0
CHm (pg)	34.9	1.3	32.5−37.2
CHr (pg)	35.6	1.3	33.1−38.6
%HYPOm (%)	3.0	3.0	0.4−9.9
%HYPOr (%)	42.0	15.6	18.3−76.8
SERUM MEASUREMENTS			
TfR (mg/l)	2.0	0.7	1.2−4.0
Ferritin (μg/l)	198	137	45−636
TfR-F index	0.95	0.43	0.49−2.1
Iron (μmol/l)	27.4	7.7	12.2−42.1
Transferrin (g/l)	2.0	0.4	1.2−2.9
TfSat (%)	55	19	21−111

[a]For reference range calculations, only samples in which Hb was greater than 146 g/l were included.
CHm, cellular hemoglobin in red blood cells; CHr, cellular hemoglobin in reticulocytes; Hb, hemoglobin; HCT, hematocrit; %HYPOm, percentage of hypochromic red blood cells; %HYPOr, percentage of hypochromic reticulocytes; IRF-H, high immature reticulocyte fraction; MCH, mean cell hemoglobin; MCHC, mean cell hemoglobin concentration; MCV, mean cell volume of red blood cells; MCVr, mean cell volume of reticulocytes; %Retic, proportion of reticulocytes; TfR, transferrin receptor; TfR-F index, transferrin receptor/log (ferritin); TfSat, transferrin saturation.
From Ervasti, M., Kotisaari, S., Sankilampi, U., et al., 2007. The relationship between red blood cell and reticulocyte indices and serum markers of iron status in the cord blood of newborns. Clin. Chem. Lab. Med. 45, 1000−1003. Hematologic data obtained in 199 full-term newborn infants with an ADVIA 120 analyzer (Siemens Diagnostic Solutions, Terrytown, NY).

RED CELL VALUES AND RELATED SERUM VALUES

TABLE A1.5   Hemoglobin Concentrations (g/dl) for Iron-Sufficient Preterm Infants[a]

		Birth weight (g)	
Age	Number	1000—1500	1501—2000
2 weeks	17,39	16.3 (11.7—18.4)	14.8 (11.8—19.6)
1 month	15,42	10.9 (8.7—15.2)	11.5 (8.2—15.0)
2 months	17,47	8.8 (7.1—11.5)	9.4 (8.0—11.4)
3 months	16,41	9.8 (8.9—11.2)	10.2 (9.3—11.8)
4 months	13,37	11.3 (9.1—13.1)	11.3 (9.1—13.1)
5 months	8,21	11.6 (10.2—14.3)	11.8 (10.4—13.0)
6 months	9,21	12.0 (9.4—13.8)	11.8 (10.7—12.6)

[a]These infants were admitted to the Helsinki Children's Hospital during a 15-month period. None had a complicated course during the first 2 weeks of life or had undergone an exchange transfusion. All infants were iron sufficient, as indicated by a serum ferritin greater than 10 ng/ml.
From Lundstrom, U., Siimes, M.A., Dallman, P.R., 1997. At what age does iron supplementation become necessary in low-birth weight infants. J Pediatr. 91, 878, with permission.

TABLE A1.6   Red Cell Values on First Postnatal Day

	Gestational age (weeks)							
	24—25 (7)[a]	26—27 (11)	28—29 (7)	30—31 (25)	32—33 (23)	34—35 (23)	36—37 (20)	Term (19)
RBC $\times 10^6$	4.65[b] ± 0.43	4.73 ± 0.45	4.62 ± 0.75	4.79 ± 0.74	5.0 ± 0.76	5.09 ± 0.5	5.27 ± 0.68	5.14 ± 0.7
Hb (g/dl)	19.4 ± 1.5	19.0 ± 2.5	19.3 ± 1.8	19.1 ± 2.2	18.5 ± 2.0	19.6 ± 2.1	19.2 ± 1.7	19.3 ± 2.2
Hematocrit (%)	63 ± 4	62 ± 8	60 ± 7	60 ± 8	60 ± 8	61 ± 7	64 ± 7	61 ± 7.4
MCV (fl)	135 ± 0.2	132 ± 14.4	131 ± 13.5	127 ± 12.7	123 ± 15.7	122 ± 10.0	121 ± 12.5	119 ± 9.4
Reticulocytes (%)	6.0 ± 0.5	9.6 ± 3.2	7.5 ± 2.5	5.8 ± 2.0	5.0 ± 1.9	3.9 ± 1.6	4.2 ± 1.8	3.2 ± 1.4
Weight (g)	725 ± 185	993 ± 194	1174 ± 128	1450 ± 232	1816 ± 192	1957 ± 291	2245 ± 213	

[a]Number of infants.
[b]Mean values ± SD.
From Zaizov, R., Matoth, Y., 1976. Red cell values on the first postnatal day during the last 16 weeks of gestation. Am. J. Hematol. 1, 276, with permission.

TABLE A1.7   Mean Hematological Values in the First 2 Weeks of Life in the Term Infant

Hematological value	Cord blood	Day 1	Day 3	Day 7	Day 14
Hb (g/dl)	16.8	18.4	17.8	17.0	16.8
Hematocrit	0.53	0.58	0.55	0.54	0.52
Red cells (310^{12}/l)	5.25	5.8	5.6	5.2	5.1
MCV (fl)	107	108	99.0	98.0	96.0
MCH (pg)	34	35	33	32.5	31.5
MCHC (%)	31.7	32.5	33	33	33
Reticulocytes (%)	3—7	3—7	1—3	0—1	0—1
Nucleated RBC (mm^3)	500	200	0—5	0	0
Platelets (310^9/l)	290	192	213	248	252

From Oski, F.A., Naiman, J.L., 1982. Hematologic Problems in the Newborn, third ed. Saunders, Philadelphia, PA, with permission.

TABLE A1.8   Red Cell Values at Various Ages: Mean and Lower Limit of Normal (22 SD)[a]

Age	Hemoglobin (g/dl)		Hematocrit (%)		Red cell count (10^{12}/l)		MCV (fl)		MCH (pg)		MCHC (g/dl)		Reticulocytes	
	Mean	22 SD	Mean	22 SD	Mean	22 SD	Mean	22 SD	Mean	22 SD	Mean	22 SD	Mean	22 SD
Birth (cord blood)	16.5	13.5	51	42	4.7	3.9	108	98	34	31	33	30	3.2	1.8
1–3 days (capillary)	18.5	14.5	56	45	5.3	4.0	108	95	34	31	33	29	3.0	1.5
1 week	17.5	13.5	54	42	5.1	3.9	107	88	34	28	33	28	0.5	0.1
2 weeks	16.5	12.5	51	39	4.9	3.6	105	86	34	28	33	28	0.5	0.2
1 month	14.0	10.0	43	31	4.2	3.0	104	85	34	28	33	29	0.8	0.4
2 months	11.5	9.0	35	28	3.8	2.7	96	77	30	26	33	29	1.6	0.9
3–6 months	11.5	9.5	35	29	3.8	3.1	91	74	30	25	33	30	0.7	0.4
0.5–2 years	12.0	10.5	36	33	4.5	3.7	78	70	27	23	33	30	1.0	0.2
2–6 years	12.5	11.5	37	34	4.6	3.9	81	75	27	24	34	31	1.0	0.2
6–12 years	13.5	11.5	40	35	4.6	4.0	86	77	29	25	34	31	1.0	0.2
12–18 Years														
Female	14.0	12.0	41	36	4.6	4.1	90	78	30	25	34	31	1.0	0.2
Male	14.5	13.0	43	37	4.9	4.5	88	78	30	25	34	31	1.0	0.2
18–49 Years														
Female	14.0	12.0	41	36	4.6	4.0	90	80	30	26	34	31	1.0	0.2
Male	15.5	13.5	47	41	5.2	4.5	90	80	30	26	34	31	1.0	0.2

[a]*These data have been compiled from several sources. Emphasis is given to studies employing electronic counters and to the selection of populations that are likely to exclude individuals with iron deficiency. The mean ± 2 SD can be expected to include 95% of the observations in a normal population.*
From Dallman, P.R., 1997. Blood and blood-forming tissue. In: Rudolph, A. (Ed.), Pediatrics, sixteenth ed. Appleton-Cernuary-Croles, Norwalk, CT, with permission.

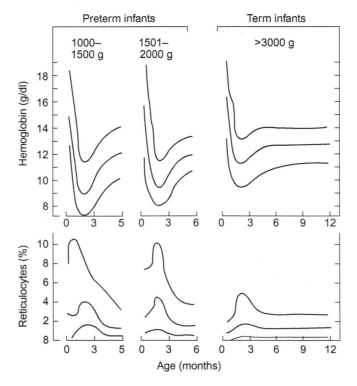

FIGURE A1.1   Physiologic nadir for term and preterm infants. The mean and range of normal hemoglobin and reticulocyte values for term and preterm infants are shown. Premature infants reach a nadir of erythrocyte production sooner and require longer to recover than their term infant counterparts. *Source: From Dallman, P.R., 1981. Anemia of prematurity. Ann. Rev. Med. 32, 143.*

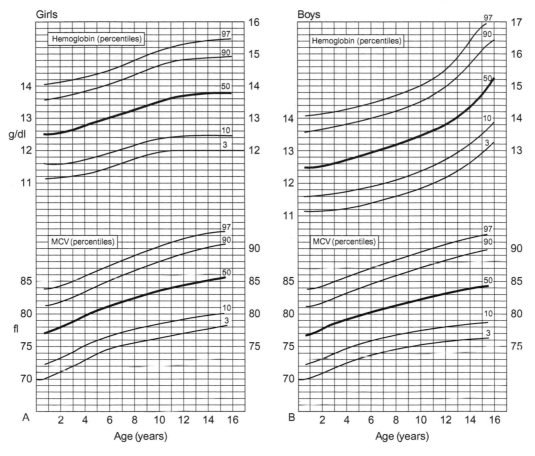

FIGURE A1.2 (A) Hemoglobin and MCV percentile curves for girls. (B) Hemoglobin and MCV percentile curves for boys. *Source: From Dallman, P.R., Siimes, M.A., 1979. Percentile curves for hemoglobin and red cell volume in infancy and childhood. J. Pediatr. 94, 28, with permission.*

TABLE A1.9 Serum Ferritin Values

Age	ng/ml
Newborn	25–200
1 month	200–600
2–5 months	50–200
6 months–15 years	7–140
Adult	
Male	15–200
Female	12–150

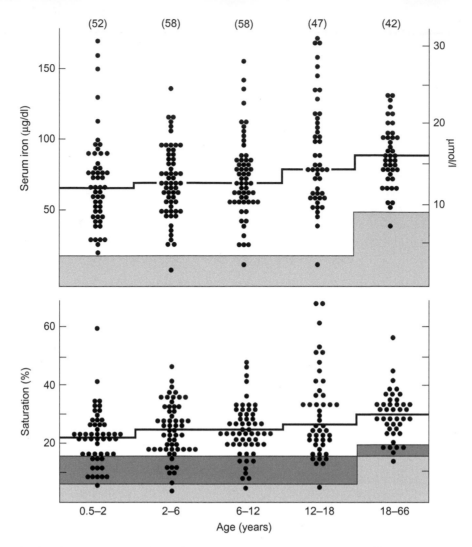

FIGURE A1.3 Normal values for serum iron and transferrin saturation. *From Koerper, M.A., Dallman, P., 1977. Serum iron concentration and transferrin saturation in the diagnosis of iron deficiency in children: normal developmental changes. J. Pediatr. 91, 870, with permission.*

TABLE A1.10 Values of Serum Iron (SI), Total Iron-Binding Capacity (TIBC), and Transferrin Saturation (S%) from Infants During the First Year of Life

		Age (months)						
		0.5	1	2	4	6	9	12
SI								
Median	μmol/l	22	22	16	15	14	15	14
		11−36	10−31	3−29	3−29	5−24	6−24	6−28
95% range	μg/dl	120	125	87	84	77	84	78
		63−201	58−172	15−159	18−164	28−135	34−135	35−155
TIBC								
Mean ± SD	μmol/l	34 ± 8	36 ± 8	44 ± 10	54 ± 7	58 ± 9	61 ± 7	64 ± 7
	μg/dl	191 ± 43	199 ± 43	246 ± 55	300 ± 39	321 ± 51	341 ± 42	358 ± 38
S%								
Median		68	63	34	27	23	25	23
95% range		30−99	35−94	21−63	7−53	10−43	10−39	10−47

Note: These data were obtained from a group of healthy, full-term infants who were born at the Helsinki University Central Hospital. Infants received iron supplementation in formula and cereal throughout the 12-month period. Infants with hemoglobin below 10 g/dl, mean corpuscular volume of red blood cells below 71 fl, or serum ferritin below 10 ng/ml were excluded from the study. The 95% range of the transferrin saturation values indicates that the lower limit of normal is about 10% after 4 months of age.

From Saarinen, U.M., Siimes, M.A., 1977. Serum iron and transferrin in iron deficiency. J. Pediatr. 91, 876, with permission.

TABLE A1.11   Mean Serum Iron and Iron Saturation Percentage

Age (years)	Serum iron (mg/dl)	Saturation (%)
0.5−2	68 ± 3.6 (16−120)	22 ± 1.1 (6−38)
2−6	72 ± 3.4 (20−124)	25 ± 1.2 (7−43)
6−12	73 ± 3.4 (23−123)	25 ± 1.2 (7−43)
181	92 ± 3.8 (48−136)	30 ± 1.1 (18−46)

From Koerper, M.A., Dallman, P.R., 1977. Serum iron concentration and transferrin saturation are lower in normal children than in adults. J. Pediatr. Res. 11, 473, with permission.

TABLE A1.12   Normal Serum Folic Acid Levels (ng/ml)

Folate	Age	Range	Mean ± SD
Serum folate	*Normal premature infants*		
	1−4 days	7.17−52.00	29.54 ± 0.98
	2−3 weeks	4.12−15.62	8.61 ± 0.55
	1−2 months	2.81−11.25	5.84 ± 0.35
	2−3 months	3.56−11.82	6.95 ± 0.50
	3−5 months	3.85−16.50	8.92 ± 0.86
	5−7 months	6.00−12.25	9.02 ± 0.74
	Normal children		
	1 year	3.0−35	9.3
	1−6 years	4.12−21.15	11.37 ± 0.82
	1−10 years	6.5−16.5	10.3
	Normal adults		
	20−45 years	4.50−28.00	10.29 ± 1.14
Red cell folate	Infants, 1 year	74−995	277
	Children, 1−11 years	96−364	215
	Adults	160−640	316
Whole blood folate	Infants, 1 year	20−160	87
	Infants, 1 year	31−400	86
	Infants, 2−24 months	34−160	96
	Children, 1−11 years	52−164	97
	Adults	50−400	195

From Shojania, A., Gross, S., 1964. Folic acid deficiency and prematurity. J. Pediatr. 64, 323, with permission.

TABLE A1.13   Percentage of Hemoglobin F (HbF) in the First Year of Life[a]

Age	Number tested	Mean	2 SD	Range
1–7 days	10	74.7	5.4	61–79.6
2 weeks	13	74.9	5.7	66–88.5
1 month	11	60.2	6.3	45.7–67.3
2 months	10	45.6	10.1	29.4–60.8
3 months	10	26.6	14.5	14.8–55.9
4 months	10	17.7	6.1	9.4–28.5
5 months	10	10.4	6.7	2.3–22.4
6 months	15	6.5	3.0	2.7–13.0
8 months	11	5.1	3.6	2.3–11.9
10 months	10	2.1	0.7	1.5–3.5
12 months	10	2.6	1.5	1.3–5.0
1–14 years and adults	100	0.6	0.4	–

[a]HbF measured by alkali denaturation.
From Schröter, W., Nafz, C., 1981. Diagnostic significance of hemoglobin F and A$_2$ levels in homo- and heterozygous beta-thalassemia during infancy. Helv. Paediatr. Acta. 36, 519.

TABLE A1.14   Percentage of Hemoglobin F and A$_2$ in Newborn and Adult

	Hemoglobin F (%)	Hemoglobin A$_2$ (%)
Newborn	60–90	1.0
Adult	1.0	1.6–3.5

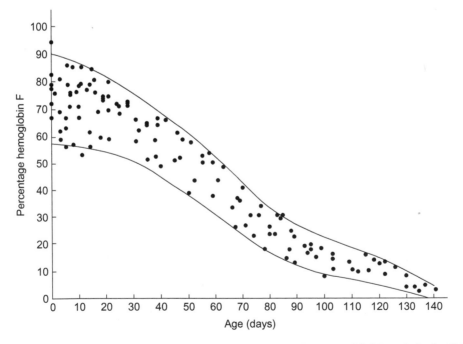

FIGURE A1.4   Relative concentration of hemoglobin F in infants and variation with age. *Modified from Garby, L., Sjolin, S., Vuille, J.C., 1962. Studies on erythro-kinetics in infancy. II The relative rate of synthesis of haemoglobin F and haemoglobin A during the first months of life. Acta Paediatr. 51, 245, with permission.*

TABLE A1.15   Estimated Blood Volumes

Age	Plasma volume (PV) (ml/kg)	Red cell mass (RCM) (ml/kg)	Total blood volume (ml/kg)	
			(From PV)	(From RCM)
Newborn	43.6	43.1	80.0	85.4
1–7 days	51–54	37.9	82.9	77.8
1–12 months	46.1	25.5	78.1	72.8
1–3 years	45.8	24.9	77.8	69.1
4–6 years	49.6	25.5	82.8	67.5
7–9 years	50.6	24.3	88.6	67.5
10–12 years	49.0	26.3	85.4	67.5
13–15 years	51.2		88.3	
16–18 years	50.1		90.2	
Adult	39–44	25–30	68–88	55–75

Modified from Price, D.C., Ries, C. In: Handmaker, H., Lowenstein, J.M. (Eds.), Nuclear Medicine in Clinical Pediatrics. Society of Nuclear Medicine, New York, p. 279.

TABLE A1.16   Methemoglobin Levels in Normal Children[a]

	Number of cases	Number of determinations	Methemoglobin (g/dl)			Number of cases	Number det.	Methemoglobin as percentage of total hemoglobin		
			Mean	Range	SD			Mean	Range	SD
Prematures (birth–7 days)	29	34	0.43	(0.02–0.83)	± 0.07	24	28	2.3	(0.08–4.4)	± 1.26
Prematures (7–72 days)	21	29	0.31	(0.02–0.78)	± 0.19	18	23	2.2	(0.02–4.7)	± 1.07
Prematures (total)	50	63	0.38	(0.02–0.83)	± 0.10	42	51	2.2	(0.08–4.7)	± 1.10
Cook Country Hospital prematures (1–14 days)	8	8	0.52	(0.18–0.83)	± 0.08	–	–	–	–	–
Newborns (1–10 days)	39	39	0.22	(0.00–0.58)	± 0.17	25	30	1.5	(0.00–2.8)	± 0.81
Infants (1 month–1 year)	8	8	0.14	(0.02–0.29)	± 0.09	8	8	1.2	(0.17–2.4)	± 0.78
Children (1–14 years)	35	35	0.11	(0.00–0.33)	± 0.09	35	35	0.79	(0.00–2.4)	± 0.62
Adults (14–78 years)	30	30	0.11	(0.00–0.28)	± 0.09	27	27	0.82	(0.00–1.9)	± 0.63

[a]*The premature and full-term infants were free of known disease. None had respiratory distress or cyanosis. Analysis of milk and water ingested by these infants revealed the nitrate level less than 0.027 ppm. The premature infants routinely received vitamin C orally each day from the seventh day of life.*
From Kravitz, H., Elegant, L.D., et al., 1956. Methemoglobin values in premature and mature infants and children. Am. J. Dis. Child 91, 2, with permission.

TABLE A1.17   Serum Erythropoietin Levels

RIA method	5–20 mU/ml
Hemagglutination method	25–125
Bioassay method	5–18

TABLE A1.18   Comparison of Enzyme Activities and Glutathione Content in Newborn and Adult Red Blood Cells

Enzyme	Activity in normal adult RBC in IU/g Hb (mean ± 1 SD at 37°C)	Mean activity in newborn RBC as percentage of mean (100%) activity in normal adult RBC
Aldolase	3.19 ± 0.86	140
Enolase	5.39 ± 0.83	250
Glucose phosphate isomerase	60.8 ± 11.0	162
Glucose-6-phosphate dehydrogenase	8.34 ± 1.59	
WHO method	12.1 ± 2.09	174
Glutathione peroxidase	30.82 ± 4.65	56
Glyceraldehyde phosphate dehydrogenase	226 ± 41.9	170
Hexokinase	1.78 ± 0.38	239
Lactate dehydrogenase	200 ± 26.5	132
NADH-methemoglobin reductase	19.2 ± 3.85 (at 30°C)	Increased
Phosphofructokinase	11.01 ± 2.33	97
Phosphoglycerate kinase	320 ± 36.1	165
Pyruvate kinase	15.0 ± 1.99	160
6-Phosphogluconate dehydrogenase	8.78 ± 0.78	150
Triose phosphate isomerase	211 ± 397	101
Glutathione	6570 ± 1040 nmol/g Hb	156

Note: The percentage activity in newborn RBC compared to mean adult (100%) values is presented with quantitative data from studies on adult RBC. Newborn data from Konrad et al. (1972); quantitative data from Beutler (1984).
From Hinchliffe, R.F., Lilleyman, J.S. (Eds.), 1987. Practical paediatric haematology: a laboratory worker's guide to blood disorders in children. Wiley, New York, with permission.

TABLE A1.19   Polymorphonuclear Leukocyte and Band Counts in the Newborn During the First 2 Days of Life[a]

Age (h)	Absolute neutrophil count (mm^3)	Absolute band count (mm^3)	B/N ratio
0	3,500–6,000	1,300	0.14
12	8,000–15,000	1,300	0.14
24	7,000–13,000	1,300	0.14
36	5,000–9,000	700	0.11
48	3,500–5,200	700	0.11

[a]*Normal values were obtained from the assessment of 3100 separate white blood cell counts obtained from 965 infants; 513 counts were from infants considered to be completely normal at the time the count was obtained and for the preceding and subsequent 48 h. There was no difference in the normal ranges when infants were compared by either birth weight (∼2500 g) or gestational age.*
From Manroe, B.L., Browne, R., et al., 1976. Normal leukocyte (WBC) values in neonates. Pediatr. Res. 10, 428, with permission.

WHITE CELL VALUES

TABLE A1.20 Normal Leukocyte Counts[a]

Age	Total leukocytes Mean	(Range)	Neutrophils Mean	(Range)	%	Lymphocytes Mean	Range	%	Monocytes Mean	%	Eosinophils Mean	%
Birth	18.1	(9.0–30.0)	11.0	(6.0–26.0)	61	5.5	(2.0–11.0)	31	1.1	6	0.4	2
12 hs	22.8	(13.0–38.0)	15.5	(6.0–28.0)	68	5.5	(2.0–11.0)	24	1.2	5	0.5	2
24 h	18.9	(9.4–34.0)	11.5	(5.0–21.0)	61	5.8	(2.0–11.5)	31	1.1	6	0.5	2
1 week	12.2	(5.0–21.0)	5.5	(1.5–10.0)	45	5.0	(2.0–17.0)	41	1.1	9	0.5	4
2 weeks	11.4	(5.0–20.0)	4.5	(1.0–9.5)	40	5.5	(2.0–17.0)	48	1.0	9	0.4	3
1 month	10.8	(5.0–19.5)	3.8	(1.0–9.0)	35	6.0	(2.5–16.5)	56	0.7	7	0.3	3
6 months	11.9	(6.0–17.5)	3.8	(1.0–8.5)	32	7.3	(4.0–13.5)	61	0.6	5	0.3	3
1 year	11.4	(6.0–17.5)	3.5	(1.5–8.5)	31	7.0	(4.0–10.5)	61	0.6	5	0.3	3
2 years	10.6	(6.0–17.0)	3.5	(1.5–8.5)	33	6.3	(3.0–9.5)	59	0.5	5	0.3	3
4 years	9.1	(5.5–15.5)	3.8	(1.5–8.5)	42	4.5	(2.0–8.0)	50	0.5	5	0.3	3
6 years	8.5	(5.0–14.5)	4.3	(1.5–8.0)	51	3.5	(1.5–7.0)	42	0.4	5	0.2	3
8 years	8.3	(4.5–13.5)	4.4	(1.5–8.0)	53	3.3	(1.5–6.8)	39	0.4	4	0.2	2
10 years	8.1	(4.5–13.5)	4.4	(1.8–8.0)	54	3.1	(1.5–6.5)	38	0.4	4	0.2	2
16 years	7.8	(4.5–13.0)	4.4	(1.8–8.0)	57	2.8	(1.2–5.2)	35	0.4	5	0.2	3
21 years	7.4	(4.5–11.0)	4.4	(1.8–7.7)	59	2.5	(1.0–4.8)	34	0.3	4	0.2	3

[a]*Numbers of leukocytes are in thousands per mm³, ranges are estimates of 95% confidence limits and percentages refer to differential counts. Neutrophils include band cells at all ages and a small number of metamyelocytes and myelocytes in the first few days of life.*

From Dallman, P.R., 1977. Blood and blood-forming tissues. In: Rudolpg, A.M. (Ed.), Pediatrics, sixteenth ed. Appleton-Century-Crofts, Norwalk, CT, with permission.

TABLE A1.21 Lymphocyte Subsets (25th, 50th, and 75th centiles) in Children and Adults

Subset	Percentage T cells			Percentage B cells			Percentage natural killer (NK) cells		
CD Numbers Reagent	CD3 Leu 4			CD19 Leu 12			CD16[1]/CD56[1], CD3[2] Leu 11[1]/19[1], Leu 4[2]		
Percentile	P_{25}	P_{50}	P_{75}	P_{25}	P_{50}	P_{75}	P_{25}	P_{50}	P_{75}
Age group	Number								
Cord blood	24	49	55	62	14	20	23	14	19.5	30
1 day–11 months	31	55	60	67	19	25	29	11	15	19
1–6 years	54	62	64	69	21	25	28	8	11	15
7–17 years	31	64	70	74	12	16	23	8.5	11	15.5
18–70 years	300	67	72	76	11	13	16	10	14	19

Subset	Percentage T helper cells			Percentage T suppressor cells			Helper/suppressor (CD4/CD8) ratio		
CD number Reagent	CD4 Leu 3			CD8 Leu 2			CD4 Leu 3	CD8 Leu 2	
Percentile	P_{25}	P_{50}	P_{75}	P_{25}	P_{50}	P_{75}	P_{25}	P_{50}	P_{75}
Age group	Number								
Cord blood	24	28	35	42	26	29	33	0.80	1.15	1.75
1 day–11 months	31	35	41	48	18	23	28	1.35	1.80	2.25
1–6 years	54	32	37	40	25	32	36	1.00	1.20	1.60
7–17 years	31	33	36	40	28	31	36	1.05	1.20	1.40
18–70 years	300	38	42	46	31	35	40	1.00	1.20	1.50

Data obtained using flow cytometry. Kindly supplied by Dr F Hulstaert, Becton Dickinson Immunocytometry Systems Medical Department Cockeysville, MD.

PLATELET VALUES

TABLE A1.22 Normal Platelet Counts

Subject	Platelet count/mm^3 (mean $\pm$ 1 SD)
Preterm, 27–31 weeks	275,000 $\pm$ 60,000
Preterm, 32–36 weeks	290,000 $\pm$ 70,000
Term infants	310,000 $\pm$ 68,000
Normal adult or child	300,000 $\pm$ 50,000

From Oski, F.A., Naiman, J.L., 1982. Normal blood values in the newborn period. In: Hematologic Problems in the Newborn. Saunders, Philadelphia, PA, with permission.

COAGULATION VALUES

TABLE A1.23 Reference Values for Coagulation Tests in Healthy Full-Term Infant During First 6 Months of Life

	Day 1		Day 5		Day 30		Day 90		Day 180		Adult	
	M	B	M	B	M	B	M	B	M	B	M	B
PT (s)	13.0	(10.1–15.9)[a]	12.4	(10.0–15.3)[a]	11.8	(10.0–14.3)[a]	11.9	(10.0–14.2)[a]	12.3	(10.7–13.9)[a]	12.4	(10.8–13.9)
INR	1.00	(0.53–1.62)	0.89	(0.53–1.48)	0.79	(0.53–1.26)	0.81	(0.53–1.26)	0.88	(0.61–1.17)	0.89	(0.64–1.17)
APTT (s)	42.9	(31.3–54.5)	42.6	(25.4–59.8)	40.4	(32.0–55.2)	37.1	(29.0–50.1)[a]	35.5	(28.1–42.9)[a]	33.5	(26.6–10.3)
TCT (s)	23.5	(19.0–28.3)[a]	23.1	(18.0–29.2)	24.3	(19.4–29.2)[a]	25.1	(20.5–29.7)[a]	25.5	(19.8–31.2)[a]	25.0	(19.7–30.3)
Fibrinogen (g/dl)	2.83	(1.67–3.99)[a]	3.12	(1.62–4.62)[a]	2.70	(1.62–3.78)[a]	2.43	(1.50–3.79)[a]	2.51	(1.50–3.87)[a]	2.78	(1.56–4.00)
II (units/ml)	0.48	(0.26–0.70)	0.63	(0.33–0.93)	0.68	(0.34–1.02)	0.75	(0.45–1.05)	0.88	(0.60–1.16)	1.08	(0.70–1.46)
V (units/ml)	0.72	(0.34–1.08)	0.95	(0.45–1.45)	0.98	(0.62–1.34)	0.90	(0.48–1.32)	0.91	(0.55–1.27)	1.06	(0.62–1.50)
VII (units/ ml)	0.66	(0.28–1.04)	0.89	(0.35–1.43)	0.90	(0.42–1.38)	0.91	(0.39–1.43)	0.87	(0.47–1.27)	1.05	(0.67–1.43)
VIII (units /ml)	1.00	(0.50–1.78)[a]	0.88	(0.50–1.54)[a]	0.91	(0.50–1.57)[a]	0.79	(0.50–1.25)[a]	0.73	(0.50–1.09)	0.99	(0.50–1.49)
vWF (units/ml)	1.53	(0.50–2.87)	1.40	(0.50–2.54)	1.28	(0.50–2.46)	1.18	(0.50–2.06)	1.07	(0.50–1.97)	0.92	(0.50–1.58)
IX (units/ml)	0.53	(0.15–0.91)	0.53	(0.15–0.91)	0.51	(0.21–0.81)	0.67	(0.21–1.13)	0.86	(0.36–1.36)	1.09	(0.55–1.63)
X (units/ml)	0.40	(0.12–0.68)	0.49	(0.19–0.79)	0.59	(0.31–0.87)	0.71	(0.35–1.07)	0.78	(0.38–1.18)	1.06	(0.70–1.52)
XI (units/ml)	0.38	(0.10–0.66)	0.55	(0.23–0.87)	0.53	(0.27–0.79)	0.69	(0.41–0.97)	0.86	(0.49–1.34)	0.97	(0.67–1.27)
XII (units/ml)	0.53	(0.13–0.93)	0.47	(0.11–0.83)	0.49	(0.17–0.81)	0.67	(0.25–1.09)	0.77	(0.39–1.15)	1.08	(0.52–1.64)
PK (units/ml)	0.37	(0.18–0.69)	0.48	(0.20–0.76)	0.57	(0.23–0.91)	0.73	(0.41–1.05)	0.86	(0.56–1.16)	1.12	(0.62–1.62)
HMW-K (units/ml)	0.54	(0.06–1.02)	0.74	(0.16–1.32)	0.77	(0.33–1.21)	0.82	(0.30–1.46)[a]	0.82	(0.36–1.28)[a]	0.92	(0.50–1.36)
XIII[a] (units/ml)	0.79	(0.27–1.31)	0.94	(0.44–1.44)[a]	0.93	(0.39–1.47)[a]	1.04	(0.36–1.72)[a]	1.04	(0.46–1.62)[a]	1.05	(0.55–1.55)
XIII (units/ml)	0.76	(0.30–1.22)	1.06	(0.32–1.80)	1.11	(0.39–1.73)[a]	1.16	(0.48–1.84)[a]	1.10	(0.50–1.70)[a]	0.97	(0.57–1.37)

[a]*Values are indistinguishable from the adult.*
PT, prothrombin time; INR, international normalized ratio; APTT, activated partial thromboplastin time; TCT, thrombin clotting time; VII, factor VIII procoagulant; vWF, von Willebrand's factor; PK, prekallikrein; HMW-K, high molecular weight kininogen.
All factors except fibrinogen are expressed as units per milliliter (units/ml) where pooled plasma contains 1.0 unit/ml. All values are expressed as mean (M) followed by the lower and upper boundaries (B) encompassing 95% of the population. Between 40 and 77 samples were assayed for each value for the newborn. Some measurements were skewed due to a disproportionate number of high values. The lower limit, which excludes the lower 2.5% of the population, has been given.
From Andrew, M., 1991. Bailliere's Clin. Hematol. 4, 251, with permission.

TABLE A1.24 Reference Values for Coagulation Tests in Healthy Premature Infants (30–36 Weeks' Gestation) During First 6 Months

	Day 1		Day 5		Day 30		Day 90		Day 180		Adult	
	M	B	M	B	M	B	M	B	M	B	M	B
PT (s)	13.0	(10.6–16.2)[b]	12.5	(10.0–15.3)[b,c]	11.8	(10.0–13.6)[b]	12.3	(10.0–14.6)[b]	12.5	(10.0–15.0)[b]	12.4	(10.8–13.9)
APTT (s)	53.6	(27.5–79.4)[c]	50.5	(26.9–74.1)[c]	44.7	(26.9–62.5)	39.5	(28.3–50.7)	37.5	(21.7–53.3)[b]	33.5	(26.6–40.3)
TCT (s)	24.8	(19.2–30.4)[b]	24.1	(18.8–29.4)[b,c,d]	24.4	(18.8–29.9)[b]	25.1	(19.4–30.8)[b]	25.2	(18.9–31.5)[b]	25.0	(19.7–30.3)
Fibrinogen (g/l)	2.43	(1.50–3.73)[b,c,d]	2.80	(1.60–4.18)[c]	2.54	(1.50–4.14)[b,c]	2.46	(1.50–3.52)[b,c]	2.28	(1.50–3.60)[d]	2.78	(1.56–4.00)
II (units/ml)	0.45	(0.20–0.77)[c]	0.57	(0.29–0.85)	0.57	(0.36–0.95)[c,d]	0.68	(0.30–1.06)	0.87	(0.51–1.23)	1.08	(0.70–1.46)
V (units/ml)	0.88	(0.41–1.44)[b,c,d]	1.00	(0.46–1.54)	1.02	(0.48–1.56)	0.99	(0.59–1.39)	1.02	(0.58–1.46)	1.06	(0.62–1.50)
VII (units/ml)	0.67	(0.21–1.13)	0.84	(0.30–1.38)	0.83	(0.21–1.45)	0.87	(0.31–1.43)	0.99	(0.47–1.51)[b]	1.05	(0.67–1.43)
VIII (units/ml)	1.11	(0.50–2.13)[b,c]	1.15	(0.53–2.05)[b,c,d]	1.11	(0.50–1.99)[b,c,d]	1.06	(0.58–1.88)[b,c,d]	0.99	(0.50–1.87)[b,c,d]	0.99	(0.50–1.49)
vWF (units/ml)	1.36	(0.78–2.10)[b]	1.33	(0.72–2.19)[c]	1.36	(0.66–2.16)[c]	1.12	(0.75–1.84)[b,c]	0.98	(0.54–1.58)[b,c]	0.92	(0.50–1.58)
IX (units/ml)	0.35	(0.19–0.65)[b,c,d]	0.42	(0.14–0.74)[c,d]	0.44	(0.13–0.80)[c]	0.59	(0.25–0.93)	0.81	(0.50–1.20)[c]	1.09	(0.55–1.63)
X (units/ml)	0.41	(0.11–0.71)	0.51	(0.19–0.83)	0.56	(0.20–0.92)	0.67	(0.35–0.99)	0.77	(0.35–1.19)	1.06	(0.70–1.52)
XI (units/ml)	0.30	(0.08–0.52)[c,d]	0.41	(0.13–0.69)[d]	0.43	(0.15–0.71)[d]	0.59	(0.25–0.93)[d]	0.78	(0.46–1.10)	0.97	(0.67–1.27)
XII (units/ml)	0.38	(0.10–0.66)[d]	0.39	(0.09–0.69)[d]	0.43	(0.11–0.75)	0.61	(0.15–1.07)	0.82	(0.22–1.42)	1.08	(0.52–1.64)
PK (units/ml)	0.33	(0.09–0.57)	0.45	(0.28–0.75)[c]	0.59	(0.31–0.87)	0.79	(0.37–1.21)	0.78	(0.40–1.16)	1.12	(0.62–1.62)
HMWK (units/ml)	0.49	(0.09–0.89)	0.62	(0.24–1.00)[d]	0.64	(0.16–1.12)[d]	0.78	(0.32–1.24)	0.83	(0.41–1.25)[b]	0.92	(0.50–1.36)
XIII[a] (units/ml)	0.70	(0.32–1.08)	1.01	(0.57–1.45)[b]	0.99	(0.51–1.47)[b]	1.13	(0.71–1.55)[b]	1.13	(0.65–1.61)[b]	1.05	(0.55–1.55)
XIII[b] (units/ml)	0.81	(0.35–1.27)	1.10	(0.68–1.58)[b]	1.07	(0.57–1.57)[b]	1.21	(0.75–1.67)	1.15	(0.67–1.63)	0.97	(0.57–1.37)
Plasminogen (CTPA, units/ml)	1.70	(1.12–2.48)[c,d]	1.91	(1.21–2.61)[d]	1.81	(1.09–2.53)	2.38	(1.58–3.18)	2.75	(1.91–3.59)[d]	3.36	(2.48–4.24)

[a] All values are given as a mean (M) followed by the lower and upper boundaries (B) encompassing 95% of the population. Between 40 and 96 samples were assayed for each value for newborns.
[b] Values indistinguishable from the adult.
[c] Measurements are skewed owing to a disproportionate number of high values. Lower limit that excludes the lower 2.5% of the population is given (B).
[d] Values differ from those of full-term infants.

From Andrew, M., Paes, B., Milner, R., et al., 1988. Development of the human coagulation system in the healthy premature infant. Blood 72, 1651, with permission.

TABLE A1.25   Reference Values for the Inhibitors of Coagulation in the Healthy Full-Term Infant During the First 6 Months of Life[a]

Inhibitors	Day 1 (n)	Day 5 (n)	Day 30 (n)	Day 90 (n)	Day 180 (n)	Adult (n)
AT-III (units/ml)	0.63 ± 0.12 (58)	0.67 ± 0.13 (74)	0.78 ± 0.15 (66)	0.97 ± 0.12 (60)[b]	1.04 ± 0.10 (56)[b]	1.05 ± 0.13 (28)
α_2-M (units/ml)	1.39 ± 0.22 (54)	1.48 ± 0.25 (73)	1.50 ± 0.22 (61)	1.76 ± 0.25 (55)	1.91 ± 0.21 (55)	0.86 ± 0.17 (29)
α_2-AP (units/ml)	0.85 ± 0.15 (55)	1.00 ± 0.15 (75)[b]	1.00 ± 0.12 (62)[b]	1.08 ± 0.16 (55)[b]	1.11 ± 0.14 (53)	1.02 ± 0.17 (29)
C_1E-INH (units/ml)	0.72 ± 0.18 (59)	0.90 ± 0.15 (76)[b]	0.89 ± 0.21 (63)	1.15 ± 0.22 (55)	1.41 ± 0.26 (55)[b]	1.01 ± 0.15 (29)
α_1-AT (units/ml)	0.93 ± 0.22 (57)[b]	0.89 ± 0.20 (75)[b]	0.62 ± 0.13 (61)	0.72 ± 0.15 (56)	0.77 ± 0.15 (55)	0.93 ± 0.19 (29)
HCII (units/ml)	0.43 ± 0.25 (56)	0.48 ± 0.24 (72)	0.47 ± 0.20 (58)	0.72 ± 0.37 (58)	1.20 ± 0.35 (55)	0.96 ± 0.15 (29)
Protein C (units/ml)	0.35 ± 0.09 (41)	0.42 ± 0.11 (44)	0.43 ± 0.11 (43)	0.54 ± 0.13 (44)	0.59 ± 0.11 (52)	0.96 ± 0.16 (28)
Protein S (units/ml)	0.36 ± 0.12 (40)	0.50 ± 0.14 (48)	0.63 ± 0.15 (41)	0.86 ± 0.16 (46)[b]	0.87 ± 0.16 (49)[b]	0.92 ± 0.16 (29)

[a]All values expressed in units/ml as mean ± 1 SD; n, numbers studied.
[b]Values indistinguishable from those of adults.
AT-III, antithrombin III; α_2-M, alpha$_2$-macroglobulin; α_2-AP, alpha$_2$-antiplasmin; C_1 E-INH, C_1-esterase inhibitor; α_1-AT, alpha$_1$-antitrypsin; HCII, heparin cofactor II.
From Andrew, M., Paes, B., Milner, R., et al., 1988. Development of the human coagulation system in the healthy premature infant. Blood 72, 1651, with permission.

TABLE A1.26   Reference Values for Inhibitors of Coagulation Healthy Premature Infants During First 6 Months of Life[a]

	Day 1		Day 5		Day 30		Day 90		Day 180		Adult	
	M	B	M	B	M	B	M	B	M	B	M	B
AT-III (units/ml)	0.38	(0.14−0.62)[d]	0.56	(0.30−0.82)[b]	0.59	(0.37−0.81)[d]	0.83	(0.45−1.21)[d]	0.90	(0.52−1.28)[d]	1.05	(0.79−1.31)
α_2-M (units/ml)	1.10	(0.56−1.82)[c,d]	1.25	(0.71−1.77)[b]	1.38	(0.72−2.04)	1.80	(1.20−2.66)[c]	2.09	(1.10−3.21)[c]	0.86	(0.52−1.20)
α_2-AP (units/ml)	0.78	(0.40−1.16)	0.81	(0.49−1.13)[b]	0.89	(0.55−1.23)[d]	1.06	(0.64−1.48)[b]	1.15	(0.77−1.53)	1.02	(0.68−1.36)
C_1E-INH (units/ml)	0.65	(0.31−0.99)	0.83	(0.45−1.21)	0.74	(0.40−1.24)[c,d]	1.14	(0.60−1.68)	1.40	(0.96−2.04)[c]	1.01	(0.71−1.31)
α_1-AT (units/ml)	0.90	(0.36−1.44)[b]	0.94	(0.42−1.46)[d]	0.76	(0.38−1.12)[d]	0.81	(0.49−1.13)[b,d]	0.82	(0.48−1.16)[b]	0.93	(0.55−1.31)
HCII (units/ml)	0.32	(0.00−0.80)[d]	0.34	(0.00−0.69)	0.43	(0.15−0.71)	0.61	(0.20−1.11)[c]	0.89	(0.45−1.40)[b,c,d]	0.96	(0.66−1.26)
Protein C (units/ml)	0.28	(0.12−0.44)[c]	0.31	(0.11−0.51)	0.37	(0.15−0.59)[d]	0.45	(0.23−0.67)[d]	0.57	(0.31−0.83)	0.96	(0.64−1.28)
Protein S (units/ml)	0.26	(0.14−0.38)[d]	0.37	(0.13−0.61)	0.56	(0.22−0.90)	0.76	(0.40−1.12)[d]	0.82	(0.44−1.20)	0.92	(0.60−1.24)

[a]All values expressed in units/ml were pooled plasma contains 1.0 unit/ml. All values are given as mean (M) followed by the lower and upper boundaries (B) encompassing 95% of the population. Between 40 and 75 samples were assayed for each value for the newborn.
[b]Values are indistinguishable from the adults.
[c]Measurements are skewed owing to a disproportionate number of high values. Lower limit which excludes the lower 2.5% of the population is given (B).
[d]Values differ from those of full-term infants.
From Andrew, M., Paes, B., Milner, R., et al., 1988. Development of the human coagulation system in the healthy premature infant. Blood 72, 1651, with permission.

TABLE A1.27 Reference Values for the Components of the Fibrinolytic System in the Healthy Full-Term Infant and the Healthy Premature infant During the First 6 Months of Life

Fibrinolytic system	Day 1 M	Day 1 B	Day 5 M	Day 5 B	Day 30 M	Day 30 B	Day 90 M	Day 90 B	Day 180 M	Day 180 B	Adult M	Adult B
HEALTHY FULL-TERM INFANT												
Plasminogen (units/ml)	1.95	(1.25−2.65)	2.17	(1.41−2.93)	1.98	(1.26−2.70)	2.48	(1.74−3.22)	3.01	(2.21−3.81)	3.36	(2.48−4.24)
t-PA (ng/ml)	9.60	(5.00−18.9)	5.60	(4.00−10.0)[a]	4.10	(1.00−6.00)[a]	2.10	(1.00−5.00)[a]	2.80	(1.00−6.00)[a]	4.90	(1.40−8.40)
α_2 AP (units/ml)	0.85	(0.55−1.15)	1.00	(0.70−1.30)[a]	1.00	(0.76−1.24)[a]	1.08	(0.76−1.40)[a]	1.11	(0.83−1.39)[a]	1.02	(0.68−1.36)
PAI (units/ml)	6.40	(2.00−15.1)	2.30	(0.00−8.10)[a]	3.4	(0.00−8.80)[a]	7.20	(1.00−15.3)	8.10	(6.00−13.0)	3.6	(0.00−11.0)
HEALTHY PREMATURE INFANT												
Plasminogen (units/ml)	1.70	(1.12−2.48)[b]	1.91	(1.21−2.61)[b]	1.81	(1.09−2.53)	2.38	(1.58−3.18)	2.75	(1.91−3.59)[b]	3.36	(2.48−4.24)
t-PA (ng/ml)	8.48	(3.00−16.70)	3.97	(2.00−6.93)[a]	4.13	(2.00−7.79)[a]	3.31	(2.00−5.07)[a]	3.48	(2.00−5.85)[a]	4.96	(1.46−8.46)
α_2 AP (units/ml)	0.78	(0.40−1.16)	0.81	(0.49−1.13)[b]	0.89	(0.55−1.23)[b]	1.06	(0.64−1.48)[a]	1.15	(0.77−1.53)	1.02	(0.68−1.36)
PAI (units/ml)	5.40	(0.00−12.2)[a,b]	2.50	(0.00−7.10)[a]	4.30	(0.00−11.8)[a]	4.80	(1.00−10.2)[a,b]	4.90	(1.00−10.2)[a,b]	3.60	(0.00−11.0)

[a]Values that are indistinguishable from those of the adult.
[b]Values that are different from those of the full-term infant.
For α_2 AP, values are expressed as units per milliliter (units/ml) were pooled plasma contains 1.0 unit/ml. Plasminogen units are those recommended by the Committee on Thrombolytic Agents. Values for t-PA are given as nanograms per milliliter. Values for PAI are given as units per milliliter; one unit of PAI activity is defined as the amount of PAI that inhibits one international unit of human single-chain TA. All values are given a mean (M) followed by the lower and upper boundaries (B) encompassing 95% of the population.
t-PA, tissue plasminogen activator; α_2 AP, α_2 antiplasmin; PAI, plasminogen activator inhibitor.
From Andrew, M., Paes, B., Johnston, M., et al., 1990. Development of the hemostatic system in the neonate and young infant. Am. J. Pediatr. Hematol. Oncol. 12, 95−104.

TABLE A1.28 Bleeding Time (min) in Children and Adults[a]

Subjects	Number tested	Mean	SD	Range
Children	36	4.6	1.4	2.5−8.5
Adults	48	4.6	1.2	2.5−6.5

[a]Bleeding time performed using a template technique, with incision 6 mm long and 1 mm deep and sphygmomanometer pressure 40 mmHg. Using this technique, bleeding times up to 11.5 min have been observed in apparently healthy children tested in the author's laboratory.
From Buchanan, G.R., Holtkamp, C.A., 1980. Prolonged bleeding time in children and young adults with hemophilia. Pediatrics 66, 951.

TABLE A1.29 Bleeding Time (min) in Newborns and Children[a]

Subjects	Number tested	Sphygmomanometer pressure (mmHg)	Mean	Range
Term newborn	30	30	3.4	1.9−5.8
Preterm Newborn				
1,000 g	6	20	3.3	2.6−4.0
1000−2000 g	15	25	3.9	2.0−5.6
2000 g	5	30	3.2	2.3−5.0
Children	17	30	3.4	1.0−5.5
Adults	20	30	2.8	0.5−5.5

[a]Bleeding time performed using a template technique, with incision 5 mm long and 0.5 mm deep. All subjects had normal platelet counts.
From Feusner, J.H., 1980. Normal and abnormal bleeding times in neonates and young children utilizing a fully standardized template technique. Am. J. Clin. Pathol. 74, 73.

TABLE A1.30 Coagulation Factor Assays (Mean ± 1 SD) and Screening Tests in the Fetus and Neonate[a]

Assays of coagulation factors	Normal adult values	28–31 weeks' gestation	32–36 weeks' gestation	Term	Time at which values attain adult norms
Fibrinogen (mg/dl)	150–400	215 ± 28 (SE)	226 ± 23 (SE)	246 ± 18 (SE)	b
		270 ± 85	244 ± 55	246 ± 55	2–12 months
II (%)	100	30 ± 10	35 ± 12	45 ± 15	2–12 months
V (%)	100	76 ± 7 (SE)	84 ± 9 (SE)	100 ± 5 (SE)	b
		90 ± 26	72 ± 23	98 ± 40	
VII and X (%)	100	38 ± 14	40 ± 15	56 ± 16	2–12 months
VIII (%)	100	90 ± 15 (SE)	140 ± 10 (SE)	168 ± 12 (SE)	b
		70 ± 30	98 ± 40	105 ± 34	
IX (%)	100	27 ± 10	NA	28 ± 8	3–9 months
XI (%)	100	5–18	NA	29–70	1–2 months
XII (%)	100	NA	30 ±	51 (25–70)	9–14 days
XIII	100	100	100	100	b
Bioassay (%)					
Quantitative (units/ml)	21 ± 5.6	5 ± 3.5	NA	11 ± 3.4	3 weeks
Prothrombin time (s)[c]	12–14	23 ±	17 (12–21)	16 (13–20)	1 week
Activated partial thromboplastin time (s)[c]	44	NA	70 ±	55 ± 10	2–9 months
Thrombin time (s)[c]	10	16–28	14 (11–17)	12 (10–16)	Few days

[a] Assays quoted are biologic unless otherwise specified.
[b] Adult levels attained prenatally.
[c] Values vary among laboratories, depending on reagents employed.
SE, standard error; NA, not available.
From Hathaway, W.E., 1975. The bleeding newborn. Semin. Hematol. 12, 175 and from Gross, S.J., Stuart, M.J., 1977. Hemostasis in the premature infant. Clin. Perinatol. 4, 260, with permission.

TABLE A1.31 Vascular—Platelet Interactions in the Fetus and Neonate

Vascular—platelet interactions	Normal adult values	27—31 weeks' gestation	32—36 weeks' gestation	Term	Term infant, 1—2 months
Capillary fragility	N	Increased	N	N	N
Platelet count (10^3/mm^3)	300 ± 50	275 ± 60	290 ± 70	310 ± 68	280 ± 56
Platelet retention (%)	N	NA	NA	N or decreased	N
Platelet aggregation with ADP, epinephrine collagen	N	Abn	Abn	Abn	
Platelet aggregation with ristocetin	N	NA	NA	N or increased	N
Platelet release I (adenide nucleotides)	N	Abn	Abn	Abn	
Platelet factor 3	N	Abn	Abn	Abn	
Platelet factor 4	N	NA	NA	N	N
Bleeding time (min)	4.0 ± 1.5		4 ± 1.5	4 ± 1.5	N

N, normal; Abn, abnormal; NA, not available.
From Hathaway, W.E., 1975. The bleeding newborn. Semin. Hematol. 12, 175 and from Gross, S.J., Stuart, M.J., 1977. Hemostasis in the premature infant. Clin. Perinatol. 4, 260, with permission.

BONE MARROW CELLS

TABLE A1.32 Bone Marrow Cell Populations of Normal Infants[a]

Cell type	Month				
	0 (n557)[b]	1 (n571)	2 (n548)	3 (n524)	4 (n519)
Small lymphocytes	14.42 ± 5.54	47.05 ± 9.24	42.68 ± 7.90	43.63 ± 11.83	47.06 ± 8.77
Transitional cells	1.18 ± 1.13	1.95 ± 0.94	2.38 ± 1.35	2.17 ± 1.64	1.64 ± 1.01
Proerythroblasts	0.02 ± 0.06	0.10 ± 0.14	0.13 ± 0.19	0.10 ± 0.13	0.05 ± 0.10
Basophilic erythroblasts	0.24 ± 0.25	0.34 ± 0.33	0.57 ± 0.41	0.40 ± 0.33	0.24 ± 0.24
Early erythroblasts	0.27 ± 0.26	0.44 ± 0.42	0.71 ± 0.51	0.50 ± 0.38	0.28 ± 0.30
Polychromatic erythroblasts	13.06 ± 6.78	6.90 ± 4.45	13.06 ± 3.48	10.51 ± 3.39	6.84 ± 2.58
Orthochromatic erythroblasts	0.69 ± 0.73	0.54 ± 1.88	0.66 ± 0.82	0.70 ± 0.87	0.34 ± 0.30
Extruded nuclei	0.47 ± 0.46	0.16 ± 0.17	0.26 ± 0.22	0.19 ± 0.12	0.16 ± 0.17
Late erythroblasts	14.22 ± 7.14	7.60 ± 4.84	13.99 ± 3.82	11.40 ± 3.43	7.34 ± 2.54
Early/late erythroblasts ratio[c]	1:50	1:15	1:18	1:22	1:23
Fetal erythroblasts	14.48 ± 7.24	8.04 ± 5.00	14.70 ± 3.86	11.90 ± 3.52	7.62 ± 2.56
Blood reticulocytes	4.18 ± 1.46	1.06 ± 1.13	3.39 ± 1.22	2.90 ± 0.91	1.65 ± 0.73

(Continued)

TABLE A1.32 *(Continued)*

Cell type	Month 0 (*n*557)[b]	1 (*n*571)	2 (*n*548)	3 (*n*524)	4 (*n*519)
NEUTROPHILS					
Promyelocytes	0.79 ± 0.91	0.76 ± 0.65	0.78 ± 0.68	0.76 ± 0.80	0.59 ± 0.51
Myelocytes	3.95 ± 2.93	2.50 ± 1.48	2.03 ± 1.14	2.24 ± 1.70	2.32 ± 1.59
Early neutrophils	4.74 ± 3.43	3.27 ± 1.94	2.81 ± 1.62	3.00 ± 2.18	2.91 ± 2.01
Metamyelocytes	19.37 ± 4.84	11.34 ± 3.59	11.27 ± 3.38	11.93 ± 13.09	6.04 ± 3.63
Bands	28.89 ± 7.56	14.10 ± 4.63	13.15 ± 4.71	14.60 ± 7.54	13.93 ± 6.13
Mature neutrophils	7.37 ± 4.64	3.64 ± 2.97	3.07 ± 2.45	3.48 ± 1.62	4.27 ± 2.69
Late neutrophils	55.63 ± 7.98	29.08 ± 6.79	27.50 ± 6.88	31.00 ± 11.17	31.30 ± 7.80
Early/late neutrophil ratio	1:12	1:9	1:9	1:9	1:11
Total neutrophils	60.37 ± 8.66	32.35 ± 7.68	30.31 ± 7.27	34.01 ± 11.95	34.21 ± 8.61
Total eosinophils	2.70 ± 1.27	2.61 ± 1.40	2.50 ± 1.22	2.54 ± 1.46	2.37 ± 4.13
Total basophils	0.12 ± 0.20	0.07 ± 0.16	0.08 ± 0.10	0.09 ± 0.09	0.11 ± 0.14
Total myeloid cells	63.19 ± 9.10	35.03 ± 8.09	32.90 ± 7.85	36.64 ± 2.26	36.69 ± 8.91
Monocytes	0.88 ± 0.85	1.01 ± 0.89	0.91 ± 0.83	0.68 ± 0.56	0.75 ± 0.75
MISCELLANEOUS					
Megakaryocytes	0.06 ± 0.15	0.05 ± 0.09	0.10 ± 0.13	0.06 ± 0.09	0.06 ± 0.06
Plasma cells	0.00 ± 0.02	0.02 ± 0.06	0.02 ± 0.05	0.00 ± 0.02	0.01 ± 0.03
Unknown blasts	0.31 ± 0.31	0.62 ± 0.50	0.58 ± 0.50	0.63 ± 0.60	0.56 ± 0.53
Unknown cells	0.22 ± 0.34	0.21 ± 0.25	0.16 ± 0.24	0.19 ± 0.21	0.23 ± 0.25
Damaged cells	5.79 ± 2.78	5.50 ± 2.46	5.09 ± 1.78	4.75 ± 2.30	4.80 ± 2.29
Total	6.38 ± 2.84	6.39 ± 2.63	5.94 ± 1.94	5.63 ± 2.36	5.66 ± 2.30

MONTH

5 (*n*522)	6 (*n*522)	9 (*n*516)	12 (*n*518)	15 (*n*512)	18 (*n*519)
47.19 ± 9.93	47.55 ± 7.88	48.76 ± 8.11	47.11 ± 11.32	42.77 ± 8.94	43.55 ± 8.56
1.83 ± 0.89	2.31 ± 1.16	1.92 ± 1.39	2.32 ± 1.90	1.70 ± 0.82	1.99 ± 1.00
0.07 ± 0.10	0.09 ± 0.12	0.07 ± 0.09	0.02 ± 0.04	0.07 ± 0.12	0.08 ± 0.13
0.47 ± 0.33	0.32 ± 0.24	0.31 ± 0.24	0.30 ± 0.25	0.38 ± 0.37	0.50 ± 0.34
0.55 ± 0.36	0.41 ± 0.30	0.39 ± 0.28	0.39 ± 0.27	0.46 ± 0.36	0.59 ± 0.34
7.55 ± 2.35	7.30 ± 3.60	7.73 ± 3.39	6.83 ± 3.75	6.04 ± 1.56	6.97 ± 3.56
0.46 ± 0.51	0.38 ± 0.56	0.39 ± 0.48	0.37 ± 0.51	0.50 ± 0.65	0.44 ± 0.49
0.14 ± 0.11	0.16 ± 0.22	0.22 ± 0.25	0.23 ± 0.25	0.17 ± 0.12	0.21 ± 0.19
8.16 ± 2.58	7.85 ± 4.11	8.34 ± 3.31	7.42 ± 4.11	6.72 ± 1.80	7.62 ± 3.63
1:15	1:17	1:19	1:17	1:15	1:10
8.70 ± 2.69	8.25 ± 4.31	8.72 ± 3.34	7.81 ± 4.26	7.18 ± 1.95	8.21 ± 37.1
1.38 ± 0.65	1.74 ± 0.80	1.67 ± 0.52	1.79 ± 0.79	2.10 ± 0.91	1.84 ± 0.46
0.87 ± 0.80	0.67 ± 0.66	0.41 ± 0.34	0.69 ± 0.71	0.67 ± 0.58	0.64 ± 0.59

(Continued)

TABLE A1.32 (Continued)

MONTH

5 (n522)	6 (n522)	9 (n516)	12 (n518)	15 (n512)	18 (n519)
2.73 ± 1.82	2.22 ± 1.25	2.07 ± 1.20	2.32 ± 1.14	2.48 ± 0.94	2.49 ± 1.39
3.60 ± 2.50	2.89 ± 1.71	2.48 ± 1.46	3.02 ± 1.52	3.16 ± 1.19	3.14 ± 1.75
11.89 ± 3.24	11.02 ± 3.12	11.80 ± 3.90	11.10 ± 3.82	12.48 ± 7.45	12.42 ± 4.15
14.07 ± 5.48	14.00 ± 4.58	14.08 ± 4.53	14.02 ± 4.88	15.17 ± 4.20	14.20 ± 5.23
3.77 ± 2.44	4.85 ± 2.69	3.97 ± 2.29	5.65 ± 3.92	6.94 ± 3.88	6.31 ± 3.91
29.73 ± 7.19	29.86 ± 6.74	29.86 ± 7.36	30.77 ± 8.69	34.60 ± 7.35	32.93 ± 7.01
1:8	1:10	1:12	1:10	1:10	1:10
33.12 ± 8.34	32.75 ± 7.03	32.33 ± 7.75	33.79 ± 8.76	37.76 ± 7.32	36.06 ± 7.40
1.98 ± 0.86	2.08 ± 1.16	1.74 ± 1.08	1.92 ± 1.09	3.39 ± 1.93	2.70 ± 2.16
0.09 ± 0.13	0.10 ± 0.13	0.11 ± 0.13	0.13 ± 0.15	0.27 ± 0.37	0.10 ± 0.12
35.40 ± 8.54	34.93 ± 7.52	34.18 ± 8.13	35.83 ± 8.84	41.42 ± 7.43	38.86 ± 7.92
1.29 ± 1.06	1.21 ± 1.01	1.17 ± 0.97	1.46 ± 1.52	1.68 ± 1.09	2.12 ± 1.59
0.08 ± 0.09	0.04 ± 0.07	0.09 ± 0.12	0.05 ± 0.08	0.00 ± 0.00	0.07 ± 0.12
0.05 ± 0.11	0.03 ± 0.07	0.01 ± 0.03	0.03 ± 0.07	0.07 ± 0.12	0.06 ± 0.08
0.50 ± 0.37	0.56 ± 0.48	0.42 ± 0.50	0.37 ± 0.33	0.46 ± 0.32	0.43 ± 0.45
0.17 ± 0.22	0.10 ± 0.15	0.14 ± 0.17	0.11 ± 0.14	0.13 ± 0.18	0.20 ± 0.23
4.86 ± 1.25	5.04 ± 1.08	4.89 ± 1.60	5.34 ± 2.19	4.99 ± 1.96	5.05 ± 2.15
5.66 ± 1.41	5.78 ± 1.16	5.55 ± 1.74	5.90 ± 2.03	5.65 ± 2.02	5.81 ± 2.16

[a]Percentages of cell types (mean ± SD) in tibial bone marrow of infants from birth to 18 months of age. Data were obtained from normal American infants of black, white, and Asian racial origin. The changes in the marrow during the first 18 months of postnatal life are based on differential counts of 1000 cells classified on stained smears on each 10 serial marrow samples aspirated from the same population of infants. Criteria for including bone marrow data in this study consisted of the absence of any clinical evidence of disease, normal rate of growth, and normal serum proteins and transferrin saturations.

[b]n, number of infants studied at each stage.

[c]Expressed in round figures for facilitating comparison. Mean ± SD were calculated from values obtained in individual infants and statistical comparisons were performed.

From Rosse, C., Kraemer, M.J., et al., 1977. Bone marrow cell populations of normal infants: the predominance of lymphocytes. J. Lab. Clin. Med. 89, 1228, with permission.

TABLE A1.33   Normal Serum Vitamin E Levels (mg/dl) in Newborns[a]

Weeks	1	2	3	4	5	6	7	8	9	10
1500 g 28−32 weeks	0.40 [0.05][a]	0.30 [0.04]	0.25 [0.03]	0.25 [0.03]	0.25 [0.03]	0.25 [0.03]	0.25 [0.03]	0.25 [0.03]	0.35 [0.04]	0.45 [0.05]
1500−2000 g 32−36 weeks	0.45 [0.05]	0.40 [0.05]	0.40 [0.05]	0.45 [0.05]	0.45 [0.05]	0.45 [0.05]	0.50 [0.05]	0.50 [0.05]	0.60 [0.06]	0.70 [0.06]
2000−2500 g 36−40 weeks	0.50 [0.05]	0.45 [0.05]	0.50 [0.05]	0.60 [0.06]	0.70 [0.06]	0.75 [0.06]	0.75 [0.60]	0.75 [0.60]	0.75 [0.60]	0.80 [0.70]
>2500 g term	0.55 [0.60]	0.55 [0.60]	0.55 [0.60]	0.60 [0.60]	0.75 [0.70]	0.80 [0.70]	0.85 [0.80]	0.85 [0.80]	0.85 [0.80]	0.85 [0.80]

[a]Mean ± 1 SD.

From Klaus, M., Fanaroff, A., 1986. Care of the High Risk Neonate, third ed. Saunders, Philadelphia, PA, with permission.

TABLE A1.34 Levels of Urinary Homovanyllic Acid and Vanillylmandelic Acid in Different Age Groups

Age	HVA (mg/mg urinary creatinine)		VMA (mg/mg urinary creatinine)	
	Range	Mean ± SD	Range	Mean ± SD
0−3 months	11.3−35.0	22.3 ± 7.5	5.0−37.0	19.5 ± 10.3
3−12 months	8.4−44.9	27.8 ± 9.6	8.4−43.8	22.8 ± 10.6
1−2 years	12.2−31.8	19.7 ± 6.0	7.9−23.0	14.9 ± 3.9
2−5 years	3.4−32.0	15.3 ± 7.7	2.9−23.0	11.3 ± 7.6
5−10 years	6.8−23.7	12.8 ± 5.1	5.8−18.7	9.3 ± 3.7
10−15 years	3.2−13.6	8.1 ± 3.9	1.6−10.6	5.2 ± 2.5
>15 years	3.2−9.6	5.7 ± 1.9	2.8−8.3	4.5 ± 1.8

HVA, homovanyllic acid; VMA, vanillylmandelic acid.
From Tuchman, M., Moriss, C.L., Ramnaraine, M.L., et al., 1985. Value of urinary homovanillic acid and vanillylmandelic acid levels in the diagnosis and management of patients with neuroblastoma: comparison with 24-hour urine collections. Pediatrics 75, 324, with permission.

TABLE A1.35   Values for Soluble Interleukin-2R (SIL-2R)

Cord blood	267−799 units/ml
2.5−9 months	580−1712 units/ml
9.5−19.5 months	341−2337 units/ml
20−60 months	322−1207 units/ml
101 years	80−600 units/ml
Adults	71−477 units/ml

TABLE A1.36   Thyroid Values

TSH (mIU/ml)	0.3−5.0	Immunoradiometric assay
Total T-3 (ng/dl)	50−190	Radioimmunoassay
Total T-4 (mg/dl)	4.0−13.0	Radioimmunoassay

2

Biological Tumor Markers

Tumor Marker	Diagnostic Significance	Normal Values
HORMONES		
Beta-human chorionic gonadotropin (beta-hCG) in serum	Choriocarcinoma	5 mIU/ml
	Hepatoblastoma	
Catecholamines urinary VMA/HVA	Neuroblastoma	Appendix 1, Table A1.34
Plasma and urinary catecholamines and urinary metanephrine and VMA	Pheochromocytoma	
Testosterone serum	Leydig cell tumor (testis)	DOM
	Sertoli–Leydig cell (ovary)	
Estrogen serum	Granulosa cell tumor (ovary)	DOM
	Sertoli cell tumor (testis)	
Cortisol, aldosterone, testosterone and estrogen serum	Adrenal hyperplasia Adrenal adenoma	DOM
	Adrenal carcinoma	
	Ectopic adrenal rests (liver, testis, ovary)	
Thyrocalcitonin serum	Medullary (C-cell) thyroid carcinoma	DOM
T3, T4 serum	Hyperthyroidism	Appendix 1, Table A1.36
	Thyroid adenoma	
	Thyroid carcinoma	
Parathormone serum	Parathyroid hyperplasia	DOM
	Parathyroid adenoma	
	Parathyroid carcinoma	
	Hepatoblastoma	
Vasoactive intestinal peptide (VIP)	Neuroblastoma	<75 pg/ml
	Vipoma	

Tumor Marker	Diagnostic Significance	Normal Values
Renin serum	Wilms' tumor	DOM
Erythropoietin serum	Wilms' tumor	Appendix 1, Table A1.17
	Adrenal carcinoma	
	Renal carcinoma	
	Hepatoma	

ENZYMES

Tumor Marker	Diagnostic Significance	Normal Values
Neurone-specific enolase in serum	Neuroblastoma	15 ng/ml
	Primitive neuroectodermal tumors	
	Medulloblastoma	
Lactic dehydrogenase (LDH) in serum	Acute leukemias	297–537 units/l (depending on laboratory)
	Non-Hodgkin's lymphoma	
	Neuroblastoma	
	Ewing's sarcoma	
	Osteosarcoma	
	Germ cell tumor	

SPECIFIC PROTEINS

Tumor Marker	Diagnostic Significance	Normal Values
Immunoglobulins	Multiple myeloma and other gammopathies	See reference values in Appendix 1
Cytokines	Acute lymphoblastic leukemia	
Soluble interleukin-2 receptor (SIL-2R) in serum	Non-Hodgkin's lymphoma	
	Hodgkin's disease	
	Hemophagocytic histiocytic syndrome	

MUCINS AND OTHER GLYCOPROTEINS

Tumor Marker	Diagnostic Significance	Normal Values
CA 125	Ovarian cancer	<35 units/ml
CA 19–9	Colon cancer, pancreatic cancer	~37 units/ml (for investigation use only)
CA 15–3	Breast cancer	<31 units/ml

OTHER TUMOR MARKERS

Tumor Marker	Diagnostic Significance	Normal Values
5-Hydroxyindoleacetic acid in urine (5-HIAA)	Carcinoid tumor	24 h urine: 2–8 mg/dl
Neopterins in urine	Non-Hodgkin lymphoma	7–150 µg/l
	Hemophagocytic lymphohistiocytosis	
Ferritin	Neuroblastoma	6 months–15 years: 12–113 ng/ml; 15–49 years: 12–156 ng/ml
	Hodgkin disease	
	Hepatocellular carcinoma	
	Germ cell tumor	

ONCOFETAL PROTEINS

α-Fetoprotein in serum	Hepatoblastoma	Age	Mean	± SD (ng/ml)
	Hepatocellular carcinoma	Premature	134,734	± 41,444
	Endodermal sinus tumor	Newborn	48,406	± 34,718
	Embryonal carcinoma	Newborn–2 weeks	33,113	± 32,503
	Pancreatoblastoma	2 weeks–1 month	9452	± 12,610
		2 months	323	± 278
		3 months	88	± 87
		4 months	74	± 56
		5 months	46.5	± 19
		6 months	12.5	± 9.8
		7 months	9.7	± 7.1
		8 months	8.5	± 5.5

Notes. SD, standard deviation. From Wu, J.T., Book, L., Sudar, K., 1981. Serum α-fetoprotein (AFP) levels in normal infants. Pediatr. Res. 15, 50–52, with permission.

Carcinoembryonic antigen (CEA) in serum	Adenocarcinoma of colon	2.5 ng/ml
	Wilms' tumor	
	Hepatoblastoma	
	Hepatocellular carcinoma	
	Germ cell tumor	
	Pulmonary blastoma	

DOM, dependent on method. The reader should consult reference laboratory or standard textbook for normal values.

Index

Note: Page numbers followed by "*f*" and "*t*" refer to figures and tables, respectively.